Neutron and X-ray Optics

Neutron and X-ray Optics

Jay Theodore Cremer, Jr.

Chief Scientist
Adelphi Technology, Inc.
2003 E. Bayshore Road
Redwood City, CA 94063
USA

ELSEVIER

AMSTERDAM • BOSTON • HEIDELBERG • LONDON • NEW YORK • OXFORD
PARIS • SAN DIEGO • SAN FRANCISCO • SINGAPORE • SYDNEY • TOKYO

Elsevier
32 Jamestown Road London NW1 7BY
225 Wyman Street, Waltham, MA 02451, USA

First edition 2013

British Library Cataloguing-in-Publication Data
A catalogue record for this book is available from the British Library

Library of Congress Cataloging-in-Publication Data
A catalog record for this book is available from the Library of Congress

ISBN: 978-0-12-407164-3

For information on all Elsevier publications visit our website at store.elsevier.com

This book has been manufactured using Print On Demand technology. Each copy is produced to order and is limited to black ink. The online version of this book will show color figures where appropriate.

Transferred to Digital Printing in 2013

Contents

1 Introduction

This Insights book is a continuation of the "Advances in Imaging and Electron Physics" volumes (AIEP) 172, 173, and 175 of the work on neutron and X-ray microscopy and optics by this author. This material is intended for a wide international readership of varying backgrounds, specialties, and educations, in which English may be a second language. The Insights book attempts to provide rapid assimilation of the presented topics that include neutron and X-ray scatter, refraction, diffraction, and reflection and their potential application.

AIEP volume 172 (Cremer, 2012b) first introduces experiments with thermal neutron and X-ray geometric, refractive, reflective, and diffractive optics, as well as neutron and X-ray microscopy, phase contrast, and thermal and fast neutron radiographic imaging. What follows is a presentation of compound refractive lenses (CRLs) and prisms followed by the monochromatic and geometric aberrations of geometric X-ray and neutron optics. This Insights volume will cover the diffractive neutron and X-ray optics and, using the Huygens–Fresnel approach, the Kirchhoff equation solutions for compound refractive X-ray and neutron lenses, which include the geometric, diffractive, monochromatic, and chromatic aberrations.

Next, the X-ray complex scatter factor and refractive index are derived, and presented are the scatter cross sections for X-rays, including X-ray attenuation processes. In addition, X-ray reflection and transmission at interfaces are included. Next, this volume derives the Schrödinger wave equation for neutron plane waves, neutron phase and group velocity, neutron reflection and transmission at interfaces, and neutron reflectometry, as well as X-ray and neutron interferometry, perfect crystals, and the Fizeau effect. Finally, the general refractive index for neutrons and X-rays in materials is derived, and useful formulas and the comparison of the scatter and attenuation process of neutrons and X-rays, multilayer mirrors, capillary optics, and Kumakhov lenses for neutrons and X-rays are presented, and then there is a discussion of bright-field, dark-field, and phase contrast microscopy for neutrons and X-rays.

AIEP volume 173 (Cremer, 2012c) starts with coverage on sources, scattering, detection, and radiographic imaging of fast neutrons, as well as the use of fast neutrons in nuclear resonant detection of isotopes. Next comes a detailed presentation of thermal neutron scatter, which includes neutron complex scatter lengths and scatter amplitudes, and refractive index in materials, and then onto a review of the relevant quantum mechanics, such as the Heisenberg, Schrödinger, and interaction pictures, creation and annihilation operators of the harmonic oscillator, and the derivation of the Breit–Wigner equation and Fermi's Golden Rule. The Eikonal equation for neutron optics is presented, and the general solution of the Schrödinger wave

Neutron and X-ray Optics. DOI: http://dx.doi.org/10.1016/B978-0-12-407164-3.00019-X

equation for plane wave neutron scatter from nuclei is derived in detail and includes derivation of the complex scatter amplitudes, neutron scatter lengths, phase shift angles, and the optical theorem.

Next, is derived the total, differential, and partial differential cross sections for coherent and incoherent, elastic and inelastic neutron scatter in materials. Then, the crystal and its unit cell, and the regular and reciprocal lattices for Bravais and non-Bravais lattice crystals, the Ewald sphere, the Wigner–Seitz cell and the first Brillouin zone, and the Bragg and Laue diffraction conditions are presented prior to coverage of neutron diffraction from crystals. The partial differential, differential, and total cross sections, are derived for coherent and incoherent, elastic and inelastic neutron scatter in Bravais and non-Bravais crystals with 1-D and 3-D, which is based on the kinematic model. These crystal neutron scatter cross sections are first derived without thermal effects and then derived with the inclusion of thermal effects via the Debye–Waller factor.

The coherent and incoherent inelastic scatter and phonon–neutron scatter are also covered. The 1-D and 3-D thermal vibration modes in crystals are covered, and the cubic crystal lattice is examined in particular, which also includes Hooke's law and the derivation of the cubic crystal stress tensor and elastic waves. Neutron residual stress analysis is discussed, along with the topics of coupled Ewald equations, extinction, and broadening of diffraction peaks in perfect crystals. Next, the dynamic model of neutron and X-ray scatter in perfect crystals is presented. The average nuclear potential and the time-independent Schrödinger equation and the solution for dynamic neutron scatter in a perfect crystal, which includes anomalous absorption, primary extinction, and Pendellösung interference, are detailed.

Next, the neutron transport equation for neutron scatter in mosaic crystals in terms of the neutron scatter and absorption cross section per unit volume is derived and then solved to yield the generalized Darwin equation. Single Bragg reflection applied to the generalized Darwin equations yields the ordinary Darwin equations, which are solved for the one-beam (off-Bragg) and two-beam (Bragg) cases for a planar mosaic crystal. Finally, the partial differential, differential, and total cross sections for kinematic neutron scatter from averaged nuclear potential in crystals are derived.

AIEP volume 175, chapter 1, starts with the coverage of the correlation, scatter, and intermediate functions of small-angle neutron scatter. Small-angle X-ray and neutron scatter from general samples are covered, followed by the Rayleigh–Gans theories, Babinets' principle, and the differential cross section of X-ray or neutron small-angle scattering from a solute–solvent sample. The resolution of the scatter vector for a small-angle neutron scatter (SANS) instrument for X-rays or neutrons is derived. Next, we present the neutron scatter length density, particle structure factor, and scatter amplitudes and intensity. Also covered are random variables, correlation, and independence, followed by derivation of the macroscopic differential cross section for neutron scatter, which involves convolution and cross-correlation. Next, the coherent and incoherent, elastic and inelastic components of the pair correlation function, intermediate function, and scatter function, the relationships between these functions, and the measured neutron small-angle scatter intensities from neutron scattering sample are discussed.

AIEP volume 175, chapter 1 then covers the Guinier, Intermediate, and Porod regimes of the sample-averaged intermediate function, the method of contrast variation, and Porod's law. Coherent neutron scatter measurements are shown to yield the solute particle size and shape in the Guinier regime, and incoherent neutron scatter measurements are shown to yield the incoherent scatter function, which give particle diffusion information. Also derived is the principle of detailed balance and covered are the static approximation, the particle number density operator and pair correlation function, and the moments of the neutron scatter function. The neutron coherent differential cross section in crystals is shown to be expressed by particle density operators, and neutron elastic scatter is shown to be only in the forward direction for liquids and gases by the coherent intermediate and scatter functions.

In the second chapter of AIEP volume 175, are derived the partial differential cross sections for neutron scatter from a nucleus, which accounts for the neutron spin and the nuclear spin. Covered first are the preliminary background topics of angular momentum vectors, spin vectors, and vector operators, the Heisenberg uncertainty principle and commutation of operators, the neutron spin operator, and the neutron spin lowering and raising operators. Next are derived, the partial differential cross section for nuclear scatter of the neutron spin-up and spin-down states, and derived are polarized neutron scatter and the partial differential cross section, which includes both the neutron spin state and nuclear spin state, via the combined neutron spin operator and nuclear spin operators. Next, the neutron nuclear scatter length, which accounts for the neutron spin states, is covered. Thermal averaging is then taken into account, and the total partial differential cross section for neutron spin state scatter is derived, as well as the neutron spin state scatter lengths for an ensemble of nuclear spins and isotopes. Finally, derived are the partial differential, differential, and total cross sections for neutron coherent and incoherent scatter from an ensemble of atoms of varying nuclear spins and isotopes, which accounts for neutron spin states.

The Insights book augments, supplements, and continues coverage of the neutron and X-ray topics in AIEP volumes 172, 173, and 175, and extends this coverage to relevant topics from a variety of branches of physics. The Insights book also presents magnetic neutron scatter, neutron refractive index in materials and fields, electric field neutron scatter, and neutron magnetic optics, X-ray optics and gamma optics, and neutron and X-ray refractive lenses and microscopy. Also included in this book is a detailed discussion of associated material, which includes (1) neutron spin echo, tomography, and holography; (2) diffractive X-ray and neutron optics; (3) Kirchoff equation solutions for X-ray and neutron refractive lenses; (4) the Klein Gordon and Dirac equations applied to magnetic neutron scatter; (5) electrodynamics with particular coverage of Lienard–Wiechert potentials and electric and magnetic fields from moving and accelerated charges, magnetic dipole moments, and dipole interaction; (6) Hamiltonians and least action; (7) special relativity transformation of electric and magnetic fields between moving inertial frames; (8) a general relativity primer for X-ray and neutron optics applied to astronomy; (9) novel X-ray sources, such as transition, channel and parametric sources for X-ray optics; (10) neutron and charged particle beam focusing and imaging optics; and (11) radiation imaging systems and performance.

This Insights book and the AIEP volumes 172, 173, and 175 use MKS units, unless specified otherwise. The extensive mathematical physics that is presented in this book may require in many cases a computer to solve, simulate, model, or calculate. For scientific programming and numerical computation in the FORTRAN computer language, see Press et al. (1989), and for the rhea C computer language, see Press et al. (1992). For a general survey of scientific programming and numerical computation, see Burden and Faires (1997) and Gerard and Wheatley (1994).

This first chapter covers the basics of refractive material lenses for X-rays, neutrons, and gammas, and covers parabolic and spherical, biconcave and biconvex CRL. In particular, the focal lengths, limiting aperture diameters due to absorption, and parabolic focusing are derived for CRLs for neutrons, X-rays, and gammas. Also presented is a 3-D array of CRLs, which allow imaging and magnification of extended objects, which is based on the 3-D microlens arrays which are used for visible light in fax machines and photocopiers. The complex refractive index and scatter cross section for X-rays are derived using the electromagnetic wave equation and the atomic electron oscillator equation. Maxwell's equations are presented and are used to derive the electromagnetic wave equation and the energy balance equation for the electromagnetic wave. A review of integration on the complex plane, which is specific to deriving the complex refractive index for photons, is presented.

In Chapter 1, the optical theorem, also known as the shadow theorem, is derived. The optical theorem allows one to calculate the total cross section for scatter and absorption of an X-ray or neutron wave from the scatter factor or scatter amplitude. Also, the Kramers–Kronig relation is derived. The optical theorem allows one to calculate the real refractive part of the complex refractive index from the imaginary attenuation component. The attenuation component of the complex refractive index is more readily measured than the attenuating component. For example, the very tiny total cross section for attenuation of neutrinos in theory would allow the calculation of the very tiny refractive part of the complex refractive index of neutrinos as a function of neutrino energy.

Chapter 1 closes with the application of the optical theorem and the Kramers–Kronig relations in gamma ray refractive optics (gamma CRLs). The complex refractive index for gammas, which is based on the Delbrück effect of photon–photon scatter, is specifically gamma scatter by the nuclear electric field. Finally, we present the first experimental results of the refractive deflection of gammas with a prism, which use an intense Compton backscatter gamma source.

This section closes with images taken with (1) neutron CRL, (2) X-ray CRL, (3) X-ray Fresnel zone plates, (4) thermal neutron radiography, (5) fast neutron radiography, (6) neutron phase contrast imaging, and (7) X-ray phase contrast imaging. The experiments and details regarding these images can be found in Chapter 1 of AIEP volume 172. Figure 1.1 shows refractive imaging of very cold, cold, and thermal neutrons with a CRL, which acts as the objective lens in a simple neutron microscope set up for obtaining magnified images (Mag $\times$). Refractive imaging of cold neutrons in Figure 1.1 can be found in Eskildsen et al. (1998), Choi et al. (2000), Beguiristain et al. (2002), Cremer et al. (2004a, 2004b, 2005, 2010), and Park et al. (2006). Refractive imaging with very cold neutrons in Lal et al. (2007) and thermal

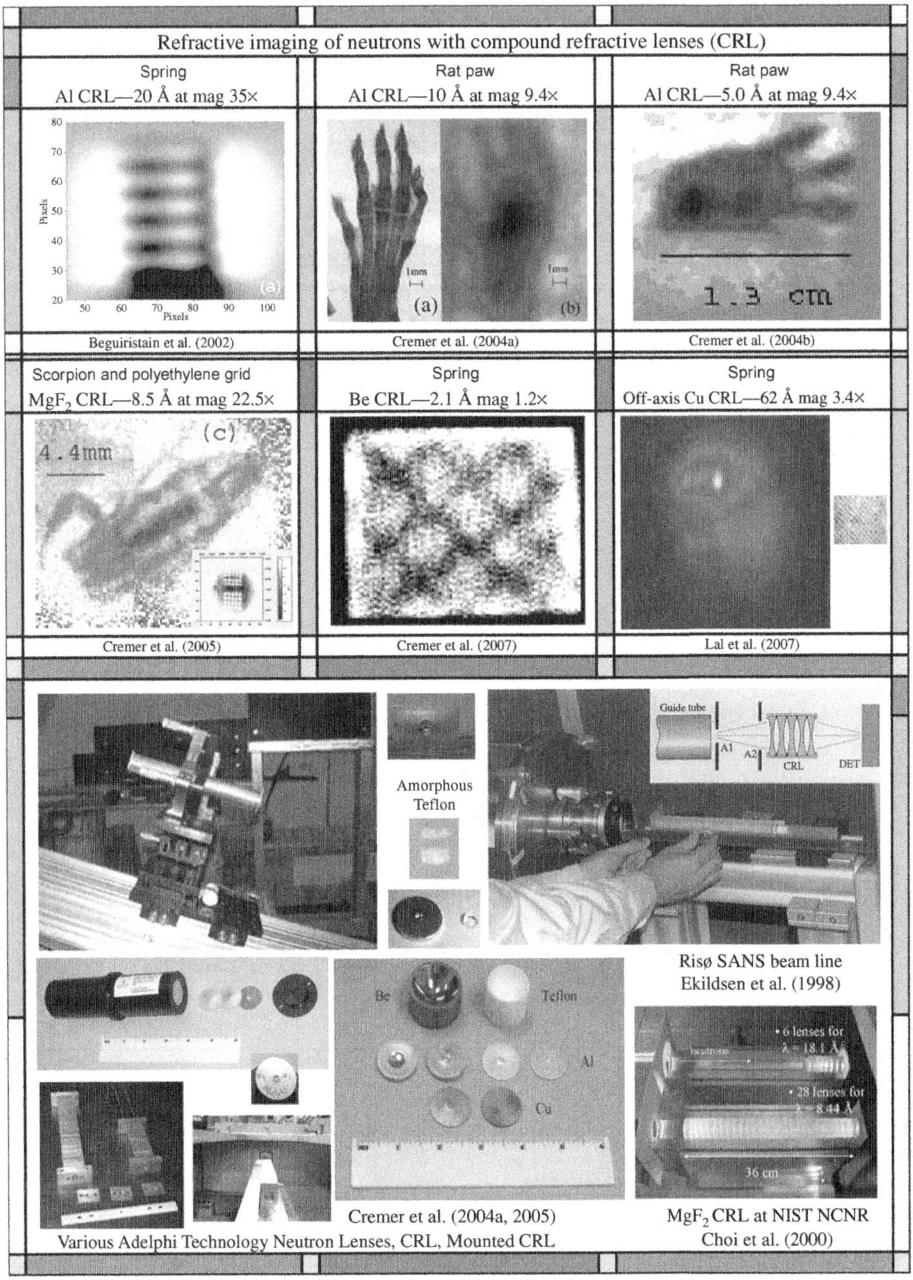

Figure 1.1 A collection of neutron CRLs, microscopes, and their magnified images.

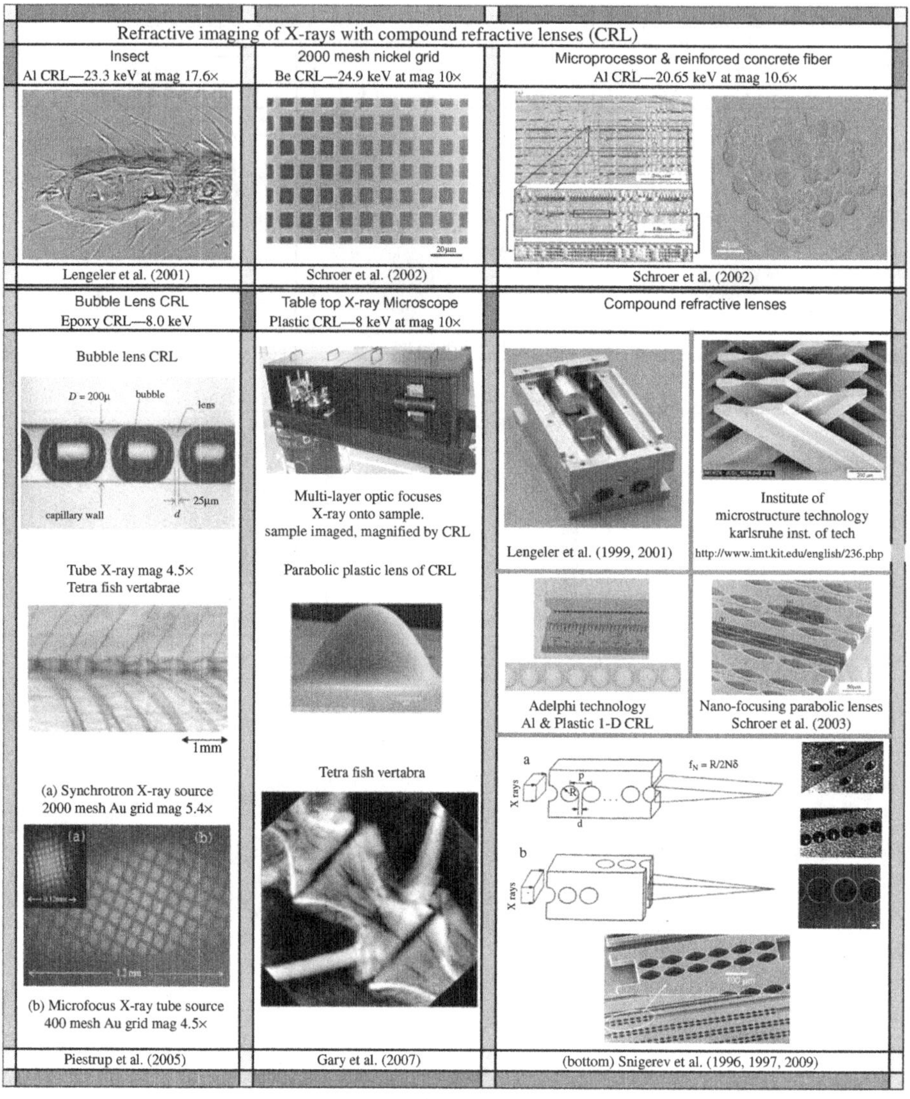

Figure 1.2 A collection of X-ray CRLs, microscopes, and their magnified images.

neutrons in Cremer et al. (2007, 2007a) are covered in section 2 of chapter 1 of Cremer (2012b).

Figure 1.2 shows examples of refractive imaging with X-ray CRLs as well as photos of the various individual X-ray lenses stacked to form a CRL. The refractive imaging of X-rays in Figure 1.2 can be found in Snigirev et al. (1996, 1997, 2009), Lengeler et al. (1999, 2001), Schroer et al. (2000, 2002, 2003), Piestrup et al. (2005), Gary et al. (2007), and section 3 of chapter 1 in Cremer (2012b).

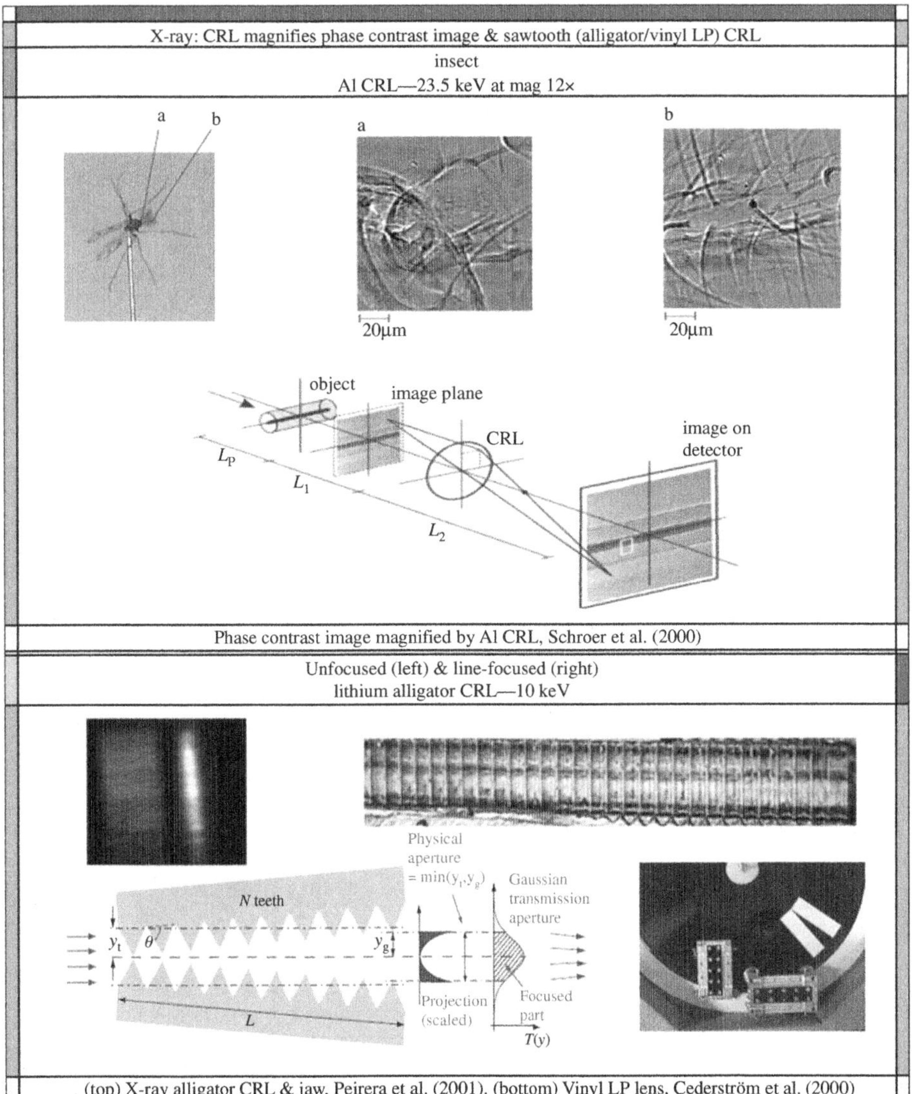

Figure 1.3 The use of CRL in phase contrast imaging of X-rays and the alligator (vinyl LP) CRL.

Figure 1.3 shows the use of a CRL to magnify a phase contrast image, produced by a synchrotron X-ray source (Schroer et al., 2000), the alligator CRL (Pereira et al., 2001), and the vinyl LP CRL (Cederström et al., 2000). Also see section 3 of chapter 1 in Cremer (2012b).

Examples of diffractive imaging of X-rays are shown in Figure 1.4. Also see section 7 of chapter 1 in Cremer (2012b). In Figure 1.4 (left panel), the differential phase contrast (DPC) imaging with a grating are shown. Conventional transmission

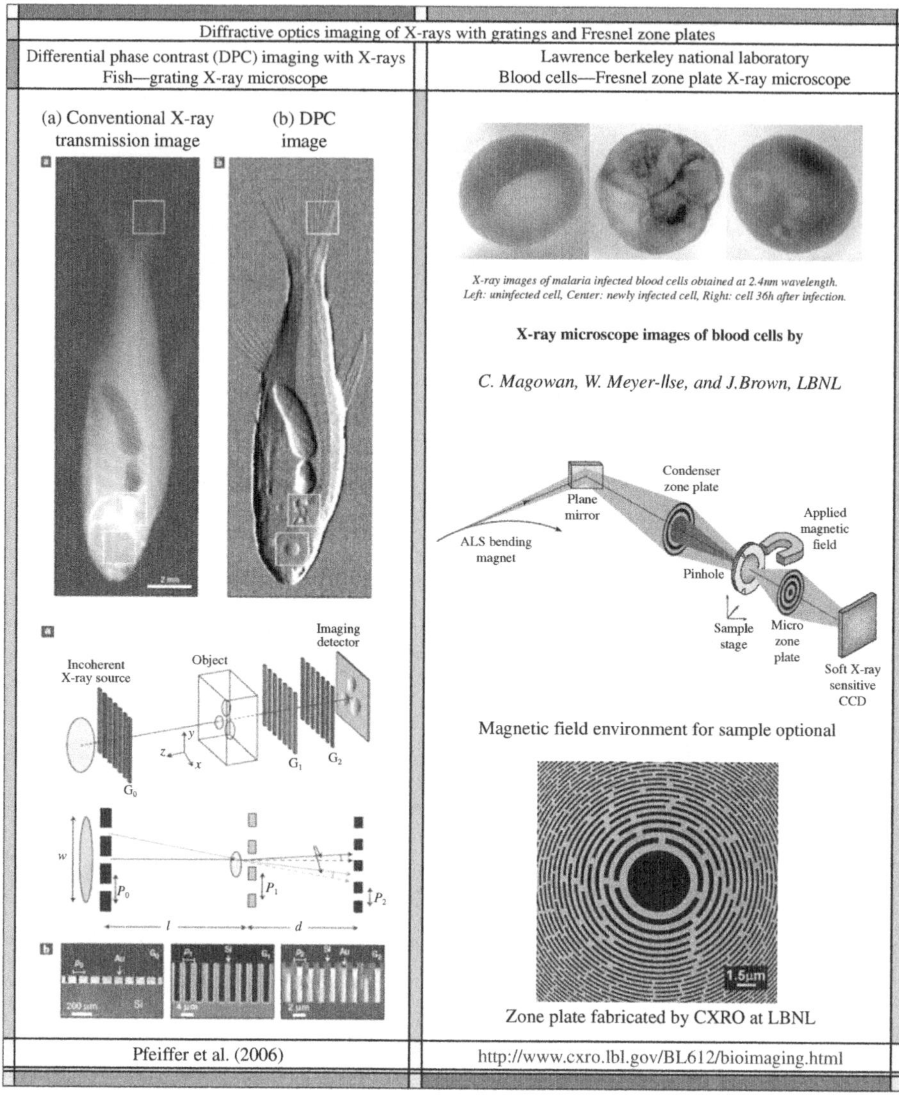

Figure 1.4 Examples of diffractive imaging of X-rays with gratings and Fresnel zone plates.

X-ray image (left fish) and DPC image (right fish) were obtained by Pfeiffer et al. (2006). In Figure 1.4 (right panel), the X-ray microscopy with Fresnel zone plates by the XM-1 bright-field, soft X-ray microscope at Advanced Light Source at Lawrence Berkeley National Laboratory (LBNL) is shown. The blood cell images are from C. Magowan, W. Meyer-Llse, and J. Brown of LBNL; see http://www.cxro.lbl.gov/zoneplate and http://www.cxro.lbl.gov/BL612/bioimaging.html.

In Figure 1.5, examples of phase contrast images with X-rays are shown. The X-ray phase contrast diagram and fish images in Figure 1.5 (left panel) are from the

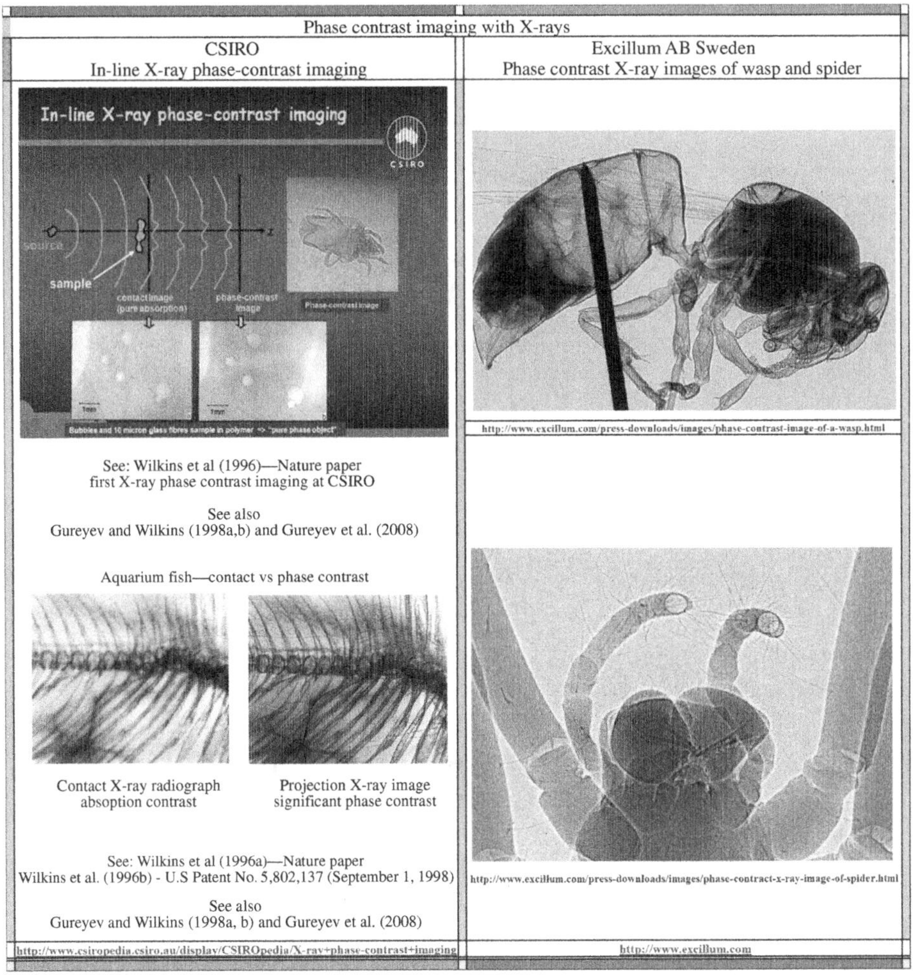

Figure 1.5 Examples of phase contrast imaging with X-rays.

CSIROpedia Web site of the Commonwealth Scientific and Industrial Research Organization (CSIRO) in Australia (http://www.csiropedia.csiro.au/display/CSIROpedia/X-ray + phase-contrast + imaging). For X-ray phase contrast imaging, see Wilkins et al. (1996), Wilkins (1996), Gureyev and Wilkins (1998a,1998b), Gureyev et al. (2008), and section 7 of chapter 1 in Cremer (2012b). The wasp and spider images in Figure 1.5 (right panel) are from the Web site of Excillum AB of Kista, Sweden (http://www.excillum.com/contact.html).

In Figure 1.6, examples of cold and thermal neutron phase contrast imaging are shown. Also see section 4 of chapter 1 in Cremer (2012b). In Figure 1.6 (left panel), the diagram and images are taken from a 2000 NIST Research Highlights article reporting on first-phase contrast imaging experiments with cold neutrons by

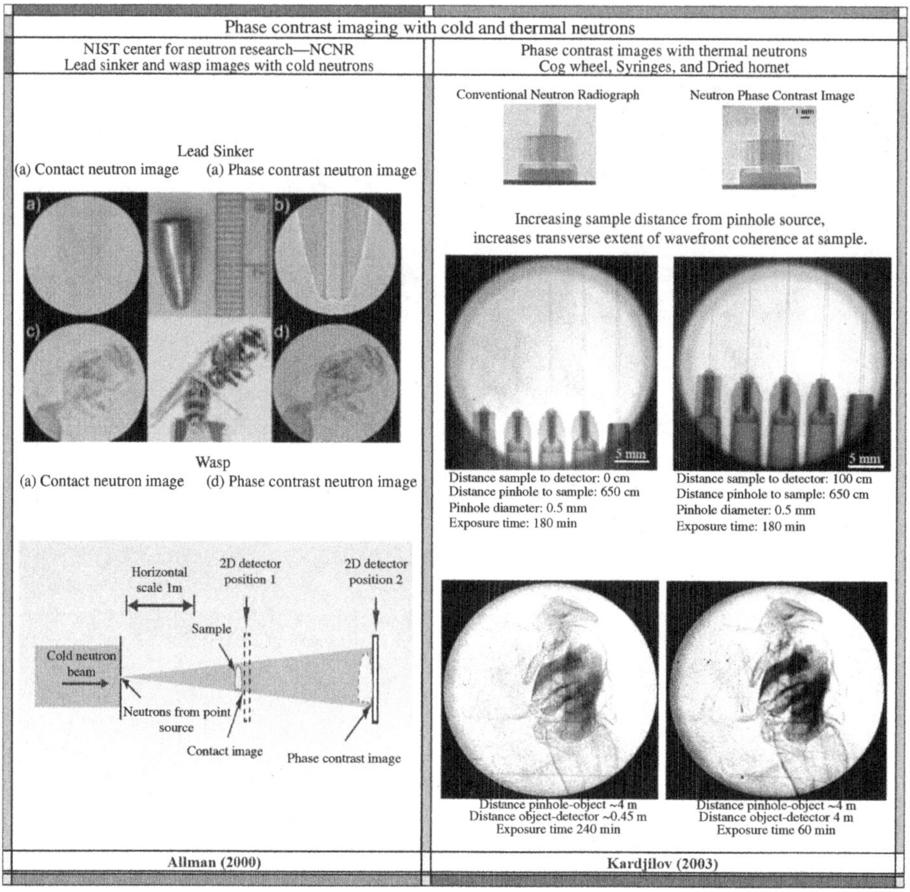

Figure 1.6 Examples of cold and thermal neutron, phase contrast imaging.

Allman et al. (2000). In Figure 1.6 (right panel), the phase contrast images are contrasted with contact images Kardjilov (2003). The neutron wave is refracted by the edge portion of the sample toward the detector in directions that are at a small angle relative to the incident direction of the wave front before sample incidence. Refraction of neutrons by the strong refractive gradient of the edge feature causes an amplitude shadow of the edge feature at the detector. The interference of the amplitudes of the edge-refracted wave front and the transmitted wave front on either side of the edge feature leads to amplitude modulation and a modulated intensity at the detector, that is, a phase contrast image of the edge feature.

In Figure 1.7, examples of thermal neutron radiography are shown. The images in Figure 1.7 (left panel) are from the McClellan Nuclear Radiation Center (MNRC) at the University of California at Davis (UC Davis; http://mnrc.ucdavis.edu). They show the comparison between a neutron contact radiograph (contrasts hydrogen-based fluid) and X-ray contact radiograph (contrasts high-Z materials) of a lighter. The lower left

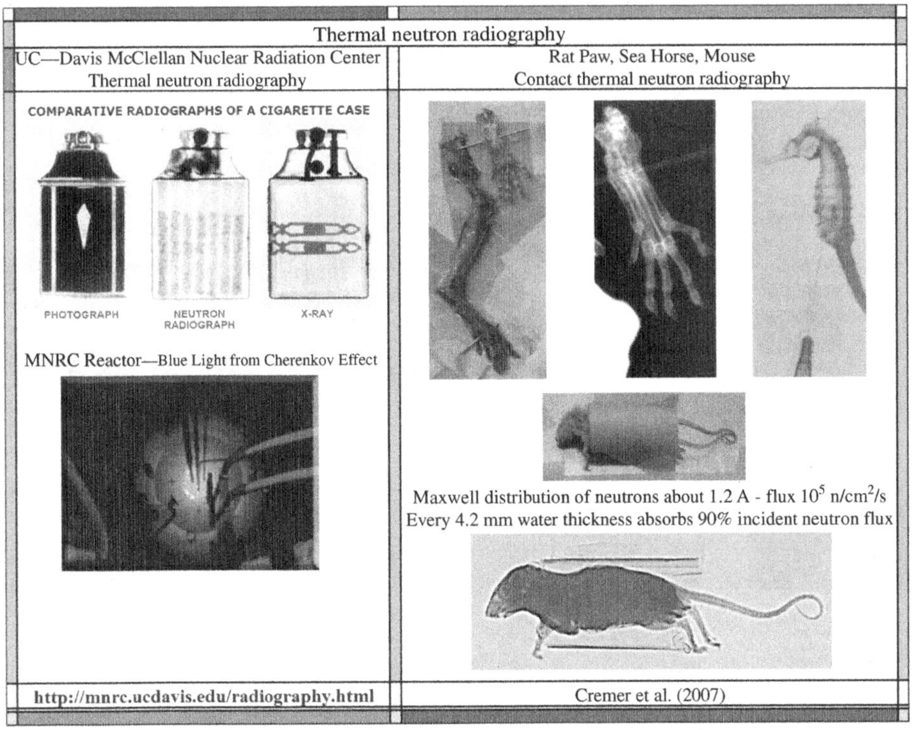

Figure 1.7 Examples of thermal neutron radiography.

photo shows the MNRC TRIGA reactor core with a blue glow from the Cherenkov radiation "sonic boom or bow wave" of charged particles, which exceed the phase velocity of light. Thermal neutron radiography can find missing glue in aircraft parts and stopped-up cooling channels in the aircraft turbine blade. In Figure 1.7 (right panel), contact thermal neutron radiography images of a rat paw, sea horse, and mouse are shown; they were taken at Bay 4 of the MNRC by Cremer et al. (2007), which is presented in section 4 of chapter 1 in Cremer (2012b). The thermal neutrons from the reactor have a Maxwell–Boltzmann distribution centered about 1.2 Å, where a sapphire crystal filters out gammas from the reactor and helps moderate the thermal neutrons to the room-temperature wavelength of 1.8 Å. The thermal neutrons in the reactor water are above room temperature, and the incident flux of 10^5 n/cm^2/s at the radiographic sample, with L/D = 270 collimation (distance L/ source_size D), are slightly warmer with distribution center wavelength 1.2 Å. Hydrogen has an 80.26 barn scatter cross section, and a 0.33 barn absorption cross section; hence, the thermal neutrons are strongly attenuated by scatter from hydrogen in the mouse.

Fast neutron imaging examples are shown in Figure 1.8; see section 5 in chapter 1 of Cremer (2012b). Disdier et al. (2003, 2004) used fast neutron imaging with liquid scintillator–filled capillary arrays to image the shape of imploding DT capsules in inertial confinement fusion experiments at the Laboratory for Laser Energetics at

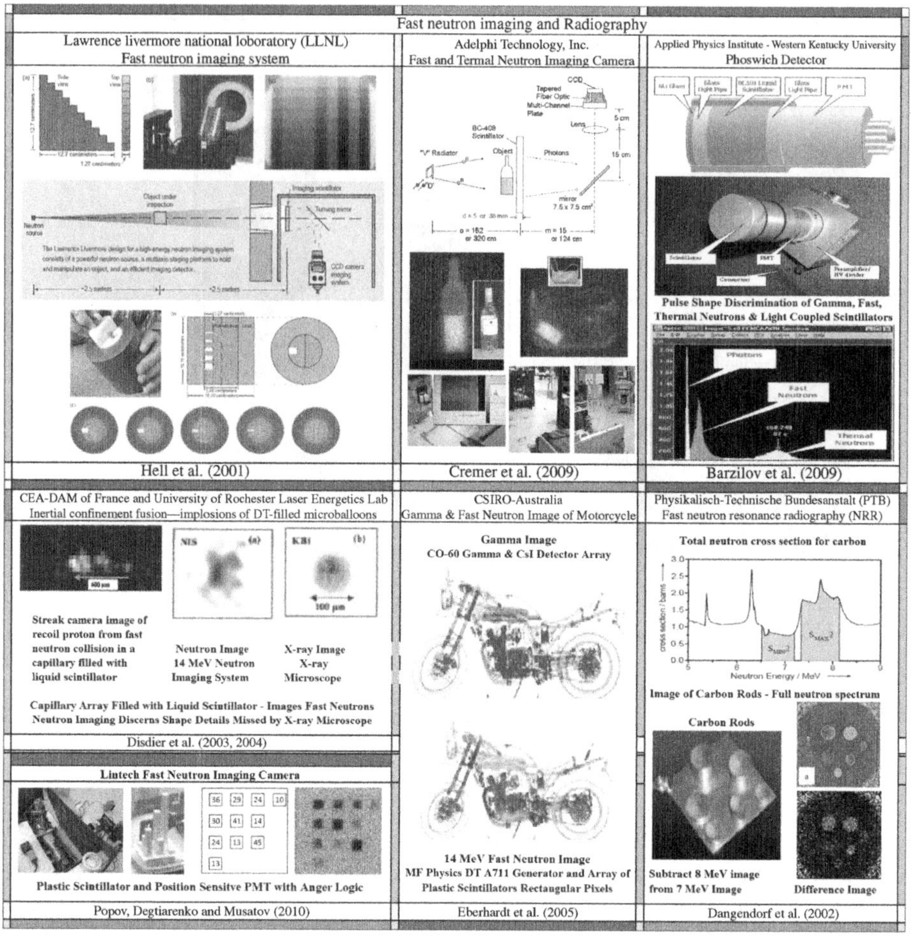

Figure 1.8 Examples of fast neutron imaging and radiography.

Rochester University with the Omega laser. Using the neutron source at the Ohio University Accelerator Laboratory (OUAL) in Athens, OH, Hall (2001) and Hall et al. (2001) obtained 10 MeV neutron radiographic and tomographic images of various phantoms, and Eberhardt et al. (2005) at CSIRO developed a combined gamma and fast neutron interrogation system for inspecting air cargo at the Brisbane airport in Australia. Dangendorf et al. (2002) exploited the fast neutron energy bands of higher and lower cross sections to obtain difference images of carbon rods. Lintech and Adelphi Technology have developed fast and thermal neutron imaging cameras see Popov et al. (2011) and Cremer et al. (2009, 2012). The phoswich detector developed by the Applied Physics Institute of Western Kentucky University by Barzilov et al. (2009) can discriminate, detect, and count the gamma, thermal, and fast neutron components from a radiation source, via pulse shape discrimination, and a plastic and lithium glass scintillator, optically coupled to a photomultiplier tube.

1.1 Refractive Index for Neutrons and X-rays

For derivation of the general refractive index for X-rays, gammas, and neutrons, see section 1 of Chapter 6 of AIEP volume 172, Cremer (2012b). The complex refractive index n of a material for gammas (or neutrons) is

$$n = 1 - \delta = 1 - (\delta_r + i\delta_i) = 1 - (\delta_r - i\beta) \tag{1.1}$$

The dimensionless, complex decrement δ is the sum of the real part δ_r and the imaginary part δ_i, where

$$\delta = \delta_r + i\delta_i \tag{1.2}$$

In X-ray refraction, the complex refractive index n of a material is expressed by convention by a real, dimensionless decrement δ for refraction and an imaginary, dimensionless attenuation component β, where

$$\delta = \delta_r \qquad \beta = -\delta_i \tag{1.3}$$

and

$$n = 1 - (\delta_r + i\delta_r) = 1 - \delta + i\beta \tag{1.4}$$

The amplitude of the electric field for X-rays or gammas or the wave-function amplitude $E(\mathbf{r}, t)$ for the neutron, inside a material medium, as a function of vector position $\mathbf{r}$ and time t, given the incident amplitude E_0 from the adjoining vacuum, is then

$$E(\mathbf{r}, t) = E_0 \exp[-i(wt - \mathbf{k} \cdot \mathbf{r})] \tag{1.5}$$

The wave vector $\mathbf{k}$ is normal to the plane wave front, where the unit wave vector $\hat{\mathbf{k}}$ is the unit length ray that points in the direction of the plane wave propagation. The magnitude of the wave vector $\mathbf{k}$ is the wave number $k = |\mathbf{k}|$, and the wave number k is the number of wavelengths per unit distance in the direction of ray propagation, where

$$k = \frac{2\pi}{\lambda} \tag{1.6}$$

In a medium with refractive index n for the X-ray, neutron, gamma, or light wave, the phase velocity v_p [cm/s] of the X-ray, neutron, gamma, or light wave propagation is related to the universal limiting, vacuum speed of light c [cm/s], where

$$v_p = \frac{w}{k} = \frac{c}{n} \tag{1.7}$$

The vacuum speed of light c is the assumed speed limit of propagation of modulated waves, which carry energy at the group velocity v_g, where

$$c \geq v_g = \frac{dw}{dk} \tag{1.8}$$

Given $n = 1 - \delta$, the refractive index is greater than 1 ($n > 1$) for negative refractive decrements $\delta < 0$, and the phase velocity $v_p = c/n$ of the wave is less than the speed of light $v_p < c$. Conversely, if the refractive index is less than 1 ($n < 1$) for positive refractive decrements $\delta > 0$, then the phase velocity of the wave is greater than the speed of light ($v_p > c$), where

$$c \geq v_p \quad \text{for } \delta < 0 \Rightarrow n > 1 \tag{1.9}$$

$$c \leq v_p \quad \text{for } \delta > 0 \Rightarrow n < 1 \tag{1.10}$$

The wave phase velocity v_p is not the speed at which energy is transported by the wave. Rather, the group velocity v_g is the speed at which the wave transports energy, and the wave phase velocity v_p is the speed of the phase of the wave. Hence, the wave phase velocity can be less than or greater than the vacuum speed of light.

The frequency f [s^{-1}] is the number of wavelengths (or cycles) per unit time in the direction of propagation and is related to the phase velocity of the wave v_p [cm/s], wavelength λ [cm] by

$$w = 2\pi \cdot f = 2\pi \frac{v_p}{\lambda} \tag{1.11}$$

The angular frequency w [rad/s] is simply the temporal frequency f [s^{-1}] multiplied by 2π [rad/cycle]. Substitution of Eq. (1.7) into Eq. (1.11) gives

$$w = 2\pi \frac{c}{n\lambda} = \frac{c}{n} k \tag{1.12}$$

For propagation in the z-direction, the vector dot product of the wave vector $\mathbf{k}$ and position vector $\mathbf{r}$ becomes

$$\mathbf{k} \cdot \mathbf{r} = kz \tag{1.13}$$

With substitution of Eqs. (1.12) and (1.13) into Eq. (1.5), the electric field amplitude $E(\mathbf{r}, t)$ of the gamma or X-ray [or probability density amplitude $E(\mathbf{r}, t)$ for a neutron] wave inside the material medium becomes

$$E(\mathbf{r}, t) = E_0 \exp[-\mathrm{i}(wt - \mathbf{k} \cdot \mathbf{r})] = E_0 \exp\left[-\mathrm{i}k\left(\frac{c}{n}t - z\right)\right] \tag{1.14}$$

One plugs into Eq. (1.14) the expression for the refractive index n of Eq. (1.4) and uses the binomial approximation, via

$$-\delta + \mathrm{i}\beta \ll 1 \tag{1.15}$$

The electric field amplitude $E(\mathbf{r}, t)$ of Eq. (1.14) becomes

$$E(\mathbf{r},t) = E_0 \exp\left[-\mathrm{i}k\left(\frac{c}{n}t - z\right)\right] \cong E_0 \exp\left[-\mathrm{i}w\left(t - \frac{z}{c}\right)\right] \exp\left(-\mathrm{i}z\frac{2\pi}{\lambda}\delta\right) \exp\left(-z\frac{2\pi}{\lambda}\beta\right) \tag{1.16}$$

The argument of the attenuation exponential of Eq. (1.16) suggests the linear attenuation coefficient μ [cm^{-1}], which has units of inverse length and is expressed in terms of dimensionless attenuation component β, and gamma (X-ray or neutron) wavelength λ, where

$$\mu = 4\pi\frac{\beta}{\lambda} \tag{1.17}$$

From section 3 of Chapter 6 in AIEP volume 172, Cremer (2012b), the linear attenuation coefficient μ [cm^{-1}] is the product of the number density ρ [cm^{-3}] of scatter centers (atoms) and the cross section for attenuation σ_{at} [cm^2], which is the sum of the cross sections for absorption σ_{abs} [cm^2], and incoherent scatter σ_{inc} [cm^2] (both elastic and inelastic), where

$$\mu = \rho\sigma_{at} = \rho(\sigma_{abs} + \sigma_{inc}) \tag{1.18}$$

Using Eq. (1.17), the electric field amplitude $E(\mathbf{r}, t)$ of Eq. (1.16) becomes

$$E(\mathbf{r},t) = E_0 \exp\left[-ik\left(\frac{c}{n}t - z\right)\right] \cong E_0 \exp\left[-iw\left(t - \frac{z}{c}\right)\right]\exp\left(-iz\frac{2\pi}{\lambda}\delta\right)\exp\left(-\frac{z\mu}{2}\right) \tag{1.19}$$

The intensity $I(\mathbf{r}, t)$ is the squared magnitude of the wave amplitude $E(\mathbf{r}, t)$. The linear attenuation coefficient μ is the wave propagation distance z in the material medium, which causes a $1/e$ attenuation of the wave intensity $I(\mathbf{r}, t)$, as compared to its incident intensity I_0, where

$$I(\mathbf{r}, t) = |E(\mathbf{r}, t)|^2 = I_0 \exp(-z\mu) \tag{1.20}$$

The intensity magnitude I_0 is the square of the magnitude of the electric field amplitude E_0, where

$$I_0 = |E_0|^2 \tag{1.21}$$

From section 2 of Chapter 6 of AIEP volume 172, Cremer (2012b), the significance of the ratio of the decrement δ divided by the linear attenuation μ, that is, δ/μ is

$$\frac{\delta}{\mu} = \frac{\Delta\lambda/\Delta z}{\Delta E/\Delta z} = \frac{\Delta\lambda}{\Delta E} \tag{1.22}$$

That is, the decrement δ is the incremental phase shift in wavelengths $\Delta\lambda$ per incremental propagation distance Δz, and linear attenuation μ is the incremental decrease in wave amplitude ΔE per incremental propagation distance Δz. Hence, δ/μ is the incremental phase shift in wavelengths $\Delta\lambda$ per incremental decrease in amplitude ΔE.

Recall from Eq. (1.462) of section 22 in Chapter 2 of AIEP volume 173, Cremer (2012c), that the complex scatter length b [fm] has units of length and is given by

$$b = b_{\mathrm{r}} - \mathrm{i}b_{\mathrm{i}} \tag{1.23}$$

Recall from Eq. (1.518) of section 24 in Chapter 2 of AIEP volume 173, Cremer (2012c), that the complex decrement is δ for neutron scatter from a single isotope Z_A of atomic number Z and atomic weight A. The complex decrement δ is proportional to the complex scatter length b, the number density ρ of the scatter centers (nuclei number density), and the square of the incident wavelength λ of the incident neutron plane waves, where

$$\delta_{\mathrm{r}} + \mathrm{i}\delta_{\mathrm{i}} = \frac{\lambda^2}{2\pi}\rho b = \frac{\lambda^2}{2\pi}\rho(b_{\mathrm{r}} - \mathrm{i}b_{\mathrm{i}}) \tag{1.24}$$

The complex index of refraction n is then

$$n = 1 - \delta_{\mathrm{r}} - \mathrm{i}\delta_{\mathrm{i}} = 1 - \frac{\lambda^2}{2\pi}\rho(b_{\mathrm{r}} - \mathrm{i}b_{\mathrm{i}}) \tag{1.25}$$

The decrement δ for X-rays, gammas, and neutrons is small compared to that of visible light. As a result, one requires the use of a stack of N lenses; that is, a compound refractive lens (CRL) to provide adequate refraction. A CRL is needed to provide focal lengths f within the range of a few centimeters to a few meters or more, depending on the wavelength λ and refractive decrement δ. Figure 1.9(a) shows a series of N individual thin unit lenses, which share a common optical axis and are used to form a compound refractive X-ray lens.

Each lens has focal length f_0, which is proportional to the lens radius of curvature R or radius for spherical lenses. For parabolic lenses, the single-lens focal length f_0 is proportional to the on-axis radius of curvature R, which is the radius of curvature at the vertex of the paraboloid lens surface, where

$$f_0 = \frac{R}{2\delta} \tag{1.26}$$

Incident plane waves of X-rays, gammas, visible light, or neutrons are focused by the CRLs to a focal point on the optical axis. The closely spaced series of N biconcave unit lenses combine to produce the focal length f of the CRL, where

$$f = \frac{f_0}{N} = \frac{R}{2N\delta} \tag{1.27}$$

The total focal length of the CRL is shorter than a single constituent lens by a factor of $1/N$. Depending on the wavelength λ of the incident plane wave, the focal length f_0 of a single unit lens could be extremely long (e.g., 100 m), but using 100 of such unit lenses would result in a CRL focal length f of only 1 m. Thus, the use of a stack of N lenses makes the focusing, collecting, and imaging of objects practical with X-rays, neutrons, or gammas, which have very small refractive decrements. See Figs. 1.9 and 1.10.

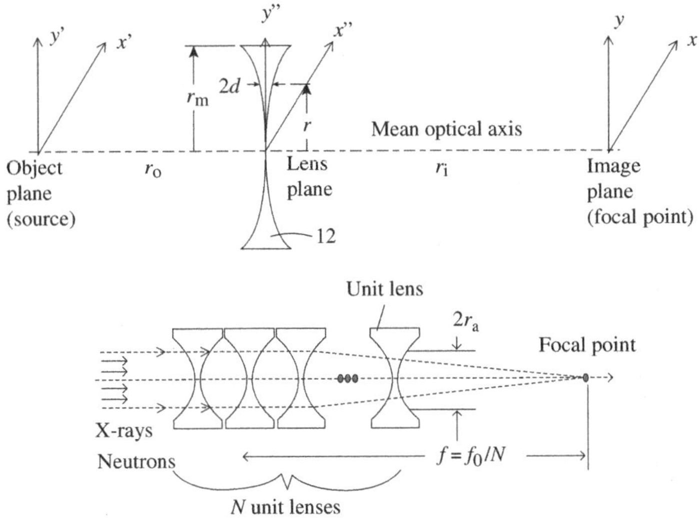

Figure 1.9 These biconcave lenses focus not only on X-rays but, depending on the lens refractive index, also focus neutrons and gammas. (a) The CRL is a linear array of N biconcave unit lenses, which reduce the overall focal length to $f = f_0/N$, where f_0 is the focal length of a unit lens. (b) A unit lens has a mechanical aperture $2R$, minimum thickness d, and limiting absorption aperture diameter $2r_a$. The increasing thickness of the biconcave lens, with increasing radial distance r from the lens center axis, then causes increasing attenuation. Hence, absorption of X-rays, neutrons, and gammas limit the effective lens aperture. At the absorption aperture radius, the transmitted intensity is $1/e^2$ of the transmitted intensity along the center lens (optical) axis (Piestrup et al., 2001).

The de Broglie wave-particle duality assumes that the neutron particle momentum p and rest mass m are related to the neutron plane wave wavelength λ by Planck's constant h, where

$$p = mv = \frac{h}{\lambda} \tag{1.28}$$

The nonrelativistic energy E of the neutron is proportional to its rest mass m and the square of its velocity v. By the de Broglie wave-particle duality, the neutron energy E is inversely proportional to its wavelength λ, where

$$E = \frac{h^2}{2m\lambda^2} = \frac{mv^2}{2} \tag{1.29}$$

The wavelength λ is inversely proportional to the square root of the neutron energy E, where

$$\lambda\left[\mathring{A}\right] = \frac{9.045}{\sqrt{E\,[\text{meV}]}} \tag{1.30}$$

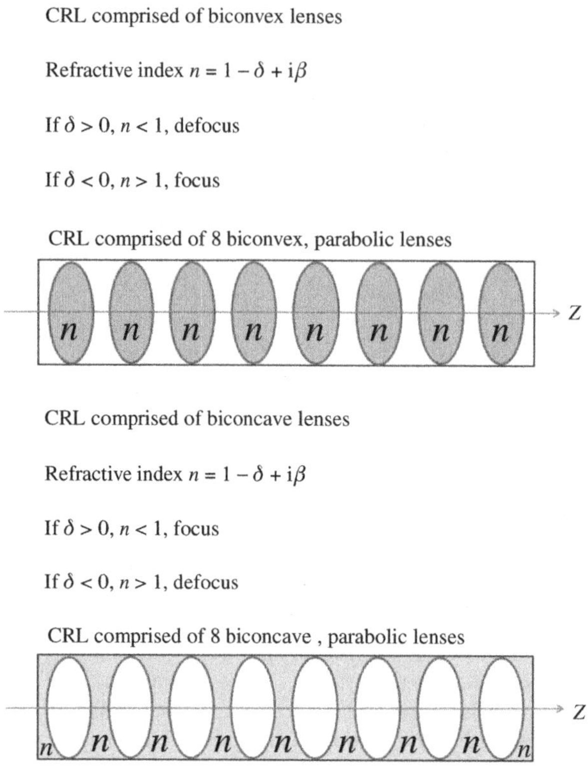

Figure 1.10 For X-rays and gammas in the X-ray energy range, all isotopes (lens materials) have $n < 1$, which require biconcave lenses to focus. For neutrons and some gamma energies, most isotopes (lens materials) have $n < 1$, which require biconcave lenses to focus. However, for neutrons and some gamma energies, some isotopes (lens materials) have $n > 1$ and require biconvex lenses to focus.

The neutron energy E is related to the squared neutron particle velocity v^2 and the wavelength λ of the neutron de Broglie plane wave:

$$E\,[\text{meV}] = 5.227 \cdot v^2\,[\text{km}/s] \tag{1.31}$$

The neutron particle velocity v equals the group velocity v_g of the equivalent neutron de Broglie plane wave, where

$$v_g = v\,\left[\text{km}/s\right] = \frac{3.956}{\lambda\,[\mathring{A}]} \tag{1.32}$$

Here, meV stands for millielectron volts. Recall that in Eq. (1.24) the decrement $\lambda\,[\mathring{A}]$ of a lens material is proportional to the square of the neutron wavelength λ, where

$$\delta(\lambda) \propto \lambda^2 \tag{1.33}$$

In the limit of low neutron energy $E \to 0$, the only neutron–nucleus interactions are elastic neutron scattering and exothermic neutron reactions (e.g., thermal neutron absorption with release of energy and gammas). The product of the cross section for an exothermic reaction σ_{ex} and the flux of neutrons incident on a nucleus Φ equals the reaction rate per nucleus r_n, or

$$\Phi\sigma_{\text{ex}} = r_n \tag{1.34}$$

However, the incident neutron flux Φ is the product of the neutron number density n_0 in the incident beam and the neutron velocity v relative to the target nucleus, or

$$\Phi = n_0 v \tag{1.35}$$

Also, the reaction rate per nucleus r_n is proportional to the incident neutron number density ρ via proportionality constant C, or

$$r_n = Cn_0 \tag{1.36}$$

Hence, the cross section for the exothermic reaction, such as thermal neutron absorption cross section σ_{abs}, is inversely proportional to the neutron velocity v, where

$$\sigma_{\text{abs}} = \sigma_{\text{ex}} = \frac{C}{v} \tag{1.37}$$

The neutron linear attenuation coefficient μ is proportional to the product of the absorption cross section σ_{abs} and nuclei density ρ, where

$$\mu\ \left[\text{cm}^{-1}\right] = \rho\ \left[\text{nuclei/cm}^3\right] \cdot \sigma_{\text{abs}}\ [\text{cm}^2/\text{nuclei}] \propto \frac{1}{v} \tag{1.38}$$

By the de Broglie relation of Eq. (1.28), the neutron velocity is inversely proportional to its wavelength:

$$v \propto \frac{1}{\lambda} \tag{1.39}$$

The linear attenuation μ for neutron absorption is proportional to the neutron wavelength λ, where $\mu \propto \lambda$.

In review, the coherent elastic cross section σ^e_{coh} of the CRL material for neutrons or X-rays gives rise to refraction and the decrement δ. For neutrons, the absorption cross section $\sigma_{\text{abs}}(\lambda)$ is primarily due to thermal neutron capture by the nuclei, which comprise the lens material, and the absorption cross section increases linearly with neutron wavelength λ, where $\sigma_{\text{abs}}(\lambda) \propto \lambda$. For X-rays, the attenuation is primarily due to photoelectric absorption, and secondarily due to incoherent, inelastic Compton scatter. In addition to the absorption cross section, neutron attenuation includes contributions from inelastic and incoherent scatter. The neutron wavelength-dependent attenuating scatter cross section includes (1) the coherent inelastic scatter $\sigma^i_{\text{coh}}(\lambda)$ due to collective or coherent thermal vibrations of lattice atoms in the lens material that result in phonon exchange between the incident neutrons and lens scatter centers; (2) the incoherent, elastic scatter $\sigma^{e}_{inc}(\lambda)$, where the lens scatter centers have no fixed phase relationships, but no energy exchange occurs between lens and neutrons; and (3) incoherent, inelastic scatter $\sigma^{i}_{inc}(\lambda)$, where the lens scatter centers have no fixed phase relation, and energy is exchanged between neutrons and lens scatter centers, via phonons. The diffraction cross section $\sigma_{dif}(\lambda)$ for unwanted coherent, elastic scatter is caused by neutron diffraction from the poly- or single crystals, which comprise the CRLs, and acts as another source of neutron attenuation. The total attenuation cross section $\sigma_{att}(\lambda)$ of the ith isotope nuclei that comprise the lens material is then

$$\sigma_{att}(\lambda) = \sigma_{abs}(\lambda) + \sigma^{i}_{coh}(\lambda) + \sigma^{e}_{inc}(\lambda) + \sigma^{i}_{inc}(\lambda) + \sigma_{dif}(\lambda) \tag{1.40}$$

Since the neutron or X-ray cross sections can vary widely for the same material, depending on the material structure and atom density nonuniformity, one should measure the candidate lens material scatter and absorption on a small-angle scatter beam line for neutrons or X-rays. For details of neutron attenuation cross sections for neutron monochromators and filters, see Freund (1983).

For thermal and cold neutrons at other wavelengths $\lambda\,[\mathring{A}]$, one can express the linear attenuation coefficient for neutron absorption $\mu(\lambda)$ in a material, as a function of wavelength λ, and the material's linear attenuation coefficient μ_{th}, evaluated for $\lambda = 1.8\,\mathring{A}$ neutrons. The linear attenuation for a single element isotope is

$$\mu\left[\mathrm{cm}^{-1}\right] = \frac{\mu_{th}\,[\mathrm{cm}^{-1}]}{1.8\,\mathring{A}} \cdot \lambda\,[\mathring{A}] \tag{1.41}$$

The decrement δ for the single element isotope is calculated from the isotope scatter length b and nuclei number density ρ, where

$$\delta = 1.59 \times 10^{-30}\,[\mathrm{cm}^3/\mathring{A}^2\,\mathrm{fm}] \cdot \rho\,[\mathrm{nuclei/cm}^3] \cdot b\,[\mathrm{fm}] \cdot \lambda^2\,[\mathring{A}]. \tag{1.42}$$

From Eqs. (1.41) and (1.42), the ratio δ/μ is thus proportional to the neutron wave length λ, or

$$\frac{\delta}{\mu} \propto \lambda b \tag{1.43}$$

The ratio δ/μ is proportional to the product of the neutron wavelength $\lambda\,[\mathring{A}]$ and scattering length $b\,[\mathrm{fm}]$. However the ratio δ/μ is inversely proportional to the thermal neutron absorption cross section σ_{abs} [barn] as a function of neutron wavelength λ. Hence,

$$\frac{\delta}{\mu}[\mathrm{cm}] = 2.86 \times 10^{-6}\,[\mathrm{barn\ cm}/\mathring{A}\,\mathrm{fm}] \cdot \lambda\,[\mathring{A}] \cdot \frac{b\,[\mathrm{fm}]}{\sigma_{abs}\,[\mathrm{barn}]} \tag{1.44}$$

In Table 1.1, the decrement δ, the linear attenuation coefficient μ, and the ratio b/σ_{abs} are evaluated at room temperature, where the thermal neutron

Table 1.1 Listed are Neutron Bound Scattering Lengths and Absorption Cross Sections for Some Specimen- and CRL-Relevant Elements in Their Natural States

Element in Natural State		Bound Coherent Scatter Length, *b* (fm)	Bound Coherent Scatter Cross at 1.8 Å σ_{coh} (barn)	Bound Incoherent Scatter Cross at 1.8 Å σ_{incoh} (barn)	Thermal Neutron Absorption Cross at 1.8 Å σ_{abs} (barn)	b/σ_{abs} (fm/barn)	Mass Density (g/cm^3)	Decrement at 1.8 Å δ_{th}	Linear Attenuate at 1.8 Å μ_{th} (cm^{-1})
Hydrogen	H	−3.739	1.7568	80.26	0.3326	11			
Deuterium	D	6.671	5.592	2.05	5.19×10^{-4}	12,854			
Lithium	Li	−1.9	0.454	0.92	70.5	0.0270	0.5300	-4.50×10^{-7}	3.24
Boron	B	0.213i	3.54	1.7	767	0.0003	2.3400		99.9
Beryllium	Be	7.79	7.63	0.0018	0.0076	1025	1.848	4.96×10^{-6}	9.38×10^{-4}
Carbon	C	6.646	5.551	0.001	0.0035	1899	2.260	3.88×10^{-6}	3.96×10^{-4}
Nitrogen	N	9.36	11.01	0.5	1.91	5			
Oxygen	O	5.803	4.232	8.00×10^{-4}	1.90×10^{-4}	30,542			
Fluorine	F	5.654	4.017	8.00×10^{-4}	9.60×10^{-3}	589			
Magnesium	Mg	5.375	3.631	0.08	0.063	85	1.738	1.19×10^{-6}	2.71×10^{-3}
MgF_2		5.55	3.866	0.032	0.082	68	3.180	2.87×10^{-6}	2.71×10^{-3}
Aluminum	Al	3.449	1.495	0.0082	0.231	15	2.700	1.07×10^{-6}	1.39×10^{-2}
Titanium	Ti	−3.438	1.485	2.87	6.09	1	4.506	-1.00×10^{-6}	3.45×10^{-1}
Manganese	Mn	−3.73	1.75	0.4	13.3	0.3	7.300	-1.54×10^{-6}	1.06
Iron	Fe	9.45	11.22	0.4	2.56	4	7.870	4.13×10^{-6}	2.17×10^{-1}
Nickel	Ni	10.3	13.3	5.2	4.49	2	8.908	4.85×10^{-6}	4.10×10^{-1}
Copper	Cu	7.718	7.485	0.55	3.78	2	8.940	3.37×10^{-6}	3.20×10^{-1}
Cadmium	Cd	4.87−0.70i	3.04	3.46	2520	0.0020	8.650		1.17×10^{2}
Gadolinium	Gd	6.5−13.82i	29.3	151	49,700	0.0003	7.900		1.50×10^{3}
Bismuth	Bi	8.532	9.148	0.0084	0.0338	252	9.790	1.24×10^{-6}	9.53×10^{-4}

Deuterium and naturally occurring beryllium, carbon, oxygen, and fluorine have large b/σ_{abs} ratios of refraction to attenuation for scatter length b divided by absorption cross section σ_{abs} at room-temperature wavelength $\lambda = 1.8$ Å. Neutron lenses, which are comprised of these elements, provide relatively strong refraction with minimal attenuation.

wavelength is $\lambda = 1.8\,\mathring{A}$, which corresponds to thermal neutron energy $E = 0.025$ eV, and thermal neutron velocity of $v = 2.2$ km/s at 293 K, respectively. Note that 1 barn $= 10^{-24}$ cm^2 and 1 fm $= 10^{-15}$ m.

An element or compound of mass density ρ_M [g/cm^3] consists of N isotopes, such that each ith isotope has weight fraction w_i, atomic weight A_i, and absorption cross section σ_i. The atomic weight A_i is the number of grams per mole of atoms A_i [gram/mole] of the element isotope and Avogadro's number N is the number of atoms per mole, or $N = 6.02 \times 10^{23}$ [atom/mole]. The wavelength-dependent, dimensionless, decrement $\delta[\lambda]$ for a mixture is

$$\begin{aligned}\delta(\lambda) &= 9.57 \times 10^{-7}\,[\text{cm}^3/\mathring{A}^2\text{ fm mole}] \\ &\times \sum_{i=1}^{N} w_i \cdot \rho_M[\text{g/cm}^3] \cdot b\,[\text{fm}] \cdot \lambda^2\,[\mathring{A}] \cdot \frac{1}{A_i\,[\text{g/mole}]}\end{aligned} \tag{1.45}$$

The linear attenuation coefficient $\mu(\lambda)$ is proportional to wavelength $\mu(\lambda) \propto \lambda$, via the thermal neutron absorption cross section $\sigma(\lambda)$, which is proportional to wavelength $\sigma(\lambda) \propto \lambda$. The wavelength-dependent linear neutron attenuation μ for a mixture is

$$\mu\,[\text{cm}^{-1}] = 0.602 \sum_{i=1}^{N} w_i \cdot \sigma_i\,[\text{barn}] \cdot \rho_M\,[\text{g/cm}^3] \cdot \frac{1}{A_i\,[\text{g/mole}]} \tag{1.46}$$

Table 1.1 lists neutron bound scattering lengths b and absorption cross sections σ_{abs} for many common elements in their natural occurring state. These elements are candidate lens materials, and also comprise the chemical composition of many of the potential specimens to be imaged. Each isotope of atomic weight A of an element of atomic number Z has its own set of bound scattering lengths b and cross sections σ_{abs}. Thus, these tabulated numbers in Table 1.1 are the weighted averages of all the natural occurring elements' constituent stable isotopes.

In Figure 1.11, the magnitude of the refraction-to-attenuation ratio $|\delta/\mu|$, or "delta over mu ratio" magnitude measured in centimeters, is plotted as a function of neutron energy for various materials. For neutrons, the magnitude of the refraction-to-attenuation ratio $|\delta/\mu|$ in all materials increases linearly with wavelength λ because of (1) the increase of dimensionless δ with increasing λ^2 and (2) the increase of linear attenuation μ [cm^{-1}] with increasing λ. Most materials have a positive decrement $\delta > 0$ with a refractive index $n = 1 - \delta < 1$ and thus require spherical or parabolic biconcave lenses for focusing and imaging. The exceptions are titanium, vanadium, and manganese, which have a negative decrement $\delta < 0$, with a refractive index $n = 1 - \delta > 1$, and thus require spherical or parabolic biconvex lenses for focusing and imaging. Of the materials plotted, titanium, vanadium, and manganese have the lowest $|\delta/\mu|$ and beryllium, carbon, and aluminum have the highest $|\delta/\mu|$, where beryllium, carbon, and aluminum are also good lens materials for X-rays.

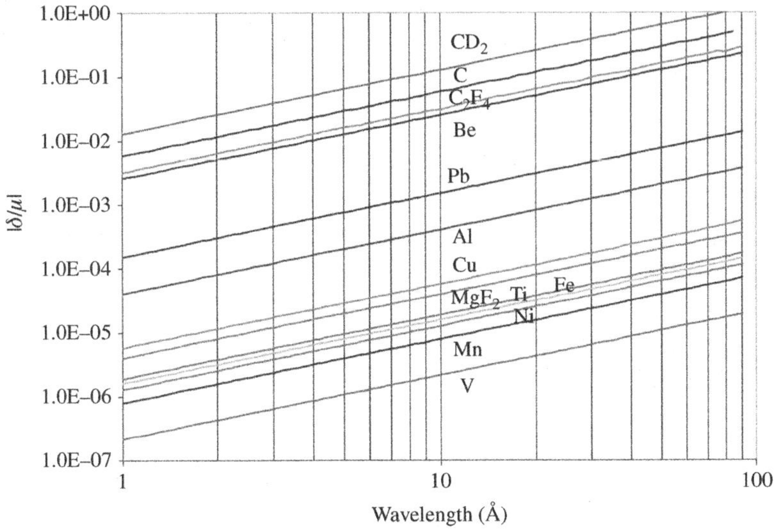

Figure 1.11 The magnitude of the refraction-to-attenuation ratio $|\delta/\mu|$, measured in centimeters, is plotted as a function of neutron energy for various materials, where most materials have a positive decrement $\delta > 0$ with a refractive index less than 1, where $n = 1 - \delta < 1$. The exceptions include titanium, vanadium, and manganese, which have a negative decrement $\delta < 0$ with a refractive index greater than 1, where $n = 1 - \delta > 1$. For neutrons, $|\delta/\mu|$ in all materials increases linearly with wavelength λ because of (1) the increase of δ with increasing λ^2 and (2) the increase of linear attenuation μ with increasing λ.

1.2 CRLs—Thin-Lens Approximation: Focal Length, Ray Path Lengths, and Attenuation

In the thin-lens approximation, the ray radial position is approximated as constant throughout the ray trajectory through the lens; however, the refraction by the lens has changed the ray angle, relative to the ray incident direction by angle $\Delta\theta$ at the input air/lens interface, and by another equal $\Delta\theta$ at the output lens/air interface, which results in a total angle change of $2\Delta\theta$. For the case of a biconcave lens, see the schematic in Figure 1.12, and for the case of a biconvex lens, see the schematic in Figure 1.13.

The X-ray or neutron ray is refracted at a lens surface by a very small angle $\Delta\theta$ toward the lens center axis, from incident angle θ_1 to refracted angle θ_2 with respect to the biconcave or biconvex, parabolic or spherical lens surface normal at radius r, where

$$\theta_2 = \theta_1 - \Delta\theta \quad \Delta\theta \ll 1 \tag{1.47}$$

On the lens left surface, the incident angle θ_1 is measured with respect to the left surface normal (pointing out of the lens), and refracted angle $(\theta_1 - \Delta\theta)$ is measured with respect to the left surface normal (pointing into lens). Snell's law applied to a ray at the left lens surface is then

$$\sin\theta_1 = (1 - \delta)\sin(\theta_1 - \Delta\theta) \tag{1.48}$$

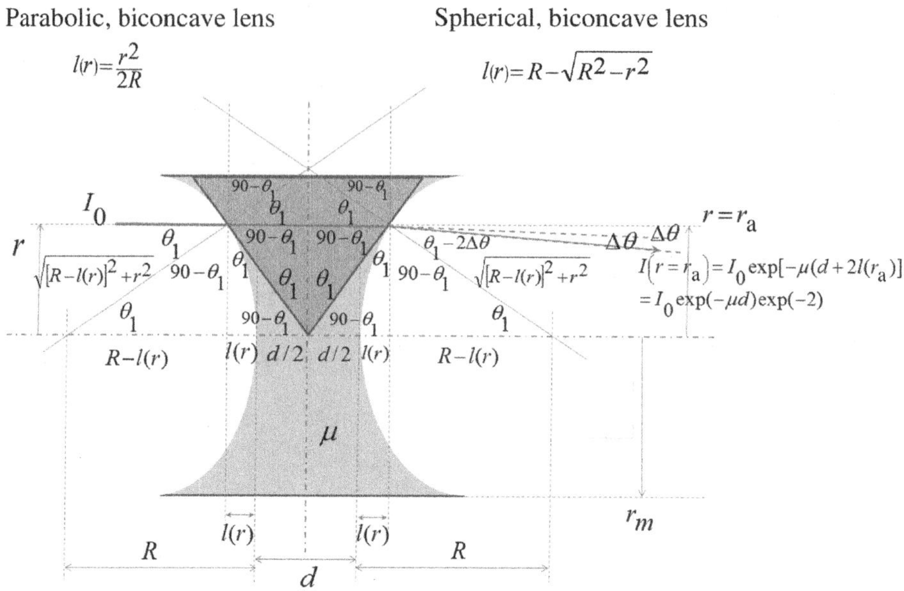

Figure 1.12 A diagrammatic representation refraction of a plane wave (X-rays, neutrons, and gammas) by a biconcave, spherical, or parabolic lens, with superimposed inverted prism to emphasis Snell's law at the left (input) and right (output) surfaces of the single lens. The thin-lens approximation is applied, where the ray is modeled as having a constant radial distance r from the lens center axis, but changes direction by $2\Delta\theta$, due to the refractive bending of the ray by $\Delta\theta$ at the input and output lens surfaces. A ray, which propagates along the absorption aperture radius r_a, has intensity $I(r_a)$, which is smaller by a factor of e^2 than the transmitted ray intensity $I(0)$ along the lens center axis. The thickness of the lens is the sum of (1) the fixed, minimum, center thickness d, and (2) the two radial-dependent thicknesses $l(r)$ of the left and right portions of the lens. Here R is the radius of curvature for a spherical lens and the on-axis radius of curvature for a parabolic lens, and r_m is the physical radius of the lens.

The sine trigonometry identity is applied to the right side of Eq. (1.48), followed by application of the small angle approximation with $\Delta\theta \ll 1$, and then the second order term $\delta\Delta\theta \cos\theta_1$, is neglected, which yields

$$(1-\delta)\sin(\theta_1-\Delta\theta) \cong \sin\theta_1 - \Delta\theta\cos\theta_1 - \delta\sin\theta_1 \tag{1.49}$$

With substitution of Eq. (1.49) into Eq. (1.48), one obtains the ray angle change $\Delta\theta$ at the lens left surface, where

$$\Delta\theta = -\delta\tan\theta_1 \tag{1.50}$$

On the lens right surface, the incident angle $\theta_1 - \Delta\theta$ is measured with respect to the right surface normal (pointing into the lens), and refracted angle $[(\theta_1-\Delta\theta)+\Delta\theta]$ is measured with respect to the right surface normal (pointing out of lens). Snell's law applied to a ray at the right lens surface is then

$$(1-\delta)\sin(\theta_1-\Delta\theta) = \sin[(\theta_1-\Delta\theta)+\Delta\theta] \tag{1.51}$$

The sine trigonometry identity is applied to both sides of Eq. (1.51), followed by application of the small angle approximation with $\Delta\theta \ll 1$, and then the second order term $\delta\Delta\theta \cos\theta_1$, is neglected, which yields

$$\sin\theta_1 - \Delta\theta\cos\theta_1 - \delta\sin\theta_1 = \sin\theta_1 \tag{1.52}$$

Solving for $\Delta\theta$ in Eq. (1.52) for the right lens surface gives

$$\Delta\theta = -\delta\tan\theta_1 \tag{1.53}$$

The left and right surfaces of the biconcave or biconvex lens cause ray angular deviation $\Delta\theta$ toward the center lens axis. Hence, a single biconcave or biconvex lens, with a small refractive decrement δ, refracts an incident ray by deflection angle $2\Delta\theta$, which is measured relative to the incident ray direction, where

$$2\Delta\theta = -2\delta\tan\theta_1 \quad \delta \ll 1 \tag{1.54}$$

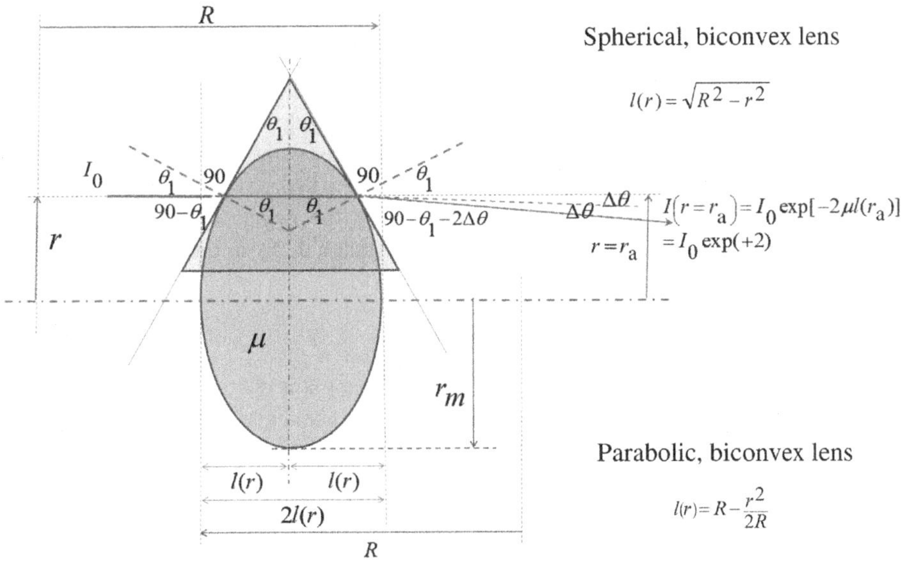

Figure 1.13 The diagrammatic representation refraction of a plane wave (neutron or gamma) by a biconvex, spherical or parabolic lens, with superimposed upright prism to emphasis Snell's law at the left (input) and right (output) surfaces of the single lens. The thin-lens approximation is applied, where the ray is modeled as having a constant radial distance r from the lens center axis, but changes direction by $2\Delta\theta$, due to the refractive bending of the ray by $\Delta\theta$ at the input and output lens surfaces. A ray, which propagates along the absorption aperture radius r_a, has intensity $I(r_a)$, which is larger by a factor of e^2 than the transmitted ray intensity $I(0)$ along the lens center axis. The thickness of the lens is the sum of (1) the fixed, minimum center thickness d, and (2) the two radial-dependent thicknesses $l(r)$ of the left and right portions of the lens. Here, R is the radius of curvature for a spherical lens and the on-axis radius of curvature for a parabolic lens, and r_m is the physical radius of the lens.

If the change in the incident ray direction by angle $2\Delta\theta$ is toward the center optical or z-axis, then the ray propagates diagonally from the off-axis, radial position r at the center, transverse plane of the lens at $z=0$, and intercepts the z-axis at downstream, axial distance $z=f_1$. The angle of the ray, which is measured relative to the lens center axis (z-axis), and is assumed to be sufficiently small for the application of the small-angle approximation, where

$$2\Delta\theta = \tan^{-1}\left(\frac{r}{f_1}\right) \cong \frac{r}{f_1} \tag{1.55}$$

Transposing, one finds a parallel ray, which is incident at the lens at radial distance r, and is refracted, and intersects the center lens or z-axis at downstream focal distance f_1, where

$$f_1 = \frac{r}{2\Delta\theta} \tag{1.56}$$

For an N-lens CRL, the focal distance f_{crl} is then

$$f_{\text{crl}} = \frac{r}{2N\Delta\theta} \tag{1.57}$$

The product of the slopes m_1 and m_2 of two perpendicular lines is equal to -1, where $m_1 m_2 = -1$. For a lens, the slope of the line, which is tangent to a point L on the lens surface, is $\mathrm{d}r/\mathrm{d}\tau$, where $\mathrm{d}r/\mathrm{d}\tau = \cot\theta_1$. Hence, the slope of the perpendicular, surface normal line at point L on the lens is $-\mathrm{d}\tau/\mathrm{d}r$, where $-\mathrm{d}\tau/\mathrm{dr} = \tan\theta_1$. The tangent of incident angle θ_1 at an interface is the negative differential rate of lens thickness change $\mathrm{d}\tau$ with change in radial distance $\mathrm{d}r$ from the lens center axis:

$$\tan\theta_1 = -\frac{\mathrm{d}\tau}{\mathrm{d}r} \tag{1.58}$$

With substitution of Eq. (1.58) into Eq. (1.54), the angular change $2\Delta\theta$ of the X-ray or neutron wave vector $\mathbf{k}$ by refraction with a single biconcave or biconvex lens is

$$2\Delta\theta = 2\delta\frac{\mathrm{d}\tau}{\mathrm{d}r} \tag{1.59}$$

The angular deviation of the output ray, compared to the input ray in a single biconcave or biconvex lens, thus depends on the product of the decrement δ and the negative differential change of the ray path length $\mathrm{d}\tau$ through the lens. The differential path length change $\mathrm{d}\tau$ of the ray, per differential change in the ray radial position $\mathrm{d}r$, depends on the radial distance r of the ray from the lens center axis ($r=0$). By substitution of Eq. (1.59) into Eq. (1.56), the focal length f_1 for a single biconcave or biconvex, parabolic, or spherical lens, which includes both lens surfaces, is then

$$f_1 = r\left[\delta\frac{\mathrm{d}\tau}{\mathrm{d}r}\right]^{-1} \tag{1.60}$$

The ray deflection angle $2\Delta\theta$ of a refracted-ray single lens is the ratio of the radial distance r on the incident ray from the center optical axis, divided by the lens length f, where

$$\tan 2\Delta\theta \cong 2\Delta\theta = \frac{r}{f} \tag{1.61}$$

If rays incident at all radial distances r from the center z-axis of the lens are to be focused at a single point on the z-axis at the downstream focal distance f from the center transverse plane of the thin lens, then ray deflection angle $2\Delta\theta$ must be proportional to the ray's radial distance r from the center optical axis. The proportionality constant is the reciprocal of the focal length $1/f$, where substitution into Eq. (1.61) from Eqs. (1.54) and (1.58) yields

$$\text{constant} = f = \frac{r}{2\Delta\theta} = -\frac{r}{2\delta \tan \theta_1} = \frac{\mathrm{d}r}{\mathrm{d}\tau}\frac{r}{2} \tag{1.62}$$

Transposing, one obtains the relation between the differential lens thickness $\mathrm{d}\tau$ and the differential radial distance $\mathrm{d}r$, where radial distance r varies but the focal distance f is fixed:

$$\mathrm{d}\tau = \frac{r}{2f}\mathrm{d}r \tag{1.63}$$

Integration of Eq. (1.63), demonstrates the thickness τ of the lens as the square r^2 of the radial distance r from the center optical axis, which requires a parabolic lens that is either biconcave for positive decrement $\delta > 0$ (and $n < 1$), or biconvex for negative decrement $\delta > 0$ (and $n > 1$), where

$$\tau(r) = \frac{r^2}{f} \tag{1.64}$$

The expression for the phase shift $\phi(r)$ of a photon wave through a refractive lens, as a function of the lens thickness $\tau(r)$, is

$$\phi(r) = 2\pi\frac{\tau(r)}{\lambda}\delta \tag{1.65}$$

With substitution of Eq. (1.64) for parabolic lens thickness $\tau(r)$ into Eq. (1.65), the phase shift $\phi(r)$ of a photon wave through a single parabolic, biconcave, or spherical lens is then

$$\phi(r) = 2\pi\frac{r^2}{f\lambda}\delta \tag{1.66}$$

We now consider the relation between the ray deflection angle $\Delta\theta$ and ray phase shift $\phi(r)$ in material, electric, gravity, or magnetic lenses of arbitrary shape. The ray deflection angle $\Delta\theta$ upon refraction, which can be related to the ray phase shift angle $\phi(r)$ as a function of distance r from the center lens axis, is obtained from the differential expression

$$\mathrm{d}\phi(r) = 2\pi \frac{\mathrm{d}\tau(r)}{\lambda}\delta \tag{1.67}$$

Equation (1.59), which relates the assumed small-ray deflection angle $2\Delta\theta$ from the two surfaces of a single biconcave or biconvex, parabolic, or spherical lens, applies to the surface of the lens of arbitrary shape, where

$$\delta = \frac{\mathrm{d}r}{\mathrm{d}\tau}\Delta\theta \tag{1.68}$$

Recall that the derivative $\mathrm{d}\tau/dr$ is the change of lens thickness per change of the ray distance r from the lens center axis. Substitution of Eq. (1.68) into Eq. (1.67) yields

$$\mathrm{d}\phi(r) = 2\pi \frac{\Delta\theta(r)}{\lambda}\mathrm{d}r \tag{1.69}$$

With the integration of Eq. (1.69) and assuming that the initial phase shift is zero, where $\phi(r=0)=0$, the phase shift angle $\phi(r)$ becomes

$$\phi(r) = \frac{2\pi}{\lambda}\int_0^r \Delta\theta(r)\mathrm{d}r \tag{1.70}$$

For a photon, the assumed small deflection angle $\Delta\theta$ is given by the ratio of the radial phase velocity v_r, divided by the axial phase velocity v_z, where

$$\tan\ \Delta\theta \cong \Delta\theta = \frac{v_r}{v_z} \qquad \Delta\theta \ll 1 \tag{1.71}$$

The axial velocity v_z is the incident photon phase velocity v_{ph}, where, given refractive index n and speed of light c,

$$v_z = v_{\mathrm{ph}} = \frac{c}{n} \tag{1.72}$$

The small deflection angle allows one to approximate the photon radial velocity v_r, which is acquired by refraction at a lens surface, as the axial velocity v_z, and then multiply it by the deflection angle $\Delta\theta$, where

$$v_r \cong v_z \tan\ \Delta\theta \cong v_{\mathrm{ph}}\Delta\theta = \frac{c}{n}\Delta\theta \tag{1.73}$$

Plugging Eq. (1.59) into Eq. (1.73) gives the radial velocity v_r, which is acquired by a photon at a refracting surface, where

$$v_r \cong -\frac{c\delta}{n}\frac{\mathrm{d}\tau}{\mathrm{d}r} = -\frac{c\delta}{1-\delta}\frac{\mathrm{d}\tau}{\mathrm{d}r} \tag{1.74}$$

The differential change of lens thickness with lens radius $\mathrm{d}\tau/dr$ for parabolic lens of Eq. (1.64) is plugged into Eq. (1.74). The resulting photon radial velocity v_r, which is acquired upon refraction at a parabolic lens surface interface, is

$$v_r \cong -\frac{c}{n}\frac{\delta\, r}{2f} \tag{1.75}$$

where

$$\Delta\theta = \frac{v_r}{v_z} = -\frac{\delta\, r}{2f} \tag{1.76}$$

Now consider a nonrelativistic particle with nonzero rest mass m_0 and energy E, such as a neutron in a magnetic field, electron in a magnetic or electric field, or a charged or neutral particle in a gravity field. The axial velocity v_z of the particle in the field-free space is assumed to be constant, where

$$v_z = \sqrt{\frac{2E}{m_0}} \tag{1.77}$$

When the particle passes through the interface between the field-free space and the field, it experiences a force due to the spatial gradient of the field. Given the potential energy Φ, which is acquired by the particle on passing from the field-free region to the field region, the force $\mathbf{F}$ acting on the particle is

$$\mathbf{F} = m_0\frac{\mathrm{d}\mathbf{v}}{\mathrm{d}t} = -\nabla\Phi \tag{1.78}$$

Given unit radius vector $\hat{\mathbf{r}}$, the radial acceleration component $\mathrm{d}v_r/dt$ is

$$\frac{\mathrm{d}v_r}{\mathrm{d}t} = -\frac{1}{m_0}\hat{\mathbf{r}}\cdot\nabla\Phi \tag{1.79}$$

Assume that the particle has zero initial radial velocity, where $v_r\,(r=0)=0$ at initial time $t=0$, when the particle passes from the field-free region into the field region. The radial velocity v_r, which is acquired by the matter wave whose particle constituent has rest mass m_0, is then

$$v_r = -\frac{1}{m_0}\hat{\mathbf{r}}\cdot\int_0^t \nabla\Phi\, \mathrm{d}t \tag{1.80}$$

For a particle, the assumed small deflection angle $\Delta\theta$ is given by the ratio of the radial phase velocity v_r divided by the axial phase velocity v_z where

$$\tan \Delta\theta \cong \Delta\theta = \frac{v_r}{v_z} \quad \Delta\theta \ll 1 \tag{1.81}$$

With the substitution of Eq. (1.77) and Eq. (1.80) into Eq. (1.81), the deflection angle $\Delta\theta$, which is acquired by the particle at time t, after crossing into the field region from the field-free region at initial time $t = 0$ is then

$$\tan \Delta\theta \cong \Delta\theta = \frac{v_r}{v_z} = -\frac{1}{\sqrt{2Em_0}}\hat{\mathbf{r}}\cdot \int_0^t \nabla\Phi \, dt \quad \Delta\theta \ll 1 \tag{1.82}$$

Consider a magnetic field **B** and a charged particle (electron) or neutral particle (neutron, He-3) with magnetic dipole moment $\boldsymbol{\mu}$ and axial component $\mu_{\parallel}$, which is parallel or antiparallel with the magnetic field direction. The potential energy $\Phi(x, y, z)$, which is acquired by the particle with magnetic dipole moment $\boldsymbol{\mu}$ on its passage from the field-free region to the field region, is the vector dot product of $\boldsymbol{\mu}$ and $\mathbf{B}(x, y, z)$, where

$$\Phi = -\boldsymbol{\mu} \cdot \mathbf{B} = -\mu_{\parallel} B \tag{1.83}$$

The deflection angle $\Delta\theta$ of the particle with magnetic dipole moment $\boldsymbol{\mu}$ in magnetic field **B** at time t after crossing from the field-free region to the field region is

$$\tan \Delta\theta \cong \Delta\theta = \frac{v_r}{v_z} = -\frac{\mu_{\parallel}}{\sqrt{2Em_0}}\hat{\mathbf{r}}\cdot \int_0^t \nabla B \, dt \quad \Delta\theta \ll 1 \tag{1.84}$$

The thickness $\tau(r)$ of the biconcave or biconvex, spherical or parabolic lens, varies with the radial distance r from the center optical axis (z-axis). The thin lens approximation is assumed, where the radius r of a ray path through the lens is approximated as constant, but the ray changes direction. The incident plane wave amplitude E_0 is assumed constant over the input face of the lens. However, the transmitted wave amplitude at the lens output face, depends on the ray path radius r. At radius r at the lens output face, the transmitted wave amplitude $E(r, \lambda)$ at wavelength λ is

$$E(r, \lambda) = E_0 \exp[-\tau(r)\mu(\lambda)] \tag{1.85}$$

The transmitted wave intensity $I(r, \lambda)$ is the squared magnitude of the transmitted wave amplitude $E(r, \lambda)$, where

$$I(r, \lambda) = |E(r, \lambda)|^2 = |E_0|^2 \exp[-2\tau(r)\mu(\lambda)] = I_0 \exp[-2\tau(r)\mu(\lambda)] \tag{1.86}$$

The intensity transmission fraction $T(r, \lambda)$ depends on the wavelength λ, via the wavelength-dependent linear attenuation $\mu(\lambda)$, and the radial distance r, via the radial-dependent lens thickness $\tau(r)$. The intensity transmission fraction $T(r, \lambda)$ is the ratio of the transmitted intensity $I(r, \lambda)$ along a lens ray path at radial distance r from the lens center axis, divided by the incident intensity I_0. The radial-dependent transmission of X-rays, gammas, and neutrons through a CRL is represented by the transmission fraction $T(r, \lambda)$, given by

$$T(r, \lambda) = \frac{I(r, \lambda)}{I_0} = \exp[-2\tau(r)\mu(\lambda)] \tag{1.87}$$

The absorption aperture radius r_a for a biconvex, spherical, or parabolic lens is the radial distance $r = r_a$ from the lens center axis, where the transmitted intensity $T(r_a)$ increases by a factor of e^2, compared to the transmitted intensity $T(0)$ along the center axis. The ratio of the transmission fraction $T(r_a)$ at the absorption aperture radius $r = r_a$ divided by the transmission fraction $T(0)$ along the center axis of the biconvex lens at $r = 0$ is

$$\frac{T(r_a)}{T(0)} = \frac{\exp[-\mu\tau(r_a)]}{\exp[-\mu\tau(0)]} = \exp(+2) \quad \text{biconvex} \tag{1.88}$$

Conversely, the absorption aperture radius r_a from the lens center axis for a biconcave, spherical, or parabolic lens, is defined by the decrease of the wave intensity $T(r_a)$ by a factor of e^{-2} at lens radius $r = r_a$, which is measured relative to the maximum transmitted wave intensity $T(0)$ along the lens center axis at $r = 0$. The ratio of the transmission fraction $T(r_a)$ at the absorption aperture radius $r = r_a$, divided by the transmission fraction $T(0)$ along the center axis of the biconcave lens at $r = 0$ is

$$\frac{T(r_a)}{T(0)} = \frac{\exp[-\mu\tau(r_a)]}{\exp[-\mu\tau(0)]} = \exp(-2) \quad \text{biconcave} \tag{1.89}$$

Equating the arguments of the exponentials in Eq. (1.88) for the biconvex lens, yields

$$\tau(r_a) = \tau(0) - \frac{2}{\mu} \tag{1.90}$$

Equating the arguments of the exponentials in Eq. (1.89) for the biconcave lens, yields

$$\tau(r_a) = \tau(0) + \frac{2}{\mu} \tag{1.91}$$

The average transmitted intensity $\bar{I}$ through CRL, which is comprised of N biconvex or biconcave, spherical, or parabolic lenses, is the average of the squared magnitude of the wave amplitude transmission through the lens, where

$$\bar{I} = \frac{1}{r_m}\int_0^{r_m} \exp[-2N\tau(r)\mu(\lambda)]\mathrm{d}r \tag{1.92}$$

With the paraxial approximation, one places a limiting iris in front of the lens, so the ray paths are confined to travel through the center portion of the spherical biconvex or biconcave lens, whose surface is approximately parabolic. The thin-lens approximation can be applied to the X-ray, gamma, and neutron lenses, because the very small refractive decrement δ, and the radial distance r of a ray from the lens center axis, are both much less than the lens radius of curvature $r \ll R$. With the thin-lens approximation, the radial distance r of a ray from the lens center axis, is approximately constant along the ray path through the lens. At the lens exit, the ray is assumed to change direction by angle $2\Delta\theta$, according to Snell's law applied at the two lens surfaces. Figure 1.14 shows a spherical and parabolic biconvex lens superimposed, and Figure 1.15 shows a spherical and parabolic biconvex lens superimposed. The transverse or radial extent of the lenses is $-r_m \leq r \leq r_m$, where the biconvex or biconcave, parabolic lens has physical radius $r_m = R\sqrt{2}$, and the biconvex or biconcave, spherical lens has physical radius $r_m = R$.

Consider the superimposed biconvex lenses in Figure 1.14. As a function of radial distance r from the center lens axis, the ray path lengths through the biconvex, spherical, and parabolic lenses are

$$\tau(r) = \tau_{\mathrm{sph}}^{\mathrm{vex}}(r) = d + 2\sqrt{R^2 - r^2} \quad \text{biconvex, spherical} \tag{1.93}$$

$$\tau(r) = \tau_{\mathrm{par}}^{\mathrm{vex}}(r) = d + 2R - 2\left(\frac{r^2}{2R}\right) = d + 2R - \frac{r^2}{R} \quad \text{biconvex, parabolic} \tag{1.94}$$

The biconvex lenses have minimum thickness $\tau(r_m) = d$, both at the spherical lens perimeter $r_m = R$ and at the parabolic lens perimeter $r_m = R\sqrt{2}$. The minimum thickness $\tau(r_m) = d$ at the lens perimeter is approximated as zero, where

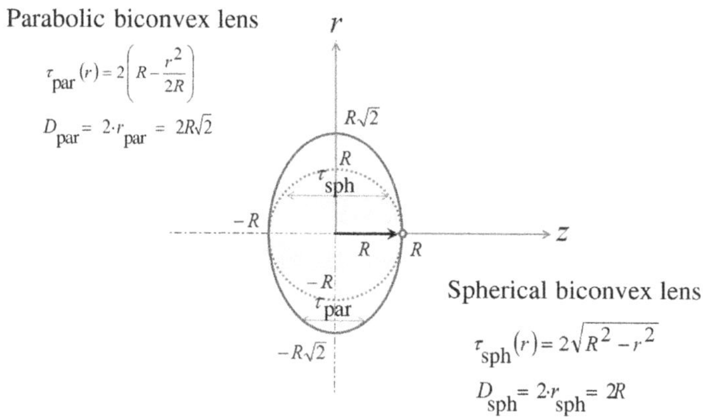

Figure 1.14 The radial-dependent thickness and physical aperture radius of a spherical and parabolic biconvex lens, which share the same on-axis radius of curvature R for the left and right lens surfaces, is compared. At the perimeter or outer edge of the biconvex lens, the lens minimum thickness is drawn as $d = 0$. However, the physical lens may require a very small minimum thickness d to provide a flat perimeter surface to aid in lens alignment and mounting.

$\tau(r_m) = d = 0$ in Figure 1.14. However, a fabricated biconvex lens is likely to have a small nonzero thickness $\tau(r_m) = d > 0$ at the lens perimeter to facilitate attachment and alignment and reduce the likelihood of lens fracture.

The biconvex, spherical, and parabolic lenses have maximum thickness $\tau(0) = 2R$ along the lens center axis $r = 0$. The lens perimeter radius r_m, and thus the radial distance r of a ray from the lens center axis from the lens center axis, are assumed to be much less than the lens radius of curvature R, where $r \le r_m \ll R$. Hence, the binomial approximation can be applied to Eq. (1.93) for the center portion of the spherical, biconvex lens. The binomial approximation shows the center portion of the spherical lens is approximately parabolic for lens radius r less than the parabolic aperture radius r_p, where $r < r_p$, so that

$$\tau_{\text{sph}}^{\text{vex}}(r) = d + 2\sqrt{R^2 - r^2} \cong d + 2R - \frac{r^2}{R} = \tau_{\text{par}}^{\text{vex}}(r) \quad r \ll R \tag{1.95}$$

Now consider the superimposed biconcave lenses in Figure 1.15. As a function of radial distance r from the center lens axis, the ray paths through the biconcave, spherical, and parabolic lenses are

$$\tau(r) = t_{\text{sph}}^{\text{ave}}(r) = d + 2R - 2\sqrt{R^2 - r^2} \quad \text{biconcave, spherical} \tag{1.96}$$

$$\tau(r) = t_{\text{par}}^{\text{ave}}(r) = d + \frac{r^2}{R} \quad \text{biconcave, parabolic} \tag{1.97}$$

The biconcave lenses have minimum thickness $\tau(r_m) = d$ along the lens center axis $r = 0$, and have maximum thickness $\tau(r_m) = 2R + d$ at the lens perimeter $r_m = R$ for spherical lenses, and at lens perimeter $r_m = R\sqrt{2}$ for parabolic lenses. With $r \le r_m \ll R$, the binomial approximation is applied to Eq. (1.96) for the center

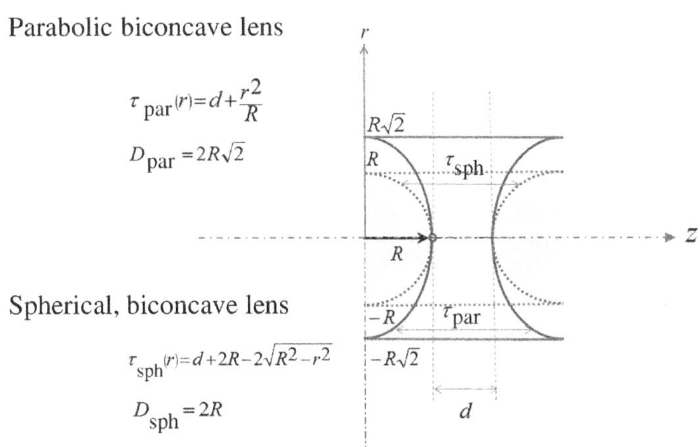

Figure 1.15 The radial-dependent thickness and physical aperture radius of a spherical and parabolic biconcave lens, which share the same on-axis radius of curvature R for the left and right lens surfaces, is compared.

portion of the spherical, biconvex lens, which again shows that the center portion of the spherical lens is approximately parabolic, where

$$\tau_{\text{sph}}^{\text{ave}}(r) = d + 2R - 2\sqrt{R^2 - r^2} \cong d + \frac{r^2}{R} = \tau_{\text{par}}^{\text{ave}}(r) \qquad r \ll R \tag{1.98}$$

A single lens can be characterized by (1) the deflection angle $2\Delta\theta$ of an incident ray, (2) the focal length f_1, (3) the ray path length $\tau(r_{\text{a}})$ through the lens at the absorption aperture radius r_{a}, and (4) the average intensity transmission $\bar{I}$ of the plane waves of X-rays, gammas, or neutrons through the lens. These expressions are contained in Eq. (1.59), Eq. (1.56), Eqs. (1.90–1.91), and Eq. (1.92), which are repeated below

$$\Delta\theta = \delta \frac{\mathrm{d}\tau}{\mathrm{d}r} \tag{1.99}$$

$$f_1 = \frac{r}{2\Delta\theta} \tag{1.100}$$

$$\tau(r_a) = \tau(0) \pm \frac{2}{\mu} \quad \text{biconcave } (+) \quad \text{biconvex } (-) \tag{1.101}$$

$$\bar{I} = \frac{1}{r_m} \int_0^{r_m} \exp[-N\tau(r)\mu(\lambda)]\mathrm{d}r \tag{1.102}$$

Visible light has a large refractive decrement δ; only a single lens is needed for refraction, and with the very small attenuation μ of the optical glass lens material, the absorption of light in a lens is ignored. However, if an antireflection coating is not used, reflection from the lens surface can lead to a 5–10% loss per surface.

We first consider the ray (wave vector **k**) deflection angles $2\Delta\theta$ of the single, biconvex, and biconcave lenses. With the substitution of Eqs. (1.93) and (1.94) into Eq. (1.99), the angular change of the wave vector **k** by refraction, for the single biconvex, spherical and parabolic lens, are

$$\Delta\theta_{\text{sph}}^{\text{vex}} = -\delta \frac{r}{\sqrt{R^2 - r^2}} \quad \text{biconvex, spherical} \tag{1.103}$$

$$\Delta\theta_{\text{par}}^{\text{vex}} = -\delta \frac{r}{R} \quad \text{biconvex, parabolic} \tag{1.104}$$

With the substitution of Eqs. (1.96) and (1.97) into Eq. (1.99), the angular change of the wave vector **k** by refraction, for the single biconcave, spherical and parabolic lens, are

$$\Delta\theta_{\text{sph}}^{\text{ave}} = \delta \frac{r}{\sqrt{R^2 - r^2}} \quad \text{biconcave, spherical} \tag{1.105}$$

$$\Delta\theta_{\text{par}}^{\text{ave}} = \delta \frac{r}{R} \quad \text{biconcave, parabolic} \tag{1.106}$$

The deflection angle $2\Delta\theta$ of the exit ray is measured relative to the incident ray direction. The deflection angle is negative ($2\Delta\theta < 0$) for counterclockwise rotation, and positive ($2\Delta\theta > 0$) for clockwise rotation. For incident plane waves, the incident rays in the positive z-direction are parallel to the positive z-axis, which lies along the center lens axis. If the deflection angle is negative ($2\Delta\theta < 0$) for counterclockwise rotation, with respect to the positive z-direction, then the ray is diverged (defocused) away from the center lens axis by angle $2\Delta\theta$. Conversely, if the deflection angle is positive ($2\Delta\theta > 0$) for clockwise rotation, with respect to the positive z direction, then the ray is converged (focused) toward the center lens axis by angle $2\Delta\theta$.

The negative deflection angle ($2\Delta\theta < 0$) for defocusing and the positive deflection angle ($2\Delta\theta > 0$) for focusing is determined by the polarity of the refractive decrement δ of the refractive index n, where $n = 1 - \delta$. Earlier in this chapter, Figure 1.10 (top) shows CRLs comprised of spherical or parabolic, biconvex lenses. The biconvex lenses focus X-rays, gammas, neutrons, or light waves, if the refractive decrement is negative ($\delta < 0$ and $n > 1$), and the biconvex lenses defocus these waves, if the refractive decrement is positive ($\delta > 0$ and $n < 1$). Figure 1.10 (bottom) shows CRLs comprised of spherical or parabolic, biconcave lenses. The biconcave lenses focus X-rays, gammas, neutrons, or light waves, if the refractive decrement is positive ($\delta > 0$ and $n < 1$), and defocus these waves, if the refractive decrement is negative ($\delta < 0$ and $n > 1$).

If the material refractive index is greater than 1 ($n > 1$) for the photon or particle wave, then the wave phase velocity v_{ph} in the material medium is less than the vacuum speed of light c, where $v_{\text{ph}} < c$. Hence, for $n > 1$, the phase of the photon or particle wave in the material medium lags behind the phase of the photon or particle wave in a vacuum. Conversely, if the material refractive index is less than 1 ($n < 1$) for the photon or particle wave, then the wave phase velocity v_{ph} in the material medium is greater than the vacuum speed of light c, where $v_{\text{ph}} > c$. Hence, for $n < 1$, the phase of the photon or particle wave in the material medium leads the phase of the same photon or particle wave in a vacuum.

We now consider the focal lengths of the single biconvex and biconcave lenses. With substitution of Eqs. (1.103–1.106) for $2\Delta\theta$ into Eq. (1.100), the focal lengths f_1 of a single biconvex or biconcave, spherical, or parabolic lens are

$$ {}^{vex}_{sph}f_1 = -\frac{\sqrt{R^2 - r^2}}{2\delta} \quad \text{biconvex, spherical} \tag{1.107} $$

$$ {}^{vex}_{par}f_1 = -\frac{R}{2\delta} \quad \text{biconvex, parabolic} \tag{1.108} $$

$$ {}^{ave}_{sph}f_1 = \frac{\sqrt{R^2 - r^2}}{2\delta} \quad \text{biconcave, spherical} \tag{1.109} $$

$$ {}^{ave}_{par}f_1 = \frac{R}{2\delta} \quad \text{biconcave, parabolic} \tag{1.110} $$

A negative focal length ($f_1 < 0$) indicates that the diverging rays from the lens converge to a focal point upstream (on the source side) of the lens. A positive focal length ($f_1 > 0$) indicates that the converging rays from the lens converge to a focal point downstream of the lens (on the image side). If the decrement of the lens material for the photon or particle wave is positive ($\delta > 0$) with refractive index $n = 1 - \delta < 1$, then the biconcave lenses focus and the biconvex lenses defocus. Conversely, if the decrement of the lens material for the photon or particle wave is negative ($\delta < 0$) with refractive index $n > 1$, then the biconvex lenses focus and the biconcave lenses defocus.

The radius of curvature R of the spherical, biconvex, or biconcave lenses with physical radius r_m is much larger than the radial distance r of the ray from the lens center axis, where $r \le r_m \ll R$, so that the focal lengths of the spherical and parabolic, convex lens are approximately equal, and the focal lengths of the spherical and parabolic, concave lens are approximately equal, where

$$ {}^{vex}_{sph}f_1 = -\frac{\sqrt{R^2 - r^2}}{2\delta} \cong -\frac{R}{2\delta} = {}^{vex}_{par}f_1 \tag{1.111} $$

$$ {}^{ave}_{sph}f_1 = \frac{\sqrt{R^2 - r^2}}{2\delta} \cong \frac{R}{2\delta} = {}^{ave}_{par}f_1 \tag{1.112} $$

Hence, for the CRL shape and material, which focus X-rays, gammas, neutrons, or light and are comprised of N biconvex or biconcave, spherical or parabolic lenses, the CRL focal length f is approximately

$$ f = \frac{R}{2N\delta} \tag{1.113} $$

We now consider the absorption apertures r_a of the biconvex and biconcave, spherical, and parabolic lenses. The absorption aperture radius r_a from the lens center axis for biconvex lenses is defined by the increase of the wave intensity $T(r_a)$ by a factor of e^2 at lens radius $r = r_a$, which is measured relative to the minimum, transmitted wave intensity $T(0)$ along the lens center axis at $r = 0$, where

$$ \frac{T(r_a)}{T(0)} = \frac{\exp[-\mu\tau(r_a)]}{\exp[-\mu\tau(0)]} = \exp(+2) \tag{1.114} $$

Equating the arguments of the exponentials in Eq. (1.114) for a biconvex lens, yields

$$ \tau(r_a) = \tau(0) - \frac{2}{\mu} \tag{1.115} $$

The absorption aperture radius r_a for a single, spherical and parabolic, biconvex lens, are obtained by substitution of Eqs. (1.93–1.94) into Eq. (1.101), where

$$ d + 2\sqrt{R^2 - r_a^2} = d + 2R - \frac{2}{\mu} \quad \text{biconvex, spherical} \tag{1.116} $$

$$ d + 2R - \frac{r_a^2}{R} = d + 2R - \frac{2}{\mu} \quad \text{biconvex, parabolic} \tag{1.117} $$

Solving Eqs. (1.116–1.117) for the absorption aperture radius r_a for biconvex, spherical, and parabolic cases, where the transmitted intensity of the X-rays, gammas, or neutrons occurs primarily outside the absorption aperture radius $r > r_a$, one obtains

$$ {}^{vex}_{sph}r_a = \sqrt{\frac{1}{\mu}\left(2R - \frac{1}{\mu}\right)} \quad \text{biconvex, spherical} \tag{1.118} $$

$$ {}^{vex}_{par}r_a = \sqrt{\frac{2R}{\mu}} \quad \text{biconvex, parabolic} \tag{1.119} $$

If twice the radius of curvature $2R$ of the convex, spherical lens is much greater than the reciprocal of the linear attenuation μ for the lens material (i.e., $2R \gg 1/\mu$), then the absorption aperture radius of the spherical and parabolic, biconvex lenses is equal, where

$$ {}^{vex}_{sph}r_a = \sqrt{\frac{1}{\mu}\left(2R - \frac{1}{\mu}\right)} \cong \sqrt{\frac{2R}{\mu}} = {}^{vex}_{par}r_a \quad 2R \gg \frac{1}{\mu} \tag{1.120} $$

The absorption aperture radius r_a for a single, spherical and parabolic, biconcave lens is obtained by substituting Eqs. (1.96–1.97) into Eq. (1.101), where

$$ d + 2R - 2\sqrt{R^2 - r_a^2} = d + 2R - 2R + \frac{2}{\mu} \quad \text{biconcave, spherical} \tag{1.121} $$

$$ d + \frac{r_a^2}{R} = d + \frac{2}{\mu} \quad \text{biconcave, parabolic} \tag{1.122} $$

Solving Eqs. (1.121–1.122) for the absorption aperture radius r_a for biconvex, spherical and parabolic lenses, yield

$$ {}^{ave}_{sph}r_a = \sqrt{\frac{1}{\mu}\left(2R + \frac{1}{\mu}\right)} \quad \text{biconcave, spherical} \tag{1.123} $$

$$ {}^{ave}_{par}r_a = \sqrt{\frac{2R}{\mu}} \quad \text{biconcave, parabolic} \tag{1.124} $$

Again, the absorption aperture radius of the spherical and parabolic biconcave lenses is equal if $2R \gg 1/\mu$, where

$$ {}^{ave}_{sph}r_a = \sqrt{\frac{1}{\mu}\left(2R + \frac{1}{\mu}\right)} \cong \sqrt{\frac{2R}{\mu}} = {}^{ave}_{par}r_a \quad 2R \gg \frac{1}{\mu} \tag{1.125} $$

With a CRL with N biconvex or biconcave, parabolic or spherical lenses with $2R \gg 1/\mu$, one replaces μ with $N\mu$ to obtain the CRL absorption aperture radius r_a, where

$$r_a = \sqrt{\frac{2R}{N\mu}} \tag{1.126}$$

Substituting R into Eq. (1.126) from Eq. (1.113) yields the alternative CRL absorption aperture radius r_a, which is expressed by the product of the δ/μ ratio, and the CRL focal length f for the N parabolic, biconvex lenses, where

$$r_a = 2\sqrt{f\frac{\delta}{\mu}} \tag{1.127}$$

One can determine the focal length of a biconvex lens, which is based on a minimum fraction of transmitted intensity through the center, thickest part of the lens, relative to the incident intensity. For a biconvex, spherical, or parabolic lens, the fraction of transmission of the incident wave amplitude is 100% at the lens perimeter $r = r_m$, assuming zero edge thickness. The transmitted wave intensity $I(r_m)$ through the zero-thickness perimeter equals the incident wave intensity I_0 at the convex lens input surface, where

$$I(r_m) = I_0 \tag{1.128}$$

However, the transmitted wave intensity $I(0)$ along the thickest part of the biconvex lens along the lens center axis $r = 0$ can be strongly attenuated to far less than 100% transmission.

One can design the single biconvex lens, or stack of biconvex lenses, in a CRL using a predetermined transmission fraction $T(0)$ of intensity through the center, thickest part of the lens. The transmission fraction $T(0)$ is the ratio of the center-transmitted intensity $I(0)$ divided by the intensity $I(r_m)$ along the lens perimeter, where $I(r_m)$ is also the incident intensity. The transmission fraction $T(0)$ is set equal to $1/e^2$, where

$$T(0) = \frac{I(0)}{I(r_m)} = \exp[-N\mu\tau(0)] = \exp(-N\mu 2R) = \exp(-2) \tag{1.129}$$

Equating the arguments of the exponentials of the last two expressions in Eq. (1.129), one then obtains the convex lens radius of curvature R, which is required to meet the $1/e^2$ center/perimeter intensity ratio, where

$$R = \frac{1}{N\mu} \tag{1.130}$$

The substitution of Eq. (1.130) into Eq. (1.113) yields the focal length f of CRL, where

$$f = \frac{1}{2N^2\mu\delta} \tag{1.131}$$

We now consider the average transmitted intensity through biconvex lenses. Consider a biconvex, N-lens CRL with parabolic lenses, or with spherical lenses, which are approximated as parabolic lenses. One substitutes the parabolic lens thickness $\tau(r)$ of Eq. (1.94) or Eq. (1.95) into Eq. (1.102), and the average transmitted intensity $\bar{I}$ through the lens is

$$\bar{I} = \frac{\exp(-4R\mu N)\exp(-2N\mu d)}{r_m}\int_0^{r_m} \exp\left[\frac{r^2 N\mu}{R}\right] dr \quad \text{biconvex} \tag{1.132}$$

The perimeter radius r_m is the upper limit of integration. The solution of the integral is obtained from the general integral solution, which uses the error function $\text{erf}(r)$. The error function along the $r = 0$ lens center axis is $\text{erf}(0) = 0$, and for $x > 2.5$, $\text{erf}(x) = 1$, so that

$$\int_0^{r_m} \exp(-ax^2)dx = \sqrt{\frac{\pi}{a}}\frac{\text{erf}(r_m\sqrt{a})}{2} \quad \text{with} \quad a = \frac{N\mu}{R} \tag{1.133}$$

The error function $\text{erf}(p)$ is related to the imaginary error function $\text{erfi}(p)$, via $\text{erfi}(p) = -\text{i}\,\text{erf}(\text{i}p)$, where

$$\int_0^{r_m} \exp(ax^2)dx = -\text{i}\sqrt{\frac{\pi}{a}}\frac{\text{erf}(\text{i}r_m\sqrt{a})}{2} = \sqrt{\frac{\pi}{a}}\frac{\text{erfi}(r_m\sqrt{a})}{2} \quad \text{with} \quad a = \frac{N\mu}{R} \tag{1.134}$$

Application of Eq. (1.134) to Eq. (1.132) gives the average transmitted intensity $\bar{I}$ of waves of X-rays, gammas, or neutrons through a biconvex, spherical lens with perimeter radius $r_m = R$, or a biconvex, parabolic lens with perimeter radius $r_m = R\sqrt{2}$, where

$$\bar{I} = \frac{I(0)}{2}\sqrt{\frac{\pi}{N\mu}}\text{erfi}\left(\sqrt{RN\mu}\right) \quad \text{biconvex, spherical} \tag{1.135}$$

$$\bar{I} = \frac{I(0)}{2\sqrt{2}}\sqrt{\frac{\pi}{N\mu}}\text{erfi}\left(\sqrt{2RN\mu}\right) \quad \text{biconvex, parabolic} \tag{1.136}$$

Along the center axis, each biconvex lens has thickness $\tau(0) = N(2R + d)$, and recalling Eq. (1.86), the transmitted intensity $I(0)$ along the center axis of the CRL of N biconvex lenses is

$$I(0) = \exp(-2N\mu d)\exp(-4R\mu N) \quad \text{biconvex} \tag{1.137}$$

Given $\text{erfi}(p) \cong 2p/\sqrt{\pi}$ for $-1 < p < 1$, then for $R\mu N < 1$ in Eq. (135–136), the average transmission $\bar{I}$ through CRLs with biconvex, spherical and parabolic lenses, are identical, where

$$\bar{I} = I(0)\sqrt{R} = \exp(-2N\mu d)\exp(-4R\mu N)\sqrt{R} \quad \text{biconvex, spherical, and } R\mu N < 1 \tag{1.138}$$

$$\bar{I} = I(0)\sqrt{R} = \exp(-2N\mu d)\exp(-4R\mu N)\sqrt{R} \quad \text{biconvex, parabolic, and } R\mu N < 1 \tag{1.139}$$

If one designs the CRL to have a center-transmitted intensity, as $1/e^2$ of the incident or lens perimeter intensity, one invokes Eq. (1.130) for $R = 1/N\mu$ in Eqs. (1.135–1.137), so that the average transmitted intensity through a CRL with biconvex spherical and parabolic lenses, are again identical, where

$$\bar{I} = I(0)\sqrt{\frac{1}{N\mu}} = \exp(-2N\mu d)\exp(-4)\sqrt{\frac{1}{N\mu}} \quad \text{biconvex, spherical} \tag{1.140}$$

$$\bar{I} = I(0)\sqrt{\frac{1}{N\mu}} = \exp(-2N\mu d)\exp(-4)\sqrt{\frac{1}{N\mu}} \quad \text{biconvex, parabolic} \tag{1.141}$$

We now consider the radial-dependent transmission profile of the CRL, which is comprised of a stack of N biconvex, parabolic lenses. In Figure 1.16, the radial-dependent transmission fraction $T(r)$ of room-temperature thermal neutrons (0.025 eV or 1.8 A) by a titanium CRL is plotted. The CRL transmission profile $T(r)$ is represented by the transmission fraction $T(r) = I(r)/I_0$ of Eq. (1.87), where the absorption aperture radius r_a is based on the criteria of Eq. (1.88), and one uses Eq. (1.94) for the thickness of a biconvex, parabolic lens. Titanium is one of the few elements that has a negative decrement ($\delta < 0$) for neutrons and a resulting refractive index greater than 1, where $n = 1 - \delta > 1$. This requires a biconvex, titanium lens for the focusing and imaging of neutrons. The plot in Figure 1.16 is an example of the relatively larger neutron attenuation by the center, thicker portion of the CRL, compared to the outer radial portion of the CRL. The CRL is comprised of $N = 70$ biconvex, parabolic lenses of titanium, with an on-axis radius of curvature $R = 0.6$ mm and a physical aperture radius $r_m = 0.8$ mm. For the 0.025-eV thermal neutrons, the titanium decrement is $\delta = -1.0 \times 10^{-6}$ and the titanium linear attenuation is $\mu = 0.35\ \text{cm}^{-1}$. This results in a titanium CRL absorption aperture radius of $r_a = 0.7$ mm and focal length of $f = 4.0$ m.

We now consider the average transmitted intensity through biconcave lenses. The transmitted wave intensity $I(r, \lambda)$ for the biconcave, parabolic lens, or the transmitted intensity $I(r, \lambda)$ through the center portion of the spherical lens, within the parabolic radius $r < r_p$, is obtained from Eqs. (1.86) and (1.98), where

$$I(r) = I_0 \exp(-2N\mu d)\exp\left(-N\mu\frac{r^2}{R}\right) \tag{1.142}$$

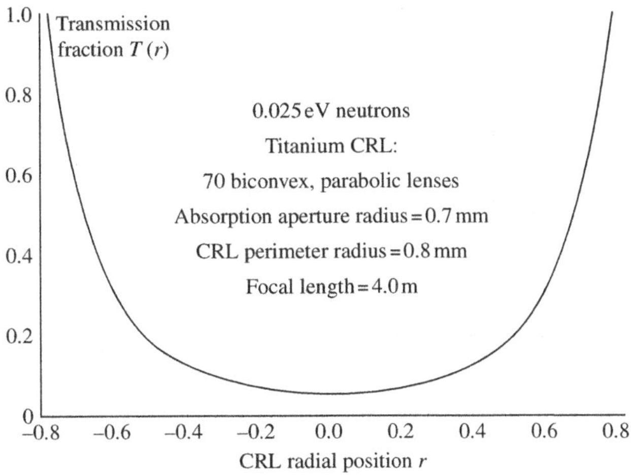

Figure 1.16 The radial-dependent transmission fraction $T(r) = I(r)/I_0$ of room-temperature thermal neutrons (0.025 eV or 1.8 A) by a CRL, which is comprised of biconvex lenses of titanium, is plotted. The CRL focuses and images the neutrons. However, the shape of the biconvex lenses of the CRL causes a relatively stronger attenuation at the thicker, center portion of the CRL relative to the outer radial portion of the CRL.

The radial-dependent transmission fraction $T(r, \lambda)$ is the ratio of the transmitted intensity $I(r, \lambda)$, divided by the incident intensity I_0, where

$$T(r, \lambda) = \frac{I(r, \lambda)}{I_0} = \exp(-2N\mu d)\exp\left(-N\mu\frac{r^2}{R}\right) \tag{1.143}$$

At the absorption aperture radius $r_a = \sqrt{2R/\mu N}$, the ratio of the transmission fraction $T(r_a)$, divided by the transmission fraction $T(0)$ along the center axis of the biconcave lens, is

$$\frac{T(\sqrt{2}r_a)}{T(0)} = \exp\left(-N\mu\frac{2r_a^2}{R}\right) = \exp(-4) = 0.018 \quad \text{biconcave} \tag{1.144}$$

Now, consider a biconcave, N-lens CRL with parabolic lenses, or with spherical lenses, which are approximated by parabolic lenses. One substitutes the parabolic lens thickness $\tau(r)$ of Eq. (1.97) or Eq. (1.98) into Eq. (1.102), and the average transmitted intensity $\bar{I}$ through the CRL is then

$$\bar{I} = \frac{\exp(-2N\mu d)}{r_m}\int_0^{r_m}\exp\left[-\frac{r^2N\mu}{R}\right]\mathrm{d}r \quad \text{biconcave} \tag{1.145}$$

The physical lens perimeter radius r_m is used as the upper limit of integration. If the CRL, with N biconcave lenses, transmits the incident wave intensity with relatively little attenuation, then the CRL absorption aperture radius r_a is much larger than the lens perimeter radius r_m, where $r_a \gg r_m$. Here, $r_m = R$ for spherical, biconcave lenses, and $r_m = R\sqrt{2}$ for parabolic, biconcave lenses. Application of the integral rule of Eq. (1.133), yields the average transmitted intensity $\bar{I}$ by CRLs with biconcave, spherical and parabolic lenses, where

$$\bar{I} = \frac{I(0)}{2}\sqrt{\frac{\pi}{N\mu}} \cdot \mathrm{erf}\left(\sqrt{RN\mu}\right) \quad \text{biconcave, spherical} \tag{1.146}$$

$$\bar{I} = \frac{I(0)}{2\sqrt{2}}\sqrt{\frac{\pi}{N\mu}} \cdot \mathrm{erf}\left(\sqrt{2RN\mu}\right) \quad \text{biconcave, parabolic} \tag{1.147}$$

The transmitted intensity along the center axis of the CRL of N biconcave lenses is

$$I(0) = \exp(-2N\mu d) \quad \text{biconcave} \tag{1.148}$$

The average transmission $\bar{I}$ through CRLs, which have a sufficiently large product $RN\mu$ in Eqs. (1.143) and (1.144) are then

$$\bar{I} = \frac{I(0)}{2}\sqrt{\frac{\pi}{N\mu}} = \frac{0.89 \cdot I(0)}{\sqrt{N\mu}} \quad \text{biconcave, spherical, and } \sqrt{RN\mu} > 2.5 \tag{1.149}$$

$$\bar{I} = \frac{I(0)}{2\sqrt{2}}\sqrt{\frac{\pi}{N\mu}} = \frac{0.63 \cdot I(0)}{\sqrt{N\mu}} \quad \text{biconcave, parabolic, and } \sqrt{2RN\mu} > 2.5 \tag{1.150}$$

If the CRL with N biconcave lenses transmits the incident wave intensity with relatively strong attenuation, then the CRL absorption aperture radius r_a is much smaller than the lens perimeter radius r_m, where $r_a \ll r_m$. The average transmitted intensity $\bar{I}$ by the CRL of biconcave lenses becomes

$$\bar{I} = \frac{\exp(-2N\mu d)}{r_m}\int_0^{r_m} \exp\left[-\frac{r^2 N\mu}{R}\right] \mathrm{d}r \quad \text{biconcave} \tag{1.151}$$

Most of the integrand contribution to the integral occurs for values of r from 0 to $\sqrt{2}r_a$ and approximately zero contribution for values of r from $\sqrt{2}r_a$ to ∞, so that

$$\begin{aligned}\bar{I} &= \frac{\exp(-2N\mu d)}{r_m}\int_0^{r_m} \exp\left(-\frac{r^2 N\mu}{R}\right) \mathrm{d}r \cong \frac{\exp(-2N\mu d)}{r_a\sqrt{2}}\int_0^{r_a\sqrt{2}} \exp\left(-\frac{2}{r_a^2}r^2\right) \mathrm{d}r \\ &\cong \frac{\exp(-2N\mu d)}{r_a\sqrt{2}}\int_0^{\infty} \exp\left(-\frac{2}{r_a^2}r^2\right) \mathrm{d}r\end{aligned} \tag{1.152}$$

The solution of Eq. (1.152) is obtained from the general integral solution:

$$\int_0^{\infty} \exp(-ar^2)\mathrm{d}r = \frac{1}{2}\sqrt{\frac{\pi}{a}} \tag{1.153}$$

with

$$a = \frac{N\mu}{R} = \frac{2}{r_\mathrm{a}^2} \tag{1.154}$$

Hence, evaluating Eq. (1.152) gives the average transmitted intensity $\bar{I}$ through a strongly attenuating CRL, where

$$\begin{aligned}\bar{I} &\cong \frac{\exp(-2N\mu d)}{r_\mathrm{a}\sqrt{2}} \int_0^{\infty} \exp\left(-\frac{2}{r_\mathrm{a}^2} r^2\right)\mathrm{d}r \\ &= \frac{\sqrt{\pi}}{4}\exp(-2N\mu d) = 0.44 \cdot \exp(-2N\mu d) \quad \text{biconcave}\end{aligned} \tag{1.155}$$

We now consider the radial-dependent transmission profile of the CRL, which is comprised of a stack of N biconcave lenses. In Figure 1.17, the radial-dependent transmission profile $T(r)$ of thermal neutrons (0.025 eV or 1.8 A) by a copper CRL is plotted. The CRL transmission profile $T(r)$ is represented by the transmission fraction $T(r) = I(r)/I_0$ of Eq. (1.87), where the absorption aperture radius r_a is based on the criteria of Eq. (1.89). Copper is one of the many elements that have a

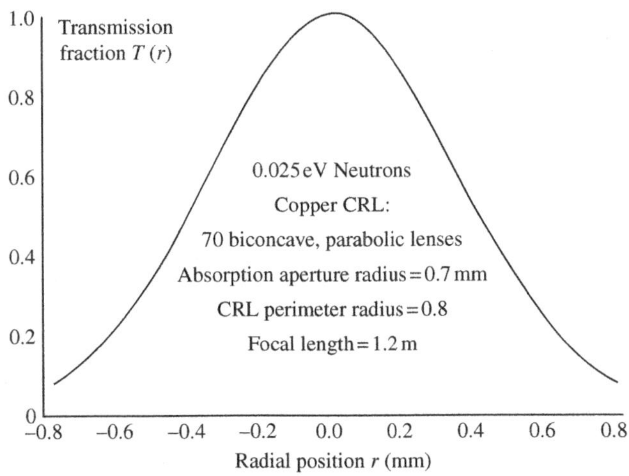

Figure 1.17 Plotted is the radial-dependent transmission fraction $T(r)$ of room-temperature thermal neutrons (0.025 eV or 1.8 A) by the CRL, which is comprised of biconcave lenses of copper that focus and image the neutrons. However, the biconcave shape of the copper lenses of the CRL causes a relatively stronger attenuation at the outer radial, thicker portion of the CRL, relative to the center portion of the CRL.

positive decrement ($\delta > 0$), and the resulting refractive index is less than 1 ($n = 1 - \delta < 1$). This requires a biconcave lens for focusing and imaging of neutrons, as well as X-rays. This is an example of the relatively larger neutron attenuation by the thicker, outer portion of the CRL, compared to the center portion of the CRL. The copper CRL is comprised of $N = 70$ biconcave, parabolic lenses of copper, with an on-axis radius of curvature $R = 0.6$ mm, and physical aperture radius $r_m = 0.8$ mm. For 0.025 eV neutrons, the copper decrement is $\delta = 3.4 \times 10^{-6}$, and the copper linear attenuation is $\mu = 0.32\ \mathrm{cm}^{-1}$. This results in a copper CRL aperture radius of $r_a = 0.7$ mm, and the focal length $f = 1.2$ m.

Comparing Figs. 1.16 and 1.17, the titanium and copper CRLs have the same number N of lenses, radius of curvature R, and physical aperture radius r_m. Titanium and copper have approximately the same neutron linear attenuation $\mu = 0.32\ \mathrm{cm}^{-1}$ for 0.025 eV thermal neutrons. The titanium biconvex CRL absorption aperture radius and the copper biconcave CRL absorption aperture radius are thus equal, where $r_a = 0.7$ mm. However, for 0.025 eV neutrons, the copper refractive decrement magnitude of $|\delta| = 3.4 \times 10^{-6}$ is larger than the magnitude of titanium decrement of $|\delta| = 1.0 \times 10^{-6}$. Hence, the copper biconcave CRL has a shorter focal length of $f = 1.2$ m than the biconvex titanium CRL focal length of $f = 4.0$ m.

In Figure 1.18, the CRL X-ray absorption aperture diameter $2r_a$ is plotted for a variety of lens materials as a function of photon energy from 5 to 30 keV, which corresponds to 2.3–0.4 Å. The CRLs are comprised of biconcave, spherical, or parabolic lenses and have a common focal length of $f = 1$ m for the X-rays, where CRL absorption aperture diameter $2r_a = 2\sqrt{f\delta/\mu}$ of Eq. (1.127) is expressed in terms of the CRL focal length f and the X-ray decrement to linear attenuation ratio δ/μ.

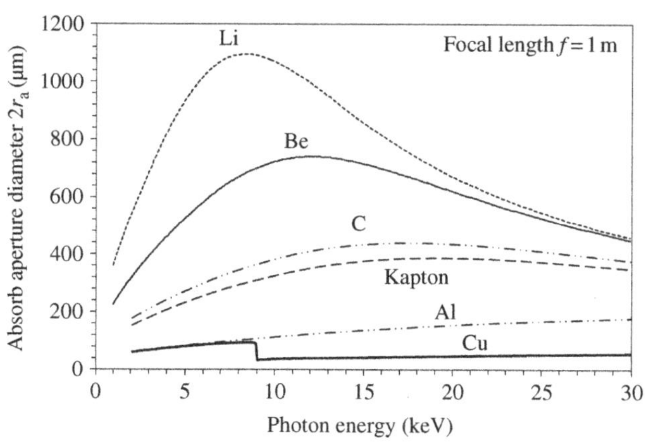

Figure 1.18 Approximate CRL absorption aperture diameter $2r_a = 2\sqrt{f\delta/\mu}$, expressed in microns, for a CRL focal length $f = 1$ m, as a function of X-ray energy, and a range of 5–30 keV, which corresponds to wavelengths over the range 2.3–0.4 Å for various materials. For the low-atomic-number materials, the absorption aperture diameter $2r_a$ of an X-ray CRL for a set focal length, such as $f = 1$ m, achieves a maximum diameter at a particular X-ray energy. The higher atomic number elements are strongly absorbing, but $2r_a$ increases slowly with increasing energy after a K, L, M, or N edge, as seen with copper.

In the low-atomic-number materials, the absorption aperture diameter $2r_a$ of X-ray CRLs for a set focal length, such as $f = 1$ m, achieves a maximum diameter for particular X-ray energy. CRLs with lithium, beryllium, and carbon have the largest absorption aperture diameters $2r_a$ for X-rays. For neutrons, lithium is strongly absorbing. However, beryllium and copper are excellent lens materials for neutrons as well as X-rays. For X-rays, the higher atomic number elements are strongly absorbing, but the CRL absorption aperture diameter $2r_a$ increases slowly, with increasing X-ray energy after a K, L, M, or N edge, as seen with copper.

In comparison, for a variety of lens materials, the CRL neutron absorption aperture diameter $2r_a$ is plotted in Figure 1.19 as a function of neutron wavelengths 1–90 Å, which corresponds to 10^{-5}–1.0 eV energy range. The CRLs all share the same focal length ($f = 1$), where the CRL absorption aperture diameter is $2r_a = 2\sqrt{f\delta/\mu}$, and from Eq. (1.43), for neutrons, $\delta/\mu \propto \lambda b$. For the $f = 1$ m focal lengths, the neutron absorption aperture diameters tend to be larger than 1 mm, where $2r_a > 1.0$ mm; the X-ray absorption aperture diameters tend to be smaller than 1 mm, where $2r_a < 1.0$ mm. The larger absorption aperture diameter for neutrons, results from the larger ratio of the refraction to attenuation δ/μ for neutrons, as compared to X-rays. This is clearly seen in the 1.0–2.3 Å common wavelength range of the X-ray and neutron absorption aperture diameter plots of Figs. 1.18 and 1.19 and δ/μ plots of Figure 1.11 for neutrons and Figure 1.26 (later in this chapter) for X-rays.

For design purposes, the CRL physical aperture radius r_m can be set equal to the absorption aperture radius r_a, via a limiting iris. Usually, the CRL absorption aperture radius is significantly larger than the parabolic aperture radius r_p, where $r_m > r_a > r_p$.

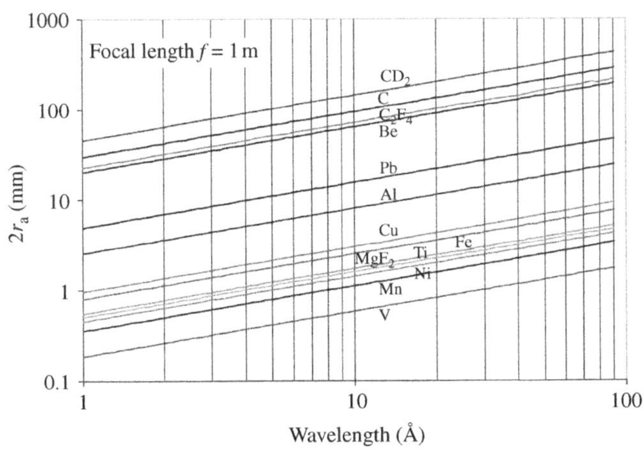

Figure 1.19 Approximate CRL absorption aperture diameter $2r_a = 2\sqrt{f\delta/\mu}$ for a CRL focal length $f = 1$ m, as a function of neutron wavelengths over the wave length range 1–90 Å or corresponding 1.0–10^{-5} eV energy range for various materials. Most materials have a neutron refractive index less than 1 ($n < 1$), which requires biconcave, spherical, or parabolic lenses for CRL focusing and imaging. Natural occurring titanium, vanadium, and manganese have refractive indexes greater than 1 ($n > 1$) and require biconvex, spherical or parabolic lenses for CRL focusing and imaging.

To eliminate spherical aberration in a spherical CRL, the CRL physical aperture r_m is stopped down farther to the CRL parabolic aperture radius r_p, via a smaller diameter iris, where $r_a > r_m = r_p$. However, for a wide-band source of X-rays or neutrons, the CRL iris radius r_m can be set to the CRL absorption aperture radius r_a, where $r_m = r_a > r_p$. In this case, the spherical aberration occurs with rays incident in the lens radial portion between the inner parabolic aperture radius r_p and outer absorption aperture radius r_a. The spherical aberration can be counteracted by the CLR chromatic aberration, where the longer wavelengths are attenuated more than the shorter wavelengths as the radius r increases from the parabolic aperture radius r_p to the limiting aperture radius r_m, which is set to the absorption aperture radius, where $r_m = r_a$. The effective aperture radius r_e of an X-ray, gamma, or neutron CRL, which is comprised of spherical lenses is the minimum of the physical or perimeter aperture radius r_m, absorption aperture radius r_a, and parabolic aperture radius r_p, where

$$r_e = \min(r_m, r_a, r_p) \tag{1.156}$$

We now derive the parabolic aperture diameter $2r_p$ of a spherical lens. For radial distances less than the parabolic aperture radius ($r \le r_p$), the shape of the spherical biconvex or biconcave lens (sphere) is approximately parabolic. For radial distances larger than the parabolic aperture radius ($r > r_p$), the spherical lens shape departs, increasing from the parabolic shape as the radial distance r increases. As a result, X-rays, gammas, or neutrons incident in the middle to outer radius of the lens undergo spherical aberration. Rays, which are incident upon the inner radial portion of the spherical, biconcave lens, do not undergo spherical aberration, and are focused without spherical aberration.

The inner radius of the biconcave, spherical lens, which is effectively a parabolic, biconcave lens, is termed the parabolic aperture radius r_p. Spherical aberration occurs outside the center, parabolic portion of a spherical lens, and causes loss of image resolution, but not a loss of intensity. At radial distance r from the lens center axis, an additional lens thickness $\Delta\tau$, is encountered by rays propagating in the spherical lens, as compared to a parabolic lens. From Eqs. (1.93) and (1.94) and Eqs. (1.96) and (1.97), is obtained the additional ray path length $\Delta\tau$ in a spherical lens,

$$\Delta\tau = \tau_{\text{sph}} - \tau_{\text{par}} = \left(d + 2R - 2\sqrt{R^2 - r^2}\right) - \left(d + \frac{r^2}{R}\right) \tag{1.157}$$

Since the radius of curvature of a single biconcave, spherical lens is much greater than the effective aperture radius r_e of the lens (where $R \gg r_e$ with $r \le r_e$, and thus $R \gg r$), the binomial expansion of the square root in Eq. (1.157) as a sum of the linear and quadratic terms in $(r/R)^2$ gives

$$\Delta\tau = \frac{r^4}{4R^3} \tag{1.158}$$

The additional thickness $\Delta\tau$ of the spherical lens, compared to the parabolic lens, causes phase shift $\Delta\phi$, where

$$\Delta\phi = 2\pi \cdot \frac{\Delta\tau}{\lambda} \cdot \delta \leq \pi \quad \text{amplitude summation} \tag{1.159}$$

Summation of ray amplitudes occurs at a single focal point, if the phase shift angles $\Delta\phi$ between the ray paths from an on-axis object point O to an on-axis conjugate image point I are less than π, where substitution of $\Delta\tau$ gives

$$\Delta\phi = 2\pi \cdot \frac{r^4}{4R^3\lambda} \cdot \delta \leq \pi \quad \text{amplitude summation} \tag{1.160}$$

Solving Eq. (1.160), for the radial distance r from the lens center axis, gives the parabolic aperture radius r_p, which contains ray paths from object point O to image point I, whose amplitudes have relative phase shifts less than 180 degree, and thus sum constructively, where

$$r_p = \left(\frac{2R^3\lambda}{\delta}\right)^{1/4} \tag{1.161}$$

For a biconcave, spherical N-lens CRL, the parabolic aperture radius r_p with path length difference $N\Delta\tau$ is then

$$r_p = \left(\frac{2R^3\lambda}{N\delta}\right)^{1/4} \tag{1.162}$$

Ray paths from on-axis object point O, which fall outside the parabolic aperture radius $r > r_p$, are focused in a locus of on-axis image points upstream and downstream of on-axis image point I. The outer rays beyond the spherical CRL parabolic aperture radius $r > r_p$ have focal length f_{sp}, which decreases with radial distance r from the lens center axis, where one recalls from Eqs. (1.107) and (1.109),

$$f_{sp} = \frac{\sqrt{R^2 - r^2}}{2N\delta} \tag{1.163}$$

1.3 CRL Arrays

Imaging techniques such as radiography, computed tomography (CT), nuclear medicine, and magnetic resonance has been used for large-area structural and anatomical imaging; however, these methods are generally not capable of using magnification to improve resolution. Electron and optical microscopy are limited to in vitro methods of imaging due to high absorption of electrons and light in soft tissue. Imaging of light with compact 3-D arrays of microlenses has long been exploited for numerous applications including photocopies, optical fibers, and CCD arrays, see Daly (2001). One can modify and apply the light microlens array technology to X-rays and neutron optics, and in particular to stacked planar arrays of microlenses for large area imaging.

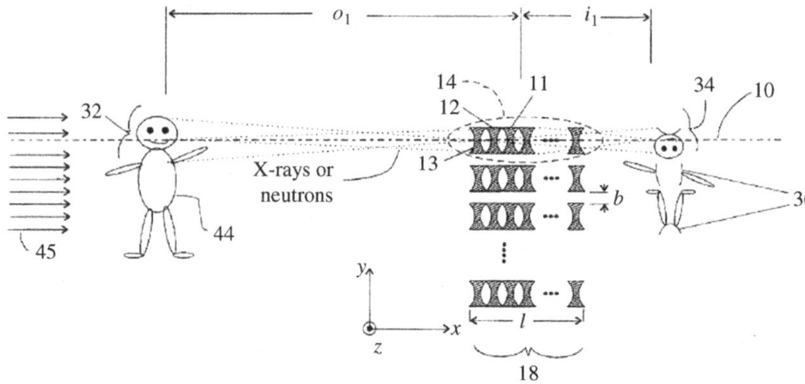

Figure 1.20 A side view of a single 3-D lens array, which gives partial inverted images of an object using X-rays. The unit lenses are in a cubic array. For a single 3-D array, the final image is scrambled (Piestrup, 2006).

These large arrays of microlenses permit the magnified imaging of large objects, overcoming the problem of currently manufactured refractive optics, which are highly limited by their apertures and, hence, by their field of view (FOV). Since the absorption length of X-rays and gamma rays is much longer than that of visible light or electrons (as in light or electron microscopy), such rays can see structures embedded in soft tissue at desired depths. Refractive lenses can be designed to have limited depth-of-fields permitting imaging of a preselected depth, whereas conventional radiography produces only shadow images of the entire thickness of the object. Thus, for refractive lenses, in vivo imaging can take place without excision.

The large area 3-D lens array for X-rays and neutrons by Piestrup (2004, 2006), is now described. A 3-D array of unit biconcave or biconvex lenses, which are spherical, but preferably parabolic, forms one optical element, and is shown in Figure 1.20. The arrays of biconcave unit lenses are aligned in rows and columns, where the optical axis of each unit lens is coaxial with succeeding unit biconcave lenses, forming a CRL. Identical CRLs are repeated in the y- and z-directions. In Figure 1.20, an object (a little man) is illuminated by X-rays, where the CRL images part of the object (i.e., the head of the man, 32) to a point on the other side of the CRL. The partial image (34) of the head is inverted and has been demagnified, as shown in the figure. Each CRL obeys the lens formula, which relates object distance o_1 to image distance i_1 and focal length f:

$$\frac{1}{o_1} + \frac{1}{i_1} = \frac{1}{f} \tag{1.164}$$

Large-area arrays of biconcave or biconvex, spherical or parabolic microlenses are capable of imaging of large-area objects, which are illuminated by angularly divergent X-ray sources. First, consider the use of planar arrays of unit lenses stacked, such that common optical axes are formed between the planes. As shown in Figure 1.21, a minimum of two sets of lenses will be used to form the image: one 3-D array to collect the

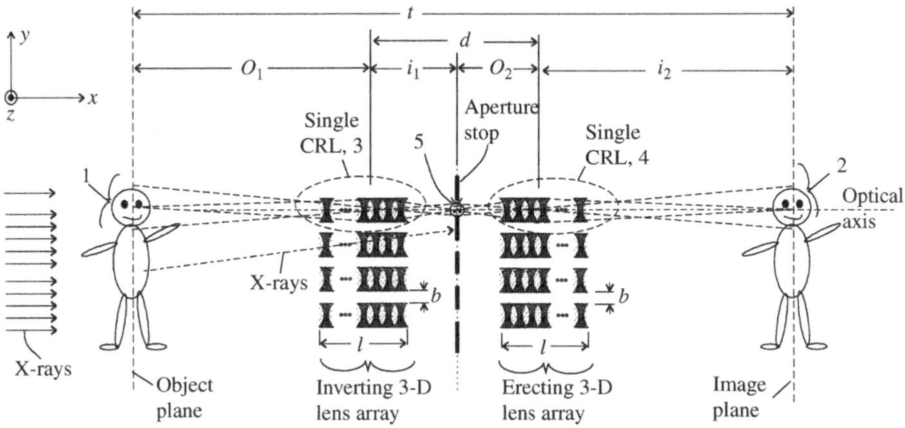

Figure 1.21 A side view of a 3-D lens array system that gives a one-to-one image of an object using X-rays (Piestrup, 2006).

object's image and, in the process, invert and fragment it; another 3-D array will reconstitute the image into a composite form. Inverting and erecting lens arrays using visible-optics microlenses has been used for various imaging systems and cameras, see Daly (2001). One-to-one imaging, magnification, and demagnification (i.e., collection and focusing) of X-rays can be accomplished using such lens systems.

1.3.1 One-to-One Imaging

In Figure 1.21, the unit lenses are aligned so that the optical axis of each is coaxial with succeeding unit convex lenses, forming a CRL. Identical CRLs are repeated in the y- and z-directions. An object (a little man) is illuminated by X-rays. The CRL images part of the object (i.e., the head of the man, 1) to a point on the other side of the CRL. The partial image (2) is inverted and has been demagnified, as shown in Figure 1.21. Each CRL obeys the lens formula

$$\frac{1}{o_1} + \frac{1}{i_1} = \frac{1}{f} \tag{1.165}$$

Here, o_1 is the object distance and i_1 is the image distance as shown in Figure 1.21. The image is demagnified by

$$M = \frac{i_1}{o_1} \tag{1.166}$$

Each succeeding CRL in the y- and z-directions will image a small section of the object (e.g., the head of the man, 1) However, each partial image (2) is inverted and overlaps other images (scrambled image). Thus, using a single 3-D lens array will produce only multiple, inverted, overlapping images (i.e., the total image is scrambled and blurred).

As shown in Figure 1.21, to produce a complete composite image of the object, we must use another 3-D lens array (called the *erecting 3-D lens array*). A one-to-one image is produced (i.e., the object is the same size as the produced image). The erecting 3-D lens array matches those of the inverting 3-D lens array. Each inverting CRL (3) and erecting CRL (4) of the two 3-D lens arrays has a common optical axis. CRL (3) forms an inverted, intermediate partial image (5) of part of object 1 (i.e., the head of the man in Figure 1.21) at an aperture stop. The second CRL (4) erects the intermediate image and places the resulting final image in overlapped registration with adjacent images, forming an erect uniform composite image in the image plane of the whole object. For this system, distances

$$i_1 = o_2 \tag{1.167}$$

$$i_2 = o_1 \tag{1.168}$$

The separation between the erecting 3-D lens array (3) and the inverting 3-D lens array (4) is

$$d = i_1 + o_2 = 2i_1 \tag{1.169}$$

In Figure 1.21, the lengths l of 3-D lens arrays are very small compared to the 3-D lens array focal lengths f. This permits the use of the simple thin-lens formulae of Eqs. (1.165) and (1.166).

To form a single complete image, the FOV of each of the adjacent CRL must overlap on the object plane. This enables the image portions, which are transmitted through adjacent compound lens pairs with partially overlapping areas. The multiple image points, which are formed from a common object point, are required to coincide at the image plane, where they are in focus. The FOV of the object for each pair of CRL arrays, and the field of coverage of the resulting image portion, extend approximately to the optic axes of the adjacent lens pairs. Since multiple images of each object point are formed, there is overlapping in the composite image.

When an object is imaged through two or more parallel CRLs, several conditions must be satisfied in order to obtain coincidence of the multiple images of each object point in the final image. In the image plane, the merging to a single point of multiple image points, which are produced by a single object point, require the following three merge conditions:

1. A correspondence between each point in the object plane must be made to each point in the image plane.
2. Brightness uniformity of the final image is obtained by:
   - (a) Having a large amount of overlap of the individual image fields of adjacent CRL pairs;
   - (b) Providing a gradual tapering off of the field's brightness toward the edge of each image by vignetting, so that sharply defined image field edges do not appear.
3. The two CRLs are spaced such that there is an intermediate image between them.

As shown in Figure 1.21, each of the 3-D lens arrays (3 and 4) is composed of multiple biconcave lenses, which form 2-D arrays of CRLs. The CRLs can be spaced apart in the transverse y- and z-directions by approximately the same distance b. The transverse spacing distance b between the CRLs can be set approximately equal to the effective aperture radius r_a of one CRL, where, from Eq. (1.126),

$$b = r_a = \sqrt{\frac{2R}{N\mu}} \tag{1.170}$$

That is, one calculates transverse spacing of the CRLs from the radius of curvature R and number of lenses N, and the lens material linear attenuation $\mu(\lambda)$ of the individual lenses that comprise the identical CRLs. The imaging array is set for a particular center neutron or X-ray wavelength λ within the X-ray or neutron source bandwidth. This provides uniform coverage of the object plane. The best selection of the effective aperture radius r_e is the absorption aperture radius r_a, where

$$r_e \cong r_a = \sqrt{\frac{2R}{N\mu}} \tag{1.171}$$

The separation b between CRLs in the transverse y- and z-directions can also vary appreciably without undue loss of image quality. To achieve a uniform brightness and a clear total image, the important design principles are primarily merge conditions (1) and (2).

1.3.2 Magnified Imaging

To obtain a magnified image, two different inverting and erecting 3-D lens arrays are used, as shown in Figure 1.22. Unlike the one-to-one imaging system, the two 3-D lenses have differing separations b_1 between the CRLs of the inverting 3-D lens array and b_2 between the CRLs of the erecting 3-D lens arrays. These separations b_1 and b_2 are repeated in the CRL arrays in the transverse y- and z-directions. As with one-to-one imaging, the inverting CRL (e.g., 3) and erecting CRL (e.g., 4) have a common optical axis AF, which is parallel to the axial x-direction. Again, the lengths l of the CRLs (e.g., 3, 4, 6, 7 in Figure 1.22), which form the 3-D lens arrays, are much smaller than their respective focal lengths f. In the transverse y- and z-directions, which are perpendicular to AF, the CRLs form a 3-D array that has the same focal length and mechanical and absorption apertures. The inverting 3-D lens array positioned closest to an object plane AE, which transmits the image of the object plane onto a central image plane, as the inverted intermediate image portions.

The erecting 3-D lens array transmits the inverted intermediate image in focus onto an image plane FJ, as an erected final image, which is in the same orientation as the image of the object. Thus, a CRL (3) on the inverting 3-D lens array and a CRL (4) on the erecting 3-D lens array form a CRL pair, which acts together as a lens train. This lens train transmits a portion of the image of an object, which appears on the

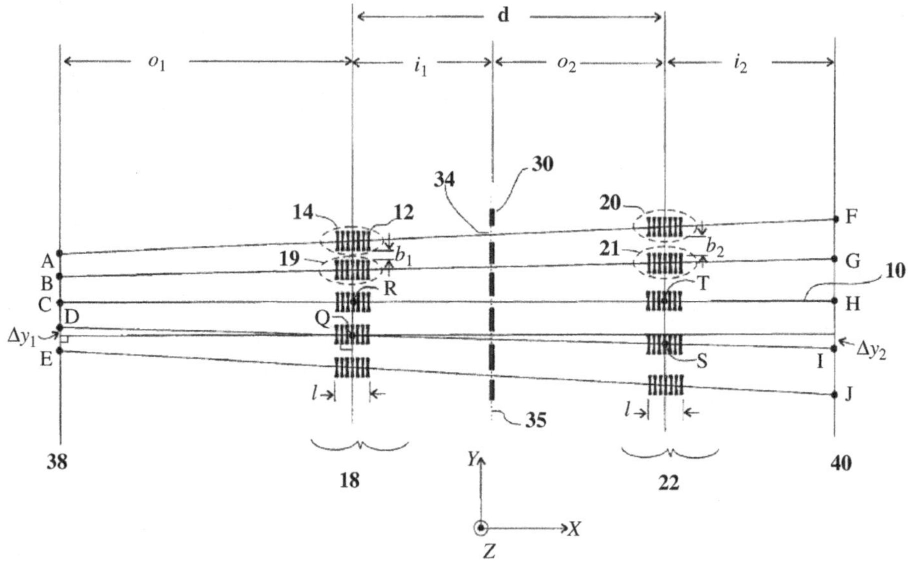

Figure 1.22 A side view of a 3-D lens array system using thin CRLs that produces a magnified image of an object (Piestrup, 2006).

object plane, onto the image plane. The final image can be magnified or demagnified depending on the selection of spacing and focal length of the lens system.

The rules for determining the dimensional relationship between the two 3-D arrays are the same for the other cases of either converging or parallel optical axes, which result in demagnification or lack of magnification, respectively. The total magnification of the overall 3-D lens system must equal the product of the magnification of the two inverting and erecting 3-D CRLs (e.g., 3 and 4). This enables coincidence of overlapping image portions and focuses each portion in the same plane, which is necessary to form a complete image.

A grid of imaginary lines is formed, such that each intersection or node of the grid lies on an optical axis of one of the CRLs. In Figure 1.22, points A, B, C, D, and E represent nodes in the object plane. The grid or latticework in the object plane is imaged onto the image plane. Points A, B, C, D, and E are imaged on to points F, G, H, I, and J in the image plane. Lattice magnification M_L is defined as the ratio of (1) the size of the object in the object plane, divided by (2) the size of its image in the image plane. Thus, the length AE will be lattice magnified to FJ, where

$$M_L = \frac{FJ}{AE} \tag{1.172}$$

Nodal magnification M_n is defined as the total magnification of a lens pair for the limited object portion covered by a particular CRL pair. This nodal magnification is usually the product of the magnification of the two CRLs (e.g., 3, 4), which lie along the same optical axis (e.g., AF).

For the complete 3-D lens system to work (that is, for complete image formation without distortion), requires the nodal magnification M_n to equal the lattice magnification M_L, where

$$M_n = M_L \tag{1.173}$$

Equal nodal and lattice magnifications $M_n = M_L$, require that separate image portions will coincide and overlap, if the size of each image portion has been magnified to equal the spacing between the image portions. The nodal magnification is given by a product of the two CRLs, where

$$M_n = M_1 M_2 \tag{1.174}$$

Using Eqs. (1.173) and (1.174), and using nodal magnifications of CRLs (10) and (12), without losing generality, the lattice magnification M_L becomes

$$M_L = M_n = M_1 M_2 = \frac{i_1}{o_1} \cdot \frac{i_2}{o_2} = \frac{UV}{CD} \cdot \frac{HI}{UV} = \frac{RQ + RH((TS - RQ)/RT)}{RQ - CR((TS - RQ)/RT)} \tag{1.175}$$

The axial separation d between the inverting and erecting CRL 3-D array lenses is set equal to the sum of the image distance i_1 of the upstream inverting CRL array and the downstream object distance o_2 of the erecting CRL array, where

$$d = i_1 + o_2 \tag{1.176}$$

The axial separation d between the inverting and erecting CRL 3-D arrays, then sets ratio r of the transverse distance RQ of the upstream inverting array and the corresponding transverse distance TS of the downstream erecting CRLs, where

$$r = \frac{TS}{RQ} \tag{1.177}$$

This ratio r then simplifies Eq. (1.175) for the lattice magnification M_L. Substituting r for TS/RQ in the expression of the lattice magnification M_L of Eq. (1.175) and solving for ratio r yields

$$r = 1 + \frac{M_n - 1}{d + i_2 - M_n o_1} \tag{1.178}$$

Equation (1.175) or (1.178) can be used to design the 3-D lens systems for X-ray or neutron, large-area imaging, and collection. These two equations give the lens spacing, which must be satisfied, to form the complete reconstructed image by means of the CRL system, as given in Figure 1.22. Equation (1.178), shows that the ratio r of the CRL separations is given by the nodal magnification M_n, lens separation d, image distance i_2 from the second lens, and object distance o_1 from the first lens.

In between the 3-D lens arrays is an accurately aligned array of apertures, which are capable of passing X-rays only at the aperture of the iris. Known as a *field stop* in visible optics, this array of irises limits the cross talk between images and prevents multiple images. A single central aperture stop is used in Figs. 1.21 and 1.22. Other aperture stops can be strategically placed on either side of the 3-D lens arrays to prevent stray rays from entering the wrong field.

The CRLs have apertures, which limit the useful collection area. There are three possible apertures, all of which limit the lenses' size. These are the ordinary mechanical aperture, the absorption aperture, and the parabolic aperture. The absorption and mechanical apertures can be useful because these limiting apertures can attenuate the scattered X-rays. This prevents skewed rays from passing through any two CRLs, which are not on the same optical axis. The field-stop iris array and the absorption and mechanical aperture are designed to prevent unintended combinations of CRLs from transmitting X-rays (neutrons), which result in spurious images, multiple images, or stray X-rays (neutrons).

1.4 Integration on the Complex Plane—Cauchy–Riemann Theorem, Cauchy Integration, and Residues

We begin this section by deriving Green's theorem. Consider first the area integral, where one integrates a function $f(z)$ over a surface, where the orientation of the surface at each point is described by the surface normal, unit vector $\hat{\mathbf{n}}$. Consider the simple case of a differential rectangle, that is, an infinitesimally small rectangle, where one calculates

$$\iint f(x,y)\mathrm{d}x\ \mathrm{d}y = \iint (\nabla \times \mathbf{p}) \cdot \hat{\mathbf{n}}\ \mathrm{d}x\ \mathrm{d}y = \left(\frac{\partial v}{\partial x} - \frac{\partial u}{\partial y}\right)\mathrm{d}x\ \mathrm{d}y \tag{1.179}$$

with

$$f(z) = (\nabla \times \mathbf{p}) \cdot \hat{\mathbf{n}} \tag{1.180}$$

$$\mathbf{p} = u\hat{\mathbf{x}} + v\hat{\mathbf{y}} \tag{1.181}$$

$$\nabla = \hat{\mathbf{x}}\frac{\partial}{\partial x} + \hat{\mathbf{y}}\frac{\partial}{\partial y} \tag{1.182}$$

$$\nabla \times \mathbf{p} = \left(\frac{\partial v}{\partial x} - \frac{\partial u}{\partial y}\right)\hat{\mathbf{n}} \tag{1.183}$$

Consider the contour integral around the infinitesimally small, closed, rectangular loop, which encloses the differential rectangular area over which the surface integral of Eq. (1.179) was performed. The orientation of the rectangular, closed-loop path of the line integral is described by the differential tangent vector $\mathbf{dl}$. In the case of the simple rectangle, one computes $\mathbf{p} \cdot \mathbf{dl}$ for the four differential lengths of the infinitesimal rectangle. Moving in a counterclockwise direction, at each corner point, the function $\mathbf{p} = u\hat{\mathbf{x}} + v\hat{\mathbf{y}}$ is evaluated, and then multiplied by the differential length segment (rectangle side) that follows the corner point. The four corners of the infinitesimal rectangle are

$$(x, y) \quad (x + \mathrm{d}x, y) \quad (x + \mathrm{d}x, y + \mathrm{d}y) \quad (x, y + \mathrm{d}y) \tag{1.184}$$

The contour integral around the infinitesimal rectangle is

$$\oint \mathbf{p} \cdot \mathbf{dl} = u(x,y)\mathrm{d}x + v(x+\mathrm{d}x,y)\mathrm{d}y - u(x+\mathrm{d}x,y+\mathrm{d}y)\mathrm{d}x - v(x,y+\mathrm{d}y)\mathrm{d}y \tag{1.185}$$

Rearrangement of Eq. (1.185) yields

$$\oint \mathbf{p} \cdot \mathbf{dl} = [u(x,y) - u(x+\mathrm{d}x,y+\mathrm{d}y)]\mathrm{d}x + [v(x+\mathrm{d}x,y) - v(x,y+\mathrm{d}y)]\mathrm{d}y, \tag{1.186}$$

where

$$[u(x,y) - u(x+\mathrm{d}x,y+\mathrm{d}y)]\mathrm{d}x = \left[\frac{u(x,y) - u(x+\mathrm{d}x,y+\mathrm{d}y)}{\mathrm{d}y}\right]\mathrm{d}x\,\mathrm{d}y = -\frac{\partial u}{\partial y}\mathrm{d}x\,\mathrm{d}y \tag{1.187}$$

$$[v(x+\mathrm{d}x,y) - v(x,y+\mathrm{d}y)]\mathrm{d}y = \left[\frac{[v(x+\mathrm{d}x,y) - v(x,y+\mathrm{d}y)]\mathrm{d}y}{\mathrm{d}x}\right]\mathrm{d}x\,\mathrm{d}y = \frac{\partial v}{\partial x}\mathrm{d}x\,\mathrm{d}y \tag{1.188}$$

With substitution of Eqs. (1.187–1.188) into Eq. (1.186), the line integral around an infinitesimal rectangle becomes

$$\oint \mathbf{p} \cdot \mathbf{dl} = \left(\frac{\partial v}{\partial x} - \frac{\partial u}{\partial y}\right)\mathrm{d}x\,\mathrm{d}y \tag{1.189}$$

Comparison of the surface integral of the infinitesimal rectangle of Eq. (1.179), with the contour integral (closed-loop line integral) of Eq. (1.189) along the four rectangle sides of differential lengths, thus yields Green's theorem, where

$$\iint (\nabla \times \mathbf{p}) \cdot \hat{\mathbf{n}}\,\mathrm{d}x\,\mathrm{d}y = \oint \mathbf{a} \cdot \mathbf{dl} \tag{1.190}$$

The previous 2-D vector $\mathbf{p} = u\hat{\mathbf{x}} + v\hat{\mathbf{y}}$ has components in two spatial directions. In comparison, the complex function $f(z)$ is a 2-D function, which has a component along the real axis and a component along the imaginary axis of the complex plane.

That is, the complex function $f(z)$ has a real part $\mathrm{Re}[f(z)]$, and an imaginary part $\mathrm{Im}[f(z)]$, where

$$f(z) = \mathrm{Re}[f(z)] + \mathrm{i}\,\mathrm{Im}[f(z)] \quad \text{and} \quad z = x + \mathrm{i}y \tag{1.191}$$

Performing the derivative of the complex function $f(z)$, requires one point z to approach another point $z + \Delta z$ in any direction in the complex plane, as their separation goes to zero ($\Delta z \to 0$). Thus, as the separation distance $\Delta z \to 0$ goes to zero, the two points, z and $z + \Delta z$, which approach each other along either the real x-axis, or the imaginary y-axis, must produce the same derivative, where

$$\lim_{\mathrm{d}z\to 0}\left[\frac{f(z+\mathrm{d}z)-f(z)}{\mathrm{d}z}\right]_{\mathrm{d}z=\mathrm{d}x} = \lim_{\mathrm{d}z\to 0}\left[\frac{f(z+\mathrm{d}z)-f(z)}{\mathrm{d}z}\right]_{\mathrm{d}z=\mathrm{i}\,\mathrm{d}y} \tag{1.192}$$

Assume that the complex function $f(z)$ is given by

$$f(z) = u(x, y) + \mathrm{i}v(x, y) \tag{1.193}$$

Substitution of Eq. (1.193) into Eq. (1.192) yields

$$\begin{aligned}\frac{\mathrm{d}f}{\mathrm{d}z} &= \lim_{\mathrm{d}x\to 0}\frac{u(x+\mathrm{d}x, y) + \mathrm{i}v(x+\mathrm{d}x, y) - u(x, y) - \mathrm{i}v(x, y)}{\mathrm{d}x}\\ &= \lim_{\mathrm{d}y\to 0}\frac{u(x, y+\mathrm{d}y) + \mathrm{i}v(x, y+\mathrm{d}y) - u(x, y) - \mathrm{i}v(x, y)}{\mathrm{i}\mathrm{d}y}\end{aligned} \tag{1.194}$$

Taking the limits along the real and imaginary axis in Eq. (1.194) yields

$$\frac{\mathrm{d}f}{\mathrm{d}z} = \frac{\mathrm{d}u}{\mathrm{d}x} + \mathrm{i}\frac{\mathrm{d}v}{\mathrm{d}x} = \frac{1}{\mathrm{i}}\left(\frac{\mathrm{d}u}{\mathrm{d}y} + \mathrm{i}\frac{\mathrm{d}v}{\mathrm{d}y}\right) \tag{1.195}$$

The complex derivative $\mathrm{d}f/dz$ of the complex function $f(z)$, with respect to the complex variable z, becomes

$$\frac{\mathrm{d}f}{\mathrm{d}z} = \frac{\mathrm{d}u}{\mathrm{d}x} + \mathrm{i}\frac{\mathrm{d}v}{\mathrm{d}x} = \frac{\mathrm{d}v}{\mathrm{d}y} - \mathrm{i}\frac{\mathrm{d}u}{\mathrm{d}y} \tag{1.196}$$

One equates the real and imaginary parts in Eq. (1.196), which yields the Cauchy–Riemann equations, where

$$\frac{\partial u}{\partial x} = \frac{\partial v}{\partial y} \tag{1.197}$$

$$\frac{\partial v}{\partial x} = -\frac{\partial u}{\partial y} \tag{1.198}$$

A function is analytic ($z = a$), if function $f(z)$ is differentiable at point $z = a$, and a small region of radius ε around $z = a$, where $|z - a| \leq \varepsilon$, for $\varepsilon > 0$. A point z_p, where the function $f(z)$ becomes infinite and $f(z_p) = \infty$, is a singularity called a *pole.* A simply connected region is a region of space where a loop can be shrunk to a single point without crossing a pole z_p. That is, a closed loop in a simply connected region contains no poles.

We next consider integration on the complex, 2-D plane, where the integral of the complex function $f(z)$ with respect to complex variable $f(z)$ is given by

$$\int_{z_1}^{z_2} f(z)\mathrm{d}z = \int_{z_1}^{z_2} (u + \mathrm{i}v)(\mathrm{d}x + \mathrm{i}\ \mathrm{d}y) = \int_{z_1}^{z_2} [u\ \mathrm{d}x - v\ \mathrm{d}y + \mathrm{i}(u\ \mathrm{d}y + v\ \mathrm{d}x)] \quad (1.199)$$

Given unit vector $\hat{\mathbf{x}}$ along the real x-axis and unit vector $\hat{\mathbf{y}}$ along the imaginary y-axis and the differential length vector $\mathbf{dl}$, one defines the vector $\mathbf{a}$ with component magnitudes (u, v), and the vector $\mathbf{b}$ with component magnitudes (v, u), where

$$\mathbf{a} = u\hat{\mathbf{x}} - v\hat{\mathbf{y}} \quad (1.200)$$

$$\mathbf{b} = v\hat{\mathbf{x}} + u\hat{\mathbf{y}} \quad (1.201)$$

$$\mathbf{dl} = \mathrm{d}x\hat{\mathbf{x}} + \mathrm{d}y\hat{\mathbf{y}} \quad (1.202)$$

Using Eqs. (1.200) and (1.201) to express the integrand of Eq. (1.199), one obtains the sum of two line integrals:

$$\int_{z_1}^{z_2} f(z)\mathrm{d}z = \int_{z_1}^{z_2} [u\ \mathrm{d}x - v\ \mathrm{d}y + \mathrm{i}(u\ \mathrm{d}y + v\ \mathrm{d}x)] = \int_{z_1}^{z_2} \mathbf{a}\cdot\mathbf{dl} + \mathrm{i}\int_{z_1}^{z_2} \mathbf{b}\cdot\mathbf{dl} \quad (1.203)$$

Equation (1.203) can be applied to the clockwise integration along a closed-loop path, or a closed-loop line integral, which is termed a *contour integral,* where

$$\oint f(z)\mathrm{d}z = \oint \mathbf{a}\cdot\mathbf{dl} + \mathrm{i}\oint \mathbf{b}\cdot\mathbf{dl} \quad (1.204)$$

Assume that someone, acting as a numerical integrator of the contour integral, walks counterclockwise along the closed-loop path of integration, where the interior region of the loop is always to the left of the numerical integrator. Each of the N points, z_i for $i = 1, 2, \ldots, N$, on the closed-loop curve is separated by incremental distance Δz, where the start point z_1 and the end point z_N are identical, with $z_1 = z_N$. The numerical integrator measures $f(z_i)$ computes the product $f(z_i)\Delta z$ and adds the point-by-point contributions $f(z_i)\Delta z$. After one full revolution around the loop, the numerical integrator has computed the contour integral as the integral sum $\sum_{i=1}^{N} f(z_i)\Delta z$.

We now derive the Cauchy integral theorem, which applies to a function $f(z)$, which is analytic at every point in a simply connected region that contains no poles. One starts with the contour integrals of Eq. (1.204), where

$$\oint f(z)\mathrm{d}z = \oint [u\ \mathrm{d}x - v\ \mathrm{d}y + \mathrm{i}(u\ \mathrm{d}y + v\ \mathrm{d}x)] = \oint \mathbf{a}\cdot\mathrm{d}\mathbf{l} + \mathrm{i}\oint \mathbf{b}\cdot\mathrm{d}\mathbf{l} \qquad (1.205)$$

Application of Green's theorem of Eq. (1.190) to Eq. (1.205) gives

$$\oint f(z)\mathrm{d}z = \oint \mathbf{a}\cdot\mathrm{d}\mathbf{l} + \mathrm{i}\oint \mathbf{b}\cdot\mathrm{d}\mathbf{l} = \iint(\nabla\times\mathbf{a})\cdot\hat{\mathbf{n}}\,\mathrm{d}x\,\mathrm{d}y + \mathrm{i}\iint(\nabla\times\mathbf{b})\cdot\hat{\mathbf{n}}\,\mathrm{d}x\,\mathrm{d}y \qquad (1.206)$$

The unit surface normal vector $\hat{\mathbf{n}}$, perpendicular to the plane, which contains the vectors **a** and **b**, where

$$\nabla\times\mathbf{a} = \left(\hat{\mathbf{x}}\frac{\partial}{\partial x} + \hat{\mathbf{y}}\frac{\partial}{\partial y}\right)\times(u\hat{\mathbf{x}} - v\hat{\mathbf{y}}) = -\left(\frac{\partial v}{\partial x} + \frac{\partial u}{\partial y}\right)\hat{\mathbf{n}} \qquad (1.207)$$

$$\nabla\times\mathbf{b} = i\left(\hat{\mathbf{x}}\frac{\partial}{\partial x} + \hat{\mathbf{y}}\frac{\partial}{\partial y}\right)\times(v\hat{\mathbf{x}} + u\hat{\mathbf{y}}) = \left(\frac{\partial u}{\partial x} - \frac{\partial v}{\partial y}\right)\hat{\mathbf{n}} \qquad (1.208)$$

Substitution of Eqs. (1.207) and (1.208) into Eq. (1.206) yields

$$\oint f(z)\mathrm{d}z = \iint\left[-\left(\frac{\partial v}{\partial x} + \frac{\partial u}{\partial y}\right) - \mathrm{i}\left(\frac{\partial u}{\partial x} - \frac{\partial v}{\partial y}\right)\right]\mathrm{d}x\,\mathrm{d}y \qquad (1.209)$$

One applies the Cauchy–Riemann equations of Eqs. (1.197) and (1.198) to the real and imaginary parts of the integrand, each of which is zero:

$$\frac{\partial v}{\partial x} = -\frac{\partial u}{\partial y} \Rightarrow -\left(\frac{\partial v}{\partial x} + \frac{\partial u}{\partial y}\right) = 0 \qquad (1.210)$$

$$\frac{\partial u}{\partial x} = \frac{\partial v}{\partial y} \Rightarrow -\mathrm{i}\left(\frac{\partial u}{\partial x} - \frac{\partial v}{\partial y}\right) = 0 \qquad (1.211)$$

Hence, one obtains the Cauchy theorem, where if the complex function $f(z)$ on the complex plane is analytic on a closed loop and in the interior of the closed loop, then

$$\oint f(z)\mathrm{d}z = 0 \qquad (1.212)$$

The integral on the right side of Eq. (1.213) is a Cauchy contour integral, which one can solve in the complex plane as shown in Figure 1.23, where we assign the function $f(k)$, so that

$$\int_{-\infty}^{\infty} \frac{\exp(ikr)}{(k-(w/c))(k+(w/c))} k \, dk = \int_{-\infty}^{\infty} \frac{f(k)}{k-(w/c)} dk \tag{1.213}$$

where

$$f(k) = \frac{k \exp(ikr)}{(k+(w/c))} \tag{1.214}$$

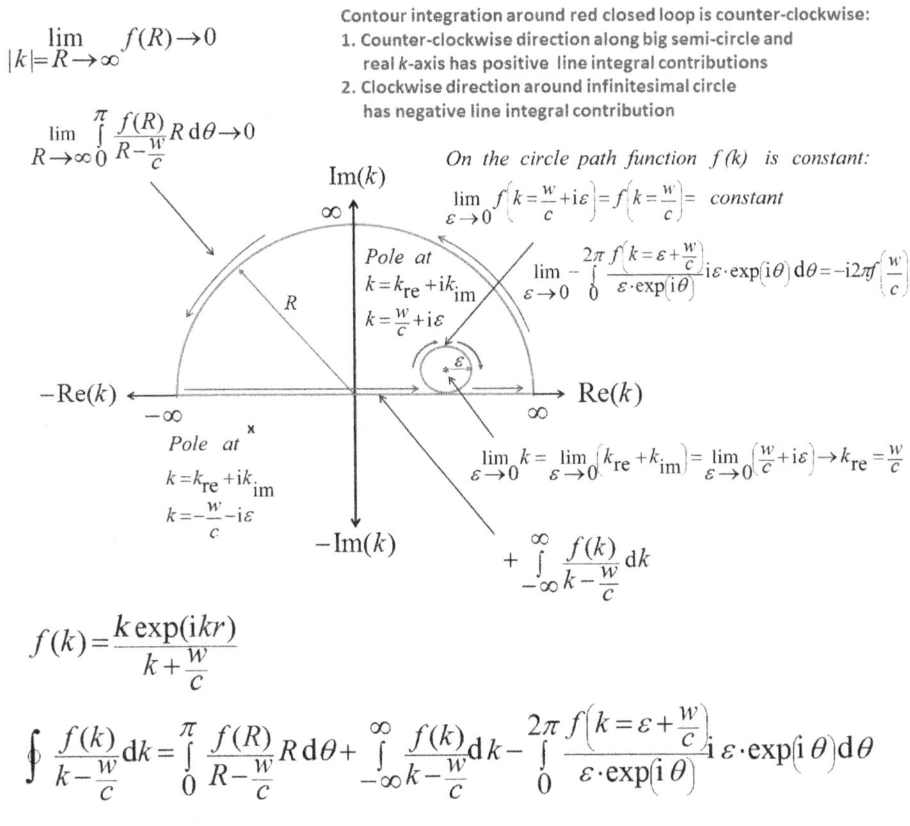

Figure 1.23 A Cauchy contour integration along a closed path in the upper half of the complex wave vector k-plane, which excludes the pole or singularity at $k = w/c$. The poles occur on the real k-axis at $k = w/c$ and $k = -w/c$. However, the poles are shifted off the real k-axis by $+\varepsilon$ and $-\varepsilon$ to facilitate the exclusion of the pole in the contour integration around the closed semicircle loops in either (1) the upper half-plane for the pole at $k = w/c$, where $f(k) \to 0$ for wave number $k = k_{re} + ik_{im}$ as $k_{im} \to \infty$, or (2) the lower half-plane for the pole at $k = -w/c$, where $f(k) \to 0$ for wave number $k = k_{re} + ik_{im}$ as $k_{im} \to -\infty$. The function $f(k)$ acquires constant value $f(k = w/c)$ along the infinitesimally small circle path as circle radius $\varepsilon \to 0$. Each contour integration around the entire closed loop contributes $i2\pi f(k = w/c)$ to the counterclockwise line integral along the real k-axis ($-\infty \le k \le \infty$). This contribution is a result of the clockwise line integration around the infinitesimally small circle over polar angle range $0 \le \theta \le 2\pi$.

That is, given the complete integrand function $g(k)$, one extracts the offensive singularity (pole) term $1/(k-a)$, which an infinite integrand $g(a)=\infty$ at $k=a$. The remaining function after extraction of the cavity (pole) is the pristine, analytic function $f(k)$, which is finite at every point of the contour and area within the contour.

In the integration over a particular region of the complex plane, this method of pristine trading allows the bad function $g(k)$ at $k=a$ to be traded for the extracted, good function $f(k)$, which is multiplied by the remaining bad term $1/(k-a)$ at $k=a$. That is, this extraction from the complete integrand function $g(k)$ of the analytic function $f(k)$, allows one to form the Cauchy integral or contour integral on the complex plane, where

$$\int_{-\infty}^{\infty} g(k)\mathrm{d}k = \int_{-\infty}^{\infty} \frac{f(k)}{k-a}\mathrm{d}k \Rightarrow \oint g(k)\mathrm{d}k = \oint \frac{f(k)}{k-a}\mathrm{d}k \tag{1.215}$$

and

$$f(k) = \frac{g(k)}{k-a} \tag{1.216}$$

The 1-D integration along the real k-axis, is converted to integration along a closed loop on the 2-D complex plane, which includes the real k-axis line integral as part of the loop. With integration on the complex plane, we can now choose a closed path or loop, which excludes the point $k=a$ from the interior of the closed-loop path. Then, along this closed curve, the function $f(k)$ is integrated. The integral of the complete, nonanalytic function $g(k)$ of Eq. (1.215), which has been converted to a Cauchy integral, via extraction of the infinity pole term, will be shown to be equal to the residue $f(k=a)$ multiplied by $\mathrm{i}2\pi$, where

$$P_{\mathrm{C}}\int_{-\infty}^{\infty} g(k)\mathrm{d}k = P_{\mathrm{C}}\int_{-\infty}^{\infty} \frac{f(k)}{k-a}\mathrm{d}k = \mathrm{i}2\pi\cdot f(a) \tag{1.217}$$

Here, P_{C} indicates the taking of only the nondivergent Cauchy principal part of the integral. The residue $\mathrm{RES}[g(k=a)]$ of the nonanalytic function $g(k)$ in the integrand of Eq. (1.217) is obtained by multiplying $g(k)$ by the factor $(k-a)$, which forms, in the limit as $k\to a$, the analytic function $f(a)$, where

$$\mathrm{RES}[g(k=a)] = \lim_{k\to a}(k-a)g(k) = f(a) \tag{1.218}$$

One useful trick for expressing the complete integrand function $g(k)$, which contains trigonometric functions such as $\cos\theta$ and $\sin\theta$, is to convert these functions of θ to functions of k, where

$$\cos\theta = \frac{\exp(\mathrm{i}\theta)+\exp(-\mathrm{i}\theta)}{2} = \frac{1}{2}\left(k+\frac{1}{k}\right) \tag{1.219}$$

$$\sin\theta = \frac{\exp(i\theta)+\exp(-i\theta)}{2i} = \frac{1}{2i}\left(k-\frac{1}{k}\right) \tag{1.220}$$

with

$$k = \exp(i\theta) \tag{1.221}$$

$$dk = i\exp(i\theta)d\theta \Rightarrow d\theta = \frac{dk}{iik} \tag{1.222}$$

Consider the case of a nonanalytic function $g(k)$ with an mth-order pole $(z-a)^m$, for positive pole integer index $m = 1, 2, 3, \ldots, N$. The goal is then to extract from $g(k)$ the mth pole $1/(z-a)^m$ and the associated analytic function $f_m(k)$. Hence, one obtains the N poles and the corresponding N analytic functions $f_n(k)$ from $g(k)$, where

$$g(k) = \frac{f(k)}{(k-a)^m} \tag{1.223}$$

One means of extracting the mth-order pole $(z-a)^m$ and associated analytic function $f_m(k)$ from $g(k)$ is to express the nonanalytic function $g(k)$ as an infinite power series of $(k-a)^p$, where power series integer index $p = -m, -m+1, \ldots, 1, 2, 3, \ldots, \infty$ and

$$g(k) = \sum_{p=-m}^{\infty} c_p(k-a)^p \tag{1.224}$$

For the pole integer index $r = 1, 2, 3, \ldots, N$, one obtains the mth analytic function, which is associated with the extraction of the mth-order pole $(z-a)^m$ from the nonanalytic function $g(k)$, where

$$f_m(a) = \lim_{k\to a}(z-a)^m g(k) = \lim_{k\to a}(z-a)^m \sum_{p=-m}^{\infty} c_p(k-a_p)^p \tag{1.225}$$

The residue of the complete, nonanalytic function $g(k)$ at $z = a$, that is RES[$g(a)$], where $g(k)$ has an mth-order pole $(z-a)^m$, is obtained by

$$\text{RES}[g(a)] = f_m(a) = \lim_{k\to a}\sum_{p=0}^{\infty} c_p(k-a_p)^p \tag{1.226}$$

Based on Eq. (1.226), for the mth-order pole $(z-a)^m$, the RES[$g(a)$] can be obtained from the mth-order derivative in the limit $z \to a$, where

$$\text{RES}[g(a)] = f_m(a) = \lim_{k\to a}\frac{1}{(m-1)!}\frac{d^{m-1}}{dk^{m-1}}[(z-a)^m g(z)] \tag{1.227}$$

We now proceed with the derivation of the Cauchy integral result, which is expressed as a summation of residues that are contained within the region bounded by the closed-loop line integral. With integration on the complex plane, the wave number radius k becomes a complex number, where

$$k = \mathrm{Re}(k) + \mathrm{iIm}(k) = k_{\mathrm{re}} + \mathrm{i}k_{\mathrm{im}} \quad (1.228)$$

Substitution into Eq. (1.214) for the complex k of Eq. (1.228) then shows directly the phase and attenuation components of function $f(k)$, where

$$f(k) = \frac{\exp(\mathrm{i}k_{\mathrm{re}}r)\exp(-k_{\mathrm{im}}r)}{k + w/c}(k_{\mathrm{re}} + \mathrm{i}k_{\mathrm{im}}) \quad (1.229)$$

In the upper half-plane of the complex plane, where the imaginary part of the complex wave number is positive $k_{\mathrm{im}} \geq 0$, the function $f(k)$ is bounded and goes to zero as $k_{\mathrm{im}} \rightarrow +\infty$, where

$$\lim_{k_{\mathrm{im}} \rightarrow +\infty} f(k) = \lim_{k_{\mathrm{im}} \rightarrow +\infty} \frac{\exp(\mathrm{i}k_{\mathrm{re}}r)\exp(-k_{\mathrm{im}}r)}{k + w/c}(k_{\mathrm{re}} + \mathrm{i}k_{\mathrm{im}}) \rightarrow 0 \quad (1.230)$$

However, in the lower half-plane of the complex plane, where the imaginary part of the complex wave number is negative $k_{\mathrm{im}} < 0$, the function $f(k)$ is not bounded and goes to infinity as $k_{\mathrm{im}} \rightarrow -\infty$, where

$$\lim_{k_{\mathrm{im}} \rightarrow -\infty} f(k) = \lim_{k_{\mathrm{im}} \rightarrow -\infty} \frac{\exp(\mathrm{i}k_{\mathrm{re}}r)\exp(-k_{\mathrm{im}}r)}{k + w/c}(k_{\mathrm{re}} + \mathrm{i}k_{\mathrm{im}}) \rightarrow \infty \quad (1.231)$$

Hence, integration of $f(k)$ in the complex plane is along a contour that is contained within the upper half-plane, where $k_{\mathrm{im}} \geq 0$.

The poles or singularities occur on the real k-axis at $k = w/c$ and $k = -w/c$. However, in Figure 1.23, the poles have been shifted off the real k-axis by $+\mathrm{i}\varepsilon$ and $-\mathrm{i}\varepsilon$, respectively, parallel to the imaginary k-axis. The electromagnetic wave, via $f(k)$, is attenuated to zero $f(k) \rightarrow 0$ for wave number $k = k_{\mathrm{re}} + \mathrm{i}k_{\mathrm{im}}$ as $k_{\mathrm{im}} \rightarrow \infty$, whereas $f(k)$ is unbounded in the lower half-plane, where $f(k) \rightarrow -\infty$, as $k_{\mathrm{im}} \rightarrow -\infty$. That is, the shifting of the pole $k = w/c$ above the real k-axis by $+\mathrm{i}\varepsilon$ allows exclusion of the $k = w/c$ pole in contour integration around the closed semicircle loop in the upper half-plane. Hence, contour integration and exclusion of the pole at $k = w/c$ occurs in the upper half-plane, as shown in Figure 1.23. However, if (1) the electromagnetic wave, via $f(k)$, is attenuated to zero $f(k) \rightarrow 0$, for wave number $k = k_{\mathrm{re}} + \mathrm{i}k_{\mathrm{im}}$, as $k_{\mathrm{im}} \rightarrow -\infty$, and (2) the function $f(k)$ is unbounded in the upper half-plane, where $f(k) \rightarrow \infty$, as $k_{\mathrm{im}} \rightarrow \infty$, then the contour integration would be around a closed semicircle loop in the lower half-plane for the pole at $k = -w/c$.

As shown in Figure 1.23, in the upper half-plane of the complex plane, the contour encloses the pole or singularity at $(k_{re}, k_{im}) = (w/c, \varepsilon)$, where $\varepsilon \to 0$, and

$$\lim_{\varepsilon\to 0} k = \lim_{\varepsilon\to 0}(k_{re} + k_{im}) = \lim_{\varepsilon\to 0}\left(\frac{w}{c} + \mathrm{i}\varepsilon\right) \to k_{re} = \frac{w}{c} \tag{1.232}$$

The integral along the real k-axis of Eq. (1.213) becomes an integration along the contour in the complex plane, as shown in Figure 1.23, where

$$\lim_{\varepsilon\to 0}\int_{-\infty}^{\infty}\frac{\exp(\mathrm{i}kr)}{(k-w/c)(k+w/c)}k\,\mathrm{d}k = \lim_{\varepsilon\to 0}\int_{-\infty}^{\infty}\frac{f(k)}{k-w/c\frac{w}{c}}\mathrm{d}k = \oint\frac{f(k)}{k-w/c}\mathrm{d}k \tag{1.233}$$

The contour integral in Eq. (1.233) is the integration along the closed curve in the upper half-plane, where

$$\begin{aligned}\oint\frac{f(k)}{k-w/c}\mathrm{d}k = &\int_0^{\pi}\frac{f(R)}{R-w/c}R\,\mathrm{d}\theta + \int_{-\infty}^{\infty}\frac{f(k)}{k-w/c}\mathrm{d}k\\ &-\int_0^{2\pi}\frac{f(k=\varepsilon+w/c)}{\varepsilon\cdot\exp(\mathrm{i}\theta)}\mathrm{i}\varepsilon\cdot\exp(\mathrm{i}\theta)\mathrm{d}\theta\end{aligned} \tag{1.234}$$

The distance of a point on the infinitesimal circle of radius ε, which is centered on the singularity at $k = w/c$, is a constant distance ε, where

$$k - \frac{w}{c} = \varepsilon\exp(\mathrm{i}\theta) = \varepsilon\cos\theta + \mathrm{i}\varepsilon\sin\theta \tag{1.235}$$

Since $\mathrm{d}(w/c)/dk = 0$, then

$$\mathrm{d}k = \mathrm{i}\varepsilon\cdot\exp(\mathrm{i}\theta) \tag{1.236}$$

The interior of the closed curve of the contour integral in the upper half-plane excludes the singularity or pole, and the integration around one cycle of the closed path in the upper half-plane is thus zero:

$$\oint\frac{f(k)}{k-w/c}\mathrm{d}k = 0 \tag{1.237}$$

That is, the pole has been shifted an incremental distance ε from $(k_{re}, k_{im}) = (w/c, 0)$ to $(k_{re}, k_{im}) = (w/c, \varepsilon)$. The pole shifted to $(k_{re}, k_{im}) = (w/c, \varepsilon)$ creates a circular path, which is traversed clockwise for 360° at infinitesimal radial distance ε from the excluded pole. In this manner, the pole has been excluded from the interior of the closed integration path of the contour integral, so that integral around the entire closed path is zero, where

$$\oint\frac{f(k)}{k-w/c}\mathrm{d}k = 0 = \int_0^{\pi}\frac{f(R)}{R-w/c}R\,\mathrm{d}\theta + \int_{-\infty}^{\infty}\frac{f(k)}{k-w/c}\mathrm{d}k - \int_0^{2\pi}\frac{f(k=\varepsilon+w/c)}{\varepsilon}\mathrm{i}\varepsilon\,\mathrm{d}\theta \tag{1.238}$$

The contour integral of zero is the sum of the following three component path integrals:

1. The counterclockwise path around the semicircle with a large radius R, equals the magnitude of the wave number, where $R = |k|$ and $R \to \infty$, where

$$\lim_{R\to\infty} f(R) \to 0 \tag{1.239}$$

and

$$\lim_{R\to\infty}\int_0^{\pi}\frac{f(R)}{R-w/c}R\,\mathrm{d}\theta = \lim_{R\to\infty}\int_0^{\pi}\frac{f(R)}{R}R\,\mathrm{d}\theta = \lim_{R\to\infty}\int_0^{\pi}f(R)\mathrm{d}\theta \to 0 \tag{1.240}$$

2. The counterclockwise path along the real k-axis from negative infinity to positive infinity, which is the sum of the real k-axis segments $-\infty < k_{\mathrm{re}} \le w/c$ and $w/c \le k_{\mathrm{re}} \le \infty$, which is thus $-\infty \le k_{\mathrm{re}} \le \infty$, where

$$\int_{-\infty}^{\infty}\frac{f(k)}{k-w/c}\mathrm{d}k = \int_{-\infty}^{w/c}\frac{f(k)}{k-w/c}\mathrm{d}k + \int_{w/c}^{\infty}\frac{f(k)}{k-w/c}\mathrm{d}k \tag{1.241}$$

3. The clockwise 360° path around the tiny circle centered on the singularity or pole at $(k_{\mathrm{re}}, k_{\mathrm{im}}) = (w/c, \varepsilon)$, which has an infinitesimal small radius ε, where $\varepsilon \to 0$. The function $f(k)$, which is evaluated on the circle path, acquires a constant value $f(k = w/c)$ as $\varepsilon \to 0$.

This constant value of $f(k = w/c)$ is effectively multiplied by $-\mathrm{i}2\pi$, via integration over the polar angle θ around the circle path from $\theta = 0$ to 2π. The counterclockwise direction of paths around the great semicircle and real k-axis is opposite to the clockwise path around the infinitesimal circle. As a result, the clockwise path produces a negative contribution (minus sign for the last term of Eq. (1.238), and the two counterclockwise paths produce positive contributions, where

$$\lim_{\varepsilon\to 0} f\left(k = \varepsilon + \frac{w}{c}\right) \to f\left(k = \frac{w}{c}\right) \tag{1.242}$$

and

$$\lim_{\varepsilon\to 0}\int_0^{2\pi}\frac{f(k=\varepsilon+w/c)}{\varepsilon}\mathrm{i}\varepsilon\,\mathrm{d}\theta = \lim_{\varepsilon\to 0}\mathrm{i}\int_0^{2\pi} f\left(k=\varepsilon+\frac{w}{c}\right)\mathrm{d}\theta = \mathrm{i}\cdot 2\pi\cdot f\left(\frac{w}{c}\right) \tag{1.243}$$

The function $f(k)$, which is integrated around the full 360° circular path with infinitesimal distance ε from $k = w/c + \mathrm{i}\varepsilon$, has an approximately constant value of $f(k = w/c)$ as $\varepsilon \to 0$ and is thus multiplied by $\mathrm{i}2\pi$.

The resulting sum of these three component integrals of Eqs. (1.240), (1.241), and (1.243), which are substituted into Eq. (1.238), then yields the general contour integral result, termed the *Cauchy's integral formula,* where

$$P_{\mathrm{C}}\int_{-\infty}^{\infty} g(k)\mathrm{d}k = P_{\mathrm{C}}\int_{-\infty}^{\infty}\frac{f(k)}{k-w/c}\mathrm{d}k = \mathrm{i}\cdot 2\pi\cdot f\left(\frac{w}{c}\right) \tag{1.244}$$

Substituting Eq. (1.214) for $f(k)$ into Eq. (1.244) yields

$$P_C \int_{-\infty}^{\infty} \frac{k \exp(ikr)}{(k - w/c)(k + w/c)} dk = i \cdot \pi \cdot \exp\left(i \frac{w}{c} r\right) \tag{1.245}$$

Now consider a singularity or pole, which is situated in the upper complex half-plane above the real k-axis (see Figure 1.24). For example, the singularity is at $k = w/c + i\sigma$, where σ is the conductivity, which accounts for the attenuation of the electromagnetic wave in a metal. In this case, one excludes the singularity in closed-loop integration by a detour, infinitesimal circle of radius ε about the point $k = w/c + i\sigma$. The function $f(k)$ acquires the value $f(k = w/c + i\sigma + i\varepsilon)$ on the circle, which converges to a constant value $f(k = w/c + i\sigma)$ around the infinitesimal circle as $\varepsilon \to 0$. In the closed-loop integration, one requires two additional line integrals along (1) the line segment of imaginary lengths $i\sigma - i\varepsilon$ from $k = w/c$ on the real k-axis to the point above on the infinitesimal circle at $k = w/c + i\sigma - i\varepsilon$ and (2) from the point on the circle $k = w/c + i\sigma - i\varepsilon$ back down to the point on the real k-axis at $k = w/c$. These two line segment integrations (1) and (2) occur in opposite directions and thus cancel. Hence, a singularity with an imaginary component, which places the singularity off the real k-axis, results in the line integral along the real k-axis for $-\infty < k < \infty$, being equated to $i2\pi f(k = w/c + i\sigma)$.

As a remark, if one were to perform the path integration around the full contour N-times, which also includes N-times around the infinitesimal circle, the value of the integration for winding the function N times around, but excluding the pole at $k = w/c$, would be $i2\pi N f(w/c)$. Each full rotation or winding around the infinitesimal circular path contributes $i2\pi f(w/c)$ to the path integral.

As another remark, recall that a circle whose radius r is multiplied by 2π to obtain the circumference distance $2\pi \cdot r$. Around the path on the infinitesimal circle, the function $f(w/c, i\varepsilon)$ acquires a constant value $f(w/c)$ because of the infinitesimally close proximity distance ε to the singularity at $k = w/c$, where $\varepsilon \to 0$. On the infinitesimal circle, the constant valued $f(k = w/c)$ acts as a radius $r = f(w/c)$ of a circle, whose imaginary circumference distance is $i \cdot 2\pi \cdot r$ or $2\pi \cdot i \cdot f(w/c)$, via the line integral around the singularity at $k = w/c$. If the radius $f(w/c)$ is a real number "length," such as a phase shift term, the refractive decrement δ, inductance L, or capacitance C of energy storage, then the circumference "length" is an imaginary $2\pi \cdot i \cdot f(w/c)$. However, if the radius $f(w/c)$ is an imaginary number "length," such as the attenuation factor $i\beta$, or resistance R of energy attenuation, then the circumference "length" $-2\pi \cdot f(w/c)$ is a negative number.

The real inductive L current energy storage and capacitive C voltage energy storage, and the imaginary resistive R energy loss terms arise from the second-order differential equation for a RLC oscillator circuit, which has a driving source. That is, the RLC and electromagnetic wave second-order differential equations have the same form because the two equations are equivalent expressions of the same physical picture of an oscillating electron charge. In closing, one can try to avoid singularities in mathematics or in life, but the singularity imposes its presence periodically by the cyclic detour that one takes to avoid the singularity.

Recall the Cauchy integral of Eq. (1.244), which is repeated below, where

$$P_C \int_{-\infty}^{\infty} g(k)\mathrm{d}k = P_C \int_{-\infty}^{\infty} \frac{f(k)}{k-a}\mathrm{d}k = \mathrm{i}2\pi \cdot f(a) \qquad (1.246)$$

Also, recall that the residue RES[$g(k=a)$] of the nonanalytic function $g(k)$ in the integrand of Eq. (1.246) is obtained by

$$\mathrm{RES}[g(k=a)] = \lim_{k\to a}(k-a)g(k) = f(a) \qquad (1.247)$$

$$\oint \frac{f(k)}{\left(k+\frac{w_1}{c}-\mathrm{i}\sigma_1\right)\left(k-\frac{w_2}{c}-\mathrm{i}\sigma_2\right)}\mathrm{d}k = 0$$

1. A closed path is chosen to excludes poles (casinos) from the interior of the closed loop.
2. Integration around the closed curve yields zero, because the start and endpoints of the integral along the curve are the same.

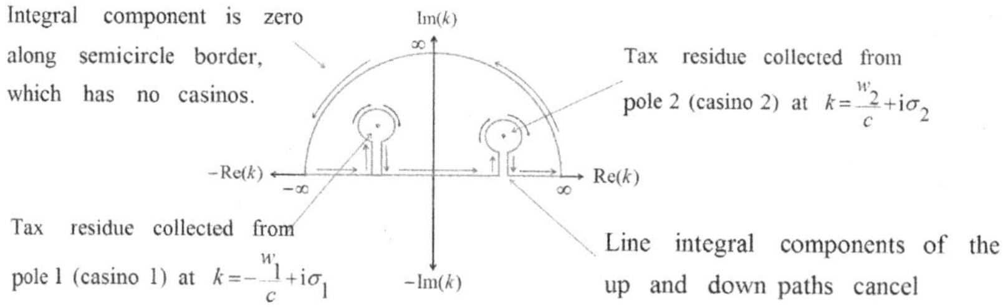

$$0 = \oint \frac{f(k)}{\left(k+\frac{w_1}{c}-\mathrm{i}\sigma_1\right)\left(k-\frac{w_2}{c}-\mathrm{i}\sigma_2\right)}dk =$$

$$= \int_{-\infty}^{\infty} \frac{f(k)}{\left(k+\frac{w_1}{c}-\mathrm{i}\sigma_1\right)\left(k-\frac{w_2}{c}-\mathrm{i}\sigma_2\right)}dk - \mathrm{i}2\pi\left[f\left(k=-\frac{w_1}{c}+\mathrm{i}\sigma_1\right)+f\left(k=\frac{w_2}{c}+\mathrm{i}\sigma_2\right)\right]$$

A tale of a two casinos and a city in the complex plane of taxes and politics:
Gambling was voted illegal in a city with two gambling casino. The city recieved real tax revenue from the casinos, and the city did not want to lose casino revenue, but the city needed to exclude the casinos from the interior of the city boundaries. So the city redrew the city boundaries to exclude the two gambling casinos. The city formally could not tax the casinos, but since the city provided water, gas, and electric to the casinos, the casinos were obliged to pay the city revene to keep operating. Hence, the casinos paid real money as an "imaginary tax" to a city's bank account in another county. Once a month the city tax collector would drive along the city limits road on border patrol, and also collect the imaginary tax revenue (residue) by driving around the two casinos, where a residue is collected of \$ $2\pi i f(-(w_1/c)+\mathrm{i}\sigma_1)$ from casino 1 and \$ $2\pi i f((w_2/c)+\mathrm{i}\sigma_2)$ from casino 2.

Figure 1.24 Illustrates the Residue theorem, which is used to calculate a variety of important integrals in the various fields of science and engineering. These integrals have poles or singularities, which yield infinity, when the integrand is evaluated at these poles. To avoid an infinite integrand, a closed path of integration is chosen to exclude the poles from the interior of the closed loop. The resulting closed-loop integral is then zero by the Cauchy theorem. However, the component line segment path along the real k-axis from Re[z] $= -\infty$ to ∞ equals the sum of the residues. The two poles at $z_1 = -w_1/c + \mathrm{i}\sigma_1$ and $z_2 = -w_2/c + \mathrm{i}\sigma_2$ contribute residues $\mathrm{i}2\pi f(z_1)$ and $\mathrm{i}2\pi f(z_2)$, which sum to equal the line integral $\int_{-\infty}^{\infty} f(z)\mathrm{d}z$. Function is assumed to be zero at infinity $f(|z| = \infty) = 0$, so the line integral along the semicircle is zero. Path integrals along parallel opposite path directions cancel. Thus, after finding a closed path to exclude the poles and evaluating the line integrals along the component curves, the line integral along the real z-axis is left with a residue.

Now, consider a nonanalytic function $g(k)$, which has $n = 1, 2, 3, \ldots, N$ poles a_n, which can be extracted via forming the analytic function $f_n(a_n)$. The residue theorem states that the nondivergent principal, Cauchy principal part of the integration of nonanalytic $g(k)$, with respect to k, is the sum of the residues multiplied by $2\pi \cdot \mathrm{i}$, where

$$\oint g(k)\mathrm{d}k = P_\mathrm{C}\int_{-\infty}^{\infty} g(k)\mathrm{d}k = \mathrm{i} \cdot 2\pi \cdot \sum_{n=1}^{N} \mathrm{RES}[f_n(a_n)] \tag{1.248}$$

The summed residues are a result of contour integration on the complex plane along a closed loop, which excludes the N poles a_n, where the function $f_n(a_n)$ is analytic on the closed loop and analytic in the loop interior. See Figure 1.24 which illustrates the residue theorem.

1.5 Derivation of the Complex Refractive Index of Material Medium (e.g., Lenses) Based on the Rayleigh Scatter of X-rays and Gammas

Rayleigh scatter by the bound atomic electrons of nuclei, which comprise a material medium, or the electrons of a plasma, accounts for the refraction, reflection, and diffraction of X-rays and gammas. Gammas can also be refracted by the polarized vacuum created by an extremely large electric field, via the Delbrück effect, which will be discussed later in this chapter. To characterize, describe, and determine the refractive index and scatter cross section of a material medium for X-rays and gamma, which is due to coherent, elastic Rayleigh scatter, the following seven steps will be done in the seven subsequent subsections, namely:

1. Derive an electromagnetic wave equation in a vacuum or dielectric medium.
2. Derive an electromagnetic field produced by an accelerated charge.
3. Derive the acceleration of a bound atomic electron by an imposed electromagnetic field.
4. Extract the complex refractive index from electromagnetic wave equation.
5. Derive the scatter, absorption, and total cross section for electromagnetic wave scatter by medium.
6. Derive the optical theorem.
7. Derive the Kramers–Kronig relation and calculate the refractive decrement from the measured attenuation cross section.

1.5.1 The Electromagnetic Wave Equation in a Vacuum or Dielectric Medium

One informally refers to **D** and **E** as the electric field and to **B** and **H** as the magnetic field. Formally, however, one differentiates between (1) the term flux density vector for the magnetic field vector **B** and electric field displacement vector **D**, which have dimensions of flux lines per area (flux lines/m^2) and (2) the magnetic field intensity **H**, which has units of current ampere-turns per meter or number of ampere current loops per meter (ampere-turns/meter), and the electric field intensity **E**, which has units of energy per charge per unit length, or simply volts per unit length volts/meter. The electric field units were derived via measurements of the force and displacement between two charged spheres or plates, and the magnetic field units were historically derived via measurements of the force and displacement between two current-carrying wires; see Maxwell (1954, 1991).

We begin with the derivation of the continuity equation or conservation of charge. Maxwell's equation of Gauss's law for an electric field for the charge density ρ and the electric displacement vector **D** is

$$\nabla \cdot \mathbf{D} = \rho \quad \text{Gauss's law for electric field} \tag{1.249}$$

The vacuum (or air) constitutive relation between the electric field intensity vector **E** and the electric field displacement vector or electric field flux density vector **D** are related linearly by the vacuum permittivity ε_0, where

$$\mathbf{D} = \varepsilon_0 \mathbf{E} \quad \text{vacuum electric field constitutive relation} \tag{1.250}$$

In MKS units, the vacuum electrical permittivity ε_0 is expressed in units of farads/meter. The farad unit (F) for electrical capacitance for vacuum and materials (such as an electronic capacitor) is expressed in units of coulomb/meters, where charge q is in units of coulombs. The charge q [coulombs] is measured in terms of the force F (N) between two charged spheres or the force between two charged plates of a capacitor. The magnitude of the repelling (attractive) force F between two identical (opposite) charges q [coulombs], which are separated by distance r [m], is expressed by Coulomb's law, where

$$\begin{aligned}\text{Electric force} = F\ \left[\text{kg}\ m/s^2\right] &= \frac{q^2}{4\pi\varepsilon_0 r^2} \Rightarrow \varepsilon_0\ \left[\text{F/m}\right] = \frac{q^2}{4\pi F r^2} \\ &\equiv [\text{coulomb}]^2 \frac{[\text{s}]^2}{[\text{kg m}]} \cdot \frac{1}{[\text{m}]^2} = \frac{[\text{coulomb}]^2 \cdot [\text{s}]^2}{[\text{kg}] \cdot [\text{m}]^3}\end{aligned} \tag{1.251}$$

Thus from Eq. (1.251), the vacuum space permittivity ε_0 is measured as

$$\varepsilon_0\ \left[\text{F/m}\right] = \frac{10^7}{c^2 4\pi} \cdot \left[\frac{\text{coulomb}^2}{\text{newton m}^2}\right] = \frac{10^7}{c^2 4\pi} \cdot \left[\frac{\text{coulomb}^2}{\text{joule m}}\right] \tag{1.252}$$

Since volt = joule/coulomb, then from this expression for ε_0, the units of the capacitance farad is expressed as

$$\text{farad} = \left[\frac{\text{coulomb}^2}{\text{joule}}\right] = \left[\frac{\text{coulomb}}{\text{volt}}\right] \tag{1.253}$$

Maxwell's equation for Gauss's law for a magnetic field, which assumes no magnetic monopoles, is

$$\nabla \cdot \mathbf{B} = 0 \quad \text{Gauss's law for magnetic field} \tag{1.254}$$

The constitutive relation between the magnetic field intensity vector **H** and the magnetic flux density or magnetic induction vector **B** is related linearly by the vacuum permeability μ_0, where

$$\mathbf{B} = \mu_0 \mathbf{H} \quad \text{vacuum magnetic field constitutive relation} \tag{1.255}$$

The vacuum magnetic permeability μ_0, in MKS units, is expressed in units of henrys per meter. The henry unit [H] for electrical inductance for vacuum and materials (such as an electronic inductor) is expressed in units of (N m/A^2), or energy per meter (J/m), where ampere [A] is in units of charge per second (coulomb/s). The current I [coulomb$/s$] or I [A] is measured in terms of force F (newton units) between two parallel wires, each with current I [A], which repel for parallel currents (identical charges moving in the same direction) or attract (opposite charges moving in the same direction). The magnitude of the repelling (attractive) force F [N] per unit length L [m], that is, F_{L} $[N/m]$ between the parallel (antiparallel) currents, which are separated by distance r, is expressed by Ampere's force law, where

$$\begin{aligned} \frac{\text{Magnetic force}}{\text{Length}} &= F_{\mathrm{L}}\left[\frac{\mathrm{kg}\cdot\mathrm{m}}{\mathrm{s}^2}\cdot\frac{1}{\mathrm{m}}\right] = \frac{\mu_0}{2\pi}\cdot\frac{I^2}{r} \Rightarrow \mu_0\left[\frac{\text{henry}}{\mathrm{m}}\right] = 2\pi\frac{Fr}{I^2} \\ &\equiv \left[\frac{1}{\text{ampere}}\right]^2\cdot\left[\frac{\text{newton}}{\mathrm{m}}\right]\cdot[\mathrm{m}] = \left[\frac{\text{newton}}{\text{ampere}^2}\right] \end{aligned} \tag{1.256}$$

Thus from Eq. (1.256), the vacuum space permittivity μ_0 is measured as

$$\mu_0\left[\frac{\text{henry}}{\mathrm{m}}\right] = 4\pi\times 10^{-7}\left[\frac{\text{newton s}^2}{\text{coulomb}^2}\right] = 4\pi\times 10^{-7}\left[\frac{\text{newton}}{\text{ampere}^2}\right] \tag{1.257}$$

Since volt = joule/coulomb and amp = coulomb/s, then from this expression for μ_0 the units of the inductance henry is expressed as

$$\text{henry} = \left[\frac{\text{newton m}}{\text{ampere}^2}\right] = \left[\frac{\text{joule}}{\text{ampere}^2}\right] = \left[\frac{\text{volt coulomb}}{\text{ampere coulomb}/s}\right] = \left[\frac{\text{volt s}}{\text{ampere}}\right] \tag{1.258}$$

If the vacuum space electric permittivity ε_0 and permeability μ_0 are multiplied, one obtains the vacuum speed of light c, where

$$\begin{aligned} \mu_0\left[\frac{\text{henry}}{\mathrm{m}}\right]\varepsilon_0\left[\frac{\text{farad}}{\mathrm{m}}\right] &= \mu_0\varepsilon_0\left[\frac{\text{henry farad}}{\mathrm{m}^2}\right] \\ &= 4\pi\times 10^{-7}\left[\frac{\text{newton s}^2}{\text{coulomb}^2}\right]\times\frac{10^7}{c^2 4\pi}\cdot\left[\frac{\text{coulomb}^2}{\text{newton m}^2}\right] \\ &= \frac{1}{c^2}\left[\frac{\mathrm{s}^2}{\mathrm{m}^2}\right] \end{aligned} \tag{1.259}$$

That is,

$$c = \frac{1}{\sqrt{\mu_0 \varepsilon_0}} \tag{1.260}$$

The ratio ε_0/μ_0 is the impedance of vacuum space, which is not dissipation, but rather expresses the relation between the component energies of the electric and the magnetic fields contained in the electromagnetic wave, where

$$\sqrt{\frac{\mu_0}{\varepsilon_0}}\left[\frac{\text{farad}}{\text{henry}}\right] = \sqrt{\frac{c^2(4\pi)^2}{10^{14}}}\left[\frac{\text{newton}^2\ \text{m}^2\ \text{s}^2}{\text{coulomb}^4}\right] = 376.7\left[\frac{\text{joule/coulomb}}{\text{ampere}}\right] \tag{1.261}$$

Or simply,

$$\sqrt{\frac{\mu_0}{\varepsilon_0}}\left[\frac{\text{farad}}{\text{henry}}\right] = \left[\frac{\text{volt}}{\text{ampere}}\right] = 376.7\ \text{ohms} \tag{1.262}$$

In a circuit, the resistance of a resistor R equals the current or charges per second that flow for an applied voltage, which is applied energy per charge, where

$$R\,[\text{ohms}] = \frac{V\ [\text{joule/coulomb}]}{I\ [\text{coulomb}/s]} = \frac{V}{I}\left[\frac{\text{volt}}{\text{ampere}}\right] \tag{1.263}$$

The vacuum impedance ε_0/μ_0 is associated with the relation of component energies where no dissipation is involved. The time-varying electric field induces a magnetic field and the time-varying magnetic field induces an electric field. Hence, the energy of the electromagnetic wave is unchanged in a vacuum; however, energy is transferred back and forth without loss between the magnetic and electric field components in a manner determined by the vacuum impedance ε_0/μ_0.

Energy is exchanged between the electric and magnetic fields of an electromagnetic wave in the same manner, as energy is exchanged between the stored potential energies of charge in capacitor and the kinetic energy of moving charge or current in an inductor of a lossless LC circuit. To avoid infinite power, energy cannot instantaneously change. Energy does not change in a zero time interval, whether the energy is that of the electric field of an electromagnetic wave, the electric field of stored electric charge in a capacitor, the energy of the magnetic field flux density **B** of an electromagnetic wave, or the magnetic field flux density **B** of moving charge (current) through an inductor. The finite time required for energy change is enforced by the spatial and time derivatives and speed of light, which are contained within Maxwell's equations.

Maxwell's equation of Ampere's law, which equates the curl or rotational distribution of the magnetic field vector $\nabla \times \mathbf{H}$ to the sum of the time variation of the electric displacement vector **D** and the current density vector **J**, is

$$\nabla \times \mathbf{H} = \frac{\partial \mathbf{D}}{\partial t} + \mathbf{J} \quad \text{Ampere's law} \tag{1.264}$$

Suppose that we now take the divergence of both sides of Eq. (1.264):

$$\nabla \cdot \nabla \times \mathbf{H} = \nabla \cdot \left(\frac{\partial \mathbf{D}}{\partial t} + \mathbf{J} \right) \tag{1.265}$$

The vector identity, which states the divergence of a curl or pure rotation is zero, is applied to Eq. (1.265), where

$$\nabla \cdot \nabla \times \mathbf{H} = 0 \tag{1.266}$$

The continuity equation which expresses the conservation of charge, is obtained from Eq. (1.265), via Eq. (1.266), and the substitution of Gauss's law for the electric field of Eq. (1.249), where

$$\frac{\partial \rho}{\partial t} + \nabla \cdot \mathbf{J} = 0 \quad \text{charge continuity equation} \tag{1.267}$$

The current density $\mathbf{J}(\mathbf{r}, t)$ is a function of vector position $\mathbf{r}$ and time t, where $\mathbf{J}(\mathbf{r}, t)$ is the product of the spatial- and time-dependent, scalar charge density $\rho(\mathbf{r}, t)$, and the spatial- and time-dependent vector velocity $\mathbf{v}(\mathbf{r}, t)$ of the charges, where

$$\mathbf{J}(\mathbf{r}, t) = \rho(\mathbf{r}, t)\mathbf{v}(\mathbf{r}, t) \tag{1.268}$$

The charge density $\rho(\mathbf{r}, t) = en(\mathbf{r}, t)$ is a product of the elementary charge e and the spatial- and time-dependent, number density $n(\mathbf{r}, t)$ of the elementary charges, where

$$\rho(\mathbf{r}, t) = en(\mathbf{r}, t) \tag{1.269}$$

Now consider an electromagnetic wave passing through a material such as a beryllium lens, where the oscillating electric field $\mathbf{E}(\mathbf{r}, t)$ is perpendicular to the oscillating magnetic field $\mathbf{H}(\mathbf{r}, t)$, and both fields are perpendicular to the direction of wave propagation, which is the ray vector or wave vector $\mathbf{k}(\mathbf{r}, t)$. The wave vector is the product of the unit wave vector $\hat{\mathbf{k}}(r, t)$ and the scalar wave number k, where

$$\mathbf{k}(\mathbf{r}, t) = k\hat{\mathbf{k}}(r, t) = \frac{2\pi}{\lambda}\hat{\mathbf{k}}(r, t) \tag{1.270}$$

The wave number k is the reciprocal of the wavelength λ of the electromagnetic wave multiplied by 2π, where

$$k\ [\text{rad/cm}] = \frac{2\pi\ [\text{rad/wavelength}]}{\lambda\ [\text{cm/wavelength}]} \tag{1.271}$$

The wave number k has units of radians per distance, where 2π radians is the angular rotation for one cycle or oscillation time period T, and the time period of the wave is the reciprocal of the angular frequency w, so that

$$T\left[\mathrm{s/wavelength}\right] = \frac{2\pi\,[\mathrm{rad/wavelength}]}{w\,[\mathrm{rad/s}]} \tag{1.272}$$

The angular frequency domain w [rad/s], which via the unit circle of trigonometry, is related to the temporal frequency domain f [s] by

$$w\,[\mathrm{rad/s}] = 2\pi\,[\mathrm{rad/wavelength}]\cdot f\,[\mathrm{wavelength/s}] \tag{1.273}$$

The electromagnetic wave phase velocity w/k in a vacuum is c, where

$$c = \frac{1}{\sqrt{\mu_0\varepsilon_0}} = \frac{w}{k} \tag{1.274}$$

The Maxwell equation of Faraday's law equates the curl or rotational distribution of the electric field $\nabla\times\mathbf{E}$ and the magnetic induction vector $\mathbf{B}$:

$$\nabla\times\mathbf{E} = -\frac{\partial\mathbf{B}}{\partial t} \quad \text{Faraday's law} \tag{1.275}$$

We now take the curl of both sides of Eq. (1.275), where

$$\nabla\times(\nabla\times\mathbf{E}) = \nabla\times\left(-\frac{\partial\mathbf{B}}{\partial t}\right) \tag{1.276}$$

Recall the vector identity for arbitrary vector $\mathbf{A}$, where

$$\nabla\times\nabla\times\mathbf{A} = \nabla(\nabla\cdot\mathbf{A}) - \nabla^2\mathbf{A} \tag{1.277}$$

Application of the vector identity of Eq. (1.277) to Eq. (1.276) yields

$$\nabla(\nabla\cdot\mathbf{E}) - \nabla^2\mathbf{E} = \nabla\times\left(-\frac{\partial\mathbf{B}}{\partial t}\right) \tag{1.278}$$

With the application of the constitutive relation between the magnetic field vector $\mathbf{H}$ and the magnetic density vector $\mathbf{B}$, where in vacuum space, $\mathbf{H} = \mathbf{B}/\mu_0$, Eq. (1.278) becomes

$$\nabla(\nabla\cdot\mathbf{E}) - \nabla^2\mathbf{E} = \nabla\times\left(-\frac{\partial\mathbf{B}}{\partial t}\right) \tag{1.279}$$

Examining the left side of Eq. (1.279), one can apply Gauss's law of Eq. (1.249) with the electric field constitutive relation in a vacuum of Eq. (1.250), where

$$\varepsilon_0 \nabla \cdot \mathbf{E} = \nabla \cdot \mathbf{D} = \rho \tag{1.280}$$

With substitution of Eq. (1.280) into Eq. (1.279), one obtains

$$\frac{1}{\varepsilon_0}\nabla\rho - \nabla^2\mathbf{E} = \nabla \times \left(-\frac{\partial \mathbf{B}}{\partial t}\right) \tag{1.281}$$

The time derivative of Ampere's law of Eq. (1.264) is taken, and the vacuum magnetic field $\mathbf{H} = \mathbf{B}/\mu_0$ constitutive relation of Eq. (1.255) and the electric field $\mathbf{E} = \mathbf{D}/\mu_0$ constitutive relation of Eq. (1.250), with the vacuum permittivity ε_0 and permeability μ_0 constants, are invoked, where

$$\frac{1}{\mu_0}\nabla \times \frac{\partial \mathbf{B}}{\partial t} = \varepsilon_0 \frac{\partial^2 \mathbf{E}}{\partial t^2} + \frac{\partial \mathbf{J}}{\partial t} \tag{1.282}$$

Substitution of Eq. (1.282) into Eq. (1.281) yields

$$\frac{1}{\varepsilon_0}\nabla\rho - \nabla^2\mathbf{E} = -\mu_0\varepsilon_0 \frac{\partial^2 \mathbf{E}}{\partial t^2} - \mu_0 \frac{\partial \mathbf{J}}{\partial t} \tag{1.283}$$

Rearrangement of Eq. (1.283), with the substitution of Eq. (1.260) for the vacuum speed of light *c*, yields the electromagnetic, vector wave equation, where

$$\left(\frac{\partial^2}{\partial t^2} - c^2\nabla^2\right)\mathbf{E}(\mathbf{r},t) = -\frac{1}{\varepsilon_0}\left[\frac{\partial \mathbf{J}(\mathbf{r},t)}{\partial t} + c^2\nabla\rho(\mathbf{r},t)\right] \quad \text{wave equation} \tag{1.284}$$

For transverse electromagnetic waves, the electric field $\mathbf{E}(\mathbf{r},t)$ is perpendicular to the direction of wave propagation, that is, the direction of the wave vector $\mathbf{k}$, where $\mathbf{E}\perp\mathbf{k}$. The transverse electric field of the electromagnetic wave is denoted by $\mathbf{E}_\perp(\mathbf{r},t)$. The induced current density $\mathbf{J}(\mathbf{r},t)$ of bound electrons in a medium is produced by the action of the imposed transverse electric field $\mathbf{E}_\perp(\mathbf{r},t)$ of the incident electromagnetic wave. Hence, the induced current density $\mathbf{J}(\mathbf{r},t)$ is parallel to the imposed transverse electric field $\mathbf{E}_\perp(\mathbf{r},t)$, and thus the induced current density $\mathbf{J}(\mathbf{r},t)$ is perpendicular to the wave vector $\mathbf{k}$, where $\mathbf{J}\perp\mathbf{k}$. The induced, transverse current density is thus denoted by $\mathbf{J}_\perp(\mathbf{r},t)$. The perpendicular electric field $\mathbf{E}_\perp(\mathbf{r},t)$ and current density $\mathbf{J}_\perp(\mathbf{r},t)$ vectors are now made explicit in the wave equation, so that Eq. (1.284) becomes

$$\left(\frac{\partial^2}{\partial t^2} - c^2\nabla^2\right)\mathbf{E}_\perp(\mathbf{r},t) = -\frac{1}{\varepsilon_0}\frac{\partial \mathbf{J}_\perp(\mathbf{r},t)}{\partial t} - \frac{c^2}{\varepsilon_0}\nabla\rho(\mathbf{r},t) \tag{1.285}$$

If no charge density gradient exists in the material, then

$$\nabla \rho(\mathbf{r}, t) = 0 \tag{1.286}$$

The wave equation then becomes

$$\left(\frac{\partial^2}{\partial t^2} - c^2\nabla^2\right)\mathbf{E}_\perp(\mathbf{r}, t) = -\frac{1}{\varepsilon_0}\frac{\partial \mathbf{J}_\perp(\mathbf{r}, t)}{\partial t} \tag{1.287}$$

From Maxwell's equation, we now directly calculate the electric and magnetic field amplitudes of a simple, plane electromagnetic wave, which propagates in the z-direction with the electric field assumed to be polarized in the transverse x-direction (see Figure 1.25). Once the electric and magnetic fields are calculated, one can obtain the simplified scalar wave equations for the single polarized electric field amplitude and the perpendicular, single, polarized magnetic field amplitudes. The scalar wave equations are found by equating the spatial derivative of Ampere's law with the time derivative of Faraday's law. Of course, with the calculated specific electric and magnetic field amplitudes, one can plug the electric field amplitude into the above vector wave equation Eq. (1.287) to obtain the simplified scalar wave equations for the transverse electric field Eq. (1.303), and transverse magnetic field Eq. (1.305), which however, are now derived directly from Maxwell's equations as follows. Assume that the electromagnetic wave propagates in a vacuum, with $\rho = 0$, along the z-axis, and the electric field is polarized in only the transverse x-direction, where

$$\mathbf{E}_\perp(\mathbf{r}, t) = E_x(z, t)\hat{\mathbf{x}} \tag{1.288}$$

Application of Gauss's law from Eq. (1.280) for the electric field $\mathbf{E}$ gives

$$\varepsilon_0 \nabla \cdot \mathbf{E} = 0 = \frac{\partial E_x}{\partial x} = 0 \tag{1.289}$$

Application of Faraday's law of Eq. (1.275) gives

$$\nabla \times \mathbf{E} = -\frac{\partial \mathbf{B}}{\partial t} \Rightarrow \left(\frac{\partial E_x}{\partial z} - \frac{\partial E_z}{\partial x}\right)\hat{\mathbf{y}} = -\frac{\partial \mathbf{B}}{\partial t} \Rightarrow \frac{\partial E_x}{\partial z} = -\frac{\partial \mathbf{B}}{\partial t} \Rightarrow \mathbf{B}(z, t) = B_y(z, t)\hat{\mathbf{y}} \tag{1.290}$$

One thus finds that the magnetic field, which propagates in the axial z-direction, is polarized (or oscillates) in the transverse y-direction, where

$$\mathbf{B}(z, t) = B_y(z, t)\hat{\mathbf{y}} \tag{1.291}$$

Substitution of Eq. (1.291) for $\mathbf{B}(z, t)$ into Faraday's law of Eq. (1.290) yields

$$\frac{\partial E_x(z, t)}{\partial z} = -\frac{\partial B_y(z, t)}{\partial t} \tag{1.292}$$

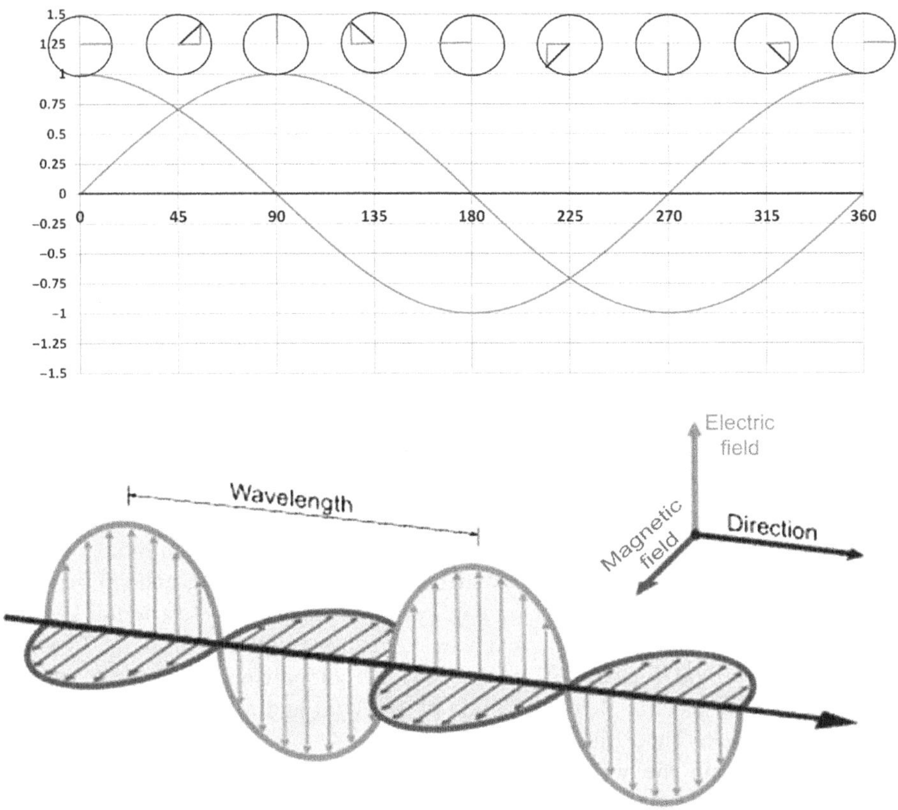

Figure 1.25 (Top) The unit circle, just like a clock, tracks the x- and y-components of the rotating clock hand (or rotating radius). Rather than plotting a sinusoidal, periodic amplitude as a function of phase angle along a line, the unit circle plots the periodic amplitude as a function polar, phase angle around a circle. The projections of the unit circle radius along the vertical y-axis, then provides the sine wave amplitude sin θ and the cosine wave amplitude cos θ along the x-axis. The unit circle clock on the 2-D real plane maps to the 2-D complex plane, via the power series for exp(iθ), cosθ, and i sin θ, and yields Euler's formula $\exp(i\theta) = \cos\theta + i\sin\theta = x + iy$. (Bottom) A schematic of the electromagnetic wave. Note the electric and magnetic fields are in phase, taken from the U.S. National Weather Service on Remote Sensing (http://www.srh.noaa.gov/jetstream/remote/remote_intro.htm).

The magnetic field can be obtained from the assumed electric field by applying the chain rule of differentiation, where

$$\frac{\partial E_x(z,t)}{\partial z} = -\frac{\partial B_y(z,t)}{\partial t} = -\frac{\partial B_y(z,t)}{\partial z}\frac{\partial z}{\partial t} \tag{1.293}$$

The wave velocity dz/dt is the speed of light in vacuum c, where

$$\frac{\partial z}{\partial t} = c \tag{1.294}$$

Substitution of Eq. (1.294) into Eq. (1.293) yields

$$\frac{\partial E_x(z,t)}{\partial z} = -c\frac{\partial B_y(z,t)}{\partial z} \Rightarrow \frac{\partial}{\partial z}(E_x(z,t) + cB_y(z,t)) = 0 \tag{1.295}$$

Hence,

$$B_y(z,t) = -\frac{1}{c}E_x(z,t) \tag{1.296}$$

From Faraday's law, we have obtained the plane wave magnetic field amplitude from the assumed electric field amplitude (polarized in the transverse x-direction), and the vacuum, axial wave propagation speed c, where (1) the magnetic field amplitude has a much smaller amplitude by $1/c$ of electric field amplitude, (2) the magnetic field has the same time dependence as the electric field, and (3) the electric and magnetic fields are perpendicular. For radio waves, the electric field amplitude of $3 \times 10^{-3}\ V/m$ produces a magnetic field flux amplitude of 10^{-11} T, where T = [weber$/m^2$], and [weber$/m^2$] = [V s/m^2]. In comparison, a focused, intense laser produces an electric field amplitude of $10^9\ V/m$ and a magnetic flux density of 3.33 T; see Bekefi and Barrett (1977).

Next, we obtain the wave equation for the transverse electric and magnetic fields. Application of Gauss's law for the magnetic field **B** gives

$$\nabla \cdot \mathbf{B} = 0 = \frac{\partial B_y}{\partial y} = 0 \tag{1.297}$$

Application of Ampere's law of Eq. (1.264), with $\mathbf{B} = B_y(z,t)\hat{\mathbf{y}}$, gives

$$\nabla \times \mathbf{B} = \frac{1}{c^2}\frac{\partial \mathbf{E}}{\partial t} \Rightarrow (\nabla \times \mathbf{B})_x = \frac{1}{c^2}\left(\frac{\partial E_x}{\partial t}\right)\hat{\mathbf{x}} \tag{1.298}$$

The result of Ampere's law of Eq. (1.298) is then

$$\frac{\partial B_y}{\partial z} = -\frac{1}{c^2}\left(\frac{\partial E_x}{\partial t}\right) \tag{1.299}$$

Recall the result of Faraday's law in Eq. (1.292):

$$\frac{\partial E_x(z,t)}{\partial z} = -\frac{\partial B_y(z,t)}{\partial t} \tag{1.300}$$

The space or z-derivative of both sides of Faraday's law of Eq. (1.300), yields

$$\frac{\partial^2 E_x(z,t)}{\partial^2 z} = -\frac{\partial B_y(z,t)}{\partial z\,\partial t} \tag{1.301}$$

The time or t-derivative of both sides of Ampere's law of Eq. (1.299) yields

$$\frac{\partial B_y}{\partial t\,\partial z} = -\frac{1}{c^2}\frac{\partial^2 E_x}{\partial t^2} \tag{1.302}$$

Comparison of Eq. (1.301) and Eq. (1.302) yields the wave equation for the transverse electric field amplitude $E_x(z,t)\hat{\mathbf{x}}$ for propagation along the z-axis as a function time t and position z, where

$$\frac{\partial^2 E_x(z,t)}{\partial t^2} = c^2\frac{\partial^2 E_x(z,t)}{\partial z^2} \tag{1.303}$$

From the result of Faraday's law in Eq. (1.296), we found

$$E_x(z,t) = -cB_y(z,t) \tag{1.304}$$

Substitution of Eq. (1.304) into Eq. (1.303) yields the wave equation for the transverse magnetic field amplitude $B_y(z,t)\hat{\mathbf{y}}$ for the propagation along the z-axis as a function of time t and position z, where

$$\frac{\partial^2 B_y(z,t)}{\partial t^2} = c^2\frac{\partial^2 B_y(z,t)}{\partial z^2} \tag{1.305}$$

and

$$c = \frac{1}{\sqrt{\mu_0\varepsilon_0}} \tag{1.306}$$

This transverse wave equation is identical to the differential equation for the wave amplitude Ψ for longitudinal wave propagation along a string at velocity v in the z-direction. That is, the amplitude of the longitudinal sound compression wave is

$$\frac{\partial^2 \Psi(z,t)}{\partial t^2} = v^2\frac{\partial^2 \Psi(z,t)}{\partial z^2} \tag{1.307}$$

For a string, the velocity v along a string is determined by the square root of the ratio string tension T and mass per unit string length μ, where

$$v = \sqrt{\frac{T}{\mu}} \tag{1.308}$$

For a sound wave, the velocity v through the fluid is determined by the square root of the ratio adiabatic bulk modulus B, divided by the ambient fluid density ρ_0, where

$$v = \sqrt{\frac{B}{\rho_0}} \tag{1.309}$$

Also, the wave equation appears in the form of the telegraph equation for the lossless voltage V and current I propagation down a transmission line with inductance L and capacitance C, with wave propagation speed v, where

$$\frac{\partial^2 V(\mathbf{r},t)}{\partial t^2} = \frac{1}{\sqrt{LC}}\nabla^2 V \tag{1.310}$$

$$\frac{\partial^2 V(\mathbf{r},t)}{\partial t^2} = \frac{1}{\sqrt{LC}}\nabla^2 V \tag{1.311}$$

$$v = \frac{1}{\sqrt{LC}} \tag{1.312}$$

The radiated power from an oscillating electromagnetic field is given by the Poynting vector $\mathbf{S}(r,t)$, which is the cross product of the electric $\mathbf{S}(r,t)$ and magnetic field intensity $\mathbf{S}(r,t)$ vectors, and gives the magnitude and direction of the energy flow per unit area per unit time, that is, the power density $\mathbf{S}$ $[J/s/m^2]$, with watt = joule/s and joule = volt · ampere, where

$$\mathbf{S}(\mathbf{r},t)\ [W/m^2] = \mathbf{E}(\mathbf{r},t)\ [V/m] \times \mathbf{H}(\mathbf{r},t)\ [A/m] \tag{1.313}$$

Recall that in a lumped circuit, the scalar power $P(\mathbf{r},t)$, expressed in watts, is the dot product of the vector current $\mathbf{I}(\mathbf{r},t)$ in A/coulomb, and vector voltage $\mathbf{V}(\mathbf{r},t)$ in J/coulomb, where

$$\mathbf{V}(\mathbf{r},t) = e\mathbf{E}(\mathbf{r},t) \tag{1.314}$$

$$P(\mathbf{r},t)\,[\mathrm{W}] = \mathbf{V}(\mathbf{r},t)\left[\frac{\text{joule}}{\text{coulomb}}\right]\cdot \mathbf{I}(\mathbf{r},t)\left[\frac{\text{coulomb}}{\text{s}}\right] = \mathbf{V}(\mathbf{r},t)\cdot\mathbf{I}(\mathbf{r},t)\left[\frac{\text{joule}}{\text{s}}\right] \tag{1.315}$$

and

$$P(\mathbf{r},t)\ [\mathrm{W}] = e\ [\text{charge}]\left(\mathbf{E}(\mathbf{r},t)\left[\frac{\text{joule}}{\text{coulomb}\cdot\text{charge}}\right]\cdot \mathbf{I}(\mathbf{r},t)\ [\text{amp}]\right) \tag{1.316}$$

Poynting's theorem, in the form of a differential or integral equation for electromagnetic waves, expresses the conservation of energy of the electromagnetic wave energy flow, storage, and attenuation. Poynting's theorem is derived by (1) taking the dot product of the magnetic field intensity **H** and the Faraday Maxwell's equation of Eq. (1.275), and (2) taking the dot product of the electric field intensity **E** and the Ampere Maxwell's equation of Eq. (1.264), where

$$\mathbf{H}\cdot(\nabla\times\mathbf{E}) = -\mathbf{H}\cdot\frac{\partial\mathbf{B}}{\partial t} \tag{1.317}$$

and

$$\mathbf{E}\cdot(\nabla\times\mathbf{H}) = \mathbf{E}\cdot\left(\frac{\partial\mathbf{D}}{\partial t}+\mathbf{J}\right) \tag{1.318}$$

Next, Eq. (1.318) is subtracted from (Eq. 1.317):

$$\mathbf{H}\cdot(\nabla\times\mathbf{E}) - \mathbf{E}\cdot(\nabla\times\mathbf{H}) = -\mathbf{H}\cdot\frac{\partial\mathbf{B}}{\partial t} - \mathbf{E}\cdot\left(\frac{\partial\mathbf{D}}{\partial t}+\mathbf{J}\right) \tag{1.319}$$

One recalls the vector identity for general vectors **A** and **B**, where

$$\nabla\cdot(\mathbf{A}\times\mathbf{B}) = \mathbf{B}\cdot(\nabla\times\mathbf{A}) - \mathbf{A}\cdot(\nabla\times\mathbf{B}) \tag{1.320}$$

With application of the vector identity of Eq. (1.320) and the vacuum constitutive relations $\mathbf{B} = \mu_0\mathbf{H}$ and $\mathbf{D} = \varepsilon_0\mathbf{E}$ to Eq. (1.319), one obtains

$$\nabla\cdot(\mathbf{E}\times\mathbf{H}) = \mu_0\mathbf{H}\cdot\frac{\partial\mathbf{H}}{\partial t} - \varepsilon_0\mathbf{E}\cdot\frac{\partial\mathbf{E}}{\partial t} - \mathbf{E}\cdot\mathbf{J} \tag{1.321}$$

The Poynting vector is defined by $\mathbf{S} = \mathbf{E}\times\mathbf{H}$, and the differential equation form of Poynting's theorem becomes

$$\nabla\cdot\mathbf{S} = \nabla\cdot(\mathbf{E}\times\mathbf{H}) = -\frac{\partial}{\partial t}\left(\frac{\mu_0 H^2}{2}+\frac{\varepsilon_0 E^2}{2}\right) - \mathbf{E}\cdot\mathbf{J} \tag{1.322}$$

The Poynting vector power flow **S** is perpendicular to the electric **E** and magnetic **B** fields, where Poynting's theorem expresses the balance of power. In the Poynting power balance equation, the power flows across area A into the enclosed volume V. The spatial change in power flow per unit area, upon crossing area A into enclosed volume V, is equal to, but opposite of, the sum of (1) the time rate of change of the stored energy per unit volume of the enclosed magnetic and electric fields, and (2) the absorbed power per unit volume within volume V, where

$$\nabla\cdot(\mathbf{E}\times\mathbf{H}) \Leftrightarrow \text{spatial change of power per area entering volume} \tag{1.323}$$

$$-\frac{\partial}{\partial t}\left(\frac{\mu_0 H^2+\varepsilon_0 E^2}{2}\right) \Leftrightarrow \text{time change of volume } E\&B \text{ field energy density} \tag{1.324}$$

$$\mathbf{E}\cdot\mathbf{J} \Leftrightarrow \text{power per volume dissipation in volume} \tag{1.325}$$

Clarification is provided by converting the Poynting differential equation to an integral equation, which is done via Gauss's theorem, which equates the volume integral of the divergence of a vector $\nabla \cdot \mathbf{C}$, with the surface integral of vector $\mathbf{C}$, where

$$\iiint \nabla \cdot \mathbf{C}\, \mathrm{d}V = \iint \mathbf{C} \cdot \hat{\mathbf{n}}\, \mathrm{d}A \tag{1.326}$$

Unit vector $\hat{\mathbf{n}}$ is the surface normal vector, which points outward from a surface. The outward-pointing, unit normal vector $\hat{\mathbf{n}}$ of the outer surface A is opposite in sign with the inward-pointing, unit normal vector $\hat{\mathbf{n}}$ of the inner surface enclosure area A of volume V. Application of Gauss's theorem of Eq. (1.326) to the differential equation form of Poynting's theorem of Eq. (1.322), thus yields the integral form of Poynting's theorem, where

$$\iint (\mathbf{E} \times \mathbf{H}) \cdot \hat{\mathbf{n}}\, \mathrm{d}A = -\frac{\partial}{\partial t} \iiint \left(\frac{\mu_0 H^2}{2} + \frac{\varepsilon_0 E^2}{2} \right) \mathrm{d}V - \iiint \mathbf{E} \cdot \mathbf{J}\, \mathrm{d}V \tag{1.327}$$

The integral Poynting's theorem, equates the power flow across volume-enclosed surface A, with the sum in volume V of (1) the power or time rate of change of the stored energy of the electric and magnetic fields, and (2) the dissipation or absorption of power, where

$$\iint (\mathbf{E} \times \mathbf{H}) \cdot \hat{\mathbf{n}}\, \mathrm{d}A \Leftrightarrow \text{power flow into volume} \tag{1.328}$$

$$-\frac{\partial}{\partial t} \iiint \left(\frac{\mu_0 H^2}{2} + \frac{\varepsilon_0 E^2}{2} \right) \mathrm{d}V \Leftrightarrow \text{time change of volume } E\&B \text{ field energy} \tag{1.329}$$

$$-\iiint \mathbf{E} \cdot \mathbf{J}\, \mathrm{d}V \Leftrightarrow \text{power dissipation in volume} \tag{1.330}$$

1.5.2 Electromagnetic Field Produced by an Accelerated Charge

Following the approach of Attwood (1999), we now calculate the electric field $\mathbf{E}$ from an accelerated electric charge e, where one starts by taking the Fourier transform of the electromagnetic vector wave equation for the general electric $\mathbf{E}(\mathbf{r}, t)$ of Eq. (1.284) and the charge continuity equation of Eq. (1.267), which are repeated here:

$$\frac{\partial^2 \mathbf{E}(\mathbf{r}, t)}{\partial t^2} - c^2 \nabla^2 \mathbf{E}(\mathbf{r}, t) + \frac{1}{\varepsilon_0} \frac{\partial \mathbf{J}(\mathbf{r}, t)}{\partial t} + \frac{c^2}{\varepsilon_0} \nabla \rho(\mathbf{r}, t) = 0 \tag{1.331}$$

and

$$\frac{\partial \rho}{\partial t} + \nabla \cdot \mathbf{J} = 0 \quad \text{charge continuity equation} \tag{1.332}$$

In the frequency domain w, one can then apply the electromagnetic wave equation and the continuity equation to obtain the electric field of an accelerating electric charge. Equivalent to the Fourier transform of these equations, one can use the phasor method of electrical engineering, which is applied to alternating voltage and current waveforms, that is, the sinusoidal waves that form the basis of the Fourier transforms (see Smith (1980)). The Fourier transforms represent the sinusoidal signals in temporal frequency space, which is based on the scalar, angular frequency w. Here, angular frequency $w = 2\pi/T$ corresponds to wave period $T = 1/f$, which is the reciprocal of the temporal frequency f. Likewise, one uses the Fourier transforms to represent the sinusoidal signals in the spatial frequency space, which is based on the vector, spatial frequency or wave vector $\mathbf{k}$. Unit wave vector $\hat{\mathbf{k}}$ points in the direction of wave propagation, where wave vector $\mathbf{k} = k\hat{\mathbf{k}}$. The angular wave number $k = 2\pi/\lambda$ for wavelength λ corresponds to the spatial wave number or the number of waves per unit distance in the direction of propagation is $f_\lambda = 1/\lambda$.

The frequency space electric field $\mathbf{E}(\mathbf{k}, w)$, current density $\mathbf{J}(\mathbf{k}, w)$, and charge density $\rho(\mathbf{k}, w)$ are obtained from the real space electric field $\mathbf{E}(\mathbf{r}, t)$, current density $\mathbf{J}(\mathbf{r}, t)$, and charge density $\rho(\mathbf{r}, t)$, via the Fourier transforms, where

$$\mathbf{E}(\mathbf{k}, w) = \int_r \mathrm{d}\mathbf{r} \int_t \mathrm{d}t\mathbf{E}(\mathbf{r}, t)\exp[\mathrm{i}(wt - \mathbf{k}\cdot\mathbf{r})] \tag{1.333}$$

$$\mathbf{J}(\mathbf{k}, w) = \int_\mathbf{r} \mathrm{d}\mathbf{r} \int_t \mathrm{d}t\mathbf{J}(\mathbf{r}, t)\exp[\mathrm{i}(wt - \mathbf{k}\cdot\mathbf{r})] \tag{1.334}$$

$$\rho(\mathbf{k}, w) = \int_\mathbf{r} \mathrm{d}\mathbf{r} \int_t \mathrm{d}t\rho(\mathbf{r}, t)\exp[\mathrm{i}(wt - \mathbf{k}\cdot\mathbf{r})] \tag{1.335}$$

Volume integration is over all position space and all time (rt-space), which is depicted by position vector $\mathbf{r}$ with rectangular components x, y, and z, and time t:

$$-\infty \le t \le \infty \tag{1.336}$$

$$\mathrm{d}\mathbf{r} = \mathrm{d}x\ \mathrm{d}y\ \mathrm{d}z \tag{1.337}$$

$$-\infty \le x \le \infty \tag{1.338}$$

$$-\infty \le y \le \infty \tag{1.339}$$

$$-\infty \le z \le \infty \tag{1.340}$$

In rectangular coordinates, the position vector $\mathbf{r}$ and wave vector $\mathbf{k}$ are

$$\mathbf{r} = \hat{\mathbf{x}}x + \hat{\mathbf{y}}y + \hat{\mathbf{z}}z \tag{1.341}$$

and

$$\mathbf{k} = \hat{\mathbf{x}}k_x + \hat{\mathbf{y}}k_y + \hat{\mathbf{z}}k_z \tag{1.342}$$

The squared magnitudes r^2 and k^2 of the position (or displacement or separation) vector **r** and wave vector **k** are obtained by the dot products of position vector **r** with itself ($\mathbf{r} \cdot \mathbf{r}$), and the dot product of wave vector **k** with itself ($\mathbf{k} \cdot \mathbf{k}$), where

$$\mathbf{r} \cdot \mathbf{r} = r^2 = x^2 + y^2 + z^2 \tag{1.343}$$

and

$$\mathbf{k} \cdot \mathbf{k} = k^2 = k_x^2 + k_y^2 + k_z^2 \tag{1.344}$$

Given polar angle θ between the position vector **r** and wave vector **k**, in rectangular coordinates, the dot product of the wave vector **k** with the radius vector **r** is, by definition,

$$\mathbf{k} \cdot \mathbf{r} \equiv kr \cos \theta = k_x x + k_y y + k_z z \tag{1.345}$$

The vector del ∇ operator and the scalar Laplacian ∇^2 operator in rectangular coordinates are

$$\nabla = \hat{\mathbf{x}} \frac{\partial}{\partial x} + \hat{\mathbf{y}} \frac{\partial}{\partial y} + \hat{\mathbf{z}} \frac{\partial}{\partial z} \tag{1.346}$$

and

$$\nabla^2 = \nabla \cdot \nabla = \frac{\partial^2}{\partial x^2} + \frac{\partial^2}{\partial y^2} + \frac{\partial^2}{\partial z^2} \tag{1.347}$$

The real space electric field $\mathbf{E}(\mathbf{r}, t)$, current density $\mathbf{J}(\mathbf{r}, t)$, and charge density $\rho(\mathbf{r}, t)$ are obtained from the frequency space electric field $\mathbf{E}(\mathbf{k}, w)$, current density $\mathbf{J}(\mathbf{k}, w)$, and charge density $\boldsymbol{\rho}(\mathbf{k}, w)$ via the inverse Fourier transforms, where

$$\mathbf{E}(\mathbf{r}, t) = \frac{1}{(2\pi)^4} \int_{\mathbf{k}} \mathrm{d}\mathbf{k} \int_{w} \mathrm{d}w \mathbf{E}(\mathbf{k}, w) \exp[-\mathrm{i}(wt - \mathbf{k} \cdot \mathbf{r})] \tag{1.348}$$

$$\mathbf{J}(\mathbf{r}, t) = \frac{1}{(2\pi)^4} \int_{\mathbf{k}} \mathrm{d}\mathbf{k} \int_{w} \mathrm{d}w \mathbf{J}(\mathbf{k}, w) \exp[-\mathrm{i}(wt - \mathbf{k} \cdot \mathbf{r})] \tag{1.349}$$

$$\rho(\mathbf{r}, t) = \frac{1}{(2\pi)^4} \int_{\mathbf{k}} \mathrm{d}\mathbf{k} \int_{w} \mathrm{d}w \rho(\mathbf{k}, w) \exp[-\mathrm{i}(wt - \mathbf{k} \cdot \mathbf{r})] \tag{1.350}$$

Volume integration is over all spatial frequency space and all temporal frequency space (*kw*-space), which is depicted by wave vector **k** with rectangular components x, y, z, and temporal frequency w, where

$$-\infty \leq w \leq \infty \tag{1.351}$$

$$\mathbf{dk} = dk_x\, dk_y\, dk_z \tag{1.352}$$

$$0 \leq k_x \leq \infty \tag{1.353}$$

$$0 \leq k_y \leq \infty \tag{1.354}$$

$$0 \leq k_z \leq \infty \tag{1.355}$$

The vector gradient ∇ operator in position space $\mathbf{r}$ (or r-space) becomes the vector multiplicative operator $-i\mathbf{k}$ in spatial frequency space $\mathbf{k}$ (or k-space), as seen by applying the vector gradient ∇ to the Fourier transform equation for the charge density $\rho(\mathbf{r}, t)$, where

$$\nabla\rho(\mathbf{r}, t) = \nabla\left\{\frac{1}{(2\pi)^4}\int_{\mathbf{k}} \mathbf{dk}\int_{w} dw\rho(\mathbf{k}, w)\exp[-i(wt - \mathbf{k}\cdot\mathbf{r})]\right\} \tag{1.356}$$

Using rectangular coordinates, Eq. (1.356) becomes

$$\begin{aligned}\nabla\rho(\mathbf{r}, t) = {} & \frac{1}{(2\pi)^4} \times \int_{-\infty}^{\infty} dt \int_{-\infty}^{\infty} dx \int_{-\infty}^{\infty} dy \\ & \times \int_{-\infty}^{\infty} dz\rho(\mathbf{k}, w)\left(\hat{\mathbf{x}}\frac{\partial}{\partial x} + \hat{\mathbf{y}}\frac{\partial}{\partial y} + \hat{\mathbf{z}}\frac{\partial}{\partial z}\right)\exp\left[-i(wt - k_x x - k_y y - k_z z)\right]\end{aligned} \tag{1.357}$$

The spatial differentiation in Eq. (1.357) demonstrates the duality between the spatial gradient (or derivative) operator ∇ of r-space, and the multiplicative vector operator $i\mathbf{k}$ of k-space. The k-space operator $\mathrm{i}\mathbf{k}$ is the wave vector $\mathbf{k}$ multiplied by $\mathrm{i} = -\sqrt{-1}$. In the complex plane, i corresponds to a 90° counterclockwise rotation of vector $\mathbf{k}$, which is then multiplied by scalar charge density $\rho(\mathbf{k}, w)$, where

$$\begin{aligned}\nabla\rho(\mathbf{r}, t) = {} & \frac{1}{(2\pi)^4} \times \int_{-\infty}^{\infty} dt \int_{-\infty}^{\infty} dx \int_{-\infty}^{\infty} dy \\ & \times \int_{-\infty}^{\infty} dz\rho(\mathbf{k}, w)(-i)\left(\hat{\mathbf{x}}k_x + \hat{\mathbf{y}}k_y + \hat{\mathbf{z}}k_z\right)\exp\left[-i(wt - k_x x - k_y y - k_z z)\right]\end{aligned} \tag{1.358}$$

or, simply,

$$\nabla\rho(\mathbf{r}, t) = \frac{1}{(2\pi)^4}\int_{\mathbf{k}} \mathbf{dk}\int_{w} dw(\mathrm{i}\mathbf{k})\rho(\mathbf{k}, w)\exp[-i(wt - \mathbf{k}\cdot\mathbf{r})] \tag{1.359}$$

The scalar Laplacian ∇^2 operator in position space $\mathbf{r}$ becomes the scalar multiplicative operator $-k^2$ in k-space, as seen by application of the scalar Laplacian ∇^2 to the Fourier transform equation for the electric field $\mathbf{E}(\mathbf{r}, t)$, where

$$\nabla^2\mathbf{E}(\mathbf{r}, t) = \nabla\left\{\frac{1}{(2\pi)^4}\int_{\mathbf{k}} \mathrm{d}\mathbf{k}\int_w \mathrm{d}w\mathbf{E}(\mathbf{k}, w)\exp[-(wt - \mathbf{k}\cdot\mathbf{r})]\right\} \tag{1.360}$$

In rectangular coordinates, substitution of the Laplacian operator in Eq. (1.360) yields

$$\begin{aligned}\nabla^2\mathbf{E}(\mathbf{r}, t) = {} & \frac{1}{(2\pi)^4} \times \int_{-\infty}^{\infty} \mathrm{d}t \int_{-\infty}^{\infty} \mathrm{d}x \int_{-\infty}^{\infty} \mathrm{d}y \int_{-\infty}^{\infty} \mathrm{d}z\mathbf{E}(\mathbf{k}, w) \\ & \times \left(\frac{\partial^2}{\partial x^2} + \hat{\mathbf{y}}\frac{\partial^2}{\partial y^2} + \hat{\mathbf{z}}\frac{\partial^2}{\partial z^2}\right)\exp\left[-\mathrm{i}(wt - k_x x - k_y y - k_z z)\right]\end{aligned} \tag{1.361}$$

The spatial differentiation in Eq. (1.361) demonstrates the duality between the real space, scalar Laplacian operator ∇^2, and the k-space multiplicative scalar operator $-k^2$, which is the squared wave number multiplied by -1. In the complex plane, this corresponds to a 180° counterclockwise rotation of vector $\mathbf{E}(\mathbf{k}, w)$ and then multiplication by k^2, where

$$\begin{aligned}\nabla^2\mathbf{E}(\mathbf{r}, t) = {} & \frac{1}{(2\pi)^4}\int_{-\infty}^{\infty} \mathrm{d}t \int_{-\infty}^{\infty} \mathrm{d}x \int_{-\infty}^{\infty} \mathrm{d}y \int_{-\infty}^{\infty} \mathrm{d}z\mathbf{E}(\mathbf{k}, w)(-1) \\ & \times (k_x^2 + k_y^2 + k_z^2)\exp\left[\mathrm{i}(wt - k_x x - k_y y - k_z z)\right]\end{aligned} \tag{1.362}$$

or, simply,

$$\nabla^2\mathbf{E}(\mathbf{r}, t) = \frac{1}{(2\pi)^4}\int_{\mathbf{k}} \mathrm{d}\mathbf{k}\int_w \mathrm{d}w(-k^2)\mathbf{E}(\mathbf{k}, w)\exp[-\mathrm{i}(wt - \mathbf{k}\cdot\mathbf{r})] \tag{1.363}$$

The application of the time derivative $\partial/\partial t$ to the Fourier transform equation for the electric field $E(\mathbf{r},\mathrm{t})$ yields

$$\frac{\partial}{\partial t}\mathbf{E}(\mathbf{r}, t) = \frac{\partial}{\partial t}\left\{\frac{1}{(2\pi)^4}\int_{\mathbf{k}} \mathrm{d}\mathbf{k}\int_w \mathrm{d}wE(\mathbf{k}, w)\exp[-\mathrm{i}(wt - \mathbf{k}\cdot\mathbf{r})]\right\} \tag{1.364}$$

The time differentiation in Eq. (1.364) demonstrates the duality between the partial time derivative $\partial/\partial t$ of t-space and the multiplication by $-\mathrm{i}w$ of w-space, where $\partial/\partial t$ and $-\mathrm{i}w$ are scalar operators. The w-space operator $-\mathrm{i}w$ is the angular frequency w multiplied by i. In the complex plane, this corresponds to a 90° counterclockwise rotation of vector $E(\mathbf{k},w)$, which is then multiplied by scalar frequency w, where

$$\frac{\partial}{\partial t}\mathbf{E}(\mathbf{r},t)=\frac{1}{(2\pi)^4}\int_{\mathbf{k}}\mathrm{d}\mathbf{k}\int_{w}\mathrm{d}w(-w)E(\mathbf{k},w)\exp[-\mathrm{i}(wt-\mathbf{k}\cdot\mathbf{r})] \tag{1.365}$$

One can apply the partial time derivative again to Eq. (1.365), where

$$\frac{\partial}{\partial t}\mathbf{E}(\mathbf{r},t)=\frac{1}{(2\pi)^4}\int_{\mathbf{k}}\mathrm{d}\mathbf{k}\int_{w}\mathrm{d}w(-w)E(\mathbf{k},w)\frac{\partial}{\partial t}\exp[-\mathrm{i}(wt-\mathbf{k}\cdot\mathbf{r})] \tag{1.366}$$

Carrying out the time differentiation in Eq. (1.366) demonstrates the duality between the partial time derivative, scalar operator $\partial^2/\partial t^2$ of t-space, and the multiplicative, scalar operator $-w^2$ of w-space. The w-space operator $-w^2$ is the squared angular frequency w multiplied by -1. In the complex plane, this corresponds to a 180° counterclockwise rotation of vector $E(\mathbf{k},w)$, which is then multiplied by the squared scalar frequency w^2, where

$$\frac{\partial^2}{\partial t^2}\mathbf{E}(\mathbf{r},t)=\frac{1}{(2\pi)^4}\int_{\mathbf{k}}\mathrm{d}\mathbf{k}\int_{w}\mathrm{d}w(-w^2)E(\mathbf{k},w)\exp[-\mathrm{i}(wt-\mathbf{k}\cdot\mathbf{r})] \tag{1.367}$$

There is a similarity of the vector operator duality of the r-space ∇ and the k-space $-\mathrm{i}\mathbf{k}$, as compared to the t-space $\partial/\partial t$ and w-space $-\mathrm{i}w$. Also, note the similarity of the scalar operator duality of the r-space ∇^2 and the k-space $-k^2$, as compared to the t-space $\partial^2/\partial t^2$ and w-space $-w^2$, where

$$\nabla=\hat{\mathbf{x}}\frac{\partial}{\partial x}+\hat{\mathbf{y}}\frac{\partial}{\partial y}+\hat{\mathbf{z}}\frac{\partial}{\partial z}\Leftrightarrow -\mathrm{i}\mathbf{k} \tag{1.368}$$

with

$$\frac{\partial}{\partial t}\Leftrightarrow -\mathrm{i}w \tag{1.369}$$

and

$$\nabla^2=\frac{\partial^2}{\partial x^2}+\frac{\partial^2}{\partial y^2}+\frac{\partial^2}{\partial z^2}\Leftrightarrow -k^2 \tag{1.370}$$

with

$$\frac{\partial^2}{\partial t^2}\Leftrightarrow -w^2 \tag{1.371}$$

Equations (1.368–1.371) are consistent with the Pythagorean theorem, updated by special and general relativity for space–time equivalence. The distance $|\mathbf{R}|=R$ in space–time remains invariant with a change of coordinates for the space–time of the separation vector, which is expressed by $\mathbf{R}=(\Delta x,\Delta y,\Delta z,\mathrm{i}c\Delta t)$ in the (x,y,z,t)

coordinate system and by $\mathbf{R} = (\Delta x', \Delta y', \Delta z', ic\Delta t')$ in the (x', y', z', t') coordinate system, where

$$|\mathbf{R}|^2 = R^2 = (\Delta x')^2 + (\Delta y')^2 + (\Delta z')^2 - (c\Delta t')^2 = (\Delta x)^2 + (\Delta y)^2 + (\Delta z) - (c\Delta t)^2 \tag{1.372}$$

Next, we obtain the spatial and temporal frequency space (*kw*) representation of the general, electromagnetic wave equation of Eq. (1.331), which is expressed in position and time–space (*rt*-space) as a second-order partial differential equation in time and space for the vacuum electric field. We start with the general wave equation of Eq. (1.331), which is transposed so that

$$\frac{\partial^2 \mathbf{E}(\mathbf{r}, t)}{\partial t^2} - c^2 \nabla^2 \mathbf{E}(\mathbf{r}, t) + \frac{1}{\varepsilon_0} \frac{\partial \mathbf{J}(\mathbf{r}, t)}{\partial t} + \frac{c^2}{\varepsilon_0} \nabla \rho(\mathbf{r}, t) = 0 \tag{1.373}$$

Also, we obtain the spatial and temporal frequency space (*kw*) representation of the charge continuity equation of Eq. (1.332), which is expressed in position and time–space (*rt*-space) as a first-order, partial differential equation in time and space, where

$$\frac{\partial \rho(\mathbf{r}, t)}{\partial t} + \nabla \cdot \mathbf{J}(\mathbf{r}, t) = 0 \tag{1.374}$$

The real space electric field $\mathbf{E}(\mathbf{r},t)$, current density $\mathbf{J}(\mathbf{r},t)$, and charge density $\rho(\mathbf{r},t)$ in the partial differential wave and charge continuity of Eqs. (1.373) and (1.374) are now expressed by their corresponding inverse Fourier transform expressions of Eqs. (1.375) and (1.376), which are integrations over spatial frequency and temporal frequency of the electric field $\mathbf{E}(\mathbf{k},w)$, current density $\mathbf{J}(\mathbf{k},w)$, and charge density $\rho(\mathbf{k},w)$ in *kw*-space, where

$$\int_w \mathrm{d}w \int_{\mathbf{k}} \mathrm{d}\mathbf{k} \left[-w^2 \mathbf{E}(\mathbf{k}, w) + k^2 c^2 \mathbf{E}(\mathbf{k}, w) + \frac{\mathrm{i}w}{\varepsilon_0} \mathbf{J}(\mathbf{k}, w) - \frac{\mathbf{i}\mathbf{k}c^2}{\varepsilon_0} \rho(\mathbf{k}, w) \right] \times \exp[-\mathrm{i}(wt - \mathbf{k} \cdot \mathbf{r})] = 0 \tag{1.375}$$

and

$$\int_w \mathrm{d}w \int_{\mathbf{k}} \mathrm{d}\mathbf{k} [-\mathrm{i}w\rho(\mathbf{k}, w) - \mathbf{i}\mathbf{k} \cdot \mathbf{J}(\mathbf{k}, w)] \exp[-\mathrm{i}(wt - \mathbf{k} \cdot \mathbf{r})] = 0 \tag{1.376}$$

These inverse Fourier transform expressions for the wave equation and charge continuity equation are integrations over $\mathbf{k}$ and w for each position and time value $(\mathbf{r}, t)$, which is present in the exponential term $\exp[-\mathrm{i}(wt - \mathbf{k} \cdot \mathbf{r})]$. The exponential term varies with $\mathbf{k}$ and w in a manner independent of the variation of the other integrand term. This other term is the inverse Fourier transform expression of the wave or charge continuity differential equation and is a function of the inverse Fourier transform variables $\mathbf{E}(\mathbf{k}, w)$, $\mathbf{J}(\mathbf{k}, w)$, $\rho(\mathbf{k}, w)$. Hence, the integrand expression with $\mathbf{E}(\mathbf{k}, w)$, $\mathbf{J}(\mathbf{k}, w)$, and $\rho(\mathbf{k}, w)$ must be zero for each $\mathbf{k}$ and w in the integrals of Eqs. (1.375) and (1.376), so that

$$-w^2\mathbf{E}(\mathbf{r}, t) + k^2c^2\mathbf{E}(\mathbf{r}, t) + \frac{\mathrm{i}w}{\varepsilon_0}\mathbf{J}(\mathbf{r}, t) - \frac{\mathrm{i}\mathbf{k}c^2}{\varepsilon_0}\rho(\mathbf{r}, t) = 0 \tag{1.377}$$

and

$$-\mathrm{i}w\rho(\mathbf{r}, t) + \mathrm{i}\mathbf{k} \cdot \mathbf{J}(\mathbf{r}, t) = 0 \tag{1.378}$$

The partial differential wave and charge continuity equations in *rt*-space have thus been converted to algebraic equations in *kw*-space.

In the phasor method of electrical engineering, one substitutes assumed sinusoidal expressions in the partial differential equations to convert the equations to algebraic equations of the frequency-dependent amplitudes of the sinusoids. To show this, one starts with the wave and charge continuity differential equations:

$$\left(\frac{\partial^2}{\partial t^2} - c^2\nabla^2\right)\mathbf{E}(\mathbf{r}, t) = -\frac{1}{\varepsilon_0}\left[\frac{\partial \mathbf{J}(\mathbf{r}, t)}{\partial t} + c^2\nabla\rho(\mathbf{r}, t)\right] \tag{1.379}$$

and

$$\frac{\partial \rho}{\partial t} + \nabla \cdot \mathbf{J} = 0 \tag{1.380}$$

In real-time–space (*rt*-space), one assumes the following assumed sinusoid solutions for the electric field $\mathbf{E}(\mathbf{r}, t)$, current density $\mathbf{J}(\mathbf{r}, t)$, and charge density $\rho(\mathbf{r}, t)$, where

$$\mathbf{E}(\mathbf{r},t) = \mathbf{E}(\mathbf{k}, w)\exp[-\mathrm{i}(wt - \mathbf{k} \cdot \mathbf{r})] \tag{1.381}$$

$$\mathbf{J}(\mathbf{r},t) = \mathbf{J}(\mathbf{k}, w)\exp[-\mathrm{i}(wt - \mathbf{k} \cdot \mathbf{r})] \tag{1.382}$$

$$\rho(\mathbf{r},t) = \rho(\mathbf{k}, w)\exp[-\mathrm{i}(wt - \mathbf{k} \cdot \mathbf{r})] \tag{1.383}$$

These assumed solutions for $\mathbf{E}(\mathbf{r}, t)$, $\mathbf{J}(\mathbf{r}, t)$, and $\rho(\mathbf{r}, t)$ are related to their inverse Fourier transform integrals by

$$\mathbf{E}(\mathbf{r}, t) = \frac{1}{(2\pi)^4}\int_{\mathbf{k}} \mathrm{d}\mathbf{k}\int_{w} \mathrm{d}w\mathbf{E}(\mathbf{k}, w)\exp[-\mathrm{i}(wt - \mathbf{k} \cdot \mathbf{r})] \tag{1.384}$$

$$\mathbf{J}(\mathbf{r}, t) = \frac{1}{(2\pi)^4}\int_{\mathbf{k}} \mathrm{d}\mathbf{k}\int_{w} \mathrm{d}w\mathbf{J}(\mathbf{k}, w)\exp[-\mathrm{i}(wt - \mathbf{k} \cdot \mathbf{r})] \tag{1.385}$$

and

$$\rho(\mathbf{r},t)=\frac{1}{(2\pi)^4}\int_{\mathbf{k}}\mathrm{d}\mathbf{k}\int_{w}\mathrm{d}w\rho(\mathbf{k},w)\exp[-\mathrm{i}(wt-\mathbf{k}\cdot\mathbf{r})] \tag{1.386}$$

The sinusoid amplitudes $\mathbf{E}(\mathbf{k},w)$, $\mathbf{J}(\mathbf{k},w)$, and $\rho(\mathbf{k},w)$, are determined by substitution of the assumed sinusoid solutions of Eqs. (1.381−1.383) into the partial differential equations of Eqs. (1.379) and (1.380). This substitution of the assumed sinusoid solutions converts the partial differential equations directly into algebraic equations of the sinusoid amplitudes $\mathbf{E}(\mathbf{k},w)$, $\mathbf{J}(\mathbf{k},w)$, and $\rho(\mathbf{k},w)$, where one again obtains Eqs. (1.377−1.378). The wave equation is then transposed to solve for electric field $\mathbf{E}(\mathbf{k},w)$, and the charge continuity equation is transposed to solve for charge density $\rho(\mathbf{k},w)$, where

$$\mathbf{E}(\mathbf{k},w)=\frac{1}{\varepsilon_0}\frac{\mathrm{i}w\mathbf{J}(\mathbf{k},w)-\mathrm{i}\mathbf{k}c^2\rho(\mathbf{k},w)}{w^2-k^2c^2} \tag{1.387}$$

and

$$\rho(\mathbf{k},w)=\frac{\mathbf{k}\cdot\mathbf{J}(\mathbf{k},w)}{w}=\frac{k}{w}\hat{\mathbf{k}}\cdot\mathbf{J}(\mathbf{k},w)=\frac{1}{c}\hat{\mathbf{k}}\cdot\mathbf{J}(\mathbf{k},w) \tag{1.388}$$

With the substitution of Eq. (1.388) for $\rho(\mathbf{k},w)$ into Eq. (1.387), the kw-space electric field $\mathbf{E}(\mathbf{k},w)$ becomes

$$\mathbf{E}(\mathbf{k},w)=\frac{1}{\varepsilon_0}\frac{\mathrm{i}w\mathbf{J}-\mathrm{i}c(\hat{\mathbf{k}}\cdot\mathbf{J})\mathbf{k}}{w^2-k^2c^2}=\frac{\mathrm{i}w}{\varepsilon_0}\cdot\frac{\mathbf{J}-(\hat{\mathbf{k}}\cdot\mathbf{J})\hat{\mathbf{k}}}{w^2-k^2c^2} \tag{1.389}$$

The current density is the sum of (1) the parallel component current density $\mathbf{J}_{\parallel}(\mathbf{k},w)$, which is parallel to the unit wave vector $\hat{\mathbf{k}}$ (wave propagation direction), and (2) the perpendicular component current density $\mathbf{J}_{\perp}(\mathbf{k},w)$, which is perpendicular to the unit wave vector $\hat{\mathbf{k}}$, where

$$\mathbf{J}(\mathbf{k},w)=\mathbf{J}_{\parallel}(\mathbf{k},w)+\mathbf{J}_{\perp}(\mathbf{k},w) \tag{1.390}$$

and

$$\mathbf{J}_{\parallel}(\mathbf{k},w)=\hat{\mathbf{k}}\cdot\mathbf{J}(\mathbf{k},w) \tag{1.391}$$

From Eqs. (1.390) and (1.391), the perpendicular component current density $\mathbf{J}_{\perp}$ is then

$$\mathbf{J}_{\perp}(\mathbf{k},w)=\mathbf{J}(\mathbf{k},w)-\mathbf{J}_{\perp}(\mathbf{k},w)=\mathbf{J}(\mathbf{k},w)-\hat{\mathbf{k}}\cdot\mathbf{J}(\mathbf{k},w) \tag{1.392}$$

With the substitution transverse current density $\mathbf{J}_{\perp}(\mathbf{k},w)$ of Eq. (1.392) into Eq. (1.389), the kw-space electric field $\mathbf{E}(\mathbf{k},w)$ becomes

$$\mathbf{E}(\mathbf{k},w)=\frac{\mathrm{i}w}{\varepsilon_0}\cdot\frac{\mathbf{J}_{\perp}(\mathbf{k},w)}{w^2-k^2c^2}=-\mathrm{i}w\frac{1}{\varepsilon_0}\frac{\mathbf{J}_{\perp}(\mathbf{k},w)}{k^2c^2-w^2} \tag{1.393}$$

The *rt*-space electric field $\mathbf{E}(\mathbf{r},t)$ is the inverse Fourier transform of $\mathbf{E}(\mathbf{k},w)$:

$$\mathbf{E}(\mathbf{r},t)=\frac{1}{(2\pi)^4}\int_{\mathbf{k}}\mathrm{d}\mathbf{k}\int_{w}\mathrm{d}w\mathbf{E}(\mathbf{k},w)\exp[-\mathrm{i}(wt-\mathbf{k}\cdot\mathbf{r})] \tag{1.394}$$

Substitution of the rightmost expression for $\mathbf{E}(\mathbf{k},w)$ of Eq. (1.393) into Eq. (1.394) gives

$$\mathbf{E}(\mathbf{r},t)=\frac{1}{\varepsilon_0}\frac{1}{(2\pi)^4}\int_{\mathbf{k}}\mathrm{d}\mathbf{k}\int_{w}\mathrm{d}w(-\mathrm{i}w)\frac{\mathbf{J}_{\perp}(\mathbf{k},w)}{k^2c^2-w^2}\exp[-\mathrm{i}(wt-\mathbf{k}\cdot\mathbf{r})] \tag{1.395}$$

One invokes $\partial/\partial t=-\mathrm{i}w$ in Eq. (1.369), so the *rt*-space electric field $\mathbf{E}(\mathbf{r},t)$ becomes

$$\mathbf{E}(\mathbf{r},t)=\frac{1}{\varepsilon_0}\frac{1}{(2\pi)^4}\int_{\mathbf{k}}\mathrm{d}\mathbf{k}\int_{w}\mathrm{d}w\frac{\partial}{\partial t}\frac{\mathbf{J}_{\perp}(\mathbf{k},w)}{k^2c^2-w^2}\exp[-\mathrm{i}(wt-\mathbf{k}\cdot\mathbf{r})] \tag{1.396}$$

For a single electron charge e, the current density $\mathbf{J}(\mathbf{r},t)$ in *rt*-space is specified by the product of (1) the electron charge e; (2) the spatial $\mathbf{r}$ and time t-dependent charge density $\rho(\mathbf{r},t)$, which is replaced by the spatial-dependent Dirac delta function $\delta(\mathbf{r})=\delta(x)\delta(y)\delta(z)$; and (3) the time-dependent electron velocity $\mathbf{v}(t)$, where

$$\mathbf{J}(\mathbf{r},t)=e\delta(\mathbf{r})\mathbf{v}(t)=e\delta(x)\delta(y)\delta(z)\mathbf{v}(t) \tag{1.397}$$

The Dirac delta function $\delta(x)$, with arbitrary function $f(x)$ and arbitrary constant a, has the following definition, normalized integration, and function sifting by integration properties:

$$\delta(x)=\begin{matrix}0 & x\neq 0\\ \infty & x=0\end{matrix} \tag{1.398}$$

$$\int_{-\infty}^{\infty}\delta(x-a)\mathrm{d}x=1 \tag{1.399}$$

$$\int_{-\infty}^{\infty}f(x)\delta(x-a)\mathrm{d}x=f(a) \tag{1.400}$$

The Fourier transform of the current density $\mathbf{J}(\mathbf{r},t)$ expression of Eq. (1.397) gives $\mathbf{J}(\mathbf{k},w)$, where

$$\begin{aligned}\mathbf{J}(\mathbf{k},w)&=e\int_{-\infty}^{\infty}\mathrm{d}t\int_{-\infty}^{\infty}\mathrm{d}x\int_{-\infty}^{\infty}\mathrm{d}y\int_{-\infty}^{\infty}\mathrm{d}z\delta(x)\delta(y)\delta(z)\mathbf{v}(t)\\&\times\exp[\mathrm{i}(wt-\mathbf{k}\cdot\mathbf{r})]=e\int_{-\infty}^{\infty}\mathbf{v}(t)\exp(-\mathrm{i}wt)\mathrm{d}t\end{aligned} \tag{1.401}$$

The kw-space perpendicular current density component $\mathbf{J}_{\perp}(\mathbf{k}, w)$ is obtained from the rt-space, perpendicular velocity component $\mathbf{v}_{\perp}(t)$, where

$$\mathbf{J}_{\perp}(\mathbf{k}, w) = e\int_{-\infty}^{\infty} \mathbf{v}_{\perp}(t)\exp(-\mathrm{i}wt)\mathrm{d}t \tag{1.402}$$

The Fourier transform of perpendicular velocity component $\mathbf{v}_{\perp}(t)$ in time−space is $v_{\perp}(w)$ in temporal frequency space, where

$$v_{\perp}(w) = e\int_{-\infty}^{\infty} \mathbf{v}_{\perp}(t)\exp(-\mathrm{i}wt)\mathrm{d}t \tag{1.403}$$

Hence,

$$\mathbf{J}_{\perp}(\mathbf{k}, w) = e\mathbf{v}_{\perp}(w) \tag{1.404}$$

With the substitution of Eq. (1.404) for $\mathbf{J}_{\perp}(\mathbf{k}, w)$ in Eq. (1.396) and changing the order of integration, the rt-space electric field $\mathbf{E}(\mathbf{r}, t)$ becomes

$$\mathbf{E}(\mathbf{r}, t) = -\frac{e}{\varepsilon_0}\frac{1}{(2\pi)^4}\int_w \mathrm{d}w\int_{\mathbf{k}} \mathrm{d}\mathbf{k}\frac{\partial}{\partial t}\frac{\mathbf{v}_{\perp}(w)}{k^2c^2 - w^2}\exp[-\mathrm{i}(wt - \mathbf{k}\cdot\mathbf{r})] \tag{1.405}$$

Next, Eq. (1.405) is integrated over k-space, where spherical coordinates are used in place of rectangular coordinates, which simplifies integration, with the wave number k as the radial variable, polar angle θ, azimuth angle ϕ, differential solid angle $\mathrm{d}\Omega$, where

$$\mathrm{d}\mathbf{k} = k^2 \sin\theta\, \mathrm{d}\theta\, \mathrm{d}\phi\, \mathrm{d}k = k^2\, \mathrm{d}\Omega\, \mathrm{d}k \tag{1.406}$$

$$\mathrm{d}\Omega = \sin\theta\, \mathrm{d}\theta\, \mathrm{d}\phi = -d(\cos\theta)\mathrm{d}\phi \tag{1.407}$$

$$0 \le k \le \infty \tag{1.408}$$

$$0 \le \theta < \pi \tag{1.409}$$

$$0 \le \phi \le 2\pi \tag{1.410}$$

$$\mathbf{k}\cdot\mathbf{r} \equiv kr\cos\theta \tag{1.411}$$

With the substitution of Eqs. (1.406−1.411) into the integral of Eq. (1.405), the rt-space electric field $\mathbf{E}(\mathbf{r}, t)$ is

$$\begin{aligned}\mathbf{E}(\mathbf{r}, t) = &-\frac{e}{\varepsilon_0}\frac{1}{(2\pi)^4} \times \int_{-\infty}^{\infty} \exp(-\mathrm{i}wt)\mathrm{d}w\int_0^{\infty} k^2\mathrm{d}k\int_0^{\pi} \mathrm{d}\theta \\ &\times \int_0^{2\pi} \mathrm{d}\phi\frac{\partial}{\partial t}\frac{\mathbf{v}_{\perp}(w)}{k^2c^2 - w^2}\exp(\mathrm{i}kr\cos\theta)\end{aligned} \tag{1.412}$$

where

$$\int_0^{\pi} \exp(ikr\cos\theta)\mathrm{d}\theta = -\int_1^{-1} \exp(ikr\cos\theta)d(\cos\theta) = -\frac{1}{ikr}[\exp(-ikr) - \exp(ikr)] \tag{1.413}$$

Integration of Eq. (1.412) over azimuth angle ϕ and polar angle θ, via $\mathrm{d}\theta = -\mathrm{d}(\cos\theta)$, gives

$$\mathbf{E}(\mathbf{r},t) = \frac{e}{\varepsilon_0}\frac{1}{(2\pi)^3}\int_{-\infty}^{\infty} \mathrm{d}w\left(\exp(-\mathrm{i}wt)\frac{\partial}{\partial t}\frac{\mathbf{v}_\perp(w)}{\mathrm{i}r}\int_0^{\infty}\frac{\exp(ikr)-\exp(-ikr)}{k^2c^2-w^2}k\,\mathrm{d}k\right) \tag{1.414}$$

where

$$\int_0^{\infty}\frac{\exp(ikr)-\exp(-ikr)}{k^2c^2-w^2}k\,\mathrm{d}k = \int_0^{\infty}\frac{\exp(ikr)}{(kc-w)(kc+w)}k\,\mathrm{d}k - \int_0^{\infty}\frac{\exp(-ikr)}{(kc-w)(kc+w)}k\,\mathrm{d}k \tag{1.415}$$

and

$$\int_0^{\infty}\frac{\exp(-ikr)}{(kc-w)(kc+w)}k\,\mathrm{d}k = \int_{-\infty}^{0}\frac{\exp(ikr)}{(kc-w)(kc+w)}k\,\mathrm{d}k \tag{1.416}$$

Substitution of Eqs. (1.415) and (1.416) into Eq. (1.414) yields

$$\int_0^{\infty}\frac{\exp(ikr)-\exp(-ikr)}{k^2c^2-w^2}k\,\mathrm{d}k = \frac{1}{c^2}\int_{-\infty}^{\infty}\frac{\exp(ikr)}{(k-w/c)(k+w/c)}k\,\mathrm{d}k \tag{1.417}$$

And with the substitution of Eq. (1.417) into Eq. (1.414), rt-space electric field $\mathbf{E}(\mathbf{r},t)$ becomes

$$\mathbf{E}(\mathbf{r},t) = \frac{e}{c^2\varepsilon_0}\frac{1}{(2\pi)^3}\int_{-\infty}^{\infty}\mathrm{d}w\left(\exp(-\mathrm{i}wt)\frac{\partial}{\partial t}\frac{\mathbf{v}_\perp(w)}{\mathrm{i}r}\int_{-\infty}^{\infty}\frac{\exp(ikr)}{(k-w/c)(k+w/c)}k\,\mathrm{d}k\right) \tag{1.418}$$

The integral on the right side of Eq. (1.418) is a Cauchy contour integral, which one can solve in the complex plane as shown in Figure 1.23, where we assign the function $f(k)$, so that

$$\int_{-\infty}^{\infty}\frac{\exp(ikr)}{(k-w/c)(k+w/c)}k\,\mathrm{d}k = \int_{-\infty}^{\infty}\frac{f(k)}{k-w/c}\mathrm{d}k \tag{1.419}$$

where

$$f(k) = \frac{k\,\exp(ikr)}{k + w/c} \tag{1.420}$$

The Cauchy integral formula then gives

$$\int_{-\infty}^{\infty} \frac{f(k)}{k - w/c}\mathrm{d}k = \mathrm{i}\cdot 2\pi \cdot f\left(\frac{w}{c}\right) \tag{1.421}$$

Substitution of Eq. (1.420) for $f(k)$ into Eq. (1.421) yields

$$\int_{-\infty}^{\infty} \frac{k\,\exp(ikr)}{(k - w/c)(k + w/c)}\mathrm{d}k = i\cdot\pi\cdot\exp\left(\mathrm{i}\frac{w}{c}r\right) \tag{1.422}$$

With the substitution of Eq. (1.422) into Eq. (1.414), the *rt*-space electric field $\mathbf{E}(\mathbf{r}, t)$ becomes

$$\mathbf{E}(\mathbf{r}, t) = \frac{e}{c^2\varepsilon_0 r}\frac{\pi}{(2\pi)^3}\frac{\partial}{\partial t}\int_{-\infty}^{\infty} \mathbf{v}_\perp(w)\exp\left[-\mathrm{i}w\left(t - \frac{r}{c}\right)\right]\mathrm{d}w \tag{1.423}$$

The electron transverse velocity in space–time $\mathbf{v}_\perp(t)$ is obtained from the electron transverse velocity in space–time $\mathbf{v}_\perp(w)$ by the inverse Fourier transform, where

$$\mathbf{v}_\perp(t) = \frac{1}{2\pi}\int_{-\infty}^{\infty} \mathbf{v}_\perp(w)\exp(-\mathrm{i}wt)\mathrm{d}w \tag{1.424}$$

Suppose that one translates time t by r/c to $t - (r/c)$ in this inverse Fourier transform of $\mathbf{v}_\perp(w)$, where

$$t \to t - \frac{r}{c} \tag{1.425}$$

One then obtains

$$\mathbf{v}_\perp\left(t - \frac{r}{c}\right) = \frac{1}{2\pi}\int_{-\infty}^{\infty} \mathbf{v}_\perp(w)\exp\left[-\mathrm{i}w\left(t - \frac{r}{c}\right)\right]\mathrm{d}w \tag{1.426}$$

With the substitution of Eq. (1.426) into Eq. (1.423), the *rt*-space electric field $\mathbf{E}(\mathbf{r}, t)$, which is emitted from an oscillating electron charge, is

$$\mathbf{E}(\mathbf{r}, t) = \frac{e}{4\pi\varepsilon_0 c^2 r}\frac{\partial}{\partial t}\left[\mathbf{v}_\perp\left(t - \frac{r}{c}\right)\right] \tag{1.427}$$

The transverse acceleration $\mathbf{a}_\perp(t - r/c)$ of the oscillating electron charge is the time derivative of the electron transverse velocity $\mathbf{v}_\perp(t - r/c)$ at retarded time $t - r/c$, where

$$\mathbf{a}_{\perp}\left(t-\frac{r}{c}\right)=\frac{\partial}{\partial t}\left[\mathbf{v}_{\perp}\left(t-\frac{r}{c}\right)\right] \tag{1.428}$$

With the substitution of Eq. (1.428) into Eq. (1.427), the position–time–space electric field $\mathbf{E}(\mathbf{r}, t)$, which is emitted from an oscillating electron charge, becomes

$$\mathbf{E}(\mathbf{r}, t)=\frac{e\mathbf{a}_{\perp}(t-(r/c))}{4\pi\varepsilon_0 c^2 r} \tag{1.429}$$

Equation (1.429) can be interpreted as follows. An observer is at distance r from an oscillating electron charge, which is undergoing periodic acceleration or undergoing an arbitrary nonperiodic acceleration. The acceleration of the electron charge causes the electron to emit an electromagnetic wave at an earlier, retarded time $t-r/c$, which travels distance r to the observer at the speed of light c for a time of flight of r/c. The observer at the present time t measures the electric field $\mathbf{E}(\mathbf{r}, t)$, which was produced by the accelerating electron at earlier time $t-r/c$.

We now calculate the magnetic field $\mathbf{H}$ of the oscillating electron charge via the Maxwell's equation of Faraday's law:

$$\nabla\times\mathbf{E}=-\frac{\partial\mathbf{B}}{\partial t}=-\mu_0\frac{\partial\mathbf{H}}{\partial t} \tag{1.430}$$

The magnetic induction vector $\mathbf{B}$ has been replaced by magnetic field vector $\mu_0\mathbf{H}$ via the constitutive relation with vacuum permeability μ_0. In real-time–space (*rt*-space), one assumes the following sinusoid solutions for the electric field $\mathbf{E}(\mathbf{r}, t)$ and magnetic field $\mathbf{H}(\mathbf{r}, t)$:

$$\mathbf{E}(\mathbf{r},t)=\mathbf{E}(\mathbf{k}, w)\exp[-\mathrm{i}(wt-\mathbf{k}\cdot\mathbf{r})] \tag{1.431}$$

$$\mathbf{H}(\mathbf{r},t)=\mathbf{H}(\mathbf{k}, w)\exp[-\mathrm{i}(wt-\mathbf{k}\cdot\mathbf{r})] \tag{1.432}$$

That is, one replaces the $\mathbf{E}(\mathbf{r}, t)$ and magnetic field intensity $\mathbf{H}(\mathbf{r}, t)$ by their Fourier transform expressions and use the *wk*-space expressions of the gradient and partial time derivative, where

$$\nabla\leftrightarrow\mathrm{i}\mathbf{k} \quad \text{and} \quad \frac{\partial}{\partial t}\leftrightarrow-\mathrm{i}w \tag{1.433}$$

Given the unit wave vector $\hat{\mathbf{k}}$ of the wave vector $\mathbf{k}=k\hat{\mathbf{k}}$, whose magnitude is the wave number k, and with the substitution of Eq. (1.433) into Faraday's law of Eq. (1.430), the magnetic field intensity $\mathbf{H}(\mathbf{r}, t)$ is

$$\mathbf{H}(k, w)=\frac{\mathbf{k}\times\mathbf{E}(k, w)}{w\mu_0}=\frac{k\hat{\mathbf{k}}\times\mathbf{E}(k, w)}{w\mu_0}=\frac{\hat{\mathbf{k}}\times\mathbf{E}(k, w)}{c\mu_0}=\sqrt{\frac{\varepsilon_0}{\mu_0}}\hat{\mathbf{k}}\times\mathbf{E}(k, w) \tag{1.434}$$

Replacing kw-space electric $\mathbf{E}(\mathbf{k}, w)$ and magnetic $\mathbf{H}(\mathbf{k}, w)$ fields by their rt-space expressions $\mathbf{E}(\mathbf{r}, t)$ and $\mathbf{H}(\mathbf{r}, t)$, the magnetic field expression of Eq. (1.434) becomes

$$\mathbf{H}(\mathbf{r}, t) = \sqrt{\frac{\varepsilon_0}{\mu_0}}\hat{\mathbf{k}} \times \mathbf{E}(\mathbf{r}, t) \tag{1.435}$$

The radiated power, Poynting vector $\mathbf{S}(\mathbf{r}, t)$, which arrives at the observer at position $\mathbf{r}$ at the observer's present time t is then

$$\mathbf{S}(\mathbf{r}, t) = \mathbf{E}(\mathbf{r}, t) \times \mathbf{H}(\mathbf{r}, t) = \sqrt{\frac{\varepsilon_0}{\mu_0}}\mathbf{E}(\mathbf{r}, t) \times \hat{\mathbf{k}} \times \mathbf{E}(\mathbf{r}, t) \tag{1.436}$$

One now employs the BAC−CAB vector identity where for vectors **A**, **B**, and **C** is

$$\mathbf{A} \times (\mathbf{B} \times \mathbf{C}) = \mathbf{B}(\mathbf{A} \cdot \mathbf{C}) - \mathbf{C}(\mathbf{A} \cdot \mathbf{B}) \tag{1.437}$$

The unit wave vector $\hat{\mathbf{k}}$ in the direction of propagation is perpendicular to both the electric $\mathbf{E}(\mathbf{r}, t)$ and magnetic $\mathbf{B}(\mathbf{r}, t)$ fields in the plane electromagnetic wave, where

$$\hat{\mathbf{k}} \cdot \mathbf{E}(\mathbf{r}, t) = \hat{\mathbf{k}} \cdot \mathbf{B}(\mathbf{r}, t) = 0 \tag{1.438}$$

With application of the BAC−CAB rule of Eqs. (1.437) and (1.438) in (1.436), the Poynting vector $\mathbf{S}(\mathbf{r}, t)$, becomes

$$\mathbf{S}(\mathbf{r}, t) = \mathbf{E}(\mathbf{r}, t) \times \mathbf{H}(\mathbf{r}, t) = \sqrt{\frac{\varepsilon_0}{\mu_0}}|\mathbf{E}(\mathbf{r}, t)|^2\hat{\mathbf{k}} = Z_0|\mathbf{E}(\mathbf{r}, t)|^2\hat{\mathbf{k}} \tag{1.439}$$

Recall that Z_0 is the impedance of a vacuum for the electromagnetic wave, where

$$Z_0 = \sqrt{\frac{\varepsilon_0}{\mu_0}} \tag{1.440}$$

The electron's retarded acceleration component $\mathbf{a}_\perp(t - r/c)$, which is perpendicular to the line connecting the electron at retarded time $t - r/c$ to the observer at present time t, is

$$\mathbf{a}_\perp\left(t - \frac{r}{c}\right) = \mathbf{a}\left(t - \frac{r}{c}\right)\sin\theta \tag{1.441}$$

The polar angle θ is between the electron charge acceleration vector $\mathbf{a}(t - r/c)$ at the retarded time $t - r/c$ and the line joining the charge at the retarded time $t - r/c$ to the observer at far-field distance r at the present time t. With the substitution into

Eq. (1.429) of the electron perpendicular acceleration component $\mathbf{a}_\perp(t-r/c)$ in terms of the total electron acceleration component $\mathbf{a}(t-r/c)$ of Eq. (1.441), the electric field $\mathbf{E}(\mathbf{r},t)$ at the observer position $\mathbf{r}$ at the present time t becomes

$$\mathbf{E}(\mathbf{r},t) = \frac{e}{4\pi\varepsilon_0 c^2 r}\mathbf{a}\left(t-\frac{r}{c}\right)\sin\theta = \frac{e}{\varepsilon_0 c^2 r}\frac{\sin\theta}{4\pi}\mathbf{a}\left(t-\frac{r}{c}\right) \tag{1.442}$$

With the substitution of the expression for the observer's electric field $\mathbf{E}(\mathbf{r},t)$ of Eq. (1.442) into Eq. (1.439), the Poynting vector $\mathbf{S}(\mathbf{r},t)$ or power flux received by the observer at the present time t from the electron at distance r, which underwent acceleration at the earlier or retarded time $t-r/c$, becomes

$$\mathbf{S}(\mathbf{r},t) = \mathbf{E}(\mathbf{r},t)\times\mathbf{H}(\mathbf{r},t) = \frac{e^2\sin^2\theta}{16\pi^2\varepsilon_0 c^3 r^2}\left|\mathbf{a}\left(t-\frac{r}{c}\right)\right|^2\hat{\mathbf{k}} \tag{1.443}$$

The differential power dP, which is radiated per differential solid angle dΩ through a differential area dA, where at distance r from the accelerating electron

$$\frac{\mathrm{d}P}{\mathrm{d}A} = \frac{\mathrm{d}P}{r^2\,\mathrm{d}\Omega} = |\mathbf{S}(\mathbf{r},t)| \tag{1.444}$$

with

$$\mathrm{d}A = r^2\,\mathrm{d}\Omega \tag{1.445}$$

Given polar angle $\theta\phi$ and azimuth angle ϕ,

$$\mathrm{d}\Omega = \sin\theta\,\mathrm{d}\theta\,\mathrm{d}\phi \tag{1.446}$$

Transposing Eq. (1.444) and substituting Eq. (1.443) for $\mathbf{S}(\mathbf{r},t)$, one obtains the power emitted per solid angle $\mathrm{d}P/d\Omega$ from the accelerated electron, where

$$\frac{\mathrm{d}P}{\mathrm{d}\Omega} = \frac{e^2\sin^2\theta}{16\pi^2\varepsilon_0 c^3}\left|\mathbf{a}\left(t-\frac{r}{c}\right)\right|^2 \tag{1.447}$$

The total power P, which is emitted from the accelerated electron, is obtained by integration over the solid angle dΩ, where

$$P = \int_0^{2\pi}\mathrm{d}\phi\int_0^{\pi}\mathrm{d}\theta\frac{e^2\sin^2\theta}{16\pi^2\varepsilon_0 c^3}\left|\mathbf{a}\left(t-\frac{r}{c}\right)\right|^2\sin\theta \tag{1.448}$$

where

$$\int_0^{\pi}\mathrm{d}\theta\sin^2\theta\sin\theta = \int_0^{\pi}\mathrm{d}\theta(\sin\theta-\sin\theta\cos^2\theta) = 2-\frac{1}{3}[\cos^3\theta]_0^{\pi} = 2-\frac{2}{3} = \frac{4}{3} \tag{1.449}$$

The total power P, which is emitted from the electon undergoing acceleration with magnitude a, is then

$$P = \frac{8\pi}{3}\left(\frac{e^2a^2}{16\pi^2\varepsilon_0 c^3}\right) \tag{1.450}$$

1.5.3 Acceleration of a Bound Atomic Electron by an Imposed Electromagnetic Field

The acceleration of an atomic electron of charge e by the source electric field, can be calculated from the equation of motion of the electron of rest mass m_0. The electron acceleration is caused by the Lorentz force, via the incident electromagnetic field, which is comprised of electric $\mathbf{E}_i$ and magnetic $\mathbf{B}_i$ field components. The second-order differential equation of driven electron oscillation is

$$m_0\frac{d^2\mathbf{x}_s}{dt^2} + m_0\gamma\frac{d\mathbf{x}_s}{dt} + m_0 w_s^2\mathbf{x}_s = -e(\mathbf{E}_i + \mathbf{v}_s \times \mathbf{B}_i) \tag{1.451}$$

The first term is the acceleration term for the electron oscillation. The second term is the dissipative force term, which inhibits electron oscillation, and accounts for energy loss, via the damping factor γ, with assumption of small dissipation, as compared to the driving frequency w of the incident electromagnetic wave, where

$$\frac{\gamma}{w} \ll 1 \tag{1.452}$$

The third term is due to the restoring force of an oscillator with resonant frequency w_s of the bound atomic electrons of the material. Here, x_s is the electron displacement from equilibrium in the direction, which is (1) parallel to the incident electric field, and (2) perpendicular to the incident magnetic field, and (3) perpendicular to the electron-to-observer direction.

The plane wave magnetic field amplitude $\mathbf{B}_i = \mu_0\mathbf{H}_i$ is much smaller than the electric field amplitude $\mathbf{E}_i$ by a factor of the speed of light c, where

$$\mathbf{E}_i = c\mathbf{B}_i = c\mu_0\mathbf{H}_i = \sqrt{\frac{\mu_0}{\varepsilon_0}}\mathbf{H}_i = 377\mathbf{H}_i \tag{1.453}$$

The electric field–induced electron velocity $\mathbf{v}_s$ follows the direction of the electric field $\mathbf{E}$, but $\mathbf{v}_s$ is perpendicular to the magnetic field $\mathbf{B}$. Hence, the force $e\mathbf{v}_s \times \mathbf{B}$ acts on the electron in a direction that is perpendicular to both the electric and magnetic fields of the incident plane wave. This $e\mathbf{v}_s \times \mathbf{B}$ force of the magnetic field, and associated acceleration, is much smaller than the electric field force $e\mathbf{E}$ and associated acceleration. Hence, on the right side of Eq. (1.451), the second term is much smaller than the first term; and thus the second term can be approximated to zero, where

$$\mathbf{v}_s \times \mathbf{B}_i \cong 0 \tag{1.454}$$

The periodic, incident electric field E_i, which is imposed on the electron, causes electron displacement x_s, velocity dx_s/dt, and acceleration d^2x_s/dt^2 with the same time, or reciprocal time (frequency) dependence.

With vector electric field amplitude $\mathbf{E}_0$, wave vector $\mathbf{k}_0$, frequency w, the incident electric field vector $\mathbf{E}_i(\mathbf{r}_s, t)$ at the electron at position $\mathbf{r}_s$ and time t is

$$\mathbf{E}_i(\mathbf{r}_s, t) = \mathbf{E}_0 \exp[-i(wt - \mathbf{k}_0 \cdot \mathbf{r}_s)] \tag{1.455}$$

The electron displacement or position vector $\mathbf{x}_s(\mathbf{r}_s, t)$ follows the direction of the imposed electric field $\mathbf{E}_i(\mathbf{r}_s, t)$, where

$$\mathbf{x}_s(\mathbf{r}_s, t) = \mathbf{x}_{os} \exp[-i(wt - \mathbf{k}_0 \cdot \mathbf{r}_s)] \tag{1.456}$$

Substitution for $\mathbf{x}_s$ and $\mathbf{E}_i$ yields

$$-w^2\mathbf{x}_{0s} - w\gamma\mathbf{x}_{0s} + w_s^2\mathbf{x}_{0s} = -\frac{e\mathbf{E}_0}{m_0} \tag{1.457}$$

Transposing Eq. (1.457), one solves for the amplitude of the electron displacement vector $\mathbf{x}_{os}$, where

$$\mathbf{x}_{os} = \frac{1}{-w^2 - iw\gamma + w_s^2}\frac{e\mathbf{E}_0}{m_0} \tag{1.458}$$

With the substitution of Eq. (1.458) for $\mathbf{x}_{os}$ into Eq. (1.457), the electron displacement vector $\mathbf{x}_s$ becomes

$$\mathbf{x}_s = \frac{\exp[-i(wt - \mathbf{k}_0 \cdot \mathbf{r}_s)]}{-w^2 - iw\gamma + w_s^2}\frac{e\mathbf{E}_0}{m_0} \tag{1.459}$$

The magnitude of the electron velocity $\mathbf{v}_s$ is the time derivative of the electron position vector, where

$$\mathbf{v}_s = \frac{d\mathbf{x}_s}{dt} = w\frac{\exp[-i(wt - \mathbf{k}_0 \cdot \mathbf{r}_s)]}{-w^2 - iw\gamma + w_s^2}\frac{e\mathbf{E}_0}{m_0} \tag{1.460}$$

The electron acceleration vector $\mathbf{a}_s$ is the time derivative of the electron velocity vector, where

$$\mathbf{a}_s = \frac{d\mathbf{v}_s}{dt} = \frac{w^2 \exp[-i(wt - \mathbf{k}_0 \cdot \mathbf{r}_s)]}{w^2 + iw\gamma - w_s^2}\frac{e\mathbf{E}_0}{m_0} \tag{1.461}$$

The perpendicular acceleration component $\mathbf{a}_{s\perp}$ of the electron acceleration vector $\mathbf{a}_s$ is the electron acceleration component $\mathbf{a}_{s\perp}$, which is perpendicular to the line that extends from the electron to the observer, where

$$\mathbf{a}_{s\perp} = \mathbf{a}_s \sin\theta \tag{1.462}$$

Here, θ is the angle between the direction of the electric field electron acceleration vector $\mathbf{a}_s$ and the scattered wave direction $\mathbf{k}$, which points toward the observer at position vector $\mathbf{r}$ from the scattering electron at position vector $\mathbf{r}_s$.

The electron acceleration $\mathbf{a}_s$ is in the direction of the applied, incident electric field $\mathbf{E}_i$. The electron acceleration vector component $\mathbf{a}_{s\perp}(t_r)$ is perpendicular to the direction from electron to observer and is evaluated at retarded time t_{r}, where

$$t_{\mathrm{r}} = t - \frac{r - r_s}{c} \tag{1.463}$$

The electron acceleration by the incident electromagnetic wave occurs before the observer receives the electron-scattered wave. The separation distance $(\mathbf{r} - \mathbf{r}_s)$ is between the electron position $\mathbf{r}_s$ at the earlier time t_{r} of wave scatter and the observer position $\mathbf{r}$ at the present time t of detection. The finite speed of light c and separation distance $(r - r_s)$ cause a time lag $(\mathbf{r} - \mathbf{r}_s)/c$ to occur between the wave scatter by the electron and the detection of this scattered wave by the distant observer. Hence, the expression for the electron acceleration vector component $\mathbf{a}_{s\perp}(t_r)$, which is perpendicular to the direction $\mathbf{r} - \mathbf{r}_s$ between the electron at $\mathbf{r}_s$ and observer at $\mathbf{r}$, must be evaluated at the previous or retarded time t_{r} or $(\mathbf{r} - \mathbf{r}_s)/c$, so that

$$\mathbf{a}_{s\perp}(t_{\mathrm{r}}, w) = \frac{-w^2}{w^2 - w_s^2 + \mathrm{i}w\gamma} \frac{e\mathbf{E}_0 \sin\,\theta \exp[-\mathrm{i}(wt_{\mathrm{r}} - \mathbf{k}_0 \cdot \mathbf{r}_s)]}{m_0} \tag{1.464}$$

Elastic scatter is assumed, where the scattered wave number k equals the incident wave number k_0. That is, the direction of the incident wave vector $\mathbf{k}_0$ changes direction by polar angle θ upon scatter. The scattered wave vector $\mathbf{k}$ and incident wave vector $\mathbf{k}_0$ have identical magnitude or wave numbers, where

$$k = k_0 = \frac{w}{c} \tag{1.465}$$

The difference wave vector or the resulting change in direction of the incident wave vector upon scatter from the electron is then

$$\Delta\mathbf{k} = \mathbf{k} - \mathbf{k}_0 \tag{1.466}$$

The magnitude of the change of wave vector direction is

$$|\Delta\mathbf{k}| = 2k_0 \sin\theta \tag{1.467}$$

The retarded time expression of Eq. (1.463) is substituted into Eq. (1.464) for $\mathbf{a}_{s\perp}(t_{\mathrm{r}})$. Next, w/c in the resulting expression is replaced by k using Eq. (1.465), which then allows substitution for the difference wave vector $\Delta\mathbf{k} = \mathbf{k} - \mathbf{k}_0$ of Eq. (1.466). The expression for the electron acceleration vector component, which is perpendicular to the direction from electron to observer, becomes

$$\mathbf{a}_{s\perp}(t_{\mathrm{r}}) = \frac{-w^2}{w^2 - w_s^2 + \mathrm{i}w\gamma} \frac{e\mathbf{E}_0 \sin\theta \exp\left[-\mathrm{i}w(t-(r/c))\right]\exp(-\mathrm{i}\Delta\mathbf{k}\cdot\mathbf{r}_s)}{m_0} \tag{1.468}$$

The electric field $\mathbf{E}$, which is emitted from a single charged particle (an electron in this case) of charge e and undergoes acceleration $\mathbf{a}$, is obtained from the Lienard–Wiechert potentials. The Lienard–Wiechert potentials and the resulting electromagnetic field are derived later in this book, in Chapters 26 and 27, and can also be found in Attwood (1999), Eyges (1972), Feynman (1964), Jackson (1975), Landau & Lifshitz (1975), Marion (1965), Panofsky and Phillips (2005), Schwartz (1972), Shadowitz (1975), Sommerfeld (1964), and Stratton (1941).

To recap, the electric field $\mathbf{E}$, emitted from the electron of charge e undergoing acceleration $\mathbf{a}_s(t_{\mathrm{r}})$, is proportional to the electron's perpendicular acceleration vector component $\mathbf{a}_{s\perp}(t_{\mathrm{r}})$. The electron's perpendicular acceleration component $\mathbf{a}_{s\perp}(t_{\mathrm{r}})$ is the component of the electron acceleration $\mathbf{a}_s(t_{\mathrm{r}})$, which occurs at the retarded time t_{r}, and is perpendicular to the propagation direction of the electric field to the observer, who measures the electric field at a later time t. The time lag $t - t_{\mathrm{r}}$ is due to the finite separation distance between the charged particle and observer, and the finite speed of light c. The electric field $\mathbf{E}(\mathbf{r}, t)$ at the present observer position $\mathbf{r}$ and time t is related to the perpendicular electron acceleration component $\mathbf{a}_s(t_{\mathrm{r}})$, which is evaluated at the earlier or retarded time t_{r} and electron position $\mathbf{r}_s$ when the electron acceleration occurred. For vacuum electric permittivity ε_0, the observer electric field $\mathbf{E}(\mathbf{r}, t)$, due to a distant accelerating electron, is

$$\mathbf{E}(\mathbf{r}, t) = \frac{e}{4\pi\varepsilon_0 c^2 r}\mathbf{a}_{s\perp}(t_{\mathrm{r}}) \tag{1.469}$$

Only the component of the charged particle acceleration, which is perpendicular to the direction of the observer, produces an electric field at the observer. Hence, the observer-measured electric field amplitude has an angular dependence that is relative to the direction of charged particle acceleration.

To recap, an incident electric field wave, which has vector amplitude $\mathbf{E}_0$, wave vector $\mathbf{k}_0$, and frequency w, causes oscillation of the bound $s = 1, 2, 3, \ldots, Z$ electrons of an atom to oscillate. The oscillating electrons of the atom then emit a scattered electric field amplitude $\mathbf{E}$, wave vector $\mathbf{k}$, and frequency w. The change $\Delta\mathbf{k}$ of the incident wave vector $\mathbf{k}_0$ to scattered wave vector $\mathbf{k}$ is

$$\Delta\mathbf{k} = \mathbf{k} - \mathbf{k}_0 \tag{1.470}$$

The damping factor for each of the sth electrons at position $\mathbf{r}_s$ is given by γ, where one assumes a small dissipation compared to the driving frequency w of the incident electromagnetic wave. The expression of Eq. (1.468) for electron acceleration component $\mathbf{a}_{s\perp}(t_r)$, which is perpendicular to the direction between the electron and observer in the observer electric field, is substituted into the expression of Eq. (1.469) for the observer electric field $\mathbf{E}(\mathbf{r}, t)$ of the accelerated electron, and then yields the scattered far-field electric field $\mathbf{E}_{s\perp}(t_r)$ from the sth bound electron of the atom of atomic number Z, where

$$\mathbf{E}_{s\perp}(t_r) = \frac{e\mathbf{a}_{s\perp}(t_r)}{4\pi\varepsilon_0 c^2 r} = -\left[\frac{e}{4\pi\varepsilon_0 c^2 r}\right] \cdot \left[\frac{w^2}{w^2 - w_s^2 + iw\gamma}\right] \cdot \left[\frac{e\mathbf{E}_0 \sin\theta \exp\left[-iw(t-(r/c))\right]\exp(-i\Delta\mathbf{k}\cdot\mathbf{r}_s)}{m_0}\right] \tag{1.471}$$

Here, θ is the angle between the direction of the electron acceleration a of each of the bound atomic electrons, which follows the direction of the incident electric field direction and the direction of the scattered wave $\mathbf{k}$ toward the observer.

The scattered far-field electric field vector $\mathbf{E}_{\perp}(t)$ from all the Z bound electrons of the atom to the observer point, can be expressed in terms of the complex atomic scattering factor $f(\Delta\mathbf{k}, w)$, where

$$\mathbf{E}_{\perp}(t) = -\frac{e^2\mathbf{E}_0 \sin\theta}{4\pi\varepsilon_o r c^2 m_0}\exp\left[-iw\left(t - \frac{r}{c}\right)\right] f(\Delta\mathbf{k}, w) \tag{1.472}$$

The dimensionless, complex atomic scattering factor $f(\Delta\mathbf{k}, w)$ is the ratio of the scatter of an incident electric field, due to a multielectron atom, divided by the electric field scatter due to a single free electron, where

$$f(\Delta\mathbf{k}, w) = w^2 \sum_{s=1}^{Z} \frac{1}{w^2 - w_s^2 + iw\gamma}\exp[-i(\Delta\mathbf{k}\cdot\mathbf{r}_s)] \tag{1.473}$$

To adjust for the discrete resonant energy states of the atomic electrons, one needs to adjust the scattering factor with a weighting of the number of atomic electrons per atomic energy level, namely, the oscillator strength g_s for each sth electron. Hence, in the numerator of Eq. (1.473), one inserts the oscillator strength g_s for each sth electron, where

$$f(\Delta\mathbf{k}, w) = w^2 \sum_{s=1}^{Z} \frac{g_s}{w^2 - w_s^2 + iw\gamma}\exp[-i(\Delta\mathbf{k}\cdot\mathbf{r}_s)] \tag{1.474}$$

The sum of the oscillator strengths for the Z electrons of an atom equals the atomic number Z, i.e.,

$$\sum_{s=1}^{Z} g_s = Z \tag{1.475}$$

The distance of the atomic electron position $\mathbf{r}_s$ from its nucleus is contained within approximately the Bohr radius a_0. In the long-wavelength λ limit, the X-ray wavelength is much larger than the Bohr radius a_0, where

$$\frac{a_0}{\lambda} \ll 1 \tag{1.476}$$

In the forward-scattering limit, the angle θ, which is measured relative to the incident wave vector $\mathbf{k}_0$ direction along the positive z-axis, is nearly zero or very small, so the scattered wave vector $\mathbf{k}$ is in the forward direction

$$\theta \ll 1 \tag{1.477}$$

In the long-wavelength limit, $a_0/\lambda \ll 1$, and/or in the forward-scatter limit of $\theta \ll 1$, the argument of the exponential term in Eq. (1.474) goes to zero, where

$$|\Delta\mathbf{k}\cdot\mathbf{r}_s| \to 0 \tag{1.478}$$

Finally, in either the long-wavelength or forward-scattering limit, the complex atomic scattering factor $f^0(w)$ becomes

$$f^0(w) = w^2 \sum_{s=1}^{Z} \frac{g_s}{w^2 - w_s^2 + \mathrm{i}w\gamma} \quad \frac{a_0}{\lambda} \ll 1 \text{ or } \theta \ll 1 \tag{1.479}$$

That is, $f(\Delta\mathbf{k}, w)$ becomes $f^0(w)$ in the long-wavelength limit and/or the forward-scatter limit. If in addition the long-wavelength limit and/or forward-scatter direction, the frequency w of the incident X-ray exceeds the oscillator with resonant frequency w_s, then the atomic scattering factor $f(\Delta\mathbf{k}, w)$ equals the atomic number Z of the atom, where for $\theta \ll 1$,

$$f^0(w) = \sum_{s=1}^{Z} g_s = Z \quad w^2 \gg w_s^2 \text{ and } w \gg \gamma \tag{1.480}$$

Consider a medium that is comprised of $q = 1, 2, 3, \ldots, Q$ elements. Then the dimensionless, complex atomic scattering factor in the forward direction $f_q^0(w)$ can be written in terms of real $f_1^0(w)$ and imaginary $f_2^0(w)$ parts, where

$$f_q^0(w) = f_{1q}^0(w) - \mathrm{i}f_{2q}^0(w) \tag{1.481}$$

The dimensionless, real part $f_{1q}^0(w)$ of the complex atomic scattering factor $f_q^0(w)$ is

$$f_{1q}^0(w) = w^2 \sum_{s=1}^{Z} \frac{g_{s,q}(w^2 - w_{s,q}^2)}{(w^2 - w_{s,q}^2)^2 + w^2\gamma_q^2} \tag{1.482}$$

The dimensionless, imaginary part $f_{2q}^0(w)$ of the complex atomic scattering factor $f_q^0(w)$ is

$$f_{2q}^0(w) = -w^3\gamma_q \sum_{s=1}^{Z} \frac{g_{s,q}}{(w^2 - w_{s,q}^2)^2 + w^2\gamma_q^2} \tag{1.483}$$

In either the long-wavelength or forward-scattering limit, the complex atomic scattering factor $f^0(w)$ is

$$f^0(w) = w^2 \sum_{q=1}^{Q}\sum_{s=1}^{Z} \frac{g_{s,q}}{w^2 - w_{s,q}^2 + \mathrm{i}w\gamma_q} \qquad \frac{a_0}{\lambda} \ll 1 \text{ or } \theta \ll 1 \tag{1.484}$$

In a material medium of mass density ρ_m, the qth atom density n_q can be calculated, given Avogadro's number N_{A}, the qth element weigh fraction w_q, atomic weight A_q, and atomic number Z_q, where

$$n_q = N_{\text{A}} \frac{\rho_m w_q}{A_q} \tag{1.485}$$

Hence, for a material medium of $q = 1, 2, 3, \ldots, Q$ elements, the complex, dimensionless, forward-scatter factor $f^0(w)$, as a function of angular frequency w, for $a_0 \ll \lambda$ or $\theta \ll 1$, is

$$f^0(w) = w^2 \sum_{q=1}^{Q} n_q \sum_{s=1}^{Z} \frac{g_{s,q}}{w^2 - w_{s,q}^2 + \mathrm{i}w\gamma_q} \qquad \frac{a_0}{\lambda} \ll 1 \text{ or } \theta \ll 1 \tag{1.486}$$

In the long-wavelength limit and/or forward-scatter direction, the forward-scattered, far-field, transverse electric field vector $\mathbf{E}_\perp(t)$, which is produced by the oscillating Z_q bound atomic electrons of the qth element for a material medium of Q elements and is received by the observer at distance r from the region electrons, is

$$\mathbf{E}_\perp(t) = -\frac{e^2}{4\pi\varepsilon_0 m_0 c^2} \frac{\mathbf{E}_0 \sin\theta}{r} \exp\left[-\mathrm{i}w\left(t - \frac{r}{c}\right)\right] f^0(w) \tag{1.487}$$

The observer distance r is assumed to be much larger than the region of the oscillating atomic electrons, that is, the far field. In MKS units, the classical electron radius r_e is

$$r_e = \frac{e^2}{4\pi\varepsilon_0 m_0 c^2} = 2.817 \times 10^{-15} \text{ m} \tag{1.488}$$

With the substitution of Eq. (1.488) into Eq. (1.487), the transverse electric field vector at the observer $\mathbf{E}_\perp(t)$ becomes

$$\mathbf{E}_{\perp}(t) = -\frac{r_e}{r}E_0 \sin\theta \exp\left[-\mathrm{i}w\left(t-\frac{r}{c}\right)\right]f^0(w) \tag{1.489}$$

The transverse electric field $\mathbf{E}_{\perp}(r,t)$ is the sum of the forward-scattered radiation from the bound electrons of all the material atoms and the incident wave. This sum, or the interference of the forward-scattered waves from the material atomic electrons and the incident wave, produce a resultant wave, which propagates differently than a resultant wave in a vacuum.

1.5.4 *Extraction of the Complex Refractive Index from the Electromagnetic Wave Equation*

Now consider an electromagnetic wave passing through a material medium such as a beryllium lens. For transverse electromagnetic waves, the electric field is perpendicular to the wave number vector, i.e., $\mathbf{E}\perp\mathbf{k}$, and assuming no charge density gradient in the material,

$$\nabla\rho(\mathbf{r}_s,t) = 0 \tag{1.490}$$

In the medium, the electromagnetic wave is incident at time t at the position $\mathbf{r}_s$ of the charge density $\rho(\mathbf{r}_s,t)$. Only the perpendicular component $\mathbf{J}_{\perp}(\mathbf{r}_s,t)$ of the current density $\mathbf{J}(\mathbf{r}_s,t)$ contributes to the transverse wave. The wave equation for a transverse electromagnetic wave, which is Eq. (1.285), is repeated here:

$$\left(\frac{\partial^2}{\partial t^2} - c^2\nabla^2\right)\mathbf{E}_{\perp}(\mathbf{r}_s,t) = -\frac{1}{\varepsilon_0}\frac{\partial \mathbf{J}_{\perp}(\mathbf{r}_s,t)}{\partial t} \tag{1.491}$$

Here, $\mathbf{E}_{\perp}(\mathbf{r}_s,t)$ and $\mathbf{J}_{\perp}(\mathbf{r}_s,t)$ are the electric field and current density component vectors, which are perpendicular to the direction of wave propagation. The atomic electrons of the material, which are induced to oscillate by the imposed electric field, are the source of the transverse current density vector $\mathbf{J}_{\perp}(\mathbf{r}_s,t)$.

In Eq. (1.491) for an electromagnetic wave in a material, we now express the material-dependent term via the refractive index n of the material, which is expressed in terms of the complex atomic scatter factor $f(\Delta\mathbf{k},w)$. Consider as a function of time t, the component of the electron displacement vector $\mathbf{x}_{\perp}(\mathbf{r}_s,t)$ from its equilibrium position vector $\mathbf{r}$. The equilibrium position $\mathbf{r}$ of the bound electron in an atom of the material medium is measured relative to the system origin. The electron displacement vector $\mathbf{x}_{\perp}(\mathbf{r}_s,t)$) is parallel to the imposed electric field $\mathbf{E}_{\perp}(\mathbf{r}_s,t)$ and perpendicular to the direction vector $\mathbf{r}_{\mathrm{obs}} - \mathbf{r}$ pointing from the electron position $\mathbf{r}_s$ to the observer point $\mathbf{r}$, where Eq. (1.459) is repeated here:

$$\mathbf{x}(\mathbf{r}_s,t) = \frac{e}{m_0}\cdot\frac{1}{w^2 - w_0^2 + \mathrm{i}\gamma w}\mathbf{E}_{\perp}(\mathbf{r}_s,t) \tag{1.492}$$

The time variation of the transverse electric field thus induces electron velocity $\mathbf{v}(\mathbf{r}_s, t)$, where the time derivative of Eq. (1.492) gives

$$\mathbf{v}(\mathbf{r}_s, t) = \frac{e}{m_0} \cdot \frac{1}{w^2 - w_0^2 + i\gamma w} \frac{\partial \mathbf{E}_\perp(\mathbf{r}_s, t)}{\partial t} \tag{1.493}$$

The transverse current density vector $\mathbf{J}_\perp(\mathbf{r}_s, t)$ includes contributions of the Z electrons of each single atom of an element (Z, A). In a material medium of mass density ρ_m [g/cm^3], the qth atom density n_q [atom/cm^3] can be calculated, given Avogadro's number N_{A} [atom/mole], the qth element, dimensionless weight fraction w_q, gram atomic weight A_q [grams/mole], and atomic number Z_q, where

$$n_q = N_{\text{A}} \frac{\rho_m w_q}{A_q} \tag{1.494}$$

The transverse current density vector $\mathbf{J}_\perp(\mathbf{r}_s, t)$ includes the oscillator strength g_s contribution of each sth atomic electron. The imposed electric field induces in the material medium of Q elements, an atomic electron current density vector $\mathbf{J}_\perp(\mathbf{r}_s, t)$. The induced, total electron current density $\mathbf{J}_\perp(\mathbf{r}_s, t)$ is the sum of the product of (1) the elementary electron charge e, (2) the atom number density n_q of the qth element, (3) the induced velocity $\mathbf{v}_{s,q}$ of the sth atomic electron of the qth element, and (4) the oscillator strength of the sth atomic electron of the qth element, where

$$\mathbf{J}_\perp(\mathbf{r}_s, t) = -e \sum_{q=1}^{Q} n_q \sum_{s=1}^{Z} g_{s,q} \mathbf{v}_{s,q}(\mathbf{r}_s, t) \tag{1.495}$$

The summed oscillator strength of the atomic electron cloud of the qth element of atomic number Z_q is

$$Z_q = \sum_{s=1}^{Z_q} g_{s,q} \tag{1.496}$$

For neutrons, the scatter length b_q [fm] of the qth isotope corresponds to the X-ray scatter length $r_e Z_q$ [fm].

The velocity $\mathbf{v}_{s,q}$ of the qth element's s^{th} electron, which is situated at position $\mathbf{r}_s$, relative to the system origin at time t, is

$$\mathbf{v}_{s,q}(\mathbf{r}_s, t) = \frac{e}{m_0} \cdot \frac{1}{w^2 - w_{s,q}^2 + i\gamma_q w} \frac{\partial \mathbf{E}_\perp(\mathbf{r}_s, t)}{\partial t} \tag{1.497}$$

Here, $\mathbf{v}_{s,q}$ is the velocity and $w_{s,q}$ is the resonant frequency of the sth electron, and γ_q is the electron oscillation damping constant in the atom of the qth element in the material of Q elements. Substitution of the $\mathbf{v}_{s,q}$ expression of Eq. (1.497) into the transverse current density $\mathbf{J}_\perp(\mathbf{r}_s, t)$ of Eq. (1.495) yields

$$\mathbf{J}_{\perp}(\mathbf{r}_s,t) = -\frac{e^2}{m_0}\sum_{q=1}^{Q} n_q \sum_{s=1}^{Z} \frac{g_{s,q}}{w^2 - w_{s,q}^2 + \mathrm{i}\gamma_q w} \frac{\partial \mathbf{E}_{\perp}(\mathbf{r}_s,t)}{\partial t} \tag{1.498}$$

One applies Eq. (1.486) for $f^0(w)$ to Eq. (1.498) and the transverse current density $\mathbf{J}_{\perp}(\mathbf{r}_s,t)$, which arises from the induced movement of charges in a medium, becomes

$$\mathbf{J}_{\perp}(\mathbf{r}_s,t) = -\frac{e^2}{m_0}\cdot\frac{f^0(w)}{w^2}\cdot\frac{\partial \mathbf{E}_{\perp}(\mathbf{r}_s,t)}{\partial t} \tag{1.499}$$

Next, substitution of transverse current density $\mathbf{J}_{\perp}$ into the wave equation gives

$$\left(\frac{\partial^2}{\partial t^2} - c^2\nabla^2\right)\mathbf{E}_{\perp}(\mathbf{r}_s,t) = \frac{e^2}{\varepsilon_0 m_0}\sum_{q=1}^{Q} n_q \sum_{s=1}^{Z} \frac{g_{s,q}}{w^2 - w_{s,q}^2 + \mathrm{i}\gamma_q w} \frac{\partial^2 \mathbf{E}_{\perp}(\mathbf{r}_s,t)}{\partial t^2} \tag{1.500}$$

Rearrangement yields

$$\left[\left(1 - \frac{e^2}{\varepsilon_0 m_0}\sum_{q=1}^{Q} n_q \sum_{s=1}^{Z} \frac{g_{s,q}}{w^2 - w_{s,q}^2 + \mathrm{i}\gamma_q w}\right)\frac{\partial^2}{\partial t^2} - c^2\nabla^2\right]\mathbf{E}_{\perp}(\mathbf{r}_s,t) = 0 \tag{1.501}$$

The forward-scatter amplitude of Eq. (1.486) is contained within Eq. (1.501), where

$$\frac{f^0(w)}{w^2} = \sum_{q=1}^{Q} n_q \sum_{s=1}^{Z} \frac{g_{s,q}}{w^2 - w_{s,q}^2 + \mathrm{i}w\gamma_q} \qquad \frac{a_0}{\lambda} \ll 1 \text{ or } \theta \ll 1 \tag{1.502}$$

Substitution of Eq. (1.502) into Eq. (1.501) yields

$$\left[\left(1 - \frac{e^2}{\varepsilon_0 m_0}\frac{f^0(w)}{w^2}\right)\frac{\partial^2}{\partial t^2} - c^2\nabla^2\right]\mathbf{E}_{\perp}(\mathbf{r}_s,t) = 0 \tag{1.503}$$

where

$$\frac{e^2}{\varepsilon_0 m_0} = 4\pi\cdot c^2 r_e \tag{1.504}$$

Substitution of Eq. (1.504) into Eq. (1.503) yields

$$\left[\left(1 - \frac{4\pi\cdot c^2 r_e}{w^2}f^0(w)\right)\frac{\partial^2}{\partial t^2} - c^2\nabla^2\right]\mathbf{E}_{\perp}(\mathbf{r}_s,t) = 0 \tag{1.505}$$

This expression can be simplified with frequency parameter $n(w)$, which we will see is the dimensionless index of refraction, where

$$n(w) = \sqrt{1 - \frac{4\pi \cdot c^2 r_e}{w^2} f^0(w)} \tag{1.506}$$

With the index of refraction $n(w)$, the wave equation becomes

$$\left(\frac{\partial^2}{\partial t^2} - \frac{c^2}{n^2(w)}\nabla^2\right)\mathbf{E}_\perp(\mathbf{r}_s, t) = 0 \tag{1.507}$$

Since the electromagnetic wave phase velocity v_p is modified from c in a vacuum or air to c/n, the parameter $n(w)$ is indeed the index of refraction, where

$$v_\mathrm{p} = \frac{c}{n} \tag{1.508}$$

The scattering process has both elastic (lossless) and inelastic (dissipative) components, and hence the resultant index of refraction $n(w)$ is complex. The index of refraction $n(w)$ describes both the modified phase velocity and attenuation of the wave electric (and magnetic) field amplitude in the material. The second term in the radical of the index of refraction $n(w)$ is much less than 1, so the index of refraction is binomially approximated as

$$n(w) = 1 - 2\pi \frac{c^2 r_e}{w^2} f^0(w) \tag{1.509}$$

The wavelength λ is related to the angular frequency w by

$$w^2 = \frac{4\pi^2 c^2}{\lambda^2} \tag{1.510}$$

The dimensionless index of refraction $n(\lambda)$, as a function of wavelength λ, is

$$n(\lambda) = 1 - \frac{\lambda^2 r_e}{2\pi} f^0(\lambda) \tag{1.511}$$

With the substitution for the dimensionless forward-scatter factor $f^0(w)$ of Eq. (1.486) into Eq. (1.511), the complex refractive index of the material medium for the electromagnetic wave photons is

$$n(w) = 1 - 2\pi \cdot c^2 r_e \sum_{q=1}^{Q} n_q \sum_{s=1}^{Z} \frac{g_{s,q}}{w^2 - w_{s,q}^2 + \mathrm{i} w \gamma_q} \tag{1.512}$$

One can transpose Eq. (1.509) and solve for the forward-scatter factor $f^0(w)$ in terms of the refractive index $n(w)$, where

$$f^0(w) = \frac{w^2}{2\pi \cdot r_e c^2}[1 - n(w)] \tag{1.513}$$

One substitutes Eq. (1.513) into Eq. (1.489), and the far-field, transverse electric field vector $\mathbf{E}_\perp(t)$ at the distant observer, which is produced by the oscillating atomic electrons in a material medium, becomes

$$\mathbf{E}_\perp(t) = \frac{w^2}{2\pi \cdot c^2}\frac{1 - n(w)}{r}E_0 \sin\theta \exp\left[-\mathrm{i}w\left(t - \frac{r}{c}\right)\right] \tag{1.514}$$

We have thus calculated transverse electric field $\mathbf{E}_\perp(\mathbf{r}_s, t)$ as a function of the material index of refraction $n(w)$, where $\mathbf{E}_\perp(\mathbf{r}_s, t)$ is the summation of the incident wave with material-scattered waves in the forward direction ($\theta \ll 1$). In the forward direction $\theta \ll 1$, the scattering factor $f_q^0(w)$ is independent of the positions of the individual electrons $\mathbf{r}_{s,q}$. Hence, in the forward-scattered direction, all electrons of the same resonant frequency produce forward-scattered waves with the same phase, which leads to constructive interference and complete summation in the forward direction. This is similar to Huygen's principle, in that the secondary waves produced by points on a primary wave front, have the same phase only in the forward direction. The constructive interference, and thus complete summation of the secondary wavelets, occur only in the forward direction, whereas the secondary waves undergo destructive interference in the backward direction. The net effect is the creation of a secondary wave front in the forward direction and hence a forward-propagating wave.

The dimensionless, complex atomic scattering factor in the forward direction $f^0(w)$ can be written in terms of real $f_1^0(w)$ and imaginary $f_2^0(w)$ parts, so that

$$f_q^0(w) = f_{1q}^0(w) - \mathrm{i}f_{2q}^0(w) \tag{1.515}$$

The dimensionless, real part $f_1^0(w)$ of the complex atomic scattering factor $f^0(w)$ is

$$f_1^0(w) = \mathrm{Re}\left[w^2 \sum_{q=1}^{Q} n_q \sum_{s=1}^{Z} \frac{g_{s,q}}{w^2 - w_{s,q}^2 + \mathrm{i}w\gamma_q}\right] = \sum_{q=1}^{Q} n_q \sum_{s=1}^{Z} \frac{g_{s,q} w^2 (w^2 - w_{s,q}^2)}{(w^2 - w_{s,q}^2)^2 + w^2\gamma_q^2} \tag{1.516}$$

The dimensionless, imaginary part $f_2^0(w)$ of the complex atomic scattering factor $f^0(w)$ is

$$f_2^0(w) = \mathrm{Im}\left[w^2 \sum_{q=1}^{Q} n_q \sum_{s=1}^{Z} \frac{g_{s,q}}{w^2 - w_{s,q}^2 + \mathrm{i}w\gamma_q}\right] = -\sum_{q=1}^{Q} n_q \sum_{s=1}^{Z} \frac{g_{s,q} w^3 \gamma}{(w^2 - w_{s,q}^2)^2 + w^2\gamma_q^2} \tag{1.517}$$

With the substitution of Eq. (1.515) into Eq. (1.509), the index of refraction $n(w)$, as a function of angular frequency w, becomes

$$n(w) = 1 - 2\pi \frac{c^2 r_e}{w^2} f^0(w) = 1 - 2\pi \frac{c^2 r_e}{w^2} \left[f_1^0(w) - \mathrm{i} f_2^0(w) \right] \tag{1.518}$$

The index of refraction $n(\lambda)$, as a function of wavelength λ, is

$$n(\lambda) = 1 - \frac{r_e \lambda^2}{2\pi} f^0(\lambda) = 1 - \frac{r_e \lambda^2}{2\pi} \left[f_1^0(\lambda) - \mathrm{i} f_2^0(\lambda) \right] \tag{1.519}$$

For more information, see Als-Nielsen and McMorrow (2011), Attwood (1999), Henke, Gullickson, and Davis (1993), and Hubbell, Gimm, and Overbo (1975).

The dimensionless, complex refractive index $n(w)$ of the material medium for X-rays, as with other electromagnetic waves, as well as neutron and other particle waves, can be expressed by

$$n = 1 - \delta + \mathrm{i}\beta = 1 - (\delta_r + \mathrm{i}\delta_i) = 1 - \delta_r + \mathrm{i}\beta \tag{1.520}$$

We consider X-rays in particular and can express the refractive decrement δ or δ_r and the attenuation factor β or $\delta_i = -\beta$ in terms of the respective real $f_1^0(\lambda)$ and imaginary $f_2^0(\lambda)$ components of forward-scatter amplitude $f^0(\lambda)$, using the results of these derivations. The real decrement δ_q for X-ray scatter from a single q^{th} element of atomic number Z and atomic weight A is proportional to (1) the real component of the forward-scatter amplitude $f_1^0(\lambda)$, (2) the number density n_q of the scatter centers (atom number density) of the qth element, and (3) the square of the incident wavelength λ of the incident gamma (or neutron) plane waves. The dimensionless, decrement $\delta_q(\lambda)$ for the qth element of a medium of Q elements is proportional to the real component of the forward-scatter amplitude $f_{1q}^0(\lambda)$, where

$$\delta_q(\lambda) = \frac{r_e n_q \lambda^2}{2\pi} f_{1q}^0(\lambda) \tag{1.521}$$

The qth element's dimensionless, attenuation factor $\beta_q(\lambda)$ is proportional to the imaginary component of the forward-scatter amplitude $f_{2q}^0(\lambda)$, where

$$\beta_q(\lambda) = \frac{r_e n_q \lambda^2}{2\pi} f_{2q}^0(\lambda) \tag{1.522}$$

Based on Rayleigh scatter, plotted in Figure 1.26, as a function of X-ray and gamma photon energies, is the "delta over mu ratio" or δ/μ measured in centimeters for various lens materials. The top δ/μ (delta/mu) plot is over a wide range of energies 2–430 keV. The data for the plot is from the National Institute of Standards and Technology (NIST) database (Chantler et al., 2005), which tabulates, as a function of photon energy E, the real $f_1^0(E)$ and imaginary $f_2^0(E)$ components of the photon

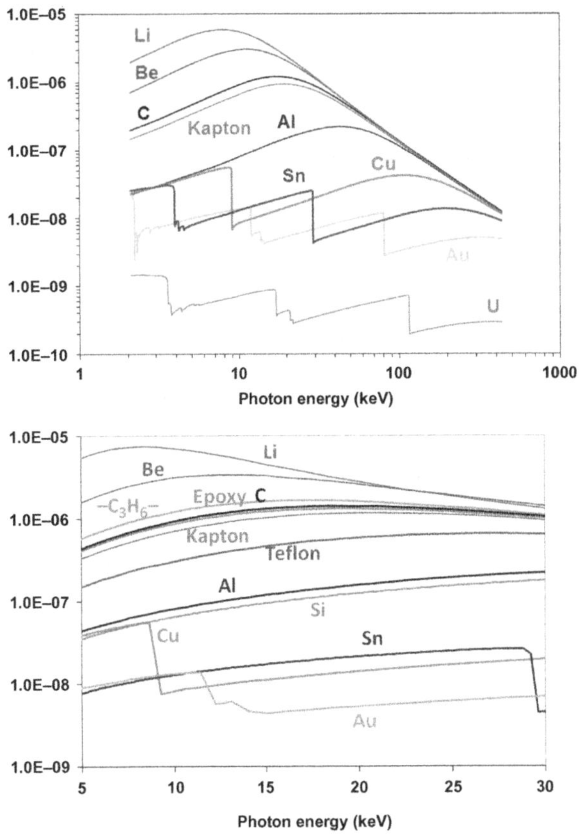

Figure 1.26 Based on Rayleigh scatter, the "delta over mu ratio" or δ/μ in units of [cm] for various lens materials, is plotted as a function of X-ray and gamma photon energies (top) over a wide range of 2–430 keV from NIST database (Chantler et al., 2005), and (bottom) over a narrow range 5–30 keV from CXRO database (Henke et al., 1993).

forward-scatter amplitude and the photon total attenuation cross section $\sigma_{\text{tot}}(E)$ in units of [cm^2/g]. The total attenuation cross section $\sigma_{\text{tot}}(E)$ is the sum of the individual cross sections for photoelectric absorption, Compton scatter, and pair production. The decrement $\delta(E)$ was calculated from Eq. (1.521) using the tabulated $f_1^0(E)$ for each element component of the material. In Figure 1.26 (top panel), the K edges of copper, the K and L edges of tin, and the K, L, and M edges of gold and uranium appear.

Each qth element in the medium of Q elements has atom number density n_q [atom/cm^3], which is multiplied by the atom cross section σ_q [cm^2/atom] for X-ray attenuation, and yields the medium's q^{th} element linear attenuation component μ_q [cm^{-1}], where

$$\mu_q\,[\text{cm}^{-1}] = n_q\,[\text{atom/cm}^3]\cdot\sigma_q\,[\text{cm}^2/\text{atom}] \tag{1.523}$$

The qth element linear attenuation component μ_q [cm^{-1}] is related to the dimensionless attenuation factor β_q via

$$\mu_q = 4\pi\frac{\beta_q}{\lambda} \tag{1.524}$$

From Eq. (1.524) for μ_q [cm^{-1}] and Eq. (1.522) for $\beta_q(\lambda)$, one obtains the absorption cross section $\sigma_q(\lambda)$, in units of area [m^2], or the more typical cross section unit of barns, where 1 barn = 10^{-28} m^2:

$$\sigma_q(\lambda) = 2r_e\lambda \cdot f_{2q}^0(\lambda) \tag{1.525}$$

With Eq. (1.524), the total linear attenuation μ for the medium is then the sum of the component linear attenuations of the Q elements, where

$$\mu = \sum_{q=1}^{Q} \mu_q = \sum_{q=1}^{Q} n_q\sigma_q = N_A\rho_m \sum_{q=1}^{Q} \frac{w_q}{A_q}\sigma_q \tag{1.526}$$

Given the medium mass density ρ_m [g/cm^3], Avogadro's number N_A [atom/mole], the qth element's gram atomic weight A_q [g/mole] and dimensionless weight fraction w_q, one obtains the qth element atom number density n_q, where

$$n_q = N_A \frac{\rho_m w_q}{A_q} \tag{1.527}$$

With the substitution of Eq. (1.525) into Eq. (1.526), the total linear attenuation μ for the medium is

$$\begin{aligned} \mu &= \sum_{q=1}^{Q} \mu_q = \frac{4\pi}{\lambda}\beta = \frac{4\pi}{\lambda}\sum_{q=1}^{Q} \beta_q = 2r_e\lambda f_2^0(\lambda) \\ &= 2r_e\lambda \sum_{q=1}^{Q} n_q f_{2q}^0(\lambda) = 2r_e\lambda N_A \rho_m \sum_{q=1}^{Q} \frac{w_q}{A_q} f_{2q}^0(\lambda) \end{aligned} \tag{1.528}$$

From Eq. (1.521), the sum of the Q element component decrements δ_q yields the total material decrement δ, where

$$\delta = \sum_{q=1}^{Q} \delta_q = \frac{r_e\lambda^2}{2\pi}\sum_{q=1}^{Q} n_q f_{1q}^0(\lambda) = \frac{r_e\lambda^2\rho_m N_A}{2\pi}\sum_{q=1}^{Q} \frac{w_q}{A_q} f_{1q}^0(\lambda) \tag{1.529}$$

Many of the materials plotted are comprised of a single material, where the weight fraction is unity, $w_q = 1$.

In Figure 1.26, the bottom curve of the δ/μ ratio is plotted over the 5–30 keV X-ray energy range for the same set of elements as the top curves. Only the K-edge of copper occurs in the 5–30 keV energy range, which is commonly available at synchrotron radiation beam lines. The data for the bottom plots of Figure 1.26 was obtained from the Center for X-Ray Optics (CXRO) database (Henke et al., 1993), which tabulates $\delta(E)$ and $\beta(E)$ as a function of photon energy E, and requires $\mu = 4\pi(\beta/\lambda)$ of Eq. (1.524) to calculate μ from β.

For X-rays, one can neglect the conductivity or effects of the free electrons in materials for the calculation of the complex refractive index, because the X-ray frequencies are much larger than the electron plasma frequency in metal conductors. For X-rays, the material's conductivity is completely neglected, and for X-rays, the material behaves as a dielectric. The phase velocity v_p of the electromagnetic X-ray wave is given by

$$v_p = \frac{w}{k} = \frac{c}{n} = \frac{c}{1 - \delta + i\beta} \tag{1.530}$$

Transposing to solve for the wave number k, one obtains the wave vector $\mathbf{k}$, where $\hat{\mathbf{k}}$ is the unit wave vector and

$$\mathbf{k} = k\hat{\mathbf{k}} = \frac{w}{c}(1 - \delta + i\beta)\hat{\mathbf{k}} \tag{1.531}$$

The phase velocity v_p of the electromagnetic wave is also given by

$$v_p = \frac{w}{k} = \frac{c}{n} = \frac{1}{n\sqrt{\mu_0 \varepsilon_0}} \tag{1.532}$$

The wave number k can be expressed as

$$k = wn\sqrt{\mu_0 \varepsilon_0} = \frac{w}{c}(1 - \delta + i\beta) \tag{1.533}$$

The phase velocity v_p can be expressed as

$$v_p = \frac{w}{k} = \frac{c}{n} = \frac{1}{\sqrt{n^2 \mu_0 \varepsilon_0}} = \frac{1}{\sqrt{\mu\varepsilon}} \tag{1.534}$$

The index of refraction n can be expressed by

$$n = c\sqrt{\mu\varepsilon} = 1 - \delta + i\beta \tag{1.535}$$

The electrical permittivity or dielectric constant is denoted by ε, and the dimensionless relative permittivity is denoted by ε_r, where

$$\varepsilon = \varepsilon_r \varepsilon_0 \tag{1.536}$$

The dimensionless relative permittivity ε_r can also be expressed in terms of the dimensionless electrical susceptibility χ_e:

$$\varepsilon_r = 1 + \chi_e \tag{1.537}$$

The magnetic permeability of a material is denoted by μ, and the dimensionless relative permeability is denoted by μ_r, where

$$\mu = \mu_r \mu_0 \tag{1.538}$$

The dimensionless relative permeability μ_r can also be expressed in terms of the dimensionless magnetic susceptibility χ_m:

$$\mu_r = 1 + \chi_m \tag{1.539}$$

For X-rays, the conductivity of a material is neglected, and for X-ray electromagnetic waves, the material impedance Z is determined by the dielectric constant ε and the magnetic permeability μ, where

$$Z = \sqrt{\frac{\mu}{\varepsilon}} \tag{1.540}$$

The magnetic flux density of the X-ray in a material medium equals the product of (1) the reciprocal of the phase velocity $1/v_p$, which is $\sqrt{\mu\varepsilon}$, and (2) the cross product of the unit wave vector $\hat{\mathbf{k}}$ and the electric field intensity $\mathbf{E}$, where

$$\mathbf{B} = \sqrt{\mu\varepsilon}\hat{\mathbf{k}} \times \mathbf{E} = \frac{\mathbf{k} \times \mathbf{E}}{w} \tag{1.541}$$

Given position vector $r\hat{\mathbf{r}}$, assume that the electric field $\mathbf{E}(\mathbf{r}, t)$ is given by

$$\mathbf{E}(\mathbf{r}, t) = \mathbf{E}_0 \exp[\mathrm{i}(\mathbf{k} \cdot \mathbf{r} - wt)] \tag{1.542}$$

With the substitution of Eq. (1.542) into Eq. (1.531), one obtains

$$\mathbf{E}(\mathbf{r}, t) = \mathbf{E}_0 \exp\left[-\mathrm{i}w\left(t - \frac{r}{c}\right)\right] \exp\left[-\mathrm{i}2\pi\delta\left(\frac{r}{\lambda}\right)\right] \exp\left[-2\pi\beta\left(\frac{r}{\lambda}\right)\right] \tag{1.543}$$

The first exponential term in Eq. (1.543), is the vacuum propagation, which represents the wave phase advance in a vacuum:

$$\exp\left[-\mathrm{i}w\left(t - \frac{r}{c}\right)\right] \Leftrightarrow \text{vacuum propagation} \tag{1.544}$$

The second exponential term in Eq. (1.543), is the phase shift of the wave as it propagates through a medium:

$$\exp\left[-\mathrm{i}2\pi\delta\left(\frac{r}{\lambda}\right)\right] \Leftrightarrow \text{phase shift in medium} \tag{1.545}$$

The third exponential term in Eq. (1.543), is the phase shift of the wave as it propagates through a medium:

$$\exp\left[-2\pi\beta\left(\frac{r}{\lambda}\right)\right] \Leftrightarrow \text{attenuation in medium} \tag{1.546}$$

For X-rays, recall the wave equation of Eq. (1.285) for the transverse electric field vector $\mathbf{E}_\perp(\mathbf{r},t)$ in a vacuum medium, which is occupied by atomic electrons, where

$$\left(\frac{\partial^2}{\partial t^2} - c^2\nabla^2\right)\mathbf{E}_\perp(\mathbf{r},t) = -\frac{1}{\varepsilon_0}\frac{\partial \mathbf{J}_\perp(\mathbf{r},t)}{\partial t} - \frac{c^2}{\varepsilon_0}\nabla\rho(\mathbf{r},t) \tag{1.547}$$

For longer wavelength electromagnetic waves such as microwaves or light, where the frequency is below the plasma frequency of the conduction electrons in a conducting medium, Eq. (1.547) is modified for a conducting medium, where

$$\left(\frac{\partial^2}{\partial t^2} - \frac{c^2}{n^2}\nabla^2\right)\mathbf{E}_\perp(\mathbf{r},t) = -\frac{1}{\varepsilon}\frac{\partial \mathbf{J}_\perp(\mathbf{r},t)}{\partial t} - \frac{c^2}{n^2\varepsilon}\nabla\rho(\mathbf{r},t) \tag{1.548}$$

The conducting medium has conductivity σ, and the transverse current density vector $\mathbf{J}_\perp(\mathbf{r},t)$ is proportional to the transverse electric field vector $\mathbf{E}_\perp(\mathbf{r},t)$ via the conductivity σ proportionality constant, where

$$\mathbf{E}_\perp(\mathbf{r},t) = \sigma\mathbf{E}_\perp(\mathbf{r},t) \tag{1.549}$$

Assume that the gradient of the free charge density is zero, where

$$\nabla\rho(\mathbf{r},t) = 0 \tag{1.550}$$

With the substitution of Eqs. (1.549) and (1.550) into Eq. (1.548) and the use of the relation $c^2/n^2 = 1/\mu\varepsilon$, the wave equation for a conducting medium for microwaves or light becomes

$$\left(\mu\varepsilon\frac{\partial^2}{\partial t^2} - \nabla^2\right)\mathbf{E}_\perp(\mathbf{r},t) = -\mu\sigma\frac{\partial \mathbf{E}_\perp(\mathbf{r},t)}{\partial t} \tag{1.551}$$

Assume that the transverse electric field vector $\mathbf{E}_\perp(\mathbf{r},t)$ for the microwave or light is

$$\mathbf{E}_\perp(\mathbf{r},t) = \mathbf{E}_0\exp[\mathrm{i}(\mathbf{k}\cdot\mathbf{r} - wt)] \tag{1.552}$$

Substitution of Eq. (1.552) into the wave equation for the transverse electric field yields

$$(-\mu\varepsilon w^2 + k^2)\mathbf{E}_\perp(\mathbf{r},t) = \mathrm{i}w\mu\sigma\mathbf{E}_\perp(\mathbf{r},t) \tag{1.553}$$

Solving for the wave number k for the microwaves or light yields

$$k = w\sqrt{\frac{\mathrm{i}\mu\sigma}{w} + \mu\varepsilon} = w\sqrt{\mu\varepsilon}\sqrt{1 + \mathrm{i}\frac{\sigma}{\varepsilon w}} = \frac{w}{v_{\mathrm{ph}}}\sqrt{1 + \mathrm{i}\frac{\sigma}{\varepsilon w}} \tag{1.554}$$

With the substitution of Eq. (1.554) into Eq. (1.552), the transverse electric field vector $\mathbf{E}_{\perp}(\mathbf{r}, t)$ becomes

$$\mathbf{E}_{\perp}(\mathbf{r}, t) = \mathbf{E}_0 \exp\left[-\mathrm{i}w\left(t - r\sqrt{\mu\varepsilon}\sqrt{1 + \mathrm{i}\frac{\sigma}{\varepsilon w}}\right)\right] \tag{1.555}$$

For sufficiently high frequencies, such as light, and/or low conductivities, the binomial approximation can be applied to the radical in Eq. (1.555), so that the transverse electric field vector $\mathbf{E}_{\perp}(\mathbf{r}, t)$ has a material phase velocity propagation component and an attenuation component, where

$$\mathbf{E}_{\perp}(\mathbf{r}, t) = \mathbf{E}_0 \exp[-\mathrm{i}w(t - r\sqrt{\mu\varepsilon})]\exp\left(-r\sqrt{\frac{\mu}{\varepsilon}}\frac{\sigma}{2}\right) \quad 1 \gg \frac{\sigma}{\varepsilon w} \tag{1.556}$$

For sufficiently low frequencies, such as microwaves, and/or high conductivities, the transverse electric field vector $\mathbf{E}_{\perp}(\mathbf{r}, t)$ has a material phase velocity propagation component and an attenuation component, where

$$\mathbf{E}_{\perp}(\mathbf{r}, t) = \mathbf{E}_0 \exp\left[-\mathrm{i}w\left(t - r\sqrt{\frac{w\mu\sigma}{2}}\right)\right]\exp\left(-r\sqrt{\frac{w\mu\sigma}{2}}\right) \quad 1 \ll \frac{\mu\sigma}{\varepsilon w} \tag{1.557}$$

and

$$\sqrt{\mathrm{i}} = \sqrt{\exp\left(\mathrm{i}\frac{\pi}{2}\right)} = \exp\left(\mathrm{i}\frac{\pi}{4}\right) = \cos\frac{\pi}{4} + \mathrm{i}\sin\frac{\pi}{4} = \frac{1}{\sqrt{2}}(1 + \mathrm{i}) \tag{1.558}$$

1.5.5 Scatter, Absorption, Total Cross Section for Electromagnetic Waves (X-rays)

We now consider the scatter of an incident X-ray electromagnetic wave from a single s^{th} atomic electron of an atom with atomic number Z, where the atom belongs to one of the qth elements, which comprise the scattering and absorbing medium of $q = 1, 2, \ldots, Q$ elements. From the calculated scattered electric field vector $\mathbf{E}(t)$ and incident electric field vector $\mathbf{E}_i(t)$, one can calculate the differential scattering cross section $d\sigma/dw$ and the total scattering cross section σ for X-rays. The scatter cross section $\sigma_s(w)$, as a function of X-ray frequency w, is defined as the ratio of the time-averaged, radiated power $P_{\mathrm{avg}}(w)$, divided by the incident, root-mean-squared (RMS) power per unit area S^i_{avg}, where

$$\sigma_s(w) = \frac{P_{\text{avg}}(w)}{S^i_{\text{avg}}} \tag{1.559}$$

To derive the scatter cross section $\sigma_s(w)$, the incident, RMS power per unit area S^i_{avg} in the denominator of Eq. (1.559) is first determined as follows. Varying with time t and frequency w, and the incident plane electromagnetic wave, the electric field intensity $\mathbf{E}_i(w,t)$ is perpendicular to the magnetic flux density $\mathbf{B}_i(w,t)$. The instantaneous time Poynting vector $\mathbf{S}_i(w,t)$ of the incident electromagnetic wave power density, or the Poynting vector $\mathbf{S}_i(w,t)$, propagates in the unit wave vector direction $\hat{\mathbf{k}}$, where

$$\mathbf{S}_i(w,t) = S_i(w,t)\hat{\mathbf{k}} = \mathbf{E}_i(w,t) \times \mathbf{H}_i(w,t) = \frac{1}{\mu_0} E_i(w,t) B_i(w,t)\hat{\mathbf{k}} \tag{1.560}$$

In the plane electromagnetic wave, the magnetic field flux density $\mathbf{B}_i(w,t)$ is obtained from the electric field intensity $\mathbf{E}_i(w,t)$, via dividing it by the speed of light c in a vacuum, where

$$\mathbf{B}_i(w,t) = \frac{\mathbf{E}_i(w,t)}{c} \tag{1.561}$$

Hence, the amplitude of the incident electromagnetic wave power density $\mathrm{S}_i(w,t)$, as a function of time t and frequency w, becomes

$$\mathrm{S}_i(w,t) = \frac{1}{\mu_0 c} E_i^2(w,t) \tag{1.562}$$

The incident plane wave electric field intensity vector $\mathbf{E}_i(w,t)$, which is incident at the position $\mathbf{r}_s$ of the scattering electron at time t, is given by

$$\mathbf{E}_i(w,t) = \mathbf{E}_0 \exp[-\mathrm{i}(wt - \mathbf{k}_0 \cdot \mathbf{r}_s)] \tag{1.563}$$

From Eq. (1.561), the corresponding, instantaneous time magnetic field $\mathbf{B}_i(w,t)$ of the incident vacuum plane wave is given by

$$\mathbf{B}_i(w,t) = \frac{\mathbf{E}_i(w,t)}{c} = \frac{\mathbf{E}_0 \exp[-\mathrm{i}(wt - \mathbf{k}_0 \cdot \mathbf{r}_s)]}{c} \tag{1.564}$$

The mean of the electric field amplitude $\mathbf{E}^i_{\text{avg}}(w,t)$, which is zero, is obtained by integrating the real part of the time-varying, electric field $\mathbf{E}_i(w,t)$ over one oscillation period T, where

$$\mathbf{E}^i_{\text{avg}}(w) = \frac{1}{T}\int_0^T \mathrm{Re}[\mathbf{E}_i(w,t)]\mathrm{d}t = \frac{\exp(\mathbf{k}_0 \cdot \mathbf{r}_s)}{T}\int_0^T \cos(wt)\mathrm{d}t = 0 \tag{1.565}$$

The RMS, electric field E^i_{rms} amplitude, which is the standard deviation of the electric field amplitude $\mathbf{E}_i(w,t)$, is defined as the square root of the average over one period T of the real part of the squared electric field $\mathbf{E}_i(w,t)$. The squared electric field $E_i^2(w,t)$ scalar quantity is the dot product of the real part of the electric field vector $\mathbf{E}_i(w)$ with itself, where

$$\mathrm{Re}[\mathbf{E}_i(w,t)]\cdot\mathrm{Re}[\mathbf{E}_i(w,t)]=(E_0)^2\cos^2[-2\mathrm{i}(wt-\mathbf{k}_0\cdot\mathbf{r}_s)] \tag{1.566}$$

With $w=2\pi/T$, the RMS, electric field E^i_{rms} amplitude, or standard deviation of the electric field amplitude, is then

$$E^i_{\mathrm{rms}}=\sqrt{\frac{1}{T}\int_0^T\mathrm{Re}[\mathbf{E}_i(w,t)]\cdot\mathrm{Re}[\mathbf{E}_i(w,t)]\mathrm{d}t}=\sqrt{\frac{E_0^2}{T}\int_0^T\cos^2\left[2\left(\frac{2\pi}{T}t-\mathbf{k}_0\cdot\mathbf{r}_s\right)\right]\mathrm{d}t}=\frac{E_0}{\sqrt{2}} \tag{1.567}$$

In Eq. (1.567), one uses the trigonometric identity

$$\cos^2\phi=\frac{1-\cos 2\phi}{2} \tag{1.568}$$

The time-averaged, power density $S^i_{\mathrm{avg}}(w)$ is defined as the average power over one oscillation period, where

$$S^i_{\mathrm{avg}}(w)=\frac{1}{T}\int_0^T S_i(w,t)\mathrm{d}t=\frac{1}{\mu_0 c}\frac{1}{T}\int_0^T E_i^2(w,t)\mathrm{d}t=\frac{1}{\mu_0 c}(E^i_{\mathrm{rms}})^2 \tag{1.569}$$

With the substitution of Eq. (1.566) for E^i_{rms} in Eq. (1.569), the incident, time-averaged power density $S^i_{\mathrm{avg}}(w)$ becomes

$$S^i_{\mathrm{avg}}(w)=\frac{E_0^2}{2\mu_0 c}=\frac{E_0^2}{2}\sqrt{\frac{\varepsilon_0}{\mu_0}} \tag{1.570}$$

From Eq. (1.570), one finds that the time-averaged Poynting vector $\mathbf{S}^i_{\mathrm{avg}}(w)$ for power density (power/area), which is transported by an electromagnetic plane wave in a vacuum, is also given by multiplying one-half by the cross product of the electric field $\mathbf{E}$ and complex conjugate of the magnetic field intensity $\mathbf{H}^*$, where

$$\mathbf{S}^i_{\mathrm{avg}}(w)=\frac{1}{2}\mathrm{Re}[\mathbf{E}\times\mathbf{H}^*]=\frac{1}{2\mu_0}\mathrm{Re}[\mathbf{E}\times\mathbf{B}^*] \tag{1.571}$$

Recall that the electric field intensity $\mathbf{E}$ is in units of V/m and the complex conjugate of the magnetic field intensity $\mathbf{H}^*$ is in units of current/m or charge/s/m, which yields the time-averaged Poynting vector $\mathbf{S}^i_{\mathrm{avg}}(w)$ in units of W/m^2.

The time-averaged power-radiated $P_{\mathrm{avg}}(w)$ in the numerator of the scatter cross section $\sigma_s(w)$ of Eq. (1.559) is next determined as follows. The scattered electric field $\mathbf{E}(t)$, which is emitted from a bound atomic electron undergoing acceleration

$\mathbf{a}_s(w,t)$, carries away the instantaneous power $P(w,t)$ of Eq. (1.450). The instantaneous power $P(w,t)$ of Eq. (1.450) is now expressed in terms of the classical electron radius r_e of Eq. (1.488), where

$$P(w,t) = \frac{8\pi}{3}\left(\frac{e^2 a_s^2(w,t)}{16\pi^2 \varepsilon_0 c^3}\right) = \frac{8\pi}{3}\frac{m_0^2 r_e^2}{e^2}\sqrt{\frac{\varepsilon_0}{\mu_0}} a_s^2(w,t) \tag{1.572}$$

The physical, time-averaged power $P_{\mathrm{avg}}(w)$, which is emitted from the oscillating bound electron, is expressed in terms of the RMS power, which is obtained from the squared magnitude of the RMS acceleration $a_{\mathrm{rms}}(w)$ of the oscillating bound electron, where

$$P_{\mathrm{avg}}(w) = \frac{8\pi}{3}\frac{m_0^2 r_e^2}{e^2}\sqrt{\frac{\varepsilon_0}{\mu_0}} |a_{\mathrm{rms}}(w)|^2 \tag{1.573}$$

That is, the physical, time-averaged power $P_{\mathrm{avg}}(w)$, which is emitted from the bound atomic electron, is expressed by the squared magnitude of the standard deviation of the electron acceleration $a_{\mathrm{rms}}^s(w)$. Recall that the incident plane wave electric field $\mathbf{E}_i(t)$ of frequency w and wave vector $\mathbf{k}_0$ is incident on a bound atomic electron, which causes the electron at position $\mathbf{r}_s$ to undergo acceleration $\mathbf{a}_s(w,t)$, which is given by Eq. (1.461) and is repeated here:

$$\mathbf{a}_s(w,t) = \frac{w^2}{w^2 + \mathrm{i}w\gamma - w_s^2}\frac{e}{m_0}\mathbf{E}_i(w,t) \tag{1.574}$$

The sinusoidal, incident electric field $\mathbf{E}_i(w,t)$ of frequency w and wave vector $\mathbf{k}_0$, which causes the bound atomic electron to undergo a sinusoidal acceleration $\mathbf{a}_s(w,t)$, is

$$\mathbf{E}_i(w,t) = \mathbf{E}_0 \exp[-\mathrm{i}(wt - \mathbf{k}_0 \cdot \mathbf{r}_s)] \tag{1.575}$$

One removes the time dependence of the bound atomic electron acceleration $\mathbf{a}_s(w,t)$, by using the scalar magnitude of the RMS acceleration $|a_{\mathrm{rms}}(w)|$ in place of the magnitude of the time-varying acceleration $|\mathbf{a}_s(w,t)|$. This requires replacing the time-dependent driving or incident electric field $\mathbf{E}_i(w,t)$ by its RMS value, namely $\mathbf{E}_{\mathrm{rms}}^i$ of Eq. (1.567). The magnitude of the RMS acceleration of the bound atomic electron $\mathbf{a}_{\mathrm{rms}}(w)$ is then

$$|\mathbf{a}_{\mathrm{rms}}(w)| = \left|\frac{w^2}{w^2 + \mathrm{i}w\gamma - w_s^2}\frac{e}{m_0}E_{\mathrm{rms}}^i(w)\right| \tag{1.576}$$

The squared magnitude of the acceleration $|\mathbf{a}_{\mathrm{rms}}(w)|^2$ or $\mathrm{a}_{\mathrm{rms}}^2(w)$ of the bound atomic electron is then

$$a_{\mathrm{rms}}^2 = |\mathbf{a}_{\mathrm{rms}}(w)|^2 = \frac{w^4}{(w^2 - w_s^2) + w^2\gamma^2}\left(\frac{eE_{\mathrm{rms}}^i}{m_0}\right)^2 \tag{1.577}$$

Substituting of Eq. (1.577) for $|\mathbf{a}_{\mathrm{rms}}(w)|^2$ into Eq. (1.573), the physical, time-averaged power $P_{\mathrm{avg}}(w)$ becomes

$$P_{\mathrm{avg}}(w) = \frac{8\pi}{3} r_e^2 \sqrt{\frac{\varepsilon_0}{\mu_0}} \frac{E_0^2}{2} \frac{w^4}{(w^2 - w_s^2) + w^2\gamma^2} \tag{1.578}$$

With the substitution of Eq. (1.570) for the time-averaged power density S^i_{avg} and Eq. (1.578) for $P_{\mathrm{avg}}(w)$ into Eq. (1.559), the complex scatter cross section $\sigma_s(w)$ for the scatter of an electromagnetic wave from a single, bound atomic electron becomes

$$\sigma_s(w) = \frac{P_{\mathrm{avg}}(w)}{\mathbf{S}^i_{\mathrm{avg}}(w)} = \frac{8\pi}{3} r_e^2 \frac{w^4}{(w^2 - w_s^2)^2 + w^2\gamma^2} \tag{1.579}$$

Recall from Eq. (1.479) that for either the long-wavelength or forward-scatter limit (toward the observer), the complex atomic scattering factor $f_s^0(w)$ for a single bound, s^{th} atomic electron in the medium is

$$f_s^0(w) = w^2 \frac{g_s}{w^2 - w_s^2 + iw\gamma} \tag{1.580}$$

With the substitution in Eq. (1.580) for $f^0(w)$ of Eq. (1.579), the complex scatter cross section $\sigma_s(w)$ for a single, bound atomic electron becomes

$$\sigma_s(w) = \frac{8\pi}{3} r_e^2 |f_s^0(w)|^2 = \frac{8\pi}{3} r_e^2 [f_s^0(w)][f_s^0(w)]^* \tag{1.581}$$

The asterisk * denotes the complex conjugate operator. If the driving frequency w of the incident electromagnetic wave is much larger than the resonant frequency w_s of the bound electron, where $w \gg w_s$, and the resonant frequency w_s is much larger than the damping frequency γ, where $w \gg \gamma$, then the bound electron becomes effectively a free electron. In this case, the incident electric field overcomes the influence of the electric field of the electron's nucleus, and one obtains the Thomson scattering cross section for a single, free electron, where

$$\sigma_s = \frac{P}{\langle S_i \rangle} = \frac{8\pi}{3} r_e^2 = 0.665 \text{ barn} \quad w \gg w_s \text{ and } w \gg \gamma \tag{1.582}$$

We now consider the scatter $\sigma_s(w)$, absorption $\sigma_a(w)$, and total $\sigma_t(w)$ cross sections for X-ray interaction by scatter and absorption in a material medium. The scatter cross section σ_s for material, which is comprised of Q elements, is obtained by replacing Eq. (1.580), the damped-resonant term for a single bound electron, by the term

$$\frac{w^4}{(w^2 - w_{s,q}^2)^2 + w^2\gamma_q^2} \tag{1.583}$$

The damped-resonant term for a single, bound atomic electron is now replaced by the squared magnitude of the forward-scatter factor $|f^0(w)|^2$ for all the bound atomic electrons for a scattering medium. From the forward-scatter factor $f^0(w)$ of Eq. (1.486) for a medium, one obtains

$$|f^0(w)|^2 = \left| w^2 \sum_{q=1}^{Q} n_q \sum_{s=1}^{Z} \frac{g_{s,q}}{w^2 - w_{s,q}^2 + \mathrm{i}w\gamma_q} \right|^2 \quad \frac{a_0}{\lambda} \ll 1 \text{ or } \theta \ll 1 \tag{1.584}$$

The scatter cross section $\sigma_s(w)$ for the single, bound atomic electron of Eq. (1.581), as a function of frequency w, now becomes the scatter cross section $\sigma_s(w)$ for a multielement medium, which is comprised of Q elements, each with Z bound electrons for q^{th} element (Z, A), where

$$\sigma_s(w) = \frac{8\pi}{3} r_e^2 |f^0(w)|^2 \tag{1.585}$$

With the substitution of the forward-scatter factor $f^0(w)$ of Eq. (1.486) for a medium into Eq. (1.585), the scatter cross section $\sigma_s(w)$ for the medium becomes

$$\sigma_s(w) = \frac{8\pi}{3} r_e^2 w^4 \left| \sum_{q=1}^{Q} n_q \sum_{s=1}^{Z} \frac{g_{s,q}}{w^2 - w_{s,q}^2 + \mathrm{i}w\gamma_q} \right|^2 \tag{1.586}$$

In Eq. (1.586), the forward-scatter factors or amplitudes from each of the electrons are first summed, and then the squared magnitude of this amplitude sum is taken.

We now determine the time-averaged power absorbed in a medium. Recall that the transverse current density $\mathbf{J}_\perp(\mathbf{r}, t)$ of Eq. (1.499) arises from the induced movement of charges in a medium by action of the incident, transverse electric field $\mathbf{E}_\perp(\mathbf{r}, t)$ and is repeated here:

$$\mathbf{J}_\perp(\mathbf{r}_s, t) = -\frac{e^2}{m_0} \cdot \frac{f^0(w)}{w^2} \cdot \frac{\partial \mathbf{E}_\perp(\mathbf{r}_s, t)}{\partial t} \tag{1.587}$$

The incident, transverse electric field vector $\mathbf{E}_\perp(\mathbf{r}_s, t)$ at the medium is given by

$$\mathbf{E}_\perp(\mathbf{r}_s, t) = \mathbf{E}_0 \exp[-\mathrm{i}(wt - \mathbf{k}_0 \cdot \mathbf{r}_s)] \tag{1.588}$$

The substitution of Eq. (1.588) into Eq. (1.587) yields the induced transverse current, vector density $\mathbf{J}_\perp(\mathbf{r}_s, t)$ by the incident electric field vector $\mathbf{E}_\perp(\mathbf{r}_s, t)$, where

$$\mathbf{J}_\perp(\mathbf{r}_s, t) = \left(\mathrm{i} \frac{e^2 f^0(w)}{m_0 w} \right) \mathbf{E}_\perp(\mathbf{r}_s, t) \tag{1.589}$$

Recalling Eq. (1.567), the standard deviation of the transverse electric field vector or the RMS $\mathbf{E}_{\perp \mathrm{rms}}(w)$ electric field amplitude of the incident plane wave is

$$\mathbf{E}_{\perp \mathrm{rms}} = \frac{\mathbf{E}_0}{\sqrt{2}} \tag{1.590}$$

The medium's atomic electron current density $\mathbf{J}_{\perp \mathrm{rms}}(w)$ vector is obtained from the RMS $\mathbf{E}_{\perp \mathrm{rms}}(w)$ electric field vector via Eq. (1.589), where

$$\mathbf{J}_{\perp \mathrm{rms}}(w) = \left(\mathrm{i}\frac{e^2 f^0(w)}{m_0 w}\right)\mathbf{E}_{\perp \mathrm{rms}} = \left(\mathrm{i}\frac{e^2 f^0(w)}{m_0 w}\right)\frac{\mathbf{E}_0}{\sqrt{2}} \tag{1.591}$$

The RMS absorbed power per volume $W_{\mathrm{rms}}(w)$, expressed in W/m^3, is the average work per unit volume and per unit time, which is performed by the RMS transverse electric field $\mathbf{E}_{\perp \mathrm{rms}}(w)$ on the movement of the medium's bound atomic electrons. That is, the RMS absorbed power per volume $W_{\mathrm{rms}}(w)$ by the atomic electrons of a medium loses energy from the incident electromagnetic wave.

The RMS absorbed power per volume $W_{\mathrm{rms}}(w)$ is the vector dot product of the RMS, applied electric field $\mathbf{E}_{\perp \mathrm{rms}}(w)$ in units of V/m, and the resulting RMS, transverse, current density $\mathbf{J}_{\perp \mathrm{rms}}(w)$ in units of A/m^2. The RMS real physical power $W_{\mathrm{rms}}(w)$, which is absorbed by the medium, is obtained by taking the real part of the vector dot product of $\mathbf{E}_{\perp \mathrm{rms}}(w)$ and $\mathbf{J}_{\perp \mathrm{rms}}(w)$, where

$$W_{\mathrm{rms}}(w) = \mathrm{Re}[\mathbf{E}_{\perp \mathrm{rms}}(w) \cdot \mathbf{J}_{\perp \mathrm{rms}}(w)] \tag{1.592}$$

With the substitution of Eq. (1.590) and Eq. (1.591) into Eq. (1.592), the RMS absorbed power per volume $W_{\mathrm{rms}}(w)$ of the incident electromagnetic wave becomes

$$W_{\mathrm{rms}}(w) = \mathrm{Re}\left[\mathrm{i}\frac{e^2 f^0(w)}{m_0 w}\mathbf{E}_{\perp \mathrm{rms}}(w) \cdot \mathbf{E}_{\perp \mathrm{rms}}(w)\right] = \frac{e^2}{2m_0 w}E_0^2 \mathrm{Im}\left[f^0(w)\right] \tag{1.593}$$

In terms of the classical electron radius r_e, the RMS absorbed power per volume $W_{\mathrm{rms}}(w)$ is

$$W_{\mathrm{rms}}(w) = 2\pi \frac{E_0^2 c^2 \varepsilon_0 r_e}{w} \mathrm{Im}\left[f^0(w)\right] \tag{1.594}$$

The time-averaged, electromagnetic wave power per unit area $S^i_{\mathrm{avg}}(w)$ of Eq. (1.570), which is incident at the partly absorbing sample, is repeated here:

$$S^i_{\mathrm{avg}}(w) = \frac{E_0^2}{2}\sqrt{\frac{\varepsilon_0}{\mu_0}} \tag{1.595}$$

This ratio of $W_{\text{rms}}(w)$, divided by $S^i_{\text{avg}}(w)$, is the absorption cross section $\sigma_a(w)$, where

$$\sigma_a(w) = \frac{W_{\text{rms}}(w)}{S^i_{\text{avg}}(w)} = 2\lambda r_e \text{Im}\left[f^0(w)\right] \tag{1.596}$$

With the substitution of Eq. (1.486) for forward-scatter factor $f^0(w)$ into Eq. (1.596), the X-ray absorption cross section in a material medium, which is comprised of Q elements, is

$$\sigma_a(w) = 2r_e\lambda \text{Im}|f^0(w)| = 2r_e\lambda f_2^0(w) = 2r_e\lambda \sum_{q=1}^{Q} n_q \sum_{s=1}^{Z} \frac{w^3\gamma_q g_{s,q}}{(w^2 - w_{s,q}^2)^2 + (w\gamma_q)^2} \tag{1.597}$$

The X-ray total cross section $\sigma_t(w)$, which is the sum of the absorption cross section $\sigma_a(w)$ of Eq. (1.596) and the scatter cross section $\sigma_s(w)$ of Eq. (1.585), is given by

$$\sigma_t(w) = \sigma_a(w) + \sigma_s(w) = 2\lambda r_e \text{Im}\left[f^0(w)\right] + \frac{8\pi}{3} r_e^2 |f^0(w)|^2 \tag{1.598}$$

With the substitution of $\sigma_a(w)$ of Eq. (1.597) and $\sigma_s(w)$ of Eq. (1.586) into Eq. (1.598), the total cross section $\sigma_t(w)$ for a material medium of Q elements, is

$$\begin{aligned}\sigma_t(w) = \sigma_a(w) + \sigma_s(w) &= 2r_e\lambda \sum_{q=1}^{Q} n_q \sum_{s=1}^{Z} \frac{w^3\gamma_q g_{s,q}}{(w^2 - w_{s,q}^2)^2 + (w\gamma_q)^2} \\ &+ \frac{8\pi}{3} r_e^2 w^4 \left| \sum_{q=1}^{Q} n_q \sum_{s=1}^{Z} \frac{g_{s,q}}{w^2 - w_{s,q}^2 + iw\gamma_q} \right|^2\end{aligned} \tag{1.599}$$

The total interaction cross section of neutron or X-rays, which strike a target sample, can be measured with a detector. The incident X-rays or neutrons strike a sample, and a portion of the incident flux is absorbed by the sample; another portion is scattered and recorded by the detector, and the remaining portion is transmitted through the sample without interaction. The measured interaction cross section $\sigma_{\text{exp}}(w)$, which is a function of frequency w [rad/s], has units of area (m^2). The experimental measured cross section $\sigma_{\text{exp}}(w)$ is the ratio of the measured, summed wave intensity I_{exp} [W], which is recorded in all directions by an ideal spherical detector enclosing the target sample, divided by the incident intensity $I_0(w)$ per sample area A [m^2], which is in units of (W/m^2), where

$$\sigma_t(w) = \sigma_{\text{exp}}(w) = \frac{I_{\text{exp}}(w)}{I_0(w)} A \tag{1.600}$$

The measured intensity $I_{\exp}(w)$ is the scattered intensity, which is measured after the absorption of some of the incident intensity by the sample. That is, the incident intensity, which is not absorbed by the target and is not transmitted by the target, is the remaining scattered intensity $I_{\exp}(w)$. This remaining scattered intensity $I_{\exp}(w)$ is measured by the ideal spherical detector, which records the full solid angle (4π sterradian) of scattered particles or photons. Hence, as a function of the X-ray or neutron wave frequency, the measured interaction cross section $\sigma_{\exp}(w)$ is also the total cross section $\sigma_t(w)$, which is the sum of the absorption cross section $\sigma_a(w)$ and scatter cross section $\sigma_s(w)$ by the sample, where

$$\sigma_t(w) = \sigma_{\exp}(w) = \sigma_a(w) + \sigma_s(w) = \frac{I_{\exp}(w)}{I_0(w)}A \tag{1.601}$$

For a material of known elemental composition, the scatter cross section $\sigma_s(w)$ can be calculated theoretically from Eq. (1.586), where

$$\sigma_s(w) = \sigma_s^{\text{cal}}(w) = \frac{8\pi}{3}r_e^2|f^0(w)|^2 \tag{1.602}$$

Hence, one can obtain the absorption cross section for the target material as a function of X-ray or neutron wave frequency w, by subtracting the theoretical, calculated scatter cross section $\sigma_s^{\text{th}}(w)$ from the measured, total cross section $\sigma_t(w)$. In the case of electromagnetic wave scatter and the theoretical calculated scatter cross section of Eq. (1.602), the absorption cross section is then

$$\sigma_a(w) = \sigma_t(w) - \sigma_s(w) = \sigma_{\exp}(w) - \sigma_s^{\text{th}}(w) = \frac{I_{\exp}(w)}{I_0(w)}A - \frac{8\pi}{3}r_e^2|f^0(w)|^2 \tag{1.603}$$

1.5.6 Derivation of the Optical Theorem

In this section, we show that the total cross section $\sigma_t(w)$ can be obtained directly from the imaginary part of the forward-scatter factor $\text{Im}[f^0(w)]$, via the optical theorem. And in the next section, we will show that the real part $\text{Re}[f^0(w)]$ of the forward-scatter factor $f^0(w)$ can be obtained from the imaginary part $\text{Im}[f^0(w)]$, and vice versa, via the Kramers–Kronig equations.

The optical theorem, also known as the shadow theorem, was first developed by Rayleigh and Sellmeier independently for light in 1871 (see Newton (1976)) and then by Feenberg (1932) for particle waves. Later, the term *optical theorem* was coined by Schweber, Bethe, and de Hoffman (1955). The optical theorem is based on the conservation of energy for photons and the conservation of probability density for matter waves, such as neutrons. An incident plane wave propagates in the z-direction and is scattered by an object that is centered on the z-axis. In the forward direction, along the z-axis in the shadow of the object, the incident plane wave intensity (squared amplitude magnitude) has been decreased by object absorption and object scatter of the incident wave in directions other than the forward direction. That is, the squared magnitude of the sum of the non-scattered, transmitted beam amplitude and the forward-scattered beam amplitude in the shadow of the object at the detector, should be less than the squared amplitude magnitude of the plane wave, which is incident at the object.

One starts with the total wave amplitude $\Psi(\mathbf{r}, w)$, as a function of vector position $\mathbf{r}$ and frequency w, where the incident, z-directed plane wave intercepts the scatter object. The scatter object is centered on the z-axis, and the incident plane wave amplitude $\Psi_{in}(z)$ is

$$\Psi_{in}(z, w) = \Psi_0(w)\exp(ikz)\exp(-iwt) \tag{1.604}$$

A fraction of the incident plane wave $f(\theta, w)$ is scattered as a function of wave frequency w and polar angle θ. The polar angle θ is measured relative to the positive z-axis of incident wave propagation. For matter waves, such as neutrons, the amplitude $\Psi_0(w)$ is the probability density amplitude, which is a function of the frequency w of the neutron wave, where

$$\iiint_{vol} \Psi\Psi^* = \iiint_{vol} \Psi_0\Psi_0^* = 1 \tag{1.605}$$

For electromagnetic waves, such as X-rays, Ψ_0 is the amplitude of the electric field, which is measured in units of V/m, where

$$\text{Wave energy} = E(w) = \varepsilon_0 \iiint_{vol} \Psi\Psi^* = \varepsilon_0 \iiint_{vol} \Psi_0\Psi_0^* \tag{1.606}$$

The scatter amplitude has units of length $f(\theta, w)$ because the scattered portion of the incident wave is scattered in all directions at distance r from the scatter object, where the scattered wave amplitude $\Psi_{scat}(w)$ is

$$\Psi_{scat}(w) = \frac{f(\theta, w)}{r}\Psi_0\exp(ikr)\exp(-iwt) \tag{1.607}$$

For photon or neutron wave scatter, the ratio of scatter amplitude $f(\theta, w)$ divided by the distance r of the observer from the scatter object [that is, $f(\theta, w)/r$] must be dimensionless, so that

$$\Psi(\mathbf{r}, w) = \Psi_0 \exp(ikz)\exp(-iwt) + \frac{f(\theta)}{r}\Psi_0 \exp(ikr)\exp(-iwt) \tag{1.608}$$

The frequency w of the incident wave is unchanged upon scatter. The squared magnitude of the total wave amplitude $\Psi(\mathbf{r}, w)$ is then

$$|\Psi(\mathbf{r}, w)|^2 = \left|\Psi_0(w)\exp(ikz)\exp(-iwt) + \frac{f(\theta, w)}{r}\Psi_0(w)\exp(ikr)\exp(-iwt)\right|^2 \tag{1.609}$$

We are interested in the total intensity in the forward direction in the image plane within the shadow of the object. The object and its shadow at the image plane are centered on the z-axis. To simplify matters, we will examine the intensity $|\Psi(\mathbf{r}, w)|^2$ of the object shadows on image (detector) planes, which are set at a large distance along the z-axis from the object. This axial distance z, where the detector is placed in the object's shadow, is set to be much larger than the transverse dimensions of the object and shadow. The binomial approximation can then be applied to the distance r from the object center point to any point on the transverse plane, which contains the shadow, where

$$r = \sqrt{x^2 + y^2 + z^2} = z\sqrt{1 + \frac{x^2 + y^2}{z^2}} \cong z + \frac{x^2 + y^2}{2z} \tag{1.610}$$

and

$$\exp(ikr) \cong \exp(ikz)\exp\left(ik\frac{x^2 + y^2}{2z}\right) \tag{1.611}$$

The position vector $\mathbf{r} = r\hat{\mathbf{r}}$ connects the object center point to a point on the shadow plane at distance r, where vector $\mathbf{r} = r\hat{\mathbf{r}}$ points in a direction of scatter to the shadow plane. Position vector $\mathbf{r} = r\hat{\mathbf{r}}$ is points along polar scatter angle θ, which is measured relative to the z-axis. The shadow is assumed to have a small transverse extent, as compared to the axial distance z of the shadow plane from the center of the object. Hence, the small-angle approximation can be applied, where

$$\theta \cong \tan\theta = \frac{\sqrt{x^2 + y^2}}{z} \cong 0 \tag{1.612}$$

Furthermore, the scatter amplitude $f(\theta)$ can be approximated by the forward-scatter factor $f(0)$. One substitutes Eqs. (1.610–1.612) into Eq. (1.609). The squared magnitude of the complex amplitude sum in Eq. (1.609), is the product of the complex amplitude sum and its complex conjugate, which yields

$$\frac{|\Psi(\mathbf{r}, w)|^2}{|\Psi_0(w)|^2} = 1 + \frac{f(0, w)}{r}\left[\exp\left(-ik\frac{x^2 + y^2}{2z}\right) + \exp\left(ik\frac{x^2 + y^2}{2z}\right)\right] + \frac{|f(0, w)|^2}{r^2} \tag{1.613}$$

To Eq. (1.613), one applies the following rule: twice the real part of complex quantity C or $2\mathrm{Re}[C]$ equals the sum of complex quantity C and its complex conjugate C^*. Also, one approximates the total, diagonal distance r from the object center point to a point on the shadow by the axial distance z between the object and shadow planes. The $1/r^2$ term is small compared to the $1/r$ term, and thus the $1/r^2$ term is dropped, where

$$\frac{|\Psi(\mathbf{r}, w)|^2}{|\Psi_0(w)|^2} = 1 + 2\mathrm{Re}\left[\frac{f(0, w)}{z}\exp\left(ik\frac{x^2 + y^2}{2z}\right)\right] \tag{1.614}$$

Next, integration of Eq. (1.614) is performed over area A, which occupies the significant shadow area in the shadow plane, where

$$\text{Scatter object area} = A = (2x_0)(2y_0) \tag{1.615}$$

The extent of the x integration limits $[-x_0, x_0]$ and y integration limits $[-x_0, x_0]$ are set to produce rapid oscillations of exponential terms in x and y, where

$$\frac{2\pi}{\lambda}\frac{x_0^2}{2z} \geq N2\pi \quad \text{where integer } N \gg 1 \tag{1.616}$$

$$\frac{2\pi}{\lambda}\frac{x_0^2}{2z} \geq M2\pi \quad \text{where integer } M \gg 1 \tag{1.617}$$

Rearrangement of Eqs. (1.616−1.617), shows

$$x_0 \geq \sqrt{2\lambda z N} \quad \text{where integer } N \gg 1 \tag{1.618}$$

$$y_0 \geq \sqrt{2\lambda z M} \quad \text{where integer } M \gg 1 \tag{1.619}$$

Integration of Eq. (1.614) is performed over the selected portion of the shadow in the shadow plane, which is below the spatial limits of Eqs. (1.618) and (1.619), for large oscillation of the exponential terms, where

$$\begin{aligned}\int_{-x_0}^{x_0} \mathrm{d}x \int_{-y_0}^{y_0} \mathrm{d}y \frac{|\Psi(\mathbf{r}, w)|^2}{|\Psi_0(w)|^2} = \int_{-x_0}^{x_0} \mathrm{d}x \int_{-y_0}^{y_0} \mathrm{d}y \\ + 2\mathrm{Re}\left[\frac{f(0,w)}{z} \int_{-x_0}^{x_0} \mathrm{d}x \exp\left(\mathrm{i}k\frac{x^2}{2z}\right) \int_{-y_0}^{y_0} \mathrm{d}y \exp\left(\mathrm{i}k\frac{y^2}{2z}\right)\right]\end{aligned} \tag{1.620}$$

The rapidly oscillating exponentials in the integrands of Eq. (1.620), make a negligible contribution to the integral outside the area A of significant shadow. Hence, the limits of the integral in Eq. (1.620), can be extended from area A to infinity, where

$$\int_{-x_0}^{x_0} \mathrm{d}x \exp\left(\mathrm{i}k\frac{x^2}{2z}\right) \int_{-y_0}^{y_0} \mathrm{d}y \exp\left(\mathrm{i}k\frac{y^2}{2z}\right) \cong \int_{-\infty}^{\infty} \mathrm{d}x \exp\left(\mathrm{i}k\frac{x^2}{2z}\right) \int_{-\infty}^{\infty} \mathrm{d}y \exp\left(\mathrm{i}k\frac{y^2}{2z}\right) \tag{1.621}$$

As a result, the integrations over x and y of Eq. (1.620) is simplified by Eq. (1.621) to the standard integration of a Gaussian function over the entire infinite range of x and y, where

$$\int_{-\infty}^{\infty} \mathrm{d}x \exp\left(\mathrm{i}k\frac{x^2}{2z}\right) \int_{-\infty}^{\infty} \mathrm{d}y \exp\left(\mathrm{i}k\frac{y^2}{2z}\right) = \mathrm{i}\frac{2z\pi}{k} \tag{1.622}$$

With the substitution of Eqs. (1.621) and (1.622) into Eq. (1.620), one obtains

$$\int_{-x_0}^{x_0} \mathrm{d}x \int_{-y_0}^{y_0} \mathrm{d}y \frac{|\Psi(\mathbf{r}, w)|^2}{|\Psi_0(w)|^2} = A + 2\mathrm{Re}\left[\mathrm{i}\frac{2z\pi}{k}\frac{f(0, w)}{z}\right] \tag{1.623}$$

where

$$2\mathrm{Re}\left[\mathrm{i}\frac{2z\pi}{k}\frac{f(0, w)}{z}\right] = \frac{4\pi}{k}\mathrm{Re}[jf(0, w)] = -\frac{4\pi}{k}\mathrm{Im}[f(0, w)] \tag{1.624}$$

Rearrangement of Eq. (1.623), using Eq. (1.624), yields the relation

$$\frac{1}{A}\int_{-x_0}^{x_0} \mathrm{d}x \int_{-y_0}^{y_0} \mathrm{d}y \frac{|\Psi(\mathbf{r}, w)|^2}{|\Psi_0(w)|^2} = 1 - \frac{1}{A}\frac{4\pi}{k}\mathrm{Im}[f(0, w)] \tag{1.625}$$

One can identify the separate terms in Eq. (1.625) as probability components, where

$$\frac{1}{A}\int_{-x_0}^{x_0} \mathrm{d}x \int_{-y_0}^{y_0} \mathrm{d}y \frac{|\Psi(\mathbf{r}, w)|^2}{|\Psi_0(w)|^2} = P_{\mathrm{No}}(w) = \text{probability of no interaction} \tag{1.626}$$

$$\frac{1}{A}\frac{4\pi}{k}\mathrm{Im}[f(0, w)] = P_{\mathrm{Yes}}(w) = \text{probability of interaction} \tag{1.627}$$

The unity probability of 1 in Eq. (1.625) is the sum of the probabilities of Eqs. (1.626) and (1.627) and is given by

$$1 = P_{\mathrm{Yes}}(w) + P_{\mathrm{No}}(w) = \text{probability of interaction} + \text{probability of no interaction} \tag{1.628}$$

As a function of frequency w, the total cross section $\sigma_t(w)$ is the product of the interaction probability $P_{\mathrm{Yes}}(w)$ of Eq. (1.627) for both scatter and absorption, and the area A of the scatter object, where

$$\sigma_t(w) = AP_{\mathrm{Yes}}(w) = \frac{4\pi}{k}\mathrm{Im}[f(0, w)] \tag{1.629}$$

As a function of wavelength λ, the total cross section $\sigma_t(\lambda)$ is the sum of the cross sections for scatter $\sigma_s(\lambda)$ and absorption $\sigma_a(\lambda)$. The total cross section $\sigma_t(\lambda)$ is then the product of the neutron or X-ray wave half-wavelength $\lambda/2$, and the imaginary component of the forward-scatter amplitude $f(0, \lambda)$, which has dimension of length, where

$$\sigma_t(\lambda) = \sigma_s(\lambda) + \sigma_a(\lambda) = 2\lambda\mathrm{Im}[f(0, \lambda)] \tag{1.630}$$

The optical theorem applies in general to waves such as photons and particles (neutrons). For the electromagnetic waves, the optical theorem is derived for electromagnetic waves by Jackson (1975) and Panofsky and Phillips (2005), using the dimensionless, forward-scatter factor $f^0(\lambda)$ as a function of wavelength $f^0(\lambda)$, where

$$\sigma_t(\lambda) = \sigma_s(\lambda) + \sigma_a(\lambda) = \frac{\lambda^2}{\pi}\mathrm{Im}\left[f^0(\lambda)\right] \tag{1.631}$$

The optical theorem for neutrons is derived in Chapter 2, section 19 of AIEP volume 173 (Cremer, 2012c) as Eq. (1.391), using the forward-scatter amplitude $f(\theta = 0)$; see Sears (1989). The forward-scatter amplitude $f(\theta = 0)$ is also used for electromagnetic waves; see Schwartz (1972). Using the forward-scatter amplitude $f(\theta = 0)$, which has units of length (m), the optical theorem, for electromagnetic waves and neutrons, and waves in general, is

$$\sigma_t(\lambda) = \sigma_s(\lambda) + \sigma_a(\lambda) = 2\lambda\mathrm{Im}[f(\theta = 0)] \tag{1.632}$$

One obtains units of area for the total cross section $\sigma_t(\lambda)$ in the application of the optical theorem, which uses either the dimensionless forward-scatter factor $f^0(\lambda)$ as in Eq. (1.631), or the forward-scatter amplitude $f(\theta = 0)$ with dimensions of length as in Eq. (1.632).

1.5.7 Derivation of the Kramers–Kronig Relation and Calculation of the Refractive Decrement from the Measured Attenuation Cross Section

We now derive the Kramers–Kronig equations, which were developed by Kramers and Kronig around 1926 to 1927. Applying the general arguments of linearity, causality, and analyticity to models of atomic gas, but not to the model-independent, Kramers (1927) derived the pair of relations; on the other hand, Kronig (1926) derived only one relation. See also Bohren (2010), Gorter and Kronig (1942), Kronig (1942), and Toll (1952, 1956).

Consider as a function of time t a driving function $d(t)$ and a response function $r(t)$, which are linked or connected by a transfer or system function $s(t)$. In the angular frequency domain (or simply the frequency domain), the time-dependent driving or input function $d(t)$ and its corresponding response or output function $o(t)$, via their relationship function $s(t)$, are each expressed by a summed spectrum of sinusoidal waves or "dos, do by system s." This is a play on words: (1) the English word *does* for the reaction of the system function s to the driving input d resulting in output o, by adding e to the word *dos* and (2) the Spanish word *dos*, meaning "2," as connecting the two entities of input d and output o by the system function s.

The spectrum of waves, which comprises the driving function $d(t)$ is a sum of an infinite set of discrete set of individual sinusoidal waves, whose angular frequency w_n and phase ϕ_n are indexed by integer n, and have a unique set of frequency-dependent amplitudes $D(w_n)$ and phases ϕ_n.

The Euler equation equates (1) the infinite series expressions of the sinusoidal exponential function $\exp(iw_nt)$ with an argument w_nt multiplied by $i = \sqrt{-1}$ to the sum of (2) a real part [namely, the cosine of the argument $\cos(w_nt)$], and (3) an imaginary part [namely, the sine of the argument $\cos(w_nt)$], where

$$D(w_n)\exp[i(w_nt + \phi_n)] = D(w_n)[\cos(w_nt + \phi_n) + i\,\sin(w_nt + \phi_n)] \quad (1.633)$$

As shown in Figure 1.25 (top panel), in the unit circle of trigonometry, the unit radius $R = 1$ rotates counterclockwise about the center point of the circle and has a time-varying, positive angle w_nt, which is measured relative to the horizontal, x-axis. Each complete cycle of rotation is 360° or 2π radians, and at time t after the kth cycle, the angular position w_nt of the unit radius, given its angular position ϕ_n at initial time $t = 0$, where for integer n

$$w_nt = k2\pi + \phi_n = k2\pi \quad \phi_n = 0 \text{ for } -\infty < n < \infty \quad (1.634)$$

Without losing generality, the unit radius vector $R = 1$ begins its counterclockwise rotation from the x-axis at initial time $t = 0$, where for index integer n and $-\infty < n < \infty$, the initial phase angle ϕ_n for each of the component sinusoidal waves is zero, that is $\phi_n = 0$ at $t = 0$. As a result, the cosine function $\cos(w_nt)$ is the projection of the unit radius $R = 1$ onto the horizontal x-axis, where $x = \cos(w_nt)$. The sine function $\cos(w_nt)$ is the projection of the unit radius $R = 1$ onto the y-axis, where $y = \sin(w_nt)$ Hence,

$$R \times \exp(w_nt) = R \times [\cos(w_nt) + i\,\sin(w_nt)] = R \times [x(t) + iy(t)] \quad (1.635)$$

or

$$\exp(w_nt) = \cos(w_nt) + i\,\sin(w_nt) = x(t) + iy(t) \quad (1.636)$$

In Eq. (1.635), the variable R for the length of the unit radius was retained to show that the unit circle could be replaced by a circle of arbitrary radius R rather than unit radius $R = 1$. This shows the generality of the circle relation to any periodic time-varying, vector, which has arbitrary magnitude for the nth sinusoidal wave, where the nth circle radius is $R_n = D(w_n)$ for integer n and $-\infty < n < \infty$, so that

$$D(w_n)\exp[iw_nt] = D(w_n)[\cos(w_nt) + i\,\sin(w_nt)] = D(w_n)[x(t) + iy(t)] \quad (1.637)$$

The time-dependent driving or input function $d(t)$ and its corresponding response or output function $o(t)$, via their relationship (transfer or system) function $s(t)$, can be expressed by the summation of the sinusoidal components of their unique frequency domain spectrum, just as a musical sound, water wave, or a pastel color. Given the respective amplitudes of their n^{th} spectral components $D(w_n)$, $O(w_n)$, and $S(w_n)$ for frequency w_n, the driving function $d(t)$, response function $o(t)$, and system function $s(t)$ are

$$d(t) = \sum_{n=-\infty}^{\infty} D(w_n)\exp(\mathrm{i}w_n t) \tag{1.638}$$

$$o(t) = \sum_{n=-\infty}^{\infty} O(w_n)\exp(\mathrm{i}w_n t) \tag{1.639}$$

$$s(t) = \sum_{n=-\infty}^{\infty} S(w_n)\exp(\mathrm{i}w_n t) \tag{1.640}$$

The nth harmonic frequency w_n is an integer multiple of fundamental frequency w_1, with fundamental oscillation period T_1, where

$$w_n = nw_1 = n\frac{2\pi}{T_1} \tag{1.641}$$

and

$$\exp(w_n t) = \cos(w_n t) + \mathrm{i}\,\sin(w_n t) = \cos\left(n\frac{2\pi}{T_1}t\right) + \mathrm{i}\,\sin\left(n\frac{2\pi}{T_1}t\right) \tag{1.642}$$

The nth spectral components $D(w_n)$, $O(w_n)$, and $S(w_n)$ for frequency w_n in the frequency domain are found, via the orthogonality properties of cosine and sine waves, where

$$\int_0^{\pi} \sin(w_n t)\cos(w_k t)\mathrm{d}t = 0 \tag{1.643}$$

$$\int_0^{2\pi} \sin(w_n t)\sin(w_k t)\mathrm{d}t = \int_0^{2\pi} \cos(w_n t)\cos(w_k t)\mathrm{d}t = \begin{matrix} 0 & n \neq k \\ \pi & n = k \end{matrix} \tag{1.644}$$

Both sides of Eqs. (1.638–1.640) are multiplied by $\exp(-\mathrm{i}w_k t)$, and time integration over one fundamental oscillation period T_1 is performed, so that the application of Eqs. (1.641–1.644) yields

$$D(w_k) = \frac{1}{T_1}\int_{-T_1/2}^{T_1/2} \mathrm{d}(t)\exp(-\mathrm{i}w_k t)\mathrm{d}t = \frac{1}{T_1}\int_{-T_1/2}^{T_1/2} \mathrm{d}(t)\exp\left(-\mathrm{i}k\frac{2\pi}{T_1}t\right)\mathrm{d}t \tag{1.645}$$

$$O(w_k) = \frac{1}{T_1}\int_{-T_1/2}^{T_1/2} o(t)\exp(-\mathrm{i}w_k t)\mathrm{d}t = \frac{1}{T_1}\int_{-T_1/2}^{T_1/2} o(t)\exp\left(-\mathrm{i}k\frac{2\pi}{T_1}t\right)\mathrm{d}t \tag{1.646}$$

$$S(w_k) = \frac{1}{T_1}\int_{-T_1/2}^{T_1/2} s(t)\exp(-\mathrm{i}w_k t)\mathrm{d}t = \frac{1}{T_1}\int_{-T_1/2}^{T_1/2} s(t)\exp\left(-\mathrm{i}k\frac{2\pi}{T_1}t\right)\mathrm{d}t \tag{1.647}$$

With the substitution of the Fourier series of Eqs. (1.645–1.647) into the integrands of the inverse Fourier series of Eqs. (1.638–1.640), the driving function d(t), response function $o(t)$, and system function $s(t)$ expressions of Eqs. (1.638–1.640) are expressed by a Fourier series, where

$$d(t) = \sum_{n=-\infty}^{\infty} \left(\frac{1}{T_1} \int_{-T_1/2}^{T_1/2} d(\tau)\exp(-\mathrm{i}w_k\tau)\mathrm{d}\tau \right) \exp(\mathrm{i}w_n t) \tag{1.648}$$

$$o(t) = \sum_{n=-\infty}^{\infty} \left(\frac{1}{T_1} \int_{-T_1/2}^{T_1/2} o(\tau)\exp(-\mathrm{i}w_k\tau)\mathrm{d}\tau \right) \exp(\mathrm{i}w_n t) \tag{1.649}$$

$$s(t) = \sum_{n=-\infty}^{\infty} \left(\frac{1}{T_1} \int_{-T_1/2}^{T_1/2} s(\tau)\exp(-\mathrm{i}w_k\tau)\mathrm{d}\tau \right) \exp(\mathrm{i}w_n t) \tag{1.650}$$

These Fourier series expressions of functions $d(t)$, $o(t)$, and $s(t)$ are limited to the time duration, which is contained within one fundamental period $-T_1/2 \le t \le T_1/2$. However, if the functions $d(t)$, $o(t)$, and $s(t)$ are periodic, then they can be represented by the Fourier series over an infinite time interval $-\infty \le t \le \infty$.

However, if the time-dependent functions $d(t)$, $o(t)$, and $s(t)$ are not periodic, one can extend their expression by sinusoidal functions from the fundamental period $-T_1/2 \le t \le T_1/2$ to an infinite time period $-\infty \le t \le \infty$, where $T_1 \to \infty$, so that

$$\frac{\Delta w}{2\pi} = \frac{w_{n+1} - w_n}{2\pi} = \frac{1}{T_1} \tag{1.651}$$

With infinite period $T_1 \to \infty$ and infinite range of integer indices, $n \to \infty$ and $k \to \infty$, the discrete frequencies pass onto a continuous range of frequencies, where

$$w_n = nw_1 = n\frac{2\pi}{T_1} \to w \quad \text{and} \quad w_k = kw_1 = k\frac{2\pi}{T_1} \to w \tag{1.652}$$

With $T_1 \to \infty$, the discrete domain of the Fourier series gives way to the continuous domain of the Fourier transform, where the general functions $d(t)$, $o(t)$, and $s(t)$ can be aperiodic. Application of Eqs. (1.651) and (1.652), $T_1 \to \infty$, and $\Delta w \to 0$ to the inverse Fourier series expressions for d(t), $o(t)$, and $s(t)$ of Eqs. (1.648–1.650) gives

$$\begin{aligned} d(t) &= \lim_{T_1\to\infty,\Delta w\to 0} \sum_{n=-\infty}^{\infty} \left(\frac{\Delta w}{2\pi} \int_{-T_1/2}^{T_1/2} d(\tau)\exp(-\mathrm{i}w_k\tau)\mathrm{d}\tau \right) \exp(\mathrm{i}w_n t) \\ &= \frac{1}{2\pi} \int_{-\infty}^{\infty} \mathrm{d}w \left(\int_{-\infty}^{\infty} d(\tau)\exp(-\mathrm{i}w\tau)\mathrm{d}\tau \right) \exp(\mathrm{i}wt) \end{aligned} \tag{1.653}$$

$$
\begin{aligned}
o(t) &= \lim_{T_1\to\infty,\Delta w\to 0}\sum_{n=-\infty}^{\infty}\left(\frac{\Delta w}{2\pi}\int_{-T_1/2}^{T_1/2} o(\tau)\exp(-\mathrm{i}w_k\tau)\mathrm{d}\tau\right)\exp(\mathrm{i}w_n t)\\
&= \frac{1}{2\pi}\int_{-\infty}^{\infty}\mathrm{d}w\left(\int_{-\infty}^{\infty} o(\tau)\exp(-\mathrm{i}w\tau)\mathrm{d}\tau\right)\exp(\mathrm{i}wt)
\end{aligned}
\tag{1.654}
$$

$$
\begin{aligned}
s(t) &= \lim_{T_1\to\infty,\Delta w\to 0}\sum_{n=-\infty}^{\infty}\left(\frac{\Delta w}{2\pi}\int_{-T_1/2}^{T_1/2} s(\tau)\exp(-\mathrm{i}w_k\tau)\mathrm{d}\tau\right)\exp(\mathrm{i}w_n t)\\
&= \frac{1}{2\pi}\int_{-\infty}^{\infty}\mathrm{d}w\left(\int_{-\infty}^{\infty} s(\tau)\exp(-\mathrm{i}w\tau)\mathrm{d}\tau\right)\exp(\mathrm{i}wt)
\end{aligned}
\tag{1.655}
$$

The frequency domain Fourier transforms $D(w)$, $O(w)$, and $S(w)$ of the time domain functions $d(t)$, $o(t)$, and $s(t)$ are extracted from Eqs. (1.653–1.655), where

$$
D(w) = \int_{-\infty}^{\infty} d(t)\exp(-\mathrm{i}wt)\mathrm{d}t \tag{1.656}
$$

$$
O(w) = \int_{-\infty}^{\infty} o(t)\exp(-\mathrm{i}wt)\mathrm{d}t \tag{1.657}
$$

$$
S(w) = \int_{-\infty}^{\infty} s(t)\exp(-\mathrm{i}wt)\mathrm{d}t \tag{1.658}
$$

One takes the inverse Fourier transform of $X(w)$, which is the Fourier transform of $x(t)$ to obtain the original function $x(t)$, which provides the integral expression of the Dirac delta function, where

$$
\begin{aligned}
\int_{-\infty}^{\infty} X(w)\exp(\mathrm{i}wt')\mathrm{d}w &= \int_{-\infty}^{\infty}\left(\int_{-\infty}^{\infty} x(t)\exp(-\mathrm{i}wt)\mathrm{d}t\right)\exp(\mathrm{i}wt')\mathrm{d}w\\
&= \int_{-\infty}^{\infty}\left(\int_{-\infty}^{\infty}\exp[\mathrm{i}w(t'-t)]\mathrm{d}w\right)x(t)\mathrm{d}t\\
&= 2\pi\int_{-\infty}^{\infty} x(t)\delta(t'-t)\mathrm{d}t' = x(t)
\end{aligned}
\tag{1.659}
$$

The Dirac delta function $\delta(t'-t)$ is extracted as

$$
\frac{1}{2\pi}\int_{-\infty}^{\infty}\exp[\mathrm{i}w(t'-t)]\mathrm{d}w = \delta(t'-t) \tag{1.660}
$$

The Euler formula, the oddness of the sine function, and the evenness of the cosine function, with a zero contribution for integration of the odd sine function, show

$$
\delta(t'-t) = \lim_{T\to 0}\frac{1}{2\pi}\int_{-2\pi/T}^{2\pi/T}\exp[\mathrm{i}w(t'-t)]\mathrm{d}w \tag{1.661}
$$

Integration of Eq. (1.661) yields

$$\frac{1}{2\pi}\int_{-2\pi/T}^{2\pi/T}\exp[\mathrm{i}w(t'-t)]\mathrm{d}w=\frac{1}{2\pi}\int_{-2\pi/T}^{2\pi/T}\cos[w(t'-t)]\mathrm{d}w$$
$$+\frac{\mathrm{i}}{2\pi}\int_{-2\pi/T}^{2\pi/T}\sin[w(t'-t)]\mathrm{d}w \qquad (1.662)$$
$$=\frac{1}{2\pi}\frac{2\sin[(2\pi/T)(t'-t)]}{t'-t}$$

We now obtain some insight into the Dirac delta function as the outcome of a limiting process, which is applied to the integral of the exponential. The resulting limiting process, which is applied to the integral, yields a constant integrated area under the sinusoidal curve. However, the sinusoidal function in the integrand becomes an infinitesimally narrow function with infinite height. Let

$$y=\frac{2\pi}{T}(t'-t) \quad \text{and} \quad \mathrm{d}y=\frac{2\pi}{T}\mathrm{d}(t'-t) \qquad (1.663)$$

The change of variables of Eq. (1.663) is applied to the integration in Eq. (1.662), where the limit of a zero period (or infinite frequency) is imposed, so that the integrated area under the function becomes π, where

$$1=\lim_{T\to 0}\frac{1}{\pi}\int_{-\infty}^{\infty}\frac{\sin\,y}{y}\mathrm{d}y=\lim_{T\to 0}\frac{1}{2\pi}\int_{-\infty}^{\infty}\frac{2\sin[(2\pi/T)(t'-t)]}{t'-t}d(t'-t)$$
$$=\int_{-\infty}^{\infty}\delta(t'-t)d(t'-t) \qquad (1.664)$$

The Dirac delta function is zero for all nonzero time $t\neq 0$, where $\delta(t\neq 0)=0$ but becomes infinite at $t=0$, where $\delta(t=0)\to\infty$, so that

$$\int_{-\infty}^{\infty}\delta(t)\mathrm{d}t=1 \qquad (1.665)$$

and

$$\int_{-\infty}^{\infty}f(t-a)\delta(t)\mathrm{d}t=f(a) \qquad (1.666)$$

With Eqs. (1.659) and (1.660) applied to Eqs. (1.656–1.658), one thus obtains the inverse Fourier transforms

$$d(t) = \frac{1}{2\pi}\int_{-\infty}^{\infty} D(w)\exp(iwt)dw \tag{1.667}$$

$$o(t) = \frac{1}{2\pi}\int_{-\infty}^{\infty} O(w)\exp(iwt)dw \tag{1.668}$$

$$s(t) = \frac{1}{2\pi}\int_{-\infty}^{\infty} S(w)\exp(iwt)dw \tag{1.669}$$

Let us now guess that the Fourier transform of the output $O(w)$ equals the product of the Fourier transform of the input $D(w)$ and the Fourier transform of the system function $S(w)$, where

$$O(w) = D(w)S(w) \tag{1.670}$$

Substitution of $O(w)$ of Eq. (1.670) into Eq. (1.668) gives

$$o(t) = \frac{1}{2\pi}\int_{-\infty}^{\infty} O(w)\exp(iwt)dw = \frac{1}{2\pi}\int_{-\infty}^{\infty} D(w)S(w)\exp(iwt)dw \tag{1.671}$$

Substitution in Eq. (1.671) for $S(w)$ from Eq. (1.658) gives

$$o(t) = \frac{1}{2\pi}\int_{-\infty}^{\infty} D(w)\left(\int_{-\infty}^{\infty} s(\tau)\exp(-iw\tau)d\tau\right)\exp(iwt)dw \tag{1.672}$$

In Eq. (1.672), the exponentials are combined and the order of integration is reversed, so that

$$o(t) = \frac{1}{2\pi}\int_{-\infty}^{\infty} s(\tau)\left(\int_{-\infty}^{\infty} D(w)\exp[iw(t-\tau)]dw\right)d\tau \tag{1.673}$$

The expression in parentheses on the right side of Eq. (1.673) is $d(t-\tau)$, which has an inverse Fourier transform of $D(w)$, with time shift τ to earlier time $t-\tau$, where

$$o(t) = \frac{1}{2\pi}\int_{-\infty}^{\infty} d(\tau)s(t-\tau)d\tau \tag{1.674}$$

The assumption $O(w) = D(w)S(w)$ is correct because its implementation resulted in the correct conclusion: The response or output function $o(t)$ is indeed obtained by convolution in the time domain of driving or input function $d(t)$ with the system function $s(t)$. Hence, the convolution theorem is proved, where the inverse Fourier transform of the product of the two Fourier transforms $D(w)$ and $S(w)$ equals the convolution of their respective inverse Fourier transforms $d(t)$ and $s(t)$, where

$$o(t) = \frac{1}{2\pi}\int_{-\infty}^{\infty} D(w)S(w)\exp(iwt)dw = \frac{1}{2\pi}\int_{-\infty}^{\infty} d(\tau)s(t-\tau)d\tau \tag{1.675}$$

We now derive the Kramers–Kronig relation, via the principle of causality, which requires that the output $o(t)$ is zero until the occurrence of the input $d(t)$. That causality circumstance causes the Kramers–Kronig relation. Let the input function $d(t)$ be the Dirac delta function $\delta(t)$. Using Eq. (1.666), the Fourier transform of the delta function $d(t) = \delta(t)$, as the driving or input function, is unity over all frequencies w, where

$$D(w) = \int_{-\infty}^{\infty} \delta(t)\exp(-\mathrm{i}wt)\mathrm{d}t = 1 \tag{1.676}$$

With the substitution of $D(w) = 1$ in Eq. (1.675), the output $o(t)$ becomes

$$o(t) = \frac{1}{2\pi}\int_{-\infty}^{\infty} D(w)S(w)\exp(\mathrm{i}wt)\mathrm{d}w = \frac{1}{2\pi}\int_{-\infty}^{\infty} S(w)\exp(\mathrm{i}wt)\mathrm{d}w \tag{1.677}$$

Next, the system function's Fourier transform $S(w)$, which in general is complex, is expressed as the sum of its real $S_{\mathrm{re}}(w)$ and imaginary $S_{\mathrm{im}}(w)$ parts:

$$S(w) = S_{\mathrm{re}}(w) + \mathrm{i}S_{\mathrm{im}}(w) \tag{1.678}$$

Substitution of Eq. (1.678) for $S(w)$ into Eq. (1.677) for the output $o(t)$ and the application of the Euler formula, give

$$\begin{aligned} o(t) = {} & \frac{1}{2\pi}\int_{-\infty}^{\infty} [S_{\mathrm{re}}(w)\cos(wt) - S_{\mathrm{im}}(w)\sin(wt)]\mathrm{d}w \\ & + \frac{\mathrm{i}}{2\pi}\int_{-\infty}^{\infty} [S_{\mathrm{re}}(w)\sin(wt) + S_{\mathrm{im}}(w)\cos(wt)]\mathrm{d}w \end{aligned} \tag{1.679}$$

The input function $d(t) = \delta(t)$ is real, and hence its output response $s(t)$ is assumed to be real, so that the imaginary part $o_{\mathrm{im}}(t)$ of output $o(t)$ is zero at each instance of time t, where

$$o_{\mathrm{im}}(t) = \int_{-\infty}^{\infty} [S_{\mathrm{re}}(w)\sin(wt) + S_{\mathrm{im}}(w)\cos(wt)]\mathrm{d}w = 0 \tag{1.680}$$

The first integral in Eq. (1.680) is an odd function with respect to time, t and the second integral is an even function with respect to time t, where

$$\int_{-\infty}^{\infty} S_{\mathrm{re}}(w)\sin(wt)\mathrm{d}w \quad \text{is odd function of time } t \tag{1.681}$$

and

$$\int_{-\infty}^{\infty} S_{\mathrm{im}}(w)\cos(wt)\mathrm{d}w \quad \text{is even function of time } t \tag{1.682}$$

If the imaginary part $o_{im}(t)$ of output $o(t)$ is to be zero at each instance of time t, then these even and odd integrals must be independently zero for each instance of time t. This requires (1) the real part $S_{re}(w)$ of the system (transfer) function to be an even function of frequency w, and (2) the imaginary part $S_{im}(w)$ to be an odd function of frequency w. Hence, for all time t,

$$\int_{-\infty}^{\infty} S_{re}(w)\sin(wt)\mathrm{d}w = 0 \quad \text{if } S_{re} \text{ is even function of frequency } w \tag{1.683}$$

and

$$\int_{-\infty}^{\infty} S_{im}(w)\cos(wt)\mathrm{d}w = 0 \quad \text{if} S_{im} \text{ is odd function of frequency } w \tag{1.684}$$

As a result, Eq. (1.679) for the output response output $o(t)$ to the impulse input $d(t) = \delta(t)$ becomes

$$o(t) = \frac{1}{2\pi}\int_{-\infty}^{\infty} [S_{re}(w)\cos(wt) - S_{im}(w)\sin(wt)]\mathrm{d}w \tag{1.685}$$

By Eqs. (1.681–1.684), $S_{re}(w)$ is an even function of frequency w and $S_{im}(w)$ is an odd function of frequency w. Hence, both integrals of Eq. (1.685) are even functions of frequency w. The two integrals over the full frequency space $-\infty < w < \infty$ can thus be expressed over the half-frequency space $0 \le w < \infty$, which is compensated for by multiplying by 2, where

$$\begin{aligned} &\text{if} \int_{-\infty}^{\infty} S_{re}(w)\cos(wt)\mathrm{d}w \quad \text{is even with } w, \\ &\text{then} \int_{-\infty}^{\infty} S_{re}(w)\cos(wt)\mathrm{d}w = 2\int_{0}^{\infty} S_{re}(w)\cos(wt)\mathrm{d}w \end{aligned} \tag{1.686}$$

and

$$\begin{aligned} &\text{if} \int_{-\infty}^{\infty} S_{im}(w)\sin(wt)\mathrm{d}w \quad \text{is even with } w, \\ &\text{then} \int_{-\infty}^{\infty} S_{im}(w)\sin(wt)\mathrm{d}w = 2\int_{0}^{\infty} S_{im}(w)\sin(wt)\mathrm{d}w \end{aligned} \tag{1.687}$$

With the application of Eqs. (1.686) and (1.687) to Eq. (1.685), the response output $o(t)$ to the impulse input $\mathrm{d}(t) = \delta(t)$ becomes

$$o(t) = \frac{1}{\pi}\int_{0}^{\infty} [S_{re}(w)\cos(wt) - S_{im}(w)\sin(wt)]\mathrm{d}w \tag{1.688}$$

We now invoke causality, which requires the output $o(t)$ not to occur before the input $d(t)$, where

$$\text{if } i(t) = 0 \text{ for } t < 0 \quad \text{then } o(t) = 0 \text{ for } t < 0 \tag{1.689}$$

Given positive time dummy variable $\infty > |\tau| \geq 0$, which is applied to Eq. (1.688), causality requires a zero output ($o(t) = 0$) as a function of negative time $t = -|\tau| < 0$, for zero input $d(t) = 0$, where

$$o(-|\tau|) = 0 = \frac{1}{\pi}\int_0^\infty [S_{re}(w)\cos(-w|\tau|) - S_{im}(w)\sin(-w|\tau|)]dw \quad t = -|\tau| < 0 \tag{1.690}$$

with

$$\cos(-w|\tau|) = \cos(w|\tau|) \tag{1.692}$$

$$\sin(-w|\tau|) = -\sin(w|\tau|) \tag{1.693}$$

The output $o(t)$ of Eq. (1.690) as a function of negative time $t = -|\tau| < 0$ for zero input $d(t) = 0$, using Eqs. (1.692) and (1.693), then requires

$$\frac{1}{\pi}\int_0^\infty \left[S_{re}(w)\cos(w|\tau|) = -\frac{1}{\pi}\int_0^\infty S_{im}(w)\sin(w|\tau|)\right]dw \quad t = -|\tau| < 0 \tag{1.694}$$

In the positive time domain after the occurrence of the input, as a function of positive time, where $t = |\tau| > 0$, causality requires

$$o(|\tau|) = \frac{1}{\pi}\int_0^\infty [S_{re}(w)\cos(w|\tau|) - S_{im}(w)\sin(w|\tau|)]dw \quad t = |\tau| > 0 \tag{1.695}$$

With the substitution of Eq. (1.694) into Eq. (1.695), the resulting causality relations for positive time $t = |\tau| > 0$ are the inverse Fourier cosine and sine transforms, where

$$o(t) = \frac{2}{\pi}\int_0^\infty S_{re}(w)\cos(wt) \quad t = |\tau| > 0 \tag{1.696}$$

and

$$o(t) = -\frac{2}{\pi}\int_0^\infty S_{im}(w)\sin(wt)dw \quad t = |\tau| > 0 \tag{1.697}$$

This result of causality then allows the derivation of a direct relationship between the real $S_{re}(w)$ and imaginary $S_{im}(w)$ components of the system function $S(w)$

in frequency w space. We start with the causality condition of the output function, where

$$\text{if } i(t) = 0 \text{ for } t < 0 \quad \text{then } o(t) = 0 \text{ for } t < 0 \tag{1.698}$$

The causality condition of Eq. (1.698) is applied to the inverse Fourier transform expression of the frequency domain system function $S(w)$ of Eq. (1.677), which is obtained from the time domain output function $o(t)$, where

$$S(w) = \int_{-\infty}^{\infty} o(t)\exp(-\mathrm{i}wt)\mathrm{d}t = \int_{0}^{\infty} o(t)\exp(-\mathrm{i}wt)\mathrm{d}t \quad o(t) = 0 \text{ for } t < 0 \tag{1.699}$$

The system function $S(w)$ of Eq. (1.699) is then expressed as the sum of its real $S_{\mathrm{re}}(w)$ and imaginary parts $S_{\mathrm{im}}(w)$, where

$$S_{\mathrm{re}}(w) + \mathrm{i}S_{\mathrm{im}}(w) = \int_{0}^{\infty} o(t)\cos(wt)\mathrm{d}t - \mathrm{i}\int_{0}^{\infty} o(t)\sin(wt)\mathrm{d}t \tag{1.700}$$

Equating the real parts and imaginary parts on both sides of Eq. (1.700), one obtains the Fourier cosine and sine transforms:

$$S_{\mathrm{re}}(w) = \int_{0}^{\infty} o(t)\cos(wt)\mathrm{d}t \tag{1.701}$$

and

$$S_{\mathrm{im}}(w) = -\int_{0}^{\infty} o(t)\sin(wt)\mathrm{d}t \tag{1.702}$$

With Eq. (1.697), the function $o(t)$ in the integrand of Eq. (1.701) is expressed as the inverse sine transform of the imaginary component of the system function $S_{\mathrm{im}}(w)$, where

$$S_{\mathrm{re}}(w) = -\frac{2}{\pi}\int_{0}^{\infty}\left[\int_{0}^{\infty} S_{\mathrm{im}}(w')\sin(w't)\mathrm{d}w'\right]\cos(wt)\mathrm{d}t \tag{1.703}$$

With Eq. (1.696), the function $o(t)$ in the integrand of Eq. (1.702) is expressed as the inverse cosine transform of the real component of the system function $S_{\mathrm{re}}(w)$, where

$$S_{\mathrm{im}}(w) = -\frac{2}{\pi}\int_{0}^{\infty}\left[\int_{0}^{\infty} S_{\mathrm{re}}(w')\cos(w't)\mathrm{d}w'\right]\sin(wt)\mathrm{d}t \tag{1.704}$$

The precursor Kramers–Kronig relations are expressed by Eqs. (1.703) and (1.704), which are rearranged to prepare for the integration over time t, where

$$S_{\mathrm{re}}(w) = -\frac{2}{\pi}\int_0^\infty S_{\mathrm{im}}(w')\left[\int_0^\infty \sin(w't)\cos(wt)\mathrm{d}t\right]\mathrm{d}w' \tag{1.705}$$

and

$$S_{\mathrm{im}}(w) = -\frac{2}{\pi}\int_0^\infty S_{\mathrm{re}}(w')\left[\int_0^\infty \cos(w't)\sin(wt)\mathrm{d}t\right]\mathrm{d}w' \tag{1.706}$$

One then invokes the two integral rules:

$$\int_0^\infty \sin(w't)\cos(wt)\mathrm{d}t = \frac{w'}{w'^2 - w^2} \tag{1.707}$$

and

$$\int_0^\infty \sin(wt)\cos(w't)\mathrm{d}t = \frac{w}{w^2 - w'^2} \tag{1.708}$$

Applying the integral rule of Eq. (1.707) to Eq. (1.705) and Eq. (1.708) to Eq. (1.706), one then obtains the Kramers–Kronig relations between the real S_{re} and imaginary S_{im} parts of the system function S, where

$$S_{\mathrm{re}}(w) = \frac{2}{\pi}\int_0^\infty S_{\mathrm{im}}(w')\frac{w'}{w'^2 - w^2}\mathrm{d}w' \tag{1.709}$$

and

$$S_{\mathrm{im}}(w) = \frac{2w}{\pi}\int_0^\infty S_{\mathrm{re}}(w')\frac{1}{w'^2 - w^2}\mathrm{d}w' \tag{1.710}$$

The Kramers–Kronig integrals in the frequency w representation have a singularity at $w' = w$. As a result, one must take the Cauchy principal values of these integrals. For the Kramers–Kronig integrals to converge to finite values, it requires the system function real $S_{\mathrm{re}}(w)$ and imaginary parts $S_{\mathrm{im}}(w)$ to become zero for infinite frequency w, where

$$S_{\mathrm{re}}(w \to \infty) \to 0 \quad \text{and} \quad S_{\mathrm{im}}(w \to \infty) \to 0 \tag{1.711}$$

In electrical engineering, the Kramers–Kronig integrals are used extensively. As an example, the Kramers–Kronig relation can be applied to finding the imaginary part $S_{\mathrm{im}}(w)$ of the electrical impedance of a network from the measured real part $S_{\mathrm{re}}(w)$ of the network impedance $S(w)$, which takes the role of the system or transfer function. Consider the system (transfer) function, which is a delay function, where

$$S(w) = \cos(wt) - \mathrm{i}\,\sin(wt), \tag{1.712}$$

One can use the Kramers–Kronig relations of Eqs. (1.709) and (1.710) to obtain the real part $S_{re}(w) = \cos(wt)$ from the imaginary part $S_{im}(w) = \sin(wt)$, and vice versa.

Now assume that the real part $S_{re}(w)$ of the system function is the sum of an frequency-dependent part $S'_{re}(w)$ and a constant C, where

$$S_{re}(w) = C + S'_{re}(w) \tag{1.713}$$

Substitution of Eq. (1.713) into the Kramers–Kronig relation of Eq. (1.710) yields

$$S_{im}(w) = \frac{2E}{\pi}\int_0^{\infty}\frac{S_{re}(w') + C}{w'^2 - w^2}\mathrm{d}w' = \frac{2w}{\pi}\int_0^{\infty}\frac{S'_{re}(w') + C}{w'^2 - w^2}\mathrm{d}w' + C\frac{2w}{\pi}\int_0^{\infty}\frac{1}{w'^2 - w^2}\mathrm{d}w' \tag{1.714}$$

However, the Cauchy principal value of the second integral, associated with the constant C, is zero, where

$$\int_0^{\infty}\frac{1}{w'^2 - w^2}\mathrm{d}w' = 0 \tag{1.715}$$

Hence, only the frequency-dependent part $S'_{re}(w)$ of the real part $S_{re}(w)$ of the scatter amplitude contributes to the imaginary part of the scatter amplitude $S_{im}(w)$:

$$S_{im}(w) = \frac{2E}{\pi}\int_0^{\infty}\frac{S'_{re}(w') + C}{w'^2 - w^2}\mathrm{d}w' = \frac{2w}{\pi}\int_0^{\infty}\frac{S'_{re}(w')}{w'^2 - w^2}\mathrm{d}w' \tag{1.716}$$

Finally, assume the range of integrated photon energies $E' = \hbar w'$, where $0 \le E' \le \infty$, and an incident photon energy $E = \hbar w$ of interest, the system function in the Kramers–Kronig relation now represents the relation between the real $S_{re}(E)$ and imaginary $S_{im}(E)$ parts of the energy-dependent scatter amplitude function $S(E)$, where

$$S_{re}(E) = \frac{2}{\pi}\int_0^{\infty} S_{im}(E')\frac{E'}{E'^2 - E^2}\mathrm{d}E' \tag{1.717}$$

and

$$S_{im}(E) = \frac{2E}{\pi}\int_0^{\infty} S_{re}(E')\frac{1}{E'^2 - E^2}\mathrm{d}E' \tag{1.718}$$

The refractive index $n = 1 - \delta + \mathrm{i}\beta$ of a material for X-rays and gammas, which is caused by Rayleigh coherent, elastic scatter of the photons by the atomic electron cloud, can be determined from the material's absorption cross section via the

Kramers–Kronig integral. The Kramers–Kronig integral allows computation of the real part of the scatter length b_r, which is responsible for refraction, diffraction, and reflection, using the imaginary part of the scatter length, which is responsible for attenuation by absorption, and incoherent, elastic and inelastic scatter. Sections 7 and 8 of Chapter 4 in AIEP volume 172 (Cremer, 2012a), which present and discuss the complex refractive index and the Kramers–Kronig relation for X-rays, are also applicable to gamma waves and the de Broglie waves of neutrons, electrons, ions, and other particles.

For X-rays, the Kramers–Kronig relation of Eq. (1.717) allows the calculation of the real part $f_1^0(w)$ of the complex forward-scatter factor $f^0(w)$ from its imaginary part $f_2^0(w)$. The real part $f_1^0(w)$ is related to the decrement δ via Eq. (1.529), and the imaginary, absorptive part $f_2^0(w)$ is related to the attenuation factor β via Eq. (1.528). For a single element of atomic number Z, the Kramers–Kronig relation yields the real refractive part $f_1^0(w)$ from the attenuating imaginary part $f_2^0(w)$, where

$$f_1^0(w) = Z - \frac{2}{\pi} P_C \int_0^{\infty} \frac{u f_2^0(u)}{u^2 - w^2} du \qquad (1.719)$$

The P_C indicates integration only over the nondivergent parts of the line along u and avoids the singular points, $u = \pm w$. The integrand is converted to the sum of single pole terms, which allows direct application of Cauchy's principal value theorem via

$$\frac{2u}{u^2 - w^2} = \frac{1}{u - w} + \frac{1}{u + w} \qquad (1.720)$$

With the substitution of Eq. (1.720) into Eq. (1.719), the real part $f_1^0(w)$ becomes

$$f_1^0(w) = Z - \frac{1}{\pi} P_C \int_0^{\infty} \frac{f_2^0(u)}{u - w} du - \frac{1}{\pi} P_C \int_0^{\infty} \frac{f_2^0(u)}{u + w} du \qquad (1.721)$$

Finally, the optical theorem of Eq. (1.629) allows the imaginary part of the forward-scatter factor $f_2^0(w)$ to be obtained from the total cross section σ_t, where

$$\sigma_t(w) = \sigma_a(w) + \sigma_s(w) = \frac{4\pi}{k} \mathrm{Im}\left[f^0(w)\right] = \frac{4\pi}{k} f_2^0(w) \qquad (1.722)$$

The X-ray attenuation cross section σ_t for a material can be substituted into Eq. (1.722) to obtain the imaginary component $f_2^0(w)$ of the complex forward-scatter factor $f^0(w)$, via the optical theorem. Next the imaginary (attenuating) component $f_2^0(w)$ is substituted into Eq. (1.721) to yield the real (refracting) component $f_1^0(w)$, via the Kramers-Kronig relation. With the real $f_1^0(w)$ and imaginary $f_2^0(w)$ components of the X-ray complex forward-scatter factor $f^0(w)$, one can then calculate the complex X-ray refractive index of the material, via Eq. (1.518). The computation of the real component $f_1^0(w)$ from the imaginary component $f_2^0(w)$ in Eq. (1.721) is facilitated by use of the Hilbert transform, which allows calculation of the X-ray refractive index for materials, see Tatchyn & Cremer (1996). The Hilbert transform (see Morse & Feshbach, 1953) is the convolution of an input function $u(t)$ with a transfer function $h(t) = 1/(t\pi)$, which yields the output function $H[u(t)] = P_C \int_{-\infty}^{\infty} u(\tau)h(t-\tau)\mathrm{d}\tau$. Inspection of Eq. (1.721), shows the Hilbert transform $H[u(t)]$ can be used to perform computation of the integrals in Eq. (1.721), which have input function $f_2^0(u)$, where $H[f_2^0(w)] = (1/\pi)P_C \int_{-\infty}^{\infty} f_2^0(u)/(u-w)\mathrm{d}u$.

1.6 Refractive of Gammas via Rayleigh and Delbrück Scatter

Recent breakthroughs in refractive deflection of gammas by a silicon prism have introduced the possibility of refractive lenses, prisms, and reflective waveguides for gammas. This breakthrough in gamma refractive optics is due to the coherent elastic scatter of gammas by the electric field of nuclei or the Delbrück effect of photon–photon scatter, which previously was overlooked. The photon–photon, coherent, elastic scatter of incident gammas by the nuclear electric field, which is also associated with the attenuation process of pair production, is a quantum electrodynamics interaction. The vacuum surrounding the nucleus is polarized by the extremely strong nuclear electric field. The strong nuclear electric field sets up virtual electron–positron pairs, which interact with an incident gamma and cause the gamma to scatter. The gamma scatter has two components: (1) a refractive coherent elastic component and (2) an attenuating component via pair production. High Z elements (that is, elements with large atomic numbers Z), which have larger nuclear electric fields than low-Z elements, are favored candidate materials for refractive lenses and prisms, and reflective waveguides. Based on the successful development of (1) gold biconvex and biconcave lenses for focusing and imaging, (2) gold prisms for deflection, and (3) gold waveguides with total internal reflection, these optical elements can be combined to produce efficiently very narrowband gamma beams with bandwidths $\Delta w/w \cong 10^{-6}$.

These narrowband gamma optics require a very intense narrowband or even broadband gamma sources, which enable practical applications of nuclear resonant excitation, nuclear resonant fluorescence, and nuclear resonant spectroscopy. Presently, intense gamma beams can be produced by (1) narrow gamma lines, which are emitted from capture (absorption) of intense reactor sources of thermal neutron such as Institut Laue-Langevin (ILL) and (2) intense gamma sources such as the HIγS at Duke University.

In thermal neutron capture, one uses materials with a large nuclei density of the isotope, which is used to absorb the thermal neutron and emit capture gamma lines. For example, chlorine-36 in the form of the solid and common polyvinyl chloride, where naturally occurring chlorine is comprised of 75.77% of chlorine-35 and 24.23% chlorine-37, absorb thermal neutrons to produce excited nuclear states of chlorine-36 and chlorine-38, respectively. Following thermal neutron capture, excited state chlorine-36 emits 384 prompt gamma lines, where the probability of thermal neutron capture and emission of these gamma lines is 43.5 barns. In comparison, chlorine-37 has a much smaller cross section of 0.43 barns to capture a thermal neutron with the subsequent emission of 71 gamma lines. The gamma emission from naturally occurring chlorine is dominated by the emission of three gamma lines from excited state chlorine-36, which have relatively large component gamma emission cross sections. There is a 43.5 barn cross section for emission of all 384 prompt gammas. The three dominant capture gammas, which are emitted from naturally occurring chlorine, are thus from excited state chlorine-36, which include a 517.030 keV line with 7.58 barn, 1164.8650 keV with 8.91 barn, and 6110.842 keV with 6.59 barn cross section. In addition, chlorine-36 emits capture gamma lines at 786.3020 keV with 3.42 barns and 788.4280 keV with 5.42 barns cross section.

Naturally occurring gadolinium, which is comprised of many isotopes, is the element that has the largest thermal neutron cross section and prompt gamma emission cross section. In particular, 15.7% gadolinium-157 has the largest thermal neutron capture cross section of 259,000 barns, of which 257,000 barns is the cross section for thermal neutron capture and the remission of 390 prompt gamma lines. These lines are very strongly dominated by 181.931 keV with 72.003 kilobarns, 79.51 keV with 40.101 kilobarns, and 944.174 keV with 3.09 kilobarns. Also, 14.8% Gd-155 of naturally occurring gadolinium has a total cross section of 60.9 kilobarns to capture a thermal neutron and emit 324 gamma lines. Gd-155 has less intense gamma lines than Gd-157, but Gd-155 does emit relatively intense gamma lines at 199.2130 with 2.02 kilobarns and 88.6970 keV with 1.38 kilobarns. For more information, see IAEA (2006).

More intense gamma sources are under construction, such as the Mono-Energetic Gamma-ray (MEGa-Ray) at Livermore National Laboratory and the ELI-NP facility under development in Bucharest, Romania. The commercial development of Compton backscatter-based intense gamma sources are also expected to be developed in the near future. These commercial intense gamma sources, combined with the gamma ray optics, could by housed in a modest-size laboratory, which could be used for a wide variety of academic, industrial, and medical R&D, as well as commercial services and applications. Habs, Günther, and Jentschel (2012) and Habs and Köster (2011) expect such applications as (1) radioactive waste management via transmuting radioactive waste into less toxic or shorter half-life waste, (2) imaging lithium-7 for lithium battery technology or biological imaging with chelated lithium-7 in biological and medical imaging, (3) the production of 50 medical isotopes for diagnostics and therapy, (4) nuclear photonics and high resolution nuclear spectroscopy with resolution of a few electron volts, (5) and gamma astronomy.

With these very narrowband optics and intense gamma sources, the nuclear structure can be probed to nuclear energy levels, which are as high as the neutron binding energy. In medical imaging and diagnosis, gamma lenses could magnify and image the distribution of medical-injected, inhaled, or ingested radionuclides via nuclear resonant excitation imaging. That is, the very narrowband gamma ray resonant absorption and fluorescent reemission by a particular isotope injected in the body allows imaging of the resulting distribution of the isotope in the body after uptake. Nuclear resonant fluorescent imaging would allow the imaging of low-density, low-Z material inside dense, high-Z material, which is presently not possible with X-ray and gamma radiography. For example, the distribution of lithium drugs in the human brain can be imaged via nuclear resonant fluorescent imaging.

1.7 Historical Introduction to Gamma Lenses—The Dirac Equation and the Delbrück Effect

Paul Dirac's search for a relativistically correct form of the Schrödinger wave equation for the electron in an imposed electromagnetic field led him in 1928 to formulate the Dirac equation, see Dirac (1928). The Dirac equation has four solutions, which consist of the pair of solutions for a negative energy electron (positron), which is in either a spin-up or spin-down state, and the pair of solutions for a positive energy electron, which is either in a spin-up or spin-down state. This led Dirac to present the "hole theory," which proposed that no electron in a vacuum can transition from a positive energy state to a negative energy state because (1) the vacuum has no vacancy and cannot accept the physically observable electrons in their positive energy states—that is, the vacuum is completely filled with electrons in negative energy states, which range from $-mc^2$ to $-\infty$, and (2) the electrons, which fill the vacuum's negative energy states, do not produce an external field, and do not contribute to the total charge, energy, and momentum of the system. A "zero point" describes the electron vacuum, where the charge, energy, and momentum of the vacuum electrons are described by negative energy states only.

Consider an external, imposed electromagnetic field, which varies frequency f rapidly enough, that is, with a minimum energy $E = hf$, which exceeds twice the electron rest mass $E > 2m_0c^2$, such as a gamma with energy greater than 1.022 MeV. Such an energetic gamma can act on the electrons in the negative, electron energy states of the vacuum and extract a vacuum electron to a positive energy state. As a result, the electron vacuum of negative energy states is left with a vacant state. The vacant negative energy state of the electron vacuum can be modeled as a positive energy state that is occupied by a positron with a positive charge, the antiparticle of the electron. H. Weyl corrected Dirac by showing the mass of the positive energy state, positive charged positron must equal that of the electron, not a proton, as Dirac originally thought. The existence of the positron and its mass equal to the electron, but with opposite charge polarity, was verified by the experiments of C. Anderson in 1932.

The coherent scattering of photons in the electric fields of atoms, via virtual electron–positron pairs (Delbrück scattering), was first investigated experimentally by Meitner and Kösters (1933) with a comment from Delbrück. The experiment of Meitner and Kösters was the first of many experiments to measure the discrepancy between the experimental and predicted differential cross sections for elastic scattering of photons by heavy atoms, which is explained by Delbrück scattering. Due to the work of Max Delbrück with Meitner and Kösters, Hans Bethe originated the term *Delbrück scattering,* which explained discrepancies between experimental and predicted data in a Compton scattering experiment on heavy atoms. Delbrück invoked the relativistic quantum mechanics or hole theory of Dirac, where the vacuum is modeled as being filled with electrons in negative energy states. Currently, the vacuum is modeled as being comprised of electron–positron pairs.

Consider the polarization effect of the intense, inhomogeneous, strong nuclear electromagnetic field upon the vacuum sea of negative energy state electrons. This is analogous in classical electromagnetics to the action of the polarization effect of an inhomogeneous, electromagnetic field, which is applied to the atoms and ions in a material medium. If one averages the nuclear electric field in a sphere, which is centered on the silicon nucleus, with a diameter equal to the electron Compton wavelength (400 fm), then the averaged silicon nuclear field equals approximately the Schwinger electric field amplitude of $E_s = 1.2 \times 10^{18}$ V/m. The intense electromagnetic fields of nuclei, especially heavy nuclei, cause coherent, elastic scatter of photons via Delbrück scattering.

Reliable theoretical predictions of Delbrück scatter awaited (1) the development of quantum electrodynamics (QED), which were based on Feynman diagrams and enabled theoretical results to be extracted via numerical computation, and (2) the development computers, which could provide sufficiently high precision to perform accurate numerical computation. Reliable experimental investigations required the development of photon detectors of high resolution and detection efficiency. Following an initial observation of Delbrück scattering, which was achieved in a high-energy, small-angle photon scattering experiment by Jarlskog et al. (1973) at DESY (Germany), the first definitive Delbrück scattering experiment was performed by Schumacher et al. (1975) at Göttingen.

1.7.1 Refractive Index and Attenuation Cross Section for the Delbrück Refraction of Gammas

We follow the treatment by Habs et al. (2012). The mechanism of refraction of gamma rays by materials, especially elements with high atomic numbers, is Delbrück scattering, which applies to gamma energies in the range of 0.7–3.0 MeV. Delbrück scattering is the coherent elastic scattering of photons in the Coulomb field of nuclei. However, Delbrück scattering is more pronounced in the larger electric fields of high-Z nuclei, such as lead, gold, and uranium. In Delbrück scattering, an incident photon encounters the strong nuclear field, where the photon fluctuates into a virtual electron–positron pair, which interacts with the nucleus, and some exchange of momentum to the nucleus but very little energy is transferred to the recoil of the massive nucleus. The virtual positron–electron pair transforms back to a single photon with nearly the same energy, but the scattered photon momentum and propagation direction have changed relative to the incident, fluctuated photon.

The relatively fixed nuclei and near-zero energy transfer to the nucleus, allow a fixed phase relation to be maintained by the various atom scatter centers, which interact with different parts of an incident gamma plane wave. Each region of the incident gamma plane wave, which is in the vicinity of the intense electromagnetic field of a nucleus, interacts with the nucleus via the formation of a virtual positron–electron pair, which recombines to produce an elastic scattered photon. And the fixed phase relation between the various portions of the incident gamma plane wave, allows the elastic scattered waves to be coherent as well. As a result, Delbrück scattering produces coherent elastic scattered waves, which produce the effects of refraction, reflection, and diffraction.

The refractive decrement δ and attenuation factor β of the complex refractive index $n = 1 - \delta + i\beta$ of a material medium for gammas, as with X-rays, are calculated with an energy-dependent, forward-scatter amplitude $A_f(E)$, which is comprised of the real, refractive part $A_{rf}(E)$ and an attenuating imaginary part $A_{if}(E)$. As a function of gamma energy E, the forward scatter amplitude $A_f(E)$, which has dimensions of length, is

$$A_f(E) = A_{rf}(E) + iA_{if}(E) \tag{1.723}$$

The real refractive part $A_{rf}(E)$ of the forward-scatter, gamma amplitude $A_f(E)$ is comprised of components from the virtual processes of (1) the coherent virtual photo effect, which is also Rayleigh scatter $A_{rf}^{ph}(E)$, (2) the virtual coherent virtual Compton effect $A_{rf}^{co}(E)$, and (3) the coherent virtual pair $A_{rf}^{pa}(E)$, where

$$A_{rf}(E) = A_{rf}^{ph}(E) + A_{rf}^{co}(E) + A_{rf}^{pa}(E) \tag{1.724}$$

These three components of the real refractive part $A_{rf}(E)$ of the forward-scatter, gamma amplitude $A_f(E)$, each produces coherent, elastic scatter.

The imaginary, attenuation part $A_{if}(E)$ of the forward-scatter, gamma amplitude $A_f(E)$ is comprised of components from the real processes of (1) photoelectric absorption $A_{if}^{ph}(E)$, (2) Compton inelastic scatter $A_{if}^{co}(E)$, and (3) pair production $A_{if}^{pa}(E)$, where

$$A_{if}(E) = A_{if}^{ph}(E) + A_{if}^{co}(E) + A_{if}^{pa}(E) \tag{1.725}$$

These three components cause absorption, and inelastic, incoherent scatter gives rise to the attenuating, imaginary part $A_{if}(E)$.

Next, consider a particular gamma interaction process in a material medium with number density ρ [#/cm^3] of the scatter centers (atom number density). The gamma interaction processes include photoelectric absorption, Rayleigh scatter, Compton scatter, and pair production in the vicinity of a nucleus. In a medium, and as a function of the incident gamma energy E and corresponding wavelength λ, the dimensionless, total refractive decrement $\delta(E)$, for a particular gamma interaction process, is directly proportional to the total real, refractive part $A_{rf}(E)$ of the forward-scatter amplitude, which has units of length, where

$$\delta(E) = \frac{\lambda^2 \rho}{2\pi} A_{rf}(E) \tag{1.726}$$

The components of the total refractive decrement $\delta(E)$ are (1) $\delta_{ph}(E)$ from the coherent virtual photo effect, which is also Rayleigh scatter, via real forward-scatter amplitude component $A_{rf}^{ph}(E)$; (2) $\delta_{co}(E)$ from the virtual coherent Compton effect, via real forward-scatter amplitude component $A_{rf}^{co}(E)$; and (3) $\delta_{pa}(E)$ from the coherent virtual pair production, via real forward-scatter amplitude component $A_{rf}^{pa}(E)$, where

$$\delta(E) = \delta_{ph}(E) + \delta_{co}(E) + \delta_{pa}(E) = \frac{\lambda^2 \rho}{2\pi}\left[A_{rf}^{ph}(E) + A_{rf}^{co}(E) + A_{rf}^{pa}(E)\right] \tag{1.727}$$

Attenuation tends to be more easily measured than refraction. One can measure the attenuation of a gamma ray by the various interaction processes of photo absorption, Compton scatter, and pair production in the vicinity of a nucleus, or in the vicinity of a free electron. Likewise, one can measure the attenuation of neutron waves, via the interaction processes of neutron absorption, inelastic scatter, elastic scatter, and nonelastic scatter (e.g., fission). From the optical theorem, as a function of the incident gamma wave energy E_γ, the differential, forward-scatter cross section $d\sigma_{scat}/d\Omega$ for a particular gamma interaction process with the atoms of the medium, then equals the squared magnitude of the real part of the forward-scatter amplitude $A_{rf}(E_\gamma)$, where

$$\frac{d\sigma_{scat}}{d\Omega}(E_\gamma) = |A_{rf}(E_\gamma)|^2 \tag{1.728}$$

The optical theorem of Eq. (1.632) relates the total imaginary component $A_{if}(E_\gamma)$ of the forward-scatter amplitude $A_f(E_\gamma)$ to the total attenuation cross section $\sigma_a(E_\gamma)$, as a function of the incident wave energy E_γ, where

$$\sigma_a(E_\gamma) = \frac{\lambda}{\pi} A_{if}(E_\gamma) \tag{1.729}$$

The total attenuation cross section $\sigma_a(E_\gamma)$ is comprised of components from the real processes of (1) photoelectric absorption $\sigma_a^{ph}(E)$, (2) Compton inelastic scatter $\sigma_a^{co}(E)$, and (3) pair production $\sigma_a^{pa}(E)$, where

$$\sigma_a(E) = \sigma_a^{ph}(E) + \sigma_a^{co}(E) + \sigma_a^{pa}(E) \tag{1.730}$$

The incident gamma wavelength λ_γ is related to its energy E_γ by

$$E_\gamma = \hbar w = \frac{h}{2\pi} \cdot 2\pi \cdot f = \frac{hc}{\lambda_\gamma} \tag{1.731}$$

Transposing Eq. (1.729), one obtains the total imaginary component of the forward-scatter amplitude $A_{if}(E)$, which is expressed in terms of the attenuation cross section $\sigma_a(E_\gamma)$, as a function of the incident gamma wave energy E_γ and wavelength λ, where

$$A_{if}(E_\gamma) = \frac{\pi}{\lambda_\gamma}\sigma_a(E_\gamma) = \pi\left(\frac{E_\gamma}{hc}\right)\sigma_a(E_\gamma) \tag{1.732}$$

The forward-scatter imaginary components for photoelectric absorption $A_{rf}^{ph}(E)$, Compton scatter $A_{rf}^{co}(E)$, and pair production $A_{rf}^{pa}(E)$, are thus obtained from the corresponding, measured, respective attenuation, component cross sections, $\sigma_a^{ph}(E)$, $\sigma_a^{co}(E)$, and $\sigma_a^{pa}(E)$:

$$A_{if}^{ph}(E_\gamma) = \pi\left(\frac{E_\gamma}{hc}\right)\sigma_a^{ph}(E_\gamma) \tag{1.733}$$

$$A_{if}^{co}(E_\gamma) = \pi\left(\frac{E_\gamma}{hc}\right)\sigma_a^{co}(E_\gamma) \tag{1.734}$$

$$A_{if}^{pa}(E_\gamma) = \pi\left(\frac{E_\gamma}{hc}\right)\sigma_a^{pa}(E_\gamma) \tag{1.735}$$

In principle, from the measured attenuation cross section, one can calculate the imaginary part of the forward-scatter amplitude $A_{\text{if}}(E)$ for a particular interaction process of the wave in question, X-ray, gamma, neutron, or other particle. The Kramers–Kronig relations can then calculate the corresponding refractive component $A_{\text{rf}}(E)$, where

$$A_{\text{rf}}(E_\gamma) = \frac{2}{\pi}\int_0^\infty \frac{EA_{\text{if}}(E)}{E^2 - E_\gamma^2}\text{d}E \tag{1.736}$$

The integrand of the Kramers–Kronig integral expresses resonance about an arbitrarily chosen gamma energy E_γ in a manner similar to the Breit–Wigner expression. The Kramers–Kronig resonance integrand is integrated over the range of gamma energies E from 0 to ∞, which then yields the real part or refractive component of the forward-scatter amplitude at arbitrarily chosen gamma energy E_γ, where E_γ acts as the energy of resonance. Via the Kramers–Kronig expression of Eq. (1.736), one finds (1) the coherent virtual photo effect or Rayleigh scatter $A_{\text{rf}}^{\text{ph}}(E)$ is obtained from the measured photoelectric absorption cross section $\sigma_a^{\text{ph}}(E)$; (2) the virtual coherent virtual Compton effect $A_{\text{rf}}^{\text{co}}(E)$ is obtained from the measured inelastic Compton scatter cross section $\sigma_a^{\text{co}}(E)$; and (3) the coherent virtual pair $A_{\text{rf}}^{\text{pa}}(E)$ is obtained from the measured absorption cross section of pair production $\sigma_a^{\text{pa}}(E)$, where

$$A_{\text{rf}}^{\text{ph}}(E_\gamma) = \frac{2}{\pi}\int_0^\infty \frac{EA_{\text{if}}^{\text{ph}}(E)}{E^2 - E_\gamma^2}\text{d}E \tag{1.737}$$

$$A_{\text{rf}}^{\text{co}}(E_\gamma) = \frac{2}{\pi}\int_0^\infty \frac{EA_{\text{if}}^{\text{co}}(E)}{E^2 - E_\gamma^2}\text{d}E \tag{1.738}$$

$$A_{\text{rf}}^{\text{pa}}(E_\gamma) = \frac{2}{\pi}\int_0^\infty \frac{EA_{\text{if}}^{\text{pa}}(E)}{E^2 - E_\gamma^2}\text{d}E \tag{1.739}$$

With the substitution of Eq. (1.732) into Eq. (1.736), one obtains the total real, refractive component of the forward-scatter amplitude $A_{\text{rf}}(E_\gamma)$ at a particular energy E_γ, which is expressed in terms of the total attenuation cross section $\sigma_a(E)$, via integration over the range of gamma energies E from 0 to ∞, where

$$A_{\text{rf}}(E_\gamma) = 2\left(\frac{E_\gamma}{hc}\right)\int_0^\infty \frac{E\sigma_a(E)}{E^2 - E_\gamma^2}\text{d}E \tag{1.740}$$

And with the substitution from Eqs. (1.733–1.735) into Eqs. (1.737–1.739), the real components of the forward-scatter amplitudes for the coherent virtual photo effect or Rayleigh scatter $A_{\text{rf}}^{\text{ph}}(E_\gamma)$, Compton scatter $A_{\text{rf}}^{\text{co}}(E_\gamma)$, and pair production $A_{\text{rf}}^{\text{pa}}(E_\gamma)$ for a specific gamma energy E_γ, are obtained from their corresponding, measured attenuation cross sections for photoelectric absorption $\sigma_a^{\rm ph}(E)$, Compton scatter $\sigma_a^{\rm co}(E)$, and pair production $\sigma_a^{\rm pa}(E)$, via integration over the range of gamma energies E from 0 to ∞, where

$$A_{\rm rf}^{\rm ph}(E_\gamma) = 2\left(\frac{E_\gamma}{hc}\right)\int_0^\infty \frac{E\sigma_a^{\rm ph}(E)}{E^2 - E_\gamma^2}\,{\rm d}E \tag{1.741}$$

$$A_{\rm rf}^{\rm co}(E_\gamma) = 2\left(\frac{E_\gamma}{hc}\right)\int_0^\infty \frac{E\sigma_a^{\rm co}(E)}{E^2 - E_\gamma^2}\,{\rm d}E \tag{1.742}$$

$$A_{\rm rf}^{\rm pa}(E_\gamma) = 2\left(\frac{E_\gamma}{hc}\right)\int_0^\infty \frac{E\sigma_a^{\rm pa}(E)}{E^2 - E_\gamma^2}\,{\rm d}E \tag{1.743}$$

Recall from Eq. (1.726) that the total refractive decrement $\delta(E_\gamma)$ is directly proportional to the total real, refractive part $A_{\rm rf}(E_\gamma)$ part of the forward-scatter amplitude, where

$$\delta(E_\gamma) = \frac{\lambda_\gamma^2 \rho}{2\pi} A_{\rm rf}(E_\gamma) = \frac{\rho}{2\pi}\left(\frac{hc}{E_\gamma}\right)^2 A_{\rm rf}(E_\gamma) \tag{1.744}$$

With the substitution of $A_{\rm rf}(E_\gamma)$ of Eq. (1.740) into Eq. (1.744), the total refractive decrement $\delta(E_\gamma)$ in a material medium for an incident gamma of energy E_γ becomes

$$\delta(E_\gamma) = \frac{\rho}{\pi}\left(\frac{hc}{E_\gamma}\right)\int_0^\infty \frac{E\sigma_a(E)}{E^2 - E_\gamma^2}\,{\rm d}E \tag{1.745}$$

From Eq. (1.745) and Eq. (1.730), one obtains the components of the total refractive decrement $\delta(E)$, which are (1) the decrement component $\delta_{\rm ph}(E)$ for the coherent virtual photo effect (Rayleigh scatter), (2) the decrement component $\delta_{\rm co}(E)$ for the virtual coherent Compton effect, and (3) the decrement component $\delta_{\rm pa}(E)$ for coherent virtual pair production, where

$$\delta_{\rm ph}(E_\gamma) = \frac{\rho}{\pi}\left(\frac{hc}{E_\gamma}\right)\int_0^\infty \frac{E\sigma_a^{\rm ph}(E)}{E^2 - E_\gamma^2}\,{\rm d}E \tag{1.746}$$

$$\delta_{\rm co}(E_\gamma) = \frac{\rho}{\pi}\left(\frac{hc}{E_\gamma}\right)\int_0^\infty \frac{E\sigma_a^{\rm co}(E)}{E^2 - E_\gamma^2}\,{\rm d}E \tag{1.747}$$

$$\delta_{\rm pa}(E_\gamma) = \frac{\rho}{\pi}\left(\frac{hc}{E_\gamma}\right)\int_0^\infty \frac{E\sigma_a^{\rm pa}(E)}{E^2 - E_\gamma^2}\,{\rm d}E \tag{1.748}$$

In Figure 1.27, as a function of photon energy E_γ, the left set of plots includes the cross sections for attenuation by photoelectric absorption $\sigma_{\rm ph}(E_\gamma)$, Compton

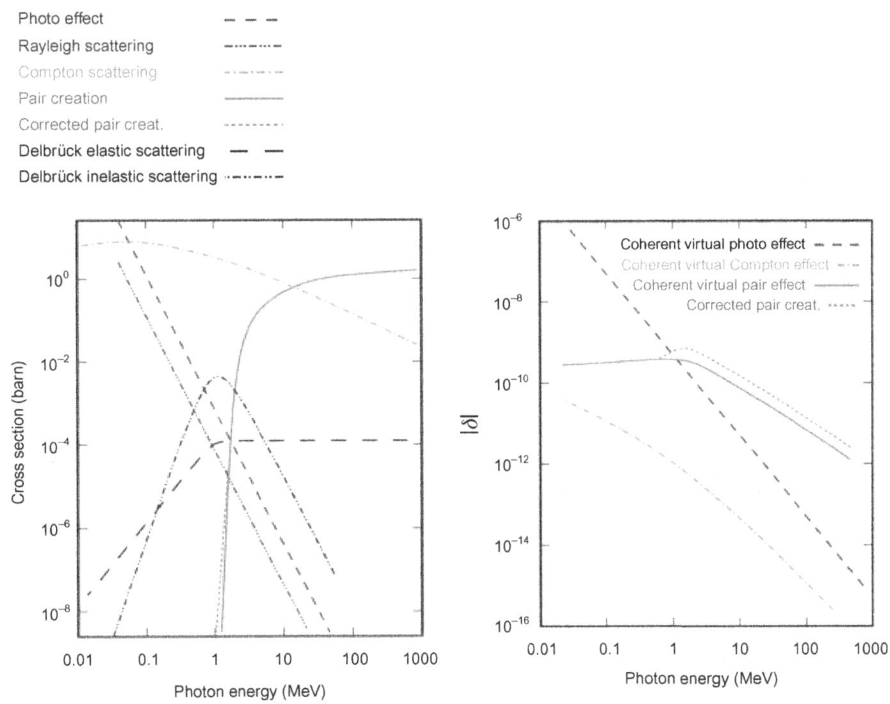

Figure 1.27 (Left) From Habs et al. (2012), as a function of photon energy E_γ, are plotted cross sections for attenuation by photoelectric absorption $\sigma_{\mathrm{ph}}(E_\gamma)$, Compton scatter $\sigma_{\mathrm{co}}(E_\gamma)$, pair production $\sigma_{\mathrm{pa}}(E_\gamma)$, and Delbrück inelastic scatter $\sigma_{\mathrm{di}}(E_\gamma)$ cross sections, and the refractive cross sections for Rayleigh $\sigma_{\mathrm{ra}}(E_\gamma)$ and Delbrück elastic scatter $\sigma_{\mathrm{de}}(E_\gamma)$. Also plotted is the corrected pair production cross section, which is increased, compared to the Bethe–Heitler cross section, via dispersion relation considerations. The component attenuation cross sections $\sigma(E_\gamma)$ of the left plots were used to calculate the corresponding decrement magnitudes $|\delta(E_\gamma)|$ of the right plots. (Right) Plotted as a function of photon energy E_γ are the magnitudes of the refractive decrement components for the coherent virtual photo (Rayleigh) $|\delta_{\mathrm{ph}}(E_\gamma)|$, Compton $|\delta_{\mathrm{co}}(E_\gamma)|$, and pair production $|\delta_{\mathrm{pa}}(E_\gamma)|$, and the corrected pair production $\infty > |\tau| \geq 0$.

scatter $\sigma_{\mathrm{co}}(E_\gamma)$, pair production $\sigma_{\mathrm{pa}}(E_\gamma)$, and Delbrück inelastic scatter $\sigma_{\mathrm{di}}(E_\gamma)$ cross sections, as well as the refractive cross sections for Rayleigh $\sigma_{\mathrm{ra}}(E_\gamma)$ and Delbrück elastic scatter $\sigma_{\mathrm{de}}(E_\gamma)$. Also in Figure 1.27, the corrected pair production cross section is plotted, which is greater than the Bethe–Heitler cross section, via dispersion relation considerations. The Kramers–Kronig relation was used to calculate component decrements for gamma refraction for the virtual coherent processes of photo (Rayleigh), Compton, and pair production from the corresponding measured and tabulated cross sections for photoelectric absorption, Compton scatter, and pair production. And in Figure 1.27, as a function of photon energy E_γ, are shown a set of plots of the magnitudes of the refractive decrement components for the coherent virtual photo (Rayleigh) $|\delta_{ph}(E_\gamma)|$, Compton $|\delta_{co}(E_\gamma)|$, and pair production $|\delta_{pa}(E_\gamma)|$, as well as the corrected pair production $|\delta_{pa}(E_\gamma)|$.

1.7.2 Gamma Refractive Optics—Experimental Results

Presently, many efforts are underway to develop highly brilliant gamma beams of low emittance, via Compton back scattering of high-power laser photons by an ultrarelativistic electron beam, which is produced by a normal conducting electron linac or a (superconducting) energy recovery linac (ERL). A renaissance in nuclear physics is occurring around the world because of a new kind of incredibly bright, gamma ray light source, which can be created with short pulse lasers and energetic electron beams.

The MEGa-ray source facility at Lawrence Livermore National Laboratory produces narrow, laser-like beams of incoherent, tunable gamma rays, which enables access to and manipulation of the nucleus with photons, and advances nuclear photonics. In the early days of the laser, photon manipulation of the valence electron structure of the atom became possible, enabling new applications and science. Similarly, nuclear photonics with laser-based gamma ray sources promises to open up wide areas of practical isotope-related materials applications and nuclear science. Very narrow bandwidths are needed for nuclear resonant fluorescents, which require chromatic filters. The chromatic filters can be fabricated from such optics as compound refractive, 2-D, biconcave, and biconvex gamma lenses and prisms, multilayer gamma mirrors, and waveguides. A CRL can be used to produce a focal spot of X-rays, neutrons, or gammas, which just passes through the limiting aperture of a tiny pinhole. The pinhole aperture thus admits a focal spot of more nearly monochromatic photons or neutrons. Provided that the pinhole has a sufficiently small diameter, an object and detector, which are placed sufficiently far downstream of the pinhole, can allow gamma phase contrast imaging.

In 2012, a breakthrough of refracting 184 keV gammas, was experimentally demonstrated by the D. Habs group of the Ludwig-Maximilians-Universität München in Garching, Germany. This led the Habs group to pursue a CRL, which is comprised of a stack of biconvex gold lenses. In 1952, J. Toll and J. A. Wheeler had predicted that the real refractive index component or decrement for 1 MeV gammas in lead, which is due to the pair production creation, would be a factor of 10^3 smaller than the refractive decrement component due to Rayleigh scatter. However, Toll and Wheeler did not account for Delbrück scatter, which in general is the scatter of photons by an external electric field (photon–photon) scatter, and particularly the radiative correction to Compton scatter of incident photons by the intense electric field of the nucleus, in addition to the normal Compton scatter by the atomic electron electric field.

The Habs group included the higher order Delbrück scatter in the calculation of the refractive decrement component from pair production, with inclusion of the inelastic Delbrück scatter in the Kramers–Kronig calculation. With the Kramers–Kronig relation, the Habs group obtained the real refractive forward-scatter amplitude, via integration over all energy, of the imaginary, attenuating, forward-scatter amplitude component. The forward scatter amplitude component was obtained from the experimentally measured real pair production attenuation cross section, and the theoretical calculated, virtual attenuation cross section of inelastic Delbrück scatter.

The Habs group theoretical calculations were in agreement with their experimental measurements of the refraction of 184 keV gammas by a single prism. As a result, they expect the refractive decrement for a stack of 17 gold, biconcave, 1-mm-diameter lenses, will be around 3×10^{-5} for 1 MeV gammas, which produces a focal length of 3 m.

The development of intense gamma sources, which are based on Compton scatter, is expected to provide the necessary gamma intensity to utilize the compound refractive gamma lenses. The Compton scatter gamma source employs a relativistic electron beam, which is directed against an intense laser light source. The Compton scatter between the relativistic electrons and photons cause the light photons to be upshifted to gamma wavelengths, via the Doppler effect. A 250-MeV, ultracompact, Compton scattering gamma ray source is under development at the Lawrence Livermore National Laboratory, which has the goal of generating tunable gamma rays in the 0.5–2.5 MeV photon range, at a repetition frequency of 120 Hz and with a peak brightness of about 10^{20} photons per s mm^2 mrad^2 0.1% BW, see Marsh et al. (2012). Also see Albert (2010), Barty (2011a, 2011b), ELI-NP (2011), and Gibson (2010).

Via the Delbrück effect, Habs et al. (2012) used a silicon wedge to refract beams of monochromatic gammas of specific energies 182, 517, and 786 keV, as well as 1.165 MeV and 1.951 MeV. Figure 1.28 shows the basic experimental configuration, and plotted are the measured prism-refracted and air-refracted 184 keV Erbium-168 gamma counts at the detector, as a function of the relative diffraction angle in nanoradians. Each diffraction angle corresponds to a gamma energy, which is selected by the pair of perfect silicon crystals that act as the spectrometer. This spectrometer compares the refraction of the upper gamma, plane wave front through the prism, and the lower gamma plane wave front through air only. See Table 1.2.

The monochromatic gammas were provided by thermal neutron capture in gadolinium-158, which provided the 182 keV gammas, and chlorine-36, which provided the other aforementioned gamma energies. The high-flux reactor at Institut Laue-Langevin (ILL) in Grenoble provided an intense beam of thermal neutrons with flux 4.0×10^{15} $n/\text{cm}^2/s$, which yielded gamma emission rates of 10^{15} g/s. The thermal neutron beam then irradiated 10-g samples of chlorine-36 or gadolinium-158, which absorbed the thermal neutrons and produced the monochromatic gammas.

These gammas were collimated over a distance of 17 m to produce a 4 mm $\times$ 20 mm rectangular plane wave front, which were incident on the input surface of two perfect crystals of silicon, which comprised the GAMS5 spectrometer at ILL. Between the two silicon crystals of the spectrometer was placed a silicon prism with an edge angle of 165°. The silicon prism intercepts only the upper portion of the gamma beam, which is diffracted by the input silicon crystal. The lower portion of the gamma beam from the input silicon crystal is transmitted through

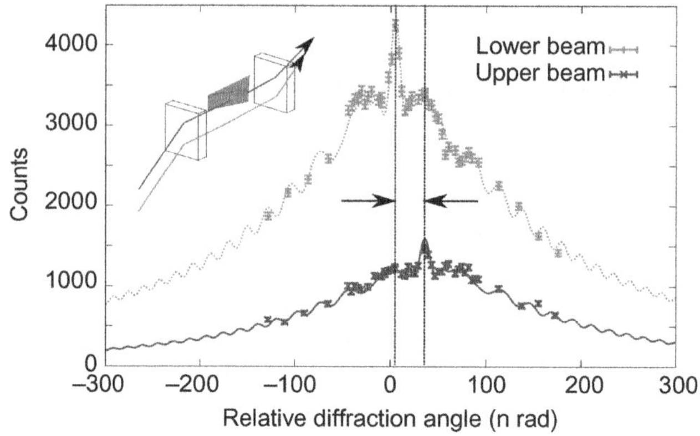

Figure 1.28 Shown from Habs et al. (2012) is their basic experimental configuration. Plotted are the measured prism-refracted and air-refracted 184 keV Erbium-168 capture gamma counts at the detector as a function of the relative diffraction angle in nanoradians. A silicon wedge (the silicon prism) was placed between two perfect silicon crystals of the GAMS5 spectrometer. The gamma beam originates from the capture of thermal neutrons by Erbium-168, where the intense thermal flux 4.0×10^{15} $n/\text{cm}^2/s$ of the Grenoble ILL high-flux reactor produced gamma emission rates of 10^{15} g/s. Each diffraction angle corresponds to a gamma energy, which is selected by the pair of perfect silicon crystals. The pair of perfect silicon crystals acts as the spectrometer and compares the refraction of the upper gamma, plane wave front (blue plot) through the prism, and the lower gamma plane wave front (red plot) through air only. The upper gamma beam (blue plot) undergoes attenuation in the prism and thus shows less relative intensity, as compared to the lower gamma beam (red plot), which passes through air. The plotted spectra have small oscillations, which are due to the Pendellösung effect. (For interpretation of the references to color in this figure legend, the reader is referred to the web version of this book.)

Table 1.2 List of decrements and linear attenuation coefficients for a few monochromatic gamma energies produced by thermal neutron capture in chlorine and gadolinium isotopes, from Habs et al. (2012)

E (keV)	Isotope Gamma Source	Wedge Silicon Decrement δ	Wedge Silicon Attenuation μ (cm^{-1})	Background Air Decrement δ	Background Air Attenuation μ (cm^{-1})
517	Cl-36	-4.6×10^{-10}	0.20		1.0×10^{-4}
786	Cl-36	1.8×10^{-10}	0.16		8.6×10^{-5}
1165	Cl-36	1.5×10^{-9}	0.13	3.2×10^{-10}	7.1×10^{-5}
1951	Cl-36	1.1×10^{-9}	0.10		5.4×10^{-5}
182	Gd-158	-1.1×10^{-8}	0.31	-6.7×10^{-10}	1.5×10^{-4}

The linear attenuation μ [cm^{-1}] for silicon and dry air at sea level was calculated using total attenuation cross sections from the NIST Web site (http://physics.nist.gov/PhysRefData), via Hubbell et al. (1980).

air, and is then diffracted to the output silicon crystal. The upper and lower portions of the gamma beam, are then diffracted by the output crystal, as parallel beams into a calibrated HP-Ge detector, which is 3 m downstream of the spectrometer. The detector measures the energy spectrum (with a 2 keV window) of the upper, refracted gamma beam and the lower, transmitted gamma beam separately. This is done via a lead shutter, which is used to block either the top or bottom half of the gamma beam.

In the plot of Figure 1.28, the measured intensity has a peak in a narrow range of diffraction angles, which corresponds to the narrow bandwidth of the Erbium-168 gamma line of 184 keV. The upper gamma beam (blue plot) undergoes attenuation in the prism and thus shows less relative intensity compared to the lower gamma beam (red plot), which passes through air. The plotted spectra of Figure 1.28 have small oscillations, which are due to the Pendellösung effect. In Chapter 7 of AIEP volume 173 (Cremer, 2012b), dynamic scatter of X-rays and neutrons by perfect crystals and the Pendellösung effect are discussed. Also, in AIEP volume 172 (Cremer, 2012a), neutron and X-ray interferometry, and the Pendellösung effect and anomalous absorption in perfect crystals are covered in more detail. The Pendellösung effect accounts for the dynamic scatter within a crystal, which is the interference between the incident X-ray or neutron plane wave and the scattered neutrons or X-rays within the perfect crystal. That is, the Pendellösung oscillations in Figure 1.28 are caused by interference effects of the multiple-scattered X-rays, neutrons and gammas by the lattice atoms of the perfect silicon crystal, which comprises the wedge (prism) and the pair of spectrometer slabs.

The first or input crystal of the pair of 2.5-mm-thick, single silicon crystals of the spectrometer determined the nanoradian acceptance angle of the spectrometer. The nanoradian acceptance angle and the silicon crystal lattice spacing $\Delta d/d = 5 \times 10^{-8}$ of the input crystal, allow the input crystal to scan through the input beam, which has a divergence angle of a few microradian, and diffract a narrow bandwidth of the gamma beam to the output crystal. The second or output crystal is set in a nondispersive geometry, where the second crystal is aligned parallel to the first crystal.

The output crystal is rocked relative to its reference position, which is parallel to the input crystal. Rocking the output crystal, relative to the position parallel to the input crystal, allows an energy scan of the refracted and transmitted gammas. The nondispersive geometry promotes the detection of the beam deflection by the silicon prism, which is placed between the two planar silicon crystals. As a result, the output silicon crystal of the spectrometer can compare angular deviations of 1 nrad between the upper and lower portions of the gamma beam.

The acquisition times were short to minimize drift problems, and 30 pairs of scans were taken and averaged for each measured energy point in the range of energies scanned by the spectrometer, while the temperature, pressure, and humidity at the spectrometer were monitored, and the time variations of the silicon lattice spacing and refractive index of air were corrected. Background measurements were done, where the upper and lower beams were measured, and the prism was removed from the upper beam.

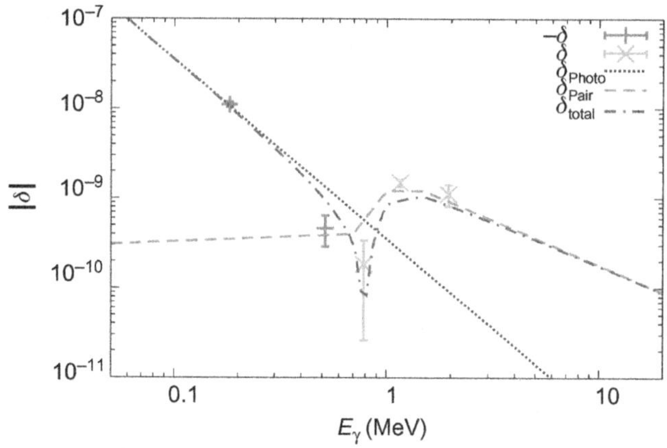

Figure 1.29 From Habs et al. (2012), as a function of gamma energies E_γ up to 2 MeV, are plotted the calculated and measured magnitudes of the total refractive index decrement $|\delta(E_\gamma)|$ (the black dashed curve), which is the sum of the component photo $|\delta_{\text{ph}}(E_\gamma)|$ decrement and pair $|\delta_{\text{pa}}(E_\gamma)|$ decrements. The blue dashed curve is the calculated negative-valued decrements $\delta < 0$ or $|\delta_{\text{ph}}(E_\gamma)|$, and the red crosses are the measured values. In comparison, the red dashed curve is the calculated positive-valued decrements $\delta > 0$ or $|\delta_{\text{pa}}(E_\gamma)|$, and the green crosses are the measured values. (For interpretation of the references to color in this figure legend, the reader is referred to the web version of this book.)

Habs et al. (2012) used the following definition of refractive index n with decrement δ and attenuation factor β, where

$$n = 1 + \delta + \mathrm{i}\beta \tag{1.749}$$

This is different from the normal convention in X-rays and neutron optics, and in this book, see, for example, Attwood (1999) and Sears (1989).

$$n = 1 - \delta + \mathrm{i}\beta \quad \text{normal convention} \tag{1.750}$$

One should thus note the use of $n = 1 + \delta$ in the discussion of refraction and choice of convex and concave lenses in the discussion and plots of Habs et al. (2012), which are discussed in this section.

In Figure 1.29, as a function of gamma energies E_γ up to 2 MeV, the calculated and measured magnitudes of the total refractive index decrement $|\delta(E_\gamma)|$, which is a sum of the component photo $|\delta_{\text{ph}}(E_\gamma)|$ decrement and pair $|\delta_{\text{pa}}(E_\gamma)|$ decrement, are plotted. The plots include (1) the black dashed curve for the total, calculated decrement $|\delta(E_\gamma)|$, (2) the blue dashed curve for the calculated negative-valued decrements $\delta < 0$, (3) the red dashed curve for the calculated, positive-valued decrements $\delta > 0$, (4) the red crosses for the measured negative decrements $\delta_{\text{ph}}(E_\gamma)$, and (5) the green crosses for the measured positive decrements $\delta > 0$. The negative

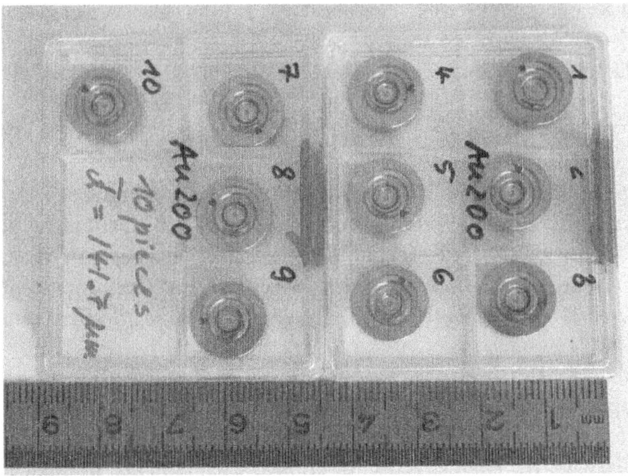

Figure 1.30 Shown are gold, biconvex lenses of Habs et al. (2012), which are expected to refract gamma radiation more strongly than silicon lenses.

decrements $\delta < 0$, or the photo decrement $\delta_{\mathrm{ph}}(E_\gamma)$, are caused by the coherent virtual photo effect, that is, Rayleigh scatter. Conversely, the positive decrements $\delta > 0$, or pair decrement $\delta_{\mathrm{pa}}(E_\gamma)$, are caused by coherent virtual pair production. The next step is to test a gamma CRL, which has been fabricated by Habs et al. (2012). Their CRL is comprised of a stack of biconvex gold lenses, which are shown in Figure 1.30.

References

Albert, F. (2010). Isotope-specific detection of low density materials with laser-based monoenergetic gamma-rays. *Optics Letters*, *35*(3), 354.

Allman, B. E., McMahon, P. J., Nugent, K. A., Paganin, D., Jacobson, D. L., & Arif, M., et al. (2000). Imaging: Phase radiography with neutrons. *Nature*, *408*, 158–159.

Attwood, D. (1999). *Soft X-rays and extreme ultraviolet radiation*. Cambridge: Cambridge University Press. Chapter 2.

Barty, C. P. J. (2011a). MEGa-rays, petawatts, and nuclear photonics. *SPIE Opt. Optoelectron.* Paper 8080B-30.

Barty, C. (2011b) (LLNL), ELI-NPmeeting. <http://www.eli-np.ro/gamma-beam = meeting-1819-august-2011/php>.

Barzilov, A., Novikov, I., Womble, P., and Hopper, L. (2009). Characterization of a pulse neutron source yield under field conditions. In *Application of accelerators in research and industry: Twentieth international conference. AIP conference proceedings*. Vol. 1099 (pp. 904–906). American Institute of Physics.

Beguiristain, H. R., Anderson, I. S., Dewhurst, C. W., Piestrup, M. A., Cremer, J. T., & Pantell, R. H. (2002). A simple neutron microscope using a compound refractive lens. *Applied Physics Letters*, *81*(22), 4290–4292.

Bekefi, G., & Barrett, A. H. (1977). *Electromagnetic vibrations, waves, and radiation*. Cambridge, MA: MIT Press.

Bohren, C. F. (2010). What did Kramers and Kronig do, and how did they do it? *European Journal of Physics, 31*, 573–577.

Burden, R. L., & Faires, J. D. (1997). *Numerical analysis*. Belmont, CA: Brooks/Cole Publishing Company.

Cederström, B., Cahn, R. N., Danielsson, M., Lundqvist, M., & Nygren, D. R. (2000). Focusing hard X-rays with old LPs. *Nature, 404*, 951.

Chantler, C. T., Olsen, K., Dragoset, R. A., Chang, J., Kishore, A. R., & Kotochigova, S. A., et al. *X-ray form factor, attenuation and scattering tables (version 2.1)*. Gaithersburg, MD: National Institute of Standards and Technology. http://physics.nist.gov/ffast. Originally published as Chantler, C. T. (2000). *Journal of Physical and Chemical Reference Data, 29*(4), 597–1048 and Chantler, C. T. (1995). *Journal of Physical and Chemical Reference Data,24*, 71–643.

Choi, S. M., Barker, J. G., Glinka, C. J., Cheng, Y. T., & Gammel, P. T. (2000). Focusing cold neutrons with multiple biconcave lenses for small-angle neutron scattering. *Journal of Applied Crystallography, 33*, 793–796.

Cremer, J. T., Piestrup, M. A., Gary, C. K., Pantell, R. H., & Glinka, C. J. (2004a). Biological imaging with a neutron microscope. *Applied Physics Letters, 85*(3), 494–496.

Cremer, J., Piestrup, M., Gary, C. K., Park, H., & Pantell, R. H. (2004b). Biological imaging with a neutron microscope. (invited paper). In F. P. Doty, R. C. Schirato, H. B. Barber, & H. Roehrig (Eds.), *Penetrating radiation systems and applications VI. Proceedings of SPIE, Vol. 5541* (pp. 75–114). Bellingham, WA: SPIE.

Cremer, J. T., Park, H., Piestrup, M. A., Gary, C. K., Pantell, R.H., Flocchini, R. G., et al. (2007). Unpublished results.

Cremer, J. T., Park, H., Piestrup, M. A., Gary, C. K., Pantell, R. H., Flocchini, R. G., et al. (2007a). Simple microscope using a compound refractive lens and a wide-bandwidth thermal neutron beam. *Applied Physics Letters, 90*, 141113.

Cremer, J. T., Piestrup, M. A., Gary, C. K., Pantell, R. H., Glinka, C. J., & Barker, J. G. (2005). Imaging hydrogenous materials with a neutron microscope. *Applied Physics Letters, 87*(16), 161913.

Cremer, J. T., Williams, D. L., Gary, C. K., Piestrup, M. A., Faber, D. R., Fuller, M. J., et al. (2009). Efficient high-resolution fast-neutron detector. Unpublished results of Phase II SBIR administered by J. Figeroa, U.S. National Science Foundation, Contract No. IIP-0724503 (2008).

Cremer, J. T., Williams, D. L., Fuller, M. J., Gary, C. K., Piestrup, M. A., Pantell, R. H., et al. (2010). Periodic magnetic field as a polarized and focusing thermal neutron spectrometer and monochromator. *Review of Scientific Instruments, 81*(1), 013902.

Cremer, J. T, Williams, D. L., Gary, C. K., Piestrup, M. A., Faber, D. R., Fuller, M. J., et al. (2012). Fast neutron imaging of hydrogenous materials using a DD fusion generator. *Nuclear Instruments and Methods A, 675*, 51–55.

Cremer, J. T. (2012a). Advances in Imaging and Electron Physics. In Hawkes P. W. (Ed.), *Neutron and X-ray Microscopy—Part 1* (Vol. 172). London, UK: Academic Press, an imprint of Elsevier.

Cremer, J. T. (2012b). Advances in Imaging and Electron Physics. In Hawkes P. W (Ed.), *Neutron and X-ray Microscopy—Part 2* (Vol. 173). London, UK: Academic Press, an imprint of Elsevier.

Cremer, J. T. (2013). Advances in Imaging and Electron Physics. In P. W. Hawkes (Ed.), *Chapters 1 and 2* (Vol. 175). London, UK: Academic Press, an imprint of Elsevier.

Daly, D. (2001). *Microlens arrays*. New York, NY: Taylor & Francis.

Dangendorf, V., Laczko, G., Kersten, C., Jagutzki, O., and Spillmann, U. (2002). Fast neutron resonance radiography in a pulsed neutron beam. In *Seventh world conference on neutron radiography*, Rome, September 2002, eprint: arXiv:nucl-ex/0301001, <http://arxiv.org/ftp/nucl-ex/papers/0301/0301001.pdf>.

Dirac, P. A. M. (1928). The quantum theory of the electron. *Proceedings of the Royal Society A: Mathematical, Physical and Engineering Sciences*, *117*(778), 610. doi:10.1098/rspa.1928.0023.

Disdier, L., Lerche, R. A., Bourgade, J. L., & Glebo, V. Y. (2004). Capillary detector with deuterated scintillator for inertial confinement fusion neutron images. *Review of Scientific Instruments*, *75*(6), 2134–2139.

Disdier, L., Rouyer, A., Fedotoff, A., Bourgade, J. L., Marshall, F. J., & Glebov, V. Y., et al. (2003). Neutron imaging of ICF target plasmas. *Review of Scientific Instruments*, *74*(3), 1832–1836.

Eberhardt, J. E., Rainey, S., Stevens, R. J., Sowerby, B. D., & Tickner, J. R. (2005). Fast neutron radiography scanner for the detection of contraband in air cargo containers. *Applied Radiation and Isotopes*, *63*(2), 179–188.

ELI-NP. (2011). Description of the extreme light infrastructure nuclear physics facility. <http://www.eli-np.ro/documents/ELI-NP-WhiteBook.pdf> Accessed 01.10.11.

Eskildsen, M. R., Gammel, P. L., Isaacs, E. D., Detlefs, C., Mortensen, K., & Bishop, D. J. (1998). Compound refractive optics for imaging and focusing of low-energy neutrons. *Nature*, *391*, 563–566.

Eyges, L. (1972). *The classical electromagnetic field*. New York, NY: Dover. Chapter 14.

Feenberg, E. (1932). The scattering of slow electrons by neutral atoms. *Physical Review*, *40*, 40.

Feynman, R. P. (1964). *The Feynman lectures on physics. Vol. II (mainly Electromagnetism and Matter)*. Reading, MA: Addison Wesley. Chapter 21.

Freund, A. K. (1983). Cross section of materials used as neutron monochromators and filters. *Nuclear Instruments and Methods*, *213*(2–3), 495–501.

Gary, C. K., Park, H., Lombardo, L. W., Piestrup, M. A., Cremer, J. T., & Pantell, R. H., et al. (2007). High resolution X-ray microscope. *Applied Physics Letters*, *90*, 181111.

Gerard, C. F., & Wheatley, P. O. (1994). *Applied numerical analysis*. New York, NY: Addison Wesley.

Gibson, D. J. (2010). Design and operation of a tunable MeV-level Compton-scattering-based (gamma-ray) source. *Physical Review Special Topics—Accelerators and Beams*, *13*(7), 070703.

Gorter, C. J., & Kronig, R. de L. (1942). On the theory of absorption and dispersion in paramagnetic and dielectric media. *Physica*, *3*, 1009–1020.

Gureyev, T. E., Nesterets, Y. I., Stevenson, A. W., Miller, P. R., Pogany, A., & Wilkins, S. W. (2008). Some simple rules for contrast, signal-to-noise and resolution in in-line X-ray phase-contrast imaging. *Optics Express*, *16*(5), 3223–3241.

Gureyev, T. E., & Wilkins, S. W. (1998a). On X-ray phase retrieval from polychromatic images. *Optics Communications*, *147*(4–6), 229–232.

Gureyev, T. E., & Wilkins, S. W. (1998b). On X-ray phase imaging with a point source. *Journal of the Optical Society of America*, *A15*(3), 579–585.

Habs, D., Günther, M. M., Jentschel, M., & Urban, W. (2012). The refractive index of silicon at γ ray energies. *Physical Review Letters*, *108*. (18)id. 184802. May 2012. Also arXiv:1111.3608v2 [physics.atom-ph]. <http://arxiv.org/abs/1111.3608>.

Habs, D., & Köster, U. (2011). Production of medical radioisotopes with large specific activity in photonuclear reactions with γ-beams of high intensity and large brilliance. *Applied Physics B*, *103*, 501.

Hall J.M. (2001). Uncovering hidden defects with neutrons. science and technology review, S&TR, May 2001. Lawrence Livermore National Laboratory (11 September 2000). <http://www.llnl.gov/str/May01/Hall.html>.

Hall, J., Dietrich, F., Logan, C., & Rusnak, B. (2001). Recent results in the development of fast neutron imaging techniques. Lawrence Livermore National Laboratory LLNL Report # UCRL-MI-140345. Available from https://e-reports-ext.llnl.gov/pdf/240834.pdf.

Henke, B. L., Gullickson, E. M., & Davis, J. C. (1993). X-ray interactions: photoabsorption, scattering, transmission, and reflection at $E = 50-30000$ eV, $Z = 1-92$. *Atomic Data and Nuclear Data Tables*, *54*(2), 181–342 http://henke.lbl.gov/optical_constants/

Hubbell, J. H., Gimm, H. A., & Overbo, I. (1980). Pair, triplet and total atomic cross sections (and mass attenuation coefficients) for 1 MeV–100 GeV photons in elements $Z = 1$ to 100. *Journal of Physical and Chemical Reference Data*, *9*, 1023–1147 http://physics.nist.gov/PhysRefData

IAEA (2006). Database of prompt gamma rays from slow neutron capture for elemental analysis. *Technical report series*. International Atomic Energy Agency (IAEA), Vienna

Jackson, J. D. (1975). *Classical electrodynamics*. New York, NY: Wiley. Chapter 14.

Jarlskog, G., Jonsson, L., Prunster, S., Schulz, H. D., Willutzki, H., & Winter, G. G. (1973). Measurement of Delbrück scattering and observation of photon splitting at high. energies. *Physical Review D*, *8*, 3813.

Kardjilov, N. (2003). Further developments and applications of radiography and tomography with thermal and cold neutrons. PhD dissertation, Technical University of Munich (April 28, 2003).

Kramers, H. A. (1927). La diffusion de la lumiére par les atomes Atti del Congresso Internazionale dei Fisici, Como-Pavia-Roma, Vol. 2 (Bologna: Nicola Zanichelli) (pp. 545–557) (English translation in Ter Haar, D. (1998). *Master of modern physics: The scientific contributions of H. A. Kramers*. Princeton University Press (Appendix D)).

Kronig, R. (1926). On the theory of dispersion of X-rays. *Journal of Optical Society of America and Review of Scientific Instruments*, *12*, 547–557.

Kronig, R. (1942). Algemeene theorie der diëlectrische en magnetische verliezen (General theory of dielectric and magnetic losses). *Ned. Tijdschr. Natuurkd*, *9*, 402–409.

Lal, J., Lang, E., Carpenter, J., Geltenbort, P., Cremer, T., Shimizu, H., et al. (2007). Unpublished results of material and magnetic CRL imaging experiment at Pf2 ILL beam line in August 2007.

Landau, L. M., & Lifshitz, E. M. (1975). *The classical theory of fields* (4th ed., Chapter 21). New York, NY: Pergamon Press. Chapter 21.

Lengeler, B., Schroer, C., Tümmler, J., Benner, B., Richwin, M., & Snigirev, A., et al. (1999). Imaging by parabolic refractive lenses in the hard X-ray range. *Journal of Synchrotron Radiation*, *6*(part 6), 1153–1167.

Lengeler, B., Schroer, C. G., Benner, B., Günzler, T. F., Kuhlmann, M., & Tummler, J., et al. (2001). Parabolic refractive X-ray lenses: a breakthrough in X-ray optics. *Nuclear Instruments and Methods in Physics Research A*, *467–468*(Part 2), 944–950.

Marion, J. B. (1965). *Classical electromagnetic radiation*. New York, NY: Academic Press. Chapter 7.

Marsh, R.A., Albert, F., Anderson, S.G., Gibson, D.J., Wu, S.S., Hartemann, F.V., et al. (2012). Ultracompact Technology for a Next-Generation Gamma-Ray Source. Lawrence Livermore National Laboratory Proceedings. LLNL-PROC-557854.

Maxwell, J. C. (1954). *A treatise on electricity and magnetism* (Vols. 1 & 2). New York, NY: Dover.

Maxwell, J. C. (1991). *Matter and motion* (Vols. 1 & 2). New York, NY: Dover.

Meitner, L., Kösters, H., (with comment by Delbrück, M). (1933). Über die Streuung kurzwelliger Gamma-Strahlen. Z. Phys., *84*, 137.

Morse, P. M., & Feshbach, H. (1953). *Methods of theoretical physics*. New York: McGraw-Hill. pp. 372, 944.

Newton, R. G. (1976). Optical theorem and beyond. *American Journal of Physics*, *44*(7), 639–642.

Park, H., Cremer, J. T., Piestrup, M. A., Gary, C. K., Hjelm, R. P., Sellyey, W. C. L. J., & Pantell, R. H. (2006). Measured operational neutron energies of compound refractive lenses. *Nuclear Instruments and Methods in Physics Research, B 251*, 507–511.

Panofsky, W. K. H., & Phillips, M. (2005). *Classical electricity and magnetism*. New York, NY: Dover. Chapter 20.

Pereira, N. R., Arms, D. A., Clarke, R., Dierker, S. B., Dufresne, E., and Foster, D. (2001). Li metal for X-ray refractive optics. In A. M. Khounsary, & C. A. MacDonald (Eds.). *Advances in laboratory-based X-ray sources and optics II. Proceedings of SPIE, Vol. 4502* (pp. 173–183). Bellingham, Washington, DC: SPIE.

Pfeiffer, F., Weitkamp, T., Bunk, O., & David, C. (2006). Phase retrieval and differential phase-contrast imaging with low-brilliance X-ray sources. *Nature Physics*, *2*, 258–261.

Piestrup, M. A. (2004). Large area X-ray and neutron imaging using three-dimensional arrays of microlenses. *Review of Scientific Instruments*, *75*(11), 4769–4774.

Piestrup, M. A. (2006). X-ray and neutron imaging. U.S. Patent No. 6,992,313 B2, January 31, 2006.

Piestrup, M. A., Gary, C. K., Park, H., Harris, J. L., Cremer, J. T., & Pantell, H., et al. (2005). Microscope using an X-ray tube and a bubble compound refractive lens. *Applied Physics Letters*, *86*, 131104.

Piestrup, M. A., Pantell, R. H., Cremer, J. T., and Beguiristain, H. R. (2001). Compound refractive lens for X-rays. U.S. Patent No. 6,269,145 B1, July 31, 2001.

Popov, V., Degtiarenko, P., & Musatov, I. (2011). New detector for use in fast neutron radiography. *Journal of Instrumentation (IOP Science)*, *6*, C01029.

Press, W. H., Flannery, B. P., Teukolsky, S. A., & Vetterling, W. T. (1989). *Numerical recipes in FORTRAN*. Cambridge: Cambridge University Press.

Press, W. H., Flannery, B. P., Teukolsky, S. A., & Vetterling, W. T. (1992). *Numerical recipes in C*. Cambridge: Cambridge University Press.

Schroer, C. G., Kuhlmann, M., Hunger, U. T., Günzler, T. F., Kurapova, O., & Feste, S., et al. (2003). Nanofocusing parabolic refractive X-ray lenses. *Applied Physics Letters*, *82*(9), 1485–1487.

Schroer, C.G., Kuhlmann, M., Lengeler, B., Günzler, T.F., Kurapova, O., Benner, B., et al. (2002). Beryllium parabolic refractive X-ray lenses. In D. C. Mancini (Ed.). *Design and microfabrication of novel X-Ray optics. (Proceedings of SPIE, Vol. 4783* (pp. 10–18). Bellingham, Washington, DC: SPIE.

Schroer, C.G., Lengeler, B., Benner, B., Günzler, F., Tümmler, J., Drakopoulos, M., et al. (2000). A microscope for hard X-rays based on parabolic compound refractive lenses. In W. Meyer-Ilse, T. Warwick, & D. Attwood (Eds.). *X-ray microscopy: Proceedings of the sixth international conference. AIP conference proceedings, Vol. 507* (pp. 340–345).

Schumacher, M., Borchert, I., Smend, F., & Rullhusen, P. (1975). Delbrück scattering of 2.75 MeV photons by lead. *Physics Letters B*, *59*(2), 134–136.

Schwartz, M. (1972). *Principles of electrodynamics*. New York, NY: Dover. Chapter 6.

Schweber, S. S., Bethe, H. A., & de Hoffman, F. (1955). *Mesons and fields*. Evanston, IL: Row, Peterson.

Sears, V. F. (1989). *Neutron optics*. UK: Oxford University Press.

Shadowitz, A. (1975). *The electromagnetic field*. New York, NY: Dover. Chapter 19.

Smith, R. J. (1980). *Circuits, devices and systems: A first course in electrical engineering*. New York, NY: Wiley.

Snigirev, A., Filseth, B., Elleaume, P., Klocke, Th., Kohn, V., Lengeler, B., et al. (1997). Refractive lenses for high-energy X-ray focusing. In A. T. Macrander, & A. M. Khounsary (Eds.). *High heat flux and synchrotron radiation beamline. Proceedings of SPIE, Vol. 3151* (pp. 164–170). Bellingham, Washington, DC: SPIE.

Snigirev, A., Kohn, V., Snigireva, I., & Lengeler, B. (1996). A compound refractive lens for focusing high-energy X-rays. *Nature*, *384*, 49–51.

Snigirev, A., Snigireva, I., Kohn, V., Yunkin, V., Kuznetsov, S., & Grigoriev, M. B., et al. (2009). X-ray nanointerferometer based on Si refractive bilenses. *Applied Physics Letters*, *103*(064801), 1–4.

Sommerfeld, A. (1964). *Electrodynamics*. New York, NY: Academic Press. Chapters 29 and 30.

Stratton, J. A. (1941). *Electromagnetic theory*. New York, NY: McGraw-Hill. Chapter 8.

Tatchyn, R., & Cremer, T. (1996). Xreflection: A new graphic database and optics simulation utility at SSRL. The Stanford Synchrotron Radiation Laboratory Users Newsletter, October 1996. Menlo, Park, CA: Stanford Synchrotron Radiation Lightsource.

Toll, J. S. (1956). Causality and the dispersion relation: logical foundations. *Physical Review*, *104*, 1760–1770.

Toll, J. S. (1952). The dispersion relation for light and the applications involving electron pairs. PhD dissertation. Princeton, New Jersey: Princeton University.

Wilkins, S. W. (1996) X-ray optics, especially for phase contrast imaging. U.S. Patent No. 5,802,137, September 1, 1998.

Wilkins, S. W., Gureyev, T. E., Gao, D., Pogany, A., & Stevenson, A. W. (1996). Phase-contrast imaging using polychromatic hard X-rays. *Nature*, *384*, 335–338.

2 Neutron Refractive Index in Materials and Fields

2.1 Calculation of General Refractive Decrement for Material or Magnetic Media

We calculate the refractive index of the neutron based on its potential energy of interaction with a material and/or applied field such as a magnetic field or gravitational field, see Cremer (2012a, 2012b). Consider a neutron with kinetic energy W_0 and mass m_n that propagates in a material or vacuum magnetic field region and acquires potential energy U. In the neutron trajectory from the source through the region of material or magnetic field to the detector, the total neutron energy E remains constant, and equal to the sum of its kinetic W and potential U energies. Hence, the neutron kinetic energy W_p in region with the material or vacuum magnetic field is less than its kinetic energy W_0 outside in a no-field vacuum due to its acquisition of potential energy U:

$$W_p < W_0 \tag{2.1}$$

Using the de Broglie relation between neutron wave momentum $\mathrm{p} = h/\lambda$ and wavelength λ, and given neutron kinetic energy $\mathrm{E} = \mathrm{p}^2/2\mathrm{m_n}$ with rest mass $\mathrm{m_n}$, the conservation of energy yields

$$E = \frac{1}{2m_n}\left(\frac{h}{\lambda_m}\right)^2 + U = \frac{1}{2m_n}\left(\frac{h}{\lambda}\right)^2 \tag{2.2}$$

The neutron wavelength λ_m in the material or magnetic field region is longer than its wavelength λ [Å] in the no-field vacuum. In the region with material or vacuum magnetic field, the refractive index is

$$n = \frac{\lambda}{\lambda_m} \tag{2.3}$$

Solving for λ/λ_m yields

$$n^2 = \left(\frac{\lambda}{\lambda_m}\right)^2 = 1 - \frac{U}{E} \tag{2.4}$$

Neutron and X-ray Optics. DOI: http://dx.doi.org/10.1016/B978-0-12-407164-3.00001-2

The neutron kinetic energy exceeds its relatively weak potential energy, where

$$U \ll E \tag{2.5}$$

Hence, the refractive index is

$$n = 1 - \delta \tag{2.6}$$

The dimensionless decrement is

$$\delta = \frac{U}{2E} \tag{2.7}$$

Conservation of energy gives

$$E = W_0 = W_p + U \tag{2.8}$$

The neutron kinetic energy W_0 [eV] outside the material or magnetic field for neutrons of wavelength λ [Å] is

$$W_0\,[\text{eV}] = \frac{0.082}{\lambda^2} \tag{2.9}$$

With $E = W_0$ from Eq. (2.8), the decrement δ is then

$$\delta = \frac{U}{2W_0} \tag{2.10}$$

Substitution of Eq. (2.2), via Eqs. (2.8–2.9) into Eq. (2.10), yields refractive index decrement δ. The decrement δ is expressed by the potential energy U [eV], which is acquired by a neutron of wavelength λ [Å] and rest mass energy $m_n c^2$ [MeV] = 939.57 in the material or magnetic medium, where

$$\delta = m_n U \left(\frac{\lambda}{h}\right)^2 = m_n c^2 U \left(\frac{\lambda}{hc}\right)^2 = 6.095\lambda^2 U \tag{2.11}$$

From Squires (1996) or Sears (1989), and derived in chapter 3 of Cremer (2012b) the scattering length b for a scattering system is obtained from the neutron interactive potential U via the Fermi pseudo potential:

$$b = \frac{m_n}{2\pi\hbar^2 \rho} U \tag{2.12}$$

The number density ρ of the scattering units produces potential U, and scattering length b has units of femtometers, where the total coherent cross section σ_{coh} of the scattering process is

$$\sigma_{\text{coh}} = 4\pi b^2 \tag{2.13}$$

2.2 Comparison of the Electron, Neutron, X-ray, and Light Refractive Index

We now compare the decrement and refractive index of the electron, X-ray photon, neutron, and visible light photon. First, we examine electrons propagating from an equipotential half-plane of zero voltage to an equipotential half-plane with an imposed positive or negative voltage. Upon passage into the half-plane region with the imposed positive or negative voltage, the electron acquires potential energy U [eV], via the imposed V [volt], which acts on the electron charge $q = -1.6 \times 10^{-19}$ coulomb, in MKS units, where

$$U_e\,[\text{eV}] = qV \tag{2.14}$$

The electron refractive index n_e is

$$n_e^2 = \left(\frac{\lambda}{\lambda_m}\right)^2 = 1 - \frac{qV}{E} \tag{2.15}$$

Consider a nonrelativistic electron of mass m_e that propagates at velocity v_e in a potential-free region with total energy E equal to its kinetic energy W_0:

$$E = W_0 = \frac{m_e v_e^2}{2} \tag{2.16}$$

The electron enters a region of constant electrostatic potential V in which the electrostatic potential energy U, which is acquired by the electron, is much less than its incident kinetic energy W_0:

$$qV \ll W_0 \tag{2.17}$$

The electron refractive index n_e is, then, related to its decrement δ_e by

$$n_e = 1 - \delta_e \quad \text{and} \quad qV \ll W_0 \tag{2.18}$$

With $E = W_0$, the decrement δ_e for the refraction of electrons of wavelength λ [Å] is

$$\delta_e = \frac{U}{2E} = m_e U \left(\frac{\lambda}{h}\right)^2 = m_e c^2 U \left(\frac{\lambda}{hc}\right)^2 \quad U \ll E \tag{2.19}$$

The electron mass is $1/1836$ that of the neutron mass. Hence, the decrement δ_e for an electron of wavelength λ [Å] in an applied voltage V [V], equals the neutron decrement δ, divided by 1836, where the neutron potential energy U is replaced by the applied voltage V, which acts on the electron, where

$$\delta_e = \frac{6.095\lambda^2 U}{1836} = -3.3 \times 10^{-3}\lambda^2 V \quad U \ll E \tag{2.20}$$

The refractive index for electrons in medium with applied voltages is birefringent in that the refractive index can be less than 1 or greater than 1, depending on the imposed voltage polarity. The electron refractive index exceeds 1 in regions with imposed positive voltages relative to regions of zero voltage:

$$n_e = 1 + 3.3 \times 10^{-3}\lambda^2 |V| > 1 \quad \text{electron } V > 0 \tag{2.21}$$

Electrons, which pass from a half-plane at zero potential to a half-plane with an applied positive voltage $V > 0$, are refracted from a larger incident angle θ_1 to a smaller refracted angle θ_2. These angles are measured relative to the surface normal vector (perpendicular vector) of the slanted (diagonal) interface, where $\theta_2 < \theta_1$. However, the electron refractive index is less than 1 in regions with imposed negative voltages relative to regions of zero voltage, where

$$n_e = 1 - 3.3 \times 10^{-3}\lambda^2 |V| < 1 \quad \text{electron } V < 0 \tag{2.22}$$

Electrons, which pass from a half-plane at zero potential to a half-plane with an applied negative voltage $V < 0$, are refracted from a smaller incident angle θ_1 to a larger refracted angle θ_2 where $\theta_2 > \theta_1$.

In contrast, consider a neutron ray or X-ray, which is incident at an angle θ_1 to the surface normal of a planar, slanted interface between vacuum, and a half-plane of material. The ray is then refracted to a larger angle θ_2 relative to the surface normal, where $\theta_2 > \theta_1$. For example, consider a beam of neutrons or X-rays propagating parallel to the base of a prism in which the beam encounters the first slanted (diagonal) face of the prism. The neutrons and X-rays are deflected in a direction away from the base of a material prism because the decrement is positive in materials $\delta > 0$, and thus the neutron refractive index and the X-ray refractive index n is less than 1 for materials, where

$$n = 1 - \delta < 1 \quad \text{neutron and } X\text{-ray} \tag{2.23}$$

However, there are exceptions for neutrons (namely, titanium, manganese, and hydrogen) for which the neutron refractive index $n > 1$ because δ is negative ($\delta < 0$, see Cremer (2012a).

The decrement magnitudes for X-rays and neutrons are less than 1 and approach 1 only except at the longest wavelengths for X-rays and neutrons. For example, the 4 Å X-ray decrement δ of beryllium is about 0.186. In comparison, the neutron decrement magnitude of beryllium is 2.5×10^{-5} at 4 Å and is about 1.25×10^{-2} at 90.5 Å.

The decrement magnitudes decrease with increasing energy (frequency) of the X-ray or photon. The decrements δ for ultracold neutrons can become significantly large, but they remain less than 1; that is, $1 \geq \delta \gg 0$. For example, Frank (1987) reported ultracold neutrons in amino acids with neutron velocities of 5 m/s (791 Å) experienced a refractive index n of 0.57 in tryptophan ($C_4NO_3H_4$) and 0.375 in deuterated tryptophan, which corresponds to decrements δ of 0.43 and 0.625.

Now we consider the refraction of visible light. Glass (SiO_2) has a refractive index n greater than 1 for visible light, where

$$n^2 = 1 + \delta > 1 \quad \text{visible light} \tag{2.24}$$

In the case of visible light, the decrement is relatively large in comparison to 1. Assume that the material has a molecule number density N. In a glass prism, a light beam is refracted from a larger incident angle θ_1 to a smaller refracted angle θ_2 relative to the normal prism surface, where $\theta_2 < \theta_1$, such that the light beam is deflected toward the base of the glass prism.

For photons in the spectrum from infrared light to X-rays, the refractive index is a result of bound atomic electron dipole oscillators, which have natural oscillation frequencies w_0 and damping constants γ. The bound atomic electron oscillators are then driven by an incident electromagnetic wave of frequency w. One can model a material as a group of molecules, which are comprised of atoms with bound atomic electron oscillators. The electrons of an atom or of the molecule as a whole can be divided into a set of J oscillators. Each jth oscillator has oscillation strength f_j, natural oscillation frequency w_{0j}, and damping constant γ_j. Given the electron mass m_e, charge q, and vacuum permittivity ε_0, the decrement for infrared light, visible light, ultraviolet light, and X-rays is then

$$\delta(w) = \frac{Nq^2}{\varepsilon_0 m_e} \sum_{j=1}^{J} \frac{f_j}{w_{0j}^2 - w^2 + i\gamma_j w} \quad \text{photons} \tag{2.25}$$

where

$$n^2(w) = 1 + \delta(w) \quad \text{photons} \tag{2.26}$$

The J oscillator strengths f_j sum to 1:

$$\sum_{j=1}^{J} f_j = 1 \tag{2.27}$$

The refractive index for photons of infrared to X-ray wavelengths is then

$$n(w) = \sqrt{1 + \frac{Nq^2}{\varepsilon_0 m_e} \sum_{j=1}^{J} \frac{f_j}{w_{0j}^2 - w^2 + i\gamma_j w}} \quad \text{photons} \tag{2.28}$$

For $\delta \ll 1$,

$$n(w) = 1 + \frac{1}{2}\frac{Nq^2}{\varepsilon_0 m_e}\sum_{j=1}^{J}\frac{f_j}{w_{0j}^2 - w^2 + \mathrm{i}\gamma_j w} \quad \text{photons} \tag{2.29}$$

As the frequency w of the incident electromagnetic radiation increases above the set of natural oscillation frequencies w_{0j}, where $\delta \ll 1$, the decrement goes from positive decrement $\delta > 0$, where $w_{0j} > w$ with positive denominators for $j \leq J$, to a negative decrement $\delta < 0$, where $w_{0j} < w$ with negative denominators for $j \leq J$. Between infrared and visible light, visible light and ultraviolet light, and ultraviolet and X-rays is a range of frequencies, termed the *anomalous dispersion region.* In the anomalous dispersion region, the decrement goes from positive to negative as the frequency increases past the center frequency or resonant frequency w_R at the center of the anomalous dispersion region.

At the resonant frequency w_R, the electron oscillators resonate, and the resulting zero denominator would cause an infinite amplitude of oscillation, but the zero denominator is prevented by the damping term $\mathrm{i}\gamma_j$. There is strong absorption of the incident photon at the resonant frequency w_R, where the damping term is dominant, so the decrement δ becomes 0, and the refractive index n becomes 1. The anomalous dispersion frequency band is thus an absorption band for the photons. Outside the anomalous region, one can ignore the absorption term $\mathrm{i}\gamma_j w$, where

$$w_{0j}^2 - w^2 \gg \mathrm{i}\gamma_j w \tag{2.30}$$

However, at the center resonant frequency w_R of an anomalous dispersion region

$$\delta(w_R) = -\frac{\mathrm{i}}{2}\frac{Nq^2}{\varepsilon_0 m_e}\sum_{j=1}^{J}\frac{f_j}{\gamma_j w} = 0 \quad \text{photons} \tag{2.31}$$

where

$$w_{0j}^2 = w_R^2 \quad j = 1, 2, \ldots, J \tag{2.32}$$

and

$$n(w_R) = 1 \tag{2.33}$$

In an anomalous dispersion region, the refractive index $n(w)$ decreases with increasing photon frequency w, where

$$\frac{\mathrm{d}n}{\mathrm{d}w} < 0 \quad \text{anomalous } w \text{ dispersion region} \tag{2.34}$$

In an anomalous dispersion band of photon absorption, the decrement δ decreases from its maximum positive value at the anomalous band minimum frequency $w_{\min}$ to $\delta = 0$ at the center, resonant frequency w_R, and then δ decreases to its maximum negative value at the anomalous band maximum frequency $w_{\max}$. The normal dispersion region is the range of frequencies between the maximum frequency $w_{\max}$ of a lower anomalous band and the minimum frequency $w_{\min}$ of the next higher frequency, upper anomalous band. In the dispersion region of photon frequencies, photon absorption is negligible compared to photon refraction.

In the normal dispersion region between a lower frequency anomalous dispersion region and a next higher frequency anomalous dispersion region, the refractive index $n(w)$ increases with frequency w, where

$$\frac{\mathrm{d}n}{\mathrm{d}w} > 0 \quad \text{normal } w \text{ dispersion region} \tag{2.35}$$

In the normal dispersion regions of visible and ultraviolet light, the decrement δ increases with increasing photon frequency w from the negative-maximum value ($\delta_- \leq \delta < 0$) through $\delta = 0$ to the positive-maximum value $\delta_+ \geq \delta \geq 0$. The corresponding refractive index $n(w)$ increases from $1 > n > 0$ to $n > 1$ as photon frequency w increases from the uppermost frequency $w_{\max}$ of the lower anomalous band to the lowermost frequency $w_{\min}$ of the upper anomalous band. For infrared and radio waves down to the zero photon frequency, the decrement δ is positive, and the index of refraction n lies between 1 and 2. There is an anomalous, absorptive, dispersion region between infrared and radio waves, where the refractive index decreases with the increasing photon frequency, $\mathrm{d}n/dw < 0$. However, the decrement remains positive ($\delta > 0$), and the refractive index is greater than 1 ($n > 1$) in the anomalous band between radio waves and infrared.

In the case of X-rays at frequencies w above the maximum frequency $w_{\max}$ of the anomalous band between ultraviolet and X-rays, $w > w_{\max}$, the positive decrement $\delta(w)$ decreases in magnitude toward zero. Correspondingly, the X-ray refractive index $n(w)$ increases and asymptotically approaches 1 as the X-ray photon frequency w increases to infinity, where

$$\frac{\mathrm{d}n}{\mathrm{d}w} > 0 \quad \text{and} \quad \lim_{w \to \infty} n(w) \to 1 \quad X\text{-rays} \tag{2.36}$$

The refractive index of neutrons has a different physical basis than that of X-rays. Neutrons are coherently scattered by the strong nuclear force as well as by the atomic magnetic and electric fields. However, as with the X-rays, the decrement magnitude $|\delta|$ of neutrons decreases toward zero with increasing neutron energy (frequency). As with X-rays, the neutron refractive index approaches 1 as the neutron frequency increases toward infinity; that is, the neutron wavelength decreases toward zero.

2.3 Neutron Decrement for Composite Materials, and Neutron Refraction Due to Decrement Gradient

The neutron is deflected and thus refracted by the force arising from the negative gradient of the potential energy of interaction of the neutron with a source, where the source and the neutron share a common attribute, such as mass or magnetic dipole moment. The effective force imposed on the neutron of wavelength λ [Å] by the potential energy gradient ∇U of the refractive medium is

$$\mathbf{F} = -\nabla U \tag{2.37}$$

or

$$\mathbf{F} = -\frac{\nabla \delta}{6.095\lambda^2} \tag{2.38}$$

Consider mediums, which are comprised of magnetized or nonmagnetized materials or of applied magnetic fields, a nonuniform medium, such as a gradient index (GRIN) lens or a mirage in visible light optics, or more traditionally, the interface between two uniform media. These medium gradients produce a decrement gradient $\nabla\delta$, from which arises a wavelength-dependent, deflection force $\mathbf{F}$, which refracts the passing neutrons. Gravity, acting upon the neutron, pulls it downward in the particle view. In the wave view, gravity creates a nonuniform refractive gradient, which causes downward refraction of the passing neutron wave.

The decrement delta δ can be expressed in terms of scattering length b [m] and the number density of scattering units ρ [#/m^3] for a specified neutron wavelength λ [m], where

$$\delta = \frac{6.095\lambda^2 \rho b}{2\pi} \tag{2.39}$$

In a material prism comprised of Q isotopes, each jth isotope with nuclei number density ρ_j [#/m^3] and nuclei bound coherent scattering length b_j [m] bestows potential energy U_j upon the neutron of mass m, where

$$U_j = \eta \rho_j b_j \tag{2.40}$$

and

$$\eta = \frac{h^2}{2\pi \cdot m} \tag{2.41}$$

The total nuclear potential energy U acquired by the neutron is

$$U = \sum_{j=1}^{Q} U_j \tag{2.42}$$

Hence, the decrement δ for materials is

$$\delta = 6.095\lambda^2 \sum_{j=1}^{Q} \frac{\rho_j b_j}{2\pi} \tag{2.43}$$

The neutron refractive index n for the composite material is expressed in terms of its decrement δ, where

$$n = 1 - \delta \tag{2.44}$$

The total decrement of the refractive index for the neutron is the sum of the decrement contributions from each of the potential energies, which is acquired by the neutron, such as the gravitational potential energy, and in materials, the material nuclear potential and material magnetization potentials, and in vacuum (air), the applied magnetic field potential.

2.4 Neutron Decrement and Refractive Index in a Gravitational Field

We now calculate the refractive index for the neutron in a gravitational field. The gradient of gravitational potential energy U_g provides a small downward force $\mathbf{F}_g$ on the neutron of mass m_n, which can impart significant downward deflections of neutrons at the longer wavelengths, but this effect is easily compensated for by using a compound material or magnetic prism, where

$$\mathbf{F}_g = -\nabla U_g = -m_n g \tag{2.45}$$

The gravitational potential energy U_g at height y above the reference floor is

$$U_g = m_n g y \tag{2.46}$$

The neutron kinetic energy W_0 is

$$W_0 = \frac{m_n v_0^2}{2} \tag{2.47}$$

This allows us to calculate the decrement δ_g for gravity, where from Eq. (2.7)

$$\delta_g = \frac{U_g}{2W_0} \tag{2.48}$$

A region with gravity has decrement δ_g that is proportional to neutron height y above the reference floor, where

$$\delta_g = \frac{U_g}{2E} = \frac{m_n g y}{m_n v^2} = \frac{g y}{v^2} \tag{2.49}$$

With unit vector $\hat{\mathbf{y}}$ in the y-direction, the gravity decrement gradient $\nabla\delta_g$ is

$$\nabla\delta_g = 6.095\lambda^2 m g\hat{\mathbf{y}} \tag{2.50}$$

The gravity force $\mathbf{F}_g$ causes the vertical, downward deflection of a neutron with zero initial vertical velocity from its initial height y_0 [mm] as it propagates horizontally along z [mm]:

$$\mathbf{F}_g = -mg\hat{\mathbf{y}} \tag{2.51}$$

This leads to a neutron vertical trajectory via the gravitational field vector acceleration $\mathbf{g}$ and potential energy U_g [eV], where at the Earth's surface

$$\mathbf{g}\,[\mathrm{mm}/s^2] = 9800 \tag{2.52}$$

The downward vertical displacement y [mm] of neutrons of wavelength λ [Å] and horizontal velocity v_0 [mm] by gravitational acceleration g that acts over neutron propagation distance z is given by

$$y = \frac{gt^2}{2} \tag{2.53}$$

where

$$t = \frac{z}{v_0} \tag{2.54}$$

The horizontal neutron velocity v_0 [mm] is

$$v_0 = \frac{3.96 \times 10^6}{\lambda} \tag{2.55}$$

The neutron vertical displacement is

$$y[\mathrm{mm}] = -3.12 \times 10^{-10} z^2 \lambda^2 \tag{2.56}$$

The neutron deflection angle ϕ, where $y \ll z$, is

$$\phi[\mathrm{rad}] = \tan\frac{y}{z} \cong \frac{y}{z} = -3.12 \times 10^{-10} z\lambda^2 \tag{2.57}$$

With $y \ll z$, its deflection angle axial gradient is

$$\frac{d\phi}{dz}\left[\mathrm{rad/mm}\right] = -3.12 \times 10^{-10}\lambda^2 \quad (2.58)$$

With substitution for y from Eq. (2.56) into Eq. (2.49), the decrement δ of the gravitational index of refraction $n = 1 - \delta$ for neutrons becomes

$$\delta = -1.95 \times 10^{-19} z^2 \lambda^4 \quad (2.59)$$

A region of gravity has a nonuniform refractive index. Gravity produces a small downward vertical deflection of the neutrons in which for $\lambda = 1.8$ Å neutrons and distance $z\,[\mathrm{mm}] = 6000$, the downward deflection of neutrons by gravity is 36.4 μm. This gravity droop can be offset by the upward neutron refraction provided a neutron prism, which can be a line of N prisms in the form of a compound refractive prism. This is similar to a line of N lenses in a compound refractive lens, which provides $2N$ refractive surfaces to compensate for the very small refractive index.

2.5 Neutron Spin and Magnetic Dipole Moment Vectors in Applied Magnetic Fields

Let us now consider the spin and the magnetic dipole moment of the neutron, which provides the basis of its refractive and scattering properties in ferromagnetic materials and applied magnetic fields. A magnetic potential energy U is acquired by a neutron with magnetic dipole moment vector $\boldsymbol{\mu}$ and spin angular momentum vector $\mathbf{S}$ in the applied vector magnetic field $\mathbf{B}$, where

$$U = -\boldsymbol{\mu} \cdot \mathbf{B} \quad (2.60)$$

The applied magnetic field is pointed in an arbitrary, reference direction, which by convention is normally the z-direction, where

$$\mathbf{B} = B_z \hat{\mathbf{z}} \quad (2.61)$$

The magnetic dipole moment vector $\boldsymbol{\mu}$ is defined as the product IA, which arises in a circular current loop of area A and radius r. A charge e of mass m moves at velocity v around the loop radius r with angular momentum L forming current I, where

$$I = \frac{ev}{2\pi \cdot r} \quad A = \pi \cdot r^2 \quad L = mvr \quad (2.62)$$

The magnitude μ of magnetic dipole moment vector $\boldsymbol{\mu}$ is, then

$$\mu \equiv IA = \frac{evr}{2} = \frac{e}{2m} \cdot mvr = \frac{e}{2m} L \tag{2.63}$$

The proportionality constant $e/2m$ between the magnetic dipole moment μ of the charged particle and its angular momentum L is the particle gyromagnetic ratio, $\gamma = e/2m$, which equals one-half the charged particle charge e to mass m ratio e/m. Current loops can be formed by (1) charged particles moving in a circle, such as electrons orbiting the nucleus that form a dipole; (2) charges spinning on axes, such as protons and electrons that form a dipole; and (3) charge-neutral particles, such as neutrons in which the equal negative and positive, spinning charge distributions do not completely overlap, thereby forming a dipole.

A magnetic dipole is formed by a current loop that gives rise to a magnetic field. The current loop magnetic dipole moment is represented by the magnetic dipole moment vector $\boldsymbol{\mu}$, which points from the south S to north N magnetic pole $\boldsymbol{\mu}(S \rightarrow N)$ of the current loop magnetic field $\mathbf{B}$. For positive current I flowing counterclockwise about the center z-axis of a circular loop, the north magnetic field N is above the loop plane along the positive z-axis, and the south magnetic pole S is below the loop plane along the negative z-axis. The magnetic dipole moment vector $\boldsymbol{\mu}$ thus points upward in the positive z-direction from the south magnetic pole S on the z-axis below the loop plane to the north N magnetic pole on the z-axis above the loop plane

The neutron has a magnetic dipole moment vector $\boldsymbol{\mu}$, which is proportional to its spin angular momentum vector $\mathbf{S}$ via the gyromagnetic ratio γ for the neutron. The gyromagnetic ratio γ proportionality constant is unique to the neutron and differs from the gyromagnetic ratio for the proton γ_p or electron γ_e, where

$$\boldsymbol{\mu} = \gamma \mathbf{S} \tag{2.64}$$

The neutron magnetic dipole moment vector $\boldsymbol{\mu}$ is antiparallel to the neutron spin vector $\mathbf{S}$, which is indicated by the negative value of the gyromagnetic ratio γ. In turn, the gyromagnetic ratio γ is a function of the Landé g factor for the neutron, where

$$\gamma = \frac{2g\mu_n}{\hbar} = -1.83 \times 10^8 \text{ rad}/(\text{T } s) \tag{2.65}$$

The negative Landé g factor g for the neutron, which is denoted by g_n, is the basic indicator of the antiparallel relation between the neutron spin $\mathbf{S}$ and magnetic dipole moment $\boldsymbol{\mu}$ vectors. The Landé g factor for the neutron g_n is

$$g_n = -1.91 \tag{2.66}$$

The other constant, μ_n, is the nuclear dipole moment or magnetron constant for proton rest mass m_p and elementary charge e. The nuclear magnetron μ_n is effectively the proton charge to mass ratio e/m_p, which is multiplied by the spin angular momentum unit $\hbar/2$. Expressed in SI units, the nuclear magnetron μ_n constant is

$$\mu_n = \frac{e\hbar}{2m_p} = 3.15 \times 10^{-8}\ \mathrm{eV/T} \tag{2.67}$$

We now examine the orientation of the neutron spin vector **S** relative to the neutron's magnetic dipole moment vector $\boldsymbol{\mu}$ under the conditions of an applied magnetic field **B**. The total neutron spin vector **S**, with total spin quantum number $s = 1/2$, has magnitude

$$|\mathbf{S}| = \hbar\sqrt{s(s+1)} = \frac{\sqrt{3}}{2}\hbar \tag{2.68}$$

In terms of the neutron spin vector **S**, or unit spin vector $\mathbf{S}/|\mathbf{S}|$, the neutron Landé g factor g_n, and nuclear magnetron μ_n constants, the neutron magnetic dipole moment vector $\boldsymbol{\mu}$ is given by

$$\boldsymbol{\mu} = \gamma\mathbf{S} = \frac{2g_n\mu_n}{\hbar}\mathbf{S} = g_n\mu_n\sqrt{3}\frac{\mathbf{S}}{|\mathbf{S}|} \tag{2.69}$$

By substituting the expression for the nuclear magnetron constant μ_n, the magnetic dipole moment vector $\boldsymbol{\mu}$ of the neutron is related to the neutron's spin vector **S** by the charge to mass ratio of the proton e/m_p multiplied by the negative-valued Landé g factor for the neutron g_n, where

$$\boldsymbol{\mu} = g_n\frac{e}{m_p}\mathbf{S} \tag{2.70}$$

Substitution of the magnitude $S = |\mathbf{S}|$ of the neutron total spin vector **S** gives the magnitude $\mu = |\boldsymbol{\mu}|$ of the neutron's moment dipole moment vector $\boldsymbol{\mu}$, where

$$|\boldsymbol{\mu}| = \mu = |g_n|\mu_n\sqrt{3} = 1.91\mu_n\sqrt{3} = \frac{\hbar|\gamma|}{2} = 1.04 \times 10^{-7}\ \mathrm{eV/T} \tag{2.71}$$

Assume that the applied magnetic field vector $\mathbf{B} = B_z\hat{\mathbf{z}}$ is pointed in the positive z-direction. The total neutron spin vector **S** has a longitudinal (azimuth) component S_z, which is either aligned in the same direction (parallel) as the magnetic field **B** for spin-up (+) neutrons, or aligned in the opposite (antiparallel) direction of the applied magnetic field **B** for spin-down (−) neutrons. The longitudinal spin component S_z has two allowed magnetic quantum numbers s_z, where

$$s_z = \pm s = \pm\frac{1}{2} \tag{2.72}$$

The two possible values of the longitudinal spin component S_z are

$$S_z = s_z\hbar = \pm\frac{\hbar}{2} \tag{2.73}$$

Relative to the magnetic field **B** direction and spin z-component S_z, the neutron total spin vector **S** is oriented at polar angle $\chi_\pm$ for spin-up (+) and spin-down (−) neutrons, where

$$\chi_\pm = \cos^{-1}\frac{S_z}{|\mathbf{S}|} = \cos^{-1}\left(\pm\frac{1}{\sqrt{3}}\right) \tag{2.74}$$

Spin-up (+) neutrons have their longitudinal spin component S_z pointed in the same direction (parallel) as the applied magnetic field $\mathbf{B} = B_z\hat{\mathbf{z}}$. The orientation angle χ_+ of total spin vector **S** of the spin-up (+) neutron is then

$$\chi_+ = \cos^{-1}\left(+\frac{1}{\sqrt{3}}\right) = 54.7^\circ \tag{2.75}$$

In contrast, spin-down (−) neutrons have their longitudinal spin component S_z oriented in the *opposite direction* (i.e., antiparallel direction) as the direction of the applied magnetic field $\mathbf{B} = B_z\hat{\mathbf{z}}$. Thus, the orientation angle χ_- of the total spin vector **S** relative to applied magnetic field vector **B** for the spin-down (−) neutrons is polar angle

$$\chi_- = \cos^{-1}\left(-\frac{1}{\sqrt{3}}\right) = 125.3^\circ \tag{2.76}$$

The polar angle $\chi_+ = 54.7^\circ$ of the spin-up (+) neutrons and the polar angle $\chi_- = 125.3^\circ$ of the opposite-directed, spin-down (−) neutrons sum to 180°.

The neutron magnetic dipole moment vector $\boldsymbol{\mu}$ orientation angle $\theta_\pm$ is related to the neutron spin vector **S** orientation $\chi_\pm$ angle for spin-up (+) and spin-down (−) neutrons by

$$\theta_\pm = 180^\circ - \chi_\pm \tag{2.77}$$

The longitudinal (azimuth) component of the magnetic dipole moment vector $\boldsymbol{\mu}$ (namely,$\mu_{||}$) is aligned parallel or antiparallel to the magnetic field direction **B**. In particular, relative to the **B** direction, the orientation angle θ_+ for spin-up (+) neutrons is

$$\theta_+ = 180^\circ - \chi_+ = 125.3^\circ \tag{2.78}$$

and relative to the magnetic field **B** direction, the magnetic dipole moment vector **μ** orientation angle θ_- for spin-down (−) neutrons is

$$\theta_- = 180^\circ - \chi_- = 54.7^\circ \tag{2.79}$$

To learn more about the polar angles of precession, see Akitt and Mann (2000) and Brehm and Mullin (1989).

The magnitude μ of the total magnetic vector dipole vector **μ** and its z-component $\mu_{||}$ are each constant and are related by

$$\mu = \sqrt{\mu_{||}^2 + \mu_{\perp}^2} \tag{2.80}$$

The perpendicular component $\mu_{\perp}$ is comprised of the x- and y-components that are not individually specified, and $\mu_{\perp}$ is determined by the specified total magnetic moment μ and its z-component $\mu_{||}$. The directions of μ and $\mu_{||}$ are defined by the direction of the magnetic field **B**. The pointing direction of $\mu_{||}$ is in the same or the opposite direction of the applied magnetic field **B** lines.

Relative to the direction of the applied magnetic field vector **B**, the polar orientation angle θ of the magnetic dipole moment vector **μ** determines its nonzero, transverse component $\mu_{\perp}$, which is perpendicular to the magnetic field vector **B**, where

$$\mu_{\perp} = \mu \sin\theta \tag{2.81}$$

The magnetic dipole moment vector **μ** has x-, y-, and z-components (μ_x, μ_y, μ_z), and each varies with position z or time t within the magnetic field region, where, by convention

$$\mu_{||} = \mu_z \quad \mu_{\perp} = \sqrt{\mu_x^2 + \mu_y^2} \tag{2.82}$$

For more information, see Cottingham and Greenwood (2001), Kittel (2005), and Liboff (2003).

2.6 Potential Energy, Force, and Decrement for Neutrons in Applied Magnetic Fields

By convention, the magnetic dipole moment vector **μ** points from the magnetic south S to north N poles, and the magnetic field vector **B** points from the magnetic north N to south S poles. The neutron magnetic dipole moment vector **μ** is oriented at polar angle θ relative to the magnetic field $\mathbf{B} = B_z\hat{\mathbf{z}}$ direction, and the neutron acquires potential energy U via its magnetic dipole moment **μ**, where

$$U = -\boldsymbol{\mu}\cdot\mathbf{B} = -|\boldsymbol{\mu}||\mathbf{B}|\cos\theta = -\mu B_z \cos\theta = -\mu_z B_z \tag{2.83}$$

The longitudinal component $\mu_{||\pm}$ or z-component μ_z (i.e., $\mu \cos \theta$) of the neutron magnetic dipole moment vector $\boldsymbol{\mu}$ for the spin-up (+) and spin-down (−) neutrons is then

$$\mu_z \equiv \mu_{||\pm} = |\boldsymbol{\mu}|\cos \theta_\pm = \mu \cos \theta_\pm = \frac{2|g_n|\mu_n}{\hbar}|\mathbf{S}|\cos \theta_\pm \tag{2.84}$$

Relative to the z-directed magnetic field $\mathbf{B} = B_z\hat{\mathbf{z}}$, the neutron magnetic dipole moment vector $\boldsymbol{\mu}$ is oriented at polar angle $\theta_\pm$, and the neutron total spin vector $\mathbf{S}$ is oriented at polar angle $\chi_\pm$ for spin-up (+) and spin-down (−) neutrons. Recall that $\theta_\pm = 180 - \chi_\pm$, so

$$\cos \theta_\pm = \cos(180 - \chi_\pm) = -\cos\chi_\pm \tag{2.85}$$

Hence, the respective z-components $\mu_{||}$ and S_z of the neutron magnetic dipole vector $\boldsymbol{\mu}$ and spin vector $\mathbf{S}$ are related by

$$\mu_z \equiv \mu_{||\pm} = -\frac{2|g_n|\mu_n}{\hbar}|\mathbf{S}|\cos\chi_\pm = -\frac{2|g_n|\mu_n}{\hbar}S_z \tag{2.86}$$

Recall that the z-component S_z of the neutron spin vector $\mathbf{S}$ is

$$S_z = s_z\hbar = \pm\frac{\hbar}{2} \tag{2.87}$$

The longitudinal component $\mu_{||\pm}$ of the magnetic dipole moment vector $\boldsymbol{\mu}$ for the spin-up (+) and spin-down (−) neutrons then becomes

$$\mu_{||\pm} = -\frac{2|g_n|\mu_n}{\hbar}S_z = \mp |g_n|\mu_n \tag{2.88}$$

With substitution, the longitudinal component $\mu_{||\pm}$ of the magnetic dipole moment vector $\boldsymbol{\mu}$ for spin-down (+) and spin-up (−) neutrons becomes

$$\mu_{||\pm} = \mp |g_n|\mu_n = \mp 6.01 \times 10^{-8} \text{ eV/T} \tag{2.89}$$

That is, the longitudinal component $\mu_{||+}$ of the magnetic dipole moment vector $\boldsymbol{\mu}$ for the spin-up (+) neutron is

$$\mu_{||+} = -|g_n|\mu_n = -6.01 \times 10^{-8} \text{ eV/T} \tag{2.90}$$

and the longitudinal component $\mu_{||-}$ of the magnetic dipole moment vector $\boldsymbol{\mu}$ for the spin-down (−) neutron is

$$\mu_{||-} = +|g_n|\mu_n = +6.01 \times 10^{-8} \text{ eV/T} \tag{2.91}$$

The potential energy $U_{\pm}$ of the spin-up (+) and spin-down (−) neutrons in the applied magnetic field **B** is

$$U_{\pm} = -\mu_{||\pm}|\mathbf{B}| = -(\mp|g_n|\mu_n)|\mathbf{B}| = \pm|g_n|\mu_n|\mathbf{B}| \tag{2.92}$$

The lower state, magnetic potential energy U_- [eV] is acquired by the spin-down (−) neutron for $\mu_{||}$ aligned parallel to **B** [T], where

$$U_- = -|g_n|\mu_n|\mathbf{B}| = -6.01 \times 10^{-8}|\mathbf{B}| \tag{2.93}$$

The upper state, magnetic potential energy U_+ [eV] is acquired by the spin-up (+) neutron for $\mu_{||}$ aligned antiparallel to **B** [T], where

$$U_+ = +|g_n|\mu_n|\mathbf{B}| = 6.01 \times 10^{-8}|\mathbf{B}| \tag{2.94}$$

Arising from the negative gradient of the magnetic potential energy $-\nabla U$, the magnetic gradient force $\mathbf{F}_{\pm}$ deflects the spin-up (+) and spin-down (−) neutrons in opposite directions, and substitution from the expression for $U_{\pm}$ yields

$$\mathbf{F}_{\pm} = -\nabla U_{\pm} = -(\pm|g_n|\mu_n)\nabla|\mathbf{B}| = \mp|g_n|\mu_n\nabla|\mathbf{B}| \tag{2.95}$$

The magnetic gradient force $\mathbf{F}_+$ acting on the higher energy spin-up (+) neutron is

$$\mathbf{F}_+ = -|g_n|\mu_n\nabla|\mathbf{B}| = -6.01 \times 10^{-8}\nabla|\mathbf{B}| \tag{2.96}$$

The magnetic gradient force $\mathbf{F}_-$ acting on the lower energy spin-down (−) neutron is

$$\mathbf{F}_- = +|g_n|\mu_n\nabla|\mathbf{B}| = 6.01 \times 10^{-8}\nabla|\mathbf{B}| \tag{2.97}$$

The magnetic field **B** decreases as the distance from the source of the magnetic field increases, so the magnetic field gradient is negative, where $\nabla|\mathbf{B}| < 0$. In a negative gradient magnetic field $\nabla|\mathbf{B}| < 0$, the spin-up (+) neutrons are deflected upward, and the spin-down (−) neutrons are deflected downward. However, in a positive gradient magnetic field $\nabla|\mathbf{B}| > 0$, the spin-up (+) neutrons are deflected downward, and the spin-down (−) neutrons are deflected upward.

In Eq. (2.95) for magnetic gradient force $\mathbf{F}_{\pm}$ acting on the neutron, the unit for $\mathbf{F}_{\pm}$ is energy in electron volts per spatial interval. As a result, the eV energy units must be converted to erg energy units (1.0 eV is replaced by 1.6×10^{-12} erg) to obtain the force $\mathbf{F}_{\pm}$ in CGS dyne units, or the eV energy units must be converted to joule energy units (1.0 eV is replaced by 1.6×10^{-19} J) to obtain the force $\mathbf{F}_{\pm}$ in MKS Newton units.

Given magnetic potential energy $U_\pm$ [eV], the decrement $\delta_\pm$ for spin-up (+) and spin-down (−) neutrons of wavelength λ [Å] in an applied magnetic field **B** is

$$\delta_\pm = 6.01\lambda^2 U_\pm = 6.01\lambda^2 \mu_{||\pm} B \tag{2.98}$$

The index of refraction $n_\pm$ for neutrons in an applied magnetic field is thus birefringent, where

$$n_\pm = 1 - \delta_\pm \tag{2.99}$$

In an applied magnetic field **B** [T], the higher energy spin-up (+) neutrons of wavelength λ [Å] have a positive decrement δ_+ and refractive index n_+ less than 1, where

$$\delta_+ = 3.67 \times 10^{-7} \lambda^2 B \tag{2.100}$$

and

$$n_+ = 1 - 3.67 \times 10^{-7} \lambda^2 B \tag{2.101}$$

The lower energy spin-down (−) neutrons of wavelength λ [Å] in an applied magnetic field **B** [T] have a negative decrement δ_- and refractive index n_- greater than 1, where

$$\delta = -3.67 \times 10^{-7} \lambda^2 B \tag{2.102}$$

and

$$n_- = 1 + 3.67 \times 10^{-7} \lambda^2 B \tag{2.103}$$

A 1.0 T applied magnetic field produces approximately the same decrement δ and refractive index $n = 1 - \delta$ for a 1.0 Å neutron as the nuclei of aluminum ($\delta_{Al} = 3.81 \times 10^{-7}$).

By comparison, both spin-up (+) and spin-down (−) neutrons in nonmagnetic materials have a positive decrement and index of refraction less than 1. However, exceptions for neutron refraction in materials include natural occurring titanium (via 73.8% Ti-48), manganese (100% Mn-55), and hydrogen (via 99.985% H-1), which have a negative decrement and refractive index greater than 1. Also, X-rays with or without an applied magnetic field have a positive decrement and an index of refraction less than 1. In contrast, the lower energy spin-down (−) neutrons have a negative decrement and a refractive index greater than 1, which is also the case for visible light.

The spin-up (+) and spin-down (−) neutrons are refracted in equal and opposite directions in an applied magnetic field. For example, if a beam of incident neutrons are propagating parallel to the base of a prism, then the higher energy spin-up neutrons are deflected upward, away from the prism base; on the other hand, the lower energy spin-down neutrons are deflected downward, toward the prism base. By comparison, both spin-up and spin-down neutrons are deflected upward in nonmagnetic material prism. With or without applied magnetic fields, X-rays are deflected upward, away from the base for magnetic or nonmagnetic material prisms. In contrast, visible light is deflected downward toward the prism base.

2.7 The Bloch Equation and Neutron Precession in an Applied Magnetic Field

In an applied magnetic field, the neutron spin vector **S** and opposite-facing magnetic dipole moment vector $\boldsymbol{\mu}$ rotates (precession) about the lines of the magnetic field direction in an applied magnetic field **B**. A neutron enters a magnetic field **B**, and its magnetic dipole moment vector $\boldsymbol{\mu}$ begins azimuth precession at the Larmor frequency Ω at polar orientation angle θ about the magnetic field direction, independent of its wavelength λ [Å] and polar angle θ, where

$$\Omega = \gamma B \tag{2.104}$$

The neutron's initial longitudinal magnetic dipole moment ($\mu_{||} = \mu \cos \theta$) is either parallel or antiparallel to the direction of the magnetic field. A spin-up (spin-down) neutron maintains its longitudinal magnetic dipole moment component $\mu_{||}$ in a fixed antiparallel (parallel) alignment with respect to the varying direction of the magnetic field lines. And in so doing, the neutron magnetic dipole moment vector $\boldsymbol{\mu}$ is kept at a constant polar orientation angle θ, that is, 125.3° for spin-up (+) and 54.7° for spin-down (−) neutrons, measured relative to the direction of the applied magnetic field vector **B**. The constant polar angle θ is maintained, provided that the temporal variation of the magnetic field w is much less than the azimuth Larmor precession frequency Ω, where this condition for adiabatic precession is given by

$$\Omega \gg w \tag{2.105}$$

A periodic magnetic field B [T], which has a spatial period p oriented in the direction of neutron propagation, presents a magnetic field that changes direction periodically with time, and produces a temporal half-period $t_{1/2}$ in the neutron rest frame. That is, in the neutron rest frame, the magnetic field changes direction every half-period $t_{1/2}$ as the neutron propagates through a periodic magnetic field with period p [mm]. The neutron realignment can occur within a Larmor precession period t_Ω [s], provided that the ratio of the magnetic field half-period $t_{1/2}$, divided by the Larmor precession period t_Ω [s], is much greater than 1, where

$$\frac{t_{1/2}}{t_\Omega} = 3.67p\lambda B \gg 1 \tag{2.106}$$

The Larmor precession period t_Ω [s] is the reciprocal of the Larmor precession frequency Ω:

$$t_\Omega\ [\mathrm{s}] = \frac{2\pi}{\Omega} \tag{2.107}$$

For more information in general about the precession of magnetic dipole vectors in an applied magnetic field, see Siegman (1971).

The Bloch equation for the magnetic dipole moment vector $\boldsymbol{\mu}$ for a neutron or other particle with a magnetic dipole moment is

$$\frac{\mathrm{d}\boldsymbol{\mu}}{\mathrm{d}t} = \gamma \mathbf{B} \times \boldsymbol{\mu} \tag{2.108}$$

or

$$\frac{\mathrm{d}\boldsymbol{\mu}}{\mathrm{d}t} = (\gamma|\mathbf{B}||\boldsymbol{\mu}|\sin\theta)\hat{\boldsymbol{\varphi}} \tag{2.109}$$

The unit vector $\hat{\boldsymbol{\varphi}}$ is the cross-product of the unit vectors of the magnetic field $\mathbf{B}/|\mathbf{B}|$ and neutron magnetic dipole moment $\boldsymbol{\mu}/|\boldsymbol{\mu}|$. The neutron magnetic dipole moment vector $\boldsymbol{\mu}$ rotates or precesses in the azimuth direction about the lines of a z-directed magnetic field $\mathbf{B} = B_z\hat{\mathbf{z}}$ at the Larmor precession frequency w_L, where

$$w_\mathrm{L}\ [\mathrm{rad/s}] = \pm\gamma|\mathbf{B}| = \mp 1.83 \times 10^8 |\mathbf{B}\ [\mathrm{T}]| \tag{2.110}$$

The $\pm$ refers to precession in the positive azimuth direction $+\hat{\boldsymbol{\varphi}}$ for $\sin\theta < 0$ and the negative azimuth direction $-\hat{\boldsymbol{\varphi}}$ for $\sin\theta > 0$. The precession or rotation direction of the magnetic dipole moment vector $\boldsymbol{\mu}$ about the magnetic field lines of $\mathbf{B}$ reverses ($w_\mathrm{L} \to -w_\mathrm{L}$) when the direction (polarity) of the magnetic field $\mathbf{B}$ reverses with ($B \to -B$).

The longitudinal component $\mu_{||}$ of the magnetic dipole moment $\boldsymbol{\mu}$ of the spin-up (+) neutron is pointed at polar angle $\theta = 125.3^\circ$ with respect to the applied magnetic field $\mathbf{B} = B_z\hat{\mathbf{z}}$. Hence, the spin-up (+) neutron magnetic dipole moment vector $\boldsymbol{\mu}$ undergoes counterclockwise precession about lines of $\mathbf{B} = B_z\hat{\mathbf{z}}$ in the positive azimuth angle direction $+\hat{\boldsymbol{\varphi}}$. Conversely, the longitudinal component $\mu_{|}|$ of the magnetic dipole moment $\boldsymbol{\mu}$ of the spin-down (−) neutron is pointed at polar angle $\theta = 54.7^\circ$ with respect to the direction of the applied magnetic field $\mathbf{B} = B_z\hat{\mathbf{z}}$. Hence, the spin-down (−) neutron magnetic dipole moment vector $\boldsymbol{\mu}$ undergoes clockwise precession about lines of $\mathbf{B} = B_z\hat{\mathbf{z}}$ in the negative azimuth angle direction $-\hat{\boldsymbol{\varphi}}$.

2.8 Temperature Effect on Neutron Spin and Magnetic Dipole Moment Orientation in an Applied Magnetic Field

The thermal energy or temperature of the neutrons determines the relative population of spin-up and spin-down neutrons in an applied magnetic field. Independent of angle θ between μ and **B**, as first demonstrated by the 1922 Stern–Gerlach experiment, a beam of N neutrons is quantized into one of two states with N_1 and N_2 neutrons, respectively, where

$$N = N_1 + N_2 \tag{2.111}$$

The lower energy state has N_1 spin-down (−) neutrons at lower energy U_1 with $\mu_{||}$ aligned in a direction parallel to the magnetic field **B** [T], where

$$U_1\ [\text{eV}] = -6.01 \times 10^{-8} |\mathbf{B}| \tag{2.112}$$

The higher energy state has N_2 spin-up (+) neutrons at higher energy U_2 with $\mu_{||}$ aligned antiparallel to **B**, where

$$U_2\ [\text{eV}] = 6.01 \times 10^{-8} |\mathbf{B}| \tag{2.113}$$

In thermal equilibrium at temperature T [K], the neutrons have thermal energy kT with the Boltzmann constant k given in convenient units:

$$k = 8.62 \times 10^{-5}\ \text{eV/K} \tag{2.114}$$

The ratio $r_{||}$ of the number of higher energy spin-up (+) neutrons N_2, divided by the number of lower energy spin-down (−) neutrons N_1, is obtained from the Boltzmann equation

$$r_{||} = \frac{N_2}{N_1} = \exp\left(-\frac{U_{21}}{kT}\right) \tag{2.115}$$

The energy difference between the spin-up (+) to spin-down (−) neutron states is

$$U_{21} = U_2 - U_1 = 1.2 \times 10^{-7} |\mathbf{B}| \tag{2.116}$$

Substitution gives the ratio $r_{||}$ of spin-up (+) to spin-down (−) neutron populations, where

$$r_{||} = \exp\left(-1.39 \times 10^{-3} \frac{B}{T}\right) \tag{2.117}$$

For laboratory magnetic fields (1−10 T) and room temperature $T = 298$ K, the neutron beam in the magnetic field is almost evenly populated by spin-down (−) and spin-up (+) neutrons, with population ratio $r_{||}$ near unity, where

$$r_{||} = 0.99999 \quad T = 298\ \text{K} \quad B < 10\ \text{T} \tag{2.118}$$

At sufficiently low temperatures and high magnetic fields, almost all the neutrons will be in the lower energy spin-down (−) state, with

$$r_{||} \cong 0 \quad T \cong 0\ \text{K} \quad B < 10\ \text{T} \tag{2.119}$$

2.9 The Bloch Equation and the Lorentz Force Equation

We now compare (1) the Bloch equation for the trajectory of the neutron spin vector in its precession in an applied magnetic field, and (2) the Lorentz force equation for the trajectory of a charged particle in an applied electric and magnetic field. One finds they are similar differential equations. The Bloch equation for precession of the magnetic dipole moment vector $\boldsymbol{\mu}$ of a neutron or other particle, with a magnetic dipole moment in an applied magnetic field $\mathbf{B}$, is

$$\frac{d\boldsymbol{\mu}}{dt} = \gamma \mathbf{B} \times \boldsymbol{\mu} \tag{2.120}$$

The Lorentz force equation for the velocity vector $\mathbf{v}$ of a charged particle is

$$\frac{d\mathbf{v}}{dt} = \frac{e}{n}(\mathbf{v} \times \mathbf{B}) \tag{2.121}$$

Interestingly, the Bloch equation solution for the magnetic dipole moment vector $\boldsymbol{\mu}$ and the Lorentz force equation for velocity vector $\mathbf{v}$ of a charged particle are not just similar—these are identical.

Consider a charged particle of mass m with charge e and gyromagnetic ratio γ. The particle gyromagnetic ratio γ is proportional to the particle's Landé g factor. In the Lorentz force equation, the charge to mass ratio e/m of the charged particle is set equal to the charged particle gyromagnetic ratio γ, and then the particle velocity vector variable $\mathbf{v}$ is replaced by the particle magnetic dipole moment vector $\boldsymbol{\mu}$. That is, in the Lorentz force equation,

$$\frac{e}{m} \to -\gamma \quad \text{and} \quad \mathbf{v} \to \boldsymbol{\mu} \tag{2.122}$$

The trajectory solution of the Lorentz force equation is now the solution of the particle magnetic moment vector $\boldsymbol{\mu}$ rather than the particle velocity $\mathbf{v}$. Thus, if there are existing Lorentz force equation solutions for the charge particle vector $\mathbf{v}$

trajectory in an applied magnetic field optic, then one can immediately obtain the trajectory of the magnetic dipole moment vector $\boldsymbol{\mu}$ of the charged particle. For example, the ratio e/m in the Lorentz force equation solution for the velocity vector $\mathbf{v}$ of a ^{3}He nucleus through the magnetic field can be replaced by the negative of the ^{3}He gyromagnetic ratio $(-\gamma)$ to obtain the trajectory of the ^{3}He magnetic dipole moment vector $\boldsymbol{\mu}$.

The proton and neutron gyromagnetic ratios γ differ due to the differing Landé g factors, where

$$\gamma = \frac{2g\mu_n}{\hbar} = \frac{ge}{m_p} \quad g = -1.91 \text{ neutrons} \quad g = 2.79 \text{ protons} \tag{2.123}$$

The magnetic dipole moment vector $\boldsymbol{\mu}$ is, then, related to the spin vector $\mathbf{S}$ by

$$\boldsymbol{\mu} = \gamma\mathbf{S} = \frac{2g\mu_n}{\hbar}\mathbf{S} \tag{2.124}$$

The nuclear dipole moment or magnetron μ_n is given by the proton rest mass m_p and elementary charge e.

The electron magnetic dipole moment vector $\boldsymbol{\mu}_e$ is directed opposite (180°) to the direction of the electron spin vector $\mathbf{S}$, and has a similar expression to that of the proton, neutron, and nuclei. However, the (electron) Bohr magnetron μ_b is used in place of the nuclear magnetron or nuclear dipole moment or magnetron μ_n so that

$$\boldsymbol{\mu}_e = \frac{g_e\mu_\mathrm{b}}{\hbar}\mathbf{S} = -\frac{g_e}{2}\frac{e}{m_e}\mathbf{S} = -\frac{e}{m_e}\mathbf{S} \tag{2.125}$$

The Bohr magnetron μ_b for elementary charge e and electron rest mass m_e and the Landé g factor g_e for the electron are

$$g_e = -2 \quad \text{and} \quad \mu_\mathrm{b} = \frac{\hbar}{2}\frac{e}{m_e} \tag{2.126}$$

Hence, the gyromagnetic ratio γ_e for the electron is

$$\gamma_e = -\frac{e}{m_e} \tag{2.127}$$

The electromagnetic energy loss from neutron magnetic dipole precession in 1–10 T applied magnetic fields is negligible; see Stump and Pollack (1998). Also, the neutron dipole–lattice T1 longitudinal energy decay with surrounding structures and the T2 dipole–dipole transverse dephasing decay between neutrons in the beam are negligible. This is due to the relatively low, attainable neutron beam fluxes and relatively weak, charge-neutral, neutron–neutron interactions, as compared to the large, attainable electron and proton beam current densities, which have electric charge interaction.

2.10 Average Spin Polarization of a Neutron in an Applied Magnetic Field

A neutron beam can be considered as a mixed state of spin-up (+) and spin-down (−) neutrons, which is represented by the spin component wave function ψ_s of the total neutron wave function amplitude ψ. The wave function spin component ψ_s is the superposition of the wave function spin components ψ_{s+} and ψ_{s-} of the spin-up (+) and spin-down (−) neutrons.

The two neutron spin states, also known as spinors, are given by

$$\psi_{s+} = u\begin{bmatrix} 1 \\ 0 \end{bmatrix} \quad \text{and} \quad \psi_{s-} = d\begin{bmatrix} 0 \\ 1 \end{bmatrix} \tag{2.128}$$

The neutron spin wave function ψ_s represented in its spinor form is

$$\psi_s = \begin{bmatrix} \psi_{s+} \\ \psi_{s-} \end{bmatrix} = u\begin{bmatrix} 1 \\ 0 \end{bmatrix} + d\begin{bmatrix} 0 \\ 1 \end{bmatrix} \tag{2.129}$$

The spin states ψ_{s+} and ψ_{s-} have respective amplitudes u and d given by

$$u = \cos\theta \quad \text{and} \quad d = \sin\theta\,\exp(\mathrm{i}\phi) \tag{2.130}$$

The normalized wave function spin component ψ_s is then

$$\psi_s = \frac{u\psi_{s+} + d\psi_{s-}}{|\psi_s|^2} = \cos(\theta)\psi_{s+} + \sin(\theta)\exp(\mathrm{i}\phi)\psi_{s-} \tag{2.131}$$

Substitution for ψ_{s+} and ψ_{s-} gives

$$\psi_s = \cos\theta\begin{bmatrix} 1 \\ 0 \end{bmatrix} + \sin\theta\,\exp(\mathrm{i}\phi)\begin{bmatrix} 0 \\ 1 \end{bmatrix} = \begin{bmatrix} \cos\theta \\ \sin\theta\,\exp(\mathrm{i}\phi) \end{bmatrix} \tag{2.132}$$

The wave function spin component ψ_s normalizes to unity, where

$$|\psi_s|^2 = \psi_s\psi_s^* \begin{bmatrix} \cos\theta & \sin\theta\,\exp(\mathrm{i}\phi) \end{bmatrix}\begin{bmatrix} \cos\theta \\ \sin\theta\,\exp(-\mathrm{i}\phi) \end{bmatrix} = 1 \tag{2.133}$$

The probability of spin-up (+) and spin-down (−) neutrons in the neutron beam are P_+ and P_-, respectively, where

$$P_+ = \frac{u^*u}{|\psi_s|^2} = |u|^2 = \cos^2\theta \quad \text{and} \quad P_- = \frac{d^*d}{|\psi_s|^2} = |d|^2 = \sin^2\theta \tag{2.134}$$

The Bloch sphere representation of the general spin state of a neutron allows representation of the expected value of the neutron spin state to be described by the neutron spin polarization vector $\mathbf{P}_s$. The spin polarization vector $\mathbf{P}_s$ of a neutron spin state ψ_s equals the expectation value $\langle\hat{\mathbf{S}}\rangle$ of the neutron spin vector $\hat{\mathbf{S}}$, which is shown as follows. The expectation value $\langle\hat{\mathbf{S}}\rangle$ of the neutron spin vector operator $\hat{\mathbf{S}}$ of the beam of spin-up (+) and spin-down (−) neutrons can be expressed in terms of the Pauli spin matrix, vector operator $\hat{\boldsymbol{\sigma}}$. Given neutron spin quantum number $s = 1/2$, the neutron spin vector matrix operator $\hat{\mathbf{S}}$ and Pauli spin vector, matrix operators $\hat{\boldsymbol{\sigma}}$ are related by

$$\hat{\mathbf{S}} = \frac{\hbar}{2}\hat{\boldsymbol{\sigma}} \quad \text{and} \quad \hat{S}\psi_s = (\hat{\mathbf{S}}\cdot\hat{\mathbf{S}})\psi_s = s(s+1)\begin{bmatrix} 1 & 0 \\ 0 & 1 \end{bmatrix}\psi_s \tag{2.135}$$

where

$$\hat{S}^2\psi_s = \frac{\hbar}{2}\hat{\sigma}^2\psi_s = \frac{\hbar}{2}(\hat{\boldsymbol{\sigma}}\cdot\hat{\boldsymbol{\sigma}})\psi_s = \frac{\hbar}{2}(\hat{\sigma}_x^2 + \hat{\sigma}_y^2 + \hat{\sigma}_z^2)\psi_s = \frac{3}{4}\hbar^2\begin{bmatrix} 1 & 0 \\ 0 & 1 \end{bmatrix} \tag{2.136}$$

The Pauli spin matrix operator $\hat{\boldsymbol{\sigma}}$ and its x-, y-, and z-component operators $\hat{\sigma}_x$, $\hat{\sigma}_y$, and $\hat{\sigma}_z$ for the x, y, and z directions, which are denoted by unit vectors $\mathbf{i}$, $\mathbf{j}$, and $\mathbf{k}$, are

$$\hat{\boldsymbol{\sigma}} = \hat{\sigma}_x\mathbf{i} + \hat{\sigma}_y\mathbf{j} + \hat{\sigma}_z\mathbf{k} = \begin{bmatrix} 0 & 1 \\ 1 & 0 \end{bmatrix}\mathbf{i} + \begin{bmatrix} 0 & -\mathrm{i} \\ \mathrm{i} & 0 \end{bmatrix}\mathbf{j} + \begin{bmatrix} 1 & 0 \\ 0 & -1 \end{bmatrix}\mathbf{k} \tag{2.137}$$

The general spin state ψ_s is a vector from the origin to a general point on a Bloch sphere of unit radius with polar angle θ. The polar angle θ is measured relative to the north pole. The azimuth angle ϕ rotates about the axis connecting the north and south poles. The neutron spin-up (+) state [1, 0] is the north pole N and the spin-down (−) state [0, 1] is the south pole S. The expected value $\langle S_z\rangle$ of the azimuth (longitudinal) spin component S_z in the z-direction is obtained via the Pauli spin matrix σ_z, where

$$\langle S_z\rangle = \langle\psi|\sigma_x|\psi\rangle = \begin{bmatrix} u^* & d^* \end{bmatrix}\begin{bmatrix} 1 & 0 \\ 0 & -1 \end{bmatrix}\begin{bmatrix} u \\ d \end{bmatrix} = \frac{|u|^2 - |d|^2}{u^2 + d^2} \tag{2.138}$$

The expectation value $\langle\hat{\boldsymbol{\sigma}}\rangle$ of the Pauli spin matrix operator $\hat{\boldsymbol{\sigma}}$ is the spin polarization vector $\mathbf{P}_s$, which points from the origin to a particular point on the Bloch sphere. That is, the polarization vector $\mathbf{P}_s$ represents expected spin state $\langle\hat{\boldsymbol{\sigma}}\rangle$ of the neutron given its two possible spin-up (+) state ψ_{s+} and spin-down (−) state ψ_{s-}, where

$$\mathbf{P}_s = \langle\hat{\boldsymbol{\sigma}}\rangle = \langle\psi_s|\hat{\boldsymbol{\sigma}}|\psi_s\rangle \tag{2.139}$$

where

$$\langle\psi_s|\hat{\boldsymbol{\sigma}}|\psi_s\rangle = \mathbf{i}\langle\psi_s|\hat{\sigma}_x|\psi_s\rangle + \mathbf{j}\langle\psi_s|\hat{\sigma}_y|\psi_s\rangle + \mathbf{k}\langle\psi_s|\hat{\sigma}_z|\psi_s\rangle \tag{2.140}$$

or

$$\langle\psi_s|\hat{\boldsymbol{\sigma}}|\psi_s\rangle = \int_{-\infty}^{\infty} \mathrm{d}z \int_{-\infty}^{\infty} \mathrm{d}y \int_{-\infty}^{\infty} \mathrm{d}x[\psi_s^*(\mathbf{i}\hat{\sigma}_x + \mathbf{j}\hat{\sigma}_y + \mathbf{k}\hat{\sigma}_z)\psi_s] \tag{2.141}$$

Integration of Eq. (2.141) yields the Pauli spin vector expected value $\langle\boldsymbol{\sigma}\rangle$ (or individual neutron polarization vector $\mathbf{P}_s$), which is measured relative to an arbitrary, reference z-axis of the Bloch sphere via polar angle θ and azimuth angle ϕ, where

$$\mathbf{P}_s = \langle\boldsymbol{\sigma}\rangle = \frac{2}{\hbar}\langle\mathbf{S}\rangle = \mathbf{i}\sin\theta\cos\phi + \mathbf{j}\sin\theta\sin\phi + \mathbf{k}\cos\theta \tag{2.142}$$

where

$$|\mathbf{P}_s| = 1 \tag{2.143}$$

The neutron spin polarization vector $\mathbf{P}_s$ vector is the unit radius vector of the Bloch sphere.

2.11 Equation of Motion of the Expected Value of the Neutron Spin Vector in an Applied Magnetic Field

One can obtain the equation of motion of the expected value $\langle\mathbf{S}\rangle$ of the neutron spin vector via the Hermitian density matrix operator $\hat{\rho}$ for this two-state neutron spin wave function ψ_s. The density matrix for a wave function ψ with N possible eigenstates ψ_n is defined by

$$\hat{\rho} \equiv \sum_{n=1}^{N} P_n|\psi_n\rangle\langle\psi_n| \tag{2.144}$$

where

$$\mathrm{Trace}[\rho] = 1 \tag{2.145}$$

The expected value $\langle A\rangle$ for wave function ψ is given by the trace of the product of the density matrix operator $\hat{\rho}$ and matrix operator $\hat{A}$, where

$$\langle A\rangle = \mathrm{Trace}(\hat{\rho}\hat{A}) = \mathrm{Trace}(\hat{A}\hat{\rho}) \tag{2.146}$$

Differentiating the density operator $\hat{\rho}$ with respect to time t gives

$$\frac{\partial\hat{\rho}}{\partial t} = \sum_{n=1}^{N} P_n \left[\frac{\partial|\psi_n\rangle}{\partial t}\langle\psi_n| + |\psi_n\rangle\frac{\partial\langle\psi_n|}{\partial t}\right] \tag{2.147}$$

The time derivatives of the eigenstates $|\psi_n\rangle$ on the right side of this equation can be expressed in terms of the Hamiltonian operator $\hat{H}$ via the time-dependent Schrödinger wave equations, where

$$\mathrm{i}\hbar\frac{\mathrm{d}|\psi_n\rangle}{\mathrm{d}t} = \hat{H}|\psi_n\rangle \quad \text{and} \quad -\mathrm{i}\hbar\frac{\mathrm{d}\langle\psi_n|}{\mathrm{d}t} = \langle\psi_n|\hat{H} \tag{2.148}$$

Substitution gives

$$\mathrm{i}\hbar\frac{\mathrm{d}\hat{\rho}}{\mathrm{d}t} = \sum_{n=1}^{N} P_n\left[\hat{H}|\psi_n\rangle\langle\psi_n| + |\psi_n\rangle\langle\psi_n|\hat{H}\right] \tag{2.149}$$

The right side of this equation is simply the commutation of the Hamiltonian operator $\hat{H}$ and the density operator $\hat{\rho}$, where

$$\mathrm{i}\hbar\frac{\partial\hat{\rho}}{\partial t} = \left[\hat{H},\hat{\rho}\right] = \hat{H}\hat{\rho} - \hat{\rho}\hat{H} \tag{2.150}$$

Next, the time derivative of the expected value $\mathrm{i}\hbar\langle A\rangle$ of the operator $\mathrm{i}\hbar\hat{A}$ can be expressed as

$$\mathrm{i}\hbar\frac{\mathrm{d}\langle A\rangle}{\mathrm{d}t} = \mathrm{i}h\left[\mathrm{Trace}\left(\frac{\partial\hat{\rho}}{\partial t}\hat{A}\right) + \mathrm{Trace}\left(\hat{\rho}\frac{\partial\hat{A}}{\partial t}\right)\right] \tag{2.151}$$

Substitution for $\mathrm{d}\hat{\rho}/dt$ gives

$$\mathrm{i}\hbar\frac{\mathrm{d}\langle A\rangle}{\mathrm{d}t} = \mathrm{Trace}((\hat{H}\hat{\rho} - \hat{\rho}\hat{H})\hat{A}) + \mathrm{i}h\left\langle\frac{\partial\hat{A}}{\partial t}\right\rangle = \mathrm{Trace}(\hat{\rho}(\hat{A}\hat{H} - \hat{H}\hat{A})) + \mathrm{i}h\left\langle\frac{\partial\hat{A}}{\partial t}\right\rangle \tag{2.152}$$

Hence,

$$\mathrm{i}\hbar\frac{\mathrm{d}\langle A\rangle}{\mathrm{d}t} = \langle\hat{A}\hat{H} - \hat{H}\hat{A}\rangle + \mathrm{i}h\left\langle\frac{\partial\hat{A}}{\partial t}\right\rangle \tag{2.153}$$

The individual neutron spin polarization vector $\mathbf{P}_s$ is given by the expected value $\langle \boldsymbol{\sigma} \rangle$ of the Pauli spin vector, which is independent of time, where

$$\frac{\partial \mathbf{P}_s}{\partial t} = \left\langle \frac{\partial \boldsymbol{\sigma}}{\partial t} \right\rangle = 0 \tag{2.154}$$

The equation of the motion-expected value $\langle \boldsymbol{\sigma} \rangle$ of the Pauli spin vector is then

$$\mathrm{i}\hbar \frac{\mathrm{d}\langle \boldsymbol{\sigma} \rangle}{\mathrm{d}t} = \langle \hat{\boldsymbol{\sigma}} \hat{H} - \hat{H} \hat{\boldsymbol{\sigma}} \rangle \tag{2.155}$$

The neutron acquires magnetic potential energy U due to the interaction of its magnetic moment vector with the magnetic field, where U is the dot product of the magnetic moment vector $\boldsymbol{\mu}$ with the applied magnetic field vector $\mathbf{B}$. And the neutron Hamiltonian operator $\hat{H}$ for the interaction of the neutron magnetic moment vector $\boldsymbol{\mu}$, with the applied magnetic field vector $\mathbf{B}$, is equal to the neutron magnetic potential energy U, where

$$\hat{H} = U = -\boldsymbol{\mu} \cdot \mathbf{B} \tag{2.156}$$

The neutron magnetic dipole moment vector $\boldsymbol{\mu}$ is related to the neutron spin vector $\mathbf{S}$ and Pauli spin vector $\boldsymbol{\sigma}$ by

$$\boldsymbol{\mu} = \gamma \mathbf{S} = \gamma \frac{\hbar}{2} \boldsymbol{\sigma} = \gamma \frac{\hbar}{2} \begin{bmatrix} 0 & 1 \\ 1 & 0 \end{bmatrix} \mathbf{i} + \gamma \frac{\hbar}{2} \begin{bmatrix} 0 & -\mathrm{i} \\ \mathrm{i} & 0 \end{bmatrix} \mathbf{j} + \gamma \frac{\hbar}{2} \begin{bmatrix} 1 & 0 \\ 0 & -1 \end{bmatrix} \mathbf{k} \tag{2.157}$$

and

$$\hat{H} = U = -\gamma \frac{\hbar}{2} \boldsymbol{\sigma} \cdot \mathbf{B} \tag{2.158}$$

The equation of the motion-expected value $\langle \boldsymbol{\sigma} \rangle$ of the Pauli spin vector with $\hat{\boldsymbol{\sigma}} = \boldsymbol{\sigma}$ becomes

$$\mathrm{i}\hbar \frac{\mathrm{d}\langle \boldsymbol{\sigma} \rangle}{\mathrm{d}t} = \langle \hat{\boldsymbol{\sigma}} \hat{H} - \hat{H} \hat{\boldsymbol{\sigma}} \rangle = \gamma \frac{\hbar}{2} \langle \boldsymbol{\sigma}(\boldsymbol{\sigma} \cdot \mathbf{B}) - (\boldsymbol{\sigma} \cdot \mathbf{B})\boldsymbol{\sigma} \rangle \tag{2.159}$$

The BAC−CAB vector rule says that the vector expression inside the brackets on the right side of the equation is double the cross-product $\mathbf{B} \times \boldsymbol{\sigma} \times \boldsymbol{\sigma}$, where

$$\frac{\mathrm{d}\langle \boldsymbol{\sigma} \rangle}{\mathrm{d}t} = \gamma \frac{1}{2\mathrm{i}} \langle \mathbf{B} \times \boldsymbol{\sigma} \times \boldsymbol{\sigma} \rangle \tag{2.160}$$

The cross-product of the Pauli spin vector with itself ($\boldsymbol{\sigma} \times \boldsymbol{\sigma}$) equals the Pauli spin vector $\boldsymbol{\sigma}$ multiplied by 2i, where

$$\boldsymbol{\sigma} \times \boldsymbol{\sigma} = 2\mathrm{i}\boldsymbol{\sigma} \tag{2.161}$$

With substitution,

$$\frac{d\langle\boldsymbol{\sigma}\rangle}{dt} = \gamma\langle\mathbf{B}\times\boldsymbol{\sigma}\rangle = \gamma\mathbf{B}\times\langle\boldsymbol{\sigma}\rangle \tag{2.162}$$

where

$$\mathbf{P}_s = \langle\boldsymbol{\sigma}\rangle \tag{2.163}$$

The equation of motion of expected value $\langle\boldsymbol{\sigma}\rangle$ of the Pauli spin vector $\boldsymbol{\sigma}$, that is, the equation of motion of the neutron spin polarization vector $\mathbf{P}_s$ is simply the classical equation of motion of a spinning top, where

$$\frac{d\mathbf{P}_s}{dt} = \gamma\mathbf{B}\times\mathbf{P}_s \tag{2.164}$$

The spin polarization vector maintains constant length $|\mathbf{P}_s|$ because

$$\frac{d}{dt}|\mathbf{P}_s|^2 = \frac{d}{dt}(P_s^2) = \frac{d}{dt}(\mathbf{P}_s\cdot\mathbf{P}_s) = 2\mathbf{P}_s\cdot\frac{d\mathbf{P}_s}{dt} = \gamma\mathbf{P}_s\cdot(\mathbf{B}\times\mathbf{P}_s) = 0 \tag{2.165}$$

The expected value $\langle\boldsymbol{\sigma}\rangle$ of the neutron Pauli spin vector can be expressed in terms of the expected values of its spin vector $\langle\mathbf{S}\rangle$ and magnetic dipole moment vector $\langle\boldsymbol{\mu}\rangle$, where

$$\mathbf{P}_s = \langle\boldsymbol{\sigma}\rangle = \frac{2}{\hbar}\langle\mathbf{S}\rangle = \frac{2}{\hbar\gamma}\langle\boldsymbol{\mu}\rangle \tag{2.166}$$

Hence, the expected value $\langle\boldsymbol{\mu}\rangle$ of the neutron magnetic dipole moment vector $\boldsymbol{\mu}$ obeys the classical precession equation of a magnetic dipole about an applied magnetic field $\mathbf{B}$, where

$$\frac{d\langle\boldsymbol{\mu}\rangle}{dt} = \gamma(\mathbf{B}\times\langle\boldsymbol{\mu}\rangle) \tag{2.167}$$

The x-, y-, and z-components in the solutions of $\langle\boldsymbol{\mu}\rangle$ in the classical trajectory equation for the expected value of the neutron magnetic dipole moment vector $\langle\boldsymbol{\mu}\rangle$ in the Bloch equation for the case of a z-directed, applied magnetic field $\mathbf{B} = B_z\hat{\mathbf{z}}$ are

$$\langle\mu_x(t)\rangle = \cos(w_{\mathrm{L}}t)\langle\mu_x(0)\rangle - \sin(w_{\mathrm{L}}t)\langle\mu_y(0)\rangle \tag{2.168}$$

$$\langle\mu_y(t)\rangle = \sin(w_{\mathrm{L}}t)\langle\mu_x(0)\rangle + \cos(w_{\mathrm{L}}t)\langle\mu_y(0)\rangle \tag{2.169}$$

$$\langle\mu_z(t)\rangle = \langle\mu_z(0)\rangle \tag{2.170}$$

The Larmor frequency w_L precession of $\langle \boldsymbol{\mu} \rangle$ about $\mathbf{B} = B_z \hat{\mathbf{z}}$ is

$$w_\mathrm{L} = \gamma B_z \tag{2.171}$$

This classical precession trajectory of the expected value of the neutron magnetic dipole moment vector $\langle \boldsymbol{\mu} \rangle$ is a result of Ehrenfest's theorem, which states the equations of motion of the *expected values* of quantum mechanical quantities (e.g., position and momentum) are identical to the classical mechanics equation of motion of the corresponding quantity.

The precession of magnetic dipole moment vector $\boldsymbol{\mu}$ about an applied magnetic field $\mathbf{B}$ in classical mechanics is identical to the precession of the expected value of the neutron magnetic dipole moment vector $\langle \boldsymbol{\mu} \rangle$ in quantum mechanics in that one can simultaneously specify the x-, y-, and z-components $\langle \mu_x(0) \rangle$, $\langle \mu_y(0) \rangle$, and $\langle \mu_z(0) \rangle$ at initial time $t = 0$. Note, however, in the case of a single-neutron magnetic dipole moment vector $\boldsymbol{\mu}$, one can only specify simultaneously the instantaneous total vector magnitude $\mu(0)$ and the instantaneous z-component $\mu_z(0)$ at initial time $t = 0$.

2.12 Expected Values of Quantum Mechanical Quantities Follow Classical Trajectories

Let us consider Ehrenfest's theorem and the relation of classical mechanical neutron spin trajectories to the expected values (measured values) of the quantum mechanical neutron spin trajectories. First, one needs to define the Poisson brackets used in the Hamiltonian of classical mechanics. Given the canonical coordinates of position q_i and momentum p_i in phase space and two functions $f(p_i, q_i, t)$ and $g(p_i, q_i, t)$, the Poisson bracket $\{f, g\}$ is defined by

$$\{f, g\} = \sum_{i=1}^{N} \left[\frac{\partial f}{\partial q_i}\frac{\partial g}{\partial p_i} - \frac{\partial f}{\partial p_i}\frac{\partial g}{\partial q_i} \right] \tag{2.172}$$

The Hamilton equations of motion in classical mechanics can be expressed in terms of Poisson brackets { }. Consider the function $f(p, q, t)$, which has a time derivative of

$$\frac{\mathrm{d}f}{\mathrm{d}t} = \frac{\partial f}{\partial q}\frac{\mathrm{d}q}{\mathrm{d}t} + \frac{\partial f}{\partial p}\frac{\mathrm{d}p}{\mathrm{d}t} + \frac{\partial f}{\partial t} \tag{2.173}$$

One then assumes that $p(t)$ and $q(t)$ are solutions to the Hamiltonian equations:

$$\frac{\mathrm{d}q}{\mathrm{d}t} = \frac{\partial H}{\partial p} \quad \text{and} \quad \frac{\mathrm{d}p}{\mathrm{d}t} = -\frac{\partial H}{\partial q} \tag{2.174}$$

Substitution and the definition of the Poisson bracket { } gives the equation of motion of the object that is represented by the function $f(p, q, t)$, where

$$\frac{\mathrm{d}f}{\mathrm{d}t} = \frac{\partial f}{\partial q}\frac{\partial H}{\partial p} - \frac{\partial f}{\partial p}\frac{\partial H}{\partial q} + \frac{\partial f}{\partial t} = \{f, H\} + \frac{\partial f}{\partial t} \tag{2.175}$$

Associated with the object trajectory can be constants of motion, such as energy, linear momentum, or angular momentum. Consider a constant of motion of the object, which is represented by function $f(p, q)$, with no explicit dependence on time t but with implicit dependence on time via $p(t)$ and $q(t)$. Function $f(p, q)$ is a constant of motion and thus commutes with the Hamiltonian, which yields Lioville's theorem, where

$$\frac{\mathrm{d}f}{\mathrm{d}t} = 0 = \{f, h\} + \frac{\partial f}{\partial t} \tag{2.176}$$

This equation describes the time evolution of function $f(p, q)$, which can, for example, represent the phase space distribution of momentum-position pairs $p(t)$ and $q(t)$, where $f(p, q)\mathrm{d}p\,\mathrm{d}q$ is the probability that a particle will be found in the infinitesimal volume $\mathrm{d}p\,\mathrm{d}q$. The classical mechanical expression for the time derivative of a classical quantity f is, then expressed by the Poisson brackets { }, where

$$\frac{\mathrm{d}f}{\mathrm{d}t} = \left(\frac{\partial f}{\partial q}\frac{\partial H}{\partial p} - \frac{\partial f}{\partial p}\frac{\partial H}{\partial q}\right) + \frac{\partial f}{\partial t} = \{f, h\} + \frac{\partial f}{\partial t} \quad \text{classical mechanics} \tag{2.177}$$

This equation is the classical mechanics trajectory equation for the quantity f.

Let us now consider the quantum mechanical trajectory of an object such as the neutron spin vector $\mathbf{S}$, which is characterized by wave function Ψ for the neutron spin wave amplitude. The time derivative of the expected value of the spin vector $(\mathrm{d}/\mathrm{d}t)\langle\mathbf{S}\rangle$ is obtained by the integration of $\Psi^*\hat{\mathbf{S}}\Psi$ overall space in which the neutron spin vector operator $\hat{\mathbf{S}}$ acts on the neutron wave function Ψ for the neutron spin wave amplitude, where

$$\frac{\mathrm{d}}{\mathrm{d}t}\langle\mathbf{S}\rangle = \frac{\mathrm{d}}{\mathrm{d}t}\int_{-\infty}^{\infty}\mathrm{d}z\int_{-\infty}^{\infty}\mathrm{d}y\int_{-\infty}^{\infty}\mathrm{d}x\Psi^*\hat{\mathbf{S}}\Psi \tag{2.178}$$

Carrying out the time derivatives on the right side of this equation yields

$$\frac{\mathrm{d}}{\mathrm{d}t}\langle\mathbf{S}\rangle = \int_{-\infty}^{\infty}\mathrm{d}z\int_{-\infty}^{\infty}\mathrm{d}y\int_{-\infty}^{\infty}\mathrm{d}x\frac{d\Psi^*}{dt}\hat{\mathbf{S}}\Psi + \int_{-\infty}^{\infty}\mathrm{d}z\int_{-\infty}^{\infty}\mathrm{d}y\int_{-\infty}^{\infty}\mathrm{d}x\Psi^*\frac{\mathrm{d}\hat{\mathbf{S}}}{\mathrm{d}t}\Psi$$
$$+ \int_{-\infty}^{\infty}\mathrm{d}z\int_{-\infty}^{\infty}\mathrm{d}y\int_{-\infty}^{\infty}\mathrm{d}x\Psi^*\hat{\mathbf{S}}\frac{\mathrm{d}\Psi}{\mathrm{d}t} \tag{2.179}$$

One then applies the Schrödinger time-dependent equation:

$$\frac{d\Psi}{dt} = \frac{1}{i\hbar}\hat{H}\Psi \quad \text{and} \quad \frac{d\Psi^*}{dt} = -\frac{1}{i\hbar}\Psi^*\hat{H}^* = -\frac{1}{i\hbar}\Psi^*\hat{H} \tag{2.180}$$

The Hamiltonian $\hat{H}$ equals its complex conjugate $\hat{H}^*$, where

$$\hat{H}^* = \hat{H} \tag{2.181}$$

One substitutes these expressions in the time-derivative equation for the expected value $\langle \mathbf{S} \rangle$ of the neutron spin vector to obtain

$$\frac{d}{dt}\langle \mathbf{S} \rangle = \frac{1}{i\hbar}\int_{-\infty}^{\infty} dz \int_{-\infty}^{\infty} dy \int_{-\infty}^{\infty} dx \Psi^*(\hat{\mathbf{S}}\hat{H} - \hat{H}\hat{\mathbf{S}})\Psi + \int_{-\infty}^{\infty} dz \int_{-\infty}^{\infty} dy \int_{-\infty}^{\infty} dx \Psi^* \frac{d\hat{\mathbf{S}}}{dt}\Psi \tag{2.182}$$

Given the integral definition of the expected value, an expression with commutation brackets [ ], the time-derivative equation of the expected value $\langle \mathbf{S} \rangle$ of the neutron spin vector is

$$\frac{d}{dt}\langle \mathbf{S} \rangle = \frac{1}{i\hbar}\langle (\hat{\mathbf{S}}\hat{H} - \hat{H}\hat{\mathbf{S}}) \rangle + \left\langle \frac{d\hat{\mathbf{S}}}{dt} \right\rangle = \frac{1}{i\hbar}\langle \left[\hat{\mathbf{S}} \cdot \hat{H}\right] \rangle + \left\langle \frac{d\hat{\mathbf{S}}}{dt} \right\rangle \tag{2.183}$$

For a generalized quantum mechanical quantity f and its corresponding quantum mechanical operator $\hat{f}$, the time-derivative equation of its expected value $\langle f \rangle$ is expressed with the commutator brackets [ ], where

$$\frac{d}{dt}\langle f \rangle = \frac{1}{i\hbar}\langle (\hat{f}\hat{H} - \hat{H}\hat{f}) \rangle + \left\langle \frac{d\hat{f}}{dt} \right\rangle = \frac{1}{i\hbar}\langle [\hat{f} \cdot \hat{H}] \rangle + \left\langle \frac{d\hat{f}}{dt} \right\rangle \quad \text{quantum mechanics} \tag{2.184}$$

Here, f is a scalar quantity, which can be the x-, y-, or z-component of a vector quantity $\mathbf{F}$. This is the quantum mechanics trajectory equation for the expected value $\langle f \rangle$ of quantity f.

An example of Ehrenfest's theorem is shown in the trajectory of the expected value of the time variation of particle momentum $(d/dt)\langle \mathbf{p} \rangle$, which is the expected value of the force $\langle \mathbf{F} \rangle$ that acts on the particle. This is the classical mechanical result of Newton's second law of motion with expected values of quantum mechanical quantities. Consider the quantum mechanical momentum $\mathbf{p}$ and the corresponding momentum operator $\hat{\mathbf{p}}$ and the Hamiltonian operator $\hat{H}$ for a particle of mass m

moving through a field region in which the particle acquires potential energy $V(\mathbf{r}, t)$, where the particle momentum $\mathbf{p}$ does not have explicit time dependence and

$$\hat{\mathbf{p}} = -\mathrm{i}\hbar\nabla \quad \left\langle \frac{\mathrm{d}\hat{\mathbf{p}}}{\mathrm{d}t} \right\rangle = 0 \quad \hat{H} = \frac{p^2}{2m} + V = -\frac{\hbar^2}{2m}\nabla^2 + V \tag{2.185}$$

The quantum mechanics trajectory equation for the expected value of the particle momentum $\langle \mathbf{p} \rangle$ then gives

$$\frac{\mathrm{d}}{\mathrm{d}t}\langle \mathbf{p} \rangle = \frac{1}{\mathrm{i}\hbar}\langle [\hat{\mathbf{p}}, .\hat{H}] \rangle + \left\langle \frac{\mathrm{d}\hat{\mathbf{p}}}{\mathrm{d}t} \right\rangle = \frac{1}{\mathrm{i}\hbar}\langle [\hat{\mathbf{p}}, .V] \rangle, \tag{2.186}$$

where

$$\frac{1}{\mathrm{i}\hbar}\langle [\hat{\mathbf{p}}, .V] \rangle = \frac{1}{\mathrm{i}\hbar}\int_{-\infty}^{\infty} \mathrm{d}z \int_{-\infty}^{\infty} \mathrm{d}y \int_{-\infty}^{\infty} \mathrm{d}x \Psi^*(\hat{\mathbf{p}}V - V\hat{\mathbf{p}})\Psi \tag{2.187}$$

Substitution of $\hat{\mathbf{p}} = -\mathrm{i}\hbar\nabla$ gives

$$\frac{1}{\mathrm{i}\hbar}\langle [\hat{\mathbf{p}}, .V] \rangle = \int_{-\infty}^{\infty} \mathrm{d}z \int_{-\infty}^{\infty} \mathrm{d}y \int_{-\infty}^{\infty} \mathrm{d}x(\Psi^* V\nabla\Psi - \Psi^*\Psi\nabla V - \Psi^* V\nabla\Psi) \tag{2.188}$$

Thus, one obtains Newton's second law of motion from the trajectory of the quantum mechanical expected value of the time rate of change of the particle momentum $(\mathrm{d}/\mathrm{d}t)\langle \mathbf{p} \rangle$, which equals the expected value of the force $\langle \boldsymbol{F} \rangle$ that acts on the particle. And $\langle \boldsymbol{F} \rangle$ equals the negative expected value of the gradient $-\langle \nabla V \rangle$ of potential energy $V(\mathbf{r}, t)$, which is by the particle moving through the field region, where

$$\frac{\mathrm{d}}{\mathrm{d}t}\langle \mathbf{p} \rangle = \frac{1}{\mathrm{i}\hbar}\langle [\hat{\mathbf{p}}, .V] \rangle = -\int_{-\infty}^{\infty} \mathrm{d}z \int_{-\infty}^{\infty} \mathrm{d}y \int_{-\infty}^{\infty} \mathrm{d}x \Psi^*(\nabla V)\Psi = -\langle \nabla V \rangle = \langle \mathbf{F} \rangle \tag{2.189}$$

In quantum mechanics, the expectation values of position $\langle \mathbf{r} \rangle$, momentum $\langle \mathbf{p} \rangle$, and the neutron magnetic dipole moment $\langle \boldsymbol{\mu} \rangle$ obey time evolution or trajectory equations, which are analogs of the classical mechanic time evolution or trajectory equations. That is, the expectation values of quantum mechanical quantities of position $\langle \mathbf{r} \rangle$, momentum $\langle \mathbf{p} \rangle$, and the neutron magnetic dipole moment $\langle \boldsymbol{\mu} \rangle$ follow classical mechanical trajectories. The spatial extent of the wave function of the quantum mechanical quantity is usually very small and on the order of the de Broglie wavelength. Thus the expected value of the quantum mechanical quantity is found within the spatial extent of the tiny wave function. As a result, the trajectory of the expected value of a quantity in wave mechanics follows the trajectory of the corresponding quantity as a pointlike object in classical mechanics.

The quantum mechanical localized wave trajectory and corresponding classical mechanical, pointlike particle trajectory differ in that the spatial gradient of a potential function $V(\mathbf{r}, t)$ can vary within the extent of the localized wave function. Consequently, the different spatial positions within the localized wave packet experience a different spatial gradient of $V(\mathbf{r}, t)$, and thus, different forces act on the various parts of the wave packet. This causes a modulation of the wave packet, which is effectively the interference of various parts of the wave packet with itself. The trajectory of the wave packet is a result of the interference of various wave trajectories within the wave packet.

One can now relate the classical and quantum mechanical results:

$$\frac{\mathrm{d}f}{\mathrm{d}t} = \left(\frac{\partial f}{\partial q}\frac{\partial H}{\partial p} - \frac{\partial f}{\partial p}\frac{\partial H}{\partial q}\right) + \frac{\partial f}{\partial t} \quad \text{and} \quad \frac{\mathrm{d}}{\mathrm{d}t}\langle f\rangle = \frac{1}{\mathrm{i}\hbar}\langle(\hat{f}\hat{H} - \hat{H}\hat{f})\rangle + \left\langle\frac{\mathrm{d}\hat{f}}{\mathrm{d}t}\right\rangle \tag{2.190}$$

By comparison with the quantum mechanical trajectory of the expected value $\langle f\rangle$ for quantum mechanical object (quantity)f and classical mechanical trajectory of f for classical mechanical object (quantity) f, one finds the following rule. An equation in quantum mechanics, which contains a commutator bracket [ ], can be turned into an equation in classical mechanics by changing the commutator bracket [ ] into a Poisson bracket { } and multiplying by $\mathrm{i}\hbar$.

To describe a system quantum mechanically, one can begin with a classical model of the system, which can be described mathematically by the Hamiltonian equation of energy and motion. Next, one replaces the classical position coordinates $\mathbf{q}$ and conjugate momentum $\mathbf{p}$ by the corresponding quantum mechanical position $\hat{\mathbf{q}}$ and momentum $\hat{\mathbf{p}}$ operators. Some quantities, such as electron or neutron spin, have no classical analogs. One must then create a model of atomic electron and nuclear spin with a Hamiltonian wave function whose mathematical results and solutions match experimental observations, such as the Stern–Gerlach experiment, the Zeeman effect, and the detailed structure of atomic spectral lines.

The resulting quantum mechanical equations may have products of noncommutating operators, such as position $\hat{\mathbf{q}}$ and momentum $\hat{\mathbf{p}}$, where $[\mathbf{r}, \mathbf{p}] = \mathrm{i}\hbar$, which yield imaginary quantities for measured physical quantities. With noncommutating operators $\hat{\mathbf{q}} \cdot \hat{\mathbf{p}}$, the order of applying the position $\hat{\mathbf{q}}$ and momentum $\hat{\mathbf{p}}$ operators to a wave function Ψ give differing results, where

$$(\hat{\mathbf{q}} \cdot \hat{\mathbf{p}})\Psi \neq (\hat{\mathbf{q}} \cdot \hat{\mathbf{p}})\Psi \tag{2.191}$$

The same physical result is expected, regardless of the order of application of the position $\hat{\mathbf{q}}$ and momentum $\hat{\mathbf{p}}$ operators on the wave function Ψ. The summation of the two possible order of applications $\hat{\mathbf{q}} \cdot \hat{\mathbf{p}}$ and $\hat{\mathbf{q}} \cdot \hat{\mathbf{p}}$, each with equal likelihood, creates an operator that allows independent application of the position $\hat{\mathbf{q}}$ and momentum $\hat{\mathbf{p}}$ operators, where

$$\frac{1}{2}(\hat{\mathbf{q}} \cdot \hat{\mathbf{p}} + \hat{\mathbf{p}} \cdot \hat{\mathbf{q}})\Psi = \frac{1}{2}(\hat{\mathbf{p}} \cdot \hat{\mathbf{q}} + \hat{\mathbf{q}} \cdot \hat{\mathbf{p}})\Psi \tag{2.192}$$

Transposition shows

$$(\hat{\mathbf{q}} \cdot \hat{\mathbf{p}} - \hat{\mathbf{p}} \cdot \hat{\mathbf{q}})\Psi = (\hat{\mathbf{q}} \cdot \hat{\mathbf{p}} - \hat{\mathbf{p}} \cdot \hat{\mathbf{q}})\Psi \tag{2.193}$$

Hence, the order of the operators does not change the outcome of the measurement in the mathematical representation of the physical system, where

$$[\hat{\mathbf{q}}, \hat{\mathbf{p}}]\Psi - [\hat{\mathbf{q}}, \hat{\mathbf{p}}]\Psi = (i\hbar - i\hbar)\Psi = 0 \tag{2.194}$$

Furthermore, the measured physical quantities, which are given by expected values of quantum mechanical representation of the physical quantity, must be real. The expected value of $\langle \mathbf{q} \cdot \mathbf{p} + \mathbf{p} \cdot \mathbf{q} \rangle$ is required to be real, which requires it to be equal to its complex conjugate:

$$\langle \mathbf{q} \cdot \mathbf{p} + \mathbf{p} \cdot \mathbf{q} \rangle = \langle \mathbf{q} \cdot \mathbf{p} + \mathbf{p} \cdot \mathbf{q} \rangle^* \tag{2.195}$$

By definition, the expected value is

$$\begin{aligned} \langle \mathbf{q} \cdot \mathbf{p} + \mathbf{p} \cdot \mathbf{q} \rangle &= \int_{-\infty}^{\infty} dz \int_{-\infty}^{\infty} dy \int_{-\infty}^{\infty} dx \Psi^* (\hat{\mathbf{q}} \cdot \hat{\mathbf{p}} + \hat{\mathbf{p}} \cdot \hat{\mathbf{q}})\Psi \\ &= \left(\int_{-\infty}^{\infty} dz \int_{-\infty}^{\infty} dy \int_{-\infty}^{\infty} dx \Psi^* (\hat{\mathbf{q}} \cdot \hat{\mathbf{p}} + \hat{\mathbf{p}} \cdot \hat{\mathbf{q}})\Psi \right)^* \end{aligned} \tag{2.196}$$

The complex conjugate on the right side is evaluated as

$$\left(\int_{-\infty}^{\infty} dz \int_{-\infty}^{\infty} dy \int_{-\infty}^{\infty} dx \Psi^* (\hat{\mathbf{q}} \cdot \hat{\mathbf{p}} + \hat{\mathbf{p}} \cdot \hat{\mathbf{q}})\Psi \right)^* = \int_{-\infty}^{\infty} dz \int_{-\infty}^{\infty} dy \int_{-\infty}^{\infty} dx \Psi (\hat{\mathbf{q}} \cdot \hat{\mathbf{p}} + \hat{\mathbf{p}} \cdot \hat{\mathbf{q}})^* \Psi^* \tag{2.197}$$

As a result, the operator $(\hat{\mathbf{q}} \cdot \hat{\mathbf{p}} + \hat{\mathbf{p}} \cdot \hat{\mathbf{q}})$ is defined to be a Hermitian operator if

$$\begin{aligned} \langle \mathbf{q} \cdot \mathbf{p} + \mathbf{p} \cdot \mathbf{q} \rangle &= \int_{-\infty}^{\infty} dz \int_{-\infty}^{\infty} dy \int_{-\infty}^{\infty} dx \Psi^* (\hat{\mathbf{q}} \cdot \hat{\mathbf{p}} + \hat{\mathbf{p}} \cdot \hat{\mathbf{q}})\Psi \\ &= \int_{-\infty}^{\infty} dz \int_{-\infty}^{\infty} dy \int_{-\infty}^{\infty} dx \Psi (\hat{\mathbf{q}} \cdot \hat{\mathbf{p}} + \hat{\mathbf{p}} \cdot \hat{\mathbf{q}})^* \Psi^* \end{aligned} \tag{2.198}$$

Since the position $\hat{\mathbf{q}}$ and momentum $\hat{\mathbf{p}}$ operators are each Hermitian, the combination operator is also Hermitian $(\hat{\mathbf{q}} \cdot \hat{\mathbf{p}} + \hat{\mathbf{p}} \cdot \hat{\mathbf{q}})$; see Pauli (1973). One requires Hermitian operators to extract the real, expected values or physical measurements from the quantum mechanical wave function representation of a physical quantity.

2.13 Average Spin Polarization of a Beam of Neutrons in an Applied Magnetic Field

Next, let us consider a generalized beam of spin-up (+) and spin-down (−) neutrons. The polarization vector **P** of the neutron beam is defined in terms of the number N_+ of spin-up (+) and the number N_- spin-down (−) neutrons in the beam, where

$$\mathbf{P} = \frac{N_+ - N_-}{N_+ + N_-}\hat{\mathbf{z}} \tag{2.199}$$

The total number N of neutrons in the beam is the sum of the number N_+ of spin-up (+) and the number N_- spin-down (−) neutrons, where

$$N = N_+ + N_- \tag{2.200}$$

Here, $\hat{\mathbf{z}}$ is the unit vector in the reference z-direction, which is the direction of neutron beam propagation in the z-directed magnetic field $\mathbf{B} = B_z\hat{\mathbf{z}}$. A beam of spin-up (+) neutrons has a polarization $P = 1$, a beam of spin-down neutrons has polarization $P = -1$, and an unpolarized beam with equal numbers of spin-up (+) and spin-down (−) neutrons has polarization $P = 0$.

For a beam of N neutrons, the beam average polarization vector **P** is the average of the expected values $\langle \mathbf{S}_n \rangle$ of each of the nth individual neutron spin vectors, that is, the beam average of the expected spin polarization vectors $(\mathbf{P}_s)_n$, where

$$\mathbf{P} = \frac{1}{N}\sum_{n=1}^{N}(\mathbf{P}_s)_n = \frac{1}{N}\frac{2}{\hbar}\sum_{n=1}^{N}\langle \mathbf{S}_n \rangle \tag{2.201}$$

The equation of motion of beam average spin polarization vector **P** also obeys the classical equation of motion of the spinning top:

$$\frac{d\mathbf{P}}{dt} = \gamma \mathbf{B} \times \mathbf{P} \tag{2.202}$$

For the beam of N neutrons, the beam average magnetic dipole moment vector **M** is the average of the expected values $\langle \mu_n \rangle$ of each of the nth individual magnetic dipole moment vector, where

$$\mathbf{M} = \frac{1}{N}\sum_{n=1}^{N}\langle \mu_n \rangle \tag{2.203}$$

and

$$\langle \mu_n \rangle = \gamma \langle \mathbf{S}_n \rangle \tag{2.204}$$

The neutron beam average magnetic dipole moment vector **M** points in the opposite direction of the beam average polarization vector **P**, where

$$\mathbf{P} = -\frac{2}{\hbar\gamma}\mathbf{M} \tag{2.205}$$

The beam average magnetization vector **M**, which obeys the classical precession equation of a magnetic dipole vector **m** about an applied magnetic field **B**, where

$$\frac{d\mathbf{M}}{dt} = \gamma \mathbf{B} \times \mathbf{M} \tag{2.206}$$

The x-, y-, and z-components P_x, P_y, and P_z of the beam polarization vector **P** for a general beam of spin-up (+) and spin-down (−) neutrons obey the same classical equation of motion as the x-, y-, and z-component expected values $\langle S_x \rangle$, $\langle S_y \rangle$, and $\langle S_z \rangle$ of the individual neutron spin vector expected value $\langle \mathbf{S} \rangle$. Relative to the z-direction of the applied magnetic field $\mathbf{B} = B_z\hat{\mathbf{z}}$, the average precession angle $\langle \phi_S \rangle$ of the expected value of the individual neutron spin vector expected value $\langle \mathbf{S} \rangle$ is related to the z-component $\langle S_z \rangle$ by

$$\langle S_z \rangle = \langle |\mathbf{S}| \rangle \cos\langle \phi_s \rangle \tag{2.207}$$

Now consider a general beam of spin-up (+) and spin-down (−) neutrons with beam average spin polarization vector **P**. Relative to the z-direction of the applied magnetic field $\mathbf{B} = B_z\hat{\mathbf{z}}$, the beam average spin polarization vector **P** rotates to the beam average precession angle $\langle \phi_P \rangle$ relative to the positive z-direction. The beam average polarization vector component $\langle P_z \rangle$ relative to the positive z-direction is

$$\langle P_z \rangle = \langle |\mathbf{P}| \rangle \cos\langle \phi_P \rangle \tag{2.208}$$

2.14 Adiabatic and Nonadiabatic Polarization Rotation About Magnetic Field Lines That Change Direction

We now examine adiabatic polarization rotation of the beam average spin polarization vector **P**, which represents an average of the expected values of the individual neutron spin vectors $(\mathbf{P}_s)_n$ that comprise a beam of N total neutrons, where, for the nth neutron

$$(\mathbf{P}_s)_n = \langle \sigma_n \rangle = \frac{2}{\hbar}\langle \mathbf{S}_n \rangle = \frac{2}{\hbar\gamma}\langle \mu_n \rangle \tag{2.209}$$

The beam average spin polarization vector **P** maintains the same parallel or antiparallel alignment along a magnetic field line as the magnetic field line changes direction sufficiently slowly. That is, the magnetic field **B** must change in a time period Δt_B that is greater than the period ΔT_L for the Larmor precession of the average (expected value) neutron spin vectors $\langle \mathbf{S} \rangle$ about the applied magnetic field line, where

$$\Delta t_B > \Delta T_\mathrm{L} \tag{2.210}$$

This relation is the condition for adiabatic spin polarization in the expected values of the individual neutron spin vectors $\langle \mathbf{S} \rangle$, and thus the expected values of the oppositely directed expected values of the individual neutron magnetic dipole moment vectors $\langle \boldsymbol{\mu} \rangle$ of the spin-up (+) and spin-down (−) neutrons.

If the neutron enters a magnetic field as a spin-down (−) neutron with its expected value magnetic dipole moment vector $\langle \boldsymbol{\mu} \rangle$ aligned parallel (pointing in the same direction) to the applied magnetic field **B**, then the spin-down (−) neutron maintains the same parallel alignment with **B** as the magnetic field **B** changes direction, where

$$\langle \boldsymbol{\mu} \rangle \uparrow \uparrow \mathbf{B} \quad \text{and} \quad \langle \mathbf{S} \rangle \downarrow \uparrow \mathbf{B} \quad \text{spin-down } (-) \text{ neutron} \tag{2.211}$$

This is so provided that the magnetic field **B** changes direction at a lesser rate than the neutron Larmor precession frequency w_L. The expected value of the neutron spin vector $\langle \mathbf{S} \rangle$ points in the opposite direction of the expected value of the neutron magnetic dipole moment vector $\langle \boldsymbol{\mu} \rangle$. The expected value $\langle \mathbf{S} \rangle$ of the spin-down (−) vector **S** points in the opposite direction of the applied magnetic field **B** (antiparallel).

Similarly, if the neutron enters a magnetic field as a spin-up (+) neutron with its expected value magnetic dipole moment vector $\langle \boldsymbol{\mu} \rangle$ aligned antiparallel (pointing in the opposite direction) to the applied magnetic field **B**, then the spin-up (+) neutron maintains the same antiparallel alignment with **B** as the magnetic field **B** changes direction, where

$$\langle \boldsymbol{\mu} \rangle \downarrow \uparrow \mathbf{B} \quad \text{and} \quad \langle \mathbf{S} \rangle \uparrow \uparrow \mathbf{B} \quad \text{spin-up } (+) \text{ neutron} \tag{2.212}$$

Again, this is provided that the magnetic field **B** changes direction at a lesser rate than the neutron Larmor precession frequency w_L. The expected value $\langle \mathbf{S} \rangle$ of the spin-up (+) neutron spin vector **S** points in the same (parallel) direction as the applied magnetic field **B**.

As a result, the direction of precession of the expected value neutron magnetic dipole moment vector $\langle \boldsymbol{\mu} \rangle$ about the lines magnetic field line **B** reverses if the magnetic field lines reverse direction in a sufficiently slow, adiabatic manner. Adiabatic refers to no significant change in energy of magnetic field because of its sufficiently slow variation.

For example, the expected value of the spin-up (+) neutron magnetic dipole moment vector $\langle \boldsymbol{\mu} \rangle$ undergoes a counterclockwise precession about lines of $\mathbf{B} = B_z\hat{\mathbf{z}}$ in the positive azimuth direction $+\hat{\varphi}$ in the same manner as the classical mechanic precession of the magnetic dipole moment vector **m** of a current loop about the applied magnetic field **B**. Conversely, the expected value of the magnetic dipole moment $\langle \boldsymbol{\mu} \rangle$ of the spin-up (−) neutron undergoes clockwise precession about lines of $\mathbf{B} = B_z\hat{\mathbf{z}}$ in the negative azimuth direction $-\hat{\varphi}$.

The condition for the adiabatic rotation of the expected value of the neutron spin vector is met by the following condition. The temporal frequency w of the change in direction of the applied magnetic field lines must be less than the Larmor precession frequency w_{L} of the expected value of the magnetic dipole moment $\langle \boldsymbol{\mu} \rangle$ about the magnetic field line, where

$$w_{\mathrm{L}} > w \tag{2.213}$$

If the change in the magnetic field **B** occurs slowly in temporal frequency w in comparison to the Larmor precession frequency w_{L} of the expected value of the magnetic dipole moment vector $\langle \boldsymbol{\mu} \rangle$, then $\langle \boldsymbol{\mu} \rangle$ maintains the same parallel or anti-parallel alignment in relation to the direction of the applied magnetic field **B**. For example from Eq. (2.110), a magnetic field $\mathbf{B}\,[\mathrm{T}] = 10^{-3}\hat{\mathbf{y}}$ gives the Larmor precession frequency $w_{\mathrm{L}} = 1.83 \times 10^5$ rad/s, and with $\mathbf{B}\,[\mathrm{T}] = -10^{-3}\hat{\mathbf{y}}$, Larmor precession frequency is $w_{\mathrm{L}} = -1.83 \times 10^5$ rad/s.

Nonadiabatic transitions occur when a magnetic field **B** suddenly changes direction thereby causing the expected value of the magnetic dipole moment $\langle \boldsymbol{\mu} \rangle$ to begin precession about the new direction of the magnetic field. Consider the precession of $\langle \boldsymbol{\mu} \rangle$ about the positive z-axis. In a $\pi/2$ flipper, a RF magnetic field is suddenly applied in the positive x-direction, which is perpendicular to the y-directed, DC guide magnetic field. The sudden x-directed magnetic field causes the expected value of the neutron spin vector $\langle \mathbf{S} \rangle$ to begin precession about the x-axis. Similarly, in a π flipper, a RF magnetic field is suddenly applied in the negative x-direction perpendicular to the y-directed, DC guide magnetic field. The sudden magnetic field directed in the negative x-direction causes the expected value of the neutron spin vector $\langle \mathbf{S} \rangle$ to begin precession about the negative x-axis.

2.15 Magnetic Resonance

In this section, we examine the magnetic resonance of a particle with a two-spin state of $s = \pm 1/2$ and the associated magnetic dipole moment, which applies to a neutron, a proton, an electron, or a fluorine-19 nucleus. The two-spin-state neutron with magnetic dipole moment vector $\boldsymbol{\mu}$ enters an applied magnetic field **B**. The magnetic field is comprised of a steady, z-directed (axial) magnetic field $\mathbf{B}_{||}$ and a smaller, transverse, oscillating magnetic field $\mathbf{B}_{\perp}$ in the xy-plane, where

$$\mathbf{B} = B_x\hat{\mathbf{x}} + B_y\hat{\mathbf{y}} + B_z\hat{\mathbf{z}} = \mathbf{B}_{||} + \mathbf{B}_{\perp} \tag{2.214}$$

The axial $\mathbf{B}_{||}$ and transverse $\mathbf{B}_{\perp}$ magnetic field components of $\mathbf{B}$ are

$$\mathbf{B}_{||} = B_z\hat{\mathbf{z}} \quad \text{and} \quad \mathbf{B}_{\perp} = B_x\hat{\mathbf{x}} + B_y\hat{\mathbf{y}} \tag{2.215}$$

The Bloch equation for precession of the neutron magnetic dipole moment vector $\boldsymbol{\mu}$ about the applied magnetic field $\mathbf{B}$ is

$$\frac{\mathrm{d}\boldsymbol{\mu}}{\mathrm{d}t} = \gamma\mathbf{B} \times \boldsymbol{\mu} \tag{2.216}$$

The neutron has a negative gyromagnetic ratio, where $\gamma < 0$, and thus the Bloch equation shows the clockwise precession of $\boldsymbol{\mu}$ at the Larmor frequency $w_{\mathrm{L}} = \gamma B_{||}$ about the lines of the steady magnetic field component $\mathbf{B}_{||}$ that are pointed in the positive z-direction. The smaller, transverse magnetic field $\mathbf{B}_{\perp}$ is set to rotate at frequency w in the same clockwise direction as the neutron magnetic dipole moment vector $\boldsymbol{\mu}$. Clockwise rotation corresponds to a negative azimuth angle $\phi = -wt$ with time t, and thus the clockwise rotation $\mathbf{B}_{\perp}$ gives rise to the sinusoidal x- and y-components B_x and B_y, in which B_y has the minus sign and lags behind B_x by 90°, where

$$B_x = B_{\perp} \cos wt \quad \text{and} \quad B_y = -B_{\perp} \sin wt \tag{2.217}$$

The two-state $\pm 1/2$ spin particle is described by the Pauli spin matrix operator $\hat{\boldsymbol{\sigma}}$ and its x-, y-, and z-component operators $\hat{\sigma}_x$, $\hat{\sigma}_y$, and $\hat{\sigma}_z$ for the x, y, and z directions. The Hamiltonian interaction energy operator $\hat{H}$ of the particle with the applied magnetic field $\mathbf{B}$ is

$$\hat{H} = -\hat{\boldsymbol{\mu}} \cdot \mathbf{B} = -|\hat{\boldsymbol{\mu}}| \cdot |\mathbf{B}|\cos\theta = -\mu_{||}B \tag{2.218}$$

where

$$\hat{\boldsymbol{\mu}} = \gamma\hat{\mathbf{S}} = \gamma\frac{\hbar}{2}\hat{\boldsymbol{\sigma}} = \gamma\frac{\hbar}{2}\begin{bmatrix} 0 & 1 \\ 1 & 0 \end{bmatrix}\mathbf{i} + \gamma\frac{\hbar}{2}\begin{bmatrix} 0 & -\mathrm{i} \\ \mathrm{i} & 0 \end{bmatrix}\mathbf{j} + \gamma\frac{\hbar}{2}\begin{bmatrix} 1 & 0 \\ 0 & -1 \end{bmatrix}\mathbf{k} \tag{2.219}$$

with

$$\mu_{||} = \pm g\mu_n \quad \text{and} \quad \gamma = \frac{2g\mu_n}{\hbar} \tag{2.220}$$

The time-dependent Schrödinger wave equation describes the spin-state motion of the two-state particle spin wave function $\psi_{\pm}$ in the applied magnetic field, via the interaction Hamiltonian, where

$$\mathrm{i}\hbar\frac{\partial}{\partial t}\begin{bmatrix} \psi_+ \\ \psi_- \end{bmatrix} = \hat{H}\begin{bmatrix} \psi_+ \\ \psi_- \end{bmatrix} = -\gamma\frac{\hbar}{2}\left(\begin{bmatrix} 0 & 1 \\ 1 & 0 \end{bmatrix}B_x + \begin{bmatrix} 0 & -\mathrm{i} \\ \mathrm{i} & 0 \end{bmatrix}B_y + \begin{bmatrix} 1 & 0 \\ 0 & -1 \end{bmatrix}B_z\right)\begin{bmatrix} \psi_+ \\ \psi_- \end{bmatrix} \tag{2.221}$$

The time-dependent neutron spin wave function ψ_s represented in its spinor form is

$$\psi_s(t) = \begin{bmatrix} \psi_{s+}(t) \\ \psi_{s-}(t) \end{bmatrix} = u(t)\begin{bmatrix} 1 \\ 0 \end{bmatrix} + d(t)\begin{bmatrix} 0 \\ 1 \end{bmatrix} = \begin{bmatrix} u(t) \\ d(t) \end{bmatrix} \tag{2.222}$$

Substitution gives

$$\frac{\partial}{\partial t}\begin{bmatrix} u \\ d \end{bmatrix} = \Omega_{\perp}\begin{bmatrix} 0 & \cos wt \\ \cos wt & 0 \end{bmatrix}\begin{bmatrix} u \\ d \end{bmatrix} - \Omega_{\perp}\begin{bmatrix} 0 & -\mathrm{i}\sin wt \\ \mathrm{i}\sin wt & 0 \end{bmatrix}\begin{bmatrix} u \\ d \end{bmatrix} + \Omega_{\|}\begin{bmatrix} 1 & 0 \\ 0 & -1 \end{bmatrix}\begin{bmatrix} u \\ d \end{bmatrix} \tag{2.223}$$

where

$$\Omega_{\perp} = \frac{\gamma}{2}B_{\perp} \quad \text{and} \quad \Omega_{\|} = \frac{\gamma}{2}B_{\|} \tag{2.224}$$

Application of the Euler formula yields the coupled equations for the spin-up (+) state amplitude $u(t)$ and the spin-down (−) state amplitude $d(t)$, where

$$\frac{\partial u}{\partial t} = \mathrm{i}\Omega_{\perp}\exp(\mathrm{i}wt)d + \mathrm{i}\Omega_{\|}u \tag{2.225}$$

$$\frac{\partial d}{\partial t} = \mathrm{i}\Omega_{\perp}\exp(-\mathrm{i}wt)u + \mathrm{i}\Omega_{\|}d \tag{2.226}$$

One assumes that the initial ($t = 0$) neutron spin state is its upper spin state (+), where

$$\psi_s(0) = \begin{bmatrix} u(0) \\ d(0) \end{bmatrix} = \begin{bmatrix} 1 \\ 0 \end{bmatrix} \quad u(0) = 1 \quad d(0) = 0 \tag{2.227}$$

One assumes sinusoid solutions for $u(t)$ and $d(t)$, where

$$u(t) = u_0 \exp(\mathrm{i}w_u t) \tag{2.228}$$

$$d(t) = d_0 \exp(\mathrm{i}w_u t) \tag{2.229}$$

Here, u_0 and d_0 are constant amplitudes. Substitution of $u(t)$ and $d(t)$ into the coupled differential equations gives, in matrix form,

$$\begin{bmatrix} w_u - \Omega_{\|} & -\Omega_{\perp}\exp[\mathrm{i}t(w_u - w_d - w)] \\ -\Omega_{\perp}\exp[\mathrm{i}t(w_u - w_d - w)] & w_d + \Omega_{\|} \end{bmatrix}\begin{bmatrix} u_0 \\ d_0 \end{bmatrix} = \begin{bmatrix} 0 \\ 0 \end{bmatrix} \tag{2.230}$$

The solution for constant amplitudes u_0 and d_0 must be independent of time t, so that

$$w_u - w_d - w = 0 \tag{2.231}$$

Substitution gives

$$\begin{bmatrix} w_d + w - \Omega_{||} & -\Omega_{\perp} \\ -\Omega_{\perp} & w_d + \Omega_{||} \end{bmatrix} \begin{bmatrix} u_0 \\ d_0 \end{bmatrix} = \begin{bmatrix} 0 \\ 0 \end{bmatrix} \tag{2.232}$$

For nonzero constant amplitudes u_0 and d_0, one then solves this matrix representation of the simultaneous equations for the spin-up state frequency w_u and spin-down state frequency w_d by setting the determinant of this coefficient matrix to zero:

$$(w_d + w - \Omega_{||})(w_d + \Omega_{||}) - \Omega_{\perp}^2 = 0 \tag{2.233}$$

Multiplication yields

$$w_d^2 + ww_d + \Omega_{||}w - \Omega_{||}^2 - \Omega_{\perp}^2 = 0 \tag{2.234}$$

Solution of this quadratic equation for the spin-down (−) neutron frequency w_d gives

$$w_d = -\frac{w}{2} \pm \Gamma \tag{2.235}$$

where

$$\Gamma = \sqrt{\left(\Omega_{||} - \frac{w}{2}\right)^2 + \Omega_{\perp}^2} \tag{2.236}$$

One then obtains the spin-up (+) neutron frequency w_u from the frequency relation that was required to make the amplitude magnitudes u_0 and d_0 of the spin-up (+) and spin-down (−) neutrons time independent, where

$$w_u = w_d + w = \frac{w}{2} \pm \Gamma \tag{2.237}$$

The two frequency values for w_d suggests a general solution for the spin-down (−) state amplitude $d(t)$, which is the sum of two terms with amplitude coefficients d_1 and d_2

$$d(t) = d_1 \exp\left[-it\left(\frac{w}{2} - \Gamma\right)\right] + d_2 \exp\left[-it\left(\frac{w}{2} + \Gamma\right)\right] \tag{2.238}$$

With initial condition $d(0) = 0$, one conveniently finds the value as follows:

$$d_1 = -d_2 = \frac{D}{2\mathrm{i}} \tag{2.239}$$

Hence,

$$d(t) = D\sin(\Gamma t)\exp\left(-\mathrm{i}\frac{wt}{2}\right) \tag{2.240}$$

To determine constant D, one plugs $d(t)$ into the second coupled differential equation for $d(t)$ and evaluates the resulting equation at the initial condition $t = 0$, where $u(0) = 1$, so that

$$D = \frac{\mathrm{i}\Omega_\perp}{\Gamma} \tag{2.241}$$

The solution for the amplitude $d(t)$ of the spin-down (−) neutron state is

$$d(t) = \mathrm{i}\Omega_\perp \frac{\sin(\Gamma t)}{\Gamma}\exp\left(-\mathrm{i}\frac{wt}{2}\right) \tag{2.242}$$

Plugging the solution for $d(t)$ in the second coupled differential equation for $d(t)$ then allows the solving of the amplitude $u(t)$ of the spin-up (−) neutron state, where

$$u(t) = \frac{\sin(\Gamma t)}{\Gamma}\left[\mathrm{i}\left(\Omega_{||} - \frac{w}{2}\right) + \Gamma\cot(\Gamma t)\right]\exp\left(\mathrm{i}\frac{wt}{2}\right) \tag{2.243}$$

In spinor representation, the two-spin-state wave function $\psi_s(t)$ of the neutron is

$$\psi_s(t) = \begin{bmatrix} u(t) \\ d(t) \end{bmatrix} = \frac{\sin(\Gamma t)}{\Gamma}\begin{bmatrix} \left[\mathrm{i}\left(\Omega_{||} - \frac{w}{2}\right) + \Gamma\cot(\Gamma t)\right]\exp\left(\mathrm{i}\frac{wt}{2}\right) \\ \mathrm{i}\Omega_\perp \exp\left(-\mathrm{i}\frac{wt}{2}\right) \end{bmatrix} \tag{2.244}$$

The initial state $\psi_s(0)$ was normalized:

$$|\psi_s(0)|^2 = |u(0)|^2 + |d(0)|^2 = 1 + 0 = 0 \tag{2.245}$$

We now show the wave function $\psi_s(t)$ is normalized for all time t given the initial state $\psi_s(0)$ was normalized. Given

$$|\psi_s(t)|^2 = |u(t)|^2 + |d(t)|^2 \tag{2.246}$$

with

$$|u(t)|^2 = \frac{\sin^2(\Gamma t)}{\Gamma^2}\left[\left(\Omega_{||} - \frac{w}{2}\right)^2 + \Gamma^2 \cot^2(\Gamma t)\right] = \left(\Omega_{||} - \frac{w}{2}\right)^2 \frac{\sin^2(\Gamma t)}{\Gamma^2} + \cos^2(\Gamma t) \tag{2.247}$$

and

$$|d(t)|^2 = \frac{\sin^2(\Gamma t)}{\Gamma^2}\Omega_\perp^2 \tag{2.248}$$

Substitution of Eq. (2.236) for Γ in Eqs. (2.247–248), and inserting the result into Eq. (2.246) for $\psi_s(t)$, then yields

$$|\psi_s(t)|^2 = \frac{\sin^2(\Gamma t)}{\Gamma^2}\left[\left(\Omega_{||} - \frac{w}{2}\right)^2 + \Omega_\perp^2\right] + \cos^2(\Gamma t) = \sin^2(\Gamma t) + \cos^2(\Gamma t) = 1 \tag{2.249}$$

Let us further examine the squared magnitude $|d(t)|^2$ of the spin-down (−) neutron state by plugging in for Γ from Eq. (2.236), which shows a resonance condition at $w = 2\Omega_{||}$, where

$$|d(t)|^2 = \frac{\sin^2(\Gamma t)}{\Gamma^2}\Omega_\perp^2 = \frac{\Omega_\perp^2}{(\Omega_{||} - (w/2))^2 + \Omega_\perp^2} \cdot \sin^2(\Gamma t) \tag{2.250}$$

If the transverse rotating magnetic field amplitude $B_\perp$ is small compared to the steady axial magnetic field amplitude $B_{||}$, where $B_\perp \ll B_{||}$, then the transverse precession frequency is much smaller than the axial precession frequency, where $\Gamma_\perp \ll \Gamma_{||}$. With $\Gamma_\perp \ll \Gamma_{||}$, the resonance is more pronounced at the resonance condition $w = 2\Omega_{||}$.

The flip of the neutron spin states from the higher energy spin-up (+) state to the lower energy spin-down (−) state transfers energy from the neutron to the rotating, transverse magnetic field $B_\perp$. If the neutron spin state flips from the lower energy spin-down (−) state to the higher energy spin-up (+) state, then energy is transferred from the rotating, transverse magnetic field $B_\perp$ to the neutron. In a thermal equilibrium, there are slightly more neutrons in the lower energy spin-down (−) state than in the higher energy spin-up (+), and as a result, there is a net absorption of energy by the neutrons from the transverse, rotating magnetic field $B_\perp$. The exchange of energy between spin-1/2 state neutrons, protons, electrons, carbon-13, fluorine-19, or phosphorous-31 and the RF coils of the transverse, rotating magnetic fields provides a RF signal that serves as the basis for magnetic resonance imaging.

If the initial neutron ensemble is comprised of only neutrons in the spin-up (+) with $u(0) = 1$ and $d(0) = 0$, then the number of neutron spin-up (+) states that flip to spin-down (−) states after the initial time $t > 0$ is maximized if the following condition is met. The transverse magnetic rotation frequency w must be set to the Larmor frequency w_L of neutron precession about the steady, z-directed magnetic field $B_{||}$, where

$$w = 2\Omega_{||} = \gamma B_{||} = \frac{2g\mu_n}{\hbar} B_{||} = w_L \tag{2.251}$$

Given (1) the z-directed, large, steady, DC magnetic field amplitude $B_{||}$, and (2) measurement of the rotation frequency w of the small transverse magnetic field $B_{\perp}$, where occurs the maximum number of neutron spin state flips, one can then measure g, the neutron Landé g factor. In general, the Landé g factor can be measured in terms of other two-spin-state particles, such as the proton and electron, by this resonant measurement method

2.16 Ferromagnetic Materials—Domains, Magnetization, Permeability, Susceptibility

We now examine the domain structure and the relation of the magnetization vector $\mathbf{M}$ to the magnetic flux density vector $\mathbf{B}$ and the applied magnetic field intensity vector $\mathbf{H}$ in ferromagnetic materials. The tangential component $\mathbf{M}_t$ of the magnetization vector $\mathbf{M}$ allows one to calculate the neutron refractive index of ferromagnetic materials. And the domains play a key role in ferromagnetic properties, such as hysteresis and maximum permanent magnetic field strength, as well as the refraction and scatter of neutrons.

In a ferromagnetic material, the net spin effects of orbit electrons constitutes a charge in motion, causing magnetic effects to occur on the atomic level. In most materials, the up and down electron spins of the atoms cancel each other out, so there is no net electron spin current or magnetic field. In ferromagnetic materials, there are unbalanced electron spins and a tendency of neighboring atoms to align such that their magnetic fields sum. In a specimen of unmagnetized ferromagnetic material, there are small domains of magnetic material in which all the atomic spins are aligned to produce a net magnetic field. Each domain, which is on the order of 25 μm in extent, is a region of intense magnetization; however, the domains are randomly ordered, and thus the material exhibits no net external magnetic field.

The magnetic flux density vector $\mathbf{B}$ [T] in a ferromagnetic material is the sum of the effects due to the applied magnetic field intensity vector $\mathbf{H}$ [A/m] and the magnetic polarization produced within the material leading to magnetization vector $\mathbf{M}$ [A/m]. With the vacuum permeability constant μ_0, this relation is expressed by

$$\mathbf{B} = \mu_0(\mathbf{H} + \mathbf{M}) \tag{2.252}$$

where

$$\mu_0 = 4\pi \times 10^{-7}\ \text{H/m} \quad (2.253)$$

The Henry or H unit for magnetic permeability can be expressed in terms of the basic units of length L, mass M, time T, and electric current A as

$$\text{H} = \text{MLT}^{-2}\text{A}^{-2} \quad (2.254)$$

The tesla unit (usually denoted by T) for magnetic flux density in terms of basic units is

$$\text{tesla} = \text{MT}^{-2}\text{I}^{-1} \quad (2.255)$$

In linear magnetic materials, the magnetization vector **M** is proportional to the magnetic field intensity vector **H**, via the proportionality constant, the magnetic susceptibility of the material χ_m (dimensionless number), where

$$\mathbf{M} = \chi_m \mathbf{H} \quad (2.256)$$

Hence, the magnetic flux density **B** [T] can be expressed as

$$\mathbf{B} = \mu_0(1 + \chi_m)\mathbf{H} \quad (2.257)$$

The relative permeability of the material μ_r relating **B** and **H** is denoted by

$$\mu_r = 1 + \chi_m \quad (2.258)$$

The relative permeability and susceptibility values for a vacuum is

$$\mu_r = 1 \quad \text{and} \quad \chi_m = 0 \quad (2.259)$$

The relative permeability and susceptibility values for air is

$$\mu_r \cong 1 \quad \text{and} \quad \chi_m = 3.6 \times 10^{-7} \quad (2.260)$$

For ferromagnetic elements, the relative permeability values are

$$\mu_r = 250 \quad \text{cobalt} \quad (2.261)$$

$$\mu_r = 600 \quad \text{nickel} \quad (2.262)$$

$$\mu_r = 5000 \quad \text{soft iron} \quad (2.263)$$

The relative permeability μ_r can be less than 1 for material such as copper and bismuth, which have completely compensated atomic magnetic moments such that they behave as superconductive current loops with no currents, as follows:

$$\mu_r = 0.99999 \quad \text{copper} \tag{2.264}$$

$$\mu_r = 0.99983 \quad \text{bismuth} \tag{2.265}$$

Substances with a relative permeability less than unity, $\mu_r < 1$, are called diamagnetic (μ_r is usually close to unity). Substances with a relative permeability greater than unity, $\mu_r > 1$, but with μ_r close to unity are called *paramagnetic.* Substances in which the relative permeability is very large ($\mu_r \gg 1$) are called ferromagnetic, that is,

$$\mu_r < 1 \quad \text{diamagnetic} \tag{2.266}$$

$$\mu_r > 1 \quad \text{paramagnetic} \tag{2.267}$$

$$\mu_r \gg 1 \quad \text{ferromagnetic} \tag{2.268}$$

The material permeability is denoted by

$$\mu = \mu_0 \mu_r \tag{2.269}$$

or

$$\mu = \mu_0(1 + \chi_m) \tag{2.270}$$

And the magnetic flux density vector **B** [T] is related to the applied magnetic field intensity vector **H** [A/m] via

$$\mathbf{B} = \mu \mathbf{H} \tag{2.271}$$

The magnetization vector **M** in the material can be calculated from the relative permeability and applied magnetic field intensity **H** via

$$\mathbf{M} = \left(\frac{\mu}{\mu_0} - 1 \right) \mathbf{H} \tag{2.272}$$

See Popović (1971) for more information on this topic.

2.17 Law of Refraction of Magnetic Field Lines

We now derive the law of refraction of magnetic field lines, which is useful in the design of magnet structures for neutron optics and allows quick visualization of the magnetic field lines. The lines of the magnetic field intensity vector **H** have sources and sinks or points where the **H** line begins and ends because **H** can have nonzero divergence since

$$\nabla \cdot \mathbf{H} = -\nabla \cdot \mathbf{M} \tag{2.273}$$

However, the lines of the magnetic flux density **B** are continuous by Gauss's Law:

$$\nabla \cdot \mathbf{B} = 0 \tag{2.274}$$

The boundary conditions for the normal components of the magnetic flux density vector **B** can be obtained by applying the conservation of magnetic flux law to a very thin cylinder or pillbox with its planar surfaces parallel to the boundary surface. Since the net magnetic flux through the closed surface is zero ($\nabla \cdot \mathbf{B} = 0$), one finds the components of **B** that are normal to the bottom surface B_{1n} and normal to the top surface B_{2n} are continuous, where

$$B_{1n} = B_{2n} \tag{2.275}$$

For the linear magnetic media with permeability μ_1 below the bottom pillbox surface, the magnetic flux density vector B_1 is related to the magnetic field intensity H_1 by

$$B_1 = \mu_1 H \tag{2.276}$$

And for linear magnetic media with permeability μ_2 above the top pillbox surface, B_2 is related to H_2 by

$$B_2 = \mu_2 H_2 \tag{2.277}$$

Thus, the boundary condition for continuity of the normal components of magnetic flux density B_n across a boundary can be expressed in terms of (1) the media permeability below and above the pillbox (μ_1 and μ_2, respectively), and (2) the normal component of magnetic field intensity at the pillbox bottom and top surfaces (H_{1n} and H_{2n}, respectively), where

$$\mu_1 H_{1n} = \mu_2 H_{2n} \tag{2.278}$$

The boundary condition for the tangential components of the magnetic field intensity H_t can be derived from Ampere's Law for a static magnetic field only and current density J, where

$$\nabla \times \mathbf{H} = \mathbf{J} \tag{2.279}$$

Assume that there are no surface currents ($\mathbf{J} = 0$) on the planar interface between two media. Consider a rectangular contour loop which, is split in half by the interface. The plane of the loop is perpendicular to the interface plane. From this contour, assuming that no surface currents cut through the loop, one finds that the tangential components of the magnetic field intensity are continuous across the interface. The magnetic field intensity components tangential to the interface, just above and below the interface, H_{1t} and H_{2t}, have

$$H_{1t} = H_{2t} \tag{2.280}$$

In the case of linear magnetic media, this can be expressed in another way by the tangential components of the magnetic flux density just above and below the interface, B_{1t} and B_{2t}, where

$$\frac{B_{1t}}{\mu_1} = \frac{B_{2t}}{\mu_2} \tag{2.281}$$

In the case of two linear media, a simple law of refraction of the lines of the magnetic flux density $\mathbf{B}$ can be derived as follows. The magnetic flux density vector $\mathbf{B}_1$ from the bottom media is "refracted" to $\mathbf{B}_2$ in the top media at the interface. Vector $\mathbf{B}_1$ makes angle α_1 measured from the surface normal of the interface bottom surface, and vector $\mathbf{B}_2$ makes angle α_2 measured from the surface normal of the interface top surface. From the boundary conditions for the normal components of $\mathbf{B}$ and tangential components of $\mathbf{H}$, one finds

$$\tan \alpha_1 = \frac{B_{1t}}{B_{1n}} \tag{2.282}$$

and

$$\tan \alpha_2 = \frac{B_{2t}}{B_{2n}} \tag{2.283}$$

since

$$\frac{B_{1t}}{B_{2t}} = \frac{\mu_1}{\mu_2} \tag{2.284}$$

The law of refraction of magnetic flux $\mathbf{B}$ lines is

$$\frac{\tan \alpha_1}{\tan \alpha_2} = \frac{\mu_1}{\mu_2} \tag{2.285}$$

An important application of this law of magnetic flux line refraction is the case where media 1 is a vacuum or air with $\mu_1 = \mu_0$, and media 2 is a ferromagnetic substance, in which μ is not constant and **B** and **H** have a nonlinear relation, where

$$\mu_2 \gg \mu_0 \tag{2.286}$$

Since the permeability of the ferromagnetic material is much larger than air or a vacuum,

$$\frac{\tan \alpha_1}{\tan \alpha_2} \cong 0 \tag{2.287}$$

The line of magnetic flux density **B** in air or a vacuum is approximately normal to the surfaces of the ferromagnetic material.

2.18 Ferromagnetic Materials with Applied Magnetic Fields and the Hysteresis Loop

The role of domains in the response of ferromagnetic materials to applied magnetic field is now examined. If an external magnetic field is applied to the material along the z-direction, for example, there is a tendency for the electron spin dipole magnetic to align with the applied magnetic field. For a small field intensity **H** applied in the positive z-direction, the domains are nearly aligned with **H** but grow at the expense of adjacent, less favorably aligned domains in an elastic (reversible) process. This results in an increase in the magnetic flux density **B** [T] over that in a vacuum. As **H** increases, the direction of the misaligned domains switches, causing an increase in **B**, which is rapid, but lags **H**. With this domain realignment go associated friction and transformation of magnetic energy into heat that causes the realignment to be an inelastic (irreversible) process. At higher values of magnetic field intensity **H**, the domains become increasingly aligned until the magnetic field intensity reaches H_{max}. At the maximum field intensity $\mathbf{H}_{max}$, no further domain alignment can occur and the magnetic flux density **B** in the material attains its maximum value B_{max}. The magnetic field flux density remains at B_{max} when the magnetic field intensity **H** is increased further above H_{max}. For commercial magnetic steel, the maximum magnetic flux density B_{max} at a north or south pole surface is about 1–2 T.

When the external field is removed, the ferromagnetic material does not return to the original state because in the magnetization, most of the domain realignment along **H** was inelastic. Rather, the magnetic flux density **B** decreases to the residual or remanent magnetic flux density B_R when $\mathbf{H} = 0$. The residual flux density is responsible for the existence of permanent magnets.

If the magnetic field intensity **H** is applied in the opposite direction of the previous application, that is, along the negative z-direction, the domains remain pointing in the positive z-direction until **H** becomes equal to the coercive magnetic field intensity H_c. At this point, the magnetic flux density of the domain $\mathbf{B}_D$ with material magnetization **M** exactly counters the magnetic flux density of the applied magnetic field B_A, where

$$\mathbf{B}_D = \mu_0 \mathbf{M} \tag{2.288}$$

$$\mathbf{B}_A = \mu_0 \mathbf{H} \tag{2.289}$$

where

$$\mathbf{B}_D = -\mathbf{B}_A \tag{2.290}$$

Hence, the total magnetic field **B** in the material is zero, where

$$\mathbf{B} = \mathbf{B}_A + \mathbf{B}_D = 0 \tag{2.291}$$

The increase of **H** in the negative z-direction elastically flips some of the domains. With even further increase of **H**, the domains flip inelastically until all the domains are realigned along the negative z-direction in which the magnetic field intensity reaches saturation $-H_{max}$. In this simple argument, symmetry is assumed, which may not be the case for many ferromagnetic materials. Then, decreasing this negative z-directed applied magnetic field intensity **H** back to zero yields a residual magnetic flux density $-B_R$ of the opposite polarity from the previous residual magnetic flux density.

If the magnetic field intensity **H** is again applied in the positive z-direction, the domains remain aligned in the negative z-direction up until the coercive magnetic field intensity H_c is reached. Increasing H above H_c leads to the saturated field intensity at a slightly different value of $H = H'_{max}$ in which the flux density is at the maximum, but at a slightly different value of B'_{max}. After about 10 cycles of sinusoidal variation of the applied magnetic field intensity, where $|\mathbf{H}| > H_{max}$, the hysteresis loop retraces itself and reaches the same saturation flux density values: B_{max} for H_{max}, and $-B_{max}$ for $-H_{max}$.

2.19 Calculation of the Magnetization Vector from Unpaired Atomic Electron Magnetic Dipole Moments

The calculation of the refractive index for neutrons in a ferromagnetic material is done in four steps. The first step is the calculation of the magnetization vector **M** from the constituent unpaired, atomic electron, magnetic dipole moments. The second step is the calculation of the potential energy U of neutron magnetic dipole moment interaction with the tangential magnetization vector component $\mathbf{M}_t$. The tangential magnetization vector component $\mathbf{M}_t$ represents the refractive interface between the medium of atomic magnetic dipole moment vectors of the ferromagnetic material and the air (or vacuum) medium. The tangential magnetization vector $\mathbf{M}_t$ is shown to be the difference in the magnetic flux density vector inside and outside the ferromagnetic material, namely $\mathbf{B}_i$ and $\mathbf{B}_0$, respectively. In the third step, the neutron potential energy U and the neutron total energy E (via its velocity v_0 or wavelength λ) are used to calculate the magnetic scatter length b_M, which, when added to the nuclear scatter length b, yields the total ferromagnetic, neutron scatter length b_F. In the final step, the total scatter length b^F is used to calculate the decrement δ, given the ferromagnetic atom density ρ and neutron wavelength λ, which then yields refractive index $n = 1 - \delta$ for neutrons in a ferromagnetic material.

In almost all materials, the neutron potential has the same value for both neutron spins, so there is no spin separation. The exceptions are ferromagnetic materials and magnetic fields applied to neutrons passing through materials or space. At each surface of a ferromagnetic material–based lens or prism, there is the additional force contribution $\mathbf{F}_M$ that arises from the gradient of the neutron ferromagnetic potential energy U_M:

$$\mathbf{F}_M = -\nabla U_M \tag{2.292}$$

The ferromagnetic potential energy U_M, imposed on the neutron via its magnetic moment μ_A, is the result of an applied magnetic field intensity $\mathbf{H}$. The magnetic field intensity $\mathbf{H}$ causes magnetization of the ferromagnetic material, with magnetic dipole number density ρ, by partial alignment of these magnetic atomic dipoles.

In MKS units, the Lorentz force equation is

$$\mathbf{F} = q(\mathbf{v} \times \mathbf{B}) \tag{2.293}$$

The Lorentz force equation defines the tesla (T) magnetic field unit as

$$\mathrm{T} = \mathrm{N\,A^{-1}\,m^{-1}} \tag{2.294}$$

The magnetic field $\mathbf{B}$ [T] is proportional to the sum of the magnetizing or auxiliary field $\mathbf{H}$ [A/M] (arising from free currents) and the magnetization $\mathbf{M}$ [A/M] (arising from material-bound currents) using

$$\mathbf{B} = \mu_0(\mathbf{H} + \mathbf{M}) \tag{2.295}$$

The permeability constant μ_0 for free space, expressed in units of newtons per ampere squared, is

$$\mu_0 = 4\pi \cdot 10^{-7}\ \mathrm{N/A^2} \tag{2.296}$$

The magnetization vector $\mathbf{M}$ for a material of volume V is the volume-averaged magnetic moment $\boldsymbol{\mu}_i$ of the ferromagnetic nuclei in the material, where

$$\mathbf{M} = \frac{1}{V}\sum_i \boldsymbol{\mu}_i \tag{2.297}$$

The magnetization vector M points from magnetic south pole S to north pole N in the same manner as the individual magnetic dipole moment vector $\boldsymbol{\mu}_i$, which is opposite to that of an applied magnetic field $\mathbf{B}$. The magnetic dipole moment vector $\boldsymbol{\mu}_i$ points from the magnetic north pole N to south pole S. A fully magnetized ferromagnetic material with magnetization M can be expressed in terms of the average atomic magnetic dipole moment $\boldsymbol{\mu}_\mathrm{A}$ and atom number density ρ of the atomic magnetic dipoles, where

$$\mathbf{M} = \rho\boldsymbol{\mu}_\mathrm{A} \tag{2.298}$$

We now derive an expression for the magnetization vector $\mathbf{M} = M\hat{\mathbf{m}}$ of the ferromagnetic material that will allow the calculation of its refractive index. The atomic magnetic dipole moments $\boldsymbol{\mu}_\mathrm{A}$ $[\mathrm{eV}/T]$ of the ferromagnetic atoms are all pointed in the direction of the magnetization $\mathbf{M}$, that is, in the direction of unit vector $\hat{\mathbf{m}}$. With the Bohr magnetron μ_b as a proportionality constant, the atomic dipole moment $\boldsymbol{\mu}_\mathrm{A}$ is expressed in terms of the effective spin I_A of the atom, where

$$\boldsymbol{\mu}_\mathrm{A} = 2\mu_\mathrm{b} I_\mathrm{A}\hat{\mathbf{m}} \tag{2.299}$$

With elementary charge e, electron rest mass m_e, and reduced Planck's constant $\hbar$, the Bohr magnetron μ_b is

$$\mu_\mathrm{b} = \frac{e\hbar}{2m_e} = 5.79 \times 10^{-5}\ \mathrm{eV/T} \tag{2.300}$$

With substitution, the atomic magnetic dipole moment $\boldsymbol{\mu}_\mathrm{A}$ $[\mathrm{eV/T}]$ becomes

$$\boldsymbol{\mu}_\mathrm{A} = 2\mu_\mathrm{b} I_\mathrm{A}\hat{\mathbf{m}} = 5.79 \times 10^{-5} I_\mathrm{A}\hat{\mathbf{m}} \tag{2.301}$$

Then the magnetization vector $\mathbf{M}$ [T] is

$$\mathbf{M}\,[\mathrm{T}] = \rho\boldsymbol{\mu}_\mathrm{A} = 2\rho\mu_\mathrm{b} I_\mathrm{A}\hat{\mathbf{m}} \tag{2.302}$$

or

$$\mathbf{M}\,[\mathrm{T}] = 5.79 \times 10^{-5}\rho \cdot I_\mathrm{A}\hat{\mathbf{m}} \tag{2.303}$$

2.20 Calculation of the Tangential Component of the Magnetization Vector from Magnetic Field Boundary Conditions

The expression for the tangential component $\mathbf{M}_t$ of the magnetization vector $\mathbf{M}$ at the surface of a ferromagnetic material is now shown to be the difference between the magnetic flux density outside $\mathbf{B}_0$ and inside $\mathbf{B}_i$. This is similar to the interaction potential between the neutron magnetic dipole moment vector $\boldsymbol{\mu}$ at an interface between a vacuum medium with the respective applied magnetic flux density vectors $\mathbf{B}_i$ and $\mathbf{B}_0$. The discontinuity of the magnetic flux density vector $\mathbf{B}$ at the ferromagnetic surface interface between the interior $\mathbf{B}_i$ and exterior $\mathbf{B}_0$ magnetic flux densities is

$$\mathbf{B}_i - \mathbf{B}_0 \tag{2.304}$$

This discontinuity of the magnetic flux density $\mathbf{B}$ gives rise to the neutron ferromagnetic potential energy U_M at the surface interface between the ferromagnetic material medium and the vacuum medium or air medium whose low atom density, allows approximation as a vacuum. The neutron magnetic dipole moment vector $\boldsymbol{\mu}$ and the magnetic potential energy U_M are

$$\boldsymbol{\mu} = \gamma\mathbf{S} = \frac{2g_n\mu_n}{\hbar}\mathbf{S} = g_n\mu_n\sqrt{3}\frac{\mathbf{S}}{|\mathbf{S}|} \tag{2.305}$$

and

$$U_M = -\boldsymbol{\mu}\cdot(\mathbf{B}_i - \mathbf{B}_0) \tag{2.306}$$

Inside the ferromagnetic material, the magnetic flux density $\mathbf{B}_i$ is proportional to the sum of (1) the applied magnetic field intensity $\mathbf{H}_i$, and (2) the magnetization vector $\mathbf{M}$, via the permeability μ_0, where

$$\mathbf{B}_i = \mu_0\mathbf{H}_i + \mu_0\mathbf{M} \tag{2.307}$$

The permeability constant μ_0 for free space in units of newtons per ampere squared is

$$\mu_0 = 4\pi \cdot 10^{-7}\ \mathrm{N/A^2} \tag{2.308}$$

Outside in the air (vacuum) is magnetic flux density in MKS units, which is related to the applied magnetic field intensity $\mathbf{H}_0$ via

$$\mathbf{B}_0 = \mu_0\mathbf{H}_0 \tag{2.309}$$

Recall Gauss's Law:

$$\nabla \cdot \mathbf{B} = 0 \tag{2.310}$$

The normal components of the magnetic flux density are continuous across the surface:

$$\mathbf{B}_{i\perp} = \mathbf{B}_{0\perp} \tag{2.311}$$

With no free currents, Ampere's Law gives

$$\nabla \times \mathbf{H} = 0 \tag{2.312}$$

Application of Stokes's theorem, shows the tangential components of the magnetic field intensity are continuous across the surface:

$$\mathbf{H}_{it} = \mathbf{H}_{0t} \tag{2.313}$$

Thus, at the surface, the tangential component of the magnetization $\mathbf{M}_t$ is found by subtracting the outside magnetic field $\mathbf{B}_0$ from the magnetic field inside the ferromagnetic material $\mathbf{B}_i$:

$$\mathbf{B}_i - \mathbf{B}_0 = \mu_0 \mathbf{M}_t \tag{2.314}$$

2.21 Calculation of the Neutron Potential Energy and Magnetic Scatter Length from the Tangential Component of the Magnetization Vector

With the calculated tangential component $\mathbf{M}_t$ of the magnetization vector $\mathbf{M}$, one can now calculate the potential energy of interaction between the neutron magnetic dipole moment and the summed magnetic dipole moments of the unpaired electrons in the ferromagnetic material. The potential energy U_M acquired by a neutron in a magnetized ferromagnet becomes

$$U_M = -\mu_0 \boldsymbol{\mu} \cdot \mathbf{M}_t \tag{2.315}$$

The component of the magnetization vector $\mathbf{M}$ [T] tangential to the surface is

$$\mathbf{M}_t\,[\mathrm{T}] = \rho(\boldsymbol{\mu}_\mathrm{A})_t = 2\rho\mu_\mathrm{b} I_\mathrm{A} \hat{\mathbf{m}}_t \tag{2.316}$$

or

$$\mathbf{M}_t\,[\mathrm{T}] = 5.79 \times 10^{-5} \rho \cdot I_\mathrm{A} \hat{\mathbf{m}}_t \tag{2.317}$$

With substitution, the ferromagnetic potential U_M of the neutron becomes

$$U_M = -4\rho\mu_n g\mu_0\mu_b I_A(\mathbf{S}\cdot\hat{\mathbf{m}}_t) \tag{2.318}$$

or

$$U_M = -4\pi\rho\frac{\hbar^2 r_0 g}{m_n}I_A(\mathbf{S}\cdot\hat{\mathbf{m}}_t) \tag{2.319}$$

The classical electron radius r_0 is

$$r_0 = \frac{\mu_0}{4\pi}\frac{e^2}{2m_n} = 2.818\times 10^{-15}\ \text{m} \tag{2.320}$$

The magnetic scattering length b_M for a scattering system is obtained from the neutron interactive potential U_M via the Fermi pseudo potential, where

$$b_M = \frac{m_n}{2\pi\hbar^2\rho}U_M \tag{2.321}$$

With substitution for U_M, the magnetic scatter length b_M becomes

$$b_M = -2r_0 g I_A(\mathbf{S}\cdot\hat{\mathbf{m}}_t) \tag{2.322}$$

with the Landé g factor, where

$$g = -1.91 \tag{2.323}$$

The longitudinal spin component S_z of the spin angular momentum $\mathbf{S}$ has two allowed magnetic quantum numbers:

$$S_z = \pm 1/2 \tag{2.324}$$

Recall that

$$S_z = S\cos\chi \tag{2.325}$$

Substitution gives

$$b_M = -2r_0 g I_A S_z \tag{2.326}$$

or

$$b_M = 5.38 I_A S_z \tag{2.327}$$

A spin-up neutron (+) is in the higher energy spin state $S_z = +1/2$. The spin-up neutron has (1) a longitudinal spin component S_z, which is parallel to $\mathbf{M}_t$, and (2) a longitudinal magnetic dipole moment component $\mu_{||}$, which is antiparallel to $\mathbf{M}_t$. The longitudinal magnetization $\mathbf{M}_t$ which is the magnetization vector component is tangential to the ferromagnetic material surface. The spin-up neutron has a positive ferromagnetic scatter length b_M, where

$$b_M \text{ [fm]} = 5.38 I_A \tag{2.328}$$

A spin-down neutron (−) is in the lower energy spin state $S_z = -1/2$. The spin-down neutron has (1) a longitudinal spin component S_z, which is antiparallel to $\mathbf{M}_t$, and (2) a longitudinal magnetic dipole moment component $\mu_{||}$, which is parallel to $\mathbf{M}_t$. The spin-down neutron has a negative ferromagnetic scatter length b_M, where

$$b_M \text{ [fm]} = -5.38 I_A \tag{2.329}$$

The average spins I_A of the ferromagnetic elements of iron, cobalt, and nickel are

$$1.11 \quad \text{iron} \tag{2.330}$$

$$0.86 \quad \text{cobalt} \tag{2.331}$$

$$0.30 \quad \text{nickel} \tag{2.332}$$

The magnetic scatter lengths b_M for the ferromagnetic elements of iron, cobalt, and nickel are

$$\pm 5.98 \quad \text{iron} \tag{2.333}$$

$$\pm 4.64 \quad \text{cobalt} \tag{2.334}$$

$$\pm 1.62 \quad \text{nickel} \tag{2.335}$$

See Sears (1989) for more information.

2.22 Refractive Decrement and Index for a Neutron in a Ferromagnetic Material

The neutron refractive index decrement δ for a ferromagnetic material is now obtained from the neutron scatter length sum b^F of the neutron magnetic scatter length b_M and nuclear scatter length b. A fully magnetized ferromagnetic material is birefringent to neutrons and uses scattering length b^F in place of nuclear scatter length b. The total bound coherent scattering length of each ferromagnetic isotope b^F is the sum of its material-bound coherent scattering length b and its magnetic scattering length b_M:

$$b^F = b \pm b_M \tag{2.336}$$

The spin-up neutrons with the higher spin-state energy have their azimuth spin component S_z parallel to the magnetization vector **M** and their magnetic dipole moment vector $\boldsymbol{\mu}$ antiparallel to **M**, where

$$b^F = b + b_M \tag{2.337}$$

The spin-down neutrons with the lower spin-state energy have their azimuth spin component S_z antiparallel to the magnetization vector **M** and their magnetic dipole moment vector $\boldsymbol{\mu}$ parallel to **M**, where

$$b^F = b - b_M \tag{2.338}$$

The higher energy spin-up neutrons with their azimuth spin component S_z parallel to the magnetization vector **M** are deflected in the same direction by nuclear and ferromagnetic scatter. However, the lower energy spin-down neutrons with S_z antiparallel to **M** are deflected in opposite directions by the nuclear and magnetic scatter. For example, in a ferromagnetic material prism, a beam of incident neutrons propagating parallel to the prism base, the higher energy spin-up neutrons are deflected upward, away from the prism base, by both the ferromagnetic field and the nuclear strong force. In contrast, the lower energy spin-down neutrons are deflected downward to the prism base by the ferromagnetic field but have an upward deflection component due to the nuclear strong force.

For a specified neutron wavelength λ [Å], the decrement delta δ of the refractive index $n = 1 - \delta$ can be expressed in terms of the neutron scattering length b^F [fm] and the nuclei number density ρ [#/cm^3] for a ferromagnetic material comprised of a single ferromagnetic isotope, where

$$\delta = \frac{6.01 \times 10^{-29} \lambda^2 \rho b^F}{2\pi} \tag{2.339}$$

The neutron refractive index in a ferromagnetic material comprised of one isotope for the higher spin-up neutrons, with S_z parallel to **M** and $\mu_{||}$ antiparallel to **M**, is

$$n = 1 - \frac{6.01 \times 10^{-29} \lambda^2 \rho}{2\pi}(b + b_M) \tag{2.340}$$

And for spin-down neutrons with their azimuthal spin component S_z antiparallel to **M** and the neutron azimuthal magnetic dipole moment component $\mu_{||}$ parallel to **M**, the refractive index for spin-down neutrons in the ferromagnetic material is

$$n = 1 - \frac{6.01 \times 10^{-29} \lambda^2 \rho}{2\pi}(b - b_M) \tag{2.341}$$

With incident, horizontal propagation parallel to a prism base, the higher energy state, spin-up neutrons are deflected upward from the base of a ferromagnetic prism, with spin S_z parallel to the magnetization vector **M**, and the prism-shaped, applied magnetic field **B**. In contrast, the lower energy state, spin-down neutrons are deflected toward the base of a ferromagnetic prism, with spin S_z anti-parallel to magnetization vector **M**, and the prism-shaped, applied magnetic field **B**. X-rays are deflected upward and visible light downward in material, ferromagnetic, and magnetic-field-shaped prisms, while spin-up and spin-down neutrons are both deflected upward from the prism base in nonferromagnetic, material prisms.

The magnetic scatter depends on the degree of magnetization of the ferromagnetic material. A ferromagnetic material with saturation magnetic field B_s can have internal magnetic field flux density B up to the saturation magnetic field flux density B_s, where

$$B \leq B_s \tag{2.342}$$

If the ferromagnetic material is not fully magnetized (saturated) and has a magnetic field flux density B that is less than the saturated magnetic field flux density B_s, then the magnetic scatter length b_M [fm] is multiplied by the ratio B/B_s. Thus, the refractive index for spin-up (+) and spin-down (−) neutrons of wavelength λ [Å] in a partly magnetic ferromagnetic material comprised of N isotopes with the ith isotope number density ρ_i [#/cm^3], nuclear scatter length b_i [fm], and magnetic scatter length $b_{M,i}$ [fm] becomes

$$n = 1 - \frac{6.01 \times 10^{-29} \lambda^2}{2\pi} \sum_{i=1}^{N} \rho_i \left(b_i \pm \frac{B}{B_s} b_{M,i} \right) \tag{2.343}$$

The decrement $\delta_\pm$ for spin-up (+) and spin-down (−) neutrons is then

$$\delta_\pm = \frac{6.01 \times 10^{-29} \lambda^2}{2\pi} \sum_{i=1}^{N} \rho_i \left(b_i \pm \frac{B}{B_s} b_{M,i} \right) \tag{2.344}$$

In the composite ferromagnetic material, the nonferromagnetic isotopes have zero magnetic scatter length and contribute only a nuclear scatter length.

References

Akitt, J. W., & Mann, B. E. (2000). *NMR and chemistry: An introduction to modern NMR spectroscopy*. Cheltenham: Stanley Thornes.

Brehm, J. J., & Mullin, W. J. (1989). *Introduction to the structure of matter*. New York, NY: Wiley.

Cottingham, W. N., & Greenwood, D. A. (2001). *An introduction to nuclear physics*. Cambridge: Cambridge University Press.

Cremer, J. T. (2012a). *Neutron and X-ray microscopy - part 1*. In P. W. Hawkes (Ed.), *Advances in imaging and electron physics* (172). London, UK: Academic Press, an imprint of Elsevier Press.

Cremer, J. T. (2012b). *Neutron and X-ray microscopy - part 2*. In P. W. Hawkes (Ed.), *Advances in imaging and electron physics* (173). London, UK: Academic Press, an imprint of Elsevier Press.

Frank, A. I. (1987). Optics of ultracold neutrons and the neutron microscope problem. (Usp. *Fiziol. Nauk* **151**) *Soviet Physics Uspekhi*, *30*(2), 110–132.

Kittel, C. (2005). *Introduction to solid state physics*. New York, NY: Wiley.

Liboff, R. L. (2003). *Introduction to quantum mechanics*. Reading, MA: Addison-Wesley.

Pauli, W. (1973). *Wave mechanics*. New York, NY: Dover.

Popović, B. D. (1971). *Introductory engineering electromagnetics*. Reading, MA: Addison-Wesley.

Sears, V. F. (1989). *Neutron optics*. Oxford, UK: Oxford University Press, Oxford, UK.

Siegman, A. G. (1971). *An introduction to lasers and masers*. New York, NY: McGraw-Hill.

Squires, G. L. (1996). *Introduction to the theory of thermal neutron scattering*. New York: Dover.

Stump, D. R., & Pollack, G. L. (1998). Radiation by a neutron in a magnetic field. *European Journal of Physics*, *19*(1), 59.

3 Magnetic Neutron Scatter from Magnetic Materials

3.1 Partial Differential Cross Section for Neutron Scatter in Magnetic Materials

In this chapter, we derive the partial differential cross section for neutron scatter from magnetic materials due to the interaction of the neutron magnetic dipole moment vector and the total magnetic dipole moment of the unpaired atomic electrons in a group of atoms. The scatter interaction has two components: (1) the interaction of the neutron magnetic dipole moment with the magnetic dipole moment associated with the unpaired atomic electron orbital motion and (2) the neutron magnetic dipole moment vector interaction with the magnetic dipole moment associated with the unpaired atomic electron spin. The partial differential cross section for these two component interactions add up to yield the total partial differential cross section for magnetic scatter of neutrons, which includes the thermal motion of the unpaired atomic electrons. Also derived is the differential coherent cross section for neutron elastic scatter in a magnetic material, which is expressed by the material magnetization vector M. Next, the coherent, elastic differential cross section, which is expressed by electron spin density, is derived. Finally, the thermal-averaged magnetization vector of a magnetic crystal is shown to be determined from the measured Bragg intensity. This chapter continues the neutron scatter presentation in AIEP Vols. 173 & 175 by Cremer (2012b, 2013), which follows Squires (1996).

For the group of N scattering nuclei, the partial differential cross section $(\mathrm{d}^2\sigma/d\Omega\,\mathrm{d}E)^{\mathrm{mag}}_{i\to m}$ accounts for the magnetic scatter of an incident neutron of mass m_n by the eth orbital electron. The electron integer index number e addresses all the orbital electrons belonging to the scattering system, which is the total number of electrons T, where $e = 1,2,3,\ldots,T$. If the number of unpaired electrons per atom is Z_{u}, then the total number of orbital electrons T in the scattering system of nuclei and their atomic electrons is

$$T = NZ_{\mathrm{u}} \tag{3.1}$$

Eq. (3.2) below is the partial differential cross section $(\mathrm{d}^2\sigma/d\Omega\,\mathrm{d}E)^{\mathrm{mag}}_{i,0s\to m,s}$ for (1) transition from initial scatter nuclei scatter system state i to final state m, and (2) transition from initial neutron spin state $0s$ to final state s. Without the neutron spin transition, Eq. (3.2) was derived in Eq. (483) & (456) in Ch. 3 of Cremer (2012b), and as Eq. (2.15)

Neutron and X-ray Optics. DOI: http://dx.doi.org/10.1016/B978-0-12-407164-3.00002-4

in Squires (1996), and with the neutron spin transition, as Eq. (149) in the chapter on nuclear scatter of neutron spin states in Cremer (2013), and Eq. (7.15) in Squires (1996):

$$\left(\frac{d^2\sigma}{d\Omega\, dE}\right)^{mag}_{i,0s\to m,s} = \left(\frac{m_n}{2\pi\hbar^2}\right)^2 \frac{k}{k_0} \sum_{e=1}^{T} |<\phi_i \psi_0 \psi_{0s} | \hat{U}_e | \phi_m \psi \psi_s>|^2 \delta(E_0 - E + E_m - E_i) \tag{3.2}$$

The magnetic potential U_e of the eth orbital electron is replaced by its multiplicative operator version $\hat{U}_e$, which will facilitate extracting the eigenvalues of the eigenfunction wave amplitudes for the neutron and electron spins. The magnetic potential energy operator $\hat{U}_e$, the neutron incident amplitude $\psi_0(\mathbf{r})$, and the scattered neutron amplitude $\psi(\mathbf{r})$ form the transition matrix element $\langle \hat{U}_e \rangle$, which is calculated first and yields the magnetic scatter length, where

$$\langle \hat{U}_e \rangle = \langle \psi | \hat{U}_e | \psi_0 \rangle \tag{3.3}$$

where

$$\psi_0 = \exp(\mathbf{i r} \cdot \mathbf{k}_0) \tag{3.4}$$

and

$$\psi = \exp(i\mathbf{r} \cdot \mathbf{k}) \tag{3.5}$$

The neutron with position vector $\mathbf{r}$ has incident wave vector $\mathbf{k}_0$ and scattered wave vector $\mathbf{k}$.

The interaction of the neutron magnetic dipole moment with the magnetic fields of the orbital unpaired electrons of the nuclei must account for the initial spin state of the neutron prior to scatter and the neutron final spin state after scatter. The initial neutron spin state ψ_{0s} has its spin-up (+) and spin-down (−) states denoted by ψ_{0+} and ψ_{0-}, and the final neutron spin state ψ_s has its spin-up (+) and spin-down (−) states denoted by ψ_{s+} and ψ_{s-}.

The ith initial ϕ_i and the mth final ϕ_m eigenstates of the scattering system nuclei and their unpaired atomic electrons form the third and final transition matrix element $\langle U_3 \rangle$ with the result of the second transition matrix element with operator $\langle \hat{U}_2 \rangle$, where

$$\langle U_3 \rangle = \langle \phi_i | \langle \hat{U}_2 \rangle | \psi_s \rangle \tag{3.6}$$

The third transition matrix element $\langle U_3 \rangle$ produces the resulting eigenvalue U_3 of the ith initial eigenstate ϕ_i and the mth final ϕ_m eigenstate for each of the J initial states ϕ_i and each of the M final eigenstates ϕ_m. The third transition matrix element $\langle U_3 \rangle$ is then weighted by the probability p_i of the ith initial eigenstate ϕ_i of the scattering system. The weighted sum of the resulting eigenvalues over the I initial eigenstates ϕ_i are weighted by the probability p_i, which is summed as $p_i U_3$ and then yields the probability amplitude of neutron magnetic scatter. Finally, the squared magnitude of this amplitude is the partial differential cross section.

The second transition matrix element $\langle \hat{U}_2 \rangle$ is formed with the neutron initial $\psi_{0s}(\mathbf{r})$ and the final $\psi_s(\mathbf{r})$ spin eigenstates, which results in the first transition matrix element $\langle \hat{U}_1 \rangle$, where

$$\langle \hat{U}_2 \rangle = \langle \psi_{0s} | \hat{U}_1 | \phi_m \psi \rangle \tag{3.7}$$

And the first transition matrix element $\langle \hat{U}_1 \rangle$ is the magnetic potential energy operator $\hat{U}_e$ with the neutron initial $\psi_0(\mathbf{r})$ and final $\psi(\mathbf{r})$ spatial eigenstates, where

$$\langle \hat{U}_1 \rangle = \langle \psi_0 | \hat{U}_e | \psi \rangle \tag{3.8}$$

From $\langle \hat{U}_1 \rangle$, one can obtain the scatter length for a single scattering element, such as the nuclear scatter length b of a single nucleus for nuclear neutron scatter, and the magnetic scatter length b_{m} for a single electron or the combined unpaired electrons of a single unit cell or atom for magnetic neutron scatter.

The potential energy U_B of magnetic interaction between the neutron and the eth unpaired electron of the T unpaired atomic electrons of the magnetic scatter system is represented by the potential energy operator $\hat{U}_e$. By virtue of the neutron magnetic dipole moment $\boldsymbol{\mu}$, the neutron acquires magnetic potential energy U_B in the electron magnetic field due to the eth electron spin $\mathbf{B}_S$ and the magnetic field due to orbital motion $\mathbf{B}_L$, where

$$U_B = -\boldsymbol{\mu} \cdot \mathbf{B} = -\boldsymbol{\mu} \cdot (\mathbf{B}_S + \mathbf{B}_L) \tag{3.9}$$

The total magnetic field $\mathbf{B}$ acting on the neutron is the sum of the magnetic fields $\mathbf{B}_S$ and $\mathbf{B}_L$ produced by the electron magnetic dipole moments arising from the unpaired electron spin and orbital angular momentum, respectively.

The Dirac delta function in Eq. (3.2) can be replaced by its integral equivalent. Recall that with incident neutron energy E_0 and scattered E neutron energy

$$w = \frac{E - E_0}{\hbar} \tag{3.10}$$

The Dirac delta function in Eq. (3.2) expresses the conservation of energy of the neutron and scattering atoms before and after scatter, where

$$\delta(E_0 - E + E_m - E_i) = \frac{1}{2\pi\hbar} \int_{-\infty}^{\infty} \mathrm{d}t \exp(-\mathrm{i}wt) \exp\left(\mathrm{i}\frac{E_m - E_i}{\hbar} t\right) \tag{3.11}$$

Plugging the integral equivalent for the Dirac delta function of Eq. (3.11) into Eq. (3.2), the partial differential cross section $(\mathrm{d}^2\sigma/\mathrm{d}\Omega\,\mathrm{d}E)^{\mathrm{mag}}_{i,0s\rightarrow m,s}$ for neutron scatter for the $i \rightarrow m$ scatter transition from the initial i^{th} to the mth final state becomes

$$\left(\frac{\mathrm{d}^2\sigma}{\mathrm{d}\Omega\,\mathrm{d}E}\right)^{\mathrm{mag}}_{i,0s\rightarrow m,s} = \left(\frac{m_n}{2\pi\hbar^2}\right)^2 \frac{k}{k_0} \sum_{e=1}^{T} \int_{-\infty}^{\infty} \mathrm{d}t \frac{\exp(-\mathrm{i}wt)}{2\pi\hbar} \left| <\phi_i \psi_0 \psi_{0s} | \hat{U}_e | \exp\left(-\mathrm{i}\frac{E_m}{\hbar} t\right) \phi_m \psi \psi_s> \right|^2 \tag{3.12}$$

The summation in Eq. (3.12) is over the $e = 1,2,3,\ldots T$ unpaired, atomic electrons of the scatter system nuclei, where the T unpaired atomic electrons have net magnetic moments, which interact with the magnetic moments of the incident neutrons. Eq. (3.12) for the partial differential cross section for the transition from initial to final nuclei scatter system state, $i \rightarrow m$, and transition from initial to final neutron spin state $0\,s \rightarrow s$ transition, is equivalently given by

$$\left(\frac{\mathrm{d}^2\sigma}{\mathrm{d}\Omega \mathrm{d}E}\right)^{\mathrm{mag}}_{i,0s\rightarrow m,s} = \left(\frac{m_n}{2\pi\hbar^2}\right)^2 \frac{k}{k_0}\sum_{e=1}^{T}\sum_{f=1}^{T}\int_{-\infty}^{\infty} \mathrm{d}t \frac{\exp(-\mathrm{i}wt)}{2\pi\hbar}$$

$$<\phi_i\psi_0\psi_{0s}|\hat{U}_e^*|\exp\left(-\mathrm{i}\frac{E_m}{\hbar}t\right)\phi_m\psi\psi_s><\phi_m\psi\psi_s|\hat{U}_f|\exp\left(\mathrm{i}\frac{E_i}{\hbar}t\right)\phi_i\psi_0\psi_{0s}> \tag{3.13}$$

One should not confuse the integer index i for the initial nuclei eigenstate energies E_i and eigenstates ϕ_i with the imaginary number $\mathrm{i} = \sqrt{-1}$, where

$$\exp(iE_it/\hbar) = \exp\left(\sqrt{-1}E_it/\hbar\right) \tag{3.14}$$

The JM pairs of J initial i^{th} eigenstates ϕ_i and the M final m^{th} eigenstate ϕ_m, both depend on the spatial positions $\mathbf{R}_n$ ($n = 1,2,\ldots,N$) of the N atoms to which are attached the unpaired electron scattering elements, where

$$\phi_i = \phi_i(\mathbf{R}_1, \mathbf{R}_2, \ldots, \mathbf{R}_N) \tag{3.15}$$

and

$$\phi_m = \phi_m(\mathbf{R}_1, \mathbf{R}_2, \ldots, \mathbf{R}_N) \tag{3.16}$$

The nth integration is performed for the nth nucleus position coordinate $\mathbf{R}_n$, where the nth differential volume element $\mathrm{d}V_n$ for integration over $\mathbf{R}_n$ in rectangular coordinates is

$$\mathrm{d}V_n = \mathrm{d}x_n\,\mathrm{d}y_n\,\mathrm{d}z_n \tag{3.17}$$

In spherical coordinates n^{th} differential volume element $\mathrm{d}V_n$ for integration over $\mathbf{R}_n$ for polar angle θ_n, azimuthal angle ϕ_n, and radius R_n is

$$\mathrm{d}V_n = R_n^2 \sin\theta_n\,\mathrm{d}\theta_n\,\mathrm{d}\phi_n\,\mathrm{d}R_n \tag{3.18}$$

These N-fold integrations must be done for each fth electron magnetic potential energy U_f contribution. The range of integer index f addresses for the magnetic potential energies U_f of the T unpaired, atomic E electrons for $f = 1,2,\ldots,T$. These N-fold integrations must be performed again for each eth electron magnetic potential energy U_e.

The resulting partial differential cross section for all transitions is found by summing the above partial differential cross sections for transition $i\rightarrow m$ over all the final states m for fixed initial state i and then taking the average over all the initial states i in which each initial state i occurs with probability p_i. With probability p_i^{mag} for the ith initial magnetic eigenstate of the scattering system, the total partial differential scatter cross section $(\mathrm{d}^2\sigma/d\Omega\,\mathrm{d}E)^{\text{mag}}$ for magnetic neutron scatter is obtained from Eq. (3.13), where

$$\left(\frac{\mathrm{d}^2\sigma}{\mathrm{d}\Omega\,\mathrm{d}E}\right)^{\text{mag}} = \sum_{i=1}^{J} p_i^{\text{mag}} \sum_{m=1}^{\mathrm{M}} \left(\frac{\mathrm{d}^2\sigma}{\mathrm{d}\Omega\,\mathrm{d}E}\right)^{\text{mag}}_{i,0s\to m,s} \tag{3.19}$$

3.2 The Transition Matrix Element for Neutron Magnetic Scatter

In this section, we begin the derivation of the total partial differential cross section $(\mathrm{d}^2\sigma/\mathrm{d}\Omega\,\mathrm{d}E)^{\text{mag}}$ for neutron scatter. First, one determines the transition matrix element for magnetic neutron scatter as follows. The incident neutron wave function $\psi_0(\mathbf{r})$ and wave vector $\mathbf{k}_0$ is scattered to the final-state neutron wave function $\psi(\mathbf{r})$ and wave vector $\mathbf{k}$ by the scattering system of unpaired atomic electrons. The transition matrix elements $\langle\hat{U}_e\rangle$ and $\langle\hat{U}_f\rangle$ represent the magnetic potential energy contributions from the eth and the fth electrons for the transition from the initial $\psi_0(\mathbf{r})$ to final $\psi(\mathbf{r})$ neutron states, where

$$\langle\hat{U}_e\rangle = \int_{-\infty}^{\infty}\mathrm{d}z\int_{\infty}^{\infty}\mathrm{d}y\int_{\infty}^{\infty}\mathrm{d}x\cdot\psi_0^*(\mathbf{r})\hat{U}_e\psi(\mathbf{r}) \tag{3.20}$$

and

$$\langle\hat{U}_f\rangle = \int_{-\infty}^{\infty}\mathrm{d}z\int_{\infty}^{\infty}\mathrm{d}y\int_{\infty}^{\infty}\mathrm{d}x\cdot\psi_0^*(\mathbf{r})\hat{U}_f\psi(\mathbf{r}) \tag{3.21}$$

In the integrals, the neutron position coordinate $\mathbf{r}$ is

$$\mathbf{r} = x\hat{\mathbf{x}} + y\hat{\mathbf{y}} + z\hat{\mathbf{z}} \tag{3.22}$$

The differential volume element $\mathrm{d}v$ in the integration of $\psi_0^* U_e\psi$ and $\psi_0^* U_f\psi$ over all space with respect to the neutron position (x,y,z) is

$$\mathrm{d}v = \mathrm{d}x\,\mathrm{d}y\,\mathrm{d}z \tag{3.23}$$

In the expression for the magnetic partial differential cross section, the incident total neutron wave amplitude $\psi_{0s}\psi_0(\mathbf{r})$ is the product of initial neutron spin amplitude ψ_{0s} multiplied by the incident neutron wave amplitude $\psi_0(\mathbf{r})$, which is a function of neutron position $\mathbf{r}$, where ψ_{0s} is not a function of $\mathbf{r}$. Likewise, the final total neutron wave amplitude after neutron scatter $\psi_s\psi(\mathbf{r})$ is the product of the final, position-independent neutron spin amplitude ψ_s and the position-dependent scattered neutron wave amplitude $\psi(\mathbf{r})$. With substitution for incident and scattered neutron wave amplitude $\psi_0(\mathbf{r})$ and $\psi(\mathbf{r})$, one gets

$$\langle \hat{U}_e \rangle = \int_{-\infty}^{\infty} \mathrm{d}z \int_{\infty}^{\infty} \mathrm{d}y \int_{\infty}^{\infty} \mathrm{d}x \cdot \exp(i\mathbf{r} \cdot \mathbf{k}_0) \hat{U}_e \exp(-i\mathbf{r} \cdot \mathbf{k}) \tag{3.24}$$

and

$$\langle \hat{U}_f \rangle = \int_{-\infty}^{\infty} \mathrm{d}z \int_{\infty}^{\infty} \mathrm{d}y \int_{\infty}^{\infty} \mathrm{d}x \cdot \exp(i\mathbf{r} \cdot \mathbf{k}_0) \hat{U}_f \exp(-i\mathbf{r} \cdot \mathbf{k}) \tag{3.25}$$

As stated before, the form of these expressions' expectation values $\langle \hat{U}_e \rangle$ and $\langle \hat{U}_f \rangle$ for the magnetic potential energy contributions from e^{th} and the fth electrons of the scattering system is identical to those of other scatter processes. In the case of a different scatter process, such as nuclear scattering, $\langle \hat{U}_e \rangle$ and $\langle \hat{U}_f \rangle$ then become the expectation values for the nuclear potential energy contributions from e^{th} and the fth nuclei of the scattering system.

3.3 Boltzmann Thermal Distribution of Initial Scatter System States

The total partial differential cross section $(\mathrm{d}^2\sigma/\mathrm{d}\Omega\, \mathrm{d}E)^{\text{mag}}$ for magnetic scatter of neutrons of Eq. (3.19) with Eq. (3.13), is now modified to include the scatter transition of the neutron spin state, which requires summation over the initial $0s = 1,2$ neutron spin states, which have probabilities p_{0s}^{mag}, and also requires summation over the scattered neutron spin states $s = 1,2$, where

$$\begin{aligned} \left(\frac{\mathrm{d}^2\sigma}{\mathrm{d}\Omega\, \mathrm{d}E}\right)^{\text{mag}} = \left(\frac{m_n}{2\pi\hbar^2}\right)^2 \frac{k}{k_0} \sum_{0s=1}^{2} p_{0s}^{\text{mag}} \sum_{i=1}^{J} p_i^{\text{mag}} \sum_{m=1}^{\text{M}} \sum_{s=1}^{2} \sum_{e=1}^{T} \sum_{f=1}^{T} \int_{-\infty}^{\infty} \mathrm{d}t \frac{\exp(-\mathrm{i}wt)}{2\pi\hbar} \\ \left\langle \phi_i \psi_{0s} \middle| \langle \hat{U}_e \rangle^* \middle| \exp\left(-\mathrm{i}\frac{E_m}{\hbar} t\right) \phi_m \psi_s \right\rangle \\ \left\langle \phi_m \psi_s \middle| \langle \hat{U}_f \rangle \middle| \exp\left(\mathrm{i}\frac{E_i}{\hbar} t\right) \phi_i \psi_{0s} \right\rangle \end{aligned} \tag{3.26}$$

The temperature dependence for the occurrence of the initial eigenstates of the magnetic scattering system is now examined. The temperature T determines the Boltzmann distribution of the initial nuclei states via the Boltzmann constant k_{b}, where the number of initial nuclear states is J. The probability of the initial i^{th} eigenstate of the magnetic scatter system is

$$p_i^{\text{mag}} = \frac{\exp\left(-E_i/k_{\text{b}}T\right)}{\sum_{i=1}^{J} \exp\left(-E_i/k_{\text{b}}T\right)} \tag{3.27}$$

We now examine the final magnetic scatter eigenstates of the scattering system. Recall that application of the Hamiltonian operator $\hat{H}$ gives

$$\exp\left(-\mathrm{i}\frac{\hat{H}}{\hbar}t\right)|\phi_m\rangle = \exp\left(-\mathrm{i}\frac{E_m}{\hbar}t\right)|\phi_m\rangle \tag{3.28}$$

and

$$\exp\left(-\mathrm{i}\frac{\hat{H}}{\hbar}t\right)|\phi_i\rangle = \exp\left(-\mathrm{i}\frac{E_i}{\hbar}t\right)|\phi_i\rangle \tag{3.29}$$

Substitution gives

$$\left(\frac{\mathrm{d}^2\sigma}{\mathrm{d}\Omega\,\mathrm{d}E}\right)^{\mathrm{mag}} = \left(\frac{m_n}{2\pi\hbar^2}\right)^2 \frac{k}{k_0} \sum_{0s=1}^{2} p_{0s}^{\mathrm{mag}} \sum_{i=1}^{J} p_i^{\mathrm{mag}} \sum_{m=1}^{M}\sum_{s=1}^{2}\sum_{e=1}^{T}\sum_{f=1}^{T} \int_{-\infty}^{\infty} \mathrm{d}t \frac{\exp(-\mathrm{i}wt)}{2\pi\hbar}$$
$$\left\langle \phi_i \psi_{0s} | \langle \hat{U}_e \rangle^* | \exp\left(-\mathrm{i}\frac{\hat{H}}{\hbar}t\right) \phi_m \psi_s \right\rangle$$
$$\left\langle \phi_m \psi_s | \langle \hat{U}_f \rangle | \exp\left(\mathrm{i}\frac{\hat{H}}{\hbar}t\right) \phi_i \psi_{0s} \right\rangle \tag{3.30}$$

Consider now the two possible spin states of the incident neutron: spin-up and spin-down. The summation over the two initial spin states of the neutron is indexed by $0s = 1,2$ and the summation over the final two neutron spin states is indexed by $s = 1,2$. And p_{0s}^{mag} is the probability of the initial $0s$th neutron spin state. For an unpolarized neutron beam, the probability for spin-up and spin-down neutrons is one-half, where $0s = 1,2$ and

$$p_{0s}^{\mathrm{mag}} = \frac{1}{2} \tag{3.31}$$

Next, we determine the effect of temperature and an applied magnetic field upon the initial neutron spin states. If an external magnetic field B is applied to the neutron beam with temperature T_{neut}, then the number of higher energy spin-up neutrons N_2, divided by the number N_1 of lower energy spin-down neutrons, is obtained from the Boltzmann equation:

$$\frac{N_2}{N_1} = \exp\left(-1.39 \times 10^{-3}\frac{B}{T_{\mathrm{neut}}}\right) \tag{3.32}$$

The probability of the occurrence of incident spin-up neutrons $p_{0s=+}^{\mathrm{mag}}$ and incident spin-down neutrons $p_{0s=-}^{\mathrm{mag}}$ is related by

$$p_{0s=-}^{\mathrm{mag}} = 1 - p_{0s=+}^{\mathrm{mag}} \tag{3.33}$$

where

$$p_{0s=+}^{\mathrm{mag}} = \frac{N_2}{N_1 + N_2} = \frac{1}{1 + \exp(1.39 \times 10^{-3}(B/T_{\mathrm{neut}}))} \tag{3.34}$$

Application of the closure relation and moving the summation signs adjacent to their indexed arguments in Eq. (3.30) for the partial differential cross section for magnetic neutron scattering by the scattering system of electron magnetic dipole moments give

$$\left(\frac{\mathrm{d}^2\sigma}{\mathrm{d}\Omega\,\mathrm{d}E}\right)^{\mathrm{mag}} = \left(\frac{m_n}{2\pi\hbar^2}\right)^2 \frac{k}{k_0} \sum_{0s=1}^{2} p_{0s}^{\mathrm{mag}} \sum_{i=1}^{J} p_i^{\mathrm{mag}} \int_{-\infty}^{\infty} \mathrm{d}t \frac{\exp(-\mathrm{i}wt)}{2\pi\hbar}$$
$$<\phi_i\psi_{0s}| \left[\sum_{e=1}^{E} <\hat{U}_e>^*\right] \exp\left(-\mathrm{i}\frac{\hat{H}}{\hbar}t\right) \left[\sum_{f=1}^{E} <\hat{U}_f>\right] \exp\left(\mathrm{i}\frac{\hat{H}}{\hbar}t\right) |\phi_i\psi_{0s}> \tag{3.35}$$

The relationship between Heisenberg operators $\langle \hat{U}_e(0) \rangle$ and $\langle \hat{U}_f(t) \rangle$ and Schrödinger operators $\langle \hat{U}_e \rangle^*$ and $\langle \hat{U}_f \rangle$ are

$$\langle \hat{U}_e(0) \rangle^* = \langle \hat{U}_e \rangle^* \tag{3.36}$$

and

$$\langle \hat{U}_f(t) \rangle = \exp\left(-\mathrm{i}\frac{\hat{H}}{\hbar}t\right) \langle \hat{U}_f \rangle \exp\left(\mathrm{i}\frac{\hat{H}}{\hbar}t\right) \tag{3.37}$$

Substitution yields

$$\left(\frac{\mathrm{d}^2\sigma}{\mathrm{d}\Omega\,\mathrm{d}E}\right)^{\mathrm{mag}} = \left(\frac{m_n}{2\pi\hbar^2}\right)^2 \frac{k}{k_0} \sum_{0s=1}^{2} p_{0s}^{\mathrm{mag}} \sum_{i=1}^{J} p_i^{\mathrm{mag}}$$
$$\int_{-\infty}^{\infty} \mathrm{d}t \frac{\exp(-\mathrm{i}wt)}{2\pi\hbar} \left\langle \phi_i\psi_{0s} \middle| \left[\sum_{e=1}^{T} \langle \hat{U}_e(0) \rangle^*\right] \left[\sum_{f=1}^{T} \langle \hat{U}_f(t) \rangle\right] \middle| \phi_i\psi_{0s} \right\rangle \tag{3.38}$$

For the scattering system of nuclei and their orbital electrons, the magnetic potential energy U_e for a neutron wave interaction with the eth electron corresponds to Heisenberg operator $\hat{U}_e(0)$ at the initial time $t = 0$. The magnetic potential energy U_f for the neutron wave interacting with the fth electron at the later time $t > 0$ corresponds to Heisenberg operator $\hat{U}_f(t)$. That is, the incident neutron wave scatters from the eth electron magnetic field at the initial time $t = 0$ and from the fth electron magnetic field at a later time t. The phase difference between the magnetic scattered waves depends on the position of fth electron at time t and the eth electron position at time $t = 0$.

Neutron coherent scatter requires fixed relative phase shift differences between the position of the same electron at different times and the fixed relative phase shift differences between different electrons at different times, which thus give rise to interference effects. Neutron incoherent scatter has no fixed relative phase differences between different electrons at different times. The incoherent scatter depends only on the correlation between the same electron position at different times, in which case there are no interference effects.

3.4 Magnetic Fields of Unpaired Atomic Electrons in Magnetic Materials

We now derive the potential energy expressions for the interaction of the neutron magnetic dipole moment with the magnetic dipole moment of the unpaired atomic electrons of the magnetic scattering system, which arise from the unpaired electron spin and orbital angular momentum. The scattering system is comprised of a lattice of atoms and their atomic electrons. The unpaired electrons have a net spin and orbital angular momentum, which produces a net electron magnetic dipole moment. Thus, the unpaired atomic electron magnetic field interacts with the neutron magnetic dipole moment.

The general magnetic potential energy Schrödinger operator expression $\hat{U}_e$ for an arbitrary e^{th} electron of the scattering system are first derived, from which one can obtain the corresponding expressions for the Heisenberg operators $\hat{U}_e(0)$ and $\hat{U}_f(t)$. Magnetic scatter of neutrons with matter occurs due to the interaction of the neutron magnetic dipole moment with magnetic fields of the electrons associated with the nuclei.

In particular, the magnetic field $\mathbf{B}_S$ of the electron dipole moment arising from the electron spin that acts on the neutron magnetic dipole moment $\boldsymbol{\mu}_n$ is given by

$$\mathbf{B}_S = \nabla \times \mathbf{A} \tag{3.39}$$

The neutron is position $\mathbf{R}$, which has a distance R with unit vector direction $\hat{\mathbf{R}}$ from the electron position. The magnetic vector potential $\mathbf{A}$ at the neutron position $\mathbf{R}$, due to the magnetic spin dipole moment of the eth orbital electron $\boldsymbol{\mu}_e$ with vacuum permeability μ_0, is

$$\mathbf{A} = \frac{\mu_0}{4\pi} \frac{\boldsymbol{\mu}_e \times \hat{\mathbf{R}}}{R^2} \tag{3.40}$$

The magnetic field $\mathbf{B}_S$ at the neutron position $\mathbf{R}$ is

$$\mathbf{B}_S = \nabla \times \mathbf{A} \tag{3.41}$$

or

$$\mathbf{B}_S = \frac{\mu_0}{4\pi}\frac{\nabla \times (\boldsymbol{\mu}_e \times \mathbf{R})}{R^3} \tag{3.42}$$

Next, we consider the unpaired electron magnetic dipole moment arising from the angular momentum of the effective current loop of the orbiting electron. The Biot–Savart Law gives the magnetic field arising from the eth electron orbital current I that acts on the neutron magnetic dipole moment in which the magnetic field $\mathbf{B}_L$ at distance $R = |\mathbf{R}|$. At neutron position $\mathbf{R}$, the magnetic field arising from the electron current element $I\,\mathbf{dI}$ along a differential length $\mathbf{dI}$ in its orbital path is given by

$$\mathbf{B}_L = \frac{\mu_0 I}{4\pi}\frac{\mathrm{d}\mathbf{l} \times \mathbf{R}}{R^3} \tag{3.43}$$

The momentum $\mathbf{p}$ of the electron of charge e and rest mass m_e gives rise to the current element $I\,\mathbf{dI}$ in its orbital motion about its nucleus. The Bohr magnetron μ_b for the electron is given by

$$\mu_\text{b} = \frac{e\hbar}{2m_e} \tag{3.44}$$

Substitution gives

$$I\,\mathrm{d}\mathbf{l} = -\frac{e}{m_e}\mathbf{p} = -\frac{2\mu_\text{b}}{\hbar}\mathbf{p} \tag{3.45}$$

Substitution gives the magnetic field contribution $\mathbf{B}_L$ at the neutron position $\mathbf{R}$ arising from the lattice e^{th} electron orbital motion:

$$\mathbf{B}_L = -\frac{\mu_0}{4\pi}\frac{2\mu_\text{b}}{\hbar}\frac{\mathbf{p} \times \mathbf{R}}{R^3} \tag{3.46}$$

The magnetic field $\mathbf{B}_e$ at the neutron position $\mathbf{R}$ is the sum of the contributions from the electron magnetic dipole moment arising from the eth electron spin and orbital motion, where

$$\mathbf{B}_e = \mathbf{B}_{S,e} + \mathbf{B}_{L,e} \tag{3.47}$$

Substitution of Eq. (3.42) and 3.46 into Eq. (3.47), gives the total e^{th} electron magnetic field at the neutron position $\mathbf{R}$ in terms of the eth electron magnetic dipole $\boldsymbol{\mu}_e$ and electron momentum $\mathbf{p}_e$:

$$\mathbf{B}_e = \frac{\mu_0}{4\pi}\left[\frac{\nabla \times (\boldsymbol{\mu}_e \times \mathbf{R})}{R^3} - \frac{2\mu_b}{\hbar}\frac{\mathbf{p}_e \times \mathbf{R}}{R^3}\right] \tag{3.48}$$

where

$$\nabla = \hat{\mathbf{x}}\frac{\partial}{\partial x} + \hat{\mathbf{y}}\frac{\partial}{\partial y} + \hat{\mathbf{z}}\frac{\partial}{\partial z} \tag{3.49}$$

and

$$\mathbf{R} = \hat{\mathbf{x}}x + \hat{\mathbf{y}}y + \hat{\mathbf{z}}z \tag{3.50}$$

3.5 Neutron Magnetic Potential Energy due to the Total Electron Magnetic Dipole Moment

Via its magnetic dipole moment $\boldsymbol{\mu}$, the neutron acquires magnetic potential energy U_e in the electron magnetic field due to the eth electron spin and orbital motion, where

$$U_e = -\boldsymbol{\mu} \cdot \mathbf{B}_e \tag{3.51}$$

With Eq. (3.48) Eq. (3.51) becomes

$$U_e = \frac{\mu_0}{4\pi}\boldsymbol{\mu} \cdot \left[\frac{\nabla \times (\boldsymbol{\mu}_e \times \hat{\mathbf{R}})}{R^2} - \frac{2\mu_b}{\hbar}\frac{\mathbf{p}_e \times \hat{\mathbf{R}}}{R^2}\right] \tag{3.52}$$

The neutron position unit vector $\hat{\mathbf{R}}$ that points from the eth electron to the neutron situated at distance R from the eth electron is

$$\hat{\mathbf{R}} = \frac{\mathbf{R}}{R} \tag{3.53}$$

The eth electron magnetic dipole moment $\boldsymbol{\mu}_e$ is given by

$$\boldsymbol{\mu}_e = 2\mu_b \mathbf{s}_e \tag{3.54}$$

Here, $\mathbf{s}_e$ is the eth electron spin angular momentum, which is expressed in units of the reduced Planck's constant $\hbar = h/2\pi$ and is multiplied by the Bohr magnetron μ_b for the electron given electron rest mass m_e and charge e to obtain magnetic dipole moment $\boldsymbol{\mu}_e$. The eth electron spin angular momentum $\mathbf{s}_e$ of $\pm$ 1/2 are in units of the reduced Planck's constant $\hbar = h/2\pi$. Corresponding to the eth electron spin $\mathbf{s}_e$ and magnetic dipole moment $\boldsymbol{\mu}_e$ are the eth electron operators $\hat{\mathbf{s}}_e$ and $\hat{\boldsymbol{\mu}}_e$, where

$$\hat{\boldsymbol{\mu}}_e = 2\mu_b \hat{\mathbf{s}}_e \tag{3.55}$$

The wave amplitude of the initial φ_i and final φ_m eigenstate of the group of nuclei are the eigenfunctions of the eth electron spin operator $\hat{\mathbf{s}}_e$, and $\pm\hbar/2$ are its eigenvalues with the eigenequations

$$\hat{\mathbf{s}}_e \phi_i = \pm\frac{\hbar}{2}\phi_i \tag{3.56}$$

and

$$\hat{\mathbf{s}}_e \phi_m = \pm\frac{\hbar}{2}\phi_m \tag{3.57}$$

The neutron spin $\mathbf{S}_n$, which is expressed in units of the reduced Planck's constant $\hbar = h/2\pi$, gives rise to neutron magnetic dipole moment $\boldsymbol{\mu}$, where

$$\boldsymbol{\mu} = 2g\mu_n \mathbf{S}_n \tag{3.58}$$

Here, g is the neutron Landé g factor, where

$$g = -1.91 \tag{3.59}$$

For elementary charge e and proton rest mass m_p, the nuclear magnetron μ_n is

$$\mu_n = \frac{e\hbar}{2m_p} \tag{3.60}$$

Corresponding to the neutron spin $\mathbf{S}_n$ and neutron magnetic dipole moment $\boldsymbol{\mu}$ are the neutron spin operator $\hat{\mathbf{S}}_n$ and neutron magnetic dipole moment operator $\hat{\boldsymbol{\mu}}$, where

$$\hat{\boldsymbol{\mu}} = 2g\mu_n \hat{\mathbf{S}}_n \tag{3.61}$$

The magnetic potential energy U_e, which is acquired by the neutron in the magnetic field, and arises from the eth electron magnetic dipole moment of Eq. (3.52) becomes

$$U_e = \frac{\mu_0}{4\pi} g\mu_n \hat{\mathbf{S}}_n \cdot \left[\frac{\nabla \times (\boldsymbol{\mu}_e \times \hat{\mathbf{R}})}{R^2} - \frac{2\mu_b}{\hbar}\frac{\mathbf{p}_e \times \hat{\mathbf{R}}}{R^2}\right] \tag{3.62}$$

The neutron position unit vector $\hat{\mathbf{R}}$ can be replaced by its operator $\hat{\mathbf{R}}$, in which the hat denotes both unit vector and operator status. Likewise, the electron momentum $\mathbf{p}_e$ and magnetic dipole moment $\boldsymbol{\mu}_e$ can be replaced by the operators $\hat{\mathbf{p}}_e$ and $\hat{\boldsymbol{\mu}}_e$. The magnetic field vector operator $\hat{\mathbf{B}}$ replaces the magnetic field $\mathbf{B}$. As a result, the magnetic potential energy U_e interaction between the neutron and e^{th} electron can be replaced by its scalar operator $\hat{U}_e$, which is the dot product of the neutron magnetic dipole vector operator $\hat{\boldsymbol{\mu}}$ with the magnetic field vector operator $\hat{\mathbf{B}}_e$, where

$$\hat{U}_e = -\hat{\boldsymbol{\mu}}\cdot\hat{\mathbf{B}}_e \tag{3.63}$$

and

$$\hat{U}_e = -\frac{\mu_0}{4\pi}g\mu_n\hat{\mathbf{S}}_n\cdot\left[-\frac{\nabla\times(\hat{\boldsymbol{\mu}}_e\times\hat{\mathbf{R}})}{R^2}+\frac{2\mu_b}{\hbar}\frac{\hat{\mathbf{p}}_e\times\hat{\mathbf{R}}}{R^2}\right] \tag{3.64}$$

The electron magnetic dipole moment vector operator $\hat{\boldsymbol{\mu}}_e$, which is proportional to the eth electron spin operator $\hat{\mathbf{s}}_e$via the constant $-2\mu_b$, is derived as follows:

$$\hat{\boldsymbol{\mu}}_e = -2\mu_b\hat{\mathbf{s}}_e \tag{3.65}$$

Substitution of the eth electron magnetic dipole moment vector operator $\hat{\boldsymbol{\mu}}_e$ into the scalar operator $\hat{U}_e$ for the eth electron magnetic field contribution to the neutron potential energy gives

$$\hat{U}_e = -\frac{\mu_0 g\mu_n\mu_b}{2\pi}\hat{\mathbf{S}}_n\cdot\left[\frac{\nabla\times(\hat{\mathbf{s}}_e\times\hat{\mathbf{R}})}{R^2}+\frac{1}{\hbar}\frac{\hat{\mathbf{p}}_e\times\hat{\mathbf{R}}}{R^2}\right] \tag{3.66}$$

To facilitate physical understanding and ultimately simplify the mathematical expression for the cross sections, the first and second terms of the eth electron magnetic potential operator $\hat{U}_e$ are further developed as discussed next.

3.6 Neutron Magnetic Potential Energy Due to the Electron Spin Magnetic Dipole Moment

The first term in Eq. (3.66) for the magnetic potential operator $\hat{U}_e$ arises from the interaction of the neutron magnetic dipole moment and the spin of the eth lattice electron of the T electrons of the scattering system, and is evaluated as follows:

$$\frac{\nabla\times\hat{\mathbf{s}}_e\times\hat{\mathbf{R}}}{R^2} = \nabla\times\left(\hat{\mathbf{s}}_e\times\nabla\frac{1}{R}\right) \tag{3.67}$$

where

$$\frac{\hat{\mathbf{R}}}{R^2} = -\nabla\frac{1}{R} \tag{3.68}$$

Now we will replace $1/R$ by a useful expression that is obtained from an integral transform. This integral transform equation relates the arbitrary vector $\mathbf{q}$ with magnitude $q = |\mathbf{q}|$ and neutron position vector $\mathbf{R}$, which yields the reciprocal of the neutron distance $1/R$ from the eth electron. Integration is performed with respect to $\mathbf{q}$ over an infinite spherical volume in a spherically symmetric q-space with differential volume element $\mathbf{dq}$. For polar θ and azimuthal ϕ angles, the differential volume element $\mathbf{dq}$ is

$$\mathbf{dq} = q^2 \sin\theta \, \mathrm{d}\theta \, \mathrm{d}\phi \tag{3.69}$$

Only the component of $\mathbf{q}$ in the direction of $\mathbf{R}$ contributes to the integration over azimuthal angle ϕ, in which

$$\sin\theta \, \mathrm{d}\theta = -\mathrm{d}(\cos\theta) \tag{3.70}$$

The useful integral transform equation is

$$\int \mathbf{dq} \frac{\exp(i\mathbf{q}\cdot\mathbf{R})}{q^2} = \int_0^{2\pi} \mathrm{d}\phi \int_0^{\infty} \mathrm{d}q \int_{-1}^{1} \exp(\mathrm{i}q\,R\cos\theta)\mathrm{d}(\cos\theta) \tag{3.71}$$

Integration over the polar θ and azimuth ϕ angles gives

$$\int_0^{2\pi} \mathrm{d}\phi \int_0^{\infty} \mathrm{d}q \int_{-1}^{1} \exp(\mathrm{i}q\,R\cos\theta)\mathrm{d}(\cos\theta) = 4\pi \int_0^{\infty} \frac{\sin(qR)}{qR} \mathrm{d}q \tag{3.72}$$

Integration over q produces the useful expression for $1/R$, given by

$$\int \mathbf{dq} \frac{\exp(\mathrm{i}\mathbf{q}\cdot\mathbf{R})}{q^2} = \frac{2\pi^2}{R} \tag{3.73}$$

Application of the ∇ to $1/R$ gives

$$\nabla \frac{1}{R} = \nabla \left[\frac{1}{2\pi^2} \int \mathbf{dq} \frac{\exp(\mathrm{i}\mathbf{q}\cdot\mathbf{R})}{q^2} \right] \tag{3.74}$$

Substitution for $\nabla(1/R)$ into the first term of Eq. (3.66) for the magnetic potential operator $\hat{U}_e$ gives

$$\frac{\nabla\times(\hat{\mathbf{s}}_e\times\hat{\mathbf{R}})}{R^2} = -\nabla\times\left(\hat{\mathbf{s}}_e\times\nabla\left[\frac{1}{2\pi^2}\int \mathbf{dq}\frac{\exp(\mathrm{i}\mathbf{q}\cdot\mathbf{R})}{q^2}\right]\right) \tag{3.75}$$

Vector variable $\mathbf{q}$ in rectangular coordinates is given by

$$\mathbf{q} = \hat{\mathbf{x}}q_x + \hat{\mathbf{y}}q_y + \hat{\mathbf{z}}q_z \tag{3.76}$$

The ∇ operator in Eq. (3.75) can be brought inside the integral since it operates only on variable $\mathbf{R}$:

$$\nabla\left[\frac{1}{2\pi^2}\int d\mathbf{q}\frac{\exp(i\mathbf{q}\cdot\mathbf{R})}{q^2}\right]=\frac{1}{2\pi^2}\int d\mathbf{q}\nabla\left[\frac{\exp(i\mathbf{q}\cdot\mathbf{R})}{q^2}\right] \tag{3.77}$$

where

$$\nabla[\exp(i\mathbf{q}\cdot\mathbf{R})]=\left[\hat{\mathbf{x}}\frac{\partial}{\partial x}+\hat{\mathbf{y}}\frac{\partial}{\partial y}+\hat{\mathbf{z}}\frac{\partial}{\partial z}\right]\exp(i\mathbf{q}\cdot\mathbf{R}) \tag{3.78}$$

With

$$\left[\hat{\mathbf{x}}\frac{\partial}{\partial x}+\hat{\mathbf{y}}\frac{\partial}{\partial y}+\hat{\mathbf{z}}\frac{\partial}{\partial z}\right]\exp(i\mathbf{q}\cdot\mathbf{R})=\hat{\mathbf{x}}iq_x\frac{\partial x}{\partial x}+\hat{\mathbf{y}}iq_y\frac{\partial y}{\partial y}+\hat{\mathbf{z}}iq_z\frac{\partial z}{\partial z}=i\mathbf{q} \tag{3.79}$$

one obtains

$$\nabla[\exp(i\mathbf{q}\cdot\mathbf{R})]=i\mathbf{q} \tag{3.80}$$

Hence,

$$\frac{1}{2\pi^2}\int d\mathbf{q}\nabla\left[\frac{\exp(i\mathbf{q}\cdot\mathbf{R})}{q^2}\right]=\frac{1}{2\pi^2}\int d\mathbf{q}\frac{i\mathbf{q}\exp(i\mathbf{q}\cdot\mathbf{R})}{q^2} \tag{3.81}$$

and

$$\nabla\left[\frac{1}{2\pi^2}\int d\mathbf{q}\frac{\exp(i\mathbf{q}\cdot\mathbf{R})}{q^2}\right]=\frac{1}{2\pi^2}\int d\mathbf{q}\frac{i\mathbf{q}\exp(i\mathbf{q}\cdot\mathbf{R})}{q^2} \tag{3.82}$$

The first term for the magnetic potential operator $\hat{U}_e$ of Eq. (3.66), becomes

$$\frac{\nabla\times(\hat{\mathbf{s}}_e\times\hat{\mathbf{R}})}{R^2}=-\frac{i}{2\pi^2}\int d\mathbf{q}\frac{\nabla\times(\hat{\mathbf{s}}_e\times\mathbf{q})\exp(i\mathbf{q}\cdot\mathbf{R})}{q^2} \tag{3.83}$$

The ∇ operator is independent of $\hat{\mathbf{s}}_e$ and $\mathbf{q}$, but it depends on the separation vector $\mathbf{R}$ between the eth electron and the neutron, so that

$$\int d\mathbf{q}\frac{\nabla\times(\hat{\mathbf{s}}_e\times\mathbf{q})\exp(i\mathbf{q}\cdot\mathbf{R})}{q^2}=-\int d\mathbf{q}\frac{(\hat{\mathbf{s}}_e\times\mathbf{q})\times\nabla\exp(i\mathbf{q}\cdot\mathbf{R})}{q^2} \tag{3.84}$$

where

$$\nabla\times(\hat{\mathbf{s}}_e\times\mathbf{q})=-(\hat{\mathbf{s}}_e\times\mathbf{q})\times\nabla \tag{3.85}$$

Since

$$-\int d\mathbf{q}\frac{(\hat{\mathbf{s}}_e\times\mathbf{q})\times\nabla\exp(i\mathbf{q}\cdot\mathbf{R})}{q^2}=-i\int d\mathbf{q}\frac{(\hat{\mathbf{s}}_e\times\mathbf{q})\times\mathbf{q}\exp(i\mathbf{q}\cdot\mathbf{R})}{q^2} \tag{3.86}$$

then

$$\int d\mathbf{q}\frac{\nabla\times(\hat{\mathbf{s}}_e\times\mathbf{q})\exp(i\mathbf{q}\cdot\mathbf{R})}{q^2}=-i\int d\mathbf{q}\frac{(\hat{\mathbf{s}}_e\times\mathbf{q})\times\mathbf{q}\exp(i\mathbf{q}\cdot\mathbf{R})}{q^2} \tag{3.87}$$

with

$$(\hat{\mathbf{s}}_e\times\mathbf{q})\times\mathbf{q}\exp(i\mathbf{q}\cdot\mathbf{R})=-\mathbf{q}\times(\hat{\mathbf{s}}_e\times\mathbf{q})\exp(i\mathbf{q}\cdot\mathbf{R}) \tag{3.88}$$

The first term in Eq. (3.66) for the magnetic potential operator $\hat{U}_e$, the electron spin part of the neutron and e^{th} electron interaction potential, thus becomes

$$\frac{\nabla\times(\hat{\mathbf{s}}_e\times\hat{\mathbf{R}})}{R^2}=\frac{1}{2\pi^2}\int d\mathbf{q}\frac{\mathbf{q}\times(\hat{\mathbf{s}}_e\times\mathbf{q})\exp(i\mathbf{q}\cdot\mathbf{R})}{q^2} \tag{3.89}$$

3.7 Neutron Magnetic Potential Energy Due to the Electron Orbital Magnetic Dipole Moment

In Eq. (3.66) the second term for the magnetic potential operator $\hat{U}_e$ arises from the interaction of the neutron magnetic dipole moment and the orbital motion of the lattice e^{th} electron, where

$$\hat{\mathbf{p}}_e\times\frac{\hat{\mathbf{R}}}{R^2}=-\hat{\mathbf{p}}_e\times\nabla\frac{1}{R} \tag{3.90}$$

With

$$\nabla\frac{1}{R}=\nabla\left[\frac{1}{2\pi^2}\int d\mathbf{q}\frac{\exp(i\mathbf{q}\cdot\mathbf{R})}{q^2}\right] \tag{3.91}$$

one obtains

$$\hat{\mathbf{p}}_e\times\frac{\hat{\mathbf{R}}}{R^2}=-\frac{1}{2\pi^2}\hat{\mathbf{p}}_e\times\nabla\int d\mathbf{q}\frac{\exp(i\mathbf{q}\cdot\mathbf{R})}{q^2} \tag{3.92}$$

The second term in Eq. (3.66) for the magnetic potential operator $\hat{U}_e$ thus becomes

$$\hat{\mathbf{p}}_e\times\frac{\hat{\mathbf{R}}}{R^2}=-\frac{i}{2\pi^2}\int d\mathbf{q}\frac{(\hat{\mathbf{p}}_e\times\mathbf{q})\exp(i\mathbf{q}\cdot\mathbf{R})}{q^2} \tag{3.93}$$

3.8 Evaluation of the Matrix Element for the Neutron Magnetic Potential Energy

We now evaluate the matrix element $\langle \hat{U}_e \rangle$ of the neutron magnetic potential operator with the incident ψ_0 and scattered ψneutron wave functions, where

$$\langle \hat{U}_e \rangle = \langle \psi | \hat{U}_e | \psi_0 \rangle \tag{3.94}$$

With substitution of Eqs. (3.89) and (3.93) into Eq. (3.66), the total magnetic potential energy operator $\hat{U}_e$ between the incident neutron and e^{th} electron of the scattering lattice, is then

$$\hat{U}_e = -\frac{\mu_0 g \mu_n \mu_b}{4\pi^3} \hat{\mathbf{S}}_n \cdot \int_{-\infty}^{\infty} dq_z \int_{-\infty}^{\infty} dq_y \int_{-\infty}^{\infty} dq_x \left[\frac{\mathbf{q} \times (\hat{\mathbf{s}}_e \times \mathbf{q}) - (i/\hbar)(\hat{\mathbf{p}}_e \times \mathbf{q})}{q^2} \right] \exp(i\mathbf{q} \cdot \mathbf{R}) \tag{3.95}$$

Integration over the q-space is denoted by

$$\int d\mathbf{q} \tag{3.96}$$

In rectangular coordinates, the differential volume $d\mathbf{q}$ in the q-space is

$$d\mathbf{q} = dq_x \, dq_y \, dq_z \tag{3.97}$$

Once the operator expression $\hat{U}_e$ is determined, operator $\hat{U}_e$ is plugged into the transition matrix element expression $\langle \hat{U}_e \rangle$, so that Eq. (3.94) becomes

$$\langle \hat{U}_e \rangle = -\frac{\mu_0 g \mu_n \mu_b}{4\pi^3} \left\langle \psi_0 \left| \int_{-\infty}^{\infty} dq_z \int_{-\infty}^{\infty} dq_y \int_{-\infty}^{\infty} dq_x [\hat{\mathbf{S}}_n \cdot \hat{\mathbf{q}}_{\perp e}] \exp(i\mathbf{q} \cdot \mathbf{R}) \right| \psi \right\rangle \tag{3.98}$$

where

$$\left\langle \psi_0 \left| \int_{-\infty}^{\infty} dq_z \int_{-\infty}^{\infty} dq_y \int_{-\infty}^{\infty} dq_x [\hat{\mathbf{S}}_n \cdot \hat{\mathbf{q}}_{\perp e}] \exp(i\mathbf{q} \cdot \mathbf{R}) \right| \psi \right\rangle = \int_{-\infty}^{\infty} dz \int_{\infty}^{\infty} dy \int_{\infty}^{\infty} dx \cdot \exp(i\mathbf{r} \cdot \mathbf{k}_0) \left\{ \int_{-\infty}^{\infty} dq_z \int_{-\infty}^{\infty} dq_y \int_{-\infty}^{\infty} dq_x [\hat{\mathbf{S}}_n \cdot \hat{\mathbf{q}}_{\perp e}] \exp(i\mathbf{q} \cdot \mathbf{R}) \right\} \exp(-i\mathbf{r} \cdot \mathbf{k}) \tag{3.99}$$

Vector operator $\hat{\mathbf{q}}_{\perp e}$ for the eth electron is given by

$$\hat{\mathbf{q}}_{\perp e} = \frac{\mathbf{q} \times (\hat{\mathbf{s}}_e \times \mathbf{q}) - (i/\hbar)(\hat{\mathbf{p}}_e \times \mathbf{q})}{q^2} \tag{3.100}$$

Vector operator $\hat{\mathbf{q}}_{\perp e}$ for the eth electron is perpendicular to vector $\mathbf{q}$, which is a mathematical artifice of a convenient mathematical solution. This leads to an expression for the partial differential cross section that clearly conveys its underlying physical basis. In the next step, the arbitrary vector $\mathbf{q}$ is integrated out. This leaves an expression in terms of eth electron spin, $\hat{\mathbf{s}}_e$and $\hat{\mathbf{p}}_e$ momentum operators, and the neutron scatter vector $\boldsymbol{\kappa}$. To aid integration over the neutron position $\mathbf{r}$ (relative to the lattice origin), neutron vector position $\mathbf{r}$ is replaced by

$$\mathbf{r} = \mathbf{r}_e + \mathbf{R} \tag{3.101}$$

Here, the eth electron position $\mathbf{r}_e$ relative to the lattice origin is fixed. However, the distance $R = |\mathbf{R}|$ from the eth electron to the neutron varies with neutron position $\mathbf{r}$. Neutron position $\mathbf{r}$ and electron position $\mathbf{r}_e$ are independent of each other, and the matrix element for the neutron magnetic potential operator of Eq. (3.94) becomes

$$\langle \hat{U}_e \rangle = -\frac{\mu_0 g \mu_n \mu_b}{4\pi^3} \exp(i\boldsymbol{\kappa} \cdot \mathbf{r}_e) \int_{-\infty}^{\infty} dq_z \int_{-\infty}^{\infty} dq_y \int_{-\infty}^{\infty} dq_x \left\{ \hat{\mathbf{S}}_n \cdot \hat{\mathbf{q}}_{\perp e} \right\}$$

$$\left\{ \int_{-\infty}^{\infty} dz \int_{\infty}^{\infty} dy \int_{\infty}^{\infty} dx \cdot \exp[i\mathbf{R} \cdot (\boldsymbol{\kappa} + \mathbf{q})] \right\} \tag{3.102}$$

The neutron scatter vector $\boldsymbol{\kappa}$ is:

$$\boldsymbol{\kappa} = \mathbf{k}_0 - \mathbf{k} \tag{3.103}$$

The neutron to electron separation $\mathbf{R}$ is:

$$\mathbf{R} = \mathbf{r} - \mathbf{r}_e \tag{3.104}$$

The eth constant electron position $\mathbf{r}_e$ is constant, where

$$\mathbf{r}_e = \hat{\mathbf{x}} x_e + \hat{\mathbf{y}} y_e + \hat{\mathbf{z}} z_e \tag{3.105}$$

where

$$\mathbf{R} = \hat{\mathbf{x}}(x - x_e) + \hat{\mathbf{y}}(y - y_e) + \hat{\mathbf{z}}(z - z_e) \tag{3.106}$$

The neutron position $\mathbf{r}$ is, then

$$\mathbf{r} = \mathbf{r}_e + \mathbf{R} \tag{3.107}$$

Next, the Dirac delta function is applied to the integration over the neutron position (x, y, z). Recall that the Dirac delta function $\delta(\mathbf{K})$ in K-space is the Fourier transform of the unity R-space function $f(\mathbf{R})$, where

$$f(\mathbf{R}) = 1 \tag{3.108}$$

$$\mathbf{K} = \boldsymbol{\kappa} + \boldsymbol{\tau} \tag{3.109}$$

and

$$\int_{-\infty}^{\infty} \mathrm{d}z \int_{\infty}^{\infty} \mathrm{d}y \int_{\infty}^{\infty} \mathrm{d}x \cdot \exp[\mathrm{i}\mathbf{R} \cdot (\boldsymbol{\kappa} + \mathbf{q})] = (2\pi)^3 \delta(\boldsymbol{\kappa} + \mathbf{q}) \tag{3.110}$$

The Dirac delta function and the full expression for $\hat{\mathbf{q}}_{\perp e}$ (to clarify the next integration step) are plugged into $\langle \hat{U}_e \rangle$, so that Eq. (3.102) becomes

$$\begin{aligned} \langle \hat{U}_e \rangle = & -2\mu_0 g \mu_n \mu_b \exp(\mathrm{i}\boldsymbol{\kappa} \cdot \mathbf{r}_e) \int_{-\infty}^{\infty} \mathrm{d}q_z \int_{-\infty}^{\infty} \mathrm{d}q_y \\ & \int_{-\infty}^{\infty} \mathrm{d}q_x \left[\hat{\mathbf{S}}_n \cdot \frac{\mathbf{q} \times (\hat{\mathbf{s}}_e \times \mathbf{q}) - (i/\hbar)(\hat{\mathbf{p}}_e \times \mathbf{q})}{q^2} \right] \delta(\boldsymbol{\kappa} + \mathbf{q}) \end{aligned} \tag{3.111}$$

The Dirac delta function $\delta(\boldsymbol{\kappa} + \mathbf{q})$ is nonzero only when

$$\mathbf{q} = -\boldsymbol{\kappa} \tag{3.112}$$

Performing the integration over the q-space gives

$$\langle \hat{U}_e \rangle = -2\mu_0 g \mu_n \mu_b \exp(\mathrm{i}\boldsymbol{\kappa} \cdot \mathbf{r}_e) \left[\hat{\mathbf{S}}_n \cdot \frac{\boldsymbol{\kappa} \times (\hat{\mathbf{s}}_e \times \boldsymbol{\kappa}) + (i/\hbar)(\hat{\mathbf{p}}_e \times \boldsymbol{\kappa})}{\kappa^2} \right] \tag{3.113}$$

The neutron spin operator $\hat{\mathbf{S}}_n$ is independent of electron position $\mathbf{r}_e$, and thus $\hat{\mathbf{S}}$ can be moved to the left of the term

$$\exp(\mathrm{i}\boldsymbol{\kappa} \cdot \mathbf{r}_e) \tag{3.114}$$

The eth electron position $\mathbf{r}_e$ is replaced by its operator $\hat{\mathbf{r}}_e$, and the electron momentum operator $\hat{\mathbf{p}}_e$ is switched to the right so as to be adjacent to its eigenfunction, via

$$\hat{\mathbf{p}}_e \times \boldsymbol{\kappa} = -\boldsymbol{\kappa} \times \hat{\mathbf{p}}_e \tag{3.115}$$

The matrix element $\langle \hat{U}_e \rangle$ of Eq. (3.94) for magnetic scatter for incident ψ_0 and scattered ψ neutron wave functions thus becomes

$$\langle \hat{U}_e \rangle = K\langle \psi | \hat{U}_e | \psi_0 \rangle = -2\mu_0 g \mu_n \mu_b \hat{\mathbf{S}}_n \cdot \left\{ \exp(\mathrm{i}\boldsymbol{\kappa} \cdot \hat{\mathbf{r}}_e) \frac{\boldsymbol{\kappa} \times (\hat{\mathbf{s}}_e \times \boldsymbol{\kappa}) - (i/\hbar)(\boldsymbol{\kappa} \times \hat{\mathbf{p}}_e)}{\kappa^2} \right\} \tag{3.116}$$

In comparison, for nuclear scatter of neutrons, the matrix element $\langle \hat{U}_j \rangle$ for the jth atom with neutron scatter length b_j and position $\mathbf{R}_j$ in a nonmagnetic material scatter system is

$$\langle \hat{U}_j \rangle = \langle \psi | \hat{U}_j | \psi_0 \rangle = b_j \exp(\mathrm{i}\boldsymbol{\kappa} \cdot \mathbf{R}_j) \tag{3.117}$$

The neutron scatter length b_j for the short-range nuclear strong force is constant and does not depend either on the neutron scatter vector $\boldsymbol{\kappa}$ or the neutron wavelength. In contrast, the magnetic scatter matrix element is complicated by the vector interactions of the neutron magnetic dipole moment with the magnetic dipole moment of the electron spin and magnetic dipole moment of the electron current due to the atomic electron orbital motion. The magnetic forces of the unpaired atomic electrons acts are noncentral forces and act on the neutron magnetic dipole moment over a much greater distance than the nuclear strong force interaction, which is confined to the nucleus radius.

3.9 Electron Magnetic Dipole Moment Operator for Unpaired Atomic Electrons

Now we define the perpendicular component vector operator $\hat{\mathbf{Q}}_{\perp}(\boldsymbol{\kappa})$ of the magnetic dipole moment, vector operator $\hat{\mathbf{Q}}(\boldsymbol{\kappa})$ for the scattering system electrons in reciprocal lattice space, which is the summation, indexed by e, over the E electrons of the scattering system. The dot product of the neutron spin vector operator $\hat{\mathbf{S}}_n$ and the perpendicular component vector operator $\hat{\mathbf{Q}}_{\perp}(\boldsymbol{\kappa})$ is

$$\hat{\mathbf{S}}_n \cdot \hat{\mathbf{Q}}_{\perp}(\boldsymbol{\kappa}) = \sum_{e=1}^{E} \exp(\mathrm{i}\boldsymbol{\kappa} \cdot \hat{\mathbf{r}}_e) \left\{ \frac{\boldsymbol{\kappa} \times (\hat{\mathbf{s}}_e \times \boldsymbol{\kappa}) - (i/\hbar)(\boldsymbol{\kappa} \times \hat{\mathbf{p}}_e)}{\kappa^2} \right\} \tag{3.118}$$

where

$$\mu_c \hat{\mathbf{S}}_n \cdot \hat{\mathbf{Q}}_{\perp}(\boldsymbol{\kappa}) = \sum_{e=1}^{E} \langle \hat{U}_e(0) \rangle = \sum_{e=1}^{E} \langle \hat{U}_e \rangle \tag{3.119}$$

and

$$\mu_c [\hat{\mathbf{S}}_n \cdot \hat{\mathbf{Q}}_{\perp}(\boldsymbol{\kappa})]^* = \mu_c \hat{\mathbf{Q}}_{\perp}{}^*(\boldsymbol{\kappa}) \cdot \hat{\mathbf{S}}_n^* = \sum_{e=1}^{E} \langle \hat{U}_e(0) \rangle^* = \sum_{e=1}^{E} \langle \hat{U}_e \rangle^* \tag{3.120}$$

with

$$\mu_c = -2\mu_0 g \mu_n \mu_b \tag{3.121}$$

Since the summation of f and e are over the same set of E scattering system electrons, then

$$\mu_c \hat{\mathbf{S}}_n \cdot \hat{\mathbf{Q}}_\perp(\boldsymbol{\kappa}) = \mu_c \hat{\mathbf{S}}_n(0) \cdot \hat{\mathbf{Q}}_\perp(\boldsymbol{\kappa}, 0) = \sum_{f=1}^{E} \langle \hat{U}_f(0) \rangle = \sum_{f=1}^{E} \langle \hat{U}_f \rangle \tag{3.122}$$

and

$$\mu_c \hat{\mathbf{S}}_n(t) \cdot \hat{\mathbf{Q}}_\perp(\boldsymbol{\kappa}, t) = \sum_{f=1}^{E} \langle \hat{U}_f(t) \rangle \tag{3.123}$$

The time-independent Schrödinger scalar $\hat{U}$ and vector operators $\hat{\mathbf{r}}_e$, $\hat{\mathbf{S}}_n$, $\hat{\mathbf{s}}_e$, and $\hat{\mathbf{p}}_e$ are replaced by their equivalently time-dependent, Heisenberg, scalar $\hat{U}_f(t)$, and vector operators $\mathbf{r}_e(t)$, $\hat{\mathbf{S}}_n(t)$, $\hat{\mathbf{s}}_e(t)$, and $\hat{\mathbf{p}}_e(t)$. That is, with

$$\hat{T} = \exp\left(-\frac{\mathrm{i}\hat{H}t}{\hbar}\right) \tag{3.124}$$

and

$$\hat{T}^{-1} = \exp\left(\frac{\mathrm{i}\hat{H}t}{\hbar}\right) \tag{3.125}$$

The transformation between Heisenberg $\hat{A}(\mathrm{t})$ and Schrodinger $\hat{A}$ operators is covered in Ch. 3 of Cremer (2012b), where

$$\hat{A}(t) = \hat{T}\hat{A}\hat{T}^{-1} \tag{3.126}$$

One starts with transforming Eq. (3.118), via Eq. (3.126), where

$$\langle \hat{U}_f(t) \rangle = \mu_c \hat{T}\hat{\mathbf{S}}_n \cdot \left\{ \exp(i\boldsymbol{\kappa} \cdot \hat{\mathbf{r}}_e) \frac{\boldsymbol{\kappa} \times (\hat{\mathbf{s}}_f \times \boldsymbol{\kappa}) - (i/\hbar)(\boldsymbol{\kappa} \times \hat{\mathbf{p}}_f)}{\kappa^2} \right\} \hat{T}^{-1} \tag{3.127}$$

The inverse transform of a transform is unity and vice-versa is unity, where

$$\hat{T}^{-1}\hat{T} = \hat{T}\hat{T}^{-1} = 1 \tag{3.128}$$

Eq. (3.128) as a unity multiplier between the Schrödinger operators, causes the conversion of individual Schrödinger operators to Heisenberg operators, where

$$\langle \hat{U}_f(t) \rangle = \mu_c \hat{T}\hat{\mathbf{S}}_n \hat{T}^{-1}\hat{T} \cdot \left\{ \exp(\mathrm{i}\boldsymbol{\kappa} \cdot \hat{\mathbf{r}}_e)\hat{T}^{-1}\hat{T} \frac{\boldsymbol{\kappa} \times (\hat{\mathbf{s}}_f \times \boldsymbol{\kappa}) - (i/\hbar)(\boldsymbol{\kappa} \times \hat{\mathbf{p}}_f)}{\kappa^2} \right\} \hat{T}^{-1} \tag{3.129}$$

Grouping by $\hat{T}\hat{A}\hat{T}^{-1}$ at each of the Schrödinger operator positions in the equation yields the corresponding Heisenberg operator $\hat{A}(t)$, where

$$\langle \hat{U}_f(t)\rangle = -\mu_c \hat{\mathbf{S}}_n(t)\cdot\left\{\exp(\mathrm{i}\boldsymbol{\kappa}\cdot\hat{\mathbf{r}}_f(t))\frac{\boldsymbol{\kappa}\times(\hat{\mathbf{s}}_f(t)\times\boldsymbol{\kappa})-(i/\hbar)(\boldsymbol{\kappa}\times\hat{\mathbf{p}}_f(t))}{\kappa^2}\right\} \quad (3.130)$$

With substitution of Eq. (3.130) for $\langle \hat{U}_e\rangle$ into Eq. (3.38), using Eq. (3.120) and 3.123, the partial differential cross section for magnetic scatter of an incident neutron from the electrons of a group of nuclei, becomes

$$\left(\frac{\mathrm{d}^2\sigma}{\mathrm{d}\Omega\,\mathrm{d}E}\right)^{\mathrm{mag}} = \left(\frac{m_p\mu_c}{2\pi\hbar^2}\right)^2\frac{k}{k_0}$$
$$\sum_{0s=1}^{2} p_{0s}^{\mathrm{mag}}\sum_{i=1}^{J} p_i^{\mathrm{mag}}\int_{-\infty}^{\infty}\mathrm{d}t\frac{\exp(-\mathrm{i}wt)}{2\pi\hbar} \quad (3.131)$$
$$\langle\phi_i\psi_{0s}|\left[\hat{\mathbf{S}}_n\cdot\hat{\mathbf{Q}}_\perp(\boldsymbol{\kappa})\right]\left[\hat{\mathbf{S}}_n(t)\cdot\hat{\mathbf{Q}}_\perp(\boldsymbol{\kappa},t)\right]|\phi_i\psi_{0s}\rangle$$

The constants in this magnetic cross section expression are now expanded and simplified in which $g=-1.91$ is the neutron Landé g factor:

$$\frac{m_p\mu_c}{2\pi\hbar^2} = -2\mu_0 g\mu_n\mu_b\frac{m_p}{2\pi\hbar^2} = -2\mu_0 g\frac{e\hbar}{2m_p}\frac{e\hbar}{2m_e}\frac{m_p}{2\pi\hbar^2} = -r_0 g \quad (3.132)$$

Here, the constant r_0 is the classical radius of the electron (2.818 fm), where

$$r_0 = \frac{\mu_0}{4\pi}\frac{e^2}{m_e} = 2.818\times10^{-15}\ \mathrm{m} \quad (3.133)$$

The partial differential cross section for magnetic scatter of Eq. (3.131), becomes

$$\left(\frac{\mathrm{d}^2\sigma}{\mathrm{d}\Omega\,\mathrm{d}E}\right)^{\mathrm{mag}} = (r_0 g)^2\frac{k}{k_0}$$
$$\sum_{0s=1}^{2} p_{0s}^{\mathrm{mag}}\sum_{i=1}^{J} p_i^{\mathrm{mag}}\int_{-\infty}^{\infty}\mathrm{d}t\frac{\exp(-\mathrm{i}wt)}{2\pi\hbar} \quad (3.134)$$
$$\langle\phi_i\psi_{0s}|\left[\hat{\mathbf{S}}_n\cdot\hat{\mathbf{Q}}_\perp(\boldsymbol{\kappa})\right]\left[\hat{\mathbf{S}}_n(t)\cdot\hat{\mathbf{Q}}_\perp(\boldsymbol{\kappa},t)\right]|\phi_i\psi_{0s}\rangle$$

where

$$\hat{\mathbf{S}}_n(t)\cdot\hat{\mathbf{Q}}_\perp(\boldsymbol{\kappa},t) = \sum_{f=1}^{E}\hat{\mathbf{S}}_n(t)\cdot\left\{\exp(\mathrm{i}\boldsymbol{\kappa}\cdot\hat{\mathbf{r}}_f(t))\frac{\boldsymbol{\kappa}\times(\hat{\mathbf{s}}_f(t)\times\boldsymbol{\kappa})-(i/\hbar)(\boldsymbol{\kappa}\times\hat{\mathbf{p}}_f(t))}{\kappa^2}\right\}$$
$$(3.135)$$

3.10 Magnetic Dipole Moment Operator and Magnetization Vector—the Spin Component

We show operator $\hat{\mathbf{Q}}(\boldsymbol{\kappa})$ is directly proportional to the reciprocal lattice magnetization vector $\mathbf{M}(\boldsymbol{\kappa})$ of the scattering system as follows. The perpendicular electron magnetic dipole moment dipole operator $\hat{\mathbf{Q}}_{\perp}(\boldsymbol{\kappa})$ is the sum of the perpendicular electron spin component $\hat{\mathbf{Q}}_{\perp s}(\boldsymbol{\kappa})$ and orbital component $\hat{\mathbf{Q}}_{\perp l}(\boldsymbol{\kappa})$,

$$\hat{\mathbf{Q}}_{\perp} = \hat{\mathbf{Q}}_{\perp s} + \hat{\mathbf{Q}}_{\perp l} \tag{3.136}$$

where

$$\hat{\mathbf{Q}}_{\perp s} = \sum_{e=1}^{E} \exp(i\boldsymbol{\kappa} \cdot \hat{\mathbf{r}}_e)\left\{\frac{\boldsymbol{\kappa} \times (\hat{\mathbf{s}}_e \times \boldsymbol{\kappa})}{\kappa^2}\right\} \tag{3.137}$$

and

$$\hat{\mathbf{Q}}_{\perp l} = \sum_{e=1}^{E} \exp(i\boldsymbol{\kappa} \cdot \hat{\mathbf{r}}_e)\left\{\frac{-(i/\hbar)(\boldsymbol{\kappa} \times \hat{\mathbf{p}}_e)}{\kappa^2}\right\} \tag{3.138}$$

Operator $\hat{\mathbf{Q}}_{\perp s}(\boldsymbol{\kappa})$ is the perpendicular component of the electron spin magnetic dipole moment operator $\hat{\mathbf{Q}}_s(\boldsymbol{\kappa})$, where

$$\hat{\mathbf{Q}}_{\perp s}(\boldsymbol{\kappa}) = \frac{\boldsymbol{\kappa} \times (\hat{\mathbf{Q}}_s \times \boldsymbol{\kappa})}{\kappa^2} \tag{3.139}$$

The neutron scatter vector $\boldsymbol{\kappa}$ can be represented by the unit vector neutron scatter operator $\hat{\boldsymbol{\kappa}}_{\mathrm{u}}$, given by

$$\boldsymbol{\kappa}_{\mathrm{u}} = \frac{\boldsymbol{\kappa}}{\kappa} = \frac{\mathbf{k} - \mathbf{k}_0}{|\mathbf{k} - \mathbf{k}_0|} \tag{3.140}$$

One can express $\hat{\mathbf{Q}}_{\perp s}(\boldsymbol{\kappa})$ in terms of the unit neutron scatter operator $\hat{\boldsymbol{\kappa}}_{\mathrm{u}}$, where

$$\hat{\mathbf{Q}}_{\perp s} = \hat{\boldsymbol{\kappa}}_u \times \hat{\mathbf{Q}}_s \times \hat{\boldsymbol{\kappa}}_{\mathrm{u}} \tag{3.141}$$

Comparison of Eqs. (3.137) and (3.141) for perpendicular component $\hat{\mathbf{Q}}_{\perp s}$ of operator $\hat{\mathbf{Q}}_s$ yields the expression for the electron spin magnetic dipole moment vector operator $\hat{\mathbf{Q}}_s(\boldsymbol{\kappa})$:

$$\hat{\mathbf{Q}}_s = \sum_{e=1}^{E} \exp(i\boldsymbol{\kappa} \cdot \hat{\mathbf{r}}_e)\hat{\mathbf{s}}_e \tag{3.142}$$

The electron spin density is given by the electron spin density operator $\hat{\boldsymbol{\rho}}_s(\mathbf{r})$, which is expressed as the sum of Dirac delta functions. Each delta function designates the point location of the spin $\mathbf{s}_e$ amplitude (via spin operator $\hat{\mathbf{s}}_e$) of the eth orbital electron in the scattering system of nuclei in which $\delta(\mathbf{r}-\mathbf{r}_e)$ is nonzero only at electron location $\mathbf{r}=\mathbf{r}_e$. The electron spin density vector operator $\hat{\boldsymbol{\rho}}_s(\mathbf{r})$ is, then

$$\hat{\boldsymbol{\rho}}_s(\mathbf{r}) = \sum_{e=1}^{E} \delta(\mathbf{r}-\mathbf{r}_e)\ \hat{\mathbf{s}}_e \tag{3.143}$$

Comparing Eq. (3.142) and 3.143 for vector operators $\hat{\mathbf{Q}}_s$ and $\hat{\boldsymbol{\rho}}_s(\mathbf{r})$, one finds the electron spin operator $\hat{\mathbf{Q}}_s(\boldsymbol{\kappa})$ is the κ-space Fourier transform of the direct-space (r-space) electron spin density operator $\hat{\boldsymbol{\rho}}_s(\mathbf{r})$, where

$$\hat{\mathbf{Q}}_s(\boldsymbol{\kappa}) = \int_{-\infty}^{\infty} \mathrm{d}z \int_{-\infty}^{\infty} \mathrm{d}y \int_{-\infty}^{\infty} \mathrm{d}x \hat{\boldsymbol{\rho}}_s(\mathbf{r}) \exp(\mathrm{i}\boldsymbol{\kappa}\cdot\mathbf{r}) \tag{3.144}$$

Magnetization vector $\mathbf{M}$ is the magnetic dipole moment per unit volume of the scattering system. Vector $\mathbf{M}$ is the product of the magnetic dipole moment vector of each electron $\boldsymbol{\mu}_e$ multiplied by electron number densityn_e, where

$$\mathbf{M} = n_e \boldsymbol{\mu}_e \tag{3.145}$$

and

$$\boldsymbol{\mu}_e = -2\mu_\mathrm{b}\mathbf{s} \tag{3.146}$$

This suggests an electron spin magnetization vector operator $\hat{\mathbf{M}}_s(\mathbf{r})$ that is defined in terms of the Bohr magnetron μ_b and electron spin density vector operator $\hat{\boldsymbol{\rho}}_s(\mathbf{r})$:

$$\hat{\mathbf{M}}_s(\mathbf{r}) = -2\mu_\mathrm{b}\hat{\boldsymbol{\rho}}_s(\mathbf{r}) \tag{3.147}$$

where

$$\mu_\mathrm{b} = \frac{e\hbar}{2m_e} \tag{3.148}$$

Substitution gives

$$\hat{\mathbf{Q}}_s(\boldsymbol{\kappa}) = -\frac{1}{2\mu_\mathrm{b}} \int_{-\infty}^{\infty} \mathrm{d}z \int_{-\infty}^{\infty} \mathrm{d}y \int_{-\infty}^{\infty} \mathrm{d}x \hat{\mathbf{M}}_s(\mathbf{r}) \exp(\mathrm{i}\boldsymbol{\kappa}\cdot\mathbf{r}) \tag{3.149}$$

where

$$\hat{\mathbf{M}}_s(\boldsymbol{\kappa}) = \int_{-\infty}^{\infty} \mathrm{d}z \int_{-\infty}^{\infty} \mathrm{d}y \int_{-\infty}^{\infty} \mathrm{d}x \hat{\mathbf{M}}_s(\mathbf{r}) \exp(\mathrm{i}\boldsymbol{\kappa}\cdot\mathbf{r}) \tag{3.150}$$

or, equivalently,

$$\hat{\mathbf{M}}_s(\boldsymbol{\kappa}) = -\int_{-\infty}^{\infty} dz \int_{-\infty}^{\infty} dy \int_{-\infty}^{\infty} dx 2\mu_b \hat{\rho}_s(\mathbf{r}) \exp(i\boldsymbol{\kappa}\cdot\mathbf{r}) \tag{3.151}$$

Comparison of the integral expressions for $\hat{\mathbf{Q}}_s(\boldsymbol{\kappa})$ and $\hat{\mathbf{M}}_s(\boldsymbol{\kappa})$ gives

$$\hat{\mathbf{Q}}_s(\boldsymbol{\kappa}) = -\frac{\hat{\mathbf{M}}_s(\boldsymbol{\kappa})}{2\mu_b} \tag{3.152}$$

Then, with

$$\hat{\mathbf{Q}}_{\perp s}(\boldsymbol{\kappa}) = \frac{\boldsymbol{\kappa}\times\hat{\mathbf{Q}}_s(\boldsymbol{\kappa})\times\boldsymbol{\kappa}}{\kappa^2} \tag{3.153}$$

one obtains

$$\hat{\mathbf{Q}}_{\perp s}(\boldsymbol{\kappa}) = -\frac{1}{2\mu_b}\frac{\boldsymbol{\kappa}\times\hat{\mathbf{M}}_s(\boldsymbol{\kappa})\times\boldsymbol{\kappa}}{\kappa^2} \tag{3.154}$$

3.11 Magnetic Dipole Moment Operator and Magnetization Vector—the Orbital Component

In this section, we examine the perpendicular component vector operator $\hat{\mathbf{Q}}_{\perp l}(\boldsymbol{\kappa})$ of the electron orbital magnetic dipole moment vector operator $\hat{\mathbf{Q}}_l(\boldsymbol{\kappa})$ in reciprocal lattice space. Like the case of the electron spin operator $\hat{\mathbf{Q}}_s(\boldsymbol{\kappa})$, the electron orbital operator $\hat{\mathbf{Q}}_l(\boldsymbol{\kappa})$ now will be related to the magnetization of the scattering system. One starts with the perpendicular component operator $\hat{\mathbf{Q}}_{\perp l}(\boldsymbol{\kappa})$ expression. In this expression, the eth electron, position-dependent term $\exp(i\boldsymbol{\kappa}\cdot\hat{\mathbf{r}}_e)$, and momentum operator $\hat{\mathbf{p}}_e$ do not commute. With

$$\hat{\mathbf{Q}}_{\perp l}(\boldsymbol{\kappa}) = \sum_{e=1}^{E} \exp(i\boldsymbol{\kappa}\cdot\hat{\mathbf{r}}_e)\left\{\frac{-(i/\hbar)(\boldsymbol{\kappa}\times\hat{\mathbf{p}}_e)}{\kappa^2}\right\} \tag{3.155}$$

one obtains an equivalent expression, which is remedied by the independent order of $\exp(i\boldsymbol{\kappa}\cdot\hat{\mathbf{r}}_e)$ and $\hat{\mathbf{p}}_e$ as follows:

$$\hat{\mathbf{Q}}_{\perp l}(\boldsymbol{\kappa}) = \frac{i}{2\kappa^2\hbar}\left[\sum_{e=1}^{E} \exp(i\boldsymbol{\kappa}\cdot\mathbf{r}_e)\hat{\mathbf{p}}_e + \hat{\mathbf{p}}_e \exp(i\boldsymbol{\kappa}\cdot\hat{\mathbf{r}}_e)\right]\times\boldsymbol{\kappa} \tag{3.156}$$

That is, the order dependence of $\exp(\mathrm{i}\boldsymbol{\kappa}\cdot\hat{\mathbf{r}}_e)$ and $\hat{\mathbf{p}}_e$ is remedied by taking the sum of the operator product and the reverse operator product and then dividing by 2, where

$$\exp(\mathrm{i}\boldsymbol{\kappa}\cdot\hat{\mathbf{r}}_e)\hat{\mathbf{p}}_e \quad \text{and} \quad \hat{\mathbf{p}}_e\exp(\mathrm{i}\boldsymbol{\kappa}\cdot\hat{\mathbf{r}}_e) \tag{3.157}$$

This is justified as described next.

Operator product $\hat{\mathbf{R}}\hat{\mathbf{p}}$ of noncommuting, position-dependent operator $\hat{\mathbf{R}}$, momentum $\hat{\mathbf{p}}$ operator, and reverse product operator $\hat{\mathbf{p}}\hat{\mathbf{R}}$ extract different eigenvalues when applied to an eigenfunction ϕ, where

$$\hat{\mathbf{R}}\hat{\mathbf{p}}\phi = c\phi \tag{3.158}$$

and

$$\hat{\mathbf{p}}\hat{\mathbf{R}}\phi = \mathrm{d}\phi \tag{3.159}$$

With real c and d, assuming that

$$c \neq d \quad \text{and} \quad c - d = -\mathrm{i}\hbar \tag{3.160}$$

one obtains

$$\frac{1}{2}(\hat{\mathbf{p}}\hat{\mathbf{R}} + \hat{\mathbf{R}}\hat{\mathbf{p}})\phi = \left(\frac{c+d}{2}\right)\phi = \left(\frac{2d - \mathrm{i}\hbar}{2}\right)\phi \tag{3.161}$$

where

$$\frac{1}{2}(\hat{\mathbf{p}}\hat{\mathbf{R}} + \hat{\mathbf{R}}\hat{\mathbf{p}})^*\phi = \left(\frac{c+d}{2}\right)^*\phi = \left(\frac{2d - \mathrm{i}\hbar}{2}\right)^*\phi = \left(\frac{2d + \mathrm{i}\hbar}{2}\right)\phi \tag{3.162}$$

In Eq. (3.13) with Eqs. (3.120) and (3.122), the magnetic partial differential cross section $(\mathrm{d}\sigma/d\Omega\,\mathrm{d}E)^{\mathrm{mag}}_{i\to m}$, which is a real number, requires its transition term W below to be real, where

$$W = \langle\phi_i\psi_{0s}|(\hat{\mathbf{Q}}_\perp{}^* \cdot \hat{\mathbf{S}}_n^* \cdot)|\psi_s\phi_m\rangle\langle\phi_m\psi_s|(\hat{\mathbf{S}}_n\cdot\hat{\mathbf{Q}}_\perp)|\psi_{0s}\phi_i\rangle \tag{3.163}$$

In the analogous expression of this example, substitution and the closure relation gives

$$\begin{aligned}
&\sum\nolimits_m \left\langle \phi_i \left| \left(\frac{\hat{\mathbf{p}}\hat{\mathbf{R}} + \hat{\mathbf{R}}\hat{\mathbf{p}}}{2}\right)^* \right| \phi_m \right\rangle \left\langle \phi_m \left| \left(\frac{\hat{\mathbf{p}}\hat{\mathbf{R}} + \hat{\mathbf{R}}\hat{\mathbf{p}}}{2}\right) \right| \phi_i \right\rangle \\
&= \left\langle \phi_i \left| \left(\frac{\hat{\mathbf{p}}\hat{\mathbf{R}} + \hat{\mathbf{R}}\hat{\mathbf{p}}}{2}\right)^* \left(\frac{\hat{\mathbf{p}}\hat{\mathbf{R}} + \hat{\mathbf{R}}\hat{\mathbf{p}}}{2}\right) \right| \phi_i \right\rangle
\end{aligned} \tag{3.164}$$

Application of position $\hat{\mathbf{R}}$ and momentum $\hat{\mathbf{p}}$ to sum product and reversed product operators on eigenstate ϕ_i then gives

$$\left\langle \phi_i \left| \left(\frac{\hat{\mathbf{p}}\hat{\mathbf{R}} + \hat{\mathbf{R}}\hat{\mathbf{p}}}{2} \right)^* \left(\frac{\hat{\mathbf{p}}\hat{\mathbf{R}} + \hat{\mathbf{R}}\hat{\mathbf{p}}}{2} \right) \right| \phi_i \right\rangle = \left\langle \phi_i \left| \left(\frac{2d + \mathrm{i}\hbar}{2} \right) \left(\frac{2d - \mathrm{i}\hbar}{2} \right) \phi_i \right. \right\rangle = \langle \phi_i | d^2 \phi_i \rangle \tag{3.165}$$

And thus real expectation value d^2 is obtained, which is required for the real-valued, partial differential cross section expression, where

$$\sum_m \left\langle \phi_i \left| \left(\frac{\hat{\mathbf{p}}\hat{\mathbf{R}} + \hat{\mathbf{R}}\hat{\mathbf{p}}}{2} \right)^* \right| \phi_m \right\rangle \left\langle \phi_m \left| \left(\frac{\hat{\mathbf{p}}\hat{\mathbf{R}} + \hat{\mathbf{R}}\hat{\mathbf{p}}}{2} \right) \right| \phi_i \right\rangle = d^2 \tag{3.166}$$

The motion of the eth orbital electron with mass m_e, charge q, and velocity $\mathbf{v}_e$ gives rise to current density

$$\mathbf{j}_e = -q\mathbf{v}_e \tag{3.167}$$

Here, $\mathbf{j}_e$ and $\mathbf{v}_e$ can be replaced by their respective operator equivalents $\hat{\mathbf{j}}_e$ and $\hat{\mathbf{v}}_e$, where

$$\hat{\mathbf{j}}_e = -q\hat{\mathbf{v}}_e \tag{3.168}$$

The direction of current is defined by convention as the direction of positive charge flow, and thus the direction of the negative electron current is opposite that of the electron velocity direction (hence the minus sign). The eth electron momentum operator is, then

$$\hat{\mathbf{p}}_e = -\frac{m_e \hat{\mathbf{j}}_e}{q} \tag{3.169}$$

Substitution yields

$$\sum_{e=1}^{E} [\exp(\mathrm{i}\boldsymbol{\kappa} \cdot \hat{\mathbf{r}}_e)\hat{\mathbf{p}}_e + \hat{\mathbf{p}}_e \exp(\mathrm{i}\boldsymbol{\kappa} \cdot \hat{\mathbf{r}}_e)] = -\frac{m_e}{q} \sum_{e=1}^{E} [\exp(\mathrm{i}\boldsymbol{\kappa} \cdot \hat{\mathbf{r}}_e)\hat{\mathbf{j}}_e + \hat{\mathbf{j}}_e \exp(\mathrm{i}\boldsymbol{\kappa} \cdot \hat{\mathbf{r}}_e)] \tag{3.170}$$

The sifting or sampling property of the Dirac delta function allows

$$\sum_{e=1}^{E} [\exp(\mathrm{i}\boldsymbol{\kappa} \cdot \hat{\mathbf{r}}_e)\hat{\mathbf{j}}_e + \hat{\mathbf{j}}_e \exp(\mathrm{i}\boldsymbol{\kappa} \cdot \hat{\mathbf{r}}_e)] = \int_{-\infty}^{\infty} \mathrm{d}z \int_{-\infty}^{\infty} \mathrm{d}y \int_{-\infty}^{\infty} \mathrm{d}x \hat{\mathbf{j}}(\mathbf{r}) \exp(\mathrm{i}\boldsymbol{\kappa} \cdot \hat{\mathbf{r}}_e) \tag{3.171}$$

The total electron current density operator $\hat{\mathbf{j}}(\mathbf{r})$ is, then

$$2\hat{\mathbf{j}}(\mathbf{r}) = \sum_{e=1}^{E}[\delta(\mathbf{r}-\mathbf{r}_e)\hat{\mathbf{j}}_e + \hat{\mathbf{j}}_e\delta(\mathbf{r}-\mathbf{r}_e)] \tag{3.172}$$

Summation of the current density operator contributions $\hat{\mathbf{j}}_e$ from all the eth electrons in the scattering system is done in the first term and repeated again in the second term for the summation over e. This yields $2\hat{\mathbf{j}}(\mathbf{r})$ or twice the actual current density operator $\hat{\mathbf{j}}(\mathbf{r})$.
Substitution gives perpendicular component $\hat{\mathbf{Q}}_{\perp l}(\boldsymbol{\kappa})$ of the electron orbital magnetic dipole moment vector operator $\hat{\mathbf{Q}}_l(\boldsymbol{\kappa})$, where

$$\hat{\mathbf{Q}}_{\perp l}(\boldsymbol{\kappa}) = \frac{\mathrm{i}}{\kappa^2\hbar}\frac{m_e}{q}\left[\int_{-\infty}^{\infty}\mathrm{d}z\int_{-\infty}^{\infty}\mathrm{d}y\int_{-\infty}^{\infty}\mathrm{d}x\hat{\mathbf{j}}(\mathbf{r})\exp(\mathrm{i}\boldsymbol{\kappa}\cdot\mathbf{r})\right]\times\boldsymbol{\kappa} \tag{3.173}$$

The orbital electron current density $\mathbf{j}(\mathbf{r})$ of the scattering system is the sum of the longitudinal or conduction electron current density $\nabla\phi(\mathbf{r})$ from the net motion of the electrons in a particular direction added to the net circular motion of the electrons $\nabla\times\mathbf{M}_l(\mathbf{r})$. The term $\nabla\times\mathbf{M}_l(\mathbf{r})$ is given by the curl of the direct-space (r-space) magnetization vector $\mathbf{M}_l(\mathbf{r})$, which arises from the magnetic dipoles created by the electron circular motion, where

$$\mathbf{j}(\mathbf{r}) = \nabla\times\mathbf{M}_l(\mathbf{r}) + \nabla\phi(\mathbf{r}) \tag{3.174}$$

The operator equivalent expressions for these position-dependent, magnetic vectors are

$$\hat{\mathbf{j}}(\mathbf{r}) = \nabla\times\hat{\mathbf{M}}_l(\mathbf{r}) + \nabla\phi(\mathbf{r}) \tag{3.175}$$

In Eq. (3.173), if

$$\mu_\mathrm{b} = \frac{q\hbar}{2m_e} \tag{3.176}$$

is substituted for $\hat{\mathbf{Q}}_{\perp l}$, one obtains

$$\hat{\mathbf{Q}}_{\perp l} = -\frac{\mathrm{i}}{2\mu_\mathrm{b}\kappa^2}\left[\int_{-\infty}^{\infty}\mathrm{d}z\int_{-\infty}^{\infty}\mathrm{d}y\int_{-\infty}^{\infty}\mathrm{d}x\{\nabla\times\hat{\mathbf{M}}_l(\mathbf{r}) + \nabla\phi(\mathbf{r})\}\exp(\mathrm{i}\boldsymbol{\kappa}\cdot\mathbf{r})\right]\times\boldsymbol{\kappa} \tag{3.177}$$

The first integral term of $\hat{\mathbf{Q}}_{\perp l}$ can be solved via the following vector relations:

$$\nabla x[\hat{\mathbf{M}}_l(\mathbf{r})\exp(\mathrm{i}\boldsymbol{\kappa}\cdot\mathbf{r})] = [\nabla\times\hat{\mathbf{M}}_l(\mathbf{r})]\exp(\mathrm{i}\boldsymbol{\kappa}\cdot\mathbf{r}) + \nabla[\exp(\mathrm{i}\boldsymbol{\kappa}\cdot\mathbf{r})]\times\hat{\mathbf{M}}_l(\mathbf{r}) \tag{3.178}$$

and

$$\nabla[\exp(\mathrm{i}\boldsymbol{\kappa}\cdot\mathbf{r})]\times\hat{\mathbf{M}}_l(\mathbf{r}) = \mathrm{i}\boldsymbol{\kappa}\times\hat{\mathbf{M}}_l(\mathbf{r}) \tag{3.179}$$

The integration of $\nabla\times f(\mathbf{r})$ with respect to $\mathbf{r}$ over the volume is defined by integral limits, yielding arbitrary function$f(\mathbf{r})$ at the volume boundary, which in this case is

$$\hat{\mathbf{M}}_l(\mathbf{r}) = 0 \quad \text{at } \mathbf{r} = \pm\infty \tag{3.180}$$

One obtains

$$\int_{-\infty}^{\infty}\mathrm{d}z\int_{-\infty}^{\infty}\mathrm{d}y\int_{-\infty}^{\infty}\mathrm{d}x\nabla\times[\hat{\mathbf{M}}_l(\mathbf{r})\exp(\mathrm{i}\boldsymbol{\kappa}\cdot\mathbf{r})] = 0 \tag{3.181}$$

Hence, the first integral term of $\hat{\mathbf{Q}}_{\perp l}$ becomes

$$\begin{aligned}&\int_{-\infty}^{\infty}\mathrm{d}z\int_{-\infty}^{\infty}\mathrm{d}y\int_{-\infty}^{\infty}\mathrm{d}x\{\nabla\times\hat{\mathbf{M}}_l(\mathbf{r})\}\exp(\mathrm{i}\boldsymbol{\kappa}\cdot\mathbf{r})\\ &= -\mathrm{i}\boldsymbol{\kappa}\times\int_{-\infty}^{\infty}\mathrm{d}z\int_{-\infty}^{\infty}\mathrm{d}y\int_{-\infty}^{\infty}\mathrm{d}x\hat{\mathbf{M}}_l(\mathbf{r})\end{aligned} \tag{3.182}$$

The second integral term of $\hat{\mathbf{Q}}_{\perp l}$ can be solved via the following vector relations:

$$\nabla[\phi(\mathbf{r})\exp(\mathrm{i}\boldsymbol{\kappa}\cdot\mathbf{r})] = \phi(\mathbf{r})\nabla[\exp(\mathrm{i}\boldsymbol{\kappa}\cdot\mathbf{r})] + \exp(\mathrm{i}\boldsymbol{\kappa}\cdot\mathbf{r})\nabla\phi(\mathbf{r}) \tag{3.183}$$

where

$$\nabla[\exp(\mathrm{i}\boldsymbol{\kappa}\cdot\mathbf{r})] = \mathrm{i}\boldsymbol{\kappa}\ \exp(\mathrm{i}\boldsymbol{\kappa}\cdot\mathbf{r}) \tag{3.184}$$

With

$$\phi(\mathbf{r} = \pm\infty) = 0 \tag{3.185}$$

one obtains

$$\int_{-\infty}^{\infty}\mathrm{d}z\int_{-\infty}^{\infty}\mathrm{d}y\int_{-\infty}^{\infty}\mathrm{d}x\nabla[\phi(\mathbf{r})\exp(\mathrm{i}\boldsymbol{\kappa}\cdot\mathbf{r})] = 0 \tag{3.186}$$

Hence, the second integral term of $\hat{\mathbf{Q}}_{\perp l}$ becomes

$$\int_{-\infty}^{\infty} dz \int_{-\infty}^{\infty} dy \int_{-\infty}^{\infty} dx\{\nabla\phi(\mathbf{r})\}\exp(i\boldsymbol{\kappa}\cdot\mathbf{r}) = -i\boldsymbol{\kappa}\times\int_{-\infty}^{\infty} dz \int_{-\infty}^{\infty} dy \int_{-\infty}^{\infty} dx\phi(\mathbf{r})\exp(i\boldsymbol{\kappa}\cdot\mathbf{r}) \tag{3.187}$$

The perpendicular component $\hat{\mathbf{Q}}_{\perp l}$ from Eq. (1.7) for the electron orbital magnetic dipole moment vector operator $\hat{\mathbf{Q}}_l$ becomes

$$\hat{\mathbf{Q}}_{\perp l} = -\frac{1}{2\mu_b\kappa^2}\boldsymbol{\kappa}\times\left[\int_{-\infty}^{\infty} dz \int_{-\infty}^{\infty} dy \int_{-\infty}^{\infty} dx\{\hat{\mathbf{M}}_l(\mathbf{r})+\phi(\mathbf{r})\}\exp(i\boldsymbol{\kappa}\cdot\mathbf{r})\right]\times\boldsymbol{\kappa} \tag{3.188}$$

Since

$$\boldsymbol{\kappa}\times\boldsymbol{\kappa} = 0 \tag{3.189}$$

one obtains $\hat{\mathbf{Q}}_{\perp l}(\boldsymbol{\kappa})$ where

$$\hat{\mathbf{Q}}_{\perp l} = -\frac{1}{2\mu_b}\frac{\boldsymbol{\kappa}\times\hat{\mathbf{M}}_l(\boldsymbol{\kappa})\times\boldsymbol{\kappa}}{\kappa^2} \tag{3.190}$$

The magnetization $\hat{\mathbf{M}}_l(\mathbf{r})$ of the scattering system produced by the orbital electron motion is the inverse Fourier transform of the reciprocal lattice magnetization $\hat{\mathbf{M}}_l(\boldsymbol{\kappa})$ in the κ-space from the orbital electron motion. That is, the orbital electron magnetization vector operator $\hat{\mathbf{M}}_l(\boldsymbol{\kappa})$ in the κ-space is the Fourier transform of the direct-space (r-space), orbital electron magnetization vector operator $\hat{\mathbf{M}}_l(\mathbf{r})$, where

$$\hat{\mathbf{M}}_l(\boldsymbol{\kappa}) = \int_{-\infty}^{\infty} dz \int_{-\infty}^{\infty} dy \int_{-\infty}^{\infty} dx\hat{\mathbf{M}}_l(\mathbf{r})\exp(i\boldsymbol{\kappa}\cdot\mathbf{r}) \tag{3.191}$$

The perpendicular component $\hat{\mathbf{Q}}_{\perp l}$ can be extracted from orbital $\hat{\mathbf{Q}}_l$ in the same manner as the perpendicular component $\hat{\mathbf{Q}}_{\perp s}$ from spin $\hat{\mathbf{Q}}_s$ in Eq. (3.153), so that

$$\hat{\mathbf{Q}}_{\perp l}(\boldsymbol{\kappa}) = \frac{\boldsymbol{\kappa}\times\hat{\mathbf{Q}}_l(\boldsymbol{\kappa})\times\boldsymbol{\kappa}}{\kappa^2} \tag{3.192}$$

where

$$\hat{\mathbf{Q}}_l(\boldsymbol{\kappa}) = -\frac{\hat{\mathbf{M}}_l(\boldsymbol{\kappa})}{2\mu_b} \tag{3.193}$$

In terms of the unit neutron scatter vector $\boldsymbol{\kappa}_u = \boldsymbol{\kappa}/\kappa$, substitution gives

$$\hat{\mathbf{Q}}_{\perp l}(\boldsymbol{\kappa}) = -\frac{1}{2\mu_b}\hat{\boldsymbol{\kappa}}_u \times \hat{\mathbf{M}}_l(\boldsymbol{\kappa}) \times \hat{\boldsymbol{\kappa}}_u \tag{3.194}$$

3.12 Magnetic Dipole Moment Operator Relation with Magnetization Vector

The resulting electron magnetic dipole moment operator $\hat{\mathbf{Q}}(\boldsymbol{\kappa})$ in the κ-space for the entire scattering system is the sum of its spin magnetic dipole moment operator $\hat{\mathbf{Q}}_s(\boldsymbol{\kappa})$ portion and the orbital magnetic dipole moment operator $\hat{\mathbf{Q}}_l(\boldsymbol{\kappa})$ portion. Electron magnetic dipole moment operator $\hat{\mathbf{Q}}(\boldsymbol{\kappa})$ in the κ-space is then proportional to the κ-space magnetization vector $\hat{\mathbf{M}}(\boldsymbol{\kappa})$ of the scattering system via proportionality constant $-1/2\mu_b$, where

$$\hat{\mathbf{Q}} = \hat{\mathbf{Q}}_s + \hat{\mathbf{Q}}_l = -\frac{\hat{\mathbf{M}}(\boldsymbol{\kappa})}{2\mu_b} \tag{3.195}$$

Magnetization operator $\hat{\mathbf{M}}(\boldsymbol{\kappa})$ in the κ-space is

$$\hat{\mathbf{M}}(\boldsymbol{\kappa}) = \hat{\mathbf{M}}_s(\boldsymbol{\kappa}) + \hat{\mathbf{M}}_l(\boldsymbol{\kappa}) \tag{3.196}$$

Magnetization operator $\hat{\mathbf{M}}(\boldsymbol{\kappa})$ is the κ-space Fourier transform of the direct lattice magnetization operator, where

$$\hat{\mathbf{M}}(\boldsymbol{\kappa}) = \int_{-\infty}^{\infty} dz \int_{-\infty}^{\infty} dy \int_{-\infty}^{\infty} dx \hat{\mathbf{M}}(\mathbf{r})\exp(i\boldsymbol{\kappa}\cdot\mathbf{r}) \tag{3.197}$$

In the reciprocal lattice space of the scattering system, the magnetic dipole moment vector operator $\hat{\mathbf{Q}}(\boldsymbol{\kappa})$ of all the electrons in the scattering system is the product of constant $-1/2\mu_b$ and the Fourier transform $\hat{\mathbf{M}}(\boldsymbol{\kappa})$ of the direct-space, scattering system, magnetization vector operator $\hat{\mathbf{M}}(\mathbf{r})$, which is a function of neutron position $\mathbf{r}$. Magnetization vector operator $\hat{\mathbf{M}}(\mathbf{r})$ is thus a Hermitian operator that acts on its eigenfunction (namely, the prescatter $\psi_{s0}\phi_i$ or postscatter $\psi_s\phi_m$ system eigenstate wave amplitudes). Multiplicative operator $\hat{\mathbf{M}}(\mathbf{r})$ extracts its eigenvalue (namely, the magnetization vector $\mathbf{M}(\mathbf{r})$ at the neutron position $\mathbf{r}$) via the multiplication of its eigenfunctions $\psi_{s0}\phi_i$ and $\psi_s\phi_m$ by the magnetization vector $\mathbf{M}(\mathbf{r})$ at $\mathbf{r}$.

The perpendicular component $\hat{\mathbf{Q}}_\perp(\boldsymbol{\kappa})$ of the electron magnetic dipole moment operator $\hat{\mathbf{Q}}(\boldsymbol{\kappa})$ is the sum of its spin $\hat{\mathbf{Q}}_{\perp s}(\boldsymbol{\kappa})$ and orbital components $\hat{\mathbf{Q}}_{\perp l}(\boldsymbol{\kappa})$:

$$\hat{\mathbf{Q}}_\perp(\boldsymbol{\kappa}) = \hat{\mathbf{Q}}_{\perp s}(\boldsymbol{\kappa}) + \hat{\mathbf{Q}}_{\perp l}(\boldsymbol{\kappa}) \tag{3.198}$$

where

$$\hat{\mathbf{Q}}_{\perp s} = \frac{\boldsymbol{\kappa} \times \hat{\mathbf{Q}}_s \times \boldsymbol{\kappa}}{\kappa^2} \tag{3.199}$$

and

$$\hat{\mathbf{Q}}_{\perp l}(\boldsymbol{\kappa}) = \frac{\boldsymbol{\kappa} \times \hat{\mathbf{Q}}_l(\boldsymbol{\kappa}) \times \boldsymbol{\kappa}}{\kappa^2} \tag{3.200}$$

Thus, the perpendicular component $\hat{\mathbf{Q}}_{\perp}(\boldsymbol{\kappa})$ of the electron magnetic dipole moment operator $\hat{\mathbf{Q}}(\boldsymbol{\kappa})$ is obtained from $\hat{\mathbf{Q}}(\boldsymbol{\kappa})$ via

$$\hat{\mathbf{Q}}_{\perp}(\boldsymbol{\kappa}) = \frac{\boldsymbol{\kappa} \times \hat{\mathbf{Q}}(\boldsymbol{\kappa}) \times \boldsymbol{\kappa}}{\kappa^2} \tag{3.201}$$

In terms of the unit neutron scatter vector $\boldsymbol{\kappa}_{\mathrm{u}} = \boldsymbol{\kappa}/\kappa$ and Fourier transform $\hat{\mathbf{M}}(\boldsymbol{\kappa})$ of the scattering system magnetization vector operator $\hat{\mathbf{M}}(\mathbf{r})$, the perpendicular component $\hat{\mathbf{Q}}_{\perp}(\boldsymbol{\kappa})$ of the electron magnetic dipole moment operator $\hat{\mathbf{Q}}(\boldsymbol{\kappa})$ in the reciprocal lattice space is expressed by

$$\hat{\mathbf{Q}}_{\perp}(\boldsymbol{\kappa}) = -\frac{1}{2\mu_{\mathrm{b}}} \hat{\boldsymbol{\kappa}}_{\mathrm{u}} \times \hat{\mathbf{M}}(\boldsymbol{\kappa}) \times \hat{\boldsymbol{\kappa}}_{\mathrm{u}} \tag{3.202}$$

From simple geometrical considerations, $\hat{\mathbf{Q}}_{\perp}(\boldsymbol{\kappa})$ can be obtained from $\hat{\mathbf{Q}}(\boldsymbol{\kappa})$ and the neutron scatter vector $\boldsymbol{\kappa}$, where

$$\hat{\mathbf{Q}}_{\perp} = \hat{\mathbf{Q}} - \frac{(\hat{\mathbf{Q}}\cdot\boldsymbol{\kappa})}{\kappa^2}\boldsymbol{\kappa} = \hat{\mathbf{Q}} - \hat{\boldsymbol{\kappa}}_{\mathrm{u}}(\hat{\mathbf{Q}}\cdot\hat{\boldsymbol{\kappa}}_{\mathrm{u}}) \tag{3.203}$$

And the dot product of $\hat{\mathbf{Q}}_{\perp}(\boldsymbol{\kappa})$ with its complex conjugate $\hat{\mathbf{Q}}_{\perp}^{+}(\boldsymbol{\kappa})$ is then

$$\hat{\mathbf{Q}}_{\perp}^{+} \cdot \hat{\mathbf{Q}}_{\perp} = \{\hat{\mathbf{Q}}^{+} - \hat{\boldsymbol{\kappa}}_{\mathrm{u}}(\hat{\mathbf{Q}}^{+} \cdot \hat{\boldsymbol{\kappa}}_{\mathrm{u}})\}\{\hat{\mathbf{Q}} - \hat{\boldsymbol{\kappa}}_{\mathrm{u}}(\hat{\mathbf{Q}}\cdot\hat{\boldsymbol{\kappa}}_{\mathrm{u}})\} \tag{3.204}$$

Expanding via the orthogonal properties of vector components, which is expressed by the Kronecker delta function:

$$\hat{\boldsymbol{\kappa}}_{\mathrm{u}i} \cdot \hat{\boldsymbol{\kappa}}_{\mathrm{u}j} = \delta_{ij} \tag{3.205}$$

one obtains

$$\hat{\mathbf{Q}}_{\perp}^{+} \cdot \hat{\mathbf{Q}}_{\perp} = (\hat{\mathbf{Q}}^{+} \cdot \hat{\mathbf{Q}}) - 2(\hat{\mathbf{Q}}^{+} \cdot \hat{\boldsymbol{\kappa}}_{\mathrm{u}})(\hat{\mathbf{Q}} \cdot \hat{\boldsymbol{\kappa}}_{\mathrm{u}}) + (\hat{\mathbf{Q}}^{+} \cdot \hat{\boldsymbol{\kappa}}_{\mathrm{u}})(\hat{\mathbf{Q}} \cdot \hat{\boldsymbol{\kappa}}_{\mathrm{u}}) \tag{3.206}$$

One then obtains

$$\hat{\mathbf{Q}}_{\perp}^{+} \cdot \hat{\mathbf{Q}}_{\perp} = (\hat{\mathbf{Q}}^{+} \cdot \hat{\mathbf{Q}}) - (\hat{\mathbf{Q}}^{+} \cdot \boldsymbol{\kappa}_{\mathrm{u}})(\hat{\mathbf{Q}} \cdot \boldsymbol{\kappa}_{\mathrm{u}}) \tag{3.207}$$

To simplify Eq. (3.207), dispensing with the unit neutron scatter vector $\boldsymbol{\kappa}_{\mathrm{u}}$ and returning to the scatter vector $\boldsymbol{\kappa}$, one gets

$$\hat{\mathbf{Q}}_{\perp}{}^{+} \cdot \hat{\mathbf{Q}}_{\perp} = (\hat{\mathbf{Q}}^{+} \cdot \hat{\mathbf{Q}}) - \frac{1}{\kappa^{2}}(\hat{\mathbf{Q}}^{+} \cdot \boldsymbol{\kappa})(\hat{\mathbf{Q}} \cdot \boldsymbol{\kappa}) \tag{3.208}$$

Carrying out the dot product operations, one obtains, in terms of Kronecker delta function $\delta_{\alpha\beta}$, the simplified expression

$$\hat{\mathbf{Q}}_{\perp}{}^{+} \cdot \hat{\mathbf{Q}}_{\perp} = \sum_{\alpha=1}^{3}\sum_{\beta=1}^{3}\left(\delta_{\alpha\beta} - \frac{\kappa_{\alpha}\kappa_{\beta}}{\kappa^{2}}\right) Q_{\alpha}^{+} Q_{\beta} \tag{3.209}$$

3.13 Evaluation of the Neutron Magnetic Potential Energy Operator

The magnetic scattering of the neutrons occurs via the neutron magnetic dipole moment $\boldsymbol{\mu}$, which interacts with a magnetic field **B**. The magnetic field B is produced by the magnetic dipole moments, which are associated with the scattering system electron orbital motion and spin $\boldsymbol{\mu}_e$. If the scattering system nuclei have unpaired electrons, then the system has net electron orbital motion and spin, which produce a net scattering system magnetization **M(r)** as a function of neutron position **r**. The net magnetic dipole moment of the unpaired electrons produces a magnetic field **B**, where

$$\mathbf{B} = \mu_0 \mathbf{M} \tag{3.210}$$

In addition, the gradient of the magnetic field $\nabla|\mathbf{B}|$ causes gradient ∇U to occur in the magnetic potential energy of interaction U between the neutron and the scattering system magnetic field.

From the negative gradient of the magnetic potential energy arises force **F**, which acts on the neutron and causes magnetic neutron scatter, where

$$\mathbf{F} = -\nabla U \tag{3.211}$$

with

$$U = -\boldsymbol{\mu} \cdot \mathbf{B} = -\mu_{||}|\mathbf{B}| \tag{3.212}$$

With the z-component S_{nz} of the neutron spin vector, the longitudinal component $\mu_{||}$ of the neutron magnetic dipole moment vector $\boldsymbol{\mu}$ for spin-down (+) and spin-up (–) neutrons becomes

$$\mu_{||} = 2g\mu_n S_{nz} = \pm g\mu_n = \pm 6.01 \times 10^{-8} \ \text{eV}/T \tag{3.213}$$

The magnetic field force acting on the neutron magnetic dipole moment vector $\boldsymbol{\mu}$ for spin-down (−) and spin-up (+) neutrons becomes

$$\mathbf{F} = \pm 6.01 \times 10^{-8} \ \nabla|\mathbf{B}| \tag{3.214}$$

The magnetic field gradient $\nabla|\mathbf{B}|$ is the magnetic field change per unit spatial interval. If the spatial interval is measured in units of femtometers and the magnetic field $\mathbf{B}$ in teslas, then the force acting on the spin-up (+) or spin-down (−) can be expressed in units of [eV/fm] or [N] as follows:

$$\mathbf{F} = \pm 6.01 \times 10^{-12} \ \nabla|\mathbf{B}| \, [N] \tag{3.215}$$

With Eq. (3.134) for $\hat{\mathbf{Q}}_{\perp}$, one can now evaluate the partial differential cross section for magnetic scatter of the neutron by the unpaired electrons of the scattering system, where

$$\left(\frac{d^2\sigma}{d\Omega \, dE}\right)^{\text{mag}} = (r_0 g)^2 \frac{k}{k_0} \sum_{0s=1}^{2} p_{0s}^{\text{mag}} \sum_{i=1}^{J} p_i^{\text{mag}} \int_{-\infty}^{\infty} dt \frac{\exp(-iwt)}{2\pi\hbar} \langle \phi_i \psi_{0s} | \left[\hat{\mathbf{S}}_n \cdot \hat{\mathbf{Q}}_{\perp}^{+}(\boldsymbol{\kappa})\right] \left[\hat{\mathbf{S}}_n(t) \cdot \hat{\mathbf{Q}}_{\perp}(\boldsymbol{\kappa}, t)\right] | \phi_i \psi_{0s} \rangle \tag{3.216}$$

The perpendicular component $\hat{\mathbf{Q}}_{\perp}(\boldsymbol{\kappa})$ of electron magnetic dipole moment operator $\hat{\mathbf{Q}}(\boldsymbol{\kappa})$ in the κ-space is obtained by pre- and post-cross products of operator $\hat{\mathbf{Q}}(\boldsymbol{\kappa})$ with the neutron scatter vector $\boldsymbol{\kappa}$, where

$$\hat{\mathbf{Q}}_{\perp}(\boldsymbol{\kappa}) = \frac{\boldsymbol{\kappa} \times \hat{\mathbf{Q}}(\boldsymbol{\kappa}) \times \boldsymbol{\kappa}}{\kappa^2} \tag{3.217}$$

From section 3.9 the Schrödinger, time-independent, scalar product operator $\hat{U}$ for the magnetic potential energy, which depicts the scattering interaction between the neutron wave with spin $\mathbf{S}_n$ and the scattering system magnetization $\mathbf{M}$ from its unpaired electrons at initial time $t = 0$, is

$$\hat{U} = \hat{\mathbf{S}}_n \cdot \hat{\mathbf{Q}}_{\perp}(\boldsymbol{\kappa}) \tag{3.218}$$

Substitution of Eq. (3.202) into Eq. (3.218), gives

$$\hat{U} = -\frac{1}{2\mu_b} \hat{\mathbf{S}}_n \cdot \frac{\boldsymbol{\kappa} \times \hat{\mathbf{M}}(\boldsymbol{\kappa}) \times \boldsymbol{\kappa}}{\kappa^2} \tag{3.219}$$

where

$$\hat{\mathbf{S}}_n \cdot \hat{\mathbf{Q}}_\perp(\boldsymbol{\kappa}) = \hat{S}_{nx} \cdot \hat{Q}_{\perp x}(\boldsymbol{\kappa}) + \hat{S}_{ny} \cdot \hat{Q}_{\perp y}(\boldsymbol{\kappa}) + \hat{S}_{nz} \cdot \hat{Q}_{\perp z}(\boldsymbol{\kappa}) \tag{3.220}$$

At a later time t, the magnetic scatter interaction of the neutron wave with the scattering system is depicted by the time-dependent, Heisenberg operator version of

$$\hat{U} = \hat{\mathbf{S}}_n \cdot \hat{\mathbf{Q}}_{\perp l}(\boldsymbol{\kappa}) \tag{3.221}$$

where

$$\hat{U}(\boldsymbol{\kappa}, t) = \hat{\mathbf{S}}_n(t) \cdot \hat{\mathbf{Q}}_\perp(\boldsymbol{\kappa}, t) = -\frac{1}{2\mu_b}\hat{\mathbf{S}}_n(t) \cdot \frac{\boldsymbol{\kappa} \times \hat{\mathbf{M}}(\boldsymbol{\kappa}, t) \times \boldsymbol{\kappa}}{\kappa^2} \tag{3.222}$$

and

$$\hat{\mathbf{S}}_n(t) \cdot \hat{\mathbf{Q}}_\perp(\boldsymbol{\kappa}, t) = \hat{S}_{nx}(t)\hat{Q}_{\perp x}(\boldsymbol{\kappa}, t) + \hat{S}_{ny}(t)\hat{Q}_{\perp y}(\boldsymbol{\kappa}, t) + \hat{S}_{nz}(t)\hat{Q}_{\perp z}(\boldsymbol{\kappa}, t) \tag{3.223}$$

With Eq. (3.223), one obtains the product:

$$\begin{aligned}[\hat{\mathbf{S}}_n \cdot \hat{\mathbf{Q}}_\perp^+(\boldsymbol{\kappa})][\hat{\mathbf{S}}_n(t) \cdot \hat{\mathbf{Q}}_\perp(\boldsymbol{\kappa}, t)] &= [\hat{S}_{nx} \cdot \hat{Q}_{\perp x}^+(\boldsymbol{\kappa}) + \hat{S}_{ny} \cdot \hat{Q}_{\perp y}^+(\boldsymbol{\kappa}) + \hat{S}_{nz} \cdot \hat{Q}_{\perp z}^+(\boldsymbol{\kappa})] \times \\ &[\hat{S}_{nx}(t) \cdot \hat{Q}_{\perp x}(\boldsymbol{\kappa}, t) + \hat{S}_{ny}(t) \cdot \hat{Q}_{\perp y}(\boldsymbol{\kappa}, t) + \hat{S}_{nz}(t) \cdot \hat{Q}_{\perp z}(\boldsymbol{\kappa}, t)]\end{aligned} \tag{3.224}$$

Expansion and collection gives

$$\hat{U}^+ \hat{U}(t) = [\hat{\mathbf{S}}_n \cdot \hat{\mathbf{Q}}_\perp^{\ +}]\ [\hat{\mathbf{S}}_n(t) \cdot \hat{\mathbf{Q}}_\perp(t)] = \sum_{\alpha=1}^{3}\sum_{\beta=1}^{3} \hat{S}_{n\alpha}\hat{Q}_{\perp\alpha}^+\hat{S}_\beta(t)\hat{Q}_{\perp\beta}(t) \tag{3.225}$$

The neutron position $\mathbf{r}$ and electron position $\mathbf{r}_e$ are independent of each other. The neutron spin vector operator $\hat{\mathbf{S}}_n$ depends only on the neutron spin, whereas vector operator $\hat{\mathbf{Q}}_\perp(\boldsymbol{\kappa})$ depends only on the electron position and spin. This allows separation of the neutron $\mathbf{r}$ and electron $\mathbf{r}_e$ position-based terms, where

$$\begin{aligned}\hat{U}^+ \hat{U}(t) &= \langle \phi_i \psi_{0s} | [\hat{\mathbf{S}}_n \cdot \hat{\mathbf{Q}}_\perp^+(\boldsymbol{\kappa})]\ [\hat{\mathbf{S}}_n(t) \cdot \hat{\mathbf{Q}}_\perp(\boldsymbol{\kappa}, t)] | \psi_{0s}\phi_i \rangle \\ &= \left\langle \phi_i \left| \sum_{\alpha=1}^{3}\sum_{\beta=1}^{3} \hat{Q}_{\perp\alpha}^+ Q_{\perp\beta}(t) \right| \phi_i \right\rangle \left\langle \psi_{0s} \left| \sum_{\alpha=1}^{3}\sum_{\beta=1}^{3} \hat{S}_{n\alpha}\hat{S}_{n\beta}(t) \right| \psi_{0s} \right\rangle\end{aligned} \tag{3.226}$$

3.14 Evaluation of Transition Matrix Element with Neutron Spin Eigenstates

The neutron magnetic potential energy operator is now applied to the spin component ψ_{0s} of the initial state neutron wave function. The neutron spin gives rise to the neutron magnetic dipole moment. We now determine the expected value of the neutron magnetic potential energy due to the interaction of the neutron magnetic dipole moment with the magnetic dipole moment of the unpaired electron spin and orbital angular momentum. The matrix elements of the potential energy operator and the spin component of the initial state neutron wave function are evaluated as follows.

In Eq. (3.226) the initial neutron spin state ψ_{0s} has its spin-up (+) and spin-down (−) states denoted by ψ_{0+} and ψ_{0-}, where

$$\langle\psi_{0+}|\hat{S}_{n\alpha}\hat{S}_{n\beta}|\psi_{0+}\rangle = \mathrm{i} \quad \alpha,\beta = 1,2,3 \tag{3.227}$$

$$\langle\psi_{0-}|\hat{S}_{n\alpha}\hat{S}_{n\beta}|\psi_{0-}\rangle = -\mathrm{i} \quad \alpha,\beta = 1,2,3 \tag{3.228}$$

$$\langle\psi_{0+}|\hat{S}^2_{n\alpha}|\psi_{0+}\rangle = \langle\psi_{0-}|\hat{S}^2_{n\alpha}|\psi_{0-}\rangle = 1 \quad \alpha = 1,2,3 \tag{3.229}$$

Hence,

$$\left\langle\psi_{0s}\left|\sum_{\alpha=1}^{3}\sum_{\beta=1}^{3}\hat{S}_{n\alpha}\hat{S}_{n\beta}(t)\right|\psi_{0s}\right\rangle = 2 \quad \alpha,\beta = 1,2,3 \tag{3.230}$$

With substitution of Eqs. (3.227−3.230) into Eq. (3.226) the expression of Eq. (3.216) for the partial differential cross section $(\mathrm{d}^2\sigma/d\Omega\,\mathrm{d}E)^{\mathrm{mag}}$ for magnetic neutron scatter becomes

$$\left(\frac{\mathrm{d}^2\sigma}{\mathrm{d}\Omega\,\mathrm{d}E}\right)^{\mathrm{mag}} = (r_0 g)^2\frac{k}{k_0}\sum_{0s=1}^{2}2p_{0s}^{\mathrm{mag}}\sum_{i=1}^{J}p_i^{\mathrm{mag}}\int_{-\infty}^{\infty}\mathrm{d}t\frac{\exp(-\,\mathrm{i}wt)}{2\pi\hbar} \left\langle\phi_i\left|\sum_{\alpha=1}^{3}\sum_{\beta=1}^{3}Q^{+}_{\perp\alpha}Q_{\perp\beta}(t)\right|\phi_i\right\rangle \tag{3.231}$$

If the incident neutrons are unpolarized, then the probabilities p_{0s}^{mag} for the two neutron spin states are the same and equal to 1/2, where

$$p_{0s}^{\mathrm{mag}} = 1/2 \quad 0s = 1,2 \tag{3.232}$$

Then, one obtains finally the partial differential cross section for magnetic neutron scatter $(\mathrm{d}^2\sigma/d\Omega\,\mathrm{d}E)^{\mathrm{mag}}$, where

$$\left(\frac{\mathrm{d}^2\sigma}{\mathrm{d}\Omega\,\mathrm{d}E}\right)^{\mathrm{mag}} = (r_0 g)^2\frac{k}{k_0}\sum_{i=1}^{J}p_i^{\mathrm{mag}}\int_{-\infty}^{\infty}\mathrm{d}t\frac{\exp(-\,\mathrm{i}wt)}{2\pi\hbar}\left\langle\phi_i\left|\sum_{\alpha=1}^{3}\sum_{\beta=1}^{3}Q^{+}_{\perp\alpha}Q_{\perp\beta}(t)\right|\phi_i\right\rangle \tag{3.233}$$

This cross section of Eq. (3.216) with Eq. (3.232), is expressed by $Q_\alpha^+ Q_\beta$, where Eq. (3.226) becomes

$$\hat{U}^+ \hat{U}(t) = \hat{\mathbf{Q}}_\perp^+ \cdot \hat{\mathbf{Q}}_\perp = \sum_{\alpha=1}^{3} \sum_{\beta=1}^{3} \left(\delta_{\alpha\beta} - \frac{\kappa_\alpha \kappa_\beta}{\kappa^2}\right) Q_\alpha^+ Q_\beta \tag{3.234}$$

where

$$\begin{aligned} \hat{\mathbf{Q}}_\perp^+ \cdot \hat{\mathbf{Q}}_\perp = & \left(Q_x^+ - \frac{\kappa_x \kappa_x Q_x^+ + \kappa_y \kappa_x Q_y^+ + \kappa_z \kappa_x Q_z^+}{\kappa^2} \right) Q_x \\ & + \left(Q_y^+ - \frac{\kappa_x \kappa_y Q_x^+ + \kappa_y \kappa_y Q_y^+ + \kappa_z \kappa_y Q_z^+}{\kappa^2} \right) Q_y \\ & + \left(Q_z^+ - \frac{\kappa_x \kappa_z Q_x^+ + \kappa_y \kappa_z Q_y^+ + \kappa_z \kappa_z Q_z^+}{\kappa^2} \right) Q_z \end{aligned} \tag{3.235}$$

With Eq. (3.226) replaced by Eq. (3.234) in Eq. (3.216), using Eq. (3.232), the partial differential cross section $(\mathrm{d}^2\sigma/d\Omega\, \mathrm{d}E)^{\mathrm{mag}}$ for neutron scatter by unpaired electrons in a general material, such as a solid, liquid, or gas, becomes

$$\begin{aligned} \left(\frac{\mathrm{d}^2\sigma}{\mathrm{d}\Omega\, \mathrm{d}E}\right)^{\mathrm{mag}} = & (r_0 g)^2 \frac{k}{k_0} \sum_{i=1}^{J} p_i^{\mathrm{mag}} \int_{-\infty}^{\infty} \mathrm{d}t \frac{\exp(-\,\mathrm{i}wt)}{2\pi\hbar} \\ & \left\langle \phi_i \left| \sum_{\alpha=1}^{3} \sum_{\beta=1}^{3} \left(\delta_{\alpha\beta} - \frac{\kappa_\alpha \kappa_\beta}{\kappa^2} \right) Q_\alpha^+ Q_\beta(t) \right| \phi_i \right\rangle \end{aligned} \tag{3.236}$$

The terms Q_β and Q_α^+ are the x, y, and z components of the electron magnetic dipole moment vector $\mathbf{Q}(\boldsymbol{\kappa})$ and its conjugate transpose in the κ-space. And $\mathbf{Q}(\boldsymbol{\kappa})$ is then proportional to the κ-space magnetization vector $\hat{\mathbf{M}}(\boldsymbol{\kappa})$ of the scattering system via proportionality constant $-1/2\mu_\mathrm{b}$, where

$$\mathbf{Q} = \mathbf{Q}_s + \mathbf{Q}_l = -\frac{\mathbf{M}(\boldsymbol{\kappa})}{2\mu_\mathrm{b}} \tag{3.237}$$

With substitution of Eq. (3.237) into Eq. (3.236), the magnetic partial differential cross section $(\mathrm{d}^2\sigma/d\Omega\, \mathrm{d}E)^{\mathrm{mag}}$ of a general material is expressed in terms of the material magnetization vector, where

$$\begin{aligned} \left(\frac{\mathrm{d}^2\sigma}{\mathrm{d}\Omega\, \mathrm{d}E}\right)^{\mathrm{mag}} = & (r_0 g)^2 \frac{k}{k_0} \frac{1}{2\mu_\mathrm{b}} \sum_{i=1}^{J} p_i^{\mathrm{mag}} \int_{-\infty}^{\infty} \mathrm{d}t \frac{\exp(-\,\mathrm{i}wt)}{2\pi\hbar} \\ & \left\langle \phi_i \left| \sum_{\alpha=1}^{3} \sum_{\beta=1}^{3} \left(\frac{\kappa_\alpha \kappa_\beta}{\kappa^2} - \delta_{\alpha\beta} \right) M_\alpha^+ M_\beta(t) \right| \phi_i \right\rangle \end{aligned} \tag{3.238}$$

3.15 Coherent, Elastic Differential Cross Section Expressed by a Magnetization Vector

We now determine the differential cross section $(d\sigma/d\Omega)^{mag}$ for elastic neutron scatter in magnetic material that is expressed by the material magnetization vector **M**. We can express $(d^2\sigma/d\Omega\, dE)^{mag}$ in terms of the magnetization of the material. With Eq. (3.195) and Eq. (3.217) applied to Eq. (3.234), one obtains

$$\hat{U}^+\hat{U}(t) = \sum_{\alpha=1}^{3}\sum_{\beta=1}^{3}\left(\frac{\kappa_\alpha \kappa_\beta}{\kappa^2} - \delta_{\alpha\beta}\right)M_\alpha^+ M_\beta(t) = \frac{(\boldsymbol{\kappa}\times\mathbf{M}(\boldsymbol{\kappa})\times\boldsymbol{\kappa})^+(\boldsymbol{\kappa}\times\mathbf{M}(\boldsymbol{\kappa},t)\times\boldsymbol{\kappa})}{\kappa^4} \tag{3.239}$$

With Eq. (3.226) replaced by Eq. (3.239) in Eq. (3.216), using Eq. (3.232), the partial differential cross section $(d\sigma/d\Omega\, dE)^{mag}_{coh}$ of coherent neutron scatter from a solid, liquid, or gas is expressed in terms of the material magnetization vector, where

$$\left(\frac{d^2\sigma}{d\Omega\, dE}\right)^{mag}_{coh} = \left(\frac{r_0 g}{2\mu_b}\right)^2 \frac{k}{k_0}\frac{1}{2\mu_b}\sum_{i=1}^{J} p_i^{mag}\int_{-\infty}^{\infty} dt \frac{\exp(-iwt)}{2\pi\hbar} \tag{3.240}$$
$$\langle\phi_i|(\boldsymbol{\kappa}_u\times\mathbf{M}(\boldsymbol{\kappa})\times\boldsymbol{\kappa}_u)+(\boldsymbol{\kappa}_u\times\mathbf{M}(\boldsymbol{\kappa},t)\times\boldsymbol{\kappa}_u)|\phi_i\rangle$$

Each of the ith initial J eigenstates ϕ_i of the scattering nuclei occurs with probability p_i, where

$$\phi_i = \phi_i(\mathbf{R}_1, \mathbf{R}_2, \mathbf{R}_3, \ldots, \mathbf{R}_N) \tag{3.241}$$

The brackets represent the N integrations over variables

$$\mathbf{R}_1, \mathbf{R}_2, \mathbf{R}_3, \ldots, \mathbf{R}_N \tag{3.242}$$

The unit neutron scatter vector $\boldsymbol{\kappa}_u$ is

$$\boldsymbol{\kappa}_u = \frac{\boldsymbol{\kappa}}{\kappa} \tag{3.243}$$

and

$$(\boldsymbol{\kappa}_u\times\mathbf{M}(\boldsymbol{\kappa})\times\boldsymbol{\kappa}_u)^+(\boldsymbol{\kappa}_u\times\mathbf{M}(\boldsymbol{\kappa})\times\boldsymbol{\kappa}_u) = |\boldsymbol{\kappa}_u\times\mathbf{M}(\boldsymbol{\kappa})\times\boldsymbol{\kappa}_u|^2 \tag{3.244}$$

We assume elastic scatter

$$k = k_0 \tag{3.245}$$

Elastic scatter is time independent so that

$$\langle \mathbf{M}(\boldsymbol{\kappa}, t)\rangle \cong \langle \mathbf{M}(\boldsymbol{\kappa}, 0)\rangle = \langle \mathbf{M}(\boldsymbol{\kappa})\rangle \tag{3.246}$$

One replaces the time integral by its delta function representation. Given fixed incident neutron energy E_0 and scattered neutron energy E, where

$$\hbar w = E - E_0 \quad \text{and} \quad \mathrm{d}E = -\mathrm{d}(\hbar w), \tag{3.247}$$

the time integral is

$$\int_{-\infty}^{\infty} \exp(-\mathrm{i}wt)\mathrm{d}t = 2\pi\delta(w) = 2\pi\hbar\delta(\hbar w) = 2\pi\hbar\delta(E) \tag{3.248}$$

Integration over the scatter neutron energy E yields the coherent differential cross section, where

$$\left(\frac{\mathrm{d}\sigma}{\mathrm{d}\Omega}\right)_{\mathrm{coh}}^{\mathrm{mag}} = \left(\frac{r_0 g}{2\mu_\mathrm{b}}\right)^2 \frac{k}{k_0}\frac{1}{2\mu_\mathrm{b}}\sum_{i=1}^{J} p_i^{\mathrm{mag}} \langle \phi_i|(\boldsymbol{\kappa}_\mathrm{u} \times \mathbf{M}(\boldsymbol{\kappa}) \times \boldsymbol{\kappa}_\mathrm{u}) + (\boldsymbol{\kappa}_\mathrm{u} \times \mathbf{M}(\boldsymbol{\kappa}) \times \boldsymbol{\kappa}_\mathrm{u})|\phi_i\rangle \tag{3.249}$$

The thermally averaged magnetization $\langle \mathbf{M}(\boldsymbol{\kappa})\rangle$ or expected value of the κ-space magnetization vector $\mathbf{M}(\boldsymbol{\kappa})$ is

$$\langle \mathbf{M}(\boldsymbol{\kappa})\rangle = \sum_{i=1}^{J} p_i^{\mathrm{mag}} \langle \phi_i|\mathbf{M}(\boldsymbol{\kappa})|\phi_i\rangle \tag{3.250}$$

The mean-squared magnetization vector $\langle [\mathbf{M}(\boldsymbol{\kappa})]^2\rangle$ equals the squared mean magnetization vector $(\langle \mathbf{M}(\boldsymbol{\kappa})\rangle)^2$, where

$$\langle [\mathbf{M}(\boldsymbol{\kappa})]^2\rangle = (\langle \mathbf{M}(\boldsymbol{\kappa})\rangle)^2 \tag{3.251}$$

The coherent, elastic differential magnetic cross section $(\mathrm{d}\sigma/d\Omega)_e^{\mathrm{mag}}$ for neutron scatter is

$$\left(\frac{\mathrm{d}\sigma}{\mathrm{d}\Omega}\right)_{\mathrm{coh}}^{e} = \left(\frac{r_0 g}{2\mu_\mathrm{b}}\right)^2 |\hat{U}|^2 = \left(\frac{r_0 g}{2\mu_\mathrm{b}}\right)^2 |\boldsymbol{\kappa}_\mathrm{u} \times \langle \mathbf{M}(\boldsymbol{\kappa})\rangle \times \boldsymbol{\kappa}_\mathrm{u}|^2 \tag{3.252}$$

The thermally averaged magnetization vector $\langle \mathbf{M}(\boldsymbol{\kappa})\rangle$ includes the magnetic dipole moment contributions from the unpaired electron spin and electron orbital motion. If the neutron scatter vector $\boldsymbol{\kappa}$ is parallel (or antiparallel) to the thermally averaged magnetization vector $\langle \mathbf{M} \rangle$, then the coherent differential cross section $(d\sigma/d\Omega)_e^{\text{mag}}$ for neutron elastic scatter is zero:

$$\left(\frac{d\sigma}{d\Omega}\right)_{\text{coh}}^{e} = 0 \quad \boldsymbol{\kappa} || \langle \mathbf{M} \rangle \tag{3.253}$$

However, if the neutron scatter vector $\boldsymbol{\kappa}$ is perpendicular to the thermally averaged magnetization vector $\langle \mathbf{M} \rangle$, then the coherent, elastic differential cross section is nonzero and becomes

$$\left(\frac{d\sigma}{d\Omega}\right)_{\text{coh}}^{e} = \left(\frac{r_0 g}{2\mu_b}\right)^2 |\langle \mathbf{M}(\boldsymbol{\kappa}) \rangle|^2 \quad \boldsymbol{\kappa} \perp \langle \mathbf{M} \rangle \tag{3.254}$$

3.16 Coherent, Elastic Differential Cross Section Expressed by Electron Spin Density

If the magnetic scattering of neutrons is due only to the magnetic dipole moments of unpaired electron spins and if one can ignore the magnetic dipole moments due to the electron orbits, such as in a quenched crystal (discussed in Chapter 4), then the thermally averaged, κ-space magnetization vector $\langle \mathbf{M}(\boldsymbol{\kappa}) \rangle$ can be calculated from the thermally averaged electron spin density vector $\langle \boldsymbol{\rho}_s(\boldsymbol{\kappa}) \rangle$ of unpaired electrons. By knowing the atomic electron structure, one also knows the number of unpaired electrons per atom, and knowing the atom density, one can determine the density of unpaired electrons, each with a net spin. Hence, one can determine the net electron spins per unit volume or spin density.

The electron magnetic dipole moment $\boldsymbol{\mu}_e$, which points from magnetic south to magnetic north, is antiparallel to the net spin vector $\mathbf{s}$ of the unpaired electrons, where

$$\boldsymbol{\mu}_e = -2\mu_b \mathbf{s} \tag{3.255}$$

The thermally averaged magnetization $\langle \mathbf{M}(\mathbf{r}) \rangle$ is proportional to the thermally averaged spin density vector of unpaired electron spins $\langle \boldsymbol{\rho}_s(\mathbf{r}) \rangle$ at position $\mathbf{r}$, where

$$\langle \mathbf{M}(\mathbf{r}) \rangle = -2\mu_b \langle \boldsymbol{\rho}_s(\mathbf{r}) \rangle \tag{3.256}$$

Note also that the average magnetization vector $\langle \mathbf{M}(\mathbf{r}) \rangle$, which points from magnetic south to north, is antiparallel to the average direction of the electron spin vector $\langle \boldsymbol{\rho}_s(\mathbf{r}) \rangle$ of the unpaired electrons.

The κ-space Fourier transform of the average magnetization vector $\langle \mathbf{M}(\mathbf{r}) \rangle$ gives

$$\langle \mathbf{M}(\kappa) \rangle = -2\mu_b \langle \boldsymbol{\rho}_s(\boldsymbol{\kappa}) \rangle \tag{3.257}$$

Here, $\langle \boldsymbol{\rho}_s(\boldsymbol{\kappa})\rangle$ is the thermally averaged electron spin density vector in the κ-space, which is the Fourier transform of the thermally averaged electron spin density vector $\langle \boldsymbol{\rho}_s(\boldsymbol{\kappa})\rangle$ in the magnetic material. Substitution into Eq. (3.252) of Eq. (3.257) for $\langle \mathbf{M}(\kappa)\rangle$ in $(\mathrm{d}\sigma/d\Omega)_e^{\mathrm{mag}}$ gives the partial differential cross section for neutron magnetic scatter in terms of the Fourier transform of the material electron spin density, where

$$\left(\frac{\mathrm{d}\sigma}{\mathrm{d}\Omega}\right)_{\mathrm{coh}}^{e} = (r_0 g)^2 |\boldsymbol{\kappa}_{\mathrm{u}} \times \langle \boldsymbol{\rho}_s(\boldsymbol{\kappa})\rangle \times \boldsymbol{\kappa}_{\mathrm{u}}|^2 \tag{3.258}$$

If the neutron scatter vector $\boldsymbol{\kappa}$ is parallel (or antiparallel) to the net spin $\mathbf{s}$ of unpaired electrons, then from Eq. (3.253) the coherent differential cross section $(\mathrm{d}\sigma/d\Omega)_{\mathrm{coh}}^{e}$ for neutron elastic scatter is zero:

$$\left(\frac{\mathrm{d}\sigma}{\mathrm{d}\Omega}\right)_{\mathrm{coh}}^{e} = 0 \quad \boldsymbol{\kappa} || \langle \boldsymbol{\rho}_s(\boldsymbol{\kappa})\rangle \tag{3.259}$$

However, if the neutron scatter vector $\boldsymbol{\kappa}$ is perpendicular to the net spin $\mathbf{s}$ of unpaired electrons, then from Eq. (3.254) the coherent, elastic differential cross section is nonzero and becomes

$$\left(\frac{\mathrm{d}\sigma}{\mathrm{d}\Omega}\right)_{\mathrm{coh}}^{e} = \left(\frac{r_0 g}{2\mu_{\mathrm{b}}}\right)^2 |\langle \boldsymbol{\rho}_s(\boldsymbol{\kappa})\rangle|^2 \quad \boldsymbol{\kappa} \perp \langle \boldsymbol{\rho}_s(\boldsymbol{\kappa})\rangle \tag{3.260}$$

3.17 Magnetization Determined from Measuring Bragg Peak Intensity

Next, we determine the differential cross section for elastic neutron scatter from ferromagnetic material in terms of the Bragg peak intensity $I_\tau(\boldsymbol{\tau}_\tau)$, which is the Fourier transform of the thermally averaged, expected value of the magnetization vector $\langle \mathbf{M}(\mathbf{r})\rangle$. That is, the experimentally measured intensities $I_\tau(\boldsymbol{\tau}_\tau)$ of the τth Bragg peaks for diffracted neutrons are used to calculate the thermally averaged or mean magnetization $\langle \mathbf{M}(\mathbf{r})\rangle$ of the ferromagnetic crystal. One starts with the elastic, differential magnetic cross section $(\mathrm{d}\sigma/d\Omega)_{\mathrm{coh}}^{e}$ of Eq. (3.258) for neutron scatter:

$$\left(\frac{\mathrm{d}\sigma}{\mathrm{d}\Omega}\right)_{\mathrm{coh}}^{e} = \left(\frac{r_0 g}{2\mu_{\mathrm{b}}}\right)^2 |\hat{U}|^2 = \left(\frac{r_0 g}{2\mu_{\mathrm{b}}}\right)^2 \left|\frac{\boldsymbol{\kappa} \times \langle \mathbf{M}(\boldsymbol{\kappa})\rangle \times \boldsymbol{\kappa}}{\kappa^2}\right|^2 \tag{3.261}$$

The expected value or mean magnetization $\langle \mathbf{M}_{\mathrm{do}}(\mathbf{r})\rangle$ for a single domain of a ferromagnetic crystal can be used to represent the average magnetization throughout the ferromagnetic crystal. The magnetization vector $\langle \mathbf{M}_{\mathrm{do}}(\mathbf{r})\rangle$ represents the volume average magnetic dipole moment of the unit cell that also has been thermally averaged. If the nuclei position and motion are not affected by the magnetization, then one can set

$$\int_{-\infty}^{\infty} \mathrm{d}z \int_{-\infty}^{\infty} \mathrm{d}y \int_{-\infty}^{\infty} \mathrm{d}x \langle \mathbf{M}_{\mathrm{do}}(\boldsymbol{\kappa}) \rangle \exp(\mathrm{i}\boldsymbol{\kappa} \cdot \mathbf{r}) = \overline{\mathbf{F}}_A(\boldsymbol{\kappa}) \sum_{l=0}^{N-1} \sum_{d=1}^{r} \langle \exp(\mathrm{i}\boldsymbol{\kappa} \cdot \hat{\mathbf{R}}_{ldu}(t)) \rangle \tag{3.262}$$

The κ-space vector $\overline{\mathbf{F}}_A(\boldsymbol{\kappa})$ is the thermal-averaged magnetic form factor for each crystal atom in a non-Bravais crystal lattice. In a Bravais lattice crystal (one atom per unit cell), $\overline{\mathbf{F}}_A(\boldsymbol{\kappa})$ is the thermally averaged magnetic form factor per unit cell. The corresponding mean value of the magnetization $\langle \mathbf{M}_{\mathrm{do}}(\boldsymbol{\kappa}) \rangle$ in reciprocal space can be obtained from $\langle \mathbf{M}_{\mathrm{do}}(\mathbf{r}) \rangle$ in the direct lattice by using the Fourier transform:

$$\langle \mathbf{M}(\boldsymbol{\kappa}) \rangle = \int_{-\infty}^{\infty} \mathrm{d}z \int_{-\infty}^{\infty} \mathrm{d}y \int_{-\infty}^{\infty} \mathrm{d}x \langle \mathbf{M}_{\mathrm{do}}(\mathbf{r}) \rangle \exp(\mathrm{i}\boldsymbol{\kappa} \cdot \mathbf{r}) \tag{3.263}$$

In the reciprocal lattice space, the mean magnetization vector $\langle \mathbf{M}(\boldsymbol{\kappa}) \rangle$ can be expressed as the product of the average magnetic form factor vector $\overline{\mathbf{F}}_A(\boldsymbol{\kappa})$ per atom multiplied by the sum of the scattered wave amplitudes from the Nr atoms of the crystal. These scattered amplitudes contain the pattern of atom spatial positions via the phase shifts $\mathrm{i}\boldsymbol{\kappa} \cdot \hat{\mathbf{R}}_{ldu}(t)$ of neutron waves scattered from these atoms, where

$$\langle \mathbf{M}(\boldsymbol{\kappa}) \rangle = \overline{\mathbf{F}}_A(\boldsymbol{\kappa}) \sum_{l=0}^{N-1} \sum_{d=1}^{r} \langle \exp(\mathrm{i}\boldsymbol{\kappa} \cdot \hat{\mathbf{R}}_{ldu}(t)) \rangle \tag{3.264}$$

The average magnetic form factor vector $\overline{\mathbf{F}}_A(\boldsymbol{\kappa})$ per unit cell is obtained by integrating the mean value of the magnetization $\langle \mathbf{M}_{\mathrm{do}}(\boldsymbol{\kappa}) \rangle$ over the unit cell volume v_0, which is defined by unit cell vectors $(\mathbf{a}, \mathbf{b}, \mathbf{c})$, where $v_0 = abc$ and

$$\overline{\mathbf{F}}_A(\boldsymbol{\kappa}) = \int_{-c/2}^{c/2} \mathrm{d}z \int_{-b/2}^{b/2} \mathrm{d}y \int_{-a/2}^{a/2} \mathrm{d}x \langle \mathbf{M}_{\mathrm{do}}(\mathbf{r}) \rangle \exp(\mathrm{i}\boldsymbol{\kappa} \cdot \mathbf{r}) \tag{3.265}$$

The neutron scatter vector $\boldsymbol{\kappa}$ and position vector $\mathbf{r}$ in rectangular coordinates are

$$\boldsymbol{\kappa} = \kappa_x \hat{\mathbf{x}} + \kappa_y \hat{\mathbf{y}} + \kappa_z \hat{\mathbf{z}} \quad \text{and} \quad \mathbf{r} = x\hat{\mathbf{x}} + y\hat{\mathbf{y}} + z\hat{\mathbf{z}} \tag{3.266}$$

Vector $\overline{\mathbf{F}}_A(\boldsymbol{\kappa})$ is the vector version of the scalar magnetic form factor $F(\boldsymbol{\kappa})$, which is obtained by integration of the normalized density $\rho_{ue}(x, y, z)$ of the unpaired electrons over the volume of the atom or unit cell. The normalized electron density ρ_{ue} is the unpaired electron density ρ_e in an atom, divided by the number n_{ue} of unpaired electrons, where

$$F(\boldsymbol{\kappa}) = \frac{1}{n_{ue}} \int_{-c/2}^{c/2} \mathrm{d}z \int_{-b/2}^{b/2} \mathrm{d}y \int_{-a/2}^{a/2} \mathrm{d}x \rho_e(x, y, z) \exp(\mathrm{i}\boldsymbol{\kappa} \cdot \mathbf{r}) \tag{3.267}$$

The mean domain magnetization $\langle \mathbf{M}_{\mathrm{do}}(\mathbf{r}) \rangle$ as a function of position $\mathbf{r}$ in the direct lattice can be expressed by the inverse Fourier transform of the mean magnetic form factor vector $\overline{\mathbf{F}}_A(\boldsymbol{\kappa})$ per unit cell by

$$\langle \mathbf{M}_{\mathrm{do}}(\mathbf{r}) \rangle = \frac{1}{v_0} \sum_{\tau=0}^{N-1} \overline{\mathbf{F}}_A(\boldsymbol{\tau}_\tau) \exp(-\, \mathrm{i}\boldsymbol{\tau}_\tau \cdot \mathbf{r}) \tag{3.268}$$

Here, v_0 is the direct-space unit cell volume:

$$v_0 = abc \tag{3.269}$$

This claimed inverse Fourier transform relation of Eq. (3.268) can be checked as follows. One multiplies both sides of the equation for $\langle \mathbf{M}_{\mathrm{do}}(\mathbf{r}) \rangle$ by $\exp(\mathrm{i}\boldsymbol{\tau}_{\tau'} \cdot \mathbf{r})$, and then integrates by position $\mathbf{r}$ over the volume of the direct-space unit cell, where

$$\begin{aligned} &\textstyle\int_{-c/2}^{c/2} \mathrm{d}z \int_{-b/2}^{b/2} \mathrm{d}y \int_{-a/2}^{a/2} \mathrm{d}x \langle \mathbf{M}_{\mathrm{do}}(\mathbf{r}) \rangle \exp(\mathrm{i}\boldsymbol{\tau}_{\tau'} \cdot \mathbf{r}) = \\ &\frac{1}{v_0} \sum_{\tau=0}^{N-1} \overline{\mathbf{F}}_A(\boldsymbol{\tau}_\tau) \int_{-c/2}^{c/2} \mathrm{d}z \int_{-b/2}^{b/2} \mathrm{d}y \int_{-a/2}^{a/2} \mathrm{d}x \exp[\mathrm{i}r \cdot (\boldsymbol{\tau}_{\tau'} - \boldsymbol{\tau}_\tau)] \end{aligned} \tag{3.270}$$

Recall that for Miller triplet integers (h, k, l) of the reciprocal lattice vector $\boldsymbol{\tau}_\tau$

$$\boldsymbol{\tau}_\tau = h\mathbf{a}^* + k\mathbf{b}^* + l\mathbf{c}^* \quad \text{and} \quad aa^* = bb^* = cc^* = 2\pi \tag{3.271}$$

Spatial integration gives the unit cell volume multiplied by the Kronecker delta function, see Chap. 4 in Cremer (2012b):

$$\begin{aligned} &\int_{-c/2}^{c/2} \mathrm{d}z \exp[\mathrm{i}x(\tau_{\tau' z} - \tau_{\tau z})] \int_{-b/2}^{b/2} \mathrm{d}y \exp[\mathrm{i}x(\tau_{\tau' y} - \tau_{\tau y})] \int_{-a/2}^{a/2} \mathrm{d}x \exp[\mathrm{i}x(\tau_{\tau' x} - \tau_{\tau x})] \\ &= v_0 \delta_{\boldsymbol{\tau}_{\tau'} \boldsymbol{\tau}_\tau} \end{aligned} \tag{3.272}$$

The only nonzero term occurs when $\boldsymbol{\tau}_{\tau'} = \boldsymbol{\tau}_\tau$, so that one obtains the average magnetic form factor vector $\overline{\mathbf{F}}_A(\boldsymbol{\kappa})$ per unit cell that was presented earlier in this section; namely,

$$\overline{\mathbf{F}}_A(\boldsymbol{\tau}_\tau) = \int_{-c/2}^{c/2} \mathrm{d}z \int_{-b/2}^{b/2} \mathrm{d}y \int_{-a/2}^{a/2} \mathrm{d}x \langle \mathbf{M}_{\mathrm{do}}(\mathbf{r}) \rangle \exp(\mathrm{i}\boldsymbol{\tau}_\tau \cdot \mathbf{r}) \tag{3.273}$$

As will be shown next, the intensity of the τth Bragg peak $I_\tau(\boldsymbol{\tau}_\tau)$ is in proportion to the squared magnitude of the thermally averaged magnetic form factor vector $\overline{\mathbf{F}}_A(\boldsymbol{\kappa})$ per atom (or per unit cell in a Bravais lattice), where

$$I_\tau(\boldsymbol{\tau}_\tau) \propto \left| \overline{\mathbf{F}}_A(\boldsymbol{\tau}_\tau) \right|^2 \tag{3.274}$$

Hence, one can obtain the mean domain magnetization $\langle \mathbf{M}_{\mathrm{do}}(\mathbf{r}) \rangle$ from $\overline{\mathbf{F}}_A(\boldsymbol{\kappa})$ via the measured intensity $I_\tau(\boldsymbol{\tau}_\tau)$ of the Bragg peaks from the neutron diffraction of the ferromagnetic crystal.

The nuclei in each lth unit cell of the crystal are thermally averaged. For $l = 1, 2, \ldots, N$

$$\langle \exp(\mathrm{i}\boldsymbol{\kappa} \cdot \hat{\mathbf{R}}_{ldu}(t)) \rangle = \sum_{i=1}^{r} p_i^{\mathrm{mag}} \langle \phi_i | \exp(\mathrm{i}\boldsymbol{\kappa} \cdot \hat{\mathbf{R}}_{ldu}(t)) | \phi_i \rangle \tag{3.275}$$

where

$$\langle \phi_i | \exp(\mathrm{i}\boldsymbol{\kappa} \cdot \hat{\mathbf{R}}_{ldu}(t)) | \phi_i \rangle = \int \mathrm{d}V_J \int \mathrm{d}V_{J-1} \int \mathrm{d}V_{J-2} \cdots \int \mathrm{d}V_2 \int \mathrm{d}V_1 \phi_i^* \exp(\mathrm{i}\boldsymbol{\kappa} \cdot \hat{\mathbf{R}}_{ldu}(t)) \phi_i \tag{3.276}$$

The nonequilibrium position $\mathbf{R}_{ldu}$ of a nucleus in the non-Bravais lattice crystal with r atoms per unit cell and N unit cells is the sum of the nucleus equilibrium position $\mathbf{R}_l + \mathbf{d}_d$ added to the nucleus nonequilibrium displacement vector $\mathbf{u}_{ld}$ from its equilibrium position due to thermal vibration. The vector nucleus position $\mathbf{R}_{ldu}$ of its vector, time-dependent, Heisenberg operator $\hat{\mathbf{R}}_{ldu}(t)$ is then

$$\hat{\mathbf{R}}_{ldu}(t) = \hat{\mathbf{R}}_l + \hat{\mathbf{d}}_d + \hat{\mathbf{u}}_{ld}(t) \tag{3.277}$$

The expected value of the reciprocal lattice vector magnetization $\langle \mathbf{M}(\boldsymbol{\kappa}) \rangle$ in the κ-space of Eq. (3.268) is now plugged into Eq. (3.261) for the elastic, differential magnetic cross section $(\mathrm{d}\sigma/d\Omega)_e^{\mathrm{mag}}$ for neutron scatter, so that

$$\left(\frac{\mathrm{d}\sigma}{\mathrm{d}\Omega}\right)_{\mathrm{coh}}^{e} = \left(\frac{r_0 g}{2\mu_{\mathrm{b}}}\right)^2 \frac{1}{\kappa^4} \left| \boldsymbol{\kappa} \times \overline{\mathbf{F}}_A(\boldsymbol{\kappa}) \sum_{l=0}^{N-1} \sum_{d=1}^{r} \langle \exp(\mathrm{i}\boldsymbol{\kappa} \cdot \mathbf{R}_{ldu}(t)) \rangle \times \boldsymbol{\kappa} \right|^2 \tag{3.278}$$

or

$$\left(\frac{\mathrm{d}\sigma}{\mathrm{d}\Omega}\right)_{\mathrm{coh}}^{e} = \left(\frac{r_0 g}{2\mu_{\mathrm{b}}}\right)^2 |\overline{\mathbf{F}}_A(\boldsymbol{\kappa})|^2 \left| \frac{\boldsymbol{\kappa} \times \overline{\mathbf{F}}_{Au} \times \boldsymbol{\kappa}}{\kappa^2} \right|^2 \tag{3.279}$$

The unit vector $\overline{\mathbf{F}}_{Au}(\boldsymbol{\kappa})$ for the mean magnetic form factor vector $\overline{\mathbf{F}}_A(\boldsymbol{\kappa})$ in the κ-space is

$$\overline{\mathbf{F}}_{Au}(\boldsymbol{\kappa}) = \frac{\overline{\mathbf{F}}_A(\boldsymbol{\kappa})}{|\overline{\mathbf{F}}_A(\boldsymbol{\kappa})|} \tag{3.280}$$

With substitution into Eq. (3.264) from Eq. (3.277) for $\mathbf{R}_{ldu}$, the mean of the magnetization vector $\langle \mathbf{M}(\boldsymbol{\kappa}) \rangle$ in κ-space is, then

$$\langle \mathbf{M}(\boldsymbol{\kappa}) \rangle = \overline{\mathbf{F}}_A(\boldsymbol{\kappa}) \sum_{l=0}^{N-1} \exp(\mathrm{i}\boldsymbol{\kappa} \cdot \mathbf{R}_l) \sum_{d=1}^{r} \exp(\mathrm{i}\boldsymbol{\kappa} \cdot \mathbf{d}_d) \langle \exp(\mathrm{i}\boldsymbol{\kappa} \cdot \hat{\mathbf{u}}_{ld}(t)) \rangle \tag{3.281}$$

We now evaluate the term $\langle \exp(i\boldsymbol{\kappa} \cdot \hat{\mathbf{u}}_{ld}(t)) \rangle$. Let

$$V = i\boldsymbol{\kappa} \cdot \hat{\mathbf{u}}_{ld}(t) \tag{3.282}$$

From Sect 4 and Eq. (147) in Chap. 5 of Cremer (2012b) in the derivation of the thermal average of V, one obtains the Debye–Waller factor:

$$\langle \exp(V) \rangle = \exp\left(\frac{1}{2}\langle V^2 \rangle\right) = \exp(-W_d) \tag{3.283}$$

where

$$\langle V^2 \rangle = \langle U^2 \rangle = -2W_d \tag{3.284}$$

The mean magnetization vector $\langle \mathbf{M}(\boldsymbol{\kappa}) \rangle$ in the κ-space for a non-Bravais lattice crystal of Eq. (3.281), becomes

$$\langle \mathbf{M}(\boldsymbol{\kappa}) \rangle = \overline{\mathbf{F}}_A(\boldsymbol{\kappa}) \sum_{d=1}^{r} \exp(i\boldsymbol{\kappa} \cdot \mathbf{d}_d) \exp(-W_d) \sum_{l=0}^{N-1} \exp(i\boldsymbol{\kappa} \cdot \mathbf{R}_l) \tag{3.285}$$

For each value of l', the summations over l gives the same set of values since the unit cell positions $\mathbf{R}_l$ are fixed. The position of the $l' = 0$ unit cell is at the crystal origin, where $\mathbf{R}_0 = 0$. One sets $l' = 0$ and replaces the summation over l' by N. The d nucleus displacement $\mathbf{d}_d$ and the d' nucleus displacement $\mathbf{d}_{d'}$ are not correlated; hence, one requires the double summation over d and d'. Implementation of this in Eq. (3.285), yields

$$|\langle \mathbf{M}(\boldsymbol{\kappa}) \rangle|^2 = N|\overline{\mathbf{F}}_A(\boldsymbol{\kappa})|^2 \left| \sum_{d=1}^{r} \exp(i\boldsymbol{\kappa} \cdot \mathbf{d}_d) \exp(-W_d) \right|^2 \sum_{l=0}^{N-1} \exp(i\boldsymbol{\kappa} \cdot \mathbf{R}_l) \tag{3.286}$$

The direct space amplitude S in reciprocal space can be expressed in reciprocal space by the sum of Dirac delta functions of the reciprocal lattice vectors $\boldsymbol{\tau}_\tau$, where from Eq. (3.47) in Squires (1996) and Eq. (105) in Ch. 4 of Cremer (2012b),

$$S = \sum_{l=0}^{N-1} \exp(i\boldsymbol{\kappa} \cdot \mathbf{R}_l) = \frac{(2\pi)^3}{v_0} \sum_{\tau=0}^{N-1} \delta(\boldsymbol{\kappa} - \boldsymbol{\tau}_\tau) \tag{3.287}$$

The mean magnetization vector $\langle \mathbf{M}(\boldsymbol{\kappa}) \rangle$ becomes

$$\langle \mathbf{M}(\boldsymbol{\kappa}) \rangle = \frac{(2\pi)^3}{v_0} \overline{\mathbf{F}}_A(\boldsymbol{\kappa}) \sum_{d=1}^{r} \exp(i\boldsymbol{\kappa} \cdot \mathbf{d}_d) \exp(-W_d) \sum_{\tau=0}^{N-1} \delta(\boldsymbol{\kappa} - \boldsymbol{\tau}_\tau) \tag{3.288}$$

The squared magnitude of the mean magnetization vector ($|\langle \mathbf{M}(\boldsymbol{\kappa})\rangle|^2$) becomes

$$|\langle \mathbf{M}(\boldsymbol{\kappa})\rangle|^2 = N\frac{(2\pi)^3}{v_0}|\overline{\mathbf{F}}_A(\boldsymbol{\kappa})|^2\left|\sum_{d=1}^{r}\exp(\mathrm{i}\boldsymbol{\kappa}\cdot\mathbf{d}_d)\ \exp(-W_d)\right|^2\sum_{\tau=0}^{N-1}\delta(\boldsymbol{\kappa}-\boldsymbol{\tau}_\tau) \tag{3.289}$$

where

$$\left[\sum_{l=0}^{N-1}\exp(\mathrm{i}\boldsymbol{\kappa}\cdot\mathbf{R}_l)\right]^2 = \sum_{l=0}^{N-1}\exp(\mathrm{i}\boldsymbol{\kappa}\cdot\mathbf{R}_l)\sum_{l'=0}^{N-1}\exp(\mathrm{i}\boldsymbol{\kappa}\cdot\mathbf{R}_{l'}) \tag{3.290}$$

So

$$\left[\sum_{l=0}^{N-1}\exp(\mathrm{i}\boldsymbol{\kappa}\cdot\mathbf{R}_l)\right]^2 = N\sum_{l=0}^{N-1}\exp(\mathrm{i}\boldsymbol{\kappa}\cdot\mathbf{R}_l) \tag{3.291}$$

With substitution of Eq. (3.289) into Eq. (3.261), in terms of unit vector $\boldsymbol{\kappa}_\mathrm{u}$ for the neutron scatter vector $\boldsymbol{\kappa}$, the coherent, elastic differential cross section $(\mathrm{d}\sigma/d\Omega)^e_{\mathrm{coh}}$ becomes

$$\left(\frac{\mathrm{d}\sigma}{\mathrm{d}\Omega}\right)^e_{\mathrm{coh}} = N|\overline{\mathbf{F}}_A(\boldsymbol{\kappa})|^2\left(\frac{r_0 g}{2\mu_\mathrm{b}}\right)^2\frac{(2\pi)^3}{v_0}\left|\sum_{d=1}^{r}\exp(\mathrm{i}\boldsymbol{\kappa}\cdot\mathbf{d}_d)\exp(-W_d)\right|^2 \\ |\boldsymbol{\kappa}_\mathrm{u}\times\overline{\mathbf{F}}_{Au}\times\boldsymbol{\kappa}_\mathrm{u}|^2\sum_{\tau=0}^{N-1}\delta(\boldsymbol{\kappa}-\boldsymbol{\tau}_\tau) \tag{3.292}$$

The thermally averaged magnetic scatter length $b_\mathrm{M}(\boldsymbol{\kappa})$ is, then

$$b_\mathrm{M}(\boldsymbol{\kappa}) = |\overline{\mathbf{F}}_A(\boldsymbol{\kappa})|\left(\frac{r_0 g}{2\mu_\mathrm{b}}\right)\left|\sum_{d=1}^{r}\exp(\mathrm{i}\boldsymbol{\kappa}\cdot\mathbf{d}_d)\ \exp(-W_d)\right||\boldsymbol{\kappa}_\mathrm{u}\times\overline{\mathbf{F}}_{Au}\times\boldsymbol{\kappa}_\mathrm{u}| \tag{3.293}$$

The cross section $(\mathrm{d}\sigma/d\Omega)^e_{\mathrm{coh}}$ is proportional to the squared magnitude of the thermally averaged magnetic scatter length, where

$$\left(\frac{\mathrm{d}\sigma}{\mathrm{d}\Omega}\right)^e_{\mathrm{coh}} = N[b_\mathrm{M}(\boldsymbol{\kappa})]^2\frac{(2\pi)^3}{v_0}\sum_{\tau=0}^{N-1}\delta(\boldsymbol{\kappa}-\boldsymbol{\tau}_\tau) \tag{3.294}$$

Recall Eq. (3.253). If the neutron scatter vector $\boldsymbol{\kappa}$ is parallel (or antiparallel) to the mean magnetic form factor vector per single atom $\overline{\mathbf{F}}_{Au}$, then the coherent differential cross section $(\mathrm{d}\sigma/d\Omega)^e_{\mathrm{coh}}$ for neutron elastic scatter is zero:

$$\left(\frac{\mathrm{d}\sigma}{\mathrm{d}\Omega}\right)^e_{\mathrm{coh}} = 0 \quad \boldsymbol{\kappa}||\overline{\mathbf{F}}_{Au} \tag{3.295}$$

However, recalling Eq. (3.254), if the neutron scatter vector $\boldsymbol{\kappa}$ is perpendicular to $\overline{\mathbf{F}}_{Au}$, then the coherent, elastic differential cross section is nonzero and becomes

$$\left(\frac{\mathrm{d}\sigma}{\mathrm{d}\Omega}\right)^{e}_{\mathrm{coh}} = N|\overline{\mathbf{F}}_A(\boldsymbol{\kappa})|^2\left(\frac{r_0 g}{2\mu_{\mathrm{b}}}\right)^2\frac{(2\pi)^3}{v_0}\left|\sum_{d=1}^{r}\exp(\mathrm{i}\boldsymbol{\kappa}\cdot\mathbf{d}_d)\exp(-W_d)\right|^2\sum_{\tau=0}^{N-1}\delta(\boldsymbol{\kappa}-\boldsymbol{\tau}_\tau)\quad \boldsymbol{\kappa}\perp\overline{\mathbf{F}}_{Au} \tag{3.296}$$

And in Eq. (3.296) with $\boldsymbol{\kappa}=\boldsymbol{\tau}_\tau$, the coherent, elastic neutron differential cross section of the τth Bragg peak is

$$\left(\frac{\mathrm{d}\sigma}{\mathrm{d}\Omega}\right)^{e}_{\mathrm{coh}} = N\left(\frac{r_0 g}{2\mu_{\mathrm{b}}}\right)^2\frac{(2\pi)^3}{v_0}|\overline{\mathbf{F}}_A(\boldsymbol{\tau}_\tau)|^2\left|\sum_{d=1}^{r}\exp(\mathrm{i}\boldsymbol{\kappa}\cdot\mathbf{d}_d)\exp(-W_d)\right|^2\quad \boldsymbol{\kappa}\perp\overline{\mathbf{F}}_{Au} \tag{3.297}$$

The measured ratio of the intensity of the τth Bragg peak divided by the incident neutron beam intensity at the ferromagnetic crystal gives the measured coherent elastic cross section $(\mathrm{d}\sigma/d\Omega)^e_{\mathrm{coh}}$ for the τth Bragg peak. The measured $(\mathrm{d}\sigma/d\Omega)^e_{\mathrm{coh}}$ expression then yields the squared magnitude of the unit cell magnetic form factor $|\overline{\mathbf{F}}_A(\boldsymbol{\tau}_\tau)|^2$. The measured $\overline{\mathbf{F}}_A(\boldsymbol{\tau}_\tau)$ values for n significant Bragg peaks $\boldsymbol{\tau}_\tau$ of the N possible Bragg peaks, where $n \ll N$, allows one to determine the approximate, mean magnetization $\langle\mathbf{M}_{\mathrm{do}}(\mathbf{r})\rangle$ for a single ferromagnetic domain via the Eq. (3.268), where

$$\langle\mathbf{M}_{\mathrm{do}}(\mathbf{r})\rangle = \frac{1}{v_0}\sum_{\tau=0}^{N-1}\overline{\mathbf{F}}_A(\boldsymbol{\tau}_\tau)\exp(-\mathrm{i}\boldsymbol{\tau}_\tau\cdot\mathbf{r}) \tag{3.298}$$

References

Cremer, J. T. (2012a). Neutron and x-ray microscopy - part 1. In P. W. Hawkes (Ed.), *Advances in imaging and electron physics* (172). London, UK: Academic Press, an imprint of Elsevier Press.

Cremer, J. T. (2012b). Neutron and x-ray microscopy - part 2. In P. W. Hawkes (Ed.), *Advances in imaging and electron physics* (173). London, UK: Academic Press, an imprint of Elsevier Press.

Cremer, J. T. (2013). In P. W. Hawkes (Ed.), *Advances in imaging and electron physics* (175). London, UK: Academic Press, an imprint of Elsevier Press.

Squires, G. L. (1996). *Introduction to the theory of thermal neutron scattering*. New York: Dover.

4 LS Coupling Basis for Magnetic Neutron Scatter

4.1 Summation and Coupling of Atomic Electron Spin and Orbital Angular Momentum

We now examine in detail the atomic electron environment that causes the magnetic scattering of neutrons. In an atom, the individual, unpaired atomic electrons couple their orbital angular momentums to a summed orbital angular momentum vector **L** for the atom, and the individual unpaired electrons of an atom couple their spin angular momentums to produce a net spin angular momentum vector **S** for the atom. The total atomic electron angular momentum vector **L** combines with the total atomic electron spin angular momentum vector **S** to produce the total electron angular momentum vector **J** of the atom (or ion), where

$$\mathbf{J} = \mathbf{L} + \mathbf{S} \tag{4.1}$$

This forms the basis of LS coupling of the the total spin angular momentum **S** and total orbit angular momentum **L** of an atom's unpaired electrons, whose associated magnetic dipole moments cause magnetic neutron scatter from an atom.

These angular momentums can be expressed in their operator forms: $\hat{J}^2, \hat{L}^2, \hat{S}^2, \hat{J}_z$, and $\hat{\mathbf{L}} \cdot \hat{\mathbf{S}}$, where operators $\hat{\mathbf{L}}$ and $\hat{\mathbf{S}}$ commute, and contain the LS coupling cross product—the LS dot-product term, where

$$\hat{\mathbf{J}}^2 = \hat{\mathbf{J}} \cdot \hat{\mathbf{J}} = (\hat{\mathbf{L}} + \hat{\mathbf{S}}) \cdot (\hat{\mathbf{L}} + \hat{\mathbf{S}}) = \hat{L}^2 + \hat{S}^2 + 2\hat{\mathbf{L}} \cdot \hat{\mathbf{S}} \tag{4.2}$$

Components in the x, y, and z directions are

$$\hat{J}_x = \hat{L}_x + \hat{S}_x \tag{4.3}$$

$$\hat{J}_y = \hat{L}_y + \hat{S}_y \tag{4.4}$$

$$\hat{J}_z = \hat{L}_z + \hat{S}_z \tag{4.5}$$

With the shared eigenfunction ψ_{jm_jls}, the eigenequations of $\hat{J}^2, \hat{L}^2, \hat{S}^2$ and J_z are

$$\hat{L}^2 |\psi_{jm_jls}\rangle = \hbar^2 l(l+1) |\psi_{jm_jls}\rangle \tag{4.6}$$

Neutron and X-ray Optics. DOI: http://dx.doi.org/10.1016/B978-0-12-407164-3.00003-6

$$\hat{S}^2|\psi_{jm_jls}\rangle = \hbar^2 s(s+1)|\psi_{jm_jls}\rangle \tag{4.7}$$

$$\hat{J}^2|\psi_{jm_jls}\rangle = \hbar^2 j(j+1)|\psi_{jm_jls}\rangle \tag{4.8}$$

Operators $\hat{O} = \hat{L}, \hat{S}, \hat{J}$ have quantum number eigenvalues $o = l, s, j$ and the eigenequation

$$\hat{O}^2|\psi_{jm_jls}\rangle = \hbar\sqrt{o(o+1)}|\psi_{jm_jls}\rangle \quad \hat{O} = \hat{L}, \hat{S}, \hat{J} \text{ and } o = l, s, j \tag{4.9}$$

The quantum numbers j range from the maximum $l+s$ to the minimum $|l-s|$, where

$$l+s \geq j \geq |l-s| \tag{4.10}$$

The electron spin quantum number s is

$$s = 1/2 \tag{4.11}$$

For $s < l$, there is a total or multiplicity of $2s+1$ values of j, where

$$j = l+s, l+s-1, l+s-2, \ldots, |l-s|+1, |l-s| \tag{4.12}$$

The azimuthal component operator J_z has quantum numbers

$$m_j = -j, \; -j+1, \ldots, j-1, \; j \tag{4.13}$$

where

$$J_z|\psi_{jm_jls}\rangle = \hbar m_j|\psi_{jm_jls}\rangle \tag{4.14}$$

An atom with N electrons has N individual electron orbital angular momentums $\mathbf{l}_l$, vector-summed to yield the total electron angular momentum $\mathbf{L}$, where

$$\mathbf{l}_1, \mathbf{l}_2, \ldots, \mathbf{l}_N \tag{4.15}$$

And the atom with N electrons has N individual electron orbital angular momentums $\mathbf{s}_s$, where

$$\mathbf{s}_1, \mathbf{s}_2, \ldots, \mathbf{s}_N \tag{4.16}$$

A total of n_l eigenstates can be formed that are also eigenstates of the squared total electron orbital angular momentum operator $\hat{L}^2$, the squared azimuthal

(z-component) orbital angular momentum operator $\hat{L}_z^2$, and orbital angular momentum representations

$$\hat{L}_1^2, \hat{L}_2^2, \ldots, \hat{L}_N^2 \tag{4.17}$$

with

$$\hat{\mathbf{L}} \cdot \hat{\mathbf{L}} = \hat{L}^2 \quad \text{and} \quad \hat{\mathbf{L}}_l \cdot \hat{\mathbf{L}}_l = \hat{L}_l^2 \tag{4.18}$$

Here, n_l is an N-fold product given by

$$n_l = \prod_{i=1}^{N} (2l_i + 1) \tag{4.19}$$

Likewise, a total of n_s eigenstates can be formed that are also eigenstates of the squared total electron spin angular momentum operator $\hat{S}^2$, the squared azimuthal (z-component) spin angular momentum operator $\hat{S}_z^2$, and spin angular momentum representations

$$\hat{S}_1^2, \hat{S}_2^2, \ldots, \hat{S}_N^2 \tag{4.20}$$

where

$$n_s = \prod_{i=1}^{N} (2s_i + 1) \tag{4.21}$$

and

$$\hat{\mathbf{S}} \cdot \hat{\mathbf{S}} = \hat{S}^2 \quad \text{and} \quad \hat{\mathbf{S}}_s \cdot \hat{\mathbf{S}}_s = \hat{S}_s^2 \tag{4.22}$$

4.2 Spin and Orbital Angular Momentum in Two- and Three-Electron Atoms

For a two-electron atom, the total electron orbital angular momentum operator $\hat{L}^2$ has an eigenvalue of

$$\hbar^2 l(l+1) \tag{4.23}$$

where

$$\hat{L}^2 = \hat{\mathbf{L}} \cdot \hat{\mathbf{L}} = (\hat{\mathbf{L}}_1 + \hat{\mathbf{L}}_2) \cdot (\hat{\mathbf{L}}_1 + \hat{\mathbf{L}}_2) \tag{4.24}$$

The three orbital angular momentum quantum numbers for the two-electron atoms are $l = 0, 1, 2$, where

$$l = |l_1 + l_2|, \ldots, |l_1 - l_2| \tag{4.25}$$

For a two-electron atom, the total electron spin angular momentum operator $\hat{S}^2$ has an eigenvalue of

$$\hbar^2 s(s+1) \tag{4.26}$$

where

$$\hat{S}^2 = \hat{\mathbf{S}} \cdot \hat{\mathbf{S}} = (\hat{\mathbf{S}}_1 + \hat{\mathbf{S}}_2) \cdot (\hat{\mathbf{S}}_1 + \hat{\mathbf{S}}_2) \tag{4.27}$$

The three spin angular momentum quantum numbers for the two-electron atoms are

$$s = 0, 1, 2 \tag{4.28}$$

where

$$s = |s_1 + s_2|, \ldots, |s_1 - s_2| \tag{4.29}$$

For a three-electron atom, the total electron orbital angular momentum operator $\hat{L}^2$ has an eigenvalue of

$$\hbar^2 l(l+1) \tag{4.30}$$

with

$$\hat{\mathbf{L}}_{12} = \hat{\mathbf{L}}_1 + \hat{\mathbf{L}}_2 \tag{4.31}$$

where

$$\hat{L}^2 = \hat{\mathbf{L}} \cdot \hat{\mathbf{L}} = (\hat{\mathbf{L}}_{12} + \hat{\mathbf{L}}_3) \cdot (\hat{\mathbf{L}}_{12} + \hat{\mathbf{L}}_3) \tag{4.32}$$

The four orbital angular momentum quantum numbers for the three-electron atom are

$$l = 0, 1, 2, 3 \tag{4.33}$$

where

$$l = |l_{12} + l_3|, \ldots, |l_{12} - l_3| \tag{4.34}$$

where

$$l_{12} = |l_1 + l_2|, \ldots, |l_1 - l_2| \tag{4.35}$$

For a three-electron atom, the total electron spin angular momentum operator $\hat{S}^2$ has an eigenvalue of

$$\hbar^2 s(s+1) \tag{4.36}$$

with

$$\hat{\mathbf{S}}_{12} = \hat{\mathbf{S}}_1 + \hat{\mathbf{S}}_2 \tag{4.37}$$

where

$$\hat{S}^2 = \hat{\mathbf{S}} \cdot \hat{\mathbf{S}} = (\hat{\mathbf{S}}_{12} + \hat{\mathbf{S}}_3) \cdot (\hat{\mathbf{S}}_{12} + \hat{\mathbf{S}}_3) \tag{4.38}$$

The four spin angular momentum quantum numbers for the three-electron atom are

$$s = 0, 1, 2, 3 \tag{4.39}$$

where

$$s = |s_{12} + s_3|, \ldots, |s_{12} - s_3| \tag{4.40}$$

where

$$s_{12} = |s_1 + s_2|, \ldots, |s_1 - s_2| \tag{4.41}$$

4.3 Spin and Orbital Angular Momentum for an *N*-Electron Atom

For an N-electron atom, the orbital angular momentum operator $\hat{L}^2$ has an eigenvalue of

$$\hbar^2 l(l+1) \tag{4.42}$$

The spin angular momentum operator $\hat{S}^2$ has an eigenvalue of

$$\hbar^2 s(s+1) \tag{4.43}$$

One uses iterative equations to generate the orbital and spin angular momentum operator quantum numbers l and s, where for iteration index number i,

$$i = 0, 1, 2, 3, \ldots, N - 2 \tag{4.44}$$

The iteration equation for the orbital angular momentum quantum number l and the spin angular quantum number s is

$$l_{i+1,\, i+2} = |l_{i,\, i+1} + l_{i+2}|, \ldots, |l_{i,\, i+1} - l_{i+2}| \tag{4.45}$$

and

$$s_{i+1,\, i+2} = |s_{i,\, i+1} + s_{i+2}|, \ldots, |s_{i,\, i+1} - s_{i+2}| \tag{4.46}$$

For an N-electron atom, the range of spin and orbital angular momentum numbers is determined quickly by the addition rule. As seen previously, with N electrons, this iteration generates N orbital and N spin angular quantum numbers in ascending order, where

$$l_1 \leq l_2 \leq \cdots \leq l_N \tag{4.47}$$

and

$$s_1 \leq s_2 \leq \cdots \leq s_N \tag{4.48}$$

The values of l_i and s_i may have two or more identical quantum numbers for

$$i = 1, 2, \ldots, N \tag{4.49}$$

For example, in a three-electron atom, one may have two identical orbital angular momentum quantum numbers $l_1 = 1$ and $l_2 = 1$, as well as a third one, $l_3 = 3$. With the quantum numbers arranged in ascending order via the iteration, one then finds

$$l = |l_{\max}|, |l_{\max} - 1|, \ldots, |l_{\min}| \tag{4.50}$$

and

$$s = |s_{\max}|, |s_{\max} - 1|, \ldots, |s_{\min}| \tag{4.51}$$

The maximum electron orbital quantum number $l_{\max}$ and the maximum electron spin quantum number $s_{\max}$ are obtained by summing the individual electron orbital quantum numbers l_i and spin quantum numbers s_i, respectively, where

$$l_{\max} = \sum_{i=1}^{N} l_i \tag{4.52}$$

and

$$s_{\max} = \sum_{i=1}^{N} s_i \tag{4.53}$$

If the minimum electron orbital quantum number $l_{\min}$ and the minimum electron spin quantum number $s_{\min}$ are both positive, then

$$l_{\min} > 0 \tag{4.54}$$

$$s_{\min} > 0 \tag{4.55}$$

Then, the expressions for $l_{\min}$ and $s_{\min}$ are

$$l_{\min} = l_N - \sum_{i=1}^{N-1} l_i \tag{4.56}$$

and

$$s_{\min} = s_N - \sum_{i=1}^{N-1} s_i \tag{4.57}$$

Otherwise, if $l_{\min}$ and $s_{\min}$ are not positive, then

$$l_{\min} = 0 \tag{4.58}$$

and

$$s_{\min} = 0 \tag{4.59}$$

4.4 LS Coupling and the Pauli Exclusion Principle

The dot product $\hat{\mathbf{L}} \cdot \hat{\mathbf{S}}$ of the angular momentum operator $\hat{\mathbf{L}}$ of the electron orbital motion and the electron spin $\hat{\mathbf{S}}$ is the electron spin–orbit interaction, LS coupling, or Russell–Saunders coupling. Operator $\hat{\mathbf{L}} \cdot \hat{\mathbf{S}}$ commutes with operators as follows:

$$\hat{J}^2, \hat{L}^2, \hat{S}^2 \text{ and } J_z \tag{4.60}$$

Operator $\hat{\mathbf{L}} \cdot \hat{\mathbf{S}}$ shares eigenfunction ψ_{jm_jls}, in which the eigenequation is

$$\hat{\mathbf{L}} \cdot \hat{\mathbf{S}} |\psi_{jm_jls}\rangle = \frac{1}{2}(\hat{J}^2 - \hat{L}^2 - \hat{S}^2)|\psi_{jm_jls}\rangle \tag{4.61}$$

or

$$\hat{\mathbf{L}}\cdot\hat{\mathbf{S}}|\psi_{jm_jls}\rangle = \frac{\hbar^2}{2}[j(j+1) - l(l+1) - s(s+1)]|\psi_{jm_jls}\rangle \tag{4.62}$$

The Pauli exclusion principle allows two electrons of opposite azimuthal spins to occupy a single-electron orbital spin state. The atomic electron quantum states are notated by

$$^{2s+1}X_j \Rightarrow X = S, P, D, F, G, H \;\; for \;\; l = 0, 1, 2, 3, 4, 5 \tag{4.63}$$

The electron orbital angular momentum quantum integer numbers are related to their capital letter designations by

$$l = 0, 1, 2, 3, \ldots \;\Rightarrow\; S, P, D, F, \ldots \tag{4.64}$$

A single-electron hydrogen atom with $s = 1/2$ has two states for each orbital quantum number l, where the total quantum number j varies in unit steps over the range:

$$l + s \geq j \geq |l - s| \;\Rightarrow\; j = l + s, l + s - 1, l - s - 2, \ldots, |l - s| + 1, |l - s| \tag{4.65}$$

The two states are two doublet P states, $^2P_{3/2}$ and $^2P_{1/2}$. For a two-electron helium atom, the individual electron spin quantum number $s = 1/2$ is added or subtracted, so the resulting spin is $s = 0$ or $s = 1$. For example, $l = 1$ produces a singlet series with $s = 0$, where

$$^1P_2, {}^1P_1, {}^1P_0 \quad \text{for } l = 1, \text{ s} = 0 \tag{4.66}$$

And $l = 1$ produces a triplet series with $s = 1$, where

$$^3P_2, {}^3P_1, {}^3P_0 \quad \text{for } l = 1, \text{ s} = 1 \tag{4.67}$$

4.5 Eigenfunctions and the Schrödinger Equation for a Two-Electron Atom

The eigenfunctions that describe the state (position, angular momentum, and spin momentum) of the atomic electrons can be determined by Schrödinger's equation. For example, consider a system comprised of a nucleus with two atomic electrons. The system has total energy E and is described by eigenstate ψ, where

$$\psi = \psi(\mathbf{r}_\alpha, \mathbf{s}_\alpha, \mathbf{r}_\beta, \mathbf{s}_\beta) \tag{4.68}$$

The two electrons interact via potential energy U, where

$$U = U(\mathbf{r}_\alpha, \mathbf{s}_\alpha, \mathbf{r}_\beta, \mathbf{s}_\beta) \tag{4.69}$$

With m_e as the mass of either electron, the Schrödinger equation for the system is

$$E_{\text{tot}} = -\frac{h^2}{2m_e}\left[\nabla^2_{\mathbf{r}_\alpha}\psi(\mathbf{r}_\alpha, \mathbf{s}_\alpha, \mathbf{r}_\beta, \mathbf{s}_\beta) - \nabla^2_{\mathbf{r}_\beta}\psi(\mathbf{r}_\alpha, \mathbf{s}_\alpha, \mathbf{r}_\beta, \mathbf{s}_\beta)\right] + U(\mathbf{r}_\alpha, \mathbf{s}_\alpha, \mathbf{r}_\beta, \mathbf{s}_\beta) \tag{4.70}$$

To simplify, the two electrons are assumed to move independently of each other, allowing system potential energy U to be the sum of the potential energies $U_1(\mathbf{r}_\alpha, \mathbf{s}_\alpha)$ and $U_2(\mathbf{r}_\beta, \mathbf{s}_\beta)$ of electrons 1 and 2, where

$$U(\mathbf{r}_\alpha, \mathbf{s}_\alpha, \mathbf{r}_\beta, \mathbf{s}_\beta) = U_1(\mathbf{r}_\alpha, \mathbf{s}_\alpha) + U_2(\mathbf{r}_\beta, \mathbf{s}_\beta) \tag{4.71}$$

The atomic electrons are identical particles, and thus their eigenfunctions and the physically measured eigenvalues must be independent of the identity of the electron. Electron 1 assumes the role of eigenstate $\psi_1(\mathbf{r}_\alpha, \mathbf{s}_\alpha)$, which specifies electron 1 at vector position $\mathbf{r}_\alpha$ with spin vector $\mathbf{s}_\alpha$. And electron 2 assumes the role of eigenstate $\psi_2(\mathbf{r}_\beta, \mathbf{s}_\beta)$, which specifies electron 2 at vector position $\mathbf{r}_\beta$ with spin vector $\mathbf{s}_\beta$. Separation of variables is applied to the Schrödinger equation, which allows the system eigenfunction ψ to be written as a product of the individual electron eigenstates, where

$$\psi(\mathbf{r}_\alpha, \mathbf{s}_\alpha, \mathbf{r}_\beta, \mathbf{s}_\beta) = \psi_1(\mathbf{r}_\alpha, \mathbf{s}_\alpha)\psi_2(\mathbf{r}_\beta, \mathbf{s}_\beta) \tag{4.72}$$

The eigenstate of a scattering system, which is comprised of a group of nuclei and their atomic electrons, has several independent state attributes. These state variables include the electron positions, and for each electron, the electron orbital and angular momentum. They also include the positions of the nuclei, and for each nucleus, the net angular moment of the nucleons relative to the nuclear center of mass and the net spin of the nucleons (nuclear spin) in a nucleus. One must multiply the wave amplitudes of the different attributes rather than summing the different attribute wave amplitudes. That is, one can multiply apples and flowers, but one does not add the apples in a tree to the flowers on the ground beneath the apple. However, as Newton found, the apple may fall to the ground onto the flowers.

To describe the apple–flowers scattering system, one needs to have a gravitational potential energy as well as an initial height and time of the apple stem release from the tree that allows the apple to fall, just as one specifies an incident neutron plane wave onto a neutron–atoms scattering system. Also, one specifies the location of the boundary enclosing the system and value of the wave amplitude at the boundary. And the squared magnitude of the wave amplitude or probability density of the system state integrated over the enclosed volume must be unity for a closed system. In the case of apples and flowers, the interaction potential is gravity, and in the case of neutron scatter from a group of atoms, the interaction potentials are the nuclear strong potential, electric and magnetic potentials, and gravitational potential (which is usually negligible). This interaction potential of the system is the sum of those potentials whose spatial gradients or forces act on the attributes of the system.

Simple system eigenstates, which are the product of component wave functions, are solved by the separation of variables in the Schrödinger equation. The temporal and spatial derivative, which are applied to the exponential phase components, cause the exponential term to be multiplied by the portion of the exponential argument that is constant with respect to the derivative. The magnitudes of the wave amplitude can vary spatially and temporally. However, in the time-independent Schrödinger equation, the separation of variables technique allows the equation to be separated into three differential equations, one for each of the mutually independent spatial directions.

4.6 Antisymmetric and Symmetric Eigenfunctions Describe an Identical Electron Pair

Since the electrons are assumed to be indistinguishable, exchanging their roles should not change the outcome of a physical measurable quantity, such as the probability density function, where the system probability density function commutes, so that

$$[\psi, \psi] = \psi^* \psi - \psi \psi^* = 0 \tag{4.73}$$

and

$$\psi^* \psi = \psi \psi^* \tag{4.74}$$

However, the simple product of the individual electron eigenstates violates this condition because

$$\psi^*(\mathbf{r}_\alpha, \mathbf{s}_\alpha, \mathbf{r}_\beta, \mathbf{s}_\beta)\psi(\mathbf{r}_\alpha, \mathbf{s}_\alpha, \mathbf{r}_\beta, \mathbf{s}_\beta) = \psi_1^*(\mathbf{r}_\alpha, \mathbf{s}_\alpha)\psi_2(\mathbf{r}_\beta, \mathbf{s}_\beta) \tag{4.75}$$

$$\psi(\mathbf{r}_\alpha, \mathbf{s}_\alpha, \mathbf{r}_\beta, \mathbf{s}_\beta)\psi^*(\mathbf{r}_\alpha, \mathbf{s}_\alpha, \mathbf{r}_\beta, \mathbf{s}_\beta) = \psi_1(\mathbf{r}_\alpha, \mathbf{s}_\alpha)\psi_2^*(\mathbf{r}_\beta, \mathbf{s}_\beta) \tag{4.76}$$

where

$$\psi_1^*(\mathbf{r}_\alpha, \mathbf{s}_\alpha)\psi_2(\mathbf{r}_\beta, \mathbf{s}_\beta) \neq \psi_1(\mathbf{r}_\alpha, \mathbf{s}_\alpha)\psi_2^*(\mathbf{r}_\beta, \mathbf{s}_\beta) \tag{4.77}$$

A system eigenstate is the sum or difference of the product of eigenfunctions for direct or reverse order of eigenfunction multiplication. In this case, reversing the position and spins of the two electrons does not change the summed or subtracted products.

The summed product is called the symmetric eigenstate ψ_s of the system, where

$$\psi_s = \frac{1}{\sqrt{2}}\left[\psi_1(\mathbf{r}_\alpha, \mathbf{s}_\alpha)\psi_2(\mathbf{r}_\beta, \mathbf{s}_\beta) + \psi_2(\mathbf{r}_\alpha, \mathbf{s}_\alpha)\psi_1(\mathbf{r}_\beta, \mathbf{s}_\beta)\right] \tag{4.78}$$

and

$$\psi_s = \psi_s^* \tag{4.79}$$

The symmetric wave function ψ_s has an invariant probability density of $|\psi_s|^2$, where

$$|\psi_s|^2 = \psi_s\psi_s^* = \psi_s^*\psi_s \tag{4.80}$$

The difference product is called the antisymmetric eigenstate ψ_A of the system, where

$$\psi_A = -\psi_A^* \tag{4.81}$$

The antisymmetric wave function ψ_A has an invariant probability density of $|\psi_A|^2$, where

$$|\psi_A|^2 = \psi_A\psi_A^* = \psi_A^*\psi_A \tag{4.82}$$

and

$$\psi_A = \frac{1}{\sqrt{2}}\left[\psi_1(\mathbf{r}_\alpha, \mathbf{s}_\alpha)\psi_2(\mathbf{r}_\beta, \mathbf{s}_\beta) - \psi_2(\mathbf{r}_\alpha, \mathbf{s}_\alpha)\psi_1(\mathbf{r}_\beta, \mathbf{s}_\beta)\right] \tag{4.83}$$

The symmetric and antisymmetric eigenstates are normalized by $1/\sqrt{2}$.

The Pauli exclusion principle allows only one electron per quantum state, which is expressed by the zero amplitude $\psi_A = 0$. In antisymmetric eigenstates, nuclei with two electrons have identical quantum states $\psi_1(\mathbf{r}_\rho, \mathbf{s}_\rho)$ and $\psi_2(\mathbf{r}_\rho, \mathbf{s}_\rho)$. In terms of the rotational invariance of coordinate systems and basis functions about a fixed origin, the symmetric eigenstate corresponds to a full 360° rotation, where the eigenstate is identical after full rotation. However, the antisymmetric eigenstate corresponds to a 180° rotation, where the eigenstate is multiplied by -1 after half-rotation.

The total antisymmetric eigenfunction $\psi_A(\mathbf{r}, \mathbf{s})$ of the atomic electrons and their nuclei is the product of spatial $\psi_r(\mathbf{r})$ and spin $\psi_{sp}(\mathbf{s})$ eigenfunction components, where

$$\psi_A(\mathbf{r}, \mathbf{s}) = \psi_r(\mathbf{r})\psi_{sp}(\mathbf{s}) \tag{4.84}$$

Consider the simple case of a nucleus with two orbital electrons. The total eigenfunction $\psi_A(\mathbf{r}, \mathbf{s})$ is antisymmetric for two possible conditions.

4.7 Two-Electron Atom—Symmetric Spatial and Antisymmetric Spin Components

In the first of two possible conditions, the antisymmetric total wave function $\psi_A(\mathbf{r}, \mathbf{s})$ is the product of a symmetric, spatial component eigenfunction $\psi_r^S(\mathbf{r})$ and a singlet, antisymmetric, spin component eigenfunction $\psi_{sp}^A(\mathbf{s})$, where

$$\psi_r^S(\mathbf{r}) = \frac{1}{\sqrt{2}}\left[\psi_1(\mathbf{r}_\alpha)\psi_2(\mathbf{r}_\beta) + \psi_2(\mathbf{r}_\alpha)\psi_1(\mathbf{r}_\beta)\right] \tag{4.85}$$

and

$$\psi_{sp}^A(\mathbf{s}) = \frac{1}{\sqrt{2}}\left[\psi_1(\mathbf{s}_\alpha)\psi_2(\mathbf{s}_\beta) - \psi_2(\mathbf{s}_\alpha)\psi_1(\mathbf{s}_\beta)\right] \tag{4.86}$$

In this case, the pair of electrons has opposite spins, so they have different quantum states. Thus, the electrons are allowed to be in very close proximity of each other via the symmetric spatial function, where

$$\mathbf{r}_\alpha \approx \mathbf{r}_\beta \tag{4.87}$$

For example, $\mathbf{r}_\alpha \approx \mathbf{r}_\beta$ occurs in a closed orbital or in a chemical bond, so that

$$\psi_r^S(\mathbf{r}) \cong \sqrt{2}\psi_1(\mathbf{r}_\alpha)\psi_2(\mathbf{r}_\alpha) \quad \mathbf{r}_\alpha \approx \mathbf{r}_\beta \tag{4.88}$$

The symmetric spatial eigenfunction causes the electrons to "attract each other" effectively via their opposite spins. The singlet spin state, with antiparallel spins, yields zero net spin angular momentum and no spin magnetic dipole moment. Thus, the singlet spin state produces no net spin magnetic dipole moment or magnetic field to scatter neutrons or other particles with magnetic dipole moments.

4.8 Two-Electron Atom—Antisymmetric Spatial and Symmetric Spin Components

In the second of two possible conditions, the antisymmetric $\psi_A(\mathbf{r}, \mathbf{s})$ is the product of an antisymmetric spatial eigenfunction $\psi_r^A(\mathbf{r})$ and a symmetric spin eigenfunction $\psi_{sp}^S(\mathbf{s})$, where

$$\psi_r^A(\mathbf{r}) = \frac{1}{\sqrt{2}}\left[\psi_1(\mathbf{r}_\alpha)\psi_2(\mathbf{r}_\beta) - \psi_2(\mathbf{r}_\alpha)\psi_1(\mathbf{r}_\beta)\right] \tag{4.89}$$

and

$$\psi_{sp}^S(\mathbf{s}) = \frac{1}{\sqrt{2}}\left[\psi_1(\mathbf{s}_\alpha)\psi_2(\mathbf{s}_\beta) + \psi_2(\mathbf{s}_\alpha)\psi_1(\mathbf{s}_\beta)\right] \tag{4.90}$$

In this case, if the electrons have similar spatial coordinates, where

$$\mathbf{r}_\alpha \approx \mathbf{r}_\beta \tag{4.91}$$

the antisymmetric nature of the spatial wave function $\psi_r^A(\mathbf{r})$ would yield zero amplitude:

$$\psi_r^A(\mathbf{r}) = 0 \qquad \mathbf{r}_\alpha \approx \mathbf{r}_\beta \tag{4.92}$$

Hence,

$$\psi_A(\mathbf{r}, \mathbf{s}) = \psi_r^A(\mathbf{r})\psi_{sp}^S(\mathbf{s}) = 0 \qquad \mathbf{r}_\alpha \approx \mathbf{r}_\beta \tag{4.93}$$

and thus the probability density function $|\psi_A(\mathbf{r}, \mathbf{s})|^2$ would also be zero:

$$|\psi_A(\mathbf{r}, \mathbf{s})|^2 = \psi_A^*(\mathbf{r}, \mathbf{s})\psi_A(\mathbf{r}, \mathbf{s}) = 0 \qquad \mathbf{r}_\alpha \approx \mathbf{r}_\beta \tag{4.94}$$

The probability density is nonzero when the electrons are far apart, and thus the electrons appear to repel each other. Since the electrons are well separated and thus occupy different energy quantum numbers due to their differing positions, the symmetric spin wave function $\psi_{sp}^S(\mathbf{s})$ allows for the two electron spins to be identical.

There are only two possible electron spin states, spin-up (+) and spin-down (−), where the total electron spin for each of the states α and β are

$$\mathbf{s}_\alpha = 1/2 \quad \text{and} \quad \mathbf{s}_\beta = 1/2 \tag{4.95}$$

Their azimuthal or z-component spin states are

$$s_{z\alpha} = \pm 1/2 \quad \text{and} \quad s_{z\beta} = \pm 1/2 \tag{4.96}$$

Hence, there are three possible symmetric spin eigenfunctions or triplet spin states:

$$\psi_{sp}^S(+,\ +) = \psi_1(+1/2)\psi_2(+1/2) \tag{4.97}$$

$$\psi_{sp}^S(-,\ -) = \psi_1(-1/2)\psi_2(-1/2) \tag{4.98}$$

$$\psi_{sp}^S(\pm) = \frac{1}{\sqrt{2}}\left[\psi_1(+1/2)\psi_2(-1/2) + \psi_2(1/2)\psi_1(-1/2)\right] \tag{4.99}$$

The exchange of the electrons does not change the triplet spin states. In the triplet state, the electron spins are effectively parallel. That is, the azimuthal spin (along the z-axis) of the two electrons sum to produce a net total spin $\mathbf{s}$ given by

$$\mathbf{s} = \mathbf{s}_\alpha + \mathbf{s}_\beta = 1 \tag{4.100}$$

The total z-component s_z has a net spin-up or spin-down, where

$$s_z = s_{z\alpha} + s_{z\beta} = \pm 1 \tag{4.101}$$

Hence, the average or net azimuthal spin $\langle s_z \rangle$ is zero, where

$$\langle s_z \rangle = 0 \tag{4.102}$$

In the triplet state, the net spin vector **s** rotates about the z-axis. The triplet state thus produces a net spin angular momentum and corresponding net magnetic dipole moment. The magnetic field of the electron magnetic dipole moment can scatter a neutron or other particle that has a magnetic dipole moment. The unpaired electrons with parallel spin are a common cause of magnetic neutron scatter. Thus, electrons in the same atom move as if they are under the influence of a force, called the *exchange force*, which is attractive or repulsive depending on the electron spin configuration. However, when the electrons are sufficiently far apart, the interactions of spin are negligible, as is the exchange force. The exchange force is not an actual force but rather the result of Coulomb or electric field repulsion and the Pauli exclusion principle.

4.9 *N*-Electron System Described by an Antisymmetric Total Eigenfunction

A system containing N electrons must be described by an antisymmetric total eigenfunction ϕ_A. The antisymmetric spin eigenfunction component $\psi^A_{sp}(\mathbf{s})$ or spatial eigenfunction component $\psi^A_r(\mathbf{r})$ can be obtained using the Slater determinant, with $\phi = \psi^A_{sp}(\mathbf{s})$ or $\phi = \psi^A_r(\mathbf{r})$, where

$$\phi_A = \frac{1}{\sqrt{N!}} \begin{vmatrix} \phi_1(\mathbf{r}_{\alpha_1}, \mathbf{s}_{\alpha_1}) & \cdots & \phi_N(\mathbf{r}_{\alpha_1}, \mathbf{s}_{\alpha_1}) \\ \vdots & \ddots & \cdots \\ \phi_1(\mathbf{r}_{\alpha_N}, \mathbf{s}_{\alpha_N}) & \cdots & \phi_N(\mathbf{r}_{\alpha_N}, \mathbf{s}_{\alpha_N}) \end{vmatrix} \tag{4.103}$$

The determinant gives sums and differences of the various combinations of the products of the N individual electron spin or spatial eigenfunctions. The spin eigenfunction component $\psi^S_{sp}(\mathbf{s})$ or spatial eigenfunction component $\psi^S_r(\mathbf{r})$ can be obtained from ϕ_A by setting all the minus signs to plus signs, so the result is only sums of the products of the N eigenfunction components.

The electron symmetric $\psi^S_r(\mathbf{r})$ or antisymmetric $\psi^A_r(\mathbf{r})$ spatial eigenfunctions produce expected values, which do not depend on the spatial coordinates of a position-dependent operator $\hat{A}(\mathbf{r})$. This can be shown, for example, in a two-electron atom, where

$$\phi_+ = \psi^S_r(\mathbf{r}) \quad \text{and} \quad \phi_- = \psi^A_r(\mathbf{r}) \tag{4.104}$$

One evaluates the expected value $\langle A(\mathbf{r}_\alpha)\rangle$ of the quantity $A(\mathbf{r}_\alpha)$, where

$$\langle A(\mathbf{r}_\alpha)\rangle = \langle \phi_\pm | \hat{A}(\mathbf{r}_\alpha) | \phi_\pm \rangle \tag{4.105}$$

This evaluation requires integration over all space for electron spatial coordinates $\mathbf{r}_\alpha$ and $\mathbf{r}_\beta$. With differential volumes, dv_α and dv_β for $\mathbf{r}_\alpha$ and $\mathbf{r}_\beta$, one obtains

$$\begin{aligned}\langle A(\mathbf{r}_\alpha)\rangle = \langle \phi_\pm | \hat{A}(\mathbf{r}_\alpha) | \phi_\pm \rangle = \frac{1}{2}\int dv_\beta \int dv_\alpha \\ [\psi_1(\mathbf{r}_\alpha)\psi_2(\mathbf{r}_\beta) \pm \psi_2(\mathbf{r}_\alpha)\psi_1(\mathbf{r}_\beta)]^* \hat{A}(\mathbf{r}_\alpha)[\psi_1(\mathbf{r}_\alpha)\psi_2(\mathbf{r}_\beta) \pm \psi_2(\mathbf{r}_\alpha)\psi_1(\mathbf{r}_\beta)]\end{aligned} \tag{4.106}$$

The two electrons are independent, and their wave functions are orthogonal, where $n = 1, 2$ and $m = 1, 2$, so that

$$\int dv_\alpha \psi_n^*(\mathbf{r}_\alpha)\psi_n(\mathbf{r}_\alpha) = 1 \quad n = 1, 2 \tag{4.107}$$

and

$$\int dv_\alpha \psi_n^*(\mathbf{r}_\alpha)\psi_m(\mathbf{r}_\alpha) = 0 \quad m, n = 1, 2 \text{ and } n \neq m \tag{4.108}$$

The integrated result shown here is independent of the coordinates because exchanging electron position coordinates $\mathbf{r}_\alpha$ by $\mathbf{r}_\beta$ does not change the integrated result, where

$$\langle A\rangle = \langle \phi_\pm | \hat{A}(\mathbf{r}_\alpha) | \phi_\pm \rangle = \frac{1}{2}\int dv_\alpha [|\psi_1(\mathbf{r}_\alpha)|^2 + |\psi_2(\mathbf{r}_\alpha)|^2] \tag{4.109}$$

Hence, the expected value $\langle A\rangle$ for the quantity $A(\mathbf{r}_\alpha)$ is independent of position $\mathbf{r}_\alpha$. Operator $\hat{A}(\mathbf{r}_\nu)$, which is a function of the νth electron position $\mathbf{r}_\nu$ in a scattering lattice, produces an expected value $\langle A\rangle$ that is independent of the position $\mathbf{r}_\nu$ of the νth electron. Therefore, matrix element $\langle \phi_\pm | \hat{A}(\mathbf{r}_\nu) | \phi_\pm \rangle$ is independent of $\mathbf{r}_\nu$, which is an important result that will be used later.

4.10 The Physical Basis of LS Coupling of Electron Spin and Orbital Motion

The outer valence electrons of an atom interact with the shielded electric field due to the nucleus and inner electrons. In a closed electron shell, the total electron and spin angular momentum is zero, so the orbital and spin angular momentum is determined by the unpaired valence electrons. The magnetic dipole moment of the electron spin interacts with the magnetic field, which is produced by the electron orbital motion. Consider an electron moving in a circle of radius r around a nucleus at velocity $\mathbf{v}$. In the rest frame of the electron, the charged nucleus orbits around the electron. As a result, the electron is at the center of the nucleus current loop of radius r. The electron experiences the magnetic field $\mathbf{B}$, which is produced by the circling nucleus. If the charged nucleus with atomic number Z and charge Ze moves with velocity $-\mathbf{v}$ in the electron rest frame, then the nucleus current is

$$\mathbf{J} = -Ze\mathbf{v} \tag{4.110}$$

From Ampere's law, the nucleus current $\mathbf{J}$ produces magnetic field $\mathbf{B}$ at the electron vector position in the electron rest frame, where

$$\mathbf{B} = \frac{\mu_0}{4\pi}\frac{\mathbf{J}\times\mathbf{r}}{r^3} = -\frac{Ze\mu_0}{4\pi}\frac{\mathbf{v}\times\mathbf{r}}{r^3} \tag{4.111}$$

The position vector $\mathbf{r}$ points from the orbiting nucleus to the electron at the center of the nucleus orbit. By Coulomb's law, the electric field $\mathbf{E}$ of the nucleus acting on the electron is

$$\mathbf{E} = \frac{Ze}{4\pi\varepsilon_0}\frac{\mathbf{r}}{r^3} \tag{4.112}$$

The electric field $\mathbf{E}$ of Eq. (4.112) is inserted into Eq. (4.111) via position vector $\mathbf{r}$, so

$$\mathbf{B} = -\varepsilon_0\mu_0\mathbf{v}\times\mathbf{E} = -\frac{\mathbf{v}\times\mathbf{E}}{c^2} \tag{4.113}$$

where

$$c = \frac{1}{\sqrt{\mu_0\varepsilon_0}} \tag{4.114}$$

Moving at velocity $\mathbf{v}$ relative to the nucleus, the electron in its rest frame encounters the time-varying electric field of the nucleus $\mathbf{v}\times\mathbf{E}$, which acts as a magnetic field $\mathbf{B}$ in the electron rest frame.

The orientation of the spin magnetic dipole moment $\boldsymbol{\mu}_e$ of the electron with spin $\mathbf{s}_e$ in the magnetic field $\mathbf{B}$ has an associated potential energy:

$$U = -\boldsymbol{\mu}_e\cdot\mathbf{B} \tag{4.115}$$

The magnitude s_e of the electron spin unit vector $\mathbf{s}_e$ is

$$s_e = \pm\frac{\hbar}{2} \tag{4.116}$$

The electron spin magnetic dipole moment $\boldsymbol{\mu}_e$ is

$$\boldsymbol{\mu}_e = \frac{2\mu_b}{\hbar}\mathbf{s}_e \tag{4.117}$$

In terms of electron rest mass m_e and charge e, the Bohr magnetron μ_b is

$$\mu_b = \frac{e\hbar}{2m_e} \tag{4.118}$$

Substitution for $\boldsymbol{\mu}_e$ and the magnetic field $\mathbf{B}$ in the electron rest frame, gives the orientation-dependent potential energy interaction U of the electron spin magnetic dipole moment with the magnetic field, as seen by the electron in its rest frame, where

$$U = -\frac{2\mu_b}{\hbar}\mathbf{s}_e \cdot \mathbf{B} \tag{4.119}$$

One must divide U by 2 due to the Thomas precession correction factor, which is caused by the change of direction or acceleration of the electron as it moves in a circle at constant speed v about the nucleus. Hence, the interaction potential energy U of the electron magnetic dipole moment $\boldsymbol{\mu}_e$ with the nucleus electric field $\mathbf{E}$ of the nucleus rest frame (lab frame), which becomes a magnetic field $\mathbf{B}$ in the rest frame of the electron is then

$$U = -\frac{\mu_b}{\hbar}\mathbf{s}_e \cdot \mathbf{B} \tag{4.120}$$

In terms of the Larmor precession frequency vector $\mathbf{w}_L$ of the electron magnetic dipole moment about the imposed magnetic field lines, the interaction potential energy U in the electron rest frame is

$$U = -\mathbf{s}_e \cdot \mathbf{w}_L \tag{4.121}$$

The Larmor precession frequency vector $\mathbf{w}_L$ points along the axis of precession. The precession axis is pointed along the magnetic field direction $\mathbf{B}$, about which precesses the electron magnetic dipole moment vector $\boldsymbol{\mu}_e$, where

$$\mathbf{w}_L = \frac{2\mu_b}{\hbar}\mathbf{B} \tag{4.122}$$

With substitution for magnetic field $\mathbf{B}$ of Eq. (4.113) and the Bohr magnetron μ_b of Eq. (4.118), the Larmor precession frequency $\mathbf{w}_L$ for the electron about the magnetic field line becomes

$$\mathbf{w}_L = -\frac{2\mu_b}{\hbar c^2}(\mathbf{v} \times \mathbf{E}) = -\frac{e}{m_e c^2}(\mathbf{v} \times \mathbf{E}) \tag{4.123}$$

With substitution into Eq. (4.121) of Eq. (4.123) for the Larmor precession frequency $\mathbf{w}_L$, the interaction potential energy U in the electron rest frame, becomes

$$U = \mathbf{s}_e \cdot \mathbf{w}_L = \frac{e}{m_e c^2}[\mathbf{s}_e \cdot (\mathbf{v} \times \mathbf{E})] \tag{4.124}$$

The potential energy U in the electron rest frame, which includes Thomas precession factor division by 2, is

$$U = \frac{\mathbf{s}_e \cdot \mathbf{w}_L}{2} = \frac{e}{2m_e c^2}[\mathbf{s}_e \cdot (\mathbf{v} \times \mathbf{E})] \tag{4.125}$$

The potential energy of interaction between the electron spin magnetic dipole moment and the magnetic field, which is produced by the relative velocity between the electron and nuclear electric field, can be expressed as the dot product $\mathbf{l} \cdot \mathbf{s}$ of the electron spin vector $\mathbf{s}$ and the electron angular momentum vector $\mathbf{l}$. This is shown next. In the negative radial inward direction toward the stationary nucleus is the electrostatic force $\mathbf{F}$, which acts on the moving electron in the laboratory rest frame, where

$$\mathbf{F} = -e\mathbf{E} \tag{4.126}$$

The force $\mathbf{F}$ is also the negative of the radial gradient $-\mathrm{d}V/\mathrm{d}r$ of the potential energy V of electrostatic interaction between the nucleus and the electron, where

$$\mathbf{F} = -\nabla V = -\hat{\mathbf{r}}\frac{\mathrm{d}V}{\mathrm{d}r} \tag{4.127}$$

In the lab frame, transposing Eq. (4.126), the electric field is

$$\mathbf{E} = -\frac{\mathbf{F}}{e} \tag{4.128}$$

Given radial unit vector $\hat{\mathbf{r}} = \mathbf{r}/r$, the static electric field that acts on the moving electron is

$$\mathbf{E} = \frac{\mathbf{r}}{er}\frac{\mathrm{d}V}{\mathrm{d}r} = \hat{\mathbf{r}}\frac{1}{e}\frac{\mathrm{d}V}{\mathrm{d}r} \tag{4.129}$$

Substitution of $\mathbf{E}$ of Eq. (4.129) in the magnetic field expression $\mathbf{B}$ of Eq. (4.113), gives

$$\mathbf{B} = -\frac{\mathbf{v} \times \mathbf{r}}{c^2 er}\frac{\mathrm{d}V}{\mathrm{d}r} \tag{4.130}$$

In the lab frame, the linear momentum **p** of the moving electron with mass m_e and velocity **v** relative to the stationary nucleus is

$$\mathbf{v} = \frac{\mathbf{p}}{m_e} \tag{4.131}$$

Substitution and reversing the order of the cross product in Eq. (4.130), multiplied by -1, gives the electron rest frame magnetic field **B**. The rest frame magnetic field **B** is the cross product of the electron radius vector **r** and linear momentum vector **p**, where

$$\mathbf{B} = \frac{\mathbf{r}\times\mathbf{p}}{m_e c^2 er}\frac{\mathrm{d}V}{dr} \tag{4.132}$$

The electron momentum vector **l** is the cross product of the electron radius **r** and momentum **p** vectors,

$$\mathbf{l} = \mathbf{r}\times\mathbf{p} \tag{4.133}$$

Thus, the electron rest frame magnetic field **B** is expressed by electron angular momentum **l**, where

$$\mathbf{B} = \frac{\mathbf{l}}{m_e c^2 er}\frac{\mathrm{d}V}{dr} \tag{4.134}$$

Finally, with substitution into Thomas precession Eq. (4.125) of Eq. (4.122) for $\mathbf{w}_\mathrm{L}$, via Eq. (4.134) for **B**, the interaction potential energy U between the electron magnetic dipole moment vector $\boldsymbol{\mu}_e$ and the magnetic field **B**, which is experienced by the electron in its rest frame, becomes

$$U = \frac{\mu_\mathrm{b}}{\hbar m_e c^2 er}\frac{\mathrm{d}V}{dr}(\mathbf{s}_e\cdot\mathbf{l}) \tag{4.135}$$

This electron rest frame's potential energy expression U clearly shows the LS coupling via the dot product $\mathbf{s}_e\cdot\mathbf{l}$ of the electron spin $\mathbf{s}_e$ and the electron orbital **l** angular momentum.

4.11 Derivation of Thomas Precession Factor for LS Coupling

In this section, we derive the Thomas precession factor of $1/2$, which was mentioned in Section 4.10 regarding LS coupling, and follows Eisberg and Resnick (1974). We start with the Lorentz transformation of Eq. (4.113), which is repeated below,

$$\mathbf{B} = -\varepsilon_0\mu_0\mathbf{v}\times\mathbf{E} = -\frac{\mathbf{v}\times\mathbf{E}}{c^2} \tag{4.136}$$

Recall that the charged nucleus with atomic number Z and charge Ze moves with velocity $-\mathbf{v}$ in the electron rest frame, where the nucleus current is $\mathbf{J} = -Ze\mathbf{v}$. Ampere's law relates the magnetic field $\mathbf{B}$ as a function of the nucleus current $\mathbf{J}$, and one obtains the magnetic field of the moving nuclear charge in the electron rest frame. The Lorentz transformation is applied to obtain the electron vector position in the electron rest frame, where

$$\mathbf{B} = \frac{\mu_0}{4\pi}\frac{\mathbf{J}\times\mathbf{r}}{r^3} = -\frac{Ze\mu_0}{4\pi}\frac{\mathbf{v}\times\mathbf{r}}{r^3} \tag{4.137}$$

By Coulomb's law, the electric field $\mathbf{E}$ of the nucleus, which acts on the electron, is

$$\mathbf{E} = \frac{Ze}{4\pi\varepsilon_0}\frac{\mathbf{r}}{r^3} \tag{4.138}$$

Substitution gives

$$\mathbf{B} = -\frac{\mathbf{v}\times\mathbf{E}}{c^2} \tag{4.139}$$

But there is a flaw in this derivation. The Maxwell equations are intrinsically invariant in transformations between frames moving in a constant velocity with one another, where the Lorentz transformation is used. That is, the Maxwell equations intrinsically obey special relativity, as does the Lorentz force equation when one uses relativistic mass in its expression, where

$$\mathbf{F} = \gamma m_0 \frac{\mathrm{d}^2\mathbf{r}}{\mathrm{d}t^2} = e\mathbf{E} + e\mathbf{v}\times\mathbf{B} \tag{4.140}$$

with

$$\gamma = \frac{1}{\sqrt{1-(v/c)^2}} \tag{4.141}$$

Special relativity is concerned with inertial frames in which there is no acceleration. Feynman (1964) discusses the twin paradox in which one twin remains on Earth and the other twin rockets at the speed of light to some distant planet and returns to Earth as a much older man. The paradox was that the twin on Earth is moving at the speed of light away from and toward his brother and thus would age in the frame of his brother. The Thomas precession falls in the same category as the twin paradox. In both cases, acceleration was overlooked and special relativity does not apply to frames that have relative acceleration. The first brother, who remains on earth, is never accelerated; whereas, the second brother is accelerated to the speed of light during his departure from earth. The second brother in the spaceship must then de-accelerate upon arrival at the distant planet. Next the brother in the spaceship returns to earth by accelerating toward earth, and then de-accelerates upon his arrival back to earth. Thus, the accelerated twin brother and his slower ticking clock returns to Earth to his nonaccelerated twin brother, who has become older at the pace of the faster-ticking, Earth-based clock. In this derivation, the electron rest frame is moving at a constant speed relative to the nucleus; however, since the electron is orbiting the nucleus in a circle, this simple derivation overlooks the change in direction or acceleration of the electron.

The flawed expression for the magnetic field assumed a constant velocity between the rest frames of the nucleus and orbiting electron. Since the change of direction was not taken into account, the Lorentz transformation intrinsic to the Maxwell equations produced a relation between the electric and magnetic field that was based on the electron and nucleus moving in opposite, collinear directions, rather than the electron circling the nucleus or the nucleus circling the electron. That is, the Lorentz transformation between the electric and magnetic field applies only to two frames moving along the same or opposite straight-line directions but at different speeds, so there is no relative acceleration between the two frames.

Special relativity arguments, however, can be applied to obtaining the Thomas correction by the use of instantaneous inertial frames at closely spaced points on the electron circular trajectory about the center nucleus. One then compares the rest frame of the nucleus with the rest frames of two closely spaced electrons on the circle. One assumes that the electron moves in a counterclockwise direction around the nucleus at constant speed v and radius r. Inside the circle can be inscribed a polygon with a very large number of sides, such that the polygon vertices form closely spaced points on the circle. Each polygon point on the circular electron trajectory is the origin of an instantaneous electron rest frame at that point.

Let the x-axis of each electron rest frame always be tangent to the electron circular trajectory, with the y and z axes mutually perpendicular. In the tangential x-direction at point 1, the electron trajectory has approximately the same tangential velocity v_x as the closely spaced point 2 due to the relatively small tangential component velocity change Δv_x compared to v_x. However, as the electron moves from point 1 to point 2, it experiences a radial inward nuclear Coulomb force. This force causes an acceleration a_y in the y-direction over time interval Δt, where the electron moves from position 1 to 2, and acquires y-directed velocity change,

$$\Delta v_y = a_y \Delta t \tag{4.142}$$

Relative to the nucleus rest frame, one determines the electron velocity $\mathbf{v}_2$ of the electron at position 2, given that the electron at position 2 has acquired a normal, y-directed velocity Δv_y relative to the electron position 1 rest frame. And relative to the electron position 1 rest frame, the nucleus has velocity $-v_x$ in the negative x-direction. The normal y-directed velocity component v_{2y}, divided by the tangential x-directed velocity component v_{2x}, gives the angle θ_2 between the nucleus rest frame's x-axis and the electron velocity vector $\mathbf{v}_2$ at position 2, where in the small-angle approximation,

$$\theta_2 \cong \frac{v_{2y}}{v_{2x}} = \frac{\Delta v_y \sqrt{1-(v_x/c)^2}}{v_x} \tag{4.143}$$

One obtains the tangential v_{2x} and normal v_{2y} electron velocity component for electron position 2, which are measured relative to the nucleus rest frame, via the relativistic transformation equations, where

$$v_{2x} = \frac{\Delta v_x - v_x}{1-(v_x\, \mathrm{d}v_x/c^2)} = \frac{0-(-v_x)}{1-0} = v_x \tag{4.144}$$

and

$$v_{2y} = \frac{\Delta v_{y\sqrt{1-(v_x/c)^2}}}{1-(v_x\, \mathrm{d}v_x/c^2)} = \Delta v_y \sqrt{1-(v_x/c)^2} \tag{4.145}$$

The electron position 1 rest frame is the common frame of comparison between the nucleus rest frame and the electron 2 position rest frame. One next determines the velocity of the nucleus, relative to the electron position 2 rest frame, where the velocity of electron rest frame 1 has been measured, relative to electron rest frame 2. Also the velocity of the nucleus relative to the electron rest frame 1 has been measured. That is, relative to the rest frame of the electron at position 2, one determines the nucleus velocity **V**, given that the nucleus has a negative x-directed velocity $-v_x$, which is measured relative to the electron position 1 rest frame. The tangential x-directed velocity of the electron at its position 1 rest frame, measured relative to its position 2 rest frame, is approximately zero, where $\Delta v_x = 0$. Also, the electron at its position 1 rest frame, relative to its position 2 rest frame, has a negative y-directed, normal velocity $-\Delta v_y$.

In the small-angle approximation, the nucleus normal y-directed velocity component V_y, which is divided by the nucleus tangential, x-directed component V_x, is measured relative to the electron position 2 rest frame, and gives the angle θ_1 between the x-axis of the electron position 2 rest frame and the nucleus velocity **V**, where

$$\theta_1 \cong \frac{V_y}{V_x} = \frac{\Delta v_y}{v_x \sqrt{1-(\Delta v_y/c)^2}} \tag{4.146}$$

One obtains the tangential V_x and normal V_y nucleus velocity components, which are measured relative to the electron position 2 rest frame by the relativistic transformation equations, where

$$V_x = \frac{-v_x \sqrt{1-(\Delta v_y/c)^2}}{1-(v_y \Delta v_y/c^2)} = -v_x \sqrt{1-(\Delta v_y/c)^2} \tag{4.147}$$

and

$$V_y = \frac{v_y - \Delta v_y}{1 - (v_y \Delta v_y / c^2)} = \frac{0 - \Delta v_y}{1 - 0} = -\Delta v_y \tag{4.148}$$

The angle difference $\theta_1 - \theta_2$ is the rotation of the x-axis of the electron position 2 rest frame, which is measured relative to the x-axis of the nucleus rest frame, where

$$\theta_1 - \theta_2 = \frac{\Delta v_y}{v_x \sqrt{1 - (\Delta v_y / c)^2}} - \frac{\Delta v_y \sqrt{1 - (v_x / c)^2}}{v_x} \tag{4.149}$$

One can ignore $(\Delta v_y / c)^2$ relative to 1, so that

$$\theta_1 - \theta_2 = \frac{\Delta v_y}{v_x}\left(1 - \sqrt{1 - (v_x / c)^2}\right) \tag{4.150}$$

Via the binomial approximation, one obtains

$$\Delta\theta = \theta_1 - \theta_2 = \frac{v_x \Delta v_y}{2c^2} = \frac{v_x a \Delta t}{2c^2} \tag{4.151}$$

The x-axis (and y-axis) of the electron rest frame rotate counterclockwise, relative to the nucleus, as the electron moves in its counterclockwise orbit from rest frame position 1 to 2, from position 2 to 3, and so forth.

Transposing Eq. (4.151), the Thomas precession frequency w_{T} of the rotation of the electron rest frame axes about the nucleus is then

$$w_{\mathrm{T}} = \frac{\Delta\theta}{\Delta t} = \frac{v_x a_y}{2c^2} \tag{4.152}$$

The radial inward electron acceleration and tangential electron velocity vector are perpendicular, and the precession frequency in vector form $\mathbf{w}_{\mathrm{T}}$ is then

$$\mathbf{w}_{\mathrm{T}} = w_{\mathrm{T}} \hat{\mathbf{z}} = \frac{v_x a_y}{2c^2} = \frac{\mathbf{v} \times \mathbf{a}}{2c^2} \tag{4.153}$$

The radial inward electron acceleration $\mathbf{a}$, which is the negative radial direction, arises from the force $\mathbf{F}$ of the nuclear electric field $\mathbf{E}$, where

$$\mathbf{a} = \frac{\mathbf{F}}{m_e} = -\frac{e}{m_e}\mathbf{E} \tag{4.154}$$

Substitution of Eq. (4.154) into Eq. (4.153), gives the Thomas precession frequency $\mathbf{w}_\mathrm{T}$ of the electron rest frame about the nucleus:

$$\mathbf{w}_\mathrm{T} = w_\mathrm{T}\hat{\mathbf{z}} = \frac{v_x a_y}{2c^2} = \frac{e}{2m_e c^2}(\mathbf{v} \times \mathbf{E}) \tag{4.155}$$

Along with the electron Thomas precession frequency $\mathbf{w}_\mathrm{T}$, is the electron Larmor precession frequency $\mathbf{w}_\mathrm{L}$ of Eq. (4.123). The Larmor precession frequency $\mathbf{w}_\mathrm{L}$ of the electron about the applied magnetic field lines, is measured by an observer in the electron rest frame, where

$$\mathbf{w}_\mathrm{L} = -\frac{2\mu_\mathrm{b}}{\hbar c^2}(\mathbf{v} \times \mathbf{E}) = -\frac{e}{m_e c^2}(\mathbf{v} \times \mathbf{E}) \tag{4.156}$$

The magnetic field and the nucleus are in the same rest frame, so the rest frame of the electron precesses about the nucleus at total precession frequency $\mathbf{w}$, which is the sum of the Thomas $\mathbf{w}_\mathrm{T}$ and Larmor $\mathbf{w}_\mathrm{L}$ precession frequencies, where

$$\mathbf{w} = \mathbf{w}_\mathrm{T} + \mathbf{w}_\mathrm{L} \tag{4.157}$$

With substitution of Eqs. (4.155–4.156) into Eq. (4.157), the total precession frequency $\mathbf{w}$, becomes

$$\mathbf{w} = \frac{e}{2m_e c^2}(\mathbf{v} \times \mathbf{E}) - \frac{e}{m_e c^2}(\mathbf{v} \times \mathbf{E}) = -\frac{e}{2m_e c^2}(\mathbf{v} \times \mathbf{E}) \tag{4.158}$$

Recalling Eqs. (4.115), (4.117), and (4.136), the interaction potential energy U_lab, which is measured by an observer in the laboratory or nucleus rest frame, now includes the Thomas precession, where

$$U_\mathrm{lab} = -\mathbf{s}_e \cdot w = -\mathbf{s}_e \cdot (\mathbf{w}_\mathrm{T} + \mathbf{w}_\mathrm{L}) = \frac{e}{2m_e c^2}[\mathbf{s}_e \cdot (\mathbf{v} \times \mathbf{E})] \tag{4.159}$$

4.12 An Alternative Derivation of the Thomas Precession Factor

Kroemer (2004) derived the Thomas precession factor in an alternative, simple fashion as described next. The electron moves in a circle with its velocity vector $\mathbf{v}$ tangent to the circular trajectory, and is acted upon by the perpendicular radial inward acceleration $\mathbf{a}_r$ due to the electric charge of the center nucleus. The y-component of the radial acceleration $\mathbf{a}_y$, which acts on the electron at each point on the circle, turns the electron inward toward the nucleus. If one adds a Lorentz magnetic field force in the opposite y-direction to that of the electric field force, then a zero force on the electron is produced. Zero force acting on the electron allows it to move in a straight line rather than the curved orbit path. This artifice serves the purpose of computing the magnetic field, which acts on the electron in the electron rest frame, is due to the nuclear electric force and the electron relative velocity $\mathbf{v}$ to the nucleus rest frame (lab frame). The computation is now consistent with the constant, linear electron velocity, which is required for application of the Lorentz transformation of electromagnetic fields.

One adds the magnetic Lorentz force $e\mathbf{v} \times \mathbf{B}$ to the nuclear Coulomb force $e\mathbf{E}$ to produce a net force of zero, where

$$\mathbf{F} = 0 = e\mathbf{E} + e\mathbf{v} \times \mathbf{B} \tag{4.160}$$

or

$$\mathbf{E} = -\mathbf{v} \times \mathbf{B} \tag{4.161}$$

The applied magnetic field is z-directed $B_z\hat{\mathbf{z}}$, and it is perpendicular to the plane that contains the nucleus and orbiting electron, where, in the rest frame of the nucleus,

$$E_y = v_x B_z \tag{4.162}$$

If one Lorentz-transforms electric E_y and magnetic B_z fields to the rest frame of the electron, then the resulting magnetic field B'_z co-moving with the electron is

$$B'_z = \frac{1}{\sqrt{1-(v_x/c)^2}}\left(B_z - E_y\frac{v_x}{c^2}\right) \tag{4.163}$$

For nonrelativistic electron orbital speeds about the nucleus ($v_x \ll c$), Taylor series expansion of the radical about v_x/c yields

$$B'_z \cong \left(1 + \frac{1}{2}\frac{v_x^2}{c^2} + \frac{3}{8}\frac{v_x^4}{c^4}\right)B_z - E_y\frac{v_x}{c^2}\left(1 + \frac{1}{2}\frac{v_x^2}{c^2} + \frac{3}{8}\frac{v_x^4}{c^4}\right) \tag{4.164}$$

One drops the third-power and higher terms in v_x/c to obtain

$$B'_z \cong B_z + \frac{1}{2}\frac{v_x}{c^2}(v_x B_z) - E_y\frac{v_x}{c^2} \tag{4.165}$$

With substitution of $E_y = v_x B_z$, the magnetic field B'_z in the electron rest frame then accounts for Thomas precession, via the ½ in the 2nd term of Eq. (4.165), where

$$B'_z \cong B_z - \frac{v_x E_y}{2c^2} \tag{4.166}$$

The tangential electron velocity v_x relative to the nucleus rest frame is perpendicular to the nuclear electric field E_y, which causes the electron to turn in a circle. And in vector form, the electron velocity vector $\mathbf{v}$ is tangential to its orbital motion around the nucleus, where the velocity vector $\mathbf{v}$ is perpendicular to the radial-directed, nuclear electric field $\mathbf{E}$. In vector form, the magnetic field in the electron rest frame is

$$\mathbf{B}' \cong \mathbf{B} - \frac{\mathbf{v} \times \mathbf{E}}{2c^2} = \mathbf{B} + \frac{\mathbf{E} \times \mathbf{v}}{2c^2} \tag{4.167}$$

With Eq. (4.119), recall that the interaction potential energy U of the electron magnetic dipole moment μ_e, with the nucleus electric field $\mathbf{E}$ in the nucleus rest frame (lab frame), which becomes a magnetic field $\mathbf{B}'$ in the rest frame of the electron, is

$$U = -\frac{\mu_b}{\hbar}\mathbf{s}_e \cdot \left(\mathbf{B} + \frac{\mathbf{E} \times \mathbf{v}}{2c^2}\right) \tag{4.168}$$

The spin–orbit interaction portion U_{SO} of this expression for the electron rest frame potential energy U is Thomas-precession-corrected, where

$$U_{SO} = -\frac{\mu_b}{\hbar}\mathbf{s}_e \cdot \left(\frac{\mathbf{E} \times \mathbf{v}}{2c^2}\right) \tag{4.169}$$

4.13 Quenching of an Electron Orbital Momentum in a Crystal

The electric field of a single isolated nucleus acts on its atomic electron at distance r as a central field, which varies as $1/r^2$, where one component of the electron angular momentum, conventionally along the z-axis L_z, and the square of the total electron angular momentum L^2 are constant. As a result, the plane of the electron orbital is constant. In a crystal, the collective electric fields of all the lattice nuclei act on an orbital electron as a noncentral electric field so that L_z is no longer constant along the electron orbital; however, L^2 may remain constant. When L_z varies such that its average value is zero, $\overline{L_z} = 0$, then the electron orbital momentum is quenched.

Very strong electric fields arise within the crystal lattice, and cause the orbit of an atomic electron to precess at a sufficiently high frequency, which causes its orbital angular momentum to average to zero. As a result, the overall scattering system has zero azimuthal angular momentum ($L_z = 0$), and the overall scattering system can even have zero total angular momentum ($\mathbf{L} = 0$), where

$$\mathbf{L} = \sum_{\nu=1}^{Z} \mathbf{l}_\nu = 0 \tag{4.170}$$

In this case, the angular magnetic dipole moment for each electron $\mathbf{l}_\nu$ is zero for each νth electron of the Z_d electrons in the crystal atom or ion with atomic number Z_d, where

$$\mathbf{l}_\nu = 0 \quad \nu = 1, 2, \ldots, Z \tag{4.171}$$

That is, the angular magnetic dipole moment for each electron $\mathbf{l}_\nu$ of the atom, including the unpaired electrons, averages to zero $\bar{\mathbf{l}}_\nu = 0$ so that $\mathbf{L} = 0$.

The orbital motion $\mathbf{l}_\nu$ and spin $\mathbf{s}_\nu$ angular momentum of each νth electron of the crystal d^{th} atom or ion contributes to an orbital $\boldsymbol{\mu}_{l\nu}$ magnetic moment dipole moment and a spin $\boldsymbol{\mu}_{s\nu}$ magnetic dipole moment, respectively. They are summed to produce the νth electron dipole magnetic dipole moment, where

$$\boldsymbol{\mu}_\nu = \boldsymbol{\mu}_{l\nu} + \boldsymbol{\mu}_{s\nu} = 2\mu_b(\mathbf{l}_\nu + \mathbf{s}_\nu) \tag{4.172}$$

with

$$\boldsymbol{\mu}_{l\nu} = 2\mu_b \mathbf{l}_\nu \tag{4.173}$$

and

$$\boldsymbol{\mu}_{s\nu} = 2\mu_b \mathbf{s}_\nu \tag{4.174}$$

The sum of the orbital $\boldsymbol{\mu}_{l\nu}$ and spin $\boldsymbol{\mu}_{s\nu}$ angular magnetic dipole moment contribution from the Z_d electrons of the atom yields the net spin and angular magnetic dipole moment $\boldsymbol{\mu}_d$ for the dth atom or ion of the crystal. Only the n_d unpaired electrons of the Z_d electrons will contribute significantly to $\boldsymbol{\mu}_d$, so that

$$\boldsymbol{\mu}_d = \sum_{\nu=1}^{Z_d} \boldsymbol{\mu}_\nu = 2\mu_b \sum_{\nu=1}^{Z_d} (\mathbf{l}_\nu + \mathbf{s}_\nu) = 2\mu_b \sum_{\nu=1}^{n_d} (\mathbf{l}_\nu + \mathbf{s}_\nu) \tag{4.175}$$

The material magnetization $\mathbf{M}(\mathbf{r})$ is the magnetic dipole moment per unit volume of material. Magnetization $\mathbf{M}$ is the product of the number of atoms per unit volume ρ_a and the magnetic dipole moment per atom, which is contributed by the unpaired atomic electrons. Consider the case of a crystal with N unit cells. Each unit cell has r atoms and volume v_0, which is bounded by dimensions a, b, and c:

$$v_0 = abc \tag{4.176}$$

The unit cell shape is defined by the unit cell vectors $\mathbf{a}$, $\mathbf{b}$, and $\mathbf{c}$. The crystal magnetization $\mathbf{M}(\mathbf{r})$ is the sum of the magnetization from the electron orbital $\mathbf{M}_l(\mathbf{r})$ and the spin $\mathbf{M}_s(\mathbf{r})$ magnetic dipole moments, where

$$\mathbf{M}(\mathbf{r}) = \frac{1}{v_0} \sum_{d=1}^{r} \boldsymbol{\mu}_d = \frac{1}{abc} \sum_{d=1}^{r} \sum_{\nu=1}^{n_d} \boldsymbol{\mu}_\nu \tag{4.177}$$

With substitution into Eq. (4.177) of Eq. (4.172) for $\boldsymbol{\mu}_\nu$, the crystal magnetization vector becomes

$$\mathbf{M}(\mathbf{r}) = \mathbf{M}_l(\mathbf{r}) + \mathbf{M}_s(\mathbf{r}) = \frac{2\mu_b}{abc} \sum_{d=1}^{r} \sum_{\nu=1}^{n_d} (\mathbf{l}_\nu + \mathbf{s}_\nu) \tag{4.178}$$

For a quenched crystal for each νth electron of the Z_d electrons in the atom (ion),

$$\mathbf{l}_\nu = 0 \quad \nu = 1, 2, \ldots, Z_d \tag{4.179}$$

The total z-directed magnetic dipole moment $\boldsymbol{\mu}_d$ of the dth crystal atom or ion in the quenched crystal is then

$$\boldsymbol{\mu}_d = 2\sum_{\nu=1}^{n_d} \mu_{\text{b}}\mathbf{s}_\nu \tag{4.180}$$

The quenched crystal magnetization vector $\mathbf{M}_q(\mathbf{r})$ with $\mathbf{M}_l(\mathbf{r}) = 0$, as a function of neutron position $\mathbf{r}$, is due to the dth atom n_d unpaired electron spins for the r atoms in each of the N unit cells of the crystal, where

$$\mathbf{M}_q(\mathbf{r}) = \mathbf{M}_s(\mathbf{r}) = \frac{2\mu_{\text{b}}}{abc}\sum_{d=1}^{r}\sum_{\nu=1}^{n_d} \mathbf{s}_\nu \tag{4.181}$$

Outside the crystal, the magnetization is assumed to be zero: $\mathbf{M}(\mathbf{r}) = 0$.

4.14 Paramagnetic and Ferromagnetic Materials

In a paramagnetic material, the atoms contain permanent magnetic dipole moments arising from the spin and orbital angular momentum of the electrons. An externally applied magnetic field of induction $\mathbf{B}$ is applied to the crystal, where magnetic susceptibility χ_m is the proportionality constant between the magnetization vector $\mathbf{M}$ (volume density of magnetic dipoles) and the magnetic field strength $\mathbf{H}$, so that

$$\mathbf{M} = \chi_m \mathbf{H} \tag{4.182}$$

This linear relation with χ_m applies only to linear magnetic materials. In a vacuum, $\chi_m = 0$ and, in non-magnetic materials, it is usually very small, in the range of 10^{-4} to 10^{-5}; for example, $\chi_m = -0.94 \times 10^{-5}$ for copper.

The magnetic field of induction $\mathbf{B}$ is the sum of the magnetization vector $\mathbf{M}$ of the material and the externally applied magnetic field strength $\mathbf{H}$, where the sum is multiplied by the vacuum permeability constant μ_0, where

$$\mathbf{B} = \mu_0(\mathbf{H} + \mathbf{M}) \tag{4.183}$$

The induction magnetic field $\mathbf{B}$ arises from microscopic (atomic) and macroscopic (wires) currents; whereas, the magnetic field strength $\mathbf{H}$ arises only from macroscopic currents. The induction magnetic field $\mathbf{B}$ will tend to align the dipole moments because the energy is lower for parallel dipole alignment. The induced magnetic field of the aligned magnetic dipoles $\mathbf{M}$ adds to the externally applied magnetic field $\mathbf{H}$. The magnetic field is thus

$$\mathbf{B} = \mu_0(1 + \chi_m)\mathbf{H} \tag{4.184}$$

The relative permeability of the material μ_r is defined as

$$\mu_r = 1 + \chi_m \tag{4.185}$$

Materials with relative permeability $\mu_r < 1$ are diamagnetic. Materials with $\mu_r > 1$, but with $\mu_r \cong 1$ are paramagnetic. Materials with $\mu_r \gg 1$ are ferromagnetic, and ferromagnetic materials are nonlinear, so χ_m and μ_r vary with applied magnetic field strength **H**.

The magnetic permeability μ is defined as

$$\mu = \mu_0 \mu_r \tag{4.186}$$

In linear magnetic materials, the proportionality constant between the induction magnetic field **B** and applied magnetic field strength **H**, is the magnetic permeability μ, where

$$\mathbf{B} = \mu \mathbf{H} \tag{4.187}$$

In nonlinear magnetic materials such as cobalt, nickel, and soft iron, the permeability μ varies with applied magnetic field strength **H**, where μ can be large and have representative values ranging from 250 in cobalt, 600 in nickel, and 5000 in soft iron.

The thermal motion of the atoms, which tends to randomize the dipole orientations, must be overcome by the applied magnetic field or lower temperatures. The susceptibility at low temperatures and magnetic fields obeys Curie's law, where for constant C, characteristic of the material is

$$\chi = \frac{C}{T} \tag{4.188}$$

Ferromagnetism is magnetization that arises in small regions of a material without the application of an external magnetic field. The natural magnetization in ferromagnetic materials, which arise from the unpaired electron spins, reaches its maximum at $T = 0$ K, and drops to zero at the ferromagnetic Curie temperature T_C. At temperatures higher than T_C, the ferromagnetic material becomes paramagnetic, and its magnetic susceptibility follows a modified Curie relation, in which

$$\chi = \frac{C}{T - T_C} \tag{4.189}$$

For temperatures below T_C, the ferromagnetic Curie's law is no longer applicable because the material is then ferromagnetic with permanent magnetization. A ferromagnetic material is not magnetized in the same direction throughout its volume, but rather has many smaller regions or domains. Each domain is fully magnetized in one direction, but the orientations of the domains are random so that direction of magnetization changes from one domain to the next. Random orientation of the domains occurs because random domains produce a lower energy state than fully aligned domains. An applied magnetic field somewhat aligns these domains so that their magnet fields are more in alignment, thereby increasing the magnetic field of the material as a whole. Increasing the applied magnetic field beyond the saturation magnetic field does not increase the magnetization of the material further, because the domains reach the maximum mutual alignment possible. When an external magnetic field is removed, the domains and their boundaries do not return completely to their original orientations, and thus the domains exhibit hysteresis. The only ferromagnetic elements are iron, cobalt, nickel, gadolinium, and dysprosium.

Recall that identical atomic electrons are required to be indistinguishable (hence the total wave function must be symmetric), but the component spatial and spin wave functions can be symmetric or asymmetric so long as the product of the spin and spatial component wave functions is asymmetric. In the 3-D crystal bands, where the ferromagnetic action takes place, and in other non-ferromagnetic, paramagnetic elements, the coulomb repulsion between a pair of electrons tends to move the electrons farther apart into an antisymmetric spatial state of minimum energy. This asymmetry of the total wave function, then requires a symmetric spin state, where unpaired electron spins are aligned in the same direction. Thus, the coulomb force, which pushes electrons apart to minimize energy, then leads to them aligning in parallel spins in paramagnetic and ferromagnetic materials.

The exchange force or interaction between the spatial and spin components of the wave function for two or more electrons is responsible for this coupling of electron spins to parallel alignment in paramagnetic and ferromagnetic materials. In the case of ferromagnetic materials, the 3-D band is not completely filled, the close proximity of the iron atoms allows the indistinguishability of the electrons of adjacent atoms, and the exchange interaction now extends to the electrons of adjacent atoms. To maintain a total asymmetric wave function for the set of electrons belonging to the closely spaced adjacent atoms, which have overlapping d-shells of unpaired electrons, there are two possibilities. In the ferromagnetic case (say for iron atoms), the spatial wave function can be asymmetric and the spin function is symmetric. In this case, the coulomb energy is minimized by the separation of the electrons, but the separation allows for the unpaired electrons to have their spins to be aligned parallel. The other possibility is that the spatial wave function will be symmetric, and the electrons localized together, and thus their spin wave function must be antisymmetric, which requires the electron spins to be antiparallel. The antiferromagnetic possibility does not minimize the coulomb energy in iron and does not occur. However, materials such as MnO_2 are antiferromagnetic.

References

Eisberg, R., & Resnick, R. (1974). *Quantum physics of atoms, molecules, solids, nuclei, and particles*. New York, NY: Wiley.

Feynman, R. P. (1964). *The Feynman lectures in physics* (Vol. 1). Reading, MA: Addison-Wesley.

Kroemer, H. (2004). *American Journal of Physics, 72*(1).

5 LS-Coupled, Localized Electron, Magnetic Scatter of Neutrons

5.1 Heitler–London Model for Neutron Scatter by Magnetic Materials

Following Squires (1996), is derived the neutron partial differential cross section for magnetic neutron scatter from a magnetic crystal. The magnetic neutron scatter treatment in this chapter and prior chapters, follows from the neutron spin state scatter from nuclei and correlations functions covered in Cremer (2013), and non-magnetic, nuclear scatter of neutrons covered in Cremer (2012b). Derived first is the neutron partial differential cross section for magnetic neutron scatter from a magnetic crystal, which assumes the Heitler–London model, LS coupling, and a quenched crystal. In LS coupling, one adds the total orbital angular momentum vector **L** and total spin momentum vector **S** of the unpaired electrons of an atom. Based on these three assumptions, evaluation of the thermal average of the initial state, unpaired, atomic electron positions, yields the resulting partial differential cross section for neutron scatter. This cross section is then extended to include unquenched scattering atoms. Derived next is the elastic, differential cross section, which is obtained from the partial differential cross section by energy integration over the Dirac delta function energy expression. The partial differential cross section is then expressed by the coherent intermediate correlation function, which is separated into elastic and inelastic components. This allows the partial differential cross section to be separated into the following terms: (1) nuclear elastic scatter and magnetic elastic scatter, (2) inelastic nuclear scatter and elastic magnetic scatter, (3) inelastic magnetic scatter and elastic nuclear scatter, and (4) inelastic nuclear scatter and inelastic magnetic scatter.

First, magnetic neutron scatter from a magnetic crystal with the following three assumptions is considered. The first assumption is the Heitler–London model for the crystal, which assumes that the unpaired electrons of the crystal are localized within a very close vicinity of the equilibrium positions of the lattice nuclei. The second assumption is LS coupling occurs in each lattice ion or atom of atomic number Z, where the orbital angular momentum $\mathbf{l}_\nu$ of the unpaired electrons, which is indexed by integer ν, combine and produce a resulting orbital angular momentum **L**, where

$$\nu = 1, 2, \ldots, Z \tag{5.1}$$

And in LS coupling, the individual spin angular momentums of the electrons $\mathbf{s}_\nu$ are assumed to combine to form a resulting spin angular momentum **S**. Finally, the

Neutron and X-ray Optics. DOI: http://dx.doi.org/10.1016/B978-0-12-407164-3.00004-8

third assumption is that the crystal is quenched, where the resulting orbital angular momentum **L** is zero:

$$\mathbf{L} = 0 \tag{5.2}$$

The non-Bravais crystal lattice has N unit cells, each of which is comprised of r atoms in which the dth atom position $\mathbf{d}_d$ is measured relative to a corner of the lth unit cell, and the lth unit cell is positioned at $\mathbf{R}_l$. The equilibrium position $\mathbf{R}_{ld}$ of the dth atom at the lth unit cell in the crystal lattice, which is measured relative to the crystal origin, is then

$$\mathbf{R}_{ld} = \mathbf{R}_l + \mathbf{d}_d \tag{5.3}$$

Displacement of the dth nucleus from its equilibrium position is given by vector $\mathbf{u}_{ld}(t)$ at time t. At initial time $t = 0$, the nonequilibrium position $\mathbf{R}_{ldu}(0)$ of the rth nucleus, which is measured relative to the lattice origin, is

$$\mathbf{R}_r(0) = \mathbf{R}_{ld} + \mathbf{u}_{ld}(0) \tag{5.4}$$

or

$$\mathbf{R}_{ldu}(0) = \mathbf{R}_l + \mathbf{d}_d + \mathbf{u}_{ld}(0) \tag{5.5}$$

In each unit cell, the number of atomic electrons of the dth nucleus is Z_d, where Z_d is the dth nucleus atomic number, if the dth atom is neutral. In many crystal lattices, which are comprised of ions such as quenched Fe^{3+} with $L = 0$, the number of dth ion electrons Z_d is obtained by subtracting the ion charge state from the dth ion atomic number, for example, in hematite Fe_2O_3, where

$$Z_d = 26 - (+3) = 23 \quad Fe^{3+} \text{ in } Fe_2O_3 \tag{5.6}$$

$$Z_d = 8 - (-2) = 10 \quad O^{2-} \text{ in } Fe_2O_3 \tag{5.7}$$

The N unit cell crystal has a total of E electrons, where the eth electron position $\mathbf{r}_e$, is measured relative to the crystal origin. The E crystal electrons are indexed by integer $e = 1, 2, \ldots, E$. The r atoms (ions) in each unit cell are indexed by integer $d = 1, 2, \ldots, r$, and the dth atom (ion) has Z_d electrons:

$$E = Nr \sum_{d=1}^{r} Z_d \tag{5.8}$$

The position $\mathbf{r}_{vd}$ of the vth electron of the dth nucleus for $v = 1, 2, \ldots, Z_d$ is measured relative to the position of the dth nucleus. Globally, the vth electron of the dth nucleus is the eth electron in the lattice. The eth electron vector position $\mathbf{r}_e(0)$ at initial time $t = 0$ is measured relative to the crystal lattice origin. The eth electron position vector $\mathbf{r}_e(0)$ is obtained via the triplet index integers (l, d, v), which specify that the eth electron is located in the lth unit cell and resides in the dth atom as its vth electron, where

$$\mathbf{r}_e(0) = \mathbf{R}_l + \mathbf{d}_d + \mathbf{u}_{ld}(0) + \mathbf{r}_{vd} \tag{5.9}$$

The initial electron position vectors can be replaced by their time-independent Schrödinger operator versions, see Liboff (2003), Squires (1996), and Ch. 3 in Cremer (2012b):

$$\mathbf{r}_{vd} \rightarrow \hat{\mathbf{r}}_{vd} \tag{5.10}$$

$$\mathbf{u}_{ld}(0) \rightarrow \hat{\mathbf{u}}_{ld}(0) \tag{5.11}$$

$$\mathbf{d}_d \rightarrow \hat{\mathbf{d}}_d \tag{5.12}$$

$$\mathbf{R}_l \rightarrow \hat{\mathbf{R}}_l \tag{5.13}$$

$$\mathbf{r}_e(0) \rightarrow \hat{\mathbf{r}}_e(0) \tag{5.14}$$

$$\mathbf{R}_{ldu}(0) \rightarrow \hat{\mathbf{R}}_{ldu}(0) \tag{5.15}$$

$$\mathbf{R}_{ld} \rightarrow \hat{\mathbf{R}}_{ld} \tag{5.16}$$

The nonequilibrium position for the dth nucleus is expressed by the Schrödinger operator:

$$\hat{\mathbf{R}}_{ldu}(0) = \hat{\mathbf{R}}_l + \hat{\mathbf{d}}_d + \hat{\mathbf{u}}_{ld}(0) \tag{5.17}$$

The Schrödinger operator $\hat{\mathbf{r}}_e(0)$ for the eth lattice electron position at time $t = 0$ is

$$\hat{\mathbf{r}}_e(0) = \hat{\mathbf{R}}_{ldu}(0) + \hat{\mathbf{r}}_{vd} \tag{5.18}$$

With substitution

$$\hat{\mathbf{r}}_e(0) = \hat{\mathbf{R}}_l + \hat{\mathbf{d}}_d + \hat{\mathbf{u}}_{ld}(0) + \hat{\mathbf{r}}_{vd} \tag{5.19}$$

Because the electron equilibrium position vectors $\mathbf{R}_{ld}$, $\mathbf{R}_l$, $\mathbf{d}_d$, and $\mathbf{r}_{vd}$ do not vary with time, their Schrödinger operators are equal to their Heisenberg operator at time t, where

$$\hat{\mathbf{R}}_{ld}(0) = \hat{\mathbf{R}}_{ld}(t) = \hat{\mathbf{R}}_{ld} \tag{5.20}$$

$$\hat{\mathbf{R}}_l(0) = \hat{\mathbf{R}}_l(t) = \hat{\mathbf{R}}_l \tag{5.21}$$

$$\hat{\mathbf{d}}_d(0) = \hat{\mathbf{d}}_d(t) = \hat{\mathbf{d}}_d \tag{5.22}$$

$$\hat{\mathbf{r}}_{vd}(0) = \hat{\mathbf{r}}_{vd}(t) = \hat{\mathbf{r}}_{vd} \tag{5.23}$$

The corresponding Heisenberg operator for the eth lattice electron position, where the eth electron can be a different lattice electron, or the same lattice electron at later time t, is

$$\hat{\mathbf{r}}_e(t) = \hat{\mathbf{R}}_l + \hat{\mathbf{d}}_d + \hat{\mathbf{u}}_{ld}(t) + \hat{\mathbf{r}}_{vd} \tag{5.24}$$

where

$$\hat{\mathbf{r}}_e(t) = \exp\left(\mathrm{i}t\frac{\hat{H}}{\hbar}\right)\hat{\mathbf{r}}_e(0)\exp\left(-\mathrm{i}t\frac{\hat{H}}{\hbar}\right) \tag{5.25}$$

and

$$\hat{\mathbf{u}}_{ld}(t) = \exp\left(\mathrm{i}t\frac{\hat{H}}{\hbar}\right)\hat{\mathbf{u}}_{ld}(0)\exp\left(-\mathrm{i}t\frac{\hat{H}}{\hbar}\right) \tag{5.26}$$

The eigenequation for spin operator $\hat{\mathbf{s}}_{ldv}$ for the vth electron of the dth atom in the lth unit cell of the crystal for the ith initial scattering system eigenstate is

$$\hat{\mathbf{s}}_{ldv}|\phi_i\rangle = \sqrt{s_{ldv}(s_{ldv}+1)}|\phi_i\rangle \tag{5.27}$$

The electron spin operator $\hat{\mathbf{S}}_{ld}$ for the dth atom in the lth unit cell in the crystal lattice is obtained by summing the Z_d individual v^{th} electron spin operators $\hat{\mathbf{s}}_{ldv}$, where

$$\hat{\mathbf{S}}_{ld} = \sum_{v=1}^{Z_d}\hat{\mathbf{s}}_{ldv} \tag{5.28}$$

The eigenequation for $\hat{\mathbf{S}}_{ld}$ for the ith initial scattering system eigenstate is

$$\hat{\mathbf{S}}_{ld}|\phi_i\rangle = \sqrt{s_{ld}(s_{ld}+1)}|\phi_i\rangle \tag{5.29}$$

After neutron scatter, the spin states of the lth unit cell of the dth atom in the mth final state is

$$\hat{\mathbf{S}}_{ld}|\phi_m\rangle = \sqrt{s_{ld}(s_{ld}+1)}|\phi_m\rangle \tag{5.30}$$

For an N-electron atom, one obtains s_d by iteration. For

$$i = 0, 1, 2, 3, \ldots, Z_d - 2 \tag{5.31}$$

one obtains

$$s_{i+1,i+2} = |s_{i,i+1} + s_{i+2}|, \ldots, |s_{i,i+1} - s_{i+2}| \tag{5.32}$$

After N iterations, one obtains the range of total spin quantum numbers s_d for the dth atom, with Z_d electrons in each unit cell of the crystal, where

$$s_d = |s_{\max}|, |s_{\max} - 1|, \ldots, |s_{\min}| \tag{5.33}$$

The maximum $s_{\max}$ total spin quantum number s_d of the dth atom is

$$s_{\max} = \sum_{v=1}^{N} s_{dv} \tag{5.34}$$

If the minimum $s_{\min}$ total spin quantum number s_d of the dth atom is positive, then the expression for $s_{\min}$ is

$$s_{\min} = s_{dN} - \sum_{v=1}^{N-1} s_{dv} > 0 \quad \text{if } s_{\min} > 0 \tag{5.35}$$

Otherwise, if the minimum $s_{\min}$ is negative, then $s_{\min}$ is set to 0, where

$$s_{\min} < 0 \Rightarrow s_{\min} = 0 \tag{5.36}$$

The electron orbital angular momentum of the crystal is assumed to be zero ($L = 0$) due to quenching. Hence, the magnetic dipole moment vector operator $\hat{\mathbf{Q}}(\boldsymbol{\kappa})$ of all the electrons in the scattering system has only the electron spin contribution $\hat{\mathbf{Q}}_s(\boldsymbol{\kappa})$, where

$$\hat{\mathbf{Q}}(\boldsymbol{\kappa}) = \hat{\mathbf{Q}}_s(\boldsymbol{\kappa}) \tag{5.37}$$

and

$$\hat{\mathbf{Q}}_l(\boldsymbol{\kappa}) = 0 \tag{5.38}$$

Electron spin operator $\hat{\mathbf{Q}}_s(\boldsymbol{\kappa})$ is the κ-space Fourier transform of the direct-space (r-space) electron spin density operator $\hat{\boldsymbol{\rho}}_s(\mathbf{r})$, where

$$\hat{\mathbf{Q}}_s(\boldsymbol{\kappa}) = \int_{-\infty}^{\infty} \mathrm{d}z \int_{-\infty}^{\infty} \mathrm{d}y \int_{-\infty}^{\infty} \mathrm{d}x \hat{\boldsymbol{\rho}}_s(\mathbf{r}) \exp(\mathrm{i}\boldsymbol{\kappa} \cdot \mathbf{r}) \tag{5.39}$$

The electron spin density is given by the electron spin density operator $\hat{\boldsymbol{\rho}}_s(\mathbf{r})$, which is expressed as the sum of Dirac delta functions. Each delta function designates the point location of the spin $\mathbf{s}_e$ amplitude (via spin operator $\hat{\mathbf{s}}_e$) of the eth orbital electron in the scattering system of nuclei, in which $\delta(\mathbf{r}-\mathbf{r}_e)$ is nonzero only at electron location $\mathbf{r}=\mathbf{r}_e$. The electron spin density vector operator $\hat{\boldsymbol{\rho}}_s(\mathbf{r})$ is then

$$\hat{\boldsymbol{\rho}}_s(\mathbf{r}) = \sum_{e=1}^{E} \delta(\mathbf{r}-\mathbf{r}_e)\,\hat{\mathbf{s}}_e \tag{5.40}$$

The total magnetic dipole moment vector operator $\hat{\mathbf{Q}}(\boldsymbol{\kappa},0)$ is just the electron spin magnetic dipole moment $\hat{\mathbf{Q}}_s(\boldsymbol{\kappa},0)$ in the quenched crystal. Substitution of Schrödinger operator $\hat{\mathbf{r}}_e(0)$ for the eth electron position in the Schrödinger operator expression for the electron spin magnetic dipole moment $\hat{\mathbf{Q}}_s(\boldsymbol{\kappa},0)$ gives

$$\begin{aligned}\hat{\mathbf{Q}}(\boldsymbol{\kappa},0) = \hat{\mathbf{Q}}_s(\boldsymbol{\kappa},0) &= \sum_{e=1}^{E}\left[\frac{\exp(\mathrm{i}\boldsymbol{\kappa}\cdot\hat{\mathbf{r}}_e(0))}{\kappa^2}\right]\hat{\mathbf{s}}_e(0)\\ &= \sum_{l=0}^{N-1}\exp(\mathrm{i}\boldsymbol{\kappa}\cdot\hat{\mathbf{R}}_l)\sum_{d=1}^{r}\exp(i\boldsymbol{\kappa}\cdot\hat{\mathbf{d}}_d)\exp(\mathrm{i}\boldsymbol{\kappa}\cdot\hat{\mathbf{u}}_{ld}(0))\sum_{v=1}^{Z_d}\exp(i\boldsymbol{\kappa}\cdot\hat{\mathbf{r}}_{vd})\hat{\mathbf{s}}_{ldv}\end{aligned} \tag{5.41}$$

To specify the position of the vth electron of the dth atom of every l^{th} unit cell in the crystal, one adds electron position vector operator $\hat{\mathbf{r}}_{vd}$ to the dth atom equilibrium position vector operator $\hat{\mathbf{d}}_d$ and the dth atom displacement from equilibrium vector operator $\hat{\mathbf{u}}_{ld}(0)$, which then is added to the lth unit cell vector position operator $\hat{\mathbf{R}}_l$.

The electron spin operator $\hat{\mathbf{s}}_{ldv}$ is specified for the vth electron of the dth atom of the lth unit cell to allow analysis of spin variation within a crystal. The crystal may be comprised of many domains in which the spin vectors differ from one domain to the next. The position operator $\hat{\mathbf{r}}_{vd}$ for the vth electron of the dth atom is the same in every unit cell of the crystal; however, the electron spin state operator $\hat{\mathbf{s}}_{ldv}$ may vary from one unit cell to the next. The transition matrix element for the contribution of the dth atom in the lth unit cell is

$$\begin{aligned}\langle\mathbf{Q}(\boldsymbol{\kappa},0)\rangle = \langle\hat{\mathbf{Q}}_s(\boldsymbol{\kappa},0)\rangle &= \langle\phi_m|\hat{\mathbf{Q}}_s(\boldsymbol{\kappa},0)|\phi_i\rangle\\ &= \langle\phi_m|\exp[\mathrm{i}\boldsymbol{\kappa}\cdot(\hat{\mathbf{R}}_{ld}+\hat{\mathbf{u}}_{ld}(0))]\sum_{v=1}^{Z_d}\exp(\mathrm{i}\boldsymbol{\kappa}\cdot\hat{\mathbf{r}}_{vd})\hat{\mathbf{s}}_{ldv}|\phi_i\rangle\end{aligned} \tag{5.42}$$

The vector position operator for the equilibrium d^{th} atom of the lth unit cell, is measured relative to the crystal origin where the vector sum $\hat{\mathbf{R}}_l+\hat{\mathbf{d}}_d$ is denoted by vector operator $\hat{\mathbf{R}}_{ld}$. The transition matrix element involves integrating $\hat{\mathbf{Q}}_s(\boldsymbol{\kappa},0)$

with the initial $\phi_i(\mathbf{R}_1 \ldots \mathbf{R}_N)$ and final $\phi_m(\mathbf{R}_1 \ldots \mathbf{R}_N)$ eigenstates of the atomic electron scattering system over the positions $\mathbf{R}_1 \ldots \mathbf{R}_N$ of the N unit cells, where

$$\langle \hat{\mathbf{Q}}_s(\boldsymbol{\kappa},0)\rangle = \langle \phi_m | \hat{\mathbf{Q}}_s(\boldsymbol{\kappa},0)|\phi_i\rangle = \langle \phi_m | \exp[i\boldsymbol{\kappa}\cdot(\hat{\mathbf{R}}_{ld}+\hat{\mathbf{u}}_{ld}(0))]\sum_{v=1}^{Z_d}\exp(i\boldsymbol{\kappa}\cdot\hat{\mathbf{r}}_{vd})\hat{\mathbf{s}}_{ldv}|\phi_i\rangle$$
$$= \int dV_N \int dV_{N-1} \int dV_{N-2} \ldots \int dV_2 \int dV_1 \phi_m^* \left\{ \exp[i\boldsymbol{\kappa}(\hat{\mathbf{R}}_{ld}+\hat{\mathbf{u}}_{ld}(0))]\sum_{v=1}^{Z_d}\exp(i\boldsymbol{\kappa}\cdot\hat{\mathbf{r}}_{vd})\hat{\mathbf{s}}_{ldv} \right\} \phi_i \tag{5.43}$$

5.2 Evaluation of the Unpaired, Atomic Electron, Magnetic Dipole Moment Transition Matrix Element $\langle \mathbf{Q}_s(\boldsymbol{\kappa}, 0)\rangle$

Eq. (5.43) for the total electron spin magnetic dipole moment vector operator $\langle \mathbf{Q}_s(\boldsymbol{\kappa}, 0)\rangle$ contains two operator expressions. The first of these is the position operator for the Nr atoms in the N unit cells of the non-Bravais lattice crystal with r atoms per unit cell, where

$$\sum_{l=0}^{N-1}\sum_{d=1}^{r}\langle \phi_m | \exp[i\boldsymbol{\kappa}\cdot(\hat{\mathbf{R}}_{ld}+\hat{\mathbf{u}}_{ld}(0))] \tag{5.44}$$

The second operator expression is the electron spin operator for Z_d electrons in the dth atom of every unit cell of the crystal:

$$\hat{\mathbf{S}}_d = \sum_{v=1}^{Z_d}\exp(i\boldsymbol{\kappa}\cdot\hat{\mathbf{r}}_{vd})\hat{\mathbf{s}}_{ldv} \tag{5.45}$$

The initial i and final state m eigenstates ϕ_i and ϕ_m for the scattering system atoms depends on the resulting spin angular momentum $\mathbf{S}$ and the orientation of the electron spins, i.e., the azimuthal or z-component $\mathbf{S}_z$ of $\mathbf{S}$, the positions of the electrons $\mathbf{r}_{vd}$, and the positions $\mathbf{R}_{ld} + \mathbf{u}_{ld}$ of the nuclei.

The energy of slow neutrons (less than 0.1 eV) is insufficient to alter the total electron spin angular momentum $\mathbf{S}$ and positions of the electron $\mathbf{r}_{vd}$; however, these slow neutrons change the azimuthal electron spin component $\mathbf{S}_z$ by flipping an electron spin, and the slow neutrons can alter the position of the nuclei from their equilibrium positions by displacement vector $\mathbf{u}_{ld}$ via elastic and inelastic collisions and resulting nuclei vibrations. To evaluate Eq. (5.43), one employs the closure relation, see Squires (1996), Ch. 14 in Merzbacher (1970), or Ch. 3 in Cremer (2012b), to obtain

$$\left\langle \phi_m \left| \exp[i\boldsymbol{\kappa}\cdot(\hat{\mathbf{R}}_{ld}+\hat{\mathbf{u}}_{ld}(0))]\sum_{v=1}^{Z_d}\exp(i\boldsymbol{\kappa}\cdot\hat{\mathbf{r}}_{vd})\hat{\mathbf{s}}_{ldv}\right| \phi_i \right\rangle$$
$$= \sum_{w=1}^{W}\sum_{v=1}^{Z_d}\langle \phi_m|\exp(i\boldsymbol{\kappa}\cdot\hat{\mathbf{r}}_{vd})|\phi_w\rangle\langle\phi_w|\exp[i\kappa(\hat{\mathbf{R}}_{ld}+\hat{\mathbf{u}}_{ld}(0))]\hat{\mathbf{s}}_{ldv}|\phi_i\rangle \tag{5.46}$$

In Eq. (5.46), independent of position operator $\hat{\mathbf{r}}_{vd}$, is matrix element

$$\langle \phi_m | \exp(i\boldsymbol{\kappa} \cdot \hat{\mathbf{r}}_{vd}) | \phi_w \rangle \tag{5.47}$$

That is, the symmetric or antisymmetric electron spatial eigenfunction components, which are contained within scattering system eigenfunctions ϕ_i and ϕ_m, cause this matrix element to be independent of the position $\mathbf{r}_{vd}$ of the vth electron in the dth atom of the lattice. The matrix element is then independent of the corresponding position operator $\hat{\mathbf{r}}_{vd}$. Thus, the multiplicative operator $\exp(i\boldsymbol{\kappa} \cdot \hat{\mathbf{r}}_{vd})$ depends only upon the electron position $\mathbf{r}_{vd}$.

The electron position $\mathbf{r}_{vd}$ is independent, and thus orthogonal to the electron spin states and to the positions of the nuclei. The initial eigenstate ϕ_i or the prescatter system of nuclei differs from its final eigenstate ϕ_m, in that upon neutron scatter, the nuclei positions $\mathbf{R}_{ld} + \mathbf{u}_{ld}$ and the atomic electron spin orientations (distribution of spin-up and spin-down states) are both changed.

Returning to the $i \rightarrow m$ transition matrix element of Eq. (5.46), one can move the summation over ν, the index for the vth electron in the dth atom, from the left matrix element to the right matrix element, where

$$\left\langle \phi_m \left| \exp[i\boldsymbol{\kappa} \cdot (\hat{\mathbf{R}}_{ld} + \hat{\mathbf{u}}_{ld}(0))] \sum_{v=1}^{Z_d} \exp(i\boldsymbol{\kappa} \cdot \hat{\mathbf{r}}_{vd}) \hat{\mathbf{s}}_{ldv} \right| \phi_i \right\rangle$$

$$= \sum_{w=1}^{W} \langle \phi_m | \exp(i\boldsymbol{\kappa} \cdot \hat{\mathbf{r}}_{vd}) | \phi_w \rangle \left\langle \phi_w \left| \exp[i\kappa(\hat{\mathbf{R}}_{ld} + \hat{\mathbf{u}}_{ld}(0))] \sum_{v=1}^{Z_d} \hat{\mathbf{s}}_{ldv} \right| \phi_i \right\rangle \tag{5.48}$$

By orthogonality, the matrix element for $\exp(i\boldsymbol{\kappa} \cdot \hat{\mathbf{r}}_{vd})$ in Eq. (5.48) is nonzero only if eigenstate ϕ_w is identical to the final (postscatter) eigenstate ϕ_m, which is expressed by the Kronecker delta function δ_{wm}, where

$$\langle \phi_m | \exp(i\boldsymbol{\kappa} \cdot \hat{\mathbf{r}}_{vd}) | \phi_w \rangle = \delta_{mw} \tag{5.49}$$

Evaluation of Eq. (5.49), shows

$$\langle \phi_m | \exp(i\boldsymbol{\kappa} \cdot \hat{\mathbf{r}}_{vd}) | \phi_w \rangle = \exp(i\boldsymbol{\kappa} \cdot \hat{\mathbf{r}}_{vd}) \int_{-\infty}^{\infty} dz \int_{-\infty}^{\infty} dy \int_{-\infty}^{\infty} dx \phi_m^* \phi_w = \delta_{mw} \exp(i\boldsymbol{\kappa} \cdot \hat{\mathbf{r}}_{vd}) \tag{5.50}$$

In Eq. (5.48), one inserts $\hat{\mathbf{S}}_{ld}$ for the summation of $\hat{\mathbf{s}}_{ldv}$ over v, and sets

$$\hat{\mathbf{R}}_{ldu} = \hat{\mathbf{R}}_{ld} + \hat{\mathbf{u}}_{ld}(0) \tag{5.51}$$

Eq. (5.48) becomes

$$\left\langle \phi_m \left| \exp[i\kappa \cdot (\hat{\mathbf{R}}_{ld} + \hat{\mathbf{u}}_{ld}(0))] \sum_{v=1}^{Z_d} \exp(i\kappa \cdot \hat{\mathbf{r}}_{vd}) \hat{\mathbf{s}}_{ldv} \right| \phi_i \right\rangle$$
$$= \sum_{w=1}^{W} \delta_{mw} \exp(i\kappa \cdot \hat{\mathbf{r}}_{vd}) \langle \phi_w | \exp(i\kappa \cdot \hat{\mathbf{R}}_{ldu}) \hat{\mathbf{S}}_{ld} | \phi_i \rangle \tag{5.52}$$

The only nonzero result in Eq. (5.52) is the $w = m$ term, so that with

$$\phi_w = \phi_m \tag{5.53}$$

one obtains

$$\sum_{w=1}^{W} \delta_{mw} \exp(i\kappa \cdot \hat{\mathbf{r}}_{vd}) \langle \phi_w | \exp(i\kappa \cdot \hat{\mathbf{R}}_{ldu}) \hat{\mathbf{S}}_{ld} | \phi_i \rangle = \exp(i\kappa \cdot \hat{\mathbf{r}}_{vd}) \langle \phi_m | \exp(i\kappa \cdot \hat{\mathbf{R}}_{ldu}) \hat{\mathbf{S}}_{ld} | \phi_i \rangle \tag{5.54}$$

With substitution of Eq. (5.54) into Eq. (5.52), the expected value of $\mathbf{Q}_s(\kappa, 0)$ of Eq. (5.43), becomes

$$\langle \hat{\mathbf{Q}}_s(\kappa, 0) \rangle = \langle \phi_m | \hat{\mathbf{Q}}_s(\kappa, 0) | \phi_i \rangle = \sum_{l=0}^{N-1} \sum_{d=1}^{r} \exp(i\kappa \cdot \hat{\mathbf{r}}_{vd}) \langle \phi_m | \exp(i\kappa \cdot \hat{\mathbf{R}}_{ldu}) \hat{\mathbf{S}}_{ld} | \phi_i \rangle \tag{5.55}$$

5.3 The Magnetic Form Factor

At this point, we are ready to derive the magnetic form factor $F_d(\kappa)$ for the dth atom, which is identical in every unit cell of the crystal. The position $\mathbf{r}_{vd}$ of the vth electron in the dth atom is assumed to be identical in every unit cell of the crystal, and $\mathbf{r}_{vd}$ is independent of the neutron position $\mathbf{r}$. One defines a scalar function $\rho_d(\mathbf{r})$ as the normalized density of unpaired electrons in the dth atom or ion of each unit cell of the crystal, where $\rho_d(\mathbf{r})$ is a function of neutron position $\mathbf{r}$ in the crystal lattice. For every d^{th} atom or ion, the normalized unpaired electron density $\rho_d(\mathbf{r})$ is the spatial density $\varepsilon_d(\mathbf{r})$ of the unpaired electrons, divided by the number n_d of unpaired electrons, where

$$\rho_d(\mathbf{r}) = \frac{\varepsilon_d(\mathbf{r})}{n_d} \tag{5.56}$$

so that

$$\int_{-\infty}^{\infty} dz \int_{-\infty}^{\infty} dy \int_{-\infty}^{\infty} dx \rho_d(\mathbf{r}) = \int_{-\infty}^{\infty} dz \int_{-\infty}^{\infty} dy \int_{-\infty}^{\infty} dx \frac{\varepsilon_d(\mathbf{r})}{n_d} = 1 \tag{5.57}$$

The magnetic form factor $F_d(\boldsymbol{\kappa})$ for the dth atom in each unit cell is the Fourier transform in the κ-space of the direct-lattice, normalized density $\rho_d(\mathbf{r})$ of the dth atom's unpaired electrons, where

$$F_d(\boldsymbol{\kappa}) = \exp(i\boldsymbol{\kappa} \cdot \hat{\mathbf{r}}_{vd}) = \int_{-\infty}^{\infty} dz \int_{-\infty}^{\infty} dy \int_{-\infty}^{\infty} dx \rho_d(\mathbf{r}) \exp(i\boldsymbol{\kappa} \cdot \mathbf{r}) \tag{5.58}$$

and

$$\boldsymbol{\kappa} = \mathbf{k}_0 - \mathbf{k} \tag{5.59}$$

The magnetic form factor $F_d(\boldsymbol{\kappa})$ is expressed by

$$\exp(i\boldsymbol{\kappa} \cdot \hat{\mathbf{r}}_{vd}) \tag{5.60}$$

As a result, the magnetic form factor $F_d(\boldsymbol{\kappa})$ is the amplitude of the phase shift presented by the vth electron of the dth atom. That is, magnetic form factor $F_d(\boldsymbol{\kappa})$ depends on the vth electron position $\mathbf{r}_{vd}$ in the dth atom, and $F_d(\boldsymbol{\kappa})$ depends on the change of the neutron scatter vector $\boldsymbol{\kappa}$ upon scatter. The magnetic form factor $F_d(\boldsymbol{\kappa})$ provides a set of relative positions of the unpaired electrons.

The number n_d of unpaired electrons of the dth number usually ranges from zero to the atomic number Z_d of the dth atom, where

$$0 \leq n_d \leq Z_d \tag{5.61}$$

Exceptions occur for lattices that contain negative ions whose unpaired electrons exceed the atomic number of the ion.

5.4 Partial and Differential Cross Section Expressions for a Quenched Magnetic Crystal

With substitution of Eq. (5.58) into Eq. (5.55), the expected value of the electron spin component of the magnetic dipole moment vector operator $\hat{\mathbf{Q}}_s(\boldsymbol{\kappa}, 0)$ becomes

$$\langle \hat{\mathbf{Q}}_s(\boldsymbol{\kappa}, 0) \rangle = \langle \phi_m | \hat{\mathbf{Q}}_s(\boldsymbol{\kappa}, 0) | \phi_i \rangle = \sum_{l=0}^{N-1} \sum_{d=1}^{r} F_d(\boldsymbol{\kappa}) \langle \phi_m | \exp(i\boldsymbol{\kappa} \cdot \hat{\mathbf{R}}_{ldu}) \hat{\mathbf{S}}_{ld} | \phi_i \rangle \tag{5.62}$$

or, equivalently,

$$\langle \phi_m | \hat{\mathbf{Q}}_s(\boldsymbol{\kappa}, 0) | \phi_i \rangle = \left\langle \phi_m \left| \sum_{l=0}^{N-1} \sum_{d=1}^{r} F_d(\boldsymbol{\kappa}) \exp(\mathrm{i}\boldsymbol{\kappa} \cdot \hat{\mathbf{R}}_{ldu}) \hat{\mathbf{S}}_{ld} \right| \phi_i \right\rangle \tag{5.63}$$

The electron magnetic dipole moment vector Schrödinger operator $\mathbf{Q}(\boldsymbol{\kappa}, 0)$ is, thus,

$$\mathbf{Q}(\boldsymbol{\kappa}, 0) = \hat{\mathbf{Q}}_s(\boldsymbol{\kappa}, 0) = \sum_{l=0}^{N-1} \sum_{d=1}^{r} F_d(\boldsymbol{\kappa}) \exp(\mathrm{i}\boldsymbol{\kappa} \cdot \hat{\mathbf{R}}_{ldu}) \hat{\mathbf{S}}_{ld} \tag{5.64}$$

and the electron magnetic dipole moment vector Heisenberg operator $\mathbf{Q}(\boldsymbol{\kappa}, t)$ is

$$\hat{\mathbf{Q}}(\boldsymbol{\kappa}, t) = \hat{\mathbf{Q}}_s(\boldsymbol{\kappa}, t) = \sum_{l=0}^{N-1} \sum_{d=1}^{r} F_d(\boldsymbol{\kappa}) \exp[\mathrm{i}\boldsymbol{\kappa} \cdot \hat{\mathbf{R}}_{ldu}(t)] \hat{\mathbf{S}}_{ld} \tag{5.65}$$

Recall that at time t, the Heisenberg vector operator $\hat{\mathbf{R}}_{ldu}(t)$ gives the nonequilibrium position $\mathbf{R}_{ldu}(t)$ for the dth nucleus in the lth unit cell, measured relative to the crystal center, where

$$\hat{\mathbf{R}}_{ldu}(t) = \hat{\mathbf{R}}_l + \hat{\mathbf{d}}_d + \hat{\mathbf{u}}_{ld}(t) \tag{5.66}$$

Vector operator $\hat{\mathbf{R}}_{ldu}(t)$ gives the position of the dth nucleus of each unit cell of the crystal, which has r nuclei per unit cell in the N unit cells of the crystal. Relative to the crystal center is the position vector $\mathbf{R}_l$ of the reference corner of the lth unit cell. From its equilibrium position $\mathbf{d}_d$ measured relative to the same corner in every unit cell, the dth nucleus is displaced from its equilibrium position at time t by displacement vector $\mathbf{u}_{ld}(t)$, which is expressed by Heisenberg vector operator $\hat{\mathbf{u}}_{ld}(t)$, where from Squires (1996) or Eq. (49) in Ch. 6 of Cremer (2012b),

$$\mathbf{u}_{ld}(t) = \sqrt{\frac{\hbar}{2M_d N}} \times \sum_{q=-N/2}^{N/2} \sum_{j=1}^{3r} \left(\frac{\hat{\mathbf{e}}_{qjd} a_{qj} \exp[\mathrm{i}(\mathbf{q}_q \cdot \mathbf{R}_l - w_{qj} t)] + \hat{\mathbf{e}}^*_{qjd} a^+_{qj} \exp[-\mathrm{i}(\mathbf{q}_q \cdot \mathbf{R}_l - w_{qj} t)]}{\sqrt{w_{qj}}} \right) \tag{5.67}$$

Recall from Eq. (3.19) of Ch. 3, that the generalized partial differential cross section $(\mathrm{d}^2\sigma/d\Omega\,\mathrm{d}E)^{\text{mag}}$ for the scatter of neutrons by the magnetic dipole fields produced by the unpaired electrons of the scattering system atoms is

$$\left(\frac{\mathrm{d}^2\sigma}{\mathrm{d}\Omega\,\mathrm{d}E} \right)^{\text{mag}} = \sum_{i=1}^{J} p_i \sum_{m=1}^{M} \left(\frac{\mathrm{d}^2\sigma}{\mathrm{d}\Omega\,\mathrm{d}E} \right)^{\text{mag}}_{i,0s \to m,s} \tag{5.68}$$

In thermal equilibrium at temperature T, the probability p_i of each ith initial eigenstate of J possible initial states of the scattering atoms and their unpaired electrons (i.e., atom positions and net electron spin of each atom) is given by

$$p_i = \frac{\exp(-\mathrm{i}E_i/kT)}{\sum_{i=1}^{I} \exp(-\mathrm{i}E_i/kT)} \tag{5.69}$$

The partial differential cross section for neutron scatter for the transition of the scattering system from initial state i to final state m for neutron spin vector operator $\hat{\mathbf{S}}_n$ is

$$\left(\frac{\mathrm{d}^2\sigma}{\mathrm{d}\Omega\,\mathrm{d}E}\right)^{\mathrm{mag}}_{i,0s\rightarrow m,s} = (r_0 g)^2 \frac{k}{k_0} \langle \phi_m \psi_s \psi | \hat{\mathbf{S}}_n \cdot \hat{\mathbf{Q}}_\perp(\boldsymbol{\kappa}) | \psi_0 \psi_{0s} \phi_i \rangle|^2 \delta(E_0 - E + E_m - E_i) \tag{5.70}$$

and, equivalently,

$$\left(\frac{\mathrm{d}^2\sigma}{\mathrm{d}\Omega\,\mathrm{d}E}\right)^{\mathrm{mag}}_{i,0s\rightarrow m,s} = (r_0 g)^2 \frac{k}{k_0} \delta(E_0 - E + E_m - E_i)$$
$$\langle \phi_i \psi_{0s} \psi_0 | [\hat{\mathbf{S}}_n \cdot \hat{\mathbf{Q}}_\perp(\boldsymbol{\kappa})]^+ | \psi \psi_s \phi_m \rangle \langle \phi_m \psi_s \psi | \hat{\mathbf{S}}_n \cdot \hat{\mathbf{Q}}_\perp(\boldsymbol{\kappa}) | \psi_0 \psi_{0s} \phi_i \rangle \tag{5.71}$$

The scatter system is described by (1) the initial $\psi_0(\mathbf{r})$ and final $\psi(\mathbf{r})$ state neutron position wave function components dependent on neutron position $\mathbf{r}$; (2) the initial ψ_{0s} and final ψ_s state neutron spin wave function components that have no dependence on the neutron position variable $\mathbf{r}$; and (3) the initial $\phi_i(\mathbf{r}_{vd}, \mathbf{R}_{ldu}, t)$ and final $\phi_m(\mathbf{r}_{vd}, \mathbf{R}_{ldu}, t)$ state position components of the wave function of the scattering electrons at position $\mathbf{r}_{vd}$ and nuclei at positions $\mathbf{R}_{ldu}(t)$.

The dot product $\hat{\mathbf{S}}_n \cdot \hat{\mathbf{Q}}_\perp(\boldsymbol{\kappa})$ of (1) the neutron spin vector operator $\hat{\mathbf{S}}_n$, and (2) the perpendicular component of the dipole moment vector operator $\hat{\mathbf{Q}}_\perp(\boldsymbol{\kappa})$ of the scattering system atomic electrons, describe the interaction of (1) the magnetic dipole moment of the neutron, and (2) the magnetic dipole moment of the atomic electron spin and orbital angular momentum. After application of the closure property to Eq. (5.71), the expected value of the scalar operator $\hat{U}_e$ or $\hat{\mathbf{S}}_n \cdot \hat{\mathbf{Q}}_\perp(\boldsymbol{\kappa})$ for the potential energy of magnetic interaction between the neutron and the atomic electrons of the non-Bravais crystal lattice, is obtained, via integration. Integration yields the partial differential cross section $(\mathrm{d}^2\sigma/d\Omega\,\mathrm{d}E)^{\mathrm{mag}}$ for neutron scatter due to the magnetic dipole moment of atomic electrons, where

$$\left(\frac{\mathrm{d}^2\sigma}{\mathrm{d}\Omega\,\mathrm{d}E}\right)^{\mathrm{mag}} = (r_0 g)^2 \frac{k}{k_0} \int_{-\infty}^{\infty} \frac{\exp(-\mathrm{i}wt)}{2\pi\hbar} \times \sum_{i=1}^{J} p_i \; \langle \phi_i \psi_{0s} \psi_0 | [\hat{\mathbf{S}}_n \cdot \hat{\mathbf{Q}}_\perp(\boldsymbol{\kappa})]^+ \left[\hat{\mathbf{S}}_n \cdot \hat{\mathbf{Q}}_\perp(\boldsymbol{\kappa})\right] | \psi_0 \psi_{0s} \phi_i \rangle \mathrm{d}t \tag{5.72}$$

We now obtain the differential cross section $(\mathrm{d}\sigma/d\Omega)^{\mathrm{mag}}$ for coherent elastic neutron scatter from the partial differential cross section $(\mathrm{d}^2\sigma/d\Omega\,\mathrm{d}E)^{\mathrm{mag}}$ by integration over the scattered neutron energy E, which in elastic scatter equals the incident neutron energy E_0. Hence, one expects integration involving the Dirac delta function. The time integral expresses the conservation of energy, via its equivalence to the Dirac delta function, which, in this case of elastic scatter, shows the scattered and incident neutron energy are equal, where

$$\int_{-\infty}^{\infty} \mathrm{d}t \frac{\exp(-\,\mathrm{i}wt)}{2\pi\hbar} = \delta(E - E_0) \quad \text{and} \quad E - E_0 = \hbar w \tag{5.73}$$

Using Eq. (5.73), integration in Eq. (5.72) over the scatter neutron energy E, given fixed incident neutron energy E_0, gives the differential cross section $(\mathrm{d}\sigma/d\Omega)^{\mathrm{mag}}$ for neutron scatter from the magnetic dipole moment of the unpaired atomic electron spins in a lattice:

$$\left(\frac{\mathrm{d}\sigma}{\mathrm{d}\Omega}\right)^{\mathrm{mag}} = (r_0 g)^2 \sum_{i=1}^{J} p_i \langle \phi_i \psi_{0s} \psi_0 | [\hat{\mathbf{S}}_n \cdot \hat{\mathbf{Q}}_\perp(\boldsymbol{\kappa})]^+ \left[\hat{\mathbf{S}}_n \cdot \hat{\mathbf{Q}}_\perp(\boldsymbol{\kappa})\right] | \psi_0 \psi_{0s} \phi_i \rangle \tag{5.74}$$

5.5 Evaluation of Partial and Differential Cross Section Expressions for a Quenched Magnetic Crystal—Separation of Unpaired, Atomic Electron Spatial and Spin Components

We now apply the results Ch. 3, section 3.14, and in particular Eq. (3.236) to a quenched crystal. The previously derived Heisenberg $\hat{\mathbf{Q}}_s(\boldsymbol{\kappa}, t)$ operator of Eq. (5.65) and Schrödinger $\hat{\mathbf{Q}}_s(\boldsymbol{\kappa}, 0)$ operator of Eq. (5.64) for the electron, magnetic dipole moment, which were obtained from the Heitler–London model, are now applied to a quenched crystal, and plugged into the partial differential cross section $(\mathrm{d}^2\sigma/d\Omega\ \mathrm{d}E)^{\mathrm{mag}}$ expression of Eq. (3.236) of Ch. 3. The vector portion of the operators $\hat{\mathbf{Q}}_s(\boldsymbol{\kappa}, 0)$ and $\hat{\mathbf{Q}}_s(\boldsymbol{\kappa}, t)$ of Eqs. (5.64) and 5.65, are the vector operators $\hat{\mathbf{S}}_{ld}(\kappa, 0)$ and $\hat{\mathbf{S}}_{ld}(\kappa, t)$, which correspond to the α and β components of the electron spin for l^{th} unit cell, d^{th} atom (ion) in Eq. (5.75) below, where

$$\begin{aligned}\left(\frac{\mathrm{d}^2\sigma}{\mathrm{d}\Omega\,\mathrm{d}E}\right)^{\mathrm{mag}} &= (r_0 g)^2 \frac{k}{k_0} \sum_{i=1}^{J} p_i \sum_{l=0}^{N-1} \sum_{l'=0}^{N-1} \sum_{d=1}^{r} \sum_{d'=1}^{r} \\ &\int_{-\infty}^{\infty} \frac{\exp(-\mathrm{i}wt)}{2\pi\hbar} \langle \phi_i | \exp\left[-\mathrm{i}\boldsymbol{\kappa} \cdot (\hat{\mathbf{R}}_{l'} + \hat{\mathbf{d}}_{d'} + \hat{\mathbf{u}}_{l'} d'(0))\right] \\ &\exp\left[\mathrm{i}\boldsymbol{\kappa} \cdot (\hat{\mathbf{R}}_l + \hat{\mathbf{d}}_d + \hat{\mathbf{u}}_{ld}(t))\right] \\ &F_{d'}^*(\boldsymbol{\kappa}) F_d(\boldsymbol{\kappa}) \sum_{\alpha=1}^{3} \sum_{\beta=1}^{3} \left(\delta_{\alpha\beta} - \frac{\kappa_\alpha \kappa_\beta}{\kappa^2}\right) \hat{S}_{l'd'\alpha}^{+}(0) \hat{S}_{ld\beta}(t) | \phi_i \rangle \mathrm{d}t\end{aligned} \tag{5.75}$$

In Eq. (5.75), the dot product of the neutron spin vector $\mathbf{S}_n$ with the perpendicular component of the unpaired, atomic electron magnetic dipole moment vector operator $\hat{\mathbf{Q}}_{\perp}(\boldsymbol{\kappa})$, has been performed for an assumed unpolarized incident beam of neutrons, where 50% of the initial state neutrons are spin-up, and 50% are spin-down. The result of this dot product is the vector operator $\hat{S}_{ld\beta}(t)$, which incorporates the interaction of the two equally likely, initial state directions of neutron magnetic dipole moment vector, with the initial state magnetic dipole moments of the unpaired, atomic electrons of the crystal.

The ith nucleus eigenstate wave function ϕ_i of the initial I nuclei eigenstates before neutron scatter, which describes the scattering system encountered by the incident neutron, is a function of the coordinates of positions of atoms of the crystal with N unit cells and the net spin of the unpaired electrons of each crystal atom, where

$$\phi_i = \phi_i(\mathbf{R}_0, \mathbf{u}_0, \mathbf{R}_1, \mathbf{u}_1, \mathbf{R}_2, \mathbf{u}_2, \ldots, \mathbf{R}_l, \mathbf{u}_l, \ldots, \mathbf{R}_{N-1}, \mathbf{u}_{N-1}) \tag{5.76}$$

The position of the atom at position $\hat{\mathbf{d}}_{d'} + \hat{\mathbf{u}}_{0d'}(0)$ at initial time $t = 0$, and the same or another atom at $\hat{\mathbf{d}}_d + \hat{\mathbf{u}}_{ld}(t)$ at time t, are each random and independent, and thus they are not correlated. Hence, we need the summations over both d and d'. The unit cell positions $\mathbf{R}_l$ are fixed and independent of time, so the unit cell at position $\mathbf{R}_l$ is correlated to the unit cell at position $\mathbf{R}_{l'}$. The summations over l and l' can be replaced by a single summation over l multiplied by N, where $\hat{\mathbf{R}}_0 = 0$ is the center of the crystal. With rearrangement, the partial differential cross section $(\mathrm{d}^2\sigma/d\Omega\ \mathrm{d}E)^{\text{mag}}$ of Eq. (5.75) for a non-Bravais crystal lattice, becomes

$$\begin{aligned}\left(\frac{\mathrm{d}^2\sigma}{\mathrm{d}\Omega\,\mathrm{d}E}\right)^{\text{mag}} &= N(r_0 g)^2 \frac{k}{k_0} \sum_{i=1}^{J} p_i \sum_{l=0}^{N-1} \exp(i\boldsymbol{\kappa}\cdot\hat{\mathbf{R}}_l) \int_{-\infty}^{\infty} \frac{\exp(-iwt)}{2\pi\hbar} \\ &\left\langle \phi_i \left| \sum_{d=1}^{r}\sum_{d'=1}^{r} \exp[-i\boldsymbol{\kappa}\cdot(\hat{\mathbf{d}}_{d'} + \hat{\mathbf{u}}_{0d'}(0))]\exp[i\boldsymbol{\kappa}\cdot(\hat{\mathbf{d}}_d + \hat{\mathbf{u}}_{ld}(t))] \right.\right. \\ &\left.\left. F_{d'}^{*}(\boldsymbol{\kappa})F_d(\boldsymbol{\kappa}) \sum_{\alpha=1}^{3}\sum_{\beta=1}^{3} \left(\delta_{\alpha\beta} - \frac{\kappa_\alpha \kappa_\beta}{\kappa^2}\right) \hat{S}_{0d'\alpha}^{+}(0)\hat{S}_{ld\beta}(t) \right| \phi_i \right\rangle \mathrm{d}t\end{aligned} \tag{5.77}$$

One applies the closure property to Eq. (5.77), where

$$|\phi_i\rangle\langle\phi_i| = 1 \tag{5.78}$$

and

$$\left(\frac{\mathrm{d}^2\sigma}{\mathrm{d}\Omega\,\mathrm{d}E}\right)^{\text{mag}} = (r_0 g)^2 \frac{k}{k_0} \sum_{i=1}^{J} p_i \sum_{l=0}^{N-1} \exp(\mathrm{i}\boldsymbol{\kappa}\cdot\hat{\mathbf{R}}_l) \int_{-\infty}^{\infty} \frac{\exp(-\mathrm{i}wt)}{2\pi\hbar} R_i(t) S_i(t)\mathrm{d}t \tag{5.79}$$

With probability p_i^{mag}, the time-dependent scattering atom position term $R_i(t)$ and the time-dependent net electron spin term $S_i(t)$ of the atom in the ith initial eigenstate of the scattering system of atoms are

$$R_i(t) = \left\langle \phi_i \left| \sum_{d=1}^{r} \sum_{d'=1}^{r} \exp[-\mathrm{i}\boldsymbol{\kappa} \cdot (\hat{\mathbf{d}}_{d'} + \hat{\mathbf{u}}_{0d'}(0))] \exp[\mathrm{i}\boldsymbol{\kappa} \cdot (\hat{\mathbf{d}}_d + \hat{\mathbf{u}}_{ld}(t))] \right| \phi_i \right\rangle \tag{5.80}$$

and

$$S_i(t) = \left\langle \phi_i \left| F_{d'}^*(\boldsymbol{\kappa}) F_d(\boldsymbol{\kappa}) \sum_{\alpha=1}^{3} \sum_{\beta=1}^{3} \left(\delta_{\alpha\beta} - \frac{\kappa_\alpha \kappa_\beta}{\kappa^2} \right) \hat{S}_{0d'\alpha}^{+}(0) \hat{S}_{ld\beta}(t) \right| \phi_i \right\rangle \tag{5.81}$$

The atom position term $R_i(t)$ is independent of the net electron spin term $S_i(t)$ of the atoms. That is, $R_i(t)$ depends only on position, and $S_i(t)$ depends only on the net spin of an atom. This is because the electron spins are assumed to have a negligible effect on the binding forces between the atoms in the lattice. Thus, the electron spins have a negligible effect upon the atom/electron positions. However, the atom position term $R_i(t)$, and the assumed, independent net electron spin term $S_i(t)$, each depends on temperature.

The expected set of spin states of the scattering system then depends on the temperature-dependent distribution of the spin portion of the J initial lattice eigenstates. Likewise, the expected set of lattice atom positions of the scattering system depends on the temperature-dependent distribution of the nuclei position portion of the J initial lattice eigenstates. If the random variable of position R, with temperature-dependent probability density $p(T)$ and expected value or mean $\overline{R}$, is independent of spin S with the same temperature-dependent probability density $p(T)$ with mean $\overline{S}$, then the expected value of the position and spin product $\overline{RS}$ equals the product of the means $\overline{R}$ and $\overline{S}$, where

$$\overline{RS} = \int R(T)S(T)p(T)\mathrm{d}T = \overline{R} \cdot \overline{S} = \int R(T)p(T)\mathrm{d}T \cdot \int S(T)p(T)\mathrm{d}T \tag{5.82}$$

and

$$\int p(T)\mathrm{d}T = 1 \tag{5.83}$$

With application of Eq. (5.82−5.83 to 5.79−5.81), the partial differential cross section becomes

$$\left(\frac{\mathrm{d}^2\sigma}{\mathrm{d}\Omega\, \mathrm{d}E} \right)^{\mathrm{mag}} = (r_0 g)^2 \frac{k}{k_0} \sum_{l=0}^{N-1} \exp(\mathrm{i}\boldsymbol{\kappa} \cdot \hat{\mathbf{R}}_l) \int_{-\infty}^{\infty} \frac{\exp(-\mathrm{i}wt)}{2\pi\hbar} R_T(t) S_T(t) \mathrm{d}t \tag{5.84}$$

where

$$R_T(t) = \sum_{i=1}^{J} p_i \left\langle \phi_i \left| \sum_{d=1}^{r} \sum_{d'=1}^{r} \exp[-\mathrm{i}\boldsymbol{\kappa} \cdot (\hat{\mathbf{d}}_{d'} - \hat{\mathbf{d}}_d)] \exp(-\mathrm{i}\boldsymbol{\kappa} \cdot \hat{\mathbf{u}}_{0d'}(t)) \exp[\mathrm{i}\boldsymbol{\kappa} \cdot \hat{\mathbf{u}}_{ld}(t)] \right| \phi_i \right\rangle \tag{5.85}$$

and

$$S_T(t) = \sum_{i=1}^{J} p_i \left\langle \phi_i \left| F_{d'}^*(\boldsymbol{\kappa}) F_d(\boldsymbol{\kappa}) \sum_{\alpha=1}^{3} \sum_{\beta=1}^{3} \left(\delta_{\alpha\beta} - \frac{\kappa_\alpha \kappa_\beta}{\kappa^2} \right) \hat{S}_{0d'\alpha}^{+}(0) \hat{S}_{ld\beta}(t) \right| \phi_i \right\rangle \tag{5.86}$$

5.6 Evaluation of the Thermal Average of Initial State, Unpaired, Atomic Electron Positions

Next, one examines and evaluates the temperature-dependent distribution of the electron position component of the initial eigenstates for the unpaired atomic electrons of the scatter system. Recalling section 3 in Ch. 6 of Cremer (2012b), one can simplify the spatial component of the unpaired atomic electrons $R_T(t)$, where one sets

$$U = -\mathrm{i}\boldsymbol{\kappa} \cdot \hat{\mathbf{u}}_{0d'}(0) \tag{5.87}$$

and

$$V = \mathrm{i}\boldsymbol{\kappa} \cdot \hat{\mathbf{u}}_{ld}(t) \tag{5.88}$$

The mean-squared values of the displacement vector $\mathbf{u}_l(t)$ is assumed to be independent of time t and the lth unit cell position, when averaging over time t or position $\mathbf{R}_{ldu}$ in the crystal, where for all t

$$\overline{[\mathbf{u}_{0d'}(0)]^2} = \overline{[\mathbf{u}_{ld}(t)]^2} \quad l = 0, 1, 2, \ldots, N-1 \text{ and } d, d' = 1, 2, \ldots, r \tag{5.89}$$

Hence,

$$\overline{U^2} = \overline{V^2} \tag{5.90}$$

The displacement $\mathbf{u}_l(t)$ of the atom in each lth unit cell is assumed to oscillate about its equilibrium with zero mean displacement $\overline{\mathbf{u}}_l(t)$, and one assumes no correlation between the displacement of one atom and the displacement of another atom, such that

$$\overline{\mathbf{u}}_{0d'}(t) = \overline{\mathbf{u}}_{ld}(0) = 0 \quad \text{and} \quad \overline{U} = \overline{V} = 0 \tag{5.91}$$

The summation expression in Eq. (5.85) for $R_T(t)$, can be compactly expressed by the mean or expected value of exp U exp V, which is denoted, via an overhead bar:

$$\overline{\exp U \exp V} \tag{5.92}$$

The expected value of exp U exp V is the correlation of the lth nucleus displacement $\mathbf{u}_l$ (via exp V), which is relative to the displacement of another nucleus $l = 0$ (via exp U). The partial differential cross section $(\mathrm{d}^2\sigma/d\Omega\,\mathrm{d}E)^{\text{mag}}$ of Eq. (5.84), with application of Eqs. (5.87–5.92) to Eq. (5.85) for $R_T(t)$, becomes

$$\begin{aligned}\left(\frac{\mathrm{d}^2\sigma}{\mathrm{d}\Omega\,\mathrm{d}E}\right)^{\text{mag}} &= N(r_0 g)^2\frac{k}{k_0}\sum_{l=0}^{N-1}\exp(\mathrm{i}\boldsymbol{\kappa}\cdot\hat{\mathbf{R}}_l)\\ &\times\int_{-\infty}^{\infty}\frac{\exp(-\,\mathrm{i}wt)}{2\pi\hbar}\sum_{d=1}^{r}\sum_{d'=1}^{r}\overline{\exp U \exp V}\exp\left[-\mathrm{i}\boldsymbol{\kappa}\cdot(\hat{\mathbf{d}}_{d'}-\hat{\mathbf{d}}_d)\right]\\ &\sum_{i=1}^{J}p_i\left\langle\phi_i\left|F^*_{d'}(\boldsymbol{\kappa})F_d(\boldsymbol{\kappa})\sum_{\alpha=1}^{3}\sum_{\beta=1}^{3}\left(\delta_{\alpha\beta}-\frac{\kappa_\alpha\kappa_\beta}{\kappa^2}\right)\hat{S}^{+}_{0d'\alpha}(0)\hat{S}_{ld\beta}(t)\right|\phi_i\right\rangle\mathrm{d}t\end{aligned} \tag{5.93}$$

where

$$\overline{\exp U \exp V} = \sum_{i=1}^{J}p_i\langle\phi_i\psi_{0s}|\exp(-\mathrm{i}\boldsymbol{\kappa}\cdot\hat{\mathbf{u}}_{0d'}(t))\exp[\mathrm{i}\boldsymbol{\kappa}\cdot\hat{\mathbf{u}}_{ld}(t)]|\psi_{0s}\phi_i\rangle \tag{5.94}$$

Since the initial neutron spin wave function component ψ_{0s} is independent of the atom/electron displacement from equilibrium vector operator $\hat{\mathbf{u}}_{ld}(t)$ and the initial state wave function component ϕ_i for atom/electron position, one gets

$$\overline{\exp U \exp V} = \sum_{i=1}^{J}p_i\langle\phi_i|\exp(-\mathrm{i}\boldsymbol{\kappa}\cdot\hat{\mathbf{u}}_{0d'}(t))\exp[\mathrm{i}\boldsymbol{\kappa}\cdot\hat{\mathbf{u}}_{ld}(t)]|\phi_i\rangle \tag{5.95}$$

We now simplify the expected value of the $\overline{\exp U \exp V}$ expression. Recall from section 3.4 in Squires (1996), or section 3 in Ch. 6 of Cremer (2012b), that the commutation of U and V in terms of $\hat{\mathbf{u}}_0(0)$ and $\hat{\mathbf{u}}_l(t)$, respectively, is a constant c, where

$$[U, V] = UV - VU = c \tag{5.96}$$

where

$$c = \overline{c} = \overline{UV - VU} \tag{5.97}$$

From Appendix I.1 in Squires (1996), or section 3 in Ch. 6 of Cremer (2012b), is derived he relation:

$$\exp U \exp V = \exp(U+V)\exp\left[\frac{1}{2}(UV-VU)\right] = \exp(U+V)\exp(c/2) \quad (5.98)$$

With series expansion of $\exp(U+V)$ in Eq. (5.98), one can show

$$\overline{\exp U \exp V} = \overline{\exp(U+V)}\exp\left[\frac{1}{2}(\overline{UV-VU})\right] \quad (5.99)$$

Then, recall from Appendix E.2 in Squires (1996), or Eq. (147) of section 4 in Ch. 5 of Cremer (2012b), that

$$\overline{\exp(U+V)} = \exp\left[\frac{\overline{(U+V)^2}}{2}\right] = \exp\left[\frac{1}{2}(\overline{U^2+V^2+UV+VU})\right] \quad (5.100)$$

Substitution of Eq. (5.100) into Eq. (5.99), gives

$$\overline{\exp U \exp V} = \exp\left[\frac{1}{2}(\overline{2U^2+UV+VU+UV-VU})\right] \quad (5.101)$$

Hence,

$$\overline{\exp U \exp V} = \exp \overline{U^2} \exp \overline{UV} \quad (5.102)$$

where substitution for U from 5.87 gives,

$$\overline{U^2} = \overline{(\mathrm{i}\kappa \cdot \mathbf{u}_{0d'}(0))^2} = -\overline{(\kappa \cdot \mathbf{u}_{0d'}(0))^2} \quad (5.103)$$

The Debye–Waller factor in Eq. (5.104) below for the dth atom in a unit cell, expresses the flux attenuation of scattered neutrons, and is caused by thermal vibration of the nuclei, where

$$\exp(-2W_d) \quad (5.104)$$

and

$$2W_d = -\overline{U^2} = -|U|^2 = |\kappa \cdot \mathbf{u}_{0d}(0)|^2 \quad (5.105)$$

Hence, Eq. (5.102) becomes

$$\overline{\exp U \exp V} = \exp(-2W_d)\exp \overline{UV}, \quad (5.106)$$

where

$$\overline{UV} = \sum_{i=1}^{J} p_i \langle \phi_i | [\boldsymbol{\kappa} \cdot \hat{\mathbf{u}}_{0d'}(0)][\boldsymbol{\kappa} \cdot \hat{\mathbf{u}}_{ld}(t)] | \phi_i \rangle \tag{5.107}$$

Thus, Eq. (5.94) contained within the partial differential cross section expression of Eq. (5.93), is transformed by Eqs. (5.106–5.107) to

$$\overline{\exp U \exp V} = \exp(-2W_d)\exp\left[\sum_{i=1}^{J} p_i \langle \phi_i | [\boldsymbol{\kappa} \cdot \hat{\mathbf{u}}_{0d'}(0)][\boldsymbol{\kappa} \cdot \hat{\mathbf{u}}_{ld}(t)] | \phi_i \rangle\right] \tag{5.108}$$

5.7 Partial Differential Cross Section for the LS-Coupled, Heitler–London Model in a Quenched Crystal with Unpaired, Localized Atomic Electron Spin

One finally obtains the expression for the partial differential cross section for the neutron scatter by only the magnetic dipole moments, which arise from the spins of the unpaired atomic electrons in a quenched crystal. Substitution into Eq. (5.93) of Eq. (5.108) for the thermal-averaged distribution of initial state, unpaired atomic electron positions in the crystal lattice, then yields the partial differential cross section $(\mathrm{d}^2\sigma/d\Omega\,\mathrm{d}E)^{\mathrm{mag}}$ for neutron scatter from a quenched crystal. This assumes that the spins of unpaired electrons are localized to the nuclei equilibrium positions, where

$$\left(\frac{\mathrm{d}^2\sigma}{\mathrm{d}\Omega\,\mathrm{d}E}\right)^{\mathrm{mag}} = N(r_0 g)^2 \frac{k}{k_0} \sum_{l=0}^{N-1} \exp(i\boldsymbol{\kappa} \cdot \hat{\mathbf{R}}_l) \int_{-\infty}^{\infty} \frac{\exp(-iwt)}{2\pi\hbar}$$
$$\sum_{d=1}^{r} F_d(\boldsymbol{\kappa})\exp(-2W_d)\exp(i\boldsymbol{\kappa} \cdot \hat{\mathbf{d}}_d) \sum_{d'=1}^{r} F_{d'}^*(\boldsymbol{\kappa})\exp(-i\boldsymbol{\kappa} \cdot \hat{\mathbf{d}}_{d'})$$
$$\exp\left[\sum_{i=1}^{J} p_i \langle \phi_i | [\boldsymbol{\kappa} \cdot \hat{\mathbf{u}}_{0d'}(0)][\boldsymbol{\kappa} \cdot \hat{\mathbf{u}}_{ld}(t)] | \phi_i \rangle\right]$$
$$\sum_{i=1}^{J} p_i \left\langle \phi_i \left| \sum_{\alpha=1}^{3}\sum_{\beta=1}^{3} \left(\delta_{\alpha\beta} - \frac{\kappa_\alpha \kappa_\beta}{\kappa^2}\right) \hat{S}^{+}_{0d'\alpha}(0)\hat{S}_{ld\beta}(t) \right| \phi_i \right\rangle \mathrm{d}t \tag{5.109}$$

Equation (5.109) serves as the precursor equation for magnetic neutron scatter, and is a starting point for derivations of particular magnetic scattering phenomena.

One can replace the multiplication of the complex conjugate summation terms over d and d in Eq. (5.109) by the squared magnitude of the summation term over d, where

$$\sum_{d=1}^{r} F_d(\boldsymbol{\kappa})\exp(-2W_d)\exp(\mathrm{i}\boldsymbol{\kappa}\cdot\hat{\mathbf{d}}_d)\sum_{d'=1}^{r} F_{d'}^{*}(\boldsymbol{\kappa})\exp(-\mathrm{i}\boldsymbol{\kappa}\cdot\hat{\mathbf{d}}_{d'}) = \left|\sum_{d=1}^{r} F_d(\boldsymbol{\kappa})\exp(-W_d)\exp(\mathrm{i}\boldsymbol{\kappa}\cdot\hat{\mathbf{d}}_d)\right|^2 \tag{5.110}$$

If one assumes $\overline{UV}=0$ in Eq. (5.107), then the partial differential cross section $(\mathrm{d}^2\sigma/d\Omega\,\mathrm{d}E)^{\mathrm{mag}}$ of Eq. (5.109) for a quenched crystal, with assumed LS coupling and the Heitler-London model, becomes

$$\left(\frac{\mathrm{d}^2\sigma}{\mathrm{d}\Omega\,\mathrm{d}E}\right)^{\mathrm{mag}} = N(r_0 g)^2\frac{k}{k_0} \sum_{l=0}^{N-1}\exp(\mathrm{i}\boldsymbol{\kappa}\cdot\hat{\mathbf{R}}_l)\int_{-\infty}^{\infty}\frac{\exp(-\mathrm{i}wt)}{2\pi\hbar}\left|\sum_{d=1}^{r} F_d(\boldsymbol{\kappa})\exp(-W_d)\exp(\mathrm{i}\boldsymbol{\kappa}\cdot\hat{\mathbf{d}}_d)\right|^2 \sum_{i=1}^{J} p_i\left\langle\phi_i\left|\sum_{\alpha=1}^{3}\sum_{\beta=1}^{3}\left(\delta_{\alpha\beta}-\frac{\kappa_\alpha\kappa_\beta}{\kappa^2}\right)\hat{S}^{+}_{0d'\alpha}(0)\hat{S}_{ld\beta}(t)\right|\phi_i\right\rangle \mathrm{d}t \tag{5.111}$$

If in Eq. (5.107), one estimates $\overline{UV}\neq 0$, then the expected value of the position of the dth atom in every l^{th} unit cell of the crystal is the exponential term:

$$\exp(\overline{UV}) = \exp\left[\sum_{i=1}^{J} p_i\langle\phi_i|[\boldsymbol{\kappa}\cdot\hat{\mathbf{u}}_{0d'}(0)][\boldsymbol{\kappa}\cdot\hat{\mathbf{u}}_{ld}(t)]|\phi_i\rangle\right] \tag{5.112}$$

Expansion of the exponential gives

$$\exp(\overline{UV}) = 1+\overline{UV}+\frac{1}{2!}(\overline{UV})^2+\cdots+\frac{1}{n!}(\overline{UV})^n \tag{5.113}$$

The first term of the expansion of the exponential is 1, which is the case of coherent, elastic scatter and is equivalent to $\overline{UV}=0$.

The second term $\overline{UV}$ of Eq. (5.113), is due to inelastic scatter of the neutrons from the thermal-induced, vibrating lattice atoms via a one-phonon absorption or emission:

$$\overline{UV} = \sum_{i=1}^{J} p_i\langle\phi_i|[\boldsymbol{\kappa}\cdot\hat{\mathbf{u}}_{0d'}(0)][\boldsymbol{\kappa}\cdot\hat{\mathbf{u}}_{ld}(t)]|\phi_i\rangle \tag{5.114}$$

Equation (5.114) for $\overline{UV}$, represents the sum of the J initial i^{th} scatter eigenstates ϕ_i, which occur with probability p_i. The summation yields the expected value of the squared displacement $\langle|\hat{\mathbf{u}}_{ld}|^2\rangle$ of the dth atom displacements $\hat{\mathbf{u}}_{ld}(t)$ in each lth unit cell. Substitution of the 2nd term of Eq. (5.113) into Eq. (5.112), and substitution into master Eq. (5.109), then yields the component of the partial differential cross section for inelastic neutron scatter with a one-phonon exchange, where

$$\left(\frac{\mathrm{d}^2\sigma}{\mathrm{d}\Omega\,\mathrm{d}E}\right)^{\mathrm{mag}}_{1-ph} = N(r_0 g)^2 \frac{k}{k_0}\sum_{l=0}^{N-1}\exp(\mathrm{i}\boldsymbol{\kappa}\cdot\hat{\mathbf{R}}_l)\int_{-\infty}^{\infty}\frac{\exp(-\,\mathrm{i}wt)}{2\pi\hbar}$$
$$\sum_{d=1}^{r}F_d(\boldsymbol{\kappa})\exp(-2W_d)\exp(\mathrm{i}\boldsymbol{\kappa}\cdot\hat{\mathbf{d}}_d)\sum_{d'=1}^{r}F^*_{d'}(\boldsymbol{\kappa})\exp(-\,\mathrm{i}\boldsymbol{\kappa}\cdot\hat{\mathbf{d}}_{d'})$$
$$\left[\sum_{i=1}^{J}p_i\langle\phi_i|[\boldsymbol{\kappa}\cdot\hat{\mathbf{u}}_{0d'}(0)][\boldsymbol{\kappa}\cdot\hat{\mathbf{u}}_{ld}(t)]|\phi_i\rangle\right]$$
$$\sum_{i=1}^{J}p_i\left\langle\phi_i\left|\sum_{\alpha=1}^{3}\sum_{\beta=1}^{3}\left(\delta_{\alpha\beta}-\frac{\kappa_\alpha\kappa_\beta}{\kappa^2}\right)\hat{S}^{+}_{0d'\alpha}(0)\hat{S}_{ld\beta}(t)\right|\phi_i\right\rangle\mathrm{d}t \tag{5.115}$$

This inelastic neutron scatter and one-phonon exchange with the lattice is via the force of the vibrating, atomic electron magnetic field. This is due to the time-varying displacement $\hat{\mathbf{u}}_{ld}(t)$ from equilibrium $\mathbf{R}_l+\hat{\mathbf{d}}_d$ of each dth atom of every l^{th} unit cell of the crystal. That is, the electrons are attached to the vibrating nuclei and present a vibrating magnetic field to the magnetic dipole moment of the neutron. By means of the vibrating atomic electron, a magnetic field phonon exchange occurs between the crystal lattice and the neutron. The neutron either absorbs a phonon from the lattice or emits a phonon that is absorbed by the lattice. The quantum number n_n has only a single p^{th} mode of vibration, and changes by ± 1, while all other modes of vibration have fixed quantum numbers n_n, where

$$n_n = n_n \pm \delta_{np} \quad n = 1, 2, \ldots, 3Nr \tag{5.116}$$

There are $3Nr$ modes of vibration for the r atoms per unit cell in the N unit cells. Each of the Nr atoms can vibrate in the three directions (x, y, and z), or alternatively, in the three directions that are defined by the unit cell vectors $(\mathbf{a},\mathbf{b},\mathbf{c})$. Each ith initial eigenstate ϕ_i of the J initial eigenstates of the crystal has $3Nr$ quantum numbers that correspond to the $3Nr$ modes of oscillation.

The third term $(\overline{UV})^2/2$ of Eq. (5.113), is due to inelastic scatter of the neutrons from the thermal-induced, vibrating lattice atoms, via a one-phonon absorption or emission, where

$$\frac{1}{2}\left\{\sum_{i=1}^{J}p_i\langle\phi_i|[\boldsymbol{\kappa}\cdot\hat{\mathbf{u}}_{0d'}(0)]\;[\boldsymbol{\kappa}\cdot\hat{\mathbf{u}}_{ld}(t)]|\phi_i\rangle\right\}^2 \tag{5.117}$$

In this case, the inelastic scatter of the neutron results in an exchange of two phonons between the neutron and the crystal lattice, where two of the $3Nr$ modes of vibration namely the pth and sth modes change by ± 1, so that

$$n_n = n_n \pm \delta_{np} \quad \text{and} \quad n_n = n_n \pm \delta_{ns} \quad n = 1, 2, \ldots, 3Nr \tag{5.118}$$

Substitution of the 3rd term of Eq. (5.113) into Eq. (5.112), and substitution into master Eq. (5.109), then yields the component of the partial differential cross section for inelastic neutron scatter with a two-phonon exchange, where

$$\begin{aligned}\left(\frac{d^2\sigma}{d\Omega\, dE}\right)^{\text{mag}}_{2-ph} &= N(r_0 g)^2 \frac{k}{k_0} \sum_{l=0}^{N-1} \exp(i\boldsymbol{\kappa} \cdot \hat{\mathbf{R}}_l) \int_{-\infty}^{\infty} \frac{\exp(-iwt)}{2\pi\hbar} \\ &\sum_{d=1}^{r} F_d(\boldsymbol{\kappa}) \exp(-2W_d) \exp(i\boldsymbol{\kappa} \cdot \hat{\mathbf{d}}_d) \sum_{d'=1}^{r} F^*_{d'}(\boldsymbol{\kappa}) \exp(-i\boldsymbol{\kappa} \cdot \hat{\mathbf{d}}_{d'}) \\ &\frac{1}{2}\left[\sum_{i=1}^{J} p_i \langle \phi_i | [\boldsymbol{\kappa} \cdot \hat{\mathbf{u}}_{0d'}(0)][\boldsymbol{\kappa} \cdot \hat{\mathbf{u}}_{ld}(t)] | \phi_i \rangle\right]^2 \\ &\sum_{i=1}^{J} p_i \left\langle \phi_i \left| \sum_{\alpha=1}^{3} \sum_{\beta=1}^{3} \left(\delta_{\alpha\beta} - \frac{\kappa_\alpha \kappa_\beta}{\kappa^2}\right) \hat{S}^+_{0d'\alpha}(0) \hat{S}_{ld\beta}(t) \right| \phi_i \right\rangle dt\end{aligned} \tag{5.119}$$

The fourth term of the exponential expansion of Eq. (5.113), accounts for coherent inelastic neutron scatter with a three-phonon exchange, where three normal mode quantum numbers change by ± 1 in each of the ith initial eigenstates ϕ_i. The nth term of the exponential expansion, expresses the coherent inelastic neutron scatter with an $n-1$ phonon exchange, and the change of $n-1$ normal modes of oscillation, each by ± 1.

Inelastic neutron scatter can also occur via the exchange of energy between the incident neutron and the unpaired, atomic electron spins of the crystal lattice. The spin term in the partial differential cross section expression of Eq. (5.109), which accounts for the interaction of the neutron magnetic dipole moment with the electron magnetic dipole moment of the unpaired electron spins, is repeated in Eq. (5.120) below:

$$\sum_{i=1}^{J} p_i \left\langle \phi_i \left| \sum_{\alpha=1}^{3} \sum_{\beta=1}^{3} \left(\delta_{\alpha\beta} - \frac{\kappa_\alpha \kappa_\beta}{\kappa^2}\right) \hat{S}^+_{0d'\alpha}(0) \hat{S}_{ld\beta}(t) \right| \phi_i \right\rangle \tag{5.120}$$

Eq. (5.120) yields the expected value $\langle |\hat{S}_{ld\beta}(t)|^2 \rangle$ of the squared neutron–electron magnetic dipole moment interaction via neutron–electron spins, which are represented by the neutron–electron spin interaction operator $\hat{S}_{ld\beta}(t)$. The correlation product $\hat{S}^+_{0d'\alpha}(0)\hat{S}_{ld\beta}(t)$ is integrated over time in the partial differential cross section expression of Eq. (5.109). In inelastic neutron scatter, the exchange of neutron energy can occur with the spins of the unpaired, atomic electrons of the lattice. The energy exchange occurs in discrete units or quanta of energy called *magnons*.

If the spins of one or more unpaired, atomic electrons flip in the collision with the neutron, then an energy exchange occurs between the lattice electrons and the neutron. In this case, one or more magnons of energy are exchanged between the neutron and the unpaired electron spins. The neutron spin is either unchanged or the neutron spin-flips. The kinetic energy and the spin potential energy of the neutron in the magnetic field of the unpaired, atomic electrons, sum to yield the total incident neutron energy, which is presented to these electrons in the scatter process. Usually, the kinetic energy of the neutron is much larger than the spin energy of the neutron, and the neutron kinetic energy is much larger than the spin energy of the unpaired atomic electrons.

Vibration energy quanta or phonons are exchanged between the incident neutron and the lattice atoms, via neutron collision with the thermal-induced vibrating atoms. The inelastic scatter of the neutron with an atom can occur by strong-force interaction between the neutron and the vibrating nuclei in direct scatter of the neutrons by the nucleus, where a direct exchange of phonons occurs between the neutron and vibrating nucleus. However, in magnetic scatter of the neutron with the unpaired atomic electrons, both phonon and magnon exchange can occur in the magnetic scatter from the atomic electrons that are attached to the vibrating lattice nuclei. Inelastic scatter of neutrons from an atom can thus result in a change in the vibration energy of the nucleus and a change in the spin energy of the atom's unpaired electrons, where one or more phonons and one or more magnons are exchanged.

Inelastic scatter of the neutron with the atom can occur via the magnetic force between the neutron and the attached unpaired electrons of the vibrating nuclei without a change in the spins of the unpaired electrons. In the case of elastic magnetic scatter (no magnon exchange) and inelastic magnetic scatter due to lattice vibration, phonons are exchanged between the neutron and vibrating nucleus via the magnetic force between the attached unpaired electrons of the vibrating nucleus and the neutron. This phonon exchange without electron spin-flip (and no magnon exchange) is called *magneto-vibrational scatter* or *elastic magnetic scatter.* In magneto-vibrational scatter, the neutron exchanges energy indirectly with the vibrating lattice nucleus, via the magnetic interaction of the neutron and the electrons of the nucleus. Finally, there is elastic neutron scatter in which neither phonons nor magnons are exchanged.

5.8 Partial Differential Cross Section for an LS-Coupled, Heitler–London Model in a Quenched Crystal with Unpaired, Localized Atomic Electron Spin and Orbital Current

We now extend the master equation of Eq. (5.111) to include unquenched scattering atoms, where the unpaired atomic electrons of an atom produce a net orbital angular momentum, in addition to a net spin angular momentum. The net orbital angular momentum of the unpaired electron arises from the electron orbital motion. The orbital motion of the electron can be modeled as an electron current loop, which produces a magnetic dipole moment and magnetic field. The magnetic field of the magnetic dipole moment from the electron spin and the electron current produces a magnetic field that interacts with the neutron magnetic dipole moment. The magnetic force between the magnetic dipole moments of the neutron and unpaired atomic electrons result in neutron deflection with (inelastic) or without (elastic) a change of neutron and crystal lattice energy.

As an example, ions of rare earth elements (e.g., gadolinium) in a crystal lattice have both spin and unquenched angular momentum. To proceed further, requires making the dipole approximation, where the mean radius $\bar{r}_{\mathrm{or}}$ of the orbital wave functions $\phi_i^{\mathrm{or}}(\mathbf{r}_e)$ of the unpaired electrons is assumed to be much less than $1/|\boldsymbol{\kappa}|$, so that

$$\frac{\bar{r}_{\mathrm{or}}}{|\boldsymbol{\kappa}|} \ll 1 \tag{5.121}$$

One then gets

$$\langle \phi_m^{\mathrm{or}} | f(\mathbf{r}_e) \exp(i\boldsymbol{\kappa} \cdot \mathbf{r}_e) | \phi_i^{\mathrm{or}} \rangle \cong \langle \phi_m^{\mathrm{or}} | f(\mathbf{r}_e) | \phi_i^{\mathrm{or}} \rangle \tag{5.122}$$

As discussed before, LS coupling is assumed in which the angular momentum of the unquenched d^{th} ion of the crystal unit cell is described by

$$\mathbf{J}_d = \mathbf{L}_d + \mathbf{S}_d \tag{5.123}$$

The partial differential cross section $(\mathrm{d}^2\sigma/d\Omega\, \mathrm{d}E)^{\mathrm{mag}}$ for the Heitler–London model, an LS-coupled, quenched crystal, is converted to one that applies to an unquenched crystal by the following two modifications.

The first modification sets the total spin angular momentum $\mathbf{S}_d$ of the unquenched d^{th} ion, equal to the total orbital and spin angular momentum $\mathbf{J}_d$. Next, the magnetic form factor $F_d(\boldsymbol{\kappa})$ is modified by multiplying it by the Landé splitting factor g_d, where

$$F_d(\kappa) \to \frac{g_d}{2} F_d(\kappa) \tag{5.124}$$

The Landé splitting factor g_d is real, where

$$g_d^* = g_d \tag{5.125}$$

The modified magnetic form factor is given by

$$\frac{g_d}{2} F_d(\kappa) = \frac{g_{Sd}}{2} C_{0d} + \frac{g_{Ld}}{2} (C_{0d} + C_{2d}) \tag{5.126}$$

The Landé splitting factor g_d for the dth atom in every crystal unit cell is the sum of spin g_{Sd} and orbital g_{Ld} angular momentum components, where

$$g_d = g_{Sd} + g_{Ld} \tag{5.127}$$

$$g_{Sd} = 1 + \frac{S_d(S_d+1) - L_d(L_d+1)}{J_d(J_d+1)} \tag{5.128}$$

$$g_{Ld} = \frac{1}{2} + \frac{L_d(L_d+1) - S_d(S_d+1)}{2J_d(J_d+1)} \tag{5.129}$$

The integral C_{nd} for the dth atom per unit cell integrates the product of $\rho_d(\mathbf{r})$ with $B_n(r\kappa)$, which is the spherical Bessel function $B_n(\kappa r)$ of order n and argument κr. The normalized density $\rho_d(\mathbf{r})$ of unpaired electrons in the dth atom is averaged over all spatial directions, where

$$C_{nd} = 4\pi \int_0^{\infty} B_n(\kappa r)\rho_d(\mathbf{r})\mathrm{d}r \tag{5.130}$$

With Eq. (5.124) applied to Eq. (5.109), and replacing the multiplication of complex conjugate summation terms over d and d by the squared magnitude of the summation term over d, the partial differential cross section $(\mathrm{d}^2\sigma/d\Omega\ \mathrm{d}E)^{\mathrm{mag}}$, based on the Heitler–London model with LS coupling and the dipole approximation, then becomes

$$\left(\frac{\mathrm{d}^2\sigma}{\mathrm{d}\Omega\,\mathrm{d}E}\right)^{\mathrm{mag}} = N(r_0 g)^2 \frac{k}{k_0}$$
$$\sum_{l=0}^{N-1} \exp(\mathrm{i}\boldsymbol{\kappa}\cdot\hat{\mathbf{R}}_l) \int_{-\infty}^{\infty} \frac{\exp(-\mathrm{i}wt)}{2\pi\hbar} \sum_{d=1}^{r} \left| \frac{g_d}{2} F_d(\boldsymbol{\kappa}) \exp(\mathrm{i}\boldsymbol{\kappa}\cdot\mathbf{d}_d)\exp(-W_d) \right|^2$$
$$\sum_{d'=1}^{r} \exp\left[\sum_{i=1}^{J} p_i \langle \phi_i | [\boldsymbol{\kappa}\cdot\hat{\mathbf{u}}_{0d'}(0)][\boldsymbol{\kappa}\cdot\hat{\mathbf{u}}_{ld}(t)] | \phi_i \rangle \right]$$
$$\sum_{i=1}^{J} p_i \left\langle \phi_i \left| \sum_{\alpha=1}^{3}\sum_{\beta=1}^{3} \left(\delta_{\alpha\beta} - \frac{\kappa_\alpha \kappa_\beta}{\kappa^2} \right) \hat{S}^{+}_{0d'\alpha}(0)\hat{S}_{ld\beta}(t) \right| \phi_i \right\rangle \mathrm{d}t \tag{5.131}$$

The relative positions of the equilibrium positions of atoms determine the phase differences between scattered waves. The phase shift of the incident wave, which is caused by scatter from the dth atom of the unit cell, produces an atom magnetic form factor $F_d(\boldsymbol{\kappa})$ that is multiplied by the Landé splitting factor g_d:

$$\frac{1}{2} g_d F_d(\boldsymbol{\kappa}) \tag{5.132}$$

In the case of a Bravais lattice crystal with one atom per unit cell, the partial differential cross section $(\mathrm{d}^2\sigma/d\Omega\,\mathrm{d}E)^{\mathrm{mag}}$ of Eq. (5.131), with $d = d' = 0$, becomes

$$\left(\frac{\mathrm{d}^2\sigma}{\mathrm{d}\Omega\,\mathrm{d}E}\right)^{\mathrm{mag}} = N(r_0 g)^2 \frac{k}{k_0}\sum_{l=0}^{N-1}\exp(\mathrm{i}\boldsymbol{\kappa}\cdot\hat{\mathbf{R}}_l)\int_{-\infty}^{\infty}\frac{\exp(-\mathrm{i}wt)}{2\pi\hbar}\left|\frac{g_0}{2}F_0(\boldsymbol{\kappa})\exp(-W_0)\right|^2$$
$$\exp\left[\sum_{i=1}^{J}p_i\langle\phi_i|[-\boldsymbol{\kappa}\cdot\hat{\mathbf{u}}_0(0)][\boldsymbol{\kappa}\cdot\hat{\mathbf{u}}_l(t)]|\phi_i\rangle\right]$$
$$\sum_{i=1}^{J}p_i\left\langle\phi_i\left|\sum_{\alpha=1}^{3}\sum_{\beta=1}^{3}\left(\delta_{\alpha\beta}-\frac{\kappa_\alpha\kappa_\beta}{\kappa^2}\right)\hat{S}_{0\alpha}^{+}(0)\hat{S}_{l\beta}(t)\right|\phi_i\right\rangle\mathrm{d}t \tag{5.133}$$

The partial differential cross section of Eq. (5.133), accounts for neutron scatter in a material, where the atoms have both a net electron spin and unquenched electron orbital angular momentum, which is represented by the net, effective spin operator $\hat{S}_{l\beta}(t)$. The relative positions of the equilibrium positions of atoms determine the phase differences between scattered waves. The fractional atoms sum to one atom each per unit cell. Each whole atom or unit cell is a scattering center, which provides a scattered wave with a phase shift relative to the incident wave. The overall phase shift of the incident wave due to the fractional atom scattering centers, which sum to one atom in the unit cell, is given by

$$\frac{g_0 F_0(\boldsymbol{\kappa})}{2} \tag{5.134}$$

This phase shift is the product of the magnetic form factor $F_0(\boldsymbol{\kappa})$ multiplied by the Landé splitting factor g_0 for the one-atom unit cell. And not to be confused by notation, the parameter g denotes the Landé splitting factor for the neutron.

The Debye–Waller factor, which expresses the attenuation of the flux of scattered neutrons, via a decrease in the differential cross section $(\mathrm{d}\sigma/d\Omega)^{\mathrm{mag}}$, due to the vibration of the nuclei in the Bravais lattice crystal, becomes

$$\exp(-2W_0) \tag{5.135}$$

where

$$2W_0 = -|\boldsymbol{\kappa}\cdot\mathbf{u}_0(0)|^2 \tag{5.136}$$

The intensity of neutrons that are elastic scattered at angle θ relative to their incident direction falls by $1/e^2$ if $W_0 = 1$. Given the displacement vector $\mathbf{u}_0(0)$ of the vibrating nuclei, which are oriented at angle ϕ, relative to the neutron scattering vector $\boldsymbol{\kappa}$, the $1/e^2$ attenuation with $W_0 = 1$ occurs for sufficiently large nuclear vibration amplitudes $u_0(0)$, where

$$u_0(0) = \frac{\sqrt{2}}{\kappa \cos \phi} = \frac{\lambda_0 \sqrt{2}}{4\pi \sin \theta \cos \phi} \tag{5.137}$$

with

$$\kappa = 2k_0 \sin \frac{\theta}{2} = \frac{4\pi}{\lambda_0} \sin \frac{\theta}{2} \tag{5.138}$$

and

$$\boldsymbol{\kappa} \cdot \mathbf{u}_0(0) = (\mathbf{k}_0 - \mathbf{k}) \cdot \mathbf{u}_0(0) = |\mathbf{k}_0 - \mathbf{k}||\mathbf{u}_0(0)|\cos \phi \tag{5.139}$$

5.9 Coherent, Elastic Differential Cross Section for an LS-Coupled, Heitler–London Model in a Quenched Crystal with Unpaired, Localized Atomic Electron Spin and Orbital Current

We now derive the elastic, differential cross section $(d\sigma/d\Omega)^{\text{mag}}$ from the previous section's partial differential cross section $(d^2\sigma/d\Omega\, dE)^{\text{mag}}$ of Eq. (5.133), which is based on the Heitler–London model, LS coupling, and the dipole approximation for a Bravais lattice crystal, with one atom per unit cell, where $d = d' = 0$. One considers a time-independent thermal equilibrium in which $t \to \infty$. With thermal equilibrium, the displacement of the atoms from their equilibrium positions and the net, effective spin state of each atom are independent of time, where

$$\hat{\mathbf{u}}_l(t \to \infty) = \hat{\mathbf{u}}_l(0) \tag{5.140}$$

and

$$\hat{S}_{l\beta}(t \to \infty) = \hat{S}_{l\beta}(0) \tag{5.141}$$

With $t = 0$, and $d = d = 0$, Eq. (5.67) becomes

$$\hat{\mathbf{u}}_l(0) = \sqrt{\frac{\hbar}{2MN}} \sum_{q=-N/2}^{N/2} \sum_{j=1}^{3} \left(\frac{\hat{\mathbf{e}}_{qj} a_{qj} \exp(\mathrm{i}\mathbf{q}_q \cdot \mathbf{R}_l) + \hat{\mathbf{e}}^*_{qj} a^+_{qj} \exp(-\mathrm{i}\mathbf{q}_q \cdot \mathbf{R}_l)}{\sqrt{w_{qj}}} \right) \tag{5.142}$$

The partial differential cross section of Eq. (5.133) for the time-independent case becomes

$$\left(\frac{d^2\sigma}{d\Omega\, dE}\right)^{mag} = N(r_0 g)^2 \frac{k}{k_0}\sum_{l=0}^{N-1}\exp(i\boldsymbol{\kappa}\cdot\hat{\mathbf{R}}_l)\int_{-\infty}^{\infty}\frac{\exp(-iwt)}{2\pi\hbar}\left|\frac{g_0}{2}F_0(\boldsymbol{\kappa})\exp(-W_0)\right|^2$$
$$\exp\left[\sum_{i=1}^{J}p_i\langle\phi_i|[\boldsymbol{\kappa}\cdot\hat{\mathbf{u}}_0(0)][\boldsymbol{\kappa}\cdot\hat{\mathbf{u}}_l(0)]|\phi_i\rangle\right]$$
$$\sum_{i=1}^{J}p_i\left\langle\phi_i\left|\sum_{\alpha=1}^{3}\sum_{\beta=1}^{3}\left(\delta_{\alpha\beta}-\frac{\kappa_\alpha\kappa_\beta}{\kappa^2}\right)\hat{S}^{+}_{0\alpha}(0)\hat{S}_{l\beta}(0)\right|\phi_i\right\rangle dt \tag{5.143}$$

We now assume elastic scatter, where

$$k = k_0 \tag{5.144}$$

and in Eq. (5.113)

$$\overline{UV} = \sum_{i=1}^{J}p_i\langle\phi_i|[\boldsymbol{\kappa}\cdot\hat{\mathbf{u}}_{0d'}(0)]\ [\boldsymbol{\kappa}\cdot\hat{\mathbf{u}}_{ld}(t)]|\phi_i\rangle = 0 \tag{5.145}$$

The time integral expresses the conservation of energy via its equivalence to the Dirac delta function, which, in the case of elastic scatter, shows that the scatter neutron energy equals the incident neutron energy, where

$$\int_{-\infty}^{\infty}dt\frac{\exp(-iwt)}{2\pi\hbar} = \delta(E - E_0) \tag{5.146}$$

Integration in Eq. (5.143), via Eq. (5.146), over the scatter neutron energy E, given a fixed incident neutron energy E_0, gives the differential cross section $(d\sigma/d\Omega)^{mag}$ for elastic neutron scatter from the Bravais lattice crystal based on the Heitler–London model, LS coupling, and dipole approximation, where

$$\left(\frac{d\sigma}{d\Omega}\right)^{mag} = N(r_0 g)^2\sum_{l=0}^{N-1}\exp(i\boldsymbol{\kappa}\cdot\hat{\mathbf{R}}_l)\left|\frac{g_0}{2}F_0(\boldsymbol{\kappa})\exp(-W_0)\right|^2$$
$$\sum_{i=1}^{J}p_i\left\langle\phi_i\left|\sum_{\alpha=1}^{3}\sum_{\beta=1}^{3}\left(\delta_{\alpha\beta}-\frac{\kappa_\alpha\kappa_\beta}{\kappa^2}\right)\hat{S}^{+}_{0\alpha}(0)\hat{S}_{l\beta}(0)\right|\phi_i\right\rangle \tag{5.147}$$

5.10 Expression of Partial Differential Cross Section by Intermediate Correlation Function

One manner in which cross section $(\mathrm{d}^2\sigma/d\Omega\,\mathrm{d}E)^{\mathrm{mag}}$ of Eq. (5.109) can be separated into its elastic and inelastic components is by the coherent intermediate function $I_{\mathrm{coh}}(\boldsymbol{\kappa},t)$, where, for the non-Bravais crystal lattice with the Heitler–London model, LS coupling, and the dipole approximation,

$$\left(\frac{\mathrm{d}^2\sigma}{\mathrm{d}\Omega\,\mathrm{d}E}\right)^{\mathrm{mag}} = (r_0 g)^2\frac{k}{k_0}\sum_{l=0}^{N-1}\sum_{l'=0}^{N-1}\sum_{d=1}^{r}\sum_{d'=1}^{r}\frac{1}{4}g_{d'}g_d F_{d'}^{*}(\boldsymbol{\kappa})F_d(\boldsymbol{\kappa})\int_{-\infty}^{\infty}\frac{\exp(-\mathrm{i}wt)}{2\pi\hbar}$$
$$\times\left[\sum_{i=1}^{J}p_i^r\langle\phi_i|\exp[-i\boldsymbol{\kappa}\cdot(\hat{\mathbf{R}}_{l'}+\hat{\mathbf{d}}_{d'}+\hat{\mathbf{u}}_{l'}d'(0))]\exp[i\boldsymbol{\kappa}\cdot(\hat{\mathbf{R}}_l+\hat{\mathbf{d}}_d+\hat{\mathbf{u}}_{ld}(t))]|\phi_i\rangle\right]$$
$$\times\left[\sum_{i=1}^{J}p_i^{sp}\left\langle\phi_i\left|\sum_{\alpha=1}^{3}\sum_{\beta=1}^{3}\left(\delta_{\alpha\beta}-\frac{\kappa_\alpha\kappa_\beta}{\kappa^2}\right)\hat{S}_{l'd'\alpha}^{+}(0)\hat{S}_{ld\beta}(t)\right|\phi_i\right\rangle\right]dt \tag{5.148}$$

The dimensionless, coherent intermediate function $I_{\mathrm{coh}}(\boldsymbol{\kappa},t)$ is the spatial Fourier transform of the coherent pair–correlation function $G_{\mathrm{coh}}(\mathbf{r},t)$. The coherent intermediate function $I_{\mathrm{coh}}(\boldsymbol{\kappa},t)$ provides the coherent component of the partial differential cross section, see Pynn (1990), Squires (1996), and Cremer (2013), where

$$I_{\mathrm{coh}}(\boldsymbol{\kappa},t)=\frac{1}{N}\sum_{i=1}^{I}p_i\sum_{r=1}^{N}\sum_{s=1}^{N}\int\mathrm{d}V_N\int\mathrm{d}V_{N-1}\cdots\int\mathrm{d}V_2\int\mathrm{d}V_1$$
$$\phi_i^*\exp[-\mathrm{i}\boldsymbol{\kappa}\cdot\mathbf{R}_r(0)]\exp[\mathrm{i}\boldsymbol{\kappa}\cdot\mathbf{R}_s(t)]\phi_i \tag{5.149}$$

The incoherent intermediate function $I_{\mathrm{inc}}(\boldsymbol{\kappa},t)$ is the spatial Fourier transform of the incoherent pair–correlation function $G_{\mathrm{inc}}(\mathbf{r},t)$. The incoherent intermediate function $I_{\mathrm{inc}}(\boldsymbol{\kappa},t)$ provides the incoherent component of the partial differential cross section, and it is defined by

$$I_{\mathrm{inc}}(\boldsymbol{\kappa},t)=\frac{1}{N}\sum_{i=1}^{I}p_i\sum_{r=1}^{N}\int\mathrm{d}V_N\int\mathrm{d}V_{N-1}\cdots\int\mathrm{d}V_2\int\mathrm{d}V_1$$
$$\phi_i^*\exp[-\mathrm{i}\boldsymbol{\kappa}\cdot\mathbf{R}_r(0)]\exp[\mathrm{i}\boldsymbol{\kappa}\cdot\mathbf{R}_r(t)]\phi_i \tag{5.150}$$

Recall from chapter 4 in Squires (1996), Pynn (1990), Cremer (2013), and section 43 in Ch. 3 of Cremer (2012b), that in the incoherent case, the correlation is between the positions of the same r^{th} nucleus $r=s$ at different times, which gives no interference effects. However, in the coherent case, the correlation is between the positions of the same nucleus $r=s$ at different times and the correlation between the positions of different r^{th} and s^{th} nuclei ($r \neq s$) at different times, which gives rise to interference effects.

Consider the correlation between the spin of the same electron or two different electrons at different times. And, separately and independently, one also considers the correlation between the position of a single nucleus or the positions of two different nuclei at different times. We will use the coherent intermediate function $I_{\text{coh}}(\boldsymbol{\kappa}, t)$, which tends to have the same limit as time $t \to \pm\infty$. One can write $I_{\text{coh}}(\boldsymbol{\kappa}, t)$ as the sum of its limiting value $I_{\text{coh}}(\boldsymbol{\kappa}, \infty)$ at $t = \pm\infty$ and a time-dependent part $I_{t,\text{coh}}(\boldsymbol{\kappa})$, which tends to be zero at $t = \pm\infty$, where

$$I_{\text{coh}}(\boldsymbol{\kappa}, t) = I_{\text{coh}}(\boldsymbol{\kappa}, \infty) + I_{t,\text{coh}}(\boldsymbol{\kappa}) \tag{5.151}$$

and

$$I_{t \to \pm\infty,\text{coh}}(\boldsymbol{\kappa}) \to 0 \tag{5.152}$$

With the above considerations, the Heitler–London, LS-coupled, dipole-approximated, partial differential cross section $(\mathrm{d}^2\sigma/d\Omega\,\mathrm{d}E)^{\text{mag}}$ of Eq. (5.148), can be expressed as the product of two independent, coherent, intermediate functions for the nuclear positions $I^R_{ll'dd'}(\boldsymbol{\kappa}, t)$ and the electron spin orientations $I^S_{ll'dd'}(\boldsymbol{\kappa}, t)$:

$$\left(\frac{\mathrm{d}^2\sigma}{\mathrm{d}\Omega\,\mathrm{d}E}\right)^{\text{mag}} = (r_0 g)^2 \frac{k}{k_0} \int_{-\infty}^{\infty} \frac{\exp(-\,\mathrm{i}wt)}{2\pi\hbar} \sum_{l=0}^{N-1} \sum_{l'=0}^{N-1} \sum_{d=1}^{r} \sum_{d'=1}^{r} \frac{1}{4} g_{d'} g_d F^*_{d'}(\boldsymbol{\kappa}) F_d(\boldsymbol{\kappa}) \left[I^R_{ll'dd'}(\boldsymbol{\kappa}, t) I^S_{ll'dd'}(\boldsymbol{\kappa}, t) \right] \mathrm{d}t \tag{5.153}$$

The intermediate function for the nuclear position is

$$I^R_{ll'dd'}(\boldsymbol{\kappa},t) = \sum_{i=1}^{J} p^r_i \langle \phi_i | \exp[-\,\mathrm{i}\boldsymbol{\kappa}\cdot(\hat{\mathbf{R}}_{l'} + \hat{\mathbf{d}}_{d'} + \hat{\mathbf{u}}_{l'}d'(0))] \exp[\mathrm{i}\boldsymbol{\kappa}\cdot(\hat{\mathbf{R}}_l + \hat{\mathbf{d}}_d + \hat{\mathbf{u}}_{ld}(t))] | \phi_i \rangle \tag{5.154}$$

The nuclear position's coherent intermediate $I^R_{ll'dd'}(\boldsymbol{\kappa}, t)$ function is the sum of the $t = \infty$ term $I^R_{ll'dd'}(\boldsymbol{\kappa}, \infty)$ and a time-varying term $I^R_{ll'dd,t}(t)$, where

$$I^R_{ll'dd'}(\boldsymbol{\kappa}, t) = I^R_{ll'dd'}(\boldsymbol{\kappa}, \infty) + I^R_{ll'dd,t}(t) \tag{5.155}$$

The intermediate function for the electron azimuthal spin or spin orientation is

$$I^S_{ll'dd'\alpha\beta}(\boldsymbol{\kappa}, t) = \sum_{i=1}^{J} p^{sp}_i \left\langle \phi_i \left| \sum_{\alpha=1}^{3} \sum_{\beta=1}^{3} \left(\delta_{\alpha\beta} - \frac{\kappa_\alpha \kappa_\beta}{\kappa^2} \right) \hat{S}^{+}_{l'd'\alpha}(0) \hat{S}_{ld\beta}(t) \right| \phi_i \right\rangle \tag{5.156}$$

The spin-orientation, coherent intermediate function $I^S_{ll'dd'\alpha\beta}(\mathbf{\kappa}, t)$ is the sum of the $t = \infty$ term $I^S_{ll'dd'\alpha\beta}(\mathbf{\kappa}, \infty)$ and a time-varying term $I^S_{ll'dd'\alpha\beta,t}(t)$, where

$$I^S_{ll'dd'\alpha\beta}(\mathbf{\kappa}, t) = I^S_{ll'dd'\alpha\beta}(\mathbf{\kappa}, \infty) + I^S_{ll'dd'\alpha\beta,t}(t) \tag{5.157}$$

With substitution of Eqs. (5.154−157) into Eq. (5.153), the partial differential cross section $(\mathrm{d}^2\sigma/d\Omega\ \mathrm{d}E)^{\mathrm{mag}}$, becomes

$$\left(\frac{\mathrm{d}^2\sigma}{\mathrm{d}\Omega\ \mathrm{d}E}\right)^{\mathrm{mag}} = (r_0 g)^2 \frac{k}{k_0} \sum_{l=0}^{N-1} \sum_{l'=0}^{N-1} \sum_{d=1}^{r} \sum_{d'=1}^{r} \frac{1}{4} g_{d'} g_d F^*_{d'}(\mathbf{\kappa}) F_d(\mathbf{\kappa})$$

$$\int_{-\infty}^{\infty} \frac{\exp(-\,\mathrm{i}wt)}{2\pi\hbar} \left[I^R_{ll'dd'}(\mathbf{\kappa}, \infty) + I^R_{ll'dd,t}(t)\right] \tag{5.158}$$

$$\left[I^S_{ll'dd'\alpha\beta}(\mathbf{\kappa}, \infty) + I^S_{ll'dd'\alpha\beta,t}(t)\right] \mathrm{d}t$$

Expansion by multiplication of terms in Eq. (5.158), leads to the following four components of the partial differential cross section:

1. Elastic nuclear scatter and elastic magnetic scatter term:

$$I^R_{ll'dd'}(\mathbf{\kappa}, \infty) I^S_{ll'dd'\alpha\beta}(\mathbf{\kappa}, \infty) \tag{5.159}$$

The neutron scatters elastically from the nuclei, and the electron spin orientations are not changed in the course of neutron scatter.

2. Inelastic nuclear scatter and elastic magnetic scatter term:

$$I^R_{ll'dd,t}(t) I^S_{ll'dd'\alpha\beta}(\mathbf{\kappa}, \infty) \tag{5.160}$$

In this case of magneto-vibrational scatter, the neutron scatters inelastically from the nuclei with an exchange of phonons; however, the orientations of the electron spins remain unchanged. The magnetic scatter of the neutron causes transitions between the discrete lattice vibration energy states, and causes the interchange of discrete packets of energy (phonons) between the neutron and the nuclei lattice. Consider a single domain of a ferromagnetic Bravais crystal, where the unpaired electron spins are aligned in the z-direction, so that

$$I^S_{ll'dd'\alpha\beta}(\mathbf{\kappa}, \infty) = \begin{matrix} \langle S_z \rangle^2 & \alpha = \beta = z \\ 0 & \text{otherwise} \end{matrix} \tag{5.161}$$

In this case, the magneto-vibrational cross section is the same as the cross section for coherent inelastic nuclear scatter, where the factor $\sigma_{\text{coh}}/4\pi$ is replaced by the magnetic scatter counterpart, so that

$$\frac{\sigma_{\text{coh}}}{4\pi} \rightarrow \frac{1}{4}(r_0 g)^2 g_0^2 F_0^2 \left(1 - \frac{\kappa_z^2}{\kappa^2}\right) \cdot \left[\sum_{i=1}^{J} p_i^{sp} \langle \phi_i | S_z^+ S_z | \phi_i \rangle\right] \tag{5.162}$$

Eq. (5.162) is obtained by inspecting Eq. (5.148) for the partial differential cross section, prior to plugging in the intermediate functions. If an applied magnetic field is large enough to align the domains in the direction of the neutron wave transition vector $\boldsymbol{\kappa}$, then

$$\kappa = \kappa_z \tag{5.163}$$

Hence,

$$1 - \frac{\kappa_z^2}{\kappa^2} = 0 \tag{5.164}$$

In this case, the magneto-vibrational cross section is zero. Nuclear scatter is not affected by magnetic fields, and thus the contributions of inelastic scatter from the nuclei and magnetic scatter can be measured and identified separately.

3. Inelastic magnetic scatter and elastic nuclear scatter term:

$$I_{ll'dd'}^{R}(\boldsymbol{\kappa}, \infty) I_{ll'dd'\alpha\beta,t}^{S}(t) \tag{5.165}$$

The neutron is elastically scattered from the nuclei. However, inelastic scatter of the neutrons occurs with respect to the spin orientations of unpaired atomic electrons, where one or more electron spins are flipped during the neutron inelastic scatter process. In a single electron spin-flip, the energy transition between the electron spin-up state with energy $\hbar w_q/2$ and the electron spin-down state with energy $-\hbar w_q/2$, produces absorbed or emitted net energy $\pm\hbar w_q$, which is defined as 1 magnon of energy. There arise forward and backward waves of deviated electron spin precessions, which leads to the normal modes of spin wave frequencies and their associated energies in the spin lattice. This is very similar to the forward- and backward-traveling nuclei oscillation waves, which are formed from nuclei vibrations about their equilibrium positions.

4. Inelastic nuclear and magnetic scatter term:

$$I_{ll'dd,t}^{R}(t) I_{ll'dd'\alpha\beta,t}^{S}(t) \tag{5.166}$$

In this case, the neutron exchanges phonons of vibration wave energy with the lattice atoms, which leads to either increased or decreased atom vibration. In addition, the magnetic interaction of the neutron magnetic dipole moment, and the unpaired electron magnetic dipole moments, causes electron spin-flips. The electron spin-flips lead to an exchange of magnons of spin-flip energy, which results in either increased or decreased amplitudes of electron spin precession waves. Both the lattice and neutron temperature affect the nuclear and magnetic scatter.

References

Cremer, J. T. (2012b). In P. W. Hawkes (Ed.), Neutron and X-ray microscopy—Part 2. *Advances in imaging and electron physics* (Vol. 173). London, UK: Academic Press, an imprint of Elsevier Press.

Cremer, J. T. (2013). In P. W. Hawkes (Ed.), *Advances in imaging and electron physics* (Vol. 175). London, UK: Academic Press, an imprint of Elsevier Press.

Liboff, R. L. (2003). *Introductory quantum mechanics*. Reading, MA: Addison-Wesley.

Merzbacher, E. (1970). *Quantum mechanics*. New York: John Wiley & Sons.

Pynn, R. (1990). *Neutron scattering: A primer*. Los Alamos Science website: http://www.mrl.ucsb/www.mrl.ucsb.edu/~pynn.

Squires, G. L. (1996). *Introduction to the theory of thermal neutron scattering*. New York, NY: Dover.

6 Magnetic Scatter of Neutrons in Paramagnetic Materials

6.1 General Expression for Coherent, Elastic Differential Cross Section for Paramagnetic Material in an Applied Magnetic Field

Following Squires (1996), we now determine the differential cross section for coherent, elastic scatter of neutrons from a paramagnetic Bravais lattice crystal with localized electrons based on the Heitler–London model, LS coupling, and dipole approximation. First considered is the case of no applied magnetic field, where the net, effective electron spin and electron orbital angular momentum, and resulting net magnetic dipole moment of the atoms, are randomly arrayed, which yield zero internal magnetic field. Without an internal and external magnetic field, a change in the net magnetic dipole moment orientation of an atom does not change the energy of the atom or scattering system of paramagnetic crystal atoms.

As a result, the effective spin operator S_l of the dth atom of the N atoms (and unit cells) of the Bravais crystal and the Hamiltonian operator are independent and commuting. Since the Hamiltonian operator represents the time-independent, constant, and total energy of the scattering system, the effective spin operators are time independent. The paramagnetic crystal can be comprised of metallic bonding as in tungsten, ionic bond compounds as in manganese fluoride MnF_2, or a covalent bond compound as in myoglobin. The covalent bond compound myoglobin is an oxygen-binding protein in vertebrate muscle tissue, which is readily crystallized. Myoglobin contains a paramagnetic porphyrin (heme) group with iron, which is held in a coordinate covalent bond at the protein center.

Now consider a very large laboratory magnetic field B of 5 T, which is applied to the paramagnetic crystal in the negative z-direction. The scatter of a neutron from an atom causes the net effective spin of an atom to flip, and yields a change in the energy ΔE of the scattering system. With Bohr magnetron μ_b, the change in energy ΔE of the scattering system is due to the net magnetic dipole moment, which is associated with the atom's net electron spin and electron orbital angular momentum. A single electron spin flip in a 5 tesla (T) applied magnetic field,

Neutron and X-ray Optics. DOI: http://dx.doi.org/10.1016/B978-0-12-407164-3.00005-X

which causes the atom (ion) spin to flip, causes an energy change ΔE of approximately

$$\Delta E = -\mu_b B = 2.8 \times 10^{-4}\ \text{eV} \ll 2.5 \times 10^{-2}\ \text{eV} \quad B = 5.0\ \text{T} \tag{6.1}$$

The Bohr magnetron for electron charge e and rest mass m_e is

$$\mu_b = \frac{e\hbar}{2m_e} = 5.8 \times 10^{-5}\ \text{eV}/T \tag{6.2}$$

At room temperature of 300 K, the average energy of the thermal neutron is 0.025 eV, which is much larger than ΔE. For a sufficiently "small" magnetic field, the change in energy ΔE of the scattering system is much less than the average 0.025 eV thermal energy of the incident neutron. Thus, for a sufficiently small magnetic field, the energy change ΔE is ignored, and the N atomic effective spin operators are independent of time.

The partial differential cross section expression of Eq (5.147) for elastic neutron scatter from the magnetic dipole moment of the atomic electron spin and orbital angular momentum in the Bravais lattice crystal, which is based on the Heitler–London model, LS coupling, and dipole approximation, is repeated below in Eq. (6.3)

$$\begin{aligned}\left(\frac{d\sigma}{d\Omega}\right)^{\text{mag}} &= N(r_0 g)^2 \sum_{l=0}^{N-1} \exp(i\boldsymbol{\kappa}\cdot\hat{\mathbf{R}}_l)\left|\frac{g_0}{2}F_0(\boldsymbol{\kappa})\exp(-W_0)\right|^2 \\ &\sum_{i=1}^{J} p_i \left\langle \phi_i \left| \sum_{\alpha=1}^{3}\sum_{\beta=1}^{3}\left(\delta_{\alpha\beta} - \frac{\kappa_\alpha \kappa_\beta}{\kappa^2}\right)\hat{S}^{+}_{0\alpha}(0)\hat{S}_{l\beta}(0)\right| \phi_i \right\rangle\end{aligned} \tag{6.3}$$

The effective electron spin is comprised of the net electron spins and electron orbital angular momentum of each dth atom (ions) in the x, y, and z directions. The x, y, and z directions are addressed (indexed) by integers α or β, where

$$\alpha, \beta = x, y, z \tag{6.4}$$

The Bravais lattice crystal considered here has only one atom per unit cell. In a paramagnetic Bravais crystal with no applied magnetic field or sufficiently small z-directed magnetic field, there is no correlation between the effective electron spin from one unit cell to the next. For all crystal atoms—that is, for all l with $\alpha \neq \beta$—;

$$\sum_{i=1}^{J} p_i \left\langle \phi_i \left| \sum_{\alpha=1}^{3}\sum_{\beta=1}^{3}\left(\delta_{\alpha\beta} - \frac{\kappa_\alpha \kappa_\beta}{\kappa^2}\right)\hat{S}^{+}_{0\alpha}(0)\hat{S}_{l\beta}(0)\right| \phi_i \right\rangle = 0 \quad \text{all } l \text{ and } \alpha \neq \beta \tag{6.5}$$

That is, the thermally averaged mean component of the effective electron spin in the x, y, and z directions is zero for each unit cell of the crystal, as well as for the entire crystal, because the spins are subject to random thermal effects.

For the case of an applied magnetic field $\mathbf{B} = B\hat{\mathbf{z}}$ in the z-direction, which is sufficiently small, but also sufficiently large to overcome the random thermal effects, the x and y directions are equivalent. For the z-directed magnetic field, the thermally averaged mean-squared effective electron spin component in the x-direction $\overline{S}_x$ equals the mean-squared effective electron spin component in the y-direction $\overline{S}_y$. Both are nonzero, where

$$\overline{(S_x)^2} = \overline{(S_y)^2} = \frac{1}{2}\left[[s(s+1) - \overline{(S_z)^2}\right] \quad \mathbf{B} = B\hat{\mathbf{z}} \tag{6.6}$$

In an applied z-directed magnetic field B_z, the squared mean z-directed effective electron spin component is correlated from unit cell to unit cell ($l \neq 0$) due to the applied magnetic field, where

$$\sum_{i=1}^{J} p_i \left\langle \phi_i \middle| \hat{S}_{0z}^{+}(0)\hat{S}_{lz}(0) \middle| \phi_i \right\rangle = (\overline{S}_z)^2 \quad l \neq 0 \text{ and } \alpha = \beta = z \tag{6.7}$$

For the entire crystal, relative to a single, reference unit cell, which is chosen as the unit cell at the crystal center $l = 0$, the mean-squared, z-component of the effective electron spin is

$$\sum_{i=1}^{J} p_i \left\langle \phi_i \middle| \hat{S}_{0z}^{+}(0)S_{lz}(0) \middle| \phi_i \right\rangle = \overline{(S_z)^2} \quad l = 0 \text{ and } \alpha = \beta = z \tag{6.8}$$

For the entire crystal, the mean-squared x and y components of the effective spin arise due to a z-directed magnetic field, which imposes a net mean-squared z component of the effective electron spin for the entire crystal, where

$$\sum_{i=1}^{J} p_i \langle \phi_i | \hat{S}_{0x}^{+}(0)\hat{S}_{0x}(0) | \phi_i \rangle = \overline{(S_x)^2} = \frac{1}{2}\left[s(s+1) - \overline{(S_z)^2}\right] \quad l = 0 \text{ and } \alpha = \beta = x \tag{6.9}$$

$$\sum_{i=1}^{J} p_i \langle \phi_i | \hat{S}_{0y}^{+}(0)\hat{S}_{0y}(0) | \phi_i \rangle = \overline{(S_y)^2} = \frac{1}{2}\left[s(s+1) - \overline{(S_z)^2}\right] \quad l = 0 \text{ and } \alpha = \beta = y \tag{6.10}$$

From one unit cell to the next unit cell, the x-directed components of the effective electron spin of the atoms are uncorrelated, and the y-directed effective electron spin components are uncorrelated from one unit cell to the next, with or without a z-directed applied magnetic field, where

$$\sum_{i=1}^{J} p_i \left\langle \phi_i \left| \sum_{\alpha=1}^{3} \sum_{\beta=1}^{3} \left(\delta_{\alpha\beta} - \frac{\kappa_\alpha \kappa_\beta}{\kappa^2} \right) \hat{S}^{+}_{0\alpha}(0) \hat{S}_{l\beta}(0) \right| \phi_i \right\rangle = 0 \quad l \neq 0 \; \alpha = \beta \neq z \tag{6.11}$$

One replaces the direct lattice vector operator $\hat{\mathbf{R}}_l$ by the direct lattice vector $\mathbf{R}_l$, and with substitution of Eqs. (6.5–6.11) into Eq. (6.3), the differential cross section $(d\sigma/d\Omega)^{mag}$ becomes

$$\left(\frac{d\sigma}{d\Omega}\right)^{mag} = N(r_0 g)^2 \left| \frac{g_0}{2} F_0(\boldsymbol{\kappa}) \exp(-W_0) \right|^2 \times$$
$$\left\{ \begin{array}{l} \sum_{l=0}^{N-1} \exp(i\boldsymbol{\kappa}\cdot\mathbf{R}_l) \left(1 - \frac{\kappa_z^2}{\kappa^2}\right) (\bar{S}_z)^2 - \left(1 - \frac{\kappa_z^2}{\kappa^2}\right) (\bar{S}_z)^2 + \frac{1}{2}\left(1 - \frac{\kappa_x^2}{\kappa^2}\right) \left[s(s+1) - \overline{(S_z)^2} \right] \\ \frac{1}{2}\left(1 - \frac{\kappa_y^2}{\kappa^2}\right) \left[s(s+1) - \overline{(S_z)^2} \right] + \left(1 - \frac{\kappa_y^2}{\kappa^2}\right) \overline{(S_z)^2} \end{array} \right\} \tag{6.12}$$

For $l \neq 0$ and $\alpha = \beta = z$, the squared mean of the z-directed effective spin is

$$(\bar{S}_z)^2 \tag{6.13}$$

For $l = 0$, the mean-squared z-directed spin component $\overline{(S_z)^2}$ equals the squared mean of the z-directed effective spin $(\bar{S}_z)^2$, where

$$\overline{(S_z)^2} = (\bar{S}_z)^2 \tag{6.14}$$

Next, the exponential expression in Eq. (6.12), whose argument contains with the direct lattice vectors $\mathbf{R}_l$ is replaced by the Dirac delta function expression with the reciprocal lattice $\boldsymbol{\tau}_\tau$ and neutron scatter vector $\boldsymbol{\kappa}$. To facilitate this, one maintains the summation from $l = 0$ to $N - 1$ by subtracting the $l = 0$ term. Note that

$$\frac{\kappa_z^2}{\kappa^2} = 1 - \frac{\kappa_x^2}{\kappa^2} - \frac{\kappa_y^2}{\kappa^2} \tag{6.15}$$

The differential cross section expression of Eq. (6.12) is then simplified to

$$\left(\frac{d\sigma}{d\Omega}\right)^{mag} = N(r_0 g)^2 \left|\frac{g_0}{2} F_0(\kappa)\exp(-W_0)\right|^2 \times \left\{ \begin{array}{l} \sum_{l=0}^{N-1} \exp(i\kappa \cdot \mathbf{R}_l) \left(1 - \frac{\kappa_z^2}{\kappa^2}\right)(\bar{S}_z)^2 + \frac{1}{2}\left(1 + \frac{\kappa_z^2}{\kappa^2}\right)\left[s(s+1) - \overline{(S_z)^2}\right] \\ + \left(1 - \frac{\kappa_z^2}{\kappa^2}\right)\left[\overline{(S_z)^2} - (\bar{S}_z)^2\right] \end{array} \right\} \tag{6.16}$$

The relation between the direct lattice vector $\mathbf{R}_l$ and reciprocal lattice vectors $\boldsymbol{\tau}_l$, given unit cell volume v_0, is

$$\sum_{l=0}^{N-1} \exp(i\kappa \cdot \mathbf{R}_l) = \frac{(2\pi)^3}{v_0} \sum_{\tau=0}^{N-1} \delta(\kappa - \boldsymbol{\tau}_\tau) \tag{6.17}$$

The differential cross section $(d\sigma/d\Omega)^{mag}$ for elastic neutron scatter from the atomic electron magnetic dipole moment in a Bravais lattice paramagnetic crystal with an applied z-directed magnetic field B_z, which is based on the Heitler–London model, LS coupling, and dipole approximation, becomes

$$\left(\frac{d\sigma}{d\Omega}\right)^{mag} = N(r_0 g)^2 \left|\frac{g_0}{2} F_0(\kappa)\exp(-W_0)\right|^2 \times \left\{ \begin{array}{l} \frac{(2\pi)^3}{v_0} \sum_{\tau=0}^{N-1} \delta(\kappa - \boldsymbol{\tau}_\tau) \left(1 - \frac{\kappa_z^2}{\kappa^2}\right)(\bar{S}_z)^2 + \frac{\kappa_z^2}{\kappa^2}\left[\frac{1}{2}s(s+1) - \frac{3}{2}\overline{(S_z)^2} + (\bar{S}_z)^2\right] \\ + \frac{1}{2}s(s+1) + \frac{1}{2}\overline{(S_z)^2} - (\bar{S}_z)^2 \end{array} \right\} \tag{6.18}$$

6.2 Coherent, Elastic Differential Cross Section for Paramagnetic Material in an Applied Magnetic Field Expressed by the Total Spin Quantum Number for a Paramagnetic Atom

To calculate the coherent, elastic differential cross section $(d\sigma/d\Omega)^{mag}$ of the paramagnetic material in an applied magnetic field, one needs to evaluate the average

z-directed component of the effective spin $\overline{S}_z$ and the thermally averaged, mean-squared, z-component of the effective spin $\overline{(S_z)^2}$ in the paramagnetic crystal with an applied, z-directed magnetic field B_z. Prior to neutron scatter, the set of scattering atoms is described by the ith initial state wave function ϕ_i.

Each paramagnetic atom with unpaired electrons has a total or net or effective spin vector **S** with total spin quantum number s. This net spin is the sum of the net electron spin and the electron orbital angular momentum. The total spin vector **S** has a z-directed azimuthal or magnetic component S_z, which has magnetic quantum number m. The range of the magnetic quantum number m is bounded by the total spin quantum number s, where

$$m = s, s-1, \ldots, -s+1, \ -s \tag{6.19}$$

The eigenequation for the z-component of the effective spin for an atom is

$$S_z\langle\ |\phi_i = m\hbar|\phi_i\rangle \tag{6.20}$$

The mean z-component of all the effective spins prior to neutron scatter is obtained by the thermal averaging over all the J initial scattering system states ϕ_i prior to neutron scatter, where

$$\overline{S}_z = \sum_{i=1}^{J} p_i\langle\phi_i|S_z|\phi_i\rangle = \sum_{i=1}^{J} p_i m\hbar\langle\phi_i|\phi_i\rangle = \hbar\sum_{i=1}^{J} p_i m \tag{6.21}$$

and

$$\overline{(S_z)^2} = \sum_{i=1}^{J} p_i\langle\phi_i|S_z^2|\phi_i\rangle = \sum_{i=1}^{J} p_i(m\hbar)^2\langle\phi_i|\phi_i\rangle = \hbar^2\sum_{i=1}^{J} p_i m^2 \tag{6.22}$$

The effective spin of a paramagnetic atom causes the atom to acquire magnetic potential energy E_m for the z-directed, applied magnetic field B, where

$$E_m = mg_0\mu_b B \tag{6.23}$$

In the ith initial state, the probability p_m for the effective spin of the atom to be in the mth spin state depends on the atom magnetic potential energy E_m and temperature T, where

$$p_m = \frac{1}{Z}\exp\left(\frac{E_m}{k_B T}\right) = \frac{\exp(mr)}{Z} \tag{6.24}$$

With Landé splitting factor g_0, the ratio r is defined by the magnetic potential energy $g_0\mu_b B$ for a single atom with total spin quantum number $s = 1$ divided by the thermal energy $k_B T$:

$$r = \frac{g_0 \mu_b B}{k_B T} \tag{6.25}$$

The partition function Z is

$$Z = \sum_{m=-s}^{s} \exp\left(\frac{E_m}{k_B T}\right) \tag{6.26}$$

where

$$Z = \sum_{m=-s}^{s} \exp(mr) = \frac{\sinh\left[r(s + (1/2))\right]}{\sinh((1/2)r)} \tag{6.27}$$

The thermally averaged, mean z-component of the effective spin $\overline{S}_z$ and the mean-squared z-component of the effective spin $\overline{(S_z)^2}$ are then

$$\overline{S}_z = \sum_{m=-s}^{s} p_m m = \frac{\sum_{m=-s}^{s} m \exp(mr)}{Z} = \frac{1}{Z}\frac{\partial Z}{\partial r} \tag{6.28}$$

and

$$\overline{(S_z)^2} = \sum_{m=-s}^{s} m^2 p_m = \frac{\sum_{m=-s}^{s} m^2 \exp(mr)}{Z} = \frac{1}{Z}\frac{\partial^2 Z}{\partial r^2} \tag{6.29}$$

Evaluation of $\partial Z/\partial r$ and $\partial^2 Z/\partial r^2$, using Eq. (6.27), and substitution into Eqs. (6.28–6.29), gives

$$\overline{S}_z = \left(s + \frac{1}{2}\right)\coth\left[r\left(s + \frac{1}{2}\right)\right] - \frac{1}{2}\coth\left(\frac{r}{2}\right) \tag{6.30}$$

and

$$\overline{(S_z)^2} = s(s+1) - \overline{S}_z \coth\left(\frac{r}{2}\right) \tag{6.31}$$

Substitution of $\overline{S}_z$ in $\overline{(S_z)^2}$ gives

$$\overline{(S_z)^2} = s(s+1) - \left\{\left(s + \frac{1}{2}\right)\coth\left[r\left(s + \frac{1}{2}\right)\right] - \frac{1}{2}\coth\left(\frac{r}{2}\right)\right\}\coth\left(\frac{r}{2}\right) \tag{6.32}$$

With substitution of Eqs. (6.30) and (6.32) into Eq. (6.18), the differential cross section $(d\sigma/d\Omega)^{mag}$ for elastic neutron scatter from a Bravais lattice paramagnetic crystal, which has applied, z-directed magnetic field B_z, is given by

$$\left(\frac{d\sigma}{d\Omega}\right)^{mag} = N(r_0 g)^2 \left|\frac{g_0}{2} F_0(\kappa)\exp(-W_0)\right|^2$$
$$\times \left\{ \begin{array}{l} \dfrac{(2\pi)^3}{v_0}\displaystyle\sum_{\tau=0}^{N-1}\delta(\kappa-\tau_\tau)\left(1-\frac{\kappa_z^2}{\kappa^2}\right)(\overline{S}_z)^2 + \frac{\kappa_z^2}{\kappa^2}\left[\frac{1}{2}s(s+1) - \frac{3}{2}\overline{(S_z)^2} + (\overline{S}_z)^2\right] \\ + \dfrac{1}{2}s(s+1) + \dfrac{1}{2}\overline{(S_z)^2} - (\overline{S}_z)^2 \end{array} \right\} \tag{6.33}$$

where

$$\overline{S}_z = \left(s+\frac{1}{2}\right)\coth\left[r\left(s+\frac{1}{2}\right)\right] - \frac{1}{2}\coth\left(\frac{r}{2}\right) \tag{6.34}$$

and

$$\overline{(S_z)^2} = s(s+1) - \overline{S}_z \coth\left(\frac{r}{2}\right) \tag{6.35}$$

The simplifying assumption of quenched orbital angular momentum for the atomic electrons in the paramagnetic lattice, is implemented by setting $g_0 = 1$ in Eq. (6.25) for ratio r. Then the ratio r is the magnetic potential for an atom, with net spin quantum number $s = 1$, divided by the thermal energy of the atom, where

$$r = \frac{\mu_b B}{kT} \tag{6.36}$$

6.3 Coherent, Differential Cross Section for Elastic Neutron Scatter in Paramagnetic Material—With an Applied Magnetic Field at Low and High Temperatures

We now calculate the coherent, differential cross section for elastic neutron scatter in paramagnetic materials with an applied magnetic field for the cases of low and high temperatures. If the temperature goes to zero, where $T \to 0$ and $r \to \infty$, then for applied z-directed magnetic field B_z,

$$\coth\left(\frac{r}{2}\right) \to 1 \quad \text{and} \quad \coth\left[r\left(s+\frac{1}{2}\right)\right] \to 1 \quad T \to 0 \text{ and } r \to \infty \tag{6.37}$$

In the case of zero temperature, one expects all the paramagnetic atoms to be in the same state, where all the atoms are vibration free, including their atomic electrons. With alignment, all the atoms have identical z-component effective spins and z-component magnetic dipole moments. The quantum number for all the atoms is identical, with a minimum spin quantum number (spin-down) that corresponds to the lowest energy state for the combined electron spin and orbital angular momentum, where

$$m = -s \tag{6.38}$$

Hence,

$$\overline{S}_z = s \quad T \rightarrow 0 \text{ and } r \rightarrow \infty \tag{6.39}$$

and

$$\overline{(S_z)^2} = s^2 \quad T \rightarrow 0 \text{ and } r \rightarrow \infty \tag{6.40}$$

Thus, for very low or zero temperatures $T \rightarrow 0$, with substitution of Eqs. (6.39–6.40) into Eq. (6.33) the differential cross section $(d\sigma/d\Omega)^{mag}$ for elastic neutron scatter from the Bravais lattice paramagnetic crystal, with an applied, z-directed magnetic field B_z, which is based on the Heitler–London model, LS coupling, and dipole approximation, becomes

$$\left(\frac{d\sigma}{d\Omega}\right)^{mag} = N(r_0 g)^2 \left|\frac{g_0}{2} F_0(\boldsymbol{\kappa}) \exp(-W_0)\right|^2 \left\{\frac{(2\pi)^3}{v_0} \sum_{\tau=0}^{N-1} \delta(\boldsymbol{\kappa} - \boldsymbol{\tau}_\tau)\left(1 - \frac{\kappa_z^2}{\kappa^2}\right)s^2 + \frac{s^2}{2}\left(1 + \frac{\kappa_z^2}{\kappa^2}\right)\right\} \tag{6.41}$$

Now we consider the paramagnetic material with applied z-directed magnetic field B_z at high temperatures. If the temperature becomes large, where $T \rightarrow \infty$ and $r \rightarrow 0$, then

$$\coth\left(\frac{r}{2}\right) \cong \frac{2}{r} + \frac{r}{6} \quad \text{for } T \rightarrow \infty \text{ and } r \rightarrow 0 \tag{6.42}$$

and

$$\coth\left[r\left(s + \frac{1}{2}\right)\right] \cong \frac{1}{r}\left(s + \frac{1}{2}\right)^{-1} + \frac{r}{3}\left(s + \frac{1}{2}\right) \quad \text{for } T \rightarrow \infty \text{ and } r \rightarrow 0 \tag{6.43}$$

Hence,

$$\overline{S_z} = \frac{1}{3}s(s+1)r = \frac{1}{3}s(s+1)\frac{g_0\mu_b B}{kT} \quad \text{for } T \to \infty \text{ and } r \to 0 \tag{6.44}$$

and

$$\overline{(S_z)^2} = \frac{1}{3}s(s+1) \quad \text{for } T \to \infty \text{ and } r \to 0 \tag{6.45}$$

6.4 Coherent, Differential Cross Section for Elastic Neutron Scatter in Paramagnetic Material—No Applied Magnetic Field

Next considered is the paramagnetic material with no applied magnetic field ($\mathbf{B}=0$), where we determine the coherent, elastic differential cross section for neutrons. One starts with Eq. (6.3) for the general expression for the coherent, elastic differential cross section:

$$\begin{aligned}\left(\frac{d\sigma}{d\Omega}\right)^{\text{mag}} &= N(r_0 g)^2 \sum_{l=0}^{N-1} \exp(i\boldsymbol{\kappa}\cdot\hat{\mathbf{R}}_l) \left|\frac{g_0}{2}F_0(\boldsymbol{\kappa})\exp(-W_0)\right|^2 \\ &\sum_{i=1}^{J} p_i \left\langle \phi_i \left| \sum_{\alpha=1}^{3}\sum_{\beta=1}^{3}\left(\delta_{\alpha\beta} - \frac{\kappa_\alpha\kappa_\beta}{\kappa^2}\right)\hat{S}_{0\alpha}^{+}(0)\hat{S}_{l\beta}(0)\right| \phi_i \right\rangle\end{aligned} \tag{6.46}$$

In the case of $B_z = 0$, with no magnetic field, there is no correlation between the x, y, and z components of the effective spins, which are caused by net atomic electron spin and orbital angular momentum from one unit cell to the next, so that

$$\sum_{i=1}^{J} p_i \left\langle \phi_i \left| \sum_{\alpha=1}^{3}\sum_{\beta=1}^{3}\left(\delta_{\alpha\beta} - \frac{\kappa_\alpha\kappa_\beta}{\kappa^2}\right)\hat{S}_{0\alpha}^{+}(0)\hat{S}_{l\beta}(0)\right| \phi_i \right\rangle = 0 \quad l \neq 0 \tag{6.47}$$

Relative to the reference unit cell of the crystal center at $l = 0$, in each direction α, the average component of the effective electron spin for the entire crystal is

$$\begin{aligned}&\sum_{i=1}^{J} p_i \left\langle \phi_i \left| \sum_{\alpha=1}^{3}\sum_{\beta=1}^{3}\left(\delta_{\alpha\beta} - \frac{\kappa_\alpha\kappa_\beta}{\kappa^2}\right)\hat{S}_{0a}^{+}(0)\hat{S}_{l\beta}(0)\right| \phi_i \right\rangle \\ &= \sum_{\alpha=1}^{3}\left(1 - \frac{\kappa_\alpha^2}{\kappa^2}\right)\sum_{i=1}^{J} p_i \langle\phi_i|\hat{S}_{0\alpha}^{+}(0)\hat{S}_{0\alpha}(0)|\phi_i\rangle \quad l = 0 \text{ and } \alpha = \beta\end{aligned} \tag{6.48}$$

With only three possible initial states I=3, which correspond to spins of equal probability $p_i = 1/3$ in the x, y, and z directions, one obtains

$$\sum_{i=1}^{J} p_i\langle\phi_i|\hat{S}_{0\alpha}^{+}(0)\hat{S}_{0\alpha}(0)|\phi_i\rangle = \frac{1}{3}\langle\phi_i|S_x^2 + S_y^2 + S_z^2|\phi_i\rangle = \frac{1}{3}\langle\phi_i|\mathbf{S}\cdot\mathbf{S}|\phi_i\rangle \tag{6.49}$$

Hence,

$$\sum_{i=1}^{J} p_i\langle\phi_i|\hat{S}_{0\alpha}^{+}(0)\hat{S}_{0\alpha}(0)|\phi_i\rangle = \frac{1}{3}s(s+1) \tag{6.50}$$

where

$$\sum_{\alpha=1}^{3}\left(1-\frac{\kappa_\alpha^2}{\kappa^2}\right) = 1-\frac{\kappa_x^2}{\kappa^2}+1-\frac{\kappa_y^2}{\kappa^2}+1-\frac{\kappa_z^2}{\kappa^2} = 2 \tag{6.51}$$

With application of Eqs. (6.47–6.51) to Eq. (6.46), the differential cross section $(\mathrm{d}\sigma/\mathrm{d}\Omega)^{\mathrm{mag}}$ for elastic neutron scatter from the Bravais lattice paramagnetic crystal, which has no applied magnetic field, but does have a net magnetic dipole moment (due to the net electron spin and orbital angular momentum), is then

$$\left(\frac{\mathrm{d}\sigma}{\mathrm{d}\Omega}\right)^{\mathrm{mag}} = \frac{2}{3}N(r_0 g)^2\left|\frac{g_0}{2}F_0(\boldsymbol{\kappa})\exp(-W_0)\right|^2 s(s+1) \tag{6.52}$$

For zero applied magnetic field, the scatter of neutrons is spread in all directions. When a magnetic field B_Z is applied, there is diffuse scatter in all directions, but in addition, there are Bragg scatter peaks of neutron scatter intensity, which is proportional to $(\overline{S}_z)^2$.

6.5 Coherent, Elastic Differential Cross Section for Scatter of Neutron Spin States from Localized Electrons in Paramagnetic Materials—No Applied Magnetic Field

In the previous sections of this chapter, we did not consider the transition of the neutron spin states in the scatter from paramagnetic materials, with or without an applied magnetic field. We now derive the coherent, elastic differential cross section for paramagnetic materials for the case of no applied magnetic field $\mathbf{B} = 0$, where the neutron spin states and their transition upon magnetic scatter are considered. The LS coupling and the Heitler model are assumed in which the unpaired electrons are localized to their atom. Nuclear scatter of neutrons is not included in this derivation of neutron magnetic scatter.

In the process of deriving the coherent, elastic differential cross section in this section, we also obtain the magnetic scatter length b_M of a neutron by the net effective electron magnetic dipole moment of a paramagnetic atom. The magnetic dipole moment of the paramagnetic atom arises from its net electron spin and orbital angular momentum. With the magnetic scatter length for each paramagnetic atom, one can determine the coherent, elastic differential cross section, and the refractive index for magnetic scatter of the two neutron spin states.

The derivation of the magnetic scattering length starts with Eq. (3.2) in Ch. 3 for the partial differential cross section $(\mathrm{d}^2\sigma/d\Omega\,\mathrm{d}E)^{\mathrm{mag}}_{i\to m}$ for the scattering system (atoms and neutron) transition from the initial, prescatter state i to the final, postscatter state m, where

$$\left(\frac{d^2\sigma}{d\Omega\,dE}\right)^{\mathrm{mag}}_{i,0s\to m,s} = \left(\frac{m_n}{2\pi\hbar^2}\right)^2 \frac{k}{k_0}\left|\langle\phi_m\psi_s\psi|\sum_{e=1}^{T}\hat{U}_e|\psi_0\psi_{0s}\phi_i\rangle\right|^2 \delta(E_0 - E + E_m - E_i) \tag{6.53}$$

Recall from section 3.9 of Ch 3, that

$$\mu_c\hat{\mathbf{S}}\cdot\hat{\mathbf{Q}}_\perp(\boldsymbol{\kappa}) = \sum_{e=1}^{T}\langle\hat{U}_e(0)\rangle = \sum_{e=1}^{T}\langle\hat{U}_e\rangle = \sum_{e=1}^{T}\int_{-\infty}^{\infty}\mathrm{d}z\int_{\infty}^{\infty}\mathrm{d}y\int_{\infty}^{\infty}\mathrm{d}x\psi_0^*(\mathbf{r})\hat{U}_e\psi(\mathbf{r}), \tag{6.54}$$

where

$$\mu_c = -2\mu_0 g\mu_n\mu_\mathrm{b} \tag{6.55}$$

The constant μ_c is expanded and simplified, where $g = -1.91$ is the neutron Landé g factor, and $r_0 g$ is 5.38 fm, so that

$$\mu_c\frac{m_p}{2\pi\hbar^2} = -2\mu_0 g\mu_n\mu_\mathrm{b}\frac{m_p}{2\pi\hbar^2} = -2\mu_0 g\frac{e\hbar}{2m_p}\frac{e\hbar}{2m_e}\frac{m_p}{2\pi\hbar^2} = -r_0 g = 5.38\times 10^{-15}\ \mathrm{m} \tag{6.56}$$

With substitution of Eq. (6.56) into Eq. (6.53), the partial differential cross section for magnetic neutron scatter, where the scattering system transitions from initial state i to final state m, is then

$$\left(\frac{d^2\sigma}{d\Omega\,dE}\right)^{\mathrm{mag}}_{i,0s\to m,s} = (r_0 g)^2\frac{k}{k_0}|\langle\phi_i\psi_{0s}|\hat{\mathbf{S}}_n\cdot\hat{\mathbf{Q}}_\perp(\boldsymbol{\kappa})|\psi_s\phi_m\rangle|^2\delta(E_0 - E + E_m - E_i) \tag{6.57}$$

Referring to Eq. (6.53), recall neutron scatter lengths, and see Squires (1996), and Cremer (2012b, 2013). The scatter length $\hat{b}_M$ for the magnetic scatter of the neutrons from the unpaired electrons of the nonmagnetized paramagnet, is expressed by

$$\hat{b}_M = \hat{\mathbf{S}}_n \cdot \hat{\mathbf{Q}}_\perp(\boldsymbol{\kappa}) = \left\langle \psi \middle| \sum_{e=1}^{T} \hat{U}_e \middle| \psi_0 \right\rangle \tag{6.58}$$

Recall that the scattered neutron vector $\boldsymbol{\kappa} = \mathbf{k}_0 - \mathbf{k}$ is the change of the neutron wave vector from initial wave vector $\mathbf{k}_0$ before scatter to final wave vector $\mathbf{k}$ upon scatter by the unpaired electrons of the lattice nuclei. The electron magnetic dipole moment operator $\hat{\mathbf{Q}}(\boldsymbol{\kappa})$ in the κ-space has perpendicular component $\hat{\mathbf{Q}}_\perp(\boldsymbol{\kappa})$ and parallel component $\hat{\mathbf{Q}}_\parallel(\boldsymbol{\kappa})$. The perpendicular component vector $\hat{\mathbf{Q}}_\perp(\boldsymbol{\kappa})$ is perpendicular to the neutron transition wave vector $\boldsymbol{\kappa}$, and the parallel component $\hat{\mathbf{Q}}_{|}|(\boldsymbol{\kappa})$ is parallel to $\boldsymbol{\kappa}$, where

$$\hat{\mathbf{Q}}_\parallel(\boldsymbol{\kappa}) \parallel \boldsymbol{\kappa} \tag{6.59}$$

and

$$\hat{\mathbf{Q}}_\perp(\boldsymbol{\kappa}) \perp \boldsymbol{\kappa} \tag{6.60}$$

Now we consider the effect of neutron spin and the spin of the nuclei of the paramagnetic crystal lattice on the differential cross section for elastic neutron scatter. In the scatter of the neutrons by the nuclear strong force, one invokes the nuclear scatter length operator, which is derived in section 9.2 of Squires (1996), and in the chapter on neutron scatter by nuclear spin states in Cremer (2013):

$$\hat{b} = 2B(\hat{\mathbf{I}} \cdot \hat{\mathbf{S}}_n) \tag{6.61}$$

With $B = 1/2$, the nuclear scatter operator $\hat{b}$ has the same form as the neutron magnetic scatter operator $\hat{\mathbf{S}}_n \cdot \hat{\mathbf{Q}}_\perp(\boldsymbol{\kappa})$, which suggests a neutron magnetic scatter operator $\hat{b}_M$, where

$$\hat{b}_M = \hat{\mathbf{S}}_n \cdot \hat{\mathbf{Q}}_\perp(\boldsymbol{\kappa}) = \hat{\mathbf{Q}}_\perp(\boldsymbol{\kappa}) \cdot \hat{\mathbf{S}}_n \tag{6.62}$$

where

$$2B(\hat{\mathbf{I}} \cdot \hat{\mathbf{S}}_n) \rightarrow \hat{\mathbf{Q}}_\perp(\boldsymbol{\kappa}) \cdot \hat{\mathbf{S}}_n \tag{6.63}$$

The dot product of the neutron spin vector operator $\hat{\mathbf{S}}_n$ and the perpendicular component of the unpaired, atomic electron magnetic dipole vector operator $\hat{\mathbf{Q}}_\perp$ is the magnetic potential energy operator $\hat{U}_e$ for the magnetic interaction between the neutron magnetic dipole moment and the unpaired, atomic electron magnetic dipole moment:

$$\hat{U}_e = \hat{\mathbf{Q}}_\perp \cdot \hat{\mathbf{S}}_n = \hat{Q}_{\perp x}\hat{S}_{nx} + \hat{Q}_{\perp y}\hat{S}_{ny} + \hat{Q}_{\perp z}\hat{S}_{nz} \tag{6.64}$$

"Note that Squires (1996) uses $\boldsymbol{\sigma}$ for the neutron spin vector and vector operator with x, y, and z component, neutron spin operators σ_x, σ_y, and σ_z. In this book, and in Cremer (2012b, 2013), is used the neutron spin vector $\mathbf{S}_n$ and the neutron spin vector operator $\hat{\mathbf{S}}_n$ with x, y, and z components $\hat{S}_{nx}$, $\hat{S}_{ny}$, and $\hat{S}_{nz}$. The eigenequations for the x, y, and z neutron spin operators $\hat{S}_{nx}$, $\hat{S}_{ny}$, and $\hat{S}_{nz}$ and their eigenfunctions, the initial, spin-up ψ_{0+} and spin-down ψ_{0-} neutron eigenstates yields respective neutron spin eigenvalues $\hbar/2$ and $-\hbar/2$, where

$$\hat{S}_{nx}|\psi_{0+}\rangle = \frac{\hbar}{2}\psi_{0-} \tag{6.65}$$

$$\hat{S}_{nx}|\psi_{0-}\rangle = \frac{\hbar}{2}\psi_{0+} \tag{6.66}$$

$$\hat{S}_{ny}|\psi_{0+}\rangle = \mathrm{i}\frac{\hbar}{2}\psi_{0-} \tag{6.67}$$

$$\hat{S}_{ny}|\psi_{0-}\rangle = -\,\mathrm{i}\frac{\hbar}{2}\psi_{0+} \tag{6.68}$$

and

$$\hat{S}_{nz}|\psi_{0+}\rangle = \frac{\hbar}{2}\psi_{0+} \tag{6.69}$$

$$\hat{S}_{nz}|\psi_{0-}\rangle = -\frac{\hbar}{2}\psi_{0-} \tag{6.70}$$

The transition matrix element of the potential energy operator $\hat{U}_e$ and the initial, spin-up ψ_{0+}, and spin-down ψ_{0-} neutron eigenstates are used to obtain the magnetic scatter lengths, via the orthogonality of the spin-up and spin-down neutron eigenstates:

$$\langle\psi_\alpha|\psi_\beta\rangle = \delta_{\alpha\beta} \tag{6.71}$$

The magnetic scatter lengths b_{M++} and b_{M--} for the case of neutron scatter and the no-spin-flip transitions $(+\to+)$ and $(-\to-)$, given neutron spin-up ψ_{0+} and spin-down ψ_{0-} eigenstates, are

$$b_{M++} = \langle\psi_{0+}|\hat{U}_e|\psi_{0+}|\rangle = \hat{Q}_{\perp z} \tag{6.72}$$

$$b_{M--} = \langle\psi_{0-}|\hat{U}_e|\psi_{0-}\rangle = -\hat{Q}_{\perp z} \tag{6.73}$$

The magnetic scatter lengths b_{M-+} and b_{M+-} for the case of neutron scatter and spin flip transitions $(- \rightarrow +)$ and $(+ \rightarrow -)$ are

$$b_{M-+} = \langle \psi_{0-} | \hat{U}_e | \psi_{0+} \rangle = \hat{Q}_{\perp x} + \mathrm{i}\hat{Q}_{\perp y} \tag{6.74}$$

$$b_{M+-} = \langle \psi_{0+} | \hat{U}_e | \psi_{0-} \rangle = \hat{Q}_{\perp x} - \mathrm{i}\hat{Q}_{\perp y} \tag{6.75}$$

The refractive index $n = 1 - \delta$ is expressed by the decrement δ, where $\delta \ll 1$. The decrements for the refraction in each of the four possible neutron spin-state transitions are obtained from the corresponding magnetic scatter lengths. Given neutron wavelength λ and the number density of the paramagnetic atoms ρ, the refractive index for the four possible neutron spin transitions, the two no-spin-flip cases $(+ \rightarrow +)$ and $(- \rightarrow -)$, and the two spin-flip cases $(- \rightarrow +)$ and $(+ \rightarrow -)$ are

$$n_{++} = 1 - \frac{\rho}{2\pi} \lambda^2 b_{M++} \tag{6.76}$$

$$n_{--} = 1 - \frac{\rho}{2\pi} \lambda^2 b_{M--} \tag{6.77}$$

$$n_{-+} = 1 - \frac{\rho}{2\pi} \lambda^2 b_{M-+} \tag{6.78}$$

$$n_{+-} = 1 - \frac{\rho}{2\pi} \lambda^2 b_{M+-} \tag{6.79}$$

The total refractive index n for the magnetic scatter of neutrons in the paramagnetic material, which includes all four possible neutron spin transitions, are

$$n = 1 - \frac{\rho}{2\pi} \lambda^2 (b_{M++} + b_{M--} + b_{M-+} + b_{M+-}) \tag{6.80}$$

These scatter lengths are only for magnetic scatter of neutrons by the unpaired atomic electrons of the paramagnetic atoms. Thus, these magnetic scatter lengths do not require averaging over the nuclei isotopes or spin states of the nuclei.

We consider the case of the no-spin-flip (nsf) transitions $(+ \rightarrow +)$ and $(- \rightarrow -)$. The partial differential cross section of Eq. (6.57) is integrated over the scattered neutron energy E, which yields the coherent, elastic differential cross section $(\mathrm{d}\sigma/d\Omega)^{\mathrm{mag}}_{\mathrm{nsf}}$ for magnetic neutron scatter, where

$$\left(\frac{\mathrm{d}\sigma}{\mathrm{d}\Omega}\right)^{\mathrm{mag}}_{\mathrm{nsf}} = (r_0 g)^2 \sum_{i=1}^{J} p_i \left\langle \phi_i \psi_{0s} | [(\hat{\mathbf{S}}_n \cdot \hat{\mathbf{Q}}_\perp(\boldsymbol{\kappa}))_z]^+ [(\hat{\mathbf{S}}_n \cdot \hat{\mathbf{Q}}_\perp(\boldsymbol{\kappa}))_z] | \psi_{0s} \phi_i \right\rangle \tag{6.81}$$

Application of Eq. (6.58) and Eqs. (6.64–6.75) to Eq. (6.81), yields the no-spin-flip $(d\sigma/d\Omega)^{mag}_{nsf}$ coherent, elastic differential cross section:

$$\begin{aligned}\left(\frac{d\sigma}{d\Omega}\right)^{mag}_{nsf} &= N(r_0 g)^2\left|\frac{g_0 F_0(\kappa)}{2}\right|^2 \\ &\exp(-2W_0)\left(1-\frac{\kappa_z^2}{\kappa^2}\right)\sum_{l=0}^{N-1}\exp(i\boldsymbol{\kappa}\cdot\hat{\mathbf{R}}_l)\sum_{i=1}^{J}p_i\langle\phi_i|\hat{S}^{+}_{0z}(0)\hat{S}_{lz}(0)|\phi_i\rangle\end{aligned} \tag{6.82}$$

The differential cross section $(d\sigma/d\Omega)^{mag}$ is nonzero because the mean-squared spin component in each direction for each lth unit cell for the N unit cells of the crystal is nonzero. One specifies the unit cell at the crystal center with index address $l = 0$, and multiplies by N with the mean-squared x, y, or z electron spin component, which is the same in every unit cell, and thus is represented by the $l = 0$ unit cell. Hence, the mean-squared z-component of the effective electron spin in a single unit cell of the unmagnetized paramagnetic crystal is

$$\sum_{i=1}^{J}p_i\langle\phi_i|\hat{S}^{+}_{0z}(0)\hat{S}_{0z}(t)|\phi_i\rangle = \frac{1}{3}s(s+1) \tag{6.83}$$

The differential cross section $(d\sigma/d\Omega)^{mag}$ for neutron elastic scatter from the nonmagnetized, paramagnetic material due to the magnetic dipole moment of unpaired electrons is identical for the $(+\rightarrow+)$ and $(-\rightarrow-)$ nsf neutron transitions. With substitution of Eq. (6.83) into Eq. (6.82) the no-spin-flip coherent, elastic differential cross section for either $(+\rightarrow+)$ or $(-\rightarrow-)$ is

$$\left(\frac{d\sigma}{d\Omega}\right)^{mag}_{nsf} = \frac{1}{3}N(r_0 g)^2\left|\frac{g_0 F_0(\boldsymbol{\kappa})}{2}\right|^2\exp(-2W_0)\left(1-\frac{\kappa_z^2}{\kappa^2}\right)s(s+1) \tag{6.84}$$

From Eqs. (6.72–6.73) and recalling both Eq. (1.21) and section 9.2 in Squires (1996), as well as Cremer (2012b, 2013) for nuclear scatter of neutron spin states, the neutron scatter lengths b_{M++} and b_{M--} for the no-spin-flip $(+\rightarrow+)$ or $(-\rightarrow-)$ neutron scatter in a paramagnetic material, with no applied magnetic field, are

$$b_{M++} = b_{M--} = (r_0 g)\left|\frac{g_0 F_0(\boldsymbol{\kappa})}{2}\right|\exp(-W_0)\sqrt{\frac{1}{3}\left(1-\frac{\kappa_z^2}{\kappa^2}\right)s(s+1)} \tag{6.85}$$

The coherent, elastic differential cross section for no-spin-flip is simply the number of unit cells N, multiplied by the squared scatter length per unit cell or per atom, where there is one atom per unit cell in the Bravais lattice crystal:

$$\left(\frac{d\sigma}{d\Omega}\right)^{mag}_{nsf++} = N(b_{M++})^2 = \left(\frac{d\sigma}{d\Omega}\right)^{mag}_{nsf--} = N(b_{M--})^2 \tag{6.86}$$

Similarly, using Eqs. (6.74–6.75) in Eq. (6.81), the coherent, elastic, differential cross section $(d\sigma/d\Omega)^{mag}_{sf}$ for the neutron spin-flip transitions $(- \to +)$ and $(+ \to -)$, which are identical, is

$$\left(\frac{d\sigma}{d\Omega}\right)^{mag}_{sf} = (r_0 g)^2 \sum_{i=1}^{I} p_i \langle \phi_i | (\hat{Q}_{\perp x} + i\hat{Q}_{\perp y})^+ (\hat{Q}_{\perp x} + i\hat{Q}_{\perp y}) | \phi_i \rangle \tag{6.87}$$

where

$$\begin{aligned}(\hat{Q}_{\perp x} + i\hat{Q}_{\perp y})^+ (\hat{Q}_{\perp x} + i\hat{Q}_{\perp y}) &= (\hat{Q}^+_{\perp x} - i\hat{Q}_{\perp y})(\hat{Q}^+_{\perp y} + i\hat{Q}_{\perp y}) \\ &= \hat{Q}^+_{\perp x}\hat{Q}_{\perp x} + \hat{Q}^+_{\perp y}\hat{Q}_{\perp y} = \hat{Q}^2_{\perp x} + \hat{Q}^2_{\perp y}\end{aligned} \tag{6.88}$$

and

$$\hat{Q}^2_{\perp x} + \hat{Q}^2_{\perp y} = \hat{Q}^2_{\perp} - \hat{Q}^2_{\perp z} \tag{6.89}$$

We now consider the case of the spin-flip (sf) transitions $(- \to +)$ and $(+ \to -)$ in which the coherent, elastic differential cross section $(d\sigma/d\Omega)^{mag}_{sf}$ for magnetic neutron scatter is

$$\begin{aligned}\left(\frac{d\sigma}{d\Omega}\right)^{mag}_{sf} = (r_0 g)^2 \sum_{i=1}^{J} p_i \langle \phi_i \psi_{0s} | [(\hat{\mathbf{S}}_n \cdot \hat{\mathbf{Q}}_{\perp}(\boldsymbol{\kappa}))_x + (\hat{\mathbf{S}}_n \cdot \hat{\mathbf{Q}}_{\perp}(\boldsymbol{\kappa}))_y]^+ \\ \times [(\hat{\mathbf{S}}_n \cdot \hat{\mathbf{Q}}_{\perp}(\boldsymbol{\kappa}))_x + (\hat{\mathbf{S}}_n \cdot \hat{\mathbf{Q}}_{\perp}(\boldsymbol{\kappa}))_y] | \psi_{0s} \phi_i \rangle\end{aligned} \tag{6.90}$$

As shown by Eqs. (6.88–6.89), the x and y component expression of Eq. (6.90), can be expressed equivalently by the z-component expression,

$$\left(\frac{d\sigma}{d\Omega}\right)^{mag}_{sf} = (r_0 g)^2$$

$$\sum_{i=1}^{J} p_i \langle \phi_i \psi_{0s} | [\hat{\mathbf{S}}_n \cdot \hat{\mathbf{Q}}_{\perp}(\boldsymbol{\kappa}) - (\hat{\mathbf{S}}_n \cdot \hat{\mathbf{Q}}_{\perp}(\boldsymbol{\kappa}))_z]^+ \left[\hat{\mathbf{S}}_n \cdot \hat{\mathbf{Q}}_{\perp}(\boldsymbol{\kappa}) - (\hat{\mathbf{S}}_n \cdot \hat{\mathbf{Q}}_{\perp}(\boldsymbol{\kappa}))_z\right] | \psi_{0s} \phi_i \rangle \tag{6.91}$$

The differential cross section for spin-flip $(d\sigma/d\Omega)^{mag}_{sp}$ of Eq. (6.91) (based on z-components), is equal to the coherent, elastic differential cross section for all cases of no-spin-flip and spin-flip $(d\sigma/d\Omega\, dE)^{mag}$, minus the differential cross section for no-spin-flip $(d\sigma/d\Omega)^{mag}_{nsf}$ of Eq. (6.81) (also based on z-components).

$$\left(\frac{\mathrm{d}\sigma}{\mathrm{d}\Omega}\right)_{\mathrm{sf}}^{\mathrm{mag}} = \left(\frac{\mathrm{d}\sigma}{\mathrm{d}\Omega}\right)^{\mathrm{mag}} - \left(\frac{\mathrm{d}\sigma}{\mathrm{d}\Omega}\right)_{\mathrm{nsf}}^{\mathrm{mag}} \tag{6.92}$$

The coherent, elastic differential cross section $(\mathrm{d}\sigma/d\Omega)_{\mathrm{sf}}^{\mathrm{mag}}$ for neutron elastic scatter from the nonmagnetized, paramagnetic material, which is due to the magnetic dipole moment of unpaired electrons, is identical for the $(- \rightarrow +)$ and $(+ \rightarrow -)$ spin-flip neutron transitions. Hence, the spin-flip, coherent, elastic differential cross section for either $(- \rightarrow +)$ or $(+ \rightarrow -)$ is

$$\left(\frac{\mathrm{d}\sigma}{\mathrm{d}\Omega}\right)_{\mathrm{sf}}^{\mathrm{mag}} = \frac{1}{3}N(r_0 g)^2\left|\frac{g_0 F_0(\boldsymbol{\kappa})}{2}\right|^2 \exp(-2W_0)\left(1 + \frac{\kappa_z^2}{\kappa^2}\right)s(s+1) \tag{6.93}$$

The neutron scatter lengths b_{M-+} and b_{M+-} for the spin-flip $(- \rightarrow +)$ or $(+ \rightarrow -)$ neutron scatter in a paramagnetic material with no applied magnetic field, are calculated as follows:

$$b_{M-+} = b_{M+-} = (r_0 g)\left|\frac{g_0 F_0(\boldsymbol{\kappa})}{2}\right| \exp(-W_0)\sqrt{\frac{1}{3}\left(1 + \frac{\kappa_z^2}{\kappa^2}\right)s(s+1)} \tag{6.94}$$

The coherent, elastic differential cross section for no-spin-flip transition is simply the number of unit cells N multiplied by the squared scatter length per unit cell or per atom, where there is one atom per unit cell in the Bravais lattice crystal:

$$\left(\frac{\mathrm{d}\sigma}{\mathrm{d}\Omega}\right)_{\mathrm{nsf}-+}^{\mathrm{mag}} = N(b_{M-+})^2 = \left(\frac{\mathrm{d}\sigma}{\mathrm{d}\Omega}\right)_{\mathrm{nsf}+-}^{\mathrm{mag}} = N(b_{M+-})^2 \tag{6.95}$$

Now assume the spin polarization $\mathbf{P}$ of the neutron beam is parallel to the direction of the positive z-axis as a spin-up $(+)$ beam, and the positive z-axis is also the direction of the neutron scatter vector $\boldsymbol{\kappa}$ in the paramagnetic material, with no applied magnetic field $\mathbf{B} = 0$. In this case, the neutron scatter vector is nonzero only in the z-direction ($\kappa = \kappa_z$), and the neutron spin polarization is parallel to the neutron scatter vector ($\mathbf{P}||\boldsymbol{\kappa}$). All spin-up $(+)$ neutrons in the polarized beam are flipped to the spin-down $(-)$ state. In this case, the coherent, elastic differential cross sections are

$$\left(\frac{\mathrm{d}\sigma}{\mathrm{d}\Omega}\right)_{+\rightarrow-}^{\mathrm{mag}} = \frac{2}{3}N(r_0 g)^2\left|\frac{g_0 F_0(\boldsymbol{\kappa})}{2}\right|^2 \exp(-2W_0)s(s+1) \quad \kappa = \kappa_z \text{ and } \mathbf{P}||\boldsymbol{\kappa} \tag{6.96}$$

$$\left(\frac{\mathrm{d}\sigma}{\mathrm{d}\Omega}\right)_{+\rightarrow+}^{\mathrm{mag}} = \left(\frac{\mathrm{d}\sigma}{\mathrm{d}\Omega}\right)_{-\rightarrow-}^{\mathrm{mag}} = 0 \quad \kappa = \kappa_z \text{ and } \mathbf{P}||\boldsymbol{\kappa} \tag{6.97}$$

Alternatively, assume the spin polarization **P** of the neutron beam is perpendicular to the z-axis as a spin-up (+) beam, and the z-axis is perpendicular to the neutron scatter vector $\boldsymbol{\kappa}$ in the paramagnetic material, with no applied magnetic field ($\mathbf{B} = 0$). The z-component of the neutron scatter vector is zero ($\kappa_z = 0$), and the neutron spin polarization is perpendicular to the neutron scatter vector ($\mathbf{P} \perp \boldsymbol{\kappa}$). Half of the spin-up (+) neutrons are flipped to spin-down (−), and half are not. In this case, the coherent, elastic differential cross sections are

$$\left(\frac{d\sigma}{d\Omega}\right)^{mag}_{+\to-} = \frac{1}{3}N(r_0 g)^2 \left|\frac{g_0 F_0(\boldsymbol{\kappa})}{2}\right|^2 \exp(-2W_0)s(s+1) \quad \kappa_z = 0 \text{ and } \mathbf{P}\perp\boldsymbol{\kappa} \tag{6.98}$$

$$\left(\frac{d\sigma}{d\Omega}\right)^{mag}_{+\to+} = \frac{1}{3}N(r_0 g)^2 \left|\frac{g_0 F_0(\boldsymbol{\kappa})}{2}\right|^2 \exp(-2W_0)s(s+1) \quad \kappa_z = 0 \text{ and } \mathbf{P}\perp\boldsymbol{\kappa} \tag{6.99}$$

A polarized neutron beam of only spin-up (+) neutrons, which is directed parallel and then antiparallel to the neutron scatter vector ($\boldsymbol{\kappa} = \mathbf{k}_0 - \mathbf{k}$), can be used to the contributions of elastic neutron scatter by the nuclei and the unpaired atomic electrons of the nuclei in paramagnetic materials. With polarized neutron beams in paramagnetic materials, one can thus separate the paramagnetic scatter contribution from other scatter contributions, such as multiple Bragg scatter, phonon scatter, and incoherent scatter, which is due to the variation of isotopes and variation of nuclear spin in the material.

References

Cremer, J. T. (2012b). In P. W. Hawkes (Ed.), *Neutron and X-ray microscopy – Part 2. Advances in imaging and electron physics* (Vol. 173). London, UK: Academic Press, an imprint of Elsevier Press.

Cremer, J. T. (2013). In P. W. Hawkes (Ed.), *Advances in imaging and electron physics* (Vol. 175). London, UK: Academic Press, an imprint of Elsevier Press.

Squires, G. L. (1996). *Introduction to the theory of thermal neutron scattering*. New York, NY: Dover.

7 Neutron Scatter in Ferromagnetic, Antiferromagnetic, and Helical Magnetic Materials

7.1 Coherent, Elastic Differential Cross Section for Magnetic Neutron Scatter in Ferromagnetic Materials—Localized Unpaired Electrons

In this chapter, which is a continuation of Cremer (2012b, 2013), we cover the coherent differential cross section of neutron elastic scatter in ferromagnetic materials. Following Squires (1996), first we derive the coherent differential cross section for elastic neutron scatter in a Bravais-lattice ferromagnetic crystal without regard to neutron spin or nuclear spin. This is based on the Heitler–London model, LS coupling, and dipole approximation. The coherent, elastic differential cross section is also derived for neutron magnetic scatter in antiferromagnetic materials. Next, the coherent, elastic differential cross section for magnetic and nuclear scatter of neutron spin states in a Bravais-lattice and then a non-Bravais-lattice ferromagnetic crystal is derived. Included in this discussion is the production and measurement of polarized neutrons by ferromagnetic materials. Derived next is a general expression for the coherent differential cross section for nuclear and magnetic elastic scatter of neutron spin states in ferromagnetic materials due to localized or delocalized unpaired electrons. Also, the production of polarized neutrons by grazing incidence reflection via nuclear scatter of neutron spin states in ferromagnetic materials is described. Finally, the coherent, elastic differential cross section for the scatter of neutron spin states from magnetic materials with helical-oriented, localized, unpaired electron spins is derived.

We now determine the differential cross section of ferromagnetic materials for neutron elastic scatter without regard to neutron spin or nuclear spin. One starts with Eq. (5.147) of chapter 5 for the coherent differential cross section $(d\sigma/d\Omega)^{mag}_{coh,el}$ for elastic neutron scatter from a Bravais-lattice crystal. The cross section of Eq. (5.147) is based on the Heitler–London model, LS coupling, and dipole approximation, where the electrons are assumed to be localized to their respective atoms, rather than the electrons wandering between the atoms within energy bands as itinerant

Neutron and X-ray Optics. DOI: http://dx.doi.org/10.1016/B978-0-12-407164-3.00006-1

electrons. Based on these assumptions, the coherent, elastic differential cross section $(d\sigma/d\Omega)^{\text{mag}}_{\text{coh,el}}$ of Eq. (5.147) for a Bravais lattice crystal, is repeated below,

$$\left(\frac{d\sigma}{d\Omega}\right)^{\text{mag}}_{\text{coh,el}} = N(r_0 g)^2 \sum_{l=0}^{N-1} \exp(i\boldsymbol{\kappa}\cdot\hat{\mathbf{R}}_l) \left|\frac{g_0}{2}F_0(\boldsymbol{\kappa})\exp(-W_0)\right|^2 \sum_{i=1}^{J} p_i \left\langle \phi_i \left| \sum_{\alpha=1}^{3}\sum_{\beta=1}^{3} \left(\delta_{\alpha\beta} - \frac{\kappa_\alpha \kappa_\beta}{\kappa^2}\right) \hat{S}^{+}_{0\alpha}(0)\hat{S}_{l\beta}(0) \right| \phi_i \right\rangle \tag{7.1}$$

Recall that the classical electron radius is r_0, the neutron Landé g factor is $g = -1.91$, the scatter atom Landé splitting factor is g_0, the magnetic scatter factor is $F_d(\boldsymbol{\kappa})$, and the Debye–Waller factor is $\exp(-W_0)$. For unit cell volume v_0 and reciprocal-lattice vector $\boldsymbol{\tau}_\tau$, where from Eq. (3.47) in Squires (1996) and Eq. 105 in Ch. 4 of Cremer (2012b),

$$\sum_{l=0}^{N-1} \exp(i\boldsymbol{\kappa}\cdot\hat{\mathbf{R}}_l) = \frac{(2\pi)^3}{v_0}\sum_{\tau=1}^{N}\delta(\kappa - \boldsymbol{\tau}_\tau) \tag{7.2}$$

With substitution of Eq. (7.2) into Eq. (7.1), coherent, elastic differential cross section, becomes

$$\left(\frac{d\sigma}{d\Omega}\right)^{\text{mag}}_{\text{coh,el}} = N(r_0 g)^2 \frac{(2\pi)^3}{v_0}\sum_{\tau=1}^{N}\delta(\kappa - \boldsymbol{\tau}_\tau) \left|\frac{g_0}{2}F_0(\boldsymbol{\kappa})\exp(-W_0)\right|^2 \sum_{i=1}^{J} p_i \left\langle \phi_i \left| \sum_{\alpha=1}^{3}\sum_{\beta=1}^{3} \left(\delta_{\alpha\beta} - \frac{\kappa_\alpha \kappa_\beta}{\kappa^2}\right) \hat{S}^{+}_{0\alpha}(0)\hat{S}_{l\beta}(0) \right| \phi_i \right\rangle \tag{7.3}$$

Without an applied magnetic field, unlike the paramagnet, the ferromagnetic crystal is comprised of tiny regions or domains, where the longitudinal (azimuthal) component of the unpaired electron spins point in the same direction, and create a net magnetization in the domain. The direction of the magnetization varies from one domain to the next, and application of an applied magnetic field tends to align the magnetization directions of the domains.

If one considers a single domain with a magnetization aligned in the positive z-direction, then the mean x- and y-components of the effective electron spin in Eq. (7.3) are each zero:

$$\sum_{i=1}^{I} p_i \langle \phi_i | \hat{S}^{+}_{0\alpha}(0)\hat{S}_{l\beta}(0)|\phi_i\rangle = 0 \quad \alpha = x, y \text{ and } \beta = x, y \tag{7.4}$$

In the domain, the squared mean (thermally averaged) z-component of the effective electron spin $\overline{S}_z$ is uniform and independent of unit cell position l, where

$$\sum_{i=1}^{J} p_i \left\langle \phi_i \left| \left(1 - \frac{\kappa_z^2}{\kappa^2}\right) \hat{S}^{+}_{0z}(0)\hat{S}_{lz}(0) \right| \phi_i \right\rangle = \left(1 - \frac{\kappa_z^2}{\kappa^2}\right)(\overline{S}_z)^2 \tag{7.5}$$

With substitution of Eqs. (7.4–7.5) into Eq. (7.3), the differential cross section $(d\sigma/d\Omega)^{\text{mag}}_{\text{coh,el}}$ for coherent, elastic neutron scatter from the domain of a Bravais-lattice ferromagnetic crystal, which is based on the Heitler–London model, LS coupling, and dipole approximation, is expressed in terms of the magnetic scatter length b_M:

$$\left(\frac{d\sigma}{d\Omega}\right)^{\text{mag}}_{\text{coh,el}} = N(2\pi)^3 \frac{b_M^2}{v_0} \exp(-2W_0) \sum_{\tau=1}^{N} \delta(\kappa - \boldsymbol{\tau}_\tau) \tag{7.6}$$

The magnetic scatter length b_M is

$$b_M\,[\text{fm}] = (r_0 g)\left(\frac{g_0}{2}F_0(\boldsymbol{\kappa})\right)\sqrt{1-\frac{\kappa_z^2}{\kappa^2}} = 2.69 g_0 F_0(\boldsymbol{\kappa})\sqrt{1-\frac{\kappa_z^2}{\kappa^2}} \tag{7.7}$$

Scatter lengths (measured in femtometers) are appropriate for nuclear dimensions, for example, the proton is 0.88 fm, a deuteron is 2.4 fm, and an oxygen nucleus is 6 fm.

Let vector $\mathbf{s}_u$ be a unit vector pointing in the direction of the longitudinal (azimuthal) component of the unpaired electron spins of the ferromagnetic crystal domain, and let $\boldsymbol{\tau}_{u\tau}$ be the unit vector of the τth reciprocal-lattice unit vector $\boldsymbol{\tau}_\tau$. The magnetic neutron scatter length is modified to

$$b_M\,[\text{fm}] = (r_0 g)\left(\frac{g_0}{2}F_0(\boldsymbol{\kappa})\right)\bar{S}_{\hat{s}}\sqrt{1-\overline{(\boldsymbol{\tau}_{u\tau}\cdot\mathbf{s}_u)^2}} \tag{7.8}$$

Here $(\bar{S}_{\hat{s}})^2$ is the squared mean value of the electron spin component in the direction $\mathbf{s}_u$ of the longitudinal (azimuthal) component of the effective electron spin in each domain of the ferromagnetic crystal.

If one assumes that the domain spin directions $\mathbf{s}_u$ point in all directions with equal probability, then the average of $(\boldsymbol{\tau}_{u\tau}\cdot\mathbf{s}_u)^2$ over all the domains is

$$\overline{(\boldsymbol{\tau}_{u\tau}\cdot\mathbf{s}_u)^2} = \frac{1}{3} \tag{7.9}$$

where

$$1-\overline{(\boldsymbol{\tau}_{u\tau}\cdot\mathbf{s}_u)^2} = \frac{2}{3} \tag{7.10}$$

Eq. (7.6) for the coherent, elastic scatter differential cross section in the ferromagnetic crystal averaged over the domains is then written as

$$\left(\frac{d\sigma}{d\Omega}\right)^{\text{mag}}_{\text{avg}} = \frac{2}{3}N(r_0 g)^2\frac{(2\pi)^3}{v_0}(\bar{S}_{\hat{s}})^2\sum_{\tau=1}^{N}\delta(\kappa-\boldsymbol{\tau}_\tau) \tag{7.11}$$

The modified magnetic scatter length is then

$$b_M\ [\text{fm}] = \sqrt{\frac{2}{3}}(r_0 g)\left(\frac{g_0}{2}F_0(\boldsymbol{\kappa})\right)\overline{S}_{\hat{s}} \tag{7.12}$$

For a non-Bravais-lattice ferromagnetic crystal with r atoms per unit cell, the coherent, elastic scatter differential cross section is obtained by simple substitution:

$$\left|\frac{g_0}{2}F_0(\boldsymbol{\kappa})\exp(-W_0)\right|^2 \Rightarrow \left|\sum_{d=1}^{r}\frac{g_d}{2}F_d(\boldsymbol{\kappa})\exp(i\boldsymbol{\kappa}\cdot\mathbf{d}_d)\exp(-W_d)\right|^2 \tag{7.13}$$

With Eq. (7.13) applied to Eq. (7.1), and using Eqs. (7.2), (7.4), (7.7–7.10), the averaged, coherent elastic differential cross section for a non-Bravais ferromagnetic lattice is

$$\left(\frac{d\sigma}{d\Omega}\right)_{\text{avg}}^{\text{mag}} = \frac{2}{3}N(r_0 g)^2\frac{(2\pi)^3}{v_0}\left|\sum_{d=1}^{r}\overline{S}_{\hat{s}}\frac{g_d}{2}F_d(\boldsymbol{\kappa})\exp(i\boldsymbol{\kappa}\cdot\mathbf{d}_d)\exp(-W_d)\right|^2\sum_{\tau=1}^{N}\delta(\kappa-\boldsymbol{\tau}_\tau) \tag{7.14}$$

Rather than a magnetic scatter length per atom b_M, one defines a magnetic scatter length per unit cell or magnetic unit cell structure factor $F_M(\boldsymbol{\kappa})$, where

$$F_M\ [\text{fm}] = (r_0 g)\sum_{d=1}^{r}\frac{g_d}{2}F_d(\boldsymbol{\kappa})\exp(i\boldsymbol{\kappa}\cdot\mathbf{d}_d)\exp(-W_d) \tag{7.15}$$

With substitution of Eq. (7.15) into Eq. (7.14), the average coherent elastic differential cross section for magnetic scatter of a neutron in a non-Bravais crystal lattice is then

$$\left(\frac{d\sigma}{d\Omega}\right)_{\text{avg}}^{\text{mag}} = \frac{2}{3}(2\pi)^3 N\frac{F_M^2(\boldsymbol{\kappa})}{v_0}\sum_{\tau=1}^{N}\delta(\kappa-\boldsymbol{\tau}_\tau) \tag{7.16}$$

In the differential cross section expression for neutron elastic scatter from a ferromagnetic crystal, the term $\delta(\kappa-\boldsymbol{\tau}_\tau)$ shows that the magnetic Bragg peaks of the unpaired atomic electrons occur at the same points in the reciprocal lattice as the nuclear Bragg peaks of their nuclei. However, they differ in terms of temperature dependence and the dependence on their form factors in reciprocal space on the neutron scatter vector $\boldsymbol{\kappa}$. The magnetic elastic differential scatter cross section is proportional to the squared mean z-component spin $(\overline{S}_{\hat{s}})^2$.

As expected, the squared mean z-component spin $(\overline{S}_{\hat{s}})^2$ decreases rapidly as spin randomizes with increasing temperature, and the neutron elastic scatter from electron spin interaction falls to zero as the temperature rises to the Curie temperature. However, the nuclear scatter is relatively independent of temperature, but does decrease with increasing temperature due to the Debye–Waller factor $\exp(-2W_0)$, where

$$\exp(-2W_0) = \exp(-2|\boldsymbol{\kappa} \cdot \mathbf{u}_0|^2) \tag{7.17}$$

That is, the increase of the displacement distance $\mathbf{u}_0$ of an atom from its equilibrium position causes the Debye–Waller factor $\exp(-2W_0)$ to decrease, and the differential cross section for elastic neutron scatter then decreases proportionally to the decrease in the Debye–Waller factor.

In nuclear scatter, the nuclear unit cell structure factor $F_N(\boldsymbol{\kappa})$ provides the unit cell amplitude arising from the atoms within the unit cell as a function in the change of the neutron wave vector upon scatter $\boldsymbol{\kappa}$, where

$$F_N(\boldsymbol{\kappa}) = \sum_{d=1}^{r} \overline{b}_d \exp(i\boldsymbol{\kappa} \cdot \mathbf{d}_d)\exp(-W_d) \tag{7.18}$$

The nuclear unit structure factor $F_N(\boldsymbol{\kappa})$ is the Fourier transform of the mean nuclear scattering length $\overline{b}_d$, which is localized to the pointlike nucleus. Hence, $F_N(\boldsymbol{\kappa})$, the Fourier transform of the mean scattering lengths $\overline{b}_d$ assigned to the r point positions $\mathbf{d}_d$, is somewhat constant over κ, except for a slight decrease with increasing $|\kappa|$ due to the Debye–Waller factor. That is, the short-range distance, over which a nuclear force acts in direct space, results in a nearly constant nuclear unit structure factor $F_N(\boldsymbol{\kappa})$ in reciprocal space.

However, this is not the case for the magnetic form factor. The magnetic form factor $F_d(\boldsymbol{\kappa})$ for the dth atom in each unit cell is the Fourier transform in κ-space of the direct-lattice, normalized-density $\rho_d(\mathbf{r})$ of the dth atom's unpaired electrons:

$$F_d(\boldsymbol{\kappa}) = \exp(i\boldsymbol{\kappa} \cdot \hat{\mathbf{r}}_{vd}) = \int_{-\infty}^{\infty} \mathrm{d}z \int_{-\infty}^{\infty} \mathrm{d}y \int_{-\infty}^{\infty} \mathrm{d}x \rho_d(\mathbf{r})\exp(i\boldsymbol{\kappa} \cdot \mathbf{r}) \tag{7.19}$$

Given the incident, neutron wave-number vector $\mathbf{k}_0$, and the scattered, wave-number vector $\mathbf{k}$, the incident, wave-number vector $\mathbf{k}_0$ changes by $\boldsymbol{\kappa} = \mathbf{k}_0 - \mathbf{k}$ upon scatter. The neutron scatter vector κ is also denoted by $\mathbf{q}$, where

$$\mathbf{q} \equiv \boldsymbol{\kappa} = \mathbf{k}_0 - \mathbf{k} \tag{7.20}$$

The magnetic field of the unpaired electron spin and orbital angular momentum acts over a much larger distance than the nuclear strong force. That is, the electron density function of the dth atom of the unit cell is spread over the much larger extent of the atom, which dwarfs the pointlike nucleus. The Fourier transform of the large extent of the dth atom electron corresponds to a small range of $F_d(\boldsymbol{\kappa})$ in the reciprocal lattice. Consequently, the magnetic form factor $F_d(\boldsymbol{\kappa})$ decreases rapidly with increasing $|\boldsymbol{\kappa}|$. Hence, the nuclear and magnetic scatter can be distinguished by the decrease of magnetic scatter with increasing temperature and increasing $|\boldsymbol{\kappa}|$ as compared to nuclear scatter.

Suppose that an external magnetic field **B** is applied in the same direction of reciprocal-lattice vector $\boldsymbol{\tau}_\tau$, which corresponds to the τth Bragg peak. That is, **B** is parallel to $\boldsymbol{\tau}_\tau$, where $\mathbf{B}||\boldsymbol{\tau}_\tau$. Then the longitudinal components of the unpaired electron spins $\mathbf{s}_u$ align antiparallel to $\boldsymbol{\tau}_{u\tau}$, where

$$\mathbf{s}_u = -\boldsymbol{\tau}_{u\tau} \tag{7.21}$$

with

$$\mathbf{s}_u \cdot \boldsymbol{\tau}_{u\tau} = -1 \tag{7.22}$$

and

$$1 - \overline{(\boldsymbol{\tau}_\tau \cdot \hat{\mathbf{s}})^2} = 0 \tag{7.23}$$

Thus, with $\mathbf{B}||\boldsymbol{\tau}_\tau$, the magnetic scatter goes to zero, where

$$\left(\frac{d\sigma}{d\Omega}\right)^{\text{mag}}_{\text{coh,el}} = 0 \quad \mathbf{B}||\boldsymbol{\tau}_\tau \tag{7.24}$$

The nuclear and magnetic scatter Bragg peaks overlap. Thus, without an externally applied magnetic field, the scattered neutron flux for the τth Bragg peak is due to nuclear and magnetic scatter, and with an applied magnetic field $\mathbf{B}||\boldsymbol{\tau}_\tau$, the scattered neutron flux for the τth Bragg peak is entirely due to nuclear scattering. The difference in the intensity of the τth Bragg peak with and without the applied magnetic field $\mathbf{B}||\boldsymbol{\tau}_\tau$, then allows one to determine the differential cross section for the τth Bragg peak that is due only to magnetic scatter.

7.2 Antiferromagnetic Materials—Coherent, Elastic Differential Cross Section for Neutron Scatter from Localized Unpaired Electrons

Each domain of antiferromagnetic material, such as $KMnF_3$, is comprised of alternating, stacked planes of atoms A and B, where the total spin of atom A is antiparallel to the total spin of atom B. The coherent, elastic differential cross section for magnetic neutron scatter in a single domain of an antiferromagnetic crystal is a simple modification of Eq. (7.14) for the cross section of a non-Bravais ferromagnetic crystal, where Eq. (7.14) is repeated here:

$$\left(\frac{d\sigma}{d\Omega}\right)^{\text{mag}}_{\text{avg}} = \frac{2}{3}N(r_0 g)^2\frac{(2\pi)^3}{v_0}\left|\sum_{d=1}^{r}\overline{S}_{\hat{s}}\frac{g_d}{2}F_d(\boldsymbol{\kappa})\exp(i\boldsymbol{\kappa}\cdot\mathbf{d}_d)\exp(-W_d)\right|^2\sum_{\tau=1}^{N}\delta(\kappa-\boldsymbol{\tau}_\tau) \tag{7.25}$$

The antiferromagnetic crystal can be treated as a non-Bravais crystal, where each unit cell is comprised of atoms A and B with their oppositely directed spins of unpaired electrons. As a result, the direction of the mean value of the domain electron spin z-component $\overline{S}_{\hat{s}}$ in atom A is pointed in the opposite direction of atom B. This can be implemented by multiplying $\overline{S}_{\hat{s}}$ for atom A by $+1/2$, and multiplying $\overline{S}_{\hat{s}}$ for atom B by $-1/2$, where half the atoms are atom A and the other half are atom B. The coherent, elastic differential cross section for magnetic scatter in the antiferromagnetic crystal domain becomes

$$\left(\frac{d\sigma}{d\Omega}\right)_{avg}^{mag} = \frac{2}{3}N(r_0 g)^2\frac{(2\pi)^3}{v_0}(\overline{S}_{\hat{s}})^2\sum_{\tau=1}^{N}\delta(\kappa-\boldsymbol{\tau}_\tau)$$
$$\left|\frac{g_A}{2}\frac{F_A(\boldsymbol{\kappa})}{2}\exp(i\boldsymbol{\kappa}\cdot\mathbf{d}_A)\exp(-W_A)-\frac{g_B}{2}\frac{F_B(\boldsymbol{\kappa})}{2}\exp(i\boldsymbol{\kappa}\cdot\mathbf{d}_B)\exp(-W_B)\right|^2 \tag{7.26}$$

One can obtain the coherent, elastic differential cross section for nuclear scatter, from the coherent, elastic differential cross section for nuclear scatter for a ferromagnetic domain, by substituting the magnetic unit cell structure factor $F_M(\boldsymbol{\kappa})$ of Eq. (7.15) by nuclear unit cell structure factor $F_N(\boldsymbol{\kappa})$ of Eq. (7.18), where

$$F_M(\boldsymbol{\kappa}) \rightarrow F_N(\boldsymbol{\kappa}) \tag{7.27}$$

The coherent, elastic differential cross section for nuclear scatter for the antiferromagnetic crystal domain is then

$$\left(\frac{d\sigma}{d\Omega}\right)_{avg}^{nuc} = \frac{2}{3}N\frac{(2\pi)^3}{v_0}|\overline{b}_A\exp(i\boldsymbol{\kappa}\cdot\mathbf{d}_A)\exp(-W_A)$$
$$+\overline{b}_B\exp(i\boldsymbol{\kappa}\cdot\mathbf{d}_B)\exp(-W_B)|^2\sum_{\tau=1}^{N}\delta(\kappa-\boldsymbol{\tau}_\tau) \tag{7.28}$$

If one assumes that the product of the Landé splitting factor with the magnetic form factor is identical, and the Debye–Waller factors for atom A and B are identical, then

$$g_A F_A(\boldsymbol{\kappa}) = g_B F_B(\boldsymbol{\kappa}) = g_0 F_0(\boldsymbol{\kappa}) \tag{7.29}$$

and

$$\exp(-W_A) = \exp(-W_A) = \exp(-W_0) \tag{7.30}$$

Hence, the coherent, differential cross section for magnetic scatter in the antiferromagnetic domain becomes

$$\left(\frac{d\sigma}{d\Omega}\right)_{avg}^{mag} = \frac{2}{3}N(r_0 g)^2 \frac{(2\pi)^3}{v_0}\left(\frac{\overline{S}_{\hat{s}} g_0 F_0(\boldsymbol{\kappa})}{4}\right)^2$$
$$\exp(-2W_0)|\exp(i\boldsymbol{\kappa}\cdot\mathbf{d}_A) - \exp(i\boldsymbol{\kappa}\cdot\mathbf{d}_B)|^2 \sum_{\tau=1}^{N}\delta(\kappa - \boldsymbol{\tau}_\tau) \tag{7.31}$$

And the coherent, differential cross section for nuclear scatter in the antiferromagnetic domain becomes

$$\left(\frac{d\sigma}{d\Omega}\right)_{avg}^{nuc} = \frac{2}{3}N\frac{(2\pi)^3}{v_0}\exp(-2W_0)|\overline{b}_A\exp(i\boldsymbol{\kappa}\cdot\mathbf{d}_A) + \overline{b}_B\exp(i\boldsymbol{\kappa}\cdot\mathbf{d}_B)|^2 \sum_{\tau=1}^{N}\delta(\kappa - \boldsymbol{\tau}_\tau) \tag{7.32}$$

The modulo operation is useful in this discussion and is described next. Given two numbers, the dividend a and the divisor n, the division operation a/n yields the quotient c, which is the sum of an integer m and a remainder r, where

$$\frac{a}{n} = c = m + r \tag{7.33}$$

and

$$|m| > |r| \tag{7.34}$$

The remainder r of the division operation a/n can be specified by the modulo operation

$$a \bmod n = r \tag{7.35}$$

For example,

$$4 \bmod 2 = 0, \quad 3 \bmod 2 = 1, \quad \text{and} \quad 14 \bmod 4 = 2 \tag{7.36}$$

If the sum of triplet Miller integers $h + k + l$ is an even number, see Ch. 4 in Cremer (2012b), then

$$(h + k + l)\bmod 2 = 0 \tag{7.37}$$

If the sum of triplet Miller integers $h + k + l$ is an odd number, then

$$(h + k + l)\bmod 2 = 1 \tag{7.38}$$

Nuclear scatter and magnetic scatter can occur at different reciprocal-lattice points in an antiferromagnetic crystal. Consider the positions $\mathbf{d}_\mathrm{A}$ and $\mathbf{d}_\mathrm{B}$ of the A and B atom:

$$\mathbf{d}_\mathrm{A} = 0 \tag{7.39}$$

$$\mathbf{d}_\mathrm{B} = \frac{1}{2}\mathbf{a} + \frac{1}{2}\mathbf{b} + \frac{1}{2}\mathbf{c} \tag{7.40}$$

Consider the neutron scatter vector $\boldsymbol{\kappa}$ to be equal to the τth reciprocal-lattice vector $\boldsymbol{\tau}_\tau$, which is specified by triplet integers (h, k, l), where

$$\boldsymbol{\kappa} = \boldsymbol{\tau}_\tau = h\mathbf{a}^* + k\mathbf{b}^* + l\mathbf{c}^* \tag{7.41}$$

and

$$\mathbf{a} \cdot \mathbf{a}^* = \mathbf{b} \cdot \mathbf{b}^* = \mathbf{c} \cdot \mathbf{c}^* = 2\pi \tag{7.42}$$

The magnetic neutron scatter is zero if the sum $h + k + l$ of the triplet Miller integers (h, k, l) is an even integer and nonzero if the sum $h + k + l$ is an odd integer, where

$$\left|1 - \exp[\mathrm{i}\pi(h+k+l)]\right|^2_{\mathrm{mag}} = 0 \Rightarrow \left(\frac{\mathrm{d}\sigma}{\mathrm{d}\Omega}\right)^{\mathrm{mag}}_{\mathrm{avg}} = 0 \quad (h + k + l)\mathrm{mod}\, 2 = 0 \tag{7.43}$$

and

$$\left|1 - \exp[\mathrm{i}\pi(h+k+l)]\right|^2_{\mathrm{mag}} = 2 \Rightarrow \left(\frac{\mathrm{d}\sigma}{\mathrm{d}\Omega}\right)^{\mathrm{mag}}_{\mathrm{avg}} \neq 0 \quad (h + k + l)\mathrm{mod}\, 2 = 1 \tag{7.44}$$

However, the nuclear scatter is nonzero for all triplet Miller integers(h, k, l):

$$|\overline{b}_\mathrm{A} + \overline{b}_\mathrm{B} \exp[\mathrm{i}\pi(h + k + l)]|^2_{\mathrm{nuc}} \neq 0 \Rightarrow \left(\frac{\mathrm{d}\sigma}{\mathrm{d}\Omega}\right)^{\mathrm{mag}}_{\mathrm{avg}} \neq 0 \quad \text{all } h, k, l \text{ and } \overline{b}_\mathrm{A} \neq \overline{b}_\mathrm{B} \tag{7.45}$$

If the neutron scatter vector $\boldsymbol{\kappa}$ equals the reciprocal-lattice vector $\boldsymbol{\tau}_\tau$, whose Miller indices $h + k + l$ add up to an even integer, then there is nuclear scatter without magnetic scatter, and the Bragg peak at (h, k, l) is entirely from nuclear scatter. However, if $\boldsymbol{\kappa}$ equals a different reciprocal-lattice vector $\boldsymbol{\tau}_{\tau'}$ whose Miller indices $h' + k' + l'$ add up to an odd integer, then there is both nuclear and magnetic scatter, and the Bragg peak at (h', k', l') is the sum of nuclear and magnetic scatter. The nuclear scatter is zero for Miller indices $h' + k' + l'$, which add up to an odd integer, where the nuclear scatter lengths of atoms A and B are identical ($\overline{b}_\mathrm{A} = \overline{b}_\mathrm{B}$).

7.3 Coherent, Elastic Differential Cross Section for Magnetic and Nuclear Scatter of Neutron Spin States in a Bravais-Lattice Ferromagnetic Crystal—Localized Unpaired Electrons

We now derive the differential cross section for elastic scatter of neutrons from ferromagnetic materials, which includes nuclear scattering, magnetic scatter from the unpaired electrons, and the interference between them. Following section 2.3 in Squires (1996), and Eq. 483 with Eq. 456 in Ch. 3 of Cremer (2012b) was modified to include neutron spin transitions. This led to Eq. 149 in the spin-state chapter of Cremer (2013), and Eq. (3.2) in this book, which is Eq. (7.15) in Squires (1996), and is repeated below. Eq. (7.46) is the partial differential cross section for (1) transition from initial scatter nuclei scatter system state to final state, and (2) transition from initial neutron spin state to final state, where

$$\left(\frac{d^2\sigma}{d\Omega\, dE}\right)^{mag}_{i,0s\rightarrow m,s} = \frac{k}{k_0}\left|\langle\phi_m\psi_s\psi|\hat{U}_f|\psi_0\psi_{0s}\phi_i\rangle\right|^2\delta(E_i - E_m + E_0 - E) \tag{7.46}$$

The transition matrix element $\langle\hat{U}_f\rangle$ is obtained by integrating over the neutron position $\mathbf{r}$ with the incident ψ_0 and scattered ψ, and the result can be expressed as the sum of the N unit cell scatter length operator $\hat{T}_l$ for neutron magnetic and nuclear scatter, where

$$\left(\frac{m_n}{2\pi\hbar^2}\right)\langle\psi|\hat{U}_f|\psi_0\rangle = \sum_{l=1}^{N}\hat{T}_l \tag{7.47}$$

The scatter length operator $\hat{T}_l$ for the lth unit cell of the Bravais-lattice ferromagnetic crystal includes the neutron interaction with nuclear spin and scatter length. Scalar, scatter length operator $\hat{T}_l$ includes the dot product of the neutron spin vector operator $\hat{\mathbf{S}}_n$ with the nuclear spin vector operator $\hat{\mathbf{I}}_l$ and the magnetic scatter length vector operator $\hat{\mathbf{C}}_l$ of the lth unit cell, where

$$\hat{T}_l = (A_l + 2B_l\hat{\mathbf{S}}_n\cdot\hat{\mathbf{I}}_l) - \hat{\mathbf{S}}_n\cdot\hat{\mathbf{C}}_l \tag{7.48}$$

In coherent scatter, the average over isotopes and spin states is

$$A_l = \bar{b} \quad \text{and} \quad B_l = 0 \tag{7.49}$$

The interaction of neutron spin vector operator $\hat{\mathbf{S}}_n$ with nuclear spin vector operator $\hat{\mathbf{I}}_l$ is then absent in coherent scatter in the Bravais lattice, where

$$2B_l\hat{\mathbf{S}}_n\cdot\hat{\mathbf{I}}_l = 0 \tag{7.50}$$

Hence, scatter length operator $\hat{T}_l$ for coherent nuclear and magnetic scatter becomes

$$\hat{T}_l = \overline{b} - \hat{\mathbf{S}}_n \cdot \hat{\mathbf{C}}_l \tag{7.51}$$

The mean neutron scatter length $\overline{b}$ is the randomly oriented nuclear spins and unpolarized neutrons, averaged over the isotopes of the N unit cells of the Bravais-lattice crystal with 1 atom per unit cell, where

$$\overline{b} = \left(\frac{b^+(I+1)+b^-I}{2I+1}\right)_{\text{sp,iso}} \tag{7.52}$$

The nucleus spin quantum number is I.

The magnetic scatter length vector operator $\hat{\mathbf{C}}_l$ for the lth unit cell of the Bravais lattice is

$$\hat{\mathbf{C}}_l = \frac{1}{2} r_0 g F_0(\boldsymbol{\kappa}) \frac{\boldsymbol{\kappa} \times \langle \mathbf{S}_l \rangle \times \boldsymbol{\kappa}}{\kappa^2} \tag{7.53}$$

The magnetic form factor $F_0(\boldsymbol{\kappa})$ is the Fourier transform in the κ-space of the direct-lattice, normalized-density $\rho(\mathbf{r})$ unpaired electrons, where

$$F_0(\boldsymbol{\kappa}) = \int_{-\infty}^{\infty} \mathrm{d}z \int_{-\infty}^{\infty} \mathrm{d}y \int_{-\infty}^{\infty} \mathrm{d}x \rho(\mathbf{r}) \exp(\mathrm{i}\boldsymbol{\kappa} \cdot \mathbf{r}) \tag{7.54}$$

The thermally-based, expected value of the effective, electron spin vector operator $\langle \hat{\mathbf{S}}_l \rangle$ of the lth unit cell is

$$\langle \hat{\mathbf{S}}_l \rangle = \sum_{i=1}^{J} p_i \langle \phi_i | \hat{\mathbf{S}}_l | \phi_i \rangle \tag{7.55}$$

In a ferromagnet, the mean, effective, electron spin vector operator $\hat{\mathbf{S}}_l$ is the same in each of the N unit cells:

$$\langle \hat{\mathbf{S}}_l \rangle = \langle \hat{\mathbf{S}} \rangle \quad \phi_i | \hat{\mathbf{S}}_l | \phi_i \tag{7.56}$$

Hence,

$$\hat{\mathbf{C}}_l = \frac{1}{2} r_0 g F_0(\boldsymbol{\kappa}) \frac{\boldsymbol{\kappa} \times \langle \hat{\mathbf{S}} \rangle \times \boldsymbol{\kappa}}{\kappa^2} \tag{7.57}$$

The spin-up neutron nuclear scattering length is b^+, and the nuclear scattering length for spin-down neutrons is b^-. One obtains the expected values of the scatter lengths b_{f++} and b_{f--} for the case of neutron scatter, and the no-spin-flip transitions $(+ \to +)$ and $(- \to -)$, given neutron spin-up ψ_{0+} and spin-down ψ_{0-} eigenstates, are

$$b_{f++} = \langle\psi_{0+}|\hat{T}_l|\psi_{0+}\rangle = \overline{b} - C_{lz} \tag{7.58}$$

$$b_{f--} = \langle\psi_{0-}|\hat{T}_l|\psi_{0-}\rangle = \overline{b} + C_{lz} \tag{7.59}$$

Here, C_{lz} is the z-component of magnetic scatter length vector $\mathbf{C}_l$ for the lth unit cell. The expected values of the neutron scatter lengths b_{f-+} and b_{f+-} for the case of neutron scatter spin-flip transitions $(- \to +)$ and $(+ \to -)$ are

$$b_{f-+} = \langle\psi_{0-}|\hat{T}_l|\psi_{0+}\rangle = -(C_{lx} + \mathrm{i}C_{ly}) \tag{7.60}$$

$$b_{f+-} = \langle\psi_{0+}|\hat{T}_l|\psi_{0-}\rangle = -(C_{lx} - \mathrm{i}C_{ly}) \tag{7.61}$$

Here, C_{lx} and C_{ly} are the x- and y-components of magnetic scatter length vector $\mathbf{C}_l$ for the lth unit cell.

Recall that the coherent elastic differential scatter cross section for nuclear scatter of neutrons is

$$\left(\frac{\mathrm{d}\sigma}{\mathrm{d}\Omega}\right)^{\text{nuc}}_{\text{coh,el}} = \frac{\sigma_{\text{coh}}}{4\pi} N \frac{(2\pi)^3}{v_0} \exp(-2W) \sum_{\tau=0}^{N-1} \delta(\boldsymbol{\kappa} - \boldsymbol{\tau}_\tau) \tag{7.62}$$

where

$$2W = -\langle U^2\rangle = \langle(\kappa \cdot \mathbf{u}_0)^2\rangle \tag{7.63}$$

and for the mean scatter length $\overline{b}$, it is

$$\sigma_{\text{coh}} = 4\pi\overline{b}^2 \tag{7.64}$$

One replaces $\sigma/4\pi$ by the squared magnitude of this scatter length for the particular neutron scatter spin-state transition. The coherent, elastic differential cross sections for neutron scatter from a Bravais-lattice ferromagnetic crystal becomes

$$\left(\frac{\mathrm{d}\sigma}{\mathrm{d}\Omega}\right)^{++}_{\text{coh,el}} = N|\overline{b} - C_{lz}|^2 \frac{(2\pi)^3}{v_0} \exp(-2W) \sum_{\tau=1}^{N} \delta(\boldsymbol{\kappa} - \boldsymbol{\tau}_\tau) \tag{7.65}$$

$$\left(\frac{\mathrm{d}\sigma}{\mathrm{d}\Omega}\right)^{--}_{\text{coh,el}} = N|\overline{b} + C_{lz}|^2 \frac{(2\pi)^3}{v_0} \exp(-2W) \sum_{\tau=1}^{N} \delta(\boldsymbol{\kappa} - \boldsymbol{\tau}_\tau) \tag{7.66}$$

$$\left(\frac{\mathrm{d}\sigma}{\mathrm{d}\Omega}\right)^{-+}_{\text{coh,el}} = N|C_{lx} + \mathrm{i}C_{ly}|^2 \frac{(2\pi)^3}{v_0} \exp(-2W) \sum_{\tau=1}^{N} \delta(\boldsymbol{\kappa} - \boldsymbol{\tau}_\tau) \tag{7.67}$$

and

$$\left(\frac{\mathrm{d}\sigma}{\mathrm{d}\Omega}\right)^{+-}_{\mathrm{coh,el}} = N|C_{lx} - \mathrm{i}C_{ly}|^2 \frac{(2\pi)^3}{v_0} \exp(-2W) \sum_{\tau=1}^{N} \delta(\boldsymbol{\kappa} - \boldsymbol{\tau}_\tau) \tag{7.68}$$

7.4 Coherent, Elastic Differential Cross Section for Magnetic and Nuclear Scatter of Neutron Spin States in a Non-Bravais-Lattice Ferromagnetic Crystal—Localized Unpaired Electrons

One can obtain the partial differential cross sections of the non-Bravais-lattice crystal from the Bravais-lattice crystal by simple modification of the scatter length operator $\hat{T}_l$ via the following method: (1) replace the mean neutron scatter length $\overline{b}$ by the nuclear unit cell structure factor $F_N(\boldsymbol{\kappa})$, and (2) in the magnetic scatter length vector $\mathbf{C}_l$ for the lth unit cell, replace the product $F_0(\boldsymbol{\kappa})\langle\hat{\mathbf{S}}_l\rangle$ by

$$\sum_{d=1}^{r} g_d F_d(\boldsymbol{\kappa}) \exp(\mathrm{i}\boldsymbol{\kappa}\cdot\mathbf{d}_d)\exp(-W_d)\langle\mathbf{S}_d\rangle \tag{7.69}$$

The summation over the r atoms in each lth unit cell is the product of the following five terms for each dth atom; namely (1) the Landé splitting factor g_d, (2) the magnetic form factor $F_d(\boldsymbol{\kappa})$, (3) the phase shift $\exp(\mathrm{i}\boldsymbol{\kappa}\cdot\mathbf{d}_d)$, (4) the Debye–Waller factor $\exp(-W_d)$, and (5) the effective, mean, electron spin vector operator $\langle\hat{\mathbf{S}}_{ld}\rangle$. The magnetic form factor $F_d(\boldsymbol{\kappa})$ for normalized-density $\rho_d(\mathbf{r})$ of the dth atom's unpaired electrons in every unit cell of the ferromagnetic crystal is

$$F_d(\boldsymbol{\kappa}) = \int_{-\infty}^{\infty} \mathrm{d}z \int_{-\infty}^{\infty} \mathrm{d}y \int_{-\infty}^{\infty} \mathrm{d}x \rho_d(\mathbf{r})\exp(\mathrm{i}\boldsymbol{\kappa}\cdot\mathbf{r}) \tag{7.70}$$

The magnetic scatter length vector $\mathbf{C}_l$ for the lth unit cell in a non-Bravais lattice is then

$$\hat{\mathbf{C}}_l = \frac{1}{2}\frac{r_0 g}{\kappa^2}\left\{\boldsymbol{\kappa}\times\left[\sum_{d=1}^{r} F_d(\boldsymbol{\kappa})\exp(\mathrm{i}\boldsymbol{\kappa}\cdot\mathbf{d}_d)\exp(-W_d)\langle\hat{\mathbf{S}}_{ld}\rangle\right]\times\boldsymbol{\kappa}\right\} \tag{7.71}$$

where

$$\langle\mathbf{S}_{ld}\rangle = \sum_{i=1}^{J} p_i\langle\phi_i|\mathbf{S}_{ld}|\phi_i\rangle \tag{7.72}$$

The thermally averaged mean effective electron spin for the dth atom is the same in each of the N unit cells:

$$\langle\mathbf{S}_{ld}\rangle = \langle\mathbf{S}_d\rangle \quad l = 0, 1, 2, 3, \ldots, N-1 \tag{7.73}$$

Hence, the magnetic scatter length vector $\hat{\mathbf{C}}_{ld}$ becomes

$$\hat{\mathbf{C}}_l = \frac{1}{2}\frac{r_0 g}{\kappa^2}\left\{\boldsymbol{\kappa}\times\left[\sum_{d=1}^{r} F_d(\boldsymbol{\kappa})\exp(i\boldsymbol{\kappa}\cdot\mathbf{d}_d)\exp(-W_d)\langle\mathbf{S}_d\rangle\right]\times\boldsymbol{\kappa}\right\} \tag{7.74}$$

In coherent scatter for the non-Bravais-lattice ferromagnetic crystal, the average over isotopes and spin states gives

$$A_l = F_N(\boldsymbol{\kappa}) \quad \text{and} \quad B_l = 0 \tag{7.75}$$

The interaction of the neutron spin vector operator $\hat{\mathbf{S}}_n$ with the nuclear spin vector operator $\hat{\mathbf{I}}_l$ is then absent in the coherent scatter in the non-Bravais lattice as with the Bravais lattice, where

$$2B_l\hat{\mathbf{S}}_n\cdot\hat{\mathbf{I}}_l = 0 \tag{7.76}$$

The scatter length $\hat{T}_l$ for every l^{th} unit cell of the non-Bravais-lattice ferromagnetic crystal for nuclear and magnetic neutron scatter is

$$\hat{T}_l = F_N - \hat{\mathbf{S}}_n\cdot\hat{\mathbf{C}}_l \tag{7.77}$$

The nuclear unit cell structure factor $F_N(\boldsymbol{\kappa})$ provides the unit cell amplitude, which arises from the atoms within the unit cell, and is a function of the change of the neutron wave vector upon scatter $\boldsymbol{\kappa}$, where

$$F_N(\boldsymbol{\kappa}) = \sum_{d=1}^{r}\overline{b}_d\exp(i\boldsymbol{\kappa}\cdot\mathbf{d}_d)\exp(-W_d) \tag{7.78}$$

The nuclear unit cell structure factor $F_N(\boldsymbol{\kappa})$ is the nuclear scatter length for each unit cell of the crystal. The mean neutron scatter length $\overline{b}_d$ is the average scatter length for the dth atom in each of the N unit cells of the non-Bravais-lattice crystal with r atoms per unit cell. The average is taken over the randomly oriented nuclear spins and unpolarized neutrons, and then over the isotopes of the particular d^{th} nucleus position in each unit cell, where

$$\overline{b}_d = \left(\frac{b_d^+(I+1)+b_d^- I}{2I+1}\right)_{\text{iso}} \tag{7.79}$$

The nucleus spin quantum number is I. In each unit cell for the dth nucleus, the spin-up neutron nuclear scattering length is b_d^+ and the nuclear scattering length for spin-down neutrons is b_d^-. The Debye–Waller factor $\exp(-W_d)$ for the dth nucleus is

$$W_d = \frac{1}{2}\langle|\boldsymbol{\kappa}\cdot\hat{\mathbf{u}}_{0d}(0)|^2\rangle \tag{7.80}$$

For a non-Bravais-lattice crystal, the expected values of the scatter lengths b_{f++}^{nb} and b_{f--}^{nb} for neutron no-spin-flip transitions $(+ \rightarrow +)$ and $(- \rightarrow -)$ and the expected values of the scatter lengths b_{f-+}^{nb} and b_{f+-}^{nb} for neutron spin-flip transitions $(- \rightarrow +)$ and $(+ \rightarrow -)$, given neutron spin-up ψ_{0+} and spin-down ψ_{0-} eigenstates, are

$$b_{f++}^{\mathrm{nb}} = \langle\psi_{0+}|\hat{T}_l|\psi_{0+}\rangle = F_N(\boldsymbol{\kappa}) - C_{lz} \tag{7.81}$$

$$b_{f--}^{\mathrm{nb}} = \langle\psi_{0-}|\hat{T}_l|\psi_{0-}\rangle = F_N(\boldsymbol{\kappa}) + C_{lz} \tag{7.82}$$

$$b_{f-+}^{\mathrm{nb}} = \langle\psi_{0-}|\hat{T}_l|\psi_{0+}\rangle = -\left(C_{lx} + \mathrm{i}C_{ly}\right) \tag{7.83}$$

$$b_{f+-}^{\mathrm{nb}} = \langle\psi_{0+}|\hat{T}_l|\psi_{0-}\rangle = -\left(C_{lx} - \mathrm{i}C_{ly}\right) \tag{7.84}$$

The refractive index $n = 1 - \delta$ is expressed by the decrement δ, where $\delta \ll 1$. The decrements for the refraction in each of the four possible neutron spin-state transitions are obtained from the corresponding ferromagnetic nuclear/magnetic scatter lengths. Given neutron wavelength λ and the number density of the ferromagnetic atoms ρ, the refractive index for the four possible neutron spin transitions, the two no-spin-flip cases $(+ \rightarrow +)$ and $(- \rightarrow -)$, and the two spin-flip cases $(- \rightarrow +)$ and $(+ \rightarrow -)$ are

$$n_{++} = 1 - \frac{\rho}{2\pi}\lambda^2 b_{f++}^{\mathrm{nb}} \tag{7.85}$$

$$n_{--} = 1 - \frac{\rho}{2\pi}\lambda^2 b_{f--}^{\mathrm{nb}} \tag{7.86}$$

$$n_{-+} = 1 - \frac{\rho}{2\pi}\lambda^2 b_{f-+}^{\mathrm{nb}} \tag{7.87}$$

and

$$n_{+-} = 1 - \frac{\rho}{2\pi}\lambda^2 b_{f+-}^{\mathrm{nb}} \tag{7.88}$$

The total refractive index n for the magnetic scatter of neutrons in the ferromagnetic material, which includes all four possible neutron spin transitions, are

$$n = 1 - \frac{\rho}{2\pi}\lambda^2\left(b_{f++}^{\mathrm{nb}} + b_{f--}^{\mathrm{nb}} + b_{f-+}^{\mathrm{nb}} + b_{f+-}^{\mathrm{nb}}\right) \tag{7.89}$$

These scatter lengths include both nuclear and magnetic scatter of neutrons by the unpaired atomic electrons of the ferromagnetic atoms. Thus, these magnetic scatter lengths have been averaged over the nuclei isotopes and spin states of the nuclei.

We now recall Eq. 275 in Ch. 6 of Cremer (2012b) or Eq. (3.75) in Squires (1996) for the the differential cross section for neutron coherent, elastic scatter from a non-Bravais lattice crystal, which is given in Eq. (7.90) below. The nuclear coherent elastic scatter cross section $(\mathrm{d}\sigma/d\Omega)^{\mathrm{nuc}}_{\mathrm{coh,el}}$ for a non-Bravais-lattice crystal is given in terms of the nuclear unit cell structure factor $F_N(\boldsymbol{\kappa})$, where

$$\left(\frac{\mathrm{d}\sigma}{\mathrm{d}\Omega}\right)^{\mathrm{nuc}}_{\mathrm{coh,el}} = N|F_N(\boldsymbol{\kappa})|^2\frac{(2\pi)^3}{v_0}\sum_{\tau=0}^{N-1}\delta(\boldsymbol{\kappa}-\boldsymbol{\tau}_\tau) \tag{7.90}$$

In the case of combined nuclear and magnetic scatter, one replaces the nuclear scatter length $F_N(\boldsymbol{\kappa})$ for a unit cell by the summed nuclear and magnetic scatter length. Each of the four neutron transitions has its own combined nuclear and magnetic scatter length, where

$$F_N(\boldsymbol{\kappa}) \Rightarrow b^{\mathrm{nb}}_{f++} = F_N(\boldsymbol{\kappa}) - C_{lz} \quad + \to + \tag{7.91}$$

$$F_N(\boldsymbol{\kappa}) \Rightarrow b^{\mathrm{nb}}_{f--} = F_N(\boldsymbol{\kappa}) + C_{lz} \quad - \to - \tag{7.92}$$

$$F_N(\boldsymbol{\kappa}) \Rightarrow b^{\mathrm{nb}}_{f-+} = -(C_{lx} + \mathrm{i}C_{ly}) \quad - \to + \tag{7.93}$$

$$F_N(\boldsymbol{\kappa}) \Rightarrow b^{\mathrm{nb}}_{f+-} = -(C_{lx} - \mathrm{i}C_{ly}) \quad + \to - \tag{7.94}$$

The coherent, elastic differential cross sections for the four possible neutron spin transitions for both nuclear and magnetic neutron scatter from a non-Bravais-lattice ferromagnetic crystal is

$$\left(\frac{\mathrm{d}\sigma}{\mathrm{d}\Omega}\right)^{++}_{\mathrm{coh,el}} = N|F_N(\boldsymbol{\kappa}) - C_{lz}|^2\frac{(2\pi)^3}{v_0}\sum_{\tau=1}^{N}\delta(\boldsymbol{\kappa}-\boldsymbol{\tau}_\tau) \tag{7.95}$$

$$\left(\frac{\mathrm{d}\sigma}{\mathrm{d}\Omega}\right)^{--}_{\mathrm{coh,el}} = N|F_N(\boldsymbol{\kappa}) + C_{lz}|^2\frac{(2\pi)^3}{v_0}\sum_{\tau=1}^{N}\delta(\boldsymbol{\kappa}-\boldsymbol{\tau}_\tau) \tag{7.96}$$

$$\left(\frac{\mathrm{d}\sigma}{\mathrm{d}\Omega}\right)^{-+}_{\mathrm{coh,el}} = N|C_{lx} + \mathrm{i}C_{ly}|^2\frac{(2\pi)^3}{v_0}\sum_{\tau=1}^{N}\delta(\boldsymbol{\kappa}-\boldsymbol{\tau}_\tau) \tag{7.97}$$

$$\left(\frac{\mathrm{d}\sigma}{\mathrm{d}\Omega}\right)^{+-}_{\mathrm{coh,el}} = N|C_{lx} - \mathrm{i}C_{ly}|^2\frac{(2\pi)^3}{v_0}\sum_{\tau=1}^{N}\delta(\boldsymbol{\kappa}-\boldsymbol{\tau}_\tau) \tag{7.98}$$

The total magnetic coherent, elastic differential scatter cross section $(\mathrm{d}\sigma/d\Omega)^{\mathrm{tot}}_{\mathrm{coh,el}}$ is the sum of the contributions from the four possible spin transitions, where

$$\left(\frac{\mathrm{d}\sigma}{\mathrm{d}\Omega}\right)^{\mathrm{tot}}_{\mathrm{coh,el}} = \left(\frac{\mathrm{d}\sigma}{\mathrm{d}\Omega}\right)^{++}_{\mathrm{coh,el}} + \left(\frac{\mathrm{d}\sigma}{\mathrm{d}\Omega}\right)^{--}_{\mathrm{coh,el}} + \left(\frac{\mathrm{d}\sigma}{\mathrm{d}\Omega}\right)^{-+}_{\mathrm{coh,el}} + \left(\frac{\mathrm{d}\sigma}{\mathrm{d}\Omega}\right)^{+-}_{\mathrm{coh,el}} \tag{7.99}$$

When the neutrons encounter the magnetic field at the face of the ferromagnetic slab, the azimuth components of the neutron spins will be directed either parallel or antiparallel to the magnetic field vector **B** of the ferromagnetic slab. Assume that the magnetic field **B** is directed in the negative z-direction, where

$$\mathbf{B} = -B_z\hat{\mathbf{z}} \tag{7.100}$$

The thermally averaged mean effective electron spin $\langle \mathbf{S}_d \rangle$ of each dth atom of the r atoms in each unit cell are aligned antiparallel to the magnetic field vector, where

$$\langle \mathbf{S}_d \rangle = \langle S_{dz} \rangle \hat{\mathbf{z}} \tag{7.101}$$

We now evaluate the double cross product of the neutron scatter vector $\boldsymbol{\kappa}$ with the thermally averaged mean z-directed electron spin vector $\langle \mathbf{S}_d \rangle$ of the effective electron spin for the dth atom of r atoms in every unit cell of a ferromagnetic crystal, where

$$\hat{\mathbf{C}}_l = \left[\sum_{d=1}^{r} \hat{\mathbf{C}}_{ld}\right] \tag{7.102}$$

where

$$\hat{\mathbf{C}}_{ld} = G_d(\boldsymbol{\kappa} \times \langle \mathbf{S}_d \rangle \times \boldsymbol{\kappa}) \tag{7.103}$$

and

$$G_d = \frac{1}{2}\frac{r_0 g}{\kappa^2} g_d F_d(\boldsymbol{\kappa}) \exp(i\boldsymbol{\kappa} \cdot \mathbf{d}_d) \exp(-W_d) \tag{7.104}$$

The first cross product of $\hat{\mathbf{C}}_{ld}$ is

$$\langle \mathbf{S}_d \rangle \times \boldsymbol{\kappa} = \begin{vmatrix} \hat{\mathbf{x}} & \hat{\mathbf{y}} & \hat{\mathbf{z}} \\ 0 & 0 & \langle S_{dz} \rangle \\ \kappa_x & \kappa_y & \kappa_z \end{vmatrix} = -\kappa_y \langle S_{dz} \rangle \hat{\mathbf{x}} + \kappa_x \langle S_{dz} \rangle \hat{\mathbf{y}} \tag{7.105}$$

and the second cross product of $\hat{\mathbf{C}}_{ld}$ is

$$\hat{\mathbf{C}}_{ld} = G_d(\boldsymbol{\kappa} \times \langle \mathbf{S}_d \rangle \times \boldsymbol{\kappa}) = G_d \begin{vmatrix} \hat{\mathbf{x}} & \hat{\mathbf{y}} & \hat{\mathbf{z}} \\ \kappa_x & \kappa_y & \kappa_z \\ -\kappa_y \langle S_{dz} \rangle & \kappa_x \langle S_{dz} \rangle & 0 \end{vmatrix} \tag{7.106}$$

Expanding, one obtains the double cross product, which yields $\hat{\mathbf{C}}_{ld}$, where

$$\hat{\mathbf{C}}_{ld} = G_d\{-\kappa_x\kappa_z\langle S_{dz} \rangle \hat{\mathbf{x}} - \kappa_z\kappa_y\langle S_{dz} \rangle \hat{\mathbf{y}} + [\kappa_x^2(\langle S_{dz} \rangle) + \kappa_y^2(\langle S_{dz} \rangle)]\hat{\mathbf{z}}\} \tag{7.107}$$

If $\boldsymbol{\kappa}$ is parallel or antiparallel to the negative z-directed magnetic field B_z ($\boldsymbol{\kappa}||\mathbf{B}$), then $\boldsymbol{\kappa}$ is antiparallel or parallel, respectively, to the positive z-directed, the thermally averaged mean z-component $\langle S_{dz}\rangle$ of the total effective spin of the unpaired electron spin and electron orbital motion in the ferromagnetic material, where

$$\mathbf{B} = -B_z\hat{\mathbf{z}} \tag{7.108}$$

and

$$\boldsymbol{\kappa}||\mathbf{B} \rightarrow \boldsymbol{\kappa}||\langle \mathbf{S}_{ld}\rangle \tag{7.109}$$

With $\boldsymbol{\kappa}$ parallel or antiparallel to magnetic field $\mathbf{B}$ and $\langle S_{dz}\rangle$, the magnetic scatter length vector $\hat{\mathbf{C}}_{ld}$ is zero because

$$\kappa_x = \kappa_y = 0 \quad \text{and} \quad \boldsymbol{\kappa} = \kappa_z\hat{\mathbf{z}} \quad \boldsymbol{\kappa}||\mathbf{B} \tag{7.110}$$

where

$$\hat{\mathbf{C}}_{ld} = 0 \quad \boldsymbol{\kappa}||\mathbf{B} \tag{7.111}$$

Hence, the magnetic component of the differential cross section for each of the four possible neutron spin-flip transitions is zero, and one is left only with nuclear scatter, where

$$(\mathrm{d}\sigma/d\Omega)^{\mathrm{mag}}_{\mathrm{coh,el}} = 0 \quad \boldsymbol{\kappa}||\mathbf{B} \tag{7.112}$$

and

$$\left(\frac{\mathrm{d}\sigma}{\mathrm{d}\Omega}\right)^{++}_{\mathrm{coh,el}} = N|F_N(\boldsymbol{\kappa})|^2\frac{(2\pi)^3}{v_0}\sum_{\tau=1}^{N}\delta(\boldsymbol{\kappa}-\boldsymbol{\tau}_\tau) \quad \boldsymbol{\kappa}||\mathbf{B} \tag{7.113}$$

$$\left(\frac{\mathrm{d}\sigma}{\mathrm{d}\Omega}\right)^{--}_{\mathrm{coh,el}} = N|F_N(\boldsymbol{\kappa})|^2\frac{(2\pi)^3}{v_0}\sum_{\tau=1}^{N}\delta(\boldsymbol{\kappa}-\boldsymbol{\tau}_\tau) \quad \boldsymbol{\kappa}||\mathbf{B} \tag{7.114}$$

$$\left(\frac{\mathrm{d}\sigma}{\mathrm{d}\Omega}\right)^{-+}_{\mathrm{coh,el}} = 0 \quad \boldsymbol{\kappa}||\mathbf{B} \tag{7.115}$$

$$\left(\frac{\mathrm{d}\sigma}{\mathrm{d}\Omega}\right)^{+-}_{\mathrm{coh,el}} = 0 \quad \boldsymbol{\kappa}||\mathbf{B} \tag{7.116}$$

The total combined nuclear and magnetic scatter cross section $(\mathrm{d}\sigma/d\Omega)^{\kappa||B}_{\mathrm{coh,el}}$ for scatter of neutrons with neutron scatter vector parallel to the magnetic field is

$$\left(\frac{\mathrm{d}\sigma}{\mathrm{d}\Omega}\right)^{\kappa||B}_{\mathrm{coh,el}} = \left(\frac{\mathrm{d}\sigma}{\mathrm{d}\Omega}\right)^{++}_{\mathrm{coh,el}} + \left(\frac{\mathrm{d}\sigma}{\mathrm{d}\Omega}\right)^{--}_{\mathrm{coh,el}} \quad \boldsymbol{\kappa}||\mathbf{B} \tag{7.117}$$

or

$$\left(\frac{d\sigma}{d\Omega}\right)^{\kappa||B}_{\text{coh,el}} = 2N|F_N(\boldsymbol{\kappa})|^2 \frac{(2\pi)^3}{v_0} \sum_{\tau=1}^{N} \delta(\boldsymbol{\kappa} - \boldsymbol{\tau}_\tau) \quad \boldsymbol{\kappa}||\mathbf{B} \tag{7.118}$$

If $\boldsymbol{\kappa}$ is perpendicular to the negative z-directed magnetic field $\mathbf{B}$ ($\boldsymbol{\kappa}\perp\mathbf{B}$), where

$$\mathbf{B} = -B_z\hat{\mathbf{z}} \tag{7.119}$$

then $\boldsymbol{\kappa}$ is perpendicular to the positive z-directed, $\langle S_{dz}\rangle$ ferromagnetic electron spins

$$\kappa_z = 0 \quad \text{and} \quad \boldsymbol{\kappa} = \kappa_x\hat{\mathbf{x}} + \kappa_y\hat{\mathbf{y}} \quad \boldsymbol{\kappa}\perp\mathbf{B} \tag{7.120}$$

The magnetic scatter length vector $\hat{\mathbf{C}}_{ld}$ has only a z-component, and no x- or y-components:

$$\hat{\mathbf{C}}_{ld} = G_d[\kappa_x^2(\langle S_{dz}\rangle) + \kappa_y^2(\langle S_{dz}\rangle)]\hat{\mathbf{z}} \quad \boldsymbol{\kappa}\perp\mathbf{B} \tag{7.121}$$

If the neutron scatter vector $\boldsymbol{\kappa}$ is perpendicular to the magnetic field $\mathbf{B}$ ($\boldsymbol{\kappa}\perp\mathbf{B}$), then $\boldsymbol{\kappa}$ is also perpendicular to the thermally averaged spin vector $|\mathbf{S}_{ld}\rangle$ of the unpaired electrons of the ferromagnetic material. Thus, there are no neutron spin-flip transitions, where

$$\hat{\mathbf{C}}_{ld} = C_{ld}\hat{\mathbf{z}} \quad \boldsymbol{\kappa}\perp\mathbf{B} \tag{7.122}$$

and

$$\boldsymbol{\kappa}\perp\mathbf{B} \rightarrow \boldsymbol{\kappa}\perp\langle\mathbf{S}_{ld}\rangle \tag{7.123}$$

Hence,

$$\left(\frac{d\sigma}{d\Omega}\right)^{-+}_{\text{coh}} = 0 \quad \boldsymbol{\kappa}\perp\mathbf{B} \tag{7.124}$$

$$\left(\frac{d\sigma}{d\Omega}\right)^{+-}_{\text{coh}} = 0 \quad \boldsymbol{\kappa}\perp\mathbf{B} \tag{7.125}$$

The two no-spin-flip, neutron coherent, elastic differential cross sections become

$$\left(\frac{d\sigma}{d\Omega}\right)^{++}_{\text{coh}} = N|F_N(\boldsymbol{\kappa}) - C_{lz}|^2 \frac{(2\pi)^3}{v_0} \sum_{\tau=1}^{N} \delta(\boldsymbol{\kappa} - \boldsymbol{\tau}_\tau) \quad \boldsymbol{\kappa}\perp\mathbf{B} \tag{7.126}$$

$$\left(\frac{d\sigma}{d\Omega}\right)^{--}_{\text{coh}} = N|F_N(\boldsymbol{\kappa}) + C_{lz}|^2 \frac{(2\pi)^3}{v_0} \sum_{\tau=1}^{N} \delta(\boldsymbol{\kappa} - \boldsymbol{\tau}_\tau) \quad \boldsymbol{\kappa}\perp\mathbf{B} \tag{7.127}$$

The total combined nuclear and magnetic scatter cross section $(\mathrm{d}\sigma/d\Omega)^{\kappa\perp B}_{\mathrm{coh,el}}$ for scatter of neutrons with the neutron scatter vector perpendicular to the magnetic field is

$$\left(\frac{\mathrm{d}\sigma}{\mathrm{d}\Omega}\right)^{\kappa\perp B}_{\mathrm{coh,el}} = \left(\frac{\mathrm{d}\sigma}{\mathrm{d}\Omega}\right)^{++}_{\mathrm{coh,el}} + \left(\frac{\mathrm{d}\sigma}{\mathrm{d}\Omega}\right)^{--}_{\mathrm{coh,el}} \quad \boldsymbol{\kappa}\perp\mathbf{B} \tag{7.128}$$

or

$$\left(\frac{\mathrm{d}\sigma}{\mathrm{d}\Omega}\right)^{\kappa\perp B}_{\mathrm{coh,el}} = N(|F_N(\boldsymbol{\kappa}) - C_{lz}|^2 + |F_N(\boldsymbol{\kappa}) + C_{lz}||^2)\frac{(2\pi)^3}{v_0}\sum_{\tau=1}^{N}\delta(\boldsymbol{\kappa} - \boldsymbol{\tau}_\tau) \quad \boldsymbol{\kappa}\perp\mathbf{B} \tag{7.129}$$

The $\pm 2F_N C_{lz}$ terms in the $+ \to +$ and $- \to -$ no-spin-flip transitions are called the *nuclear-magnetic interference terms*, which are nonzero for only both nonzero nuclear scatter length F_N and magnetic scatter length C_{lz}. In ferromagnets, $\pm 2F_N C_{lz}$ are nonzero, and the magnetic and nuclear scatter Bragg peaks occur at the same point.

7.5 Production and Measurement of Polarized Neutrons by Ferromagnetic Materials

The $\boldsymbol{\kappa}\perp\mathbf{B}$ case has important applications in the production of polarized neutrons and in the detection and measurement of polarized neutron beams. If the neutron scatter vector is perpendicular to the ferromagnetic field $\boldsymbol{\kappa}\perp\mathbf{B}$, then the scattered neutrons retain their incident spin. In an incident beam of unpolarized neutrons, one obtains an unpolarized beam of scattered neutrons.

However, in certain fortunate circumstances, such as the (1, 1, 1) Bragg reflection in a cobalt–iron alloy that is 92% cobalt and 8% iron, one can obtain polarized scattered neutrons. In this case, the z-component of the magnetic scatter length C_{lz} equals the nuclear scatter length component $F_N(\boldsymbol{\kappa})$, where

$$F_N(\boldsymbol{\kappa}) = C_{lz} \tag{7.130}$$

Consider the case of the neutron scatter vector $\boldsymbol{\kappa}$, which is perpendicular to the ferromagnetic field $\mathbf{B}$, where the nuclear scatter length $F_N(\boldsymbol{\kappa})$ is equal or nearly equal to the z-component magnetic scatter length C_{lz}. Then for an incident, unpolarized beam of neutrons, one obtains scatter, spin-down polarized neutrons, where

$$\left(\frac{\mathrm{d}\sigma}{\mathrm{d}\Omega}\right)^{++}_{\mathrm{coh}} = N|F_N(\boldsymbol{\kappa}) - C_{lz}|^2\frac{(2\pi)^3}{v_0}\sum_{\tau=1}^{N}\delta(\boldsymbol{\kappa} - \boldsymbol{\tau}_\tau) = 0 \quad \boldsymbol{\kappa}\perp\mathbf{B} \text{ and } F_N(\boldsymbol{\kappa}) = C_{lz} \tag{7.131}$$

and

$$\left(\frac{d\sigma}{d\Omega}\right)_{coh}^{--} = N|F_N(\boldsymbol{\kappa}) + C_{lz}|^2 \frac{(2\pi)^3}{v_0} \sum_{\tau=1}^{N} \delta(\boldsymbol{\kappa} - \boldsymbol{\tau}_\tau) \quad \boldsymbol{\kappa}\perp\mathbf{B} \text{ and } F_N(\boldsymbol{\kappa}) = C_{lz} \tag{7.132}$$

The neutron does not spin-flip if $\boldsymbol{\kappa}\perp\mathbf{B}$; hence

$$\left(\frac{d\sigma}{d\Omega}\right)_{coh}^{-+} = \left(\frac{d\sigma}{d\Omega}\right)_{coh}^{+-} = 0 \quad \boldsymbol{\kappa}\perp\mathbf{B} \text{ and } F_N(\boldsymbol{\kappa}) = C_{lz} \tag{7.133}$$

Thus, spin-down polarized neutrons for the case $(- \to -)$ in Eq. (7.132), are obtained from unpolarized neutrons, via cancellation of nuclear scatter by the magnetic scatter for spin-up neutrons in Eq. (7.131), where

$$\left(\frac{d\sigma}{d\Omega}\right)_{coh}^{tot} = \left(\frac{d\sigma}{d\Omega}\right)_{coh}^{--} \quad \boldsymbol{\kappa}\perp\mathbf{B} \text{ and } F_N(\boldsymbol{\kappa}) = C_{lz} \tag{7.134}$$

In each lth unit cell, the magnetic form factor $F_l(\boldsymbol{\kappa})$ is related to the sum of the r atoms of unpaired electron density in the unit cell. With the dth atom electron density $\rho_d(\mathbf{r})$, the lth unit cell magnetic form factor $F_l(\boldsymbol{\kappa})$ is

$$F_l(\boldsymbol{\kappa}) = \sum_{d=1}^{r} F_d(\boldsymbol{\kappa}) = \sum_{d=1}^{l} \int_{-\infty}^{\infty} dz \int_{-\infty}^{\infty} dy \int_{-\infty}^{\infty} dx \rho_d(\mathbf{r}) \exp(i\boldsymbol{\kappa}\cdot\mathbf{r}) \tag{7.135}$$

The magnetic form factor $F_l(\boldsymbol{\kappa})$ for each lth unit cell is identical for the N unit cells of the ferromagnetic crystal and is now denoted by $F(\boldsymbol{\kappa})$, where

$$F(\boldsymbol{\kappa}) = F_l(\boldsymbol{\kappa}) \quad l = 1, 2, 3, \ldots, N \tag{7.136}$$

The measured intensity I_τ of the diffraction for the τth reciprocal-lattice vector of the ferromagnetic crystal is proportional to the squared magnitude of magnetic form factor $F(\boldsymbol{\kappa})$:

$$I_\tau \propto |F(\boldsymbol{\kappa})|^2 \tag{7.137}$$

If the diffraction measurements are taken with unpolarized neutrons, then the coherent differential cross section $(d\sigma/d\Omega)^{tot}_{coh,el}$ is

$$\left(\frac{d\sigma}{d\Omega}\right)^{tot}_{coh,el} = \left(\frac{d\sigma}{d\Omega}\right)^{++}_{coh,el} + \left(\frac{d\sigma}{d\Omega}\right)^{--}_{coh,el} = N\frac{(2\pi)^3}{v_0}\sum_{\tau=1}^{N}\delta(\boldsymbol{\kappa}-\boldsymbol{\tau}_\tau)$$
$$\left\{|F_N(\boldsymbol{\kappa}) - C_{lz}|^2 + |F_N(\boldsymbol{\kappa}) - C_{lz}|^2\right\} \quad \boldsymbol{\kappa}\perp\mathbf{B} \tag{7.138}$$

Expanding and collecting terms in Eq. (7.138), yields

$$\left(\frac{d\sigma}{d\Omega}\right)^{tot}_{coh,el} = N\frac{(2\pi)^3}{v_0}\left[F_N^2(\boldsymbol{\kappa}) + C_{lz}^2\right]\sum_{\tau=1}^{N}\delta(\boldsymbol{\kappa}-\boldsymbol{\tau}_\tau) \quad \boldsymbol{\kappa}\perp\mathbf{B} \tag{7.139}$$

In most cases, the nuclear scatter length $F_N(\boldsymbol{\kappa})$ is much larger than the z-component magnetic scatter length C_{lz}. The ratio of the magnetic scatter length, divided by the nuclear scatter length r, is defined as

$$r = \frac{C_{lz}}{F_N(\boldsymbol{\kappa})} \tag{7.140}$$

Typically,

$$r \ll 1 \tag{7.141}$$

Substitution of Eq. (7.140) into Eq. (7.139) gives

$$\left(\frac{d\sigma}{d\Omega}\right)^{tot}_{coh,el} = N\frac{(2\pi)^3}{v_0}F_N^2(\boldsymbol{\kappa})\left[1 + r^2\right]\sum_{\tau=1}^{N}\delta(\boldsymbol{\kappa}-\boldsymbol{\tau}_\tau) \quad \boldsymbol{\kappa}\perp\mathbf{B} \tag{7.142}$$

Thus, with $r \ll 1$, the difference in the diffraction intensity resulting from unpolarized neutron beam diffraction with a nonmagnetized iron crystal and a fully magnetized iron crystal would be very small, and thus determining the effect of magnetization on neutron elastic scatter would be difficult.

Now consider the experiment, where one compares the diffraction intensity of a polarized spin-up neutron beam to that of a polarized spin-down neutron beam in a fully magnetized iron crystal. In this case, one defines a flipping ratio R as the differential cross section for coherent scatter of spin-up neutrons, divided by that of spin-down neutrons, where

$$R = \frac{(d\sigma/d\Omega)^{++}_{coh,el}}{(d\sigma/d\Omega)^{--}_{coh,el}} = \frac{|F_N(\boldsymbol{\kappa}) - C_{lz}|^2}{|F_N(\boldsymbol{\kappa}) + C_{lz}|^2} = \left(\frac{1-r}{1+r}\right)^2 \quad \boldsymbol{\kappa}\perp\mathbf{B} \tag{7.143}$$

For small magnetic scatter to nuclear scatter length ratio $r \ll 1$,

$$R \cong 1 - 4r \quad \boldsymbol{\kappa} \perp \mathbf{B} \tag{7.144}$$

The magnetic scatter contribution $4r$ with use of polarized neutron beams is much larger than the magnetic scatter contribution r^2 with unpolarized beams. The incident-polarized beam of spin-up or spin-down neutrons is not flipped upon scatter, so one need only measure the intensity of the diffracted neutrons without the need of a polarization sensitive detector.

7.6 General Expression for the Coherent Differential Cross Section for Nuclear and Magnetic Elastic Scatter of Neutron Spin States in Ferromagnetic Materials—Localized or Delocalized Unpaired Electrons

Let us now consider a ferromagnetic material with its thermally averaged magnetization vector $\langle \mathbf{M}(\boldsymbol{\kappa}) \rangle$. The unpaired electron spin and orbital motion are assumed to produce the magnetization, and the unpaired electrons can be in a band, which is not localized to a nucleus, or the electrons can be localized to an atom. That is, the unpaired electrons can be itinerant, where the unpaired electrons are not localized to a single atom. Now we will derive the general coherent elastic differential cross section for combined nuclear and magnetic scatter of neutron spin states, where the electrons can be localized or nonlocalized in a non-Bravais-lattice ferromagnetic crystal.

We start by recalling Eq. (7.90), which is the nuclear coherent elastic scatter cross section $(\mathrm{d}\sigma/d\Omega)^{\mathrm{nuc}}_{\mathrm{coh,el}}$ for a non-Bravais-lattice crystal, and is expressed in terms of the thermally averaged nuclear unit cell structure factor or nuclear scatter length $F_N(\boldsymbol{\kappa})$, where Eq. (7.90) is repeated below:

$$\left(\frac{\mathrm{d}\sigma}{\mathrm{d}\Omega}\right)^{\mathrm{nuc}}_{\mathrm{coh,el}} = N|F_N(\boldsymbol{\kappa})|^2 \frac{(2\pi)^3}{v_0} \sum_{\tau=0}^{N-1} \delta(\boldsymbol{\kappa} - \boldsymbol{\tau}_\tau) \tag{7.145}$$

Recall from Eq. (7.78), the thermally averaged nuclear scatter length $F_N(\boldsymbol{\kappa})$ per unit cell is obtained by summing over the r atoms of the unit cell:

$$F_N(\boldsymbol{\kappa}) = \sum_{d=1}^{r} \bar{b}_d \exp(\mathrm{i}\boldsymbol{\kappa} \cdot \mathbf{d}_d) \exp(-W_d) \tag{7.146}$$

We now recall Eq. (3.294) from section 3.17, which is the coherent, elastic differential cross section $(\mathrm{d}\sigma/d\Omega)^{\mathrm{mag}}_{\mathrm{coh,el}}$ for magnetic scatter of neutrons, and is expressed in terms of the thermally averaged magnetic scatter length $b_M(\boldsymbol{\kappa})$. Eq. (3.294) is repeated below,

$$\left(\frac{d\sigma}{d\Omega}\right)^{\text{mag}}_{\text{coh,el}} = N[b_M(\boldsymbol{\kappa})]^2 \frac{(2\pi)^3}{v_0} \sum_{\tau=0}^{N-1} \delta(\boldsymbol{\kappa} - \boldsymbol{\tau}_\tau) \tag{7.147}$$

Recall from Eq. (3.293) of section 3.17, the thermally averaged magnetic scatter length $b_M(\boldsymbol{\kappa})$ per unit cell is

$$b_M(\boldsymbol{\kappa}) = |\overline{\mathbf{F}}_A(\boldsymbol{\kappa})| \left(\frac{r_0 g}{2\mu_b}\right) \left| \sum_{d=1}^{r} \exp(i\boldsymbol{\kappa} \cdot \mathbf{d}_d) \exp(-W_d) \right| |\boldsymbol{\kappa}_u \times \overline{\mathbf{F}}_{Au} \times \boldsymbol{\kappa}_u| \tag{7.148}$$

And recall that $\overline{\mathbf{F}}_{Au}$ is the mean magnetic form factor vector per single atom. The x-, y-, and z-components of magnetic scatter length $b_M(\boldsymbol{\kappa})$ are

$$b_{Mx}(\boldsymbol{\kappa}) = |\overline{\mathbf{F}}_A(\boldsymbol{\kappa})| \left(\frac{r_0 g}{2\mu_b}\right) \left| \sum_{d=1}^{r} \exp(i\boldsymbol{\kappa} \cdot \mathbf{d}_d) \exp(-W_d) \right| (\boldsymbol{\kappa}_u \times \overline{\mathbf{F}}_{Au} \times \boldsymbol{\kappa}_u)_x \tag{7.149}$$

$$b_{My}(\boldsymbol{\kappa}) = |\overline{\mathbf{F}}_A(\boldsymbol{\kappa})| \left(\frac{r_0 g}{2\mu_b}\right) \left| \sum_{d=1}^{r} \exp(i\boldsymbol{\kappa} \cdot \mathbf{d}_d) \exp(-W_d) \right| (\boldsymbol{\kappa}_u \times \overline{\mathbf{F}}_{Au} \times \boldsymbol{\kappa}_u)_y \tag{7.150}$$

and

$$b_{Mz}(\boldsymbol{\kappa}) = |\overline{\mathbf{F}}_A(\boldsymbol{\kappa})| \left(\frac{r_0 g}{2\mu_b}\right) \left| \sum_{d=1}^{r} \exp(i\boldsymbol{\kappa} \cdot \mathbf{d}_d) \exp(-W_d) \right| (\boldsymbol{\kappa}_u \times \overline{\mathbf{F}}_{Au} \times \boldsymbol{\kappa}_u)_z \tag{7.151}$$

In the case of combined nuclear and magnetic scatter, one replaces the nuclear scatter length $F_N(\boldsymbol{\kappa})$ for a unit cell by the summed nuclear and magnetic scatter lengths. Each of the four neutron transitions has its own combined nuclear and magnetic scatter length, where

$$F_N(\boldsymbol{\kappa}) \Rightarrow b^{\text{nb}}_{f++} = F_N(\boldsymbol{\kappa}) - b_{Mz} \quad + \rightarrow + \tag{7.152}$$

$$F_N(\boldsymbol{\kappa}) \Rightarrow b^{\text{nb}}_{f--} = F_N(\boldsymbol{\kappa}) + b_{Mz} \quad - \rightarrow - \tag{7.153}$$

$$F_N(\boldsymbol{\kappa}) \Rightarrow b^{\text{nb}}_{f-+} = -(b_{Mx} + ib_{My}) \quad - \rightarrow + \tag{7.154}$$

and

$$F_N(\boldsymbol{\kappa}) \Rightarrow b^{\text{nb}}_{f+-} = -(b_{Mx} - ib_{My}) \quad + \rightarrow - \tag{7.155}$$

The coherent, elastic differential cross sections for the four possible neutron spin transitions for both nuclear and magnetic neutron scatter from a non-Bravais-lattice ferromagnetic crystal is

$$\left(\frac{\mathrm{d}\sigma}{\mathrm{d}\Omega}\right)^{++}_{\mathrm{coh,el}} = N|F_N(\boldsymbol{\kappa}) - b_{Mz}|^2 \frac{(2\pi)^3}{v_0} \sum_{\tau=1}^{N} \delta(\boldsymbol{\kappa} - \boldsymbol{\tau}_\tau) \tag{7.156}$$

$$\left(\frac{\mathrm{d}\sigma}{\mathrm{d}\Omega}\right)^{--}_{\mathrm{coh,el}} = N|F_N(\boldsymbol{\kappa}) + b_{Mz}|^2 \frac{(2\pi)^3}{v_0} \sum_{\tau=1}^{N} \delta(\boldsymbol{\kappa} - \boldsymbol{\tau}_\tau) \tag{7.157}$$

$$\left(\frac{\mathrm{d}\sigma}{\mathrm{d}\Omega}\right)^{-+}_{\mathrm{coh,el}} = N|b_{Mx} + \mathrm{i}b_{My}|^2 \frac{(2\pi)^3}{v_0} \sum_{\tau=1}^{N} \delta(\boldsymbol{\kappa} - \boldsymbol{\tau}_\tau) \tag{7.158}$$

and

$$\left(\frac{\mathrm{d}\sigma}{\mathrm{d}\Omega}\right)^{+-}_{\mathrm{coh,el}} = N|b_{Mx} - \mathrm{i}b_{My}|^2 \frac{(2\pi)^3}{v_0} \sum_{\tau=1}^{N} \delta(\boldsymbol{\kappa} - \boldsymbol{\tau}_\tau) \tag{7.159}$$

The total magnetic coherent, elastic differential scatter cross section $(\mathrm{d}\sigma/d\Omega)^{\mathrm{tot}}_{\mathrm{coh,el}}$ is the sum of the contributions from the four possible spin transitions, where

$$\left(\frac{\mathrm{d}\sigma}{\mathrm{d}\Omega}\right)^{\mathrm{tot}}_{\mathrm{coh,el}} = \left(\frac{\mathrm{d}\sigma}{\mathrm{d}\Omega}\right)^{++}_{\mathrm{coh,el}} + \left(\frac{\mathrm{d}\sigma}{\mathrm{d}\Omega}\right)^{--}_{\mathrm{coh,el}} + \left(\frac{\mathrm{d}\sigma}{\mathrm{d}\Omega}\right)^{-+}_{\mathrm{coh,el}} + \left(\frac{\mathrm{d}\sigma}{\mathrm{d}\Omega}\right)^{+-}_{\mathrm{coh,el}} \tag{7.160}$$

This derivation for the general case, as with the case of localized electrons, shows that one can produce polarized neutrons from an unpolarized beam of neutrons. This requires (1) the neutron scatter vector $\boldsymbol{\kappa}$ is perpendicular to ferromagnetic field **B**, and (2) the z-component of the magnetic scatter length b_{Mz} equals the nuclear scatter length component $F_N(\boldsymbol{\kappa})$, where

$$F_N(\boldsymbol{\kappa}) = b_{Mz} \tag{7.161}$$

7.7 Polarized Neutrons by Grazing Incidence Reflection via Nuclear Scatter of Neutron Spin States in Ferromagnetic Materials

As discussed in sections 7.5 and 7.6, polarized neutrons can be produced by reflection from ferromagnetic mirrors. The differential nuclear scatter of neutron spin states by ferromagnetic materials occurs, because the nuclear scatter lengths for spin-up b^+ and spin-down b^- neutrons, differ for each ferromagnetic isotope with nuclei density ρ. The product $\rho b^{\pm}$ is the neutron scatter length density for the particular ferromagnetic isotope.

The different neutron scatter lengths b^+ and b^- yield decrements δ^+ and δ^- of the refractive index $n^{\pm} = 1 - \delta^{\pm}$ for spin-up (+) and spin-down (–) neutrons, which lead to different critical angles $\gamma^{\pm}$ (measured relative to the mirror surface tangent plane) of neutron-grazing incidence reflection, where

$$\gamma_c^{\pm} = \sqrt{2\delta^{\pm}} \tag{7.162}$$

and

$$\delta^{\pm} = \frac{\rho\lambda^2 b^{\pm}}{2\pi} \tag{7.163}$$

Variable A is the isotope atomic weight, which is expressed in units of grams per mole. The isotope atom number density ρ is obtained from ferromagnetic material mass density ρ_m, Avogadro number N_A, and weight fraction w_A of the isotope in the ferromagnetic material, where

$$\rho = N_A w_A \frac{\rho_m}{A} \quad \text{with} \quad N_A = 6.02 \times 10^{23}\ \text{atom/mole} \tag{7.164}$$

The molar density M [mole/cm^3] is obtained from the atom number density ρ, via division by Avogadro's number N_A.

Table 7.1 shows the decrement and critical reflection angles for spin-up (+) and spin-down (–) neutrons at wavelength $\lambda = 1$ Å for the ferromagnetic elements iron, cobalt, and nickel. These elements are represented by isotopes Fe-56, Co-59, and Ni-58, whose natural abundance are 91.72%, 100%, and 68%, respectively. See Sears (1989) and the National Institute of Standards and Technology (NIST) Web site (http://www.ncnr.nist.gov/resources/n-lengths). If the scatter length and decrement are negative, so that the refractive index is greater than 1, then there is incomplete reflection, where the reflectivity $R < 1$ for all incident angles. Listed for the three ferromagnetic isotopes is the total effective, electron spin quantum number for the unpaired electrons of a single atom, and the nuclear scatter length b_N, which is the same for spin-up and spin-down neutrons.

Also listed is the neutron magnetic scatter length b_M, which is added to the nuclear scatter length b_N for spin-up (+) neutrons, and is subtracted from b_N for spin-down (–) neutrons, to obtain the total scatter length for spin-up b^+ and spin-down b^-neutrons, where

$$b^+ = b_N + b_M \tag{7.165}$$

$$b^- = b_N - b_M \tag{7.166}$$

Table 7.1 Ferromagnetic Elements: Nuclear and Magnetic Scatter Lengths, Decrements, and Critical Angle of Reflections for 1 Å Spin-up (+) and Spin-down (−) Neutrons

		Iron	Cobalt	Nickel
Parameter		**Fe-56**	**Co-59**	**Ni-58**
Weight fraction	w_A	1	1	1
Total electron spin number for atom	S	2.22	1.72	0.6
Magnetic scatter length [fm]	b_M	5.98	4.64	1.62
Nuclear scatter length [fm]	b_N	9.54	2.5	10.3
Spin-up neutron scatter length [fm]	b^+	15.52	7.14	11.92
Spin-down neutron scatter length [fm]	b^-	3.56	−2.14	8.68
Mass density [g/cm^3]	ρ_m	7.87	8.9	8.91
Gram atomic weight [g/mole]	A	56	59	58
Molar atom number density [moles/cm^3]	ρ	0.141	0.151	0.154
Spin-up neutron decrement at 1 Å	δ^+	2.09×10^{-6}	1.03×10^{-6}	1.75×10^{-6}
Spin-down neutron decrement at 1 Å	δ^-	4.79×10^{-7}	-3.09×10^{-7}	1.28×10^{-6}
Spin-up neutron critical angle at 1 Å [rad]	γ_c^+	2.04×10^{-3}	1.44×10^{-3}	1.87×10^{-3}
Spin-down neutron critical angle at 1 Å [rad]	γ_c^-	9.79×10^{-4}	$R < 1$*	1.60×10^{-3}

*Reflectivity R is less than 1 for all angles; neutron scatter length is negative, $b < 0$.

The weight fraction W_A of each isotope is 1, because the parameters are for an assumed single-isotope sample. The mass density ρ_m and gram atomic weight A are used to obtain the isotope atom density ρ. With ρ and the total scatter length (combined nuclear and magnetic) for the spin-up b^+ and spin-down b^- neutrons, one calculates the decrements δ^+ and δ^-, and then the critical angles of reflection γ_c^+ and γ_c^- for the spin-up (+) and spin-down (−) neutrons for the 1-Å wavelength. The decrement has a quadratic dependence, and the critical angle has a linear dependence on wavelength, which allows easy calculation of decrements and critical reflection angles at other wavelengths, given the results at 1 Å.

7.8 Coherent, Elastic Differential Cross Section for Scatter of Neutron Spin States from Magnetic Materials with Helical-Oriented, Localized, Unpaired Electron Spins

We now consider the class of magnetic materials, which are comprised of helical arrays of unpaired, localized electron spins. The coherent elastic differential cross section will be derived for the combined magnetic and nuclear scatter of the neutron spin states by helical magnetic Bravais-lattice crystals. The crystals MnO_2 and Au_2Mn as well as many rare-earth-based crystals exhibit helical polarization of the spins of the unpaired electrons, localized to their atoms. We first derive the coherent elastic differential cross section for magnetic scatter, and obtain the magnetic scatter length for the helical Bravais-lattice crystal. Then, the nuclear scatter length is added to the magnetic scatter length to produce a combined neutron scatter length. From the combined scatter length, one obtains the total coherent elastic differential cross section for the two neutron spin states.

In helical magnetic materials, the unpaired atomic electron spins lie in horizontal xy planes, where the parallel xy planes are stacked perpendicular to the vertical z-axis. In a horizontal plane the electron spins all point in the same direction. From one horizontal plane to the next, the spin direction rotates slightly. As one proceeds vertically along the perpendicular z-axis, the spins rotate helically. That is, the spins all point in one direction in a horizontal plane, and then they rotate about the symmetric z-axis by azimuth angle ϕ in the neighboring horizontal plane. Thus, the spin direction rotates by azimuth angle ϕ about the z-axis, as one proceeds from one horizontal plane to the next. And as one proceeds vertically along the z-axis through several parallel planes, the spin vectors rotate a full revolution in the axial distance of pitch length P of the helix. The thermally averaged mean x-, y-, and z-components for the lth unit cell in a Bravais-lattice crystal are expressed in terms of a helix pitch number vector **Q**, which points in the direction of the helix axis, and the position vector $\mathbf{R}_l$ of the crystal l^{th} unit cell, where

$$\langle S_l^x \rangle = \langle S \rangle \cos(\mathbf{Q} \cdot \mathbf{R}_l) \tag{7.167}$$

$$\langle S_l^y \rangle = \langle S \rangle \sin(\mathbf{Q} \cdot \mathbf{R}_l) \tag{7.168}$$

and

$$\langle S_l^z \rangle = 0 \tag{7.169}$$

For the lth unit cell, the unpaired, localized electron spin vector $\mathbf{S}_l$ of the atom comprising the unit cell is

$$\mathbf{S}_l = \langle S \rangle \exp(i\boldsymbol{\kappa} \cdot \mathbf{R}_l)[\hat{\mathbf{x}} \cos(\mathbf{Q} \cdot \mathbf{R}_l) + \hat{\mathbf{y}} \sin(\mathbf{Q} \cdot \mathbf{R}_l)] \tag{7.170}$$

The pitch number vector **Q** points in the direction of the vertical z-axis of the helix, where **Q** has a magnitude of 2π, divided by the axial pitch length P of the helix, where

$$\mathbf{Q} = \frac{2\pi}{P}\hat{\mathbf{z}} \tag{7.171}$$

In one axial pitch length, the spin vectors rotate a full 360°. The total or net spin of the unpaired electron of an atom (i.e., unit cell) is specified by the total quantum number S, and the thermally averaged, total spin quantum number is $\langle S \rangle$.

The position of the lth unit cell (i.e., atom) is given by position vector $\mathbf{R}_l$, measured relative to the center of the crystal specified by $l = 0$, where $\mathbf{R}_0 = 0$. The direct crystal lattice unit cell $(\mathbf{a}, \mathbf{a}, \mathbf{c})$ and reciprocal-lattice unit cell $(\mathbf{a}^*, \mathbf{a}^*, \mathbf{c}^*)$ are symmetric about the z-axis. The direct-lattice unit cell $(\mathbf{a}, \mathbf{a}, \mathbf{c})$ is rectangular volume with its three dimensions in the perpendicular x-, y-, and z-directions. The unit cell position vector $\mathbf{R}_l$ is addressed by triplet integers (m, n, o), where

$$\mathbf{R}_l = ma\hat{\mathbf{x}} + na\hat{\mathbf{y}} + oc\hat{\mathbf{z}} \tag{7.172}$$

Recall from Ch. 4 in Cremer (2012b), the reciprocal-lattice vector $\boldsymbol{\tau}_\tau$ is addressed with triplet Miller integers (h, k, l), where

$$\boldsymbol{\tau}_\tau = h\frac{2\pi}{a}\hat{\mathbf{x}} + k\frac{2\pi}{a}\hat{\mathbf{y}} + l\frac{2\pi}{c}\hat{\mathbf{z}} \tag{7.173}$$

We now determine the coherent elastic differential cross section $(\mathrm{d}\sigma/d\Omega)^{\mathrm{mag}}_{\mathrm{hel}}$ for magnetic scatter of the helical array of electron spins. One starts with Eq. (7.1) or Eq. (5.147) for a Bravais lattice crystal, which is repeated below in Eq. (7.174), where Eq. (7.174) is the coherent, elastic differential cross section $(\mathrm{d}\sigma/d\Omega)^{\mathrm{mag}}_{\mathrm{coh,el}}$ for magnetic scatter of the neutrons by only the spins of the LS-coupled unpaired, localized electrons from prior derivations. The Heitler–London model, which includes the effective spin due to the orbital motion of the electrons (i.e., magnetic moment from electron current loops), is not used here because we are considering only the electron spins, where

$$\left(\frac{\mathrm{d}\sigma}{\mathrm{d}\Omega}\right)^{\mathrm{mag}}_{\mathrm{coh,el}} = N(r_0 g)^2 \sum_{l=0}^{N-1} \exp(i\boldsymbol{\kappa}\cdot\hat{\mathbf{R}}_l) \left|\frac{g_0}{2} F_0(\boldsymbol{\kappa}) \exp(-W_0)\right|^2 \sum_{i=1}^{J} p_i \left\langle \phi_i \left| \sum_{\alpha=1}^{3}\sum_{\beta=1}^{3} \left(\delta_{\alpha\beta} - \frac{\kappa_\alpha \kappa_\beta}{\kappa^2}\right) \hat{S}^{+}_{0\alpha}(0)\hat{S}_{l\beta}(0) \right| \phi_i \right\rangle \tag{7.174}$$

In the case of the helical spins, substitution of Eqs. (7.167–7.169) into Eq. (7.174), yields

$$\sum_{l=0}^{N-1}\sum_{\alpha=1}^{3}\sum_{\beta=1}^{3}\left(\delta_{\alpha\beta} - \frac{\kappa_\alpha\kappa_\beta}{\kappa^2}\right)\hat{S}^{+}_{0\alpha}(0)\hat{S}_{l\beta}(0) = S^2 \sum_{l=0}^{N-1} \exp(i\boldsymbol{\kappa}\cdot\mathbf{R}_l) \left\{\left(1 - \frac{\kappa_x^2}{\kappa^2}\right)\cos(\mathbf{Q}\cdot\mathbf{R}_l) - \frac{\kappa_x\kappa_y}{\kappa^2}\sin(\mathbf{Q}\cdot\mathbf{R}_l)\right\} \tag{7.175}$$

The range 0 to $N-1$ of the index integer address l for the N unit cells is changed to the equivalent l range of $-(N-1)/2$ to $(N-1)/2$, which is symmetric about $l=0$, where

$$\sum_{l=0}^{N-1}\sum_{\alpha=1}^{3}\sum_{\beta=1}^{3}\left(\delta_{\alpha\beta} - \frac{\kappa_\alpha\kappa_\beta}{\kappa^2}\right)\hat{S}^{+}_{0\alpha}(0)\hat{S}_{l\beta}(0) = S^2 \sum_{l=-(N-1)/2}^{(N-1)/2} \exp(i\boldsymbol{\kappa}\cdot\mathbf{R}_l) \left\{\left(1 - \frac{\kappa_x^2}{\kappa^2}\right)\cos(\mathbf{Q}\cdot\mathbf{R}_l) - \frac{\kappa_x\kappa_y}{\kappa^2}\sin(\mathbf{Q}\cdot\mathbf{R}_l)\right\} \tag{7.176}$$

The second term with the sine function sums to zero, which is shown next. With the symmetric summation range over l, the position vectors $\mathbf{R}_l$ of the crystal are now centered about the crystal origin at $l = 0$:

$$\mathbf{R}_{-(N-1)/2} = -\mathbf{R}_{(N-1)/2} \quad \text{and} \quad \mathbf{R}_0 = 0 \tag{7.177}$$

The sine function is antisymmetric in unit cell integer index l and corresponding position vector $\mathbf{R}_l$, where

$$\sin(\mathbf{Q}\cdot\mathbf{R}_{-(N-1)/2}) = -\sin(\mathbf{Q}\cdot\mathbf{R}_{(N-1)/2}) \tag{7.178}$$

Application of the Euler identity in the second term of the summation of Eq. (7.176), gives

$$\sum_{l=-(N-1)/2}^{(N-1)/2} \exp(\mathrm{i}\boldsymbol{\kappa}\cdot\mathbf{R}_l)\sin(\mathbf{Q}\cdot\mathbf{R}_l) = \frac{1}{2\mathrm{i}}\sum_{l=-(N-1)/2}^{(N-1)/2} \left\{\exp[\mathrm{i}\mathbf{R}_l\cdot(\boldsymbol{\kappa}+\mathbf{Q})] - \exp[\mathrm{i}\mathbf{R}_l\cdot(\boldsymbol{\kappa}-\mathbf{Q})]\right\} \tag{7.179}$$

The neutron scatter vector $\boldsymbol{\kappa} = \mathbf{k}_0 - \mathbf{k}$ is

$$\boldsymbol{\kappa} = \kappa_x\hat{\mathbf{x}} + \kappa_y\hat{\mathbf{y}} + \kappa_z\hat{\mathbf{z}} \tag{7.180}$$

The z-directed vector $\mathbf{Q}$ has magnitude $\mathbf{Q} = 2\pi/P$, where P is the pitch of the helix. Let p be the ratio of the direct-lattice unit cell dimension c, divided by the pitch of the helix P:

$$p = \frac{c}{P} \tag{7.181}$$

The z-directed vector $\mathbf{Q}$ of Eq. (7.171), becomes

$$\mathbf{Q} = \frac{2\pi}{P}\hat{\mathbf{z}} = 2\pi\frac{p}{c}\hat{\mathbf{z}} \tag{7.182}$$

The integers N_m, N_n, and N_o give the crystal dimensions in the x-, y-, and z-directions in units of the respective unit cell dimensions in the x-, y-, and z-directions. The rectangular crystal is comprised of N unit cells with dimensions in the x-, y-, and z-directions:

$$-(N_m-1)/2 \le m \le (N_m-1)/2 \quad \text{and} \quad -(N_m-1)m/2 \le x \le (N_m-1)m/2 \tag{7.183}$$

$$-(N_n-1/2 \le n \le (N_n-1)/2 \quad \text{and} \quad -(N_n-1)n/2 \le y \le (N_n-1)n/2 \tag{7.184}$$

$$-(N_o-1)/2 \le o \le (N_o-1)/2 \quad \text{and} \quad -(N_o-1)o/2 \le z \le (N_o-1)o/2 \tag{7.185}$$

The direct-lattice l^{th} unit cell position vector $\mathbf{R}_l$ is specified by integer l (not to be confused with reciprocal-lattice vector Miller index l), where

$$\mathbf{R}_l = \mathbf{R}_l(x, y, z) \quad -(m+n+o) \le l \le m+n+o \tag{7.186}$$

and

$$N = N_m N_n N_o \tag{7.187}$$

Substitution into Eq. (7.179) of Eqs. (7.186), (7.180), and (7.182) respectively for $\mathbf{R}_l$, $\boldsymbol{\kappa}$, and $\mathbf{Q}$, gives

$$\begin{aligned} l \;&= \sum_{-(N-1/2)}^{(N-1)/2} \exp[\mathrm{i}\mathbf{R}_l \cdot (\boldsymbol{\kappa} \pm \mathbf{Q})] = \sum_{m=-(N_m-1)/2}^{(N_m-1)/2} \exp(\mathrm{i}2\pi\kappa_x m) \\ &\times \sum_{-(N_n-1)/2}^{(N_n-1)/2} \exp(\mathrm{i}2\pi\kappa_y n) \sum_{o=-(N_o-1)/2}^{(N_o-1)/2} \exp[\mathrm{i}2\pi o(\kappa_z \pm p)] \end{aligned} \tag{7.188}$$

where

$$\mathbf{Q} \cdot \mathbf{R}_l = 2\pi \frac{p}{c} \hat{\mathbf{z}} \cdot oc\hat{\mathbf{z}} = 2\pi \cdot po \tag{7.189}$$

Each summation in Eq. (7.188) equals a series expansion of the ratio of two sine functions, where

$$\sum_{l=-(N-1)/2}^{(N-1)/2} \exp[\mathrm{i}\mathbf{R}_l \cdot (\boldsymbol{\kappa} + \mathbf{Q})] = \frac{\sin(N_m\kappa_x\pi)}{\sin(\kappa_x\pi)} \frac{\sin(N_n\kappa_y\pi)}{\sin(\kappa_y\pi)} \frac{\sin(N_o\kappa_z\pi)}{\sin(\kappa_z\pi)} \frac{\sin(N_o p\pi)}{\sin(p\pi)} \tag{7.190}$$

and

$$\sum_{l=-(N-1)/2}^{(N-1)/2} \exp[\mathrm{i}\mathbf{R}_l \cdot (\boldsymbol{\kappa} - \mathbf{Q})] = \frac{\sin(N_m\kappa_x\pi)}{\sin(\kappa_x\pi)} \frac{\sin(N_n\kappa_y\pi)}{\sin(\kappa_y\pi)} \frac{\sin(N_o\kappa_z\pi)}{\sin(\kappa_z\pi)} \frac{\sin(N_o p\pi)}{\sin(p\pi)} \tag{7.191}$$

where

$$\frac{\sin(-N_o p\pi)}{\sin(-p\pi)} = \frac{-\sin(N_o p\pi)}{-\sin(p\pi)} = \frac{\sin(N_o p\pi)}{\sin(p\pi)} \tag{7.192}$$

Hence,

$$\sum_{l=-(N-1)/2}^{(N-1)/2} \exp[\mathrm{i}\mathbf{R}_l \cdot (\boldsymbol{\kappa} + \mathbf{Q})] = \sum_{l=-(N-1)/2}^{(N-1)/2} \exp[\mathrm{i}\mathbf{R}_l \cdot (\boldsymbol{\kappa} - \mathbf{Q})] \tag{7.193}$$

and

$$\sum_{l=-(N-1)/2}^{(N-1)/2} \exp(\mathrm{i}\boldsymbol{\kappa} \cdot \mathbf{R}_l)\sin(\mathbf{Q} \cdot \mathbf{R}_l) = \frac{1}{2\mathrm{i}} \sum_{l=-(N-1)/2}^{(N-1)/2} \{\exp[\mathrm{i}\mathbf{R}_l \cdot (\boldsymbol{\kappa} + \mathbf{Q})] - \exp[\mathrm{i}\mathbf{R}_l \cdot (\boldsymbol{\kappa} - \mathbf{Q})]\} = 0 \tag{7.194}$$

The spin-dependent portion of the coherent elastic differential cross section of Eq. (7.176) becomes

$$\sum_{l=0}^{N-1} \sum_{\alpha=1}^{3} \sum_{\beta=1}^{3} \left(\delta_{\alpha\beta} - \frac{\kappa_\alpha \kappa_\beta}{\kappa^2}\right) \hat{S}_{0\alpha}^{+}(0)\hat{S}_{l\beta}(0) = S^2 \left(1 - \frac{\kappa_x^2}{\kappa^2}\right) \sum_{l=-(N-1)/2}^{(N-1)/2} \exp(\mathrm{i}\boldsymbol{\kappa} \cdot \mathbf{R}_l)\cos(\mathbf{Q} \cdot \mathbf{R}_l) \tag{7.195}$$

In Eq. (7.195), the cosine is symmetric in unit cell integer index l and corresponding position vector $\mathbf{R}_l$:

$$\cos(\mathbf{Q} \cdot \mathbf{R}_{-(N-1)/2}) = \cos(\mathbf{Q} \cdot \mathbf{R}_{(N-1)/2}) \tag{7.196}$$

The cosine can be expressed as a sum of exponentials by the Euler identity, so that with Eq. (7.196), Eq. (7.195) becomes

$$\sum_{l=0}^{N-1} \sum_{\alpha=1}^{3} \sum_{\beta=1}^{3} \left(\delta_{\alpha\beta} - \frac{\kappa_\alpha \kappa_\beta}{\kappa^2}\right) \hat{S}_{0\alpha}^{+}(0)\hat{S}_{l\beta}(0) = S^2 \left(1 - \frac{\kappa_x^2}{\kappa^2}\right) \sum_{l=-(N-1)/2}^{(N-1)/2} \frac{\exp[\mathrm{i}\mathbf{R}_l \cdot (\boldsymbol{\kappa} + \mathbf{Q})] + \exp[\mathrm{i}\mathbf{R}_l \cdot (\boldsymbol{\kappa} - \mathbf{Q})]}{2} \tag{7.197}$$

The direct- and reciprocal-lattice unit cells are assumed to be symmetric about the z-axis. Thus, the direct-lattice vector $\mathbf{R}_l$ and reciprocal-lattice vector $\boldsymbol{\tau}_\tau$ are also symmetric about the z-axis. Bragg reflection thus requires the neutron scatter vector $\boldsymbol{\kappa}$ to be symmetric about the z-axis, so that

$$\kappa_x = \kappa_y \tag{7.198}$$

where

$$1-\frac{\kappa_x^2}{\kappa^2}=\left(\frac{1}{2}-\frac{1}{2}\frac{\kappa_x^2}{\kappa^2}\right)+\left(\frac{1}{2}-\frac{1}{2}\frac{\kappa_x^2}{\kappa^2}\right)=\left(\frac{1}{2}-\frac{1}{2}\frac{\kappa_x^2}{\kappa^2}\right)+\left(\frac{1}{2}-\frac{1}{2}\frac{\kappa_y^2}{\kappa^2}\right)=\frac{1}{2}+\frac{1}{2}\frac{\kappa_z^2}{\kappa^2} \tag{7.199}$$

In Eq. (7.197), one converts from the summation of exponentials in the direct lattice to a summation of Dirac delta functions in the reciprocal-lattice space, so that with Eq. (7.199),

$$\sum_{l=0}^{N-1}\sum_{\alpha=1}^{3}\sum_{\beta=1}^{3}\left(\delta_{\alpha\beta}-\frac{\kappa_\alpha\kappa_\beta}{\kappa^2}\right)\hat{S}_{0\alpha}^{+}(0)\hat{S}_{l\beta}(0)=\frac{1}{4}\frac{(2\pi)^3}{v_0}S^2\left(1+\frac{\kappa_z^2}{\kappa^2}\right)$$
$$\sum_{\tau=0}^{N-1}[\delta(\boldsymbol{\kappa}+\mathbf{Q}-\boldsymbol{\tau}_\tau)+\delta(\boldsymbol{\kappa}-\mathbf{Q}-\boldsymbol{\tau}_\tau)] \tag{7.200}$$

With substitution into Eq. (7.174) of Eq. (7.200) for the helical spin terms, the coherent elastic differential cross section $(\mathrm{d}\sigma/d\Omega)_{\mathrm{coh,el}}^{\mathrm{mag}}$, then becomes the coherent elastic differential cross section $(\mathrm{d}\sigma/d\Omega)_{\mathrm{hel}}^{\mathrm{mag}}$ for magnetic scatter of neutrons in helical magnetic material:

$$\left(\frac{\mathrm{d}\sigma}{\mathrm{d}\Omega}\right)_{\mathrm{hel}}^{\mathrm{mag}}=N\frac{1}{4}\frac{(2\pi)^3}{v_0}\left(1+\frac{\kappa_z^2}{\kappa^2}\right)(r_0 g)^2\left|\frac{g_0}{2}F_0(\boldsymbol{\kappa})\exp(-W_0)\right|^2$$
$$\sum_{\tau=0}^{N-1}[\delta(\boldsymbol{\kappa}+\mathbf{Q}-\boldsymbol{\tau}_\tau)+\delta(\boldsymbol{\kappa}-\mathbf{Q}-\boldsymbol{\tau}_\tau)]\sum_{i=1}^{J}p_i\langle\phi_i|S^2|\phi_i\rangle \tag{7.201}$$

In Eq. (7.201), one performs thermal averaging on the squared total electron spin per unit cell, where

$$\langle S^2\rangle=\sum_{i=1}^{J}p_i\langle\phi_i|S^2|\phi_i\rangle \tag{7.202}$$

With substitution of Eq. (7.202) into Eq (7.201), the coherent elastic differential cross section $(\mathrm{d}\sigma/d\Omega)_{\mathrm{hel}}^{\mathrm{mag}}$ for magnetic neutron scatter, becomes

$$\left(\frac{\mathrm{d}\sigma}{\mathrm{d}\Omega}\right)_{\mathrm{hel}}^{\mathrm{mag}}=\frac{N}{4}\frac{(2\pi)^3}{v_0}\langle S^2\rangle\left(1+\frac{\kappa_z^2}{\kappa^2}\right)(r_0 g)^2\left|\frac{g_0}{2}F_0(\boldsymbol{\kappa})\exp(-W_0)\right|^2$$
$$\sum_{\tau=0}^{N-1}[\delta(\boldsymbol{\kappa}+\mathbf{Q}-\boldsymbol{\tau}_\tau)+\delta(\boldsymbol{\kappa}-\mathbf{Q}-\boldsymbol{\tau}_\tau)] \tag{7.203}$$

Magnetic Bragg scatter of neutrons occurs for neutron spin vectors $\boldsymbol{\kappa}$:

$$\boldsymbol{\kappa} = \boldsymbol{\tau}_\tau \pm \mathbf{Q} \tag{7.204}$$

Extracted from Eq. (7.203), is the magnetic scatter length b_M for the helical magnetic material, where

$$b_M = \frac{1}{2}\sqrt{\langle S^2\rangle\left(1+\frac{\kappa_z^2}{\kappa^2}\right)}(r_0 g)\left|\frac{g_0}{2}F_0(\boldsymbol{\kappa})\exp(-W_0)\right| \tag{7.205}$$

and the x, y, and z components are

$$b_{Mx} = \frac{1}{2}\sqrt{\langle S^2\rangle\left(1+\frac{\kappa_z^2}{\kappa^2}\right)}(r_0 g)\left|\frac{g_0}{2}F_0(\boldsymbol{\kappa})\exp(-W_0)\right|\cos(\mathbf{Q}\cdot\mathbf{R}_l) \tag{7.206}$$

$$b_{My} = \frac{1}{2}\sqrt{\langle S^2\rangle\left(1+\frac{\kappa_z^2}{\kappa^2}\right)}(r_0 g)\left|\frac{g_0}{2}F_0(\boldsymbol{\kappa})\exp(-W_0)\right|\sin(\mathbf{Q}\cdot\mathbf{R}_l) \tag{7.207}$$

$$b_{Mz} = 0 \tag{7.208}$$

By comparison, the nuclear unit cell structure factor $F_N(\boldsymbol{\kappa})$ for the Bravais-lattice crystal is simply the mean nuclear scatter length $\overline{b}$ of the single atom in the unit cell, where

$$F_N(\boldsymbol{\kappa}) = \overline{b} \tag{7.209}$$

Recalling section 2.4 in Squires (1996), as well as section 28 in Ch. 2 of Cremer (2012b) and Ch. 2 in Cremer (2013), the average of the nuclear scatter length $\overline{b}$ is taken over the randomly oriented nuclear spins and unpolarized neutrons, and it is averaged over the isotopes of the single atom in each unit cell, where

$$\overline{b} = \left(\frac{b^+(I+1)+b^-I}{2I+1}\right)_{\text{iso}} \tag{7.210}$$

The nucleus spin quantum number is I, see section 9.2 in Squires (1996), and Ch. 2 in Cremer (2013). In each unit cell, the spin-up neutron nuclear scattering length is b^+ and the nuclear scattering length for spin-down neutrons is b^-.

Associated with the nuclear scatter and magnetic scatter is the Debye–Waller factor for each unit cell that accounts for attenuation of the coherent differential cross section for nuclear scatter due to lattice atom vibrations, see section 3.6 in Squires (1996), and section 4 in Ch. 4 of Cremer (2012b). The intensity of the nuclear and magnetic components of the Bragg scattered neutron wave and thus the cross sections for nuclear and magnetic scatter are attenuated by the Debye–Waller factor:

$$\exp(-2W_0) \tag{7.211}$$

The amplitudes of the nuclear and magnetic components of the scattered neutron wave are attenuated by the square root of the Debye–Waller factor:

$$\exp(-W_0) \tag{7.212}$$

In the case of combined nuclear and magnetic scatter, each of the four neutron transitions has its own combined nuclear and magnetic scatter length, where

$$b_{f++}^{\mathrm{nb}} = \overline{b} \quad + \rightarrow + \tag{7.213}$$

$$b_{f--}^{\mathrm{nb}} = \overline{b} \quad - \rightarrow - \tag{7.214}$$

$$b_{f-+}^{\mathrm{nb}} = -(b_{Mx} + \mathrm{i}b_{My}) \quad - \rightarrow + \tag{7.215}$$

and

$$b_{f+-}^{\mathrm{nb}} = -(b_{Mx} - \mathrm{i}b_{My}) \quad + \rightarrow - \tag{7.216}$$

The coherent, elastic differential cross sections for the two possible neutron no-spin-flip cases for both nuclear and magnetic neutron scatter from a Bravais-lattice ferromagnetic crystal are identical, where

$$\left(\frac{\mathrm{d}\sigma}{\mathrm{d}\Omega}\right)_{\mathrm{coh,el}}^{++} = \frac{1}{2}N|\overline{b}|^2\frac{(2\pi)^3}{v_0}\exp(-2W_0)\sum_{\tau=1}^{N}\delta(\boldsymbol{\kappa} - \boldsymbol{\tau}_\tau) \tag{7.217}$$

and

$$\left(\frac{\mathrm{d}\sigma}{\mathrm{d}\Omega}\right)_{\mathrm{coh,el}}^{--} = \frac{1}{2}N|\overline{b}|^2\frac{(2\pi)^3}{v_0}\exp(-2W_0)\sum_{\tau=1}^{N}\delta(\boldsymbol{\kappa} - \boldsymbol{\tau}_\tau) \tag{7.218}$$

In an unpolarized incident neutron beam, the probability of spin-up and spin-down neutrons are each $1/2$, and thus these two cross sections for the no-spin-flip cases ($+ \rightarrow +$ and $- \rightarrow -$) are each multiplied by $1/2$. The combined nuclear and magnetic cross sections for the two possible neutron no-spin-flip cases are identical and have no magnetic component:

$$\left(\frac{\mathrm{d}\sigma}{\mathrm{d}\Omega}\right)_{\mathrm{coh,el}}^{++} = \left(\frac{\mathrm{d}\sigma}{\mathrm{d}\Omega}\right)_{\mathrm{coh,el}}^{--} \tag{7.219}$$

The coherent, elastic differential cross sections for the two possible neutron spin-flip transitions for both nuclear and magnetic scatter of unpolarized neutrons from a Bravais-lattice ferromagnetic crystal are given by

$$\left(\frac{d\sigma}{d\Omega}\right)^{-+}_{\text{coh,el}} = N|b_{Mx} + ib_{My}|^2 \frac{(2\pi)^3}{v_0} \sum_{\tau=0}^{N-1} [\delta(\boldsymbol{\kappa} + \mathbf{Q} - \boldsymbol{\tau}_\tau) + \delta(\boldsymbol{\kappa} - \mathbf{Q} - \boldsymbol{\tau}_\tau)] \tag{7.220}$$

and

$$\left(\frac{d\sigma}{d\Omega}\right)^{+-}_{\text{coh,el}} = N|b_{Mx} - ib_{My}|^2 \frac{(2\pi)^3}{v_0} \sum_{\tau=0}^{N-1} [\delta(\boldsymbol{\kappa} + \mathbf{Q} - \boldsymbol{\tau}_\tau) + \delta(\boldsymbol{\kappa} - \mathbf{Q} - \boldsymbol{\tau}_\tau)] \tag{7.221}$$

From Eq. (7.220) and (7.221), the squared magnitudes of the magnetic scatter lengths for the two neutron spin-flip transitions ($- \to +$ and $+ \to -$) are identical, and equal to the squared magnetic scatter length b_M for the helical magnetic material, where

$$b_M^2 = |b_{Mx} + ib_{My}|^2 = |b_{Mx} - ib_{My}|^2 \tag{7.222}$$

where

$$b_M^2 = \frac{1}{4}\langle S^2 \rangle \left(1 + \frac{\kappa_z^2}{\kappa^2}\right)^2 (r_0 g)^2 \left|\frac{g_0}{2} F_0(\boldsymbol{\kappa}) \exp(-W_0)\right|^2 \tag{7.223}$$

The coherent elastic differential cross section for the two neutron spin-flip transitions of Eqs. (7.220) and (7.221), become

$$\begin{aligned}\left(\frac{d\sigma}{d\Omega}\right)^{-+}_{\text{coh,el}} &= \frac{N}{8}\frac{(2\pi)^3}{v_0} \langle S^2 \rangle \left(1 + \frac{\kappa_z^2}{\kappa^2}\right)^2 (r_0 g)^2 \left|\frac{g_0}{2} F_0(\boldsymbol{\kappa}) \exp(-W_0)\right|^2 \\ &\times \sum_{\tau=0}^{N-1} [\delta(\boldsymbol{\kappa} + \mathbf{Q} - \boldsymbol{\tau}_\tau) + \delta(\boldsymbol{\kappa} - \mathbf{Q} - \boldsymbol{\tau}_\tau)]\end{aligned} \tag{7.224}$$

and

$$\begin{aligned}\left(\frac{d\sigma}{d\Omega}\right)^{+-}_{\text{coh,el}} &= \frac{N}{8}\frac{(2\pi)^3}{v_0} \langle S^2 \rangle \left(1 + \frac{\kappa_z^2}{\kappa^2}\right)^2 (r_0 g)^2 \left|\frac{g_0}{2} F_0(\boldsymbol{\kappa}) \exp(-W_0)\right|^2 \\ &\times \sum_{\tau=0}^{N-1} [\delta(\boldsymbol{\kappa} + \mathbf{Q} - \boldsymbol{\tau}_\tau) + \delta(\boldsymbol{\kappa} - \mathbf{Q} - \boldsymbol{\tau}_\tau)]\end{aligned} \tag{7.225}$$

These two cross sections for the neutron spin-flip transitions ($- \rightarrow +$ and $+ \rightarrow -$) are each multiplied by 1/2 because the unpolarized beam of spin-up and spin-down neutrons occurs with a probability of 1/2. The combined nuclear and magnetic cross sections for the two possible neutron spin-flip transitions are identical and caused only by magnetic scatter:

$$\left(\frac{d\sigma}{d\Omega}\right)_{\text{coh,el}}^{-+} = \left(\frac{d\sigma}{d\Omega}\right)_{\text{coh,el}}^{+-} \tag{7.226}$$

The total magnetic coherent, elastic differential scatter cross section $(d\sigma/d\Omega)_{\text{coh,el}}^{\text{tot}}$ is the sum of the contributions from the four possible spin transitions, where

$$\left(\frac{d\sigma}{d\Omega}\right)_{\text{coh,el}}^{\text{tot}} = \left(\frac{d\sigma}{d\Omega}\right)_{\text{coh,el}}^{++} + \left(\frac{d\sigma}{d\Omega}\right)_{\text{coh,el}}^{--} + \left(\frac{d\sigma}{d\Omega}\right)_{\text{coh,el}}^{-+} + \left(\frac{d\sigma}{d\Omega}\right)_{\text{coh,el}}^{+-} \tag{7.227}$$

Substitution into Eq. (7.227) of Eqs. (7.217), (7.218), (7.224), and (7.225), gives

$$\begin{aligned}\left(\frac{d\sigma}{d\Omega}\right)_{\text{coh,el}}^{\text{tot}} &= N|\bar{b}|^2\frac{(2\pi)^3}{v_0}\exp(-2W_0)\sum_{\tau=1}^{N}\delta(\boldsymbol{\kappa}-\boldsymbol{\tau}_\tau)\\ &+\frac{N}{4}\frac{(2\pi)^3}{v_0}\langle S^2\rangle\left(1+\frac{\kappa_z^2}{\kappa^2}\right)^2(r_0 g)^2\left|\frac{g_0}{2}F_0(\boldsymbol{\kappa})\exp(-W_0)\right|^2\\ &\times\sum_{\tau=0}^{N-1}[\delta(\boldsymbol{\kappa}+\mathbf{Q}-\boldsymbol{\tau}_\tau)+\delta(\boldsymbol{\kappa}-\mathbf{Q}-\boldsymbol{\tau}_\tau)]\end{aligned} \tag{7.228}$$

Each nuclear Bragg peak at $\boldsymbol{\kappa} = \boldsymbol{\tau}_\tau$ has a magnetic Bragg peak on each side, where $\boldsymbol{\kappa} = \boldsymbol{\tau}_\tau \pm \mathbf{Q}$. Thus, the nuclear and the two magnetic Bragg peaks occur at three separate locations in the diffraction pattern for a helical magnetic crystal that is produced by an unpolarized neutron beam.

The electron spins rotate about the z-axis by azimuth angle ϕ from one xy-plane to the next in the positive z-direction, and rotate by $-\phi$ from one xy-plane to the next in the negative z-direction. Therefore, for magnetic scatter,

$$\boldsymbol{\kappa} = \boldsymbol{\tau}_\tau \pm \frac{\phi}{\pi}\hat{\mathbf{z}} = h\frac{2\pi}{a}\hat{\mathbf{x}} + k\frac{2\pi}{a}\hat{\mathbf{y}} + \left(l\frac{2\pi}{c} \pm \frac{\phi}{\pi}\right)\hat{\mathbf{z}} \tag{7.229}$$

Also,

$$\boldsymbol{\kappa} = \boldsymbol{\tau}_\tau \pm \frac{2\pi}{P}\hat{\mathbf{z}} = h\frac{2\pi}{a}\hat{\mathbf{x}} + k\frac{2\pi}{a}\hat{\mathbf{y}} + \frac{2\pi}{c}(l \pm p)\hat{\mathbf{z}} \tag{7.230}$$

The magnitude of the helix pitch number vector **Q** is expressed in terms of the helix pitch P, pitch integer p, direct-lattice unit cell dimension c, and azimuth angle ϕ by

$$Q = \frac{2\pi}{P} = \frac{\phi}{\pi} \quad \text{and} \quad p = \frac{c}{P} \tag{7.231}$$

Then the ratio of azimuth angle ϕ, divided by π, is given by

$$\frac{\phi}{n} = \frac{2\pi}{P} = 2\pi\frac{p}{c} \tag{7.232}$$

That is, the ratio ϕ/π is the fractional rotation of the spin direction from one xy-plane to the next in the positive z-direction, and $-\phi/\pi$ is the fractional rotation of the spin direction from one xy-plane to the next in the negative z-direction. If the crystal temperature T is higher than the Neel temperature T_N, then the thermal vibrations randomize the electron spins, thereby eliminating magnetic Bragg scatter. That is, the angle ϕ between the uniform direction of electron spins in one xy-plane and its neighboring upper and lower xy-planes becomes randomized. Thus, if the crystal temperature T is above the Neel temperature T_N, the mean spin azimuthal rotation angle $\langle\phi\rangle$ from one xy-plane to the next is zero, where

$$\langle\phi\rangle = 0 \quad \text{and} \quad \boldsymbol{\kappa} = \boldsymbol{\tau}_\tau \quad T > T_N \tag{7.233}$$

Thus, if the crystal temperature T is above the Neel temperature T_N, then only nuclear scatter of the neutron occurs, and one obtains only the nuclear Bragg peak, whose neutron intensity is thermally attenuated by the Debye–Waller factor $\exp(-2W_0)$.

If one has a polarized beam of only spin-up (+) neutrons, then the total magnetic coherent, elastic differential scatter cross section $(d\sigma/d\Omega)^{\text{tot},+}_{\text{coh,el}}$ for the incident spin-up (+) neutrons is

$$\left(\frac{d\sigma}{d\Omega}\right)^{\text{tot},+}_{\text{coh,el}} = \left(\frac{d\sigma}{d\Omega}\right)^{++}_{\text{coh,el}} + \left(\frac{d\sigma}{d\Omega}\right)^{+-}_{\text{coh,el}} \tag{7.234}$$

The magnetic peak occurs at $\boldsymbol{\kappa} = \boldsymbol{\tau}_\tau - \mathbf{Q}$, along with nuclear Bragg peak $\boldsymbol{\kappa} = \boldsymbol{\tau}_\tau$, where

$$\left(\frac{d\sigma}{d\Omega}\right)^{\text{tot},+}_{\text{coh,el}} = N|\bar{b}|^2\frac{(2\pi)^3}{v_0}\exp(-2W_0)\sum_{\tau=1}^{N}\delta(\boldsymbol{\kappa} - \boldsymbol{\tau}_\tau) + \frac{N}{4}\frac{(2\pi)^3}{v_0}\langle S^2\rangle\left(1 + \frac{\kappa_z^2}{\kappa^2}\right)^2$$
$$(r_0 g)^2\left|\frac{g_0}{2}F_0(\boldsymbol{\kappa})\exp(-W_0)\right|^2\sum_{\tau=0}^{N-1}\delta(\boldsymbol{\kappa} + \mathbf{Q} - \boldsymbol{\tau}_\tau) \tag{7.235}$$

Conversely, if one has a polarized beam of only spin-down (–) neutrons, then the total magnetic coherent, elastic differential scatter cross section $(d\sigma/d\Omega)^{tot,-}_{coh,el}$ for the incident spin-down neutrons (–) is

$$\left(\frac{d\sigma}{d\Omega}\right)^{tot,-}_{coh,el} = \left(\frac{d\sigma}{d\Omega}\right)^{--}_{coh,el} + \left(\frac{d\sigma}{d\Omega}\right)^{-+}_{coh,el} \tag{7.236}$$

The magnetic peak occurs at $\boldsymbol{\kappa} = \boldsymbol{\tau}_\tau + \mathbf{Q}$, along with nuclear Bragg peak $\boldsymbol{\kappa} = \boldsymbol{\tau}_\tau$, where

$$\left(\frac{d\sigma}{d\Omega}\right)^{tot,-}_{coh,el} = N|\bar{b}|^2 \frac{(2\pi)^3}{v_0} \exp(-2W_0) \sum_{\tau=1}^{N} \delta(\boldsymbol{\kappa} - \boldsymbol{\tau}_\tau) + \frac{N}{4}\frac{(2\pi)^3}{v_0} \langle S^2 \rangle \left(1 + \frac{\kappa_z^2}{\kappa^2}\right)^2 (r_0 g)^2 \left|\frac{g_0}{2} F_0(\boldsymbol{\kappa}) \exp(-W_0)\right|^2 \sum_{\tau=0}^{N-1} \delta(\boldsymbol{\kappa} - \mathbf{Q} - \boldsymbol{\tau}_\tau) \tag{7.237}$$

In this discussion of helical magnetic material, we have assumed a right-handed helix. In the right-handed helix, the electron spin direction in one xy-plane is rotated by angle ϕ in the adjacent xy-plane in the positive z-direction and by angle $-\phi$ in the adjacent xy-plane in the negative z-direction. In the case of a right-handed helical array of electron spins, the incident-polarized beam of spin-up $(+)$ neutrons has a magnetic Bragg peak at neutron scatter vector $\boldsymbol{\kappa} = \boldsymbol{\tau}_\tau - \mathbf{Q}$, and an incident-polarized spin-down $(-)$ beam of neutrons has a magnetic Bragg peak at neutron scatter vector $\boldsymbol{\kappa} = \boldsymbol{\tau}_\tau + \mathbf{Q}$.

On the other hand, in a left-handed helix, the electron spin direction in one xy-plane is rotated by angle $-\phi$ in the adjacent xy-plane in the positive z-direction, and by angle ϕ in the adjacent xy-plane in the negative z-direction. In the left-handed helix, one has $\mp \mathbf{Q}$ as opposed to the right-handed helix, which has $\pm \mathbf{Q}$. Hence, for a left-handed helix for an incident-polarized beam of spin-up $(+)$ neutrons, the magnetic Bragg peak occurs at neutron scatter vector $\boldsymbol{\kappa} = \boldsymbol{\tau}_\tau + \mathbf{Q}$, and for an incident-polarized beam of spin-down neutrons, the magnetic Bragg peak occurs at $\boldsymbol{\kappa} = \boldsymbol{\tau}_\tau - \mathbf{Q}$. To observe the spin-flip of an incident-polarized beam of only spin-up $(+)$ or only spin-down $(-)$ neutrons by magnetic scatter in a helical material, requires the domains in this material to be comprised either of all right-handed helical electron spin arrays, or of all left-handed helical electron spin arrays.

References

Cremer, J. T. (2012b). In P. W. Hawkes (Ed.), Neutron and x-ray microscopy – Part 2. *Advances in imaging and electron physics* (Vol. 173). London, UK: Academic Press, an imprint of Elsevier Press.

Cremer, J. T. (2013). In P. W. Hawkes (Ed.), *Advances in imaging and electron physics* (Vol. 175). London, UK: Academic Press, an imprint of Elsevier Press.
Sears, V. F. (1989). *Neutron optics*. Oxford: Oxford University Press.
Squires, G. L. (1996). *Introduction to the theory of thermal neutron scattering*. New York, NY: Dover.

8 Coherent, Inelastic, Magnetic Neutron Scatter, Spin Waves, and Magnons

8.1 Electron Spin, Magnetic Dipole Moment, and Precession in Applied Magnetic Field

In this chapter, we cover the topics of coherent inelastic scatter of neutrons in magnetic materials, electron spin waves, and magnons in magnetic materials. Specific topics discussed include the electron spin and magnetic dipole moment, electron spin precession in an applied magnetic field, and the Heisenberg model of unpaired electron spin in magnetic materials. Next, we describe the physical basis of the exchange integral in the Heisenberg model, the expression of Heisenberg Hamiltonian by spin operators, and we examine spin waves, the dispersion relation, and magnons in ferromagnetic and antiferromagnetic materials. The exchange and anisotropy energy and domain formation in magnetic materials and the Hamiltonian eigenequation for 1-D ferromagnetic spin lattice is then explored, followed by spin and spin deviation operators, creation and annihilation operators, Holstein–Primakoff transformations, and the linear approximation of Heisenberg Hamiltonian. We also cover the application of the Bloch theorem to express creation and annihilation operators, and the Heisenberg Hamiltonian expressed as a sum of harmonic oscillators. Derived is the coherent, inelastic partial differential cross section for one-magnon absorption or emission for neutron scatter in ferromagnetic crystals. Next, the magnon forward scatter is examined. Finally, the spontaneous magnetization in a ferromagnetic crystal as a function of temperature T in the low-temperature limit, which is based on the Planck distribution, and the $T^{3/2}$ dependence in a ferromagnetic cubic crystal are derived. This chapter follows the treatment of topics in Grosso and Parravicini (2000), Kittel (2005), and Squires (1996), and is a continuation of the neutron scatter presentation in Cremer (2012b, 2013).

We first examine the relation between the electron spin vector $\mathbf{S}$ and the magnetic dipole moment vector of an electron $\boldsymbol{\mu}_e$ and the precession of the electron spin $\mathbf{S}$ and magnetic dipole moment $\boldsymbol{\mu}_e$ in an applied magnetic field $\mathbf{B}$, such as the magnetization vector $\mathbf{M}$ in a magnetized ferromagnet, where

$$\mathbf{B} = \mu_0 \mathbf{M} \tag{8.1}$$

Neutron and X-ray Optics. DOI: http://dx.doi.org/10.1016/B978-0-12-407164-3.00007-3

The total spin of the electron is

$$s = 1/2 \tag{8.2}$$

The electron total angular momentum S and electron angular momentum components along the z-axis S_z are, respectively,

$$S = \hbar\sqrt{s(s+1)} = \sqrt{\frac{3}{4}} \quad s = 1/2 \tag{8.3}$$

and

$$S_z = \pm\hbar s = \pm\frac{1}{2}\hbar \quad s = 1/2 \tag{8.4}$$

The electron spin vector $\mathbf{s}$ is tilted at polar tilt angle θ_s, where

$$S_z = s\cos\theta_s \tag{8.5}$$

and

$$\theta_s = \cos^{-1}\frac{S_z}{S} = \cos^{-1}\pm\sqrt{\frac{1}{3}} = \begin{matrix} 54.77^\circ & S_z = 1/2 \\ 125.33^\circ & S_z = -1/2 \end{matrix} \tag{8.6}$$

Parallel electron spins refer to parallel longitudinal or z-component spins S_z. The electron spin vectors are all directed along the positive z-direction, or the electron spin vectors are all directed along the negative z-direction. An antiparallel pair of electron spins refers to one longitudinal electron spin component pointed in the positive z-direction and the other longitudinal electron spin component pointed in the negative z-direction. The total spin vector is titled at equilibrium polar tilt angle θ_s.

The electron magnetic dipole moment vector directed opposite (180°) to the direction of the electron spin vector, where

$$\boldsymbol{\mu}_e = \frac{g\mu_b}{\hbar}\mathbf{S} = -\frac{2\mu_b}{\hbar}\mathbf{S} = -\frac{e}{m_e}\mathbf{S} \tag{8.7}$$

Here, the electron Landé g factor g_e is

$$g_e = -2 \tag{8.8}$$

The Bohr magnetron μ_b for elementary charge e and electron rest mass m_e is

$$\mu_b = \frac{e\hbar}{2m_e} = 5.788\times10^{-11}\ \mathrm{MeV/T} \tag{8.9}$$

The reduced Planck's constant $\hbar$ is

$$\hbar = 6.582\times10^{-22}\ \mathrm{MeV\ s} \tag{8.10}$$

and

$$\frac{\mu_b}{\hbar} = \frac{e}{2m_e} = 8.79 \times 10^{10} \text{ rad/T s} \tag{8.11}$$

The electron spin vector **S** with its unit vector $\hat{\mathbf{S}}_u$ and half-integer spin quantum number s is

$$\mathbf{S} = \hbar\sqrt{s(s+1)}\hat{\mathbf{S}} = \frac{\hbar\sqrt{3}}{2}\hat{\mathbf{S}}_u \quad s = 1/2 \tag{8.12}$$

The electron spin magnetic dipole moment vector $\boldsymbol{\mu}_e$ becomes

$$\boldsymbol{\mu}_e = -5.788 \times 10^{-11}\sqrt{3} \cdot \hat{\mathbf{S}}_u \text{ MeV/T} \tag{8.13}$$

The polar tilt angle of the electron spin vector θ_s is oriented at 180° from the polar tilt angle θ_μ of the electron magnetic dipole vector, where

$$\theta_\mu = -\theta_s \tag{8.14}$$

Hence, the polar tilt angle θ_μ of the electron magnetic dipole vector, measured relative to the direction of the positive z-axis, is

$$\theta_\mu = \cos^{-1}\frac{\mu_z}{\mu} = \cos^{-1}\frac{S_z}{S} = \cos^{-1} \mp \sqrt{\frac{1}{3}} = \begin{matrix} 125.33^\circ & S_z = 1/2 \\ 54.77^\circ & S_z = -1/2 \end{matrix} \tag{8.15}$$

The electron potential energy U in the ferromagnetic material depends on applied magnetic field **B** at the electron and the polar tilt angle θ of the electron magnetic dipole moment vector relative to the positive z-axis direction. If the initial polar tilt angle θ of the electron magnetic dipole moment vector relative to the positive z-axis increases by increment angle $\Delta\theta$, then the potential energy U of the electron increases by ΔU, where

$$\Delta U = -\Delta(\boldsymbol{\mu} \cdot \mathbf{B}) = -|\boldsymbol{\mu}||\mathbf{B}|\Delta(\cos\theta) \tag{8.16}$$

One then obtains

$$\Delta U = \mu B \sin\theta\Delta\theta \tag{8.17}$$

The electron magnetic dipole moment vector $\boldsymbol{\mu}_e$ will precess about a magnetic field **B** directed along the positive z-axis at the Lamor angular frequency $\mathbf{w}_L$ [rad/s] where

$$\mathbf{w}_L \text{ [rad/s]} = \frac{g_s\mu_b}{\hbar}\mathbf{B} = 1.76 \times 10^{11}\mathbf{B}\ [T] \tag{8.18}$$

The electron magnetic dipole moment vector $\boldsymbol{\mu}_e$ precession is counterclockwise for positive Lamor frequency $\mathbf{w}_L > 0$; for a positive magnetic field pointed in the positive z-direction:

$$\mathbf{w}_L > 0 \quad \mathbf{B} = |B_0|\hat{\mathbf{z}} \tag{8.19}$$

The precession of μ_e is counterclockwise for negative Lamor frequency $\mathbf{w}_L < 0$ for a negative magnetic field pointed in the positive z-direction:

$$\mathbf{w}_L < 0 \quad \mathbf{B} = -|B_0|\hat{\mathbf{z}} \tag{8.20}$$

That is, the electron spin precession frequency f_L in the lab rest frame is 28 GHz/T of applied magnetic field, where

$$f_L\,[\mathrm{s}^{-1}] = \pm 2.8 \times 10^{10} B_0\,[\mathrm{T}] \tag{8.21}$$

8.2 No Magnetic Field—Unpaired Electron Spins Tend to Align in the Same Direction

In the case of no applied magnetic field ($\mathbf{B} = 0$), the electron spin direction can be pointed in any direction with equal probability. The spin–orbit coupling or LS coupling of unpaired electrons localized in an atom or delocalized (itinerant), tends to align the spin vectors **S** in the same direction, because the positions of the electrons are separated. That is, the spatial component of the wave function for the state of a group of electrons is asymmetric, which then allows a symmetric spin component of the wave function for the electron group. In this manner, the total wave function of the group of electrons is asymmetric as required by the strong condition of the Pauli exclusion principle. The weaker condition of the Pauli exclusion principle in a multi-electron atom is that the quantum state of each electron must be different from the other electrons. The stronger Pauli exclusion principle states that a system containing several electrons must be described by an asymmetric total wave function, see chapter 9 in Eisberg, R. and Resnick, R. (1974).

A group of electron localized to an atom or nonlocalized and distributed within an energy band of the crystal (itinerant) have separate positions (antisymmetric spatial eigenfunction component), and thus the electrons can have aligned spins (symmetric spin eigenfunction component), which yields their required total asymmetric eigenstates. In modeling spin waves, one can approximate the z-component spin S_z as equaling the magnitude of the total spin vector S at sufficiently low temperatures T below the Curie temperature T_C in ferromagnetic materials and below the Néel temperature T_N in antiferromagnetic materials, where

$$S_z \cong S \quad T < T_C \text{ ferromagnet} \tag{8.22}$$

$$S_z \cong S \quad T < T_N \text{ antiferromagnet} \tag{8.23}$$

In ferromagnetic material, the spins of parallel atomic planes are aligned in the same direction; that is, all spins point in the positive z-direction. However, in ferromagnetic materials, the spin direction alternates from one atomic plane to the next. That is, the spins alternate from pointing in the positive z-direction in one atomic plane to pointing in the negative z-direction in the neighboring left and right parallel atomic planes.

Without a magnetic field ($B = 0$), the electrons all have the same total energy, but they are in different quantum states. That is, without a magnetic field, the different electron quantum states have the same total quantized energy, which are termed *degenerate quantum states.* The Zeeman effect showed that with the application of a magnetic field ($B \neq 0$), the different quantum states of the electrons acquired different quantized total energies.

In an applied magnetic field ($B \neq 0$), the electron z-component was now either parallel or antiparallel to the magnetic field direction. In the applied magnetic field ($B \neq 0$), the total electron spin and the z-component of the electron spin commute and can thus be specified or measured simultaneously. However, the x and y spin components S_x and S_y do not commute with the z-spin component S_z nor with the total electron spin S with or without a magnetic field ($B = 0$ or $B \neq 0$). Therefore, the x and y spin components S_x and S_y cannot be specified or measured simultaneously with the total spin S or z-component S_z of the electron spin.

With an applied magnetic field ($B \neq 0$), since the z-component of the electron spin is $S_z = \pm 1/2$ and the total electron spin is $S = \sqrt{3}/2$, the polar angle θ_s between the total spin S and z-component S_z of the electron spin is

$$\theta_s = \begin{matrix} 54.77^\circ & S_z = 1/2 \\ 125.33^\circ & S_z = -1/2 \end{matrix} \quad B \neq 0 \tag{8.24}$$

With nonzero magnetic field, $B \neq 0$, the electron total spin vector $\mathbf{S}$ precesses about the positive or negative z-directed spin component S_z at polar tilt angle θ_s at the Lamor frequency f_L [s^{-1}], which is proportional to the applied magnetic field amplitude B_0:

$$f_L\,[\mathrm{s}^{-1}] = \pm 2.8 \times 10^{10} B_0\,[\mathrm{T}] \tag{8.25}$$

For a constant applied magnetic field $\mathbf{B}$, the change in the potential energy of the electron ΔU is associated with a change in the polar tilt $\Delta\theta_s$, where

$$\Delta U = -|\mu_e||\mathbf{B}|\Delta(\cos\theta_s) = \pm|\mu||\mathbf{B}|\sin\theta_s\Delta\theta_s \tag{8.26}$$

8.3 Heisenberg Model of Unpaired Electron Spin in Magnetic Materials

The electric field repulsion between electrons and the Pauli exclusion principle, which does not allow electrons of the same quantum state to be at the same position at the same time, causes the exchange force. The Pauli exclusion principle requires two electrons of the same spin to be relatively far apart; however, it does allow two electrons of opposite spin to be close to each other. The electrostatic energy of a system of electron spins thus depends on the relative orientation of the electron spins. The energy difference between the different spin orientations is termed the *exchange energy.* The exchange energy U_{ex} arises from the spin interaction energy between the total $\mathbf{S}_i$ and $\mathbf{S}_j$ electron spin vectors of the ith and j^{th} atoms, where J is the exchange integral, and in the Heisenberg model,

$$U_{ex} = -2J\mathbf{S}_i\mathbf{S}_j \tag{8.27}$$

The exchange integral J arises from the overlap of the charge distribution of two neighboring atoms, atom 1 and atom 2. Consider the exchange energy, which arises between electron 1 of atom 1, and electron 2 of atom 2. In this simple case, the exchange energy of electron 1 of atom 1 and electron 2 of atom 2, depends on the direction of the spin $\mathbf{s}_1$ of electron 1, relative to the direction of the spin $\mathbf{s}_2$ of electron 2, where:

$$U_{ex} = -2J\mathbf{s}_1 \cdot \mathbf{s}_2 \tag{8.28}$$

The lower state exchange energy U_{ex} occurs for parallel electron spins $\mathbf{s}_1$ and $\mathbf{s}_2$, where

$$U_{ex} = -2J \tag{8.29}$$

and

$$\mathbf{s}_1 \cdot \mathbf{s}_2 = 1 \tag{8.30}$$

The higher state exchange energy U_{ex} occurs for antiparallel electron spins $\mathbf{s}_1$ and $\mathbf{s}_2$, where

$$U_{ex} = 2J \tag{8.31}$$

and

$$\mathbf{s}_1 \cdot \mathbf{s}_2 = -1 \tag{8.32}$$

The increase of energy ΔU_{ex} of a pair of electrons, which flip from the lower energy state $U_{ex} = -2J$ of parallel spins to the upper energy state $U_{ex} = 2J$ of antiparallel spins, is then

$$\Delta U_{ex} = 4J \tag{8.33}$$

In the mean field theory, the exchange integral J for a center atom is related to the number of nearest atom neighbors, the Curie temperature T_C, total electron spin magnitude for the atom S, and the Boltzmann constant k_B, where

$$J = \frac{3k_B T_C}{2nS(S+1)} \tag{8.34}$$

The number of nearest neighbors is $n = 8$ in a body-centered cubic (bcc) crystal such as iron. Iron has a total quantum number of $s = 1$ due to unpaired electrons.

Hence, for iron and other bcc crystals with a total spin quantum number $s = 1$ per atom, the exchange integral energy J is

$$J = \frac{1}{8} k_B T_C \quad k_B = 8.617 \times 10^{-5}\ \text{eV/K} \tag{8.35}$$

At the Curie temperature T_C for iron at 1043 K, the exchange integral is $J \cong 1.1 \times 10^{-2}$ eV. At room temperature (298 K), the exchange integral for iron is $J \cong 3.2 \times 10^{-3}$ eV.

In nickel, the net magnetic moment per atom arises because the 3d band of 5 spin-up electrons is filled, but the 3d band of spin-down electrons is not completely filled and contains only 4.46 electrons. The net magnetic dipole moment of nickel is due to an excess of 0.54 spin-up electrons per atom. However, in the hole–electron model, the net magnetic dipole moment of the nickel atom is expressed as arising from the 0.54 hole in the spin-down electron band. In the filled 3d band of copper, the five spin-up electrons are matched by five spin-down electrons. Hence, there is no net unpaired electron spin and hence no magnetic dipole moment for copper.

8.4 Physical Basis of Exchange Integral in the Heisenberg Model

The Heisenberg model applies to atoms with no net orbital angular momentum, where the electrons are localized, and one considers only the net unpaired electron spin per atom. The physical basis of the exchange integral J in the Heisenberg model for a two-electron system, as well as an expression for J for a two-electron system, is derived as follows. One starts with the Hamiltonian, which describes a two-electron system, where each electron is bound to a neighboring atom:

$$\hat{H} = \hat{H}_0(\mathbf{r}_1) + \hat{H}_0(\mathbf{r}_2) + \frac{e^2}{|\mathbf{r}_1 - \mathbf{r}_2|} \tag{8.36}$$

The one-electron Hamiltonian $\hat{H}_0$ operator is

$$H_0(\mathbf{r}) = \frac{p^2}{2m} + V(\mathbf{r}) \tag{8.37}$$

The electron–electron repulsion term is

$$\frac{e^2}{r_{12}} \tag{8.38}$$

The core electron potential is $V(\mathbf{r})$, and r_{12} is the separation distance of the two electrons:

$$r_{12} = |\mathbf{r}_1 - \mathbf{r}_2| \tag{8.39}$$

The total electron system wave function must be antisymmetric, which allows (1) the spatial component of the wave function to be symmetric and the spin component to be antisymmetric in a singlet state, or (2) the spatial component of the wave function to be antisymmetric and the spin component to be symmetric in a triplet state. When the wave function is symmetric, the electrons can be close to each other, so the electron–electron repulsion energy is higher than in the case of the antisymmetric spatial wave component. In the antisymmetric wave function, the electrons are relatively far apart and the electron–electron repulsion is thus smaller.

As a result, the antisymmetric spatial wave function with triplet spin state with parallel electron spins is the ground state, and the symmetric spatial wave function with singlet spin state with antiparallel electron spins is the higher energy state of the two-electron system. The ground, triplet spin state, with the pair of parallel electron spins, is the magnetic state for a ferromagnet like iron, where the exchange energy interaction can model the ferromagnetic interaction between parallel electron spins. The exchange energy of this two-electron system interaction, is also used to model two hydrogen atoms, which approach each other (two electrons in the same atom, such as *para*-helium and *ortho*-helium). Hydrogen and helium are the starting point for the Hartree theory for multielectron atoms.

The total wave function Ψ_S of the higher energy singlet state is the product of its symmetric spatial wave function component $\phi_S(\mathbf{r}_1, \mathbf{r}_2)$ and its antisymmetric spin component $\chi_S(\mathbf{s}_1, \mathbf{s}_2)$, where

$$\Psi_S(\mathbf{r}_1, \mathbf{r}_2, \mathbf{s}_1, \mathbf{s}_2) = \phi_S(\mathbf{r}_1, \mathbf{r}_2)\chi_S(\mathbf{s}_1, \mathbf{s}_2) \tag{8.40}$$

The symmetric spatial component $\phi_S(\mathbf{r}_1, \mathbf{r}_2)$ of the two-electron wave function is expressed as the sum of the products of the coordinate-swapped individual electron spatial wave function components $\phi_1(\mathbf{r})$ and $\phi_2(\mathbf{r})$ where

$$\phi_S(\mathbf{r}_1, \mathbf{r}_2) = \frac{1}{\sqrt{2}}[\phi_1(\mathbf{r}_1)\phi_2(\mathbf{r}_2) + \phi_1(\mathbf{r}_2)\phi_2(\mathbf{r}_1)] \tag{8.41}$$

The antisymmetric electron spin wave function component $\chi_S(\mathbf{s}_1, \mathbf{s}_2)$ is

$$\chi_S(\mathbf{s}_1, \mathbf{s}_2) = \frac{1}{\sqrt{2}}(\chi_{1+}\chi_{2+} - \chi_{1-}\chi_{2-}) \tag{8.42}$$

Here, + indicates a spin-up electron, and − indicates a spin-down electron.

The total wave function Ψ_T of the lower energy triplet state is the product of its antisymmetric spatial wave function component $\phi_T(\mathbf{r}_1, \mathbf{r}_2)$ and its symmetric spin component $\chi_T(\mathbf{s}_1, \mathbf{s}_2)$, where

$$\Psi_T(\mathbf{r}_1, \mathbf{r}_2, \mathbf{s}_1, \mathbf{s}_2) = \phi_T(\mathbf{r}_1, \mathbf{r}_2)\chi_T(\mathbf{s}_1, \mathbf{s}_2) \tag{8.43}$$

The antisymmetric spatial component $\phi_T(\mathbf{r}_1, \mathbf{r}_2)$ of the two-electron wave function is expressed as the difference of the products of the coordinate-swapped individual electron spatial wave function components $\phi_1(\mathbf{r})$ and $\phi_2(\mathbf{r})$, where

$$\phi_T(\mathbf{r}_1, \mathbf{r}_2) = \frac{1}{\sqrt{2}}[\phi_1(\mathbf{r}_1)\phi_2(\mathbf{r}_2) - \phi_1(\mathbf{r}_2)\phi_2(\mathbf{r}_1)] \tag{8.44}$$

The symmetric electron spin wave function component $\chi_T(\mathbf{s}_1, \mathbf{s}_2)$ can be in one of three symmetric spin states:

$$\chi_T(\mathbf{s}_1, \mathbf{s}_2) = \frac{1}{\sqrt{2}} \begin{pmatrix} \chi_{1+}\chi_{2+} \\ \chi_{1+}\chi_{2+} + \chi_{1-}\chi_{2-} \\ \chi_{1-}\chi_{2-} \end{pmatrix} \tag{8.45}$$

The Schrödinger equation for the Heisenberg model of a two-electron system, which has a symmetric spatial wave function and an antisymmetric singlet spin state, is

$$\hat{H}\Psi_S = \left[\hat{H}_0(\mathbf{r}_1) + \hat{H}_0(\mathbf{r}_2) + \frac{e^2}{|\mathbf{r}_1 - \mathbf{r}_2|}\right]\Psi_S = E\Psi_S \tag{8.46}$$

And the Schrödinger equation for the Heisenberg model of a two-electron system, which has an antisymmetric spatial wave function and a symmetric triplet spin state, is

$$\hat{H}\Psi_T = \left[\hat{H}_0(\mathbf{r}_1) + \hat{H}_0(\mathbf{r}_2) + \frac{e^2}{|\mathbf{r}_1 - \mathbf{r}_2|}\right]\Psi_T = E\Psi_T \tag{8.47}$$

An unperturbed solution of the two-electron system, where each electron belongs to a separate atom, is the sum of the electron wave functions of each atom. In the unperturbed solution, the electron–electron interaction is ignored. This is similar to the planetary trajectory calculations in classical mechanics in which the trajectory of each planet in the solar system is determined by the planer's interaction with the sun only. However, between the pair of electrons and between the planets, there exist small forces of interaction. In the case of electrons, there is the electrostatic repulsion and their spin interactions, and the planets' gravity. The relatively small change in momentum, force, energy, and trajectory of the electrons and planets, via their relatively small interaction, as compared to the undisturbed state, can be calculated by application of the disturbing interaction to the undisturbed system state. One adds these tiny changes to the unperturbed state to arrive at the solution for the perturbed state.

This first-order perturbation of the undisturbed state added to the unperturbed state is similar to the sum of the first two terms of the Taylor series. The first term of the Taylor series is the undisturbed state function, which is evaluated at its equilibrium point. Added to the unperturbed first term of the Taylor series is a second, perturbed term, which is the derivative of the function at its equilibrium point multiplied by a small change in position from the equilibrium point (i.e., a small distance from equilibrium). The resulting two terms of the Taylor series thus evaluates the perturbed state function at a position relatively close to its unperturbed position.

Consider the interaction energy ΔE of the two electrons, where we neglect the electron spin effects and the indistinguishable nature of the two electrons. One ignores the inability to identify and follow the trajectory of each electron individually. The interaction energy ΔE of the two electrons is approximated by multiplying the interaction potential energy e^2/r_{12} by the probability p_{12} that the electrons have a particular separation distance r_{12}. The probability p_{12} requires the integration of probability density or squared magnitude of the wave function amplitude $\varphi(\mathbf{r}_1)\varphi(\mathbf{r}_2)$ of the pair of independent electrons over a range of possible electron positions $\mathbf{r}_1$ with $\mathbf{r}_2$, which are respectively surrounded by associated differential volumes d^3r_1 and d^3r_2, where

$$\Delta E = \int \mathrm{d}^3r_2 \int \mathrm{d}^3r_1 \frac{e^2}{r_{12}} |\varphi(\mathbf{r}_1)\varphi(\mathbf{r}_2)|^2 \tag{8.48}$$

We now consider the effects of spin and electrostatic interaction together with the inability to distinguish the two electrons. The two identical electrons share a total system state, which is based on two single electron states. As a result of the Pauli exclusion principle, the total system state is antisymmetric, where the lower energy state has an antisymmetric spatial component and symmetric spin component (triplet state), and the upper energy state has a symmetric spatial component and antisymmetric spin component (singlet state).

One cannot follow the individual trajectories of the electrons, which is implied by the representation of the total antisymmetric wave function as a product of the individual electron spatial wave functions and individual electron spin states, where the electrons can interact via their spin states. The total wave function now allows swapping the coordinates of the spin and spatial components between the pair of identical electrons, via the use of symmetric and antisymmetric wave functions for representation of the spin and spatial components of the total state function Ψ_S or Ψ_T.

The interaction energy ΔE_T of the two electrons now includes the electron spin effects and the indistinguishable nature of the two electrons for the lower energy triplet spin state, which has an antisymmetric spatial component and symmetric spin component, so that

$$\Delta E_T = \int \mathrm{d}^3r_2 \int \mathrm{d}^3r_1 \frac{e^2}{r_{12}} |\Psi_T(\mathbf{r}_1)\psi_T(\mathbf{r}_2)|^2 \tag{8.49}$$

As before, the spin component of the wave function, which can be in one of its three possible configurations, has no spatial coordinate dependence and can be taken outside the integral, so that substitution gives

$$\Delta E_T = |\chi_T(\mathbf{s}_1, \mathbf{s}_2)|^2 \int \mathrm{d}^3r_2 \int \mathrm{d}^3r_1 \frac{e^2}{r_{12}} |\phi_T(\mathbf{r}_1)\phi_T(\mathbf{r}_2)|^2 \tag{8.50}$$

The triplet symmetric spin component is normalized in each of its three possible states, where

$$|\chi_T(\mathbf{s}_1, \mathbf{s}_2)|^2 = 1 \tag{8.51}$$

Substitution for $\phi_T(\mathbf{r}_1)$ and $\phi_T(\mathbf{r}_2)$ in ΔE_T gives

$$\Delta E_T = \Delta E_1 - \Delta E_2 \tag{8.52}$$

The energy ΔE_1 is the increase of the electron interaction energy due to the electron–electron repulsion as r_{12} decreases and the electrons are brought closer together:

$$\Delta E_1 = \int_0^{2\pi} \mathrm{d}\varphi_2 \int_0^{\pi} \mathrm{d}\theta_2 \sin\theta_2 \int_0^{\infty} \mathrm{d}r_2 r_2^2 \int_0^{2\pi} \mathrm{d}\varphi_1 \int_0^{\pi} \mathrm{d}\theta_1 \int_0^{\infty} \mathrm{d}r_1 r_1^2 \sin\theta_1 \\ \times \frac{e^2}{2} \frac{|\phi_1(\mathbf{r}_1)|^2|\phi_2(\mathbf{r}_2)|^2 + |\phi_1(\mathbf{r}_2)|^2|\phi_2(\mathbf{r}_1)|^2}{\sqrt{r_1^2 + r_2^2 - 2r_1r_2[\cos_1 \cos\theta_2 + \sin\theta_1 \sin\theta_2 \cos(\varphi_1 - \varphi_2)]}} \tag{8.53}$$

The exchange energy ΔE_2, which accounts for each electron's interaction with each of the one-electron states, is then

$$\Delta E_2 = \int_0^{2\pi} \mathrm{d}\varphi_2 \int_0^{\pi} \mathrm{d}\theta_2 \sin\theta_2 \int_0^{\infty} \mathrm{d}r_2 r_2^2 \int_0^{2\pi} \mathrm{d}\varphi_1 \int_0^{\pi} \mathrm{d}\theta_1 \int_0^{\infty} \mathrm{d}r_1 r_1^2 \sin\theta_1 \\ \times \frac{e^2}{2} \frac{\phi_1^*(\mathbf{r}_1)\phi_2(\mathbf{r}_1)\phi_2^*(\mathbf{r}_2)\phi_1(\mathbf{r}_2) + \phi_2^*(\mathbf{r}_1)\phi_1(\mathbf{r}_1)\phi_1^*(\mathbf{r}_2)\phi_2(\mathbf{r}_2)}{\sqrt{r_1^2 + r_2^2 - 2r_1r_2[\cos_1 \cos\theta_2 + \sin\theta_1 \sin\theta_2 \cos(\varphi_1 - \varphi_2)]}} \tag{8.54}$$

For spherical coordinates, the law of cosines was applied to $|\mathbf{r}_1 - \mathbf{r}_2|$ where

$$|\mathbf{r}_1 - \mathbf{r}_2| = \sqrt{r_1^2 + r_2^2 - 2r_1r_2 \cos\gamma} \tag{8.55}$$

In spherical coordinates, the angle γ is between the two electron position vectors $\mathbf{r}_1(r_1, \theta_1, \phi_1)$ and $\mathbf{r}_2(r_2, \theta_2, \phi_2)$. Here $\mathbf{r}_1$ and $\mathbf{r}_2$ are oriented at respective polar angles θ_1 and θ_2, which are measured relative to the z-axis, and ϕ_1 and ϕ_2 are their respective azimuth angles in this right-handed, rectangular (x, y, z) coordinate system, where

$$\cos\gamma = \cos_1 \cos\theta_2 + \sin\theta_1 \sin\theta_2 \cos(\varphi_1 - \varphi_2) \tag{8.56}$$

In a similar fashion, one determines the interaction energy for the upper energy singlet spin state, which has a symmetric spatial component and antisymmetric spin component. The interaction energy ΔE_S of the two electrons, where one includes the electron spin effects, and the lack of separate identity of the two electrons, is

$$\Delta E_S = \int \mathrm{d}^3r_2 \int \mathrm{d}^3r_1 \frac{e^2}{r_{12}} |\Psi_S(\mathbf{r}_1)\psi_S(\mathbf{r}_2)|^2 \tag{8.57}$$

The singlet antisymmetric spin component is normalized in its one possible state, where

$$|\chi_S(\mathbf{s}_1, \mathbf{s}_2)|^2 = 1 \tag{8.58}$$

Substitution for $\phi_S(\mathbf{r}_1)$ and $\phi_S(\mathbf{r}_2)$ in ΔE_S gives

$$\Delta E_S = \Delta E_1 + \Delta E_2 \tag{8.59}$$

The exchange energy is $\Delta E_S - \Delta E_T$ in the lower energy, two-electron system state, which has the symmetric, spatial component and the antisymmetric, singlet spin state Ψ_T of two antiparallel electron spins. The exchange energy is $\Delta E_S + \Delta E_T$ in the upper energy, two-electron system state, which has the antisymmetric, spatial component and the symmetric, triplet spin state Ψ_S of two parallel electron spins. The difference in the exchange energy between the upper and lower states is the exchange integral J, where

$$J = \Delta E_S - \Delta E_T = (\Delta E_1 + \Delta E_2) - (\Delta E_1 - \Delta E_2) = 2\Delta E_2 \tag{8.60}$$

A pair of electrons with antiparallel spins can be close together because they do not violate the Pauli exclusion principle. However, the coulomb repulsion between the electrons increases the energy of the antiparallel configuration, as compared to the parallel spin configuration. In the parallel spin configuration, the two electrons must move apart to different quantum states, avoiding violation of the Pauli exclusion principle, and thereby the electrostatic repulsion energy decreases the interaction energy below that of antiparallel spins.

The coulomb energy contribution ΔE_1 to the total interaction energy ΔE_T of the two electrons is positive definite, and the exchange energy ΔE_2 contribution (which usually is smaller) is also positive definite. Hence, the nonnegative exchange energy ΔE_2 is subtracted from the larger nonnegative coulomb energy contribution ΔE_1 to yield a nonnegative total interaction energy ΔE_T, where

$$\Delta E_1 \geq \Delta E_2 \geq 0 \tag{8.61}$$

and

$$\Delta E_T \geq 0 \tag{8.62}$$

Thus, the exchange integral J is zero or positive:

$$J \geq 0 \tag{8.63}$$

8.5 Expression of the Heisenberg Hamiltonian by Spin Operators

The low-energy modulation of the electron spin precession in the domain of a ferromagnet, leads to perturbed electron spin polar tilt angles, and the formation of quantized spin waves in the periodic lattice of unpaired atomic electrons. A magnon is a quantized spin wave similar to the phonon, which is a quantized atomic displacement wave in lattices of vibrating atoms. The quantization of the spin wave, in terms of magnons, is based on electron spin reversal, which is caused by thermal heating (phonons) or collisions of particles (e.g., neutrons) or photons. Spin waves and magnons also apply to other types of magnetic ordered materials, such as antiferromagnets, ferrimagnets, and helical magnets.

Following Squires (1996), the ferromagnetic interaction between the net electron spins of each atom in a ferromagnetic material, is assumed to be limited to interactions between each ferromagnetic atom and its nearest neighbor ferromagnetic atoms. The domain of a ferromagnetic crystal is modeled as having volume V, which contains N atoms, where the lth ferromagnetic atom position vector $\mathbf{R}_l$ in the lattice is indexed by integers

$$l = 1, 2, 3, \ldots, N \tag{8.64}$$

Each lth ferromagnetic atom has n_c nearest neighbor ferromagnetic atoms. The position vector $\mathbf{R}_{li}$ connects the lth ferromagnetic atom with its n_c nearest neighbor ferromagnetic i^{th} atoms, where

$$i = 1, 2, \ldots, n_c \tag{8.65}$$

The net electron spin of the unpaired electrons of the lth ferromagnetic atom is S_l, and the net electron spin of the ith nearest neighbor ferromagnetic atom to the lth atom is S_{l+i}.

The net electron spins of the ferromagnetic atoms of the lattice and the spin waves, which arise from thermal and collision effects, is described by the Heisenberg Hamiltonian with positive definite exchange integral $J > 0$, where

$$\hat{H} = -J \sum_{l=1}^{N} \sum_{i=1}^{n_c} \hat{S}_l \hat{S}_{l+i} \tag{8.66}$$

The net, unpaired, electron spins of the lth ferromagnetic atom S_l and the net unpaired, electron spins S_{l+i} of the n_c nearest neighbor $(l + i)^{\text{th}}$ ferromagnetic atoms in Eq. (8.66), are represented by their respective, associated spin operators $\hat{S}_l$ and $\hat{S}_{l+i}$. The eigenequation for the Heisenberg Hamiltonian has a spin-state wave function $\psi_{S,m}$ and spin-state eigenvalue-based energy E_m that is based on the sum of the net unpaired electron spin quantum number for each ferromagnetic atom in the domain lattice, where

$$\hat{H}\psi_{S,m} = \left(-J \sum_{l=1}^{N} \sum_{i=1}^{n_c} \hat{S}_l \hat{S}_{l+i} \right) \cdot \psi_{S,m} = E_m \psi_{S,m} \tag{8.67}$$

The total spin quantum number (integer) S is the total number of unpaired electron spins in every unit cell of the ferromagnetic crystal. The azimuth or z-component spin quantum numbers m range from the S electron spins all aligned in the same direction up, to all S electron spins aligned in the same direction down, where

$$m = -S, \ -S + 1, \ -S + 2, \ldots, S - 2, S - 1, S \tag{8.68}$$

The ground-state eigenfunction $\psi_{S,0}$ of the spin system in the domain of the ferromagnet, occurs when the longitudinal components of the net spin vector of the unpaired atomic electrons of each ferromagnetic atom, all point in the same positive z-direction. This state is described by the eigenequation

$$\hat{H}\psi_{S,0} = -JS^2Nn_c\psi_{S,0} = E_0\psi_{S,0} \tag{8.69}$$

When an electron spin flips from parallel to antiparallel, the system energy increases due to the interaction of the electron with is nearest electron spin neighbors. In a single line of electron spins, one-electron spin flips, and the system energy increases by $4J$ with respect to each of the two neighboring electron spins, so the electron spin-flip increases the energy of the spin system by $8J$. If the electrons all exist in a single atom, then the z-component angular momentum of the atom for a single electron spin-flip becomes $m = S - 1$. At zero temperature, all the electron spins in the domain are aligned to be parallel in the lowest state, where $m = S$.

We now derive a more mathematically convenient and physically transparent form of the Heisenberg Hamiltonian, which will lead to the magnon dispersion relation, and specifically identify the discrete magnon energy units. This simplification is done by converting the net atomic electron spin operators to equivalent ladder operators. One starts with the definition of the total vector spin operator $\hat{\mathbf{S}}$, which represents the summation of the unpaired electron spins of a single ferromagnetic atom, where with associated total spin quantum numbers S,

$$(\hat{\mathbf{S}} \cdot \hat{\mathbf{S}})\psi_{S,m} = (\hat{S}_x^2 + \hat{S}_y^2 + \hat{S}_z^2)\psi_{S,m} = S(S+1)\hbar^2\psi_{S,m} \tag{8.70}$$

The raising S^+ and lowering S^- ladder operators are defined by

$$\hat{S}^+ = \hat{S}^x + i\hat{S}^y \tag{8.71}$$

and

$$\hat{S}^- = \hat{S}^x - i\hat{S}^y \tag{8.72}$$

The raising operator S^+ converts state $\psi_{S,m}$ to $\psi_{S,m+1}$, where

$$\hat{S}^+\psi_{S,m} = \hbar\sqrt{(S-m)(S+1+m)}\,\psi_{S,m+1} \qquad -S \le m \le S \tag{8.73}$$

And the lowering operator S^- converts state $\psi_{S,m}$ to $\psi_{S,m-1}$, where

$$\hat{S}^-\psi_{S,m} = \hbar\sqrt{(S+m)(S+1-m)}\,\psi_{S,m-1} \qquad -S \le m \le S \tag{8.74}$$

The commutation of raising S^+ and S^- lowering operators yields the longitudinal spin component operator $\hat{S}_z$ of the total spin operator $\hat{S}$, where

$$\hat{S}_z = \frac{1}{2}\left[\hat{S}^+, \hat{S}^-\right] = \frac{1}{2}\left(\hat{S}^+\hat{S}^- - \hat{S}^-\hat{S}^+\right) \tag{8.75}$$

The Heisenberg Hamiltonian in terms of the x-, y-, and z-component spin operators ($\hat{S}_j^x$, $\hat{S}_j^y$, and $\hat{S}_j^z$) is

$$\hat{H} = -J\sum_{l=1}^{N}\sum_{i=1}^{n_c}\hat{\mathbf{S}}_l \cdot \hat{\mathbf{S}}_{l+i} = -J\sum_{l=1}^{N}\sum_{i=1}^{n_c}(\hat{S}_l^z\hat{S}_{l+i}^z + \hat{S}_l^y\hat{S}_{l+i}^y + \hat{S}_l^x\hat{S}_{l+i}^x) \tag{8.76}$$

This Hamiltonian represents the isotropic Heisenberg model. Alternatively, the Heisenberg Hamiltonian can be expressed in terms of the raising $\hat{S}^+$ and lowering $\hat{S}^-$ operators and longitudinal spin component operator $\hat{S}_z$. Here one substitutes for $\hat{S}_j^x$ and $\hat{S}_j^y$ in terms of $\hat{S}^+$ and $\hat{S}^-$ and collects terms to obtain

$$\hat{H} = -J\sum_{l=1}^{N}\sum_{i=1}^{n_c}\left(\hat{S}_l^z\hat{S}_{l+i}^z + \frac{1}{2}\hat{S}_l^+\hat{S}_{l+i}^- + \frac{1}{2}\hat{S}_l^-\hat{S}_{l+i}^+\right) \tag{8.77}$$

The Ising Heisenberg model has only the z-component $\hat{S}_l^z\hat{S}_{l+i}^z$ term. For ferromagnetic material, as shown earlier, the exchange integral is positive ($J > 0$). However, for antiferromagnetic materials, the exchange integral is negative ($J < 0$).

8.6 Ferromagnetic Materials—Spin Waves, Dispersion Relation, and Magnons

We now obtain expressions for the quantized traveling spin waves, which are produced as a result of the deviated precession of the unpaired electrons in a ferromagnetic crystal. One starts with the potential energy U of interaction between the lth atom spin vector $\mathbf{S}_l$ with the spin vector $\mathbf{S}_{l+i}$ of the nearest i^{th} neighbor, where

$$U = -2J\sum_{l=1}^{N}\mathbf{S}_l \cdot \mathbf{S}_{l+i} \tag{8.78}$$

For simplicity and clarity, we now examine the 1-D case of a line of atoms equally spaced with separation distance a.

The potential energy U_l of interaction between the center l^{th} atom spin vector $\mathbf{S}_l$ and its left $l-1$ neighbor atom spin vector $\mathbf{S}_{l-1}$ and right $l+1$ neighbor atom spin vector $\mathbf{S}_{l+1}$ is

$$U_l = -2J\mathbf{S}_l \cdot (\mathbf{S}_{l-1} + \mathbf{S}_{l+1}) \tag{8.79}$$

The magnetic moment of the lth atom is

$$\boldsymbol{\mu}_l = -|g\mu_{\text{b}}|\mathbf{S}_l \tag{8.80}$$

One substitutes for the lth atom spin vector $\mathbf{S}_l$ in Eq. (8.79), via Eq. (8.80), and obtains the magnetic potential energy U_l, which is expressed by the dot product of the magnetic moment $\boldsymbol{\mu}_l$ of the lth atom, with the effective magnetic field $\mathbf{B}_l$. The effective magnetic field $\mathbf{B}_l$ or exchange field, acts on the lth atom spin vector, where

$$U_l = -\boldsymbol{\mu}_l \cdot \mathbf{B}_l, \tag{8.81}$$

where

$$\mathbf{B}_l = -\frac{2J}{|g\mu_b|}(\mathbf{S}_{l-1} + \mathbf{S}_{l+1}) \tag{8.82}$$

The torque $\mathbf{T}_l$ of the effective magnetic field $\mathbf{B}_l$, which acts on the lth atom magnetic dipole moment vector $\boldsymbol{\mu}_l$, equals the time rate of change of the lth atom angular momentum vector $\mathbf{L}_l$, where

$$\frac{d\mathbf{L}_l}{dt} = \mathbf{T}_l = \boldsymbol{\mu}_l \times \mathbf{B}_l, \tag{8.83}$$

where

$$\mathbf{L}_l = \hbar\mathbf{S}_l \tag{8.84}$$

and

$$\hbar\frac{d\mathbf{S}_l}{dt} = \boldsymbol{\mu}_l \times \mathbf{B}_l \tag{8.85}$$

The lth atom's angular momentum vector $\mathbf{L}_l$, is simply the atom's spin angular momentum vector $\mathbf{S}_l$, multiplied by the reduced Planck's constant $\hbar$, where substitution into Eq. (8.85) of Eq. (8.82) for $\mathbf{B}_l$ gives

$$\frac{d\mathbf{S}_l}{dt} = -\frac{2J}{\hbar|g\mu_b|}\boldsymbol{\mu}_l \times (\mathbf{S}_{l-1} + \mathbf{S}_{l+1}) \tag{8.86}$$

Substitution of Eq. (8.80) into Eq. (8.86), yields

$$\frac{d\mathbf{S}_l}{dt} = \frac{2J}{\hbar}\mathbf{S}_l \times (\mathbf{S}_{l-1} + \mathbf{S}_{l+1}) \tag{8.87}$$

Expansion of Eq. (8.87) in rectangular coordinates, gives

$$\frac{dS_l^x}{dt} = \frac{2J}{\hbar}[S_l^y(S_{l-1}^z + S_{l+1}^z) - S_l^z(S_{l-1}^y + S_{l+1}^y)] \tag{8.88}$$

$$\frac{dS_l^y}{dt} = \frac{2J}{\hbar}[S_l^z(S_{l-1}^x + S_{l+1}^x) - S_l^x(S_{l-1}^z + S_{l+1}^z)] \tag{8.89}$$

$$\frac{dS_l^z}{dt} = \frac{2J}{\hbar}[S_l^x(S_{l-1}^y + S_{l+1}^y) - S_l^y(S_{l-1}^x + S_{l+1}^x)] \tag{8.90}$$

One assumes that the inelastic interaction with the lattice is sufficiently small so as to obtain a simple solution, which is a perturbation of the spin system. For each

l^{th} atom, the z-component of its unpaired electron spin S_l^z is approximated to be essentially constant and equal to the total unpaired electron spin S. The transverse spin components S_l^x and S_l^y are very small compared to S, and one ignores cross terms of the x- and y-components of the spin so that

$$S_l^x, S_l^y \ll S_l^z \cong S \quad \text{and} \quad S_l^x S_{l-1}^y = S_l^x S_{l+1}^y \cong 0 \tag{8.91}$$

The spin equations of precession in Eq. (8.88–8.90), are simplified to

$$\frac{\mathrm{d}S_l^x}{\mathrm{d}t} = \frac{2JS}{\hbar}[2S_l^y - S_{l-1}^y - S_{l+1}^y] \tag{8.92}$$

$$\frac{\mathrm{d}S_l^y}{\mathrm{d}t} = -\frac{2JS}{\hbar}[2S_l^x - S_{l-1}^x - S_{l+1}^x] \tag{8.93}$$

$$\frac{\mathrm{d}S_l^z}{\mathrm{d}t} = 0 \tag{8.94}$$

As a further simplification, one assumes a cubic crystal, whose unit cell dimension is a. The solution for S_l^x and S_l^y are then assumed to be traveling waves of the form

$$S_l^x = S_{0x} \exp[\mathrm{i}(lqa - wt)] \tag{8.95}$$

and

$$S_l^y = S_{0y} \exp[\mathrm{i}(lqa - wt)] \tag{8.96}$$

The amplitude magnitudes S_{0x} and S_{0y} are constants, and q is the wave number of the spin wave. The x and y spin wave solutions of Eqs. (8.95–8.96) are plugged into the spin vector precession Eqs. (8.92–8.93), so

$$-\mathrm{i}wS_{0x} = \frac{2JS}{\hbar}[2 - \exp(-\mathrm{i}qa) - \exp(\mathrm{i}qa)]S_{0y} \tag{8.97}$$

$$-\mathrm{i}wS_{0y} = \frac{2JS}{\hbar}[2 - \exp(-\mathrm{i}qa) - \exp(\mathrm{i}qa)]S_{0x} \tag{8.98}$$

With the Euler identity applied to Eqs. (8.97–8.98), one obtains

$$S_{0x} = -\frac{4JS}{\hbar \mathrm{i}w}[1 - \cos(\mathrm{i}qa)]S_{0y} \tag{8.99}$$

$$S_{0y} = -\frac{4JS}{\hbar \mathrm{i}w}[1 - \cos(\mathrm{i}qa)]S_{0x} \tag{8.100}$$

Substitution of S_{0x} of Eq. (8.99) into Eq. (8.100) for S_{0y} eliminates S_{0y}, and one obtains the dispersion relation for the spin wave:

$$\hbar w = -4JS[1 - \cos(iqa)] \tag{8.101}$$

The energy of a spin wave mode of frequency w_q and wave number q_q, which is comprised of n_q magnons, is then

$$E_q = \left(n_q + \frac{1}{2}\right)\hbar w_q \tag{8.102}$$

The excitation of one magnon corresponds to the reversal of one-electron spin of 1/2. For small wave numbers (q) or long wavelengths ($2\pi/q$), which are measured relative to the unit cell dimension a, the dispersion relation of Eq. (8.101) acquires quadratic dependence, so that

$$\hbar w \cong 2JSa^2q^2 \quad qa \ll 1 \tag{8.103}$$

where in Eq. (8.101),

$$\cos(qa) \cong 1 - \frac{1}{2}(qa)^2 \tag{8.104}$$

Recalling a spring constant k for the potential energy $E = kx^2$ of a spring, with a stiffness constant D, the long-wavelength dispersion relation, can be expressed by

$$\hbar w \cong Dq^2 \quad qa \ll 1 \tag{8.105}$$

Lattice stiffness constant D [meV Å^2] is found by comparison of Eqs. (8.103) and (8.105), where

$$D = 2JSa^2 \tag{8.106}$$

From Kittel (2005), at 295 K, the stiffness constant D [meV Å^2] is 281, 500, and 364 for iron, cobalt, and nickel lattices, respectively, where

$$w_{\mathrm{Fe}}\ [\mathrm{rad/s}] = \frac{281}{6.58 \times 10^{-13}} q_{\mathrm{Fe}}^2 \quad qa \ll 1 \tag{8.107}$$

$$w_{\mathrm{Co}}\ [\mathrm{rad/s}] = \frac{500}{6.58 \times 10^{-13}} q_{\mathrm{Co}}^2 \quad qa \ll 1 \tag{8.108}$$

and

$$w_{\mathrm{Co}}\ [\mathrm{rad/s}] = \frac{364}{6.58 \times 10^{-13}} q_{\mathrm{Co}}^2 \quad qa \ll 1 \tag{8.109}$$

8.7 Antiferromagnetic Materials—Spin Wave Dispersion Relation

We now derive the dispersion relation for a 1-D antiferromagnetic material. In antiferromagnetic materials, the directions of the magnetization and z-component spin vector alternate from one atomic plane to the next. The spins are ordered in an antiparallel arrangement, so long as the crystal temperature T is below the Néel temperature T_N. The periodic direction of magnetization can be modeled by two alternating planes of atoms: atom A with spins and magnetization pointed in the positive z-direction, and atom B with spin and magnetization pointed in the negative z-direction. One can assume a non-Bravais lattice, where each lth unit cell is comprised of atoms A and B. Atom A is addressed by even index integer $2l$, and atom B is addressed by odd integer index $2l+1$. Antiferromagnetic materials have a negative exchange integral ($J<0$).

The spin precession equations for the ferromagnetic material, which were derived in the prior section, can be applied to the alternating, magnetic polarity bilayers of the antiferromagnetic material. The layer of atom A is modeled with one set of the previously derived ferromagnetic equations, and the layer of atom B by a second set of the ferromagnetic equations with a reversal of both the polarity of J and the spin z-component spin quantum number S_z (i.e., S). Atom A with index $2l$ has z-component spin quantum numbers $\mathbf{S}_{2l}$, whereas the neighbor atoms $2l-1$ and $2l+1$ have opposite-directed spins $-\mathbf{S}_{2l-1}$ and $-\mathbf{S}_{2l+1}$, and the exchange integral J is replaced by $-J$, where recalling Eq. (8.86),

$$\frac{\mathrm{d}\mathbf{S}_{2l}}{\mathrm{d}t} = -\frac{2(-J)}{\hbar|g\mu_b|}\boldsymbol{\mu}_{2l}\times(-\mathbf{S}_{2l-1}-\mathbf{S}_{2l+1}) \tag{8.110}$$

Into Eq. (8.110), one substitutes

$$\boldsymbol{\mu}_{2l} = -|g\mu_b|\mathbf{S}_{2l} \tag{8.111}$$

Hence, Eq. (8.110) becomes

$$\frac{\mathrm{d}\mathbf{S}_{2l}}{\mathrm{d}t} = \frac{2J}{\hbar}\mathbf{S}_{2l}\times(\mathbf{S}_{2l-1}+\mathbf{S}_{2l+1}) \tag{8.112}$$

Expansion and approximation of Eq. (8.112), gives the following x and y precession equations for the unpaired electrons of atom A:

$$\frac{\mathrm{d}S_{2l}^x}{\mathrm{d}t} = \frac{2JS}{\hbar}[-2S_{2l}^y + S_{2l-1}^y + S_{2l+1}^y] \tag{8.113}$$

$$\frac{\mathrm{d}S_{2l}^y}{\mathrm{d}t} = \frac{2JS}{\hbar}[-2S_{2l}^x + S_{2l-1}^x + S_{2l+1}^x] \tag{8.114}$$

One can use S^+ to combine these two equations, where

$$S^+ = S^x + iS^y \tag{8.115}$$

Multiplication of Eqs. (8.113) and (8.114) by i, gives

$$i\frac{dS_{2l}^x}{dt} = \frac{2JS}{\hbar}\left[-i2S_{2l}^y + iS_{2l-1}^y + iS_{2l+1}^y\right] \tag{8.116}$$

and

$$i\frac{dS_{2l}^y}{dt} = i\frac{2JS}{\hbar}\left[-2S_{2l}^x + S_{2l-1}^x + S_{2l+1}^x\right] \tag{8.117}$$

With Eqs. (8.116) and (8.117), application of Eq. (8.115), then yields

$$\frac{dS_{2l}^+}{dt} = i\frac{dS_{2l}^x}{dt} + i\frac{dS_{2l}^y}{dt} = i\frac{2JS}{\hbar}\left[-2S_{2l}^+ + S_{2l-1}^+ + S_{2l+1}^+\right] \tag{8.118}$$

Atom B has the same precession equation, but $2l$ is replaced by $2l + 1$, and because the spin direction of atom A at position $2l$ is opposite to that of atom B at adjacent position $2l + 1$, one replaces S by $-S$. For atom B, via Eq. (8.118), one obtains

$$\frac{dS_{2l+1}^+}{dt} = i\frac{dS_{2l+1}^x}{dt} + i\frac{dS_{2l+1}^y}{dt} = -i\frac{2JS}{\hbar}[-2S_{2l+1}^+ + S_{2l}^+ + S_{2l+2}^+] \tag{8.119}$$

As a further simplification, one assumes a cubic crystal whose unit cell dimension is a. The solution for S_{2l}^+ and S_{2l+1}^+ are then assumed to be traveling waves. Atoms A are at positions $2l$ and $2l + 2$. These two A atoms contribute traveling waves with a positive constant S_A as their amplitude magnitude:

$$S_{2l}^+ = S_A \exp[i(2lqa - wt)] \tag{8.120}$$

$$S_{2l+2}^+ = S_A \exp[i((2l + 2)qa - wt)] \tag{8.121}$$

Atoms B are at positions $2l + 1$ and $2l - 1$, and these two B atoms contribute traveling waves with opposite polarity to that of atoms A. To be relative to atoms A, the two atom B traveling waves thus have positive constants S_B as their amplitude magnitude, which are multiplied by -1:

$$S_{2l+1}^+ = -S_B \exp[i((2l + 1)qa - wt)] \tag{8.122}$$

$$S_{2l-1}^+ = -S_B \exp[i((2l - 1)qa - wt)] \tag{8.123}$$

Here, q is the wave number of the spin wave, which receives contributions from the two A atoms and two B atoms in this periodic ABAB...ABAB line of atoms A and B. The atoms A, B, A, and B at positions $2l-1$, $2l$, $2l+1$, and $2l+2$, establish the repeat period of the spin wave of wave number q and temporal frequency w.

The S_{2l}^{+} and S_{2l+1}^{+} spin wave solutions of Eqs. (8.120) and (8.122), are plugged into the spin vector precession equations of Eqs. (8.118) and (8.119) for S_{2l}^{+} and S_{2l+1}^{+}, and yield the equation pair

$$-wS_A = \frac{2JS}{\hbar}[-2S_A - S_B\exp(-iqa) - S_B\exp(iqa)] \tag{8.124}$$

$$wS_B = -\frac{2JS}{\hbar}[2S_B + S_A\exp(-iqa) + S_A\exp(iqa)] \tag{8.125}$$

Let

$$w_{\text{ex}} = \frac{4JS}{\hbar} \tag{8.126}$$

With Eq. (8.126), simplification of Eqs. (8.124–8.125), via the Euler formula, gives

$$S_A(w - w_{\text{ex}}) = S_B w_{\text{ex}}\cos(qa) \tag{8.127}$$

$$S_B(w + w_{\text{ex}}) = -S_A w_{\text{ex}}\cos(qa) \tag{8.128}$$

One solves the simultaneous Eqs. (8.127–8.128) for w^2 to obtain

$$w^2 = w_{\text{ex}}^2[1 - \cos^2(qa)] \tag{8.129}$$

The dispersion relation of Eq. (8.129) for magnons in an antiferromagnetic crystal, becomes

$$w = w_{\text{ex}}|\sin(qa)| \tag{8.130}$$

If the wavelength of the spin wave is larger than the cubic unit cell dimension a, where $qa \ll 1$, then the dispersion relation for the antiferromagnetic material is linear, where

$$w \cong w_{\text{ex}}|qa| \quad qa \ll 1 \tag{8.131}$$

Substitution into Eq. (8.131) from Eq. (8.126) for w_{ex}, gives

$$w \cong \frac{4JS}{\hbar}|qa| \quad qa \ll 1 \tag{8.132}$$

In comparison, the long-wavelength dispersion relation of Eq. (8.105) with Eq. (8.106) for a ferromagnetic material, is quadratic:

$$w \cong \frac{2JS}{\hbar}(qa)^2 \quad qa \ll 1 \tag{8.133}$$

8.8 Exchange and Anisotropy Energy and Domain Formation in Magnetic Materials

The Bloch wall in a crystal is the transition layer that separates adjacent domains of the magnetic material; see Kittel (2005). The direction of magnetization varies from one domain to the next. The change in the unpaired electron spin direction of the atoms, and the resulting magnetic dipole moment per unit volume or magnetization in a domain, then transitions to that of the neighboring domain over the spatial distance of many atomic planes. These many atomic planes form the Bloch wall, which prevents the discontinuous jump of magnetization change between neighboring domains. One can estimate the thickness of the Bloch walls between domains as follows.

If the magnetization vector at the center of the wall is perpendicular to the surface of the magnetic material, then the wall is termed a *Bloch wall.* However, if the magnetization vector at the center of the wall is pointed parallel to the surface of the magnetic material, then the wall is termed a *Néel wall.* For example, in one domain, the magnetization vector may point perpendicular to the material surface in the positive z-direction and then pass through a Néel wall; in the next domain, the magnetization vector direction changed to point in the negative z-direction.

We assume a cubic Bravais-lattice crystal (one atom per unit cell) with a unit cell that has dimension a. The potential energy of interaction between the unpaired electron spins of each atom and the anisotropy energy together will determine the Bloch wall thickness. The Heisenberg potential energy U_{H} [eV] of interaction between the lth atom's spin vector $\mathbf{S}_l$ and the spin vector $\mathbf{S}_{l+i}$ of the nearest i^{th} neighbor, which is at angle θ relative to $\mathbf{S}_l$, is

$$U_{\mathrm{H}} = -2J\mathbf{S}_l \cdot \mathbf{S}_{l+i} = -2JS_lS_{l+i}\cos\theta \tag{8.134}$$

The Heisenberg exchange energy U_{H} [eV] is proportional to the exchange integral energy J [eV] and the product of the lth atom spin vector $\mathbf{S}_l$ and the $l+1$ atom spin vector $\mathbf{S}_{l+i}$, via the angle dependence $\cos\theta$ between the two adjacent spin vectors.

If one imagines a line of atoms, which cross the domain wall, where the angle θ between one spin vector and the next is small, then one can use the small angle approximation for $\cos\theta$. The Heisenberg exchange energy U_{H} [eV] is proportional to the exchange integral energy J [eV], and in the small θ angle approximation,

$$U_{\mathrm{H}} = -2JS_lS_{l+i}\left(1-\frac{\theta^2}{2}\right) \tag{8.135}$$

The direction of the magnetization vector is assumed to flip by 180° from one domain through the wall to the next domain. The wall formation is governed by minimizing the angle θ from one atom to the next across the wall between domains. Minimization of U_{H} favors angles θ approaching zero between one atomic spin vector and the next. That is, minimization of the Heisenberg potential energy favors the formation of thick walls with a large number of atomic layers, where θ

between spin directions in adjacent atomic layers approaches zero. The wall thickness $N_L a$ between two domains, is the number of unit cells N_L, multiplied by the unit cell width a.

For a line of atoms, the exchange potential energy U_{ex} is the second term of the Heisenberg potential energy U_H. Since the 180° change in direction of magnetization in one domain, occurs through a wall of N_L atomic layers, the exchange energy, via the spin vector direction change from one atom to the next, occurs through angle θ, given by

$$\theta = \frac{\pi}{N_L} \tag{8.136}$$

The exchange energy $U_{ex}^{l,l+1}$ [eV] for the spin direction change from one l^{th} atom to the neighboring $l+1$ atom is, thus,

$$U_{ex}^{l,l+1} = JS^2\left(\frac{\pi}{N_L}\right)^2 \tag{8.137}$$

The magnitude of the lth atom's spin S_l vectors is set to equal the total unpaired electron spin quantum number S for each atom.

There are $1/a^2$ lines of atoms per unit area of the wall face, and each line is comprised of a line of N_L atoms, which cross the domain wall. The resultant exchange energy U_{ex} [erg/cm^2] per unit wall area for the domain wall volume, which is comprised of a wall area of a^2 and wall thickness of N_L atomic layers, is then

$$U_{ex} = \frac{N_L}{a^2} U_{ex}^{l,l+1} = JS^2 \frac{\pi^2}{a^2 N_L} \tag{8.138}$$

Lower energy states favor a long spatial transition, or thick wall separating adjacent domains. However, this tendency for thick walls and lower energy states is offset by the spin–orbit coupling between the unpaired electrons, which tend to align the spins of the electrons in the same direction. The magnetization vector of the crystal follows the orbital overlap of electrons and the aligned spin directions, which are caused by LS coupling. If it were not for spin–orbit or LS coupling, the walls between domains would thicken without limit, so as to move in the direction of the lowest energy state of the crystal.

There is a potential energy U_K per unit area that arises as a consequence of the electron spin–orbit coupling and tendency for electron spins to align in the same direction. This causes the magnetization vector to follow crystallographic axes that do not hinder electron spin alignment. This energy is termed the *magnetocrystalline energy* or *anisotropy energy* per unit area U_K [erg/cm^2], which is assumed to be proportional to the domain wall thickness $N_L a$, via the energy density proportionality constant K_A [erg/cm^3], where

$$U_K = K_A N_L a \tag{8.139}$$

The unit cell dimension a is somewhat fixed. Thus, minimization of the anisotropy energy is accomplished by decreasing the number N_L of wall atomic layers.

The tendency of unpaired electron spins to align in the same direction, causes a decrease in the wall thickness, where more spins are aligned inside the domain and fewer are associated with the domain wall.

The minimization of the anisotropy energy is accomplished by decreasing the number N_L of wall atomic layers, which is juxtaposed to the minimization of the exchange energy by increasing the number N_L of wall atomic layers. The total potential energy of wall formation per unit area U_W [erg/cm^2] is the sum of the Heisenberg and anisotropy potential energies, where

$$U_W = U_{ex} + U_K = JS^2 \frac{\pi^2}{a^2 N_L} + K_A N_L a \tag{8.140}$$

One can find the minimum total wall energy U_W with respect to the wall thickness in terms of the number of wall atomic layers N_L, where

$$\frac{dU_W}{dN_L} = -JS^2 \frac{\pi^2}{a^2 N_L^2} + K_A a = 0 \tag{8.141}$$

Solving for N_L, one obtains the number N_L^{min} of atomic layers that minimize the wall energy:

$$N_L^{min} = \frac{S\pi}{a}\sqrt{\frac{J}{K_A a}} \tag{8.142}$$

The total minimum wall energy U_W^{min} [erg/cm^2] per unit area is then

$$U_W^{min} = 2\pi\sqrt{\frac{S^2 J K_A}{a}} \tag{8.143}$$

The magnetic materials of the number of atomic layers N_L can range from 40 to 10,000 atomic layers. The bcc crystal of iron has cubic unit cell dimension $a = 2.87$ Å, and the net unpaired electron spin per iron atom yields a spin quantum number of $S = 1$.

At room temperature (298 K), the exchange integral for iron is $J \cong 3.2 \times 10^{-3}$ eV or $J \cong 5.1 \times 10^{-15}$ erg. If one estimates the bulk anisotropy constant $K_A = 2 \times 10^4$ erg/cm^3, then the estimated number of atomic layers to minimize the wall energy is $N_L^{min} = 327$. The anisotropy energy per unit wall area is $U_K^{min} = 0.19$ erg/cm^2, the exchange energy per unit wall area is $U_{ex} = 0.38$ erg/cm^2, and the total energy per unit wall area is $U_W^{min} = 0.56$ erg/cm^2.

In the process of the crystal transitioning to the least energy, the crystal forms domains, which reduce the spatial volume of the magnetic field outside the crystal. A magnet with a single domain produces external field lines, which pass from the north pole (N) through a relatively large volume of space outside the domain, and terminate at the south pole (S) on the opposite end of the magnet. Now consider the magnet, which has two parallel domains, where a pair of adjacent north (N) and south (S) poles, are at each end of the magnet. On each end of the magnet, the magnetic field lines now pass mainly from the north pole (N) half of the end face to the adjacent south pole (S) half. This path between N and S on a single end face is much shorter than the path between a north pole (N) on one end face and a south pole (S) on the opposite end face of the magnet. As a result, the spatial volume and energy of the magnetic field is reduced. For a nonmagnetized or magnetized magnet, there is generally a multitude of domains between the pole end faces of the magnet, which are arrayed in various directions and have various shapes and sizes, usually on the order of about 10 μm.

The reduction of the spatial extent of the field lines, especially the flux of field lines that pass from one end of the magnet to the other, thereby reduces the magnetic energy E_B. The magnetic energy E_B is given by the volume integral for the squared magnetic field B^2 for differential volume dV, where

$$E_B = \frac{1}{8\pi}\int B^2\, \mathrm{d}V \tag{8.144}$$

If the magnet is divided into n domains, the magnetic energy E_B is reduced by approximately $1/n$ because of the n-fold reduced spatial extent of the field. That is, if one divides the magnet into n parallel domains running the length of the magnet, then the magnet has n alternating poles (N–S–N–S...) on each magnet end. The magnetic field lines from a north pole (N) will have a much stronger tendency pass through the small amount of external space and terminate on the adjacent south pole (S), rather than passing through a large volume of space to a south pole (S) on the opposite end of the magnet.

The magnet transitions toward the lowest energy state by forming more and more domains until the energy to produce one more domain wall equals the decrease in the magnetic field energy external to the crystal. The result of this energy minimization process is the division of the crystal into a multitude of domains that are each about 10 μm in height, width, and thickness. The shapes and sizes of the domains can vary depending on the magnetic material composition, temperature, pressure, applied magnetic fields, and whether the magnetic material has been magnetized fully, partially, or not at all, and its history of magnetization (hysteresis).

A magnet can transition to an even lower energy state by the formation of prism domains with 45° triangular boundaries near the magnetic poles. Consider a set of parallel long, 45° prism-tipped rectangular bar-shaped domains, which are adjoined on each end to smaller 45° prism-shaped domains, and form 45° interfaces. The 45° interfaces allow the magnetic flux lines to circulate in complete loops within the interior of the magnet, which then minimizes the magnetic field external to the magnet. A set of two 45° prism-tipped rectangle domains adjoined to two 45° prism domains form four 45° interfaces, which enable a closed internal path for the magnetization vector in the magnetic material. This set of rectangle and prism domains are called *domains of closure.*

Let us now examine these domains of closure and the boundary conditions of the magnetization vector at a 45° interface between adjacent prism-tipped rectangle and prism domains. Consider a nonmagnetized magnetic crystal, such as a single-crystal iron whisker, where every pair of long, rectangular bar-shaped domains has antiparallel magnetization in the positive and negative z-directions. The ends of the rectangular domains are shaped like an arrowhead with 45° faces; that is, the tips of the rectangular domains are prism-shaped with 45° faces. A prism domain is fit into the 45° v-groove at each end of the pair of rectangular domains. The prism domain magnetization vector is directed in the xy-plane, which is perpendicular to the positive and negative z-directed magnetization vectors of the two rectangular domains.

The positive or negative z-directed magnetization vector of each rectangular domain has equal components perpendicular and parallel to the 45° interface between the prism-tipped rectangle domain and the prism domain. The magnetization vector of the prism and rectangular domains have the same magnitude. However, the magnetization vector in the prism domain is directed in the xy-plane and is thus perpendicular to the rectangular domain's z-directed magnetization vector. The adjacent prism domain also has equal magnetization vector components perpendicular and parallel to the 45° interface that are shared with the 45° prism-tipped rectangle domain.

The normal component of the z-directed magnetization vector in the rectangular domain at the 45° interface equals the normal component of the xy-plane directed magnetization vector in the prism domain. That is, the normal component of the magnetization vector is continuous across the 45° boundary between the rectangle and prism domains. The tangential component of the z-directed magnetization vector in the rectangular domain at the 45° interface is equal to but in the opposite direction of the tangential component of the xy-plane directed magnetization vector in the prism domain. At the 45° interface, the tangential components of the rectangular and prism domains cancel each other out.

The magnetic field thus can circulate around an internal track comprised of (1) a positive, z-directed magnetization vector in the first rectangular domain with 45° prism tips, (2) a positive, x- and/or y-directed magnetization vector in the first 45° prism domain, (3) a negative, z-directed magnetization vector in the second rectangular domain with 45° prism tips, and (4) a negative x- and/or y-directed magnetization vector in the second 45° prism domain. This is a clockwise circulation of the magnetization vector, up through the first rectangle domain, across the top prism domain, down the second rectangular domain, and back across the bottom prism domain, to the first rectangle domain.

The component of the magnetization vector, which is normal to the 45° interfaces between the prism-shaped and -tipped rectangular domains, is continuous. The net tangential component of the magnetization vector is zero at each 45° interface by cancelling the tangential components of the prism-tipped rectangle domain and adjoining the prism domain. In this manner, the magnetic field can circulate in loops inside the magnet without spatial extension into the regions external to the magnet. These internal loop pathways for the magnetic field then greatly reduce the spatial extent of the magnetic field outside the magnet and thus minimize magnetic field energy.

8.9 Hamiltonian Eigenequation for 1-D Ferromagnetic Spin Lattice

We now derive the eigenequation for the Hamiltonian for a single row of N uniformly spaced ferromagnetic atoms, where each atom has S unpaired electrons, which produce a net spin vector $\mathbf{S}_j$ for the jth atom. The total electron spin of the jth atom is then given by total spin quantum number S for the S unpaired electrons per atom, and the total spin operator is $\hat{S}$. The x-, y-, and z-component spin operators are $\hat{S}_j^x$, $\hat{S}_j^y$, and $\hat{S}_j^z$, respectively, and the raising and lowering operators are $\hat{S}_j^+$ and $\hat{S}_j^-$.

We model a 1-D ferromagnet, where the Heisenberg nearest neighbor interaction is comprised of N atoms, each with total spin $S = 1/2$ and azimuthal spin component $S_z = \pm 1/2$. The separation between neighboring atoms is linear distance a. This treatment follows Squires (1996).

Consider the eigenstate wave function ψ_S as the weighted sum of eigenfunctions ϕ_n, via expansion coefficients c_n, where

$$\psi_S = \sum_{n=1}^{N} c_n \phi_n \tag{8.145}$$

The eigenfunction ϕ_n has all N atom spins pointing in the same direction except the oppositely directed n^{th} atom. The eigenfunction ϕ_n is a function of the N atom spin states, where the jth atom spin state is indicated by s_j, so that

$$s_j = \begin{matrix} u_j & \text{spin-up} \\ d_j & \text{spin-down} \end{matrix} \qquad j = 1, 2, \ldots, N \tag{8.146}$$

In the present case, $N - 1$ atoms have a spin-up net electron spin, indicated by u_j, and the single n^{th} atom has a spin-down net electron spin, indicated by d_n. The eigenfunction ϕ_n is then

$$\phi_n = |u_1 u_2 u_3 \cdots u_{n-1} d_n u_{n+1} \cdots u_N\rangle \tag{8.147}$$

or

$$\phi_n = |\phi_{n,1}\phi_{n,2}\phi_{n,3} \cdots \phi_{n,n-1}\phi_{n,n}\phi_{n,n+1} \cdots \phi_{n,N}\rangle \tag{8.148}$$

where

$$\phi_{n,m} = \begin{matrix} u_m & m \neq n \\ d_m & m = n \end{matrix} \tag{8.149}$$

and

$$\hat{S}_j^+ u_j = 0 \qquad \hat{S}_j^- u_j = \hbar d_j \qquad \hat{S}_j^z u_j = \frac{\hbar}{2} u_j \tag{8.150}$$

$$\hat{S}_j^+ d_j = \hbar u_j \qquad \hat{S}_j^- d_j = 0 \qquad \hat{S}_j^z d_j = -\frac{\hbar}{2} d_j \tag{8.151}$$

That is,

$$\phi_n = |u_1 u_2 u_3 \cdots u_{n-1} d_n u_{n+1} \cdots u_N\rangle \tag{8.152}$$

The Heisenberg Hamiltonian for this 1-D case considers that the spin interaction of the lth atom and its left and right neighbors is

$$\begin{aligned}\hat{H} &= -J\sum_{l=1}^{N}\left\{\hat{\mathbf{S}}_{l}\cdot\hat{\mathbf{S}}_{l+1} + \hat{\mathbf{S}}_{l}\cdot\hat{\mathbf{S}}_{l-1}\right\}\\ &= -J\sum_{l=1}^{N}\left\{\left(\hat{S}_l^z\hat{S}_{l+1}^z + \hat{S}_l^y\hat{S}_{l+1}^y + \hat{S}_l^x\hat{S}_{l+1}^x\right) + \left(\hat{S}_l^z\hat{S}_{l-1}^z + \hat{S}_l^y\hat{S}_{l-1}^y + \hat{S}_l^x\hat{S}_{l-1}^x\right)\right\}\end{aligned} \tag{8.153}$$

In terms of the z-component operator $\hat{S}_l^z$ and the raising $\hat{S}_l^+$ and lowering operators $\hat{S}_l^-$ for the lth atom and its left and right nearest neighbors, the Heisenberg Hamiltonian is

$$\hat{H} = -J\sum_{l=1}^{N}\left[\left(\hat{S}_l^z\hat{S}_{l+1}^z + \frac{1}{2}\hat{S}_l^+\hat{S}_{l+1}^- + \frac{1}{2}\hat{S}_l^-\hat{S}_{l+1}^+\right) + \left(\hat{S}_l^z\hat{S}_{l-1}^z + \frac{1}{2}\hat{S}_l^+\hat{S}_{l-1}^- + \frac{1}{2}\hat{S}_l^-\hat{S}_{l-1}^+\right)\right] \tag{8.154}$$

The results of the leading or lagging z-component operator $\hat{S}_{l\pm 1}^z$, followed by the z-component operator $\hat{S}_l^z$ tandem application, gives the same result:

$$\begin{aligned}&\sum_{l=1}^{N}\hat{S}_l^z\hat{S}_{l\pm 1}^z|u_1u_2u_3\cdots u_{n-1}d_nu_{n+1}\cdots u_N\rangle\\ &\quad= \frac{\hbar^2}{4}(N-2)|u_1u_2u_3\cdots u_{n-1}d_nu_{n+1}\cdots u_N\rangle\\ &\quad\quad- \frac{\hbar^2}{4}|u_1u_2u_3\cdots u_{n-1}d_nu_{n+1}\cdots u_N\rangle\\ &\quad\quad- \frac{\hbar^2}{4}|u_1u_2u_3\cdots u_{n-1}d_nu_{n+1}\cdots u_N\rangle\end{aligned} \tag{8.155}$$

or

$$\sum_{l=1}^{N}\hat{S}_l^z\hat{S}_{l\pm 1}^z|u_1u_2u_3\cdots u_{n-1}d_nu_{n+1}\cdots u_N\rangle = \frac{\hbar^2}{4}(n-2)\phi_n - \frac{\hbar^2}{4}(2\phi_n) \tag{8.156}$$

The result of the leading, lowering $\hat{S}_{l+1}^-$ operator, followed by the raising $\hat{S}_l^+$ operator tandem application, gives

$$\sum_{l=1}^{N}\frac{1}{2}\hat{S}_l^+\hat{S}_{l+1}^-|u_1u_2u_3\cdots u_{n-1}d_nu_{n+1}\cdots u_N\rangle = \frac{\hbar^2}{2}|u_1u_2u_3\cdots u_{n-1}u_nd_{n+1}\cdots u_N\rangle \tag{8.157}$$

or

$$\sum_{l=1}^{N}\frac{1}{2}\hat{S}_l^{+}\hat{S}_{l+1}^{-}|u_1u_2u_3\cdots u_{n-1}d_nu_{n+1}\cdots u_N\rangle=\frac{\hbar^2}{2}\phi_{n+1} \tag{8.158}$$

The result of the lagging, lowering $\hat{S}_{l-1}^{-}$ operator, followed by the raising $\hat{S}_l^{+}$ operator tandem application, gives

$$\sum_{l=1}^{N}\frac{1}{2}\hat{S}_l^{+}\hat{S}_{l-1}^{-}|u_1u_2u_3\cdots u_{n-1}d_nu_{n+1}\cdots u_N\rangle=\frac{\hbar^2}{2}|u_1u_2u_3\cdots d_{n-1}u_nu_{n+1}\cdots u_N\rangle \tag{8.159}$$

or

$$\sum_{l=1}^{N}\frac{1}{2}\hat{S}_l^{+}\hat{S}_{l-1}^{-}|u_1u_2u_3\cdots u_{n-1}d_nu_{n+1}\cdots u_N\rangle=\frac{\hbar^2}{2}\phi_{n-1} \tag{8.160}$$

The result of the leading, raising $\hat{S}_{l+1}^{+}$ operator, followed by the lowering $\hat{S}_l^{-}$ operator tandem application, gives

$$\sum_{l=1}^{N}\frac{1}{2}\hat{S}_l^{-}\hat{S}_{l+1}^{+}|u_1u_2u_3\cdots u_{n-1}d_nu_{n+1}\cdots u_N\rangle=\frac{\hbar^2}{2}|u_1u_2u_3\cdots d_{n-1}u_nu_{n+1}\cdots u_N\rangle \tag{8.161}$$

or

$$\sum_{l=1}^{N}\frac{1}{2}\hat{S}_l^{-}\hat{S}_{l+1}^{+}|u_1u_2u_3\cdots u_{n-1}d_nu_{n+1}\cdots u_N\rangle=\frac{\hbar^2}{2}\phi_{n-1} \tag{8.162}$$

The result of the lagging, raising $\hat{S}_{l-1}^{+}$ operator, followed by the lowering $\hat{S}_l^{-}$ operator tandem application, gives

$$\sum_{l=1}^{N}\frac{1}{2}\hat{S}_l^{-}\hat{S}_{l-1}^{+}|u_1u_2u_3\cdots u_{n-1}d_nu_{n+1}\cdots u_N\rangle=\frac{\hbar^2}{2}|u_1u_2u_3\cdots u_{n-1}u_nd_{n+1}\cdots u_N\rangle \tag{8.163}$$

or

$$\sum_{l=1}^{N}\frac{1}{2}\hat{S}_l^{-}\hat{S}_{l-1}^{+}|u_1u_2u_3\cdots u_{n-1}u_nd_{n+1}\cdots u_N\rangle=\frac{\hbar^2}{2}\phi_{n+1} \tag{8.164}$$

The contributions are tallied for the summed, tandem application of the unpaired electron spin operator of each atom with the unpaired electron spin operators in the left and right, nearest atom neighbors. One thus obtains the Hamiltonian eigenequation for one eigenstate of the 1-D array of ferromagnetic atoms:

$$\hat{H}\phi_n = -J\hbar^2\left\{\frac{1}{2}(N-2)\phi_n - \phi_n + \phi_{n+1} + \phi_{n-1}\right\} \tag{8.165}$$

The Hamiltonian eigenequation for the summed eigenstates of the 1-D array of ferromagnetic atoms is then

$$\hat{H}\psi_S = \hat{H}\sum_{n=1}^{N} c_n\phi_n = \sum_{n=1}^{N} c_n\hat{H}\phi_n \tag{8.166}$$

Substitution of Eq. (8.165) into Eq. (8.166), yields the Hamiltonian eigenequation

$$\hat{H}\psi_S = -J\hbar^2\sum_{n=1}^{N} c_n\left\{\frac{1}{2}(N-2)\phi_n - \phi_n + \phi_{n+1} + \phi_{n-1}\right\} \tag{8.167}$$

The Hamiltonian eigenequation is also expressed in terms of eigenenergy E, where

$$\hat{H}\psi_S = E\sum_{n=1}^{N} c_n\phi_n \tag{8.168}$$

The two equivalent expressions of Eqs. (8.167) and (8.168) for the Hamiltonian eigenequation are equated:

$$E\sum_{n=1}^{N} c_n\phi_n = -J\hbar^2\sum_{n=1}^{N} c_n\left\{\frac{1}{2}(N-2)\phi_n - \phi_n + \phi_{n+1} + \phi_{n-1}\right\} \tag{8.169}$$

In Eq. (8.169), collecting terms and equating the coefficients of ϕ_n give

$$Ec_n\phi_n = -J\hbar^2\left\{\frac{1}{2}(N-2)c_n - c_n + c_{n-1} + c_{n+1}\right\}\phi_n \tag{8.170}$$

With atom spacing a and spin wave number q, one defines c_n as the amplitude of the nth spin wave normal mode for the 1-D lattice, where

$$c_n = \exp(iqna) \tag{8.171}$$

Substitution of Eq. (8.171) into Eq. (8.170), gives

$$Ec_n\phi_n = -J\hbar^2\left\{\frac{1}{2}(N-2)c_n - c_n + c_{n-1} + c_{n+1}\right\}\phi_n \tag{8.172}$$

Multiplying both sides of Eq. (8.172) by c_{-n}, gives

$$E = -2J\hbar^2\left\{\frac{N}{4} - 1 + \frac{c_{-1} + c_1}{2}\right\}\phi_n \tag{8.173}$$

One obtains the eigenenergy E of the Heisenberg Hamiltonian for the unpaired electron spin interactions in the ferromagnetic 1-D spin lattice:

$$E = 2J\hbar^2[1 - \cos(qa)] - \frac{1}{2}J\hbar^2N \tag{8.174}$$

The ground-state E_0 occurs when there are no spin waves, and the N atoms (each with total spin $S = 1/2$ and z-component spin $S_z = \pm 1/2$) are all aligned in the same direction, where $q = 0$ and

$$E_0 = -\frac{1}{2}J\hbar^2 N \tag{8.175}$$

8.10 Spin and Spin Deviation Operators, Creation and Annihilation Operators, Holstein–Primakoff Transformations, and Linear Approximation of Heisenberg Hamiltonian

We now simplify the Heisenberg Hamiltonian operator, eigenfunction, and eigenvalues of the eigenequation, which describes spin waves in the domain of a ferromagnetic material, and follows the treatment by Squires (1996). Mathematical convenience causes one to translate from spin quantum number m, with $-S \le m \le S$, to spin deviation quantum number n, with $2S \ge n \ge 0$, where S is the number of unpaired electron spins per unit cell and thus the total spin number (integer) quantum number. One can define the spin deviation quantum number n as

$$n = S - m \tag{8.176}$$

The spin deviation operator $\hat{n}$ is defined as longitudinal spin component operator $\hat{S}_z$ subtracted from the total spin operator $\hat{S}$, where

$$\hat{n} = \hat{S} - \hat{S}_z \tag{8.177}$$

The spin states are translated to spin deviation states $\psi_{S,n}$ so that

$$\{\psi_{S,S}, \psi_{S,S-1}, \ldots, \psi_{S,-S+1}, \psi_{S,-S}\} \rightarrow \{\psi_{S,0}, \psi_{S,1}, \ldots, \psi_{S,2S-1}, \psi_{S,2S}\} \tag{8.178}$$

The eigenequations for the raising S^+ and S^- lowering operators with the spin deviation eigenstates become

$$\hat{S}^+ \psi_{S,n} = \hbar\sqrt{n}\sqrt{2S-(n-1)}\,\psi_{S,n-1} \tag{8.179}$$

and

$$\hat{S}^- \psi_{S,n} = \hbar\sqrt{n+1}\sqrt{2S-n}\,\psi_{S,n+1} \tag{8.180}$$

With $2S \ge n \ge 0$, the arguments of the square root are positive or zero. Factoring Eqs. (8.179) and (8.180), one obtains

$$\hat{S}^+ \psi_{S,n} = \hbar\sqrt{2Sn}\sqrt{1-\frac{n-1}{2S}}\,\psi_{S,n-1} \tag{8.181}$$

and

$$\hat{S}^{-}\psi_{S,n} = \hbar\sqrt{2S(n+1)}\sqrt{1-\frac{n}{2S}}\,\psi_{S,n+1} \tag{8.182}$$

We now further simplify the Heisenberg Hamiltonian operator $\hat{H}$ and its spin deviation eigenfunction $\psi_{S,n}$ that describes the spin waves in the domain of a ferromagnetic material. The spin deviation operator $\hat{n}$ is analogous to the creation (raising) a^+ and annihilation (lowering) a operators for the harmonic oscillator with eigenfunction ψ_n, see Liboff (2003), Merzbacher (1970), Squires (1996), and Cremer (2012b), where

$$a^+\psi_n = \sqrt{n+1}\,\psi_{n+1} \tag{8.183}$$

$$a\psi_n = \sqrt{n}\,\psi_{n-1} \tag{8.184}$$

with

$$a^+a\psi_n = n\psi_n \tag{8.185}$$

$$aa^+\psi_n = (n+1)\psi_n \tag{8.186}$$

$$a^+a = n \tag{8.187}$$

The creation a^+ and annihilation a operators, which were applied to thermal and collision induced, vibrating lattice atoms, can be also applied, via their mathematical properties, to "vibrating" spin lattices. Also due to thermal and collision effects, the spin vectors vibrate about their equilibrium polar tilt angles θ_l. The eigenequations for the raising S^+ and S^- lowering operators, which are associated with the spin deviation operator $\hat{n}$ and eigenstates, can be expressed in terms of the creation a^+ and annihilation a operators for the harmonic oscillator:

$$a^+ \leftrightarrow \sqrt{n+1} \tag{8.188}$$

and

$$a \leftrightarrow \sqrt{n} \tag{8.189}$$

With application of Eqs. (8.183–8.187) to Eqs. (8.181) and (8.182), one obtains the Holstein–Primakoff transformations for operators S^+, S^-, and $\hat{S}_z$:

$$\hat{S}^{+}\psi_{S,n} = \hbar\sqrt{2S}\,a\sqrt{1-\frac{a^+a-1}{2S}}\,\psi_{S,n-1} \tag{8.190}$$

and

$$\hat{S}^{-}\psi_{S,n} = \hbar\sqrt{2S}\,a^+\sqrt{1-\frac{a^+a}{2S}}\,\psi_{S,n+1} \tag{8.191}$$

Via the commutation relation of Eq. (8.75), one obtains the z-component operator $\hat{S}_z$,

$$\hat{S}_z = \frac{1}{2}[\hat{S}^+, \hat{S}^-] = \frac{1}{2}[\sqrt{2S - a^+a}\, aa^+ \sqrt{2S - a^+a} - a^+(2S - a^+a)a] \tag{8.192}$$

One multiplies and expands terms in Eq. (8.192)

$$\hat{S}_z = \frac{1}{2}[(2S - a^+a) + (2S - a^+a)a^+a - 2Sa^+a + a^+a^+aa] \tag{8.193}$$

Summation of terms in Eq. (8.193), and required use of $a^+a = 1 - aa^+$, then yields

$$\hat{S}_z = S - a^+a \tag{8.194}$$

The vibration amplitudes $\mathbf{u}_{ld}$ of the lattice of N atoms in the three directions about their equilibrium positions $\mathbf{R}_{ld}$ produce $3N$ normal modes of oscillation for the lattice. Similarly, for the N ferromagnetic atoms of a domain, where the unpaired atomic electrons form the N spin vector, the vibration of the polar tilt angle θ_l of the electron spin vectors about their equilibrium polar tilt angle θ_{l0}, then form N modes of oscillation for the spin waves. There are only N modes of oscillations for spin waves, because the vibration involves only the polar tilt angle θ, but does not involve the vibration of the atom in 3 independent directions, where the 3 vibration directions are defined by the unit cell vectors $(\mathbf{a}, \mathbf{b}, \mathbf{c})$.

The Heisenberg Hamiltonian is further simplified by a linear approximation, which applies for low-temperature magnetic materials and cold neutron collisions. A linear approximation can be applied, where one assumes that the spin deviation quantum number n is small compared to total spin quantum number (integer) S, which is the number of unpaired electron spins in every unit cell of the crystal, so that

$$2S \gg n \tag{8.195}$$

That is, one considers only spin waves, which arise by a few spin vector deviations that produce just lower energy excited spin states in the ferromagnetic domain. This linear approximation applies to the low-temperature case or collisions with low-temperature neutrons. With the linear approximation, the Holstein–Primakoff transformations of Eqs. (8.190), (8.191), and (8.194), become

$$\hat{S}^+ = \sqrt{2S}\hbar a \tag{8.196}$$

$$\hat{S}^- = \sqrt{2S}\hbar a^+ \tag{8.197}$$

$$\hat{S}_z = S - a^+a \tag{8.198}$$

The Heisenberg Hamiltonian can then be expressed in terms of the raising S^+ and lowering S^- operators and longitudinal spin component operator $\hat{S}_z$. One starts with the Heisenberg Hamiltonian of Eq. (8.77):

$$\hat{H} = -J\sum_{l=1}^{N}\sum_{i=1}^{n_c}\left(S_l^z S_{l+i}^z + \frac{1}{2}S_l^+ S_{l+i}^- + \frac{1}{2}S_l^- S_{l+i}^+\right) \tag{8.199}$$

With substitution into Eq. (8.199) of Eqs. (8.196–8.198) for operators S^+, S^-, and $\hat{S}_z$, the Heisenberg Hamiltonian with the linear approximation for the low-temperature case, is expressed in terms of the creation a_l^+ and annihilation a_l operators, where

$$\hat{H} = -J\sum_{l=1}^{N}\sum_{i=1}^{n_c}[(S - a_l^+ a_l)(S - a_{l+i}^+ a_{l+i}) + S a_l a_{l+i}^+ + S a_l^+ a_{l+i}] \tag{8.200}$$

In view of the low-temperature-based linear approximation, one can expand the Heisenberg Hamiltonian, and drop the terms containing four operators, so that the Hamiltonian becomes

$$\hat{H} = E_0 + 2Jn_c S\sum_{l=1}^{N} a_l^+ a_l - 2JS\sum_{l=1}^{N}\sum_{i=1}^{n_c} a_l^+ a_{l+i} \tag{8.201}$$

In Eq (8.201), the ground-state energy of the magnetized ferromagnetic material E_0 is

$$E_0 = -JS^2 N n_c \tag{8.202}$$

8.11 Application of the Bloch Theorem to Express Creation and Annihilation Operators

In this section, the Heisenberg Hamiltonian is expressed as a sum of simple harmonic oscillators, which characterize spin waves in the domain of a ferromagnetic lattice, in a manner similar to the Hamiltonian expression of atom vibrations in a lattice by a sum of simple harmonic oscillators. To obtain such an expression, one applies the Bloch theorem to a ferromagnetic spin lattice in a single domain, which has a periodic spin-based potential energy that results from the periodic positions of the unpaired atomic electrons of the ferromagnetic atoms. This spin-based potential energy can be expressed by the Heisenberg Hamiltonian. A physical solution of the Schrödinger equation for a periodic potential of the electron spin lattice is then a traveling plane wave, which is modulated on the atomic scale by a function that contains the periodicity of the spin lattice potential; that is, the lattice periodicity.

In the 1-D case of N ferromagnetic atoms, which are equally spaced by Δy in the y-direction, each atom with net unpaired electron spin vector $\mathbf{S}$, precesses at polar tilt angle θ about the z-direction of magnetization $\mathbf{B}$, where

$$\mathbf{B} = \mu_0 \mathbf{M} \tag{8.203}$$

The Schrödinger solution $\psi_{S,n}(y)$ for a 1-D spin lattice, is the product of a traveling plane wave $\exp(iqy)$, multiplied by a function $u_q(y)$, which has the same periodicity b as the spin lattice potential, so with wave number q,

$$\psi_{S,n}(y) = u_q(y)\exp(iqy), \tag{8.204}$$

where

$$u_q(y+b) = u_q(y) \tag{8.205}$$

For $y + y_l$, the wave function $\psi_{S,n}(y)$ becomes

$$\psi_{S,n}(y+y_l) = u_q(y+y_l)\exp[iq(y+y_l)] \quad \text{for } l = 1,2,3,\ldots,N \tag{8.206}$$

where

$$y_l = lb \quad \text{for } l = 1,2,3,\ldots,N \tag{8.207}$$

The Bloch theorem is thus illustrated via the result

$$\psi_{S,n}(y+y_l) = u_q(y)\exp(iqy)\exp(iqy_l) = \exp(iqy_l)\psi_{S,n}(y) \tag{8.208}$$

If the number N of spin vectors (or atoms) in the 1-D lattice is large, one can apply cyclic or Born–von Karman boundary conditions, where the spin wave function at points $y + Nb$ and y are equal, so that

$$\psi_{S,n}(y+Nb) = \psi_{S,n}(y) \tag{8.209}$$

The cyclic boundary condition applied to large N requires

$$\exp(iqNb) = 1 \tag{8.210}$$

As a result, the allowed values of q are discrete and expressed via integers n by

$$q = \frac{2\pi}{Nb}n \quad \text{for } n = 0, \pm 1, \pm 2, \ldots \tag{8.211}$$

The density of states q is $N/2\pi$, which is proportional to the length $L = Nb$ of the 1-D lattice. The first Brillouin zone in the reciprocal lattice, contains N uniformly distributed values of q within the interval $[-2/b, 2/b]$, which is centered about the center point of the corresponding 1-D reciprocal lattice.

Inspection of the linearly approximated Heisenberg Hamiltonian, in view of the Bloch theorem, then suggests a Fourier expansion of the creation a_l^+ and annihilation a_l operators. This would allow one to express the Heisenberg Hamiltonian in terms of spin wave oscillations with the qth associated wave vector $\mathbf{q}_q$ and thus yield a dispersion relation for the spin waves. With Fourier expansion, creation a_l^+ and annihilation a_l operators are derived from the corresponding creation a_q^+ and annihilation a_q operators in the reciprocal lattice or q-space. The direct-lattice creation and annihilation operators a_l^+ and a_l are obtained from their reciprocal lattice operator creation and annihilation operators a_q^+ and a_q, via inverse Fourier transforms, where one sums over the N spin wave vectors $\mathbf{q}_q$ of the lattice. Following Squires (1996), the annihilation operator is given by

$$a_l = \frac{1}{\sqrt{N}} \sum_{q=-N/2}^{N/2} \exp(\mathrm{i}\mathbf{q}_q \cdot \mathbf{R}_l) a_q \tag{8.212}$$

The creation operator is given by

$$a_l^+ = \frac{1}{\sqrt{N}} \sum_{q=-N/2}^{N/2} \exp(-\,\mathrm{i}\mathbf{q}_q \cdot \mathbf{R}_l) a_q^+ \tag{8.213}$$

These a_l^+ and a_l are Schrödinger time-independent operators. Recall that the lth ferromagnetic atom position vector $\mathbf{R}_l$ in the direct lattice of the ferromagnetic domain is indexed by integers

$$l = 1, 2, 3, \ldots, N \tag{8.214}$$

The q-space corresponds to the 3-D direct spin lattice, which is produced by the ferromagnetic atoms in a single domain of the magnetized ferromagnetic material. The reciprocal lattice space creation and annihilation operators a_q^+ and a_q are obtained by Fourier transform of their direct-lattice counterpart operators a_l^+ and a_l, where

$$a_q = \frac{1}{\sqrt{N}} \sum_{q=-N/2}^{N/2} \exp(-\,\mathrm{i}\mathbf{q} \cdot \mathbf{R}_l) a_l \tag{8.215}$$

and

$$a_q^+ = \frac{1}{\sqrt{N}} \sum_{q=-N/2}^{N/2} \exp(\mathrm{i}\mathbf{q} \cdot \mathbf{R}_l) a_l^+ \tag{8.216}$$

The N allowed spin wave vectors $\mathbf{q}$ are contained within the first Brillouin zone of the reciprocal lattice q-space, and the commutation relation between the reciprocal lattice space creation a_q^+ and annihilation a_q operators is expressed by the Kronecker delta function, where

$$[a_q, a_{q'}^+] = \delta_{q,q'} \tag{8.217}$$

There is a 1:1 correspondence between the position vectors $\mathbf{R}_l$ in the direct lattice and their corresponding "wave vectors" $\mathbf{q}_q$ in the reciprocal lattices, where

$$\mathbf{q}_q = \frac{2\pi}{\mathbf{R}_l} \quad l = 1, 2, \ldots, N \text{ and } q = 1, 2, \ldots, N \tag{8.218}$$

Using Eqs. (8.217–8.218), and substitution into Eq. (8.201) of the inverse Fourier transform expressions of Eqs. (8.212–8.213) for the direct-lattice creation and annihilation operators a_l^+ and a_l, yields the Heisenberg Hamiltonian, which is expressed by the reciprocal lattice creation and annihilation operators a_q^+ and a_q, so that

$$\hat{H} = E_0 + 2Jn_cS \sum_{q=-N/2}^{N/2} [1 - \gamma(\mathbf{q}_q)]a_q^+ a_q \tag{8.219}$$

In Eq. (8.219), the nearest neighbor factor $\gamma(\mathbf{q}_q)$ is

$$\gamma(\mathbf{q}_q) = \frac{1}{n_c} \sum_{j=1}^{n_c} \exp(i\mathbf{q}_q \cdot \mathbf{R}_{lj}) \tag{8.220}$$

Recall that the position vector $\mathbf{R}_{li}$ connects the lth ferromagnetic atom with its nearest j^{th} neighbor ferromagnetic atoms of n_c neighboring atoms, where $j = 1, 2, \ldots, n_c$.

8.12 The Heisenberg Hamiltonian Expressed as a Sum of Harmonic Oscillators

Following Squires (1996), the Heisenberg Hamiltonian $\hat{H}$ is the total energy E of the spin system in the ferromagnetic domain, where we start with Eq. (8.219), repeated below,

$$\hat{H} = E = E_0 + 2Jn_cS \sum_{q=-N/2}^{N/2} [1 - \gamma(\mathbf{q}_q)]a_q^+ a_q \tag{8.221}$$

The integer energy level n_q of a wave function, which has wave number q in the reciprocal lattice space, corresponds to an integer number n of modulated or deviated spin vectors at positions $\mathbf{R}_l$ in the direct-lattice space:

$$a_l^+ a_l = n = n_q = a_q^+ a_q \tag{8.222}$$

Substitution of Eq. (8.222) into Eq. (8.221) for $a_q^+ a_q$ in the Heisenberg Hamiltonian $\hat{H}$ expression gives

$$\hat{H} = H_0 + \sum_{q=-N/2}^{N/2} H_q n_q \tag{8.223}$$

In Eq. (8.223), the ground-state, unperturbed Heisenberg Hamiltonian operator $\hat{H}_0$ is

$$\hat{H}_0 = E_0 \tag{8.224}$$

And in Eq. (8.223), the perturbed Heisenberg Hamiltonian operator H_q is

$$H_q = 2Jn_cS[1 - \gamma(\mathbf{q}_q)] = 2Jn_cS\left[1 - \frac{1}{n_c}\sum_{i=1}^{n_c}\exp(\mathrm{i}\mathbf{q}_q \cdot \mathbf{R}_{li})\right] \tag{8.225}$$

In Eq. (8.225) for the perturbed Heisenberg Hamiltonian expression $\hat{H}_q$, one now invokes the Planck relation between the harmonic oscillator energy E_q and frequency $w(\mathbf{q}_q)$ via Planck's reduced constant $\hbar$:

$$E_q = hw(\mathbf{q}_q) = 2Jn_cS\left[1 - \frac{1}{n_c}\sum_{j=1}^{n_c}\exp(\mathrm{i}\mathbf{q}_q \cdot \mathbf{R}_{lj})\right] \tag{8.226}$$

where

$$\hat{H}_q = \sum_{q=-N/2}^{N/2} E_q n_q = \sum_{q=-N/2}^{N/2} hw(\mathbf{q}_q)n_q \tag{8.227}$$

The dependence of the oscillator (spin wave) energy $\hbar w(\mathbf{q}_q)$ on the oscillator wave vector $\mathbf{q}_q$, is the frequency dispersion relation between $w(\mathbf{q}_q)$ and $\mathbf{q}_q$, where

$$w_q = w(\mathbf{q}_q) = \frac{2}{\hbar}Jn_cS\left[1 - \frac{1}{n_c}\sum_{i=1}^{n_c}\exp(\mathrm{i}\mathbf{q}_q \cdot \mathbf{R}_{li})\right] \tag{8.228}$$

The position vector $\mathbf{R}_{li}$ connects the lth ferromagnetic atom with its nearest i^{th} neighbor ferromagnetic atom of n_c nearest neighbor atoms.

If the spin wavelength $2\pi/q$ is longer than the distance R_{li} between the lth ferromagnetic atom and its nearest i^{th} neighbor, then $\mathbf{q} \cdot \mathbf{R}_{li} \ll 1$. In this case, one can expand the exponential with small argument $\mathbf{q}_q \cdot \mathbf{R}_{li}$ in Eq. (8.228), so that

$$w(\mathbf{q}_q) \cong \frac{2}{\hbar}Jn_cS\left[1 - \frac{1}{n_c}\sum_{i=1}^{n_c}\left(1 + \mathrm{i}\mathbf{q}_q \cdot \mathbf{R}_{li} - \frac{1}{2}(\mathbf{q}_q \cdot \mathbf{R}_{li})^2\right)\right] \quad \mathbf{q}_q \cdot \mathbf{R}_{li} \ll 1 \tag{8.229}$$

In Eq. (8.229), one invokes the antisymmetry of the positions of the nearest neighbor i^{th} atoms to each lth atom, where

$$\sum_{i=1}^{n_c}\mathrm{i}\mathbf{q}_q \cdot \mathbf{R}_{li} = \sum_{i=-(n_c-1)/2}^{(n_c-1)/2}\mathrm{i}\mathbf{q}_q \cdot \mathbf{R}_{li} \quad \text{and} \quad \mathbf{R}_{l,+i} = -\mathbf{R}_{l,-i} \tag{8.230}$$

Hence in Eq. (8.229),

$$\sum_{i=1}^{n_c} i\mathbf{q}_q \cdot \mathbf{R}_{li} = 0 \tag{8.231}$$

In Eq. (8.229), the frequency dispersion relation between w_q and $\mathbf{q}_q$ for spin wavelengths longer than the spacing between the atoms in the lattice becomes

$$w_q \cong \frac{2JS}{\hbar}(\mathbf{q}_q \cdot \mathbf{R}_{li})^2 \quad \mathbf{q}_q \cdot \mathbf{R}_{li} \ll 1 \tag{8.232}$$

Consider a simple cubic ferromagnetic crystal with unit cell vectors $(\mathbf{a}, \mathbf{a}, \mathbf{a})$. The vectors $\mathbf{R}_{li}$ between the lth ferromagnetic atom and its six nearest i^{th} neighbors along the positive and negative x-, y-, and z-directions are

$$\mathbf{R}_{li} = \pm a\hat{\mathbf{x}} \quad \mathbf{R}_{li} = \pm a\hat{\mathbf{y}} \quad \mathbf{R}_{li} = \pm a\hat{\mathbf{z}} \tag{8.233}$$

The spin wave whose wavelength is longer than the atom spacing has wave vector

$$\mathbf{q}_q = q_{qx}\hat{\mathbf{x}} + q_{qy}\hat{\mathbf{y}} + q_{qz}\hat{\mathbf{z}} \tag{8.234}$$

The dispersion relationship for the long-wavelength approximation of Eq. (8.232) becomes

$$w_q \cong \frac{2JS}{\hbar}(q_q a)^2 \quad \mathbf{q}_q \cdot \mathbf{R}_{li} \ll 1 \tag{8.235}$$

In simple, face-centered (fcc), and bcc lattices, the energy dispersion relation for electron spin waves, which are longer than the atom spacing, is obtained from Eq. (8.235), where

$$\hbar w_q = Dq_q^2 \tag{8.236}$$

In Eq. (8.236), the lattice stiffness constant D is defined as

$$D = 2JSa^2 \tag{8.237}$$

The perturbed Heisenberg Hamiltonian $\hat{H}_q$ is thus expressed as a set of harmonic oscillators (spin waves), each with frequency $w(\mathbf{q}_q)$, where

$$w_q = w(\mathbf{q}_q) \tag{8.238}$$

The minimum quanta of harmonic oscillator (spin wave) energy E_q is called the *magnon*, whose energy is

$$E_q = \hbar w_q \tag{8.239}$$

The eigenequation for the perturbed Heisenberg Hamiltonian operator $\hat{H}_q$ is then

$$\hat{H}_q \psi_{S,n} = n_q \hbar w_q \psi_{S,n} \tag{8.240}$$

Recall, the number of deviated spin states equals the number of spin wave vectors ($n = n_q$). The eigenequation for the spin waves in the domain of the ferromagnetic material is expressed by (1) the perturbed Heisenberg Hamiltonian operator $\hat{H}_q$, and (2) the spin wave wavefunction $\psi_{S,n}$ for the ferromagnetic domain, which is comprised of atoms, each with net electron spin S, where

$$\hat{H}_q \psi_{S,n} = n\hbar w_q \psi_{S,n} \tag{8.241}$$

For a harmonic oscillator or set of N harmonic oscillators, the energy of a normal mode, whether it be photon, phonon, or magnon, has a total Hamiltonian operator $\hat{H}$ expressed in terms of the reciprocal lattice creation a_q^+ and annihilation a_q operators as

$$\hat{H} = \sum_{q=-N/2}^{N/2} \left(a_q^+ a_q + \frac{1}{2} \right) \hbar w_q \tag{8.242}$$

Since

$$a_q^+ a_q = n_q = n \tag{8.243}$$

the total Hamiltonian operator $\hat{H}$ is the sum of the energies of all the spin wave oscillation modes in which n is the number of spin deviations, that is, spin flips within the set of N spin vectors $\mathbf{S}$ within the ferromagnetic domain:

$$\hat{H} = \sum_{n=1}^{N} \left(n + \frac{1}{2} \right) \hbar w_n = \sum_{n=1}^{N} E_n \tag{8.244}$$

The energy E_n of the nth excited spin wave state results from n deviated (flipped) spins in the ferromagnetic domain lattice. This energy is that of a harmonic oscillator in its nth excited state with frequency mode w_n, where energy E_n is the energy of n magnons, each of energy $\hbar w_n$, so that

$$E_n = \left(n + \frac{1}{2} \right) \hbar w_n \tag{8.245}$$

One spin reversal in the lattice produces one magnon of energy $\hbar w_n$.

8.13 Coherent, Inelastic Partial Differential Cross Section for One-Magnon Absorption or Emission for Neutron Scatter in a Ferromagnetic Crystal

In this section, we derive the coherent, inelastic, partial differential cross section for one-magnon absorption or emission, which is caused by to neutron scatter in a ferromagnetic crystal, and follows the treatment by Squires (1996). We start with the time-dependent Heisenberg direct lattice creation $a_l^+(t)$ and annihilation $a_l(t)$ operators, which are obtained by adding time dependence to the Schrödinger operators of Eqs. (8.212–8.213). The annihilation operator is given by

$$a_l(t) = \frac{1}{\sqrt{N}} \sum_{q=-N/2}^{N/2} \exp[\mathrm{i}(\mathbf{q}_q \cdot \mathbf{R}_l - w_q t)] a_q \tag{8.246}$$

The creation operator is given by

$$a_l^+(t) = \frac{1}{\sqrt{N}} \sum_{q=-N/2}^{N/2} \exp[-\mathrm{i}(\mathbf{q}_q \cdot \mathbf{R}_l - w_q t)] a_q^+ \tag{8.247}$$

Energy E_{n_q} of the $n_q{}^{\text{th}}$ excited spin wave, with wave vector $\mathbf{q}$, results from n_q deviated (flipped) spins in the ferromagnetic domain lattice. A single spin-flip is one magnon of energy $\hbar w_n$, and E_n is the energy of n_q magnons, where

$$E_{n_q} = \left(n_q + \frac{1}{2}\right) \hbar w_{n_q} \tag{8.248}$$

and

$$a_q^+ a_q |\phi_q\rangle = n_q |\phi_q\rangle \tag{8.249}$$

The Heisenberg versions of the spin raising $\hat{S}_l^+(t)$ and lowering $\hat{S}_l^-(t)$ operator, as well as the spin z-component operator $\hat{S}_l^z(t)$ for the lth atom at position $\mathbf{R}_l$, which has net electron spin vector $\mathbf{S}_l$ within a single ferromagnetic domain, are determined from their corresponding time-independent Schrödinger operator versions, and given by the Holstein–Primakoff transformations in the low-temperature, linear approximation. With substitution of Eqs (8.246–8.247) for the creation and annihilation operators into Eqs. (8.196–8.197) for the spin raising and lowering operators, one obtains

$$\hat{S}_l^+(t) = \sqrt{\frac{2S}{N}} \sum_{q=-N/2}^{N/2} \exp\left[\mathrm{i}\left(\mathbf{q}_q \cdot \mathbf{R}_l - w_q t\right)\right] a_q \tag{8.250}$$

$$\hat{S}_l^-(t) = \sqrt{\frac{2S}{N}} \sum_{q=-N/2}^{N/2} \exp\left[-\mathrm{i}\left(\mathbf{q}_q \cdot \mathbf{R}_l - w_q t\right)\right] a_q^+ \tag{8.251}$$

With spin quantum number S, or number of electron spins per l^{th} unit cell, the temperature-averaged, mean z-component of the net electron spin of the one-atom unit cell is obtained by recalling Eq. (8.198), where

$$\hat{S}_l^z(t) = S - a_l^+(t) a_l(t) \tag{8.252}$$

For the lth spin vector $\mathbf{S}_l$, the time-dependent Heisenberg x-component spin $\hat{S}_l^x(t)$ and y-component spin $\hat{S}_l^y(t)$ operators are obtained from the Heisenberg spin raising $\hat{S}_l^+(t)$ and lowering $\hat{S}_l^-(t)$ operators. Solving for the x and y component spin operators in the pair of simultaneous equations for the spin rasing and lowering operators of Eqs. (8.71–8.72), one obtains

$$\hat{S}_l^x(t) = \frac{1}{2}[\hat{S}_l^+(t) + \hat{S}_l^-(t)] \tag{8.253}$$

and

$$\hat{S}_l^y(t) = -\frac{\mathrm{i}}{2}[\hat{S}_l^+(t) - \hat{S}_l^-(t)] \tag{8.254}$$

We now recall Eq. (5.109) in section 5.7 of Ch. 5 for a non-Bravais lattice crystal, which is modified for one atom per unit cell, where d = 0. The resulting Eq. (8.255) below is based on the Heitler–London model, LS coupling, and dipole approximation for a Bravais lattice crystal with one atom per unit cell, the general partial differential cross section, where

$$\begin{aligned}\left(\frac{\mathrm{d}\sigma}{\mathrm{d}\Omega\,\mathrm{d}E}\right)^{\mathrm{mag}} &= N(r_0 g)^2 \frac{k}{k_0}\sum_{l=0}^{N-1}\exp(\mathrm{i}\boldsymbol{\kappa}\cdot\hat{\mathbf{R}}_l)\int_{-\infty}^{\infty}\frac{\exp(-\mathrm{i}wt)}{2\pi\hbar}\left|\frac{g_0}{2}F_0(\boldsymbol{\kappa})\exp(-W_0)\right|^2 \\ &\times\exp\left[\sum_{i=1}^{J}p_i\langle\phi_i|[-\boldsymbol{\kappa}\cdot\hat{\mathbf{u}}_0(0)][\boldsymbol{\kappa}\cdot\hat{\mathbf{u}}_l(t)]|\phi_i\rangle\right] \\ &\times\sum_{i=1}^{J}p_i\langle\phi_i|\sum_{\alpha=1}^{3}\sum_{\beta=1}^{3}\left(\delta_{\alpha\beta}-\frac{\kappa_\alpha\kappa_\beta}{\kappa^2}\right)\hat{S}_{0\alpha}^{+}(0)\hat{S}_{l\beta}(t)|\phi_i\rangle\mathrm{d}t\end{aligned} \tag{8.255}$$

As a simplifying approximation in Eq. (8.255), one assumes that the lth nucleus displacement vector $\hat{\mathbf{u}}_l(t)$ from its equilibrium position $\mathbf{R}_l$ is the same as that of any other unit cell or itself at a later time t, so that

$$\hat{\mathbf{u}}_0(0) = \hat{\mathbf{u}}_l(t) \tag{8.256}$$

and

$$\exp\left[\sum_{i=1}^{J}p_i\langle\phi_i|[-\boldsymbol{\kappa}\cdot\hat{\mathbf{u}}_0(0)]\,[\boldsymbol{\kappa}\cdot\hat{\mathbf{u}}_l(t)]|\phi_i\rangle\right] = \exp\sum_{i=1}^{J}p_i\langle\phi_i|\phi_i\rangle = 1 \tag{8.257}$$

With substitution of Eqs. (8.256–8.257) into Eq. (8.255), the coherent, inelastic, partial differential cross section for magnon absorption or emission, due to neutron scatter, becomes

$$\begin{aligned}\left(\frac{\mathrm{d}\sigma}{\mathrm{d}\Omega\,\mathrm{d}E}\right)^{\mathrm{mag}} &= N(r_0 g)^2 \frac{k}{k_0}\sum_{l=0}^{N-1}\exp(\mathrm{i}\boldsymbol{\kappa}\cdot\hat{\mathbf{R}}_l)\int_{-\infty}^{\infty}\frac{\exp(-\mathrm{i}wt)}{2\pi\hbar}\left|\frac{g_0}{2}F_0(\boldsymbol{\kappa})\exp(-W_0)\right|^2 \\ &\times\sum_{i=1}^{J}p_i\langle\phi_i|\sum_{\alpha=1}^{3}\sum_{\beta=1}^{3}\left(\delta_{\alpha\beta}-\frac{\kappa_\alpha\kappa_\beta}{\kappa^2}\right)\hat{S}_{0\alpha}^{+}(0)\hat{S}_{l\beta}(t)|\phi_i\rangle\mathrm{d}t\end{aligned} \tag{8.258}$$

Next, one evaluates the expected or thermally averaged values $\langle \hat{S}_{0\alpha}(0)\hat{S}_{l\beta}(t)\rangle$ of the electron spin product operator $\hat{S}_{0\alpha}(0)\hat{S}_{l\beta}(t)$ for the various combinations of the component $\alpha = x, y, z$ and $\beta = x, y, z$ directions:

$$\left\langle \hat{S}^{+}_{0\alpha}(0)\hat{S}_{l\beta}(t)\right\rangle = \sum_{i=1}^{J} p_i \left\langle \phi_i \left| \sum_{\alpha=1}^{3}\sum_{\beta=1}^{3}\left(\delta_{\alpha\beta} - \frac{\kappa_\alpha\kappa_\beta}{\kappa^2}\right)\hat{S}^{+}_{0\alpha}(0)\hat{S}_{l\beta}(t)\right| \phi_i \right\rangle \tag{8.259}$$

The $\langle\rangle$ in Eq. (8.259), indicates thermal averaging of the initial states of the scattering system at temperature T, where the initial states occur with probability p_i. See Eq. (3.27) in section 3.3 of chapter 3, and in Cremer (2012b), see chapters 3–7. The ith eigenfunction ϕ_i of the J unpaired electron spin states in each unit cell of the crystal, is an eigenstate of the lowering (annihilation) a and raising (creation) a^+ operators, where

$$a|\phi_i\rangle = \sqrt{\mathrm{i}}|\phi_{i-1}\rangle \tag{8.260}$$

$$a^+|\phi_i\rangle = \sqrt{\mathrm{i}+1}|\phi_{i+1}\rangle \tag{8.261}$$

$$a^+a|\phi_i\rangle = i|\phi_i\rangle \tag{8.262}$$

$$aa^+|\phi_i\rangle = (i+1)|\phi_i\rangle \tag{8.263}$$

The product operator is of the time-independent Schrödinger electron spin operator $\hat{S}_{0\beta}(0)$ for the $l=0$ reference unit cell at initial time $t=0$, with the time-dependent Heisenberg electron spin operator $\hat{S}_{l\beta}(t)$ in the β direction for the lth unit cell at later time t. Evaluation of Eqs. (8.259) will be done in four steps, and is explained as follows.

Step 1: In Eq. (8.259), the expected value for the product of the z-component or longitudinal electron spin operators is evaluated, where

$$\left\langle \hat{S}^{+}_{0z}(0)\hat{S}_{lz}(t)\right\rangle = \left(1 - \frac{\kappa_z^2}{\kappa^2}\right)\sum_{i=1}^{J} p_i \left\langle \phi_i \left| \hat{S}^{+}_{0z}(0)\hat{S}_{lz}(t)\right| \phi_i \right\rangle \tag{8.264}$$

From Eqs. (8.198) and (8.252), the z-component of the spin operator, is expressed by the total spin quantum number (integer) S, or the number of unpaired electron spins per unit cell:

$$\hat{S}_{lz}(t) = \hat{S}^{z}_{l}(t) = S - a^+_l(t)a_l(t) \tag{8.265}$$

With substitution into Eq. (8.264) of Eq. (8.265) for $\hat{S}_{lz}(t)$ and $\hat{S}_{0z}(t)$, one drops the 4th power term $a^+_l(0)a_l(0)a^+_l(t)a_l(t)$, and obtains

$$\left\langle \hat{S}^{+}_{0z}(0)\hat{S}_{lz}(t)\right\rangle \cong \left(1 - \frac{\kappa_z^2}{\kappa^2}\right)\sum_{i=1}^{J} p_i \left\langle \phi_i \left| S^2 - Sa^+_l(0)a_l(0) - Sa^+_l(t)a_l(t)\right| \phi_i \right|\right\rangle \tag{8.266}$$

Dropping the fourth term produces a linear approximation, which will limit the derivation of neutron inelastic scatter by electron spin wave interaction absorption or emission of only one magnon. With the linear approximation, and expression of the separate products in Eq. (8.266), the expected value of the longitudinal spin product, becomes

$$\left\langle \hat{S}^{+}_{0z}(0)\hat{S}_{lz}(t)\right\rangle \cong \left(1-\frac{\kappa_z^2}{\kappa^2}\right)\left(S^2 - S\left\langle a_l^+(0)a_l(0)\right\rangle - S\left\langle a_l^+(t)a_l(t)\right\rangle\right) \tag{8.267}$$

The thermal averaged product of the time-dependent and time-independent creation and annihilation operators for the lth unit cell are equal, and given by

$$\langle a_l^+(0)a_l(0)\rangle = \langle a_l^+(t)a_l(t)\rangle = \sum_{i=1}^{J} p_i\langle\phi_i|a_l^+(t)a_l(t)|\phi_i|\rangle \tag{8.268}$$

Recall S is a constant; that is, S is the total, spin quantum number per atom, or the number of un-paired electron spins per unit cell. Substitution for $a_l^+(t)$ and $a_l(t)$ from Eqs. (8.246–8.247) into the term $\langle a_l^+(t)a_l(t)\rangle$ in Eq. (8.268), yields

$$\langle a_l^+(t)a_l(t)\rangle = \frac{1}{N}\sum_{q=-N/2}^{N/2}\sum_{q'=-N/2}^{N/2} \exp[-\mathrm{i}\{(\mathbf{q}_{q'}-\mathbf{q}_q)\cdot\mathbf{R}_l - (w_{q'}-w_q)t\}]\langle a_q^+ a_{q'}\rangle \tag{8.269}$$

In Eq. (8.269), the only nonzero terms are for $q = q'$, so that one is left with only the summation over q, where

$$\langle a_l^+(t)a_l(t)\rangle = \frac{1}{N}\sum_{q=-N/2}^{N/2}\langle a_q^+ a_q\rangle \tag{8.270}$$

For spin wave vector $\mathbf{q}_q$, the expected number of magnons $\langle n_q\rangle$, each with energy $\hbar w_q$, is

$$\langle n_q\rangle = \langle a_q^+ a_q\rangle = \sum_{i=1}^{J} p_i\langle\phi_i|a_q^+ a_q|\phi_i|\rangle \tag{8.271}$$

Substitution of Eq. (8.271) into Eq. (8.270), gives

$$\langle a_l^+(t)a_l(t)\rangle = \frac{1}{N}\sum_{q=-N/2}^{N/2}\langle n_q\rangle \tag{8.272}$$

With $\langle n_l(t)\rangle = \langle a_l^+(t)a_l(t)\rangle$ substituted into Eq. (8.272), one obtains

$$\langle n_l(t)\rangle = \frac{1}{N}\sum_{q=-N/2}^{N/2}\langle n_q\rangle = \frac{1}{N}\sum_{q=-N/2}^{N/2}\sum_{i=1}^{J} p_i\langle\phi_i|a_q^+ a_q|\phi_i|\rangle \tag{8.273}$$

With substitution of Eq. (8.270) into Eq. (8.268), one then obtains the time-independent result for the expected value of the product of the z-component electron spin operators, which is called the *longitudinal term.* Due to this time independence, the longitudinal term is responsible for coherent elastic scatter of the neutrons by the electron spins of the ferromagnet, where

$$\langle \hat{S}_{0z}^{+}(0)\hat{S}_{lz}(t)\rangle \cong \left(1-\frac{\kappa_z^2}{\kappa^2}\right)\left(S^2-\frac{2S}{N}\sum_{q=-N/2}^{N/2}\langle a_q^+ a_{q'}\rangle\right) \tag{8.274}$$

With substitution of Eq. (8.271) into Eq. (8.274), one obtains

$$\langle \hat{S}_{0z}^{+}(0)\hat{S}_{lz}(t)\rangle \cong \left(1-\frac{\kappa_z^2}{\kappa^2}\right)\left(S^2-\frac{2S}{N}\sum_{q=-N/2}^{N/2}\langle n_q\rangle\right) \tag{8.275}$$

Step 2: In Eq, (8.259), the expected value of the product of the x- and y-components of the electron spin operators $\langle \hat{S}_{0x}(0)\hat{S}_{lx}(t)\rangle$ and $\langle \hat{S}_{0y}(0)\hat{S}_{ly}(t)\rangle$, which are called *transverse terms*, are evaluated using the raising $\hat{S}_l^{+}(t)$ and lowering $\hat{S}_l^{-}(t)$ operator expressions, where recalling Eqs. (8.250–8.254),

$$\begin{aligned}\hat{S}_{lx}(t)=\hat{S}_l^{x}(t)&=\frac{1}{2}\left[\hat{S}_l^{+}(t)+\hat{S}_l^{-}(t)\right]\\&=\frac{1}{2}\sqrt{\frac{2S}{N}}\sum_{q=-N/2}^{N/2}\{\exp[\mathrm{i}(\mathbf{q}_q\cdot\mathbf{R}_l-w_q t)]a_q+\exp[-\mathrm{i}(\mathbf{q}_q\cdot\mathbf{R}_l-w_q t)]a_q^+\}\end{aligned} \tag{8.276}$$

and

$$\begin{aligned}\hat{S}_{ly}(t)=\hat{S}_l^{y}(t)&=-\frac{\mathrm{i}}{2}[\hat{S}_l^{+}(t)-\hat{S}_l^{-}(t)]\\&=-\frac{\mathrm{i}}{2}\sqrt{\frac{2S}{N}}\sum_{q=-N/2}^{N/2}\{\exp[\mathrm{i}(\mathbf{q}_q\cdot\mathbf{R}_l-w_q t)]a_q-\exp[-\mathrm{i}(\mathbf{q}_q\cdot\mathbf{R}_l-w_q t)]a_q^+\}\end{aligned} \tag{8.277}$$

To determine expected values of cross products of x and y spin operator components in Eq. (8.259), one first evaluates the expected values $\langle \hat{S}_l^{+}(0)\hat{S}_l^{-}(t)\rangle$ and $\langle \hat{S}_l^{-}(0)\hat{S}_l^{+}(t)\rangle$, where

$$\langle \hat{S}_l^{+}(0)\hat{S}_l^{-}(t)\rangle=\left\{\frac{2S}{N}\sum_{q=-N/2}^{N/2}\exp[-\mathrm{i}(\mathbf{q}_q\cdot\mathbf{R}_l)]a_q^+\right\}\left\{\sum_{q'=-N/2}^{N/2}\exp[\mathrm{i}(\mathbf{q}'_q\cdot\mathbf{R}_l-w_{q'}t)]a_{q'}\right\} \tag{8.278}$$

and

$$\langle \hat{S}_l^{-}(0)\hat{S}_l^{+}(t)\rangle=\frac{2S}{N}\left\{\sum_{q=-N/2}^{N/2}\exp[\mathrm{i}(\mathbf{q}_q\cdot\mathbf{R}_l)]a_q^+\right\}\left\{\sum_{q'=-N/2}^{N/2}\exp[-\mathrm{i}(\mathbf{q}'_q\cdot\mathbf{R}_l-w_{q'}t)]a_{q'}\right\} \tag{8.279}$$

In Eqs. (8.278–8.279), the only nonzero terms are for $q = q'$, so that

$$\langle \hat{S}_l^+(0)\hat{S}_l^-(t)\rangle = \frac{2S}{N}\sum_{q=-N/2}^{N/2} \exp[-\mathrm{i}(\mathbf{q}_q \cdot \mathbf{R}_l - w_q t)]\langle n_q + 1\rangle \tag{8.280}$$

with

$$a_q a_q^+ |\phi_q\rangle = (n_q + 1)|\phi_q\rangle \tag{8.281}$$

and

$$\langle \hat{S}_l^-(0)\hat{S}_l^+(t)\rangle = \frac{2S}{N}\sum_{q=-N/2}^{N/2} \exp[\mathrm{i}(\mathbf{q}_q \cdot \mathbf{R}_l - w_q t)]\langle n_q\rangle \tag{8.282}$$

Substitution of Eqs. (8.280) and (8.282) with Eqs. (8.276–8.277) in Eq. (8.259) gives the x- and y-directed transverse terms, which contribute to the coherent, inelastic neutron scatter in Eq. (8.258), where

$$\langle \hat{S}_{0x}^+(0)\hat{S}_{lx}(t)\rangle = \frac{S}{2N}\left(1 - \frac{\kappa_x^2}{\kappa^2}\right)\sum_{q=-N/2}^{N/2} \{\exp[-\mathrm{i}(\mathbf{q}_q \cdot \mathbf{R}_l - w_q t)]\langle n_q + 1\rangle + \exp[\mathrm{i}(\mathbf{q}_q \cdot \mathbf{R}_l - w_q t)]\langle n_q\rangle\} \tag{8.283}$$

and

$$\langle \hat{S}_{0y}^+(0)\hat{S}_{ly}(t)\rangle = \frac{S}{2N}\left(1 - \frac{\kappa_y^2}{\kappa^2}\right)\sum_{q=-N/2}^{N/2} \{\exp[-\mathrm{i}(\mathbf{q}_q \cdot \mathbf{R}_l - w_q t)]\langle n_q + 1\rangle + \exp[\mathrm{i}(\mathbf{q}_q \cdot \mathbf{R}_l - w_q t)]\langle n_q\rangle\} \tag{8.284}$$

Step 3: In Eq, (8.259), one evaluates the expected value of the product of the x- and y-components of the electron spin operators or cross terms $\langle \hat{S}_{0x}(0)\hat{S}_{ly}(t)\rangle$ and $\langle \hat{S}_{0y}(0)\hat{S}_{lx}(t)\rangle$, which are also called *transverse terms*. Evaluation is done in the same manner as described in step 2, so that one obtains the transverse cross terms that also contribute to coherent, inelastic neutron scatter, where

$$\langle \hat{S}_{0x}^+(0)\hat{S}_{ly}(t)\rangle = -\frac{\mathrm{i}S}{2N}\frac{\kappa_x\kappa_y}{\kappa^2}\sum_{q=-N/2}^{N/2} \{\exp[-\mathrm{i}(\mathbf{q}_q \cdot \mathbf{R}_l - w_q t)]\langle n_q + 1\rangle - \exp[\mathrm{i}(\mathbf{q}_q \cdot \mathbf{R}_l - w_q t)]\langle n_q\rangle\} \tag{8.285}$$

and

$$\langle \hat{S}^{+}_{0y}(0)\hat{S}_{lx}(t)\rangle = -\frac{iS}{2N}\frac{\kappa_x\kappa_y}{\kappa^2}\sum_{q=-N/2}^{N/2}\{\exp[-i(\mathbf{q}_q\cdot\mathbf{R}_l - w_q t)]\langle n_q + 1\rangle - \exp[i(\mathbf{q}_q\cdot\mathbf{R}_l - w_q t)]\langle n_q\rangle\} \tag{8.286}$$

Step 4: In Eq, (8.259), one evaluates $\langle \hat{S}_{0x}(0)\hat{S}_{lz}(t)\rangle$, $\langle \hat{S}_{0z}(0)\hat{S}_{lx}(t)\rangle$, $\langle \hat{S}_{0y}(0)\hat{S}_{lz}(t)\rangle$, and $\langle \hat{S}_{0z}(0)\hat{S}_{ly}(t)\rangle$, where each result will be found to be zero. Operator $\hat{S}_{lx}(t)$ is the sum and $\hat{S}_{ly}(t)$ is the difference of two terms, one term with $a_l^+(t)$ and one term with $a_l(t)$. Operator $\hat{S}_{lz}(t)$ is $a_l^+a_l$ subtracted from the total, electron spin quantum number S. The product of z-component operator $\hat{S}_{lz}(t)$ with $\hat{S}_{lx}(t)$, and the product of $\hat{S}_{lz}(t)$ with $\hat{S}_{ly}(t)$, both yield expressions, which are a sum of products. Each product contains a single raising $a_l^+(t)$ or lowering $a_l(t)$ operator, or a triple product of a combination of $a_l^+(t)$ and $a_l(t)$, where

$$\langle\phi_i|a|\phi_i\rangle = \sqrt{i}\langle\phi_i|\phi_{i-1}\rangle = 0 \tag{8.287}$$

$$\langle\phi_i|a^+|\phi_i\rangle = \sqrt{i+1}\langle\phi_i|\phi_{i+1}\rangle = 0 \tag{8.288}$$

$$\langle\phi_i|aa^+a|\phi_i\rangle = i\langle\phi_i|\phi_{i-1}\rangle = 0 \tag{8.289}$$

$$\langle\phi_i|a^+a^+a|\phi_i\rangle = \sqrt{i(i+1)}\langle\phi_i|\phi_{i+1}\rangle = 0 \tag{8.290}$$

$$\langle\phi_i|aaa^+|\phi_i\rangle = \sqrt{i(i+1)}\langle\phi_i|\phi_{i-1}\rangle = 0 \tag{8.291}$$

$$\langle\phi_i|a^+aa^+|\phi_i\rangle = (i+1)\langle\phi_i|\phi_{i+1}\rangle = 0 \tag{8.292}$$

Hence, the cross products between x and z and y and z are both zero:

$$\langle \hat{S}^{+}_{0x}(0)\hat{S}_{lz}(t)\rangle = \langle \hat{S}^{+}_{0z}(0)\hat{S}_{lx}(t)\rangle = 0 \tag{8.293}$$

and

$$\langle \hat{S}^{+}_{0y}(0)\hat{S}_{lz}(t)\rangle = \langle \hat{S}^{+}_{0z}(0)\hat{S}_{ly}(t)\rangle = 0 \tag{8.294}$$

Once the four steps are completed, one has determined the nonzero expected values of the products of the electron spin operator x-, y-, and z-components. Next, one applies the following relation of the neutron scatter vector components to obtain the resultant sum of these products:

$$1 - \frac{\kappa_x^2}{\kappa^2} + 1 - \frac{\kappa_y^2}{\kappa^2} = 1 + \frac{\kappa_z^2}{\kappa^2} \tag{8.295}$$

Substitution of the expected values of the spin products of Eqs. (8.275), (8.283–8.286), (8.293–8.294), with Eq. (8.295) into Eq. (8.258–8.259), then yields the partial differential cross section $(\mathrm{d}\sigma/\mathrm{d}\Omega\,\mathrm{d}E)^{\mathrm{inel}}_{\mathrm{coh,avg}}$ for coherent, inelastic neutron scatter for single magnon absorption or emission:

$$\left(\frac{\mathrm{d}\sigma}{\mathrm{d}\Omega\mathrm{d}E}\right)^{\mathrm{inel}}_{\mathrm{coh}} = (r_0 g)^2 \frac{k}{k_0}\frac{S}{2}\left(1+\frac{\kappa_z^2}{\kappa^2}\right)\sum_{l=0}^{N-1}\exp(\mathrm{i}\boldsymbol{\kappa}\cdot\hat{\mathbf{R}}_l)$$
$$\times\left|\frac{g_0}{2}F_0(\boldsymbol{\kappa})\exp(-W_0)\right|^2 \frac{1}{2\pi\hbar}\int_{-\infty}^{\infty}\exp(-\mathrm{i}wt)$$
$$\times\sum_{q=-N/2}^{N/2}\{\exp[-\mathrm{i}(\mathbf{q}_q\cdot\mathbf{R}_l - w_q t)]\langle n_q+1\rangle + \exp[\mathrm{i}(\mathbf{q}_q\cdot\mathbf{R}_l - w_q t)]\langle n_q\rangle\}\mathrm{d}t \tag{8.296}$$

Recall from Appendix A.4 in Squires (1996), and Eq. 105 in Chapter 4 of Cremer (2012b),

$$\sum_{l=0}^{N-1}\exp[\mathrm{i}\hat{\mathbf{R}}_l\cdot(\boldsymbol{\kappa}-\mathbf{q}_q\cdot)] = \frac{(2\pi)^3}{v_0}\sum_{\tau=0}^{N-1}\delta(\boldsymbol{\kappa}-\mathbf{q}_q-\boldsymbol{\tau}_\tau) \tag{8.297}$$

and

$$\sum_{l=0}^{N-1}\exp[\mathrm{i}\hat{\mathbf{R}}_l\cdot(\boldsymbol{\kappa}+\mathbf{q}_q\cdot)] = \frac{(2\pi)^3}{v_0}\sum_{\tau=0}^{N-1}\delta(\boldsymbol{\kappa}-\mathbf{q}_q+\boldsymbol{\tau}_\tau) \tag{8.298}$$

With substitution of Eqs. (8.297–8.298) into Eq. (8.296), the partial differential cross section, becomes

$$\left(\frac{\mathrm{d}\sigma}{\mathrm{d}\Omega\,\mathrm{d}E}\right)^{\mathrm{inel}}_{\mathrm{coh}} = (r_0 g)^2 \frac{k}{k_0}\frac{S}{2}\left(1+\frac{\kappa_z^2}{\kappa^2}\right)\frac{(2\pi)^3}{v_0}\left|\frac{g_0}{2}F_0(\boldsymbol{\kappa})\exp(-W_0)\right|^2$$
$$\times\sum_{\tau=1}^{N-1}\sum_{q=-N/2}^{N/2}\left\{\langle n_q+1\rangle\delta(\boldsymbol{\kappa}-\mathbf{q}_q-\boldsymbol{\tau}_\tau)\frac{1}{2\pi\hbar}\int_{-\infty}^{\infty}\mathrm{d}t\exp[-\mathrm{i}t(w-w_q)]\right.$$
$$\left.+\langle n_q\rangle\delta(\boldsymbol{\kappa}+\mathbf{q}_q-\boldsymbol{\tau}_\tau)\frac{1}{2\pi\hbar}\int_{-\infty}^{\infty}\mathrm{d}t\exp[-\mathrm{i}t(w+w_q)]\right\} \tag{8.299}$$

Then into Eq. (8.299), one substitutes

$$\frac{1}{2\pi\hbar}\int_{-\infty}^{\infty}\mathrm{d}t\exp[-\mathrm{i}t(w-w_q)] = \delta(\hbar w-\hbar w_q) \tag{8.300}$$

and

$$\int_{-\infty}^{\infty} \mathrm{d}t \exp[-\mathrm{i}t(w + w_q)] = \delta(\hbar w_q + \hbar w) \tag{8.301}$$

Thus, one finally obtains the coherent, inelastic, magnetic partial differential cross section for neutron scatter by spin waves in a single domain of a ferromagnetic crystal:

$$\left(\frac{\mathrm{d}\sigma}{\mathrm{d}\Omega\,\mathrm{d}E}\right)_{\mathrm{coh}}^{\mathrm{inel}} = (r_0 g)^2 \frac{k}{k_0}\frac{S}{2}\left(1 + \frac{\kappa_z^2}{\kappa^2}\right)\frac{(2\pi)^3}{v_0}\left|\frac{g_0}{2}F_0(\boldsymbol{\kappa})\exp(-W_0)\right|^2$$
$$\times \sum_{\tau=0}^{N-1}\sum_{q=-N/2}^{N/2}\{\langle n_q + 1\rangle\delta(\boldsymbol{\kappa} - \mathbf{q}_q - \boldsymbol{\tau}_\tau)\delta(\hbar w - \hbar w_q)$$
$$+ \langle n_q\rangle\delta(\boldsymbol{\kappa} + \mathbf{q}_q - \boldsymbol{\tau}_\tau)\delta(\hbar w_q + \hbar w)\} \tag{8.302}$$

The partial differential cross section $(\mathrm{d}\sigma/\mathrm{d}\Omega\,\mathrm{d}E)_{\mathrm{coh,avg}}^{\mathrm{inel}}$ is the sum of two terms, where the first term is the creation of a magnon, and the second term is the annihilation of a magnon. In the first term, the quantum number of the spin wave vector $\mathbf{q}_q$ increases by 1, and in the second term, the quantum number decreases by 1. The creation of a magnon is described by the first term, where the spin wave magnon gains energy $\hbar w_q$, and the neutron loses energy $\hbar w_q$, where

$$\frac{\hbar^2}{2m_n}(k_0^2 - k^2) = \hbar w_q \tag{8.303}$$

and

$$\mathbf{k}_0 - \mathbf{k} = \boldsymbol{\tau}_\tau + \mathbf{q}_q \tag{8.304}$$

The annihilation of a magnon is described by the second term, where the spin wave magnon loses energy $\hbar w_q$, and the neutron gains energy $\hbar w_q$, where

$$\frac{\hbar^2}{2m_n}(k_0^2 - k^2) = -\hbar w_q \tag{8.305}$$

and

$$\mathbf{k}_0 - \mathbf{k} = \boldsymbol{\tau}_\tau - \mathbf{q}_q \tag{8.306}$$

8.14 Coherent Inelastic Neutron Scatter—One-Magnon Exchange in Ferromagnetic Material

Following Squires (1996), we now consider one-magnon scattering in the forward direction in a ferromagnetic crystal. In elastic forward scatter, the neutron scattering vector is zero ($\boldsymbol{\kappa} = 0$), where the scattering vector equals a corresponding reciprocal lattice vector $\boldsymbol{\tau}_\tau$, where $\boldsymbol{\kappa} = \boldsymbol{\tau}_\tau$, so that $\boldsymbol{\tau}_\tau = 0$. However, in inelastic, one-magnon neutron scatter in forward-scattered neutrons, where the lattice spin wave gains energy from the neutron (+) or loses energy to the neutron (−), the momentum conservation and energy conservation relations are

$$\mathbf{k}_0 - \mathbf{k} = \pm\,\mathbf{q}_q \tag{8.307}$$

and

$$\frac{\hbar^2}{2m_n}(k_0^2 - k^2) = \pm\,\hbar w_q \tag{8.308}$$

where

$$\boldsymbol{\tau}_\tau = 0 \tag{8.309}$$

Given neutron scatter polar angle θ for the scattered neutron wave vector $\mathbf{k}$, measured relative to the direction of the incident neutron wave vector $\mathbf{k}_0$, the squared magnitude of the momentum conservation relation gives

$$(\mathbf{k}_0 - \mathbf{k})\cdot(\mathbf{k}_0 - \mathbf{k}) = \pm\,\mathbf{q}_q \cdot \pm\,\mathbf{q}_q \tag{8.310}$$

or

$$k_0^2 + k^2 - 2k_0 k \cos\theta = q_q^2 \tag{8.311}$$

Azimuthal symmetry is assumed.

If the wavelengths of the spin waves are much larger than the unit cell dimensions, then for lattice stiffness constant D, the dispersion relation between the spin wave frequency w_q and the magnitude of the wave vector $\mathbf{q}_q$ (i.e., wave number q_q for the qth spin wave oscillation in the long-wavelength approximation) is

$$\hbar w_q = D q_q^2 \tag{8.312}$$

Substitution of Eq. (8.312) into the energy conservation relation of Eq. (8.308), gives

$$\frac{\hbar^2}{2m_n}(k_0^2 - k^2) = \pm\,D q_q^2 \tag{8.313}$$

The energy and momentum conservation relations of Eqs. (8.311) and (8.313), can now be equated and combined, via their common dependence on q_q^2, so that

$$k_0^2 + k^2 - 2k_0 k \cos\theta = \pm \frac{\hbar^2}{2Dm_n}(k_0^2 - k^2) \tag{8.314}$$

Let ratio r be

$$r = \frac{\hbar^2}{2Dm_n} = \frac{\hbar^2 c^2}{2Dm_n c^2} = \frac{20.72}{D}\ \text{MeV fm}^2 \tag{8.315}$$

or

$$r = \frac{2.072}{D}\ \text{meV A}^2 \tag{8.316}$$

Here, the conversion constant $\hbar c$ is the product of speed of light c and reduced Planck's constant $\hbar = h/2\pi$, and the neutron rest mass energy is $m_n c^2$, where

$$\hbar c = 197.326\ \text{MeV fm} \quad \text{and} \quad m_n c^2 = 939.565\ \text{MeV} \tag{8.317}$$

The quadratic equation of Eq. (8.314) is solved in terms of the ratio of the scatter neutron wave number k, divided by the incident wave number k_0. From Eq. (8.314), one obtains the quadratic equation in k/k_0:

$$\frac{k^2}{k_0^2}(1 \pm r) - 2\frac{k}{k_0}\cos\theta - (1 \mp r) = 0 \tag{8.318}$$

Solving for k/k_0, one obtains

$$\frac{k}{k_0} = \frac{\cos\theta \pm \sqrt{r^2 - \sin^2\theta}}{1 \pm r} \tag{8.319}$$

One-magnon inelastic neutron scatter occurs if the wave vector k of the scattered neutron is real:

$$\theta \le \sin^{-1} r \tag{8.320}$$

If the forward-scattered neutron falls within a cone with half angle θ, which is less than the one-magnon limiting polar angle θ_0, then the neutron can exchange one magnon with the spin lattice, where

$$\theta \le \theta_0 = \sin^{-1}\left(\frac{\hbar^2}{2Dm_n}\right) \tag{8.321}$$

The one-magnon limiting angle θ_0 is independent of neutron wavelength λ, and the assumption of forward scatter requires θ_0 to be small. For example, in iron and nickel, at temperatures below the Curie temperature, the stiffness constant can range from 100 to 400 [meV/Å^2], so that r is in the range of 0.005–0.02. The forward-inelastic-scatter neutrons, which undergo a one-magnon exchange in the lattice, scatter within a half-cone angle of about 0.005 to 0.02 rad about the incident neutron direction.

The energy–momentum relations of Eqs. (8.307–8.308) for the neutron inelastic scatter, which are caused by spin waves of the unpaired electrons in a lattice, result in the absorption or emission of a quantized unit of spin-flip energy or magnon. The energy-momentum relations of Eqs. (8.307–8.308) for magnons are identical to as the energy–momentum relations for neutron inelastic scatter, which is caused by to lattice nucleus vibration waves, and result in the absorption or emission of a quantized unit of thermal vibration or phonon. The energy spectrum of a scattered beam of incident, monoenergetic neutrons, are similar for the cases of coherent, inelastic neutron scatter from (1) thermal induced nuclear vibrations about equilibrium (causing nuclear vibration waves), and (2) thermal-induced deviations of unpaired, electron spins about equilibrium (causing spin waves).

In magnetic scatter, monoenergetic neutrons impinge on a single crystal and produce a velocity spectrum with peaks. Measurement of the incident $\mathbf{k}_0$ and scattered $\mathbf{k}_0$ wave vectors for these peaks, then allows one to calculate the corresponding spin wave frequency w_q and wave vector $\mathbf{q}_q$. The measured w_q and $\mathbf{q}_q$ allow one to determine a dispersion relation for the spin wave for the particular ferromagnetic crystal.

Without a magnetic field, the mean spin direction $\boldsymbol{\eta}_\eta$ varies from one domain to the next, and in a cubic crystal, where one averages over many domains,

$$\left(1+\frac{\kappa_\eta^2}{\kappa^2}\right)_{\text{avg}} = \frac{4}{3} \tag{8.322}$$

With substitution of Eq. (8.322) into Eq. (8.302), the domain-averaged, coherent, inelastic differential cross section neutron scatter with one-magnon absorption or emission in a nonmagnetized ferrromagnet, becomes

$$\begin{aligned}\left(\frac{\mathrm{d}\sigma}{\mathrm{d}\Omega\,\mathrm{d}E}\right)^{\text{inel}}_{\text{coh,avg}} &= (r_0 g)^2 \frac{k}{k_0}\frac{4}{3}\frac{S(2\pi)^3}{2\,v_0}\left|\frac{g_0}{2}F_0(\boldsymbol{\kappa})\exp(-W_0)\right|^2 \\ &\times \sum_{\tau=0}^{N-1}\sum_{q=-N/2}^{N/2}\{\langle n_q+1\rangle\delta(\boldsymbol{\kappa}-\mathbf{q}_q-\boldsymbol{\tau}_\tau)\delta(\hbar w-\hbar w_q) \\ &+\langle n_q\rangle\delta(\boldsymbol{\kappa}+\mathbf{q}_q-\boldsymbol{\tau}_\tau)\delta(\hbar w_q+\hbar w)\}\end{aligned} \tag{8.323}$$

Now suppose that one applies a strong magnetic field $\mathbf{B}$ to the ferromagnetic crystal, which is sufficient to align the spins of the various domains, so that $\boldsymbol{\eta}_\eta$

points in the same approximate direction from one domain to another. If the neutron scatter vector $\boldsymbol{\kappa}$ is perpendicular to $\mathbf{B}$, then in Eq. (8.302),

$$\left(1+\frac{\kappa_\eta^2}{\kappa^2}\right)_{\text{avg}} = 1 \quad \boldsymbol{\kappa}\perp\mathbf{B} \rightarrow \kappa_\eta = 0 \tag{8.324}$$

If the neutron scatter vector $\boldsymbol{\kappa}$ is parallel to $\mathbf{B}$, then in Eq. (8.302),

$$\left(1+\frac{\kappa_\eta^2}{\kappa^2}\right)_{\text{avg}} = 2 \quad \boldsymbol{\kappa}||\mathbf{B} \rightarrow \kappa_\eta = \kappa \tag{8.325}$$

With a sufficiently strong magnetic field $\mathbf{B}$ applied to the ferromagnetic crystal, the partial differential cross section $(\mathrm{d}\sigma/d\Omega\,\mathrm{d}E)^{\text{inel}}_{\text{coh,avg}}$ for one-magnon neutron inelastic scatter is almost twice what it would be without an applied magnetic field.

An applied magnetic field $\mathbf{B}$, and measurements with the neutron scatter vector $\boldsymbol{\kappa}$ parallel to the magnetic field ($\boldsymbol{\kappa}||\mathbf{B}$), can allow separate determination of the partial differential cross section contributions to inelastic neutron scatter from one-phonon and one-magnon scatter. The interference effect of magnetic and nuclear scatter is zero for $\boldsymbol{\kappa}||\mathbf{B}$ because the magneto-vibration term is proportional to $(1-\kappa_\eta^2/\kappa^2)$, which is zero when $\kappa_\eta = \kappa$. Hence, one can easily distinguish one-magnon inelastic scatter from both nuclear scatter and the interference term between nuclear and magnetic scatter. See Eq. (5.164) in section 5.10 of this book, and Eq. (8.85) in Squires (1996).

8.15 Integral Expression for Temperature Dependence of Spin and Magnetization in a Ferromagnetic Crystal Based on the Planck Distribution

Following Squires (1996), we now determine the spontaneous magnetization in a ferromagnetic material as a function of temperature T in the low-temperature limit, where one can apply the linear spin wave theory discussed previously. The average number $\langle n_q\rangle$ of magnons in the qth spin wave normal mode with spin wave vector $\mathbf{q}_q$ and frequency w_q, is given by the Planck distribution. See Eq. 18 in chapter 5 of Cremer (2012b), or Eq. (8.80) in Squires (1996), or Eq. 7 in chapter 5 of Kittel (2005). In each of the x-, y-, and z-directions, the x-, y-, and z-spin wave vector components q_{qx}, q_{qy}, and q_{qz} and frequency components w_{qx}, w_{qy}, and w_{qz} for the qth spin wave mode, are given by the product of the x-, y-, and z-component Planck distributions, where

$$\begin{aligned}\langle n_q\rangle = \langle n_{qx}\rangle\langle n_{qy}\rangle\langle n_{qz}\rangle &= [\exp(hw_{qx}/k_BT)-1]^{-1}[\exp(hw_{qy}/k_BT)-1]^{-1}\\ &\quad\times[\exp(hw_{qz}/k_BT)-1]^{-1}\end{aligned} \tag{8.326}$$

These statistics of the qth spin wave mode for the average number of magnons also describes those of the vibration wave phonons with wave vector $\mathbf{q}_q$ and frequency w_q, where the energy levels belong to a set of harmonic oscillators.

Recall that the z-component $\hat{S}_l^z(t)$ of the total unpaired electron spin Heisenberg operator $\hat{S}_l(t)$ for each lth unit cell is obtained by subtracting the product of the creation $a_l^+(t)$ and annihilation $a_l(t)$ operators from the lth unit cell's total unpaired electron spin quantum spin number S, which is the number of unpaired electron spins per l^{th} unit cell, where Eq. (8.252) is repeated below,

$$\hat{S}_l^z(t) = S - a_l^+(t)a_l(t) \tag{8.327}$$

Also recall that the thermal-averaged product of the creation $a_l^+(t)$ and annihilation $a_l(t)$ Heisenberg operators for the net unpaired electron spin in the lth unit cell, equals the thermal-averaged product of the creation a_q^+ and annihilation a_q Schrödinger operators for the crystal q^{th} mode spin wave vector $\mathbf{q}_q$ per unit cell (i.e., the division by N, the number of crystal unit cells), where

$$\begin{aligned}&\langle a_l^+(t)a_l(t)\rangle = \\ &\frac{1}{N}\sum_{q=-N/2}^{N/2}\langle a_q^+ a_q\rangle = \frac{1}{N}\sum_{q=-N/2}^{N/2}\langle n_q\rangle = \frac{1}{N}\sum_{q=-N/2}^{N/2}\sum_{i=1}^{J} p_i\langle\phi_i|a_q^+ a_q|\phi_i|\rangle\end{aligned} \tag{8.328}$$

With substitution of Eq. (8.328) into Eq. (8.327) one obtains the thermal-averaged z-component spin $\langle S_l^z(t)\rangle$, where

$$\langle S_l^z(t)\rangle = S - \langle a_l^+(t)a_l(t)\rangle = S - \frac{1}{N}\sum_{q=-N/2}^{N/2}\langle n_q\rangle \tag{8.329}$$

Substitution into Eq. (8.239) of Eq. (8.326) for $\langle n_q\rangle$, which is expressed by the Planck distribution, is then

$$\begin{aligned}\langle S_l^z(t)\rangle = S - &\frac{1}{N_x}\sum_{q_{qx}=-N_x/2}^{N_x/2}[\exp(hw_{qx}/k_BT)-1]^{-1} \\ &\times\frac{1}{N_y}\sum_{q_{qy}=-N_y/2}^{N_y/2}[\exp(hw_{qy}/k_BT)-1]^{-1} \\ &\times\frac{1}{N_z}\sum_{q_{qz}=-N_z/2}^{N_z/2}[\exp(hw_{qz}/k_BT)-1]^{-1}\end{aligned} \tag{8.330}$$

The large number of unit cells and spin wave oscillation modes, where the oscillation modes are closely spaced, can allow one to replace the summation expressions for thermal-averaged z-component spin $\langle S_l^z(t)\rangle$ by equivalent integrals. The integrals will allow one to obtain a simple direct temperature dependence of the thermal-averaged magnetization vector $\mathbf{M}(T)$ and the associated thermal-averaged z-component $\langle S_l^z(t)\rangle$ of the unpaired electron spin quantum number per atom. The summation expressions are replaced by integrals as follows:

$$\frac{1}{N_i}\sum_{q_{qi}=-N_i/2}^{N_i/2}\frac{1}{\exp(hw_{qi}/k_\mathrm{B}T)-1}=\frac{1}{N_i}\sum_{q_{qi}=-N_i/2}^{N_i/2}\frac{1}{\Delta q}\frac{\Delta q}{\exp(hw_{qi}/k_\mathrm{B}T)-1}\qquad i=x,y,z \tag{8.331}$$

Recalling Chapter 4 in Cremer (2012b), the first Brillouin zone occupies the volume v_r of a single reciprocal lattice unit cell, where

$$v_r = N_xN_yN_z(\Delta q)^3 \tag{8.332}$$

The x-, y-, and z-dimensions q_{0x}, q_{0y}, and q_{0z} of the reciprocal lattice unit cell are:

$$q_{0i} = N_i\Delta q \quad i=x,y,z \tag{8.333}$$

The direct unit cell volume v_0 is related to the reciprocal lattice unit cell volume v_r by

$$v_0=\frac{(2\pi)^3}{v_r}=\frac{(2\pi)^3}{(N_x\Delta q)(N_y\Delta q)(N_z\Delta q)}=\frac{(2\pi)^3}{q_{0x}q_{0y}q_{0z}} \tag{8.334}$$

In the limit of an infinitesimally small spin wave number increment Δq, and the limit of a infinitely large number N of crystal normal spin wave modes, one integrates the Planck distribution of spin wave normal modes over the x, y, and z bounds $\pm q_{0x}/2$, $\pm q_{0y}/2$, and $\pm q_{0z}/2$ of the first Brillouin zone, which also is the dimension of the unit cell, where

$$\frac{1}{q_{0i}}\int_{-q_{0i}/2}^{q_{0i}/2}\frac{\mathrm{d}q_{qi}}{\exp(hw_{qi}/k_\mathrm{B}T)-1}=\lim_{\Delta q\to 0}\lim_{N_i\to\infty}\frac{1}{N_i\Delta q}\sum_{q_{qi}=-N_i/2}^{N_i/2}\frac{\Delta q}{\exp(hw_{qi}/k_\mathrm{B}T)-1}$$
$$i=x,y,z \tag{8.335}$$

The thermal-averaged z-component $S_l^z(t)$ of Eq. (8.330) for the total unpaired electron spin $S_l(t)$ for the lth unit cell becomes

$$\langle S_l^z(t)\rangle = S-\frac{v_0}{(2\pi)^3}\int_0^{2\pi/c}\mathrm{d}q_{qz}\int_0^{2\pi/b}\mathrm{d}q_{qy}\int_0^{2\pi/a}\mathrm{d}q_{qx}\frac{1}{\exp(hw_q/k_\mathrm{B}T)-1} \tag{8.336}$$

Now consider a ferromagnetic crystal domain of volume V, which is comprised of N net spin vectors $\mathbf{S}$ that arise from the unpaired electrons of the domain's

N ferromagnetic atoms. Corresponding to the N net electron spin vectors, are the N net electron magnetic dipole moment vectors $\boldsymbol{\mu}_e$, such that the summed total magnetic dipole moment vector of the domain $\boldsymbol{\mu}_T$ is

$$\boldsymbol{\mu}_T = N\boldsymbol{\mu}_e \tag{8.337}$$

The net spin of the unpaired electron spin of an atom, and the net electron spin magnetic dipole moment of the atom, is determined by the Landé g factor for the ferromagnetic atom (ion), where

$$\boldsymbol{\mu}_e = \frac{g\mu_b}{\hbar}\mathbf{S} \tag{8.338}$$

The magnetization vector $\mathbf{M}(T)$ of the domain of volume V, as a function of temperature T, is defined as the magnetic dipole moment per unit volume. At low temperatures, one can express the magnetization (magnetic dipole moment per unit volume) in the domain at temperature T by

$$M(T) = M(0) - \frac{\mu_T}{Nv_0}\sum_{q=1}^{N}\langle n_q\rangle \tag{8.339}$$

Transposing, one obtains

$$M(0) - M(T) = \frac{\mu_T}{Nv_0}\sum_{q=1}^{N}\langle n_q\rangle \tag{8.340}$$

The domain of volume V is comprised of N unit cells, each of volume v_0, and substitution into Eq. (8.340) of Eq. (8.326) for the average number $\langle n_q\rangle$ of magnons in the qth spin wave normal mode with spin wave vector $\mathbf{q}_q$ and frequency w_q gives

$$\begin{aligned} M(0) - M(T) = {} & \frac{\mu_T}{Nv_0}[\exp(hw_{qx}/k_B T) - 1]^{-1} \\ & \times [\exp(hw_{qy}/k_B T) - 1]^{-1}[\exp(hw_{qz}/k_B T) - 1]^{-1} \end{aligned} \tag{8.341}$$

The domain of volume V is comprised of N unit cells, each of volume v_0. At zero temperature ($T = 0$), all the net spin vectors of the atoms (ions) in the ferromagnetic domain are aligned in the minimum potential energy configuration, where

$$M(0) = \frac{\mu_T}{V} = \frac{\mu_T}{Nv_0} \tag{8.342}$$

One then converts the summation of Eq. (8.340) to an integral over the volume of a reciprocal lattice unit cell, which comprises the first Brillouin zone, where

$$M(0) - M(T) = \frac{\mu_T}{(2\pi)^3} \int_{-q_{0z}/2}^{q_{0z}/2} dq_z \int_{-q_{0y}/2}^{q_{0y}/2} dq_y \int_{-q_{0z}/2}^{q_{0z}/2} dq_x \frac{1}{\exp(hw_q/k_B T) - 1} \tag{8.343}$$

and

$$\frac{1}{q_{0x}q_{0y}q_{0z}v_0} = \frac{1}{v_r v_0} = \frac{1}{(2\pi)^3} \tag{8.344}$$

8.16 Evaluation of Low-Temperature Spin and Magnetization of the Integral Expression for a Cubic Ferromagnetic Crystal—$T^{3/2}$ Dependence

In this section, we derive Bloch's $T^{3/2}$ law for magnetization of ferromagnetic materials, following Kittel (2005) and Squires (1996). We will evaluate the integral expression of Eq. (8.343) for the simple case of a cubic-unit-cell crystal. For cubic crystals, also see section 5 in chapter 6 of Cremer (2012b). In a cubic crystal, one can directly show the decrease of the thermal-averaged, z-directed crystal magnetization vector magnitude M, and spin component $\langle S_z \rangle$ per crystal unit cell, as the crystal temperature T increases.

We now evaluate the thermal-averaged magnetization and the thermal-averaged spin components per unit cell in a cubic-unit-cell ferromagnetic crystal. If one assumes a cubic unit cell lattice $(\mathbf{a}, \mathbf{a}, \mathbf{a})$ at sufficiently small temperatures, then one can use the quadratic dispersion relation

$$hw_q = Dq^2 \tag{8.345}$$

Substitution of Eq. (8.345) into Eq. (8.343), gives

$$M(0) - M(T) = \frac{\mu_T}{(2\pi)^3} \int_{-\pi/a}^{\pi/a} dq_z \int_{-\pi/a}^{\pi/a} dq_y \int_{-\pi/a}^{\pi/a} dq_x \frac{1}{\exp(Dq^2/k_B T) - 1} \tag{8.346}$$

The thermal-averaged z-component $S_l^z(t)$ of the total, unpaired electron spin $S_l(t)$, is subtracted from the total spin quantum number S of the lth cubic unit cell, where

$$S - \langle \hat{S}_l^z(t) \rangle = \frac{v_0}{(2\pi)^3} \int_{-\pi/a}^{\pi/a} dq_z \int_{-\pi/a}^{\pi/a} dq_y \int_{-\pi/a}^{\pi/a} dq_x \frac{1}{\exp(Dq^2/k_B T) - 1} \tag{8.347}$$

The q^2 term in Eq. (8.347) suggests a change to spherical coordinates. One then approximates the first Brillouin zone as a sphere of radius π/a, and assumes spherical symmetry, where 4π is obtained from integration over the polar and azimuthal angles, where

$$M(0) - M(T) = \frac{\mu_T}{2\pi^2} \int_0^{\pi/a} dq_r \frac{q_r^2}{\exp(Dq_r^2/k_B T) - 1} \tag{8.348}$$

and

$$S - \langle \hat{S}_l^z(t) \rangle = \frac{v_0}{2\pi^2} \int_0^{\pi/a} dq_r \frac{q_r^2}{\exp(Dq_r^2/k_B T) - 1} \tag{8.349}$$

A change of variables simplifies Eqs. (8.348–8.349):

$$x = \frac{Dq_r^2}{k_B T} \quad \text{and} \quad x_a = \frac{D}{k_B T}\left(\frac{\pi}{a}\right)^2 \tag{8.350}$$

where

$$q_r^2 = x\frac{k_B T}{D} \quad \text{and} \quad dq_r = \frac{x^{-1/2}}{2}\sqrt{\frac{k_B T}{D}} \tag{8.351}$$

The temperature-averaged magnetization expression then becomes

$$M(0) - M(T) = \frac{\mu_T}{4\pi^2}\left(\frac{k_B T}{D}\right)^{3/2} \int_0^{x_a} \mathrm{d}x \frac{\sqrt{x}}{\exp(x) - 1} \tag{8.352}$$

where

$$x_a = \frac{D}{k_B T}\left(\frac{\pi}{a}\right)^2 \tag{8.353}$$

The temperature-averaged expression for the xy-plane component spin $\langle S_l^{xy}(t)\rangle$ per the lth cubic unit cell is perpendicular to the thermal-averaged z-component spin $\langle S_l^z(t)\rangle$, where

$$\langle S_l^{xy}(t)\rangle = S - \langle S_l^z(t)\rangle = \frac{v_0}{4\pi^2}\left(\frac{k_B T}{D}\right)^{3/2} \int_0^{x_a} \mathrm{d}x \frac{\sqrt{x}}{\exp(x) - 1} \tag{8.354}$$

Because of the singularity of the integral at or near $x = 0$, the integral is dominated by values of x near zero. As a result, one can let $x_a \to \infty$, and thus the integration extends over the entire reciprocal lattice and beyond, yet the result of the integral with $x_a = \infty$ is nearly the same as integral over the first Brillouin zone. Definite integral tables give

$$\int_0^\infty dx \frac{\sqrt{x}}{\exp(x) - 1} = 0.0587 \cdot 4\pi^2 \tag{8.355}$$

One assumes sufficiently long spin wavelengths $2\pi/q$ or small spin wave numbers q relative to the dimension a of the cubic unit cell, so the lattice stiffness constant D (recall Eq. (8.106)) is approximated by

$$D = 2JSa^2 \quad qa \ll 1 \tag{8.356}$$

One thus obtains Bloch's $T^{3/2}$ law for magnetization of ferromagnetic materials at low temperatures for a cubic-unit-cell crystal. The reduction of magnetization is thus proportional to the 3/2 power of the crystal temperature:

$$M(0) - M(T) = 0.0587\mu_T \left(\frac{k_B T}{2JSa^2}\right)^{3/2} \quad qa \ll 1 \tag{8.357}$$

The thermal-averaged xy-plane spin component for each lth cubic unit cell is also proportional to the 3/2 power of the crystal temperature, where

$$\langle S_l^{xy}(t)\rangle = S - \langle \hat{S}_l^{z}(t)\rangle = 0.0587 v_0 \left(\frac{k_B T}{2JSa^2}\right)^{3/2} \quad qa \ll 1 \tag{8.358}$$

The magnetization decreases with the decreasing z-component spin. As the z-component spin decreases, the xy-component spin increases. The decrease in thermal-averaged magnetization, and the decrease in the thermal-averaged z-component of the unpaired electron spin per unit cell are each proportional to $T^{3/2}$.

8.17 Magnon Population Low-Temperature Dependence in a Ferromagnetic Cubic Crystal

We now derive the $T^{3/2}$ temperature dependence of the number of magnons in the domain of a ferromagnetic crystal at low temperatures. The number of magnons $n_{\text{mag}}(T)$ excited at temperature T at sufficiently low temperatures is

$$n_{\text{mag}}(T) = \sum_{q=1}^{N} n_q = \int_0^\infty dw \rho(w)\langle n(w)\rangle \tag{8.359}$$

The number of magnon modes per frequency range dw at frequency w is $\rho(w)$, where

$$\rho(w)dw = \frac{4\pi q^2}{(2\pi)^3}\frac{\mathrm{d}q}{\mathrm{d}w}\mathrm{d}w \tag{8.360}$$

The integral is actually over the range of spin wave frequencies w allowed by the corresponding spin wave vectors in the first Brillouin zone. However, at sufficiently low temperatures, as $T \to 0$, the average number of high-frequency magnon modes tend to become zero, where

$$\lim_{w \to \infty} \langle n(w) \rangle \to 0 \tag{8.361}$$

In terms of the lattice stiffness constant D, the quadratic magnon dispersion relation at low temperatures T (i.e., temperatures below the Curie temperature T_{C}) is

$$\hbar w = Dq^2 \quad T < T_{\mathrm{C}} \tag{8.362}$$

One assumes spin wavelengths that are much longer than the unit cell dimensions, where $qa \ll 1$, so the lattice stiffness constant D is

$$D = 2JSa^2 \quad qa \ll 1 \tag{8.363}$$

The magnon frequency for sufficiently low temperatures is then

$$\hbar w = \frac{Dq^2}{\hbar} = \frac{2JSa^2q^2}{\hbar} \quad qa \ll 1 \text{ and } T < T_c \tag{8.364}$$

The group velocity dw/dq of the magnon can be obtained from its dispersion relation, where for a cubic lattice with unit cell $(\mathbf{a}, \mathbf{a}, \mathbf{a})$,

$$\frac{\mathrm{d}w}{\mathrm{d}q} = \frac{\mathrm{d}}{\mathrm{d}q}\left(\frac{D}{\hbar}q^2\right) = \frac{2D}{\hbar}q = \frac{4JSa^2}{\hbar}q \quad qa \ll 1 \text{ and } T < T_{\mathrm{C}} \tag{8.365}$$

Substitution of Eq. (8.365) into Eq. (8.360), using Eqs. (8.362–8.363), then gives the magnon mode density $\rho(w)$ as

$$\rho(w) = \frac{\sqrt{w}}{4\pi^2}\left(\frac{\hbar}{2JSa^2}\right)^{3/2} \quad qa \ll 1 \text{ and } T < T_{\mathrm{C}} \tag{8.366}$$

For spin wave lengths longer than the unit cell dimension, and at temperatures below the Curie temperature, and with substitution of Eqs. (8.326) and (8.366) into Eq. (8.359), the number of magnons $n_{\mathrm{mag}}(T)$ is approximately

$$n_{\mathrm{mag}}(T) = \sum_{q=1}^{N} n_q = \frac{1}{4\pi^2}\left(\frac{\hbar}{2JSa^2}\right)^{3/2} \int_0^{\infty} \mathrm{d}w \frac{\sqrt{w}}{\exp(hw/k_{\mathrm{B}}T) - 1} \quad qa \ll 1 \text{ and } T < T_{\mathrm{C}} \tag{8.367}$$

With integration of Eq. (8.367), via tables of definite integrals, one obtains the Bloch $T^{3/2}$ temperature dependence, where

$$n_{\mathrm{mag}}(T) = 0.0587\left(\frac{k_{\mathrm{B}}T}{2JSa^2}\right)^{3/2} \quad qa \ll 1 \text{ and } T < T_{\mathrm{C}} \tag{8.368}$$

In the ferromagnetic domain with a cubic lattice unit cell with dimension a, the number N_v of atoms per unit volume is

$$N_v = \frac{r}{a^3} \tag{8.369}$$

In a simple cubic (sc) lattice unit cell, there is one atom ($r = 1$); in a bcc lattice unit cell, there are two atoms ($r = 2$); and in an fcc lattice unit cell, there are four atoms ($r = 4$). The total spin quantum number S of spins per unit cell is multiplied by the number N_v of atoms per unit volume (i.e., the unit cell per unit volume of one atom per unit cell) to obtain the spin density ρ_{sp}, where

$$\rho_{sp} = N_v S \tag{8.370}$$

The number of magnons per unpaired electron spin density in the domain $n_{\mathrm{mag}}(T)/\rho_{sp}$, is equal to the fractional change in the domain magnetization $\Delta M(T)/M(0)$. At low temperatures and long spin wavelengths, these ratios follow the Bloch $T^{3/2}$ law, where

$$\frac{n_{\mathrm{mag}}(T)}{\rho_{sp}} = \frac{\Delta M(T)}{M(0)} = \frac{0.0587}{N_v S}\left(\frac{k_{\mathrm{B}}T}{2JS}\right)^{3/2} \quad qa \ll 1 \text{ and } T < T_{\mathrm{C}} \tag{8.371}$$

As the number of magnons increase, an increased exchange of magnons occurs between the crystal lattice and its environment, which indicates inelastic magnetic interactions between the lattice, and leads to a change in temperature. A temperature change causes a change in the alignment of unpaired electron spins in the z-direction, which changes the magnetization.

References

Cremer, J. T. (2012b). In W. Peter (Ed.), Neutron and X-ray microscopy—Part 2. *Advances in imaging and electron physics* (173). London, UK: Hawkes, Academic Press, an imprint of Elsevier Press.

Cremer, J. T. (2013). In W. Peter (Ed.), *Advances in imaging and electron physics* (175). London, UK: Hawkes, Academic Press, an imprint of Elsevier Press.

Eisberg, R., & Resnick, R. (1974). *Quantum physics of atoms, molecules, solids, nuclei, and particles*. New York, NY: Wiley.

Grosso, G., & Parravicini, G. P. (2000). *Solid state physics*. New York, NY: Academic Press.

Kittel, C. (2005). *Introduction to solid state physics*. New York, NY: Wiley.

Liboff, R. L. (2003). *Introductory quantum mechanics*. Reading, MA: Addison-Wesley.

Merzbacher, E. (1970). *Quantum mechanics*. New York, NY: John Wiley & Sons.

Squires, G. L. (1996). *Introduction to the theory of thermal neutron scattering*. New York, NY: Dover.

9 Coherent, Elastic Scatter of Neutrons by Atomic Electric Field

9.1 Spin–Orbit Electric Field Scatter of Neutrons

We now consider scattering of a neutron by electric fields, which in the rest frame of the neutron appear as a magnetic field, as seen directly by Maxwell's equations or by relativistic transformation of the electric field from the rest frame of the source of the electric field to the neutron rest frame. Maxwell's equations were found to be invariant under the Lorentz transform by Lorentz, which led to their application in mechanics. Hence, Maxwell's equations and special relativity are equivalent means of calculating the magnetic field, which acts on the neutron magnetic dipole moment, or acts on the electron charge and magnetic dipole moment, when the neutron or electron is moving at velocity **v** relative to the electric field source. In the derivation of neutron scatter by the net magnetic field of unpaired atomic electrons, via LS coupling, the magnetic field, which is created by the electron orbital motion, interacts with the electron's spin magnetic dipole moment. In LS coupling, one adds the total orbital angular momentum vector **L** and total spin momentum vector **S** of the unpaired electrons of an atom, see Merzbacher (1970), Eisberg and Resnick (1974), Liboff (2003), and Ch. 4.

Following section 9.5 in Squires (1996), we first consider neutron scatter by the net atomic electric field **E**(**r**) (nucleus and electrons) when neutron moves though the atomic electric field **E**(**r**) at constant velocity **v**, which produces magnetic field **B**(**r**) in the neutron rest frame, as a function of neutron position **r**, where

$$\mathbf{B}(\mathbf{r}) = \frac{\mathbf{E}(\mathbf{r}) \times \mathbf{v}}{c^2} \tag{9.1}$$

In the lab frame, the linear momentum **p** of the moving neutron with mass m_n and velocity **v** relative to the stationary nucleus is

$$\mathbf{v} = \frac{\mathbf{p}}{m_n} \tag{9.2}$$

The electric field **E**(**r**) and magnetic field **B**(**r**) are replaced by their position-dependent vector operators $\hat{\mathbf{E}}(\mathbf{r})$ and $\hat{\mathbf{B}}(\mathbf{r})$, and momentum vector **p** by the momentum vector operator $\hat{\mathbf{p}}$, so the magnetic field vector operator $\hat{\mathbf{B}}$ is, then,

$$\hat{\mathbf{B}} = \frac{\hat{\mathbf{E}} \times \hat{\mathbf{p}}}{m_n c^2} \tag{9.3}$$

Neutron and X-ray Optics. DOI: http://dx.doi.org/10.1016/B978-0-12-407164-3.00008-5

The momentum operator $\hat{\mathbf{p}}$ and the position-dependent electric field operator $\hat{\mathbf{E}}$ do not commute; hence, one must express $\hat{\mathbf{E}} \times \hat{\mathbf{p}}$ as

$$\hat{\mathbf{E}} \times \hat{\mathbf{p}} = \frac{1}{2}\left[(\hat{\mathbf{E}} \times \hat{\mathbf{p}}) + (\hat{\mathbf{E}} \times \hat{\mathbf{p}})\right] = \frac{1}{2}\left[(\hat{\mathbf{E}} \times \hat{\mathbf{p}}) - (\hat{\mathbf{p}} \times \hat{\mathbf{E}})\right] \tag{9.4}$$

However, one assumes that only an atomic electrostatic electric field is produced, in which there are no average currents or time-varying electric fields, so

$$\nabla \times \mathbf{E} = 0 \tag{9.5}$$

and

$$\hat{\mathbf{p}} \times \hat{\mathbf{E}} = -i\hbar\nabla \times \hat{\mathbf{E}} = 0 \tag{9.6}$$

Hence, it is sufficient to use operator $\hat{\mathbf{E}} \times \hat{\mathbf{p}}$ because $\hat{\mathbf{p}} \times \hat{\mathbf{E}} = 0$.
The electric field $\mathbf{E}$ is obtained via

$$\mathbf{E} = -\nabla\Phi \tag{9.7}$$

The electric potential $\Phi(r)$ is calculated by integrating the charge density $n(\mathbf{r})$ over the box volume that contains the scattering system, where

$$\Phi(r) = \frac{1}{4\pi\varepsilon_0}\int_{-\infty}^{\infty} \mathrm{d}z \int_{-\infty}^{\infty} \mathrm{d}y \int_{-\infty}^{\infty} \mathrm{d}x \frac{n(\mathbf{r}')}{|\mathbf{r}-\mathbf{r}'|} \mathrm{d}x\,\mathrm{d}y\,\mathrm{d}z \tag{9.8}$$

The scattering system is assumed to be electrically neutral, where

$$\iiint n(\mathbf{r})\mathrm{d}x\,\mathrm{d}y\,\mathrm{d}z = 0 \tag{9.9}$$

The κ-space Fourier transform $n(\kappa)$ of direct or r-space charge density $n(\mathbf{r})$ is given by

$$n(\boldsymbol{\kappa}) = \iiint n(\mathbf{r}) \exp(-\mathrm{i}\boldsymbol{\kappa} \cdot \mathbf{r})\mathrm{d}x\,\mathrm{d}y\,\mathrm{d}z \tag{9.10}$$

The κ-space charge density $n(\kappa)$ is zero for $\kappa = 0$, where

$$n(\boldsymbol{\kappa} = 0) = 0 \tag{9.11}$$

Charge density $n(\mathbf{r})$ in direct r-space is obtained by the Fourier transform of $n(\kappa)$:

$$n(\mathbf{r}) = (2\pi)^{-3}\int_{-\infty}^{\infty} \mathrm{d}\kappa_z \int_{-\infty}^{\infty} \mathrm{d}\kappa_y \int_{-\infty}^{\infty} d\kappa_x \; n(\boldsymbol{\kappa})\exp(\mathrm{i}\boldsymbol{\kappa} \cdot \mathbf{r}) \tag{9.12}$$

Substitution of Eq. (9.12) for $n(\mathbf{r})$ in the electric potential $\Phi(r)$ of Eq. 9.8 gives

$$\Phi(r) = \frac{1}{4\pi\varepsilon_0}\iiint \frac{\iiint n(\boldsymbol{\kappa})\exp(\mathrm{i}\boldsymbol{\kappa}\cdot\mathbf{r}')\mathrm{d}\kappa_x\mathrm{d}\kappa_y\mathrm{d}\kappa_z}{(2\pi)^3|\mathbf{r}-\mathbf{r}'|}\mathrm{d}x\,\mathrm{d}y\,\mathrm{d}z \tag{9.13}$$

Substitute the unity relation of Eq. 9.14 into Eq. (9.13) for electric potential $\Phi(r)$:

$$\exp(-\mathrm{i}\boldsymbol{\kappa}\cdot\mathbf{r})\exp(\mathrm{i}\boldsymbol{\kappa}\cdot\mathbf{r}) = 1 \tag{9.14}$$

The electric potential becomes

$$\Phi(r) = \frac{1}{4\pi\varepsilon_0}\iiint \frac{\iiint n(\boldsymbol{\kappa})\exp[\mathrm{i}\boldsymbol{\kappa}\cdot(\mathbf{r}'-\mathbf{r})]\exp(\mathrm{i}\boldsymbol{\kappa}\cdot\mathbf{r})\mathrm{d}\kappa_x\mathrm{d}\kappa_y\mathrm{d}\kappa_z}{(2\pi)^3|\mathbf{r}'-\mathbf{r}|}\mathrm{d}x\,\mathrm{d}y\,\mathrm{d}z \tag{9.15}$$

One employs the integration result shown in Eq. (9.16) to Eq. (9.15):

$$\iiint \frac{\exp(\mathrm{i}\kappa\cdot\mathbf{r})}{\mathbf{r}}\mathrm{d}x\,\mathrm{d}y\,\mathrm{d}z = \frac{4\pi}{\kappa^2} \tag{9.16}$$

One obtains

$$\Phi(r) = \frac{1}{(2\pi)^3\varepsilon_0}\iiint \frac{n(\boldsymbol{\kappa})\exp(\mathrm{i}\boldsymbol{\kappa}\cdot\mathbf{r})}{\kappa^2}\mathrm{d}\kappa_x\,\mathrm{d}\kappa_y\,\mathrm{d}\kappa_z \tag{9.17}$$

Substitution for potential Φ in the electric field expression $\mathbf{E}$ gives

$$\mathbf{E} = -\nabla\Phi = -\frac{1}{(2\pi)^3\varepsilon_0}\iiint \frac{n(\boldsymbol{\kappa})\nabla[\exp(\mathrm{i}\boldsymbol{\kappa}\cdot\mathbf{r})]}{\kappa^2}\mathrm{d}\kappa_x\,\mathrm{d}\kappa_y\,\mathrm{d}\kappa_z \tag{9.18}$$

With

$$\nabla[\exp(\mathrm{i}\boldsymbol{\kappa}\cdot\mathbf{r})] = \mathrm{i}\boldsymbol{\kappa}\exp(\mathrm{i}\boldsymbol{\kappa}\cdot\mathbf{r}) \tag{9.19}$$

one obtains

$$\mathbf{E} = -\frac{\mathrm{i}}{(2\pi)^3\varepsilon_0}\int_{-\infty}^{\infty}\mathrm{d}\kappa_z\int_{-\infty}^{\infty}\mathrm{d}\kappa_y\int_{-\infty}^{\infty}\mathrm{d}\kappa_x\frac{\boldsymbol{\kappa}n(\boldsymbol{\kappa})\exp(\mathrm{i}\boldsymbol{\kappa}\cdot\mathbf{r})}{\kappa^2} \tag{9.20}$$

The atomic electric field $\mathbf{E}(\mathbf{r})$ produces magnetic field $\mathbf{B}(\mathbf{r})$ in the neutron rest frame, which acts on the neutron magnetic dipole moment $\boldsymbol{\mu}$ with magnetic interaction potential energy operator $U_{so}(\mathbf{r})$ for spin–orbit electric field interaction:

$$U_{so}(\mathbf{r}) = -\boldsymbol{\mu}\cdot\mathbf{B} \tag{9.21}$$

and

$$\boldsymbol{\mu} = 2g\mu_n \mathbf{S} \tag{9.22}$$

The magnetic field $\mathbf{B}(\mathbf{r})$ is directed along the z-axis, where

$$\mathbf{B}(\mathbf{r}) = B(\mathbf{r})\hat{\mathbf{z}} \tag{9.23}$$

Recall that the Landé g factor for the neutron is $g = -1.91$, and the nuclear magneton μ_n for elementary charge e and proton rest mass m_p is

$$\mu_n = \frac{e\hbar}{2m_p} \tag{9.24}$$

We now move to the operator format. The potential energy operator $\hat{U}_{so}$ is the dot product of the neutron magnetic dipole moment $\hat{\boldsymbol{\mu}}$ and the magnetic field vector operator $\hat{\mathbf{B}}(\hat{\mathbf{r}})$, which depends on the neutron position operator $\hat{\mathbf{r}}$, resulting in potential energy operator $\hat{U}_{so}$ that depends on the azimuthal neutron spin operator $\hat{S}_z$, where

$$\hat{U}_{so} = -\hat{\boldsymbol{\mu}} \cdot \hat{\mathbf{B}} = -2g\mu_n \hat{\mathbf{S}} \cdot \frac{\hat{\mathbf{E}} \times \hat{\mathbf{p}}}{m_p c^2} \tag{9.25}$$

One obtains the potential energy operator $\hat{U}_{so}$ in terms of the z-component S_z of the neutron spin vector $\mathbf{S}$ and magnetic field magnitude $\hat{B}(r)$, where

$$\hat{U}_{so} = -2g\mu_n \hat{B}(r)\hat{S}_z \tag{9.26}$$

Expression of the magnetic field vector operator $\mathbf{B}$ in terms of vector operator $\hat{\mathbf{E}} \times \hat{\mathbf{p}}$ is

$$\hat{U}_{so} = -2g\mu_n \frac{\hat{\mathbf{E}} \times \hat{\mathbf{p}}}{m_p c^2} \hat{S}_z \tag{9.27}$$

The momentum operator $\hat{\mathbf{p}}$ is given by

$$\hat{\mathbf{p}} = \mathrm{i}\hbar\nabla \tag{9.28}$$

The magnetic field operator $\hat{\mathbf{E}} \times \hat{\mathbf{p}}$ is not a multiplicative operator; hence, one cannot move $\hat{U}_{so}$ about the integrand until $\hat{U}_{so}$ has acted on the initial neutron wave function ψ_0. The neutron transition matrix element of potential energy operator $\hat{U}_{so}$ for incident neutron wave function $\psi_0 = \exp(i\mathbf{k}_0 \cdot \mathbf{r})$ and scattered neutron wave function $\psi = \exp(i\mathbf{k} \cdot \mathbf{r})$, is expressed as $\langle \psi_0 | \hat{U}_{so} | \psi \rangle$ or equivalently

$$\langle \psi_0 | \hat{U}_{so} | \psi \rangle = \frac{2g\mu_n}{m_p c^2} \int_{-\infty}^{\infty} \mathrm{d}z \int_{-\infty}^{\infty} \mathrm{d}y \int_{-\infty}^{\infty} \mathrm{d}x \exp(-\mathrm{i}\mathbf{k} \cdot \mathbf{r}) \hat{\mathbf{S}} \cdot (\hat{\mathbf{E}} \times \hat{\mathbf{p}}) \exp(\mathrm{i}\mathbf{k}_0 \cdot \mathbf{r}) \tag{9.29}$$

Substitution into Eq. 9.29 for the electric field $\hat{\mathbf{E}}$ operator from Eq. 9.20 and momentum $\hat{\mathbf{p}}$ operator of Eq. 9.28 gives

$$\langle\psi_0|\hat{U}_{so}|\psi\rangle = -\frac{2g\mu_n}{m_pc^2}\frac{\hbar}{(2\pi)^3\varepsilon_0}\int_{-\infty}^{\infty}\mathrm{d}z\int_{-\infty}^{\infty}\mathrm{d}y\int_{-\infty}^{\infty}\mathrm{d}x\,\exp(-i\mathbf{k}\cdot\mathbf{r})\hat{\mathbf{S}}$$
$$\left[\int_{-\infty}^{\infty}\mathrm{d}\kappa_z'\int_{-\infty}^{\infty}\mathrm{d}\kappa_y'\int_{-\infty}^{\infty}\mathrm{d}\kappa_x'\frac{\boldsymbol{\kappa}'n(\boldsymbol{\kappa}')\exp(i\boldsymbol{\kappa}'\cdot\mathbf{r})}{\kappa'^2}\right]\times\nabla\exp(i\mathbf{k}_0\cdot\mathbf{r}) \tag{9.30}$$

The κ'-space integrals operate on $\boldsymbol{\kappa}'$, and not $\mathbf{k}$, which are independent of the r-space variables and integration; hence, the integrals can be moved to the outer position, and the ∇ operates on the exponential to bring down $i\mathbf{k}_0$, where

$$\langle\psi_0|\hat{U}_{so}|\psi\rangle = -\frac{2g\mu_n}{m_pc^2}\frac{\hbar\mathrm{i}}{(2\pi)^3\varepsilon_0}\int_{-\infty}^{\infty}\mathrm{d}\kappa_z'\int_{-\infty}^{\infty}\mathrm{d}\kappa_y'$$
$$\times\int_{-\infty}^{\infty}\mathrm{d}\kappa_x'\left[\int_{-\infty}^{\infty}\mathrm{d}z\int_{-\infty}^{\infty}\mathrm{d}y\int_{-\infty}^{\infty}\mathrm{d}x\,\exp(\mathrm{i}\boldsymbol{\kappa}\cdot\mathbf{r})\exp(\mathrm{i}\boldsymbol{\kappa}'\cdot\mathbf{r})\right]\hat{\mathbf{S}}$$
$$\left[\frac{\boldsymbol{\kappa}'n(\boldsymbol{\kappa}')}{\kappa'^2}\right]\times\mathbf{k}_0 \tag{9.31}$$

Since

$$\int_{-\infty}^{\infty}\mathrm{d}z\int_{-\infty}^{\infty}\mathrm{d}y\int_{-\infty}^{\infty}\mathrm{d}x\,\exp(\mathrm{i}\boldsymbol{\kappa}'\cdot\mathbf{r})\exp(\mathrm{i}\boldsymbol{\kappa}\cdot\mathbf{r}) = (2\pi)^3\delta(\boldsymbol{\kappa}+\boldsymbol{\kappa}') \tag{9.32}$$

one obtains the magnetic, potential energy, and transition element expression

$$\langle\psi_0|\hat{U}_{so}|\psi\rangle = -\frac{2g\mu_n}{m_pc^2}\frac{\hbar\mathrm{i}}{\varepsilon_0}\int_{-\infty}^{\infty}\mathrm{d}\kappa_z\int_{-\infty}^{\infty}\mathrm{d}\kappa_y\int_{-\infty}^{\infty}\mathrm{d}\kappa_x\delta(\boldsymbol{\kappa}+\boldsymbol{\kappa}')\hat{\mathbf{S}}\cdot\left[\frac{\boldsymbol{\kappa}'n(\boldsymbol{\kappa}')}{\kappa'^2}\right]\times\mathbf{k}_0 \tag{9.33}$$

In Eq. 9.33 the performance of the κ-space integration, gives the transition element of the interactive magnetic potential energy, which is measured in the rest frame of the neutron, and arises from the electric field of the nucleus and atomic electrons. The κ-space charge density $n(-\boldsymbol{\kappa})$ is replaced by its operator form $\hat{n}(-\boldsymbol{\kappa})$, where

$$\langle\psi_0|\hat{U}_{so}|\psi\rangle = \mathrm{i}\frac{2g\mu_n}{m_pc^2}\frac{\hbar}{\varepsilon_0}\hat{\mathbf{S}}\cdot\left[\frac{\boldsymbol{\kappa}\hat{n}(-\boldsymbol{\kappa})}{\kappa^2}\right]\times\mathbf{k}_0 \tag{9.34}$$

The proton classical radius r_p is

$$r_p = \frac{\mu_0}{4\pi}\frac{e^2}{m_p} \tag{9.35}$$

Recall that the nuclear magneton μ_n and speed of light c are

$$\mu_n = \frac{e\hbar}{2m_p} \tag{9.36}$$

and

$$c^2 = \frac{1}{\mu_0 \varepsilon_0} \tag{9.37}$$

With expressions for μ_n, c^2, and r_p, one obtains the combined expression:

$$\frac{g\mu_n}{\hbar}\frac{\mu_0}{4\pi} = \frac{r_p g}{2e} \tag{9.38}$$

With substitution of Eq. 9.38 into Eq. 9.34, the transition element expression becomes

$$\langle \psi_0 | \hat{U}_{so} | \psi \rangle = \mathrm{i}\frac{4\pi\hbar^2}{m_p}\frac{r_p g}{e}\hat{\mathbf{S}} \cdot \left[\hat{n}(-\boldsymbol{\kappa})\frac{\boldsymbol{\kappa}}{\kappa^2}\right] \times \mathbf{k}_0 \tag{9.39}$$

The neutron scatter is assumed to be elastic because the neutron energy is insufficient to significantly change the positions of the nuclei and electron sources of charge, and thus the neutron does not affect the neutron and atomic electron spatial wave functions. For elastic neutron scatter, the magnitude of the neutron wave vector $\boldsymbol{\kappa}$ is constant, but the neutron wave vector changes direction by angle θ, where

$$\boldsymbol{\kappa} = \mathbf{k}_0 - \mathbf{k} \quad \text{and} \quad |\mathbf{k}| = |\mathbf{k}_0| \tag{9.40}$$

so that

$$\kappa = 2k_0 \sin\frac{\theta}{2} \tag{9.41}$$

Unit vector $\mathbf{u}_{\perp\kappa}$ is in the direction of $\mathbf{k} \times \mathbf{k}_0$, which is perpendicular to $\boldsymbol{\kappa} = \mathbf{k}_0 - \mathbf{k}$, where

$$\mathbf{k} \times \mathbf{k}_0 = \mathbf{u}_{\perp\kappa}\kappa^2 \sin\theta \tag{9.42}$$

One invokes trigonometric identity:

$$\sin\left(2 \cdot \frac{\theta}{2}\right) = 2\sin\left(\frac{\theta}{2}\right)\cos\left(\frac{\theta}{2}\right) \tag{9.43}$$

Application to the $\mathbf{k} \times \mathbf{k}_0$ expression yields

$$\frac{\boldsymbol{\kappa} \times \mathbf{k}_0}{\kappa^2} = -\frac{\mathbf{k} \times \mathbf{k}_0}{\kappa^2} = -\frac{1}{2}\mathbf{u}_{\perp\kappa} \cot\frac{\theta}{2} \tag{9.44}$$

With substitution of Eq. 9.44 into Eq. 9.39, the transition element expression becomes

$$\langle\psi_0|\hat{U}_{so}|\psi\rangle = -\mathrm{i}\frac{2\pi\hbar^2}{m_p}\frac{r_p g}{e}\hat{n}(-\boldsymbol{\kappa})(\hat{\mathbf{S}}\cdot\mathbf{u}_{\perp\kappa})\cot\frac{\theta}{2} \tag{9.45}$$

Recalling Eqs. 2.26 and 2.31 in Squires (1996) and Eq. 4.55 in Ch. 3 of Cremer (2012b), the neutron spin−orbit scattering length b_{so} is obtained from the neutron spin−orbit transition matrix element. The rest mass of the neutron m_n and proton m_p are approximately equal ($m_n \cong m_p$), so

$$b_{so} = \frac{m_n}{2\pi\hbar^2}\langle\psi_0|\hat{U}_{so}|\psi\rangle = -\mathrm{i}\frac{r_p g}{2e}\hat{n}(-\boldsymbol{\kappa})(\hat{\mathbf{S}}\cdot\mathbf{u}_{\perp\kappa})\cot\frac{\theta}{2} \tag{9.46}$$

The number of charges per unit area in the κ-space is

$$\frac{\hat{n}(-\boldsymbol{\kappa})}{e} \tag{9.47}$$

The scatter length per charge b_{so}^0 is a constant, where

$$b_{so}^0 = -\mathrm{i}\frac{r_p g}{2} \tag{9.48}$$

The assumed overall charge neutrality of the atom causes the κ-space transform of the charge density to be zero $\hat{n}(-\boldsymbol{\kappa})$ in the direction of forward scatter $\boldsymbol{\kappa} = 0$, where $\theta = 0$, and

$$\hat{n}(-\boldsymbol{\kappa}) = 0 \quad \boldsymbol{\kappa} = 0 \tag{9.49}$$

Recalling Eq. 2.15 in Squires (1996), Eq. 1.49 in the spin-state chapter in Cremer (2013), as well as Eq. 4.83 in Ch. 3 and Eq. 3.53 in Ch. 7 of Cremer (2012b), the partial differential cross section $(\mathrm{d}^2\sigma/d\Omega\,\mathrm{d}E)_{U,i\to m}$ for the $i \to m$ transition is

$$\left(\frac{\mathrm{d}^2\sigma}{\mathrm{d}\Omega\,\mathrm{d}E}\right)_{U,i\to m} = \left(\frac{m_n}{2\pi\hbar^2}\right)^2\frac{k}{k_0}|\langle\phi_m\psi_s\psi|\hat{U}|\psi_0\psi_{0s}\phi_i\rangle|^2\delta(E_0 - E + E_m - E_i) \tag{9.50}$$

With substitution of Eq. 9.46 into Eq. 9.50, the partial differential cross sections $(\mathrm{d}^2\sigma/d\Omega\,\mathrm{d}E)_{so,i\to m}$ for the spin−orbit electric field scatter of neutrons become

$$\left(\frac{\mathrm{d}^2\sigma}{\mathrm{d}\Omega\,\mathrm{d}E}\right)_{so,i\to m} = \frac{k}{k_0}\langle\phi_i\psi_{0s}|b_{so}^*|\psi_s\phi_m\rangle\langle\phi_m\psi_s|b_{so}|\psi_{0s}\phi_i\rangle\delta(E_0 - E + E_m - E_i) \tag{9.51}$$

9.2 Foldy Electric Field Scatter of Neutrons

Following section 9.5 in Squires (1996), we now consider the Foldy interaction of the neutron with the atomic electric field due to the slight separation of the positive and negative charge distributions within the charge-neutral neutron. The Foldy interaction is the electrostatic interaction between the nuclei and atomic electrons with the two electric-charge components of the charge-neutral neutron. The charge-neutral neutron can be considered to be a combination of a time-varying disassociation into a positive-charged proton and a negative-charged π^{-1} meson that are in close proximity. The time-varying position of the proton and π^{-1} meson relative to each other gives rise to proton π^{-1} meson current loops. The proton and π^{-1} meson current loops thus are a pair of interacting magnetic dipoles, whose sum gives rise to the net neutron spin $\mathbf{S}$ and magnetic dipole moment $\boldsymbol{\mu}$, where

$$\boldsymbol{\mu} = 2g\mu_n\mathbf{S} \tag{9.52}$$

However, in the rest frame of the neutron, the separated proton charge and π^{-1} meson experience the electric fields of the nucleus and its electrons.

The potential energy for this Foldy electrostatic interaction is proportional to the net neutron magnetic moment $\boldsymbol{\mu}$, but it is independent of the neutron spin $\mathbf{S}$ and velocity $\mathbf{v}$. That is, the Foldy potential U_F is proportional to the position-dependent charge density $n(\mathbf{r})$ of the nuclei and electrons, and by Gauss's law, the Foldy potential is proportional to the divergence of the total electric field of the nuclei and electrons:

$$\nabla \cdot \mathbf{E} = \frac{n(\mathbf{r})}{\varepsilon_0} \tag{9.53}$$

Charge density $n(\mathbf{r})$, Foldy potential energy U_F replaced by their operator forms $\hat{n}(\mathbf{r})$ and $\hat{U}_F$, and the Foldy potential energy operator are, then,

$$\hat{U}_F = \frac{\hbar g\mu_n}{2m_nc^2}\nabla \cdot \hat{\mathbf{E}} = \frac{\hbar g\mu_n}{2m_nc^2}\frac{\hat{n}(\mathbf{r})}{\varepsilon_0} \tag{9.54}$$

The Foldy neutron scatter length b_F is obtained from the neutron spin–orbit transition matrix element, where b_F is zero in the forward direction, and b_F is non-zero and uniform in all other directions. The neutron rest mass m_n is approximated as the proton rest mass m_p, where $m_n \cong m_p$, so that

$$b_F = \frac{m_n}{2\pi\hbar^2}\langle\psi_0|\hat{U}_F|\psi\rangle = \frac{r_pg}{2e}\hat{n}(-\boldsymbol{\kappa}) \tag{9.55}$$

The number of charges per unit area in the κ-space is given by

$$\frac{\hat{n}(-\boldsymbol{\kappa})}{e} \tag{9.56}$$

Hence, the scatter length per charge b_F^0 is

$$b_F^0 = \frac{r_pg}{2} \tag{9.57}$$

Recall that for $\boldsymbol{\kappa} = 0$ forward scatter, the κ-space transform $\hat{n}(-\boldsymbol{\kappa})$ of the charge density $\hat{n}(\mathbf{r})$ is zero because of the assumed overall charge neutrality of the atom, where

$$\hat{n}(-\boldsymbol{\kappa}) = 0 \tag{9.58}$$

Similar to the derivation of Eqs. 9.50-9.51, the partial differential cross sections $(\mathrm{d}^2\sigma/d\Omega\,\mathrm{d}E)_{F,i\rightarrow m}$ for the $i \rightarrow m$ transition in the Foldy electric field scatter of neutrons becomes

$$\left(\frac{\mathrm{d}^2\sigma}{\mathrm{d}\Omega\,\mathrm{d}E}\right)_{F,i\rightarrow m} = \frac{k}{k_0}\langle\phi_i\psi_{0s}|b_F^*|\psi_s\phi_m\rangle\langle\phi_m\psi_s|b_F|\psi_{0s}\phi_i\rangle\delta(E_0 - E + E_m - E_i) \tag{9.59}$$

Both the spin–orbit and Foldy electric field scatter of the neutron are minor compared to the magnetic field scatter. In magnetic field neutron scatter, the partial differential cross section has the scattering length factor $r_0 g$, whereas the electric field neutron scatter has the scattering length factor $r_p g$, in which the ratio of the electric field scatter length b_E to magnetic field scatter length b_M is calculated as follows:

$$\frac{b_E}{b_M} = \frac{r_0 g}{r_p g} = \frac{m_e}{m_p} \tag{9.60}$$

The ratio of the electric scatter elastic cross section σ_E divided by the magnetic scatter elastic cross section σ_M is the square of the ratio of the electron to proton mass $(m_e/m_p)^2$, where

$$\sigma_E = 4\pi b_e^2 \tag{9.61}$$

and

$$\sigma_M = 4\pi b_M^2 \tag{9.62}$$

so that

$$\frac{\sigma_E}{\sigma_M} = \left(\frac{b_E}{b_M}\right)^2 = \left(\frac{m_e}{m_p}\right)^2 = \left(\frac{1}{1836}\right)^2 \tag{9.63}$$

9.3 Scatter of Neutrons by Nuclear and Electric Field Interactions in a Non-Bravais Lattice Crystal

Following section 9.5 in Squires (1996), we now consider the scatter of neutrons from a non-Bravais lattice crystal, where the neutron and nuclear spin is taken into account, as well as scatter of the neutron by the atomic electric fields. The total partial differential cross section $d^2\sigma/d\Omega\, dE$ for the Bragg scatter of neutrons, which is due to the nuclear and electric field interactions for a non-Bravais crystal, is derived from the general partial differential cross section expression of Eq. 3.19 with Eq. 3.2, and is repeated in Eq. 9.64 below,

$$\frac{d^2\sigma}{d\Omega\, dE} = \sum_{i=1}^{I} p_i \sum_{m=1}^{M} \left(\frac{d\sigma}{d\Omega\, dE}\right)_{i,0s\rightarrow m,s} \tag{9.64}$$

where

$$\left(\frac{d^2\sigma}{d\Omega\, dE}\right)_{i,0s\rightarrow m,s} = \left(\frac{m_n}{2\pi\hbar^2}\right)^2 \frac{k}{k_0} |\langle \phi_m \psi_s \psi | \hat{U} | \psi_0 \psi_{0s} \phi_i \rangle|^2 \delta(E_0 - E + E_m - E_i) \tag{9.65}$$

and

$$|\langle \phi_i \psi_{0s} \psi_0 | \hat{U} | \psi \psi_s \phi_m \rangle|^2 = \langle \phi_i \psi_{0s} \psi_0 | \hat{U}^* | \psi \psi_s \phi_m \rangle \langle \phi_m \psi_s \psi | \hat{U} | \psi_0 \psi_{0s} \phi_i \rangle \tag{9.66}$$

The potential energy operator $\hat{U}$ is the sum of (1) the spin–orbit magnetic field potential energy $\hat{U}_{so}$ between the electron magnetic dipole moment, (2) the Foldy electric potential energy $\hat{U}_F$ between the atomic electric field and the unequal charge distribution of the neutron, which gives rise to the neutron magnetic dipole moment, and (3) the strong-force potential energy $\hat{U}_n$ between the nucleus and neutron, which includes the potential energy of the interaction of the neutron spin $\mathbf{S}$ and the nuclear spin $\mathbf{I}$, where

$$\hat{U} = U_{so} + \hat{U}_F + \hat{U}_n \tag{9.67}$$

and

$$U_{so} = -\boldsymbol{\mu} \cdot \mathbf{B} = -2g\mu_n \frac{\mathbf{E(r)} \times \mathbf{v}}{c^2} \hat{S}_z \tag{9.68}$$

$$\hat{U}_F = \frac{\hbar g \mu_n}{2m_n c^2} \nabla \cdot \hat{\mathbf{E}} = \frac{\hbar g \mu_n}{2m_n c^2} \frac{\hat{n}(\mathbf{r})}{\varepsilon_0} \tag{9.69}$$

$$\hat{U}_n = \hat{b}_j \exp(\mathrm{i}\hat{\mathbf{R}}_j \cdot \boldsymbol{\kappa}) = [A + 2B(\hat{\mathbf{I}} \cdot \hat{\mathbf{S}})] \exp(\mathrm{i}\hat{\mathbf{R}}_j \cdot \boldsymbol{\kappa}) \tag{9.70}$$

The Bragg scatter of neutrons due to the nuclear and electric field interactions is now derived for a non-Bravais crystal, where the nuclei vibrations are assumed to be negligible, and thus the nuclei are assumed to be static at their equilibrium positions. The sources of the electric field of an atom arise only at the expected regions of the negative electron charge and positive nuclei charge in an atom. In an atom, the negative electron charge is

$$-ef_d(\boldsymbol{\kappa}) \tag{9.71}$$

The positive nucleus charge from the Z_d protons of the dth nucleus is

$$Z_d e \tag{9.72}$$

The net charge of the dth atom is then,

$$eZ_d - ef_d(\kappa) \tag{9.73}$$

The form factor $f_d(\kappa)$ is given by

$$f_d(\kappa) = \int_{-\infty}^{\infty} \mathrm{d}z \int_{-\infty}^{\infty} \mathrm{d}y \int_{-\infty}^{\infty} \mathrm{d}x \;\rho(\mathbf{r}') \exp(\mathrm{i}\kappa \cdot \mathbf{r}')\mathrm{d}\mathbf{r}' \tag{9.74}$$

The total density $\rho(\mathbf{r}')$ of paired and unpaired electrons in the atom as a function of position $\mathbf{r}'$ is measured from the nucleus at the origin. The integrand of the form factor $f_d(\kappa)$ for the dth atom (or ion) is nonzero only at the idealized point positions of the electrons and thus represents the sum of the electron charge in the dth atom (or ion).

The position $\mathbf{R}_{ld}$ of the net charge of an atom is centered at the crystal atom positions given by:

$$\mathbf{R}_{ld} = \mathbf{R}_l + \mathbf{d}_d \tag{9.75}$$

The position $\mathbf{R}_{ld}$ of an atom is addressed by the integer pair (l, d) that specifies which l^{th} unit cell $l = 1, 2, \ldots, N$ is measured relative to the crystal center at $l = 0$, and which d^{th} in that unit cell $d = 1, 2, \ldots, r$ is measured relative to the corner point or origin of the unit cell vectors $(\mathbf{a}, \mathbf{b}, \mathbf{c})$. The net charge of the atom is assumed to be localized to the reference corner point of the unit cell. The net charge of each (l, d) atom is assumed to be a point charge situated at its equilibrium lattice position $\mathbf{R}_{ld}$. Thus, the charge density is the set of equilibrium lattice points of the crystal. The integration over the position $\mathbf{r}$ of the neutron in the crystal gives nonzero contributions only at the equilibrium lattice point positions of the atoms $\mathbf{R}_{ld}$, where

$$\mathbf{r} = \mathbf{R}_l + \mathbf{d}_d \tag{9.76}$$

The expected positions of the atoms in the crystal for an initial i^{th} eigenstate of the neutron-crystal scattering system, is multiplied by the net charge of each atom, and gives the expected crystal charge density $\langle \hat{n}(\kappa) \rangle$ for the initial d^{th} eigenstate, where

$$\langle \hat{n}(-\kappa) \rangle = \langle \phi_i | \hat{n}(-\kappa) | \phi_i \rangle = \sum_{l=1}^{N} \sum_{d=1}^{r} e[Z_d - f_d(\kappa)] \exp(\mathrm{i}\kappa \cdot \mathbf{R}_l) \exp(\mathrm{i}\kappa \cdot \mathbf{d}_d) \tag{9.77}$$

Recall that in general X-ray or neutron scatter, the crystal structure information is obtained by measuring the intensities of points in the diffraction pattern. The points in the diffraction pattern occur at the crystal reciprocal lattice points. The wave amplitude at each of these reciprocal lattice points is proportional to the structure factor $F(\kappa)$, which in the general case, f_d is the scattering factor of the single scattering unit (electron, nucleus, sum of both, etc.), and

$$F(\kappa) = \sum_{d=1}^{r} f_d \exp(i\kappa \cdot \mathbf{d}_d) \tag{9.78}$$

The measured diffraction spot intensity is then proportional to $|F(\kappa)|^2$, where

$$|F(\kappa)|^2 = \left| \sum_{d=1}^{r} f_d \exp(i\kappa \cdot \mathbf{d}_d) \right|^2 \tag{9.79}$$

An incident plane wave travels from the reference corner point of a unit cell to the position $\mathbf{d}_d$ of the dth atom, and in that distance, it undergoes phase shift $\kappa \cdot \mathbf{d}_d$. The sum of the phase shifts in the single unit cell, provides the net phase shift, which is acquired by the wave in its passage through the unit cell. This summed phase shift $\sum_{d=1}^{r} \kappa \cdot \mathbf{d}_d$ is added to the phase shift $\kappa \cdot \mathbf{R}_l$, which the plane wave acquires in passing from a reference center point of the crystal at $l = 0$ to the position $\mathbf{R}_l$ at the corner reference point of the dth unit cell.

In the case of nuclear scattering in a non-Bravais lattice crystal, the nuclear scattering factor $F_n(\kappa)$ has been derived previously as Eq. 273 in Ch. 6 of Cremer (2012b) and Eq. 3.76 in Squires (1996) where

$$F_n(\kappa) = \sum_{d=1}^{r} \overline{b}_d \exp(i\kappa \cdot \mathbf{d}_d) \exp(-W_d) \tag{9.80}$$

The crystal is cooled so that after a sufficient cooling time, the nuclear vibrations are assumed to be negligible, which allows one to neglect the Debye–Waller term, where $W_d \cong 0$. The nuclear scattering factor $F_n(\kappa)$ is, then,

$$F_n(\kappa) = \sum_{d=1}^{r} \overline{b}_d \exp(i\kappa \cdot \mathbf{d}_d) \tag{9.81}$$

Without considering neutron spin and nuclear spin, the mean nuclear scattering operator $\hat{b}_n(\kappa)$ for the crystal is

$$\hat{b}_n(\kappa) = F_n(\kappa) \tag{9.82}$$

Recalling the spin-state chapter of Cremer (2013) and section 9.2 in Squires (1996), if the neutron spin $\mathbf{S}$ and nuclear spin $\mathbf{I}$ are considered, then the neutron scatter operator becomes

$$\hat{b}_n(\kappa) = A + 2B(\hat{\mathbf{I}} \cdot \hat{\mathbf{S}}) \tag{9.83}$$

The neutron spin $\mathbf{S}$ and nuclear spin $\mathbf{I}$ dependent component of the scatter is

$$B = \frac{b^+ - b^-}{2I + 1} \tag{9.84}$$

Constant A is the nuclear mean scatter length component, which does not depend on the neutron/nucleus spin interaction. Hence, it can be set as

$$F_n(\kappa) = A \tag{9.85}$$

where

$$A = \frac{b^+(I+1) + b^- I}{2I+1} \tag{9.86}$$

The "plus" scatter length b^+ corresponds to total spin quantum number

$$t = I + \frac{1}{2} \tag{9.87}$$

The "minus" scatter length b^- corresponds to total spin quantum number

$$t = I - \frac{1}{2} \tag{9.88}$$

Operator $\hat{b}_n(\kappa)$ for the nuclear scattering length is then defined for eigenvalues b^+ and b^-, whose eigenfunctions are also ψ_{t+} and ψ_{t-}, respectively, where

$$\hat{b}_n(\kappa)\psi_{t+} = b^+\psi_{t+} \tag{9.89}$$

and

$$\hat{b}_n(\kappa)\psi_{t-} = b^-\psi_{t-} \tag{9.90}$$

Hence, if one accounts for the neutron spin **S** and nuclear spin **I**, then with substitution of Eq. 9.85 into Eq. 9.83, the mean nuclear scatter length operator becomes

$$\hat{b}_n(\kappa) = F_n(\kappa) + 2B(\hat{\mathbf{I}} \cdot \hat{\mathbf{S}}) \tag{9.91}$$

The spin–orbit and Foldy electric field scattering length $\hat{b}_e$ for a single charge e is

$$\hat{b}_e = \hat{b}_{so} + \hat{b}_F = \hat{n}(-\kappa)\frac{r_p g}{2}\left[1 - \mathrm{i}(\hat{\mathbf{S}} \cdot \mathbf{u}_{\perp\kappa})\cot\frac{\theta}{2}\right] \tag{9.92}$$

The electric field scattering length of Eq. 9.92 can be expressed by the electric field structure factor $F_e(\kappa)$, where

$$\hat{b}_e = F_e(\kappa)\left[1 - i(\hat{\mathbf{S}} \cdot \mathbf{u}_{\perp\kappa})\cot\frac{\theta}{2}\right] \tag{9.93}$$

From Eq. 9.73 and with $\hat{n}(-\boldsymbol{\kappa}) = Z_d - f_d(\boldsymbol{\kappa})$, the electric field structure factor $F_e(\boldsymbol{\kappa})$ in Eq. 9.93, is defined as

$$F_e(\boldsymbol{\kappa}) = \frac{r_p g}{2} \sum_{d=1}^{r} [Z_d - f_d(\kappa)] \exp(i\boldsymbol{\kappa} \cdot \mathbf{d}_d) \tag{9.94}$$

The measured diffraction pattern intensity is proportional to $|F_e(\boldsymbol{\kappa})|^2$ and does not include the relative positions of the nuclei, which are contained in the imaginary phase arguments of the summed exponential terms in Eq. 9.94. The phase information, which contains the relative nuclei positions, is removed upon taking the squared magnitude of summed expression of Eq. 9.94, where:

$$|F_e(\boldsymbol{\kappa})|^2 = \left| \frac{r_p g}{2} \sum_{d=1}^{r} [Z_d - f_d(\boldsymbol{\kappa})] \exp(i\boldsymbol{\kappa} \cdot \mathbf{d}_d) \right|^2 \tag{9.95}$$

If the orientations of the crystal nuclei are random, then $B = 0$ and the mean scattering length operator for the combined nuclear and electric field $\hat{b}_{ne}(\boldsymbol{\kappa})$ is

$$\hat{b}_{ne}(\boldsymbol{\kappa}) = \hat{b}_n(\boldsymbol{\kappa}) + \hat{b}_e(\boldsymbol{\kappa}) \tag{9.96}$$

With substitution of Eqs. 9.91 with B = 0 and Eq. 9.93 into Eq. 9.96,

$$\hat{b}_{ne}(\boldsymbol{\kappa}) = F_n(\boldsymbol{\kappa}) + F_e(\boldsymbol{\kappa}) - \hat{\mathbf{S}} \cdot \mathbf{u}_{\perp\kappa} i F_e(\boldsymbol{\kappa}) \cot\left(\frac{\theta}{2}\right) \tag{9.97}$$

If the neutron's spin does not change in scatter, then the expected values $b_{++}(\boldsymbol{\kappa})$ and $b_{--}(\boldsymbol{\kappa})$ of the combined atomic electric field and nuclear scattering lengths are

$$b_{++}(\boldsymbol{\kappa}) = \langle \psi_{0+} | \hat{b}_{ne}(\boldsymbol{\kappa}) | \psi_+ \rangle = F_n(\boldsymbol{\kappa}) + F_e(\boldsymbol{\kappa}) - i\frac{\hbar}{2} F_e(\boldsymbol{\kappa}) u_{\perp\kappa z} \cot\left(\frac{\theta}{2}\right) \tag{9.98}$$

and

$$b_{--}(\boldsymbol{\kappa}) = \langle \psi_{0-} | \hat{b}_{ne}(\boldsymbol{\kappa}) | \psi_- \rangle = F_n(\boldsymbol{\kappa}) + F_e(\boldsymbol{\kappa}) + i\frac{\hbar}{2} F_e(\boldsymbol{\kappa}) u_{\perp\kappa z} \cot\left(\frac{\theta}{2}\right) \tag{9.99}$$

If in scatter, the neutron spin transitions from spin-down to spin-up $(- \rightarrow +)$, with expected scatter length b_{-+}, or transitions from spin-up to spin-down $(+ \rightarrow -)$, with expected scatter length b_{+-}, then

$$b_{-+}(\boldsymbol{\kappa}) = \langle \psi_{0-} | \hat{b}_{ne}(\boldsymbol{\kappa}) | \psi_+ \rangle = -i\frac{\hbar}{2} F_e(\boldsymbol{\kappa}) (u_{\perp\kappa x} + i u_{\perp\kappa y}) \cot\left(\frac{\theta}{2}\right) \tag{9.100}$$

and

$$b_{+-}(\boldsymbol{\kappa}) = \langle \psi_{0+} | \hat{b}_{ne}(\boldsymbol{\kappa}) | \psi_- \rangle = -i\frac{\hbar}{2} F_e(\boldsymbol{\kappa}) (u_{\perp\kappa x} - i u_{\perp\kappa y}) \cot\left(\frac{\theta}{2}\right) \tag{9.101}$$

The differential cross section $(\mathrm{d}\sigma/\mathrm{d}\Omega)^e_{coh}$ can be obtained from partial differential cross section $(\mathrm{d}^2\sigma/\mathrm{d}\Omega\,\mathrm{d}E)^e_{coh}$ using these four expected scatter lengths for the four possible neutron spin transitions.

One can reduce computational effort by recalling Eq. 275 in Ch. 6 of Cremer (2012b) or Eq. 3.75 in Squires (1996) for the differential cross section for coherent, elastic scatter in a non-Bravais lattice crystal with N unit cells each, with volume v_0:

$$\left(\frac{\mathrm{d}\sigma}{\mathrm{d}\Omega}\right)^e_{coh} = N\frac{(2\pi)^3}{v_0}\sum_{\tau=1}^{N}\delta(\boldsymbol{\kappa}-\boldsymbol{\tau})|F_n(\boldsymbol{\kappa})|^2 \tag{9.102}$$

From Eq. 1.21 and Eq. 2.31 in Squires (1996), as well as Eq. 475 in Ch. 2 and Eq. 454 in Ch. 3 of Cremer (2012b), the coherent, elastic, scatter cross section for a single atom in terms of the atom's scatter length b is

$$\left(\frac{\mathrm{d}\sigma}{\mathrm{d}\Omega}\right)^e_{coh} = b^2 \tag{9.103}$$

By inspection of Eq. 9.102, the number n_{sc} of scattering lattice points is

$$n_{sc} = N\frac{(2\pi)^3}{v_0}\sum_{\tau=1}^{N}\delta(\boldsymbol{\kappa}-\boldsymbol{\tau}) \tag{9.104}$$

A polarized neutron beam is assumed to be incident in the z-direction on the crystal. A fraction f of these neutrons are in the spin-up state $(+)$, where the z-directed polarization vector $\mathbf{P}$ has magnitude

$$P = 2f - 1 \tag{9.105}$$

If one accounts for the neutron and nucleus spin in nuclear scatter, as well as the atomic electric field scatter of neutrons, then for a single atom, the coherent, elastic differential cross section is

$$\left(\frac{\mathrm{d}\sigma}{\mathrm{d}\Omega}\right)^e_{coh} = f(|b_{++}|^2 + |b_{+-}|^2) + (1-f)(|b_{--}|^2 + |b_{-+}|^2) \tag{9.106}$$

For the non-Bravais lattice crystal, the coherent, elastic differential cross section becomes

$$\left(\frac{\mathrm{d}\sigma}{\mathrm{d}\Omega}\right)^e_{coh} = N\frac{(2\pi)^3}{v_0}\sum_{\tau=1}^{N}\delta(\boldsymbol{\kappa}-\boldsymbol{\tau})\big[f(|b_{++}|^2 + |b_{+-}|^2) + (1-f)(|b_{--}|^2 + |b_{-+}|^2)\big] \tag{9.107}$$

The magnitudes of Eqs. 9.100 and 9.101 are equal, where $|b_{+-}(\boldsymbol{\kappa})| = |b_{-+}(\boldsymbol{\kappa})|$, so the sum of these terms Eq. 9.107 are

$$f|b_{+-}(\boldsymbol{\kappa})|^2 - (1-f)|b_{-+}(\boldsymbol{\kappa})|^2 = \left(\frac{\hbar}{2}F_e(\boldsymbol{\kappa})\cot\frac{\theta}{2}\right)^2\left(u^2_{\perp\kappa x} + u^2_{\perp\kappa y}\right) \tag{9.108}$$

From Eq. 9.42, one recalls the unit vector $u_{\perp k}$, which is perpendicular to the neutron scatter vector $\boldsymbol{\kappa} = \mathbf{k}_0 - \mathrm{k}$, where the magnitude of $u_{\perp k}$, and the square root of the sum of its squared x, y, and z components are equal one,

$$1 = u_{\perp\kappa}^2 = u_{\perp\kappa x}^2 + u_{\perp\kappa y}^2 + u_{\perp\kappa z}^2 \tag{9.109}$$

then Eq. 9.108 becomes

$$f|b_{+-}(\boldsymbol{\kappa})|^2 - (1-f)|b_{-+}(\boldsymbol{\kappa})|^2 = \left(\frac{\hbar}{2}F_e(\boldsymbol{\kappa})\cot\frac{\theta}{2}\right)^2 (1 - u_{\perp\kappa z}^2) \tag{9.110}$$

From Eq. 9.98,

$$\begin{aligned} |b_{++}(\boldsymbol{\kappa})|^2 = &\left[F_n(\boldsymbol{\kappa}) + F_e(\boldsymbol{\kappa}) - i\frac{\hbar}{2}F_e(\boldsymbol{\kappa})u_{\perp\kappa z}\cot\left(\frac{\theta}{2}\right)\right] \\ &\cdot \left[F_n(\boldsymbol{\kappa}) + F_e(\boldsymbol{\kappa}) - i\frac{\hbar}{2}F_e(\boldsymbol{\kappa})u_{\perp\kappa z}\cot\left(\frac{\theta}{2}\right)\right]^* \end{aligned} \tag{9.111}$$

Expanding Eq. 9.111, one obtains

$$\begin{aligned} |b_{++}(\boldsymbol{\kappa})|^2 = &|F_n(\boldsymbol{\kappa})|^2 + |F_e(\boldsymbol{\kappa})|^2 + [F_e(\boldsymbol{\kappa})F_n^*(\boldsymbol{\kappa}) + F_e^*(\boldsymbol{\kappa})F_n(\boldsymbol{\kappa})] \\ &+ |F_e(\boldsymbol{\kappa})|^2\left(\frac{\hbar}{2}\cot\frac{\theta}{2}\right)^2 u_{\perp\kappa z}^2 \\ &+ \mathrm{i}\frac{\hbar}{2}\cot\frac{\theta}{2}u_{\perp\kappa z}\left(\left[F_n(\boldsymbol{\kappa})F_e^*(\boldsymbol{\kappa})\right] - [F_n(\boldsymbol{\kappa})F_e^*(\boldsymbol{\kappa})]^*\right) \end{aligned} \tag{9.112}$$

From Eq. 9.99

$$\begin{aligned} |b_{--}(\boldsymbol{\kappa})|^2 = &\left[F_n(\boldsymbol{\kappa}) + F_e(\boldsymbol{\kappa}) + i\frac{\hbar}{2}F_e(\boldsymbol{\kappa})u_{\perp\kappa z}\cot\left(\frac{\theta}{2}\right)\right] \\ &\cdot \left[F_n(\boldsymbol{\kappa}) + F_e(\boldsymbol{\kappa}) + i\frac{\hbar}{2}F_e(\boldsymbol{\kappa})u_{\perp\kappa z}\cot\left(\frac{\theta}{2}\right)\right]^* \end{aligned} \tag{9.113}$$

Expanding Eq. 9.113, one obtains

$$\begin{aligned} |b_{--}(\boldsymbol{\kappa})|^2 = &|F_n(\boldsymbol{\kappa})|^2 + |F_e(\boldsymbol{\kappa})|^2 + [F_n^*(\boldsymbol{\kappa})F_e(\boldsymbol{\kappa}) + F_n(\boldsymbol{\kappa})F_e^*(\boldsymbol{\kappa})] \\ &+ |F_e(\boldsymbol{\kappa})|^2\left(\frac{\hbar}{2}\cot\frac{\theta}{2}\right)^2 u_{\perp\kappa z}^2 \\ &- \mathrm{i}\frac{\hbar}{2}\cot\frac{\theta}{2}u_{\perp\kappa z}\left(\left[F_n(\boldsymbol{\kappa})F_e^*(\boldsymbol{\kappa})\right] - [F_n(\boldsymbol{\kappa})F_e^*(\boldsymbol{\kappa})]^*\right) \end{aligned} \tag{9.114}$$

Then from Eqs. 9.112 and 9.114,

$$f|b_{++}(\boldsymbol{\kappa})|^2+(1-f)|b_{--}(\boldsymbol{\kappa})|^2=|F_n(\boldsymbol{\kappa})|^2+|F_e(\boldsymbol{\kappa})|^2+2\,\mathrm{Re}\left[F_n(\boldsymbol{\kappa})F_e^*(\boldsymbol{\kappa})\right] + |F_e(\boldsymbol{\kappa})|^2\left(\frac{\hbar}{2}\cot\frac{\theta}{2}\right)^2 u_{\perp\kappa z}^2 + (1-2f)\frac{\hbar}{2}\cot\frac{\theta}{2}u_{\perp\kappa z}2\,\mathrm{Im}\left[F_n(\boldsymbol{\kappa})F_e^*(\boldsymbol{\kappa})\right] \tag{9.115}$$

where

$$F_n^*(\boldsymbol{\kappa})F_e(\boldsymbol{\kappa})+F_n(\boldsymbol{\kappa})F_e^*(\boldsymbol{\kappa})=[F_n(\boldsymbol{\kappa})F_e^*(\boldsymbol{\kappa})]^*+F_n(\boldsymbol{\kappa})F_e^*(\boldsymbol{\kappa})=2\,\mathrm{Re}[F_n(\boldsymbol{\kappa})F_e^*(\boldsymbol{\kappa})] \tag{9.116}$$

and

$$F_n(\boldsymbol{\kappa})F_e^*(\boldsymbol{\kappa})-F_n^*(\boldsymbol{\kappa})F_e(\boldsymbol{\kappa})=F_n(\boldsymbol{\kappa})F_e^*(\boldsymbol{\kappa})-[F_n(\boldsymbol{\kappa})F_e^*(\boldsymbol{\kappa})]^*=2\mathrm{i}\,\mathrm{Im}[F_n(\boldsymbol{\kappa})F_e^*(\boldsymbol{\kappa})] \tag{9.117}$$

Recall that unit vector $\mathbf{u}_{\perp\kappa}$ is in the direction of $\mathbf{k}\times\mathbf{k}_0$, which is perpendicular to $\boldsymbol{\kappa}=\mathbf{k}_0-\mathbf{k}$. The z-directed polarization vector $\mathbf{P}$ has magnitude $(2f-1)$, and hence, the dot product of the polarization vector $\mathbf{P}$ and the vector $\mathbf{u}_{\perp\kappa}$ is

$$\mathbf{P}\cdot\mathbf{u}_{\perp\kappa}=(2f-1)u_{\perp\kappa z} \tag{9.118}$$

With substitution of Eqs. 9.110, 9.115-9.118 into Eq. 9.107, the coherent, elastic, differential cross section becomes

$$\left(\frac{\mathrm{d}\sigma}{\mathrm{d}\Omega}\right)_{coh}^{e}=N\frac{(2\pi)^3}{v_0}\sum_{\tau=1}^{N}\delta(\boldsymbol{\kappa}-\boldsymbol{\tau}) \times\left[\begin{array}{l}|F_n(\boldsymbol{\kappa})|^2+|F_e(\boldsymbol{\kappa})|^2+2\,\mathrm{Re}\left[F_n(\boldsymbol{\kappa})F_e^*(\boldsymbol{\kappa})\right]+|F_e(\boldsymbol{\kappa})|^2\left(\dfrac{\hbar}{2}\cot\dfrac{\theta}{2}\right)^2\\ -2(\mathbf{P}\cdot\mathbf{u}_{\perp\kappa})\left(\dfrac{\hbar}{2}\cot\dfrac{\theta}{2}\right)\mathrm{Im}\left[F_n(\boldsymbol{\kappa})F_e^*(\boldsymbol{\kappa})\right]\end{array}\right] \tag{9.119}$$

The nuclear, Foldy, and spin−orbit contributions are the first three terms of $(\mathrm{d}\sigma/d\Omega)_{coh}^{e}$; the fourth term is interference between nuclear and Foldy scatter, and the fifth term is a polarization-dependent interference between nuclear and spin−orbit scatter.

References

Cremer, J. T. (2012b). In P. W. Hawkes (Ed.), Neutron and X-ray microscopy—part 2. *Advances in imaging and electron physics* (Vol. 173). London, UK: Academic Press, an imprint of Elsevier Press.

Cremer, J. T. (2013). In P. W. Hawkes (Ed.), *Advances in imaging and electron physics* (Vol. 175). London, UK: Academic Press, an imprint of Elsevier Press.

Eisberg, R., & Resnick, R. (1974). *Quantum physics of atoms, molecules, solids, nuclei, and particles*. New York: John Wiley & Sons.

Liboff, R. L. (2003). *Introductory quantum mechanics*. Reading, MA: Addison-Wesley.

Merzbacher, E. (1970). *Quantum mechanics*. New York: John Wiley & Sons.

Squires, G. L. (1996). *Introduction to the Theory of Thermal Neutron Scattering*. New York, NY: Dover.

10 Diffractive X-ray and Neutron Optics

10.1 Derivation of Helmholtz–Kirchhoff Integral Theorem

In this chapter, we derive the Helmholtz–Kirchhoff integral theorem and Kirchhoff–Fresnel diffraction equation, with the obliquity factor, and apply the paraxial approximation to the Kirchhoff–Fresnel diffraction equation. Also covered is the Fraunhofer diffraction of X-rays or neutrons from a rectangular aperture, as well as the Fraunhofer diffraction of an X-ray or neutron line source by a parallel single slit, a parallel slit pair, and N parallel slits. Diffraction from 3D gratings is shown to model the coherent, elastic scatter of X-ray or neutrons from material lattices. The Abbe theory of imaging applied to X-rays or neutrons is then covered, followed by the derivation of the Fraunhofer diffraction of X-rays or neutrons from a circular aperture. Using the Huygens–Fresnel approach, the Kirchhoff equation is derived for a compound refractive lens (CRL) used with X-rays or neutrons. Next, the compound refractive Fresnel lens for X-rays and neutrons, followed by the Fresnel diffraction of X-rays or neutrons from a circular aperture, a rectangular aperture, and a knife edge, is discussed. Also, the Fresnel zone plates (FZPs) for X-rays or neutrons and an achromat fabricated from an FZP and a CRL are presented. Finally, the Helmholtz differential equation for X-rays and neutrons is presented, and the Helmholtz paraxial differential equation and its solution for a Gaussian, spherical, X-ray laser beam are derived. Many of the topics in this chapter follow Elmore and Heald (1969), Hecht and Zajac (1979), and Sommerfeld (1954).

The Helmholtz equation provides a description and solution of wave propagation and wave interactions with apertures, as well as neutron and X-ray optical elements. The solution of the Helmholtz differential equation is subject to boundary conditions. On the other hand, in the case of the equivalent Helmholtz integral equation, the boundary conditions are included, rather than imposed in the final stage of the solution as is the case for the differential equation. The integral equation relates the unknown function (wave amplitude, in our case) to the values of the function at neighboring points as in the differential equations. In addition, the integral equation relates the unknown function to values throughout a region, including the boundary. In many cases, the integral equation is easier and more powerful to use than the differential equation. Recall that the Helmholtz equation is

$$\nabla^2 E(\mathbf{r}) + k^2 E(\mathbf{r}) = 0 \tag{10.1}$$

Neutron and X-ray Optics. DOI: http://dx.doi.org/10.1016/B978-0-12-407164-3.00009-7

Amplitude $E(\mathbf{r})$ can represent the electric field amplitude $E(\mathbf{r})$ of an X-ray or the wave amplitude $\psi(\mathbf{r})$ of a neutron with wave vector magnitude k, where the equation can be applied to both X-rays and neutrons.

The amplitude at an image point I can be solved from the amplitude and the gradient of the amplitude on a surface that surrounds the image point I, via solving the Helmholtz equation, which is accomplished with Green's theorem. This solution of the Helmholtz equation is called the *Kirchhoff integral theorem*, and is described next. Green's theorem is derived from Gauss's theorem, which equates a surface integral of a function over a closed surface to the volume integral of the divergence of that function. Green's theorem can be applied to a volume of space, which surrounds an excluded observation point I.

Following Hecht and Zajac (1979), the volume integral is shown to vanish (contains no source). As a result, the surface integral of the inner surface, which surrounds the observation point, is equal to the surface integral of the outer surface. In the limit, where the inner surface converges to the excluded observation point I, the amplitude at the observation point I, is then equal to the outer surface integral. The integrand of the outer surface integral is the commutation of the product of (1) the amplitude solution $E_1(\mathbf{r})$ to the wave equation, and (2) the gradient of a second solution $\nabla E_2(\mathbf{r})$. This might be expected in boundary value problems of partial differential equations, where the boundary values are specified in terms of the value of a function Ψ and/or its normal derivative:

$$\frac{\mathrm{d}\Psi}{\mathrm{d}n} = \hat{\mathbf{n}}\cdot\nabla\Psi \tag{10.2}$$

For example, with Gauss's theorem, for volume V, enclosing surface S, and the outward surface normal vector $\hat{\mathbf{n}}$ for amplitude scalar solutions $E_1(\mathbf{r})$ and $E_2(\mathbf{r})$ to the Helmholtz equation, one gets

$$\oint_S E_1(\mathbf{r})(\hat{\mathbf{n}}\cdot\nabla E_2(\mathbf{r}))\mathrm{d}S = \int_V \nabla\cdot(E_1\nabla E_2)\mathrm{d}V \tag{10.3}$$

The integrand on the right-hand side is expanded, so for scalar solutions $E_1(\mathbf{r})$ and $E_2(\mathbf{r})$,

$$\oint_S E_1(\mathbf{r})(\hat{\mathbf{n}}\cdot\nabla E_2(\mathbf{r}))\mathrm{d}S = \int_V \nabla E_1\cdot\nabla E_2\,\mathrm{d}V + \int_V E_1\nabla^2 E_2\,\mathrm{d}V \tag{10.4}$$

One reverses $E_1(\mathbf{r})$ and $E_2(\mathbf{r})$ to obtain

$$\oint_S E_2(\mathbf{r})(\hat{\mathbf{n}}\cdot\nabla E_1(\mathbf{r}))\mathrm{d}S = \int_V \nabla E_2\cdot\nabla E_2\,\mathrm{d}V + \int_V E_2\nabla^2 E_1\,\mathrm{d}V \tag{10.5}$$

Subtraction of the reversed $E_1(\mathbf{r})$ and $E_2(\mathbf{r})$ in Eq. (10.5) from that of the previous nonreversed expression in Eq. (10.4), yields Green's theorem:

$$\oint_S (E_1(\mathbf{r}) \cdot \nabla E_2(\mathbf{r}) - E_2(\mathbf{r}) \cdot \nabla E_1(\mathbf{r})) \cdot \hat{\mathbf{n}}\, \mathrm{d}S = \int_V (E_1 \nabla^2 E_2 - E_2 \nabla^2 E_1) \mathrm{d}V \qquad (10.6)$$

From Green's theorem, one can derive the Helmholtz–Kirchhoff theorem by assuming the two wave solutions to the Helmholtz wave equation, each of which has wavelength λ and wave vector $\mathbf{k}$ with magnitude k. The first solution $E_1(\mathbf{r})$ is an unspecified monochromatic wave. The second solution $E_2(\mathbf{r})$ is an assumed spherically symmetric, monochromatic wave, which is measured at an infinitesimal distance r_i from the observation (image) point I, where image point I is excluded from volume V, so that

$$E_2(\mathbf{r}) = \frac{\exp(ikr_i)}{r_i} \quad \text{and} \quad k = \frac{2\pi}{\lambda} \qquad (10.7)$$

The source point P and observation (image) point I are outside the surface S that bounds volume V. That is, the image point I is surrounded by a spherical shell, where the inner surface of the spherical shell can be brought arbitrarily close to image point I. The source point P lies outside the spherical shell, and source point P can be an arbitrary distance from the outer surface of the spherical shell. Hence, inside the spherical shell of volume V and enclosed by inner and outer surface components of surface S, the solutions $E_1(\mathbf{r})$ and $E_2(\mathbf{r})$ satisfy the source-free Helmholtz equation

$$\nabla^2 E_1(\mathbf{r}) + k^2 E_1(\mathbf{r}) = \nabla^2 E_2(\mathbf{r}) + k^2 E_2(\mathbf{r}) = 0 \qquad (10.8)$$

Substitution of the solutions $E_1(\mathbf{r})$ and $E_2(\mathbf{r})$ into the Green's theorem equation of Eq. (10.6), yields

$$\oint_S (E_1(\mathbf{r}) \nabla E_2(\mathbf{r}) - E_2(\mathbf{r}) \nabla E_1(\mathbf{r})) \cdot \hat{\mathbf{n}}\, \mathrm{d}S = \int_V (E_1 \nabla^2 E_2 - E_2 \nabla^2 E_1) \mathrm{d}V \qquad (10.9)$$

Substitution for $E_1(\mathbf{r})$ and $E_2(\mathbf{r})$ of Eq. (10.7) into Eq. (10.9), and application of Eq. (10.8) to Eq. (10.9), yields the remaining surface integral:

$$\oint_S \left(E_1(\mathbf{r}) \nabla \left(\frac{\exp(ikr_i)}{r_i} \right) - \frac{\exp(ikr_i)}{r_i} \cdot \nabla E_1(\mathbf{r}) \right) \cdot \hat{\mathbf{n}} r_i^2\, \mathrm{d}\Omega = 0 \qquad (10.10)$$

The surface integral of Eq. (10.10), on the spherical shell surface S can be divided into two parts, an outer surface and an inner surface. The first surface integral is over an arbitrarily shaped outer surface of S whose distance r_i from the observation point I can vary. The second surface is an inner spherical surface that surrounds but excludes the observation point I. The inner spherical surface has infinitesimal radius $r_i = \varepsilon$ that is measured from the observation point I. The inner spherical surface radius $r_i = \varepsilon$ is set to approach zero, so that the inner surface integral S_{in} is made arbitrarily close to the observation point I.

In summary, the observation (image) point I is excluded, but surrounded by the spherical shell of volume V with inner and outer surfaces S_{in} and S_{out}. The source point P is outside the spherical shell outer surface S_{out}. The inner surface S_{in}, which encloses the observation point I, can now be shrunk in radius, and thereby approach image point I. In the limit, the value of the inner surface integral S_{in} approaches the value of the amplitude $E_1(0)$ at the observation point I, where

$$\lim_{\varepsilon \to 0}\left[\oint_{S_{\text{in}}}\left(E_1(\varepsilon)\left(-\frac{1}{\varepsilon^2}+\frac{ik}{\varepsilon}\right)\exp(ik\varepsilon)-\frac{\exp(ik\varepsilon)}{\varepsilon}\cdot\frac{\mathrm{d}E_1}{\mathrm{d}n}(\varepsilon)\right)\varepsilon^2\,\mathrm{d}\Omega\right]\to 4\pi E_1(\varepsilon) \tag{10.11}$$

The value of the amplitude $E_1(I)$ at the observation point I can then be determined by the outer surface integral S_{out}, and this result is the Helmholtz–Kirchhoff integral theorem, where

$$E_1(I) = -\frac{1}{4\pi}\oint_{S_{\text{out}}}\left(E_1(\mathbf{r})\nabla\left(\frac{\exp(ikr_i)}{r_i}\right) - \frac{\exp(ikr_i)}{r_i}\cdot\nabla E_1(\mathbf{r})\right)\cdot\hat{\mathbf{n}}\,\mathrm{d}S_{\text{out}} \tag{10.12}$$

The spherical shell can, in fact, be a random shape in which its inner surface is brought arbitrarily close to the image point I.

10.2 Derivation of the Kirchhoff–Fresnel Diffraction Equation

The Helmholtz–Kirchhoff theorem determines the amplitude at the observation point I, via the solution of the surface integral of the amplitude and amplitude gradient over the arbitrarily shaped surface S_{out}, which encloses the observation (image) point I, but excludes the source point P. That is, one could consider the image point I at the center of a sphere, which is opaque to X-rays and neutrons except for a circular or rectangular aperture S_{a} in the spherical surface S_{out}, whose radius or rectangular dimensions are larger than the wavelength of the X-rays or neutrons. Outside of the spherical surface S_{out} is the source point P, which emits X-rays or neutrons that pass only through the aperture S_{a} and onto all points inside the sphere, including the observation (image) point I at the center.

Let us now consider the Kirchhoff boundary conditions appropriate for neutron or X-ray optics. In this case, the image or observation point I is surrounded by an opaque (neutron and X-ray absorbing) wall, which comprises surface S_{out}, except for one aperture S_{a}, where the aperture S_{a} dimensions and distance to the observation point are large compared to the neutron or X-ray wavelength and the geometric optics limit. These conditions will allow the Kirchhoff assumptions that follow to be accurate. Outside the wall enclosure S_{out} is a point source P of spherical waves whose distance to the wall and observation point is also large compared to the wavelengths.

The amplitude of the emitted spherical wave at unit distance from the source point is A. The center point L of the aperture S_a is situated at source (object) distance r_0 from the source point P, and center point L is at image distance r_i from the observation point I. In the absence of the wall, the wave amplitude $E_1(r_0)$ at the wall aperture point L is

$$E_1(r_0) = \frac{A}{r_0}\exp(ikr_0) \tag{10.13}$$

The surface S_{out} is characterized by unit normal vector $\hat{\mathbf{n}}$. The unit normal vector $\hat{\mathbf{n}}$ points outward and is perpendicular to the surface S_{out} at each of its points, including the aperture surface S_a. At a point L on the aperture surface S_a, one finds

$$\hat{\mathbf{n}}\cdot\nabla E_1 = \frac{A}{r_0}\left(ik - \frac{1}{r_0}\right)\exp(ikr_0)\cos\theta_0 \tag{10.14}$$

where

$$\hat{\mathbf{n}}\cdot\nabla\left(\frac{\exp(ikr_i)}{r_i}\right) = \frac{1}{r_i}\left(-ik + \frac{1}{r_i}\right)\exp(ikr_i)\cos\theta_i \tag{10.15}$$

Observation (image) angle θ_i is measured relative to the positive, z-direction of wave propagation, and source (object) angle θ_0 is measured relative to the opposite, negative, z-direction, see Fig. 10.1. Object (source) angle θ_0 is the angle of the source point vector, which points from aperture point L on S_a to source point P, and is measured relative to the direction of the surface normal, unit vector $\hat{\mathbf{n}}$ at L, which points toward the source (object) plane. And image angle θ_i is the angle of the observation point vector, which points from aperture point L to image point I, and is measured relative to the direction of the surface S_a normal, unit vector $\hat{\mathbf{n}}$ at L which points toward the observation (image) plane.

Other than on the aperture S_a, the wave amplitude $E_1(r)$ and its gradient $dE_1(r)/dn$ are zero on the portion of the surface S_{out} that does not contain S_a. That is, the only nonzero contribution to the surface integral of the Helmholtz–Kirchhoff equation is that portion of surface S_{out}, which contains the aperture surface S_a, so with these Kirchhoff boundary conditions, Eq. (10.12) becomes

$$E_1(I) = -\frac{A}{4\pi}\oint_{S_a}\left[\left(ik - \frac{1}{r_i}\right)\cos\theta_i + \left(ik - \frac{1}{r_0}\right)\cos\theta_0\right]\frac{\exp(ik(r_i + r_0))}{r_0 r_i}\,dS_a \tag{10.16}$$

Equation (10.16) represents only the spatial variation, but the time variation can be added by multiplying the right side by the time-varying component of source point P. For a monochromatic source of angular frequency w [rad/s] and time t [s], one multiplies the right side of Eq. (10.16) by $\exp(-iwt)$.

If the object r_0 and image r_i distances are each much larger than the wavelength λ, then the Kirchhoff–Fresnel diffraction formula is obtained.

And if the time variation of a monochromatic sinusoidal source point is included, then Eq. (10.16) becomes

$$E_1(I) = -\frac{iA}{\lambda}\exp(-iwt)\oint_{S_a} \frac{\cos\theta_i + \cos\theta_0}{2}\frac{\exp(ik(r_i + r_0))}{r_0 r_i}dS_a \quad r_0 \gg \lambda \text{ and } r_i \gg \lambda \tag{10.17}$$

If source point P has anisotropic emission of neutrons or X-rays, where amplitude coefficient A varies with emission direction, then A varies on the aperture surface S_a, and A must be set inside the integrand. In this case, A is termed the *aperture illumination function*, and one assumes that the illumination function A is uniform over the aperture.

Consider the case, where the aperture S_a is not completely transparent, and has a variable transmission $T_0(x, y)$ over its surface. The aperture surface lies in the transverse xy-plane, where wave propagation is in the axial z-direction. The aperture can be comprised of a material of varying thickness, which introduces a variable phase shift $\Delta\phi(x, y)$ in the transmitted wave. In this case, the complex transmission function $T(x, y)$ is given by

$$T(x, y) = T_0\exp(i\Delta\phi) \leq 1 \tag{10.18}$$

With these generalities, the Kirchhoff–Fresnel diffraction formula of Eq. (10.17), becomes

$$E_1(I) = -\frac{iA}{\lambda}\exp(-iwt)\oint_{S_a} \frac{\cos\theta_i + \cos\theta_0}{2}\frac{T_0\exp(i\Delta\phi)\exp(ik(r_i + r_0))}{r_0 r_i}dS_a \tag{10.19}$$

However, for subsequent discussions, the complex transmission factor of the aperture is assumed to be unity. That is, the aperture is assumed to have no material, or the attenuation of the waves by the material in the aperture is ignored, so that

$$T_0\exp(i\Delta\phi) = 1 \tag{10.20}$$

where

$$\Delta\phi(x, y) = 0 \quad \text{and} \quad T_0(x, y) = 1 \tag{10.21}$$

Coefficient $-i/\lambda$ indicates the amplitude at the image point I is inversely proportional to the neutron or X-ray wavelength λ, and the observer amplitude has a phase advance of 90° relative to the source amplitude via $-i$. Without material in the aperture, the Kirchhoff–Fresnel diffraction equation of Eq. (10.19), becomes

$$E_1(I) = -\frac{iA}{\lambda}\exp(-iwt)\oint_{S_a} \frac{\cos\theta_i + \cos\theta_0}{2}\frac{\exp(ik(r_i + r_0))}{r_0 r_i}dS_a \tag{10.22}$$

10.3 The Obliquity Factor in the Kirchhoff–Fresnel Diffraction Equation

In this section, let us examine the Kirchhoff–Fresnel diffraction equation of Eq. (10.19). The surface integral over the aperture S_a contains the obliquity or inclination factor $K(\theta_i, \theta_0)$, where

$$K = \frac{\cos\theta_i + \cos\theta_0}{2} \tag{10.23}$$

In the free propagation of a spherical wave, which originates from a source point P, the Huygens–Fresnel principle states that every point on a wave front becomes an emitter of another spherical wave. One can calculate the contribution of each point on the primary wave surface W to an observer point I. However, no wave moving backward toward the source point P is observed in experimental conditions. Hence, the secondary wave, which is produced at each point on the primary spherical wave surface, is observed to be directed away from the source point P.

Fresnel added the principle the obliquity or inclination factor $K(\theta)$ to each point W on the primary wave surface. The polar angle θ is measured relative to the normal unit vector of the primary spherical wave surface at point W. At each point W on the wave surface is a unit normal vector, which defines the direction of the ray vector at that tiny portion of the wave surface. The unit normal vector is the X-ray or neutron ray at point W on the wave front. In the Huygens–Fresnel formulation, the inclination factor $K(\theta)$ is assumed in order to account for lack of backward waves:

$$K(\theta) = \frac{1 + \cos\theta}{2} \tag{10.24}$$

That is, no backward waves are observed, and the inclination factor $K(\theta)$ mathematically enforces zero amplitude for backward waves.

However, the obliquity factor $K(\theta_i, \theta_0)$ arises naturally in this derivation of the Kirchhoff–Fresnel diffraction formula by Kirchhoff. Recall, as shown in Fig. 10.1, the observation (image) angle θ_i is measured relative to the positive, z-direction of wave propagation, and source (object) angle θ_0 is measured relative to the opposite, negative, z-direction. The obliquity factor $K(\theta_i, \theta_0)$ is at its maximum in the forward direction of the wave, away from the wave source point, and in the case of the wave at aperture point L, the observation (image) angle θ_i toward the image point I is zero and the source (object) angle θ_0 toward the source (object) point P is zero:

$$\theta_i = 0 \quad \text{and} \quad \theta_0 = 0 \tag{10.25}$$

where

$$K = \frac{\cos\theta_i + \cos\theta_0}{2} = \frac{1+1}{2} = 1 \tag{10.26}$$

The obliquity factor $K(\theta_i, \theta_0)$ is minimum or zero in the backward direction from the aperture point L toward the source (object) point P. That is, the observation (image) angle θ_i toward the source (object) point P is π, and the source (object) angle θ_0 toward the source (object) point P is zero:

$$\theta_i = \pi \quad \text{and} \quad \theta_0 = 0 \tag{10.27}$$

where

$$K = \frac{\cos\theta_i + \cos\theta_0}{2} = \frac{-1+1}{2} = 0 \tag{10.28}$$

In the Kirchhoff-Fresnel diffraction formula, the symmetry of (1) the image distance and its image angle, compared to (2) the source (object) distance and its source angle, implies the reciprocity theorem. In the reciprocity theorem, (1) the wave amplitude received at source point P by wave emission from observation point I, is equal to (2) the wave amplitude received at observation point I by wave emission from source point P. Another implied theorem is Babinet's principle, where one assumes the enclosure surface is completely transparent, and then considers the amplitude at observation point I, produced by source point P. Babinet's principle asserts the amplitude at observation point I, is equal to the sum of (1) the amplitude contribution from an opaque enclosure surface, which has transparent aperture surface, and (2) the amplitude from the complementary case of a transparent enclosure surface, which is opaque at aperture surface.

10.4 The Paraxial Approximation Applied to the Kirchhoff–Fresnel Diffraction Equation

In this section, the paraxial approximation is applied to the Kirchhoff–Fresnel equation. The condition for the paraxial approximation is applied to an aperture S_a in an aperture plane, which is between a distant source or object plane and a distant observation plane. The aperture plane consists of a thin wall, with a center aperture, where the wall does not transmit X-rays or neutrons. X-ray or neutrons can be transmitted only through the aperture S_a from a source point P to an observation point I. If the largest lateral dimension $l = L_A$ of the aperture S_a is much smaller than the distance between a point L in the aperture S_a and an isotropic source (object) point P, and if the largest lateral aperture dimension l is much smaller than the distance from the aperture point L to observation (image) point I, then one can invoke the paraxial approximation (Figure 10.1).

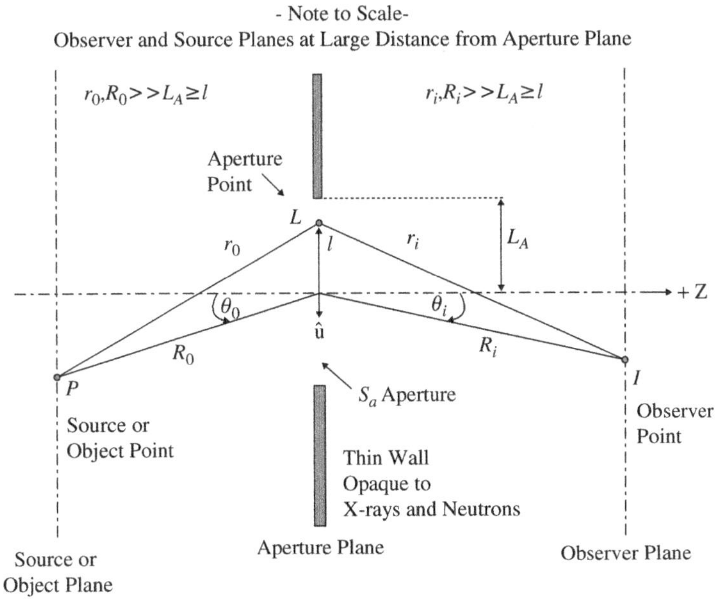

Figure 10.1 Diagram of an isotropic source or object point P, which emits X-ray or thermal neutrons that propagate to point L in an aperture. The aperture is centered in a thin, ideal wall, which absorbs X-rays and neutrons. The X-rays or neutrons are diffracted in the aperture and propagate to an observer point I in the observation plane. Source or object point P is at source or object angle θ_0, which is positive for a counterclockwise rotation, relative to the negative z-axis. Observer point I is at observer angle θ_i, which is positive for a clockwise rotation, relative to the positive z-axis. In the plane of the aperture S_a, the radial distance of point L from the center z-axis is l (or u). Vector **u**, with unit vector **û**, lies in the aperture plane, and points from the center z-axis to point L, where $\mathbf{u} = u\hat{\mathbf{u}} = l\hat{\mathbf{u}}$.

In the paraxial approximation, the source (object) distance r_0 and the image distance r_i are constant over the aperture S_a. The transverse distance of the source (object) point P from the center z-axis, which passes through the center of aperture S_a is assumed to be much smaller than the axial distance between the source plane at point P and the aperture plane at point L. And the axial distance between the aperture plane at point L and the observer (image) plane at point I, are assumed to be much larger than the transverse distance of the observer (image) point I from the center z-axis.

As a result, with the paraxial approximation, $1/r_0 r_i$ and the obliquity factor $K(\theta)$ can be taken outside the integral, and since angle θ is sufficiently small,

$$\cos\theta \cong 1 \quad \theta \ll 1 \tag{10.29}$$

Hence, the obliquity factor is unity in the paraxial case, where

$$K(\theta) = \frac{1+\cos\theta}{2} \cong 1 \quad \theta \ll 1 \tag{10.30}$$

For isotropic source point P and 100% aperture S_a transmission, the Kirchhoff–Fresnel diffraction formula of Eq. (10.22) for the paraxial case, becomes

$$E_1(I) = -\frac{iA_I}{\lambda}\cdot\frac{\exp(-iwt)}{r_0 r_i}\oint_{S_a}\exp\left[i2\pi\left(\frac{r_i + r_0}{\lambda}\right)\right]dS_a \tag{10.31}$$

Recall that A is the amplitude of the incident X-ray or neutron wave at aperture S_a, which is assumed to be uniform over the aperture S_a.

Even a small variation in the object (source) distance r_0 and/or the image (observer) distance r_i can lead to a large variation of the exponential in the integrand. Assume a fixed source position vector $\mathbf{R}_0$ from the aperture S_a center point $L = 0$ to source point P. With unit vector $\hat{\mathbf{R}}_0$, the source (object) central vector $\mathbf{R}_0$ points from center aperture point $L = 0$ to source point P, where

$$\mathbf{R}_0 = R_0\hat{\mathbf{R}}_0 \tag{10.32}$$

Assume a fixed image position vector $\mathbf{R}_i$ from aperture S_a center point $L = 0$ to the observation (image) point I. With unit vector $\hat{\mathbf{R}}_i$, the observation (image) central vector $\mathbf{R}_i$ points from aperture center point $L = 0$ to the observation (image) point I, where

$$\mathbf{R}_i = R_i\hat{\mathbf{R}}_i \tag{10.33}$$

The aperture S_a is assumed to be in a transverse xy-plane perpendicular to the center, axial z-axis. Within the xy-plane of aperture S_a, the radial vector $\mathbf{u}$ extends from the aperture center point $L = 0$ to the off-axis, aperture point L. The radial vector position $\mathbf{u}$ within the xy-plane of the aperture S_a, which is measured relative to the aperture center at $L = 0$, is

$$\mathbf{u} = u\hat{\mathbf{u}} = u_x\hat{\mathbf{x}} + u_y\hat{\mathbf{y}} \tag{10.34}$$

The radial distance of off-axis point L from the z-axis is u or l, where u = l, as shown in Fig. 10.1. The variable, source position vector $\mathbf{r}_0$ points from the off-axis point L within the aperture plane S_a to the source point P. With unit vector $\hat{\mathbf{r}}_0$, the source (object) vector $\mathbf{r}_0$ from off-axis point L within the aperture plane S_a to source point P is approximated by

$$r_0 = r_0\hat{\mathbf{r}}_0 = R_0 - \left(\hat{\mathbf{R}}_0\cdot\mathbf{u}\right) \tag{10.35}$$

The variable image position vector $\mathbf{r}_i$ points from the off-axis point L within the aperture plane S_a to the observer (image) point I. With unit vector $\hat{\mathbf{r}}_i$, the observation (image) vector $\mathbf{r}_i$ from off-axis point L within the aperture plane S_a to image point I is approximated by

$$r_i = r_i\hat{\mathbf{r}}_i = R_i - \left(\hat{\mathbf{R}}_i\cdot\mathbf{u}\right) \tag{10.36}$$

The center or optical axis is the z-axis, and X-rays or neutrons propagate from the source point P in the $z = 0$ source plane, through point L in the aperture S_a at $z = z_0$ in the aperture, to the observer (image) point in the observer (image) plane at $z = z_i$.

The neutron or X-ray wave vector $\mathbf{k}_0$ incident at the aperture S_a from the source point P to any point L on the aperture can be expressed as

$$\mathbf{k}_0 = k\hat{\mathbf{r}}_0 = \frac{2\pi}{\lambda}\hat{\mathbf{r}}_0 \tag{10.37}$$

The wave vector $\mathbf{k}_i$ that exits the aperture S_a from any aperture point L and terminates on the observation point I can be expressed by

$$\mathbf{k}_i = k\hat{\mathbf{r}}_i = \frac{2\pi}{\lambda}\hat{\mathbf{r}}_i \tag{10.38}$$

At aperture S_a, the wave function changes direction but maintains its wave vector magnitude k.

At aperture S_a, the change $\mathbf{q}$ in the neutron or X-ray wave vector from incident wave vector $\mathbf{k}_0$ to diffracted wave vector $\mathbf{k}$ at each point L on S_a is the same:

$$\mathbf{q} = k(\hat{\mathbf{r}}_0 - \hat{\mathbf{r}}_i) = \frac{2\pi}{\lambda}(\hat{\mathbf{r}}_0 - \hat{\mathbf{r}}_i) \tag{10.39}$$

The scattering vector is

$$\boldsymbol{\kappa} \equiv \mathbf{q} = \mathbf{k}_0 - \mathbf{k} \tag{10.40}$$

The neutron or X-ray scattering vector $\mathbf{q}$ is also denoted by $\boldsymbol{\kappa}$. For incident X-ray or neutron wave vector $\mathbf{k}_0$, the magnitude k of the X-ray or neutron wave vector remains constant for diffraction by aperture S_a. The diffraction of X-rays or neutrons is a form of coherent, elastic scatter. However, the diffracted X-ray or neutron wave vector $\mathbf{k}$ changes direction relative to the incident wave vector $\mathbf{k}_0$ direction by polar angle θ, which assumes azimuth symmetry in the elastic scatter process, where

$$|\boldsymbol{\kappa}| \equiv |\mathbf{q}| = 2k\sin\frac{\theta}{2} \tag{10.41}$$

In terms of the x- and y-directions, the change in the neutron or X-ray wave vector at each point L in aperture plane S_a, or scattering vector, is

$$\mathbf{q} = q_x\hat{\mathbf{x}} + q_y\hat{\mathbf{y}} \tag{10.42}$$

The phase shift χ of the neutron or X-ray, on passage from the source point P to a point L on the aperture S_a and then to the observation point I, is the dot product of the scattering vector $\mathbf{k}_0 - \mathbf{k}$ (i.e., $\mathbf{q}$) with the radial vector $\mathbf{u} = u\hat{\mathbf{u}}$ within the aperture plane, where

$$\chi = \mathbf{u} \cdot \mathbf{q} = \mathbf{u} \cdot (\mathbf{k}_0 - \mathbf{k}) \tag{10.43}$$

where

$$\chi = u_x q_x + u_y q_y \tag{10.44}$$

As before and in Fig. 10.1, the source (object) angle θ_0 of the incident ray from the source point P to the center point $L = 0$ of aperture S_a in the xy-plane is measured relative to the normal at $L = 0$, which points along the negative z-axis toward the source (object) plane with point P. A counterclockwise source (object) angle θ_0, measured relative to the negative z-axis, is positive, and a clockwise angle θ_0 is negative.

Source (object) vector $\mathbf{r}_0$ connects the generally off-axis point L on the aperture S_a to source (object) point P, which is also generally off-axis. Since the aperture S_a is small, one approximates the different source (object) angles for each point L on the aperture by a single source (object) angle θ_0, which belongs to central vector $\mathbf{R}_0$ pointing from aperture center point $L = 0$ to source (object) point P. Measured from the center z-axis, the off-axis radial (perpendicular component) distance $R_0^{\perp}$ of source (object) point P, is much smaller than the total or "diagonal" distance $|\mathbf{R}_0|$ between source (object) point P and center aperture point $L = 0$, where

$$\sin\theta_0 = \frac{R_0^{\perp}}{|\mathbf{R}_0|} \tag{10.45}$$

and

$$R_0^{\perp} \ll |\mathbf{R}_0| \tag{10.46}$$

Hence,

$$\sin\theta_0 \cong \theta_0 \tag{10.47}$$

The transverse dimensions of the source and aperture are assumed to be much smaller than the axial separation between the source and aperture planes. Thus, the distance R_0 between the source (object) point P and the center point $L = 0$ on aperture S_a, as well as the distance r_0 between the source (object) point P and the off-axis point L on aperture S_a, can each be approximated by the axial distance z_0 between the source (object) plane and the aperture plane, where

$$r_0 \cong R_0 \cong z_0 \tag{10.48}$$

The observer (image) angle θ_i of the incident ray from the center point $L = 0$ of aperture S_a in the xy-plane to the observer (image) point I is measured relative to the normal, at $L = 0$. For image angle θ_i the normal at $L = 0$ points along the positive z-axis toward the observer (image) plane with point I. A clockwise image angle θ_i relative to the positive z-axis is positive, and a counterclockwise angle θ_i is negative.

Observer (image) vector $\mathbf{r}_i$ connects observer (image) point I, which is generally off the z-axis, to a point L on the aperture S_a, which is also generally off-axis. Since the aperture S_a is small, one approximates the different observer (image) angles for each point L on the aperture by a single observer (image) angle θ_i, which belongs to central vector $\mathbf{R}_i$ pointing from aperture center point $L = 0$ to observer (image) point I. Measured form the center z-axis, the off-axis radial (perpendicular component) distance $R_i^{\perp}$ of observer (image) point I, is much smaller than the total of "diagonal" distance $|\mathbf{R}_i|$ between observer (image) point I and center aperture point $L = 0$, where

$$\sin\theta_i = \frac{R_i^{\perp}}{|\mathbf{R}_i|} \tag{10.49}$$

where

$$R_i^{\perp} \ll |\mathbf{R}_i| \tag{10.50}$$

Hence,

$$\sin\theta_i \cong \theta_i \tag{10.51}$$

Plane waves are assumed to arrive at the aperture from the distant source point P, and plane waves are assumed to arrive at the image point I from the distant aperture. The transverse dimensions of the image and aperture are assumed to be much smaller than the axial separation between the image and aperture planes. Thus, the distance R_i between the observer (image) point I and the center point $L = 0$ on aperture S_a, and the distance r_i between the source (object) point P and the off-axis point L on aperture S_a, can each be approximated by the axial distance z_i between the observer (image) plane and the aperture plane, where

$$r_i \cong R_i \cong z_i \tag{10.52}$$

The polar angle θ is the sum of the small source (object) angle θ_0 and the small observer (image) angle θ_i. As shown in Fig. 10.1, positive source angles θ_0 are measured counterclockwise from the negative z-axis, and positive image angles θ_i are measured clockwise from the positive z-axis. Thus, adding small source angle θ_0 and small image angle θ_i, actually yields their difference θ, where polar angle θ is thus also small, and

$$\theta = \theta_i + \theta_0 \ll 1 \tag{10.53}$$

Thus, in spite of the confusing sum in Eq. (10.53), the polar angle θ is the change of the ray angle from θ_0 to θ_i (i.e. difference θ), as the ray propagates from source (object) point P, through center aperture point L, and terminates on observation (image) point I. For example, in a straight-line ray path from source point P, through the aperture center point L to the image point I, the ray has no change in angle, so that $\theta = 0$, where source angle $\theta_0 = \theta_{\text{lin}}$ and image angle $\theta_i = -\theta_{\text{lin}}$. The change of the neutron or X-ray wave vector in the x-direction q_x and the y-direction q_y, on passage through the xy-plane rectangular aperture S_a, is given by

$$q_x = 2k\sin\left(\frac{\theta}{2}\right)\cos\phi \tag{10.54}$$

and

$$q_y = 2k\sin\left(\frac{\theta}{2}\right)\sin\phi \tag{10.55}$$

The azimuth angle ϕ lies in the transverse xy-plane of the aperture S_a, and is measured from the transverse x-axis. Since angle θ is small, the changes of the neutron or X-ray wave vector in the x and y directions at the aperture S_a are

$$q_x = k(\theta_0 + \theta_i)\cos\phi \tag{10.56}$$

and

$$q_y = k(\theta_0 + \theta_i)\sin\phi \tag{10.57}$$

At aperture S_a, the deviation angles θ_x in the xz-plane and θ_y in the yz-plane of diffracted wave vector (angle θ_i) at the aperture S_a are

$$\theta_x = (\theta_0 + \theta_i)\cos\phi \quad \theta_0 \ll 1 \text{ and } \theta_i \ll 1 \tag{10.58}$$

and

$$\theta_y = (\theta_0 + \theta_i)\sin\phi \quad \theta_0 \ll 1 \text{ and } \theta_i \ll 1 \tag{10.59}$$

Hence,

$$q_x = k\theta_x \quad \theta_0 \ll 1 \text{ and } \theta_i \ll 1 \tag{10.60}$$

and

$$q_y = k\theta_y \quad \theta_0 \ll 1 \text{ and } \theta_i \ll 1 \tag{10.61}$$

The phase shift χ of the neutron or X-ray wave from the source point through an aperture point L to the observation point I is

$$\chi = \mathbf{u} \cdot \mathbf{q} = u_x k\theta_x + u_y k\theta_y = u_x q_x + u_y q_y \tag{10.62}$$

The neutron or X-ray wave amplitude A_s at the source point P changes in its trajectory through aperture S_a to observation point I. At image point I, the amplitude A_I is proportional to exp(iχ), where χ is the phase shift that the wave has incurred in its trajectory from source point P to observation point I. At the image point I, the amplitude A_I of Eq. (10.31) is then

$$A_I = A_S \exp(\mathrm{i}\chi)\exp(-\mathrm{i}wt) \tag{10.63}$$

The observation point amplitude A_I of Eq. (10.31) in rectangular coordinates is then

$$A_I = A_S\exp[\mathrm{i}(u_x q_x + u_y q_y)]\exp(-\mathrm{i}wt) \tag{10.64}$$

10.5 Fraunhofer Diffraction of X-rays or Neutrons from a Rectangular Aperture

In this section, we examine the diffraction of X-rays and neutrons from a rectangular aperture. Consider a rectangular aperture S_a whose extent in the x- and y-directions are $[-x_\mathrm{a}, x_\mathrm{a}]$ and $[-y_\mathrm{a}, y_\mathrm{a}]$, respectively, whose differential areas are calculated by

$$\mathrm{d}S_\mathrm{a} = \mathrm{d}u_x\, \mathrm{d}u_y \tag{10.65}$$

As a result of Eq. (10.65), one finds the amplitude A_I at the image point I for the paraxial approximation, becomes the x- and y-direction Fourier transforms of the xy-plane aperture, and Eq. (10.31) with amplitude A_I of Eq. (10.64), becomes

$$E_1(I) = -\frac{\mathrm{i}A_S\exp(-\mathrm{i}wt)}{\lambda}\cdot\frac{\exp(\mathrm{i}k(R_i + R_0))}{r_0 r_i}\int_{-x_\mathrm{a}}^{x_\mathrm{a}}\exp(\mathrm{i}u_x q_x)\mathrm{d}u_x\int_{-y_\mathrm{a}}^{y_\mathrm{a}}\exp(\mathrm{i}u_y q_y)\mathrm{d}u_y \tag{10.66}$$

Here, A_S is the magnitude of the source point P amplitude, where

$$A_S = |A_0| \tag{10.67}$$

Integration yields the amplitude at the observer point I, where Eq. (10.66) becomes

$$E_1(I) = -\frac{\mathrm{i}A_S\exp(-\mathrm{i}wt)}{\lambda}\cdot\frac{\exp(\mathrm{i}k(R_i + R_0))}{r_0 r_i}\cdot\frac{2\sin(x_\mathrm{a}k\theta_x)}{k\theta_x}\cdot\frac{2\sin(y_\mathrm{a}k\theta_y)}{k\theta_y} \tag{10.68}$$

The intensity $I_1(I)$ at the observer point I is the squared magnitude of the amplitude $E_1(I)$ of Eq. (10.68). The far-field intensity $I_1(I)$ from the paraxial approximation is the Fraunhofer diffraction pattern at observation (image) plane I, which occurs as a result of diffraction by a rectangular aperture S_a with x and y widths $2x_\mathrm{a}$ and $2y_\mathrm{a}$, where

$$I_1(I) = |E_1(I)|^2 = \left(\frac{2A_S}{r_0 r_i \lambda}\right)^2\cdot\frac{\sin^2(x_\mathrm{a}k\theta_x)}{k^2\theta_x^2}\cdot\frac{\sin^2(y_\mathrm{a}k\theta_y)}{k^2\theta_y^2} \tag{10.69}$$

Multiplying the numerator and denominator by $x_a^2 y_a^2$ of the intensity $I_1(I)$, yields the expected Fraunhofer far-field diffraction pattern for a rectangular aperture. The far-field diffraction has x-direction and y-direction dependencies is given by

$$\frac{\sin^2(\alpha)}{\alpha^2} \quad \text{and} \quad \frac{\sin^2(\beta)}{\beta^2} \tag{10.70}$$

where

$$\alpha = x_a k\theta_x \quad \text{and} \quad \beta = y_a k\theta_y \tag{10.71}$$

The intensity $I_1(I)$ in the far-field diffraction pattern at image plane I due to rectangular aperture S_a is

$$I_1(I) = |E_1(I)|^2 = (x_a y_a)^2 \left(\frac{2A_S}{r_0 r_i \lambda}\right)^2 \cdot \frac{\sin^2(x_a k\theta_x)}{x_a^2 k^2 \theta_x^2} \cdot \frac{\sin^2(y_a k\theta_y)}{y_a^2 k^2 \theta_y^2} \tag{10.72}$$

At image plane I, the far-field or Fraunhofer diffraction of the aperture S_a can also be generated by a focusing lens, which is placed between the source point P and aperture S_a. The lens receives the diverging rays of a nearby point source P and refracts them to near parallel, an effect that is normally produced by a distant source point P. These near-parallel rays are then incident on the aperture S_a, where they are transmitted and refracted, and they produce a diffraction pattern or Fourier transform of the aperture S_a at a distant detector at observer plane I. With a nearby point source P, the focusing lens thus mimics the effect of a faraway point source P, as seen by an observer in image plane I.

10.6 Fraunhofer Diffraction of X-ray or Neutron Line Source by a Parallel Single Slit

Consider the 1-D case of a horizontal-line X-ray or neutron source P, which illuminates a horizontal slit that acts as the aperture S_a. The slit is narrow in the x-direction and wide in the y-direction. One can modify the point source P illumination of a rectangular aperture S_a and the amplitude solution at a distant observer point I in Eq. (10.72) to obtain the far-field diffraction pattern or intensity at the observer point I, which is caused by a line source P illumination of the slit. The points on the line source P are assumed to emit independently of one another, so that there is no coherence of amplitudes at the observation point I. Thus, to obtain the intensity at the observation point I, by superposition, one sums the intensity contribution to the observation point I from all the points on the line source P. A sufficiently wide slit and wide line source in the y-direction allows integration in θ_y to be approximated by the range from $-\pi/2$ to $\pi/2$:

$$I_1(I) = |E_1(I)|^2 = (x_a y_a)^2 \left(\frac{2A_S}{r_0 r_i \lambda}\right)^2 \cdot \frac{\sin^2(x_a k\theta_x)}{x_a^2 k^2 \theta_x^2} \int_{-\pi/2}^{\pi/2} \frac{\sin^2(y_a k\theta_y)}{y_a^2 k^2 \theta_y^2} \mathrm{d}\theta_y \tag{10.73}$$

With a change in variables, the integral is transformed to

$$\int_{-\infty}^{\infty} \frac{\sin^2(\beta)}{\beta^2} \mathrm{d}\beta = \pi \tag{10.74}$$

where

$$\beta = y_a k\theta_y \quad \text{and} \quad \mathrm{d}\beta = y_a k \, \mathrm{d}\theta_y \tag{10.75}$$

Integration over the points of an "infinite," incoherent X-ray or neutron line source via integration of β over $[-\infty, \infty]$ yields the far-field intensity at observation point I due to an incoherent line X-ray or neutron source. The result is the sinc function diffraction pattern of the Fraunhofer slit, where

$$I_1(I) = x_a^2 \pi \left(\frac{2A_S}{r_0 r_i \lambda}\right)^2 \cdot \frac{\sin^2(x_a k\theta_x)}{x_a^2 k^2 \theta_x^2} \tag{10.76}$$

The angular width $\Delta\theta_x$ of the central diffraction peak is found by setting

$$x_a k\theta_x = \pm \pi \tag{10.77}$$

and then solving θ_x for π and $-\pi$ to find

$$\theta_x = \pm \frac{\lambda}{2x_a} \tag{10.78}$$

Two point sources of the same amplitude are resolved by the Rayleigh criteria when their angular separation is

$$\Delta\theta_x = \frac{\lambda}{x_a} \tag{10.79}$$

The results of the diffraction by a single slit give insight into the resolution limits of a lens or CRL. The aperture dimensions of a CRL limits its resolution due to diffraction. The resolution of a parabolic lens CRL, or a spherical lens CRL, with an aperture stop set to its parabolic aperture radius r_p to eliminate spherical aberration, is determined by the angular width $\Delta\theta$ of the center diffraction peak that is produced by the lens aperture, where

$$\Delta\theta = \frac{\lambda}{r_p} \tag{10.80}$$

At object distance r_0 from the center of the CRL, and at wavelength λ, the Rayleigh-limited or diffraction-limited object resolution σ_{obj} of the CRL is

$$\sigma_{obj} = 0.61 \cdot r_0 \cdot \frac{\lambda}{r_p} \tag{10.81}$$

The image resolution σ_{im} (with image distance r_i) of the CRL is

$$\sigma_{im} = 0.61 \cdot r_i \cdot \frac{\lambda}{r_p} \tag{10.82}$$

The object resolution is also found by dividing the image resolution by the transverse magnification M_T, which is given by image distance r_i divided by object distance r_0, where

$$M_T = \frac{r_i}{r_0} \tag{10.83}$$

The Rayleigh-limited resolution allows definition of the depth of focus of the CRL of aperture radius r_m. The CRL depth of focus Δf, about focal length f, is the axial extent that the detector can be dithered about the image plane, corresponding to a fixed object plane, such that the blur diameter of image points is within twice the Rayleigh-limited resolution. The numerical aperture NA is defined by the ratio of the CRL aperture radius r_m divided by the CRL focal length f, where

$$\mathrm{NA} = \frac{r_m}{f} \tag{10.84}$$

As a result, depth of focus Δf is

$$\Delta f = 0.61 \frac{\lambda f^2}{r_m^2} = \frac{0.61\lambda}{\mathrm{NA}^2} \tag{10.85}$$

The depth of field (DOF) is related to the depth of focus Δf, via the CRL axial magnification M_A given by

$$M_A = -M_T^2 \tag{10.86}$$

Expressed in terms of the transverse magnification M_T, the depth of field (DOF) equals

$$\mathrm{DOF} = \frac{\Delta f}{M_A} = \frac{0.61\lambda}{M_T^2 \mathrm{NA}^2} \tag{10.87}$$

Image blur occurs in spherical aberration, coma, and astigmatism, whereas curvature of field and distortion cause shifting of the image, but no blurring. For monochromatic aberrations, the CRL pipelike field of view (FOV) is limited to primarily paraxial rays. Aberrations arising from sufficiently off-axis marginal rays lead to astigmatism, coma, distortion, and curvature of field occur in spherical and parabolic lens CRLs. Lateral and transverse spherical aberrations occur in spherical lens CRLs, but not in parabolic lens CRLs. For a small-aperture diameter and long CRLs, only paraxial rays are transmitted, and thus the off-axis ray aberrations (astigmatism, coma, distortion, and curvature of field) are minor. Both the parabolic and spherical lens CRLs have chromatic aberration.

10.7 Fraunhofer Diffraction of X-ray or Neutron Line Source by a Parallel Slit Pair

Next, consider the 1-D case of a horizontal-line X-ray or neutron source illumination of double horizontal slits, which are narrow in the vertical x-direction, with width $2a$, and wide in the y-direction, with slit center to slit center spaced $2b$ apart in the x-direction. The origin of the vertical x-axis is centered between the two slits, so that the center of the upper slit is distance b above the origin and the center of the lower slit is distance b below the origin. The source angle θ_0 from the slits S_a to source point P, and the image angle θ_i from the slits S_a to the observation (image) point I, are each measured relative to the normal, unit vector to the plane S_a of the slits.

With u_x as the coordinate in the x-direction in the plane of the two slits, angle θ_x in the xz-plane is given by

$$\theta_x = \theta_0 + \theta_i \quad \theta_0 \ll 1 \text{ and } \theta_i \ll 1 \tag{10.88}$$

and

$$q_x = k\theta_x \quad \theta_0 \ll 1 \text{ and } \theta_i \ll 1 \tag{10.89}$$

The center or optical axis is the z-axis as always. The distance of the source (object) point P and observation (image) point from the two-slit aperture S_a is assumed to be much larger than the transverse distance of the image point I from the center z-axis. As a result, the source angle θ_0 and image angle θ_i are each sufficiently small, so that $\sin\theta_0$ can be replaced by θ_0, and $\sin\theta_i$ can be replaced by θ_i.

One now modifies Eq. (10.66) for the case of 1-D diffraction by a pair of slits, which are narrow in the x-direction, and broad in the y-direction, so the y-directed diffraction is ignored. The amplitude $E_1(I)$ at the observer point for the paraxial approximation is then

$$E_1(I) = -\frac{\mathrm{i}A_S \exp(-\mathrm{i}wt)}{\lambda} \cdot \frac{\exp(\mathrm{i}k(R_i + R_0))}{r_0 r_i} \left(\int_{-b-a}^{-b+a} \exp(\mathrm{i}u_x q_x) \mathrm{d}u_x + \int_{b-a}^{b+a} \exp(\mathrm{i}u_x q_x) \mathrm{d}u_x \right) \tag{10.90}$$

Integration yields the amplitude. The squared magnitude of the amplitude yields intensity $I_1(I)$ at the observer point I due to the two slits, where

$$I_1(I) = \left(\frac{2aA_S}{r_0 r_i \lambda} \right)^2 \cdot \frac{\sin^2(q_x a)}{(q_x a)^2} \cdot \cos^2(q_x b) \tag{10.91}$$

The diffraction pattern of two slits of width a and separated by interval b, is the single-slit diffraction pattern of a squared function $\mathrm{sinc}^2(\alpha)$, with $\alpha = q_x a$, and modulated by the squared cosine dependece $\cos^2(\beta)$ with $\beta = q_x b$ for the slit spacing b.

The angular width $\Delta\theta_x$ of the wide center diffraction peak produced by each slit is found by setting

$$\alpha = q_x a = k\theta_x a = \frac{2\pi}{\lambda} \theta_x a = \pm \pi \tag{10.92}$$

One solves θ_x for π and $-\pi$ to find

$$\theta_x = \pm \frac{\lambda}{2a} \tag{10.93}$$

Thus, the angular width $\Delta\theta_x$ of the wide center diffraction peak produced by each slit is

$$\Delta\theta_x = \frac{\lambda}{a} \tag{10.94}$$

The angular width $\Delta\theta_{xm}$ of the narrow modulation peaks, which is caused by interference of the two slits, is found by setting

$$\beta = q_x b = k\theta_x b = \frac{2\pi}{\lambda} \theta_x b = \pm \frac{\pi}{2} \tag{10.95}$$

One solves for θ_{xm} for $\pi/2$ and $-\pi/2$ to find

$$\theta_{xm} = \pm \frac{\lambda}{2b} \tag{10.96}$$

The angular width $\Delta\theta_{xm}$ of the narrow modulation peaks is then

$$\Delta\theta_{xm} = \frac{\lambda}{b} \tag{10.97}$$

10.8 Fraunhofer Diffraction of an X-ray or a Neutron Line Source from N Parallel Slits

The Fraunhofer diffraction of an X-ray or neutron horizontal-line source from a vertical array of horizontal slits is now considered. The aperture S_a is comprised of a vertical (x-direction) array of N horizontal slits. Each horizontal slit of has a narrow vertical width $2a$ and uniform vertical spacing of $2b$ between slit centers. Let the origin of the x-axis be at the center of the bottom slit of the N-slit vertical array. The far-field or Fraunhofer diffraction of a horizontal-line source of neutrons or X-rays by the N-slit is a precursor to the Fraunhofer diffraction from a vertical diffraction grating.

The diffraction grating can represent an array of neutron or X-ray scattering atoms or molecules. For N multiple slits, the intensity of the diffracted neutron or X-ray wave is the product of a diffraction factor, which is a function of the width $2a$ of the individual slits, multiplied by the interference factor, which is a function of the uniform slit spacing $2b$ and the number of slits N. The diffraction and interference factors each depend on the angle of neutron or X-ray incidence angle θ_0 on the array and the angle of diffraction θ_i, each measured relative to the normal array.

Recall that the change of the neutron or X-ray wave amplitude from the source (object) point P through any aperture point L on S_a to the observation (image) point I, is proportional to the exponential factor whose argument is the phase shift χ, where

$$\exp(\mathrm{i}\chi) \tag{10.98}$$

where

$$\chi = \mathbf{u} \cdot \mathbf{q} = \mathbf{u} \cdot (\mathbf{k}_0 - \mathbf{k}) \tag{10.99}$$

or

$$\chi = u_x k\theta_x \quad \text{and} \quad q_x = k\theta_x \tag{10.100}$$

The image point I amplitude $E_1(I)$ for the paraxial approximation is then a modification of Eq. (10.66), where

$$E_1(I) = -\frac{\mathrm{i}A_S \exp(-\,\mathrm{i}wt)}{\lambda} \cdot \frac{\exp(\mathrm{i}k(R_i + R_0))}{r_0 r_i} \sum_{n=0}^{N-1} \left(\int_{-a+nb}^{a+nb} \exp(\mathrm{i}u_x q_x)\mathrm{d}u_x \right) \tag{10.101}$$

In Eq. (10.101), one performs a substitution of variables with

$$u'_x = u_x - nb \tag{10.102}$$

With substitution of Eq. (10.102) into Eq. (10.101), the image point I amplitude becomes:

$$E_1(I) = -\frac{\mathrm{i}A_S \exp(-\mathrm{i}wt)}{\lambda} \cdot \frac{\exp(\mathrm{i}k(R_i + R_0))}{r_0 r_i} \left(\int_{-a}^{a} \exp(\mathrm{i}u'_x q_x)\mathrm{d}u'_x \right) \sum_{n=0}^{N-1} \exp(\mathrm{i}nbq_x) \tag{10.103}$$

Performing the integration in Eq. (10.103), and substitution of q_x by $k\theta_x$ from Eq. (10.100), yields

$$E_1(I) = -2a\frac{\mathrm{i}A_S \exp(-\mathrm{i}wt)}{\lambda} \cdot \frac{\exp(\mathrm{i}k(R_i + R_0))}{r_0 r_i} \frac{\sin(k\theta_x a)}{k\theta_x a} \sum_{n=0}^{N-1} \exp(\mathrm{i}n2bk\theta_x) \tag{10.104}$$

Let

$$p = \mathrm{i}2bk\theta_x \tag{10.105}$$

In Eq. (10.104), one recognizes the geometric progression

$$\sum_{n=0}^{N-1} \exp(pn) = \frac{p^N - 1}{p - 1} \tag{10.106}$$

With substitution of Eq. (10.106) into Eq. (10.104), the image point I amplitude is then

$$E_1(I) = -2a\frac{\mathrm{i}A_S \exp(-\mathrm{i}wt)}{\lambda} \cdot \frac{\exp(\mathrm{i}k(R_i + R_0))}{r_0 r_i} \frac{\sin(k\theta_x a)}{k\theta_x a} \cdot \frac{\exp(\mathrm{i}N2bk\theta_x) - 1}{\exp(\mathrm{i}2bk\theta_x) - 1} \tag{10.107}$$

Factoring the exponential ratio into the last term of Eq. (10.107), leads to a more convenient image point I amplitude $E_1(I)$ expression:

$$E_1(I) = -2a \cdot \exp(\mathrm{i}bk\theta_x(N-1)) \cdot \frac{\mathrm{i}A_S \exp(-\mathrm{i}wt)}{\lambda} \cdot \frac{\exp(\mathrm{i}k(R_i + R_0))}{r_0 r_i} \cdot \frac{\sin(k\theta_x a)}{k\theta_x a} \cdot \frac{\sin(Nbk\theta_x)}{N \sin(bk\theta_x)} \tag{10.108}$$

For N parallel slits, the intensity $I_1(I)$ at the observation (image) point I, is the squared magnitude of the amplitude of Eq. (10.108):

$$I_1(I) = \left(\frac{2aNA_S}{r_0 r_i \lambda}\right)^2 \cdot \left(\frac{\sin(k\theta_x a)}{k\theta_x a}\right)^2 \cdot \left(\frac{\sin(Nbk\theta_x)}{N \sin(bk\theta_x)}\right)^2 \tag{10.109}$$

The numerator and denominator of Eq. (10.108) were multiplied by N^2, so the N-slit interference factor can be unity when $\theta_x = 0$ or

$$\lim_{bk\theta_x \to 0} \left(\frac{\sin(Nbk\theta_x)}{N \sin(bk\theta_x)} \right)^2 = 1 \tag{10.110}$$

If $\theta_x = 0$, the diffraction factor is also unity; or if the slit width becomes very narrow, $a \to 0$, the diffraction factor approaches unity, where

$$\lim_{k\theta_x a \to 0} \left(\frac{\sin(k\theta_x a)}{k\theta_x a} \right)^2 = 1 \tag{10.111}$$

For an image point I on the z-axis, the image intensity $I_1(0)$ has the maximum amplitude I_0, which occurs when $\theta_x = 0$, where Eqs. (10.110–10.111), applied to Eq. (10.109), yields

$$I_1(0) = I_0 = \left(\frac{2aNA_S}{r_0 r_i \lambda} \right)^2 \quad \theta_x = 0 \tag{10.112}$$

The N-slit interference factor has zeros when n is any positive or negative integer, where in Eq. (10.109),

$$Nbk\theta_x = n\pi \tag{10.113}$$

If n is not an integer multiple of integer N, then

$$bk\theta_x = \frac{n\pi}{N} \tag{10.114}$$

If n is an integer multiple of integer N, then the N-slit interference factor is unity, with principal maximum intensities determined by the diffraction factor. That is, for integers n, m, and the integer number of slits N, the N-slit interference factor becomes unity for $n = mN$, where in Eq. (10.109),

$$\lim_{bk\theta_x \to \frac{n\pi}{N}} \left[\frac{\sin(Nbk\theta_x)}{N \sin(bk\theta_x)} \right] = 1 \quad n = mN \tag{10.115}$$

The width of the principal maxima at $m = n/N$ in the image intensity distribution of Eq. (10.109), occurs between the bounding adjacent zeros at

$$m = \frac{n-1}{N} \quad \text{and} \quad m = \frac{n+1}{N} \tag{10.116}$$

with

$$\Delta m = \frac{2}{N} \tag{10.117}$$

A principal maxima width in the image intensity distribution of Eq. (10.109) for N slits, is then

$$\frac{2\pi}{N} \tag{10.118}$$

There are $N-1$ zeros that occur between pairs of adjacent principal maxima:

$$bk\theta_x = m\pi \tag{10.119}$$

and

$$bk\theta_x = (m+1)\pi \tag{10.120}$$

The $N-1$ zeros occur at

$$bk\theta_x = \frac{\pi}{N} + m\pi, \frac{2\pi}{N} + m\pi, \frac{3\pi}{N} + m\pi, \ldots, \frac{N-1}{N}\pi + m\pi \tag{10.121}$$

Also, between the adjacent principal maxima in the image intensity distribution of Eq. (10.109), are $N-2$ subsidiary maxima, which occur at

$$bk\theta_x = \frac{\pi}{2N} + m\pi, \frac{3\pi}{2N} + m\pi, \frac{5\pi}{2N} + m\pi, \ldots, \frac{2N-3}{2N}\pi + m\pi \tag{10.122}$$

The subsidiary maxima have a width of $\pi/2N$ between adjacent zeros, which is half the width of the principal maxima. At the subsidiary maxima, the interference factor is $1/N^2$ rather than unity, as is the case with the principal maxima.

In Eq. (10.109), when N is large, the subsidiary maxima are very small, and only the principal maxima have significant intensity and occur at angles given by

$$bk\theta_x = m\pi \quad m = 0, \pm 1, \pm 2, \ldots \tag{10.123}$$

One assumes small object θ_0 and image angles θ_i, and thus in the xz-plane, the change θ_x of wave vector angle upon diffraction at the aperture S_a is

$$\theta_x = \theta_0 + \theta_i \quad \theta_0 \ll 1 \text{ and } \theta_i \ll 1 \tag{10.124}$$

With substitution of Eq. (10.124) into Eq. (10.123), the 1-D case with $\phi = 0$ for the paraxial case, gives the grating equation

$$\theta_x = \theta_0 + \theta_i = \frac{m\lambda}{2b} \quad \phi = 0 \tag{10.125}$$

A 1-D array of scattering atoms can be modeled as a 1-D grating with regard to the diffraction of incident neutrons or X-rays. For more divergent source θ_0 and image

θ_i angles, the grating equation of Eq. (10.125) is written more accurately with $\sin\theta_0$ and $\sin\theta_i$, where

$$\sin\theta_i + \sin\theta_0 = \frac{m\lambda}{2b} \tag{10.126}$$

For $m = 0$, the source angle θ_0 from the grating S_a to source P, and the image angle θ_i from the grating S_a to the observation (image) point I, relative to the grating normal, are equal for all neutron or X-ray wavelengths λ. In addition, all the spectral components are superimposed on the central or zero-order maximum, where

$$\theta_0 = -\theta_i \quad m = 0 \tag{10.127}$$

For $m = 0$, the incident ray and diffracted ray lie on the same straight line from source point P through aperture point L to image point I. Recall that if source angle θ_0 is clockwise (counterclockwise) relative to the aperture normal pointing in the negative z-direction toward the source plane, then source angle θ_0 is negative (positive). However, the corresponding and equal magnitude image angle θ_i is clockwise (counterclockwise), and thus positive (negative) relative to the aperture normal pointing in the positive z-direction toward the image plane. Thus, the source θ_0 and θ_i image angles are equal but carry opposite signs for $m = 0$.

When $m = \pm 1$, the principal maxima, which correspond to each wavelength λ, form a first-order spectrum of principal maxima. Each principal maximum, is an image of the point source, which is formed by the neutrons or X-rays of a particular wavelength λ, and when $m = \pm 1$, this image is called a first-order spectral point. The dispersion or spread of the 1-D grating or angular spread of the spectral lines, with respect to wavelength, is obtained by taking the differential of the general (nonparaxial) grating equation, and defining dispersion D as the magnitude of the wavelength derivative of the angle of the observation (image) point I. The observation angle is measured relative to the normal of the grating normal:

$$D = \left|\frac{\mathrm{d}\theta_i}{\mathrm{d}\lambda}\right| = \frac{|m|}{2\,b\cos\theta_i} \tag{10.128}$$

Recall from Eq. (10.117), that the principal maxima have width $\Delta m = 2/N$ between adjacent zeros, and the angular width $\Delta\theta_i$ between principal maxima can be determined from the differential of the grating equation, where, for a monochromatic source of wavelength λ,

$$\Delta\theta_i = \frac{\lambda\Delta m}{2b\cos\theta_i} = \frac{\lambda}{Nb\cos\theta_i} \tag{10.129}$$

A pair of neutron or X-ray waves, which are emitted from the same source at the same intensity and slightly different wavelengths, are resolved by the Rayleigh criteria, if the principal maximum of the first wavelength source falls on the adjacent zero of the second wavelength source. The angular separation $\Delta\theta_i/2$ between the principal maximum of one source, and the adjacent zero of the other source, is the Rayleigh angular resolution $\Delta\theta_{ray}$, where

$$\Delta\theta_{ray} = \frac{\Delta\theta_i}{2} = \frac{\lambda}{2Nb\cos\theta_i} \tag{10.130}$$

From the dispersion relationship of Eq. (10.128), the Rayleigh minimum angular resolution corresponds to a minimum separation in wavelength, $\Delta\lambda_{ray}$, which is used to determine the grating resolving power R, defined by

$$R \equiv \frac{\lambda}{\Delta\lambda_{ray}} \tag{10.131}$$

where

$$\Delta\lambda_{ray} = \frac{\Delta\theta_{ray}}{(\Delta\theta_i/\Delta\lambda)} = \frac{\lambda}{N|m|} \tag{10.132}$$

Via Eq. (10.124), the maximum of index m is limited by the condition

$$\sin\theta_0 + \sin\theta_i \le 1 \tag{10.133}$$

In the grating equation of Eq. (10.126), this requires

$$\frac{m\lambda}{2b} \le 1 \tag{10.134}$$

With substitution of Eq. (10.134) into Eq. (10.131) with Eq. (10.132), the maximum grating resolving power is

$$R_{max} \le \frac{2b}{\lambda} \tag{10.135}$$

In summary, the intensity distribution in the Fraunhofer diffraction pattern of a 1-D grating is separated into the product of a diffraction factor and an interference factor. In this case, the source and aperture are sufficiently distant from the 1-D array to produce near-parallel rays at the 1-D array. The near-parallel rays form a wave front of a single phase, which is slightly tilted for off-axis source points, and passes through the aperture. In the aperture, this wave front is a source of secondary rays that converge to a single distant observation point. With a nearby source point, which produces divergent rays at the 1-D array, a lens can be placed between the source point and the 1-D array.

The lens refracts the divergent rays from the nearby source point to near-parallel rays incident on the 1-D array. In addition, a second lens placed between the 1-D array and a nearby observation point allows convergence of the parallel secondary rays to a nearby observation point by its refraction. Thus, for a nearby source and nearby observation point, a lens placed before and after the 1-D aperture array, can produce a Fraunhofer diffraction pattern in the observation plane.

The diffraction factor depends on the width of the individual slit (single structure), and is a broad intensity distribution. In comparison, the interference factor depends on the number N of identical slits (or number of repeated structures). The interference factor is the narrow intensity distributions of principal maxima, which are contained within the broad distribution of the diffraction factor. A linear array of identical narrow-slit sources or identical narrow slat scatter obstacles, which yield identical phases or phases that increase linearly along the array, give rise to the grating interference factor. A grating in an opaque sheet of material in the aperture plane is comprised of a linear array of alternating slits and slats. In the opaque sheet, the slits are the narrow, rectangular openings, and the slats are the narrow, rectangular, opaque bars. Each narrow-slit aperture in the grating acts as a source of secondary wavelets whose phase is set by the direction of the primary wave that originates at the neutron or X-ray source and arrives at the aperture, where the primary wave acts as the source of secondary wavelets in the Huygens view. And this applies to an array of slat obstacles, which act as scattering centers.

10.9 Fraunhofer Diffraction from Gratings Is Archetype for Coherent, Elastic Scatter of X-ray or Neutrons from Material Lattices

One can model the lattices of materials as 3-D diffraction gratings. In this case, the coherent, elastic scatter amplitude of X-rays or neutrons from a material lattice is calculated via the diffraction formula for a 3-D grating. For example, consider a 1-D array of N slits and opaque slats, where the slit or slat of width a becomes sufficiently narrow, and acts as a line source or line obstacle, as $a \to 0$. The diffraction factor of each aperture, which modulates the intensity of the interference pattern, then approaches unity. In Eq. (10.109), let

$$bk\theta_x = \frac{n\pi}{N} \tag{10.136}$$

Consider the condition

$$k\theta_x a \to 0 \tag{10.137}$$

Equation (10.137) is equivalent to an aperture width a, which is much smaller than the spacing b between apertures, where $a \ll b$ and

$$\frac{an\pi}{Nb} \to 0 \quad a \ll b \tag{10.138}$$

What remains is the interference pattern with primary maxima of uniform amplitude, surrounded by subsidiary maxima with near-zero amplitude. The intensity $I_1(I)$ at the observation point of this 1-D grating of line source or line obstacles is

$$I_1(I) = \left(\frac{2aNA_S}{r_0 r_i \lambda}\right)^2 \cdot \left(\frac{\sin(Nbk\theta_x)}{N\sin(bk\theta_x)}\right)^2 \tag{10.139}$$

Next, we consider a 2-D or 3-D array of source apertures or scattering obstacles, where the source point and observation point are quite distant compared to the spatial extent of the 2-D or 3-D array. If the source point is not sufficiently distant from the array, then a source lens is required to produce parallel rays at the array from the divergent rays, which are emitted from the nearby source point. If the observation point is not sufficiently distant from the array, then an observation lens is required to produce converging rays at the observation point from parallel secondary rays, which are emitted from the array apertures or secondary rays scattered from the array-scattering obstacles.

The 2-D array case has apertures or obstacles with near-zero transverse dimensions in x and y (namely $a_x \to 0$ and $a_y \to 0$), and the number of apertures or scattering centers in x and y are N_x and N_y. The 3-D case adds the axial direction z, so the aperture or obstacle dimension in the z-direction is also near-zero ($a_z \to 0$), and the number of apertures or scattering centers in z is N_z.

In the 3-D case, when N_x, N_y, and N_z are large, the subsidiary maxima are very small, and only the principal maxima have significant intensity and occur at angles θ_x, θ_y, and θ_z, which are measured relative to the transverse x- and y-axis, and the axial z-axis, respectively, so applying Eq. (10.123) to the x, y, and z directions,

$$b_x k\theta_x = m_x\pi \quad m_x = 0, \pm 1, \pm 2, \ldots \tag{10.140}$$

$$b_y k\theta_y = m_y\pi \quad m_y = 0, \pm 1, \pm 2, \ldots \tag{10.141}$$

$$b_z k\theta_z = m_z\pi \quad m_z = 0, \pm 1, \pm 2, \ldots \tag{10.142}$$

The angle components θ_{0x}, θ_{0y}, and θ_{0z} from the source to the array, and the angle components θ_{ix}, θ_{iy}, and θ_{iz} from the array to the observation point, each are measured relative to the x-, y-, and z-axis, respectively. Hence, the changes in direction angles in the x-, y-, and z-directions of the neutron or X-ray wave front, upon passage through and diffraction by the 3-D array, are obtained by applying 10.125 to the x, y, and z directions,

$$\theta_x = \theta_{0x} + \theta_{ix} = \frac{m_x\lambda}{2b_x} \tag{10.143}$$

$$\theta_y = \theta_{0y} + \theta_{iy} = \frac{m_y\lambda}{2b_y} \tag{10.144}$$

$$\theta_z = \theta_{0z} + \theta_{iz} = \frac{m_z\lambda}{2b_z} \tag{10.145}$$

The transverse x- and y-axis and axial z-axis can be positioned so that the origin is at one corner of the array of apertures or scattering centers, where each aperture or scattering center is narrow in spatial extent in the x-, y-, and

z-directions. We assume uniform spacing of the "point" apertures or scattering centers in x, y, and z. For more divergent x-, y-, and z-component source angles $(\theta_{0x}, \theta_{0y}, \theta_{0z})$ and x-, y-, and z-component image angles $(\theta_{ix}, \theta_{iy}, \theta_{iz})$, the small-angle approximation of the sine function no longer applies. For divergent angles, which resulted in the one-dimensional diffraction grating Eq. (10.126), the three-dimensional diffraction grating Eqs. (10.143–10.145) are modified to

$$\sin\theta_{0x} + \sin\theta_{ix} = \frac{m_x\lambda}{2b_x} \tag{10.146}$$

$$\sin\theta_{0y} + \sin\theta_{iy} = \frac{m_y\lambda}{2b_y} \tag{10.147}$$

$$\sin\theta_{0z} + \sin\theta_{iz} = \frac{m_z\lambda}{2b_z} \tag{10.148}$$

The change of the neutron or X-ray wave amplitude from the source point, through an aperture point to the observation point, is proportional to the exponential amplitude expression whose imaginary argument is the phase shift χ, where recalling Eq. (10.98),

$$\exp(\mathrm{i}\chi) \tag{10.149}$$

The phase shift χ is the dot product of the position vector $\mathbf{u}$ within the aperture plane and the X-ray or neutron scatter vector $\mathbf{q}$ (or κ), where applying 10.99 to the x, y, and z directions,

$$\chi = \mathbf{u}\cdot\mathbf{q} = u_x k\theta_x + u_y k\theta_y + u_z k\theta_z \tag{10.150}$$

Recalling Eq. (10.100), the x, y, and, z scatter vector $\mathbf{q}$ (sometimes denoted by κ) components are

$$q_x = k\theta_x, \quad q_y = k\theta_y, \quad q_z = k\theta_z \tag{10.151}$$

For the 3D grating, Eq. (10.101) is modified, and the amplitude $E_1(I)$ at the observer point for the paraxial approximation is then

$$E_1(I) = -\frac{\mathrm{i}A_S\exp(-\mathrm{i}wt)}{\lambda}\cdot\frac{\exp(\mathrm{i}k(R_i+R_0))}{r_0 r_i}\sum_{n_x=0}^{N_x-1}\left(\int_{-a_x+n_xb_x}^{a_x+n_xb_x}\exp(\mathrm{i}u_xq_x)\mathrm{d}u_x\right)$$
$$\sum_{n_y=0}^{N_y-1}\left(\int_{-a_y+n_yb_y}^{a_y+n_yb_y}\exp(\mathrm{i}u_yq_y)\mathrm{d}u_y\right)\sum_{n_z=0}^{N_z-1}\left(\int_{-a_z+n_zb_z}^{a_z+n_zb_z}\exp(\mathrm{i}u_zq_z)\mathrm{d}u_z\right) \tag{10.152}$$

A simplified expression for the observation point amplitude for the 3-D case is obtained in the same manner as the 1-D case. The dimensions of the apertures or scattering centers are assumed to be much smaller than the spacing between them, i.e.,

$$a_x \ll b_x, \quad a_y \ll b_y, \quad a_z \ll b_z \tag{10.153}$$

Substitution of variables is performed in Eq. (10.152), where

$$u'_x = u_x - n_x b_x, \quad u'_y = u_y - n_y b_y, \quad u'_z = u_z - n_z b_z \tag{10.154}$$

This is followed by integration and substitution of q_x, q_y, and q_z, where one assigns

$$p_x = \mathrm{i}2b_x k\theta_x, \quad p_y = \mathrm{i}2b_y k\theta_y, \quad p_z = \mathrm{i}2b_z k\theta_z \tag{10.155}$$

Applying the geometric progression formula of Eq. (10.106), and factoring the resulting exponential terms, allow the application of Euler's relation to Eq. (10.152). This yields the observation point I amplitude for the 3-D aperture or scatter array in terms of sine functions:

$$\begin{aligned} E_1(I) = &-\frac{\mathrm{i}2a_x a_y a_z A_S N_x N_y N_z \exp(-\mathrm{i}wt)}{\lambda} \\ &\cdot \frac{\exp[\mathrm{i}k(b_x\theta_x(N_x-1) + b_y\theta_y(N_y-1) + b_z\theta_z(N_z-1) + (R_i + R_0))]}{r_0 r_i} \cdot \\ &\frac{\sin(k\theta_x a_x)}{k\theta_x a_x} \cdot \frac{\sin(k\theta_x a_x)}{k\theta_x a_x} \cdot \frac{\sin(k\theta_x a_x)}{k\theta_x a_x} \cdot \frac{\sin(N_x b_x k\theta_x)}{N_x \sin(b_x k\theta_x)} \\ &\cdot \frac{\sin(N_y b_y k\theta_y)}{N_y \sin(b_y k\theta_y)} \cdot \frac{\sin(N_z b_z k\theta_z)}{N_z \sin(b_z k\theta_z)} \end{aligned} \tag{10.156}$$

Eq. (10.156) is further simplified by approximating the $\sin x/x$ terms as unity so that

$$\begin{aligned} E_1(I) = &-\frac{\mathrm{i}2a_x a_y a_z A_S N_x N_y N_z \exp(-\mathrm{i}wt)}{\lambda} \\ &\cdot \frac{\exp[\mathrm{i}k(b_x\theta_x(N_x-1) + b_y\theta_y(N_y-1) + b_z\theta_z(N_z-1) + (R_i + R_0))]}{r_0 r_i} \cdot \\ &\frac{\sin(N_x b_x k\theta_x)}{N_x \sin(b_x k\theta_x)} \cdot \frac{\sin(N_y b_y k\theta_y)}{N_y \sin(b_y k\theta_y)} \cdot \frac{\sin(N_z b_z k\theta_z)}{N_z \sin(b_z k\theta_z)} \end{aligned} \tag{10.157}$$

For the 3-D array of apertures or scatter centers, intensity $I_1(I)$ at the observation point I of the 3-D array of aperture or scattering centers, is the squared magnitude of the amplitude of Eq. (10.157), where

$$\begin{aligned} I_1(I) = &\left[\frac{2a_x a_y a_z A_S N_x N_y N_z}{r_0 r_i \lambda}\right]^2 \cdot \left[\frac{\sin(k\theta_x a_x)}{k\theta_x a_x} \cdot \frac{\sin(k\theta_y a_y)}{k\theta_y a_y} \cdot \frac{\sin(k\theta_z a_z)}{k\theta_z a_z}\right]^2 \\ &\cdot \left[\frac{\sin(N_x b_x k\theta_x)}{N_x \sin(b_x k\theta_x)} \cdot \frac{\sin(N_y b_y k\theta_y)}{N_y \sin(b_y k\theta_y)} \cdot \frac{\sin(N_z b_z k\theta_z)}{N_z \sin(b_z k\theta_z)}\right]^2 \end{aligned} \tag{10.158}$$

Now assume the change of direction angles in Eq. (10.158) are zero, where

$$\theta_x = \theta_{0x} + \theta_{ix} = 0 \quad \theta_y = \theta_{0y} + \theta_{iy} = 0 \quad \theta_z = \theta_{0z} + \theta_{iz} = 0 \tag{10.159}$$

Then, the x-, y-, and z-component angles from the aperture array to source $(\theta_{0x}, \theta_{0y}, \theta_{0z})$, and from the aperture array to observation points $(\theta_{ix}, \theta_{iy}, \theta_{iz})$, relative to the x-, y-, and z-axis, are

$$\theta_{0x} = -\theta_{ix}, \quad \theta_{0y} = -\theta_{iy}, \quad \theta_{0z} = -\theta_{iz} \tag{10.160}$$

Thus, for all neutron or X-ray wavelengths λ, all the spectral components are superimposed on the central or zero-order maximum. For zero change of direction angles, the ray from the source point P through the aperture point L to the image point I is a straight line. The negative signs are due to the pair of oppositely directed, outward normal, unit vectors $\hat{\mathbf{n}}$ on the aperture surface S_a, which point to the object (source) and image (observer) planes, respectively.

Recall that the x-, y-, and z-components of the source angle θ_{0x}, θ_{0y}, and θ_{0z} are negative, if clockwise from the aperture normal ($\hat{\mathbf{n}} = -\hat{\mathbf{z}}$) at point L, pointing along the negative z-axis to the source plane, and positive, if counterclockwise. However, the x-, y-, and z-components of the source angle θ_{ix}, θ_{iy}, and θ_{iz} are positive, if clockwise from aperture normal ($\hat{\mathbf{n}} = \hat{\mathbf{z}}$) at point L, pointing along the positive z-axis to the image plane, and negative if counterclockwise. Hence, the undiffracted, zero-order rays above the incident and diffracted rays lie on the same line, such that the source and image angles are equal but carry opposite signs.

The changes of direction angles $\theta_x, \theta_y, \theta_z$ in the x-, y-, and z-directions of the neutron or X-ray wave front of wavelength λ on passage through the 3-D array are measured relative to the normal array. From Eq. (10.158), the changes in direction, which are caused by array diffraction of a neutron or an X-ray wave front, where the diffraction produces principal maxima in the observation plane, occur when

$$\theta_x = \frac{m_x \lambda}{2b_x} \quad m_x = 0, \pm 1, \pm 2, \ldots \tag{10.161}$$

$$\theta_y = \frac{m_y \lambda}{2b_y} \quad m_y = 0, \pm 1, \pm 2, \ldots \tag{10.162}$$

$$\theta_z = \frac{m_z \lambda}{2b_z} \quad m_z = 0, \pm 1, \pm 2, \ldots \tag{10.163}$$

These changes in angle satisfy the Bragg condition. For larger angle changes, one dispenses with the small-angle sine approximation, and the condition of the principal maxima and the Bragg condition of Eqs. (10.161–10.163), then become

$$\sin\theta_x = \frac{m_x \lambda}{2b_x} \tag{10.164}$$

$$\sin\theta_y = \frac{m_y\lambda}{2b_y} \tag{10.165}$$

$$\sin\theta_z = \frac{m_z\lambda}{2b_z} \tag{10.166}$$

With substitution of Eqs. (10.161–10.163) into Eq. (10.158), the intensity of these maxima are modulated by the diffraction intensity, which is produced by each of the single identical apertures or scattering objects, so that

$$I_1(I) = \left[\frac{2a_xa_ya_zA_SN_xN_yN_z}{r_0r_i\lambda}\right]^2 \cdot \left[\frac{\sin(m_xa_x\pi/b_x)}{(m_xa_x\pi/b_x)} \cdot \frac{\sin(m_ya_y\pi/b_y)}{(m_xa_x\pi/b_x)} \cdot \frac{\sin(m_za_z\pi/b_z)}{(m_za_z\pi/b_z)}\right]^2 \tag{10.167}$$

The dimensions of the apertures or scattering centers are assumed to be much smaller than the spacing between them, i.e.,

$$a_x \ll b_x, \quad a_y \ll b_y, \quad a_z \ll b_z \tag{10.168}$$

Hence, the maxima intensities are approximately uniform within a region of the observation plane, where the variation of source point to array distance r_0, and the variation of the array-to-observation-point distance r_i are small, as compared to the array-to-source, and array to observation distances r_0 and r_i of the region's central observation point, so that Eq. (10.167) becomes

$$I_1(I) = \left(\frac{2a_xa_ya_zA_SN_xN_yN_z}{r_0r_i\lambda}\right)^2 \tag{10.169}$$

10.10 Abbe Theory of Imaging Applied to X-rays or Neutrons

In the Abbe theory of imaging, the undeflected (zero-order) X-rays or neutrons illuminate the observation (image) plane without amplitude modulation, i.e., there is no interference with scattered neutrons (X-rays). Hence, the resultant uniform illumination in the image plane, by only the undeviated, zero-order neutrons (X-rays), does not allow one to detect an image of the scattering object, which is much smaller than the wavelength of the incident X-rays or neutrons.

Consider the case of a specimen whose size s is larger than the neutron or X-ray wavelength λ, where

$$s \geq \lambda \tag{10.170}$$

And assume that the CRL has a large numerical aperture NA, given by

$$\mathrm{NA} = n \sin\left(\frac{\theta}{2}\right) \tag{10.171}$$

Here, n is the refractive index of the medium between the specimen and lens, and $\sin\theta$ is the ratio of lens aperture radius r_m divided by the lens focal f, where

$$\sin\theta = \frac{r_m}{f} \tag{10.172}$$

A larger numerical aperture allows the CRL to intercept more diffraction orders. Diffracted waves, with increasing diffraction orders (larger m), interfere in the image plane with the zero-order diffracted wave. This leads to an interference pattern, which is an increasingly resolved replica of the object.

An incident neutron or X-ray plane wave of wavelength λ propagates in the axial z-direction and is diffracted by a simple specimen, which is represented as a 2-D variable period grating with period Δx along the transverse x-direction and with period Δy along the transverse y-direction. The diffracted waves shown in the xz-plane emerge at direction angle α_x with respect to the center optical z-axis, where from Eq. (10.164),

$$\alpha_x = \sin^{-1}\left(\frac{m_x\lambda}{\Delta x}\right) \tag{10.173}$$

The diffracted waves shown in the yz-plane emerge at direction angle α_y with respect to the center optical z-axis, where from Eq. (10.165),

$$\alpha_y = \sin^{-1}\left(\frac{m_y\lambda}{\Delta y}\right) \tag{10.174}$$

The path lengths along the various discrete directions differ by integer wavelengths. The zero-order undiffracted wave corresponds to $m_x = 0$ and $m_y = 0$, and the nonzero-order diffracted waves, with increasing positive (negative) direction angles, correspond to increasing positive (negative) integer values of m_x and m_y.

The intensity amplitude of neutrons or X-rays emerging from the specimen is modulated by the specimen neutron optical thickness. The complex and realistic specimens, which has dimensions d_x and d_y along the x and y transverse directions, can be represented by a superposition of 2-D gratings in the object plane of varying periods along the x- and y-directions. The pattern of the intensity amplitude of the neutron or X-ray distribution, which emerge from the specimen, can be constructed of sine waves of varying spatial frequencies, amplitudes, and phase shifts. The undiffracted wave provides the DC level amplitude, and the diffracted waves at diffraction angle pair (θ_x, θ_y) represent the Fourier component sine waves of spatial frequencies $m_x/\Delta x$ along x and $m_y/\Delta y$ along y.

Consider sufficiently small specimens of size d_x and d_y in x and y, with respect to the incident wavelength λ, where

$$d_x \ll \lambda \quad \text{and} \quad d_y \ll \lambda \tag{10.175}$$

The diffraction angle for the $m_x = 1$ or $m_y = 1$ order is each near 90°, and the higher diffraction orders $m_x > 1$ or $m_y > 1$ are even closer to 90°. With 90° angles, no lens can intercept any of the diffracted nonzero orders. In this case, the lens only intercepts the zero-order $m_x = 0$ or $m_y = 0$ wave, which passes undeflected through the object or specimen. The undeflected zero-order neutrons or X-rays illuminate the observation (image) plane without amplitude modulation (via interference with scattered neutrons or X-rays). Hence, the resultant uniform illumination in the image plane, by only the undeviated, zero-order neutrons (X-rays), does not allow one to detect or see an image of the scattering object, which is much smaller than the wavelength of the incident neutrons (X-rays).

10.11 Fraunhofer Diffraction of X-rays or Neutrons from a Circular Aperture

Instead of a rectangular aperture, consider now the far-field diffraction of X-rays or neutrons from a circular aperture of radius r_a with the paraxial approximation, see Fig. 10.1. For radius u and symmetry in the azimuth angle ϕ in the circular aperture plane, a point in the circular aperture has radial position vector $\mathbf{u}$ with unit radial vector $\hat{\mathbf{u}}$ and unit azimuthal angle vector $\hat{\varphi}$, where, for circular coordinates,

$$\mathbf{u} = u\hat{\mathbf{u}} \quad \text{and} \quad \mathrm{d}\mathbf{u} = \mathrm{d}u\hat{\mathbf{u}} + u\,\mathrm{d}\phi\hat{\varphi} \tag{10.176}$$

The differential area $\mathrm{d}S_a$ in the aperture plane is

$$\mathrm{d}S_a = u\,\mathrm{d}u\,\mathrm{d}\phi \tag{10.177}$$

The change in the neutron or X-ray wave vector $\mathbf{q}$ is the same at each aperture point. The source angle θ_0 and image angle θ_i are assumed to be sufficiently small that the sine of each angle is approximately equal to the angle, where

$$\sin\theta_0 \cong \theta_0 \quad \text{and} \quad \sin\theta_i \cong \theta_i \tag{10.178}$$

Therefore, change θ in the neutron trajectory angle from source angle θ_0 to diffracted, image angle θ_i at the circular aperture is also small, where

$$\theta = \theta_0 + \theta_i \ll 1 \tag{10.179}$$

The change in the X-ray or neutron wave vector $\mathbf{q}$ for diffraction or coherent, elastic scatter is

$$\mathbf{q} = 2k \sin\left(\frac{\theta}{2}\right)\hat{\mathbf{r}} \tag{10.180}$$

Within the aperture plane, azimuth angle ϕ is defined by the dot product of the position vector $\mathbf{u}$ within the aperture plane and the neutron or X-ray scatter vector $\mathbf{q}$, where recalling Eq. (10.34), and Eqs. (10.40–10.43),

$$\mathbf{u} \cdot \mathbf{q} = 2uk \sin\left(\frac{\theta}{2}\right)(\hat{\mathbf{u}}\cdot\hat{\mathbf{r}}) = 2uk \sin\left(\frac{\theta}{2}\right)\cos \phi \tag{10.181}$$

The azimuth angle ϕ is between the radial unit vectors $\hat{\mathbf{u}}$ and $\hat{\mathbf{r}}$ is the plane of the circular aperture S_a, so that the dot product of unit vectors $\hat{\mathbf{u}}$ and $\hat{\mathbf{r}}$ is cos ϕ. The argument of the exponential within the integrand of the Kirchhoff–Fresnel diffraction formula of Eq. (10.31), is then

$$2\pi\left(\frac{r_i + r_0}{\lambda}\right) = k(R_i + R_0) + \mathbf{u} \cdot \mathbf{q} = k(R_i + R_0) + 2uk \sin(\theta/2)\cos \phi \tag{10.182}$$

With Eqs. (10.43), (10.63), and (10.182), the Kirchhoff–Fresnel diffraction formula of Eq. (10.31) is adapted to circular aperture, where

$$E_1(I) = -\frac{\mathrm{i}A_S \exp(-\mathrm{i}wt)}{\lambda} \cdot \frac{\exp(\mathrm{i}k(R_i + R_0))}{r_0 r_i} \int_0^{2\pi}\int_0^{r_a} \exp(\mathrm{i}2k \sin(\theta/2)u \cos \phi)u\, \mathrm{d}u\, \mathrm{d}\phi \tag{10.183}$$

Integration of Eq. (10.183), requires a Bessel function solution, where the zero-order Bessel function J_0 is

$$J_0(\alpha) = \frac{1}{2\pi}\int_0^{2\pi} \exp(\mathrm{i}\alpha \cos \phi)\mathrm{d}\phi \tag{10.184}$$

One makes the change of variables in the integral of Eq. (10.183), where

$$\alpha = 2ku \sin(\theta/2) \tag{10.185}$$

and

$$\mathrm{d}\alpha = 2k \sin(\theta/2)\mathrm{d}u \tag{10.186}$$

Substitution of Eqs. (10.184–10.186) into (10.183), yields

$$E_1(I) = -\frac{\mathrm{i}A_S \exp(-\mathrm{i}wt)}{[2k \sin(\theta/2)]^2 \lambda} \cdot \frac{\exp(\mathrm{i}k(R_i + R_0))}{r_0 r_i} \int_0^{r_a} \alpha J_0(\alpha)\mathrm{d}\alpha \tag{10.187}$$

To solve the integral in Eq. (10.187), the recurrence relation for the Bessel function is invoked:

$$\frac{\mathrm{d}}{\mathrm{d}\alpha}[\alpha^m J_m(\alpha)] = \alpha^m J_{m-1}(\alpha) \tag{10.188}$$

In Eqs. (10.187–10.188), with integer $m = 1$ and aperture diameter $d = 2r_\mathrm{a}$ and integration from $u = 0$ to $u = d/2$, the amplitude $E_1(I)$ at the observer point for the Fraunhofer diffraction through a circular aperture is

$$E_1(I) = -\frac{\mathrm{i}A_S d\,\exp(-\mathrm{i}wt)}{4\,\sin(\theta/2)} \cdot \frac{\exp(\mathrm{i}k(R_i + R_0))}{r_0 r_i} \cdot J_1\left(\frac{2\pi d\,\sin(\theta/2)}{\lambda}\right) \tag{10.189}$$

With the squared magnitude of Eq. (10.189), the far-field intensity from the paraxial approximation thus yields the Fraunhofer diffraction through a circular aperture of diameter d, where

$$I_1(I) = \left[\frac{A_S d}{4r_0 r_i\,\sin(\theta/2)} J_1\left(\frac{2\pi d\,\sin(\theta/2)}{\lambda}\right)\right]^2 \tag{10.190}$$

Recall that the incident neutron or X-ray current density J_in [#/cm^2] at a rectangular aperture is the product of the squared amplitude $|E_\mathrm{in}|^2$ or probability density of the incident neutrons or X-rays and their group velocity v_g, where

$$J_\mathrm{in} = v_g |E_\mathrm{in}|^2 \tag{10.191}$$

Point source P produces waves with group velocity v_{gn}, which are incident at the aperture with amplitude

$$E_\mathrm{in} = \frac{A_S\,\exp(\mathrm{i}wt)\exp(\mathrm{i}kr_0)}{r_0} \tag{10.192}$$

From section 1.1, as well as section 1 in chapter 5 of Cremer (2012a), the neutron group velocity v_{gn} is

$$v_{gn} = \frac{\hbar k}{m} \tag{10.193}$$

From section 1.1 as well as from sections 18–20 in chapter 4 of Cremer (2012a), the X-ray group velocity v_{gx} is

$$v_{gx} = \frac{\mathrm{d}w}{\mathrm{d}k} \tag{10.194}$$

The neutrons or X-rays exit the aperture with the same group velocity v_g, but with amplitude E_out. The exit current density from the aperture J_out is then

$$J_\mathrm{out} = v_g |E_\mathrm{out}|^2 \tag{10.195}$$

Recall from Eq. (10.66), for a rectangular aperture, the amplitude E_{out} of the exiting wave is

$$E_{\text{out}} = -\frac{\mathrm{i}A_S \exp(-\mathrm{i}wt)}{\lambda} \cdot \frac{\exp(\mathrm{i}kR_0)}{r_0} \int_{-x_a}^{x_a} \exp(\mathrm{i}u_x q_x)\mathrm{d}u_x \int_{-y_a}^{y_a} \exp(\mathrm{i}u_y q_y)\mathrm{d}u_y \tag{10.196}$$

In the paraxial approximation with a distant source, or a nearby source near the axis with a focusing lens placed between the source and aperture, the differential scattering cross section $\mathrm{d}\sigma/d\Omega$ of the neutrons of wavelength λ by the rectangular aperture is

$$\frac{\mathrm{d}\sigma}{\mathrm{d}\Omega} = \frac{J_{\text{out}}}{J_{\text{in}}} = \frac{|E_{\text{out}}|^2}{|E_{\text{in}}|^2} = \left(\frac{1}{\lambda}\right)^2 \left[\int_{-x_a}^{x_a} \exp(\mathrm{i}u_x q_x)\mathrm{d}u_x \int_{-y_a}^{y_a} \exp(\mathrm{i}u_y q_y)\mathrm{d}u_y\right]^2 \tag{10.197}$$

Solving the integrals yields the differential cross section $\mathrm{d}\sigma/d\Omega$ for a rectangular aperture with x and y widths $2x_a$ and $2y_a$, where

$$\frac{\mathrm{d}\sigma}{\mathrm{d}\Omega} = \left(\frac{2}{\lambda}\right)^2 \cdot \frac{\sin^2(x_a k\theta_x)}{k^2\theta_x^2} \cdot \frac{\sin^2(y_a k\theta_y)}{k^2\theta_y^2} \tag{10.198}$$

In Eq. (10.197) replaces the rectangular aperture diffraction amplitude of Eq. (10.196) by the circular aperture amplitude of Eq. (10.189), and obtains the differential cross section $\mathrm{d}\sigma/d\Omega$ for a circular aperture of diameter d, where

$$\frac{\mathrm{d}\sigma}{\mathrm{d}\Omega} = \left[\frac{d}{4\sin(\theta/2)} J_1\left(\frac{2\pi d \sin(\theta/2)}{\lambda}\right)\right]^2 \tag{10.199}$$

This equation for the differential cross section also applies to X-rays and neutrons because the group velocity v_g, via Eqs. (10.191) and (10.195), is divided out. The group velocity of the X-ray or neutron is unchanged as it passes through the circular aperture.

10.12 Huygens–Fresnel Approach: The Kirchhoff Equation for a Compound Refractive Lens with X-rays or Neutrons

We now examine the refraction of X-rays or neutrons with a CRL using the Huygens–Fresnel approach expressed by the Kirchhoff equation. Consider a circular aperture, which is formed by a stack of lenses, called a compound refractive lens (CRL), and comprised of N biconcave parabolic lenses. Each lens has on-axis radius of curvature R and center minimum thickness t. See chapters 1 and 11, as well as Cremer (2012a), regarding compound refractive lenses. The optical or center z-axis passes through the centers of the object (source), aperture, and image (observation) planes. A source point at transverse radius s from the optical axis emits uniformly the X-rays or neutron rays of wavelength λ in all directions.

One begins with the Kirchhoff–Fresnel diffraction formula for a circular aperture of diameter $d = 2r_a$. The Kirchhoff–Fresnel equation sums the wave amplitudes for the set of ray paths, which originate at the single source point P at radial distance s, and pass through any point L of the CRL aperture at radial distance u. The ray paths from the points L on the CRL aperture then converge to a single image (observation) point I at radial distance y. The X-rays or neutrons are assumed to propagate in the positive z-direction from the point source plane within the FOV of the cylindrical CRL to the image plane.

The amplitude at the image point I depends on radial image position y, via angle $\theta(y, s, r_i, r_0)$, and thus the amplitude also depends on the object point radial distance s from the z-axis, as well as the object point intensity A_S. The amplitude at the image point also depends on the object-to-CRL drift-space distance r_0 between object (source) point P at radius s and lens point L at radius u in the center transverse plane of the CRL. And the image (observer) point I amplitude depends on the ray travel image distance r_i in the CRL-to-image drift space from lens point L to the image point I at radius y from the z-axis in the image plane.

The drift-space rays are generally diagonal and, by azimuthal symmetry, propagate in the same plane. Hence, (1) the single source point at radius s, (2) the radial line of lens points at radius u for $0 < u < r_a$, and (3) the single image point at radius y all lie in the same plane. One must sum the wave amplitudes of the ray paths from all these planes centered about the center optical axis, where each plane is specified by azimuth angle ϕ measured with respect to the transverse vertical x-direction. One must integrate over all these planes, which contain the range of ray-path azimuth angles ϕ through the CRL, namely, $0 < \phi < 2\pi$.

The CRL is assumed to be a thin lens comprised of N biconcave, parabolic lenses. The resultant "thin-lens" CRL has its two principal planes merged into its center transverse plane. A ray, which travels through a thin-lens CRL, is assumed to change direction, but has no significant change of radial position. The lens material has linear attenuation μ and refractive decrement δ. The neutron or X-rays travel through the N-lens, and undergo a phase shift and attenuation, via complex phase shift ψ, where

$$\psi = \psi(\lambda, u, r_0, r_i, \delta, \mu, R, t) \tag{10.200}$$

This phase shift ψ depends on the neutron wavelength λ, radius of the path u through the CRL, the ray path distances through object r_0 and image drift space r_i, the number N of biconcave parabolic lenses, on-axis radius of curvature R, and center minimum wall thickness t of each lens, the lens material decrement δ, and linear attenuation μ.

From Eq. (10.183), the Kirchhoff–Fresnel diffraction formula for a circular aperture with embedded CRL is

$$E_1(y) = -\frac{\mathrm{i}A_S\exp(-\mathrm{i}wt)}{\lambda} \cdot \frac{\exp(\mathrm{i}k(R_i + R_0))}{r_0 r_i} \int_0^{2\pi}\int_0^{r_a} \exp(-\mathrm{i}\psi)\exp(\mathrm{i}2k\sin(\theta/2)u\cos\phi)u\,\mathrm{d}u\,\mathrm{d}\phi \tag{10.201}$$

From Eqs. (11.39–11.41) in the next chapter, in a N parabolic CRL, the neutrons or X-rays are phase-shifted by ψ_N^{par} or

$$\psi = \psi_N^{\text{par}} = -kNt\delta - k\frac{u^2}{2}\left(\frac{1}{r_0} + \frac{1}{r_i}\right) - \mathrm{i}\frac{N\mu}{2}\left(t + \frac{u^2}{R}\right) \tag{10.202}$$

The aperture radius r_a of the CRL is assumed to be much less than the radius of curvature of the lenses comprising the CRL:

$$r_a \ll R \tag{10.203}$$

Lens radius r_a is the smaller of two choices: (i) physical radius $d/2$ of the lens or (ii) the lens absorption aperture radius, where the intensity is $1/e$ of that of the center minimum thickness due to attenuation by lens absorption and scatter.

The absorption aperture radius r_{as} of a biconcave spherical lens with radius of curvature R is expressed by the parabolic lens aperture radius r_{ap} and the on-axis radius of curvature R, where from Eqs. (1.124–1.125) in chapter 1, or Eq. 55 in chapter 2 of Cremer (2012a),

$$r_{as} = r_{ap}\sqrt{1 - \left(\frac{r_{ap}}{2R}\right)^2} \tag{10.204}$$

The center portion of the spherical lens is parabolic, and the radius of this center parabolic-shaped portion of the spherical lens is the parabolic aperture radius r_{ap}, given by

$$r_{ap} = 2\sqrt{\frac{f\delta}{\mu}} = 2\sqrt{\frac{R}{\mu N}} \tag{10.205}$$

Then, the phase shift due to N-spherical biconcave lenses is approximately the same as that of N-parabolic biconcave lenses, where

$$\psi_N^{\text{sph}} = \psi_N^{\text{par}} \tag{10.206}$$

As before, the azimuthal symmetry is assumed. Integration over azimuth angle ϕ for the N-parabolic, biconcave lens CRL, gives the amplitude at the observation (image) point I at radius y, which is caused by source (object) point P at radius s, via Eq. (10.53) (or 10.179) for the change in the neutron trajectory angle $\theta(y, s, r_i, r_0)$:

$$\begin{aligned} E_1(y,s) = &-\frac{\mathrm{i}A_S\exp(-\mathrm{i}wt)}{\lambda}\cdot\frac{\exp(\mathrm{i}k(R_i + R_0 - Nt\delta))}{r_0 r_i} \\ &\cdot\exp\left(-\frac{N\mu t}{2}\right)\times\int_0^{r_a} J_0(2ku\sin(\theta/2))\exp\left(-\frac{N\mu}{2R}u^2\right) \\ &\exp\left[ik\frac{u^2}{2}\left(\frac{1}{r_0} + \frac{1}{r_i}\right)\right]udu \end{aligned} \tag{10.207}$$

The source (object) angle θ_0 of the incident ray from the source (object) point S to a point on the CRL entrance lens, and the observation (image) angle θ_i of the refracted ray from the lens point to the observation point I, are each measured with respect to the center z-axis of the CRL.

Given radial position s on the object and radial position y on the image, the incident and observation angles are

$$\sin\theta_0 = \frac{s}{r_0} \quad \text{and} \quad \sin\theta_i = \frac{y}{r_i} \tag{10.208}$$

Assume that the object and image dimensions are much less than the distance of the object (image) to the aperture plane with the embedded CRL, where

$$s \ll r_0 \quad \text{and} \quad y \ll r_i \tag{10.209}$$

One approximates

$$\theta_i \cong \frac{s}{r_0}, \quad \theta_0 \cong \frac{y}{r_i}, \quad \sin\left(\frac{\theta}{2}\right) \cong \frac{\theta}{2} \tag{10.210}$$

Angle change θ is the change of angle experienced by the incident ray as it passes through the CRL and is refracted toward the observation (image) point. The angle change θ is assumed to be small, where

$$\theta = \theta_0 + \theta_i \ll 1 \tag{10.211}$$

From Eq. (1.27) in this book, or chapters 1 and 2 in Cremer (2012a), the focal length f of a CRL with parabolic lenses is

$$f = \frac{R}{2N\delta} \tag{10.212}$$

The lens-maker equation is

$$\frac{1}{f} = \frac{1}{r_0} + \frac{1}{r_i} \tag{10.213}$$

With substitution of Eqs. (10.208), (10.210), and (10.212–10.213) into Eq. (10.207), one obtains

$$E_1(y,s) = -\frac{\mathrm{i}A_S \exp(-\mathrm{i}wt)}{\lambda} \cdot \frac{\exp\left(\mathrm{i}(2\pi/\lambda)(R_i + R_0 - Nt\delta)\right)}{r_0 r_i} \cdot \exp\left(-\frac{Nt\mu}{2}\right) \times \int_0^{r_a} J_0\left(\frac{2\pi}{\lambda}\left(\frac{y}{r_i} - \frac{s}{r_0}\right)\cdot u\right) \exp\left[-\frac{u^2 N}{2R}\left(\mu - \mathrm{i}\frac{4\pi\delta}{\lambda}\right)\right] u\, du \tag{10.214}$$

The intensity $I_1(y,s)$ at the image point at radius y is the squared magnitude of the amplitude, where

$$I_1(y,s) = |E_1(y,s)|^2 \tag{10.215}$$

The CRL produces intensity $I_1(y,s)$ at image point at radius y by focusing rays of intensity A_s^2 emitted from an object point at radius s in the object plane. Substitution of Eq. (10.214) into Eq. (10.215), gives

$$I_1(y,s)=\left(\frac{A_S}{r_0 r_i \lambda}\right)^2 \cdot \exp(-Nt\mu)\cdot\left|\int_0^{r_a} J_0\left(\frac{2\pi}{\lambda}\left(\frac{y}{r_i}-\frac{s}{r_0}\right)\cdot u\right)\exp\left[-\frac{u^2 N}{2R}\left(\mu-i\frac{4\pi\delta}{\lambda}\right)\right]u\,du\right|^2 \tag{10.216}$$

Polychromatic X-rays or neutrons have distribution $\Gamma(\lambda)$, and hence an additional integration is also performed over the range of X-ray or neutron wavelengths from λ_1 to λ_2 in the spectral distribution. Substitution of $\Gamma(\lambda)$ and rearrangement in the exponential argument of Eq. (2.6), gives

$$\begin{aligned} I_1(y,s)=&\left(\frac{A_S}{r_0 r_i}\right)^2 \int_{\lambda_1}^{\lambda_2} d\lambda \cdot \frac{\Gamma(\lambda)\exp(-Nt\mu)}{\lambda^2} \\ &\cdot\left|\int_0^{r_a} J_0\left(\frac{2\pi}{\lambda}\cdot u\cdot\left(\frac{y}{f}-\frac{s}{r_0}\right)\right)\exp\left(-\frac{u^2}{r_{ap}^2}\right)\exp\left(i\frac{\pi u^2}{2f\lambda}\right)u\,du\right|^2 \end{aligned} \tag{10.217}$$

The intensity $I_1(y,s)$ at radius y in the image plane can be simplified further. The CRL inverts and magnifies, where

$$\frac{s}{r_0}=-\frac{y}{r_i} \tag{10.218}$$

Hence,

$$\theta=\theta_0+\theta_i=\frac{2y}{f} \tag{10.219}$$

with

$$r_i \cong f \tag{10.220}$$

One obtains the intensity $I_1(y)$, which depends only on the radius y of the image point, where

$$\begin{aligned} I_1(y)=&\left(\frac{A_S}{r_0 r_i}\right)^2 \int_{\lambda_1}^{\lambda_2} d\lambda \\ &\cdot\frac{\Gamma(\lambda)\exp(-Nt\mu)}{\lambda^2}\cdot\left|\int_0^{r_a} J_0\left(\frac{4\pi}{\lambda}\cdot\frac{y}{f}\cdot u\right)\exp\left(-\frac{u^2}{r_{ap}^2}\right)\exp\left(i\frac{\pi u^2}{2f\lambda}\right)u\,du\right|^2 \end{aligned} \tag{10.221}$$

The intensity $I_{CRL}(0)$ at the center image point $y=0$ is found by integration over the CRL radius from $u=0$ to $u=\infty$. The intensity between the absorption aperture diameter $u = r_a$ of the CRL and $u = \infty$ is small compared to that from $u = 0$ to $u = \infty$ and can be neglected, thus simplifying integration. The zero order Bessel function with an argument of 0, gives

$$J_0(0) = 1 \tag{10.222}$$

With Eq. (10.221), the CRL on-axis intensity transmitted to the center of the image plane $y = 0$, is then

$$I_{\mathrm{CRL}}(0) = \left(\frac{A_S}{r_0 r_i}\right)^2 \int_{\lambda_1}^{\lambda_2} \mathrm{d}\lambda \cdot \frac{\Gamma(\lambda)\exp(-Nt\mu)}{\lambda^2} \cdot \left| \int_0^{\infty} \exp\left[-u^2\left(\frac{1}{r_{\mathrm{ap}}^2} - \mathrm{i}\frac{\pi}{2f\lambda}\right)\right] u\,\mathrm{d}u \right|^2 \tag{10.223}$$

One substitutes into Eq. (10.223) for the CRL parabolic aperture radius r_{ap} of Eq. (10.205). Integration then yields the CRL transmitted and focused intensity $I_{\mathrm{CRL}}(0)$ from the on-axis object point ($s = 0$) to the on-axis image point ($y = 0$):

$$I_{\mathrm{CRL}}(0) = \left(\frac{A_S}{2r_0 r_i}\right)^2 \int_{\lambda_1}^{\lambda_2} \mathrm{d}\lambda \cdot \frac{\Gamma(\lambda)}{\lambda^2} \cdot \frac{\exp(-Nt\mu)}{\frac{1}{8}(\mu/f\delta)^2 + (\pi/2f\lambda)^2} \tag{10.224}$$

The intensity I_0 of incident of X-rays or neutrons at the CRL principal plane at object distance r_0 (neglecting attenuation by the CRL), is

$$I_0 = \frac{A_s^2}{4\pi \cdot r_0^2} \tag{10.225}$$

For monochromatic X-rays or neutrons, the on-axis transmission T_0 through the CRL is

$$T_0 = \frac{I_{\mathrm{CRL}}(0)}{I_0} \tag{10.226}$$

Given I_0, the intensity transmitted to the center image point is

$$T_0 = \frac{I_{\mathrm{CRL}}(0)}{I_0} = \frac{\pi}{r_i^2\lambda^2} \cdot \frac{\exp(-Nt\mu)}{\frac{1}{8}(\mu/f\delta)^2 + (\pi/2f\lambda)^2} \tag{10.227}$$

The point spread function (PSF) is the image plane intensity distribution along radius y. It is azimuthally symmetric and is produced by the single source point of intensity A_s^2 at radius s_0 in the object plane, where

$$\mathrm{PSF}(y) = I_1(y, s_0) \tag{10.228}$$

The image distance r_i is approximated by the CRL focal length

$$r_i \cong f = \frac{R}{2N\delta} \tag{10.229}$$

For a single X-ray or neutron wavelength λ, the PSF for an object point at the center of the object plane, $s_0 = 0$, is obtained by substitution of Eq. (10.216) into Eq. (10.228), where

$$\mathrm{PSF}(y) = \left(\frac{A_S}{r_0 r_i}\right)^2 \cdot \frac{\exp(-Nt\mu)}{\lambda^2} \cdot \left| \int_0^{r_a} J_0\left(\frac{2\pi}{\lambda} \cdot u \cdot \frac{y}{f}\right) \exp\left(-\frac{u^2}{r_{ap}^2}\right) \exp\left(\mathrm{i}\frac{\pi u^2}{2f\lambda}\right) u \, \mathrm{d}u \right|^2 \tag{10.230}$$

The field of view (FOV) of an object is the radial portion s of an object of radius s_a that contributes neutron or X-ray intensity to an observer at radius y_0 in the image plane. That is, at the image plane, one can observe object points at radial positions s within the FOV:

$$0 < s < s_{FOV} \tag{10.231}$$

And the radius s_{FOV} of the FOV is assumed to be confined to the object radius s_a, where

$$0 \le s_{FOV} \le s_a \tag{10.232}$$

For a single X-ray or neutron wavelength λ, the FOV is given by

$$\mathrm{FOV}(s) = I_1(s, y_0) \tag{10.233}$$

With substitution of Eq. (10.216) into Eq. (10.233), the FOV of an observer, which is situated at the center of the image plane at $y_0 = 0$, is then

$$\mathrm{FOV}(s) = \left(\frac{A_S}{r_0 r_i \lambda}\right)^2 \cdot \exp(-Nt\mu) \cdot \left| \int_0^{r_a} J_0\left(-\frac{2\pi}{\lambda} \cdot \frac{us}{r_0}\right) \exp\left(-\frac{u^2}{r_{ap}^2}\right) \exp\left(\mathrm{i}\frac{\pi u^2}{2f\lambda}\right) u \, \mathrm{d}u \right|^2 \tag{10.234}$$

10.13 Compound Refractive Fresnel Lens for X-rays and Neutrons

Neutrons and X-rays can be focused by diffraction using a compound refractive Fresnel lens (CFL). In comparison, compound refractive lenses (CRL) have a limited aperture due to material absorption. One means of overcoming the limited absorption aperture of CRLs is to use a Fresnel lens shape similar to lighthouse lenses. The concept of fabricating a larger, lighter lens by separating the surface gradient of the lens into separate annular sections, which are mounted on a base, dates from the 1700s, where single Fresnel lenses were first used as lighthouse lenses in the 1800s.

The Fresnel lens subdivides a spherical or parabolic lens into a set of concentric annular sections known as *Fresnel zones.* The annular, concentric sections or Fresnel zones retain the surface contour of the parabolic or spherical lens. However, the interior portion of the lens material is removed, and the remaining parabolic or spherical surface contours are set on a thin, flat planar plate. This results in the each Fresnel zone varying from the same maximum thickness to the same minimum zero thickness. Each Fresnel zone has the same axial thickness, and since the radial variation of the lens thickness increases with radial distance from the center z-axis of a planoconcave or planoconvex lens, the radial width of the Fresnel zones decreases with increasing radius.

Only the material interfaces of the Fresnel segments deflect and focus the neutrons. The absorbing segment behind the refractive segment is of no use, resulting in increased neutron absorption. Removing material in steps does not interfere with the ability of the lens to refract the neutrons. Since the Fresnel lens can be thin and its surface follows a parabolic shape in steps, the refractive Fresnel lens can achieve a larger clear aperture and provide a larger FOV with increased image intensity and possibly better resolution. Fresnel lenses are used for visible optics when a thin lens is needed.

The compound Fresnel lens (CFL) is comprised of a stack of N aligned Fresnel lenses (Figure 10.2). In a CFL, the steps in each of the N Fresnel lenses are placed so the power absorption of the CFL is reduced by the exponential attenuation factor

$$e^{-a} \tag{10.235}$$

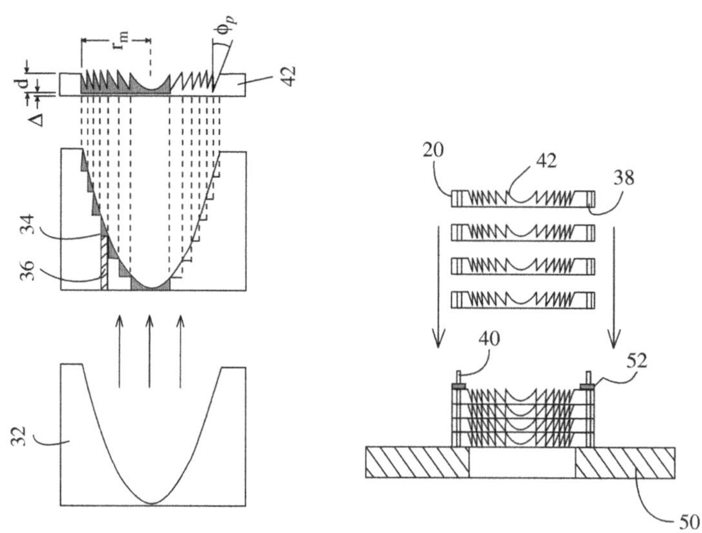

Figure 10.2 Shown at left is a diagram of the derivation of a Fresnel lens from a thick planoconcave lens. Seen at right is a diagram showing the stacking and precise alignment of N Fresnel lenses to form a CFL, which can be used as a large aperture condenser lens to focus X-rays or neutrons onto a sample. The sample can then be imaged by an objective lens CFL or CRL onto an imaging detector. (Piestrup et al., 2001).

The argument a of the exponential attenuation factor e^{-a} is the product of the linear attenuation μ of the Fresnel lens material, the number N of unit Fresnel lenses in the CFL, and the thickness T of the unit Fresnel lens, where

$$a = \mu NT \tag{10.236}$$

The unit Fresnel lens is parabolic-shaped with a parabolic lens-based thickness T, where

$$T = \frac{r_n^2 - r_{n-1}^2}{2R} \tag{10.237}$$

The scatter and absorption of neutron rays passing through the CFL are characterized by the linear attenuation coefficient $\mu\,[\mathrm{cm}^{-1}]$ of the CFL material. The thickness $t(l)$ of each Fresnel lens varies periodically from 0 to T as a function of radial distance l, measured from the lens center axis. Each Fresnel lens is mounted on a thin support base substrate of uniform thickness b comprised of material that is the same or different from the CFL material. The nth step location r_n is given by

$$r_n = \sqrt{2RnT} = \sqrt{\frac{2Rna}{N\mu}} = d_{\text{ap}}\sqrt{\frac{na}{8}} \tag{10.238}$$

In Figure 10.2, the base substrate thickness is indicated by Δ, where $b \equiv \Delta$; the maximum thickness of each lens zone is indicated by d, where $T \equiv d$; and the physical aperture radius of the CFL and each of the N Fresnel lenses is indicated by r_m.

Here, d_{ap} is the absorption aperture diameter $2r_{\text{ap}}$ for the stack of N Fresnel lenses, which is approximated by twice the absorption aperture radius r_{ap} of a parabolic-shaped lens. With Eq. (10.212) for the focal length f of a parabolic lens CRL, the parabolic lens CRL absorption aperture r_{ap} expression from Eq. (10.205), is

$$r_{\text{ap}} = 2\sqrt{\frac{f\delta}{\mu}} = \sqrt{\frac{2R}{N\mu}} \quad \text{and} \quad f = \frac{R}{2N\delta} \tag{10.239}$$

The maximum parabolic thickness $T_{\max}$ occurs at the zone edge and is termed *step thickness*, which can be conveniently expressed in terms of a and a parabolic on-axis radius of curvature R by transposing Eq. (10.238), where

$$T_{\max} = \frac{a d_{\text{ap}}^2}{16R} \tag{10.240}$$

With substitution of Eq. (10.240) into Eq. (10.238), the aperture diameter D_Z and radius r_Z of the CFL of Z-zones are then

$$D_Z = 2r_Z = \sqrt{\frac{8RZa}{N\mu}} = d_{\text{ap}}\sqrt{\frac{Za}{2}} \tag{10.241}$$

From Eq. (10.241), the outer zone radial width Δr_Z is

$$\Delta r_Z = d_{\text{ap}}\sqrt{\frac{a}{8}}\left[\sqrt{Z} - \sqrt{Z-1}\right] \tag{10.242}$$

The base thickness b, for each Fresnel lens of the CFL, contributes to an overall uniform decrease of transmission through the CFL, via $e^{-\mu b}$. A thin base, which readily transmits X-ray or neutron material, can yield a small μb product. For example, consider a 60% base transmission where

$$\mu Nb = 0.5 \Rightarrow \exp(-\mu Nb) = 0.606 \tag{10.243}$$

In this case, the overall transmission ranges from 13.5% at a radius of minimum CFL thickness to 8.2% at a radius of maximum CFL thickness.

To increase transmission, the Fresnel lens can be coated on a mounting substrate of different material whose thickness b and linear attenuation μ_b yield a $\mu_b b$ product, which is less than the Fresnel lens material μb product. By proper choice of substrate material, one can increase transmission through the CFL and improve the structural support of the Fresnel lenses in the CFL.

Consider an X-ray or neutron source, which illuminates an object imaged by a CFL onto an imaging detector. The center z-axis passes through the object center, CFL center, and image center. Neutrons with wavelength λ are emitted from an object at object point (s_x, s_y) at radial distance s from the center axis, where

$$s = \sqrt{s_x^2 + s_y^2} \tag{10.244}$$

At axial distance (object distance) r_0 downstream of the object point, the neutron ray intercepts the CFL entrance plane at point (l_x, l_y) at radial distance l from the center optical axis along z, where

$$l = \sqrt{l_x^2 + l_y^2} \tag{10.245}$$

The ray passes through the N Fresnel lenses of the CFL from the entrance plane to the exit plane. Provided that the radius of curvature R of each Fresnel lens is much larger than its thickness (thin-lens approximation), the very small decrement δ (where $\delta \ll 1$) causes the neutron ray to acquire a transverse deflection angle. However, the neutron ray does not have a significant change in its radial distance from the center axis of the CFL in the ray transit through the CFL. From the CFL exit plane, the ray travels image distance r_i to image point (h_x, h_y), which is situated in the image plane at radial distance y from the center axis, where

$$y = \sqrt{h_x^2 + h_y^2} \tag{10.246}$$

The X-ray or neutron trajectories must satisfy the focus condition for a single planoconcave parabolic lens, which allow those rays to contribute to the image for each wavelength λ. Each step of the Fresnel lens is comprised of an annular parabolic lens. For each Fresnel lens, b is the base thickness and Z is the number of zones. A stack of N Fresnel lenses comprises an N-lens CFL. Critical to the performance of the CFL is its PSF, which is equivalent to its impulse response. The image distance r_i for a single CFL planoconcave parabolic lens is approximated by its focal length f, given by

$$f = \frac{R}{\delta} \tag{10.247}$$

The focal length f for a CFL comprised of a stack of N Fresnel lenses is then

$$f = \frac{R}{N\delta} \tag{10.248}$$

An N-planoconcave, parabolic CFL is placed between the object and image planes. The CFL produces intensity $I_{\mathrm{CFL}}(y, s)$ at a single image point I at radius y in the image plane, by focusing rays from a single object point P at radius s_0 in the object plane. The object point P emits polychromatic X-rays or neutrons with distribution $\Gamma(\lambda)$. One can apply Eq. (10.217) for the CRL to the CFL, where integration is performed over the range of X-ray or neutron wavelengths from λ_1 to λ_2 of the source spectral distribution. A single zone CFL, which has a parabolic surface, is equivalent to a parabolic lens CRL, where from Eq. (10.217), the image intensity at image point I, is

$$I_{\mathrm{CFL}}(y, s) = \left(\frac{A_S}{r_0 r_i}\right)^2 \int_{\lambda_1}^{\lambda_2} \mathrm{d}\lambda \cdot \frac{\Gamma(\lambda)\exp(-Nb\mu_b)}{\lambda^2} \cdot \left| \int_0^{r_a} J_0\left(\frac{2\pi}{\lambda} \cdot u \cdot \left(\frac{y}{f} - \frac{s}{r_0}\right)\right) \exp\left(-\frac{u^2}{r_{\mathrm{ap}}^2}\right) \exp\left(i\pi\frac{u^2}{f\lambda}\right) u\,\mathrm{d}u \right|^2 \tag{10.249}$$

Integration must be performed separately for each of the Z-zones of the CFL. Integration for each zone n for $n = 1, 2, \ldots, Z$ extends over the radial interval of the zone. Each nth zone is bounded by an inner radius r_{n-1}, which has minimum thickness $T_{\min}$, and an outer radius r_n, which has maximum thickness $T_{\max}$.

The transmission through the zone maximum thickness or wall height $T_{\max}$ is a set fraction of the transmission through the zone minimum thickness $T_{\min}$, which is zero or almost zero. Commonly, one chooses an intensity transmission $F(T_{\max})$ for the maximum thickness $T_{\max}$, which is smaller than the intensity transmission $F(T_{\min})$ for minimum thickness $T_{\min}$ by a factor $e^{-2/N}$, where

$$F(T_{\max}) = F(T_{\min})\exp\left(-\frac{2}{N}\right) \quad \text{and} \quad T_{\min} \cong 0 \tag{10.250}$$

To obtain e^{-2}, the maximum allowed attenuation through the zone, one sets $a=2$ in Eq. (10.236), and the maximum thickness of Eq. (10.240), via Eq. (10.239), becomes

$$T_{\max}=\frac{a}{\mu N}=\frac{2}{\mu N} \tag{10.251}$$

For N lenses, the transmission e^{-a} through the CFL is $e^{-2}=0.135$ multiplied by the transmission F_{base} through both the N lens bases of thickness b and the base material linear attenuation μ_b (1/cm), where

$$F_{\text{base}}=\exp(-Nb\mu_b) \tag{10.252}$$

The integration over a single zone is from r_{n-1} to r_n, where

$$r_n=r_{\text{ap}}\sqrt{\frac{na}{2}}=\sqrt{2na\frac{f\delta}{\mu}} \tag{10.253}$$

For polychromatic sources, the X-ray or neutron rays transmitted from one zone have a fixed phase relationship (coherent), but the phase relationship between polychromatic rays transmitted through one zone and the other zones is not fixed (incoherent). Hence, one sums the intensity contributions from each zone at the image point. The intensity $I_{\text{CFL}}(y,s)$ at an image point at radius y in the image plane of an object point at radius s_0 in the object plane, which emits polychromatic X-rays or neutrons, is produced by an N-lens CFL that is centered in the aperture plane. Modification of Eq. (10.249) for a single zone CFL to an N zone CFL with parabolic surfaces, is then

$$I_{\text{CFL}}(y,s)=\left(\frac{A_S}{r_0 r_i}\right)^2\int_{\lambda_1}^{\lambda_2}\mathrm{d}\lambda\cdot\frac{\Gamma(\lambda)}{\lambda^2}\cdot\exp(-Nb\mu_b)\cdot$$
$$\sum_{n=0}^{Z-1}\left|\int_{r_{\text{ap}}\sqrt{an/2}}^{r_{\text{ap}}\sqrt{a(n+1)/2}}J_0\left(\frac{2\pi}{\lambda}\left(\frac{y}{f}-\frac{s}{r_0}\right)u\right)\exp\left[-u^2\left(\frac{1}{r_{ap}^2}-\mathrm{i}\frac{\pi}{f\lambda}\right)+\frac{an}{2}\left(1-\mathrm{i}\frac{r_{\text{ap}}^2\pi}{f\lambda}\right)\right]u\,\mathrm{d}u\right|^2 \tag{10.254}$$

The nonradial-dependent imaginary component of the exponential in Eq. (10.254), has a magnitude of unity, and moving the real nonradial real component of the exponential outside of the integral, one obtains

$$I_{\text{CFL}}(y,s)=\left(\frac{A_S}{r_0 r_i}\right)^2\int_{\lambda_1}^{\lambda_2}\mathrm{d}\lambda\cdot\frac{\Gamma(\lambda)}{\lambda^2}\cdot\exp(-Nb\mu_b)\cdot$$
$$\sum_{n=0}^{Z-1}\exp(an)\cdot\left|\int_{r_{\text{ap}}\sqrt{an/2}}^{r_{\text{ap}}\sqrt{a(n+1)/2}}J_0\left(\frac{2\pi}{\lambda}\left(\frac{y}{f}-\frac{s}{r_0}\right)u\right)\exp\left[-u^2\left(\frac{1}{r_{ap}^2}-\mathrm{i}\frac{\pi}{f\lambda}\right)\right]u\,\mathrm{d}u\right|^2 \tag{10.255}$$

PSF(y, s_0) or point spread function, is a radial-dependent image plane intensity distribution, which is obtained by fixing the object point radius $s = s_0$, and then plotting the intensity at an image point, as a function of its radius y in the image plane. The PSF at the image plane, produced by polychromatic X-ray or neutron emission from the center point $s_0 = 0$ in the object plane, is obtained from Eq. (10.256), where

$$\mathrm{PSF}(y,0) = I_{\mathrm{CFL}}(y,0) = \left(\frac{A_S}{r_0 r_i}\right)^2 \int_{\lambda_1}^{\lambda_2} \mathrm{d}\lambda \cdot \frac{\Gamma(\lambda)}{\lambda^2} \cdot \exp(-Nb\mu_b) \cdot \sum_{n=0}^{Z-1} \exp(an) \cdot \left| \int_{r_{\mathrm{ap}}\sqrt{an/2}}^{r_{\mathrm{ap}}\sqrt{a(n+1)/2}} J_0\left(\frac{2\pi}{\lambda}\cdot\frac{y}{f}\cdot u\right) \exp\left[-u^2\left(\frac{1}{r_{\mathrm{ap}}^2} - \mathrm{i}\frac{\pi}{f\lambda}\right)\right] u\,\mathrm{d}u \right|^2 \tag{10.256}$$

FOV(y_0, s) or field of view, is a radial-dependent object plane intensity distribution, which is obtained by fixing the image point radius $y = y_0$, and then plotting the intensity at an object point, as a function of its radius s in the object plane. The FOV of the object plane, as seen at a center point $y_0 = 0$ in the image plane, for the case of a polychromatic X-ray or neutron emission from the object points, is obtained from Eq. (10.256), where

$$\mathrm{FOV}(0,s) = I_{\mathrm{CFL}}(0,s) = \left(\frac{A_S}{r_0 r_i}\right)^2 \int_{\lambda_1}^{\lambda_2} \mathrm{d}\lambda \cdot \frac{\Gamma(\lambda)}{\lambda^2} \cdot \exp(-Nb\mu_b) \cdot \sum_{n=0}^{Z-1} \exp(an) \cdot \left| \int_{r_{\mathrm{ap}}\sqrt{an/2}}^{r_{\mathrm{ap}}\sqrt{a(n+1)/2}} J_0\left(-\frac{2\pi}{\lambda}\cdot\frac{s}{r_0}\cdot u\right) \exp\left[-u^2\left(\frac{1}{r_{\mathrm{ap}}^2} - i\frac{\pi}{f\lambda}\right)\right] u\,du \right|^2 \tag{10.257}$$

If the X-ray or neutron source is monochromatic with only one wavelength λ, then coherence, or a fixed phase relationship, exists between all the points (and the attached normal rays) along the entire wave front, which passes through the CFL. In this coherent case, the image intensity at radius y is obtained by modifying Eq. (10.255), where instead of summation of the intensities of the $Z - 1$ zones, the squared magnitude is taken of the sum amplitudes of the $Z - 1$ zones, where

$$I_{\mathrm{CFL}}^{\mathrm{coh}}(y,s) = \left(\frac{A_S}{r_0 r_i \lambda}\right)^2 \cdot \exp(-Nb\mu_b) \cdot \left| \sum_{n=0}^{Z-1} \int_{r_{\mathrm{ap}}\sqrt{an/2}}^{r_{\mathrm{ap}}\sqrt{a(n+1)/2}} J_0\left(\frac{2\pi}{\lambda}\left(\frac{y}{f} - \frac{s}{r_0}\right) u\right) \exp\left[-u^2\left(\frac{1}{r_{\mathrm{ap}}^2} - \mathrm{i}\frac{\pi}{f\lambda}\right) + \frac{an}{2}\left(1 - \mathrm{i}\frac{r_{\mathrm{ap}}^2 \pi}{f\lambda}\right)\right] u\,du \right|^2 \tag{10.258}$$

Now we return to the case of the polychromatic X-ray or neutron source and the CFL. The CFL inverts and magnifies, where

$$\frac{s}{r_0} = -\frac{y}{r_i} \tag{10.259}$$

The image distance r_i approximates the focal length f by

$$r_i \cong f \quad \text{and} \quad f = \frac{R}{N\delta} \tag{10.260}$$

With substitution of Eqs. (10.259–10.260) into Eq. (10.258), one can then obtain the image intensity $I_{\mathrm{CFL}}(y)$, which depends only on the radius y of the image point, where

$$I_{\mathrm{CFL}}(y) = \left(\frac{A_S}{r_0 r_i}\right)^2 \int_{\lambda_1}^{\lambda_2} \mathrm{d}\lambda \cdot \frac{\Gamma(\lambda)}{\lambda^2} \cdot \exp(-Nb\mu_b) \cdot$$
$$\sum_{n=0}^{Z-1} \exp(an) \left| \int_{r_{\mathrm{ap}}\sqrt{an/2}}^{r_{\mathrm{ap}}\sqrt{a(n+1)/2}} J_0\left(\frac{2\pi}{\lambda} \cdot \frac{y}{f} \cdot u\right) \exp\left[-u^2\left(\frac{1}{r_{ap}^2} - \mathrm{i}\frac{\pi}{f\lambda}\right)\right] u \, \mathrm{d}u \right|^2 \tag{10.261}$$

The zero-order Bessel function with a zero argument is unity, where

$$J_0(0) = 1 \tag{10.262}$$

With substitution of Eq. (10.262) into Eq. (10.261), the on-axis intensity $I_{\mathrm{CFL}}(0)$, which is produced by the CFL at the center point $y = 0$ in the image plane, is then

$$I_{\mathrm{CFL}}(0) = \left(\frac{A_S}{r_0 r_i}\right)^2$$
$$\int_{\lambda_1}^{\lambda_2} \mathrm{d}\lambda \cdot \frac{\Gamma(\lambda)}{\lambda^2} \cdot \exp(-Nb\mu_b) \cdot \sum_{n=0}^{Z-1} \exp(an) \left| \int_{r_{\mathrm{ap}}\sqrt{an/2}}^{r_{\mathrm{ap}}\sqrt{a(n+1)/2}} \exp\left[-u^2\left(\frac{1}{r_{\mathrm{ap}}^2} - \mathrm{i}\frac{\pi}{f\lambda}\right)\right] u \, \mathrm{d}u \right|^2 \tag{10.263}$$

Integrating and performing the summation in Eq. (10.263), yields the on-axis CFL, focused and transmitted intensity from an on-axis source point ($s = 0$) to an on-axis image point ($y = 0$):

$$I_{\mathrm{CFL}}(0) = \left(\frac{A_S}{2r_0 r_i}\right)^2 \int_{\lambda_1}^{\lambda_2} \mathrm{d}\lambda \cdot \frac{\Gamma(\lambda)}{\lambda^2} \cdot \frac{\exp(-Nb\mu_b)}{\frac{1}{8}(\mu/f\delta)^2 + (\pi/f\lambda)^2} \cdot Z\left[1 - \exp\left(-\frac{a}{2}\right)\right]^2 \tag{10.264}$$

The intensity I_0 of incident X-rays or neutrons at a point at the CFL is

$$I_0 = \frac{A_s^2}{4\pi \cdot r_0^2} \tag{10.265}$$

The on-axis transmission T_0 through the CFL is

$$T_0 = \frac{I_{\text{CFL}}(0)}{I_0} \tag{10.266}$$

Given I_0, the intensity T_0 transmitted to the center image point is

$$T_0 = \frac{I_{\text{CFL}}(0)}{I_0} = \frac{\pi}{r_i^2}\int_{\lambda_1}^{\lambda_2} \mathrm{d}\lambda \cdot \frac{\Gamma(\lambda)}{\lambda^2} \cdot \frac{\exp(-Nb\mu_b)}{\frac{1}{8}(\mu/f\delta)^2 + (\pi/f\lambda)^2} \cdot Z\left[1-\exp\left(-\frac{a}{2}\right)\right]^2 \tag{10.267}$$

The increase of focused and transmitted on-axis intensity from the center object point ($s = 0$) to the center image point ($y = 0$) by the CFL $I_{\text{CFL}}(0)$ of Eq. (10.264) over that of the CRL $I_{\text{CRL}}(0)$ of Eq. (10.224) is the ratio

$$\frac{I_{\text{CFL}}(0)}{I_{\text{CRL}}(0)} = Z \cdot \left[1-\exp\left(-\frac{a}{2}\right)\right]^2 \tag{10.268}$$

With Fresnel lenses, X-rays or neutrons can be reflected from the surfaces of the flat sides of the lenses and the radial surfaces of the zone walls. X-rays or neutrons may travel at a sufficiently large angle to move from one zone to another as they pass through the multiple unit lenses in the CRL. In this case, some neutrons will strike the zone wall. For the larger inner zones, the cross section of the wall is negligible; however, for thin outer zones, the probability of striking the wall will be larger.

As an example, for an amorphous Teflon CFL for neutrons at 1.8 Å with $N = 100$ Fresnel lenses, and with maximum attenuation per zone set at $a = 0.939$ for e^{-a} (where $a = \mu N T_{\max}$), the zone wall height is $T_{\max} = 380\ \mu m$. Given a 1.0 m focal length and FOV of 2.0 cm, the maximum ray angle through the lens is less than 10 mrad. Thus, the radial thickness of a parabolic zone, which is struck by the neutrons on its flat side wall surface (with radial outward surface normal), is about 3.8 μm. For $Z = 5$, the outer zone size is 89 μm. Thus, the radial dimensions of the side, flat wall that scatters neutrons in each nth zone is small compared to the radial dimensions of the parabolic surface of the same n^{th} zone that refracts neutrons.

An X-ray or neutron entering the central region of the lens is unlikely to strike a zone flat wall, but a neutron entering in the outer zone often has a greater than 50% probability of striking a zone flat wall. However, X-rays or neutrons, which through the flat wall surface, have a greater likelihood of being absorbed, or being transmitted to points outside the region of interest in the image plane. The critical grazing angle for total reflection of neutrons from amorphous Teflon at 1.8 Å is about 2 mrad, so many of the somewhat collimated neutrons from the source or specimen object that strike a highly polished, flat zone wall surface will be reflected. This reflection can be avoided by not polishing the zone wall, so the neutron will pass through the zone wall relatively unaffected.

If an X-ray or neutron does not strike the steep wall between zones, it will refract just as it does for a parabolic lens. Ignoring losses to the walls, each zone must be about twice the median misalignment, which preserves the transmitted intensity, and where resolution is not affected significantly by misalignment. Essentially, the effective aperture is reduced, where there is random variation of the unit lenses off the average optical axis of the lenses. If those X-rays or neutrons, which do pass through the region of the walls, and are not scattered, then alignment will be no more critical than for the parabolic case. If scattering from the boundary walls is significant, then alignment should be on the order of the radial extent of the wall, that is, on the order of 10–20 μm.

A tolerance estimate may be given for the step size for zone coherence by allowing a phase error of $\pm 0.05 \cdot (2\pi)$. Each zone of the Fresnel lens varies from zero thickness to a maximum thickness or step size T. If the errors are random, this corresponds to a thickness error, ΔT. The phase shift condition occurs between the Fresnel lens material with refractive index $n = 1 - \delta$ and the surrounding vacuum (air) with refractive index $n = 1$. The refractive index $n = v/c$ gives the relative phase velocity of the X-ray or neutron in the material compared to that in a vacuum, where v is the phase velocity in the material, and c is the phase velocity in a vacuum (air).

The difference in time Δt, which is needed to traverse the same distance interval ΔT in a vacuum (air) or material, leads to a phase shift $\Delta\phi$. Equivalently, the same phase shift $\Delta\phi$ occurs if equal time intervals $\Delta t = 0$ are used instead, which lead to a difference $\delta\,\Delta T$ in distance interval (or optical path segment) traversed. The phase shift condition, which occurs when a random step size variation occurs in the N Fresnel lenses of the CFL, is then

$$\sqrt{N}\left(2\pi\frac{\delta\Delta T}{\lambda}\right) = \pm 0.05(2\pi) \tag{10.269}$$

Hence, the tolerance on the step size is

$$\Delta T = \frac{\pm 0.05\lambda}{\delta\sqrt{N}} \tag{10.270}$$

If the error is not random, then

$$\Delta T \sim \frac{1}{N} \tag{10.271}$$

For Teflon at $\lambda = 1.8$ Å, using $N = 78$ unit lenses, the tolerance is $\Delta T = \pm 0.38$ *μm*.

10.14 Fresnel Diffraction of X-rays or Neutrons from a Circular Aperture

We now consider the diffraction of X-rays or neutrons from a circular aperture in the near-field, which is Fresnel diffraction, as compared to the previously derived far-field or Fraunhofer diffraction. In Fresnel diffraction, the diffracted X-ray or neutron amplitude at a nearby observation point I is considered for a distant or close isotropic source point P. The Fresnel or near-field condition is applied to the previously Kirchhoff–Fresnel diffraction equation. As a simplification, the source point P and observation point I are assumed to both lie on the z-axis, which passes through the aperture S_a center point L. The aperture is also assumed to be sufficiently small, so that the paraxial approximation applies, and the inclination factor $K(\theta_i, \theta_0) = 1$ is unity.

The spherical wave from the source point is approximated to be a planar wave at the aperture plane S_a. A spherical wave surface at the aperture plane S_a, or a planar wave surface at the aperture plane S_a, contain the same set of Fresnel zones, and thus the approximation of a planar wave at the aperture plane S_a simplifies the analysis. It is further simplified by having the source point P and observer point I lie on the center z-axis passing though the aperture S_a center point $L = 0$. The variable position vector $\mathbf{r}_0 = r_0\hat{\mathbf{r}}_0$ is between the on-axis source point P and any point L within the aperture plane S_a, and the variable position vector $\mathbf{r}_i = r_i\hat{\mathbf{r}}_i$ is between any point L in the aperture plane S_a and the on-axis observer point I.

Consider a circular aperture centered on the z-axis at radius $u = 0$. In the plane of the circular aperture S_a is position vector $\mathbf{u}$, where $\hat{\mathbf{u}}$ is the unit radius vector, and $\hat{\varphi}$ is the unit azimuth angle vector, so that in cylindrical coordinates with $z = 0$,

$$\mathbf{u} = u\hat{\mathbf{u}} \quad \text{and} \quad \mathrm{d}\mathbf{u} = \mathrm{d}u\hat{\mathbf{u}} + u\,\mathrm{d}\phi\hat{\boldsymbol{\varphi}} \tag{10.272}$$

One assumes azimuthal symmetry in this simplified near-field analysis. The circular aperture diameter d is assumed to be much less than the aperture distance R_0 to the source point P and the aperture distance R_i to observation point I, where

$$d \ll R_0 \quad \text{and} \quad d \ll R_i \tag{10.273}$$

Hence, the binomial approximation is used to obtain source (object) r_0 and observer (image) r_i distances:

$$r_0 = \sqrt{R_0^2 + u^2} = R_0\left(1 + \frac{u^2}{2R_0^2}\right) \tag{10.274}$$

and

$$r_i = \sqrt{R_i^2 + u^2} = R_i\left(1 + \frac{u^2}{2R_i^2}\right) \tag{10.275}$$

The Fresnel approximation is then invoked, where

$$r_0 r_i \cong R_0 R_i \tag{10.276}$$

Integration of ϕ over $[0, 2\pi]$ using the Fresnel approximation gives the amplitude $E_1(I)$ at the observer point I, which is near the circular aperture. This amplitude expression $E_1(I)$ is, then, the Kirchhoff–Fresnel diffraction formula for a circular aperture of diameter d, where

$$E_1(I) = -\frac{iA_S}{\lambda}\exp(-iwt)\frac{\exp[ik(R_i + R_0)]}{R_0 R_i}2\pi\int_0^{d/2} u\exp\left[\frac{iku^2}{2}\left(\frac{1}{R_0}+\frac{1}{R_i}\right)\right]du \tag{10.277}$$

In the near-field observer amplitude expression $E_1(I)$ for a circular aperture, one can divide the integral over the aperture surface into zones, where the distance from points L in each single zone to the observation point I, differ by half a wavelength or less. This allows amplitude contributions from the points in a single zone to constructively interfere (sum) at the observation point. The zones are indexed by integer n for a total of N zones, which just cover the circular aperture, so that

$$n\frac{\lambda}{2} \geq \frac{u^2}{2}\left(\frac{1}{R_0}+\frac{1}{R_i}\right) \quad n = 1, 2, 3\ldots, N \tag{10.278}$$

In this integral over the circular aperture, the differential area dS_a can be written as

$$dS_a = 2\pi u\, du = 2\pi u\frac{du}{dn}dn = \lambda\pi\left(\frac{1}{R_0}+\frac{1}{R_i}\right)^{-1} dn \tag{10.279}$$

Integrating dS_a over one zone for $\Delta n = 1$, yields the area, where each full zone ΔS_a is constant:

$$\Delta S_a = \int_{n-1}^{n} \lambda\pi\left(\frac{1}{R_0}+\frac{1}{R_i}\right)^{-1} dn = \lambda\pi\left(\frac{1}{R_0}+\frac{1}{R_i}\right)^{-1} \tag{10.280}$$

With substitution of Eq. (10.279) into Eq. (10.277), the contribution to the amplitude at the observation point from one full zone is then

$$E_1(I) = -iA_S\pi\exp(-iwt)\frac{\exp[ik(R_i + R_0)]}{R_0 + R_i}\int_{n-1}^{n} \exp(in\pi)dn \tag{10.281}$$

With Eq. (10.280), performing the integration in Eq. (10.281), yields the amplitude at the nearby observer point I from each nth zone, where

$$E_1(I) = 2A_S(-1)^{n-1}\exp(-iwt)\frac{\exp[ik(R_i + R_0)]}{R_0 + R_i} \tag{10.282}$$

Suppose that the circular aperture of radius $d/2$, is covered by N zones, where

$$n = \frac{u^2}{\lambda}\left(\frac{1}{R_0} + \frac{1}{R_i}\right) \qquad dn = \frac{2u}{\lambda}\left(\frac{1}{R_0} + \frac{1}{R_i}\right) \tag{10.283}$$

One substitutes Eq. (10.283), into Eq. (10.277), where integral limit $d/2$ is replaced by N, and obtains the summed amplitude at the nearby observation point I from these N zones:

$$E_1(I) = -\,\mathrm{i}A_S\pi\exp(-\,\mathrm{i}wt)\frac{\exp[\mathrm{i}k(R_i + R_0)]}{R_0 + R_i}\int_0^N \exp(\mathrm{i}n\pi)\mathrm{d}n \tag{10.284}$$

Application of the Euler formula shows:

$$\mathrm{i} = \exp\left(\frac{\mathrm{i}\pi}{2}\right) \tag{10.285}$$

With application of Eq. (10.285) to Eq. (10.284), one integrates to find the amplitude at nearby observation point I from all N zones, which just cover the circular aperture, where

$$E_1(I) = 2\mathrm{i}^{N-1}A_S\exp(-\,\mathrm{i}wt)\frac{\exp[\mathrm{i}k(R_i + R_0)]}{R_0 + R_i}\sin\frac{N\pi}{2} \tag{10.286}$$

The squared magnitude of the amplitude of Eq. (10.286), yields the corresponding intensity $I_1(I)$ at the observation point I from the entire aperture, where

$$I_1(I) = |E_1(I)|^2 = \left(\frac{2A_S}{R_0 + R_i}\right)^2 \sin^2\frac{N\pi}{2} \tag{10.287}$$

Without a dividing wall and aperture (i.e., open space in the dividing wall between source and observer), the amplitude $E_0(I)$ and intensity $I_0(I)$ at observation point I at distance $R_0 + R_i$ from the source point S are

$$E_0(I) = \frac{A_S\exp(\mathrm{i}wt)}{R_0 + R_i}\exp[\mathrm{i}k(R_0 + R_i)] \tag{10.288}$$

and

$$I_0(I) = |E_0(I)|^2 = \left(\frac{A_S}{R_0 + R_i}\right)^2 \tag{10.289}$$

Hence, the intensity $I_1(I)$ at the nearby observation point I, which is caused by the aperture in the dividing wall, can be related to the intensity $I_0(I)$ without a wall and aperture, where

$$I_1(I) = 4I_0(I)\sin^2\frac{N\pi}{2} = 2I_0(I)(1 - \cos N\pi) \tag{10.290}$$

The intensity at observation point I is a maximum $2I_0(I)$, when an odd number N of zones covers the circular aperture. The number of zones N, which cover a circular aperture, changes with aperture diameter d, and also changes with the axial distance R_0 from the aperture center point C to the source point S, and/or axial distance to the observation point R_i.

The neutron or X-ray current density J_{in}, which has group velocity v_g, and originates at a source point P, and then arrives at the observation point I in free space, without aperture and dividing wall, is

$$J_0 = v_g I_0(I) \tag{10.291}$$

However, with a dividing wall and aperture between source point S and observation point I, the current density is

$$J_1 = v_g I_1(I) \tag{10.292}$$

The group velocity is not changed by passing through the aperture. In the paraxial approximation with a nearby source near the axis, the flux ratio F of the aperture for the neutrons or X-rays, which is expressed in terms of the number N of Fresnel zones that just cover the circular aperture, is then obtained via Eqs. (10.290–10.292), where

$$F = \frac{J_1}{J_0} = \frac{|E_1(I)|^2}{|E_0(I)|^2} = 4\sin^2\frac{N\pi}{2} = 2(1 - \cos N\pi) \tag{10.293}$$

That is, the flux ratio F is the ratio of the current density at the observer position I on the center z-axis, with dividing wall and aperture, divided by the current density with no dividing wall (nor aperture). Here the flux ratio F is 4 for an odd number of Fresnel zones, and zero for an even number of Fresnel zones, where

$$F = 4 \quad \text{odd } N = 1, 3, 5, \ldots \tag{10.294}$$

$$F = 0 \quad \text{even } N = 2, 4, 6, \ldots \tag{10.295}$$

Flux ratio F is also the differential cross section for forward directed X-rays or neutrons along the z-axis, which is due to the presence of the aperture in the dividing wall between the on-axis source point P and image point I.

10.15 Fresnel Diffraction of X-rays or Neutrons from a Rectangular Aperture

Consider now the Fresnel diffraction of the z-directed, neutrons or X-rays from a rectangular aperture. As with Eq. (10.277) for the circular aperture, the source point P and image point I lie on the z-axis, which passes through the center of the rectangular aperture. The on-axis image point I is near the rectangular aperture. The rectangular x- and y-coordinates of the position vector $\mathbf{u}$ in the transverse, aperture plane, are u_x and u_y, respectively, where

$$\mathbf{u} = u_x\hat{\mathbf{x}} + u_y\hat{\mathbf{y}} = u\hat{\mathbf{r}} = \hat{\mathbf{r}}\sqrt{u_x^2 + u_y^2} \tag{10.296}$$

The rectangular aperture in the transverse, xy-plane extends from $[-x_a, x_a]$ along the x-axis for u_x, and $[-y_a, y_a]$ along the y-axis for u_y. One recalls Eqs. (10.274–10.275), where the binomial approximation is applied, thereby yielding the Fresnel approximation, where

$$2\pi\left(\frac{r_i + r_0}{\lambda}\right) = 2\pi\left(\frac{\sqrt{R_0^2 + u^2} + \sqrt{R_i^2 + u^2}}{\lambda}\right) \cong \frac{2\pi}{\lambda}\left[R_0\left(1 + \frac{u^2}{2R_0^2}\right) + R_i\left(1 + \frac{u^2}{2R_i^2}\right)\right] \tag{10.297}$$

Collecting terms in Eq. (10.297), the Fresnel approximations for the sum $r_0 + r_i$ of the object r_0 and image r_i distances, as well as the product $r_0 r_i$, are then

$$2\pi\left(\frac{r_i + r_0}{\lambda}\right) \cong \frac{2\pi}{\lambda}\left[R_0 + R_i + \left(\frac{1}{2R_i} + \frac{1}{2R_0}\right)\left(u_x^2 + u_y^2\right)\right] \quad \text{and} \quad r_0 r_i \cong R_0 R_i \tag{10.298}$$

From Eq. (10.31) with Eqs. (10.65) and (10.298), one then obtains the Kirchhoff–Fresnel diffraction formula for a rectangular aperture, which gives the amplitude $E_1(\mathrm{I})$ at an on-axis observer point I, where

$$E_1(\mathrm{I}) = -\frac{\mathrm{i}A_S}{\lambda}\exp(-\mathrm{i}wt)\frac{\exp[\mathrm{i}k(R_i + R_0)]}{R_0 R_i}\int_{-x_a}^{x_a}\exp\left(\frac{iku_x^2}{2f}\right)\mathrm{d}u_x\int_{-y_a}^{y_a}\exp\left(\frac{iku_y^2}{2f}\right)\mathrm{d}u_y \tag{10.299}$$

Here, constant f, termed "focal length," is suggested by its relation with the source distance R_0 and observation point distance R_i from the aperture center C, where

$$\frac{1}{f} = \frac{1}{R_0} + \frac{1}{R_i} \tag{10.300}$$

The integral in Eq. (10.299) is rearranged, via variable substitution, and converted into the standard Fresnel integral form, where the integral limits become

$$v_{ax} = x_a\sqrt{\frac{k}{f\pi}} \tag{10.301}$$

$$v_{ay} = y_a\sqrt{\frac{k}{f\pi}} \tag{10.302}$$

The amplitude at the on-axis observer point I becomes

$$E_1(I) = -\frac{\mathrm{i}fA_S}{2}\exp(-\mathrm{i}wt)\frac{\exp[\mathrm{i}k(R_i + R_0)]}{R_0 R_i}\int_{-v_{ax}}^{v_{ax}}\exp\left(\mathrm{i}\frac{\pi}{2}v_x^2\right)\mathrm{d}v_x\int_{-v_{ay}}^{v_{ay}}\exp\left(\mathrm{i}\frac{\pi}{2}v_y^2\right)\mathrm{d}v_y \tag{10.303}$$

The variables in the integrand of Eq. (10.303) are

$$v_x = u_x\sqrt{\frac{k}{f\pi}} \tag{10.304}$$

and

$$v_y = u_y\sqrt{\frac{k}{f\pi}} \tag{10.305}$$

With application of Euler's formula, the Fresnel integrals of Eq. (10.303) take the following form:

$$\int_0^u \exp\left(\mathrm{i}\frac{\pi}{2}u^2\right)\mathrm{d}u = C(u) + \mathrm{i}S(u) \tag{10.306}$$

where

$$C(u) = \int_0^u \cos\left(\mathrm{i}\frac{\pi}{2}u^2\right)\mathrm{d}u \tag{10.307}$$

and

$$S(u) = \int_0^u \sin\left(\mathrm{i}\frac{\pi}{2}u^2\right)\mathrm{d}u \tag{10.308}$$

Taking the difference of Eq. (10.307), which is evaluated at different upper limits u_1 and u_2, and repeating for Eq. (10.308), and then applying the results to Eq. (10.306) with lower and upper limits u_1 and u_2, yields the Fresnel integrals in the near-field diffraction formulas for the rectangular and circular apertures, where

$$\int_{u_1}^{u_2} \exp\left(\mathrm{i}\frac{\pi}{2}u^2\right)\mathrm{d}u = [C(u_2) - C(u_1)] + \mathrm{i}[S(u_2) - S(u_1)] = R\cdot\exp(\mathrm{i}\phi) \tag{10.309}$$

In the complex plane $(C(u), S(u))$ is plotted $C(u) + \mathrm{i}S(u)$ as a function of u, thereby forming the Cornu spiral, where $C(u)$ is the real axis and $S(u)$ is the imaginary axis. The magnitude R is the length of the straight line that connects point $(C(u_2), S(u_2))$ to point $(C(u_1), S(u_1))$, i.e., distance R and phase angle φ is the angle of the line measured relative to the real $C(u)$ axis:

$$R = \sqrt{[C(u_2) - C(u_1)]^2 + [S(u_2) - S(u_1)]^2} \tag{10.310}$$

and

$$\varphi = \tan^{-1}\frac{S(u_2) - S(u_1)}{C(u_2) - C(u_1)} \tag{10.311}$$

The Cornu spiral is antisymmetric, and hence

$$C(-u) = -C(u) \tag{10.312}$$

and

$$S(-u) = -S(u) \tag{10.313}$$

The z-axis connecting the source point P to the image point I is the center axis of the rectangular aperture, so that

$$u_2 = -u_1 = u \tag{10.314}$$

The distance R and phase angle φ are

$$R = 2\sqrt{C^2(u) + S^2(u)} \tag{10.315}$$

and

$$\varphi = \tan^{-1}[S(u)/C(u)] \tag{10.316}$$

Since

$$C(\pm\infty) = \pm 0.5 \quad \text{and} \quad S(\pm\infty) = \pm 0.5 \tag{10.317}$$

integration over

$$[u_1, u_2] = [-\infty, \infty] \tag{10.318}$$

gives

$$R = \frac{1}{\sqrt{2}} \tag{10.319}$$

From the above Fresnel integral expressions of Eq. (10.309), and 10.315–10.316, the amplitude $E_1(I)$ and intensity $I_1(I)$ at an on-axis observer point I, which is near the rectangular aperture, can then be determined, where

$$E_1(I) = \frac{-2\mathrm{i}fA_S}{R_0R_i}\exp(-\mathrm{i}wt)\exp[\mathrm{i}k(R_i + R_0)]\exp\left(\mathrm{i}\tan^{-1}\left[\frac{S(u)}{C(u)}\right]\right) \sqrt{C^2(v_{ax}) + S^2(v_{ax})}\sqrt{C^2(v_{ay}) + S^2(v_{ay})} \tag{10.320}$$

The intensity $I_1(I)$ at the on-axis observer point I is proportional to the square of the distance R:

$$I_1(I) \propto R^2 \tag{10.321}$$

The intensity $I_1(I)$ at the on-axis observer point I can be applied to the case of an off-axis observer point I at (x,y), if (1) the z-axis, which passes through the rectangular aperture to the observer point I, is fixed, and (2) the aperture center is shifted from the z-axis at (0,0) to the off-axis point (x,y). In this case with the aperture centered at (x,y), the values change in the integral limits of Eqs. (10.301–10.302), and the integrand variables of Eqs. (10.304–10.305). The intensity at the on-axis or off-axis image point I, is then the squared magnitude of Eq. (10.320), where

$$I_1(I) = \left(\frac{2fA_S}{R_0R_i}\right)^2 \left[C^2(v_{ax}) + S^2(v_{ax})\right] \cdot \left[C^2(v_{ay}) + S^2(v_{ay})\right] \tag{10.322}$$

10.16 Fresnel Diffraction of X-rays or Neutrons from a Knife Edge

The knife edge is a particular limiting case of a rectangular aperture, and we now consider the Fresnel diffraction of X-ray or neutron waves from a knife edge. The integral limits of Eq. (10.302) are taken to infinity in Eq. (10.303), which converts the rectangular aperture to a slit, where

$$v_{ay} \rightarrow \infty \quad \text{and} \quad -v_{ay} \rightarrow -\infty \tag{10.323}$$

In Eq. (10.303), the integration over v_y gives

$$\int_{-\infty}^{\infty} \exp\left(i\frac{\pi}{2}v_y^2\right)\mathrm{d}v_y = \frac{1}{\sqrt{2}} \tag{10.324}$$

With substitution of Eq. (10.324) into Eq. (10.303), the Fresnel diffraction by a slit produces amplitude $E_1(I)$ and intensity $I_1(I)$ at the observation point I, where

$$E_1(I) = -\frac{\mathrm{i}2fA_S}{\sqrt{2}}\exp(-\mathrm{i}wt)\frac{\exp[\mathrm{i}k(R_i + R_0)]}{R_0R_i}\int_{-v_{ax}}^{v_{ax}} \exp\left(\mathrm{i}\frac{\pi}{2}v_x^2\right)\mathrm{d}v_x \tag{10.325}$$

and

$$I_1(I) = \left(\frac{2fA_S}{R_0R_i}\right)^2\left[C^2(v_{ax}) + S^2(v_{ax})\right] \tag{10.326}$$

In the case of a knife edge, where the slit opens up one side,

$$-v_{ax} = 0 \tag{10.327}$$

and

$$v_{ax} = x_a\sqrt{\frac{k}{f\pi}} \tag{10.328}$$

The Fresnel diffraction by a knife edge produces amplitude $E_1(I)$ and intensity $I_1(I)$ at the observation point I, where

$$E_1(I) = -\frac{\mathrm{i}2fA_S}{\sqrt{2}}\exp(-\mathrm{i}wt)\frac{\exp[\mathrm{i}k(R_i + R_0)]}{R_0R_i}\int_{0}^{v_{ax}} \exp\left(\mathrm{i}\frac{\pi}{2}v_x^2\right)\mathrm{d}v_x \tag{10.329}$$

and

$$I_1(I) = \frac{1}{2}\left(\frac{2fA_S}{R_0R_i}\right)^2\left[C^2(v_{ax}) + S^2(v_{ax})\right] \tag{10.330}$$

The far-field Fraunhofer diffraction pattern is the limit of Fresnel diffraction, and this occurs when only a portion of the first Fresnel zone covers the aperture. It can be accomplished by increasing the distance of the source point and/or observation point from the aperture, as well as decreasing the aperture diameter. Classical geometric optics dominates, and diffraction becomes negligible when the source aperture dimensions are much larger than the neutron or X-ray wavelength. Likewise, classical mechanics is valid in the limit where the particle wavelength $\lambda = h/mv$ is small compared to the dimensions of the interacting objects and apertures; otherwise, quantum mechanics is required. Diffractive optics and quantum mechanics, where the photon or particle wavelength is comparable to the dimensions of interacting objects and apertures, require the summation of complex wave amplitudes of the incident and scattered waves.

10.17 Fresnel Zone Plates (FZP) for X-rays or Neutrons

Neutrons or X-rays can be focused by diffraction using a FZP such as the one shown in Figure 10.3. FZPs are alternating, flat, annular transparent, and opaque rings, where the radial thickness of the zone decreases with increasing radius. The radial thickness, which decreases as the radial distance increases from the center z-axis, allows successive flat surface rings to be 180° out of phase with the preceding flat surface annular ring. This allows constructive 360° phase shifts for each transparent ring and the preceding and succeeding transparent rings. The opaque rings between each pair of transparent rings do not contribute destructive 180° phase-shift contributions to the image point. Alternatively, the opaque rings can be replaced by transmitting rings of a different thickness than their preceding and succeeding transmitting neighbors. In this case, the phase shift of each pair of neighboring flat rings, which have different axial thickness, can transmit photons or neutrons to an image point, where the phase shifts have been set for a constructive 360°.

The radius of the annular zones varies so that the path-length difference between every other zone is a wavelength λ. The alternating zones can be either opaque or made of a given thickness of material that causes a 180° phase shift (a "phase-reversal" FZP). In both cases, which are incident perpendicular to the zone plate

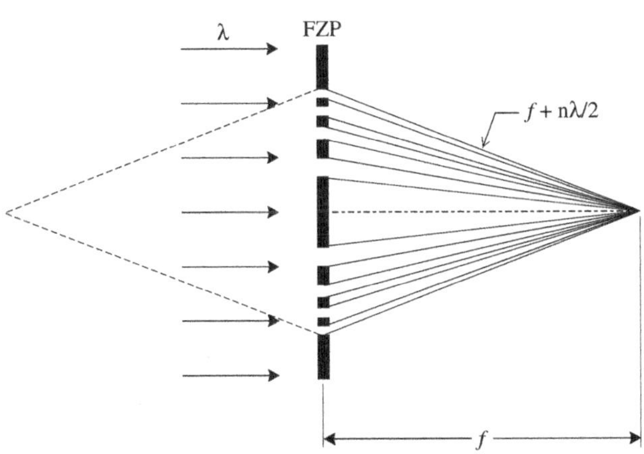

Figure 10.3 A schematic diagram of an FZP. Incident X-rays or neutrons of wavelength λ are diffracted by the radial zones. The radial width of each of the nth zones varies, so the diffracted radiation is focused to the axis at a focal distance f.

surface, are diffracted from the zones, and add constructively at the focal point. For a given design focal length f, the required radius of each zone is given by

$$r_n^2 + f^2 = \left(f + \frac{n\lambda}{2}\right)^2 \tag{10.331}$$

If the design focal length f is assumed to be much larger than any of the path differences or multiples of the wavelength λ, then

$$f \gg \frac{n\lambda}{2} \tag{10.332}$$

Then, the radius r_n of the nth zone is

$$r_n \approx \sqrt{n\lambda f} \tag{10.333}$$

The radius r_1 of the first zone is denoted by l, where

$$r_1 = l \tag{10.334}$$

Since $f \gg \lambda$, one then obtains

$$f = \frac{l^2}{\lambda} \tag{10.335}$$

The focal length is determined by the radius l of the first zone. The aperture diameter D of a zone plate is simply the diameter of the last zone:

$$D = 2r_N = 2l\sqrt{N} \tag{10.336}$$

Here, N is the total number of zones. Typically, the manufacturing limit for FZPs is the width of the outer zone:

$$r_N - r_{N-1} = \Delta r \tag{10.337}$$

For a given focal length, the first zone radius l is given, so that Δr limits the total number of zones, which can be produced, and thus limits the aperture. The limiting aperture diameter D can be calculated from Δr via

$$r_N^2 - r_{N-1}^2 = \lambda f = r_N^2 - (r_N - \Delta r)^2 = 2r_N\Delta r - (\Delta r)^2 \approx 2r_N\Delta r = D\Delta r \tag{10.338}$$

or

$$D = \frac{f\lambda}{\Delta r} \tag{10.339}$$

If we assume a minimum zone spacing of 1 μm, for a 10 m focal length for 10 Å neutrons, then the FZP aperture diameter is $D = 1$ cm, which is similar to the apertures of CRLs for neutrons. Neutron FZPs have typically been phase plates, rather than opaque, due to the low absorption of neutrons in micron thicknesses of most materials.

10.18 X-ray or Neutron Achromat Fabricated from FZPs and CRLs

Chromatic aberration or wavelength-dependent aberrations of X-rays or neutrons can be corrected by achromatic lenses. Two types of chromatic aberration need to be corrected by an achromatic lens, namely, longitudinal aberration and lateral aberration. The change in the position of the focal spot of the lens with the wavelength of the radiation is *longitudinal aberration.* The change in the magnification of objects in the image plane with wavelength is *lateral* or *transverse chromatic aberration.* Consider the following two achromatic lens designs that can apply to X-rays or neutrons.

The first achromatic design corrects for transverse chromatic aberration. This achromat design is comprised of two focusing optics of the same material (e.g., CRL pair, CFL pair, or FZP pair) separated by a distance d as follows:

$$d = \frac{1}{2}(f_1 + f_2) \tag{10.340}$$

Here, f_1 and f_2 are the focal lengths of the lenses. The second focusing intercepts the diverging rays of different wavelengths and bends them to the extent that they are separated, but parallel, thereby reducing the transverse chromatic aberration at the image plane (especially for long image distances). However, the longitudinal chromatic aberration remains, where each neutron wavelength has a different image plane. While the focal length is invariant, giving constant magnification, the locations of the principal planes in the system move according to the wavelength.

A second method of making an achromat, requires two component lenses, which have different wavelength dispersions. This method is generally referred to as a Crown and Flint, because this achromat was first produced, using two types of glass (crown and flint), which had different dispersions for visible light. This design provides both lateral and longitudinal chromatic correction. One lens is positive and the other negative.

As derived in Section 10.17, FZPs, regardless of their design, all have wavelength dispersions proportional to λ^{-1}, where the FZP focal length is

$$f = \frac{l^2}{\lambda} \tag{10.341}$$

Here l is the first zone radius. For X-rays and neutrons, the wavelength dispersion of a CRL is proportional to λ^{-2}, where a CRL comprised of N biconcave parabolic lenses, and has focal length of

$$f = \frac{R}{2N\delta} \tag{10.342}$$

Given refractive index n, the decrement δ_0 is for a specific wavelength λ_0, where

$$\delta = \delta_0 \left(\frac{\lambda}{\lambda_0}\right)^2 \tag{10.343}$$

and

$$n = 1 - \delta \tag{10.344}$$

With neutrons, the optical constants (namely the decrement δ and the linear attenuation coefficient μ) are tabulated for individual isotopes and natural occurring elements by the National Institute of Standards and Technology (NIST) for room-temperature thermal neutrons at wavelength $\lambda_0 = 1.8$ Å. With X-rays, 1.0 Å is a convenient and relevant wavelength for the tabulation and comparison of optical constants. Since CRLs and FZPs have different dispersions, an achromat can be designed from their combination, analogous to the design of crown and flint achromats for visible optics.

The general design of the achromat proceeds as follows. Since CRL and FZP powers add up to give the total achromat power, and the power of an optic is inversely proportional to its focal length, then the total focal length of the achromat is expressed by

$$\frac{1}{f_{\text{total}}} = \frac{1}{f_{\text{CRL}}} + \frac{1}{f_{\text{FZP}}} \tag{10.345}$$

This assumes that the lenses are collocated. Since the CRL has a finite thickness, the length of the CRL must also be treated. Locating the FZP midway in the CRL would eliminate any problem with translating principal planes if necessary. Using the power P of the lenses, which is inversely proportional to the focal length $1/f$, the focusing power of the combined optics is a function of the individual focal lengths of the CRL and FZP. The total power of the achromat P_T is the sum of the CRL power P_{CRL} and the FZP power P_{FZP}, where

$$P_T = P_{\text{CRL}} + P_{\text{FZP}} \tag{10.346}$$

Substitution of Eqs. (10.342–10.343) into Eq. (10.345), and taking the reciprocal to get Eq. (10.346), gives the achromat power:

$$P_T = \frac{2N\delta_0}{R}\left(\frac{\lambda}{\lambda_0}\right)^2 + \frac{\lambda_0}{l^2}\left(\frac{\lambda}{\lambda_0}\right) \tag{10.347}$$

At reference wavelength λ_0, the CRL reference power P^0_{CRL} and CFL reference power P^0_{FZP} simplify the achromat power expression to

$$P_{\mathrm{T}} = P^{0}_{\mathrm{CRL}} \frac{\lambda^{2}}{\lambda_{0}^{2}} + P^{0}_{\mathrm{FZP}} \frac{\lambda}{\lambda_{0}} \tag{10.348}$$

To form an achromat, the derivative of the focal power of Eq. (10.348), is set equal to zero at the reference or design wavelength λ_0, where

$$\frac{\mathrm{d}}{\mathrm{d}\lambda} P_{\mathrm{T}} \bigg| \lambda_0 = 0 = \left(2P^{0}_{\mathrm{CRL}} \frac{\lambda}{\lambda_{0}^{2}} + P^{0}_{\mathrm{FZP}} \frac{1}{\lambda_{0}} \right)_{\lambda=\lambda_0} = 2P^{0}_{\mathrm{CRL}} + P^{0}_{\mathrm{FZP}} \tag{10.349}$$

Let us assume

$$\frac{\mathrm{d}P_{\mathrm{T}}}{\mathrm{d}\lambda} = 0 \tag{10.350}$$

Then, the variation of power with wavelength is minimized, and evaluation at the reference wavelength λ_0 yields the minimum chromatic aberration in the vicinity of λ_0, where

$$2P^{0}_{\mathrm{CRL}} = -P^{0}_{\mathrm{FZP}} \tag{10.351}$$

and

$$P_{\mathrm{T}}(\lambda_0) = -P^{0}_{\mathrm{CRL}} = \frac{1}{2} P^{0}_{\mathrm{FZP}} \tag{10.352}$$

One could also design the achromat by setting the focal length equal to two different wavelengths.

An important result of Eq. (10.352) is that the power of the CRL must have the opposite sign of the power of the FZP, and the power of the achromat must have the same sign as the FZP. Thus, to have a positive (focusing) achromat, the FZP must be a positive lens and the CRL a negative lens. FZPs are typically used as positive lenses, so this presents no particular design challenge. However, the design of a negative CRL requires the decrement $\delta < 0$ in the refractive index $n = 1 - \delta$ for biconcave lenses for a negative power, such as titanium. The use of biconvex lenses with $\delta < 0$, suffers from the large attenuation, which is due to the maximum thickness being at the center of the lens.

To include any aberration, such as spherical, gravity, or chromatic, we need to only include the appropriate term in the Kirchhoff formula before calculating the PSF. While this is easily stated, it can be exceedingly difficult, requiring extensive numerical analysis, to perform the given integrations. One must be careful in the choice of lens material and fabrication to reduce effects such as incoherent scattering, which can result from microcrystalline structures, porous regions, or impurities in the material.

10.19 The Helmholtz Differential Equation for X-rays and Neutrons

The solutions of the electromagnetic equation for X-ray photons, and the Schrödinger equation for neutrons for propagation in a vacuum, reduce to the Helmholtz equation. In free space in which there are no free charges or currents, the electromagnetic wave equation equates the second spatial derivative of the wave amplitude E to the its second spatial derivative, via the speed of light c proportionality constant:

$$\nabla^2 E - \frac{1}{c^2}\frac{\partial^2 E}{\partial t^2} = 0 \tag{10.353}$$

In free space, where the scattering potentials are zero, the Schrödinger wave equation equates the second derivative of the wave amplitude $\Psi(\mathbf{r}, t)$ to the first derivative of the wave amplitude, via the imaginary proportionality constant $2\text{im}/\hbar$, where

$$\nabla^2 \Psi + \frac{2\text{im}}{\hbar}\frac{\partial \Psi}{\partial t} = 0 \tag{10.354}$$

These equations are used for air for sufficiently short propagation distances, where attenuation and refraction by the air can be neglected.

A wave solution $E(\mathbf{r}, t)$ to the electromagnetic equation for X-rays is assumed to take the form

$$E(\mathbf{r}, t) = E_0(\mathbf{r})\exp(-\text{i}wt) \tag{10.355}$$

Substitution of $E(\mathbf{r}, t)$ into the electromagnetic wave equation of Eq. (10.353), yields the electromagnetic Helmholtz equation:

$$\nabla^2 E + k^2 E = 0 \tag{10.356}$$

A wave solution $\Psi(\mathbf{r}, t)$ to the Schrödinger equation for neutrons is assumed to take the form

$$\Psi(\mathbf{r}, t) = \Psi_0(\mathbf{r})\exp(-\text{i}wt) \tag{10.357}$$

Substitution of $\Psi(\mathbf{r}, t)$ into the Schrödinger equatiion of Eq. (10.354), yields the neutron wave Helmholtz equation:

$$\nabla^2 \Psi + k^2 \Psi = 0 \tag{10.358}$$

In free space, the electromagnetic wave number k is the angular wave frequency w divided by the speed of light in vacuum c, where

$$k = \frac{w}{c} \tag{10.359}$$

In free space, the total X-ray energy W is related to the X-ray momentum p by

$$W = \hbar w = \frac{wp}{k} \tag{10.360}$$

where

$$p = \hbar k = \frac{h}{\lambda} = \frac{hf}{\lambda f} = \frac{W}{c} \tag{10.361}$$

In a vacuum without gravity, magnetic, electric, or nuclear strong force fields, the neutron total energy W is related to its momentum of motion by

$$W = \frac{p^2}{2m} \tag{10.362}$$

In free space, the wave amplitude of the electric field for X-rays and the wave amplitude of the wave function of the neutron thus both obey Helmholtz's equation. In a material medium with refractive index n, the electromagnetic wave number k is modified to

$$k = \frac{nw}{c} \tag{10.363}$$

The photons are not affected by electric, magnetic, strong force, or ordinary, weak gravitational fields.

Neutrons, however, are affected by a material medium and/or medium with applied magnetic, electric, and gravitational fields. On entering a material or applied field medium, the neutron maintains its incident total energy W, but the neutron acquires interaction potential energy U at the expense of reduced kinetic energy K. The neutron kinetic energy K in a vacuum is its total energy W, where $K = W$. However, in the material or applied field medium, the neutron kinetic energy is reduced to $K = W - U$. Thus, in the material or applied field medium, the neutron wave number k is modified to

$$k = \sqrt{\frac{2m(W - U)}{\hbar^2}} = \sqrt{\frac{2mK}{\hbar^2}} \tag{10.364}$$

In a material medium, the neutron can acquire nuclear potential energy U_n and material magnetization potential energy U_M, and additionally, in the material medium or vacuum medium, it can acquire gravitational potential energy U_g, applied, free-space magnetic field potential energy U_B, or electric field U_E (e.g., atomic electric field). Thus, the total neutron energy W in a material and/or applied field medium can be

$$W = \frac{p^2}{2m} + U_n + U_M + U_g + U_B + U_E \tag{10.365}$$

To summarize, the neutron wave and the electromagnetic wave both obey the Helmholtz equation in free space and in materials. The wave solutions and analysis of the neutron wave and X-ray are thus identical. Consequently, the Huygens–Fresnel–Kirchhoff approach and equation and analysis can be applied to both X-rays and neutrons. The refractive index expression for neutrons and X-rays is identical, i.e.,

$$n = 1 - \delta + \beta \tag{10.366}$$

where

$$\beta = \frac{\mu}{2k} \tag{10.367}$$

With β, the differences between neutrons and X-rays are the values of the optical constants δ and β (or μ). The difference between the value of neutron and X-ray optical constants, results from the different physical nature of the neutron and X-ray, as well as their interactions with matter and applied fields such as magnetic, electric, and gravitational fields.

10.20 Derivation of the Helmholtz Paraxial Equation for a Gaussian, Spherical X-ray Laser Beam

Following Siegman (1971) and Milonni and Eberly (1988), we now derive the paraxial form of the Helmholtz equation for the spatially dependent amplitude $E(\mathbf{r})$ of a transverse and longitudinal coherent source of photons (X-ray laser), where $E(\mathbf{r})$ is the electric field amplitude of the electromagnetic wave. In particular, we examine the propagation of a Gaussian, spherical beam of electromagnetic radiation, which applies to visible light lasers as well as the newly developed X-ray lasers. The Helmholtz equation will be solved for a Gaussian spherical beam. The Gaussian spherical beam of an expanding, spherical, electromagnetic wave, which emerges from its beam waist in the laser source cavity, has a Gaussian-distributed amplitude on the spherical wave surface about the center ray of the beam. That is, the amplitude along the spherical surface of the expanding wave, decreases in a gaussian manner, from the center ray of the spherical wave. In addition, the spherical wave emission from its beam waist in the laser cavity, is anisotropic (directed) rather than isotropic.

The Gaussian, spherical wave solution applies expanding spherical waves, whose amplitude on the wave surface is Gaussian, rather than uniform. An expanding spherical wave, with uniform amplitude on the spherical wave surface, is not a beam, because the source emits energy uniformly in all directions. However, in a Gaussian, spherical wave, the decrease of amplitude on either side of a center ray, constitutes nonisotropic energy emission, i.e., a beam of directed energy. In the case of a visible light laser or an X-ray laser, the photon beam has spatial coherence over the transverse surface of the beam, and longitudinal (temporal) coherence in the direction of beam propagation. The solution of the Helmholtz equation for a Gaussian, spherical beam, can apply to the as-yet-undeveloped coherent beam of thermal or cold neutrons. In this case, the neutron wave amplitude $\Psi(\mathbf{r})$ is used in place of electric field amplitude $E(\mathbf{r})$ in the Helmholtz equation.

We start with the Helmholtz equation:

$$\nabla^2 E(\mathbf{r}) + k^2 E(\mathbf{r}) = 0 \qquad (10.368)$$

If the source is at the origin, where the observation (image) plane at $z = R$ is tangent to the expanding spherical wave of radius R, then the distance r from the source point P to an off-axis image point I on the observation (image) plane at transverse coordinates x, y is

$$r = \sqrt{x^2 + y^2 + R^2} \qquad (10.369)$$

The distance r can be approximated by its binomial expansion or Fresnel approximation, so that

$$r \cong R\left(1 + \frac{x^2 + y^2}{2R^2}\right) = R + \frac{x^2 + y^2}{2R} \quad x, y \ll R \qquad (10.370)$$

The spherical wave amplitude $E(\mathbf{r})$, which is valid for neutrons and X-rays, then becomes

$$E(\mathbf{r}) = \frac{E_{00}}{R}\exp(-\,\mathrm{i}kR)\exp\left(-\mathrm{i}k\frac{x^2 + y^2}{2R}\right) \quad x, y \ll R \qquad (10.371)$$

Isotropic X-ray emission, which propagates from a source point P, occurs in the case of simple spherical waves. In the case of the X-ray laser, the X-rays are created in minimum focal spot of radius w_0 at the center of a resonant cavity, and the X-rays propagate in a very narrow cone as a unidirectional beam. The cone angle is defined by a center vertex point at the center of the focal spot of the resonator cavity. The cone half-angle, whose angle θ is defined by wavelength λ and beam waist radius w_0, where

$$\theta = \frac{\lambda}{\pi w_0} \tag{10.372}$$

In the case of neutrons, which are focused by a CRL with long focal length f and aperture radius r_m, the neutrons diverge from the diffraction and aberration-limited focal spot diameter into a narrow cone with cone half-angle θ, thereby forming a somewhat unidirectional beam expanding neutrons, where

$$\theta = \frac{r_m}{f} \tag{10.373}$$

In the observation plane at time t and axial position $z = R$, one observes variation with transverse position x, y in the phase angle ϕ and radius of curvature r of the crossing spherical waves, where

$$\phi = \frac{\pi\sqrt{x^2 + y^2}}{\lambda R} \tag{10.374}$$

and

$$r = \sqrt{R^2 + x^2 + y^2} \tag{10.375}$$

However, the magnitude of the spherical wave amplitude is constant as a function of position $z = R$ in the observation plane, where

$$|E(R, t)| = \frac{E_0}{R} \tag{10.376}$$

A beam, on the other hand, requires a decrease in the wave amplitude E, as the transverse distance ρ along the wave front, increases from the center z-axis, where

$$\rho = \sqrt{x^2 + y^2} \tag{10.377}$$

To fulfill the requirement of limited transverse extent of the wave front, a modified spherical wave solution is required, where the magnitude of the wave amplitude decreases with transverse distance ρ from the center z-axis of the beam. The modified spherical wave solution $E(\mathbf{r})$ is separated into axial spatial dependence and transverse spatial dependence in the form

$$E(\mathbf{r}) = E_0(\mathbf{r})\exp(-\mathrm{i}kz) \tag{10.378}$$

Substitution of $E(\mathbf{r})$ into the Helmholtz equation of 10.368, gives

$$\left(\frac{\partial^2}{\partial x^2} + \frac{\partial^2}{\partial y^2} + \frac{\partial^2}{\partial z^2}\right)E_0(\mathbf{r})\exp(-\mathrm{i}kz) + k^2 E_0(\mathbf{r})\exp(-\mathrm{i}kz) = 0 \tag{10.379}$$

Eq. (10.379) is now inspected to determine whether one or more terms, whose values are much smaller than the other terms, can thus be dropped to simplify the equation without losing the essential predictions of the equation. First, one examines the z-dependent portion of Eq. (10.379). Note in Eq. (10.379), the second derivative with respect to z is

$$\frac{\partial^2}{\partial z^2}E_0(\mathbf{r})\exp(-ikz) = \left(\frac{\partial^2 E_0(\mathbf{r})}{\partial z^2} + 2ik\frac{\partial E_0(\mathbf{r})}{\partial z} - k^2 E_0(\mathbf{r})\right)\exp(-ikz) \quad (10.380)$$

The paraxial approximation can be invoked at this point to simplify the z-dependent Eq. (10.380). That is, one assumes that the variation of the amplitude $E_0(\mathbf{r})$, and the variation of the amplitude gradient in the z-direction $\partial E_0(\mathbf{r})/\partial z$, within the distance of one wavelength λ in the z-direction, are negligible, i.e.,

$$\lambda\left|\frac{\partial E_0(\mathbf{r})}{\partial z}\right| \ll |E_0(\mathbf{r})| \quad \text{and} \quad \lambda\left|\frac{\partial E_0(\mathbf{r})}{\partial z}\right| \ll |E_0(\mathbf{r})| \quad (10.381)$$

In terms of wave number, the paraxial approximation is

$$\left|\frac{\partial E_0(\mathbf{r})}{\partial z}\right| \ll k|E_0(\mathbf{r})| \quad \text{and} \quad \left|\frac{\partial E_0(\mathbf{r})}{\partial z}\right| \ll k|E_0(\mathbf{r})| \quad (10.382)$$

The other two spatial terms in x and y in Eq. (10.379) have no simplifying approximation. Hence, the paraxial approximation, which is applied to the Helmholtz equation of Eq. (10.379), yields paraxial wave equation

$$\frac{\partial^2 E_0(\mathbf{r})}{\partial x^2} + \frac{\partial^2 E_0(\mathbf{r})}{\partial y^2} + 2ik\frac{\partial E_0(\mathbf{r})}{\partial z} = 0 \quad x, y \ll R \quad (10.383)$$

This paraxial equation version of the Helmholtz equation has some similarity to the time-dependent Schrödinger wave equation for a neutron, whose motion is limited to the xy-plane. That is, the time variable t of the Schrödinger equation is replaced by the axial dimension variable z in the paraxial equation.

10.21 Solution of the Helmholtz Paraxial Equation for a Gaussian, Spherical X-ray Laser Beam

Following Eisberg and Resnick (1974), Milonni and Eberly (1988), and Siegman (1971), we now derive a solution of the Helmholtz paraxial equation for a Gaussian, spherical, X-ray laser beam. The solution $E(\mathbf{r})$ of the above paraxial wave equation of Eq. (10.383) for an X-ray laser beam is expected to be a spherical wave having a Gaussian variation in amplitude across the wave front. The Gaussian standard deviation or the X-ray beam spot radius $w(z)$ increases with axial distance z from the Gaussian source with spot radius $w(0)$ or w_0 at $z = 0$. In the case of X-rays, the power flow into the wave front is normalized to unity, where the power flow is the squared magnitude of the electric field intensity. The magnitude of the X-ray wave amplitude is then

$$|E(x,y,z)| = \sqrt{\frac{2}{\pi}} \frac{1}{w(z)} \exp\left(-\frac{x^2+y^2}{w^2(z)} \right) \tag{10.384}$$

where

$$\int_{-\infty}^{\infty} \int_{-\infty}^{\infty} |E(x,y,z)|^2 = 1 \tag{10.385}$$

The paraxial equation of Eq. (10.383) is solved by assuming a trial solution $E_0(\mathbf{r})$ of the form

$$E_0(\mathbf{r}) = E_{00} g\left(\frac{x}{w(z)} \right) h\left(\frac{y}{w(z)} \right) \exp(\mathrm{i}P(z)) \exp\left(\mathrm{i}k \frac{x^2+y^2}{2q(z)} \right) \tag{10.386}$$

Parameter $q(z)$ is defined as the complex radius of curvature, which is analogous to the radius of curvature R of a spherical wave. The complex radius of curvature $q(z)$ increases with increasing axial distance z. With the source at $z = 0$, the complex radius of curvature $q(0)$ is a constant q_0, where

$$q(z=0) = q_0 \tag{10.387}$$

With a spherical wave, the radius of curvature R increases linearly with axial distance z, where

$$R = z \tag{10.388}$$

The amplitude $E(\mathbf{r})$ is assumed to decrease with increasing transverse distance from the center z-axis in a Gaussian fashion. In Eq. (10.386), the relation between the complex radius of curvature $q(z)$, and the standard deviation of the Gaussian $w(z)$, which increases with axial distance z from the source spot radius w_0, must be

$$\frac{1}{q(z)} = \frac{2\mathrm{i}}{kw^2(z)} = \frac{\mathrm{i}\lambda}{\pi w^2(z)} \tag{10.389}$$

The minimum beam radius w_0, which is at the beam waist in the laser cavity, is

$$w(z=0) = w_0 \tag{10.390}$$

With Eq. (10.389), the exponential term in Eq. (10.386), becomes

$$\exp\left(ik\frac{x^2+y^2}{2q(z)}\right)=\exp\left(-\frac{x^2+y^2}{w^2(z)}\right) \tag{10.391}$$

The complex radius of curvature $q(z)$ also linearly increases with axial distance z, where

$$q(z)=q_0+z \tag{10.392}$$

From Eqs. (10.389–10.390),

$$\frac{1}{q_0}=\frac{\mathrm{i}\lambda}{\pi w_0^2} \tag{10.393}$$

The complex radius of curvature $q(z)$ should also account for the phase shift across the observation plane, as expressed in Eq 10.391 in terms of the radius of curvature R of the spherical wave

$$\exp\left(-ik\frac{x^2+y^2}{2R}\right) \tag{10.394}$$

This suggests that the real component of $1/q(z)$ should be similar to $1/R$. The real radius of curvature $R(z)$ should be a function of the axial propagation distance z of the spherical, Gaussian wave. At this point, we have the complex radius of curvature $q(z)$ in Eq. (10.386) in the form

$$\frac{1}{q(z)}=\frac{1}{R(z)}+\frac{\mathrm{i}\lambda}{\pi w^2(z)} \tag{10.395}$$

Next, we determine $R(z)$ and $w(z)$ in $1/q(z)$ of Eq. (10.395), where the complex radius of curvature $q(z)$ of Eq.10.392 is repeated below,

$$q(z)=q_0+z \tag{10.396}$$

Substitution into Eq. (10.396) of Eq. (10.393), for q_0 of Eq. (10.393), gives

$$q(z)=-\mathrm{i}z_0+z \tag{10.397}$$

where

$$z_0=\frac{\pi w_0^2}{\lambda} \tag{10.398}$$

Inverting Eq. (10.397), one obtains

$$\frac{1}{q(z)} = \frac{1}{-\mathrm{i}z_0 + z} = \frac{\mathrm{i}(z_0/z) + 1}{(z_0^2/z) + z} \tag{10.399}$$

The real and imaginary parts of Eq. (10.399) for $1/q(z)$, can be compared to the Eq. (10.395) for $1/q(z)$, which then determines the real part $R(z)$ of the complex radius of curvature $q(z)$, where

$$R(z) = \frac{z_0^2}{z} + z \tag{10.400}$$

Note, the radius of curvature $R(z)$ of the beam is infinity at $z = 0$, which corresponds to a plane wave at the source. The imaginary part of the complex radius of curvature $q(z)$, is obtained by comparison of Eqs. (10.395) and (10.399) for $1/q(z)$, where

$$\frac{\mathrm{i}\lambda}{\pi w^2(z)} = \frac{\mathrm{i}(z_0/z)}{(z_0^2/z) + z} = \frac{\mathrm{i}(1/z_0)}{1 + (z/z_0)^2} \tag{10.401}$$

Recall Eq. (10.398), which is the area of the beam waist, divided by the beam wavelength, where

$$z_0 = \pi \frac{w_0^2}{\lambda} \tag{10.402}$$

One substitutes Eq. (10.402) into 10.401, and obtains

$$\frac{\mathrm{i}\lambda}{\pi w^2(z)} = \frac{\mathrm{i}\lambda}{\pi w_0^2(1 + (z/z_0)^2)} \tag{10.403}$$

With transposition of Eq. (10.403), one determines the increase of the standard deviation of the X-ray laser beam radius $w(z)$, as a function of axial propagation distance z from the beam minimum radius w_0 at the source plane $z = 0$, where

$$w(z) = w_0 \sqrt{1 + \left(\frac{z}{z_0}\right)^2} \tag{10.404}$$

With the trial solution $E_0(\mathbf{r})$, a pair of variable changes in the last exponential term of Eq. (10.386), are done as follows:

$$u = \frac{x}{w(z)} \quad \text{and} \quad v = \frac{y}{w(z)} \tag{10.405}$$

where

$$E_0(\mathbf{r}) = E_{00} \cdot g(u) \cdot h(v) \cdot \exp(\mathrm{i}P(z)) \cdot \exp\left(\mathrm{i}k \frac{u^2 + v^2}{2q(z)} w^2\right) \tag{10.406}$$

The chain rule is applied to the $g(u)$ and $h(v)$ terms in Eq. (10.406), with substitution from Eq. (10.405), where

$$\frac{\partial g}{\partial x} = \frac{\partial g}{\partial u}\frac{\partial u}{\partial x} = \frac{1}{w(z)}\frac{\partial g}{\partial u} \tag{10.407}$$

$$\frac{\partial h}{\partial y} = \frac{\partial h}{\partial v}\frac{\partial v}{\partial y} = \frac{1}{w(z)}\frac{\partial h}{\partial v} \tag{10.408}$$

$$\frac{\partial^2 g}{\partial x^2} = \frac{1}{w^2(z)}\frac{\partial^2 g}{\partial u^2} \tag{10.409}$$

$$\frac{\partial^2 h}{\partial y^2} = \frac{1}{w^2(z)}\frac{\partial^2 h}{\partial v^2} \tag{10.410}$$

and

$$\frac{\partial g}{\partial z} = \frac{\partial g}{\partial u}\frac{\partial u}{\partial z} = -\frac{1}{w^2(z)}\frac{\partial w}{\partial z}\frac{\partial g}{\partial u} \tag{10.411}$$

$$\frac{\partial h}{\partial z} = \frac{\partial h}{\partial v}\frac{\partial v}{\partial z} = -\frac{1}{w^2(z)}\frac{\partial w}{\partial z}\frac{\partial h}{\partial v} \tag{10.412}$$

Substitution of the above trial solution of Eq. (10.406) into the paraxial equation of Eq. (10.383) with the chain rule expressions of Eqs. (10.407–10.412), leads to an equation, which allows separation variables. The equation consists of an x-dependent term, a y-dependent term, and a z-dependent term. After division by g and h, one finds

$$\frac{1}{g(u)}\left(\frac{\partial^2 g}{\partial u^2} - 4u\frac{\partial g}{\partial u}\right) + \frac{1}{h(v)}\left(\frac{\partial^2 h}{\partial v^2} - 4v\frac{\partial h}{\partial v}\right) + w^2(z)\left(\frac{2ik}{q(z)} - 2k\frac{\partial P}{\partial z}\right) = 0 \tag{10.413}$$

Eq. (10.413) can be solved for all x and y for a specified, but arbitrary z if (1) the x-dependent expression of g (the first term) equals a constant $-\lambda_1$, (2) the y-dependent expression of h (the second term) equals a constant $-\lambda_2$, and (3) the third term, which depends on $q(z)$, $w(z)$, and $p(z)$, equals constant $\lambda_1 + \lambda_2$. Application of these 3 requirements in Eq. (10.413), results in three ordinary differential equations:

$$\frac{1}{g(u)}\left(\frac{\partial^2 g}{\partial u^2} - 4u\frac{\partial g}{\partial u}\right) = -\lambda_1 \tag{10.414}$$

$$\frac{1}{h(v)}\left(\frac{\partial^2 h}{\partial v^2} + 4v\frac{\partial h}{\partial v}\right) = -\lambda_2 \tag{10.415}$$

$$w^2(z)\left(\frac{2ik}{q(z)} - 2k\frac{\partial P}{\partial z}\right) = \lambda_1 + \lambda_2 \tag{10.416}$$

In Eq. (10.416), one substitutes

$$q = -\mathrm{i}z_0 + z, \quad k = \frac{2\pi}{\lambda}, \quad w(z) = w_0\sqrt{1 + \left(\frac{z}{z_0}\right)^2} \tag{10.417}$$

With substitution of Eq. (10.417) into Eq. (10.416), one obtains

$$\frac{\partial P}{\partial z} = \frac{\mathrm{i}z}{z_0^2 + z^2} - \frac{z_0}{z_0^2 + z^2} - \frac{z_0(\lambda_1 + \lambda_2)}{z_0^2 + z^2}\frac{\lambda}{4\pi w_0^2} \tag{10.418}$$

Recall from Eq. (10.398),

$$\frac{\lambda}{w_0^2\pi} = \frac{1}{z_0} \tag{10.419}$$

With substitution of Eq. (10.419) into Eq. (10.418), third equation (Eq. (10.416)), then becomes

$$\frac{\partial P}{\partial z} = \frac{\mathrm{i}z}{z_0^2 + z^2} - \frac{z_0}{z_0^2 + z^2}\left(1 + \frac{\lambda_1 + \lambda_2}{4}\right) \tag{10.420}$$

Integration of Eq. (10.420), yields

$$P(z) = \mathrm{i}\ \ln\sqrt{1 + \frac{z^2}{z_0^2}} - \left(\frac{\lambda_1 + \lambda_2}{4} + 1\right)\tan^{-1}\left(\frac{z}{z_0}\right) \tag{10.421}$$

One substitutes $P(z)$ of Eq. (10.421) into its exponential term in trial solution Eq. (10.406) for differential equation Eq. (10.416), which yields

$$\begin{aligned} \exp(iP(z)) &= \frac{\exp\left[-\mathrm{i}((\lambda_1 + \lambda_2/4) + 1)\tan^{-1}(z/z_0)\right]}{\sqrt{1 + (z^2/z_0^2)}} \\ &= \frac{w_0}{w(z)}\exp\left[-\mathrm{i}\left(\frac{\lambda_1 + \lambda_2}{4} + 1\right)\tan^{-1}\left(\frac{z}{z_0}\right)\right] \end{aligned} \tag{10.422}$$

With the third equation (Eq. (10.416)) solved, the first two equations (10.414 and 10.415) are now solved as follows. A change of variable is performed in Eqs. (10.414) and (10.415), where

$$\xi = u\sqrt{2} = \frac{x\sqrt{2}}{w} \tag{10.423}$$

$$\eta = v\sqrt{2} = \frac{y\sqrt{2}}{w} \tag{10.424}$$

This results in two Hermite differential equations in x and y, or, respectively, ξ and η, where

$$\frac{\partial^2 g}{\partial \xi^2} - 2\xi \frac{\partial g}{\partial \xi} + \frac{\lambda_1}{2} g = 0 \tag{10.425}$$

$$\frac{\partial^2 h}{\partial \eta^2} - 2\eta \frac{\partial h}{\partial \eta} + \frac{\lambda_2}{2} h = 0 \tag{10.426}$$

These two Hermite equations of Eqs. (10.425) and (10.426) can be solved, by assuming a series solution $G(\xi)$ and $H(\eta)$ of the form

$$g(\xi) = \sum_{m=0}^{\infty} a_m \xi^m \tag{10.427}$$

$$h(\eta) = \sum_{n=0}^{\infty} b_n \eta^n \tag{10.428}$$

In Eqs. (10.427–10.428), the coefficients $a_0, a_1, a_2, \ldots$ and $b_0, b_1, b_2, \ldots$ are determined by plugging the polynomial solution $g(\xi)$ and $h(\eta)$ into their respective Hermite Eqs. (10.425) and (10.426). This results in new polynomial equations for ξ and η, where

$$\sum_{m=0}^{\infty} a_m \xi^m = 0 \tag{10.429}$$

$$\sum_{n=0}^{\infty} b_n \eta^n = 0 \tag{10.430}$$

In Eq. (10.429), each mth coefficient A_m is a function of a_m and a_{m+2}, and one requires each coefficient of each power of u to be zero, where the recursion relation is

$$(m+1)(m+2)a_{m+2} + \left(\frac{\lambda_1}{2} - 2m\right)a_m = 0 \tag{10.431}$$

Likewise in Eq. (10.430), each nth coefficient B_n is a function of b_n and b_{n+2}, and one requires each coefficient of each power of v to be zero, where the recursion relation is

$$(n+1)(n+2)b_{n+2} + \left(\frac{\lambda_2}{2} - 2n\right)b_n = 0 \tag{10.432}$$

In the recursion relation of Eq. (10.431) for large m,

$$\frac{a_{m+2}}{a_m} = \frac{2}{m} \tag{10.433}$$

The series of Eq. (10.433) then converges for large m,

$$\lim_{m\to\infty} \frac{2}{m} \to 0 \tag{10.434}$$

In the recursion relation of Eq. (10.432) for large n,

$$\frac{b_{n+2}}{b_n} = \frac{2}{n} \tag{10.435}$$

The series of Eq. (10.435) then converges for large n,

$$\lim_{n\to\infty} \frac{2}{n} \to 0 \tag{10.436}$$

The two Hermite equations of Eqs. (10.429) and (10.430) have a solution, only if the following parameters are true:

$$\frac{\lambda_1}{2} = 2m \quad m = 0, 1, 2, 3\ldots \tag{10.437}$$

$$\frac{\lambda_2}{2} = 2n \quad n = 0, 1, 2, 3\ldots \tag{10.438}$$

With the recursion relationships of Eqs. (10.431–10.432), with Eqs. (10.437–10.438), the even coefficients $a_2, a_4, a_6, \ldots$ and $b_2, b_4, b_6, \ldots$ are each expressed as a function of a_0. And the odd coefficients $a_3, a_5, a_7, \ldots$ and $b_3, b_5, b_7, \ldots$ are each expressed as functions of a_1 and b_1, so that the Hermite solutions of Eqs. (10.427) and (10.428) become

$$\begin{aligned} g(\xi) = a_0 &\left(1 + \frac{a_2}{a_0}\xi^2 + \frac{a_4}{a_2}\frac{a_2}{a_0}\xi^4 + \frac{a_6}{a_4}\frac{a_4}{a_2}\frac{a_2}{a_0}\xi^6 + \ldots\right) \\ &+ a_1\left(1 + \frac{a_3}{a_1}\xi^3 + \frac{a_5}{a_3}\frac{a_3}{a_1}\xi^5 + \frac{a_7}{a_5}\frac{a_5}{a_3}\frac{a_3}{a_1}\xi^7 + \ldots\right) \end{aligned} \tag{10.439}$$

and

$$\begin{aligned} h(\eta) = b_0 &\left(1 + \frac{b_2}{b_0}\eta^2 + \frac{b_4}{b_2}\frac{b_2}{b_0}\eta^4 + \frac{b_6}{b_4}\frac{b_4}{b_2}\frac{b_2}{b_0}\eta^6 + \ldots\right) \\ &+ b_1\left(1 + \frac{b_3}{b_1}\eta^3 + \frac{b_5}{b_3}\frac{b_3}{b_1}\eta^5 + \frac{b_7}{b_5}\frac{b_5}{b_3}\frac{b_3}{b_1}\eta^7 + \ldots\right) \end{aligned} \tag{10.440}$$

To obtain a solution of each of the Hermite equations of Eq. (10.439) and 10.440, one can set

$$a_0 = b_0 = 0 \tag{10.441}$$

This forces the series in a_1 and b_1 to terminate after a finite number of terms. Alternatively in Eq. (10.439) and 10.440, one can set

$$a_1 = b_1 = 0 \tag{10.442}$$

This forces the series in a_0 and b_0 to terminate after a finite number of terms. The series termination can be seen by inspecting the recursion relations for a_m and b_n, where

$$a_{m+2} = \frac{-a_m}{(m+1)(m+2)}\left(\frac{\lambda_1}{2} - 2m\right) \tag{10.443}$$

$$b_{n+2} = \frac{-b_n}{(n+1)(n+2)}\left(\frac{\lambda_2}{2} - 2n\right) \tag{10.444}$$

Let us first assume

$$a_0 = b_0 = 0 \tag{10.445}$$

With a_0 and b_0 each zero, Eqs. (10.443–10.444) terminate for odd integer m and n, where

$$\frac{\lambda_1}{2} = 2m \quad m = 1, 3, 5, \ldots \text{ and } a_0 = b_0 = 0 \tag{10.446}$$

$$\frac{\lambda_2}{2} = 2n \quad n = 1, 3, 5, \ldots \text{ and } a_0 = b_0 = 0 \tag{10.447}$$

Next assume a_1 and b_1 are each zero. In this case, Eqs. (10.443–10.444) terminate for even integer m and n, where

$$\frac{\lambda_1}{2} = 2m \quad m = 0, 2, 4, 6, \ldots. \text{ and } a_1 = b_1 = 0 \tag{10.448}$$

$$\frac{\lambda_2}{2} = 2n \quad n = 0, 2, 4, 6, \ldots. \text{ and } a_1 = b_1 = 0 \tag{10.449}$$

Based on these series termination relationships of Eqs. (10.446–10.447) or (10.448–10.449), the two Hermite equations of Eqs. (10.425) and (10.426), have the Hermite polynomial solutions $H_m(\xi)$ and $H_n(\eta)$:

$$g(\xi) = H_m(\xi) \tag{10.450}$$

with

$$\xi = \frac{x\sqrt{2}}{w} \tag{10.451}$$

and

$$h(\eta) = H_n(\eta) \tag{10.452}$$

with

$$\eta = \frac{y\sqrt{2}}{w} \tag{10.453}$$

That is, with series termination in Eqs. (10.443–10.444), the two Hermite equations of Eqs. (10.425) and (10.426), have Hermite polynomial solutions

$$g\left(\frac{x\sqrt{2}}{w}\right) = H_m\left(\frac{x\sqrt{2}}{w}\right) \tag{10.454}$$

and

$$h\left(\frac{y\sqrt{2}}{w}\right) = H_n\left(\frac{y\sqrt{2}}{w}\right) \tag{10.455}$$

The solution $g(\xi)$ of Eq. (10.454) along the x-direction, is a Hermite polynomial solution, which, for each integer m, is obtained from the generating function

$$H_m(\xi) = (-1)^m \exp(\xi^2)\frac{\mathrm{d}^m}{\mathrm{d}\xi^m}\left[\exp(-\xi^2)\right] \tag{10.456}$$

Similarly, the solution $h(\eta)$ of Eq. (10.455) along the y-direction is a Hermite polynomial solution, which, for each integer n, is obtained from the generating function

$$H_n(\eta) = (-1)^n \exp(\eta^2)\frac{\mathrm{d}^n}{\mathrm{d}\eta^n}\left[\exp(-\eta^2)\right] \tag{10.457}$$

The particular values of the Hermite polynomial generating functions in Eqs. 10.456 for $g(\xi)$ and 10.457 for $h(\eta)$, for the cases of $m = n = 0$ and $m = n = 1$, are:

$$H_0(\xi) = H_0(\eta) = 1 \quad m = n = 0 \tag{10.458}$$

and

$$H_1(\xi) = 2\xi = \frac{2x\sqrt{2}}{w} \quad m = n = 1 \tag{10.459}$$

$$H_1(\eta) = 2\eta = \frac{2y\sqrt{2}}{w} \quad m = n = 1 \tag{10.460}$$

Recall from Eq. (10.406), the trial paraxial equation amplitude solution $E_0(\mathbf{r})$:

$$E_0(\mathbf{r}) = E_{00}g\left(\frac{x}{w(z)}\right)h\left(\frac{y}{w(z)}\right)\exp(\mathrm{i}P(z))\exp\left(\mathrm{i}k\frac{x^2+y^2}{2q(z)}\right) \tag{10.461}$$

In Eq. (10.406), and recalling Eqs. (10.423–10.424), one then substitutes Eqs. (10.422), (10.454–10.455), (10.395), and (10.437–10.438) for the following:

$$\exp(\mathrm{i}P(z)) \quad g\left(\frac{x}{w}\right) \quad h\left(\frac{y}{w}\right) \quad \frac{1}{q(z)} \quad \lambda_1 = 4m \quad \lambda_2 = 4n \tag{10.462}$$

The result of substitution in Eq. (10.461) is the paraxial amplitude solution $E_0(\mathbf{r})$, where

$$\begin{aligned} E(\mathbf{r}) = {} & \frac{E_{00}w_0}{w(z)}\exp\left(-\frac{x^2+y^2}{w^2(z)}\right)H_m\left(\frac{x\sqrt{2}}{w}\right)H_n\left(\frac{y\sqrt{2}}{w}\right) \\ & \times \exp\left[-\mathrm{i}(m+n+1)\tan^{-1}\left(\frac{z}{z_0}\right)\right]\exp\left(\mathrm{i}k\frac{x^2+y^2}{2R(z)}\right) \end{aligned} \tag{10.463}$$

The lowest order amplitude solution $E_0(\mathbf{r})$ of the paraxial equation has integers m and n, each being zero, which characterizes an open-sidewall laser cavity, where

$$E_0(\mathbf{r}) = \frac{E_{00}w_0}{w(z)}\exp\left(-\frac{x^2+y^2}{w^2(z)}\right)\exp\left[-\mathrm{i}\tan^{-1}\left(\frac{z}{z_0}\right)\right]\exp\left(\mathrm{i}k\frac{x^2+y^2}{2R(z)}\right) \quad m=n=0 \tag{10.464}$$

Recall from Eq. (10.378), which is repeated below, that the amplitude of a Gaussian beam $E(\mathbf{r})$ is

$$E(\mathbf{r}) = E_0(\mathbf{r})\exp(-\mathrm{i}kz) \tag{10.465}$$

Substitution into Eq. (10.465) of Eq. (10.464) for the lowest order beam, along with Eq. (10.404) for $W(z)$ and Eq. (10.400) for $R(z)$, yields the explicit expression of the amplitude $E(\mathbf{r})$ of a Gaussian beam in an open-sidewall laser cavity, which is applicable to an X-ray laser, where

$$\begin{aligned} E(\mathbf{r}) = {} & \frac{E_{00}}{\sqrt{1+(z/z_0)^2}}\exp\left(-\frac{x^2+y^2}{w_0^2[1+(z/z_0)^2]}\right) \\ & \times \exp\left[-\mathrm{i}\tan^{-1}(z/z_0)\right]\exp\left(\mathrm{i}\pi\frac{x^2+y^2}{\lambda z[1+(z_0/z)^2]}\right)\exp\left(-\mathrm{i}2\pi\frac{z}{\lambda}\right) \end{aligned} \tag{10.466}$$

The wave amplitude $E(\mathbf{r})$ of Eq. (10.465)−10.466 can also apply to neutron wave functions, and thus a neutron Gaussian beam in a neutron laser, if it were possible to produce a Gaussian, monchromatic, spatially and temporally coherent, neutron beam.

References

Cremer, J. T. (2012a). In W. Peter (Ed.), Neutron and X-ray microscopy—Part 1. *Advances in imaging and electron physics* (172). London, UK: Hawkes, Academic Press, an imprint of Elsevier Press.

Eisberg, R., & Resnick, R. (1974). *Quantum physics of atoms, molecules, solids, nuclei, and particles*. New York, NY: John Wiley & Sons.

Elmore, W. C., & Heald, M. A. (1969). *Physics of waves*. New York, NY: Dover.

Hecht, E., & Zajac, A. (1979). *Optics*. Reading, MA: Addison-Wesley.

Milonni, P. W., & Eberly, J. H. (1988). *Lasers*. New York, NY: John Wiley & Sons.

Piestrup, M. A., Pantell, R. H., Cremer, J. T., & Beguiristain, H. R. (2001). Compound refractive lens for X-rays, U.S. Patent No. 6,269,145 B1, July 31.

Siegman, A. E. (1971). *An introduction to lasers and masers*. New York, NY: McGraw-Hill.

Sommerfeld, A. (1954). *Optics*. New York: Academic Press.

11 Kirchhoff Equation Solution for CRL, Pinhole, and Phase Contrast Imaging

11.1 Kirchhoff Equation with a 1-D Biconcave, Parabolic, or Spherical CRL and Thin-Sample Approximation for X-rays or Neutrons with Gravity

In this chapter, we derive the Kirchhoff equation for 1-D biconcave, parabolic, or spherical compound refractive lens (CRL), which assumes a thin sample, where the X-ray or neutron-ray bundle passes from the source point through the sample without change of direction, and passes to the CRL. The X-ray or neutron bundle wave changes amplitude and/or phase but not direction on passage through the thin sample. In addition to the thin-sample approximation, the paraxial approximation and the thin-lens approximation for the CRL are applied, and gravity droop is included for the case of neutrons. Next, the thin-sample approximation is dropped and a thick sample is assumed, which can change the wave (ray) direction and the wave amplitude and/or phase.

Derived next is the general Kirchhoff equation with 2-D biconcave parabolic or spherical CRL for X-rays or neutrons with gravity, which assumes a thick sample, the thin-lens approximation, and the paraxial approximation. Next follows the derivations of the biconcave, parabolic CRL for the image amplitude distribution, point spread function (PSF), modulation transfer function (MTF), field of view (FOV), and image intensity distribution for incoherent X-rays or neutrons with gravity. Also derived is the PSF of a pinhole camera (no CRL) for an incoherent X-ray or a neutron source, including gravity droop for neutrons.

Next, we discuss phase contrast imaging without a CRL. In particular, the image intensity distribution of an object image with a point X-ray or neutron source, which provides transverse coherence at the object and includes gravity droop for neutrons, is derived. Next, the parameter settings are given for the CRL image intensity distribution for the special case of a merged object plane and source plane. Finally, the stationary phase approach is presented, which is applied to the image amplitude integral of a spherical, biconcave CRL for incoherent X-rays or neutrons with gravity droop included.

We first derive the Kirchhoff equation for a 1-D CRL, where a thin object (sample) is assumed. In the thin-object (sample) approximation, only at the sample do neutrons or X-rays pass through a sample point, either unscattered or forward

Neutron and X-ray Optics. DOI: http://dx.doi.org/10.1016/B978-0-12-407164-3.00010-3

scattered, or the X-ray or neutron is completely absorbed. Gravity is not included in this 1-D treatment, so the results apply to the electric field amplitude of an X-ray or to the neutron wave amplitude.

The Huygens–Fresnel approach is not used at the thin sample, where each sample point would have emitted a secondary wave. Instead, for this 1-D model, each point on the source acts as a cylindrical wave source in which ray paths from every source point pass in a straight line through the object (sample) plane to every point on the CRL center or principal plane. And each point on the CRL principal plane, has a ray path to every point on the image plane. Thus, each source point and each point on the CRL principal plane acts as a source of a cylindrical waves.

Next, we will derive the Kirchhoff equation for a 2-D CRL in which a thin object (sample) is assumed. In this case, every point on the source and every point on the CRL principal plane is a source of spherical waves. Neutron or X-ray propagation is in the positive z-direction, and the z-axis is a center axis or an optical axis of the source, object (sample), CRL, and image planes. Ray paths from each point of the source xy-plane pass through the object (sample) in a straight line to every point on the xy-principal plane of the CRL. And from each point on the xy-principal plane of the CRL, ray paths go to every point on the image xy-plane. The thin object (sample) is assumed to completely absorb the X-ray or neutron, or the neutron or X-ray is unscattered; or, if it is scattered, the X-ray or neutron is forward scattered along its line of incidence.

Gravity is included in the thin-sample, 2-D analysis, where gravity acts in the downward, negative y-direction. Hence, one sets the gravity term to zero in the case of X-rays. The Kirchhoff equation result applies to X-rays if one sets the gravity terms to zero. If the gravity terms are zero, then the x- and y-transverse X-ray or neutron excursion are the same. This results in azimuthal symmetry for the neutron or X-ray trajectory from the source point, through the object (sample) point, to CRL principal plane point, and onto a point on the image plane. If one retains the gravity terms for the neutron case, then in the transverse negative y-direction, along which gravity acts, the neutron then has a different trajectory component than the transverse x-component.

The Huygens–Fresnel approach to X-ray or neutron refractive optics, allows treatment of the monochromatic and chromatic aberrations and yields insight into the details of the image formation, resolution, FOV, and contrast. The Kirchhoff equation, which calculates the propagation of radiation, and includes both refractive and diffractive effects, is an expression of the Huygens–Fresnel approach. The Kirchhoff equation has an advantage over ray tracing, because the Kirchhoff equation includes both absorption and diffraction effects, which are not easily included in a ray tracing analysis. Also, the Kirchhoff equation can lead to closed-form solutions in some cases, which yields a more intuitive understanding. See Goodman (1968), Hecht and Zajac (1979), Elmore and Heald (1969), and Sommerfeld (1954).

The resolution of the lens, and the size of the source spot on the detector, can be calculated, by first determining, the image intensity distribution from a single object point (the PSF). To simply computation of the PSF, the single object point is set on the origin of the z-axis, so that the object is centered on the z-axis in the object plane. We next translate the object point, and the resulting intensity is added to the original intensity. When the distance between the two points in the object plane gives a minimum-to-maximum ratio of 0.81 between the two peaks in the image plane (the Rayleigh criterion), the distance of translation is defined as the resolution.

To include any aberration, such as spherical, gravity, or chromatic, we only need to include the appropriate term in the Kirchhoff formula before calculating the PSF. While this is stated easily, it can be exceedingly difficult, requiring extensive numerical analysis, to perform the given integrations. We will perform this analysis for lens surface and chromatic aberrations along with gravity effects, lens thickness, and source collimation, though not simultaneously.

We now derive the Kirchhoff equation for this system in Figure 11.1 using the thin-sample approximation, the thin-lens approximation, and the paraxial

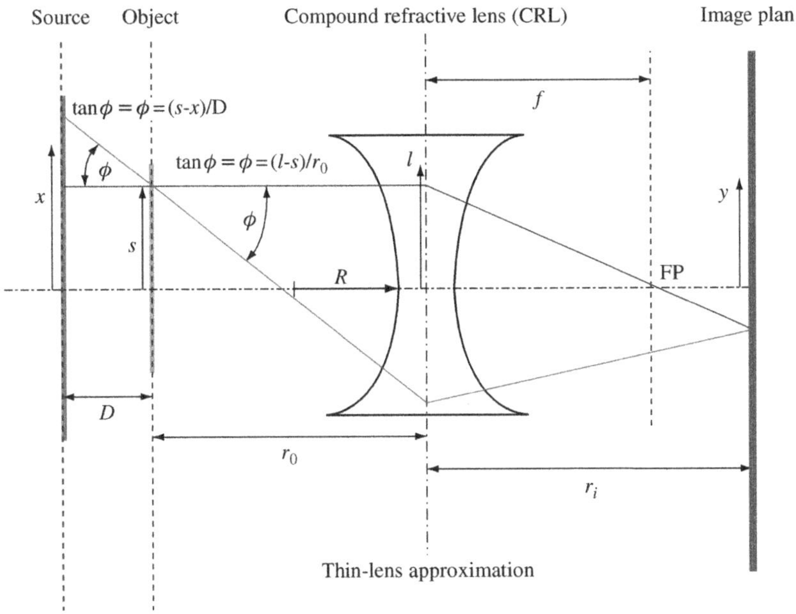

Figure 11.1 An X-ray or neutron source point x illuminates an object at point s, and the X-rays or neutrons propagate and intercept a CRL principal or lens center plane at point l, which refracts and focuses the neutrons on the image plane at point y. This setup assumes the thin-sample approximation, thin-lens approximation, and the paraxial approximation. The ray bundle passes from a source point through an object point to the lens point without changing direction. Only the phase and/or amplitude of the wave change as the X-ray or neutron wave passes through the sample, but the direction of the wave is unchanged in this thin-sample approximation. This allows application of similar triangles, which simplifies computation. The thin lens approximation assumes an incident ray maintains a constant transverse distance from the CRL center axis, but the ray changes direction at the center CRL principal plane. The paraxial approximation assumes the transverse distance of the rays in the CRL is much smaller than the axial separation distances between the CRL principal plane, and the source, object, and image planes plane. The paraxial approximation allows application of the binomial approximation to the ray trajectory equations.

approximation. In the thin-sample approximation, the X-rays or neutron rays emitted from the source plane, are assumed to pass in a straight-line trajectory through the object plane. Any X-ray or neutron scatter in the thin sample is assumed to be forward scattered in the direction of incidence. In the thin-lens approximation, the CRL is treated as a thin lens, where refraction in the CRL is assumed to take place at the principal or center plane of the CRL, and rays change angle but not position within the CRL. To qualify as a thin lens, the focal length and the object and image distances of the CRL, must be much greater than the length of the CRL.

The paraxial approximation assumes that the rays or the X-rays and neutrons have small angles relative to the center positive z-axis or optical axis, which extends through the center of the source plane, object plane, CRL principal plane, and image plane. In the paraxial approximation, the X-ray or neutron trajectories stay in close proximity to the center optical axis. The paraxial approximation assumes that the transverse dimensions of the source, object (sample), the CRL, and the image are all small compared to the axial separation distances among the source, object, CRL, and image planes. The CRL plane is assumed to be the CRL center or principal plane.

Consider the phase shift of a ray of an X-ray or neutron wave of amplitude E_s and wavelength λ, where the ray is emitted from source point x at transverse position x in the source plane at $z=0$. The ray intercepts an object point s at transverse position s in the object plane at $z=D$, where the object is a thin sample (see Figure 11.1). The ray from the object point s propagates to point l at transverse position l in the CRL principal plane at $z=D+r_0$, where the CRL principal plane is the center plane of the thin-lens CRL. The CRL refracts the ray at the principal plane, and the ray propagates from the CRL principal plane and terminates at image point y at transverse position y in the image plane at $z=D+r_0+r_i$. As shown in Figure 11.1, via the thin sample approximation, the ray from the source point x is forward-scattered at object point s by the thin-sample toward point l on the CRL principal plane, with possible attenuation, or the ray is transmitted, unscattered through object point s to point l on the CRL principal plane. That is, the ray is assumed to be a straight line from source point x through object point s to CRL principal plane point l. Hence, the source transverse position coordinate x can be expressed as a function of r_0, D, s, and l; via similar triangles, where

$$x=f(s,l,D,r_0) \tag{11.1}$$

The CRL is circular, and hence by azimuthal symmetry, the 1-D transverse source, object, lens, and image coordinate lines can be considered radial coordinate lines, which extend from the center z-axis (the optical axis).

The ray of the neutron wave at the image plane has wavelength-dependent amplitude E_y. This amplitude is determined by the total path length L_T, and the resulting phase shift ϕ_T of the local wave front, which propagates along this single path (represented by a ray). The wave amplitude for the ray at image position y is

$$E_y(\lambda,y,l,s)=E_s(\lambda,s,l)\exp(-\mathrm{i}\phi_T(\lambda,y,l,s)) \tag{11.2}$$

The total path length L_T is the sum of the path length in vacuum drift space L_{drf} and the optical path length L_{CRL} through the N-lens CRL, where $L_T = L_{\text{drf}} + L_{\text{CRL}}$. The optical path length through a single lens L_{CRL} is the product of the single lens thickness Δt along the path, and the refractive index difference Δn between the lens $n = 1 - \delta_r - \text{i}\delta_i$ and vacuum $n = 1$, where $\Delta n = -\delta_r - \text{i}\delta_i$, so that $L_{\text{CRL}} = N\Delta t\Delta n$. The total phase shift ϕ_T for a ray from source point s, through lens point l in the CRL principal plane to image point y, is then

$$\phi_T = 2\pi\frac{L_T}{\lambda} = 2\pi\frac{L_{\text{drf}} + L_{\text{CRL}}}{\lambda} = 2\pi\frac{L_{\text{drf}} - N\Delta t(\delta_r + \text{i}\delta_i)}{\lambda} \tag{11.3}$$

The 1-D image intensity profile I_y, that is, the intensity at each point y in the image plane of the CRL, is obtained as follows. At each image point y one sums the amplitude contributions of all the ray paths through the CRL from single object point s. The squared magnitude of the summed amplitude (intensity) at image point y is thus obtained for a single object point s. One then sums (integrates) the intensity contributions at image point y from all the object points s, where

$$I_y(\lambda, y) = \int_{\text{source}} \text{d}s \left| \int_{\text{lens}} E_y(\lambda, y, l, s)\text{d}l \right|^2 \tag{11.4}$$

At the image plane, one can calculate the gain in image plane intensity, using a CRL, in place of an aperture at the CRL position. The aperture diameter equals the CRL effective aperture diameter, which is a minimum of the CRL absorption aperture diameter and the CRL physical aperture diameter. The 1-D intensity gain $G_{1\text{-}D}$ at the image plane, with the use of a CRL, with intensity $I_y^{\text{CRL}}(y, \lambda)$, compared to the case of no CRL and aperture only with intensity $I_y^0(\lambda, y)$, is

$$G_{1\text{-}D}(\lambda, y) = \frac{I_y^{\text{CRL}}}{I_y^0} \tag{11.5}$$

Depending on whether the CRL is comprised of parabolic or spherical, biconcave lenses, one substitutes for I_y^{CRL} either

$$I_y^{\text{CRL}} = I_y^{\text{par}} \quad \text{or} \quad I_y^{\text{CRL}} = I_y^{\text{sph}} \tag{11.6}$$

For an azimuthally symmetric lens, the 2-D gain is the square of the 1-D gain, that is,

$$G_{2\text{-}D}(\lambda, y) = [G_{1\text{-}D}(\lambda, y)]^2 = \left[\frac{I_y^{\text{CRL}}(\lambda, y)}{I_y^0(\lambda, y)}\right]^2 \tag{11.7}$$

One can determine the image and object resolution of the CRL, by considering a specific object point s, for example, the center object point at $s = 0$, where one then calculates the image intensity distribution $I_y(\lambda, y)$ for $s = 0$, as a function of the transverse position $y$in the image plane, where

$$I_y(\lambda, y) = \left| \int_{\text{lens}} E_y(\lambda, y, l, s=0) \mathrm{d}l \right|^2 \tag{11.8}$$

The FWHM of the image intensity distribution $I_y(\lambda, y)$ of Eq. (11.8), as a function of transverse distance y from the center z-axis of the image plane, is the CRL image resolution, and division by the transverse magnification M_T, yields the CRL object resolution.

Alternatively, one can determine the field of view (FOV) of the CRL by considering a specific image point; for example, the center image point at $y = 0$. One can determine the object intensity distribution along the transverse radial line s, which contributes to the intensity of the image point at $y = 0$, that is,

$$I_y(\lambda, s) = \left| \int_{\text{lens}} E_y(\lambda, y=0, l, s) \mathrm{d}l \right|^2 \tag{11.9}$$

The FWHM of the object intensity distribution of Eq. (11.9), as a function of transverse distance s from the center z-axis of the object plane, is the field of view (FOV) of the CRL. The integral rule (see Selby and Girling, 1965) in Eq. (11.10) below, where $a > 0$, will be used to solve the integrals, which occur in the derived CRL Kirchhoff equation, where

$$\int_{-\infty}^{\infty} \exp(-ax^2 + ibx)\mathrm{d}x = \exp\left(-\frac{b^2}{4a}\right) \int_{-\infty}^{\infty} \exp\left[-a\left(x - \frac{ib}{2a}\right)^2\right] \mathrm{d}x = \sqrt{\frac{\pi}{a}} \exp\left(-\frac{b^2}{4a}\right) \tag{11.10}$$

The task remains to determine the total path length L_T and the resulting phase shift ϕ_T, for each ray path of the neutron wave, emitted from a single source point, which is focused by the CRL at its principal plane to a single image point. Assume that the neutron source has a Gaussian distribution of wavelength λ, about a mean wavelength λ_0 and standard deviation σ_λ. Further, assume that the neutron source is azimuthally symmetric and has a Gaussian amplitude distribution with respect to the transverse dimension x, with standard deviation σ_x about its center ($x = 0$) amplitude E_0. Hence, at the source plane ($z = 0$), the X-ray or neutron wave amplitude along a particular dimension, such as the transverse x axis $E_s(\lambda, x) = E(\lambda, x, z = 0)$, has a radial dependence and is given by

$$E_s(\lambda, x) = \frac{2E_0}{\pi \sigma_x \sigma_\lambda} \exp\left(-\frac{\lambda - \lambda_0}{2\sigma_\lambda^2}\right) \exp\left(-\frac{x^2}{2\sigma_x^2}\right) \tag{11.11}$$

The trajectory length of flight of the neutron ray, which is expressed by the number of neutron wavelengths, or by the phase shift of the wave, extends from the neutron source point x to an object point s. The source and object planes are separated by axial distance D. The paraxial approximation assumes the axial distance D between the source and object planes is much greater than the transverse extent of the source points x, and that D is much greater than the transverse extent of the object points s. With the paraxial approximation, where the rays have relatively small angles with respect to the z-axis, one can apply the binomial approximation, because $D \gg |s-x|$, so that

$$L_s = \sqrt{D^2 + (s-x)^2} \approx D + \frac{(s-x)^2}{2D} \quad D \gg |s-x| \tag{11.12}$$

If the sample or object is sufficiently thin that most neutrons are transmitted unscattered, or most of the scattered neutrons are forward scattered along the line of incidence, then one can use the thin-sample approximation. The thin-sample approximation assumes that the source X-ray or neutron rays pass from a source point in a straight line to a point on the principal plane of the CRL. In strict application of the Huygens–Fresnel approach, an X-ray or neutron ray from the source point x propagates to an object point s, and from the single object point s, rays go to every point l on the principal plane of the CRL. That is, every point on a wave front is the source of a new wave front. However, if most of the neutron rays are absorbed by the sample, or pass through the sample in a straight line (either unscattered or forward scattered), then one can apply the thin-sample approximation.

With the thin-sample approximation, one is able to eliminate the source transverse dimension variable x, by expressing it in terms of the object transverse dimension s and lens transverse dimension l. Assume a parallel ray and a diagonal ray are emitted from 2 different points x_1 and x_2 in the source plane, and assume the two rays converge to single object point s, and then diverge to two points l_1 and l_2 in CRL principal plane. This pair of rays form a pair of similar triangles, which have their common vertex at object point s, where

$$\tan o = \frac{s-x}{D} = \frac{l-s}{r_0} \tag{11.13}$$

One solves Eq. (11.13) for the ray transverse distance x at the source plane in terms of the ray transverse distance s at the object plane, and the ray transverse distance l at the CRL principal plane, where

$$x = s\left(1 + \frac{D}{r_0}\right) - \frac{lD}{r_0} \tag{11.14}$$

Substitution of Eq. (11.14) for x into Eq. (11.12) for L_s, then yields the ray path length from source point x to an object point s, where

$$L_s = D + \frac{D}{2r_0^2}(l-s)^2 \tag{11.15}$$

With substitution of Eq. (11.14) into Eq. (11.11), one obtains the amplitude distribution $E_s(s)$ in the source plane, where

$$E_s = E(s, z = 0) \tag{11.16}$$

and

$$\begin{aligned} E_s = {} & \frac{2E_0}{\pi\sigma_x\sigma_\lambda}\exp\left[-\frac{\lambda-\lambda_0}{2\sigma_\lambda^2}\right]\exp\left[-\frac{s^2}{2\sigma_x^2}\left(1+\frac{D}{r_0}\right)^2\right] \\ & \exp\left[-\frac{l^2}{2\sigma_x^2}\left(\frac{D}{r_0}\right)^2\right]\exp\left[\frac{ls}{\sigma_x^2}\frac{D}{r_0}\left(1+\frac{D}{r_0}\right)\right] \end{aligned} \tag{11.17}$$

The length of flight L_0 in the drift space, from illuminated object (specimen) point to CRL principal or center plane, and the length of flight L_i in the drift space between the CRL principal plane and the image point are, respectively,

$$L_0 = \sqrt{r_0^2 + (l-s)^2} \cong r_0 + \frac{(l-s)^2}{2r_0} \qquad r_0 \gg |l - s| \tag{11.18}$$

and

$$L_i = \sqrt{r_i^2 + (y-l)^2} \cong r_i + \frac{(y-l)^2}{2r_i} \qquad r_i \gg |y - l| \tag{11.19}$$

The transverse dimensions of the image, CRL, object, and source are assumed to be much less than the axial distances between them.

The total path length L_{drf} in vacuum drift space is the sum of (1) the ray path length L_s of Eq. (11.15) from the source point x to the illuminated object (specimen) point s, (2) the ray path length L_0 of Eq. (11.18) from object point s to point l in the center, principal plane of the CRL, and (3) the ray path length L_i of Eq. (11.19) from CRL principal plane point l to image point y. The axial length L_{CRL} of the CRL itself, is assumed to be very small compared to (1) the axial distance (object distance) r_0 between the CRL principal plane and object plane, where $L_{CRL} \ll r_0$, and (2) the axial distance (image distance) r_i between the CRL principal plane and image plane, where $L_{CRL} \ll r_i$. This allows the approximation of using the CRL principal plane for point l, rather than the CRL input plane as the end point for the ray path length L_0 in the object-CRL drift space, and the CRL output plane as the start point for the ray path length L_i in the CRL-image drift space. Hence,

$$L_{drf} = L_s + L_0 + L_i \tag{11.20}$$

In propagation over the total distance L_{drf} in the source-object, object-CRL, and CRL-image drift spaces, the neutron or X-ray wave undergoes phase shift ϕ_{drf}, where

$$\phi_{\mathrm{drf}} = kL_{\mathrm{drf}} \tag{11.21}$$

The wave number k of the X-ray or neutron wave in units of $[1/length]$, is the number of wavelengths per unit propagation distance, which is expressed by the number of wave cycles (2π radians) per unit propagation distance, where

$$k = \frac{2\pi}{\lambda} \tag{11.22}$$

One substitutes Eq. (11.20), with Eqs. (11.15), (11.18), and (11.19) into Eq. (11.21), and then one expands and collects terms, and neglects (drops) the constant phase-shift term $K(D + r_0 + r_i)$. This yields ϕ_{drf}, which is the sum of the X-ray or neutron wave phase shifts in the source-object, object-CRL, and CRL-image drift spaces, where

$$\phi_{\mathrm{drf}} = \left(\frac{ks^2}{2r_0}\right)\left(1+\frac{D}{r_0}\right)+\left(\frac{kl^2}{2}\right)\left[\frac{1}{r_0}\left(1+\frac{D}{r_0}\right)+\frac{1}{r_i}\right]-\left(\frac{kls}{r_0}\right)\left(1+\frac{D}{r_0}\right)-\left(\frac{kyl}{r_i}\right)+\left(\frac{ky^2}{2r_i}\right) \tag{11.23}$$

Recall the lens-maker formula:

$$\frac{1}{f} = \frac{1}{r_0} + \frac{1}{r_i} \tag{11.24}$$

Substitution of Eq. (11.24) into Eq. (11.23), yields

$$\phi_{\mathrm{drf}} = \left(\frac{ks^2}{2r_0}\right)\left(1+\frac{D}{r_0}\right)+\left(\frac{kl^2}{2f}\right)+\left(\frac{kl^2D}{2r_0^2}\right)-\left(\frac{kls}{r_0}\right)\left(1+\frac{D}{r_0}\right)-\left(\frac{kyl}{r_i}\right)+\left(\frac{ky^2}{2r_i}\right) \tag{11.25}$$

The CRL index of refraction is

$$n = 1 - \delta - \mathrm{i}\beta \tag{11.26}$$

The minus sign in front of $\mathrm{i}\beta$ in the complex refractive index $n = 1 - \delta - \mathrm{i}\beta$ in Eq. (11.26) produces a negative, real argument term in the exponential $\exp(-\mathrm{i}\phi_T)$ of Eq. (11.2) for the wave phase shift of a wave traveling in the forward, positive z-direction in the CRL, where this minus sign is needed to properly show wave attenuation in the CRL material, and

$$\beta = \frac{\lambda\mu}{4\pi} \tag{11.27}$$

The CRL index of refraction n in a vacuum is

$$n = 1 \tag{11.28}$$

The CRL index of refraction differs from the refractive index for a vacuum by Δn, where

$$\Delta n = \left(1 - \delta - \mathrm{i}\frac{\mu}{2k}\right) - 1 \tag{11.29}$$

or

$$\Delta n = -\delta - \mathrm{i}\frac{\mu}{2k} \tag{11.30}$$

In most CRL materials, the X-ray or neutron wave front moves at a larger phase velocity v_{p} in the CRL than its vacuum velocity v, where

$$v_{\mathrm{p}} = \frac{v}{1 - \delta} \tag{11.31}$$

The ray path length Δt through a single lens, can vary as a function of ray distance l from the lens center z-axis, which leads to ray refraction (change of direction), and a phase shift $\Delta\phi$ of the ray's complex amplitude, where

$$\Delta\phi = k\Delta n\Delta t \tag{11.32}$$

One substitutes Eq. (11.30) for Δn into Eq. (11.32), which yields the phase shift of the complex amplitude of a ray, which propagates through the N-lens CRL, where

$$\Delta\phi = -Nk\delta\Delta t - \mathrm{i}\frac{N\mu\Delta t}{2} \tag{11.33}$$

The thickness of a biconcave, parabolic lens $\Delta t_{\mathrm{par}}(l)$, as a function of distance l from the center axis of the lens, with on-axis radius of curvature R, is

$$\Delta t_{\mathrm{par}}(l) = d + \frac{l^2}{R} \tag{11.34}$$

In contrast, the thickness of a biconcave, spherical lens $\Delta t_{\mathrm{sph}}(l)$, as a function of distance l from the center axis of the lens, with radius of curvature R, is

$$\Delta t_{\mathrm{sph}}(l) = d + 2\left(R - \sqrt{R^2 - l^2}\right) \tag{11.35}$$

With application of the binomial approximation to Eq. (11.35), where $l \ll R$, the single lens thickness $\Delta t_{\mathrm{sph}}(l)$ of a biconcave, spherical lens is approximately equal to the thickness $\Delta t_{\mathrm{par}}(l)$ of a biconcave, parabolic lens, as a function of transverse ray position l, where

$$\Delta t(l) = \Delta t_{\mathrm{sph}}(l) = \Delta t_{\mathrm{Par}}(l) \quad l \ll R \tag{11.36}$$

The phase shift $\Delta\phi_N^{\mathrm{par}}$ of the complex amplitude of a X-ray or neutron ray, which propagates through a CRL with N parabolic, biconcave lenses is

$$\Delta\phi_N^{\mathrm{par}} = -\Delta\phi_d - kN\delta\frac{l^2}{R} - \mathrm{i}\cdot\left(\mu_d + \frac{\mu N l^2}{2R}\right) \tag{11.37}$$

For a CRL with N biconcave, parabolic lenses, the focal length f is

$$f = \frac{R}{2N\delta} \tag{11.38}$$

Substitution of Eq. (11.38) into Eq. (11.37), yields the phase-shift term $\Delta\phi_N^{\mathrm{par}}$ for the CRL with N parabolic, biconcave lenses, where

$$\Delta\phi_N^{\mathrm{par}} = -\Delta\phi_d - \frac{kl^2}{2f} - \mathrm{i}\cdot\left(\mu_d + \frac{\mu N l^2}{2R}\right) \tag{11.39}$$

The purpose of the lens is to provide equal optical paths from the object point s through the lens to the image point y. Equal optical paths allow maximum constructive interference and summing of the amplitudes at the image point y. In Eq. (11.39), the term $\Delta\phi_d$ is the path-independent phase-shift term, and the first imaginary term μ_d is the path-independent absorption. Both terms are due to the fixed, minimum wall thickness d of each biconcave lens, where

$$\Delta\phi_d = kN\delta d \tag{11.40}$$

and

$$\mu_d = \frac{N\mu d}{2} \tag{11.41}$$

The second, imaginary term in Eq. (11.39), which has a quadratic dependence l^2 on the transverse distance l of a ray from the CRL center axis, accounts for the path-dependent, wave absorption. For the case of a CRL with N spherical, biconcave lenses, the complex phase shift $\Delta\phi_N^{\mathrm{sph}}$ of the X-ray or neutron ray, which passes through the CRL is then

$$\Delta\phi_N^{\mathrm{sph}} = -\Delta\phi_d - 2kN\delta\left(R - \sqrt{R^2 - l^2}\right) - \mathrm{i}\cdot\left(\mu_d + \mu N\left(R - \sqrt{R^2 - l^2}\right)\right) \tag{11.42}$$

For $l \ll R$, and application of the binomial approximation to Eq. (11.42), the phase shift, which is due to N-spherical biconcave lenses, is approximately the same as that of N-parabolic biconcave lenses, where

$$\Delta\phi_N^{\mathrm{sph}} = \Delta\phi_N^{\mathrm{par}} \quad l \ll R \tag{11.43}$$

For a CRL with N parabolic, biconcave lenses, the total complex phase shift is

$$\phi_{\mathrm{T}}^{\mathrm{par}} = \phi_{\mathrm{drf}} + \Delta\phi_N^{\mathrm{par}} \tag{11.44}$$

The source and object planes are merged, and the ray phase shift term $\Delta\phi_N^{\mathrm{par}}$ of Eq. (11.39) for a parabolic lens CRL, is then added to the drift space phase shift term ϕ_{drf} of Eq. (11.25), with $D = 0$, to obtain ϕ_T^{par} of Eq. (11.44). The constant phase shift term $\Delta\phi_d$ from Eq. (11.40) in $\Delta\phi_N^{\mathrm{par}}$ is dropped, because $\Delta\phi_d$ is a constant, which in the amplitude expression $E_y(\lambda, y, l, s)$ in Eqs. (11.2), (11.4), (11.8), and (11.9), will become the imaginary argument of an exponential, whose magnitude is 1. In this sum of $\Delta\phi_N^{\mathrm{par}}$ and ϕ_{drf}, occurs the cancellation of the quadratic l^2 dependent terms, so that the phase shift ϕ_T^{par} of a ray from a point s in object plane, through point l in the principal plane of the parabolic lens CRL, to a point y on the image plane, becomes

$$\phi_{\mathrm{T}}^{\mathrm{par}} = \frac{k}{2}\left(\frac{s^2}{r_0}\right) - kl\left(\frac{s}{r_0} + \frac{y}{r_i}\right) + \frac{k}{2}\left(\frac{y^2}{r_i}\right) - \mathrm{i}\left(\mu_d + \frac{l^2 \mu N}{2R}\right) \tag{11.45}$$

In the case of the merged source and object planes ($D = 0$), the parabolic CRL produces an inverted image of the object with transverse magnification:

$$M_{\mathrm{T}} = -\frac{y}{s} \tag{11.46}$$

We consider the case of no lens aberration. The chief ray is a straight line ray from object point s, which extends through the z-axis at center point C at the CRL principal plane, and then extends to the corresponding image point y. The chief ray and z-axis form two similar triangles, which share a common vertex at the CRL center point C. From this pair of similar triangles, one obtains the transverse magnification of Eq. (11.46), and the following ratio relation in Eq. (11.47). If there is no lens aberration, then the ratio relation of Eq. (11.47) causes the cross terms for ls and yl to cancel in the parabolic CRL phase-shift expression of Eq. (11.45), where

$$\frac{s}{r_0} = -\frac{y}{r_i} \tag{11.47}$$

The value of the Kirchhoff equation approach is that one can account for lens aberrations. Hence, in the derivations in this chapter, we will assume lens aberrations occur, so that the condition of Eq. (11.47) does not occur. For simplicity, assume that the source and object plane are the same, where $D = 0$. The shape and index of refraction of the lens causes each ray path length from object point s, through any point l in the lens principal plane, to image point y to be identical. Hence, the phase shift ϕ_T^{par} is independent of the ray path through the lens. That is, the phase shift ϕ_T^{par} is independent of the lens transverse dimension l through which the ray from object point s passes on its way to an image point y. With substitution of Eq. (11.41), one rearranges Eq. (11.45) for the ray phase shift from object point s through principal plane point l in the parabolic lens CRL to image point y, so that

$$\phi_T^{\text{par}} = \frac{k}{2}\left(\frac{s^2}{r_0} + \frac{y^2}{r_i}\right) - kl\left(\frac{s}{r_0} + \frac{y}{r_i}\right) - \mathrm{i} \cdot \left(\frac{N\mu d}{2} + \frac{l^2 \mu N}{2R}\right) \tag{11.48}$$

With $D = 0$ in Eq. (11.25), and with Eqs. (11.41) and (11.42), the phase shift ϕ_T^{sph} of the X-ray or neutron ray from object point s through principal plane point l in a spherical lens CRL to image point y, is then

$$\begin{aligned}\phi_T^{\text{sph}} = \phi_{\text{drf}} + \Delta\phi_N^{\text{sph}} = \frac{k}{2}\left(\frac{s^2}{r_0} + \frac{y^2}{r_i} + \frac{l^2}{f}\right) - kl\left(\frac{s}{r_0} + \frac{y}{r_i}\right) \\ - \mathrm{i} \cdot \left(\frac{N\mu d}{2} + \mu N\left(R - \sqrt{R^2 - l^2}\right)\right)\end{aligned} \tag{11.49}$$

The merged source and object planes set $D = 0$ in Eq. (11.17), and with substitution of Eq. (11.17) and Eq. (11.48) into Eq. (11.2) for a CRL with parabolic, biconcave lenses, one obtains the X-ray or neutron wave amplitude $E_y^{\text{par}}(\lambda, y, l, s)$ at image point y, for an arbitrary path from object point s through CRL principal plane point l to image point y, where

$$\begin{aligned}E_y^{\text{par}} = E_s(\lambda, s, l)\,\exp\left(-\mathrm{i}\phi_T^{\text{par}}(\lambda, y, l, s)\right) = \left(\frac{2E_0\,\mathrm{e}^{-N\mu d/2}}{(r_0 + r_i)\pi\sigma_x\sigma_\lambda}\right)\exp\left(-\frac{\lambda - \lambda_0}{2\sigma_\lambda^2}\right) \\ \times \exp\left[\mathrm{i}kl\left(\frac{s}{r_0} + \frac{y}{r_i}\right)\right]\exp\left[-s^2\left(\frac{1}{2\sigma_x^2} + \frac{\mathrm{i}\pi}{r_0\lambda}\right)\right]\exp\left(-\frac{\mu N}{2R}l^2\right)\exp\left(-\frac{\mathrm{i}\pi}{r_i\lambda}y^2\right)\end{aligned} \tag{11.50}$$

The phase shifts along the various paths, from an object point s to the corresponding inverted image point y, are thus independent of the ray incidence position l at the CRL principal plane, because of the compensating thicknesses of the N biconcave, parabolic lenses of the CRL. The parabolic lens expression of Eq. (11.50) can be integrated analytically. The thin-lens approximation is assumed for the CRL, and refraction at the CRL takes place only at the CRL principal or center plane. A ray propagating through CRL is assumed to change its angle at the CRL principal plane, but the transverse position of the ray within the CRL is approximated as constant.

In comparison, with merged source and object planes ($D = 0$) in Eq. (11.17), and with substitution of Eq. (11.17) and Eq. (11.49) into Eq. (11.2) for a CRL with spherical, biconcave lenses, one obtains the X-ray or neutron wave amplitude $E_y^{\text{sph}}(\lambda, y, l, s)$ at image point y, for an arbitrary path from object point s through CRL principal plane point l to image point y, where

$$E_y^{\text{sph}} = E_s(\lambda, s, l)\exp\left(-\mathrm{i}\phi_T^{\text{sph}}(\lambda, y, l, s)\right) = \left(\frac{2E_0\,\mathrm{e}^{-N\mu d/2}}{(r_0 + r_i)\pi\sigma_x\sigma_\lambda}\right)\exp\left(-\frac{\lambda-\lambda_0}{2\sigma_\lambda^2}\right)$$
$$\times\exp\left[\mathrm{i}kl\left(\frac{s}{r_0} + \frac{y}{r_i}\right)\right]\exp\left(-\frac{s^2}{2\sigma_x^2}\right)\exp\left[-\mu N\left(R - \sqrt{R^2 - l^2}\right)\right]$$
$$\exp\left[-\mathrm{i}\frac{\pi}{\lambda}\left(\frac{s^2}{r_0} + \frac{l^2}{f} + \frac{y^2}{r_i}\right)\right] \tag{11.51}$$

This spherical lens expression of Eq. (11.51) requires numerical integration.

Approximation of the spherical lenses of a CRL by parabolic lenses, allows an analytic solution, which can be almost as accurate as that of the actual spherical lens CRL. One can approximate a spherical lens CRL by a parabolic lens CRL, provided that the CRL has an entrance pupil or aperture, which only admits rays whose incident transverse positions l on the CRL principal plane, are much less than the radius of curvature R of the lenses, where $l \ll R$. The parabolic or spherical lens CRL occupies the aperture in an X-ray/neutron absorbing sheet, and an X-ray/neutron absorbing pupil limits the CRL aperture to effective aperture radius r_e. If the CRL is comprised of parabolic lenses, then the CRL effective aperture radius r_e is the minimum of the absorption aperture radius $r_\mathrm{a} = \sqrt{2R/\mu N}$ (see Eqs. (1.126–1.127)) and the physical aperture radius r_m, where $r_\mathrm{e} = \min(r_\mathrm{a}, r_\mathrm{m})$. If the CRL is comprised of spherical lenses, then the CRL effective aperture radius r_e is a minimum of the physical r_m and absorption r_a aperture radii, as well as the CRL parabolic aperture radius $r_\mathrm{p} = (2R^3\lambda/N\delta)^{1/4}$ from Eq. (1.162), where $r_\mathrm{e} = \min(r_\mathrm{a}, r_\mathrm{p}, r_\mathrm{m})$ from Eq. (1.156).

With the source and object planes merged ($D = 0$), the neutron or X-ray wave intensity $I_y^{\text{par}}(\lambda, y, l, s)$ at image point y is obtained by substitution of Eq. (11.50) into Eq. (11.4). The integrand is nonzero only over the range of l within the CRL effective aperture radius r_e. Hence, the integral is unchanged by extension of the range of l integration to infinity, where $-\infty < l < \infty$. This allows application of the integral rule of Eq. (11.10) to the integration in Eq. (11.4), with integral parameter $a = \mu N/2R$, which is conveniently $a = 1/r_\mathrm{a}^2$, and integral parameter $b = k[(y/r_i) + (s/r_0)]$ with $k = 2\pi/\lambda$. One thus integrates the amplitude $E_y^{\text{par}}(\lambda, y, l, s)$ of Eq. (11.50) in Eq. (11.4) over CRL coordinate l, and squares the magnitude of the integration result, which then yields the image intensity distribution $I_y^{\text{par}}(\lambda, y, l, s)$ from a single source point s, where

$$I_y^{\text{par}}(s, y, \lambda) = \left| \int_{-\infty}^{\infty} E_y^{\text{par}}(s, y, l, \lambda) \mathrm{d}l \right|^2 = \frac{r_a^2}{\pi} \left(\frac{2E_0 \, \mathrm{e}^{-N\mu d/2}}{(r_0 + r_i)\pi\sigma_x\sigma_\lambda} \right)^2 \exp\left(-\frac{\lambda - \lambda_0}{\sigma_\lambda^2} \right)$$
$$\times \exp\left[-2\frac{r_a^2\pi^2}{\lambda^2} \left(\frac{s}{r_0} + \frac{y}{r_i} \right)^2 \right] \exp\left[-2s^2 \left(\frac{1}{2\sigma_x^2} + \frac{\mathrm{i}\pi}{r_0\lambda} \right) \right] \exp\left(-2\frac{\mathrm{i}\pi}{r_i\lambda} y^2 \right) \tag{11.52}$$

One then integrates Eq. (11.52) in Eq. (11.4) over infinite range $-\infty < s < \infty$ in the object plane, and thereby obtains the image plane intensity $I_y^{\text{par}}(\lambda, y)$ at each image point y from all the object points s, due to the CRL, where again one uses the integral rule of Eq. (11.10). Parabolic lenses focus rays from a single object point s to a single image point y, along any paraxial path through the CRL, if the ray path lengths are equal, within the accuracy of the binomial approximation, which was used in the ray phase-shift calculations for the drift spaces. Hence, the X-ray or neutron rays, which are emitted from an object point s, are provided compensating path lengths (and thus compensating phase shifts) by the parabolic lens CRL. The equal ray path lengths, allow the rays from object point s to arrive at the single, corresponding image point y, all with the same phase. Equal phase shifts of the rays, permit maximum constructive interference, which produces maximum intensity at image point y. However, image point y suffers intensity loss, which is caused by ray scatter and absorption in the CRL. In addition, intensity loss at image point y is caused by the limited CRL field of view (FOV), which allows the CRL to intercept only a fraction of the rays that are emitted from object point s.

The parabolic CRL produces an image distribution along y, or impulse response from a single object point, following integration of $E_y^{\text{par}}(\lambda, y, l, s = 0)$ over the lens coordinate l. The object resolution of the CRL is obtained from Eq. (11.52) by setting $s = 0$ (recall Eq. (11.8)), and inspecting the argument of the exponential in y, where one obtains the CRL object resolution as the full width half maximum (FWHM) of the image intensity distribution Δy_{par}, where

$$\Delta y_{\text{par}} = \frac{\lambda r_i}{r_a \pi} \sqrt{\frac{\log_e(2)}{2}} \tag{11.53}$$

The CRL field of view (FOV) is obtained from Eq. (11.52) by setting $y = 0$ (recall Eq. (11.9)), and inspecting the argument of the exponential in s, where one obtains the FOV as the full width half maximum (FWHM) of the source intensity distribution Δs_{par}, where

$$\Delta s_{\text{par}} = \sqrt{\left(\frac{1}{\sigma_x^2} + \frac{2\pi^2 r_a^2}{\lambda^2 r_0^2} \right)^{-1} \log_e(2)} \tag{11.54}$$

If the CRL is removed, and replaced by an aperture of the same diameter (akin to a pinhole camera), the total complex phase shift, with merged source and object planes ($D = 0$), is given by

$$\phi_T^{\text{aper}} = -kl\left(\frac{s}{r_0} + \frac{y}{r_i}\right) + \frac{k}{2}\left[\frac{s^2}{r_0} + \frac{y^2}{r_i} + l^2\left(\frac{1}{r_0} + \frac{1}{r_i}\right)\right] \tag{11.55}$$

Note that the aperture also inverts the image with a projected transverse magnification M_T without aberration, where the lack of aberration allows cancellation of cross terms in ls and yl in the first term of Eq. (11.55) for ϕ_T^{aper}, where

$$\frac{s}{r_0} = -\frac{y}{r_i} \tag{11.56}$$

Hence,

$$\phi_T^{aper} = \frac{k}{2}\left[\frac{s^2}{r_0} + \frac{y^2}{r_i} + l^2\left(\frac{1}{r_0} + \frac{1}{r_i}\right)\right] \tag{11.57}$$

The aperture exhibits phase shifts along the various paths from object point s to image point y, which are dependent on the ray position l in the aperture, except for ray positions l that are sufficiently close to the aperture center axis, where

$$l \ll s \quad \text{and} \quad l \ll y \tag{11.58}$$

This path-independent phase shift through the pinhole requires an aperture or a pinhole with a very small diameter, which results in a dramatic decrease in the image intensity profile.

One can account for the effect of gravity on the phase shift of cold and ultracold neutrons in their passage from the source to image planes. To calculate the phase shift in these drift spaces, due to the extra fall distance of the neutron due to gravity, one calculates the transit time of the neutron across each of the drift spaces, via the drift space length and the velocity v of the neutron. The neutron group velocity v, with h as Planck's constant and m as the neutron rest mass, is

$$v = \frac{h}{m\lambda} \tag{11.59}$$

The transit time t_D in the drift space between the source and illuminated object is

$$t_D = \frac{D}{v} = \frac{m\lambda D}{h} \tag{11.60}$$

The transit time t_0 between the object and CRL center or principal plane is

$$t_0 = \frac{r_0}{v} = \frac{m\lambda r_0}{h} \tag{11.61}$$

The transit time t_i between the CRL principal plane and the image plane is

$$t_i = \frac{r_i}{v} = \frac{m\lambda r_i}{h} \tag{11.62}$$

The additional distances that the neutron falls, which is caused by gravity g in the neutron transit time t_D between (1) the source and object D_g, (2) the object and the CRL principal plane o_g, and (3) the CRL principal plane and image i_g are, respectively,

$$D_g = \frac{g}{2}\left(\frac{m\lambda D}{h}\right)^2 \tag{11.63}$$

$$o_g = \frac{g}{2}\left(\frac{m\lambda r_0}{h}\right)^2 \tag{11.64}$$

$$i_g = \frac{g}{2}\left(\frac{m\lambda r_i}{h}\right)^2 \tag{11.65}$$

The total drift space phase shift ϕ_{drf}^{G}, including trajectory length increase due to gravity, is obtained by inserting Eqs. (11.63) into Eq. (11.15), and Eq. (11.64) into Eq. (11.18), and Eq. (11.65) into Eq. (11.19), and then substituting into Eq. (11.21), where

$$\phi_{\text{drf}}^{G} = k\left(D + r_0 + r_i\right) + \frac{kD}{2r_0^2}\left(l - s - D_g\right)^2 + \frac{k}{2r_0}\left(l - s - o_g\right)^2 + \frac{k}{2r_i}\left(y - l - i_g\right)^2 \tag{11.66}$$

11.2 Derivation of Kirchhoff Equation with a 2-D Biconcave Parabolic or Spherical CRL for X-rays or Neutrons with Gravity—Thick Sample

In this section and the sections that follow, we discuss 2-D models using the Kirchhoff equation in which *the thin-sample approximation is not used*. The samples can be *thick* specimens that strongly scatter, refract, and absorb. Every point on source, object, CRL, and image plane is an emitter of spherical waves, as would be expected by the Huygens's wave front and wave propagation constructions. Without the thin-sample approximation applied to the object plane, the mathematical derivation and results are more complicated, and are derived as follows. However, the thin-lens approximation and the paraxial approximation are still assumed in this 2-D analysis.

In this 2-D derivation, each point on the source, each point on the object (sample), and each point on the CRL is a spherical wave source. That is, a point on the source will have ray paths to every point on the object (sample) plane. Every point on the object (sample) plane will have a ray path to every point on the CRL principal plane. And finally, every point on the CRL principal plane will have a ray path to every point on the image plane. For neutrons, gravity is included in this analysis, so the transverse x- and y-directions are not equivalent because of the neutron gravity drift in the downward, negative y-direction. If one neglects gravity, or for the case of X-rays, which are not affected by gravity, one simply sets the gravity terms to zero.

The Huygens–Fresnel approach leads to the Kirchhoff equation, which calculates the propagation of X-rays and neutrons, including both refractive and diffractive effects. This method has an advantage over ray tracing since it includes both absorption and diffraction effects, which are not easily included in a ray tracing analysis. Also, the Kirchhoff equation can lead to closed-form solutions in some cases, which yields a more intuitive understanding.

The resolution of the lens and the size of the source spot on the detector can be calculated by first determining the image intensity distribution from a single point at the origin on the object (the PSF). We then translate the object point and the resulting intensity is added to the original intensity. When the distance between the two points in the object plane gives a minimum-to-maximum ratio of 0.81 between the two peaks in the image plane (the Rayleigh criterion), the distance of translation is defined as the resolution.

Using the paraxial approximation, the Kirchhoff equation is derived for an X-ray or neutron source, which illuminates an object that is magnified by a CRL onto an image plane. The CRL is comprised of a stack of N circular, biconcave, parabolic lenses, which are azimuthally symmetric about the z-axis that passes through the centers of the source, object, lens, and image. A ray of a neutron wave of amplitude E_s and wavelength λ is emitted from the source point at transverse position x and axial position $z = 0$. The ray intercepts an object at axial distance D and transverse object position s. The ray propagates axial distance r_0 from the object to intercept a thin-lens CRL at transverse-lens position l at the CRL principal or center plane. The CRL then refracts the ray at the CRL principal plane, and the ray propagates axial distance r_i to a point on the image plane at transverse position y.

The ray of the X-ray or neutron wave at the image plane has wavelength-dependent amplitude $E_y(\lambda, y, l, s)$. This amplitude is determined by the total path length L_T, and the resulting phase shift ϕ_T of the local wave front, which propagates along this single path (represented by a ray), is

$$\phi_T = 2\pi \frac{L_T}{\lambda} \tag{11.67}$$

Its wave amplitude at image position y is

$$E_y(\lambda, y, l, s) = E_s(\lambda, s, l)\exp(-\mathrm{i}\phi_T(\lambda, y, l, s)) \tag{11.68}$$

The 1-D image or radial image intensity profile I_y is obtained by determining the intensity at image point y from the various paths from each object point s, which is illuminated by a ray from a particular source point x, and then summing (integrating) the contributions from each of the source points x. The summed amplitude at y is

$$I_y(\lambda, y) = \int_{\text{source}} \mathrm{d}s \left| \int_{\text{lens}} E_y(\lambda, y, l, s)\mathrm{d}l \right|^2 \tag{11.69}$$

At the image plane, one can calculate the gain in image plane intensity, using a CRL, in place of only an aperture at the CRL position. The aperture diameter equals the CRL effective aperture diameter $2r_e$. The 1-D intensity gain $G_{1\text{-}D}$ at the image plane with the use of a CRL and image intensity $I_y^{\text{CRL}}(y, \lambda)$, compared to the case of no CRL and aperture only with image intensity $I_y^0(\lambda, y)$, is

$$G_{1D}(\lambda, y) = \frac{I_y^{\text{CRL}}(\lambda, y)}{I_y^0(\lambda, y)} \tag{11.70}$$

For a parabolic CRL or spherical CRL for I_y^{CRL}, one substitutes either

$$I_{\text{CRL}} = I_y^{\text{par}} \quad \text{or} \quad I_{\text{CRL}} = I_y^{\text{sph}} \tag{11.71}$$

For an azimuthally symmetric lens, the 2-D gain is the square of the 1-D gain, that is,

$$G_{2\text{-}D}(\lambda, y) = [G_{1\text{-}D}(\lambda, y)]^2 = \left[\frac{I_y^{\text{CRL}}(\lambda, y)}{I_y^0(\lambda, y)}\right]^2 \tag{11.72}$$

One can determine the image and object resolution of the CRL imaged object by considering a specific object point (e.g., the center object point at $s = 0$). One then calculates the image plane distribution from this single object point, that is,

$$I_y(\lambda, y) = \left| \int_{\text{lens}} E_y(\lambda, y, l, s = 0)\mathrm{d}l \right|^2 \tag{11.73}$$

The FWHM of the image plane distribution along transverse position y, yields the image resolution, and division of the image resolution by the CRL transverse magnification $M_T = y/s$, then determines the object resolution.

Alternatively, one can determine the FOV of the CRL by considering a specific image point, for example, the center image point at $y=0$. One can determine the intensity distribution in the object plane along transverse direction s, which contributes to the intensity of the image point at $y=0$, that is,

$$I_y(\lambda, s) = \left| \int_{\text{lens}} E_y(\lambda, y=0, l, s) \mathrm{d}l \right|^2 \tag{11.74}$$

The FWHM of this distribution of source points along s is the FOV of the CRL. The integral rule of Eq. (11.10) is repeated below, where $a>0$, and

$$\int_{-\infty}^{\infty} \exp\left(-ax^2 + ibx\right) \mathrm{d}x = \sqrt{\frac{\pi}{a}} \exp\left(-\frac{b^2}{4a}\right) \quad a>0 \tag{11.75}$$

The task remains to determine the total path length L_T and the resulting phase shift ϕ_T for each ray path of the X-ray or neutron wave, emitted from a single source point, which is focused by the CRL to a single image point. Assume that the X-ray or neutron source has a Gaussian distribution of wavelength λ, about a mean wavelength λ_0 and standard deviation σ_λ. Further, assume that the X-ray or neutron source is azimuthally symmetric and has a Gaussian amplitude distribution with respect to the transverse dimensions x, with standard deviation σ_x about its center ($x=0$) amplitude E_0. Hence, at the source plane ($z=0$), the X-ray or neutron wave amplitude along a particular dimension, such as the transverse x-axis $E_s(\lambda, x)$, has a radial dependence, and is given by

$$E_s(\lambda, x) = E(\lambda, x, z=0) = \frac{2E_0}{\pi \sigma_x \sigma_\lambda} \exp\left[-\frac{(\lambda-\lambda_0)^2}{2\sigma_\lambda^2}\right] \exp\left(-\frac{x^2}{2\sigma_x^2}\right) \tag{11.76}$$

Consider the 2-D imaging case for a parabolic CRL. The positive vertical downward y-direction points in the direction of the gravity (and the ground) and the horizontal positive x-direction points outward from the yz-plane. The right-hand rule thus has the neutrons propagating in the positive z-direction from the source at the left, through the object and through the CRL, to the image plane at the right.

For the 2-D case at the source plane ($z=0$), the X-ray or neutron wave amplitude along the two transverse source dimensions x and y, $E_s(\lambda, x, y, z=0)$, is given by

$$E_s(\lambda, x, y) = E_s^0 E_s^x E_s^y \tag{11.77}$$

The distance r is from the source point at source plane transverse coordinates x_s, y_s to an image point with transverse coordinates x, y, where the source amplitude coefficient E_s^0, is

$$E_s^0 = \frac{E_{0x} E_{0y}}{r} \tag{11.78}$$

The axial distance from the source plane at axial position $z = 0$, through the object and CRL of length L to the image plane at axial position $z = z_{im}$, where one recalls the object r_0 and image r_i distances are measured from the CRL center, principal plane, where

$$z = z_{im} = D + r_0 + r_i \tag{11.79}$$

The distance from source point $(x_s, y_s, z = 0)$ to image point $(x, y, z = z_{im})$ is

$$r = z\sqrt{1 + \left(\frac{x - x_s}{z}\right)^2 + \left(\frac{y - y_s}{z}\right)^2} \tag{11.80}$$

One assumes the axial distance z between the source plane and image plane, and thus (1) the source distance D between the source plane and object plane, (2) the object distance r_0 between the object (sample) plane and CRL principal plane, and (3) the image distance r_i between the CRL principal plane and image plane, are each much larger than (1) the transverse extent of the source points (x_s, y_s), (2) the transverse extent of the object (sample) points (s_x, s_y), (3) the transverse extent of the CRL principal plane points (l_x, l_y), and (4) the transverse extent of the image points (x, y), where

$$z, D, r_0, r_i \gg x_s, y_s, s_x, s_y, l_x, l_y, x, y \quad \text{where} \quad z = D + r_0 + r_i \tag{11.81}$$

With these assumptions, the distance r from transverse source position x_s, y_s to image transverse image position x, y becomes

$$r \cong z + \frac{(x - x_s)^2}{2z} + \frac{(y - y_s)^2}{2z} \qquad z \gg x_s, y_s, x, y \tag{11.82}$$

A single path from source point (x_s, y_s) to image point (x, y) is comprised of three rays, which generally do not form a single straight line, but instead form 3 joined, diagonal, line segments (zig-zag). The total path length r is the sum of the three diagonal ray path lengths, which are (1) from source point (x_s, y_s) to object point (s_x, s_y), (2) from object point (s_x, s_y) to CRL principal plane point (l_x, l_y), and (3) from CRL principal plane point (l_x, l_y) to image point (x, y). Since the transverse component distances in the x-direction (x_s, s_x, l_x, x) and y-direction (y_s, s_y, l_y, y), are each much less than the axial component path length z of the total path length r from source point (x_s, y_s) to image point (x, y), one can then approximate the total path length r by its axial component z, where

$$r \cong z \qquad z \gg x_s, y_s, x, y \tag{11.83}$$

From section 10.3, Eq. (10.24), the obliquity factor F_0 is

$$F_0 = \frac{1 + \cos\theta}{2} \tag{11.84}$$

where

$$\theta = \tan^{-1}\frac{\sqrt{(x-x_s)^2 + (y-y_s)^2}}{z} \tag{11.85}$$

With the small transverse source (x_s, y_s) and object (x, y) dimensions, as compared to the source plane to image distance z, the obliquity factor F_0 is approximately unity:

$$F_0 \cong 1 \qquad z \gg x_s, y_s, x, y \tag{11.86}$$

Assume that the X-rays or neutrons in the x- and y-directions have respective center wavelengths λ_{0x} and λ_{0y} with Gaussian standard deviations about these center wavelengths of $\sigma_{\lambda x}$ and $\sigma_{\lambda y}$, respectively. The X-ray or neutron source amplitude component in the x-direction $E_s^x(x_s, \lambda)$ is a function of the wavelength λ and transverse source coordinate x_s, and the source amplitude component in the y-direction $E_s^y(y_s, \lambda)$ is a function of wavelength λ and transverse source coordinate y_s, where

$$E_s^x = \exp\left[-\frac{(\lambda-\lambda_{0x})^2}{2\sigma_{\lambda x}^2}\right]\exp\left(-\frac{x_s^2}{2\sigma_x^2}\right) \tag{11.87}$$

and

$$E_s^y = \exp\left[-\frac{(\lambda-\lambda_{0y})^2}{2\sigma_{\lambda y}^2}\right]\exp\left(-\frac{y_s^2}{2\sigma_y^2}\right) \tag{11.88}$$

For neutron sources and nonlaser X-ray sources, the standard deviation or approximate dimensions of the X-ray or neutron source along the x-direction are σ_x, and the source size along the y-direction is σ_y. The source dimensions σ_x and σ_y are related to parameters W_x and W_y by

$$W_x = W_x^{\mathrm{Re}} = \frac{1}{\sigma_x^2} \qquad W_x^{\mathrm{Im}} = 0 \tag{11.89}$$

$$W_y = W_y^{\mathrm{Re}} = \frac{1}{\sigma_y^2} \qquad W_y^{\mathrm{Im}} = 0 \tag{11.90}$$

where

$$W_x = W_x^{\mathrm{Re}} + \mathrm{i}\, W_x^{\mathrm{Im}} \tag{11.91}$$

$$W_y = W_y^{\mathrm{Re}} + \mathrm{i}\, W_y^{\mathrm{Im}} \tag{11.92}$$

For X-ray lasers, the standard deviation along the x- and y-directions, $\sigma_x(z)$ and $\sigma_y(z)$, in its most general form is complex, for example, for an X-ray laser with a Gaussian amplitude profile. A minus sign occurs in the exponential phase shift term of Eq. (11.68), as compared to a positive sign in the exponential phase shift term in the paraxial equation of 10.386. Hence using Eqs. (10.386), (10.392), and (10.393) in section 10.21, one obtains the positive real and imaginary components

$$\frac{ik}{2q_x(z)} = \frac{1}{\sigma_x^2} = \frac{1}{w_{0x}^2 + \mathrm{i}\frac{\lambda z}{\pi}} \tag{11.93}$$

$$\frac{ik}{2q_y(z)} = \frac{1}{\sigma_y^2} = \frac{1}{w_{0y}^2 + \mathrm{i}\frac{\lambda z}{\pi}} \tag{11.94}$$

For a Gaussian amplitude X-ray laser, the real W_x^{Re} and imaginary W_x^{Im} x-components for X-ray laser minimum waist w_{0x} along x and the real and imaginary y-components W_y^{Re} and W_y^{Im} for minimum X-ray laser waist w_{0y} along y are given by

$$W_x^{\mathrm{Re}} = w_{0x}^2 \quad \text{and} \quad W_x^{\mathrm{Im}} = \frac{z\lambda}{\pi} \tag{11.95}$$

and

$$W_y^{\mathrm{Re}} = w_{0y}^2 \quad \text{and} \quad W_y^{\mathrm{Im}} = \frac{z\lambda}{\pi} \tag{11.96}$$

Consider neutron or X-ray wave propagation in the positive z-direction, where the wave originates at the source plane at $z = 0$ and propagates through the object, through the lens, and finally to the image plane. The 2-D image intensity distribution is

$$E_{xy}(\lambda,x,y,l_x,l_y,s_x,s_y,x_s,y_s) = E_s(\lambda,l_x,l_y,x_s,y_s)\exp\left[-\mathrm{i}\phi_{\mathrm{T}}^{2D}(\lambda,x,y,l_x,l_y,s_x,s_y,x_s,y_s)\right] \tag{11.97}$$

where

$$\phi_{\mathrm{T}}^{2D}(\lambda,x,y,l_x,l_y,s_x,s_y,x_s,y_s) = \phi_{\mathrm{Drift}}^{2D}(\lambda,x,y,l_x,l_y,s_x,s_y,x_s,y_s) + \phi_{\mathrm{CRL}}^{2D}(\lambda,l_x,l_y) \tag{11.98}$$

and

$$E_s(\lambda, l_x, l_y, x_s, y_s) = E_s^x(\lambda, l_x, x_s) \cdot E_s^y(\lambda, l_y, y_s) \tag{11.99}$$

The negative sign is used with the argument ϕ_{T}^{2D} because the propagation is in the positive z-direction.

One can account for the effect of gravity on the phase shift of cold and ultracold neutrons in their passage from the source to image planes. Normal gravity does not affect X-ray trajectories, but neutrons have a significant trajectory droop in ordinary gravitational fields. In the case of neutrons, to calculate the phase shift in these drift spaces, due to the extra fall distance of the neutron via gravity, one calculates the transit time of the neutron across each of the drift spaces, via the drift space length and the velocity v of the neutron. With Planck's constant h and neutron mass m, the neutron velocity group velocity v is

$$v = \frac{h}{m\lambda} \tag{11.100}$$

Gravity acts vertically downward in the positive y-direction. The positive y-axis points vertically downward, the perpendicular, positive, horizontal x-axis points out of the plane of the page to the viewer, and the positive, horizontal, z-axis lying in the plane of the page points to the right. The transit time t_D in the drift space between the source and illuminated object is

$$t_D = \frac{D}{v} = \frac{m\lambda D}{h} \tag{11.101}$$

The transit time t_0 between the object and CRL principal plane is

$$t_0 = \frac{r_0}{v} = \frac{m\lambda r_0}{h} \tag{11.102}$$

The transit time t_L through the CRL of length L is

$$t_L = \frac{L}{v} = \frac{m\lambda L}{h} \tag{11.103}$$

The transit time t_i between the CRL principal plane and the image plane is

$$t_i = \frac{r_i}{v} = \frac{m\lambda r_i}{h} \tag{11.104}$$

The vertical distance contributions in the positive, y-direction (transverse to the positive, z-direction of incident neutron velocity), through which the neutron falls downward, due to gravity g, is given by the following:

1. Neutron transit time t_D between the source and the object yields neutron drop distance D_g, where

$$D_g = \left(\frac{g}{2}\right)\left(\frac{m\lambda D}{h}\right)^2 \tag{11.105}$$

2. Neutron transit time t_0 between the object and the CRL principal plane yields neutron drop distance o_g, where

$$o_g = \left(\frac{g}{2}\right)\left(\frac{m\lambda r_o}{h}\right)^2 \tag{11.106}$$

3. Neutron transit time t_L through the CRL of length L yields neutron drop distance

$$L_g = \left(\frac{g}{2}\right)\left(\frac{m\lambda L}{h}\right)^2 \tag{11.107}$$

4. Neutron transit time t_i between the CRL principal plane and image yields neutron drop distance i_g, where

$$i_g = \left(\frac{g}{2}\right)\left(\frac{m\lambda r_i}{h}\right)^2 \tag{11.108}$$

The next step is to determine the phase shifts due to the drift spaces. First, one calculates the length of flight L_s of the drift space between a neutron source point and the neutron-illuminated object (specimen) point. The trajectory length of flight of the X-ray or neutron ray is expressed in terms of X-ray or neutron wavelengths, or in terms of phase shift. The transit distance L_s of the X-ray or neutron from the source point (x_s, y_s) at $z = 0$ to an illuminated object point (s_x, s_y) at downstream axial distance $z = D$ is

$$L_s = \sqrt{D^2 + (s_x - x_s)^2 + (s_y - y_s - D_g)^2} \tag{11.109}$$

One can apply the Fresnel approximation (binomial) to this drift space expression:

$$D \gg |s_x - x_s| \tag{11.110}$$

and

$$D \gg |s_y - y_s - D_g| \tag{11.111}$$

to obtain

$$L_s \cong D + \frac{1}{2D}\left[(s_x - x_s)^2 + (s_y - y_s - D_g)^2\right] \tag{11.112}$$

The length of flight L_0 in the drift space, from illuminated object point (s_x, s_y) to a point (l_x, l_y) on the CRL center plane or principal plane, is

$$L_0 = \sqrt{r_0^2 + (l_x - s_x)^2 + (l_y - s_y - o_g)^2} \tag{11.113}$$

Application of the binomial approximation yields

$$L_0 \cong r_0 + \frac{1}{2r_0}\left[(l_x - s_x)^2 + \left(l_y - s_y - o_g\right)^2\right] \tag{11.114}$$

The length of flight L_i in the drift space between a point (l_x, l_y) on the CRL principal plane and the image point (x, y) is

$$L_i = \sqrt{r_i^2 + (x - l_x)^2 + (y - l_y - i_g)^2} \tag{11.115}$$

Application of the binomial approximation yields

$$L_i \cong r_i + \frac{1}{2r_i}\left[(x - l_x)^2 + (y - l_y - i_g)^2\right] \tag{11.116}$$

The drift space path length L_{drift} from the source to the illuminated object (specimen) through the CRL to its image plane is the sum of the "diagonal" drift space transit lengths. To recap, the summed transit lengths are (1) drift space transit length L_s from source point (x_s, y_s) at source plane $z = 0$ to an object point (s_x, s_y) at object plane $z = D$; (2) drift space transit length L_0 from object point (s_x, s_y) to point (l_x, l_y) on the CRL principal plane at $z = D + r_0$; and (3) drift space transit length L_i from point (l_x, l_y) on the CRL principal plane to image point (x, y) at image plane z_{im}, where

$$L_{\text{drift}} = L_s + L_0 + L_i \quad \text{and} \quad z_{\text{im}} = D + r_0 + r_i \tag{11.117}$$

The corresponding phase shift ϕ_{drift}^{2D} of the wavelength λ neutron wave due to the entire drift space length is as follows:

$$\phi_{\text{drift}}^{2D} = kL_{\text{drift}} \tag{11.118}$$

This drift space phase shift can be expressed in terms of components that depend on the x and y transverse positions in the image plane ϕ_{drift}^{x} and ϕ_{drift}^{y}, and a constant component ϕ_k dependent only on wavelength λ and axial distance between the source and image planes, where

$$\phi_{\text{drift}}^{2D} = \phi_{\text{drift}}^{x} + \phi_{\text{drift}}^{y} + \phi_k \tag{11.119}$$

Substitution of Eq. (11.117) with Eqs. (11.112), (11.114), (11.116) into Eq. (11.118), yields

$$\phi_{\text{drift}}^{x} = \left(\frac{k}{2D}\right)(s_x - x_s)^2 + \left(\frac{k}{2r_0}\right)(l_x - s_x)^2 + \left(\frac{k}{2r_i}\right)(x - l_x)^2 \tag{11.120}$$

$$\phi_{\text{drift}}^{y} = \left(\frac{k}{2D}\right)(s_y - y_s - D_g)^2 + \left(\frac{k}{2r_0}\right)(l_y - s_y - o_g)^2 + \left(\frac{k}{2r_i}\right)(y - l_y - i_g)^2 \tag{11.121}$$

and

$$\phi_k = k(D + r_0 + r_i) \tag{11.122}$$

The constant phase-shift term ϕ_k is dropped because subsequent calculations of intensity require the squared magnitude of the neutron wave amplitude, where

$$|\exp(i\phi_k)|^2 = 1 \tag{11.123}$$

Next, we consider the phase shift of the incident X-ray neutron wave, which is caused by its passage through the CRL for the 2-D case. With $\beta = \mu/2k$, the CRL index of refraction differs from a vacuum by

$$\Delta n = -\delta - i\beta \tag{11.124}$$

The thickness of a single parabolic lens t is a function of distance l from the center axis, and depends on the lens on-axis radius R and minimum or center wall thickness d, where

$$t = d + \frac{l^2}{R} \tag{11.125}$$

An X-ray or neutron ray, which intercepts a spherical biconcave lens at transverse x- and y-distances l_x and l_y from the lens center axis, travels an approximate axial distance or length of flight L_{lens} through the lens material, where

$$L_{\text{lens}} = 2\left(R - \sqrt{R^2 - l_x^2 - l_y^2}\right) + d \tag{11.126}$$

If an X-ray or neutron travels at phase velocity v in a vacuum and velocity v/n in a material with refractive index n, then the time of flight of the neutron is identical for path length L in a vacuum and path length nL in the material. The optical path length Γ_{lens} through the lens is related to the lens refractive index n and the neutron path length L_{lens}, where

$$\Gamma_{lens} = L_{lens}n \tag{11.127}$$

The complex index of refraction n of a material consists of the dimensionless, real, refractive decrement δ, and the dimensionless, imaginary, absorptive β, where

$$n = 1 - \delta - i\beta \tag{11.128}$$

The minus sign in front of $i\beta$ in the complex refractive index $n = 1 - \delta - i\beta$ in Eq. 11.128 produces a negative, real argument term in the exponential term of eq. 11.97 for the wave phase shift of a wave traveling in the forward, positive z-direction in the CRL. This minus sign is needed to properly show wave attenuation in the CRL material. The dimensionless absorption parameter β is related to the linear attenuation coefficient of absorption of the material $\mu\,[\mathrm{cm}^{-1}]$ and the neutron wave number k by

$$\beta = \frac{\mu}{2k} = \frac{\mu\lambda}{4\pi} \tag{11.129}$$

The optical path length Γ_{lens} through the lens is, by substitution of Eq. (11.128) into Eq. (11.127),

$$\Gamma_{lens} = L_{lens}\left(1 - \delta - i\frac{\mu}{2k}\right) \tag{11.130}$$

Between the left plane and the left surface of the biconcave lens is air, and between the right surface and the right plane of the lens is air. The refractive index of air, which has a very low density, is approximated as the vacuum refractive index. The left and right planes of each biconcave lens are separated by axial distance d_{max}, which is the lens perimeter or maximum thickness. The thickness L_{air} of the air space of each spherical lens is

$$L_{air} = d_{max} - L_{lens} \tag{11.131}$$

or

$$L_{air} = d_{max} - 2\left(R - \sqrt{R^2 - l_x^2 - l_y^2}\right) - d \tag{11.132}$$

The optical path length Γ_{CRL} of a ray through a CRL with N biconcave lenses is

$$\Gamma_{CRL} = N\Gamma_{lens} \tag{11.133}$$

For the case of a spherical N biconcave CRL, with substitution of Eq. (11.126) into Eq. (11.133), the optical path length $\Gamma_{\mathrm{CRL}}^{\mathrm{sph}}$ cm^{-1} is

$$\Gamma_{\mathrm{CRL}}^{\mathrm{sph}} = N\left(d_{\max} - d\left(\delta - \mathrm{i}\frac{\mu}{2k}\right)\right) - 2N\left(\delta - \mathrm{i}\frac{\mu}{2k}\right)\left(R - \sqrt{R^2 - l_x^2 - l_y^2}\right) \quad (11.134)$$

The radius of curvature R of each lens of the CRL is assumed to be much larger than the transverse x- and y-dimensions of the lens, and R is assumed to be much larger than the x- and y-positions l_x and l_y on the CRL principal plane, where

$$R \gg l_x \quad \text{and} \quad R \gg l_y \quad (11.135)$$

Expansion of radical to the third order in Eq. (11.134), yields the optical path length $\Gamma_{\mathrm{CRL}}^{\mathrm{sph}}$ in the spherical lens CRL, where

$$\Gamma_{\mathrm{CRL}}^{\mathrm{sph}} \approx N\left(d_{\max} - d\left(\delta + \mathrm{i}\frac{\mu}{2k}\right)\right) - 2N\left(\delta + \mathrm{i}\frac{\mu}{2k}\right)\left[\frac{l_x^2 + l_y^2}{2R} + \frac{(l_x^2 + l_y^2)^2}{8R^3}\right] \quad (11.136)$$

Recalling Eqs. (11.37–11.39), the optical path length $\Gamma_{\mathrm{CRL}}^{\mathrm{par}}$ in a parabolic lens CRL, is also obtained by dropping the l^4 term in spherical lens optical path length $\Gamma_{\mathrm{CRL}}^{\mathrm{sph}}$ expression of Eq. (11.136), where

$$\Gamma_{\mathrm{CRL}}^{\mathrm{par}} = N\left(d_{\max} - d\left(\delta + \mathrm{i}\frac{\mu}{2k}\right)\right) - \frac{N}{R}\left(\delta + \mathrm{i}\frac{\mu}{2k}\right)(l_x^2 + l_y^2) \quad (11.137)$$

Rearrangement of Eq. (11.137), yields the real and imaginary parts of the optical path length in a parabolic lens CRL:

$$\Gamma_{\mathrm{CRL}}^{\mathrm{par}} = N(d_{\max} - d\delta) - \frac{N\delta}{R}(l_x^2 + l_y^2) - \mathrm{i}\cdot\frac{N\mu}{2k}\left(d + \frac{1}{R}(l_x^2 + l_y^2)\right) \quad (11.138)$$

An X-ray or neutron wave with wavelength λ or wave number k undergoes a phase shift ϕ in propagation through optical path length Γ, where

$$\phi = 2\pi\frac{\Gamma}{\lambda} = k\Gamma \quad (11.139)$$

With substitution of Eq. (11.136) into Eq. (11.139), the phase shift $\phi_{\mathrm{CRL}}^{\mathrm{sph}}$ for the case of a spherical, biconcave CRL is

$$\phi_{\mathrm{CRL}}^{\mathrm{sph}} \approx kNd_{\max} - kNd\delta - 2kN\delta\left[\frac{l_x^2 + l_y^2}{2R} + \frac{(l_x^2 + l_y^2)^2}{8R^3}\right] - \mathrm{i}N\mu\left[\frac{d}{2} + \frac{l_x^2 + l_y^2}{2R} + \frac{(l_x^2 + l_y^2)^2}{8R^3}\right] \quad (11.140)$$

With substitution of Eq. (11.138) into Eq. (11.139), in a parabolic lens CRL, the X-ray or neutron wave phase shift $\phi_{\mathrm{CRL}}^{\mathrm{par}}$ is

$$\phi_{\mathrm{CRL}}^{\mathrm{par}} = kNd_{\max} - kNd\delta - kN\delta\left(\frac{l_x^2 + l_y^2}{R}\right) - \mathrm{i}N\mu\left(\frac{d}{2} + \frac{l_x^2 + l_y^2}{2R}\right) \tag{11.141}$$

For a 2-D parabolic CRL, the X-ray or neutron wave phase shift $\phi_{\mathrm{CRL}}^{\mathrm{par}}$ expression of Eq. (11.141) can be separated into x- and y-components, where

$$\phi_{\mathrm{CRL}}^{2\text{-}D}(\lambda, l_x, l_y) = kNd_{\max} - kN\delta d - \mathrm{i}\left(\frac{N\mu d}{2}\right) + \phi_x^{\mathrm{par}} + \phi_y^{\mathrm{par}} \tag{11.142}$$

The CRL phase-shift components ϕ_x^{par} and ϕ_y^{par} in the x- and y-directions are

$$\phi_x^{\mathrm{par}} = -kN\delta\left(\frac{l_x^2}{R}\right) - \mathrm{i}\left(\frac{\mu N l_x^2}{2R}\right) \tag{11.143}$$

and

$$\phi_y^{\mathrm{par}} = -kN\delta\left(\frac{l_y^2}{R}\right) - \mathrm{i}\left(\frac{\mu N l_y^2}{2R}\right) \tag{11.144}$$

For the case of the parabolic CRL, the 2-D expression of Eq. (11.97) for the X-ray or neutron wave amplitude in the image plane is separable into the product of the 1-D dependences in x and y, so that

$$E_{xy}(\lambda, x, y, l_x, l_y, s_x, s_y) = E_0(\lambda)E_x(\lambda, l_x, s_x, x)E_y(\lambda, l_y, s_y, y) \tag{11.145}$$

In Eq. (11.145), the amplitude component perpendicular to gravity $E_x(x, \lambda, x_s, s_x, l_x)$ is

$$E_x(x, \lambda, x_s, s_x, l_x) = E_s^x \exp[-\mathrm{i}\cdot(\phi_{\mathrm{drift}}^x + \phi_x^{\mathrm{par}})] \tag{11.146}$$

Substitution of Eqs. (11.87), (11.120), (11.143) into Eq. (11.146), yields

$$E_x(x, \lambda, x_s, s_x, l_x) = \exp\left[-\frac{(\lambda - \lambda_{0x})^2}{2\sigma_{\lambda x}^2}\right]\exp\left(-\frac{x_s^2}{2\sigma_x^2}\right) \times \exp\left[-\mathrm{i}k\frac{(s_x - x_s)^2}{2D} - \mathrm{i}k\frac{(l_x - s_x)^2}{2r_0} - \mathrm{i}k\frac{(x - l_x)^2}{2r_i} + \mathrm{i}k\frac{N\delta}{R}l_x^2 - \frac{\mu N}{2R}l_x^2\right] \tag{11.147}$$

Expansion and factoring of terms in Eq. (11.147), yields the amplitude at the image plane along the x-direction:

$$E_x(x,\lambda,x_s,s_x,l_x)=E_{1x0}(\lambda)E_{1lx}(\lambda,l_x)\exp\left[\mathrm{i}xl_x\left(\frac{k}{r_i}\right)\right]\exp\left[\mathrm{i}s_xl_x\left(\frac{k}{r_0}\right)\right]E_{1x}(x,\lambda,s_x,x_s) \tag{11.148}$$

where

$$E_{1x0}(\lambda)=\exp\left[-\frac{(\lambda-\lambda_{0x})^2}{2\sigma_{\lambda x}^2}\right] \tag{11.149}$$

$$E_{1lx}(\lambda,l_x)=\exp\left[-\frac{\mu N}{2R}l_x^2\right]\exp\left[-\mathrm{i}l_x^2\left(\frac{k}{2r_0}+\frac{k}{2r_i}-\frac{kN\delta}{R}\right)\right] \tag{11.150}$$

and

$$E_{1x}(x,\lambda,s_x,x_s)=\exp\left[\mathrm{i}x_ss_x\left(\frac{k}{D}\right)\right]\exp\left[-\mathrm{i}s_x^2\left(\frac{k}{2D}+\frac{k}{2r_0}\right)\right]$$
$$\times\exp\left(-\frac{x_s^2}{2\sigma_x^2}\right)\exp\left[-\mathrm{i}x_s^2\left(\frac{k}{2D}\right)\right]\exp\left[-\mathrm{i}x^2\left(\frac{k}{2r_i}\right)\right] \tag{11.151}$$

In Eq. (11.145), the amplitude component in the direction of the gravity $E_y(y,\lambda,y_s,s_y,l_y)$ is

$$E_y(y,\lambda,y_s,s_y,l_y)=E_s^y\exp[-\mathrm{i}\cdot(\phi_{\mathrm{drift}}^y+\phi_y^{\mathrm{par}})] \tag{11.152}$$

Substitution of Eqs. (11.88), (11.121), (11.144) into Eq. (11.152), yields

$$E_y(y,\lambda,y_s,s_y,l_y)=\exp\left[-\frac{(\lambda-\lambda_{0y})^2}{2\sigma_{\lambda y}^2}\right]\exp\left(-\frac{y_s^2}{2\sigma_y^2}\right)\exp\left(-\frac{\mu N}{2R}l_y^2\right)$$
$$\times\exp\left[-\frac{ik}{2}\left\{\frac{1}{D}(s_y-y_s-D_g)^2+\frac{1}{r_0}(l_y-s_y-o_g)^2\right.\right.$$
$$\left.\left.+\frac{1}{r_i}(y-l_y-i_g)^2-2\frac{N\delta}{R}l_y^2\right\}\right] \tag{11.153}$$

The CRL axial extent is much smaller than the source plane to the CRL principal plane, as well as the distance from the CRL principal plane to the image plane.

Hence, the neutron fall due to gravity within the CRL will be ignored. To simplify integration over l_y, the term below will be dropped in Eq. (11.153), so that

$$l_y\left(\frac{ko_g}{r_0}-\frac{ki_g}{r_i}\right)\cong 0 \tag{11.154}$$

Expansion and factoring of terms in Eq. (11.153), yields the amplitude at the image plane along the y-direction:

$$E_y(y,\lambda,y_s,s_y,l_y)=E_{1y0}(\lambda)E_{1ly}(\lambda,l_y)\exp\left[\mathrm{i}s_yl_y\left(\frac{k}{r_0}\right)\right]\exp\left[\mathrm{i}yl_y\left(\frac{k}{r_i}\right)\right]E_{1y}(y,\lambda,s_y,y_s) \tag{11.155}$$

where

$$E_{1y0}(\lambda)=\exp\left[-\frac{(\lambda-\lambda_{0y})^2}{2\sigma_{\lambda y}^2}\right]\exp\left[-ik\left(\frac{D_g^2}{2D}+\frac{o_g^2}{2r_0}+\frac{i_g^2}{2r_i}\right)\right] \tag{11.156}$$

$$E_{1ly}(\lambda,l_y)=\exp\left[-\frac{\mu N}{2R}l_y^2\right]\exp\left[-\mathrm{i}l_y^2\left(\frac{k}{2r_0}+\frac{k}{2r_i}-\frac{kN\delta}{R}\right)\right] \tag{11.157}$$

and

$$\begin{aligned}E_{1y}(y,\lambda,s_y,y_s)=&\exp\left[-is_y\left(\frac{ko_g}{r_0}-\frac{kD_g}{D}\right)\right]\exp\left[-\mathrm{i}s_y^2\left(\frac{k}{2D}+\frac{k}{2r_0}\right)\right]\\&\times\exp\left[iy_ss_y\left(\frac{k}{D}\right)\right]\exp\left(-\frac{y_s^2}{2\sigma_y^2}\right)\exp\left[-\mathrm{i}y_s\left(\frac{kD_g}{D}\right)\right]\\&\times\exp\left[-iy_s^2\left(\frac{k}{2D}\right)\right]\exp\left[\mathrm{i}y\left(\frac{ki_g}{r_i}\right)\right]\exp\left[-\mathrm{i}y^2\left(\frac{k}{2r_i}\right)\right]\end{aligned} \tag{11.158}$$

Recall from Eq. (11.83), the total path length r from source point (x_s, y_s) to image point (x, y), one can then approximate the total path length r by its axial component z, where

$$r=z \tag{11.159}$$

The X-ray or neutron wave amplitude coefficient E_0 in Eq. (11.145) is

$$E_0(\lambda,z)=\left(\frac{E_{0x}E_{0y}}{z}\right)\exp\left(-\frac{N\mu d}{2}\right)\exp\left[-\mathrm{i}kN\left(d_{\max}-\delta d\right)\right] \tag{11.160}$$

The center portion of a spherical lens is parabolic. A spherical CRL with a pupil diameter set equal to the parabolic aperture diameter of the spherical CRL will allow application of the analytic parabolic CRL solution to the center parabolic portion of the spherical CRL. These wave function amplitudes apply to the case of a spherical CRL of radius of curvature R, if the aperture radius r_{stop}, which is formed by an iris over the lens, equals the parabolic aperture radius r_{p} of the spherical CRL, where recalling Eq. 1.161 in section 1.2 of chapter 1,

$$r_{\text{stop}} = r_{\text{p}} = \left(\frac{2R^3\lambda}{N\delta}\right)^{1/4} \tag{11.161}$$

The nonzero portion of the lens integration occurs only for the radial lens coordinates l that lie within lens aperture $[-r_{\text{stop}}, r_{\text{stop}}]$. If a ray intercepts the spherical lens of radius of curvature R at a radial distance l from the center lens axis with $l \ll R$, then the thickness t of the spherical CRL approximately equals that of a parabolic CRL with on-axis lens radius R, where the thickness t from Eq. (1.98) in section 1.2 of chapter 1, is

$$t = R - \sqrt{R^2 - l^2} \approx \frac{l^2}{2R} \quad l \ll R \tag{11.162}$$

A spherical CRL, whose aperture is limited to the center parabolic portion, has an intensity distribution I_{xy} in the image plane, which is separable into wavelength-dependent x- and y-components, and a component that is dependent only on wavelength, where

$$I_{xy}(\lambda, x, y, l_x, l_y, s_x, s_y) = |E_0(\lambda)|^2 |E_x(\lambda, l_x, s_x, x)|^2 |E_y(\lambda, l_y, s_y, y)|^2 \tag{11.163}$$

The PSF (Green's function), FOV, and image intensity distribution for a spherical CRL, whose aperture is set to the parabolic aperture radius, can be obtained from analytic solutions of the Kirchhoff equation for the parabolic CRL. This saves time, as the spherical CRL Kirchhoff equation must be solved numerically.

11.3 Biconcave, Parabolic CRL—Image Amplitude Distribution for Incoherent X-rays or Neutrons with Gravity

The source of neutrons and X-rays is incoherent. The gravity terms are set to zero for the case of an incoherent X-ray source. Gravity applies only to the neutrons unless the gravitational field is exceptionally large, such as that of a large star. We now determine the 2-D X-ray or neutron amplitude at the image plane which is produced by a CRL that is comprised of a stack of N biconcave, parabolic lenses. Neutrons or X-rays are emitted from a source and propagate in the positive z-direction through an object (sample plane) to the CRL. The thin-lens approximation is applied, which assumes that the length of the CRL is much less than its focal length or the object and image distances. The CRL then refracts, focuses, and images the neutrons or X-rays onto the image plane. In the thin-lens approximation, the X-rays or neutrons are refracted at the center, single principal plane of the CRL that divides the CRL into its left and right axial halves.

For a parabolic CRL, the analytic expression for the neutron wave function amplitude $E_{xy}(x, y, \lambda, x_s, y_s, s_x, s_y)$ is obtained by integration of the amplitude of Eq. 11.145 over the lens transverse coordinates l_x, l_y, where

$$E_{xy}(x, y, \lambda, x_s, y_s, s_x, s_y) = \int_{-r_m}^{r_m} \int_{-r_m}^{r_m} E_{xy}(x, y, \lambda, x_s, y_s, s_x, s_y, l_x, l_y)\, \mathrm{d}l_x\, \mathrm{d}l_y \tag{11.164}$$

The binomial approximation, which was applied to the ray path length, allows separation of the integrand into independent x- and y-expressions. Setting aperture stops, with radius r_m at the CRL entrance and exit, limits the contributions to the integral to only neutrons, which propagate within the CRL aperture radius r_m. The image amplitude distribution becomes

$$E_{xy}(x, y, \lambda, x_x, y_s, s_x, s_y) = E_0(\lambda) \int_{-\infty}^{\infty} E_x(x, \lambda, x_s, s_x, l_x)\, \mathrm{d}l_x \int_{-\infty}^{\infty} E_y(y, \lambda, y_s, s_y, l_y)\, \mathrm{d}l_y \tag{11.165}$$

11.3.1 Amplitude in the y-Direction Parallel to the Direction of Gravity

The integral of the amplitude over the parabolic CRL transverse coordinate l_y in the direction of gravity has gravity dependence, where from Eq. (11.155),

$$E_y(y, \lambda, s_y, y_s) = E_{2y}(y, \lambda, s_y, y_s) \int_{-\infty}^{\infty} \mathrm{d}l_y \exp(-Al_y^2)\exp[\mathrm{i}\cdot(-El_y^2 + Fs_y l_y + Gyl_y)] \tag{11.166}$$

where

$$\begin{aligned} E_{2y}(y,\lambda,s_y,y_s) = E_{2y0}(\lambda)&\exp\left[\mathrm{i}y\left(\frac{ki_g}{r_i}\right)\right]\exp\left[-\mathrm{i}y^2\left(\frac{k}{2r_i}\right)\right]\exp\left(-\frac{y_s^2}{2\sigma_y^2}\right) \\ &\times\exp\left[-\mathrm{i}y_s\left(\frac{kD_g}{D}\right)\right]\exp\left[-\mathrm{i}y_s^2\left(\frac{k}{2D}\right)\right]\exp\left[-\mathrm{i}s_y\left(\frac{ko_g}{r_0}-\frac{kD_g}{D}\right)\right] \\ &\times\exp\left[-\mathrm{i}s_y^2\left(\frac{k}{2D}+\frac{k}{2r_0}\right)\right]\exp\left[\mathrm{i}y_s s_y\left(\frac{k}{D}\right)\right] \end{aligned} \tag{11.167}$$

with

$$E_{2y0}(\lambda) = \exp\left[-\frac{(\lambda-\lambda_{0y})^2}{2\sigma_{\lambda y}^2}\right]\exp\left[-ik\left(\frac{D_g^2}{2D}+\frac{o_g^2}{2r_0}+\frac{i_g^2}{2r_i}\right)\right] \tag{11.168}$$

Consolidation of terms in $E_{2y}(y, \lambda, s_y, y_s)$ yields

$$E_{2y}(y, \lambda, s_y, y_s) = E_{2y0}(\lambda)\exp\left[iy\left(\frac{ki_g}{r_i}\right)\right]\exp\left(-W_y y_s^2\right)$$
$$\times\exp\left[-i\cdot\left(Cs_y^2 + Hy_s^2 + \frac{G}{2}y^2 - 2Hy_s s_y + U_y s_y + V_y y_s\right)\right] \tag{11.169}$$

The parameters for the integration over l_y are as follows. The CRL parameters A and E are

$$A = \frac{\mu N}{2R} \tag{11.170}$$

$$E = \frac{k}{2r_0} + \frac{k}{2r_i} - \frac{kN\delta}{R} \tag{11.171}$$

For a CRL parameter, E is zero, where

$$E = \frac{\pi}{\lambda}\left(\frac{1}{r_0} + \frac{1}{r_i} - \frac{2N\delta}{R}\right) = 0 \tag{11.172}$$

The distance r_0 from the object plane to the CRL principal plane and the distance D from the source to the object plane give parameters C, F, and H, where

$$C = \frac{k}{2D} + \frac{k}{2r_0} = \frac{\pi}{\lambda}\left(\frac{1}{D} + \frac{1}{r_0}\right) \tag{11.173}$$

$$F = \frac{k}{r_0} = \frac{2\pi}{r_0\lambda} \tag{11.174}$$

$$H = \frac{k}{2D} = \frac{\pi}{D\lambda} \tag{11.175}$$

The image distance r_i parameter G is

$$G = \frac{k}{r_i} = \frac{2\pi}{r_i} \tag{11.176}$$

The y-component source parameter W_y is

$$W_y = \frac{1}{2\sigma_y^2} \tag{11.177}$$

The y-directed gravity parameters U_y and V_y are

$$U_y = \frac{ko_g}{r_0} - \frac{kD_g}{D} = \frac{2\pi}{\lambda}\left(\frac{o_g}{r_0} - \frac{D_g}{D}\right) \tag{11.178}$$

$$V_y = \frac{kD_g}{D} = \frac{2\pi}{\lambda}\frac{D_g}{D} \tag{11.179}$$

The CRL has zero parameter $E = 0$ and nonzero parameter $A \neq 0$, whereas a pinhole has nonzero parameter $E \neq 0$ and zero parameter $A = 0$. This will be examined in more detail later in this chapter, in Section 11.5.

The above integral in Eq. (11.166) can be solved by using the Euler formula below, where

$$\exp\left(\frac{\mathrm{j}kls}{r_0}\right) = \cos\left(\frac{kls}{r_0}\right) + \mathrm{j}\sin\left(\frac{kls}{r_0}\right) \tag{11.180}$$

The asymmetric sine term integrates to zero, so only the symmetric cosine term is retained in the integration. Then, one applies the tabulated integral, see Selby and Girling (1965), for the condition $a \neq 0$ so that

$$\int_0^\infty \cos(bs)\exp(-a^2s^2)\,\mathrm{d}s = \frac{\sqrt{\pi}}{2a}\exp\left(-\frac{b^2}{4a^2}\right) \qquad a \neq 0 \tag{11.181}$$

Another integral encountered, which is a special case of the previous integral with $b = 0$, has the integration requirement, $a \neq 0$, as follows:

$$\int_0^\infty \exp(-a^2s^2)\mathrm{d}s = \frac{\sqrt{\pi}}{2a} \qquad a \neq 0 \tag{11.182}$$

Mathematica, a software program by the Wolfram Company, is used to solve the integral expressions encountered in this analysis.

Rather than integrating over the lens diameter, we integrate along in the lens plane in the x- and y-directions from $-\infty$ to $+\infty$, so as to obtain a simpler expression without the complications of erf function. The erf functions result from integration of l_x and l_y from $-r_m$ to $+r_m$, where r_m is the lens radius, defined by a limiting iris. The iris is made of cadmium or gadolinium in the case of thermal neutrons; and for the case of X-rays, the iris is lead or tungsten. There is little error, considering that at twice the absorption aperture, radius $2r_a$ of the lens transmission fraction is 3.3×10^{-4}. The integration from twice the absorption aperture to infinity contributes only 0.03% of the integrated result.

The solution to the integral of Eq. (11.166) over the parabolic CRL transverse coordinate l_y, which is parallel to the direction of gravity, is

$$E_y(y, \lambda, s_y, y_s) = \frac{E_{2y}(y, \lambda, s_y, y_s)\sqrt{\pi}}{\sqrt{A + iE}} \exp\left[-\frac{(Fs_y + Gy)^2}{4(A + iE)}\right] \tag{11.183}$$

Expansion and collection of terms in Eq. (11.183), yields

$$\begin{aligned} &E_y(y, \lambda, s_y, y_s) \\ &= E_{y0}(\lambda)\exp(-W_y y_s^2)\exp\left[-ik\left(\frac{D_g^2}{2D} + \frac{o_g^2}{2r_0} + \frac{i_g^2}{2r_i}\right)\right]\exp\left[iy\left(\frac{ki_g}{r_i}\right)\right] \\ &\times \exp(-Js_y^2 - Ks_y y - Ny^2)\exp\left[-\mathrm{i}\cdot\left(Cs_y^2 + Hy_s^2 + \frac{G}{2}y^2 - 2Hy_s s_y + U_y s_y + V_y y_s\right)\right] \end{aligned} \tag{11.184}$$

where

$$E_{y0}(\lambda) = \sqrt{\frac{\pi}{A + \mathrm{i}E}} \exp\left[-\frac{(\lambda - \lambda_{0y})^2}{2\sigma_{\lambda y}^2}\right] \tag{11.185}$$

The parameters J, K, and N are closely interrelated, and each depends on the CRL radius of curvature R, number of lenses N, the lens material linear attenuation μ, object to lens distance, r_0, lens to image distance r_i, and wavelength λ, where

$$J = \frac{F^2}{4(A + iE)} \tag{11.186}$$

Explicit expression of the real and imaginary parts of J yields

$$J = \frac{F^2(A - \mathrm{i}E)}{4(A^2 + E^2)} \tag{11.187}$$

Substitution yields the explicit expression for J:

$$J = \frac{(\mu N/2R) - \mathrm{i}\cdot(\pi/\lambda)(1/r_0) + (1/r_i) - (2N\delta/R)}{(\mu N/2R)^2 + (\pi/\lambda)^2(1/r_0) + (1/r_i) - (2N\delta/R)^2} \cdot \left(\frac{\pi}{r_0\lambda}\right)^2 \tag{11.188}$$

The parameter K is

$$K = \frac{FG}{2(A + iE)} \tag{11.189}$$

Explicit representation of K in terms of its real and imaginary parts is

$$K = \frac{FG(A - iE)}{2(A^2 + E^2)} \tag{11.190}$$

Substitution yields the explicit expression for K:

$$K = \frac{(\mu N/2R) - i(\pi/\lambda)(1/2r_0) + (1/2r_i) - (2kN\delta/R)}{(\mu N/2R)^2 + (\pi/\lambda)^2((1/2r_0) + (1/2r_i) - (N\delta/R))^2}\left(\frac{2}{r_0 r_i}\right)\left(\frac{\pi}{\lambda}\right)^2 \tag{11.191}$$

Comparison of K with J reveals

$$K = 2J\left(\frac{r_0}{r_i}\right) \tag{11.192}$$

The parameter N is

$$N = \frac{G^2}{4(A + iE)} \tag{11.193}$$

Explicit representation of N in terms of its real and imaginary parts is

$$N = \frac{G^2(A - iE)}{4(A^2 + E^2)} \tag{11.194}$$

Substitution yields the explicit expression for N:

$$N = \frac{(\mu N/2R) - i \cdot (\pi/\lambda)(1/r_0) + (1/r_i) - (2N\delta/R)}{(\mu N/2R)^2 - i \cdot (\pi/\lambda)^2((1/r_0) + (1/r_i) - (2N\delta/R))^2}\left(\frac{\pi}{r_i \lambda}\right)^2 \tag{11.195}$$

Comparison of N with J reveals

$$N = J\left(\frac{r_0}{r_i}\right)^2 \tag{11.196}$$

For the case of very transmissive lenses, whose absorption aperture radius r_a greatly exceeds their physical aperture radius r_m, where $r_a \gg r_m$, one must integrate over the diameter of the lens. Outside the lens diameter, one assumes complete absorption of the X-rays or absorption. In the case of X-rays, outside the lens radius is lead shielding so that it allows only X-rays passing through the lens to reach the detector. In the case of neutrons, outside of the neutron lens radius is cadmium or lithium shielding (no capture gammas), which allows only neutrons passing through the lens to reach the detector. If the $[-\infty, \infty]$ integration limits for integration over the y-dimension of the lens l_y in Eq. (11.166) are replaced by $[-r_m, r_m]$, where the physical aperture diameter of the lens is $2r_m$, then the resulting wave amplitude E_y of Eq. (11.184) is multiplied by the factor L_y, where

$$L_y = \frac{1-\mathrm{i}}{\sqrt{2}}\sqrt{\frac{A+\mathrm{i}E}{E-\mathrm{i}A}}\left[\mathrm{Erf}\left(\frac{2\mathrm{i}Ar_m - 2r_mE + Fs_y + Gy}{2\sqrt{E-\mathrm{i}A}}\sqrt{\mathrm{i}}\right)\right. \\ \left. + \mathrm{Erf}\left(\frac{2\mathrm{i}Ar_m - 2r_mE - Fs_y - Gy}{2\sqrt{E-\mathrm{i}A}}\sqrt{\mathrm{i}}\right)\right] \tag{11.197}$$

The factor L_y is the sum of two error functions, resulting from evaluation at the integration limits.

The error function Erf(u) is an integral of the Gaussian function with respect to its dependent variable u over a portion $u \in (0, t)$ of its infinite interval $u \in (0, \infty)$, where

$$\mathrm{Erf}(u) = \frac{2}{\sqrt{\pi}}\int_0^t \exp(-u^2)\mathrm{d}u \tag{11.198}$$

The complementary error function Erf$c(u)$ is simply

$$\mathrm{Erf}c(u) = 1 - \mathrm{Erf}(u) \tag{11.199}$$

The imaginary error function Erfi(u) is obtained from the error function via

$$\mathrm{Erf}(u) = \frac{\mathrm{Erf}(\mathrm{i}u)}{\mathrm{i}} \tag{11.200}$$

For large u, the asymptotic expansion of the error function is

$$\mathrm{Erf}(u) = 1 - \frac{\exp(-u^2)}{u\sqrt{\pi}}\left(1 - \frac{1}{2u^2} + \frac{1\cdot 3}{2^2u^4} - \frac{1\cdot 3\cdot 5}{2^3u^6}\right) \tag{11.201}$$

11.3.2 Amplitude in the x-Direction Perpendicular to the Direction of Gravity

By dropping gravity, the y-directed amplitude of Eq. (11.184), with replacement of y by x, then becomes the x-directed amplitude, where

$$E_x(x, \lambda, s_x, x_s) = E_{x0}(\lambda)\exp(-W_x x_s^2)\exp(-Js_x^2 - Ks_x x - Nx^2) \\ \times \exp\left[-\mathrm{i}\cdot\left(Cs_x^2 + Hx_s^2 + \frac{G}{2}x^2 - 2Hx_s s_x\right)\right] \tag{11.202}$$

where

$$E_{x0}(\lambda) = \sqrt{\frac{\pi}{A + \mathrm{i}E}} \exp\left[-\frac{(\lambda - \lambda_{0x})^2}{2\sigma_{\lambda x}^2}\right] \tag{11.203}$$

If the $[-\infty, \infty]$ integration limits for integration over the x-dimension of the lens l_x are replaced by $[-r_m, r_m]$, where the physical aperture diameter of the lens is $2r_m$, resulting wave amplitude E_x of Eq. (11.202) is multiplied by the factor L_x, where

$$L_x = \frac{1-\mathrm{i}}{\sqrt{2}}\sqrt{\frac{A+\mathrm{i}E}{E-\mathrm{i}A}}\left[\mathrm{Erf}\left(\frac{2\mathrm{i}Ar_m - 2r_mE + Fs_x + Gx}{2\sqrt{E-\mathrm{i}A}}\sqrt{\mathrm{i}}\right)\right.$$
$$\left. + \mathrm{Erf}\left(\frac{2\mathrm{i}Ar_m - 2r_mE - Fs_x - Gx}{2\sqrt{E-\mathrm{i}A}}\sqrt{\mathrm{i}}\right)\right] \tag{11.204}$$

11.3.3 Object Transmission Function

The image amplitude distribution after integration over l_x, l_y in the lens plane is

$$E_{xy}(\lambda, x, y, s_x, s_y, x_s, y_s) = E_0(\lambda)E_x(\lambda, x, s_x, x_s)E_y(\lambda, y, s_y, y_s) \tag{11.205}$$

We now integrate Eq. (11.205) over the object plane coordinates s_x, s_y. In the object plane, one has placed a specimen that may attenuate, phase shift, refract, or diffract the incident wave, which is emitted from the source plane at axial distance D upstream. The attenuation and phase shift, and thus refraction or diffraction of the incident wave through the object as a function of transverse object position s_x, s_y, can be written as the inverse Fourier transform of the spectral distribution of the object transmission function.

The object transmission function $T(s_x, s_y)$, expressed as the inverse Fourier transform of the object spatial frequency distribution $T(f_{sx}, f_{sy})$, is

$$T(s_x, s_y) = \int_{-\infty}^{\infty}\int_{-\infty}^{\infty} T(f_{sx}, f_{sy}) \exp(\mathrm{i}2\pi f_{sx}\, s_x) \exp(\mathrm{i}2\pi f_{sy}\, s_y)\, \mathrm{d}f_{sx}\, \mathrm{d}f_{sy} \tag{11.206}$$

Here, $T(f_{sx}, f_{sy})$ is the Fourier transform of the object transmission function $T(s_x, s_y)$, and f_{sx} and f_{sy} are the spatial frequencies, which correspond to the object plane coordinates s_x and s_y, respectively, where

$$T(f_{sx}, f_{sy}) = \int_{-\infty}^{\infty}\int_{-\infty}^{\infty} T(s_x, s_y) \exp(-\mathrm{i}2\pi f_{sy}\, s_y) \exp(-\mathrm{i}2\pi f_{sx}\, s_x)\, \mathrm{d}s_x\, \mathrm{d}s_y \tag{11.207}$$

The object plane transmission function is now included in the expression for the image plane amplitude, so that

$$E_{xy}(\lambda, x, y, s_x, s_y, x_s, y_s) = E_0(\lambda)E_x(\lambda, x, s_x, x_s)E_y(\lambda, y, s_y, y_s)T(s_x, s_y) \tag{11.208}$$

Substitution of Eq. (11.207) into Eq. (11.208), yields

$$E_{xy}(\lambda, x, y, s_x, s_y, x_s, y_s) = E_0(\lambda)E_x(\lambda, x, s_x, x_s)E_y(\lambda, y, s_y, y_s) \times \int_{-\infty}^{\infty}\int_{-\infty}^{\infty} T(f_{sx}, f_{sy})\exp(\mathrm{i}2\pi f_{sx}s_x)\exp(\mathrm{i}2\pi f_{sy}s_y)\mathrm{d}f_{sx}\,\mathrm{d}f_{sy} \tag{11.209}$$

In Eq. (11.209), the x- and y-amplitudes $E_x(\lambda, x, s_x, x_s)$ and $E_y(\lambda, y, s_y, y_s)$ can be inserted into the double integral over the object plane spatial frequencies f_{sx} and f_{sy}. This is done to facilitate integration over the object plane spatial variables s_x and s_y:

$$E_{xy}(\lambda, x, y, s_x, s_y, x_s, y_s) = E_0(\lambda)\int_{-\infty}^{\infty}\int_{-\infty}^{\infty} T(f_{sx}, f_{sy})E_x(\lambda, x, s_x, x_s) \exp(\mathrm{i}2\pi f_{sx}s_x)E_y(\lambda, y, s_y, y_s)\exp(\mathrm{i}2\pi f_{sy}s_y)\mathrm{d}f_{sx}\,\mathrm{d}f_{sy} \tag{11.210}$$

11.3.4 Integration in the y-Direction in the Object Plane

Next, Eq. (11.210) for the amplitude $E_{xy}(\lambda, x, y, s_x, s_y, x_s, y_s)$ is integrated over the object plane spatial variables s_x and s_y to obtain amplitude $E_{xy}(\lambda, x, y, x_s, y_s)$, where

$$E_{xy}(\lambda, x, y, x_s, y_s) = \int_{-\infty}^{\infty}\int_{-\infty}^{\infty} E_{xy}(\lambda, x, y, s_x, s_y, x_s, y_s)\mathrm{d}s_x\,\mathrm{d}s_y \tag{11.211}$$

Substitution of Eq. (11.210) into Eq. (11.211), and switching the order of integration, yields

$$E_{xy}(\lambda, x, y, x_s, y_s, f_{sx}, f_{sy}) = E_0(\lambda)\int_{-\infty}^{\infty}\int_{-\infty}^{\infty} T(f_{sx}, f_{sy})E_{sx}(\lambda, x, x_s, f_{sx}) E_{sy}(\lambda, y, y_s, f_{sy})\mathrm{d}f_{sx}\,\mathrm{d}f_{sy} \tag{11.212}$$

Recall from Eq. (11.160), the amplitude coefficient is,

$$E_0(\lambda, z) = \left(\frac{E_{0x}E_{0y}}{z}\right)\exp\left(-\frac{N\mu d}{2}\right)\exp[-\mathrm{i}kN(d_{\max} - \delta d)] \tag{11.213}$$

We begin with integration in the y-direction in the object plane. The y-directed amplitude $E_{sy}(\lambda, y, y_s, f_{sy})$ component is

$$E_{sy}(\lambda, y, y_s, f_{sy}) = \int_{-\infty}^{\infty} E_y(\lambda, y, s_y, y_s)\exp(\mathrm{i}2\pi f_{sy}s_y)\mathrm{d}s_y \tag{11.214}$$

Substitution of Eq. (11.184) for Ey into Eq. (11.214), and placement of terms, which are dependent on s_y inside the integral for s_y, yields

$$\begin{aligned}E_{sy}(\lambda, y, y_s, f_{sy}) = {} & E_{sy0}(\lambda)\exp(-Ny^2)\exp\left(\mathrm{i}\frac{yki_g}{r_i}\right)\exp(-W_y y_s^2)\\ & \exp\left[-\mathrm{i}\cdot\left(Hy_s^2 + \frac{G}{2}y^2 + V_y y_s\right)\right] \times \int_{-\infty}^{\infty}\exp(-Js_y^2 - Ks_y y)\\ & \exp[-\mathrm{i}\cdot(Cs_y^2 - 2Hy_s s_y + U_y s_y)]\exp(\mathrm{i}2\pi f_{sy}s_y)\mathrm{d}s_y\end{aligned} \tag{11.215}$$

where

$$E_{sy0}(\lambda) = \sqrt{\frac{\pi}{A + \mathrm{i}E}}\exp\left[-\frac{(\lambda - \lambda_{0y})^2}{2\sigma_{\lambda y}^2}\right]\exp\left[-ik\left(\frac{D_g^2}{2D} + \frac{o_g^2}{2r_0} + \frac{i_g^2}{2r_i}\right)\right] \tag{11.216}$$

The parameters J and C in section 11.3.1 meet the condition,

$$\mathrm{Re}[J] > \mathrm{Im}[C] \tag{11.217}$$

where

$$\mathrm{Re}[J] > 0 \quad \text{and} \quad \mathrm{Im}[C] = 0 \tag{11.218}$$

Integration over the y-direction in the object plane s_y, using Mathematica, yields $E_{sy}(\lambda, y, y_s, f_{sy})$. The resulting $E_{sy}(\lambda, y, y_s, f_{sy})$ in the y-direction includes gravity, which acts in the positive y-direction.

As a result of integration of Eq. (11.215) over the object plane in the y-direction s_y, the amplitude $E_{sy}(\lambda, y, y_s, f_{sy})$ consists of four terms, all are dependent on wavelength λ. The first term $E_{30y}(\lambda, f_{sy})$ depends on the spatial frequency f_{sy} along the y-direction in the object plane; the second term $E_{30y}(\lambda, y, f_{sy})$ depends on spatial frequency f_{sy} and the position in the y-direction at the image plane; the third term $E_{3ys}(\lambda, y_s, f_{sy})$ depends on spatial frequency f_{sy} and the position in the y-direction at the source plane; and the fourth term $E_{3yys}(\lambda, y, y_s)$ depends on both position y_s in the source plane and position y in the image plane.

The resulting amplitude expression $E_{sy}(\lambda, y, y_s, f_{sy})$ is

$$E_{sy}(\lambda, y, y_s, f_{sy}) = E_{30y}(\lambda, f_{sy}) \cdot E_{3y}(\lambda, y, f_{sy}) \cdot E_{3ys}(\lambda, y_s, f_{sy}) \cdot E_{3yys}(\lambda, y, y_s) \tag{11.219}$$

The first y-term is

$$E_{30y}(\lambda,f_{sy}) = \frac{\pi}{\sqrt{AJ - EC + i(EJ + AC)}} \exp\left[-\frac{(\lambda-\lambda_{0y})^2}{2\sigma_{\lambda y}^2}\right]$$
$$\times \exp\left[-\mathrm{i}\cdot\left(\frac{f_{sy}U_y\pi}{C-\mathrm{i}J} - \frac{f_{sy}^2\pi^2}{C-\mathrm{i}J} - \frac{U_y^2}{4(C-\mathrm{i}J)} + k\left(\frac{D_g^2}{2D} + \frac{o_g^2}{2r_0} + \frac{i_g^2}{2r_i}\right)\right)\right] \tag{11.220}$$

The second y-term is

$$E_{3y}(\lambda, y, f_{sy}) = \exp\left[-y\left(\frac{f_{sy}K\pi}{C-\mathrm{i}J} - \frac{KU_y}{2(C-\mathrm{i}J)}\right)\right]\exp(-Ny^2)$$
$$\times \exp\left(\mathrm{i}y\frac{ki_g}{r_i}\right)\exp\left[-\mathrm{i}y^2\left(\frac{G}{2} + \frac{K^2}{4(C-\mathrm{i}J)}\right)\right] \tag{11.221}$$

The third y-term is

$$E_{3ys}(\lambda, y_s, f_{sy}) = \exp(-W_y y_s^2)$$
$$\times \exp\left[-\mathrm{i}y_s\left(V_y + \frac{HU_y}{C-\mathrm{i}J} - \frac{2f_{sy}H\pi}{C-\mathrm{i}J}\right)\right]\exp\left[-\mathrm{i}y_s^2\left(H - \frac{H^2}{C-\mathrm{i}J}\right)\right] \tag{11.222}$$

The fourth y-term is

$$E_{3yys}(\lambda, y, y_s) = \exp\left[-y_s y\left(\frac{HK}{C-\mathrm{i}J}\right)\right] \tag{11.223}$$

The first term can be expressed as a product of two components. The first component is a magnitude term $E_{30y}^{\mathrm{Re}}(\lambda,f_{sy})$, where the exponentials have only a real argument, that is, $\exp(\alpha)$, in which α is real. The second component is a phase term $E_{30y}^{\mathrm{Im}}(\lambda,f_{sy})$, where the exponentials have an imaginary argument, that is, $\exp(\mathrm{i}\alpha)$, in which α is real, so that

$$E_{30y}(\lambda,f_{sy}) = E_{30y}^{\mathrm{Re}}(\lambda, f_{sy})\, E_{30y}^{\mathrm{Im}}(\lambda, f_{sy}) \tag{11.224}$$

where

$$|E_{30y}^{\mathrm{Re}}(\lambda,f_{sy})| = E_{30y}^{\mathrm{Re}}(\lambda,f_{sy}) \tag{11.225}$$

and

$$|E_{30y}^{\rm Im}(\lambda, f_{sy})| = 1 \tag{11.226}$$

The first term has the complex component

$$\frac{\pi}{\sqrt{AJ - EC + {\rm i}(EJ + AC)}} = \sqrt{\frac{(AJ - EC) - {\rm i}(EJ + AC)}{(AJ - EC)^2 + (EJ + AC)^2}} \tag{11.227}$$

with

$$Z \exp({\rm i}\theta) = X + {\rm i}Y \tag{11.228}$$

where

$$Z = \sqrt{X^2 + Y^2} \tag{11.229}$$

and

$$\theta = \tan^{-1}\left(\frac{Y}{X}\right) \tag{11.230}$$

One has the useful identity

$$\frac{1}{\sqrt{X + iY}} = \frac{1}{\sqrt{Z\, {\rm e}^{i\theta}}} = \frac{\cos(\theta/2) - i\,\sin(\theta/2)}{(X^2 + Y^2)^{1/4}} \tag{11.231}$$

Application of Eqs. (11.228–11.231) to Eq. (11.227), yields

$$\frac{\pi}{\sqrt{AJ - EC + {\rm i}\cdot(EJ + AC)}} = \pi\frac{\exp[-({\rm i}/2)\tan^{-1}((EJ + AC)/(AJ - EC))]}{\left[(AJ - EC)^2 + (EJ + AC)^2\right]^{1/4}} \tag{11.232}$$

The magnitude and phase components of the first y-term $E_{30y}(\lambda, f_{sy})$ are

$$E_{30y}^{\rm Re}(\lambda, f_{sy}) = \frac{\pi}{\left[(AJ - EC)^2 + (EJ + AC)^2\right]^{1/4}} \times \exp\left[-\frac{(\lambda - \lambda_{0y})^2}{2\sigma_{\lambda y}^2}\right]\exp\left[\frac{f_{sy}U_yJ\pi}{C^2 + J^2} - \frac{f_{sy}^2\pi^2J}{C^2 + J^2} - \frac{U_y^2J}{4(C^2 + J^2)}\right] \tag{11.233}$$

and

$$E_{30y}^{\mathrm{Im}}(\lambda,f_{sy}) = \exp\left[-\frac{\mathrm{i}}{2}\tan^{-1}\left(\frac{EJ+AC}{AJ-EC}\right)\right] \times \exp\left[-\mathrm{i}\frac{2\pi}{\lambda}\left(\frac{D_g^2}{2D}+\frac{o_g^2}{2r_0}+\frac{i_g^2}{2r_i}\right)\right] \times \exp\left[-\mathrm{i}\cdot\left(\frac{f_{sy}U_y\pi(C)}{C^2+J^2}-\frac{f_{sy}^2\pi^2(C)}{C^2+J^2}-\frac{U_y^2(C)}{4(C^2+J^2)}\right)\right] \tag{11.234}$$

The magnitude and phase components of the second y-term $E_{3y}(\lambda, y, f_{sy})$ are

$$E_{3y}^{\mathrm{Re}}(\lambda, y, f_{sy}) = \exp\left[-y\left(\frac{f_{sy}KC\pi}{C^2+J^2}-\frac{KU_yC}{2(C^2+J^2)}\right)\right]\exp\left[y^2\left(\frac{K^2J}{4(C^2+J^2)}-N\right)\right] \tag{11.235}$$

and

$$E_{3y}^{\mathrm{Im}}(\lambda, y, f_{sy}) = \exp\left[-\mathrm{i}y\left(\frac{f_{sy}KJ\pi}{C^2+J^2}-\frac{KU_yJ}{2(C^2+J^2)}\right)\right]\exp\left(\mathrm{i}y\frac{ki_g}{r_i}\right) \times \exp\left[-\mathrm{i}y^2\left(\frac{G}{2}+\frac{K^2C}{4(C^2+J^2)}\right)\right] \tag{11.236}$$

The magnitude and phase components of the third y-term $E_{3ys}(\lambda, y_s, f_{sy})$ are

$$E_{3ys}^{\mathrm{Re}}(\lambda, y_s, f_{sy}) = \exp(-W_y^{\mathrm{Re}}y_s^2) \times \exp\left[y_s\left(\frac{HJ}{C^2+J^2}\right)(U_y-2f_{sy}\pi)\right]\exp\left[-y_s^2\left(\frac{H^2J}{C^2+J^2}\right)\right] \tag{11.237}$$

and

$$E_{3ys}^{\mathrm{Im}}(\lambda, y_s, f_{sy}) = \exp\left[-\mathrm{i}y_s\left(V_y+\frac{HU_yC}{C^2+J^2}-\frac{2f_{sy}HC\pi}{C^2+J^2}\right)\right] \times \exp\left[-\mathrm{i}y_s^2\left(H-\frac{H^2C}{C^2+J^2}-W_y^{\mathrm{Im}}\right)\right] \tag{11.238}$$

The magnitude and phase components of the fourth y-term $E_{3yys}(\lambda, y, y_s)$ are

$$E_{3yys}^{\mathrm{Re}}(\lambda, y, y_s) = \exp\left[-y_s y\left(\frac{HKC}{C^2 + J^2}\right)\right] \tag{11.239}$$

and

$$E_{3yys}^{\mathrm{Im}}(\lambda, y, y_s) = \exp\left[-\mathrm{i} y_s y\left(\frac{HKJ}{C^2 + J^2}\right)\right] \tag{11.240}$$

11.3.5 Integration in the x-Direction in the Object Plane

We now perform integration in the x-direction in the object plane. In Eq. (11.212), the x-directed amplitude $E_{sx}(\lambda, x, x_s, f_{sx})$ component is

$$E_{sx}(\lambda, x, x_s, f_{sx}) = \int_{-\infty}^{\infty} E_x(\lambda, x, s_x, x_s)\exp(\mathrm{i}2\pi f_{sx} s_x)\mathrm{d}s_x \tag{11.241}$$

Substitution of Eq. (11.202) for E_x into Eq. (11.241), and placement of terms, which are dependent on s_x inside the integral for s_x, yields

$$\begin{aligned} E_{sx}(x, \lambda, x_s, f_{sx}) = {} & E_{sx0}(\lambda) \exp(-W_x x_s^2) \exp(-Nx^2)\exp\left[-\mathrm{i}\cdot\left(Hx_s^2 + \frac{G}{2}x^2\right)\right] \\ & \times \int_{-\infty}^{\infty} \exp(-Js_x^2 - Ks_x x) \exp[-\mathrm{i}\cdot(Cs_x^2 - 2Hx_s s_x)] \\ & \exp(\mathrm{i}2\pi f_{sx} s_x)\mathrm{d}s_x \end{aligned} \tag{11.242}$$

where

$$E_{sx0}(\lambda) = \sqrt{\frac{\pi}{A + \mathrm{i}E}}\exp\left[-\frac{(\lambda - \lambda_{0x})^2}{2\sigma_{\lambda x}^2}\right] \tag{11.243}$$

Performing the integration along the x-direction in the object plane, yields x-directed amplitude expression $E_{sx}(\lambda, x, x_s, f_{sx})$, which can also be obtained from the y-directed expression of Eq. (11.219), where y is set to x, and gravity is set to zero, so that

$$E_{sx}(\lambda, x, y_s, f_{sy}) = E_{30x}(\lambda, f_{sx}) \cdot E_{3x}(\lambda, x, f_{sx}) \cdot E_{3xs}(\lambda, x_s, f_{sx}) \cdot E_{3xxs}(\lambda, x, x_s) \tag{11.244}$$

11.3.6 Amplitude in the Image Plane Along the x- and y-Directions

We now determine the amplitude in the image plane along the x-direction, perpendicular to gravity, and in the y-direction, parallel to gravity. The amplitude $E_{xy}(x, y, \lambda, x_s, y_s, f_{sx}, f_{sy})$ at a single (x, y) in the image plane for wavelength λ is due to a single source point (x_s, y_s). That is, the resulting amplitude E_{xy} at a point I in the image plane begins with the illumination of all the object points O by a wave from this single source point P, which was mathematically accomplished by the previous integration over all the source plane (s_x, s_y) points P. Each of the object points O emits a wave, which passes through all the points L in the CRL principal plane. Each point L on the CRL principal plane then emits a wave that is refracted and converged by the CRL onto the single image point I. The amplitude at image point I at position (x, y) in the image plane is the sum of the waves from points (l_x, l_y) in the CRL principal plane, which was mathematically realized by the first integration (l_x, l_y) over the CRL principal plane. Following integration in source plane in the y-direction, one obtains Eq. (11.219), and following integration over the source plane in the x-direction, one obtains Eq. (11.244). Insertion of the results of Eq. (11.219) and (11.244) into Eq. (11.212), yields the amplitude at image plane point (x, y), which is due to source point (x_s, y_s) where

$$\begin{aligned}E_{xy}(\lambda,x,y,x_s,y_s,f_{sx},f_{sy}) &= E_0(\lambda)E_{3yys}(\lambda,y,y_s)E_{3xxs}(\lambda,x,x_s)\\ &\times \int_{-\infty}^{\infty} \mathrm{d}f_{sx}E_{30x}(\lambda,f_{sx}) \cdot E_{3x}(\lambda,x,f_{sx}) \cdot E_{3xs}(\lambda,x_s,f_{sx})\\ &\times \int_{-\infty}^{\infty} \mathrm{d}f_{sy}T(f_{sx},f_{sy})E_{30y}(\lambda,f_{sy}) \cdot E_{3y}(\lambda,y,f_{sy}) \cdot E_{3ys}(\lambda,y_s,f_{sy})\end{aligned} \tag{11.245}$$

Next, we list all the parameters and their resulting values for the case of a CRL with N biconcave, parabolic lenses with minimum thickness d at its center $r = 0$, which obeys the lens-maker equation. The CRL absorption aperture radius r_{a} is a useful parameter, and it is defined as the radius at which the amplitude falls to $1/e$ of its center amplitude $\exp(-N\mu d)$. For the case of a lens, which obeys the lens-maker equation $E = 0$, where

$$\frac{1}{r_0} + \frac{1}{r_i} = \frac{2N\delta}{R} \tag{11.246}$$

The CRL parameters for an incoherent source of neutrons or X-rays are the following. The x-component W_x and y-component W_y source terms are real, where

$$W_x = \frac{1}{2\sigma_x^2} \quad \text{and} \quad W_y = \frac{1}{2\sigma_y^2} \tag{11.247}$$

The absorption aperture radius r_{a} and other CRL parameters E, A, J, K, and N of section 11.3.1, are

$$E = \frac{\pi}{\lambda}\left(\frac{1}{r_0} + \frac{1}{r_i} - \frac{2N\delta}{R}\right) = 0 \quad A = \frac{\mu N}{2R} = \frac{1}{r_{\mathrm{a}}} \quad r_{\mathrm{a}} = \sqrt{\frac{2R}{\mu N}} \tag{11.248}$$

$$J = \left(\frac{\pi}{r_0\lambda}\right)^2 \left(\frac{\mu N}{2R}\right)^{-1} = \left(\frac{r_a\pi}{r_0\lambda}\right)^2 \tag{11.249}$$

$$K = \left(\frac{2}{r_0 r_i}\right)\left(\frac{\pi}{\lambda}\right)^2 \left(\frac{\mu N}{2R}\right)^{-1} = \left(\frac{2}{r_0 r_i}\right)\left(\frac{r_a\pi}{\lambda}\right)^2 \tag{11.250}$$

$$N = \left(\frac{\pi}{r_i\lambda}\right)^2 \left(\frac{\mu N}{2R}\right)^{-1} = \left(\frac{r_a\pi}{r_i\lambda}\right)^2 \tag{11.251}$$

The parameters C, F, and H for source-to-object-plane distance D and object distance r_0, as well as the parameter G for image distance r_i, from section 11.3.1, are

$$C = \frac{\pi}{\lambda}\left(\frac{1}{D} + \frac{1}{r_0}\right) \quad H = \frac{\pi}{D\lambda} \quad F = \frac{2\pi}{r_0\lambda} \quad G = \frac{2\pi}{r_i} \tag{11.252}$$

The constant amplitude factor is

$$\frac{\pi}{\sqrt{AJ - EC + i(EJ + AC)}} = \frac{\pi}{(A^2J^2 + A^2C^2)^{1/4}} \exp\left[-\frac{i}{2}\tan^{-1}\left(\frac{C}{J}\right)\right] \tag{11.253}$$

In the case of a CRL, substitution into Eq. (11.253) for A, C, and J, yields

$$\frac{\pi}{\sqrt{AJ - EC + i(EJ + AC)}} = \frac{\pi\sqrt{r_a}}{\left(\left(\frac{r_a\pi}{r_0\lambda}\right)^4 + \left(\frac{\pi}{\lambda}\right)^2\left(\frac{1}{D} + \frac{1}{r_0}\right)^2\right)^{1/4}} \exp\left[-\frac{i}{2}\tan^{-1}\left(\frac{\left(\frac{\pi}{\lambda}\right)\left(\frac{1}{D} + \frac{1}{r_0}\right)}{\left(\frac{r_a\pi}{r_0\lambda}\right)^2}\right)\right] \tag{11.254}$$

In cases where gravity is not applicable, such as X-rays, or neglecting the effect of gravity for neutrons, the gravity parameters U_y, V_y, D_g, o_g, and i_g are set to zero, where

$$U_y = \frac{2\pi}{\lambda}\left(\frac{o_g}{r_0} - \frac{D_g}{D}\right) \quad V_y = \frac{2\pi}{\lambda}\frac{D_g}{D} \tag{11.255}$$

and

$$D_g = \left(\frac{g}{2}\right)\left(\frac{m\lambda D}{h}\right)^2 \quad o_g = \left(\frac{g}{2}\right)\left(\frac{m\lambda r_0}{h}\right)^2 \quad i_g = \left(\frac{g}{2}\right)\left(\frac{m\lambda r_i}{h}\right)^2 \tag{11.256}$$

11.4 Point Spread Function (PSF) of the Biconcave, Parabolic CRL for Incoherent X-rays or Neutrons with Gravity

The PSF is the Green's function or impulse response. That is, the amplitude (intensity) PSF is the image amplitude (intensity) distribution produced by a single source point P. The PSF depends on the optical components, such as a CRL, placed between the source plane and image plane. The following is a derivation for the 2-D PSF for a CRL, which images the incident waves from an object, which has been illuminated by an incoherent source of X-rays or neutrons.

Axial (z-axis) distance D separates the source and the object planes, axial object distance r_o separates the object plane and the CRL center plane, and axial image distance r_i separates the center lens or aperture plane and the image plane. The object r_o and image r_i distances in general are unrelated, and in the specific case of the CRL, they are related by the lens-maker formula. From Eq. (11.245), the amplitude at image plane point (x, y) due to source point (x_s, y_s) is the impulse response or the PSF of the lens or aperture, where the PSF is

$$\begin{aligned} E_{xy}(\lambda,x,y,x_s,y_s,f_{sx},f_{sy}) = {} & E_0(\lambda)E_{3yys}(\lambda,y,y_s)E_{3xxs}(\lambda,x,x_s) \\ & \times \int_{-\infty}^{\infty} \mathrm{d}f_{sx} E_{30x}(\lambda,f_{sx}) \cdot E_{3x}(\lambda,x,f_{sx}) \cdot E_{3xs}(\lambda,x_s,f_{sx}) \\ & \times \int_{-\infty}^{\infty} \mathrm{d}f_{sy} T(f_{sx},f_{sy}) E_{30y}(\lambda,f_{sy}) \cdot E_{3y}(\lambda,y,f_{sy}) \cdot E_{3ys}(\lambda,y_s,f_{sy}) \end{aligned} \tag{11.257}$$

For mathematical convenience and symmetry, one typically evaluates the PSF (Green's function) or image amplitude $E_{xy}(x, y, \lambda, x_s, y_s)$ for an on-axis source point alone by setting $x_s = y_s = 0$. In this case,

$$E_{3xs}(x_s = 0, f_{sx}) = 1 \tag{11.258}$$

$$E_{3ys}(y_s = 0, f_{sy}) = 1 \tag{11.259}$$

$$E_{3xxs}(\lambda, y, x_s = 0, f_{sx}) = 1 \tag{11.260}$$

$$E_{3yys}(\lambda, y, y_s = 0, f_{sy}) = 1 \tag{11.261}$$

With substitution of Eqs. (11.258–11.261) into Eq. (11.257), the resulting expression for the PSF in terms of its real and imaginary parts is

$$\begin{aligned} E_{xy}(\lambda,x,y,x_s,y_s,f_{sx},f_{sy}) = {} & E_0^{\mathrm{Re}}(\lambda)E_0^{\mathrm{Im}}(\lambda) \\ & \times \int_{-\infty}^{\infty} \mathrm{d}f_{sx} E_{30x}^{\mathrm{Re}}(\lambda,f_{sx}) \cdot E_{30x}^{\mathrm{Im}}(\lambda,f_{sx}) \cdot E_{3x}^{\mathrm{Re}}(\lambda,x,f_{sx}) \cdot E_{3x}^{\mathrm{Im}}(\lambda,x,f_{sx}) \\ & \times \int_{-\infty}^{\infty} \mathrm{d}f_{sy} T(f_{sx},f_{sy}) E_{30y}^{\mathrm{Re}}(\lambda,f_{sy}) E_{30y}^{\mathrm{Im}}(\lambda,f_{sy}) E_{3y}^{\mathrm{Re}}(\lambda,y,f_{sy}) \\ & \quad \cdot E_{3y}^{\mathrm{Im}}(\lambda,y,f_{sy}) \end{aligned} \tag{11.262}$$

where from Eq. (11.160)

$$E_0^{\text{Re}}(\lambda, z) = \left(\frac{E_{0x}E_{0y}}{z}\right)\exp\left(-\frac{N\mu d}{2}\right) \tag{11.263}$$

and

$$E_0^{\text{Im}}(\lambda, z) = \exp[-ikN(d_{\max} - \delta d)] \tag{11.264}$$

Consider the case of an object with uniform transmission T_0 of unity, which is independent of the source plane's spatial coordinates (s_x, s_y), where

$$T(s_x, s_y) = T_0 = 1 \tag{11.265}$$

Then the transmission in the spatial frequency domain $T(f_{sx},f_{sy})$ is a Dirac delta function $\delta(f_{sx},f_{sy})$, where

$$T(f_{sx}, f_{sy}) = \delta(f_{sx}, f_{sy}) \tag{11.266}$$

This results in the amplitude distribution $E_{xy}(x, y, \lambda, 0, 0, 0, 0)$ at the image plane resulting in a single center source point, $s_x = 0$ and $s_y = 0$, for the case uniform object transmission, which corresponds to

$$T(s_x, s_y) = T_0 \tag{11.267}$$

where

$$f_{sx} = 0 \quad \text{and} \quad f_{sy} = 0 \tag{11.268}$$

The PSF amplitude distribution $E_{xy}(\lambda, x, y)$ can be written as the product of two terms. The first term, $E_{3xsys}^{\text{inc}}(\lambda, x_s, y_s)$, is the incoherent PSF amplitude, which is the contribution of the amplitude at the image point from a single source point, where there is no fixed relationship between the waves emitted from the individual source points. The second term, $E_{3xy}^{\text{coh}}(\lambda, x, y)$, accounts for coherent interactions between the source points. In the case of the PSF, there is only a single source point, and this yields a coherent component of the amplitude distribution in the image plane. Given center source point $s_x = 0$ and $s_y = 0$ and uniform object transmission,

$$f_{sx} = 0 \quad \text{and} \quad f_{sy} = 0 \tag{11.269}$$

The PSF amplitude $E_{xy}(\lambda, x, y)$ is the product of coherent and incoherent components, where in the case of incoherence, the coherent component is unity, and

$$E_{xy}(\lambda, x, y) = E_{3xy}^{\text{inc}}(\lambda, x, y)E_{3xy}^{\text{coh}}(\lambda, x, y) \tag{11.270}$$

The incoherent term $E_{3xy}^{\mathrm{inc}}(\lambda, x, y)$ is

$$E_{3xy}^{\mathrm{inc}}(\lambda, x, y) = E_0^{\mathrm{Re}}(\lambda)E_{30x}^{\mathrm{Re}}(\lambda, f_{sx} = 0)E_{3x}^{\mathrm{Re}}(\lambda, x, f_{sx} = 0) \times E_{30y}^{\mathrm{Re}}(\lambda, f_{sy} = 0)E_{3y}^{\mathrm{Re}}(\lambda, y, f_{sy} = 0) \tag{11.271}$$

This first term, $E_{3xsys}^{\mathrm{inc}}(\lambda, x_s, y_s)$, represents waves from the X-ray or neutron source point, which arrive at the image plane without any fixed phase relationship to the other source point waves. Consequently, the contribution to the intensity distribution at an image point, is the sum of each source point wave intensity. The X-ray or neutron wave intensity I is the squared magnitude of the X-ray or neutron wave amplitude E, where

$$I = |E|^2 \tag{11.272}$$

The incoherent amplitude term $E_{3xy}^{\mathrm{inc}}(\lambda, x, y)$ has no imaginary components so

$$E_{3xy}^{\mathrm{inc}}(\lambda, x, y) = \left|E_{3xy}^{\mathrm{inc}}(\lambda, x, y)\right| = E_0^{\mathrm{Re}}(\lambda)E_{30x}^{\mathrm{Re}}(\lambda, f_{sx} = 0)E_{3x}^{\mathrm{Re}}(\lambda, x, f_{sx} = 0) \times E_{30y}^{\mathrm{Re}}(\lambda, f_{sy} = 0)E_{3y}^{\mathrm{Re}}(\lambda, y, f_{sy} = 0) \tag{11.273}$$

From Eq. (11.233) for the real y-component amplitude $E_{30y}^{\mathrm{Re}}(\lambda, f_{sy} = 0)$, and Eq. (11.235) for real y-component amplitude $E_{3y}^{\mathrm{Re}}(\lambda, y, f_{sy} = 0)$, one sets the gravity terms to zero, and replaces y by x, and obtains the real x-component amplitude $E_{30x}^{\mathrm{Re}}(\lambda, f_{sx} = 0)$, and real x-component amplitude $E_{3x}^{\mathrm{Re}}(\lambda, y, f_{sx} = 0)$, where

$$E_{30x}^{\mathrm{Re}}(\lambda, f_{sx} = 0) = \frac{\pi}{2}\frac{1}{\left[(AJ - EC)^2 + (EJ + AC)^2\right]^{1/4}}\exp\left[-\frac{(\lambda - \lambda_{0x})^2}{2\sigma_{\lambda x}^2}\right] \tag{11.274}$$

$$E_{3x}^{\mathrm{Re}}(\lambda, x, f_{sx} = 0) = \exp\left[x^2\left(\frac{K^2 J}{4(C^2 + J^2)} - N\right)\right] \tag{11.275}$$

$$E_{30y}^{\mathrm{Re}}(\lambda, f_{sy} = 0) = \frac{\pi}{2}\frac{1}{\left[(AJ - EC)^2 + (EJ + AC)^2\right]^{1/4}}\exp\left[-\frac{(\lambda - \lambda_{0y})^2}{2\sigma_{\lambda y}^2}\right]\exp\left[-\frac{U_y^2 J}{4(C^2 + J^2)}\right] \tag{11.276}$$

and

$$E_{3y}^{\mathrm{Re}}(\lambda, y, f_{sy} = 0) = \exp\left[y\left(\frac{KU_y C}{2(C^2 + J^2)}\right)\right]\exp\left[y^2\left(\frac{K^2 J}{4(C^2 + J^2)} - N\right)\right] \tag{11.277}$$

The source points are coherent, if a fixed phase relationship between their emitted waves exists. The coherent amplitude term $E_{3xy}^{\mathrm{coh}}(\lambda, x, y)$ thus depends on the phase of the waves, which is based on their position in space and time. For the PSF, the single source point is coherent with itself, and thus there is a coherent component of the amplitude of the PSF. In Eq. (11.270), the coherent term $E_{3xy}^{\mathrm{coh}}(\lambda, x, y)$ consists only of exponentials, with an imaginary argument $i\phi$ of the real phase shift ϕ of the X-ray or neutron, that is,

$$E_{3xy}^{\mathrm{coh}}(\lambda, x, y) = \exp(i\phi) \quad \text{where } \phi \text{ is real} \tag{11.278}$$

The coherent term $E_{3xy}^{\mathrm{coh}}(\lambda, x, y)$ is then

$$E_{3xy}^{\mathrm{coh}}(\lambda, x, y) = E_{30x}^{\mathrm{Im}}(\lambda, f_{sx} = 0)\ E_{3x}^{\mathrm{Im}}(\lambda, x, f_{sx} = 0)\ E_{30y}^{\mathrm{Im}}(\lambda, f_{sy} = 0) E_{3y}^{\mathrm{Im}}(\lambda, y, f_{sy} = 0) \tag{11.279}$$

The magnitude of all-imaginary coherent term $E_{3xy}^{\mathrm{coh}}(\lambda, x, y)$ is unity:

$$|E_{3xy}^{\mathrm{coh}}(\lambda, x, y)| = 1 \tag{11.280}$$

From Eq. (11.264) for $E_0^{\mathrm{Im}}(\lambda)$, and the x-component version $E_{30x}^{\mathrm{Im}}(\lambda, f_{sx} = 0)$ of the y-component amplitude $E_{30y}^{\mathrm{Im}}(\lambda, f_{sy} = 0)$ of Eq. (11.234), which has gravity terms set to zero, and y replaced by x, one obtains the imaginary amplitude components for the incoherent source case, where

$$E_0^{\mathrm{Im}}(\lambda) = \exp[-ikN(d_{\max} - \delta d)] \tag{11.281}$$

$$E_{30x}^{\mathrm{Im}}(\lambda, f_{sx} = 0) = 1 \tag{11.282}$$

For the incoherent source case, the following equations are the imaginary y-component amplitudes $E_{30y}^{\mathrm{Im}}(\lambda, f_{sy} = 0)$ of Eq. (11.234) and $E_{3y}^{\mathrm{Im}}(\lambda, y, f_{sy} = 0)$ of (11.236), and the imaginary x-component version $E_{3x}^{\mathrm{Im}}(\lambda, x, f_{sx} = 0)$ of the y-component amplitude $E_{3y}^{\mathrm{Im}}(\lambda, f_{sy} = 0)$ of Eq. (11.236), which has the gravity terms set to zero, and y replaced by x, where

$$E_{3x}^{\mathrm{Im}}(\lambda, x, f_{sx} = 0) = \exp\left[-\mathrm{i}x^2 \frac{G}{2}\left(1 + \frac{K^2 C}{2G(C^2 + J^2)}\right)\right] \tag{11.283}$$

$$E_{30y}^{\mathrm{Im}}(\lambda, f_{sy} = 0) = \exp\left[-\mathrm{i}\tan^{-1}\left(\frac{EJ + AC}{AJ - EC}\right)\right] \times \exp\left[-\mathrm{i}\cdot\left(-\frac{U_y^2 C}{4(C^2 + J^2)} + \frac{2\pi}{\lambda}\left(\frac{D_g^2}{2D} + \frac{o_g^2}{2r_0} + \frac{i_g^2}{2r_i}\right)\right)\right] \tag{11.284}$$

and

$$E_{3y}^{\mathrm{Im}}(\lambda,y,f_{sy}=0)=\exp\left[-\mathrm{i}y\left(-\frac{KU_yJ}{2(C^2+J^2)}\right)\right]\exp\left(\mathrm{i}y\frac{ki_g}{r_i}\right)\exp\left[-\mathrm{i}y^2\left(\frac{G}{2}+\frac{K^2C}{4(C^2+J^2)}\right)\right] \tag{11.285}$$

Consider the case of a CRL absorption aperture radius r_{a}, which is much larger than the lens aperture radius r_m, where

$$r_{\mathrm{a}} \gg r_m \tag{11.286}$$

This amplitude was obtained by integration in the plane of the lens in the x- and y-directions, respectively, where

$$l_x \in (-\infty, \infty) \quad \text{and} \quad l_y \in (-\infty, \infty) \tag{11.287}$$

In the case of integration over only the lens aperture diameter in the x- and y-directions,

$$l_x \in (-r_m, r_m) \quad \text{and} \quad l_y \in (-r_m, r_m) \tag{11.288}$$

The amplitude $E_{xy}(\lambda, x, y)$ of Eq. (11.245), which was derived for integration over an infinite lens aperture, must be multiplied by the x- and y-component factors L_x and L_y to obtain the corrected amplitude for a finite lens aperture, where

$$L_x = \frac{1-\mathrm{i}}{\sqrt{2}}\sqrt{\frac{A+\mathrm{i}E}{E-\mathrm{i}A}}\left[\mathrm{Erf}\left(\frac{2\mathrm{i}Ar_m - 2r_mE + Fs_x + Gx}{2\sqrt{E-\mathrm{i}A}}\sqrt{\mathrm{i}}\right)\right.$$
$$\left. + \mathrm{Erf}\left(\frac{2\mathrm{i}Ar_m - 2r_mE - Fs_x - Gx}{2\sqrt{E-\mathrm{i}A}}\sqrt{\mathrm{i}}\right)\right] \tag{11.289}$$

and

$$L_y = \frac{1-\mathrm{i}}{\sqrt{2}}\sqrt{\frac{A+\mathrm{i}E}{E-\mathrm{i}A}}\left[\mathrm{Erf}\left(\frac{2\mathrm{i}Ar_m - 2r_mE + Fs_y + Gy}{2\sqrt{E-\mathrm{i}A}}\sqrt{\mathrm{i}}\right)\right.$$
$$\left. + \mathrm{Erf}\left(\frac{2\mathrm{i}Ar_m - 2r_mE - Fs_y - Gy}{2\sqrt{E-\mathrm{i}A}}\sqrt{\mathrm{i}}\right)\right] \tag{11.290}$$

These factors L_x and L_y also multiply the x- and y-component terms, respectively, of the image plane amplitude $E_{xy}(\lambda, x, y)$ in the case of a pinhole aperture of radius r_m put in place of the CRL. A CRL has parameter $E = 0$, and a pinhole has parameter $A = 0$. In either case, integration over a finite aperture pinhole or lens is equal to the integration over an infinite aperture, and then multiplication of the result by factors L_x and L_y.

This generalized $E_{xy}(\lambda, x, y)$ amplitude for a center source point and uniform transmission applies to a CRL or an empty aperture, such as a pinhole or pipe with infinite radius or finite radius. The amplitude PSF for a CRL $E_{xy0}^{\mathrm{CRL}}(\lambda, x, y)$ is obtained from the generalized image amplitude $E_{xy}(\lambda, x, y)$ for a single source center point P, and recalling section 11.3.3, uniform object transmission $T(s_s, s_y)$, where

$$(x_s, y_s) = (0, 0) \quad \text{and} \quad T(s_s, s_y) = 1 \tag{11.291}$$

We now determine the amplitude PSF for the specific case of a biconcave, parabolic CRL. The focal length f for the case of a CRL, which consists of N parabolic, biconcave CRLs, each with on-axis radius of curvature R, is

$$f = \frac{R}{2N\delta} \tag{11.292}$$

The CRL obeys the lens-maker equation:

$$\frac{1}{f} = \frac{1}{r_0} + \frac{1}{r_i} \tag{11.293}$$

In the case of the CRL, the parameter E of section 11.3.1 is then altered by the lens-maker equation, where

$$E = \frac{k}{2r_0} + \frac{k}{2r_i} - \frac{kN\delta}{R} = \frac{\pi}{\lambda}\left(\frac{1}{r_0} + \frac{1}{r_i} - \frac{2N\delta}{R}\right) = 0 \quad \text{CRL} \tag{11.294}$$

For a parabolic, biconcave CRL, the parameter $E = 0$, and the E-dependent parameters, parameter J of section 11.3.1 is

$$J = \frac{F^2}{4A} = \left(\frac{\pi}{r_0\lambda}\right)^2 \left(\frac{2R}{\mu N}\right) \tag{11.295}$$

The parameter K of section 11.3.1 is

$$K = 2J\left(\frac{r_0}{r_i}\right) \tag{11.296}$$

Substitution for J gives

$$K = \frac{FG}{2A} = \left(\frac{2}{r_0 r_i}\right)\left(\frac{\pi}{\lambda}\right)^2 \left(\frac{2R}{\mu N}\right) \tag{11.297}$$

The parameter N of section 11.3.1 is

$$N = J\left(\frac{r_0}{r_i}\right)^2 \tag{11.298}$$

Substitution for J gives

$$N = \frac{G^2}{4A} = \left(\frac{\pi}{r_i\lambda}\right)^2\left(\frac{2R}{\mu N}\right) \tag{11.299}$$

One assumes that the CRL has an absorption aperture radius r_a, which was much larger than the lens aperture radius r_m, where $r_a \gg r_m$. With this assumption, the corrective factors L_x and L_y multiply the amplitude, which is obtained by integration over infinite CRL aperture. The result is equal to integration over a finite aperture CRL, where Eq. (11.289) and (11.290), become

$$L_x = (1-\mathrm{i})\sqrt{\frac{\mathrm{i}}{2}}\left[\mathrm{Erf}\left(\frac{-2Ar_m + \mathrm{i}(Fs_x + Gx)}{2\sqrt{A}}\right) + \mathrm{Erf}\left(\frac{-2Ar_m - \mathrm{i}(Fs_x + Gx)}{2\sqrt{A}}\right)\right] \tag{11.300}$$

and

$$L_y = (1-\mathrm{i})\sqrt{\frac{\mathrm{i}}{2}}\left[\mathrm{Erf}\left(\frac{-2Ar_m + \mathrm{i}(Fs_y + Gy)}{2\sqrt{A}}\right) + \mathrm{Erf}\left(\frac{-2Ar_m - \mathrm{i}(Fs_y + Gy)}{2\sqrt{A}}\right)\right] \tag{11.301}$$

In summary, repeating Eq. (11.270), the CRL PSF amplitude $E_{xy0}^{\mathrm{CRL}}(\lambda, x, y)$ is the product of coherent and incoherent components, where in the case of an incoherent source, the coherent component is unity, and

$$E_{xy0}^{\mathrm{CRL}}(\lambda, x, y) = E_{\mathrm{CRL}}^{\mathrm{inc}}(\lambda, x_s, y_s)E_{\mathrm{CRL}}^{\mathrm{coh}}(\lambda, x, y) \tag{11.302}$$

The incoherent amplitude term $E_{\mathrm{CRL}}^{\mathrm{inc}}(\lambda, x_s, y_s)$ for the CRL, is expressed by exponentials with only real arguments, such that the magnitude of $E_{\mathrm{CRL}}^{\mathrm{inc}}(\lambda, x_s, y_s)$ is not unity. Substitution into Eq. (11.271) of Eqs. (11.273–11.277), with parameters from Eqs. (11.294–11.299), yields incoherent amplitude term

$$E_{\mathrm{CRL}}^{\mathrm{inc}}(\lambda,x,y) = E_{00}(\lambda)\exp\left[-(x^2+y^2)\left(\frac{r_a\pi}{r_i\lambda}\right)^2\left(1 - \frac{\left(\frac{r_a\pi}{r_0\lambda}\right)^4}{\left(\frac{\pi}{\lambda}\right)^2\left(\frac{1}{D}+\frac{1}{r_0}\right)^2 + \left(\frac{r_a\pi}{r_0\lambda}\right)^4}\right)\right]$$

$$\times\exp\left[y\cdot\left(\frac{\left(\frac{2r_a^2}{r_0 r_i}\right)\left(\frac{\pi}{\lambda}\right)^4\left(\frac{o_g}{D}-\frac{D_g}{D}\right)\left(\frac{1}{D}+\frac{1}{r_0}\right)}{\left(\frac{\pi}{\lambda}\right)^2\left(\frac{1}{D}+\frac{1}{r_0}\right)^2 + \left(\frac{r_a\pi}{r_0\lambda}\right)^4}\right)\right] \tag{11.303}$$

The maximum amplitude $E_{00}(\lambda)$ of the image plane distribution, which is produced from a single source point, via CRL imaging, is expressed by exponentials with both real and imaginary arguments, such that the magnitude of $E_{\mathrm{CRL}}^{\mathrm{inc}}(\lambda, x_s, y_s)$ is not unity, where

$$E_{00}(\lambda)=\left(\frac{\pi E_{0x}E_{0y}\exp\left(-\frac{N\mu d}{2}\right)}{z}\right)\exp\left[-\frac{(\lambda-\lambda_{0x})^2}{2\sigma_{\lambda x}^2}\right]\exp\left[-\frac{(\lambda-\lambda_{0y})^2}{2\sigma_{\lambda y}^2}\right]$$
$$\times\exp\left[-\mathrm{i}\frac{2\pi}{\lambda}\left(\frac{D_g^2}{2D}+\frac{o_g^2}{2r_0}+\frac{i_g^2}{2r_i}\right)\right]\sqrt{\frac{\left(\frac{2\pi}{r_0\lambda}\right)^2-\mathrm{i}\left(\frac{2\mu N\pi}{R\lambda}\right)\left(\frac{1}{D}+\frac{1}{r_0}\right)}{\left(\frac{2\pi}{r_0\lambda}\right)^4+\left(\left(\frac{2\mu N\pi}{R\lambda}\right)\left(\frac{1}{D}+\frac{1}{r_0}\right)\right)^2}}$$
$$\times\exp\left[-\mathrm{i}\left(\frac{\left(\frac{2\pi}{\lambda}\right)^2\left(\frac{o_g}{D}-\frac{D_g}{D}\right)^2}{4\left(\frac{\pi}{\lambda}\left(\frac{1}{D}+\frac{1}{r_0}\right)-\mathrm{i}\left(\frac{r_a\pi}{r_0\lambda}\right)^2\right)}\right)\right] \tag{11.304}$$

The coherent amplitude term $E_{\mathrm{CRL}}^{\mathrm{coh}}(\lambda, x_s, y_s)$ for the CRL is expressed by exponentials, which have only imaginary arguments, such that the magnitude of $E_{\mathrm{CRL}}^{\mathrm{coh}}(\lambda, x_s, y_s)$ is unity. Substitution into Eq. (11.279) of Eqs. (11.281–11.285), with parameters from Eqs. (11.294–11.299), yields coherent amplitude term

$$E_{\mathrm{CRL}}^{\mathrm{coh}}(\lambda,x,y)=\exp[-\mathrm{i}kN(d_{\max}-\delta d)]$$
$$\times\exp\left[-i(x^2+y^2)\left(\frac{\pi}{r_i}\right)\left(1+\frac{\left(\frac{r_i}{\lambda}\right)\left(\frac{1}{D}+\frac{1}{r_0}\right)\left(\frac{r_o}{r_i}\right)^2\left(\frac{r_a\pi}{r_0\lambda}\right)^4}{\left(\frac{\pi}{\lambda}\right)^2\left(\frac{1}{D}+\frac{1}{r_0}\right)^2+\left(\frac{r_a\pi}{r_0\lambda}\right)^4}\right)\right]$$
$$\times\exp\left[\mathrm{i}y\left(\frac{\left(\frac{2r_0\pi}{r_i\lambda}\right)\left(\frac{o_g}{D}-\frac{D_g}{D}\right)\left(\frac{r_a\pi}{r_0\lambda}\right)^4}{\left(\frac{\pi}{\lambda}\right)^2\left(\frac{1}{D}+\frac{1}{r_0}\right)^2+\left(\frac{r_a\pi}{r_0\lambda}\right)^4}+\frac{2i_g\pi}{r_i\lambda}\right)\right] \tag{11.305}$$

The intensity PSF for the CRL $I_{xy0}^{\mathrm{CRL}}(\lambda, x, y)$ is the squared magnitude of Eq. (11.302), where

$$I_{xy0}^{\mathrm{CRL}}(\lambda, x, y) = |E_{\mathrm{CRL}}^{\mathrm{inc}}(\lambda, x_s, y_s)E_{\mathrm{CRL}}^{\mathrm{coh}}(\lambda, x, y)|^2 = |E_{\mathrm{CRL}}^{\mathrm{inc}}(\lambda, x_s, y_s)|^2 \tag{11.306}$$

11.5 PSF of a Pinhole for Incoherent X-rays or Neutrons with Gravity

Assume an incoherent source of X-rays or neutrons. We now consider the PSF of a pinhole $E_{xy0}^{\mathrm{pin}}(\lambda, x, y)$ that is produced by the center source point P. The amplitude (intensity) PSF is the image amplitude (intensity) distribution produced by a single source point P, which emits X-rays or neutrons that pass through the pinhole to the image plane. From section 11.3.3, uniform, unity transmission $T_0 = 1$ through the object plane is assumed, where

$$T(s_x, s_y) = T_0 = 1 \tag{11.307}$$

The image amplitude $E_{xy0}^{\mathrm{pin}}(\lambda, x, y)$ for a pinhole is obtained from the generalized image amplitude $E_{xy}(\lambda, x, y)$ of a single source center point P at

$$(x_s, y_s) = (0, 0) \tag{11.308}$$

In the case of a pinhole, the refractive decrement δ and the linear attenuation coefficient μ are each zero, where

$$\delta = 0 \quad \text{and} \quad \mu = 0 \tag{11.309}$$

We now evaluate the CRL parameters of section 11.3.1, which are modified for a pinhole. The parameter A of Eq. (11.170) is zero:

$$A = 0 \tag{11.310}$$

And parameter E for a pinhole becomes

$$E = E_{\mathrm{pin}} = \frac{k}{2r_0} + \frac{k}{2r_i} = \frac{\pi}{\lambda}\left(\frac{1}{r_0} + \frac{1}{r_i}\right) \tag{11.311}$$

In the case of the pinhole, the A-dependent and E-dependent parameter J becomes

$$J = J_{\mathrm{pin}} = \frac{-\mathrm{i}F^2}{4E_{\mathrm{pin}}} = \frac{-\mathrm{i}}{r_0^2(\lambda/\pi)((1/r_0) + (1/r_i))} \tag{11.312}$$

The parameter K for a pinhole is

$$K = K_{\text{pin}} = 2J_{\text{pin}}\left(\frac{r_0}{r_i}\right) \tag{11.313}$$

Substitution for J_{pin} gives

$$K_{\text{pin}} = \frac{-\mathrm{i}FG}{2E_{\text{pin}}} = \frac{-4\mathrm{i}}{(\lambda/\pi)(r_0 + r_i)} \tag{11.314}$$

The parameter N for a pinhole is

$$N_{\text{pin}} = J_{\text{pin}}\left(\frac{r_0}{r_i}\right)^2 \tag{11.315}$$

Substitution for J_{pin} gives

$$N_{\text{pin}} = \frac{-\mathrm{i}G^2}{4E_{\text{pin}}} = \frac{i}{r_i^2(\lambda/\pi)((1/r_0) + (1/r_i))} \tag{11.316}$$

The corrective factors L_x and L_y multiply the x- and y-components of the image plane amplitude, which are obtained by integration over an infinite radius pinhole aperture, so as to obtain integration over the finite radius r_m of the pinhole with an aperture. In the case of a pinhole with Eqs. (11.309–11.316), the expressions for L_x and L_y of Eqs. (11.289) and (11.290), become

$$L_x = (1-\mathrm{i})\sqrt{\frac{\mathrm{i}}{2}}\left[\mathrm{Erf}\left(\frac{-2r_mE_{\text{pin}} + Fs_x + Gx}{2\sqrt{E_{\text{pin}}}}\sqrt{\mathrm{i}}\right) + \mathrm{Erf}\left(\frac{-2r_mE_{\text{pin}} - Fs_x - Gx}{2\sqrt{E_{\text{pin}}}}\sqrt{\mathrm{i}}\right)\right] \tag{11.317}$$

and

$$L_y = (1-\mathrm{i})\sqrt{\frac{\mathrm{i}}{2}}\left[\mathrm{Erf}\left(\frac{-2r_mE_{\text{pin}} + Fs_y + Gy}{2\sqrt{E_{\text{pin}}}}\sqrt{\mathrm{i}}\right) + \mathrm{Erf}\left(\frac{-2r_mE_{\text{pin}} - Fs_y - Gy}{2\sqrt{E_{\text{pin}}}}\sqrt{\mathrm{i}}\right)\right] \tag{11.318}$$

At the center source point, the amplitude PSF of a pinhole $E_{xy0}^{\text{pin}}(\lambda, x, y)$ is

$$E_{xy0}^{\text{pin}}(\lambda, x, y) = E_{\text{pin}}^{\text{inc}}(\lambda, x_s, y_s)E_{\text{pin}}^{\text{coh}}(\lambda, x, y) \quad x_s = y_s = 0 \tag{11.319}$$

Substitution of the above pinhole parameters into Eq. (11.271) and (11.279) for the CRL, converts the CRL to a pinhole. The pinhole incoherent amplitude contribution $E_{\text{pin}}^{\text{inc}}(\lambda, x_s, y_s)$, which is comprised of exponentials with only real terms. The magnitude of $E_{\text{pin}}^{\text{inc}}(\lambda, x_s, y_s)$ is not unity. The incoherent amplitude term for a pinhole is

$$E_{\text{pin}}^{\text{inc}}(\lambda,x,y)=\left(\frac{E_{0x}E_{0y}}{z+(x^2/2z)+(y^2/2z)}\right)\exp\left(-\frac{N\mu d}{2}\right)$$
$$\times\frac{\pi^2}{4E_{\text{pin}}\sqrt{C^2+J_{\text{pin}}^2}}\exp\left[-\frac{(\lambda-\lambda_{0x})^2}{2\sigma_{\lambda x}^2}\right]\exp\left[-\frac{(\lambda-\lambda_{0y})^2}{2\sigma_{\lambda y}^2}\right]\exp\left[-\frac{U_y^2J_{\text{pin}}}{4\left(C^2+J_{\text{pin}}^2\right)}\right]$$
$$\times\exp\left[(x^2+y^2)\left(\frac{K_{\text{pin}}^2J_{\text{pin}}}{4(C^2+J_{\text{pin}}^2)}-N_{\text{pin}}\right)\right]\exp\left[y\left(\frac{K_{\text{pin}}U_yC}{2\left(C^2+J_{\text{pin}}^2\right)}\right)\right] \tag{11.320}$$

With substitution of the above pinhole parameters into Eq. (11.279), the CRL coherent amplitude contribution is converted to the pinhole coherent amplitude contribution $E_{\text{pin}}^{\text{coh}}(\lambda,x_s,y_s)$, which is expressed by exponentials with only imaginary arguments, where the magnitude of $E_{\text{pin}}^{\text{coh}}(\lambda,x_s,y_s)$ is unity, and

$$E_{\text{pin}}^{\text{coh}}(\lambda,x,y)=\exp[-\mathrm{i}kN(d_{\max}-\delta d)]\exp\left(\mathrm{i}y\frac{ki_g}{r_i}\right)\exp\left[\mathrm{i}\tan^{-1}\left(\frac{J_{\text{pin}}}{C}\right)\right]$$
$$\times\exp\left[-\mathrm{i}\cdot\left(-\frac{U_y^2C}{4(C^2+J_{\text{pin}}^2)}+\frac{2\pi}{\lambda}\left(\frac{D_g^2}{2D}+\frac{o_g^2}{2r_0}+\frac{i_g^2}{2r_i}\right)\right)\right]$$
$$\times\exp\left[-\mathrm{i}(x^2+y^2)\frac{G}{2}\left(1+\frac{K_{\text{pin}}^2C}{2G(C^2+J_{\text{pin}}^2)}\right)\right]\exp\left[\mathrm{i}y\left(\frac{K_{\text{pin}}U_yJ_{\text{pin}}}{2(C^2+J_{\text{pin}}^2)}\right)\right] \tag{11.321}$$

From section 11.3.1, the parameter C for the source-to-object-plane distance D and object distance r_0, as well as the parameter G for image distance r_i, are

$$C=\frac{\pi}{\lambda}\left(\frac{1}{D}+\frac{1}{r_0}\right)\quad G=\frac{2\pi}{r_i} \tag{11.322}$$

The gravity parameter U_y of Eq. (11.178) is

$$U_y=\frac{2\pi}{\lambda}\left(\frac{o_g}{r_0}-\frac{D_g}{D}\right) \tag{11.323}$$

At the center source point, the intensity PSF of a pinhole $I_{xy0}^{\text{pin}}(\lambda,x,y)$ is the squared magnitude of Eq. (11.319), where

$$I_{xy0}^{\text{pin}}(\lambda,x,y)=|\,E_{\text{pin}}^{\text{inc}}(\lambda,x_s,y_s)\,E_{\text{pin}}^{\text{coh}}(\lambda,x,y)|^2=|\,E_{\text{pin}}^{\text{inc}}(\lambda,x_s,y_s)|^2\quad x_s=y_s=0 \tag{11.324}$$

11.6 The Modulation Transfer Function

We now discuss the modulation transfer function, MTF(f_x,f_y), which describes the gain of an imaging system as a function of the x- and y-component spatial frequencies f_x and f_y of the imaged object (specimen) features. For zero spatial frequency of DC gain, the MTF is maximum and normalized to unity. As the spatial frequencies f_x and f_y increase, the MTF monotonically decreases, approaching zero. The MTF provides a means of describing the decrease of gain and contrast as the feature size of the imaged object (specimen) decreases in size with the corresponding increase in the reciprocal spatial frequency.

When an X-ray or neutron point source in the object plane of the CRL is focused by the CRL onto a detector in the image plane, the conjugate image is not a point but a patch of distributed intensity. This distributed intensity in its image plane of the CRL is its impulse response to a point neutron source in its object plane. The distributed intensity arises because of the CRL diffraction, chromatic aberration, and monochromatic aberrations. The impulse response or point spread function (PSF) of the CRL, which can be expressed by a Green's function via the Kirchhoff equation, is an expression of the Huygens–Fresnel principles of optics. Alternatively and more conveniently, one can image a line source of neutrons with the CRL to obtain the line spread function.

Spatial resolution is the smallest distance that two objects can be separated and still be distinguished. *Contrast* is the relative signal difference between an object and its background. *Radiographic contrast* is the relative difference between the flux emerging from adjacent portions of a specimen, whereas *image contrast* is the relative difference in image brightness between the corresponding specimen portions on the image on the detector by the CRL. Consider a sinusoidal variation ΔI of the image intensity between $I_{\min}$ and $I_{\max}$, where

$$\Delta I = I_{\max} - I_{\min} \tag{11.325}$$

The mean intensity I_0 is the average of the maximum $I_{\max}$ and minimum intensity $I_{\min}$:

$$I_0 = \frac{I_{\max} + I_{\min}}{2} \tag{11.326}$$

Contrast C is defined by

$$C = \frac{I_{\max} - I_{\min}}{I_{\max} + I_{\min}} \tag{11.327}$$

The modulation M is defined as the ratio of the intensity variation ΔI divided by the mean intensity I_0:

$$M = \frac{\Delta I}{I_0} \tag{11.328}$$

One performance test for the CRL is how much modulation can be transferred from the object to the image, as a function of the spatial frequencies h_x and h_y in the x- and y-directions. The spatial frequencies in the x- and y-directions are the reciprocal of the spatial periods Δx and Δy of the transferred modulation, which occurs along the x- and y-directions in the planes of the object and its conjugate image plane. The modulation in the object planes $M_{\text{obj}}(h_x, h_y)$ and the corresponding modulation in the image plane $M_{\text{ima}}(h_x, h_y)$ are

$$M_{\text{ima}} = \frac{\Delta I^{\text{ima}}}{I_0^{\text{ima}}} \tag{11.329}$$

and

$$M_{\text{obj}} = \frac{\Delta I^{\text{obj}}}{I_0^{\text{obj}}} \tag{11.330}$$

The ratio of the image modulation $M_{\text{ima}}(h_x, h_y)$, divided by the object modulation $M_{\text{obj}}(h_x, h_y)$ is $\text{MTF}(h_x, h_y)$, that is,

$$\text{MTF}(h_x, h_y) = \frac{M_{\text{image}}(h_x, h_y)}{M_{\text{object}}(h_x, h_y)} \tag{11.331}$$

For zero spatial frequency, the MTF is unity, where

$$\text{MTF}(0, 0) = \frac{M_{\text{image}}(0, 0)}{M_{\text{object}}(0, 0)} = 1 \tag{11.332}$$

The spatial frequencies in the x- and y-directions (h_x, h_y) are the reciprocal of the spatial periods $(\Delta x, \Delta y)$ of the transferred modulation, which occurs along the x- and y-directions in the planes of the object and its conjugate image. The MTF of the CRL is the magnitude Fourier transform of its PSF. Aberrations of the CRL may cause some image pixels to become dislocated, as in coma and distortion. The extent of the dislocations is measured by the phase shift $\phi(h_x, h_y)$, which is a function of the spatial frequency. Hence, $\phi(h_x, h_y)$ is the phase transfer function of the object modulation, imaged by the CRL into the image modulation. Together, the magnitude of the modulation transfer given by the $\text{MTF}(h_x, h_y)$, and the phase shift accompanying the object-to-image modulation transfer $\phi(h_x, h_y)$, lead to the complex optical transfer function $\text{OTF}(h_x, h_y)$, given by

$$\text{OTF}(h_x, h_y) = \text{MTF}(h_x, h_y)\exp(\text{i}\phi(h_x, h_y)) \tag{11.333}$$

The OTF is the ratio of the Fourier transform of the image intensity distribution, divided by the Fourier transform of the object intensity distribution.

11.7 The Modulation Transfer Function (MTF) with a Biconcave, Parabolic CRL for Incoherent X-rays or Neutrons with Gravity

One assumes an incoherent source of X-rays or neutrons. The modulation transfer function (MTF) can be readily calculated from the point spread function (PSF). The MTF is the magnitude of the Fourier transform of the PSF. Recalling section 11.3.3, assume the case of an object with uniform transmission T_0, which is independent of the source plane spatial coordinates (s_x, s_y):

$$T(s_x, s_y) = T_0 = 1 \tag{11.334}$$

Then, the transmission in the spatial frequency domain $T(f_{sx},f_{sy})$ is a Dirac delta function:

$$T(f_{sx},f_{sy}) = \delta(f_{sx},f_{sy}) \tag{11.335}$$

The amplitude at image plane point (x, y), which is due to source point (x_s, y_s), is the impulse response or the PSF of the lens or aperture. The generalized PSF for a lens or aperture (pinhole or pipe) is

$$\begin{aligned} E_{xy}(\lambda,x,y,x_s,y_s,f_{sx}=0,f_{sy}=0) &= E_0(\lambda)\cdot E_{3yys}(\lambda,y,y_s)\cdot E_{3xxs}(\lambda,x,x_s) \\ &\quad \times E_{30x}(\lambda,f_{sx}=0)\cdot E_{3x}(\lambda,x,f_{sx}=0)\cdot E_{3xs}(\lambda,x_s,f_{sx}=0) \\ &\quad \times E_{30y}(\lambda,f_{sy}=0)\cdot E_{3y}(\lambda,y,f_{sy}=0)\cdot E_{3ys}(\lambda,y_s,f_{sy}=0) \end{aligned} \tag{11.336}$$

Into the above Eq. (11.336) for $E_{xy}(\lambda, x, y, x_s, y_s)$ with $f_{sx} = 0$ and $f_{sy} = 0$, one substitutes the following expressions for the y-components: Eq. (11.213) for $E_0(\lambda)$, and Eq. (11.220) for $E_{30y}(\lambda,f_{sy}=0)$, and Eq. (11.221) for $E_{3y}(\lambda, y,f_{sy}=0)$, and Eq. (11.222) for $E_{3ys}(\lambda, y_s,f_{sy}=0)$, and Eq. (11.223) for $E_{3yys}(\lambda, y, y_s)$. The gravity terms are then set to zero in these y-component terms to obtain the corresponding x-component terms of $E_{30x}(\lambda,f_{sx}=0)$, $E_{3x}(\lambda, x,f_{sx}=0)$, $E_{3xs}(\lambda, x_s,f_{sx}=0)$, and $E_{3xxs}(\lambda, x, x_s)$, which are then substituted into Eq. (11.336). These substitutions yield Eq. (11.337) below for the generalized PSF amplitude $E_{xy}(\lambda, x, y, x_s, y_s)$ with $f_{sx} = 0$ and $f_{sy} = 0$, which is in a form convenient for integration over the image plane transverse coordinates x, y, where

$$\begin{aligned} &E_{xy}(\lambda,x,y,x_s,y_s) \\ &= E_{\lambda 00}(\lambda,z)E_s(\lambda,z,x_s,y_s)\exp\left[-x_s x\left(\frac{HKC}{C^2+J^2}\right)\right] \\ &\times \exp\left[-\mathrm{i}x_s x\left(\frac{HKJ}{C^2+J^2}\right)\right]\exp\left[x^2\left(\frac{K^2 J}{4(C^2+J^2)}-N\right)\right]\exp\left[-\mathrm{i}x^2\frac{G}{2}\left(1+\frac{K^2 C}{2G(C^2+J^2)}\right)\right] \\ &\times \exp\left[-y_s y\left(\frac{HKC}{C^2+J^2}\right)\right]\exp\left[-\mathrm{i}y_s y\left(\frac{HKJ}{C^2+J^2}\right)\right]\exp\left[y^2\left(\frac{K^2 J}{4(C^2+J^2)}-N\right)\right] \\ &\times \exp\left[y\left(\frac{KU_y C}{2(C^2+J^2)}\right)\right]\exp\left[-\mathrm{i}y^2\left(\frac{G}{2}+\frac{K^2 C}{4(C^2+J^2)}\right)\right]\exp\left[\mathrm{i}y\left(\frac{KU_y J}{2(C^2+J^2)}+\frac{k\mathrm{i}_g}{r_i}\right)\right] \end{aligned} \tag{11.337}$$

where

$$E_{\lambda 00}(\lambda,z,x,y)=\left(\frac{\pi}{2}\right)^2\left(\frac{E_{0x}E_{0y}}{z}\right)\exp\left[-\frac{(\lambda-\lambda_{0x})^2}{2\sigma_{\lambda x}^2}\right]\exp\left[-\frac{(\lambda-\lambda_{0y})^2}{2\sigma_{\lambda y}^2}\right]$$
$$\times\exp\left[-\mathrm{i}\cdot\left(\frac{2\pi}{\lambda}\left(\frac{D_g^2}{2D}+\frac{o_g^2}{2r_0}+\frac{i_g^2}{2r_i}\right)-\frac{U_y^2C}{4(C^2+J^2)}\right)\right]\exp\left[-\frac{U_y^2J}{4(C^2+J^2)}\right]$$
$$\times\left(\frac{\exp\left(-\frac{N\mu d}{2}\right)\exp\left[-ikN(d_{\max}-\delta d)\right]}{\left[(AJ-EC)^2+(EJ+AC)^2\right]^{1/2}}\right)\exp\left[-i\tan^{-1}\left(\frac{EJ+AC}{AJ-EC}\right)\right] \tag{11.338}$$

and

$$E_s(\lambda,x_s,y_s)=\exp\left[-x_s^2\left(\frac{H^2J}{C^2+J^2}+W_x^{\mathrm{Re}}\right)\right]\exp\left[-\mathrm{i}x_s^2\left(H-\frac{H^2C}{C^2+J^2}-W_x^{\mathrm{Im}}\right)\right]$$
$$\times\exp\left[-y_s^2\left(\frac{H^2J}{C^2+J^2}+W_y^{\mathrm{Re}}\right)\right]\exp\left[y_s\left(\frac{HJU_y}{C^2+J^2}\right)\right]$$
$$\times\exp\left[-\mathrm{i}y_s^2\left(H-\frac{H^2C}{C^2+J^2}-W_y^{\mathrm{Im}}\right)\right]\exp\left[-\mathrm{i}y_s\left(V_y+\frac{HU_yC}{C^2+J^2}\right)\right] \tag{11.339}$$

The x-component W_x and y-component W_y source parameters of Eq. (11.247) in section 11.3.1 are real, where

$$W_x^{\mathrm{Im}}=W_y^{\mathrm{Im}}=0\quad W_x^{\mathrm{Re}}=\frac{1}{\sigma_x^2}\quad W_y^{\mathrm{Re}}=\frac{1}{\sigma_y^2} \tag{11.340}$$

The absorption aperture radius r_{a}, and E,A, J, K, and N are CRL parameters from section 11.3.1, where

$$r_{\mathrm{a}}=\sqrt{\frac{2R}{\mu N}}\quad E=\frac{\pi}{\lambda}\left(\frac{1}{r_0}+\frac{1}{r_i}-\frac{2N\delta}{R}\right)=0\quad A=\frac{\mu N}{2R}=\frac{1}{r_{\mathrm{a}}} \tag{11.341}$$

$$J=\left(\frac{r_{\mathrm{a}}\pi}{r_0\lambda}\right)^2\quad K=\left(\frac{2}{r_0r_i}\right)\left(\frac{r_{\mathrm{a}}\pi}{\lambda}\right)^2\quad N=\left(\frac{r_{\mathrm{a}}\pi}{r_i\lambda}\right)^2 \tag{11.342}$$

The parameters C and H of section 11.3.1 for source-to-object distance D and object distance r_0, as well as the parameter G for image distance r_i, are

$$C = \frac{\pi}{\lambda}\left(\frac{1}{D} + \frac{1}{r_0}\right) \quad H = \frac{\pi}{D\lambda} \quad G = \frac{2\pi}{r_i} \tag{11.343}$$

The y-directed gravity parameter U_y of Eq. (11.178), is

$$U_y = \frac{2\pi}{\lambda}\left(\frac{o_g}{r_0} - \frac{D_g}{D}\right) \tag{11.344}$$

MTF(f_x,f_y, λ), as a function of the spatial frequencies f_x,f_y along the x- and y-directions for a specific neutron wavelength λ, is the modulus of the Fourier transform of the PSF of Eq. (11.257). The MTF is given by

$$\mathrm{MTF}(g_x, g_y, \lambda) = \left| \int_{-\infty}^{\infty} \int_{-\infty}^{\infty} E_{xy}(x, y, z, \lambda, x_s, y_s) \exp(-\mathrm{j}2\pi h_y) \exp(-\mathrm{j}2\pi h_y) \mathrm{d}x\, \mathrm{d}y \right| \tag{11.345}$$

Substitution of Eqs. (11.337), with Eqs. (11.338–11.344), into Eq. (11.345), yields

$$\mathrm{MTF}(z, \lambda, x_s, y_s) = \left| \begin{array}{l} E_{\lambda 00}(\lambda, z) E_s(\lambda, z, x_s, y_s) \times \\ \displaystyle\int_{-\infty}^{\infty} \mathrm{d}x \exp(-\mathrm{i}2\pi h_x x) \exp[-\alpha_x x_s x] \exp(-\mathrm{i}\beta_x x_s x) \exp(\delta_x x^2) \exp(-\mathrm{i}\varepsilon_x x^2) \\ \displaystyle\int_{-\infty}^{\infty} \mathrm{d}y \exp(-\mathrm{i}2\pi h_y y) \exp(-\alpha_y y_s y) \exp(-\mathrm{i}\beta_y y_s y) \exp(\delta_y y^2) \\ \exp(\eta_y y) \exp(-\mathrm{i}\varepsilon_y y^2) \exp(\mathrm{i}\kappa_y y) \end{array} \right| \tag{11.346}$$

Parameters α, β, ζ, and ε in Eq. (11.346), are

$$\alpha = \frac{HKC}{C^2 + J^2} \quad \beta = \frac{HKJ}{C^2 + J^2} \tag{11.347}$$

$$\zeta = \frac{K^2 J}{4(C^2 + J^2)} - N \quad \varepsilon = \frac{K^2 C}{4(C^2 + J^2)} + \frac{G}{2} \tag{11.348}$$

The gravity terms η and κ in Eq. (11.346), are

$$\eta = \frac{KU_yC}{2(C^2+J^2)} \quad \kappa = \frac{KU_yJ}{2(C^2+J^2)} + \frac{ki_g}{r_i} \tag{11.349}$$

In the coefficient $E_{\lambda 00}(\lambda, z, x, y)$, the axial distance is set at the image plane $z = z_{\mathrm{im}}$, which assumes $z \gg x^2 + y^2$, where

$$\frac{E_{0x}E_{0y}}{z} \cong \frac{E_{0x}E_{0y}}{D + r_0 + r_i} \tag{11.350}$$

Assume that

$$\mathrm{Im}[\varepsilon] + \mathrm{Re}[\zeta] < 0 \tag{11.351}$$

Then, integration by Mathematica over x and y in the image plane yields the $\mathrm{MTF}(z, \lambda, x_s, y_s)$, and the magnitude of the Fourier transform of the PSF, which depends on source point (x_s, y_s), where

$$\begin{aligned}\mathrm{MTF}(z, \lambda, x_s, y_s) = \pi\left|\frac{E_{\lambda 00}(\lambda, z)E_s(\lambda, z, x_s, y_s)}{-\zeta + \mathrm{i}\varepsilon}\right| \times \left|\exp\left(\frac{(2h_x\pi + x_s(-\mathrm{i}\alpha + \beta))^2}{4(\zeta - \mathrm{i}\varepsilon)}\right)\right| \\ \times \left|\exp\left(\frac{(2h_y\pi + y_s(-\mathrm{i}\alpha + \beta))^2}{4(\zeta - \mathrm{i}\varepsilon)}\right)\right|\end{aligned} \tag{11.352}$$

Since $\mathrm{Im}[\varepsilon] = 0$, the condition allowing integration along x and y in the image plane is $\mathrm{Re}[\zeta] < 0$. Substitution of the parameters C, J, K, and N into ζ of Eq. (11.348), shows that the previous integration along x and y is valid if

$$\zeta = -\frac{\left(\frac{r_0}{D}+1\right)^2\left(\frac{r_0}{r_a r_i}\right)^2}{\left(\frac{r_0}{D}+1\right)^2\left(\frac{r_0\lambda}{r_a^2\pi}\right)^2 + 1} < 0 \tag{11.353}$$

Simplifying Eq. (11.353), one obtains

$$\left(\frac{r_0}{r_a r_i}\right)\left(\frac{r_0}{D}+1\right) > 0 \tag{11.354}$$

The integration is thus valid because variables r_a, r_0, r_i, and D in Eq. (11.354) are greater than zero.

If gravity is neglected, then gravity terms η, κ, U_y, and V_y are set to zero, where

$$\eta = \kappa = 0 \tag{11.355}$$

$$U_y = V_y = 0 \tag{11.356}$$

With gravity neglected, the y-directed gravity droop distances in the vertical y-direction are zero. That is, the vertical droop distance between the source and the object plane D_g, the vertical droop distance between the object plane and CRL principal or center plane o_g, and the vertical droop distance between the CRL principal plane and image plane i_g are all zero, where from Eqs. (11.105), (11.106), and (11.108)

$$o_g = i_g = D_g = 0 \tag{11.357}$$

With the gravity terms set to zero in the MTF expression of Eq. (11.352), the resulting MTF expression then applies to X-rays, where the effect of ordinary gravity is negligible. Since there is a lens, the lens-maker condition is met with parameter $E = 0$. For the case of zero gravity or gravity neglected, the MTF for X-rays or neutrons of Eq. (11.352), becomes

$$\begin{aligned}
\mathrm{MTF}(z,\lambda,x_s,y_s) = {} & \frac{\pi^3}{4}\left(\frac{E_{0x}E_{0y}}{D+r_0+r_i}\right)\left(\frac{\exp\left(-\frac{N\mu d}{2}\right)}{\left[(AJ)^2+(AC)^2\right]^{1/2}}\right) \\
& \times \exp\left[-\frac{(\lambda-\lambda_{0x})^2}{2\sigma_{\lambda x}^2}\right]\exp\left[-\frac{(\lambda-\lambda_{0y})^2}{2\sigma_{\lambda y}^2}\right]\exp\left[-x_s^2\left(\frac{H^2J}{C^2+J^2}+W_x^{\mathrm{Re}}\right)\right] \\
& \times \exp\left[-y_s^2\left(\frac{H^2J}{C^2+J^2}+W_y^{\mathrm{Re}}\right)\right]\frac{1}{(\zeta^2+\varepsilon^2)^{1/2}} \\
& \times \exp\left(\frac{4h_x^2\pi^2\zeta+4h_x x_s\pi(\beta\zeta+\alpha\varepsilon)+x_s^2(\beta^2\zeta+2\alpha\beta\varepsilon-\alpha^2\zeta)}{4(\zeta^2+\varepsilon^2)}\right) \\
& \times \exp\left(\frac{4h_y^2\pi^2\zeta+4h_y y_s\pi(\beta\zeta+\alpha\varepsilon)+y_s^2(\beta^2\zeta+2\alpha\beta\varepsilon-\alpha^2\zeta)}{4(\zeta^2+\varepsilon^2)}\right)
\end{aligned} \tag{11.358}$$

11.8 Field of View (FOV) with a Biconcave, Parabolic CRL for Incoherent X-rays or Neutrons with Gravity

An observer at a point I in the image plane can see through the CRL only a portion of the source plane, or only a portion of the object plane, which is illuminated by incoherent X-rays or neutrons. This portion of the source or source-illuminated object plane, which is observable by a point-sized observer at a point I in the image plane, is the field of view (FOV) at that image point I. This FOV varies depending on the chosen observer point I in the image plane, and it is the source amplitude distribution that contributes to the formation of single image point I by constructive and destructive interference of the source waves. The FOV depends on the optic (e.g., CRL) between the source plane and image plane.

The FOV is defined as the set of points P at positions (x_s, y_s) in the source plane at $z = 0$, which emit waves that pass through the object plane and are refracted by the CRL so as to converge to a single point I at position (x, y) in the image plane at $z = z_{\text{im}}$. The first integration over the lens plane at $z = D + r_0$ include only those CRL-emitted rays, which converge to the center image point I at $(x, y) = (0, 0)$. At each point on the CRL center principal plane, the second integration includes the waves, which are emitted from all the points P at positions (s_x, s_y) on the source plane at $z = D$.

Recall from Eq. (11.245), for a particular source point P at (x_s, y_s), the amplitude distribution $E_{xy}(\lambda, x, y, x_s, y_s)$ is the contribution of the source point P amplitude to the observer point I at (x, y) in the image plane. From Eq. (11.245), one can obtain the source amplitude distribution $E_{xy}(\lambda, x_{\text{obs}}, y_{\text{obs}}, x_s, y_s)$, which is the "visible part of the source" in the image plane at the observer point I at $(x, y) = (x_{\text{obs}}, y_{\text{obs}})$. The field of view (FOV) amplitude distribution $E_{xy}(\lambda, x_{\text{obs}}, y_{\text{obs}}, x_s, y_s)$ at observer point I is simplified for an observer at the center image plane point $(x_{\text{obs}}, y_{\text{obs}}) = (0, 0)$, where

$$
\begin{aligned}
E_{xy}(\lambda, x_{\text{obs}}, y_{\text{obs}}, x_s, y_s) = E_0(\lambda)\exp\left[-y_s y_{\text{obs}}\left(\frac{HK}{C - \mathrm{i}J}\right)\right]\exp\left[-x_s x_{\text{obs}}\left(\frac{HK}{C - \mathrm{i}J}\right)\right] \\
\times \int_{-\infty}^{\infty} \mathrm{d}f_{sx} E_{30x}(\lambda, f_{sx}) E_{3xs}(\lambda, x_s, f_{sx}) \\
\times \int_{-\infty}^{\infty} \mathrm{d}f_{sy} T(f_{sx}, f_{sy}) E_{30y}(\lambda, f_{sy}) E_{3ys}(\lambda, y_s, f_{sy})
\end{aligned}
\tag{11.359}
$$

where

$$
E_0(\lambda, z) = \left(\frac{E_{0x}\,E_{0y}}{z}\right)\exp\left(-\frac{N\mu d}{2}\right)\exp[-\mathrm{i}kN(d_{\max} - \delta d)] \tag{11.360}
$$

Here,

$$
E_{3x}(x = 0, f_{sx}) = 1 \tag{11.361}
$$

and

$$
E_{3y}(y = 0, f_{sy}) = 1 \tag{11.362}
$$

If the object has uniform transmission T_0, which is independent of the source plane spatial coordinates (s_x, s_y), where

$$T(s_x, s_y) = T_0 \tag{11.363}$$

Then recalling section 11.3.3, the transmission in the spatial frequency domain is a Dirac delta function, where

$$T(f_{sx}, f_{sy}) = \delta(f_{sx}, f_{sy}) \tag{11.364}$$

With substitution of Eqs. (11.361−11.364) into Eq. (11.359), the FOV amplitude distribution $E_{xy}(\lambda, x_s, y_s)$ then becomes

$$E_{xy}(\lambda, x_s, y_s) = E^{\text{inc}}_{3xsys}(\lambda, x_s, y_s) E^{\text{coh}}_{3xsys}(\lambda, x_s, y_s) \tag{11.365}$$

We now examine the field of view amplitude of Eq. (11.359) for an observer at image point I in the center of the image plane at $(x_{\text{obs}}, y_{\text{obs}}) = (0, 0)$. The FOV amplitude of Eq. (11.359) is then expressed as the product of an incoherent amplitude $E^{\text{inc}}_{3xsys}(\lambda, x_s, y_s)$ and a coherent amplitude $E^{\text{coh}}_{3xsys}(\lambda, x_s, y_s)$. The incoherent amplitude $E^{\text{inc}}_{3xsys}(\lambda, x_s, y_s)$ is extracted from Eq. (11.359), and is the product of real x- and y-components. The y-component of the incoherent amplitude $E^{\text{inc}}_{3xsys}(\lambda, x_s, y_s)$ is the product of real part of Eq. (11.359) for $E_0(\lambda)$, and Eq. (11.233) for the real part of Eq. (11.220) for $E_{30y}(\lambda, f_{sy} = 0)$, and Eq. (11.237) for the real part of Eq. (11.222) for $E_{3ys}(\lambda, y_s, f_{sy} = 0)$. The gravity terms are then set to zero in these y-component terms, and y is set to x, to obtain the corresponding real parts of the x-component terms of $E_{30x}(\lambda, f_{sx} = 0)$ and $E_{3xs}(\lambda, x_s, f_{sx} = 0)$. Uniform object transmission is assumed, so that the integral expressions in Eq. (11.359) for the x- and y-components, via the transmission delta function of Eq. (11.364), reduce to evaluation of the integrands at the single points $f_{sx} = 0$ and $f_{sy} = 0$. With substitution of parameters from Section 11.3.1, the incoherent amplitude $E^{\text{inc}}_{3xsys}(\lambda, x_s, y_s)$ component of the field view (FOV), for an observer at image point I in the center of the image plane at $(x_{\text{obs}}, y_{\text{obs}}) = (0, 0)$, and for an object with uniform transmission, is then

$$\begin{aligned}
E^{\text{inc}}_{3xsys}(\lambda, x_s, y_s) = {} & \left(\frac{E_{0x} E_{0y}}{z}\right) \exp\left(-\frac{N\mu d}{2}\right) \exp\left[-(W_x^{\text{Re}} x_s^2 + W_y^{\text{Re}} y_s^2)\right] \\
& \times \exp\left[-(x_s^2 + y_s^2)\left(\frac{(\pi/D\lambda)^2 (r_a \pi/r_0 \lambda)^2}{(\pi/D\lambda)^2((1/D)+(1/r_0))^2 (r_a \pi/r_0 \lambda)^4 + (r_a \pi/r_0 \lambda)^4}\right)\right] \\
& \exp\left[y_s \left(\frac{(2\pi^2/D\lambda^2)((o_g/D) - (D_g/D))(r_a \pi/r_0 \lambda)^2}{(\pi/\lambda)^2((1/D)+(1/r_0))^2 + (r_a \pi/r_0 \lambda)^4}\right)\right]
\end{aligned} \tag{11.366}$$

With Eq. (11.365), we now consider the effects of coherent interactions of the source points P at observer point I at the image plane center point $(x_{\text{obs}}, y_{\text{obs}}) = (0, 0)$, which is contained in the coherent amplitude component $E^{\text{coh}}_{3xsys}(\lambda, x_s, y_s)$ of the FOV. That is, the source points are coherent, if there is a fixed phase relationship between their emitted waves. The coherent amplitude term $E^{\text{coh}}_{3xsys}(\lambda, x_s, y_s)$ thus depends on the phase of the waves, which is based on their position in space and time.

The fixed spatial phase relationships between the waves emitted by the source points are a manifestation of the spatial arrangement of the source points. The source points are assumed to be coherent temporally; that is, the timing of the emission of waves from each source point has a fixed delay with respect to the other source points. For simplicity, we assume that the source points emit their waves at the same time. Waves diverge from each source point, and the wave front (or, equivalently, the rays perpendicular to the wave front) then pass through the object plane to the lens. For each wavelength, each part of the wave front, or each ray path, experiences a phase velocity that depends on the medium through which it propagates.

Also, the spatial distance varies along each ray path or point on the wave front. Consequently, the optical path lengths vary across the wave front. Along the incident wave front, the lens provides a compensating distance via its shape, and phase velocity change via its material refractive index $n = 1 - \delta$. The lens thus changes the radius of curvature of the wave front in both magnitude and sign, so as to converge the wave onto the single image point.

The coherent amplitude $E^{\text{coh}}_{3xsys}(\lambda, x_s, y_s)$ is extracted from Eq. (11.359), and is the product of imaginary x- and y-components. The y-component of the coherent amplitude $E^{\text{coh}}_{3xsys}(\lambda, x_s, y_s)$ is the product of imaginary part of Eq. (11.359) for $E_0(\lambda)$, and Eq. (11.234) for the imaginary part of Eq. (11.220) for $E_{30y}(\lambda, f_{sy} = 0)$, and Eq. (11.238) for the imaginary part of Eq. (11.222) for $E_{3ys}(\lambda, y_s, f_{sy} = 0)$. The gravity terms are then set to zero in these y-component terms, and y is set to x, to obtain the corresponding imaginary parts of the x-component terms of $E_{30x}(\lambda, f_{sx} = 0)$ and $E_{3xs}(\lambda, x_s, f_{sx} = 0)$. Uniform object transmission is assumed, so that the integral expressions in Eq. (11.359) for the x- and y-components, via the transmission delta function of Eq. (11.364), reduce to evaluation of the integrands at the single points $f_{sx} = 0$ and $f_{sy} = 0$. With substitution of parameters from Section 11.3.1, the coherent amplitude $E^{\text{coh}}_{3xsys}(\lambda, x_s, y_s)$ component of the field view (FOV), for an observer at image point I in the center of the image plane at $(x_{\text{obs}}, y_{\text{obs}}) = (0, 0)$, and for an object with uniform transmission, is then

$$E^{\text{coh}}_{3xsys}(\lambda, x_s, y_s) = \exp[-ikN(d_{\max} - \delta d)]\exp\left[\mathrm{i}\cdot\left(x_s^2 W_x^{\text{Im}} + y_s^2 W_y^{\text{Im}}\right)\right]$$
$$\times\exp\left[-i\left(x_s^2 + y_s^2\right)\left(\frac{\pi}{D\lambda} - \frac{\left(\frac{\pi}{D\lambda}\right)^2\left(\left(\frac{\pi}{\lambda}\right)\left(\frac{1}{D} + \frac{1}{r_0}\right)\right)}{\left(\frac{\pi}{\lambda}\right)^2\left(\frac{1}{D} + \frac{1}{r_0}\right)^2 + \left(\frac{r_a\pi}{r_0\lambda}\right)^4}\right)\right]$$
$$\times\exp\left[-\mathrm{i}y_s^2\left(\frac{2\pi D_g}{\lambda D} + \frac{\left(\frac{2\pi^2}{D\lambda^2}\right)\left(\frac{o_g}{D} - \frac{D_g}{D}\right)\left(\frac{\pi}{\lambda}\right)\left(\frac{1}{D} + \frac{1}{r_0}\right)}{\left(\frac{\pi}{\lambda}\right)^2\left(\frac{1}{D} + \frac{1}{r_0}\right)^2 + \left(\frac{r_a\pi}{r_0\lambda}\right)^4}\right)\right] \qquad (11.367)$$

The FOV intensity distribution $I_{xy}(\lambda, x_s, y_s)$ is then the squared magnitude of the amplitude expression $E_{xy}(\lambda, x_s, y_s)$ of Eq. (11.365), where

$$I_{xy}(\lambda, x_s, y_s) = |E^{\mathrm{inc}}_{3xsys}(\lambda, x_s, y_s) E^{\mathrm{coh}}_{3xsys}(\lambda, x_s, y_s)|^2 = |E^{\mathrm{inc}}_{3xsys}(\lambda, x_s, y_s)|^2 \qquad (11.368)$$

11.9 Image Intensity Distribution of a Biconcave, Parabolic CRL for Coherent X-rays or Neutrons with Gravity

We now consider the case of an object illuminated with a coherent source. This coherent source can be a short-pulsed point source of X-rays or neutrons, which are used in phase contrast imaging. Alternatively, the coherent source could be the recently developed X-ray laser. In the future, an extended, coherent source of neutrons or neutron laser may be developed. At present, pulsed, high-intensity point sources of thermal and cold neutrons with short pulse times are not available. Pinhole sources of cold or thermal neutrons can be produced with significant transverse coherence for phase contrast imaging, but at present, there are no directed, neutron laser beams with large temporal (longitudinal) and transverse (spatial) coherence.

The amplitude $E_{xy}(\lambda, x, y, z, x_s, y_s)$ at image point (x, y, z_{im}) for wavelength λ, which is to a *single* source point (x_s, y_s), is the PSF, or Green's function. The PSF is now integrated over all the source plane points (x_s, y_s) to obtain the amplitude $E^{\mathrm{coh}}_{xy}(\lambda, x, y, z)$ at image point (x, y) due to *all* the source points (x_s, y_s). This amplitude distribution at the image plane is the integration over a coherent source.

An advantage of deriving the image intensity distribution for a coherent source, is that one can divide the image intensity distribution into coherent and incoherent terms. With an incoherent source, one simply sets the coherent terms to zero, and if the source is coherent, then the image distribution solution requires both the coherent and incoherent terms. For example, consider the wave amplitude C, which is the sum of two complex wave amplitudes with magnitudes A and B for wave number k as a function of axial propagation distance z relative to two-source, axial positions z_1 and z_2, where

$$C = A \exp[ik(z - z_1)] + B \exp[ik(z - z_2)] \qquad (11.369)$$

If the source is incoherent, then the intensity I^{inc}_c of wave C is the sum of the intensities of wave A and wave B, where

$$I^{\mathrm{inc}}_c = |A|^2 + |B|^2 \qquad (11.370)$$

If the source is coherent, then the intensity I^{coh}_c of wave C is the squared magnitude of the summed amplitudes of wave A and wave B, where

$$I^{\mathrm{coh}}_c = |C|^2 = |A \exp[ik(z - z_1)] + B \exp[ik(z - z_2)]|^2 \qquad (11.371)$$

One now expands Eq. (11.371),

$$I_c^{\rm coh} = C^* C = \{A\exp[ik(z-z_1)] + B\exp[ik(z-z_2)]\}^* \{A\exp[ik(z-z_1)] + B\exp[ik(z-z_2)]\} \tag{11.372}$$

Collecting terms in Eq. (11.372), one obtains

$$I_c^{\rm coh} = C^* C = |A|^2 + |B|^2 + BA\ \exp[ik(z_2 - z_1)] + AB\ \exp[ik(z_1 - z_2)] \tag{11.373}$$

Hence, the image intensity of the coherent source is the image intensity of an incoherent source with added interference intensity terms, where

$$I_c^{\rm coh} = C^* C = I_c^{\rm inc} + I_c^{\rm cross} \tag{11.374}$$

and

$$I_c^{\rm cross} = BA\ \exp[ik(z_2 - z_1)] + AB\ \exp[ik(z_1 - z_2)] \tag{11.375}$$

With the solution for a coherent source, one has the full possible solution. If the coherent source result is applied to the case of an incoherent source, then one simply sets the cross terms to zero. One can thus anticipate future coherent sources, and solve the image intensity distributions for the coherent case. However, one can apply the coherent solution to a present-day source, which is incoherent, by simply setting the coherent cross terms to zero.

At present, there are no neutron laser sources, but there are X-ray lasers, such as the FLASH at DESY in Germany and the Stanford University Linear Coherent Light Source (LCLS), which have recently become operational. However, even prior to coherent X-ray sources, in phase contrast imaging, at a sufficient distance from a sufficiently small, but sufficiently intense, incoherent X-ray or thermal neutron source, one can obtain a coherent X-ray or neutron wave front with a significant transverse coherence length. In the future, for phase contrast imaging, there may be intense sources of short pulsed, monochromatic neutrons, which passed through a sufficiently small pinhole, can create at sufficiently large distances, substantial transverse and longitudinal coherence lengths.

We will now assume a coherent X-ray source, which is a Gaussian beam with minimum waist diameters w_{0x} and w_{0y} along x_s and y_s, respectively, at the source plane ($z = 0$). Recall Eq. (11.257) for the image amplitude of a parabolic lens CRL, which was applied to the case of a CRL imaging a coherent source, and yielded Eqs. (11.302–11.305). The image plane amplitude distribution $E_{xy}^{\rm coh}(\lambda, x, y, z)$ at $z = z_{\rm im}$, obtained by integration over the coherent source, is then

$$E_{xy}^{\rm coh}(\lambda, x, y, z) = E_0(\lambda) \int_{-\infty}^{\infty} {\rm d}y_s \int_{-\infty}^{\infty} {\rm d}x_s \left(\int_{-\infty}^{\infty} \int_{-\infty}^{\infty} T(f_{sx}, f_{sy}) \right.$$
$$\left. E_{sx}(\lambda, x, x_s, f_{sx}) E_{sy}(\lambda, y, y_s, f_{sy}) {\rm d}f_{sx}\ {\rm d}f_{sy} \right) \tag{11.376}$$

For a coherent X-ray source, recall from section 10.21 and Eq. (10.466), that the Gaussian amplitude coefficient $E_{0G}(\lambda, z)$ varies as a function of distance z from its beam waist at $z = 0$:

$$E_{0G}(\lambda, z, x_s, y_s) = -E_{0x}E_{0y}\left(\frac{w_{0x}w_{0y}\pi}{z\lambda}\right)^2 \exp\left(-\frac{2x_s^2}{\sigma_x^2}\right)\exp\left(-\frac{2y_s^2}{\sigma_y^2}\right) \quad (11.377)$$

For a coherent X-ray source, one assumes in Eq. (11.377) that distance z from the beam waist at $z = 0$, is much larger than the beam waist dimensions z_{0x} and z_{0y} in the x- and y-directions, where

$$z \gg z_{0x}, z_{0y} \quad (11.378)$$

Hence, with the Gaussian amplitude of Eq. (11.377), the amplitude coefficient of Eq. (11.376), becomes

$$E_0(\lambda, z) = -E_{0x}E_{0y}\left(\frac{w_{0x}w_{0y}\pi}{z\lambda}\right)^2 \exp\left(-\frac{N\mu d}{2}\right)\exp[-ikN(d_{\max} - \delta d)] \quad (11.379)$$

Recall from section 11.3.3, that $T(f_{sx}, f_{sy})$ is the Fourier transform of the object transmission function $T(s_x, s_y)$, and f_{sx} and f_{sy} are the spatial frequencies, which correspond respectively to the object plane coordinates s_x and s_y, where

$$T(f_{sx}, f_{sy}) = \int_{-\infty}^{\infty}\int_{-\infty}^{\infty} T(s_x, s_y)\exp(-i2\pi f_{sy}s_y)\exp(-i2\pi f_{sx}s_x)\,ds_x\,ds_y \quad (11.380)$$

We can change the order of integration in Eq. (11.380), so that

$$E_{xy}^{coh}(\lambda, x, y, z) = E_0(\lambda, z)\int_{-\infty}^{\infty}\int_{-\infty}^{\infty} df_{sx}\,df_{sy}E_x^{coh}(\lambda, x, z, f_{sx})E_y^{coh}(\lambda, y, z, f_{sy})T(f_{sx}, f_{sy}) \quad (11.381)$$

where

$$E_x^{coh}(\lambda, x, f_{sx}) = \int_{-\infty}^{\infty} dx_s E_{sx}(\lambda, x, x_s, f_{sx}) \quad (11.382)$$

and

$$E_y^{coh}(\lambda, y, f_{sy}) = \int_{-\infty}^{\infty} dy_s E_{sy}(\lambda, y, y_s, f_{sy}) \quad (11.383)$$

First, we solve for $E_y^{\rm coh}(\lambda, y, f_{sy})$ by performing the integral over y_s. Substitution for $E_{sy}(\lambda, y, y_s, f_{sy})$ yields

$$E_y^{\rm coh}(\lambda, y, f_{sy}) = E_{30y}^{\rm Re}(\lambda, f_{sy})E_{30y}^{\rm Im}(\lambda, f_{sy})E_{3y}^{\rm Re}(\lambda, y, f_{sy})E_{3y}^{\rm Im}(\lambda, y, f_{sy}) \times \int_{-\infty}^{\infty} dy_s E_{3ys}^{\rm Re}(\lambda, y_s, f_{sy})E_{3ys}^{\rm Im}(\lambda, y_s, f_{sy}) \cdot E_{3yys}^{\rm Re}(\lambda, y, y_s)E_{3yys}^{\rm Im}(\lambda, y, y_s) \tag{11.384}$$

The real part of $E_y^{\rm coh}(\lambda, y, f_{sy})$ in the integrand of Eq. (11.384), consists of $E_{3ys}^{\rm Re}(\lambda, y_s, f_{sy})$ of Eq. (11.237) and $E_{3yys}^{\rm Re}(\lambda, y, y_s)$ of Eq. (11.239), where

$$E_{3ys}^{\rm Re}(\lambda, y_s, f_{sy}) = \exp(-W_y^{\rm Re} y_s^2)\exp\left[y_s\left(\frac{HJ}{C^2+J^2}\right)(U_y - 2f_{sy}\pi)\right]\exp\left[-y_s^2\left(\frac{H^2 J}{C^2+J^2}\right)\right] \tag{11.385}$$

and

$$E_{3yys}^{\rm Re}(\lambda, y, y_s) = \exp\left[-y_s y\left(\frac{HKC}{C^2+J^2}\right)\right] \tag{11.386}$$

The imaginary part of $E_y^{\rm coh}(\lambda, y, f_{sy})$ in the integrand of Eq. (11.384), consists of $E_{3ys}^{\rm Im}(\lambda, y_s, f_{sy})$ of Eq. (11.238) and $E_{3yys}^{\rm Im}(\lambda, y, y_s)$ of Eq. (11.240), where

$$E_{3ys}^{\rm Im}\left(\lambda, y_s, f_{sy}\right) = \exp\left[-{\rm i}y_s\left(V_y + \frac{HU_yC}{C^2+J^2} - \frac{2f_{sy}HC\pi}{C^2+J^2}\right)\right] \times \exp\left[-{\rm i}y_s^2\left(H - \frac{H^2C}{C^2+J^2} - W_y^{\rm Im}\right)\right] \tag{11.387}$$

$$E_{3yys}^{\rm Im}(\lambda, y, y_s) = \exp\left[-{\rm i}y_s y\left(\frac{HKJ}{C^2+J^2}\right)\right] \tag{11.388}$$

To simplify the integration Eq. (11.384) over y_s, one simplifies the form of the integral to

$$E_y^{\rm coh}(\lambda, y, f_{sy}) = E_{30y}^{\rm Re}(\lambda, f_{sy})E_{30y}^{\rm Im}(\lambda, f_{sy})E_{3y}^{\rm Re}(\lambda, y, f_{sy})E_{3y}^{\rm Im}(\lambda, y, f_{sy}) \times \int_{-\infty}^{\infty} dy_s \exp(-Q_y y_s^2 - R_y y_s)\exp(-{\rm i}S_y y_s^2 - {\rm i}T_y y_s) \tag{11.389}$$

The parameters Q_y, R_y, S_y, and T_y in Eq. (11.389), are

$$Q_y = W_y^{\mathrm{Re}} + \frac{H^2 J}{C^2 + J^2} \tag{11.390}$$

$$R_y = \frac{2f_{sy} HJ\pi}{C^2 + J^2} - U_y + \frac{yHKC}{C^2 + J^2} \tag{11.391}$$

$$S_y = H - W_y^{\mathrm{Im}} - \frac{H^2 C}{C^2 + J^2} \tag{11.392}$$

$$T_y = V_y + \frac{HU_y C}{C^2 + J^2} - \frac{2f_{sy} HC\pi}{C^2 + J^2} + \frac{yHKJ}{C^2 + J^2} \tag{11.393}$$

Integration over y_s in Eq. (11.389), yields

$$\begin{aligned} E_y^{\mathrm{coh}}(\lambda, y, f_{sy}) &= E_{30y}^{\mathrm{Re}}(\lambda, f_{sy}) E_{30y}^{\mathrm{Im}}(\lambda, f_{sy}) E_{3y}^{\mathrm{Re}}(\lambda, y, f_{sy}) E_{3y}^{\mathrm{Im}}(\lambda, y, f_{sx}) \\ &\times \sqrt{\frac{\pi}{Q_y + \mathrm{i}S_y}} \exp\left[\frac{(R_y + \mathrm{i}T_y)^2}{4(Q_y + \mathrm{i}S_y)}\right] \end{aligned} \tag{11.394}$$

Expansion and collection of terms in Eq. (11.394), yields

$$E_y^{\mathrm{coh}}(\lambda, x, f_{sy}) = E_{30y}^{\mathrm{Re}}(\lambda, f_{sy}) E_{30y}^{\mathrm{Im}}(\lambda, f_{sy}) E_{3y}^{\mathrm{Re}}(\lambda, y, f_{sy}) E_{3y}^{\mathrm{Im}}(\lambda, y, f_{sy}) E_{40y}^{\mathrm{Re}}(\lambda, f_{sy}) E_{40y}^{\mathrm{Im}}(\lambda, f_{sy}) \tag{11.395}$$

where

$$E_{40y}^{\mathrm{Re}}(\lambda, f_{sy}) = \frac{\sqrt{\pi}}{(Q_y^2 + S_y^2)^{1/4}} \exp\left[\frac{2R_y S_y T_y + Q_y(R_y^2 - T_y^2)}{4(Q_y^2 + S_y^2)}\right] \tag{11.396}$$

and

$$E_{40y}^{\mathrm{Im}}(\lambda, f_{sy}) = \exp\left[-\frac{\mathrm{i}}{2}\tan^{-1}\left(\frac{S_y}{Q_y}\right)\right] \exp\left[\mathrm{i}\frac{S_y T_y^2 + 2Q_y R_y T_y - R_y^2 S_y}{4(Q_y^2 + S_y^2)}\right] \tag{11.397}$$

This integral is valid, providing that

$$\mathrm{Re}[Q_y] > \mathrm{Im}[S_y] \tag{11.398}$$

And since

$$\mathrm{Im}[S_y] = 0 \tag{11.399}$$

Hence, the condition of Eq. (11.398) becomes

$$\mathrm{Re}[Q_y] > 0 \tag{11.400}$$

With substitution of Eq. (11.390) into Eq. (11.400),

$$\mathrm{Re}\left[W_y^{\mathrm{Re}} + \frac{H^2 J}{C^2 + J^2}\right] > 0 \tag{11.401}$$

Substitution of the parameters C, J, and H of section 11.3.1 into ζ of Eq. (11.348), and the resulting positive values, indicate the required integration condition is met:

$$W_y^{\mathrm{Re}} + \frac{(\pi/D\lambda)^2(r_{\mathrm{a}}\pi/r_0\lambda)^2}{(\pi/\lambda)^2((1/D)+(1/r_0))^2 + (r_{\mathrm{a}}\pi/r_0\lambda)^2(r_{\mathrm{a}}\pi/r_0\lambda)^2} > 0 \tag{11.402}$$

The integration limits of the source plane in the x-direction along y_s are $-\infty$ and ∞. In this case, one assumes that the source intensity decreases rapidly with y_s, such that the two contributions to the integral outside of the beam waist, namely, between w_{0y} and ∞ and between $-w_{0x}$ and $-\infty$, are small compared to the integration across the beam waist at $z=0$, that is, between $-w_{0y}$ and w_{0y}. If the contributions outside of the beam waist are significant, but X-ray shielding outside the beam waist prevents that contribution, then one must integrate Eq. (11.389) across the beam waist at $z=0$, that is, between $-w_{0y}$ and w_{0y} only, so that

$$E_y^{\mathrm{coh}}(\lambda, y, f_{sy}) = E_{30y}^{\mathrm{Re}}(\lambda, f_{sy})E_{30y}^{\mathrm{Im}}(\lambda, f_{sy})E_{3y}^{\mathrm{Re}}(\lambda, y, f_{sy})E_{3y}^{\mathrm{Im}}(\lambda, y, f_{sy})$$
$$\times \int_{-w_{0y}}^{w_{0y}} \mathrm{d}y_s \exp(-Q_y y_s^2 - R_y y_s)\exp(-\mathrm{i}S_y y_s^2 - \mathrm{i}T_y y_s) \tag{11.403}$$

Integration of Eq. (11.403), yields the previous result for integration of y_s from $-\infty$ to ∞, with the addition of a multiplicative term $Z_{1y}(\lambda, f_{sy})$, which consists of the sum of two error functions evaluated at the $-w_{0y}$ and w_{0y} beam waist limits, where

$$E_y^{\mathrm{coh}}(\lambda, y, f_{sy}) = Z_{1y}(\lambda, f_{sy})E_{30y}^{\mathrm{Re}}(\lambda, f_{sy})E_{30y}^{\mathrm{Im}}(\lambda, f_{sy})E_{3y}^{\mathrm{Re}}(\lambda, y, f_{sy})E_{3y}^{\mathrm{Im}}(\lambda, y, f_{sy})$$
$$E_{40y}^{\mathrm{Re}}(\lambda, f_{sy})E_{40y}^{\mathrm{Im}}(\lambda, f_{sy}) \tag{11.404}$$

The corrective term $Z_{1y}(\lambda, f_{sy})$ multiplies the image amplitude, which results from integration in the source plane over an infinite source diameter in the y-direction $[x \in (-\infty, \infty)]$. This obtains the result for a finite source diameter in the y-direction $[y \in (-w_{0y}, w_{0y})]$, where

$$Z_{1y} = \frac{1}{2}\mathrm{Erf}\left[\frac{-R_y + 2w_{0y}(Q_y + \mathrm{i}S_y) - \mathrm{i}T_y}{2\sqrt{Q_y + \mathrm{i}S_y}}\right] + \frac{1}{2}\mathrm{Erf}\left[\frac{R_y + 2w_{0y}(Q_y + \mathrm{i}S_y) + \mathrm{i}T_y}{2\sqrt{Q_y + \mathrm{i}S_y}}\right] \tag{11.405}$$

Complex expansion and simplification in Eq. (11.405), yields

$$Z_{1y} = \frac{1}{2}\mathrm{Erf}\left[P_{1y}\exp\left(\mathrm{i}\frac{\theta_{1y}}{2}\right)\right] + \frac{1}{2}\mathrm{Erf}\left[P_{2x}\exp\left(\mathrm{i}\frac{\theta_{2y}}{2}\right)\right] \tag{11.406}$$

where

$$P_{1y} = \frac{1}{2}\left(\frac{(R_y^2+4w_{0y}^2(Q_y^2+S_y^2)-4w_{0y}(Q_yR_y+S_yT_y)+T_y^2)^2}{Q_y^2+S_y^2}\right)^{1/4} \tag{11.407}$$

$$\theta_{1y} = \tan^{-1}\left[\frac{-R_y^2S_y+4w_{0y}^2S_y(Q_y^2+S_y^2)+2Q_yR_yT_y-4w_{0y}T_y(Q_y^2+S_y^2)+S_yT_y^2}{4w_{0y}^2Q_y(Q_y^2+S_y^2)-4w_{0y}R_y(Q_y^2+S_y^2)+2R_yS_yT_y+Q(R_y^2-T_y^2)}\right] \tag{11.408}$$

$$P_{2y} = \frac{1}{2}\left(\frac{(R_y^2+4w_{0y}^2(Q_y^2+S_y^2)+4w_{0y}(Q_yR_y+S_yT_y)+T_y^2)^2}{Q_y^2+S_y^2}\right)^{1/4} \tag{11.409}$$

and

$$\theta_{2y} = \tan^{-1}\left[\frac{-R_y^2S_y+4w_{0y}^2S_y(Q_y^2+S_y^2)+2Q_yR_yT_y+4w_{0y}T_y(Q_y^2+S_y^2)+S_yT_y^2}{4w_{0y}^2Q_y(Q_y^2+S_y^2)+4w_{0y}R_y(Q_y^2+S_y^2)+2R_yS_yT_y+Q(R_y^2-T_y^2)}\right] \tag{11.410}$$

For a coherent beam of neutrons, the solution for the x-component of the coherent amplitude $E_x^{\mathrm{coh}}(\lambda,x,f_{sx})$ is obtained from the y-component of the coherent amplitude $E_y^{\mathrm{coh}}(\lambda,y,f_{sy})$ by setting the gravity terms to zero. For coherent X-rays, the gravity terms in the y-component of the coherent amplitude $E_y^{\mathrm{coh}}(\lambda,y,f_{sy})$ should be set to zero. And for coherent X-rays, the x-component, which is obtained from the y component, should have no gravity terms. And for X-rays or neutrons without gravity, the x- and y-components should be equal in an azimuthal symmetric optical system without an object (sample). With an object (sample), the X-rays and neutrons can be scattered differently in the x- and y-directions, so scatter by the object (sample) causes the x- and y-components $E_x^{\mathrm{coh}}(\lambda,x,f_{sx})$ and $E_y^{\mathrm{coh}}(\lambda,y,f_{sy})$ to differ.

The image plane intensity distribution is now calculated for the case, where one can integrate over the source plane from $-\infty$ to ∞, and thus ignore the integrand contributions outside of the source diameter (or beam waist for a Gaussian source). This simplifies integration without the burden of erf functions. The image plane intensity distribution $I_{xy}^{\mathrm{coh}}(\lambda,x,y)$ for a coherent source-illuminated object, which is imaged by a CRL, is the square of the magnitude of the image plane amplitude $E_{xy}^{\mathrm{coh}}(\lambda,x,y)$, where

$$I_{xy}^{\mathrm{coh}}(\lambda,x,y) = |E_{xy}^{\mathrm{coh}}(\lambda,x,y)|^2 \tag{11.411}$$

With substitution of Eqs. (11.381) into Eq. (11.411), one obtains

$$I_{xy}^{\mathrm{coh}}(\lambda,x,y,z)=|E_0(\lambda,z)|^2\left|\int_{-\infty}^{\infty}\int_{-\infty}^{\infty}\mathrm{d}f_{sx}\,\mathrm{d}f_{sy}E_x^{\mathrm{coh}}(\lambda,x,f_{sx})E_y^{\mathrm{coh}}(\lambda,y,f_{sy})T(f_{sx},f_{sy})\right|^2 \tag{11.412}$$

Recall that for the triangle inequality, for complex numbers A, B, the triangle inequality demands

$$|A+B|\leq|A|+|B| \tag{11.413}$$

However, if A, B are real and positive (or zero), then

$$|A+B|=|A|+|B| \tag{11.414}$$

One requires the integrand of Eq. (11.412) to be real and positive (or zero) at each spatial frequency coordinate pair (f_{sx},f_{sy}), so that the magnitude can be applied to the integrand term. That is, one requires the integrand to be positive and semidefinite:

$$E_x^{\mathrm{coh}}(\lambda,x,f_{sx})E_y^{\mathrm{coh}}(\lambda,y,f_{sy})T(f_{sx},f_{sy})\geq 0 \tag{11.415}$$

A positive, semidefinite integrand allows the image intensity of Eq. (11.412) to be written as:

$$I_{xy}^{\mathrm{coh}}(\lambda,x,y,z)=|E_0(\lambda,z)|^2\int_{-\infty}^{\infty}\int_{-\infty}^{\infty}\mathrm{d}f_{sx}\,\mathrm{d}f_{sy}|E_x^{\mathrm{coh}}(\lambda,x,f_{sx})E_y^{\mathrm{coh}}(\lambda,y,f_{sy})T(f_{sx},f_{sy})^2| \tag{11.416}$$

The property of modulus multiplication is applied to Eq. (11.416), where

$$|AB|=|A|\cdot|B| \tag{11.417}$$

The image intensity profile of an object illuminated by a coherent source (such as an X-ray laser) and imaged by a CRL is then

$$I_{xy}^{\mathrm{coh}}(\lambda,x,y,z)=|E_0(\lambda,z)|^2\int_{-\infty}^{\infty}\int_{-\infty}^{\infty}\mathrm{d}f_{sx}\,\mathrm{d}f_{sy}|E_x^{\mathrm{coh}}(\lambda,x,f_{sx})|^2|E_y^{\mathrm{coh}}(\lambda,y,f_{sy})|^2|T(f_{sx},f_{sy})|^2 \tag{11.418}$$

If the integrand is not positive and semidefinite, then this integrated result overestimates the image intensity distribution, and represents the upper-bound image intensity distribution.

The imaginary terms are arguments of exponentials; thus, their magnitude is 1, while the real terms are retained. With substitution into Eq. (11.418) of Eq. (11.395) for the y-component, and then substitution of the x-component version of Eq. (11.395), with the gravity terms set to zero in the parameters, one obtains image plane intensity distribution

$$I_{xy}^{\text{coh}}(\lambda, x, y, z) = |E_0(\lambda, z)|^2 \times \int_{-\infty}^{\infty} df_{sy}[E_{30y}^{\text{Re}}(\lambda, f_{sy})E_{3y}^{\text{Re}}(\lambda, y, f_{sy})E_{40y}^{\text{Re}}(\lambda, f_{sy})]^2 \times \int_{-\infty}^{\infty} df_{sx}T^2(f_{sx}, f_{sy})[E_{30x}^{\text{Re}}(\lambda, f_{sx})E_{3x}^{\text{Re}}(\lambda, x, f_{sx})E_{40x}^{\text{Re}}(\lambda, f_{sx})]^2 \tag{11.419}$$

All the intensity components depend on the X-ray or neutron wave length λ. The source intensity contribution I_{0xy} term is independent of transverse image positions x and y but can vary with propagation distance in the z-direction, where

$$I_{0xy} = |E_0(\lambda, z)|^2 \tag{11.420}$$

The intensity x-component I_x^{coh} depends on the image position x and the object (sample) spatial frequency f_{sx} in the x-direction, and the intensity y-component, and I_y^{coh} depends on the image position y and the object (sample) spatial frequency f_{sy} in the y-direction:

$$I_x^{\text{coh}} = |E_x^{\text{coh}}(\lambda, x, f_{sx})|^2 \tag{11.421}$$

and

$$I_y^{\text{coh}} = |E_y^{\text{coh}}(\lambda, y, f_{sy})|^2 \tag{11.422}$$

The expression for I_x^{coh} is obtained from I_y^{coh} by setting the gravity terms to zero. The intensity term I_{0xy} is independently given by

$$I_{0xy} = |E_0(\lambda, z)|^2 = (E_{0x}E_{0y})^2\left(\frac{w_{0x}w_{0y}\pi}{z\lambda}\right)^4 \exp(-N\mu d) \tag{11.423}$$

The y-directed intensity term I_y^{coh} is obtained from the squared magnitude of the amplitude E_y^{coh} of Eq. (11.389), where amplitude E_y^{coh} is obtained from integration along the y-direction y_s in the source plane over the infinite range $[-\infty, \infty]$, where

$$I_y^{\text{coh}} = |E_y^{\text{coh}}(\lambda, y, f_{sy})|^2 = \frac{\pi^3}{4\sqrt{(AJ-EC)^2+(EJ+AC)^2}}\frac{1}{\sqrt{Q_y^2+S_y^2}}\exp\left[-\frac{(\lambda-\lambda_{0y})^2}{\sigma_{\lambda y}^2}\right] \times \exp\left[\frac{2f_{sy}U_yJ\pi}{C^2+J^2}-\frac{2f_{sy}^2\pi^2J}{C^2+J^2}-\frac{U_y^2J}{2(C^2+J^2)}\right]\exp\left[-y\left(\frac{2f_{sy}KC\pi}{C^2+J^2}-\frac{KU_yC}{C^2+J^2}\right)\right] \times \exp\left[y^2\left(\frac{K^2J}{2(C^2+J^2)}-2N\right)\right]\exp\left[\frac{1}{2}\frac{Q_y(R_y^2-T_y^2)}{Q_y^2+S_y^2}\right]\exp\left(\frac{R_yS_yT_y}{Q_y^2+S_y^2}\right) \tag{11.424}$$

One assumes a Gaussian beam X-ray laser in which the laser minimum waist is at the source plane $z = 0$ with x- and y-waist dimension w_{0x} and w_{0y}. The real and imaginary, transverse x- and y-components of the X-ray laser beam source W_x and W_y are obtained from Eqs. (11.95) and (11.96), which are repeated below

$$W_x^{\mathrm{Re}} = w_{0x}^2 \quad \text{and} \quad W_x^{\mathrm{Im}} = \frac{z\lambda}{\pi} \tag{11.425}$$

and

$$W_y^{\mathrm{Re}} = w_{0y}^2 \quad \text{and} \quad W_y^{\mathrm{Im}} = \frac{z\lambda}{\pi} \tag{11.426}$$

Recall from Eq. (11.390–11.393), the composite parameters Q_y, R_y, S_y, and T_y are

$$Q_y = W_y^{\mathrm{Re}} + \frac{H^2 J}{C^2 + J^2} \tag{11.427}$$

$$R_y = \frac{2f_{sy}HJ\pi}{C^2 + J^2} - U_y + \frac{yHKC}{C^2 + J^2} \tag{11.428}$$

$$S_y = H - W_y^{\mathrm{Im}} - \frac{H^2 C}{C^2 + J^2} \tag{11.429}$$

and

$$T_y = V_y + \frac{HU_yC}{C^2 + J^2} - \frac{2f_{sy}HC\pi}{C^2 + J^2} + \frac{yHKJ}{C^2 + J^2} \tag{11.430}$$

The gravity parameters U_y and V_y of Eqs. (11.178) and 11.179 are set to zero:

$$U_y = \frac{2\pi}{\lambda}\left(\frac{o_g}{r_0} - \frac{D_g}{D}\right) = 0 \quad \text{and} \quad V_y = \frac{2\pi}{\lambda}\frac{D_g}{D} = 0 \tag{11.431}$$

Recall from section 11.3.1, the absorption aperture radius r_{a}, and E, A, J, K, and N are CRL parameters, where

$$r_{\mathrm{a}} = \sqrt{\frac{2R}{\mu N}} \quad E = \frac{\pi}{\lambda}\left(\frac{1}{r_0} + \frac{1}{r_i} - \frac{2N\delta}{R}\right) = 0 \quad A = \frac{\mu N}{2R} = \frac{1}{r_{\mathrm{a}}} \tag{11.432}$$

$$J = \left(\frac{r_{\mathrm{a}}\pi}{r_0\lambda}\right)^2 \quad K = \left(\frac{2}{r_0 r_i}\right)\left(\frac{r_{\mathrm{a}}\pi}{\lambda}\right)^2 \quad N = \left(\frac{r_{\mathrm{a}}\pi}{r_i\lambda}\right)^2 \tag{11.433}$$

From section 11.3.1, the parameters C and H for source-to-object distance D and object distance r_0 are

$$C = \frac{\pi}{\lambda}\left(\frac{1}{D} + \frac{1}{r_0}\right) \quad H = \frac{\pi}{D\lambda} \tag{11.434}$$

Consider the special case of uniform object transmission, where

$$T(s_x, s_y) = T_0 < 1 \tag{11.435}$$

The transmission in the spatial frequency domain is

$$T(f_{sx}, f_{sy}) = T_0\delta(f_{sx}, f_{sy}) \tag{11.436}$$

From Eq. (11.418), the simplified expression for the image intensity distribution is

$$I_{xy}^{\mathrm{coh}}(\lambda, x, y, z) = T_0^2|E_0(\lambda, z)|^2|E_x^{\mathrm{coh}}(\lambda, x, f_{sx} = 0)|^2|E_y^{\mathrm{coh}}(\lambda, y, f_{sy} = 0)|^2 \tag{11.437}$$

The image of the object is constructed from the sum of each of the spatial frequency components of the object. The zero spatial frequencies $f_{sx} = 0$ and $f_{sy} = 0$ of the object in the respective x- and y-directions indicate the existence of the object by a decrease of overall amplitude in the image plane. However, no detail or shape of the object is indicated by the zero spatial frequency amplitude components. The nonzero object spatial frequencies in the x- and y-directions provide the spatial detail or shape and size of the object. Finer object detail is provided by the higher spatial frequencies. Thus, one increasingly adds detail to the image as the successively higher partial frequency terms are added to the image, which is composed of the sum of the spatial frequency terms.

11.10 Biconcave, Parabolic CRL Image Intensity Distribution for Incoherent X-rays or Neutrons with Gravity—Coherent Amplitude Cross Terms Set to Zero

We now determine the image intensity distribution in the image plane for an incoherent source of X-rays or neutrons. The amplitude $E_{xy}(\lambda, x, y, z, x_s, y_s)$ at image point (x, y, z) for wavelength λ due to a *single* source point (x_s, y_s) is the PSF, or Green's function. In the case of an incoherent source, the squared magnitude of the PSF is obtained, which is the intensity from a single source point. For an incoherent source of X-rays or neutrons, the intensity at an image point is the sum of the intensity contributions from all the source points. The amplitude cross terms are not included, and thus the amplitude cross terms are set to zero for incoherent X-ray or neutron sources.

Recall that in the coherent source case (such as an X-ray laser or a distant, point X-ray source), the intensity at each image point was the squared magnitude of the sum of the amplitude contributions from all the source points in which the amplitude cross terms are included. The squared magnitude of the PSF is now integrated over all the source plane points (x_s, y_s) to obtain the image plane intensity $I_{xy}^{\mathrm{coh}}(\lambda, x, y, z)$ at image point (x, y, z_{im}), due to *all* the source points (x_s, y_s).

Most X-ray sources and all the thermal neutron sources are incoherent. An object illuminated by an incoherent source and imaged by a CRL produces image intensity profile $I_{xy}^{\mathrm{inc}}(\lambda, x, y)$, given by

$$I_{xy}^{\mathrm{inc}}(\lambda, x, y, z) = \int_{-\infty}^{\infty} \mathrm{d}y_s \int_{-\infty}^{\infty} \mathrm{d}x_s \left[\int_{-\infty}^{\infty} \int_{-\infty}^{\infty} |E_{xy}(\lambda, x, y, x_s, y_s, f_{sx}, f_{sy})|^2 \mathrm{d}f_{sx}\, \mathrm{d}f_{sy} \right] \tag{11.438}$$

With substitution of Eq. (11.212) into Eq. (11.438),

$$\begin{aligned} I_{xy}^{\mathrm{inc}}(\lambda, x, y, z) = |E_0(\lambda, z)|^2 \int_{-\infty}^{\infty} \mathrm{d}y_s \int_{-\infty}^{\infty} \mathrm{d}x_s \Bigg[\int_{-\infty}^{\infty} \int_{-\infty}^{\infty} |T(f_{sx}, f_{sy})|^2 \\ |E_{sx}(\lambda, x, x_s, f_{sx})|^2 |E_{sy}(\lambda, y, y_s, f_{sy})|^2 \mathrm{d}f_{sx}\, \mathrm{d}f_{sy} \Bigg] \end{aligned} \tag{11.439}$$

We can change the order of integration in Eq. (11.439), so that

$$I_{xy}^{\mathrm{inc}}(\lambda, x, y, z) = |E_0(\lambda, z)|^2 \int_{-\infty}^{\infty} \int_{-\infty}^{\infty} \mathrm{d}f_{sx}\, \mathrm{d}f_{sy} |T(f_{sx}, f_{sy})|^2 I_x^{\mathrm{inc}}(\lambda, x, z, f_{sx}) I_y^{\mathrm{inc}}(\lambda, y, z, f_{sy}) \tag{11.440}$$

where

$$|E_0(\lambda, z)|^2 = (E_{0x} E_{0y})^2 \left(\frac{w_{0x} w_{0y} \pi}{z\lambda} \right)^4 \exp(-N\mu d) \tag{11.441}$$

$$I_x^{\mathrm{inc}}(\lambda, x, f_{sx}) = \int_{-\infty}^{\infty} \mathrm{d}x_s |E_{sx}(\lambda, x, x_s, f_{sx})|^2 \tag{11.442}$$

and

$$I_y^{\mathrm{inc}}(\lambda, y, f_{sy}) = \int_{-\infty}^{\infty} \mathrm{d}y_s |E_{sy}(\lambda, y, y_s, f_{sy})|^2 \tag{11.443}$$

First, we solve for $I_y^{\mathrm{inc}}(\lambda, y, f_{sy})$. Substitution into Eq. (11.443) of Eq. (11.219) for $E_{sy}(\lambda, y, y_s, f_{sy})$ yields

$$I_y^{\mathrm{inc}}(\lambda, y, f_{sy}) = [E_{30y}^{\mathrm{Re}}(\lambda, f_{sy}) E_{3y}^{\mathrm{Re}}(\lambda, y, f_{sy})]^2 \int_{-\infty}^{\infty} \mathrm{d}y_s [E_{3ys}^{\mathrm{Re}}(\lambda, y_s, f_{sy}) E_{3yys}^{\mathrm{Re}}(\lambda, y, y_s)]^2 \tag{11.444}$$

Recall Eqs. (11.233), (11.235), (11.237), and (11.239), where

$$E_{30y}^{\mathrm{Re}}(\lambda,f_{sy}) = \frac{\pi}{\left[(AJ-EC)^2+(EJ+AC)^2\right]^{1/4}} \times \exp\left[-\frac{(\lambda-\lambda_{0y})^2}{2\sigma_{\lambda y}^2}\right] \exp\left[\frac{f_{sy}U_yJ\pi}{C^2+J^2} - \frac{f_{sy}^2\pi^2J}{C^2+J^2} - \frac{U_y^2J}{4(C^2+J^2)}\right] \tag{11.445}$$

$$E_{3y}^{\mathrm{Re}}(\lambda,y,f_{sy}) = \exp\left[-y\left(\frac{f_{sy}KC\pi}{C^2+J^2} - \frac{KU_yC}{2(C^2+J^2)}\right)\right] \exp\left[y^2\left(\frac{K^2J}{4(C^2+J^2)} - N\right)\right] \tag{11.446}$$

$$E_{3ys}^{\mathrm{Re}}(\lambda,y_s,f_{sy}) = \exp(-W_y^{\mathrm{Re}}y_s^2)\exp\left[y_s\left(\frac{HJ}{C^2+J^2}\right)(U_y-2f_{sy}\pi)\right]\exp\left[-y_s^2\left(\frac{H^2J}{C^2+J^2}\right)\right] \tag{11.447}$$

and

$$E_{3yys}^{\mathrm{Re}}(\lambda,y,y_s) = \exp\left[-y_sy\left(\frac{HKC}{C^2+J^2}\right)\right] \tag{11.448}$$

The major difference between the case of the coherent and incoherent source is that the imaginary terms are not involved in the integration over the source plane, since their squared magnitude is 1, and the minor difference is that the real terms in the integrand are squared before integration—thus the integrand semidefinite.

To simplify the integration over y_s, one simplifies the form of the integral of Eq. (11.444) to

$$I_y^{\mathrm{inc}}(\lambda,y,f_{sy}) = [E_{30y}^{\mathrm{Re}}(\lambda,f_{sy})E_{3y}^{\mathrm{Re}}(\lambda,y,f_{sy})]^2 \int_{-\infty}^{\infty} dy_s \exp(-2Q_yy_s^2 - 2R_yy_s) \tag{11.449}$$

This result takes the same form as the coherent case, except that $S_y = 0$ and $T_y = 0$, and that Q_y and R_y are multiplied by 2, due to squaring the exponential amplitudes. Hence, the incoherent, y-directed amplitude can be obtained from the coherent, y-directed amplitude with $S_y = 0$ and $T_y = 0$, and Q_y and R_y are replaced by $2Q_y$ and $2R_y$. Integration over y_s in Eq. (11.449), thus yields

$$I_y^{\mathrm{inc}}(\lambda,y,f_{sy}) = [E_{30y}^{\mathrm{Re}}(\lambda,f_{sy})E_{3y}^{\mathrm{Re}}(\lambda,y,f_{sy})]^2\sqrt{\frac{\pi}{2Q_y}}\exp\left[\frac{R_y^2}{2Q_y}\right] \tag{11.450}$$

Recalling Eqs. (11.427) and (11.428), the parameters Q_y and R_y are given by

$$Q_y = W_y^{\mathrm{Re}} + \frac{H^2 J}{C^2 + J^2} \tag{11.451}$$

$$R_y = \frac{2f_{sy} HJ\pi}{C^2 + J^2} - U_y + \frac{yHKC}{C^2 + J^2} \tag{11.452}$$

The result of integration of Eq. (11.449), the incoherent component of the image intensity $I_y^{\mathrm{inc}}(\lambda, y, f_{sy})$ is valid, if the real part of the exponential argument in the integrand of Eq. (11.449) is negative, which is true, because of positive $R_y > 0$ of Eq. (11.428) and positive $Q_y > 0$ of Eq. (11.427), via positive $W_y^{\mathrm{Re}} > 0$ of (11.426), where

$$W_y^{\mathrm{Re}} > 0 \quad H > 0 \quad C > 0 \quad J > 0 \quad K > 0 \tag{11.453}$$

Substitution of the parameters of Eqs. (11.451–11.453) into Eq. (11.450), yields

$$I_y^{\mathrm{inc}}(\lambda, y, f_{sy}) = \frac{\pi^2}{4\sqrt{2}} \sqrt{\frac{\pi}{(AJ - EC)^2 + (EJ + AC)^2}} \frac{1}{\sqrt{Q_y}} \exp\left(\frac{R_y^2}{2Q_y}\right)$$
$$\times \exp\left[-\frac{1}{\sigma_{\lambda y}^2}(\lambda - \lambda_{0y})^2\right] \exp\left[y^2\left(\frac{K^2 J}{2(C^2 + J^2)} - 2N\right)\right]$$
$$\times \exp\left[-y\left(\frac{KC(2f_{sy}\pi - U_y)}{C^2 + J^2}\right)\right] \exp\left[\frac{J\left(2f_{sy}U_y\pi - 2f_{sy}^2\pi^2 - \frac{1}{2}U_y^2\right)}{C^2 + J^2}\right] \tag{11.454}$$

The integration limits of the source plane in the y-direction along y_s are $-\infty$ and ∞. In this case, one assumes that the source intensity decreases rapidly with y_s, such that the two contributions to the integral outside of the beam waist, (namely, between w_{0y} and ∞ and between $-w_{0y}$ and $-\infty$) are small compared to the integration across the beam waist at $z = 0$, that is, between $-w_{0y}$ and w_{0y}. If the contributions outside of the beam waist are significant, but X-ray shielding outside the beam waist prevents that contribution, then one must integrate across the beam waist at $z = 0$, that is, between $-w_{0y}$ and w_{0y} only, so that

$$I_y^{\mathrm{inc}}(\lambda, y, f_{sy}) = [E_{30y}^{\mathrm{Re}}(\lambda, f_{sy}) E_{3y}^{\mathrm{Re}}(\lambda, y, f_{sy})]^2 \int_{-w_{0y}}^{w_{0y}} dy_s \exp(-2Q_y y_s^2 - 2R_y y_s) \tag{11.455}$$

Integration of Eq. (11.455), yields the previous result of Eq. (11.454) for integration of y_s from $-\infty$ to ∞, which is multiplied by $Z_{1y}(\lambda, f_{sy})$, which consists of the sum of two error functions evaluated at the $-w_{0y}$ and w_{0y} beam waist limits, where

$$I_y^{\text{inc}}(\lambda, y, f_{sy}) = Z_{1y}(\lambda, f_{sy})[E_{30y}^{\text{Re}}(\lambda, f_{sy})E_{3y}^{\text{Re}}(\lambda, y, f_{sy})]^2 \sqrt{\frac{\pi}{8Q_y}} \exp\left[\frac{R_y^2}{2Q_y}\right] \qquad (11.456)$$

and

$$Z_{1y}(\lambda, f_{sy}) = \frac{1}{2}\text{Erf}\left[\frac{2w_{0y}Q_y - R_y}{\sqrt{2Q_y}}\right] + \frac{1}{2}\text{Erf}\left[\frac{2w_{0y}Q_y + R_y}{\sqrt{2Q_y}}\right] \qquad (11.457)$$

The expression for $I_x^{\text{inc}}(\lambda, x, f_{sx})$ is obtained from $I_y^{\text{inc}}(\lambda, y, f_{sy})$, where the gravity terms are set to zero.

Next, one substitutes Eq. (11.454) for the y-component intensity into Eq. (11.440), and then substitutes Eq. (11.454) into Eq. (11.440) as the x-component intensity, where the gravity terms are set to zero, and y is replaced by x. The image intensity distribution $I_{xy}^{\text{inc}}(\lambda, x, y)$ of an object, which is illuminated by an incoherent X-ray or neutron source and imaged by a CRL or aperture, such as a pinhole or pipe, becomes

$$I_{xy}^{\text{inc}}(\lambda, x, y, z) = \frac{\pi^5}{32}(E_{0x}E_{0y})^2 \left(\frac{w_{0x}w_{0y}\pi}{z\lambda}\right)^4 \exp(-N\mu d)$$

$$\times \exp\left[-\frac{(\lambda-\lambda_{0x})^2}{\sigma_{\lambda x}^2}\right] \exp\left[-\frac{(\lambda-\lambda_{0y})^2}{\sigma_{\lambda y}^2}\right] \int_{-\infty}^{\infty}\int_{-\infty}^{\infty} \frac{|(f_{sx}, f_{sy})|^2}{[(AJ-EC)^2 + (EJ+AC)^2]\sqrt{Q_x Q_y}}$$

$$\times \exp\left[\frac{J(2f_{sy}U_y\pi - 2f_{sy}^2\pi^2 - \frac{1}{2}U_y^2)}{C^2+J^2}\right] \exp\left[-f_{sx}^2\left(\frac{2\pi^2 J}{C^2+J^2}\right)\right] \exp\left[-y\left(\frac{U_y}{C^2+J^2}\right)\right]$$

$$\times \exp\left[(x^2+y^2)\left(\frac{K^2 J}{2(C^2+J^2)} - 2N\right)\right] \exp\left[-(f_{sx}x + f_{sy}y)\left(\frac{2KC\pi}{C^2+J^2}\right)\right]$$

$$\times \exp\left[\frac{1}{2Q_y}\left(\frac{2f_{sx}HJ\pi}{C^2+J^2} + \frac{xHKC}{C^2+J^2}\right)^2\right] \exp\left[\frac{R_y^2}{2Q_y}\right] df_{sx}\, df_{sy}$$

(11.458)

The composite CRL parameters for an incoherent source of neutrons or X-rays are obtained from the CRL parameters in the previous section 11.9, where

$$Q_x = W_x^{\text{Re}} + \frac{H^2 J}{C^2 + J^2} \tag{11.459}$$

$$Q_y = W_y^{\text{Re}} + \frac{H^2 J}{C^2 + J^2} \tag{11.460}$$

$$R_y = \frac{2f_{sy} HJ\pi}{C^2 + J^2} - U_y + \frac{yHKC}{C^2 + J^2} \tag{11.461}$$

$$S_y = H - W_y^{\text{Im}} - \frac{H^2 C}{C^2 + J^2} \tag{11.462}$$

and

$$T_y = V_y + \frac{HU_y C}{C^2 + J^2} - \frac{2f_{sy} HC\pi}{C^2 + J^2} + \frac{yHKJ}{C^2 + J^2} \tag{11.463}$$

From section 11.3.1, the gravity parameters U_y and V_y are

$$U_y = \frac{2\pi}{\lambda}\left(\frac{o_g}{r_0} - \frac{D_g}{D}\right) \quad \text{and} \quad V_y = \frac{2\pi}{\lambda}\frac{D_g}{D} \tag{11.464}$$

From Eqs. (11.425−11.426), the transverse x- and y-components W_x and W_y of an incoherent source of X-rays or neutrons is real, with zero imaginary components, where

$$W_x^{\text{Re}} = \frac{1}{2\sigma_x^2} \quad W_x^{\text{Im}} = W_y^{\text{Im}} = 0 \quad W_y^{\text{Re}} = \frac{1}{2\sigma_y^2} \tag{11.465}$$

The absorption aperture radius r_{a}, and E, A, J, K, and N are CRL parameters from section 11.3.1, where

$$E = \frac{\pi}{\lambda}\left(\frac{1}{r_0} + \frac{1}{r_i} - \frac{2N\delta}{R}\right) = 0 \quad r_{\text{a}} = \sqrt{\frac{2R}{\mu N}} \quad A = \frac{\mu N}{2R} = \frac{1}{r_{\text{a}}} \tag{11.466}$$

$$J = \left(\frac{r_{\text{a}}\pi}{r_0\lambda}\right)^2 \quad K = \left(\frac{2}{r_0 r_i}\right)\left(\frac{r_{\text{a}}\pi}{\lambda}\right)^2 \quad N = \left(\frac{r_{\text{a}}\pi}{r_i\lambda}\right)^2 \tag{11.467}$$

From section 11.3.1, the parameters C, H, and F for source-to-object distance D and object distance r_0, as well as the parameter G for image distance r_i, are

$$C = \frac{\pi}{\lambda}\left(\frac{1}{D} + \frac{1}{r_0}\right) \quad H = \frac{\pi}{D\lambda} \quad F = \frac{2\pi}{r_0\lambda} \quad G = \frac{2\pi}{r_i} \tag{11.468}$$

The x-component terms are identical to the y-component terms, except that the gravity is set to zero in the x-component terms. The gravity terms are set to zero in the case of X-rays.

The X-ray or neutron image intensity distribution $I_{xy}^{\mathrm{inc}}(\lambda, x, y)$ is then constructed from the integral or summation of the contributions of the intensity terms for each object spatial frequency in the x- and y-directions. The zero or DC spatial frequency yields the presence of the object by a decrease in uniform change in intensity at the image plane. Intensity terms for the increasing object spatial frequencies in the x- and y-directions sum to add detail to the image. The largest spatial frequency included in the integral provides the finest detail of the object.

For the special case of uniform object transmission T_0 of X-rays or neutrons,

$$T(s_x, s_y) = T_0 < 1 \tag{11.469}$$

Recalling section 11.3.3, the transmission $T(f_{sx}, f_{sy})$ in spatial frequency domain (f_{sx}, f_{sy}) is

$$T(f_{sx}, f_{sy}) = T_0\delta(f_{sx}, f_{sy}) \tag{11.470}$$

The resulting image intensity distribution expression is simplified to

$$I_{xy}^{\mathrm{inc}}(\lambda, x, y) = |E_0(\lambda, z)|^2 T_0^2 I_x^{\mathrm{inc}}(\lambda, x, z, f_{sx} = 0) I_y^{\mathrm{inc}}(\lambda, y, z, f_{sy} = 0) \tag{11.471}$$

Thus we have derived the image intensity distribution $I_{xy}^{\mathrm{inc}}(\lambda, x, y, z)$ for an incoherent source-illuminated object with uniform transmission for the case of a CRL or aperture, such as a pinhole or pipe.

The X-ray or neutron image is produced by a CRL or pinhole aperture. In the case of a CRL, the parameter $E = 0$ because of the lens-maker formula, where

$$A = \frac{\mu N}{2R} = \frac{1}{r_a}, \quad \text{CRL} \tag{11.472}$$

$$E = E_{\mathrm{CRL}} = \frac{\pi}{\lambda}\left(\frac{1}{r_0} + \frac{1}{r_i} - \frac{2N\delta}{R}\right) = 0, \qquad \text{CRL} \tag{11.473}$$

and

$$\frac{1}{f} = \frac{2N\delta}{R} = \frac{1}{r_0} + \frac{1}{r_i} \quad \text{CRL} \tag{11.474}$$

In the case of a pinhole, one sets $\delta = 0$ and $\mu = 0$ so that parameter A is set to zero:

$$A = 0 \quad \text{pinhole} \tag{11.475}$$

$$E = E_{\text{pin}} = \frac{k}{2r_0} + \frac{k}{2r_i} = \frac{\pi}{\lambda}\left(\frac{1}{r_0} + \frac{1}{r_i}\right) \neq 0 \quad \text{pinhole} \tag{11.476}$$

With the uniform object transmission condition of Eq. (11.470), substitution of the parameters into Eq. (11.471), then yields the image intensity distribution $I_{xy}^{\text{inc}}(\lambda, x, y, z)$, which is applicable to a lens or a pinhole, is

$$I_{xy}^{\text{inc}}(\lambda,x,y,z)$$

$$= \left(\frac{T_0^2 \frac{\pi^5}{32}(E_{0x}E_{0y})^2 \left(\frac{w_{0x}w_{0y}\pi}{z\lambda}\right)^4 \exp(-N\mu d)\exp\left[-\frac{(\lambda-\lambda_{0x})^2}{\sigma_{\lambda x}^2}\right]\exp\left[-\frac{(\lambda-\lambda_{0y})^2}{\sigma_{\lambda y}^2}\right]}{\left[(AJ-EC)^2+(EJ+AC)^2\right]\sqrt{W_x^{\text{Re}}W_y^{\text{Re}}+\frac{H^2J}{C^2+J^2}(W_x^{\text{Re}}+W_y^{\text{Re}})+\left(\frac{H^2J}{C^2+J^2}\right)^2}} \right)$$

$$\times\exp\left[\frac{J(2f_{sy}U_y\pi-2f_{sy}^2\pi^2-\frac{1}{2}U_y^2)}{C^2+J^2}\right]\exp\left[(x^2+y^2)\left(\frac{K^2J}{2(C^2+J^2)}-2N\right)\right]$$

$$\times\exp\left[-y\left(\frac{U_y}{C^2+J^2}\right)\right]\exp\left[\frac{\frac{1}{2}\left(\frac{xHKC}{C^2+J^2}\right)^2}{W_x^{\text{Re}}+\frac{H^2J}{C^2+J^2}}\right]\exp\left[\frac{\frac{1}{2}\left(-U_y+\frac{yHKC}{C^2+J^2}\right)^2}{W_y^{\text{Re}}+\frac{H^2J}{C^2+J^2}}\right] \tag{11.477}$$

11.11 Without CRL—Phase Contrast Imaging for Incoherent X-rays or Neutrons with Gravity

A CRL is not needed for phase contrast imaging of incoherent X-rays or neutrons. Phase contrast imaging can be performed with a tiny, incoherent point X-ray or neutron source, a specimen placed at a very large distance compared to the source diameter, and a high spatial resolution imaging detector placed at a moderate distance from the specimen. Phase contrast imaging records information from the bending or refraction of X-rays or neutrons, as they pass through materials, such as soft biological and medical specimens, multilevel circuit boards, or ceramics and composite materials. X-ray or neutron refraction occurs because the denser material changes the speed of the X-rays or neutrons more than the less dense materials, producing what is known as a phase shift. The goal of the phase contrast microscope is to convert this phase shift of the X-ray or neutron illuminated specimen into a sharp and magnified image, which reveals subtle variations in its structures. For example the phase contrast microscope can be applied to a biopsy of soft tissue, which is examined for signs of breast cancer, or applied to quality control inspection of "through-wafer vias" and "holes and trenches" in semiconductor wafers, or applied to a ball and grid arrays in circuit boards.

Thin-specimen structures provide large changes in the refractive index, with the surrounding material similar to a high-curvature, strongly refracting lens. The refractive index $n = 1 - \delta$ of the material is very close to 1. Thus, the index of refraction of the material is not what strongly refracts the incident X-rays or neutrons. Rather, the change in shape and density of the object (specimen), creates large gradients in the object refractive index. That is, the change or gradient of the refractive index of the object (specimen) along a direction, which is perpendicular to the incident X-rays or neutrons, creates the relatively large refraction angles of the coherent scattered X-rays or neutrons.

The refractive, transverse gradient features of the object (specimen) create a range of refraction angles $\Delta\alpha$ of scattered X-rays or neutrons, which diverge at small angles with respect to the axial z-direction of incident X-rays or neutrons. These rays interfere with unscattered rays at the detector. This creates an interference pattern or image of the object (specimen) edges and structures that, which have large, transverse refractive gradients. The detector lateral resolution must be sufficiently small, and the specimen-detector distance sufficiently large, that the small angular difference between two rays scattered from adjacent specimen features lead to the rays intercepting the detector at different pixels. To image the edge features, the incident X-ray or neutron wave must have transverse spatial coherence over a portion of the wave front surface, which is larger than the specimen features.

The phase shift for the incident X-rays, which propagate in the z-direction, and pass though specimen thickness $\Delta z_{x,y}$ at lateral coordinates (x, y), is given by

$$\phi_{x,y}(W) = 2\pi \frac{\Delta z_{x,y}}{\lambda} \delta_{x,y}(W) \tag{11.478}$$

The change in refraction angle $\Delta\alpha$ with the gradient of the phase shift $\phi_{x,y}$, that is, the gradient $\nabla_{x,y}\phi$ in the xy-plane, as a function of transverse position (x, y) in the specimen, the change in the refraction angle per the thickness of specimen Δz is given by

$$\frac{\Delta\alpha_{x,y}}{\Delta z} = \frac{\lambda}{2\pi} |\nabla_{x,y}\phi| \tag{11.479}$$

The gradient of the phase shift $\phi_{x,y}$ is the decrement $\delta(x, y)$ divided by the wavelength λ:

$$\nabla_{x,y}\phi = \frac{2\pi}{\lambda}\delta(x, y) \tag{11.480}$$

The change in refraction angle $\Delta\alpha$, equals the gradient of the decrement $\delta(x, y)$ in the xy-plane, which is perpendicular to the z-directed rays of incident X-rays or neutrons, where

$$\frac{\Delta\alpha_{x,y}(E)}{\Delta z} = |\nabla_{x,y}\delta_{x,y}(W)| \tag{11.481}$$

The change in refraction angle $\Delta\alpha$ is measured with respect to X-ray incident direction z. Both the change in refraction angle $\Delta\alpha_{x,y}(W)$ and the decrement $\delta_{x,y}(W)$, depend on the X-ray or neutron energy W. In addition to refraction, there is reflection of the incident X-rays or neutrons by the sample. Complete reflection occurs for X-rays or neutrons at grazing incident angles θ_i, which are measured relative to the surface tangent, and are less than the critical angle $\theta_i < \theta_c$, where

$$\theta_c = \sqrt{2\delta} \quad \theta_i < \theta_c \text{ complete reflection} \tag{11.482}$$

Thus, larger grazing incidence reflection, and large angles of refraction of incident X-rays and neutrons, can occur at edges or discontinuities in specimen density (edges) and composition, thus leading to detail of structural profiles within the specimen. The edges in the object produce shadows in the image plane, and on either side of the shadow of an edge is a brighter region formed by the edge-refracted or -reflected rays.

Assume an incoherent X-ray or neutron source. If the source diameter D_{sou} is sufficiently small, then at a relatively large distance D in the object plane on a small circular region or disk of diameter d_{coh}, the wave front points have fixed spatial phase relationships, which is transverse spatial coherence. In the object plane over diameter d_{coh} of the wave front, there is a high degree of spatial coherence. The diameter d_{coh} in the object plane is termed the *transverse* or *spatial coherence length,* and as a result of the van Cittert–Zernike theorem, is usually expressed as

$$d_{coh} = \frac{1}{2\pi}\frac{\lambda D}{D_{sou}} \tag{11.483}$$

This coherence length expression for d_{coh} can also be obtained as follows. The top perimeter point of the source disk of X-rays or neutrons is on the z-axis. The bottom perimeter point of the disk is vertically below the z-axis by distance D_{sou}. The top perimeter source point is equidistant from a pair of upper and lower slits centered about the z-axis in the object plane at distance D. The upper and lower slits in the object plane have vertical separation distance d_{coh}, and each slit is at transverse distance $d_{coh}/2$ from the center z-axis. The bottom perimeter point of the source disk is not equidistant to the upper and lower slits on the distant object plane. If this difference in the distance is one wavelength λ or an integer multiple $n\lambda$ of the X-ray or neutron wavelength for integer n, then constructive interference occurs at the upper slit in the object plane.

Consider the single-wavelength difference λ with $n = 1$, which creates a central region of constructive interference at the upper slit in the object plane. Rays emitted from the source disk, propagate in the positive z-direction to the distant transverse xy-plane, on which lies the pair of upper and lower slits vertically separated by d_{coh}. The polar inclination angle θ of a ray, which is from the source disk, is measured relative to the positive z-axis. The same inclination angle θ is also the angle between the distant object plane and the wave front, which is just touching the lower slit, but is still one wavelength from making contact with the upper slit. Angle θ is assumed to be small because the diameter D_{sou} of the X-ray or neutron source disk is assumed to be much larger than the wavelength λ of the X-rays or neutrons. The dual representation of angle θ is expressed by

$$\theta \cong \tan\,\theta = \frac{D_{sou}}{D} = \frac{\lambda}{d_{coh}} \quad D_{sou} \gg \lambda \tag{11.484}$$

Rearrangement of Eq. (11.484), yields the spatial coherence diameter d_{coh}, where

$$d_{coh} = \frac{D}{D_{sou}}\lambda \quad D_{sou} \gg \lambda \tag{11.485}$$

If the wave front just touches the lower slit in the object plane, the upper slit should actually be less than one-half wavelength $\lambda/2$ from the wave front of the incident plane wave from the source disk. This allows only constructive interference to occur for all source disk emitted rays, which fall onto the upper and lower slits, and the points between the two slits. Transverse spatial coherence requires almost complete constructive interference of the waves, which are emitted from all points of the source disk, and fall on all the points within the circular region of diameter d_{coh} in the distant object plane, whose diameter is d_{coh}. Almost complete constructive interference or transverse spatial coherence occurs, if one divides Eq. (11.485) by 2π. The spatial coherence diameter expression then becomes the previously asserted spatial coherence length d_{coh}, where

$$d_{coh} = \frac{1}{2\pi}\frac{D}{D_{sou}}\lambda \quad D_{sou} \gg \lambda \tag{11.486}$$

A single point emitter of X-rays or neutrons produces a spherical wave front, where all points of the surface have the same phase, and thus a single spherical wave emitted from a point source has transverse spatial coherence over its entire spherical surface. However, the points on the spherical wave surface may not necessarily have a fixed temporal or spatial relationship with points on the other concentric spherical wave surfaces. If the single point source emits a sequence of spherical waves over time interval Δt, where the points on each wave front do not have a fixed spatial or temporal phase relationship with the points on the other concentric wave fronts, then the point source does not have temporal or longitudinal coherence. If a set of wave fronts are emitted in time interval Δt, and the points of each wave front have fixed phase relationships with the points of the other concentric wave fronts, then the point source has temporal coherence for time interval Δt.

Temporal coherence requires narrow frequency bandwidth Δw of X-ray or neutron emission from the point source, or an extended source, where the center frequency w of the bandwidth is much larger than the bandwidth Δw, so that

$$\frac{\Delta w}{w} \ll 1 \tag{11.487}$$

The temporal coherence time τ_{coh} is defined as

$$\tau_{\text{coh}} = \frac{1}{\Delta w} \tag{11.488}$$

The temporal coherence length or longitudinal coherence length l_{coh} for X-rays is obtained by multiplication of the temporal coherence time τ_{coh} by the speed of light in vacuum c, where

$$l_{\text{coh}} = c\tau_{\text{coh}} = \frac{c}{\Delta w} \tag{11.489}$$

Now consider the temporal coherence length for a point or extended source of neutrons. The neutron group velocity v, with h as Planck's constant and m as the neutron rest mass, is

$$v = \frac{h}{m\lambda} = \frac{w}{2\pi}\lambda \tag{11.490}$$

The temporal coherence length or longitudinal coherence length l_{coh} for neutrons is obtained by multiplying the temporal coherence time τ_{coh} by the neutron group velocity v, where

$$l_{\text{coh}} = v\tau_{\text{coh}} = \frac{v}{\Delta w} = \frac{\lambda}{2\pi}\frac{w}{\Delta w} \tag{11.491}$$

In monochromatic neutron sources, $w/\Delta w$ ranges from about 10 to 100, and in Boltzmann distributed reactor neutron sources, $w/\Delta w \leq 1$, so the temporal or longitudinal coherence length of present-day thermal or cold neutron sources is less than 1 Å.

By comparison, monochromatic, nonlaser light sources have temporal coherence lengths of about 0.5 m, and visible light lasers, in spite of mirror vibrations and thermal expansion, have temporal coherence lengths exceeding 10^5 m. Each Gaussian beam wave front of a visible light or X-ray laser has transverse spatial coherence and temporal coherence, where a sequence of wave fronts that are emitted over a relatively long time interval Δt have fixed phase relationships. In phase contrast imaging, with a continuous or pulsed incoherent X-ray or neutron point-like source, one can achieve transverse spatial coherence over a small portion of a wave front; however, there is no temporal coherence.

In the case of phase contrast microscopy, no lens is required, and one requires only a small source, an object, and a detector. A tiny X-ray or neutron source is required, whose diameter $D_{\rm sou}$ is much less than the required distance D to an object of interest, where $D \gg D_{\rm sou}$. The small-source $D_{\rm sou}$ and large-distance D allow a significant part of the object to be intercepted by a coherent wavelet. That is, each point on a small region of an object is intercepted by a ray whose phase is the same as the neighboring rays within the region. Effectively, each small region of the object experiences a wavelet of one phase. The wavelet is refracted and phase-shifted due to the variation of the optical thickness of the target material across the width of the wavelet.

The transverse coherence length $d_{\rm coh}$ is proportional to the wavelength λ of the wavelet, and proportional to the ratio of the source diameter $D_{\rm sou}$, divided by the axial distance D from the source plane to the object plane, where

$$d_{\rm coh} = \frac{1}{2\pi}\frac{D}{D_{\rm sou}}\lambda \quad D_{\rm sou} \gg \lambda \tag{11.492}$$

The reciprocal of $D_{\rm sou}/D$ is the X-ray or neutron beam collimation ratio $D/D_{\rm sou}$, which is measured at a point on the object (specimen), which acts as a vertex of an angle bounded by the source diameter $D_{\rm sou}$ at source–object distance D. The beam collimation ratio $D/D_{\rm sou}$ is commonly referred to as the "L/D" ratio, expressed by source-to-object distance L and source diameter D. For thermal neutron radiographic imaging at reactor neutron sources, for example, the beam collimation L/D ratio can range from 100 to 1000, or even more.

With wavelengths of 1 Å, phase contrast imaging requires very large L/D ratios above 6500, so as to have transverse coherence lengths on the order of 0.1 mm or more. The spatially coherent X-rays or neutrons within the transverse spatial coherence length $d_{\rm coh}$ are refracted by the variation of the refractive index or thickness of the object, especially at the edges. These refracted rays intercept the image plane on either side of the projection of the object edge onto the image plane. This leaves a shadow or intensity variation in the image plane that corresponds to the object edge or discontinuity.

In phase contrast microscopy, no lens is needed. The source-to-object plane distance is D, and the pinhole source of X-rays or neutrons has diameter $D_{\rm sou}$, which is centered in the source plane at $(x_s, y_s) = (0, 0)$ at $z = 0$. The object or specimen is centered at $(s_x, s_y) = (0, 0)$ in the object plane at $z = D$. The CRL plane at $z = D + r_0$ is empty space through which pass the source rays, which are transmitted or scattered by the object plane specimen to the image or detector plane at $z = D + r_0 + r_i$.

We now recall the general CRL parameters in sections 11.3.1, which can also be applied to the case of a pinhole or no CRL (phase contrast imaging without a lens).

No lens in the CRL plane is equivalent to an aperture of infinite diameter, where $\delta = 0$ and $\mu = 0$, so that parameter $A = 0$ and parameter $E = E_{no}$, or

$$E_{no} = \frac{k}{2r_0} + \frac{k}{2r_i} = \frac{\pi}{\lambda}\left(\frac{1}{r_0} + \frac{1}{r_i}\right) \tag{11.493}$$

In the no-lens case, parameter J becomes

$$J = J_{no} = \frac{-\mathrm{i}F^2}{4E_{no}} = \frac{-\mathrm{i}}{r_0^2(\lambda/\pi)((1/r_0) + (1/r_i))} \tag{11.494}$$

The parameter K for the no-lens case becomes

$$K = K_{no} = 2J_{no}\left(\frac{r_0}{r_i}\right) = \frac{-\mathrm{i}FG}{2E_{no}} = \frac{-4\mathrm{i}}{(\lambda/\pi)(r_0 + r_i)} \tag{11.495}$$

The parameter N for the no-lens case becomes

$$N = N_{no} = J_{no}\left(\frac{r_0}{r_i}\right)^2 = \frac{-\mathrm{i}G^2}{4E_{no}} = \frac{\mathrm{i}}{r_i^2(\lambda/\pi)((1/r_0) + (1/r_i))} \tag{11.496}$$

The parameter Q_y for the no-lens case becomes

$$Q_y = Q_{yno} = W_y^{Re} + \frac{H^2 J_{no}}{C^2 + J_{no}^2} \tag{11.497}$$

The parameter R_y for the no-lens case becomes

$$R_y = R_{yno} = \frac{2f_{sy}HJ_{no}\pi}{C^2 + J_{no}^2} + \frac{yHK_{no}C}{C^2 + J_{no}^2} \tag{11.498}$$

The correction factor for a finite source aperture in the y-direction for the no-lens case Z_{1y} becomes

$$Z_{1y} = Z_{1y\,no}(\lambda, f_{sy}) = \frac{1}{2}\mathrm{Erf}\left[\frac{2w_{0y}Q_{y\,no} - R_{y\,no}}{\sqrt{2Q_{y\,no}}}\right] + \frac{1}{2}\mathrm{Erf}\left[\frac{2w_{0y}Q_{y\,no} + R_{y\,no}}{\sqrt{2Q_{y\,no}}}\right] \tag{11.499}$$

The correction factor for a finite source aperture in the x-direction for the no-lens case Z_{1x} is simply the y-term $Z_{1y} = Z_{1y\,no}$, with gravity set to zero.

The phase contrast image intensity distribution $I_{xy}^{ph}(\lambda, x, y)$ of an object, which is illuminated by an incoherent, small (point) source of X-rays or neutrons, is obtained by modifying the general image intensity distribution for an incoherent source. The aperture radius r_m in the lens or aperture plane is set to infinity $r_m = \infty$ so that the integration range along the x-direction is $l_x \in (-\infty, \infty)$, and the integration direction along the y-direction is $l_y \in (-\infty, \infty)$. The decrement δ and linear attenuation μ are set to zero (no lens—only air or vacuum) so that parameter $A = 0$ and parameter E is modified, which leads to the modification of the other A- and E-dependent parameters as discussed earlier. Hence, after substitution of the modified parameters, the image intensity $I_{xy}^{\text{ph}}(\lambda, x, y)$ distribution for phase contrast imaging is

$$
\begin{aligned}
I_{xy}^{\text{ph}}(\lambda,x,y,z) = {} & Z_{1x\text{no}} Z_{1y\text{no}} \frac{\pi^5}{32}\left(E_{0x}E_{0y}\right)^2 \left(\frac{w_{0x}w_{0y}\pi}{z\lambda}\right)^4 \\
& \times \exp(-N\mu d) \exp\left[-\frac{(\lambda-\lambda_{0x})^2}{\sigma_{\lambda x}^2}\right] \exp\left[-\frac{(\lambda-\lambda_{0y})^2}{\sigma_{\lambda y}^2}\right] \\
& \times \int_{-\infty}^{\infty}\int_{-\infty}^{\infty} \frac{|T(f_{sx},f_{sy})|^2}{E_{\text{no}}^2[C^2+J_{\text{no}}^2]\sqrt{Q_{x\text{no}}Q_{y\text{no}}}} \exp\left[\frac{J\left(2f_{sy}U_y\pi - 2f_{sy}^2\pi^2 - \frac{1}{2}U_y^2\right)}{C^2+J_{\text{no}}^2}\right] \\
& \times \exp\left[(x^2+y^2)\left(\frac{K_{\text{no}}^2 J_{\text{no}}}{2(C^2+J_{\text{no}}^2)} - 2N_{\text{no}}\right)\right] \exp\left[-f_{sx}^2\left(\frac{2\pi^2 J_{\text{no}}}{C^2+J_{\text{no}}^2}\right)\right] \\
& \times \exp\left[-(f_{sx}x+f_{sy}y)\left(\frac{2K_{\text{no}}C\pi}{C^2+J_{\text{no}}^2}\right)\right] \exp\left[-y\left(\frac{U_y}{C^2+J_{\text{no}}^2}\right)\right] \\
& \times \exp\left[\frac{1}{2Q_{x\text{no}}}\left(\frac{2f_{sx}HJ_{\text{no}}\pi}{C^2+J_{\text{no}}^2} + \frac{xHK_{\text{no}}C}{C^2+J_{\text{no}}^2}\right)^2\right] \exp\left(\frac{R_{y\text{no}}^2}{2Q_{nyo}}\right) df_{sx}\, df_{sy}
\end{aligned} \tag{11.500}
$$

The composite parameter $R_{y\text{no}}$ is

$$
R_{y\text{no}} = \frac{2f_{sy}HJ_{\text{no}}\pi}{C^2+J_{\text{no}}^2} - U_y + \frac{yHK_{\text{no}}C}{C^2+J_{\text{no}}^2} \tag{11.501}
$$

The x- and y-component composite parameters $Q_{x\text{no}}$ and $Q_{y\text{no}}$, which are related to the x- and y-component source parameters W_x and W_y, are

$$
Q_{x\text{no}} = W_x^{\text{Re}} + \frac{H^2 J_{\text{no}}}{C^2+J_{\text{no}}^2} \qquad Q_{y\text{no}} = W_y^{\text{Re}} + \frac{H^2 J_{\text{no}}}{C^2+J_{\text{no}}^2} \tag{11.502}
$$

The x- and y-component source parameters W_x and W_y are real, with zero imaginary components because the X-ray or neutron source is incoherent, where

$$W_x^{\mathrm{Re}} = \frac{1}{2\sigma_x^2} \quad W_x^{\mathrm{Im}} = W_y^{\mathrm{Im}} = 0 \quad W_y^{\mathrm{Re}} = \frac{1}{2\sigma_y^2} \tag{11.503}$$

The absorption aperture radius r_a, and E, A, J, K, and N are CRL parameters, where

$$\begin{aligned} J_{\mathrm{no}} &= \frac{-\mathrm{i}}{(r_0)^2(\lambda/\pi)((1/r_0)+(1/r_i))} \qquad K_{\mathrm{no}} = \frac{-4\mathrm{i}}{(\lambda/\pi)(r_0+r_i)} \\ N_{\mathrm{no}} &= \frac{\mathrm{i}}{r_i^2(\lambda/\pi)((1/r_0)+(1/r_i))} \end{aligned} \tag{11.504}$$

The parameters C and H for source-to-object distance D and object distance r_0 are

$$C = \frac{\pi}{\lambda}\left(\frac{1}{D} + \frac{1}{r_0}\right) \quad H = \frac{\pi}{D\lambda} \tag{11.505}$$

The gravity term is

$$U_y = \frac{2\pi}{\lambda}\left(\frac{o_g}{r_0} - \frac{D_g}{D}\right) \tag{11.506}$$

The terms for the x-direction are identical to the terms for the y-direction, except that the gravity-based variables are set to zero in the x-direction terms. For X-ray phase contrast imaging, one sets the gravity terms to zero.

11.12 Special Case of a Merged Object Plane and Source Plane

With X-rays or neutrons, consider the case in which the coherent or incoherent source and object planes are the same. This section gives the parameter settings for a merged source and object planes. These are the general CRL parameters in sections 11.3.1 and 11.9, which can also be applied to the case of a pinhole or no CRL (phase contrast imaging without a lens). When the source and the object planes are merged, the source-to-object plane distance D is zero, where

$$D = 0 \tag{11.507}$$

Then the parameters H and C are changed to

$$H = 0 \tag{11.508}$$

and

$$C = \frac{k}{2r_0} = \frac{\pi}{r_0 \lambda} \tag{11.509}$$

Also, the source transverse dimension variables $(x_s.y_s)$ are now identical to the corresponding object dimension variables (s_x, s_y), where

$$x_s = s_x \quad \text{and} \quad y_s = s_y \tag{11.510}$$

The parameters, which depend on parameters H and C, the y-directed parameters Q_y, R_y, S_y, T_y, P_{1y}, P_{2y}, θ_{1y}, and θ_{2y}, and the corresponding x-directed parameters Q_x, R_x, S_x, T_x, P_{1x}, P_{2x}, θ_{1x}, and θ_{2x} (with gravity set to zero), then are accordingly altered. The y-directed, gravity-dependent parameters D_g and V_y are set to zero, where

$$D_g = 0 \quad \text{and} \quad V_y = 0 \tag{11.511}$$

The parameter U_y is altered, as are the MTF parameters, where

$$\alpha = 0 \quad \text{and} \quad \beta = 0 \tag{11.512}$$

The C-dependent MTF parameters, namely, δ, ε, η, and κ, are also altered. Thus, all the amplitude expressions are altered by this change of parameters.

The expressions for FOV, PSF, amplitudes, and intensities are greatly simplified by the case of the source and the object planes being identical with

$$D = 0 \tag{11.513}$$

and

$$x_s = s_x \quad \text{and} \quad y_s = s_y \tag{11.514}$$

Hence, the following parameters are set to zero:

$$C = D_g = V_y = \alpha = \beta = 0 \tag{11.515}$$

The terms containing D are removed, rather than setting the source-to-object plane distance D to zero.

11.13 Stationary Phase Approach Applied to an Image Amplitude Integral of a Spherical, Biconcave CRL for Incoherent X-rays or Neutrons with Gravity

We now consider the case of forming X-ray or neutron images with a spherical CRL. The absorption and phase-shift terms for the spherical lens are radicals, which do not yield an analytic solution, unless one limits the effective aperture of the spherical lens to that of the parabolic aperture radius. The radicals can then be removed via the binomial approximation, and one obtains analytic solutions with the Kirchhoff equation. In this section, however, we present a method to reduce the computation time in the numerical integration of the Kirchhoff equation for a spherical CRL that has the radical expression for phase and magnitude in the arguments of the exponentials.

The object is illuminated by an incoherent X-ray or neutron source. The source distribution is expressed in terms of the object transverse positions (s_x, s_y) and the source distribution standard deviations (σ_x, σ_y). From section 11.2, with Eq. (11.126) and Eqs. (11.155−11.160), the 2-D image intensity distribution for X-rays or neutrons for a CRL with biconcave spherical lenses, is

$$E_{xy}^{\mathrm{sph}}(\lambda, x, y, l_x, l_y, s_x, s_y) = E_0(\lambda)E_x(\lambda, l_x, s_x, x)E_y(\lambda, l_y, s_y, y)E_{\mathrm{CRL}}(\lambda, l_x, l_y) \tag{11.516}$$

where

$$E_{\mathrm{CRL}}(\lambda, l_x, l_y) = \exp[-\mu N(R-\sqrt{R^2 - l_x^2 - l_y^2}] \exp[\mathrm{i}2kN\delta(R-\sqrt{R^2 - l_x^2 - l_y^2})] \tag{11.517}$$

$$\begin{aligned} E_y(\lambda,l_y,s_y,y)=&\exp\left[-\frac{s_y}{\sigma_y^2}\left(\frac{Do_g}{r_0}-D_g\right)\left(1+\frac{D}{r_0}\right)\right]\exp\left[-\left(\frac{s_y^2}{2\sigma_y^2}\right)\left(1+\frac{D}{r_0}\right)^2\right] \\ &\times\exp\left[\left(\frac{\mathrm{i}kD}{2r_0^2}\right)\left(l_y-s_y-D_g\right)^2+\left(\frac{\mathrm{i}k}{2r_0}\right)(l_y-s_y-o_g)^2+\left(\frac{\mathrm{i}k}{2r_i}\right)(y-l_y-\mathrm{i}_g)^2\right] \\ &\times\exp\left[\left(\frac{l_y s_y}{\sigma_y^2}\right)\left(\frac{D}{r_0}\right)\left(1+\frac{D}{r_0}\right)\right]\exp\left[\frac{l_y D}{r_0\sigma_y^2}\left(\frac{Do_g}{r_0}-D_g\right)\right] \\ &\times\exp\left[-\left(\frac{\mathrm{i}kl_y^2}{2}\right)\left(\frac{1}{r_0}+\frac{1}{r_i}\right)\right]\exp\left[-\left(\frac{l_y^2}{2\sigma_y^2}\right)\left(\frac{D}{r_0}\right)^2\right] \end{aligned} \tag{11.518}$$

and

$$E_0(\lambda) = \frac{4F_0}{\pi^2 r}\left(\frac{E_{0x}}{\sigma_x\sigma_\lambda}\right)\left(\frac{E_{0y}}{\sigma_y\sigma_\lambda}\right)\exp\left(-\frac{N\mu d}{2}\right)\exp\left[-\frac{(\lambda-\lambda_0)^2}{\sigma_\lambda^2}\right]$$
$$\times\exp\left[-\frac{1}{2\sigma_y^2}\left(\frac{Do_g}{r_0}-D_g\right)^2\right]\exp\left[-ik\left(\frac{DD_g^2}{2r_0^2}+\frac{o_g^2}{2r_0}+L_g+\frac{i_g^2}{2r_i}+N(d_{\max}-\delta d)\right)\right] \tag{11.519}$$

For X-rays, the x-directed amplitude $E_x(\lambda, l_x, s_x, x)$ is obtained from the y-directed amplitude $E_y(\lambda, l_y, s_y, y)$ by setting the gravity terms to zero. The result also applies to neutrons when gravity is neglected.

The source point-to-image point distance is given by the binomial approximation as

$$r = z + \frac{x^2}{z} + \frac{y^2}{z} \cong z \quad \text{where} \quad z \gg x, y \tag{11.520}$$

The axial distance from the source plane at axial position $z = 0$ to the image plane at axial position z, including the CRL length L, is further simplified by the approximation

$$z = z_{\mathrm{im}} = D + r_0 + L + r_i \tag{11.521}$$

As before, the obliquity factor F_0 is approximated as unity:

$$F_0 \cong 1 \tag{11.522}$$

To facilitate integration, this expression along the lens transverse coordinates (l_x, l_y) is stated in the form

$$E_{xy}^{\mathrm{sph}}(x,y,\lambda,s_x,s_y,l_x,l_y) = E_0(\lambda)E_s(s_x,s_y)H(\lambda,s_x,s_y,l_x,l_y)\exp[\mathrm{im}(\lambda,x,y,s_x,s_y,l_x,l_y)] \tag{11.523}$$

where

$$H(\lambda,s_x,s_y,l_x,l_y) = \exp\left[-\mu N\left(R-\sqrt{R^2-l_x^2-l_y^2}\right)\right]$$
$$\times\exp\left[-\left(\frac{l_x^2}{2\sigma_x^2}\right)\left(\frac{D}{r_0}\right)^2\right]\exp\left[\left(\frac{l_x s_x}{\sigma_x^2}\right)\left(\frac{D}{r_0}\right)\left(1+\frac{D}{r_0}\right)\right]$$
$$\times\exp\left[-\left(\frac{l_y^2}{2\sigma_y^2}\right)\left(\frac{D}{r_0}\right)^2\right]\exp\left[\left(\frac{l_y s_y}{\sigma_y^2}\right)\left(\frac{D}{r_0}\right)\left(1+\frac{D}{r_0}\right)\right]\exp\left[\frac{l_y D}{r_0\sigma_y^2}\left(\frac{Do_g}{r_0}-D_g\right)\right] \tag{11.524}$$

$$m(\lambda,x,y,s_x,s_y,l_x,l_y)=2kN\delta(R-\sqrt{R^2-l_x^2-l_y^2})-\left(\frac{kl_x^2}{2}\right)\left(\frac{1}{r_0}+\frac{1}{r_i}\right)$$
$$+\left(\frac{kD}{2r_0^2}\right)(l_x-s_x)^2+\left(\frac{k}{2r_0}\right)(l_x-s_x)^2+\left(\frac{k}{2r_i}\right)(x-l_x)^2$$
$$-\left(\frac{kl_y^2}{2}\right)\left(\frac{1}{r_0}+\frac{1}{r_i}\right)+\left(\frac{kD}{2r_0^2}\right)(l_y-s_y-D_g)^2$$
$$+\left(\frac{k}{2r_0}\right)(l_y-s_y-o_g)^2+\left(\frac{k}{2r_i}\right)(y-l_y-i_g)^2 \qquad (11.525)$$

and

$$E_s(s_x,s_y)=\exp\left[-\left(\frac{s_x^2}{2\sigma_x^2}\right)\left(1+\frac{D}{r_0}\right)^2\right]\exp\left[-\left(\frac{s_y^2}{2\sigma_y^2}\right)\left(1+\frac{D}{r_0}\right)^2\right]$$
$$\times\exp\left[-\frac{s_y}{\sigma_y^2}\left(\frac{Do_g}{r_0}-D_g\right)\left(1+\frac{D}{r_0}\right)\right] \qquad (11.526)$$

The wave function amplitude $E_{xy}^{\mathrm{sph}}(\lambda,x,y,l_x,l_y,s_x,s_y)$ is zero outside the CRL physical aperture radius r_m, where

$$E_{xy}^{\mathrm{sph}}(x,y,\lambda,s_x,s_y,l_x,l_y)=0 \quad \text{for} \quad |l_x|>r_m,\ \ |l_y|>r_m \qquad (11.527)$$

Also, the wave amplitude E_{xy} of Eq. (11.523) is not defined for CRL principal plane points that are outside the CRL physical aperture radius R, where

$$l_x,l_y>R \qquad (11.528)$$

Inside the aperture radius r_m of spherical lens CRL, the wave function amplitude is

$$E_{xy}^{\mathrm{sph}}(x,y,\lambda,s_x,s_y)=E_0E_s\int_{-\infty}^{\infty}\int_{-\infty}^{\infty}H(l_x,l_y)\exp[\mathrm{i}m(l_x,l_y)]\mathrm{d}l_x\,\mathrm{d}l_y \qquad (11.529)$$

The zero amplitude outside the CRL physical aperture radius, and non-zero amplitude inside the CRL aperture radius is enforced by

$$H(l_x,l_y)=0 \quad \text{for} \quad l_x,l_y>r_m \qquad (11.530)$$

The wave amplitude of Eq. (11.529) is a function of the exponential, whose argument is $m(l_x, l_y)$. Argument $m(l_x, l_y)$ is a radical, which does not give a direct analytic solution of Eq. (11.529), and thus one requires numerical integration.

Numerical integration of Eq. (11.529) is time consuming because of the rapid, complex, sinusoidal variation of the integrand with l_x, l_y via

$$\exp\left[im\left(l_x, l_y\right)\right] = \cos\left[m\left(l_x, l_y\right)\right] + \mathrm{i}\sin\left[m\left(l_x, l_y\right)\right] \tag{11.531}$$

Note that $H(l_x, l_y)$ in Eq. (11.529) is not sinusoidal, and is a slowly varying function of l_x and l_y. The double integral of Eq. (11.529) can be evaluated approximately by the method of stationary phase. Since the integrand is rapidly oscillating, the integral is nearly zero, except for contributions from the immediate regions about those points, where there are no oscillations. These regions about these points l_{sx}, l_{sy} are where the phase of the integrand is constant or stationary. These points are the extreme points of the argument of the complex exponential $m(l_x, l_y)$, where its first partial derivatives are zero, that is,

$$\frac{\partial m}{\partial l_x} = 0 \quad \text{and} \quad \frac{\partial m}{\partial l_y} = 0 \tag{11.532}$$

The double integral of Eq. (11.529) can be estimated by finding places where $m(l_x, l_y)$ has a zero derivative, evaluating (approximately) the integral in the neighborhood of each of these points, and summing the contributions.

Let one assume that $m(l_x, l_y)$ has zero first partial derivatives at

$$l_x = l_{sx} \quad \text{and} \quad l_y = l_{sy} \tag{11.533}$$

That is,

$$m_x^{'} = \frac{\partial m(l_x, l_y)}{\partial l_x}\left[l_{sx}, l_{sy}\right] = 0 \quad \text{and} \quad m_y^{'} = \frac{\partial m(l_x, l_y)}{\partial l_y}\left[l_{sx}, l_{sy}\right] = 0 \tag{11.534}$$

Then, in the neighborhood of this point (l_{sx}, l_{sy}), the function $m(l_x, l_y)$ can be expanded in a Taylor series:

$$m(l_x, l_y) = m_s + \frac{(l_x - l_{sx})^2 m_x^{''}}{2} + \frac{(l_y - l_{sy})^2 m_y^{''}}{2} + \cdots \in \cdots \in \cdots \tag{11.535}$$

where m_s is a constant and

$$m_x^{''} = \frac{\partial^2 m(l_x, l_y)}{\partial l_x^2}\left[l_{sx}, l_{sy}\right] \quad \text{and} \quad m_y^{''} = \frac{\partial^2 m(l_x, l_y)}{\partial l_y^2}\left[l_{sx}, l_{sy}\right] \tag{11.536}$$

Here, the second partial derivatives $m_x^{''}$ and $m_y^{''}$ are evaluated at

$$l_x = l_{sx} \quad \text{and} \quad l_y = l_{sy} \tag{11.537}$$

Since $H(l_x, l_y)$ is slowly varying and approximated as constant, as compared to the rapidly varying exponential term, this allows removal from the integral. Also given Eq. (11.535), the double integral of Eq. (11.529) can be expressed as a product of x- and y-component integrals, where

$$E_{xy}^{\mathrm{sph}}\left(x, y, \lambda, s_x, s_y\right) = E_0 E_s \int_{-\infty}^{\infty}\int_{-\infty}^{\infty} H\left(l_x, l_y\right)\exp\left[im\left(l_x, l_y\right)\right] dl_x dl_y =$$
$$E_s H\left(l_{sx}, l_{sy}\right)\exp\left(im_s\right)$$
$$\times \int_{-\infty}^{\infty} \exp\left[i\frac{\left(l_x - l_{sx}\right)^2 m_x''}{2}\right] dl_x \int_{-\infty}^{\infty} \exp\left[i\frac{\left(l_y - l_{sy}\right)^2 m_y''}{2}\right] dl_y \tag{11.538}$$

The integrand of Eq. (11.529), via Eq. (11.535), is thus separable into a l_x-dependent portion and a l_y-dependent portion, which can be integrated separately. This leads to two single, identical integrations that can be evaluated independent of the other. To evaluate the integrals, one first makes a change of variables of l_x and l_y to A and B, where

$$\frac{\pi A^2}{2} = \frac{(l_x, l_{sx})^2 m_x''}{2} \quad \text{and} \quad \frac{\pi B^2}{2} = \frac{\partial^2 m(l_{sx}, l_{sy})}{\partial l_y^2}\frac{(l_y - l_{sy})^2}{2} \tag{11.539}$$

Since l_{sx} and l_{sy} are constants, one finds

$$\mathrm{dA} = \sqrt{\frac{\pi}{m_x''}}\mathrm{d}l_x \quad \text{and} \quad \mathrm{dB} = \sqrt{\frac{\pi}{m_y''}}\mathrm{d}l_y \tag{11.540}$$

With substitution of Eqs. (11.539) and (11.540) into Eq. (11.538), the resulting integrands are even functions of variables A and B. Consequently, the integration limits $[-\infty, \infty]$ can be replaced by $[0, \infty]$ in each integral, and the resulting integral is then multiplied by 2. Furthermore, in each integral, the exponential can be Euler-expanded into the sum of the real sine and imaginary sine, resulting in an expression that is the product of Fresnel integrals:

$$E_{xy}^{\mathrm{sph}}\left(x, y, \lambda, s_x, s_y\right) = \frac{4\pi E_i E_s H\left(l_{sx}, l_{sy}\right)\exp\left(\mathrm{i}m_s\right)}{\sqrt{m_x'' m_y''}}$$
$$\times \int_0^{\infty}\left[\cos\left(\frac{\pi A^2}{2}\right) + i\sin\left(\frac{\pi A^2}{2}\right)\right] dA$$
$$\times \int_0^{\infty}\left[\cos\left(\frac{\pi B^2}{2}\right) + i\sin\left(\frac{\pi B^2}{2}\right)\right] dB \tag{11.541}$$

Since

$$\int_0^{\infty} \cos\left(\frac{\pi X^2}{2}\right) dX = \int_0^{\infty} \sin\left(\frac{\pi X^2}{2}\right) dX = \frac{1}{2} \text{ and } \sqrt{2}\exp\left(i\frac{\pi}{4}\right) = (1 + i) \tag{11.542}$$

With substitution of Eq. (11.542) into Eq. (11.541), the 2-D image intensity distribution, which is produced by a spherical CRL, becomes

$$E_{xy}^{\text{sph}}(x, y, \lambda, s_x, s_y) = i\frac{2\pi E_i E_s H(l_{sx}, l_{sy})\exp(im_s)}{\sqrt{m_x'' m_y''}} \tag{11.543}$$

One final task remains, and that is to calculate the stationary points l_{sx}, l_{sy} in Eqs. (11.543), via (11.534), where

$$\frac{\partial m}{\partial l_x} = 0 \quad \text{and} \quad \frac{\partial m}{\partial l_y} = 0 \tag{11.544}$$

The l_x and l_y partial derivatives for $m(l_x, l_y)$ are each to zero. These partial derivative equations of Eq. (11.544) are then solved for l_x and l_y, respectively. The solutions are thus the stationary points l_{sx} and l_{sy}, where

$$\begin{aligned} 0 = \frac{\partial m(l_x, l_y)}{\partial l_x} &= 2l_x kN\delta(R^2 - l_x^2 - l_y^2)^{-1/2} - kl_x\left(\frac{1}{r_0} + \frac{1}{r_i}\right) \\ &+ \frac{kD}{r_0^2}(l_x - s_x) + \frac{k}{r_0}(l_x - s_x) - \frac{k}{r_i}(x - l_x) \end{aligned} \tag{11.545}$$

and

$$\begin{aligned} 0 = \frac{\partial m(l_x, l_y)}{\partial l_y} &= 2l_y kN\delta(R^2 - l_x^2 - l_y^2)^{-1/2} - kl_y\left(\frac{1}{r_0} + \frac{1}{r_i}\right) \\ &+ \frac{kD}{r_0^2}(l_y - s_y - D_g) + \frac{k}{r_0}(l_y - s_y - o_g) - \frac{k}{r_i}(y - l_y - i_g) \end{aligned} \tag{11.546}$$

One solves Eqs. (11.545) and (11.546) for l_x and l_y to obtain

$$l_x\left[2kN\delta(R^2 - l_x^2 - l_y^2)^{-1/2} - k\left(\frac{1}{r_0} + \frac{1}{r_i}\right) + \frac{kD}{r_0^2} + \frac{k}{r_0} + \frac{k}{r_i}\right] = \frac{kDs_x}{r_0^2} + \frac{ks_x}{r_0} + \frac{kx}{r_i} \tag{11.547}$$

and

$$l_y\left[2kN\delta(R^2-l_x^2-l_y^2)^{-1/2}-k\left(\frac{1}{r_0}+\frac{1}{r_i}\right)+\frac{kD}{r_0^2}+\frac{k}{r_0}+\frac{k}{r_i}\right]=\frac{kD}{r_0^2}(s_y+D_g)$$
$$+\frac{k}{r_0}(s_y+o_g)+\frac{k}{r_i}(y-i_g) \qquad (11.548)$$

The individual lens radius of curvature R is assumed to be larger than the radial distance of CRL principal plane points (l_x, l_y) from the center z-axis, where lens points (l_x, l_y) lie within the CRL physical aperture radius r_m, so that

$$R^2 > l_x^2 + l_y^2 \qquad (11.549)$$

The x- and y-lens position coordinates l_x and l_y multiply the same term in the left side of Eqs. (11.547) and (11.548). This common factor on the left side of Eqs. (11.547) and (11.548), allows the 2 equations to be combined to yield

$$l_x\left(\left(\frac{kD}{r_0^2}\right)(s_y+D_g)+\left(\frac{k}{r_0}\right)(s_y+o_g)+\left(\frac{k}{r_i}\right)(y-i_g)\right)=l_y\left(\frac{kDs_x}{r_0^2}+\frac{ks_x}{r_0}+\frac{kx}{r_i}\right)$$
$$(11.550)$$

If one ignores gravity, then the gravity parameters of Eqs. (11.105), (11.106), and (11.108) are set to zero, where

$$D_g = o_g = i_g = 0 \qquad (11.551)$$

Then, Eq. (11.550) simplifies to

$$l_x\left[\left(\frac{D}{r_0^2}+\frac{1}{r_0}\right)r_is_y+y\right]=l_y\left[\left(\frac{D}{r_0^2}+\frac{1}{r_0}\right)r_is_x+x\right] \qquad (11.552)$$

where

$$R^2 > l_x^2 + l_y^2 \qquad (11.553)$$

We now impose the point spread function (PSF) condition on Eq. (11.552). The expression of Eq. (11.552) is evaluated at the object point coordinates at the center of the object plane, where

$$s_x = s_y = 0 \qquad (11.554)$$

The radius of curvature R of each lens of the spherical lens CRL is much larger than the CRL physical aperture radius r_m, which is the upper bound of the x- and

y-coordinates (l_x, l_y) in the CRL principal plane, where $r_m \geq l_x, l_y$. Given Eqs. (11.553) and (11.554), one then simplifies Eq. (11.552), and obtains

$$yl_x = xl_y \quad R^2 > l_x^2 + l_y^2 \tag{11.555}$$

We now impose the field of view (FOV) condition on Eq. (11.552). Again, since the radius of curvature of the each lens of the CRL is much larger than the lens radius, when calculating the FOV of the object plane from the perspective of an image point at the center of the image plane $(x, y) = (0, 0)$, the relationship simplifies to

$$s_y l_x = s_x l_y \quad R^2 > l_x^2 + l_y^2 \tag{11.556}$$

One must iteratively solve to find the single values of l_x and l_y, which satisfies both the equality and the inequality relationships, that is, the values of l_{sx} and l_{sy}, and thus $m(l_{sx}, l_{sy})$ and $H(l_{sx}, l_{sy})$. Finally, only the second partial derivatives of $m(l_x, l_y)$; i.e., $m_x''(l_{sx})$ and $m_y''(l_{sy})$ evaluated at l_{sx} and l_{sy}, remain to be determined. With or without gravity, the second derivatives of $m(l_x, l_y)$ evaluated at l_{sx} and l_{sy} are thus

$$\begin{aligned}\frac{\partial^2 m(l_{sx}, l_{sy})}{\partial l_x^2} &= 2kN\delta(R^2 - l_{sx}^2 - l_{sy}^2)^{-1/2} + 2l_{sx}^2 kN\delta(R^2 - l_{sx}^2 - l_{sy}^2)^{-3/2} \\ &\quad - k\left(\frac{1}{r_0} + \frac{1}{r_i}\right) + \frac{kD}{r_0^2} + \frac{k}{r_0} - \frac{k}{r_i}\end{aligned} \tag{11.557}$$

and

$$\begin{aligned}\frac{\partial^2 m(l_{sx}, l_{sy})}{\partial l_y^2} &= 2kN\delta(R^2 - l_{sx}^2 - l_{sy}^2)^{-1/2} + 2l_{sy}^2 kN\delta(R^2 - l_{sx}^2 - l_{sy}^2)^{-3/2} \\ &\quad - k\left(\frac{1}{r_0} + \frac{1}{r_i}\right) + \frac{kD}{r_0^2} + \frac{k}{r_0} - \frac{k}{r_i}\end{aligned} \tag{11.558}$$

By substitution of Eqs. (11.557) and (11.558) into Eq. (11.541), the 2-D wave function amplitude $E_{xy}^{\mathrm{sph}}(x, y, \lambda, s_x, s_y)$ can be calculated.

With an incoherent neutron source illuminating an object, the 2-D image plane intensity distribution produced by a spherical CRL is

$$I_{xy}(\lambda,x,y) = \int_{-\infty}^{\infty}\int_{-\infty}^{\infty}\left|\int_{-\infty}^{\infty} E_y^{\mathrm{sph}}(\lambda,y,l_y,s_y)\mathrm{d}l_y \int_{-\infty}^{\infty} E_y^{\mathrm{sph}}(\lambda,x,l_x,s_x)\mathrm{d}l_x\right|^2 T(s_x,s_y)\mathrm{d}s_y\,\mathrm{d}s_x \tag{11.559}$$

The nominal 2-D image plane resolution, which is produced by a spherical CRL, is found by considering a specific object point (e.g., the center object point at $s = 0$), and then determining the image plane distribution from this single point. This is the PSF(λ, x, y), where

$$\mathrm{PSF}(\lambda,x,y) = \left| \int_{-\infty}^{\infty} E_y^{\mathrm{sph}}(\lambda, y, l_y, s_y{=}0)\mathrm{d}l_y \int_{-\infty}^{\infty} E_y^{\mathrm{sph}}(\lambda, x, l_x, s_x{=}0)\mathrm{d}l_x \right|^2 \quad (11.560)$$

The FWHM of the image plane distribution along transverse positions x and y yields the image resolution, and division by the CRL transverse magnification determines the corresponding 2-D object resolution in x and y. Alternatively, the spatial resolution of the spherical CRL can be characterized by MTF(f_x, f_y, λ), which is the magnitude of the Fourier transform of PSF(λ, x, y), and is a function of spatial frequencies f_x, f_y in the image plane x- and y-directions, so that

$$\mathrm{MTF}(f_x, f_y, \lambda) = \left| \int_{-\infty}^{\infty} \left| \int_{-\infty}^{\infty} E_x^{\mathrm{sph}}(\lambda, x, l_x, s_x{=}0)\mathrm{d}l_x \right|^2 \exp(-\mathrm{j}2\pi f_x x)\mathrm{d}x \right| \times \left| \int_{-\infty}^{\infty} \left| \int_{-\infty}^{\infty} E_y^{\mathrm{sph}}(\lambda, y, l_y, s_y{=}0)\mathrm{d}l_y \right|^2 \exp(-\mathrm{j}2\pi f_y y)\mathrm{d}y \right| \quad (11.561)$$

Alternatively, for an X-ray or neutron incoherent source, one can determine the FOV of the spherical CRL by considering a specific image point, for example, the center image point at $y = 0$. The intensity distribution of the illuminated object along the transverse object coordinate s, which contributes to the intensity of the image point at $y = 0$, has the FWHM. By the azimuthal symmetry of the CRL, this FWHM is a measure of the CRL FOV. The FOV for the CRL is the intensity distribution in the object plane, which contributes to the image point $y = 0$, where

$$I_{xy}(\lambda, s_x, s_y) = T(s_x, s_y) \left| \int_{-\infty}^{\infty} E_y^{\mathrm{sph}}(\lambda, y{=}0, l_y, s_y)\mathrm{d}l_y \int_{-\infty}^{\infty} E_y^{\mathrm{sph}}(\lambda, x{=}0, l_x, s_x)\mathrm{d}l_x \right|^2 \quad (11.562)$$

References

Elmore, W. C., & Heald, M. A. (1969). *Physics of waves*. New York, NY: Dover.

Goodman, J. W. (1968). *Introduction to Fourier optics*. New York, NY: McGraw-Hill.

Hecht, E., & Zajac, A. (1979). *Optics*. Reading, MA: Addison-Wesley.

Selby, S. M., & Girling, B. (1965). *Standard mathematical tables,* 14th ed. Boca Raton, Florida: CRC Press p. 346.

Sommerfeld, A. (1954). *Optics*. New York, NY: Academic Press.

12 Electromagnetic Fields of Moving Charges, Electric and Magnetic Dipoles

12.1 Maxwell's Equations and the Lienard–Wiechert Potentials

In this chapter, we cover Maxwell's equations, the Lienard–Wiechert potentials, the Lorentz gauge, and the Helmholtz theorem. Derived are the scalar and vector potentials of an oscillating electric dipole, magnetic field of an oscillating electric dipole via vector potential, and the electric field emitted from the oscillating electric dipole. The conversion of electric and magnetic fields from MKS (meter-kilogram-second) units to (centimeter-gram-second) units is presented. Next, we derive the near- and far-zone electric fields and power emission from an oscillating electric dipole, and the frequency and wave-number domain expressions for electric and magnetic fields emission and power from an oscillating electric dipole. Then, we use the binomial expansion of vector potential of a moving charge to derive the contributions from electric and magnetic dipole and electric quadrupole moments. Also, the transformation of fields and vector potential of electric dipole to magnetic dipole and the electric and magnetic fields and power emission from oscillating magnetic dipole are derived.

This discussion is followed by the derivation of the electric and magnetic fields of a charge in arbitrary motion from the Lienard–Wiechert potentials, and the Poynting vector and electromagnetic energy radiated from an arbitrary accelerated charge per solid angle per frequency interval. Next, we derive the electron trajectory in synchrotron ring insertion devices and a simplified velocity-dependent expression for a per solid angle frequency spectrum emitted by accelerated charged particles. Also covered is bremsstrahlung and electromagnetic wave polarization, and the net neutron magnetic dipole moment, which is produced in ultracold neutron population with an applied magnetic field and has the potential possibility for population inversion. The energy radiated per solid angle for the moving magnetic dipole moment of neutrons and charged particles is then derived. Finally, we relate the magnetization vector of a moving charge with the polarization vector of a stationary charge. "The topics in this chapter are drawn from books by Becker (1982), Eyges (1980), Feynman (1964), Jackson (1975), Landau and Lifshitz (1975), Marion (1965), Maxwell (1954),

Neutron and X-ray Optics. DOI: http://dx.doi.org/10.1016/B978-0-12-407164-3.00011-5

Panofsky and Phillips (2005), Popović (1971), Purcell (1965), Ramo, Whinnery & Van Duzer (1965), Schey (2005), Schwartz (1972), Shadowitz (1988a,b), Shen & Kong (1983), Sommerfeld (1964), Stratton (1941), and Zahn (1979)."

We begin with Maxwell's equations in MKS (SI) units. Given electric field intensity **E** and electric permittivity ε, Gauss's Law for the electric field states in integral form, that the free charge e, which is obtained by volume integration of the free charge density ρ, and is enclosed by a closed surface, is equal to the surface integral of the electric displacement field or electric flux density $\mathbf{D} = \varepsilon\mathbf{E}$. The surface integral integrates only the component of the electric flux density $\mathbf{D} = \varepsilon\mathbf{E}$, which is perpendicular to the closed surface that contains the free charge. The differential form of Gauss's Law for the electric field is

$$\nabla \cdot (\varepsilon\mathbf{E}) = \rho \tag{12.1}$$

In vacuum permittivity, $\varepsilon = \varepsilon_0$ measures the resistance of formation of an electric field, and in a material ε measures how the electric field is affected by the material and how the material is affected by the electric field via the formation of induced electric dipoles in the material. Permittivity measures the ability of a material to sustain a static electric field or transmit a time-varying electric field. In general, the permittivity $\varepsilon(\mathbf{r}, t)$, like the electric field $\mathbf{E}(\mathbf{r}, t)$, can vary with position $\mathbf{r}$ and time t.

Given magnetic field intensity **H** and magnetic permeability μ, Faraday's Law of induction states in integral form, states that the line integral of the electric field intensity **E** around a planar, closed loop, equals the surface integral of the negative time rate of change of the loop-enclosed component of the magnetic flux density $\mathbf{B} = \mu\mathbf{H}$, which is perpendicular to the loop plane, and integrated over the loop area. The line integral of the electric field around the closed loop integrates only the component of the electric field intensity **E**, which is tangent to the loop. Faraday's Law in differential form is

$$\nabla \times \mathbf{E} = -\frac{\partial \mathbf{B}}{\partial t} \tag{12.2}$$

Gauss's Law for the magnetic field states in integral form, that the surface integral of the magnetic flux density **B** over a closed surface is zero, which thus assumes no free magnetic charges exist. The surface integral integrates only the component of the magnetic flux density **B**, which is perpendicular to the closed surface. The differential form of Gauss's Law for the magnetic field is

$$\nabla \cdot \mathbf{B} = 0 \tag{12.3}$$

Maxwell-Ampere's Law states in integral form, that the line integral of the magnetic field intensity **H** around a planar, closed loop, equals the sum of the surface integrals of (1) the enclosed component of the electric flux density **D**, which is perpendicular to the loop plane, and integrated over the loop area, and (2) the enclosed current **I**, which is obtained by integration over the loop area of the component of the enclosed current density **J** that is perpendicular to the loop plane. The line integral of the magnetic field around the closed loop integrates only the component of the magnetic field intensity **H**, which is tangent to the loop. Ampere's Law in differential form is

$$\nabla \times \mathbf{H} = \frac{\partial(\varepsilon \mathbf{E})}{\partial t} + \mathbf{J} \tag{12.4}$$

In integral form the conservation of charge states the current **I** through a closed surface, is equal to the negative time rate of change of the charge q, which is enclosed by the surface. That is, the volume integration of the negative time rate of change of the free charge density ρ is equal the surface integral of the current density **J**, which integrates only the component of the current density **J**, which is perpendicular to the closed surface that contains $\partial q/\partial t$. The differential form of the conservation of charge is

$$\nabla \cdot \mathbf{J} = -\frac{\partial \rho}{\partial t} \tag{12.5}$$

The solutions of Maxwell's equations in terms of the scalar electric potential φ and the magnetic vector potential **A** are

$$\mathbf{E} = -\nabla \phi - \frac{\partial \mathbf{A}}{\partial t} \tag{12.6}$$

$$\mathbf{B} = \nabla \times \mathbf{A} \tag{12.7}$$

The scalar electric potential $\phi(\mathbf{r}, t)$ and the magnetic vector potential $\mathbf{A}(\mathbf{r}, t)$ at the observer position **r** and time t have integral solutions. The scalar $\phi(\mathbf{r}, t)$ and vector potential $\mathbf{A}(\mathbf{r}, t)$ are related to the charge $\rho(\mathbf{r}_s, t_{\text{ret}})$ and current density $\mathbf{J}(\mathbf{r}_s, t_{\text{ret}})$ at source position $\mathbf{r}_s$ at an earlier time or retarded time t_{ret}, where

$$\phi(\mathbf{r}, t) = \frac{1}{4\pi\varepsilon} \iiint dx\ dy\ dz \frac{1}{|\mathbf{r} - \mathbf{r}_s|} \rho\left(\mathbf{r}_s, t - \left|\frac{\mathbf{r} - \mathbf{r}_s}{c}\right|\right) \tag{12.8}$$

and

$$\mathbf{A}(\mathbf{r}, t) = \frac{\mu}{4\pi} \frac{1}{c} \iiint dx\ dy\ dz \frac{1}{|\mathbf{r} - \mathbf{r}_s|} \mathbf{J}\left(\mathbf{r}_s, t - \left|\frac{\mathbf{r} - \mathbf{r}_s}{c}\right|\right) \tag{12.9}$$

Here, ε is the permittivity and μ is the permeability of the medium in which, for a vacuum,

$$\varepsilon = \varepsilon_0 \quad \text{and} \quad \mu = \mu_0 \quad \text{vacuum} \tag{12.10}$$

Since the electromagnetic wave is limited to a maximum of the speed of light, the electromagnetic field, which arrives at the observer at time t, was produced by source charges or current densities at an earlier time or retarded time t_{ret}, where

$$t_{\text{ret}} = t - \left|\frac{\mathbf{r}-\mathbf{r}_s}{c}\right| \tag{12.11}$$

The electromagnetic field, which is emitted from the charge or current density at source position $\mathbf{r}_s$ at retarded or prior time t_{ret}, travels distance $|\mathbf{r}-\mathbf{r}_s|$ at velocity c in time interval $|\mathbf{r}-\mathbf{r}_s|/c$. This time interval $|\mathbf{r}-\mathbf{r}_s|/c$ is subtracted from the observer time t to obtain the earlier point in time t_{ret} in which electromagnetic wave emission from the source charge or current density occurred. The current density $\mathbf{J}(\mathbf{r}_s, t_{\text{ret}})$ is the product of the charge density $\rho(\mathbf{r}_s, t_{\text{ret}})$ multiplied by the velocity of the charge density $\mathbf{v}(\mathbf{r}_s, t_{\text{ret}})$, where

$$\mathbf{J}(\mathbf{r}_s, t_{\text{ret}}) = \rho(\mathbf{r}_s, t_{\text{ret}}) \cdot \mathbf{v}(\mathbf{r}_s, t_{\text{ret}}) \tag{12.12}$$

The equation for the current density $\mathbf{J}(\mathbf{r}_s, t_{\text{ret}})$ at source position $\mathbf{r}_s$ at earlier or retarded time t_{ret} becomes

$$\mathbf{A}(\mathbf{r}, t) = \frac{\mu}{4\pi}\frac{1}{c}\iiint \mathrm{d}x\ \mathrm{d}y\ \mathrm{d}z\frac{\rho(\mathbf{r}_s, t_{\text{ret}})v(\mathbf{r}_s, t_{\text{ret}})}{|\mathbf{r}-\mathbf{r}_s|} \tag{12.13}$$

Following Feynman (1964), presented now is an alternative derivation of the Lienard–Wiechert electric potential, which is thus a correction of Coulomb's law. Consider a stationary observer and a charge, which moves in the positive z-direction toward the observer. At initial time $t = 0$, the position of the charge and observer are $\mathbf{r}_s(0)$ and $\mathbf{r}(0)$, and at retarded time $t_{\text{ret}} \geq \Delta t$, the charge has moved axial distance vt_{ret} to position $\mathbf{r}_s(t_{\text{ret}})$, where

$$\mathbf{r}_s(t_{\text{ret}}) = \mathbf{r}_s(0) + vt_{\text{ret}} = \mathbf{r}_s(0) + v\frac{|\mathbf{r}(0)-\mathbf{r}_s(0)|}{c} \qquad t_{\text{ret}} \geq \Delta t \tag{12.14}$$

And the stationary observer position is fixed, where

$$\mathbf{r}(0) = \mathbf{r}(t_{\text{det}}) \tag{12.15}$$

In the static case, Coulomb's law describes the electric field at observer position $\mathbf{r}$ due to a charge e at source position $\mathbf{r}_s$, where, in a vacuum,

$$\phi(\mathbf{r}, t) = \frac{1}{4\pi\varepsilon_0}\frac{e}{|\mathbf{r}-\mathbf{r}_s|} \tag{12.16}$$

The charge e is modeled as a cube dimension a with charge density ρ, where

$$e = \rho a^3 \tag{12.17}$$

One assumes that the distance $|\mathbf{r} - \mathbf{r}_s|$ between the charge cube center position $\mathbf{r}_s$ and the observer position $\mathbf{r}$ at initial time $t = 0$ is much greater than $\mathbf{r}_s$, where

$$|\mathbf{r} - \mathbf{r}_s| \gg r_s \tag{12.18}$$

The electric field is emitted from the charge at initial $t = 0$, and travels distance $|\mathbf{r} - \mathbf{r}_s|$, and arrives at the observer at later or retarded time $t = t_{\text{ret}}$, where

$$t_{\text{ret}} = \frac{|\mathbf{r} - \mathbf{r}_s|}{c} \tag{12.19}$$

In time interval Δt after departure time $t = 0$, the charge cube moves in the positive z-direction at velocity $\mathbf{v} = v\hat{\mathbf{z}}$ (speed v) toward the observer and propagates distance $v\Delta t$. The length, which is occupied by the charge in the z-direction toward the observer, has increased from a at initial time $t = 0$ to length b at later time $t = \Delta t$, where

$$b = a + v\Delta t \tag{12.20}$$

The electric field departs at speed of light velocity c from the rear (farthest) portion of the charge cube at initial $t = 0$, and travels distance b in the direction of the observer over time interval Δt:

$$\Delta t = \frac{b}{c} \tag{12.21}$$

Substitution for Δt in the axial length b, which contains the charge over time interval Δt, gives

$$b = a + v\frac{b}{c} \tag{12.22}$$

Solving for b yields

$$b = \frac{a}{1 - (v/c)} \tag{12.23}$$

The charge density $\rho(0)$ at initial time $t = 0$ is

$$\rho(0) = \frac{e}{a^3} \tag{12.24}$$

In time interval Δt, the charge e is contained within length b with cross-sectional area a^2. Consequently, the volume, which contains the charge, has increased from a^3 at $t=0$ to volume a^2b at time $t=\Delta t$, and charge density $\rho(\Delta t)$ at later time $t=\Delta t$ is

$$\rho(\Delta t) = \frac{e}{a^2 b} = \frac{e}{a^3}\left(1 - \frac{v}{c}\right) \tag{12.25}$$

Application of Coulomb's law allows one to integrate over the axial extent of the charge, from $z=0$ to $z=b$, prior to the departure of the electric field from the leading edge of the charge cube (edge closest to the observer) at time $t=\Delta t$, where

$$\phi(\mathbf{r}, t \geq \Delta t) = \frac{1}{4\pi\varepsilon_0}\frac{1}{|\mathbf{r}-\mathbf{r}_s|}\int_0^a \mathrm{d}x \int_0^a \mathrm{d}y \int_0^b \rho(\Delta t)\mathrm{d}z \tag{12.26}$$

Substitution for charge density $\rho(\Delta t)$ at later time $t=\Delta t$ gives

$$\phi(\mathbf{r}, t \geq \Delta t) = \frac{1}{4\pi\varepsilon_0}\frac{1}{|\mathbf{r}(0)-\mathbf{r}_s(0)|}\int_0^a \mathrm{d}x \int_0^a \mathrm{d}y \int_0^b \mathrm{d}z \frac{e}{a^3}\left(1 - \frac{v}{c}\right) \tag{12.27}$$

Integration yields the corrected version of Coulomb's law, given the speed of the electric field is a finite speed of light c, instead of an infinite speed:

$$\phi(\mathbf{r}, t \geq \Delta t) = \frac{e}{4\pi\varepsilon_0}\frac{1}{|\mathbf{r}-\mathbf{r}_s|}\frac{b}{a} \tag{12.28}$$

Substitution for b gives the equation for the Lienard–Wiechert potential for electric charge e, which moves at velocity v in the direction of the observe position vector $\mathbf{r}$, where

$$\phi(\mathbf{r}, t \geq \Delta t) = \frac{e}{4\pi\varepsilon_0}\frac{1}{|\mathbf{r}-\mathbf{r}_s|}\cdot\frac{1}{1-(v/c)} \tag{12.29}$$

Next, assume the charge e moves with velocity $\mathbf{v}$ in an arbitrary direction, relative to distance or separation vector $\mathbf{R}$, which points from the charge at position $\mathbf{r}_s$ to the observer position vector $\mathbf{r}$. The component of the velocity v_r of the charge e, along the distance vector $\mathbf{R}$, is the dot product of the charge velocity $\mathbf{v}$ and $\mathbf{R}$, where

$$v_r = \mathbf{v}\cdot\mathbf{R} \tag{12.30}$$

and

$$\mathbf{R} = \mathbf{r} - \mathbf{r}_s \tag{12.31}$$

The scalar Lienard–Wiechert potential for electric charge e, which moves at arbitrary velocity $\mathbf{v}$, relative to the observer position vector $\mathbf{r}$, is then

$$\phi(\mathbf{r}, t \geq \Delta t) = \frac{e}{4\pi\varepsilon_0} \cdot \frac{1}{R - (\mathbf{v} \cdot \mathbf{R}/c)} \tag{12.32}$$

The vector Lienard–Wiechert potential is magnetic potential vector $\mathbf{A}$ for a charge e, which moves at arbitrary velocity $\mathbf{v}$ relative to the observer position vector $\mathbf{r}$, and is

$$\mathbf{A}(\mathbf{r}, t \geq \Delta t) = \frac{e}{4\pi\varepsilon_0 c^2} \cdot \frac{\mathbf{v}}{R - (\mathbf{v} \cdot \mathbf{R}/c)} \tag{12.33}$$

This derivation of the Lienard–Wiechert electric and magnetic vector potentials for a moving charge, given the finite maximum velocity c of energy transfer and field propagation (electric, magnetic, electromagnetic, gravity, strong, and weak), which has a dependence of $(1-\beta)^{-1}$, thereby gives a hint of the Lorentz factor $(1-\beta^2)^{-1}$ of special relativity.

12.2 The Lorentz Gauge and the Helmholtz Theorem

For a vector to be unique and single-valued, the Helmholtz theorem requires that the curl and divergences of the vector be specified, and that the fields approach zero far from the sources. The Lorentz gauge or condition is a particular choice of the curl and divergence of the vector potential $\mathbf{A}$, which allows $\mathbf{A}$ and the vector potential-dependent magnetic $\mathbf{B}$ and electric $\mathbf{E}$ fields to have a single value. First, the Helmholtz theorem is proved and then the Lorentz gauge is derived.

The vector potential $\mathbf{A}$ does not uniquely specify magnetic field $\mathbf{B}$ because for scalar function Φ,

$$\mathbf{B} = \nabla \times (\mathbf{A} + \nabla\Phi) = \nabla \times \mathbf{A} \tag{12.34}$$

However, vector potential $\mathbf{A}$ can uniquely specify the magnetic field $\mathbf{B}$, if $\mathbf{A}$ is specified by its curl $\nabla \times \mathbf{A}$ and divergence $\nabla \cdot \mathbf{A}$. Suppose that one adds a vector $\mathbf{C}$ to $\mathbf{A}$, where

$$\mathbf{A} \to \mathbf{A} + \mathbf{C} \tag{12.35}$$

If the magnetic field $\mathbf{B} = \nabla \times \mathbf{A}$ is to be uniquely specified by vector potential $\mathbf{A}$, then this requires $\nabla \times \mathbf{C} = 0$, where

$$\mathbf{B} = \nabla \times (\mathbf{A} + \mathbf{C}) = \nabla \times \mathbf{A} \Rightarrow \nabla \times \mathbf{C} = 0 \tag{12.36}$$

If $\nabla \times \mathbf{C} = 0$, then vector $\mathbf{C}$ must be obtained from a gradient of a scalar function ϕ because the curl of a gradient is zero, where

$$\nabla \times \nabla\phi = 0 \Rightarrow \mathbf{C} = \nabla\phi \tag{12.37}$$

Since the divergence of vector potential $\mathbf{A}$ must be uniquely specified with a single value, then if one adds a vector $\mathbf{C}$ to $\mathbf{A}$, the divergence of vector $\mathbf{C}$ must be zero, where

$$\nabla \cdot (\mathbf{A} + \mathbf{C}) = \nabla \cdot \mathbf{A} \Rightarrow \nabla \cdot \mathbf{C} = 0 \tag{12.38}$$

Since $\mathbf{C} = \nabla\phi$, then the Laplacian ∇^2 of the scalar function ϕ must be zero, where

$$\nabla \cdot \mathbf{C} = \nabla \cdot \nabla\phi = \nabla^2\phi = 0 \tag{12.39}$$

However, $\nabla^2\phi = 0$ is Laplace's equation, which applies to the case of zero charge density ($\rho = 0$). Scalar function ϕ is also the electric potential $\phi(x, y, z)$, which is measured by an observer at position (x, y, z), and is produced by charge density distribution $\rho(x_s, y_s, z_s)$. Scalar function $\phi(x, y, z)$ is obtained by solving Poisson's equation, which is

$$\nabla^2\phi(x, y, z) = -\frac{\rho(x_s, y_s, z_s)}{\varepsilon} \tag{12.40}$$

The solution of Poisson's equation is

$$\phi(x, y, z) = \int_{-\infty}^{\infty} dz_s \int_{-\infty}^{\infty} dy_s \int_{-\infty}^{\infty} dx_s \frac{\rho(x_s, y_s, z_s)}{4\pi\varepsilon\sqrt{(x - x_s)^2 + (y - y_s)^2 + (z - z_s)^2}} \tag{12.41}$$

If the charge density is zero ($\rho = 0$), then from Poisson's equation, the electric potential is zero (i.e., $\phi = 0$). Thus, vector $\mathbf{C}$, which is the gradient of ϕ, is also zero, where

$$\mathbf{C} = \nabla\phi = 0 \tag{12.42}$$

Recall that one assumed an arbitrary, nonzero vector $\mathbf{C}$ was added to the vector potential $\mathbf{A}$, where $\mathbf{A} \rightarrow \mathbf{A} + \mathbf{C}$. The vector $\mathbf{C}$, however, was shown to be zero. With $\mathbf{C} = 0$, the vector potential $\mathbf{A}$ is uniquely specified. Helmholtz's theorem was applied to specify the curl and divergence of vector potential $\mathbf{A}$, via curl $\nabla \times \mathbf{C} = 0$ and divergence $\nabla \cdot \mathbf{C} = 0$. The divergence $\nabla \cdot \mathbf{C} = 0$, with $\mathbf{C} = \nabla\phi$, requries the Poisson equation solution for ϕ to show zero electric potential ($\phi = 0$) for zero charge density ($\rho = 0$), or for infinite observer distance ($R \to \infty$) from a nonzero charge density ($\rho \neq 0$), which is consistent with the Helmholtz theorem. The zero electric potential ($\phi = 0$) for the case of zero charge density ($\rho = 0$), then confirms that $\mathbf{C} = 0$. Thus, the Helmholtz theorem has been demonstrated by indirect proof.

The Lorentz gauge applies the Helmholtz theorem to vector potential **A**, where scalar potential ϕ is specifically expressed in terms of **A**. One obtains the Lorentz gauge or condition by assuming that the magnetic field $\mathbf{B}(\mathbf{r}, t)$ can be expressed in terms of vector potential $\mathbf{A}(\mathbf{r}, t)$ via $\mathbf{B} = \nabla \times \mathbf{A}$, which in a vacuum must satisfy Faraday's law, where $\mathbf{B} = \mu_0 \mathbf{H}$ and

$$\nabla \times \mathbf{E} = -\frac{\partial \mathbf{B}}{\partial t} \tag{12.43}$$

Substitution for **B** in Faraday's law, and application of the vector identity, which states the curl of the gradient of a scalar potential is zero, then yields

$$\nabla \times \left(\mathbf{E} + \frac{\partial \mathbf{A}}{\partial t} \right) = 0 = \nabla \times (-\nabla \phi) \tag{12.44}$$

The minus sign of $-\nabla \phi$ is chosen so that in a static case (no time variation), the electric field is the negative of the gradient of the electric potential ϕ. One thus obtains the expression for the electric field $\mathbf{E}(\mathbf{r}, t)$ in terms of $\phi(\mathbf{r}, t)$ and $\mathbf{A}(\mathbf{r}, t)$:

$$\mathbf{E} = -\nabla \phi - \frac{\partial \mathbf{A}}{\partial t} \tag{12.45}$$

The electric field **E** is not changed, if the partial time derivative of another scalar function Ψ is subtracted from the scalar potential Φ, and the gradient of the same scalar function Ψ is added to vector potential **A**, where

$$\Phi \Rightarrow \Phi - \frac{\partial \Psi}{\partial t} \tag{12.46}$$

and

$$\mathbf{A} \Rightarrow \mathbf{A} + \nabla \Psi \tag{12.47}$$

The electric field is unchanged by substitution of the altered scalar and vector potentials:

$$\mathbf{E} = -\nabla \left(\Phi - \frac{\partial \Psi}{\partial t} \right) - \frac{\partial}{\partial t}(\mathbf{A} + \nabla \Psi) = -\nabla \Phi - \frac{\partial \mathbf{A}}{\partial t} \tag{12.48}$$

Next, one invokes Ampere's law for a dielectric medium:

$$\nabla \times \mathbf{H} = \mathbf{J} + \frac{\partial \mathbf{D}}{\partial t} \tag{12.49}$$

where

$$\mathbf{D} = \varepsilon\mathbf{E} \quad \text{and} \quad \mathbf{B} = \mu\mathbf{H} \tag{12.50}$$

Assume that permeability μ and permittivity ε do not vary with time, where

$$c^2 = \frac{1}{\mu\varepsilon} \tag{12.51}$$

Plugging into Eq. (12.49) for $\mathbf{E}(\mathbf{r}, t)$ of Eq. (12.48) and $\mathbf{B}(\mathbf{r}, t)$ of Eq. (12.34), which are expressed by the scalar potential ϕ and vector potential $\mathbf{A}$, gives

$$\nabla \times (\nabla \times \mathbf{A}) = \mu\mathbf{J} + \frac{1}{c^2}\left(-\nabla\left(\frac{\partial\phi}{\partial t}\right) - \frac{\partial^2\mathbf{A}}{\partial t^2}\right) \tag{12.52}$$

One invokes the BAC–CAB vector identity:

$$\nabla \times (\nabla \times \mathbf{A}) = \nabla(\nabla \cdot \mathbf{A}) - \nabla^2\mathbf{A} \tag{12.53}$$

Hence,

$$\nabla^2\mathbf{A} - \nabla\left(\nabla \cdot \mathbf{A} + \frac{1}{c^2}\frac{\partial\phi}{\partial t}\right) - \frac{1}{c^2}\frac{\partial^2\mathbf{A}}{\partial t^2} = -\mu\mathbf{J} \tag{12.54}$$

The divergence of the vector potential **A** is specified to cause the above second-order differential equation for vector **A** to become an inhomogeneous wave equation for vector **A**. This particular choice of the divergence of the vector potential **A** is the Lorentz gauge, which requires the above gradient term to become zero. That is, the Lorentz gauge is

$$\nabla \cdot \mathbf{A} + \frac{1}{c^2}\frac{\partial\phi}{\partial t} = 0 \tag{12.55}$$

As a result of the choice of the Lorentz gauge, one obtains the wave equation for vector potential **A**, which provides the experimentally correct description of electromagnetic wave propagation.

The scalar version of the wave equation is obtained from Coulomb's law and application of the Lorentz gauge of Eq. (12.55), where

$$\nabla \cdot \mathbf{E} = \frac{\rho}{\varepsilon_0} \Rightarrow \nabla \cdot \left(-\nabla\phi - \frac{\partial\mathbf{A}}{\partial t}\right) = \frac{\rho}{\varepsilon_0} \tag{12.56}$$

and

$$\nabla^2\phi + \frac{\partial}{\partial t}\nabla \cdot \mathbf{A} = -\frac{\rho}{\varepsilon_0} \tag{12.57}$$

Application of the Lorentz gauge shows

$$\nabla \cdot \mathbf{A} = -\frac{1}{c^2}\frac{\partial \phi}{\partial t} \tag{12.58}$$

One thus obtains the inhomogeneous wave equation for the scalar potential ϕ due to source charge density ρ, where

$$\nabla^2 \Phi - \frac{1}{c^2}\frac{\partial^2 \Phi}{\partial t^2} = -\frac{\rho}{\varepsilon_0} \tag{12.59}$$

12.3 Calculation of Scalar and Vector Potentials of an Oscillating Electric Dipole

Many sources of X-rays are produced by oscillating electric dipoles such as atomic electron oscillators, charge oscillations in the Earth's upper atmosphere or from space, or in the form of undulators and wigglers in synchrotron rings. In this section, we derive the vector potential $\mathbf{A}$ and scalar potential ϕ, which are produced by an oscillating electric dipole $\mathbf{p}_e$. The dipole is a pair of opposite polarity charges ($+e$ and $-e$), which are separated by a time-varying distance $s(t_s)$ or $|\mathbf{s}(t_s)|$, where the distance and orientation direction are described by the separation vector $\mathbf{s}(t_s)$, where

$$\mathbf{p}_e(t_s) = e\mathbf{s}(t_s) \tag{12.60}$$

The time-varying separation $\mathbf{s}(t_s)$ between the two charges $+e$ and $-e$ gives rise to velocity $\mathbf{v}(t_s)$, where

$$\frac{\mathrm{d}\mathbf{p}_e(t_s)}{\mathrm{d}t} = e\frac{\mathrm{d}\mathbf{s}(t_s)}{\mathrm{d}t} = e\mathbf{v}(t_s) \tag{12.61}$$

The oscillation amplitude $s(t_s)$ is assumed to be small compared to the wavelength λ of the emitted electromagnetic radiation with temporal frequency f and period $T = 1/f$, where

$$s(t_s) \ll \lambda = \frac{c}{f} = cT \tag{12.62}$$

The effective time-dependent velocity v of the positive and negative charges of the oscillating dipole is $v = s(t_s)/T$, where

$$\frac{s}{\lambda} = \frac{fs}{c} = \frac{s}{cT} = \frac{v}{c} \ll 1 \tag{12.63}$$

The dipole charges $+e$ and $-e$ oscillate about a fixed center position at the origin of a spherical coordinate system in a fixed direction, where dipole center position $\mathbf{r}_s = 0$. Since the oscillating dipole charges have nonrelativistic velocities, the scalar $\phi(\mathbf{r}, t)$ and vector $\mathbf{A}(\mathbf{r}, t)$ Lienard–Wiechert potentials for an electric charge e, which moves at arbitrary velocity $\mathbf{v}$ at fixed distance R from the fixed observer position $\mathbf{r}$, are approximated by

$$\phi(R, t) = \frac{e}{4\pi\varepsilon_0 R} \tag{12.64}$$

$$\mathbf{A}(R, t) = \frac{e\mathbf{v}(t_s)}{4\pi\varepsilon_0 c^2 R} \tag{12.65}$$

With substitution from Eq. (12.61) for $\mathbf{v}(t_s)$, the vector potential $\mathbf{A}(R, t)$ for the oscillating electric dipole becomes

$$\mathbf{A}(R, t) = \frac{1}{4\pi\varepsilon_0 c^2 R}\frac{\partial \mathbf{p}_e}{\partial t} \tag{12.66}$$

At earlier time t_s, the oscillating charges of the stationary dipole at position $\mathbf{r}_s$ emit an electromagnetic wave, which travels at the speed of light c, and arrives at present time t at the fixed observer position $\mathbf{r}(t)$ at fixed distance R from the stationary dipole, where

$$t = t_s + \frac{R}{c} \tag{12.67}$$

As a result of Eq. (12.67),

$$\frac{\partial \mathbf{p}_e(t_s)}{\mathrm{d}t} = \frac{\partial \mathbf{p}_e(t - R/c)}{\mathrm{d}t} \tag{12.68}$$

Hence,

$$\frac{\partial \mathbf{p}_e(t - R/c)}{\mathrm{d}R} = -\frac{1}{c}\frac{\partial \mathbf{p}_e(t - R/c)}{\mathrm{d}t} \tag{12.69}$$

Consider, for example, an oscillating electric dipole moment $\mathbf{p}_e(t_s)$ with amplitude $\mathbf{p}_0$ and oscillation frequency w, where

$$\mathbf{p}_e(t_s) = \mathbf{p}_e\left(t - \frac{R}{c}\right) = \mathbf{p}_0 \sin\left[w\left(t - \frac{R}{c}\right)\right] \tag{12.70}$$

A single, unique solution of the magnetic $\mathbf{B}(R,t)$ and electric $\mathbf{E}(R,t)$ fields, which is expressed in terms of the scalar and vector potentials, require that $\phi(R,t)$ and $\mathbf{A}(R,t)$ satisfy the Lorentz condition, where

$$\nabla \cdot \mathbf{A} + \frac{1}{c^2}\frac{\partial \phi}{\partial t} = 0 \tag{12.71}$$

Substitution into Eq. (12.71) of Eq. (12.66) for $\mathbf{A}(R,t)$ in the Lorentz gauge allows the determination of scalar potential $\phi(R,t)$:

$$\frac{\partial \phi}{\partial t} = -\nabla \cdot \left(\frac{1}{4\pi\varepsilon_0 R}\frac{\partial \mathbf{p}_e(t_s)}{\partial t}\right) \tag{12.72}$$

Integration over observer time t yields

$$\phi = -\frac{1}{4\pi\varepsilon_0}\left(\nabla \cdot \frac{\mathbf{p}_e}{R}\right) \tag{12.73}$$

The vector identity in Eq. (12.74) below is now applied to Eq. (12.73), where

$$\nabla \cdot \frac{\mathbf{p}_e}{R} = \frac{1}{R}(\nabla \cdot \mathbf{p}_e) + (\mathbf{p}_e \cdot \nabla)\frac{1}{R} \tag{12.74}$$

The angle θ is between the vector $\mathbf{s}(t_s)$ of the oscillating dipole and unit vector $\hat{\mathbf{R}}$, which points from the dipole to the observer, where

$$(\mathbf{p}_e \cdot \nabla)\frac{1}{R} = -\frac{\hat{\mathbf{R}}\cdot\mathbf{p}_e}{R^2} = -\frac{p_e \cos\theta}{R^2} \tag{12.75}$$

Recall (Eq. 12.69), where

$$\frac{\partial \mathbf{p}_e}{\partial R} = -\frac{1}{c}\frac{\partial \mathbf{p}_e}{\partial t} \tag{12.76}$$

One evaluates the first term on the right side of Eq. (12.74) with application of Eq. (12.76), where

$$\frac{1}{R}(\nabla \cdot \mathbf{p}_e) = \frac{1}{R}\frac{\partial(\mathbf{p}_e \cdot \hat{\mathbf{R}})}{\partial R} = -\frac{\hat{\mathbf{R}}}{Rc}\cdot\frac{\partial \mathbf{p}_e}{\partial t} = -\frac{\cos\theta}{Rc}\frac{\partial p_e}{\partial t} \tag{12.77}$$

With substitution of Eqs. (12.75) and (12.77) into Eq. (12.74), followed by substitution of Eq. (12.74) into Eq. (12.73), the scalar potential $\phi(R,t)$ for the oscillating electric dipole becomes

$$\phi = \frac{1}{4\pi\varepsilon_0}\left(\frac{\cos\theta}{Rc}\frac{\mathrm{d}p_e}{\mathrm{d}t} + \frac{p_e \cos\theta}{R^2}\right) \tag{12.78}$$

In vector form, $\phi(\mathbf{r}, t)$ for the oscillating electric dipole is obtained by application of Eqs. (12.75) and (12.77) to Eq. (12.78), where

$$\phi = \frac{1}{4\pi\varepsilon_0}\left(\frac{\hat{\mathbf{R}}}{Rc}\cdot\frac{\partial \mathbf{p}_e}{\partial t} + \frac{\hat{\mathbf{R}}\cdot\mathbf{p}_e}{R^2}\right) \tag{12.79}$$

12.4 Calculation of a Magnetic Field of an Oscillating Electric Dipole via the Vector Potential

In this section, we derive the magnetic field, whcih is produced by an oscillating electric dipole $\mathbf{p}_e$, via the vector potential $\mathbf{A}$, derived in Section 12.3. The magnetic field $\mathbf{B}$ is the curl of the vector potential $\mathbf{A}$:

$$\mathbf{B} = \nabla \times \mathbf{A} \tag{12.80}$$

With substitution of Eq. (12.66) into Eq. (12.80), the magnetic field $\mathbf{B}$ of the oscillating electric dipole becomes

$$\mathbf{B} = \frac{1}{4\pi\varepsilon_0 c^2}\nabla \times \left(\frac{1}{R}\frac{\partial \mathbf{p}_e}{\partial t}\right) \tag{12.81}$$

Next, one employs the vector identity

$$\nabla \times \left(\frac{1}{R}\frac{\partial \mathbf{p}_e}{\partial t}\right) = \nabla\frac{1}{R} \times \frac{\partial \mathbf{p}_e}{\partial t} + \frac{1}{R}\nabla \times \frac{\partial \mathbf{p}_e}{\partial t} \tag{12.82}$$

where

$$\nabla \times \frac{\partial \mathbf{p}_e}{\partial t} = \hat{\mathbf{R}} \times \left(\frac{\partial}{\partial t}\frac{\partial \mathbf{p}_e}{\partial R}\right) = -\frac{1}{c}\hat{\mathbf{R}} \times \frac{\partial^2 \mathbf{p}_e}{\partial R^2} \tag{12.83}$$

With application of Eqs. (12.82) and (12.83) to Eq. (12.81), the magnetic field $\mathbf{B}(R, t)$ for an oscillating electric dipole $\mathbf{p}_e(t - R/c)$ becomes

$$\mathbf{B}(\mathbf{r}, t) = \frac{1}{4\pi\varepsilon_0 c^2}\left(\frac{\partial \mathbf{p}_e}{\partial t} \times \frac{\hat{\mathbf{R}}}{R^2}\right) + \frac{1}{4\pi\varepsilon_0 c^3}\left(\frac{\partial^2 \mathbf{p}_e}{\partial t^2} \times \frac{\hat{\mathbf{R}}}{R}\right) \tag{12.84}$$

Angle θ is between oscillating dipole unit vector $\hat{\mathbf{p}}_e$, and the unit vector $\hat{\mathbf{R}}$ points from dipole to observer, where the dipole unit vector is

$$\hat{\mathbf{p}}_e = \frac{\mathbf{p}_e}{|\mathbf{p}_e|} \tag{12.85}$$

Alternatively, the magnetic field $\mathbf{B}(R,t)$ for an oscillating electric dipole in Eq. (12.84) can be expressed by

$$\mathbf{B}(R,t) = \frac{\sin\theta}{4\pi\varepsilon_0 c^2 R^2}\frac{\partial p_e}{\partial t}(\hat{\mathbf{p}}_e \times \hat{\mathbf{R}}) + \frac{\sin\theta}{4\pi\varepsilon_0 c^3 R}\frac{\partial^2 p_e}{\partial t^2}(\hat{\mathbf{p}}_e \times \hat{\mathbf{R}}) \tag{12.86}$$

12.5 Calculation of the Electric Field Emitted from an Oscillating Electric Dipole

Calculation of the electric field $\mathbf{E}$ for an oscillating electric dipole $\mathbf{p}_e$ is more involved. The electric $\mathbf{E}$ is obtained from the scalar ϕ and vector $\mathbf{A}$ potentials derived previously, where

$$\mathbf{E} = -\nabla\phi - \frac{\partial \mathbf{A}}{\partial t} \tag{12.87}$$

From the derivation in Section 12.4, the scalar ϕ and vector $\mathbf{A}$ potentials are

$$\phi = \frac{1}{4\pi\varepsilon_0}\left(\frac{\hat{\mathbf{R}}}{Rc}\cdot\frac{\partial \mathbf{p}_e}{\partial t} + \frac{\hat{\mathbf{R}}\cdot\mathbf{p}_e}{R^2}\right) \tag{12.88}$$

$$\mathbf{A}(R,t) = \frac{1}{4\pi\varepsilon_0 c^2 R}\frac{\partial \mathbf{p}_e}{\partial t} \tag{12.89}$$

With substitution of Eq. (12.88) and (12.89) into Eq. (12.87), the electric field of the oscillating electric dipole is

$$\mathbf{E} = -\frac{1}{4\pi\varepsilon_0}\nabla\left(\frac{\hat{\mathbf{R}}}{Rc}\cdot\frac{\partial \mathbf{p}_e}{\partial t} + \frac{\hat{\mathbf{R}}\cdot\mathbf{p}_e}{R^2}\right) - \frac{1}{4\pi\varepsilon_0 c^2 R}\frac{\partial^2 \mathbf{p}_e(t_s)}{\partial t^2} \tag{12.90}$$

Next, one applies the following vector identity to Eq. (12.90):

$$\nabla(\mathbf{F}\cdot\mathbf{G}) = (\mathbf{F}\cdot\nabla)\mathbf{G} + (\mathbf{G}\cdot\nabla)\mathbf{F} + \mathbf{F}\times(\nabla\times\mathbf{G}) + \mathbf{G}\times(\nabla\times\mathbf{F}) \tag{12.91}$$

The first two terms in parenthesis on the right side of Eq. (12.90) become

$$\nabla\left(\frac{\hat{\mathbf{R}}}{Rc}\cdot\frac{\partial \mathbf{p}_e}{\partial t}\right) = \left(\frac{\hat{\mathbf{R}}}{Rc}\cdot\nabla\right)\frac{\partial \mathbf{p}_e}{\partial t} + \left(\frac{\partial \mathbf{p}_e}{\partial t}\cdot\nabla\right)\frac{\hat{\mathbf{R}}}{Rc} + \frac{\hat{\mathbf{R}}}{Rc}\times\left(\nabla\times\frac{\partial \mathbf{p}_e}{\partial t}\right) + \frac{\partial \mathbf{p}_e}{\partial t}\times\left(\nabla\times\frac{\hat{\mathbf{R}}}{Rc}\right) \tag{12.92}$$

and

$$\nabla\left(\frac{\hat{\mathbf{R}}\cdot\mathbf{p}_e}{R^2}\right)=\nabla\left(\frac{\hat{\mathbf{R}}}{R^2}\cdot\mathbf{p}_e\right)=\left(\frac{\hat{\mathbf{R}}}{R^2}\cdot\nabla\right)\mathbf{p}_e+(\mathbf{p}_e\cdot\nabla)\frac{\hat{\mathbf{R}}}{R^2}+\frac{\hat{\mathbf{R}}}{R^2}\times(\nabla\times\mathbf{p}_e)+\mathbf{p}_e\times\left(\nabla\times\frac{\hat{\mathbf{R}}}{R^2}\right) \tag{12.93}$$

The last term of Eq. (12.92) is evaluated, where

$$\nabla\times\frac{\hat{\mathbf{R}}}{R}=\begin{vmatrix}\hat{\mathbf{R}} & \hat{\boldsymbol{\theta}} & \hat{\phi}\\ \dfrac{\partial}{\partial R} & \dfrac{1}{R}\dfrac{\partial}{\partial\theta} & \dfrac{1}{R\sin\theta}\dfrac{\partial}{\partial\phi}\\ \dfrac{\hat{\mathbf{R}}}{R} & 0 & 0\end{vmatrix}=0 \tag{12.94}$$

One then invokes the vector identity in Eq. (12.95) below to evaluate the second terms on the right sides of Eq. (12.92) and (12.93):

$$\begin{aligned}(\mathbf{C}\cdot\nabla)\hat{\mathbf{R}}=(\mathbf{C}\cdot\nabla)\frac{\mathbf{R}}{R}&=\left(C_x\frac{\partial}{\partial x}+C_y\frac{\partial}{\partial y}+C_z\frac{\partial}{\partial z}\right)\left(\frac{x\hat{\mathbf{x}}+y\hat{\mathbf{y}}+z\hat{\mathbf{z}}}{\sqrt{x^2+y^2+z^2}}\right)\\ &=\frac{C_x\hat{\mathbf{x}}+C_y\hat{\mathbf{y}}+C_z\hat{\mathbf{z}}}{\sqrt{x^2+y^2+z^2}}-\frac{C_xx+C_yy+C_zz}{(x^2+y^2+z^2)^{3/2}}(x\hat{\mathbf{x}}+y\hat{\mathbf{y}}+z\hat{\mathbf{z}})\end{aligned} \tag{12.95}$$

Collecting terms, the useful vector identity of Eq. (12.95) becomes,

$$(\mathbf{C}\cdot\nabla)\hat{\mathbf{R}}=\frac{\mathbf{C}-\hat{\mathbf{R}}(\mathbf{C}\cdot\hat{\mathbf{R}})}{R} \tag{12.96}$$

Another pair of useful vector identities invoked is

$$\begin{aligned}&(\mathbf{C}\cdot\nabla)\frac{1}{R}\\ &=\left(C_x\frac{\partial}{\partial x}+C_y\frac{\partial}{\partial y}+C_z\frac{\partial}{\partial z}\right)\left(\frac{1}{\sqrt{x^2+y^2+z^2}}\right)=-\frac{1}{2}\frac{2C_xx+2C_yy+2C_zz}{(x^2+y^2+z^2)^{3/2}}=-\mathbf{C}\cdot\frac{\hat{\mathbf{R}}}{R^2}\end{aligned} \tag{12.97}$$

and

$$(\mathbf{C}\cdot\nabla)\frac{1}{R^2} = \left(C_x\frac{\partial}{\partial x} + C_y\frac{\partial}{\partial y} + C_z\frac{\partial}{\partial z}\right)\left(\frac{1}{x^2+y^2+z^2}\right) = -\frac{2C_x x + 2C_y y + 2C_z z}{(x^2+y^2+z^2)^2}$$

$$= -2\mathbf{C}\cdot\frac{\hat{\mathbf{R}}}{R^3} \tag{12.98}$$

Application of Eqs. (12.96–12.97) to the second term on the right side of Eq. (12.92) yields

$$\left(\frac{\partial\mathbf{p}_e}{\partial t}\cdot\nabla\right)\frac{\hat{\mathbf{R}}}{Rc} = \frac{1}{R^2c}\left[\frac{\partial\mathbf{p}_e}{\partial t} - 2\hat{\mathbf{R}}\left(\frac{\partial\mathbf{p}_e}{\partial t}\cdot\hat{\mathbf{R}}\right)\right] \tag{12.99}$$

Application of Eqs. (12.96) and (12.98) to the second term on the right side of Eq. (12.93) yields

$$(\mathbf{p}_e\cdot\nabla)\frac{\hat{\mathbf{R}}}{R^2} = \frac{\mathbf{p}_e - 3\hat{\mathbf{R}}(\mathbf{p}_e\cdot\hat{\mathbf{R}})}{R^3} \tag{12.100}$$

The first term on the right side of Eq. (12.92) is evaluated, where

$$\left(\frac{\hat{\mathbf{R}}}{Rc}\cdot\nabla\right)\frac{\partial\mathbf{p}_e}{\partial t} = \left[\frac{\hat{\mathbf{R}}}{Rc}\cdot\left(\hat{\mathbf{R}}\frac{\partial}{\partial R} + \hat{\boldsymbol{\theta}}\frac{1}{R}\frac{\partial}{\partial\theta} + \hat{\varphi}\frac{1}{R\sin\theta}\frac{\partial}{\partial\phi}\right)\right]\frac{\partial\mathbf{p}_e}{\partial t} = \frac{1}{Rc}\frac{\partial}{\partial R}\left(\frac{\partial\mathbf{p}_e}{\partial t}\right)$$

$$= -\frac{1}{c^2R}\frac{\partial^2\mathbf{p}_e}{\partial t^2} \tag{12.101}$$

The first term on the right side of Eq. (12.93) is evaluated, where

$$\left(\frac{\hat{\mathbf{R}}}{R^2}\cdot\nabla\right)\mathbf{p}_e = -\frac{1}{R^2c}\frac{\partial\mathbf{p}_e}{\partial t} \tag{12.102}$$

With substitution of Eqs. (12.94), (12.99), and (12.101) into Eq. (12.92), one obtains

$$\nabla\left(\frac{\hat{\mathbf{R}}}{Rc}\cdot\frac{\partial\mathbf{p}_e}{\partial t}\right) = -\frac{1}{c^2R}\frac{\partial^2\mathbf{p}_e}{\partial t^2} + \frac{1}{R^2c}\left[\frac{\partial\mathbf{p}_e}{\partial t} - 2\hat{\mathbf{R}}\left(\frac{\partial\mathbf{p}_e}{\partial t}\cdot\hat{\mathbf{R}}\right)\right] + \frac{\hat{\mathbf{R}}}{Rc}\times\left(\nabla\times\frac{\partial\mathbf{p}_e}{\partial t}\right) \tag{12.103}$$

With substitution of Eqs. (12.100), and (12.102) into Eq. (12.93), one obtains

$$\nabla\left(\frac{\hat{\mathbf{R}}\cdot\mathbf{p}_e}{R^2}\right) = -\frac{1}{R^2c}\frac{\partial\mathbf{p}_e}{\partial t} + \frac{\mathbf{p}_e}{R^3} - \frac{3\hat{\mathbf{R}}(\mathbf{p}_e\cdot\hat{\mathbf{R}})}{R^3} + \frac{\hat{\mathbf{R}}}{R^2}\times(\nabla\times\mathbf{p}_e) \tag{12.104}$$

We now evaluate the last terms on the right sides of Eqs. (12.103) and (12.104) as follows. The dipole charges $+e$ and $-e$ oscillate in a fixed direction, so the emission of electromagnetic radiation is azimuthally symmetric and

$$\frac{\mathrm{d}\mathbf{p}_e}{\mathrm{d}\theta}=0 \quad \text{and} \quad \frac{\mathrm{d}\mathbf{p}_e}{\mathrm{d}\phi}=0 \tag{12.105}$$

The cross product $\hat{\mathbf{R}}\times\hat{\mathbf{R}}=0$, so the last term of Eq. (12.103) becomes

$$\frac{\hat{\mathbf{R}}}{Rc}\times\left(\nabla\times\frac{\partial\mathbf{p}_e}{\partial t}\right)=\frac{\hat{\mathbf{R}}}{Rc}\times\begin{vmatrix}\hat{\mathbf{R}} & \hat{\boldsymbol{\theta}} & \hat{\boldsymbol{\varphi}}\\ \dfrac{\partial}{\partial R} & \dfrac{1}{R}\dfrac{\partial}{\partial\theta} & \dfrac{1}{R\sin\theta}\dfrac{\partial}{\partial\phi}\\ \left(\dfrac{\partial p_e}{\partial t}\right)_{\hat{\mathbf{R}}} & \left(\dfrac{\partial p_e}{\partial t}\right)_{\hat{\boldsymbol{\theta}}} & \left(\dfrac{\partial p_e}{\partial t}\right)_{\hat{\boldsymbol{\varphi}}}\end{vmatrix}$$

$$=\frac{1}{Rc}\begin{vmatrix}0 & \hat{\boldsymbol{\varphi}} & -\hat{\boldsymbol{\theta}}\\ -\dfrac{1}{c}\dfrac{\partial}{\partial t} & 0 & 0\\ \left(\dfrac{\partial p_e}{\partial t}\right)_{\hat{\mathbf{R}}} & \left(\dfrac{\partial p_e}{\partial t}\right)_{\hat{\boldsymbol{\theta}}} & \left(\dfrac{\partial p_e}{\partial t}\right)_{\hat{\boldsymbol{\varphi}}}\end{vmatrix} \tag{12.106}$$

Evaluation of the determinant of Eq. (12.106) yields:

$$\frac{\hat{\mathbf{R}}}{Rc}\times\left(\nabla\times\frac{\partial\mathbf{p}_e}{\partial t}\right)=\frac{1}{c^2R}\left(\frac{\partial^2 p_e}{\partial t^2}\right)_{\hat{\boldsymbol{\varphi}}}\hat{\boldsymbol{\varphi}}+\frac{1}{c^2R}\left(\frac{\partial^2 p_e}{\partial t^2}\right)_{\hat{\boldsymbol{\theta}}}\hat{\boldsymbol{\theta}} \tag{12.107}$$

One can simplify Eq. (12.107) further, so that the last tem of Eq. (12.103) becomes:

$$\frac{\hat{\mathbf{R}}}{Rc}\times\left(\nabla\times\frac{\partial\mathbf{p}_e}{\partial t}\right)=\frac{1}{c^2R}\left[\frac{\partial^2\mathbf{p}_e}{\partial t^2}-\hat{\mathbf{R}}\left(\hat{\mathbf{R}}\cdot\frac{\partial^2\mathbf{p}_e}{\partial t^2}\right)\right] \tag{12.108}$$

The result of Eq. (12.108), without the time derivative, yields the last term of Eq. (12.104), where

$$\frac{\hat{\mathbf{R}}}{R^2}\times(\nabla\times\mathbf{p}_e)=\frac{1}{R^2c}\left[\frac{\partial\mathbf{p}_e}{\partial t}-\hat{\mathbf{R}}\left(\hat{\mathbf{R}}\cdot\frac{\partial\mathbf{p}_e}{\partial t}\right)\right] \tag{12.109}$$

Substitution of Eq. (12.108) into Eq. (12.103) yields

$$\nabla\left(\frac{\hat{\mathbf{R}}}{Rc}\cdot\frac{\partial \mathbf{p}_e}{\partial t}\right)=-\frac{1}{c^2R}\frac{\partial^2\mathbf{p}_e}{\partial t^2}+\frac{1}{R^2c}\left[\frac{\partial \mathbf{p}_e}{\partial t}-2\hat{\mathbf{R}}\left(\frac{\partial \mathbf{p}_e}{\partial t}\cdot\hat{\mathbf{R}}\right)\right]+\frac{1}{c^2R}\left[\frac{\partial^2\mathbf{p}_e}{\partial t^2}-\hat{\mathbf{R}}\left(\hat{\mathbf{R}}\cdot\frac{\partial^2\mathbf{p}_e}{\partial t^2}\right)\right] \tag{12.110}$$

Substitution of Eq. (12.109) into Eq. (12.104) yields

$$\nabla\left(\frac{\hat{\mathbf{R}}\cdot\mathbf{p}_e}{R^2}\right)=-\frac{1}{R^2c}\frac{\partial \mathbf{p}_e}{\partial t}+\frac{\mathbf{p}_e}{R^3}-\frac{3\hat{\mathbf{R}}(\mathbf{p}_e\cdot\hat{\mathbf{R}})}{R^3}+\frac{1}{cR^2}\left[\frac{\partial \mathbf{p}_e}{\partial t}-\hat{\mathbf{R}}\left(\hat{\mathbf{R}}\cdot\frac{\partial \mathbf{p}_e}{\partial t}\right)\right] \tag{12.111}$$

With substitution of Eqs. (12.110) and (12.111) into Eq. (12.90), the electric field $\mathbf{E}(R,t)$ for an oscillating electric dipole $\mathbf{p}_e(t-R/c)$ becomes

$$\mathbf{E}(R,t)=\frac{1}{4\pi\varepsilon_0}\times$$

$$\left[-\frac{1}{c^2R}\frac{\partial^2\mathbf{p}_e(t_s)}{\partial t^2}+\frac{\hat{\mathbf{R}}}{c^2R}\left(\frac{\partial^2\mathbf{p}_e}{\partial t^2}\cdot\hat{\mathbf{R}}\right)-\frac{1}{cR^2}\frac{\partial \mathbf{p}_e}{\partial t}+3\frac{\hat{\mathbf{R}}}{cR^2}\left(\frac{\partial \mathbf{p}_e}{\partial t}\cdot\hat{\mathbf{R}}\right)-\frac{\mathbf{p}_e}{R^3}+\frac{3\hat{\mathbf{R}}(\mathbf{p}_e\cdot\hat{\mathbf{R}})}{R^3}\right] \tag{12.112}$$

12.6 Conversion of Electric and Magnetic Fields from MKS Units to CGS Units

This section summarizes the electric and magnetic fields of an oscillating electric dipole and presents the conversion constants to obtain the electric and magnetic fields in CGS units given their expressions in MKS units. To recap, the electric dipole $\mathbf{p}_e$ is a pair of opposite polarity charges $+e$ and $-e$, which are separated by a time-varying distance $s(t_s)$ or $|\mathbf{s}(t_s)|$, where the distance and orientation direction are described by the separation vector $\mathbf{s}(t_s)$, so that

$$\mathbf{p}_e\left(t-\frac{R}{c}\right)=\mathbf{p}_e(t_s)=e\mathbf{s}(t_s) \tag{12.113}$$

and

$$\frac{\mathrm{d}\mathbf{p}_e(t_s)}{\mathrm{d}t}=e\frac{\mathrm{d}\mathbf{s}(t_s)}{\mathrm{d}t}=e\mathbf{v}(t_s)=e\mathbf{v}\left(t-\frac{R}{c}\right) \tag{12.114}$$

The electric field $\mathbf{E}(R,t)$ and magnetic flux density $\mathbf{B}(R,t)$ for an oscillating electric dipole $\mathbf{p}_e(t-R/c)$ in MKS units are obtained from Eqs. (12.112) and (12.85), where

$$\mathbf{E}(R,t) = \frac{1}{4\pi\varepsilon_0} \times$$

$$\left[-\frac{1}{c^2R}\frac{\partial^2\mathbf{p}_e(t_s)}{\partial t^2} + \frac{\hat{\mathbf{R}}}{c^2R}\left(\frac{\partial^2\mathbf{p}_e}{\partial t^2}\cdot\hat{\mathbf{R}}\right) - \frac{1}{cR^2}\frac{\partial\mathbf{p}_e}{\partial t} + 3\frac{\hat{\mathbf{R}}}{cR^2}\left(\frac{\partial\mathbf{p}_e}{\partial t}\cdot\hat{\mathbf{R}}\right) - \frac{\mathbf{p}_e}{R^3} + \frac{3\hat{\mathbf{R}}(\mathbf{p}_e\cdot\hat{\mathbf{R}})}{R^3}\right] \tag{12.115}$$

and

$$\mathbf{B}(R,t) = \frac{1}{4\pi\varepsilon_0c^2}\left(\frac{\partial\mathbf{p}_e}{\partial t}\times\frac{\hat{\mathbf{R}}}{R^2}\right) + \frac{1}{4\pi\varepsilon_0c^3}\left(\frac{\partial^2\mathbf{p}_e}{\partial t^2}\times\frac{\hat{\mathbf{R}}}{R}\right) \tag{12.116}$$

If the electric dipole is static, then the time-varying terms are zero (i.e., the magnetic field is zero), and one is left with the static electric produced by a stationary pair of constant charges.

The electric $\mathbf{E}_{\mathrm{mks}}(R,t)$ and magnetic flux density $\mathbf{B}_{\mathrm{mks}}(R,t)$ fields of Eqs. (12.115) and (12.116) in their MKS units are converted to electric field $\mathbf{E}_{\mathrm{cgs}}(R,t)$ and magnetic field intensity $\mathbf{H}_{\mathrm{cgs}}(R,t)$ in CGS units as follows:

1. Multiply MKS $\mathbf{B}_{\mathrm{mks}}(R,t)$ magnetic flux density by $4\pi/\mu_0c$ to obtains CGS magnetic field intensity $\mathbf{H}_{\mathrm{cgs}}(R,t)$, where vacuum permeability μ_0 and speed of light c are in MKS units:

$$\mathbf{H}_{\mathrm{cgs}}(R,t) = \frac{4\pi}{\mu_0c}\mathbf{B}_{\mathrm{mks}}(R,t) \tag{12.117}$$

2. Multiply MKS electric field $\mathbf{E}_{\mathrm{mks}}(R,t)$ by $4\pi\varepsilon_0$ to obtain CGS electric field $\mathbf{E}_{\mathrm{cgs}}(R,t)$, where vacuum permittivity ε_0 is in MKS units:

$$\mathbf{E}_{\mathrm{cgs}}(R,t) = 4\pi\varepsilon_0\mathbf{E}_{\mathrm{mks}}(R,t) \tag{12.118}$$

The electric field $\mathbf{E}_{\mathrm{cgs}}(R,t)$ and magnetic field intensity $\mathbf{H}_{\mathrm{cgs}}(R,t)$ of the electric dipole in CGS units are

$$\mathbf{E}_{\mathrm{cgs}}(R,t) = -\frac{1}{c^2R}\frac{\partial^2\mathbf{p}_e(t_s)}{\partial t^2} + \frac{\hat{\mathbf{R}}}{c^2R}\left(\frac{\partial^2\mathbf{p}_e}{\partial t^2}\cdot\hat{\mathbf{R}}\right) - \frac{1}{cR^2}\frac{\partial\mathbf{p}_e}{\partial t} + 3\frac{\hat{\mathbf{R}}}{cR^2}\left(\frac{\partial\mathbf{p}_e}{\partial t}\cdot\hat{\mathbf{R}}\right) - \frac{\mathbf{p}_e}{R^3} + \frac{3\hat{\mathbf{R}}(\mathbf{p}_e\cdot\hat{\mathbf{R}})}{R^3} \tag{12.119}$$

and

$$\mathbf{H}_{\mathrm{cgs}}(R,t) = \frac{1}{c}\left(\frac{\partial\mathbf{p}_e}{\partial t}\times\frac{\hat{\mathbf{R}}}{R^2}\right) + \frac{1}{c^2}\left(\frac{\partial^2\mathbf{p}_e}{\partial t^2}\times\frac{\hat{\mathbf{R}}}{R}\right) \tag{12.120}$$

12.7 Near- and Far-Zone Electric Fields and Power Emission from an Oscillating Electric Dipole

The electric and magnetic fields of the oscillating electric dipole are now examined for distances very close to the dipole, and distances very far from the dipole. In the near zone, the observer distance R from the dipole is small, and the $1/R^3$ terms dominate. The near-zone electric $\mathbf{E}(R,t)$ and magnetic $\mathbf{B}(R,t)$ fields of the oscillating electric dipole are obtained from Eqs. (12.115) and (12.116), where

$$\mathbf{E}(R,t) = \frac{1}{4\pi\varepsilon_0}\left[-\frac{\mathbf{p}_e}{R^3} + \frac{3\hat{\mathbf{R}}(\mathbf{p}_e \cdot \hat{\mathbf{R}})}{R^3}\right] \tag{12.121}$$

and

$$\mathbf{B}(R,t) = \frac{1}{4\pi\varepsilon_0 c^2}\left(\frac{\partial \mathbf{p}_e}{\partial t} \times \frac{\hat{\mathbf{R}}}{R^2}\right) \tag{12.122}$$

The electric $\mathbf{E}(R,t)$ and magnetic $\mathbf{B}(R,t)$ fields in the near-zone are not perpendicular because

$$\hat{\mathbf{R}} \times \mathbf{E}(R,t) = \frac{1}{4\pi\varepsilon_0}\hat{\mathbf{R}} \times \left[-\frac{\mathbf{p}_e}{R^3} + \frac{3\hat{\mathbf{R}}(\mathbf{p}_e \cdot \hat{\mathbf{R}})}{R^3}\right] = \mathbf{p}_e \times \frac{\hat{\mathbf{R}}}{4\pi\varepsilon_0 R^3} \tag{12.123}$$

and

$$c\mathbf{B}(R,t) = \frac{1}{4\pi\varepsilon_0 c}\left(\frac{\partial \mathbf{p}_e}{\partial t} \times \frac{\hat{\mathbf{R}}}{R^2}\right) \neq \hat{\mathbf{R}} \times \mathbf{E}(\mathbf{r},t) = \mathbf{p}_e \times \frac{\hat{\mathbf{R}}}{4\pi\varepsilon_0 c R^3} \tag{12.124}$$

This is because in general,

$$\frac{\partial \mathbf{p}_e}{\partial t} \times \mathbf{R} \neq \mathbf{p}_e \times \frac{\hat{\mathbf{R}}}{R} \tag{12.125}$$

A strange and short-lived exception to this rule would be

$$\mathbf{p}_e = \mathbf{p}_0 \exp\left(\frac{t}{R}\right) \tag{12.126}$$

In the far-zone or wave-zone, the observer distance R from the dipole is large, and the $1/R$ terms dominate. The near-zone electric $\mathbf{E}(R,t)$ and magnetic $\mathbf{B}(R,t)$

fields of the oscillating electric dipole are obtained from Eqs. (12.115) and (12.116), where

$$\mathbf{E}(R,t)=\frac{1}{4\pi\varepsilon_0}\left[-\frac{1}{c^2R}\frac{\partial^2\mathbf{p}_e}{\partial t^2}+\frac{\hat{\mathbf{R}}}{c^2R}\left(\hat{\mathbf{R}}\cdot\frac{\partial^2\mathbf{p}_e}{\partial t^2}\right)\right] \tag{12.127}$$

and

$$\mathbf{B}(R,t)=\frac{1}{4\pi\varepsilon_0c^3}\left(\frac{\partial^2\mathbf{p}_e}{\partial t^2}\times\frac{\hat{\mathbf{R}}}{R}\right) \tag{12.128}$$

The electric $\mathbf{E}(R,t)$ and magnetic $\mathbf{B}(R,t)$ fields of Eqs. (12.127) and (12.128) in the far-zone are perpendicular:

$$c\mathbf{B}(R,t)=\frac{1}{4\pi\varepsilon_0c^2}\left(\frac{\partial^2\mathbf{p}_e}{\partial t^2}\times\frac{\hat{\mathbf{R}}}{R}\right)=\hat{\mathbf{R}}\times\mathbf{E}(\mathbf{r},t)=\frac{1}{4\pi\varepsilon_0c^2}\left(\frac{\partial^2\mathbf{p}_e}{\partial t^2}\times\frac{\hat{\mathbf{R}}}{R}\right) \tag{12.129}$$

In CGS units, recalling section 12.6, the electric field $\mathbf{E}_{\text{cgs}}(R,t)$ and magnetic field $\mathbf{H}_{\text{cgs}}(R,t)$ intensities are

$$\mathbf{E}_{\text{cgs}}(R,t)=-\frac{1}{c^2R}\frac{\partial^2\mathbf{p}_e}{\partial t^2}+\frac{\hat{\mathbf{R}}}{c^2R}\left(\hat{\mathbf{R}}\cdot\frac{\partial^2\mathbf{p}_e}{\partial t^2}\right) \tag{12.130}$$

$$\mathbf{H}_{\text{cgs}}(R,t)=\frac{\partial^2\mathbf{p}_e}{\partial t^2}\times\frac{\hat{\mathbf{R}}}{R} \tag{12.131}$$

The Poynting vector $\mathbf{S}(\mathbf{r},t)$ is the power density transmitted by the charge in the form of the electromagnetic field, which is caused by the charge acceleration $\mathbf{a}$, where, in MKS units,

$$\mathbf{S}(\mathbf{r},t)=c^2\varepsilon_0\mathbf{E}(\mathbf{r},t)\times\mathbf{B}(\mathbf{r},t) \tag{12.132}$$

Examination of Eqs. (12.127) and (12.128), shows the magnetic field $\mathbf{B}(R,t)$ is obtained from the electric field $\mathbf{E}(\mathbf{r},t)$ by

$$\mathbf{B}(R,t)=\frac{1}{c}\hat{\mathbf{R}}\times\mathbf{E}(\mathbf{r},t) \tag{12.133}$$

The electric field $\mathbf{E}(R,t)$ is obtained from the magnetic field $\mathbf{B}(\mathbf{r},t)$, taking the cross product of both sides of Eq. (12.133) with the unit vector $\hat{\mathbf{R}}$ from dipole to observer, and then applying the BAC−CAB rule, where

$$\hat{\mathbf{R}}\times\mathbf{B}=\frac{1}{c}\hat{\mathbf{R}}\times(\hat{\mathbf{R}}\times\mathbf{E})=\frac{\hat{\mathbf{R}}}{c}(\hat{\mathbf{R}}\cdot\mathbf{E})-\frac{\mathbf{E}}{c}(\hat{\mathbf{R}}\cdot\hat{\mathbf{R}}) \tag{12.134}$$

The electric field $\mathbf{E}$ is perpendicular to the direction of wave propagation $\hat{\mathbf{R}}$, and the magnetic field $\mathbf{B}$ is perpendicular to $\hat{\mathbf{R}}$, where

$$\hat{\mathbf{R}}\cdot\mathbf{E}=0 \quad \hat{\mathbf{R}}\cdot\mathbf{B}=0 \tag{12.135}$$

With application of Eq. (12.135) to Eq. (12.134), one obtains the electric field $\mathbf{E}$ in terms of the magnetic field $\mathbf{B}$, where

$$\mathbf{E}=-c\hat{\mathbf{R}}\times\mathbf{B} \tag{12.136}$$

Application of the BAC−CAB rule to the Poynting vector equation of Eq. (12.132), after substituting for electric field $\mathbf{E}$ of Eq. (12.136), gives

$$\mathbf{S}=-c^3\varepsilon_0(\hat{\mathbf{R}}\times\mathbf{B})\times\mathbf{B}=c^3\varepsilon_0\mathbf{B}\times(\hat{\mathbf{R}}\times\mathbf{B})=c^3\varepsilon_0\left[\hat{\mathbf{R}}(\mathbf{B}\cdot\mathbf{B})-\mathbf{B}(\mathbf{B}\cdot\hat{\mathbf{R}})\right]=c^3\varepsilon_0\hat{\mathbf{R}}(\mathbf{B}\cdot\mathbf{B}) \tag{12.137}$$

With substitution of Eq. (12.128) into Eq. (12.137), the Poynting vector $\mathbf{S}(\mathbf{r},t)$ or power density (w/m^2) becomes

$$\mathbf{S}=\frac{\hat{\mathbf{R}}}{4\pi R^2}\left[\left(\frac{\partial^2\mathbf{p}_e}{\partial t^2}\times\hat{\mathbf{R}}\right)\cdot\left(\frac{\partial^2\mathbf{p}_e}{\partial t^2}\times\hat{\mathbf{R}}\right)\right] \tag{12.138}$$

Given angle θ between the electric dipole vector $\mathbf{p}_e(t-R/c)$ and the unit vector $\hat{\mathbf{R}}$ pointing from the dipole to observer, the Poynting vector $\mathbf{S}(R,t)$ of Eq. (12.138) becomes

$$\mathbf{S}(R,t)=\frac{\hat{\mathbf{R}}}{4\pi R^2}\left(\frac{\partial^2 p_e}{\partial t^2}\sin\theta\right)^2 \tag{12.139}$$

12.8 Frequency and Wave-Number Domain Expressions for Electric and Magnetic Fields Emission and Power from the Oscillating Electric Dipole

In this section, we derive the frequency and wave-number domain expressions for the electric and magnetic fields, and Poynting vector power emission from an oscillating electric dipole. We return to the general expression for the electric $\mathbf{E}(R,t)$ and magnetic $\mathbf{B}(R,t)$ fields of an electric dipole in steady-state oscillation, where

$$\mathbf{p}_e(t_s) = \mathbf{p}_e(t - R/c) = \mathbf{p}_0 \exp\left[-iw\left(t - \frac{R}{c}\right)\right] \tag{12.140}$$

$$\mathbf{E}(R,t) = \mathbf{E}_0 \exp\left[-iw\left(t - \frac{R}{c}\right)\right] \tag{12.141}$$

$$\mathbf{B}(R,t) = \mathbf{B}_0 \exp\left[-iw\left(t - \frac{R}{c}\right)\right] \tag{12.142}$$

$$w = kc = \frac{2\pi}{\lambda}c \tag{12.143}$$

From Eq. (12.139), the Poynting vector $\mathbf{S}(R,t)$ is

$$\mathbf{S}(R,t) = \mathbf{S}_0 \exp\left[-iw\left(t - \frac{R}{c}\right)\right] = \frac{\hat{\mathbf{R}}}{4\pi R^2}\left(\frac{\partial^2 p_e}{\partial t^2}\sin\theta\right)^2 \tag{12.144}$$

Substitution into Eq. (12.144) of Eq. (12.140) for p_e gives

$$\mathbf{S}_0 = \frac{\hat{\mathbf{R}}w^4}{4\pi R^2}(p_0 \sin\theta)^2 \tag{12.145}$$

The Fourier transform is performed on the dipole electric $\mathbf{E}(R,t)$ and magnetic $\mathbf{B}(R,t)$ field differential equations for $\mathbf{p}_e(t - R/c)$ to obtain the frequency domain expressions for the electric and magnetic fields, which are emitted from an oscillating electric dipole. The Fourier transform is effectively obtained by substituting $\mathbf{E}(R,t)$, $\mathbf{B}(R,t)$, and $\mathbf{p}_e(t - R/c)$ expressions of Eqs. (12.140–12.142) into the time-domain differential equations for the dipole electric $\mathbf{E}(R,t)$ and magnetic $\mathbf{B}(R,t)$ fields of Eqs. (12.115) and (12.116). With substitution of Eqs. (12.140) and (12.141) into Eq. (12.115), the electric field emitted by an oscillating electric dipole in the frequency domain $\mathbf{E}_0(R,w)$ is

$$\mathbf{E}_0(R,w) = \frac{1}{4\pi\varepsilon_0}\left[\frac{w^2\mathbf{p}_0}{c^2R} - w^2\frac{\hat{\mathbf{R}}}{c^2R}(\hat{\mathbf{R}}\cdot\mathbf{p}_0) + (3\hat{\mathbf{R}}(\hat{\mathbf{R}}\cdot\mathbf{p}_0) - \mathbf{p}_0)\left(\frac{1}{R^3} - \frac{iw}{cR^2}\right)\right] \tag{12.146}$$

With application of the BAC–CAB vector identity to the first two terms in Eq. (12.146), the frequency domain $\mathbf{E}_0(R,w)$ is also expressed by

$$\mathbf{E}_0(R,w) = \frac{1}{4\pi\varepsilon_0}\left[\frac{w^2((\hat{\mathbf{R}}\times\mathbf{p}_0)\times\hat{\mathbf{R}})}{c^2R} + (3\hat{\mathbf{R}}(\hat{\mathbf{R}}\cdot\mathbf{p}_0) - \mathbf{p}_0)\left(\frac{1}{R^3} - \frac{iw}{cR^2}\right)\right] \tag{12.147}$$

With substitution of Eqs. (12.140) and (12.142) into Eq. (12.116), the magnetic field emitted by an oscillating electric dipole in the frequency domain $\mathbf{B}_0(R, w)$ is

$$\mathbf{B}_0(R, w) = \frac{w^2}{4\pi\varepsilon_0 c^3}\left(\frac{\hat{\mathbf{R}} \times \mathbf{p}_0}{R}\right)\left(1 + \frac{\mathrm{i}}{wR}\right) \tag{12.148}$$

From Eq. (12.145), the power flow or Poynting vector $\mathbf{S}_0(R, w)$ in the frequency domain is

$$\mathbf{S}_0(R, w) = \frac{w^4}{4\pi R^2}(p_0 \sin\theta)^2\hat{\mathbf{R}} \tag{12.149}$$

The total electromagnetic power $P(w)$ in watts emitted through a spherical surface of radius R as a function of frequency w is

$$P(w) = \int_0^{2\pi} \mathrm{d}\phi \int_0^{\pi} \mathrm{d}\theta\left[\hat{\mathbf{R}}\cdot\mathbf{S}_0(R, w)\right]R^2 \sin\theta \tag{12.150}$$

Substitution of Eq. (12.149) into Eq. (12.150) and the trivial integration over the symmetric azimuthal angle yields

$$P(w) = \frac{w^4 p_0^2}{2}\int_0^{\pi} \mathrm{d}\theta \sin^3\theta \tag{12.151}$$

With integration over the polar angle, the instantaneous power $P(w)$ crossing the spherical surface becomes

$$P(w) = \frac{w^4 p_0^2}{2}\int_0^{\pi} \mathrm{d}\theta\left[\sin\theta - \sin\theta\cos^2\theta\right] = \frac{2w^4 p_0^2}{3} \tag{12.152}$$

Averaged over one oscillation period, the time-averaged power $\overline{P}(w)$ is one-half the instantaneous power $P(w)$ for sinusoidal power, so that

$$\overline{P}(w) = \frac{1}{2}P(w) = \frac{w^4 p_0^2}{3} \tag{12.153}$$

One obtains the time-averaged power $\overline{P}(w)$ via the time-averaged Poynting vector $\overline{\mathbf{S}}(R, t)$, where the Poynting vector $\mathbf{S}(R, t)$ is integrated over one oscillation period, so that

$$\overline{\mathbf{S}}(R, t) = \frac{1}{T}\int_0^{T} \mathrm{Re}[\mathbf{S}(R, t)]\mathrm{d}t = \frac{S_0}{T}\int_0^{1/f} \cos^2\left[2\pi f\left(t - \frac{R}{c}\right)\right]\mathrm{d}t = \frac{S_0}{2} \tag{12.154}$$

Now consider the wave-vector **k** domain with wave-number k. Substitution of $w = kc$ into Eqs. (12.147) and (12.148), yields the electric $\mathbf{E}_0(R,k)$ and magnetic $\mathbf{B}_0(R,k)$ fields, which are emitted from an oscillating magnetic dipole in the wave-number domain k, where

$$\mathbf{E}_0(R,k) = \frac{1}{4\pi\varepsilon_0}\left[-\frac{k^2\mathbf{p}_0}{R} + \frac{k^2\hat{\mathbf{R}}(\hat{\mathbf{R}}\cdot\mathbf{p}_0)}{R} + (3\hat{\mathbf{R}}(\mathbf{p}_0\cdot\hat{\mathbf{R}}) - \mathbf{p}_0)\left(\frac{1}{R^3} - \frac{\mathrm{i}\mathrm{k}}{R^2}\right)\right] \tag{12.155}$$

Equivalently, the k-space electric $\mathbf{E}_0(R,k)$ field is via the BAC–CAB rule:

$$\mathbf{E}_0(R,k) = \frac{1}{4\pi\varepsilon_0}\left[\frac{k^2((\hat{\mathbf{R}}\times\mathbf{p}_0)\times\hat{\mathbf{R}})}{R} + (3\hat{\mathbf{R}}(\mathbf{p}_0\cdot\hat{\mathbf{R}}) - \mathbf{p}_0)\left(\frac{1}{R^3} - \frac{ik}{R^2}\right)\right] \tag{12.156}$$

With substitution of $w = kc$ into Eq. 12.148, the magnetic field in k-space becomes

$$\mathbf{B}_0(R,k) = \frac{k^2}{4\pi\varepsilon_0 c}\left(\frac{\hat{\mathbf{R}}\times\mathbf{p}_0}{R}\right)\left(1 + \frac{\mathrm{i}}{ckR}\right) \tag{12.157}$$

The power flow or Poynting vector $\mathbf{S}_0(R,k)$ in the k-space is

$$\mathbf{S}(R,w) = \frac{c^4k^4}{4\pi R^2}(p_0 \sin\theta)^2\hat{\mathbf{R}} \tag{12.158}$$

or alternatively, in terms of wavelength, the Poynting vector is

$$\mathbf{S}(R,w) = 4\pi^5\frac{c^4}{\lambda^4R^2}(p_0 \sin\theta)^2\hat{\mathbf{R}} \tag{12.159}$$

The criteria kR for near-, intermediate-, and far-zone electric fields is the product of the wave-number k and the observer distance R from the electric dipole, where

$$kR = 2\pi\frac{R}{\lambda} \ll 1 \quad \text{near zone} \tag{12.160}$$

$$kR = 2\pi\frac{R}{\lambda} = 1 \quad \text{intermediate zone} \tag{12.161}$$

$$kR = 2\pi\frac{R}{\lambda} \gg 1 \quad \text{far-radiation zone} \tag{12.162}$$

The k-space electric $\mathbf{E}_0(R,k)$ and magnetic $\mathbf{B}_0(R,k)$ fields of Eqs. (12.156) and (12.157) are rearranged for application of criteria kR, where

$$\mathbf{E}_0(R,k) = \frac{(kR)^2((\hat{\mathbf{R}} \times \mathbf{p}_0) \times \hat{\mathbf{R}}) + (3\hat{\mathbf{R}}(\mathbf{p}_0 \cdot \hat{\mathbf{R}}) - \mathbf{p}_0)(1 - \mathrm{i}kR)}{4\pi\varepsilon_0 R^3} \tag{12.163}$$

and

$$\mathbf{B}_0(R,k) = \frac{1}{4\pi\varepsilon_0 c}\left(\frac{\hat{\mathbf{R}} \times \mathbf{p}_0}{R^3}\right)(kR)\left(kR + \frac{\mathrm{i}}{c}\right) \tag{12.164}$$

In the near (static) zone, where $kR \ll 1$,

$$\mathbf{E}_0(R,k) = \frac{1}{4\pi\varepsilon_0 R^3 c}\left[3\hat{\mathbf{R}}(\mathbf{p}_0 \cdot \hat{\mathbf{R}}) - \mathbf{p}_0\right] \tag{12.165}$$

$$\mathbf{B}_0(R,k) = \frac{\mathrm{i}kR}{4\pi\varepsilon_0 R^3 c}(\hat{\mathbf{R}} \times \mathbf{p}_0) \tag{12.166}$$

In the far (radiation) zone, where $kR \gg 1$, one obtains the plane electromagnetic wave or TEM (transverse electric and magnetic fields, which are perpendicular to direction $\hat{\mathbf{R}}$ of propagation), where

$$\mathbf{E}_0(R,k) = \frac{k^2((\hat{\mathbf{R}} \times \mathbf{p}_0) \times \hat{\mathbf{R}})}{4\pi\varepsilon_0 R} \tag{12.167}$$

$$\mathbf{B}_0(R,k) = \frac{k^2(\hat{\mathbf{R}} \times \mathbf{p}_0)}{4\pi\varepsilon_0 Rc} \tag{12.168}$$

12.9 Binomial Expansion of the Vector Potential of a Moving Charge Includes Contributions from Electric and Magnetic Dipole and Electric Quadrupole Moments

In this section, we calculate the vector potential due to a moving charge or current. The moving charge or current can form a pure magnetic dipole or produce a magnetic dipole component. Magnetic dipoles are of particular interest in neutron optics, which is due to the neutron magnetic moment and the magnetic moments of unpaired atomic electrons and other sources of magnetic moments. Thus, dipole magnetic fields, which interact with the dipole magnetic field neutrons, cause scatter, refraction, diffraction, and reflection, as well as precession.

Also, X-ray emission can also occur from laboratory-based and space-based sources of oscillating or moving magnetic dipoles. Intense and rapidly varying currents in plasmas, such as laser-produced plasmas or stellar plasmas, can produce X-rays.

The binomial expansion leads to a vector potential expression, which is the sum of the electric and magnetic dipole moments and electric quadrupole moment contributions. Of particular interest is the comparison of the magnetic and electric dipole contributions to the vector potential. As will be shown, a simple relation is established between the vector potential of the electric dipole and the magnetic dipole components of the moving charge. The relation between the vector potential, which is due to the electric dipole moment and the magnetic dipole moment, allows transformation of the electric and magnetic fields of the electric dipole to that of the magnetic dipole. As a result, the derived electric- and magnetic-field expressions for an oscillating electric dipole (as discussed in the previous sections), can be used to determine the electric and magnetic field expressions for an oscillating magnetic dipole.

The magnetic dipole field and electric dipole field are dual solutions, where the expressions for the electric and magnetic fields are swapped, if the electric dipole moment is replaced with the magnetic dipole moment. This is explained as follows. From the vector Lienard–Wiechert potential, which was derived in section 12.1, we consider the magnetic field produced by a tiny current loop that forms a magnetic dipole. The magnetic potential vector $\mathbf{A}(\mathbf{R}, t)$, which is measured by an observer at position $\mathbf{R}$ present time t, is produced at earlier time t_s by a charge e at position $\mathbf{r}(t_s)$, where the charge circulates at velocity $\mathbf{v}(t_s)$ in a tiny current loop of radius r_s, where from Eq. (12.33),

$$\mathbf{A}(\mathbf{R}, t) = \frac{e\mathbf{v}(t_s)}{4\pi\varepsilon_0 c^2 R} \cdot \left[1 - \frac{\mathbf{v}(t_s) \cdot \hat{\mathbf{R}}}{c}\right]^{-1} = \frac{e\mathbf{v}(t_s)}{4\pi\varepsilon_0 c^2 R} \cdot \left[1 - \frac{\mathrm{d}}{\mathrm{d}t_s} \frac{r(t_s) \cdot \hat{\mathbf{R}}}{c}\right]^{-1} \tag{12.169}$$

The radius r_s of the current loop, or the spatial extent $\mathbf{r}_s(t_s)$ of a moving charge distribution or current density $\mathbf{J}(t_s)$ is assumed to be approximately constant over time τ, given by

$$\tau = r(t_s) \cdot \frac{\hat{\mathbf{R}}}{c} \tag{12.170}$$

The period T of the electromagnetic radiation must be much longer than the time-of-flight variation τ, the electromagnetic radiation, which is emitted from various positions $\mathbf{r}_s(t_s)$ of the moving charges that comprise a current loop or other type of localized current or moving charge distribution, where

$$T \gg \tau = r(t_s) \cdot \frac{\hat{\mathbf{R}}}{c} \tag{12.171}$$

Rearrangement of Eq. (12.171), shows the wavelength λ of the emitted electromagnetic radiation must be larger than the spatial extent $r(t_s)$ of the charge, where

$$cT = \lambda \gg r(t_s) \tag{12.172}$$

Thus, the radius of the current loop of a magnetic dipole, or the separation of two charges of an electric dipole, must be much smaller than the wavelength of the emitted electromagnetic radiation for the following derived electric and magnetic field results for the electric and magnetic dipoles.

Starting at earlier time t_s, the electromagnetic radiation, which is emitted from the magnetic dipole, travels at speed of light c over distance R, and arrives at the observer at present time t, where

$$t = t_s + \frac{R}{c} \tag{12.173}$$

One assumes that the velocity $\mathbf{v}(t_s)$ of the electron is nonrelativistic, where

$$\mathbf{v}(t_s) \cdot \frac{\hat{\mathbf{R}}}{c} \ll 1 \tag{12.174}$$

From section 71 in Landau and Lifshitz (1975), the second term on the right side of Eq. (12.169) for the magnetic potential vector $\mathbf{A}(\mathbf{r}, t)$ is expanded in powers of $\mathbf{r}(t_s) \cdot \hat{\mathbf{R}}/c$ to the first order (binomial expansion), where

$$\mathbf{A}(\mathbf{R}, t) \cong \frac{e\mathbf{v}(t_s)}{4\pi\varepsilon_0 c^2 R} + \frac{e\mathbf{v}(t_s)[\mathbf{v}(t_s) \cdot \hat{\mathbf{R}}]}{4\pi\varepsilon_0 c^3 R} \cong \frac{e\mathbf{v}(t_s)}{4\pi\varepsilon_0 c^2 R} + \frac{\partial}{\partial t_S}\frac{e\mathbf{v}(t_s)[\mathbf{r}(t_s) \cdot \hat{\mathbf{R}}]}{4\pi\varepsilon_0 c^3 R} \tag{12.175}$$

The last expression in Eq. (12.175) is obtained by neglecting the higher order acceleration term $\mathbf{a}(t_s)[\mathbf{r}(t_s) \cdot \hat{\mathbf{R}}]$ compared to the velocity term $\mathbf{v}(t_s)[\mathbf{r}(t_s) \cdot \hat{\mathbf{R}}]$, where

$$\mathbf{v} = \mathbf{v}(t_s) = \frac{\partial \mathbf{r}_s(t_s)}{\partial t_s} \quad \mathbf{a} = \mathbf{a}(t_s) = \frac{\partial \mathbf{v}_s(t_s)}{\partial t_s} \quad \mathbf{a}[\mathbf{r} \cdot \hat{\mathbf{R}}] \gg \mathbf{v}[\mathbf{r} \cdot \hat{\mathbf{R}}] \tag{12.176}$$

We now evaluate $\mathbf{v}(t_s)(\mathbf{r}(t_s) \cdot \hat{\mathbf{R}})$ in Eq. (12.175), where one starts with

$$\frac{\partial}{\partial t_s}\mathbf{r}_s(\mathbf{r}_s \cdot \hat{\mathbf{R}}) = \mathbf{v}(\mathbf{r}_s \cdot \hat{\mathbf{R}}) + \mathbf{r}_s(\mathbf{v} \cdot \hat{\mathbf{R}}) \tag{12.177}$$

One solves Eq. (12.177) for $\mathbf{v}(\mathbf{r}_s \cdot \hat{\mathbf{R}})$ to obtain

$$\mathbf{v}(\mathbf{r}_s \cdot \hat{\mathbf{R}}) = \frac{\partial}{\partial t_s}\mathbf{r}_s(\mathbf{r}_s \cdot \hat{\mathbf{R}}) - \mathbf{r}_s(\mathbf{v} \cdot \hat{\mathbf{R}}) \tag{12.178}$$

One the right side of Eq. (12.178), $\mathbf{v}(\mathbf{r}_s\cdot\hat{\mathbf{R}})/2$ is added and subtracted, where

$$\mathbf{v}(\mathbf{r}_s\cdot\hat{\mathbf{R}}) = \frac{\partial}{\partial t_s}\mathbf{r}_s(\mathbf{r}_s\cdot\hat{\mathbf{R}}) - \mathbf{r}_s(\mathbf{v}\cdot\hat{\mathbf{R}}) + \left[\frac{\mathbf{v}(\mathbf{r}_s\cdot\hat{\mathbf{R}})}{2} - \frac{\mathbf{v}(\mathbf{r}_s\cdot\hat{\mathbf{R}})}{2}\right] \tag{12.179}$$

One simplifies Eq. (12.179) to obtain

$$\frac{\mathbf{v}(\mathbf{r}_s\cdot\hat{\mathbf{R}})}{2} = \frac{1}{2}\frac{\partial}{\partial t_s}\mathbf{r}_s(\mathbf{r}_s\cdot\hat{\mathbf{R}}) - \frac{1}{2}\mathbf{r}_s(\mathbf{v}\cdot\hat{\mathbf{R}}) \tag{12.180}$$

One adds $\mathbf{v}(\mathbf{r}_s\cdot\hat{\mathbf{R}})/2$ to both sides of Eq. (12.180), where

$$\mathbf{v}(\mathbf{r}_s\cdot\hat{\mathbf{R}}) = \frac{1}{2}\frac{\partial}{\partial t_s}\mathbf{r}_s(\mathbf{r}_s\cdot\hat{\mathbf{R}}) + \frac{1}{2}\mathbf{v}(\mathbf{r}_s\cdot\hat{\mathbf{R}}) - \frac{1}{2}\mathbf{r}_s(\mathbf{v}\cdot\hat{\mathbf{R}}) \tag{12.181}$$

Unit vector $\hat{\mathbf{R}}$, which points from the center of the current loop at the origin to the fixed observer, is independent of (1) the earlier time t_s of the dipole emission of electromagnetic radiation, and (2) the present time t when the radiation arrives at the observer, where

$$\frac{\partial\hat{\mathbf{R}}}{\partial t_s} = \frac{\partial\hat{\mathbf{R}}}{\partial t} = 0 \tag{12.182}$$

With application of the BAC−CAB identity to Eq. (12.181), $\mathbf{v}(\mathbf{r}_s\cdot\hat{\mathbf{R}})$ becomes

$$\mathbf{v}(\mathbf{r}_s\cdot\hat{\mathbf{R}}) = \frac{1}{2}\frac{\partial}{\partial t_s}\mathbf{r}_s(\mathbf{r}_s\cdot\hat{\mathbf{R}}) + \frac{1}{2}(\hat{\mathbf{R}}\times(\mathbf{v}\times\mathbf{r}_s)) \tag{12.183}$$

Reversing the order twice in the double cross product $(-1\times -1 = 1)$ leaves the sign unchanged in the double cross product, where

$$\mathbf{v}(\mathbf{r}_s\cdot\hat{\mathbf{R}}) = \frac{1}{2}\frac{\partial}{\partial t_s}\mathbf{r}_s(\mathbf{r}_s\cdot\hat{\mathbf{R}}) + \frac{1}{2}((\mathbf{r}_s\times\mathbf{v})\times\hat{\mathbf{R}}) \tag{12.184}$$

Substitution of Eq. (12.184) into (12.175) yields

$$\mathbf{A}(\mathbf{R},t) = \frac{e\mathbf{v}(t_s)}{4\pi\varepsilon_0 c^2 R} + \frac{1}{8\pi\varepsilon_0 c^3 R}\frac{\partial^2}{\partial t_s^2}\left[e\mathbf{r}_s(\mathbf{r}_s\cdot\hat{\mathbf{R}})\right] + \frac{1}{8\pi\varepsilon_0 c^3 R}\frac{\partial}{\partial t_s}\left[(\mathbf{r}_s\times e\mathbf{v})\times\hat{\mathbf{R}}\right] \tag{12.185}$$

The expression for the vector potential of Eq. (12.185) has been readied for the appropriate insertion of electric dipole and magnetic dipole moment terms, via the appropriate shifting of the charge e in the second and third terms. First, we review the relation of the electric dipole $\mathbf{p}_e(t_s)$ to charge e, charge position $\mathbf{r}_s(t_s)$, and velocity $\mathbf{v}(t_s)$, and the loop current $\mathbf{I}(t_s)$ produced by the moving charge. At earlier time t_s, the loop current $\mathbf{I}(t_s)$ is the product of the charge e and charge velocity $\mathbf{v}(t_s)$, and current $\mathbf{I}(t_s)$ can also be considered as the time derivative of an electric dipole moment $\mathbf{p}_e(t_s)$, where

$$\mathbf{I}(t_s) = e\mathbf{v}(t_s) = \frac{\mathrm{d}\mathbf{p}_e(t_s)}{\mathrm{d}t_s} \quad \text{and} \quad \mathbf{p}_e(t_s) = e\mathbf{r}_s(t_s) \tag{12.186}$$

The magnetic dipole moment $\mathbf{p}_m(t_s)$, at earlier time t_s, or expressed in terms of the present time t of the observer, is defined as

$$\mathbf{p}_m\left(t - \frac{R}{c}\right) = \mathbf{p}_m(t_s) = \frac{1}{2}(\mathbf{r}_s \times e\mathbf{v}) \tag{12.187}$$

Substitution of Eqs. (12.186) and (12.187) into magnetic vector potential $\mathbf{A}(\mathbf{r}, t)$ expression of Eq. (12.185), yields

$$\mathbf{A}(\mathbf{R}, t) = \frac{1}{4\pi\varepsilon_0 c^2 R}\frac{\partial \mathbf{p}_e(t_s)}{\partial t_s} + \frac{1}{4\pi\varepsilon_0 c^3 R}\frac{1}{2}\frac{\partial^2}{\partial t_s^2}\left[\mathbf{p}_e(t_s)(\mathbf{r}_s \cdot \hat{\mathbf{R}})\right] + \frac{1}{4\pi\varepsilon_0 c^3 R}\left(\frac{\partial \mathbf{p}_m(t_s)}{\partial t_s} \times \hat{\mathbf{R}}\right) \tag{12.188}$$

The first term of the magnetic vector potential $\mathbf{A}(\mathbf{r}, t)$ is due to the electric dipole moment vector $\mathbf{p}_e(t_s)$; the second term is due to the electric quadrupole moment vector $\mathbf{Q}_m(t_s)$; and the last term is due to the magnetic dipole moment vector $\mathbf{p}_m(t_s)$.

12.10 Transformation of the Fields and Vector Potential of the Electric Dipole to the Magnetic Dipole

In this section, the transformation of the vector potential, electric field, and magnetic field of an electric dipole to that of a magnetic dipole is demonstrated. The vector potential of the magnetic dipole is shown to be related simply to the vector potential of the electric dipole due to the duality of the electric and magnetic fields in a vacuum. From the relation between vector potentials for the electric and magnetic dipoles, one can determine the transformation of the electric and magnetic fields of the oscillating electric dipole to that of an oscillating magnetic dipole.

One starts with the magnetic vector potential $\mathbf{A}(\mathbf{r}, t)$ for the electric dipole and magnetic dipole contributions, which are

$$\mathbf{A}(\mathbf{R}, t) = \mathbf{A}_{pe}(\mathbf{R}, t) + \mathbf{A}_{pm}(\mathbf{R}, t) \tag{12.189}$$

From Eq. (12.188), the magnetic vector potential $\mathbf{A}_{pe}(\mathbf{R}, t)$ due to the electric dipole moment vector $\mathbf{p}_e(t_s)$ is

$$\mathbf{A}_{pe}(\mathbf{R}, t) = \frac{1}{4\pi\varepsilon_0 c^2 R}\frac{\partial \mathbf{p}_e(t_s)}{\partial t_s} \tag{12.190}$$

From Eq. (12.188), the magnetic vector potential $\mathbf{A}_{pm}(\mathbf{R}, t)$ due to the magnetic dipole moment vector $\mathbf{p}_m(t_s)$ is

$$\mathbf{A}_{pm}(\mathbf{R}, t) = \frac{1}{4\pi\varepsilon_0 c^3 R}\left(\frac{\partial \mathbf{p}_m(t_s)}{\partial t_s} \times \hat{\mathbf{R}}\right) \tag{12.191}$$

Comparison of electric dipole vector potential $\mathbf{A}_{pe}(\mathbf{R}, t)$ with magnetic dipole vector potential $\mathbf{A}_{pm}(\mathbf{R}, t)$ suggests that

$$\mathbf{A}_{pm}(\mathbf{R}, t) = \frac{1}{4\pi\varepsilon_0 c^2 R}\frac{\partial \mathbf{p}_m(t_s)}{\partial t_s} \times \frac{\hat{\mathbf{R}}}{c} \tag{12.192}$$

where

$$\frac{\partial \mathbf{p}_m(t_s)}{\partial t_s} \rightarrow \frac{\partial \mathbf{p}_e(t_s)}{\partial t_s} \tag{12.193}$$

With Eqs. (12.192) and (12.193), as suggested by Eqs. (12.190) and (12.191), one obtains

$$\mathbf{A}_{pm}(\mathbf{R}, t) = \mathbf{A}_{pe}(\mathbf{R}, t) \times \frac{\hat{\mathbf{R}}}{c} \tag{12.194}$$

The magnetic field of the magnetic dipole $\mathbf{B}_{pm}(r, t)$ is obtained by taking the curl of both sides of Eq. (12.194), where

$$\nabla \times \mathbf{A}_{pm}(\mathbf{R}, t) = \nabla \times \mathbf{A}_{pe}(\mathbf{R}, t) \times \frac{\hat{\mathbf{R}}}{c} \tag{12.195}$$

The curl of the magnetic dipole vector potential $\mathbf{A}_{pm}(\mathbf{R}, t)$ on the left side of the equation is the magnetic field $\mathbf{B}_{pm}(\mathbf{R}, t)$ of the magnetic dipole $\mathbf{p}_m(t_s)$, and the curl of the electric dipole vector potential $\mathbf{A}_{pe}(\mathbf{R}, t)$ on the right side of the equation is the magnetic field $\mathbf{B}_{pe}(\mathbf{R}, t)$ of the electric dipole $\mathbf{p}_e(t_s)$, where

$$\mathbf{B}_{pm}(\mathbf{R}, t) = \mathbf{B}_{pe}(\mathbf{R}, t) \times \frac{\hat{\mathbf{R}}}{c} \tag{12.196}$$

Substitution of Eq. (12.136) for the cross product on the right side of Eq. (12.196) shows

$$\mathbf{B}_{pm}(\mathbf{R},t) = \frac{1}{c^2}\mathbf{E}_{pe}(\mathbf{R},t) \tag{12.197}$$

Thus, from Eq. (12.197), the electric field $\mathbf{E}_{pe}(\mathbf{R},t)$ of an electric dipole $\mathbf{p}_e(t_s)$, divided by c^2, gives the magnetic field $\mathbf{B}_{pm}(\mathbf{R},t)$ of a magnetic dipole $\mathbf{p}_m(t_s)$, where

$$\frac{1}{c^2}\mathbf{E}_{pe}(\mathbf{R},t) \rightarrow \mathbf{B}_{pm}(\mathbf{R},t) \tag{12.198}$$

The plane wave electric field $\mathbf{E}_{pe}(\mathbf{R},t)$ of the electric dipole is obtained from its magnetic field $\mathbf{B}_{pe}(\mathbf{R},t)$, via Eq. (12.136), where

$$\mathbf{E}_{pe}(\mathbf{R},t) = -c\hat{\mathbf{R}} \times \mathbf{B}_{pe}(\mathbf{R},t) = c\mathbf{B}_{pe}(\mathbf{R},t) \times \hat{\mathbf{R}} \tag{12.199}$$

That is, in free space, the electromagnetic waves emitted from an electric dipole or magnetic dipole are plane waves, which obey Faraday's Law and Ampere's Law, where

$$\nabla \times \mathbf{E}_{pe}(\mathbf{R},t) = -\frac{\partial \mathbf{B}_{pe}(\mathbf{R},t)}{\partial t} \tag{12.200}$$

and

$$\nabla \times \mathbf{B}_{pe}(\mathbf{R},t) = \frac{1}{c^2}\frac{\partial_{pe}\mathbf{E}(\mathbf{R},t)}{\partial t} \tag{12.201}$$

The form of the electric and magnetic field components of the plane wave from an electric dipole, which obeys Maxwell equations of Eqs. (11.200) and (11.201), are

$$\mathbf{E}_{pe}(\mathbf{R},t) = f(R-ct) \tag{12.202}$$

and

$$c\mathbf{B}_{pe}(\mathbf{R},t) = f(R-ct) \tag{12.203}$$

To obtain the electric field $\mathbf{E}_{pm}(\mathbf{R},t)$ of the magnetic dipole $\mathbf{p}_m(t_s)$, one returns to the expression of Eq. (12.196),

$$\mathbf{B}_{pm}(\mathbf{R},t) = \mathbf{B}_{pe}(\mathbf{R},t) \times \frac{\hat{\mathbf{R}}}{c} \tag{12.204}$$

The plane wave electric field $\mathbf{E}_{pm}(\mathbf{R}, t)$ of the magnetic dipole is obtained from its magnetic field $\mathbf{B}_{pm}(\mathbf{R}, t)$, via Eq. (12.136), where

$$\mathbf{E}_{pm}(\mathbf{R}, t) = -c\hat{\mathbf{R}} \times \mathbf{B}_{pm}(\mathbf{R}, t) = c\mathbf{B}_{pm}(\mathbf{R}, t) \times \hat{\mathbf{R}} \tag{12.205}$$

One takes the cross product of the origin−observer unit vector $\hat{\mathbf{R}}$ with both sides of Eq. (12.204), where

$$\hat{\mathbf{R}} \times \mathbf{B}_{pm}(\mathbf{R}, t) = \hat{\mathbf{R}} \times \mathbf{B}_{pe}(\mathbf{R}, t) \times \frac{\hat{\mathbf{R}}}{c} \tag{12.206}$$

Substitution for the electric field $\mathbf{E}_{pm}(\mathbf{R}, t)$ of Eq. (12.205) for the cross product on the left side of Eq. (12.206), and application of the BAC−CAB vector identity on the right side of the resulting equation, yields

$$\mathbf{E}_{pm}(\mathbf{R}, t) = -\mathbf{B}_{pe}(\mathbf{R}, t) \tag{12.207}$$

where

$$\hat{\mathbf{R}} \cdot \mathbf{B}_{pe}(\mathbf{R}, t) = 0 \tag{12.208}$$

Examination of Eq. (12.207), shows the magnetic field $\mathbf{B}_{pe}(\mathbf{R}, t)$ of an electric dipole $\mathbf{p}_e(t_s)$ multiplied by -1 gives the electric field $\mathbf{E}_{pm}(\mathbf{R}, t)$ of a magnetic dipole $\mathbf{p}_m(t_s)$, where

$$-\mathbf{B}_{pe}(\mathbf{R}, t) \rightarrow \mathbf{E}_{pm}(\mathbf{R}, t) \tag{12.209}$$

12.11 Electric and Magnetic Fields and Power Emission from an Oscillating Magnetic Dipole

In this section, we derive the electric and magnetic fields and power emission from an oscillating magnetic dipole using the transformation results derived in Section 12.10. The expression for the electric $\mathbf{E}(R, t)$ and magnetic $\mathbf{B}(R, t)$ field vectors, and the electromagnetic power density vector; that is, the Poynting vector $\mathbf{S}(R, t)$ of an oscillating magnetic dipole $p_m(t - R/c)$, can thus be obtained from the corresponding dual expressions for an oscillating electric dipole $p_e(t - R/c)$. In the electric and magnetic field expressions for the electric dipole, one replaces the electric dipole moment $\mathbf{p}_e(t - R/c)$ by the magnetic dipole moment $\mathbf{p}_m(t - R/c)$. The electric field expression for the electric dipole $\mathbf{E}_{pe}(R, t)$ divided by c^2 becomes the magnetic field expression for the magnetic dipole $\mathbf{B}_{pm}(R, t)$. And the magnetic field expression for the electric dipole $\mathbf{B}_{pe}(R, t)$ is multiplied by -1 to become the electric field expression for the magnetic dipole $\mathbf{E}_{pm}(R,t)$. The transformations expressed in MKS units are

$$\mathbf{p}_e(t-R/c) \rightarrow \mathbf{p}_m(t-R/c) \tag{12.210}$$

$$\frac{1}{c^2}\mathbf{E}_{pe}(R,t) \rightarrow \mathbf{B}_{pm}(R,t) \tag{12.211}$$

$$-1 \times \mathbf{B}_{pe}(R,t) \rightarrow \mathbf{E}_{pm}(R,t) \tag{12.212}$$

With application of Eqs. (12.210–12.212) to Eqs. (12.115) and Eq. (12.116), the time-domain electric field $\mathbf{E}_{pm}(R,t)$ and magnetic flux density $\mathbf{B}_{pm}(R,t)$ for an oscillating magnetic dipole $\mathbf{p}_m(t-R/c)$, expressed in MKS units, are

$$\mathbf{B}_{pm}(R,t) = \frac{1}{4\pi\varepsilon_0 c^2} \times$$

$$\left[\frac{-1}{c^2 R}\frac{\partial^2 \mathbf{p}_m(t_s)}{\partial t^2} + \frac{\hat{\mathbf{R}}}{c^2 R}\left(\frac{\partial^2 \mathbf{p}_m}{\partial t^2}\cdot\hat{\mathbf{R}}\right) - \frac{1}{cR^2}\frac{\partial \mathbf{p}_m}{\partial t} + 3\frac{\hat{\mathbf{R}}}{cR^2}\left(\frac{\partial \mathbf{p}_m}{\partial t}\cdot\hat{\mathbf{R}}\right) - \frac{\mathbf{p}_m}{R^3} + \frac{3\hat{\mathbf{R}}(\mathbf{p}_m\cdot\hat{\mathbf{R}})}{R^3}\right] \tag{12.213}$$

and

$$\mathbf{E}_{pm}(R,t) = -\frac{1}{4\pi\varepsilon_0 c^2}\left(\frac{\partial \mathbf{p}_m}{\partial t}\times\frac{\hat{\mathbf{R}}}{R^2}\right) - \frac{1}{4\pi\varepsilon_0 c^3}\left(\frac{\partial^2 \mathbf{p}_m}{\partial t^2}\times\frac{\hat{\mathbf{R}}}{R}\right) \tag{12.214}$$

If the magnetic dipole is static, then the time-varying terms are zero (with zero electric field), and a static magnetic field is produced by a stationary current loop with constant current.

Next, the frequency w domain expressions for the magnetic dipole are obtained from the expressions for the electric dipole. The magnetic dipole $\mathbf{p}_m(t-R/c)$ sinusoidal expression is expressed as a phasor (See Smith, 1966), where

$$\mathbf{p}_m(t_s) = \mathbf{p}_m(t-R/c) = \mathbf{p}_{0m}\ \exp\left[-\mathrm{i}w\left(t-\frac{R}{c}\right)\right] \tag{12.215}$$

The electric $\mathbf{E}_{pm}(R,t)$ and magnetic $\mathbf{B}_{pm}(R,t)$ fields of the oscillating magnetic dipole, expressed as phasors, are

$$\mathbf{E}_{pm}(R,t) = \mathbf{E}_{0m}\ \exp\left[-\mathrm{i}w\left(t-\frac{R}{c}\right)\right] \tag{12.216}$$

$$\mathbf{B}_{pm}(R,t) = \mathbf{B}_0\ \exp\left[-\mathrm{i}w\left(\mathrm{t}-\frac{R}{c}\right)\right] \tag{12.217}$$

With substitution of Eqs. (12.215–12.217) into Eqs. (12.213) and (12.214), the frequency w domain electric field $\mathbf{E}_{0m}(R,w)$ and magnetic flux density $\mathbf{B}_{0m}(R,w)$ for an oscillating magnetic dipole $\mathbf{p}_m(t-R/c)$, expressed in MKS units, are

$$\mathbf{B}_{0m}(R,w) = \frac{1}{4\pi\varepsilon_0 c^2}\left[\frac{w^2((\hat{\mathbf{R}}\times\mathbf{p}_{0m})\times\hat{\mathbf{R}})}{c^2R} + (3\hat{\mathbf{R}}(\hat{\mathbf{R}}\cdot\mathbf{p}_{0m}) - \mathbf{p}_{0m})\left(\frac{1}{R^3} - \frac{\mathrm{i}w}{cR^2}\right)\right] \tag{12.218}$$

and

$$\mathbf{E}_{0m}(R,w) = -\frac{w^2}{4\pi\varepsilon_0 c^3}\left(\frac{\hat{\mathbf{R}}\times\mathbf{p}_{0m}}{R}\right)\left(1+\frac{\mathrm{i}}{wR}\right) \tag{12.219}$$

With substitution of Eq. (12.215) into Eq. (12.144), the power flow or Poynting vector $\mathbf{S}_{0m}(R,w)$ from the oscillating magnetic dipole in the frequency w domain is

$$\mathbf{S}_{0m}(R,w) = \frac{w^4}{4\pi R^2}(p_{0m}\ \sin\ \theta)^2\hat{\mathbf{R}} \tag{12.220}$$

Finally, the wave-number k domain expressions for the magnetic dipole are obtained from Eqs. (12.218–12.220), via $w = kc$. The wave-number k domain electric field $\mathbf{E}_{0m}(R,k)$ and magnetic flux density $\mathbf{B}_{0m}(R,k)$ for an oscillating magnetic dipole $\mathbf{p}_m(t-R/c)$, expressed in MKS units, are

$$\mathbf{B}_{0m}(R,k) = \frac{1}{4\pi\varepsilon_0 c^2}\left[\frac{k^2((\hat{\mathbf{R}}\times\mathbf{p}_{0m})\times\hat{\mathbf{R}})}{R} + (3\hat{\mathbf{R}}(\mathbf{p}_{0m}\cdot\hat{\mathbf{R}}) - \mathbf{p}_{0m})\left(\frac{1}{R^3} - \frac{\mathrm{i}k}{R^2}\right)\right] \tag{12.221}$$

and

$$\mathbf{E}_{0m}(R,k) = -\frac{k^2}{4\pi\varepsilon_0 c}\left(\frac{\hat{\mathbf{R}}\times\mathbf{p}_{0m}}{R}\right)\left(1+\frac{\mathrm{i}}{ckR}\right) \tag{12.222}$$

The power flow or Poynting vector $\mathbf{S}_{0m}(R,k)$ from the oscillating magnetic dipole in the wave-number k domain is

$$\mathbf{S}_{0m}(R,k)\frac{c^4k^4}{4\pi R^2}(p_{0m}\ \sin\ \theta)^2\hat{\mathbf{R}} \tag{12.223}$$

The Poynting vector expressed in terms of wavelength is

$$\mathbf{S}_{om}(R,k) = 4\pi^5\frac{c^4}{\lambda^4R^2}(p_{0m}\ \sin\ \theta)^2\hat{\mathbf{R}} \tag{12.224}$$

Recalling section 12.6 and Eqs. (12.210–12.212), to obtain the electric and magnetic fields for a magnetic dipole from the electric and magnetic fields for an electric dipole, one uses the following transformations, in CGS units:

$$\mathbf{p}_e(t - R/c) \rightarrow \mathbf{p}_m(t - R/c) \tag{12.225}$$

$$\mathbf{E}_{pe}(R,t) \rightarrow \mathbf{H}_{pm}(R,t) \tag{12.226}$$

$$-1 \times \mathbf{H}_{pe}(R,t) \rightarrow \mathbf{E}_{pm}(R,t) \tag{12.227}$$

The only difference between the CGS transformations of Eq. (12.225–12.227) and MKS transformations of Eqs. (12.210–12.212), is the division of the electric field of the electric dipole by c^2 in the MKS transformation, whereas in the CGS transformation, the electric field of the electric dipole is identical to the magnetic field of the magnetic dipole. With application of Eqs. (12.225–12.227) to Eqs. (12.119) and Eq. (12.120), the electric field $\mathbf{E}_{\text{cgs}}(R,t)$ and magnetic field intensity $\mathbf{H}_{\text{cgs}}(R,t)$ of the magnetic dipole, expressed in CGS units, are

$$\mathbf{E}_{\text{pm,cgs}}(R,t) = -\frac{1}{c}\left(\frac{\partial \mathbf{p}_m}{\partial t} \times \frac{\hat{\mathbf{R}}}{R^2}\right) - \frac{1}{c^2}\left(\frac{\partial^2 \mathbf{p}_m}{\partial t^2} \times \frac{\hat{\mathbf{R}}}{R}\right) \tag{12.228}$$

and

$$\mathbf{H}_{\text{pm,cgs}}(R,t) = -\frac{1}{c^2R}\frac{\partial^2 \mathbf{p}_m(t_s)}{\partial t^2} + \frac{\hat{\mathbf{R}}}{c^2R}\left(\frac{\partial^2 \mathbf{p}_m}{\partial t^2}\cdot\hat{\mathbf{R}}\right) - \frac{1}{cR^2}\frac{\partial \mathbf{p}_m}{\partial t} + \frac{3\hat{\mathbf{R}}}{cR^2}\left(\frac{\partial \mathbf{p}_m}{\partial t}\cdot\hat{\mathbf{R}}\right) - \frac{\mathbf{p}_e}{R^3} + \frac{3\hat{\mathbf{R}}(\mathbf{p}_m\cdot\hat{\mathbf{R}})}{R^3} \tag{12.229}$$

One now compares Eqs. (12.228) and (12.229). In CGS units, for a magnetic dipole, the plane wave electric field $\mathbf{E}_{\text{pm,cgs}}(R,t)$ and magnetic field intensity $\mathbf{H}_{\text{pm,cgs}}(R,t)$ in a vacuum are equal in magnitude, but they are perpendicular, and related by a simple cross product with the origin–observer unit vector $\hat{\mathbf{R}}$, where

$$\mathbf{E}_{\text{pm,cgs}}(R,t) = \mathbf{H}_{\text{pm,cgs}}(R,t) \times \hat{\mathbf{R}} \tag{12.230}$$

and

$$\mathbf{H}_{\text{pm,cgs}}(R,t) = -\mathbf{E}_{\text{pm,cgs}}(R,t) \times \hat{\mathbf{R}} \tag{12.231}$$

12.12 Derivation of the Electric Field of a Charge in Arbitrary Motion from Lienard–Wiechert Potentials

Consider now a charge moving arbitrarily at relativistic or nonrelativistic velocities and accelerations. In this section, the electric field from a charged particle undergoing arbitrary motion is derived. From the Lienard–Wiechert potentials, the electric **E** and magnetic **B** fields can be calculated for the charge e moving at velocity **v** and undergoing acceleration $d\mathbf{v}/dt$. One begins with the Lienard–Wiechert scalar and vector potentials of Eqs. (12.32) and (12.33), expressed in MKS units:

$$\phi(\mathbf{r},t) = \frac{e}{4\pi\varepsilon_0} \cdot \frac{1}{R - (\mathbf{v}\cdot\mathbf{R}/c)} \tag{12.232}$$

and

$$\mathbf{A}(\mathbf{r},t) = \frac{e}{4\pi\varepsilon_0 c^2} \cdot \frac{\mathbf{v}}{R - (\mathbf{v}\cdot\mathbf{R}/c)} \tag{12.233}$$

The distance or separation vector **R** points from the charge at position $\mathbf{r}_s$ to the observer position vector **r**. The electric **E** and magnetic **B** field solutions are obtained via the scalar electric potential ϕ and magnetic vector potential **A**:

$$\mathbf{E} = -\nabla\phi - \frac{\partial\mathbf{A}}{\partial t} \tag{12.234}$$

$$\mathbf{B} = \nabla \times \mathbf{A} \tag{12.235}$$

We begin with the time derivatives. At earlier time (retarded time) t_s, the charge is at position $\mathbf{r}_s(t_s)$ at distance R from the observer at position **r**. The charge emits the electromagnetic field at earlier time t_s, which propagates at velocity c over distance R, and arrives at the observer at later time t, where

$$\mathbf{R} = (x - x_s)\hat{\mathbf{x}} + (y - y_s)\hat{\mathbf{y}} + (z - z_s)\hat{\mathbf{z}} \tag{12.236}$$

and

$$R = c(t - t_s) = \sqrt{(x-x_s)^2 + (y-y_s)^2 + (z-z_s)^2} \tag{12.237}$$

The velocity **v** of the charge at the earlier time t_s is

$$\frac{\partial\mathbf{R}}{\partial t_s} = -\frac{\partial x_s}{\partial t_s}\hat{\mathbf{x}} - \frac{\partial y_s}{\partial t_s}\hat{\mathbf{y}} - \frac{\partial z_s}{\partial t_s}\hat{\mathbf{z}} = -\mathbf{v} \tag{12.238}$$

The unit vector $\hat{\mathbf{R}}$ that points from the charge to the observer at earlier time t_s is

$$\hat{\mathbf{R}} = \frac{\mathbf{R}}{R} = \frac{(x - x_s)\hat{\mathbf{x}} + (y - y_s)\hat{\mathbf{y}} + (z - z_s)\hat{\mathbf{z}}}{\sqrt{(x-x_s)^2 + (y-y_s)^2 + (z-z_s)^2}} \tag{12.239}$$

The earlier time derivative of the distance R between the charge and observer is

$$\frac{\partial R}{\partial t_s} = \frac{\partial R}{\partial x_s}\frac{\partial x_s}{\partial t_s} + \frac{\partial R}{\partial y_s}\frac{\partial y_s}{\partial t_s} + \frac{\partial R}{\partial z_s}\frac{\partial z_s}{\partial t_s} \tag{12.240}$$

Substitution for R of Eq. (12.237), yields the earlier time derivative $\partial R/\partial t_s$ of the distance between the observer and charge at earlier time t_s, where

$$\frac{\partial R}{\partial t_s} = -\frac{x - x_s}{R}\frac{\partial x_s}{\partial t_s} - \frac{y - y_s}{R}\frac{\partial y_s}{\partial t_s} - \frac{z - z_s}{R}\frac{\partial z_s}{\partial t_s} \tag{12.241}$$

Eq. (12.241) can be expressed as the dot product of the charge velocity at earlier time, and the the charge-observer unit vector, which points from the charge at earlier time to the observer point at the present time, where

$$\frac{\partial R}{\partial t_s} = -\frac{\mathbf{R} \cdot \mathbf{v}}{R} = -\hat{\mathbf{R}}\cdot\mathbf{v} \tag{12.242}$$

With Eq. (12.242), the observer present time derivative $\partial/\partial t$ can be obtained in terms of the earlier time derivative $\partial/\partial t_s$. One starts with application of the chain rule, which gives the first expression for the observer time derivative $\partial R/\partial t$ of the charge-to-observer distance, where

$$\frac{\partial R}{\partial t} = \frac{\partial R}{\partial t_s}\frac{\partial t_s}{\partial t} = -(\hat{\mathbf{R}}\cdot\mathbf{v})\frac{\partial t_s}{\partial t} \tag{12.243}$$

With $R = c(t - t_s)$ of Eq. (12.237), one can obtain a second, alternative expression for $\partial R/\partial t$, where

$$\frac{\partial R}{\partial t} = c\left(1 - \frac{\partial t_s}{\partial t}\right) \tag{12.244}$$

Equating Eqs. (12.243) and (12.244) for the two alternative expressions for $\partial R/\partial t$, one obtains

$$\frac{\partial R}{\partial t} = c\left(1 - \frac{\partial t_s}{\partial t}\right) = -(\hat{\mathbf{R}}\cdot\mathbf{v})\frac{\partial t_s}{\partial t} \tag{12.245}$$

Solving for $\partial t_s/\partial t$, one obtains the observer time derivative of the earlier time:

$$\frac{\partial t_s}{\partial t} = \frac{1}{1 - (\hat{\mathbf{R}}\cdot\mathbf{v}/c)} \tag{12.246}$$

The chain rule gives

$$\frac{\partial}{\partial t}=\frac{\partial}{\partial t_s}\frac{\partial t_s}{\partial t} \tag{12.247}$$

Substitution into Eq. (12.247) of Eq. (12.246) for $\partial t_s/\partial t$, then yields an expression for the observer time derivative $\partial/\partial t$ in terms of the earlier time derivative $\partial t_s/\partial t$ at the charge, and the unit vector $\hat{\mathbf{R}}$, which points from the moving charge of velocity $\mathbf{v}$ at earlier time t_s to the observer at the present time t, where

$$\frac{\partial}{\partial t}=\frac{1}{1-(\hat{\mathbf{R}}\cdot\mathbf{v}/c)}\frac{\partial}{\partial t_s}=\frac{R}{R-(\mathbf{R}\cdot\mathbf{v}/c)}\frac{\partial}{\partial t_s} \tag{12.248}$$

Next, the spatial derivatives are evaluated. The gradient ∇R of the distance $R(t_s)$ between the observer and the charge at earlier time t_s is

$$\nabla R=[\nabla R]_{t_s}+\frac{\partial R}{\partial t_s}\nabla t_s \tag{12.249}$$

From Eq. (12.242),

$$\frac{\partial R}{\partial t_s}=-\hat{\mathbf{R}}\cdot\mathbf{v} \tag{12.250}$$

The spatial derivatives of the gradient $[\nabla R]_{t_s}$ of the distance $R(t_s)$ between the charge at earlier time t_s and the observer at present time t, are evaluated at a constant earlier time t_s, where

$$[\nabla R]_{t_s}=\hat{\mathbf{R}} \tag{12.251}$$

Substitution of Eqs. (12.250) and (12.251) into Eq. (12.249), and transposing, yields the first expression for ∇R, where

$$\nabla R=\hat{\mathbf{R}}-(\hat{\mathbf{R}}\cdot\mathbf{v})\nabla t_s \tag{12.252}$$

With $R=c(t-t_s)$ of Eq. (12.237), one can obtain a second, alternative expression for ∇R, where

$$\nabla R=-c\nabla t_s \tag{12.253}$$

Equating Eqs. (12.252) and (12.253) for the two alternative expressions for ∇R, one obtains

$$\nabla R=\hat{\mathbf{R}}-(\hat{\mathbf{R}}\cdot\mathbf{v})\nabla t_s=-c\nabla t_s \tag{12.254}$$

Solving for ∇t_s then yields

$$\nabla t_s = -\frac{1}{c}\left[\frac{\hat{\mathbf{R}}}{1-\left(\mathbf{v}\cdot\hat{\mathbf{R}}/c\right)}\right] \tag{12.255}$$

Substitution of Eq. (12.255) into Eq. (12.254), yields

$$\nabla R = \frac{\hat{\mathbf{R}}}{1-\left(\mathbf{v}\cdot\hat{\mathbf{R}}/c\right)} \tag{12.256}$$

The observer gradient ∇ can be expressed in terms of the earlier time t_s, where the charge is at distance $R(t_s)$, and moves at arbitrary velocity $\mathbf{v}$, where

$$\nabla = [\nabla]_{t_s} + \nabla t_s \frac{\partial}{\partial t_s} \tag{12.257}$$

The resulting observer gradient ∇ at present time t and position $\mathbf{r}$, is expressed in terms of (1) the gradient $[\nabla]_{t_s}$, which is evaluated at constant source (earlier or retarded) time t_s and position $\mathbf{r}_s$, (2) the unit vector $\hat{\mathbf{R}}$, which points from the charge at the earlier time t_s and position $\mathbf{r}_s$ to the observer at present time t and position $\mathbf{r}$, and (3) the earlier time derivative $\partial R/\partial t_s$ of the separation distance $R(t_s)$ between the charge, which moves at arbitrary velocity $\mathbf{v}(t_s)$ at earlier time t_s and position $\mathbf{r}_s$, and the observer at present time t and position $\mathbf{r}$, where

$$\nabla = [\nabla]_{t_s} - \frac{1}{c}\left[\frac{\hat{\mathbf{R}}}{1-(\mathbf{v}\cdot\hat{\mathbf{R}}/c)}\right]\frac{\partial}{\partial t_s} \tag{12.258}$$

or

$$\nabla = [\nabla]_{t_s} - \frac{1}{c}\left[\frac{\mathbf{R}}{R-(\mathbf{v}\cdot\mathbf{R}/c)}\right]\frac{\partial}{\partial t_s} \tag{12.259}$$

The first term of electric field $\mathbf{E}$ is the observer gradient $\nabla\phi$ of the scalar potential ϕ, which is evaluated as follows:

$$\nabla\phi = \frac{e}{4\pi\varepsilon_0}\left[\nabla\left(\frac{1}{R-(\mathbf{v}\cdot\mathbf{R}/c)}\right)\right]_{t_s} - \frac{e}{4\pi\varepsilon_0}\frac{1}{c}\left[\frac{\hat{\mathbf{R}}}{1-(\mathbf{v}\cdot\hat{\mathbf{R}}/c)}\right]\frac{\partial}{\partial t_s}\left(\frac{1}{R-(\mathbf{v}\cdot\mathbf{R}/c)}\right) \tag{12.260}$$

The first term of the $\nabla\phi$ expression contains the observer gradient, which must be evaluated with the earlier time t_s kept constant:

$$\left[\nabla\left(\frac{1}{R-(\mathbf{v}\cdot\mathbf{R}/c)}\right)\right]_{t_s} = -\frac{1}{[R-(\mathbf{v}\cdot\mathbf{R}/c)]^2}\left([\nabla R]_{t_s} - \frac{1}{c}[\nabla(\mathbf{v}\cdot\mathbf{R})]_{t_s}\right) \tag{12.261}$$

For the first term on the right side of Eq. (12.261), one recalls Eq. (12.251), where

$$[\nabla R]_{t_s} = \hat{\mathbf{R}} \tag{12.262}$$

Next, one evaluates the second term on the right side of Eq. (12.261), where

$$[\nabla(\mathbf{v}\cdot\mathbf{R})]_{t_s} = (v_x\hat{\mathbf{x}}\cdot\hat{\mathbf{x}})[\nabla(x-x_s)]_{t_s} + (v_y\hat{\mathbf{y}}\cdot\hat{\mathbf{y}})[\nabla(y-y_s)]_{t_s} + (v_z\hat{\mathbf{z}}\cdot\hat{\mathbf{z}})[\nabla(z-z_s)]_{t_s} \tag{12.263}$$

with

$$[\nabla(x-x_s)]_{t_s} = \hat{\mathbf{x}}\frac{\partial}{\partial x}(x-x_s) = \hat{\mathbf{x}} \tag{12.264}$$

$$[\nabla(y-y_s)]_{t_s} = \hat{\mathbf{y}}\frac{\partial}{\partial y}(y-y_s) = \hat{\mathbf{y}} \tag{12.265}$$

$$[\nabla(z-z_s)]_{t_s} = \hat{\mathbf{z}}\frac{\partial}{\partial z}(z-z_s) = \hat{\mathbf{z}} \tag{12.266}$$

With substitution of Eqs. (12.264–12.266) into Eq. (12.263), the second term on the right side of Eq. (12.261), becomes

$$[\nabla(\mathbf{v}\cdot\mathbf{R})]_{t_s} = v_x\hat{\mathbf{x}} + v_y\hat{\mathbf{y}} + v_z\hat{\mathbf{z}} = \mathbf{v} \tag{12.267}$$

With substitution of Eqs. (12.262) and (12.267) into Eq. (12.261), the first term of the scalar potential $\nabla\phi$ becomes

$$\left[\nabla\left(\frac{1}{R-(\mathbf{v}\cdot\mathbf{R}/c)}\right)\right]_{t_s} = -\frac{1}{[R-(\mathbf{v}\cdot\mathbf{R}/c)]^2}\left(\hat{\mathbf{R}} - \frac{\mathbf{v}}{c}\right) \tag{12.268}$$

Next, one evaluates the earlier time derivative of Eq. (12.268) for the second term of Eq. (12.260) for $\nabla\phi$, where

$$\frac{\partial}{\partial t_s}\left(\frac{1}{R-(\mathbf{v}\cdot\mathbf{R}/c)}\right) = -\frac{1}{[R-(\mathbf{v}\cdot\mathbf{R}/c)]^2}\left(\frac{\partial R}{\partial t_s} - \frac{1}{c}\frac{\partial}{\partial t_s}(\mathbf{v}\cdot\mathbf{R})\right) \tag{12.269}$$

For the first term on the right side of Eq. (12.269), one recalls Eq. (12.250), where

$$\frac{\partial R}{\partial t_s} = -\hat{\mathbf{R}}\cdot\mathbf{v} = -\frac{\mathbf{R}\cdot\mathbf{v}}{R} \tag{12.270}$$

The earlier time derivative of $\mathbf{v}\cdot\mathbf{R}$ in Eq. (12.269), becomes

$$\frac{\partial}{\partial t_s}(\mathbf{v}\cdot\mathbf{R}) = \frac{\partial\mathbf{v}}{\partial t_s}\cdot\mathbf{R} + \mathbf{v}\cdot\frac{\partial\mathbf{R}}{\partial t_s} = \mathbf{a}\cdot\mathbf{R} - \mathbf{v}\cdot\mathbf{v} \tag{12.271}$$

The evaluation of Eq. (12.271) is obtained by again recalling Eq. (12.250), where

$$\frac{\partial \mathbf{R}}{\partial t_s} = -\mathbf{v} \tag{12.272}$$

$$\frac{\partial \mathbf{v}}{\partial t_s} = \frac{\partial^2 x_s}{\partial t_s^2}\hat{\mathbf{x}} + \frac{\partial y_s}{\partial t_s^2}\hat{\mathbf{y}} + \frac{\partial z_s}{\partial t_s^2}\hat{\mathbf{z}} = \mathbf{a} \tag{12.273}$$

With substitution of Eqs. (12.270) and (12.271) into Eq. (12.269), the second term of $\nabla\phi$ becomes

$$\frac{\partial}{\partial t_s}\left(\frac{1}{R-(\mathbf{v}\cdot\mathbf{R}/c)}\right) = \frac{1}{[R-(\mathbf{v}\cdot\mathbf{R}/c)]^2}\left(\frac{\mathbf{R}\cdot\mathbf{v}}{R} + \frac{\mathbf{a}\cdot\mathbf{R}-\mathbf{v}\cdot\mathbf{v}}{c}\right) \tag{12.274}$$

With substitution of Eqs. (12.268) and (12.274) into Eq. (12.260), one obtains the following expression for $\nabla\phi$:

$$\nabla\phi = -\frac{e}{4\pi\varepsilon_0}\frac{1}{[R-(\mathbf{v}\cdot\mathbf{R}/c)]^2}\left(\frac{\mathbf{R}}{R}-\frac{\mathbf{v}}{c}\right) - \frac{e}{4\pi\varepsilon_0}\frac{1}{c}\frac{\mathbf{R}}{[R-(\mathbf{v}\cdot\mathbf{R}/c)]^3}\left(\frac{\mathbf{R}\cdot\mathbf{v}}{R}+\frac{\mathbf{a}\cdot\mathbf{R}-\mathbf{v}\cdot\mathbf{v}}{c}\right) \tag{12.275}$$

The second term of electric field **E** is the observer temporal derivative $\partial\mathbf{A}/\partial t$ of the magnetic vector potential **A**, where one applies Eqs. (12.246) and (12.247), so that

$$\frac{\partial \mathbf{A}}{\partial t} = \frac{1}{1-(\hat{\mathbf{R}}\cdot\mathbf{v}/c)}\frac{\partial \mathbf{A}}{\partial t_s} = \frac{R}{R-(\mathbf{R}\cdot\mathbf{v}/c)}\frac{\partial \mathbf{A}}{\partial t_s} \tag{12.276}$$

The earlier (retarded) time derivative of the magnetic vector potential **A** of Eq. (12.233), yields

$$\frac{\partial \mathbf{A}}{\partial t_s} = \frac{e}{4\pi\varepsilon_0}\cdot\left[\frac{1}{R-(\mathbf{v}\cdot\mathbf{R}/c)}\frac{\partial \mathbf{v}}{\partial t_s} - \frac{\mathbf{v}}{[R-(\mathbf{v}\cdot\mathbf{R}/c)]^2}\left(\frac{\partial R}{\partial t_s} - \frac{1}{c}\frac{\partial}{\partial t_s}(\mathbf{v}\cdot\mathbf{R})\right)\right] \tag{12.277}$$

Substitution into Eq. (12.277) of Eqs. (12.271−12.273) for the earlier (retarded) time derivatives, gives

$$\frac{\partial \mathbf{A}}{\partial t_s} = \frac{e}{4\pi\varepsilon_0}\cdot\left[\frac{\mathbf{a}}{R-(\mathbf{v}\cdot\mathbf{R}/c)} + \frac{\mathbf{v}}{[R-(\mathbf{v}\cdot\mathbf{R}/c)]^2}\left(\frac{\mathbf{R}\cdot\mathbf{v}}{R}+\frac{\mathbf{a}\cdot\mathbf{R}-\mathbf{v}\cdot\mathbf{v}}{c}\right)\right] \tag{12.278}$$

Substitution of $\partial \mathbf{A}/\partial t_s$ into the expression for the observer temporal derivative $\partial \mathbf{A}/\partial t$ gives

$$\frac{\partial \mathbf{A}}{\partial t}=\frac{e}{4\pi\varepsilon_0}\cdot\left[\frac{\mathbf{a}R}{[R-(\mathbf{v}\cdot\mathbf{R}/c)]^2}+\frac{\mathbf{v}R}{[R-(\mathbf{v}\cdot\mathbf{R}/c)]^3}\left(\frac{\mathbf{R}\cdot\mathbf{v}}{R}+\frac{\mathbf{a}\cdot\mathbf{R}-\mathbf{v}\cdot\mathbf{v}}{c}\right)\right] \tag{12.279}$$

The electric field $\mathbf{E}(\mathbf{r},t)$, which is measured by the observer at position $\mathbf{r}$ at present time t, and arises from a charge e moving a velocity $\mathbf{v}$ at distance R at previous time t_s, is obtained by substitution of Eqs. (12.275) and (12.279) into Eq. (12.234):

$$\begin{aligned}\mathbf{E}=&\frac{e}{4\pi\varepsilon_0}\frac{1}{[R-(\mathbf{v}\cdot\mathbf{R}/c)]^2}\left(\frac{\mathbf{R}}{R}-\frac{\mathbf{v}}{c}\right)+\frac{e}{4\pi\varepsilon_0}\frac{1}{c}\frac{\mathbf{R}}{[R-(\mathbf{v}\cdot\mathbf{R}/c)]^3}\left(\frac{\mathbf{R}\cdot\mathbf{v}}{R}+\frac{\mathbf{a}\cdot\mathbf{R}-\mathbf{v}\cdot\mathbf{v}}{c}\right)\\&-\frac{e}{4\pi\varepsilon_0 c^2}\cdot\left[\frac{\mathbf{a}R}{[R-(\mathbf{v}\cdot\mathbf{R}/c)]^2}+\frac{\mathbf{v}R}{[R-(\mathbf{v}\cdot\mathbf{R}/c)]^3}\left(\frac{\mathbf{R}\cdot\mathbf{v}}{R}+\frac{\mathbf{a}\cdot\mathbf{R}-\mathbf{v}\cdot\mathbf{v}}{c}\right)\right]\end{aligned} \tag{12.280}$$

In Eq. (12.280), the observer electric field $\mathbf{E}(\mathbf{r},t)$ at the present time t is separated into acceleration terms with $\mathbf{a}$ and velocity terms with only $\mathbf{R}$ and $\boldsymbol{\beta}=\mathbf{v}/c$. All the terms on the right side of the equation are evaluated at the earlier time t_s so as to obtain on the left side of the equation the electric field $\mathbf{E}(\mathbf{r},t)$ at observer position $\mathbf{r}(t)$ at present time t, where

$$\mathbf{E}(\mathbf{r},t)=\frac{e}{4\pi\varepsilon_0}\left[\frac{R(1-\beta^2)}{[R-\boldsymbol{\beta}\cdot\mathbf{R}]^3}(\hat{\mathbf{R}}-\boldsymbol{\beta})\right]+\frac{e}{4\pi\varepsilon_0 c^2}\left\{\frac{R(\mathbf{a}\cdot\mathbf{R})}{[R-\boldsymbol{\beta}\cdot\mathbf{R}]^3}(\hat{\mathbf{R}}-\boldsymbol{\beta})-\frac{\mathbf{a}R}{[R-\boldsymbol{\beta}\cdot\mathbf{R}]^2}\right\} \tag{12.281}$$

The vector $\boldsymbol{\beta}$ is the source charge velocity $\mathbf{v}(t_s)$ at the earlier time t_s, divided by the vacuum speed of light c; that is, the source charge velocity normalized to the vacuum speed of light, where

$$\boldsymbol{\beta}=\frac{\mathbf{v}}{c} \tag{12.282}$$

Further simplification of Eq. (12.281), yields the explicit $1/R$ dependence of the acceleration terms, and the $1/R^2$ dependence of the velocity terms in the observer electric field $\mathbf{E}(\mathbf{r},t)$, where

$$\mathbf{E}=\frac{e}{4\pi\varepsilon_0}\left[\frac{1-\beta^2}{[1-\boldsymbol{\beta}\cdot\hat{\mathbf{R}}]^3}\frac{\hat{\mathbf{R}}-\boldsymbol{\beta}}{R^2}\right]+\frac{e}{4\pi\varepsilon_0 c^2}\left\{\mathbf{a}\cdot\frac{\hat{\mathbf{R}}}{[1-\boldsymbol{\beta}\cdot\hat{\mathbf{R}}]^3}\frac{\hat{\mathbf{R}}-\boldsymbol{\beta}}{R}-\frac{1}{[1-\boldsymbol{\beta}\cdot\hat{\mathbf{R}}]^2}\frac{\mathbf{a}}{R}\right\} \tag{12.283}$$

A further simplification of the electric field expression of Eq. (12.283) occurs by combining acceleration terms, where

$$\mathbf{a} \cdot \frac{\hat{\mathbf{R}}}{[1-\boldsymbol{\beta} \cdot \hat{\mathbf{R}}]^3} \frac{\hat{\mathbf{R}}-\boldsymbol{\beta}}{R} - \frac{1}{[1-\boldsymbol{\beta} \cdot \hat{\mathbf{R}}]^2} \frac{\mathbf{a}}{R} = \frac{1}{R} \frac{(\mathbf{a} \cdot \hat{\mathbf{R}})(\hat{\mathbf{R}}-\boldsymbol{\beta}) - \mathbf{a}(1-\boldsymbol{\beta} \cdot \hat{\mathbf{R}})}{[1-\boldsymbol{\beta} \cdot \hat{\mathbf{R}}]^3} \tag{12.284}$$

One applies the vector identity (BAC–CAB) to Eq. (12.284), where

$$\mathbf{A} \times \mathbf{B} \times \mathbf{C} = \mathbf{B}(\mathbf{A} \cdot \mathbf{C}) - \mathbf{C}(\mathbf{A} \cdot \mathbf{B}) \tag{12.285}$$

with

$$\mathbf{A} = \hat{\mathbf{R}}, \quad \mathbf{B} = \hat{\mathbf{R}} - \boldsymbol{\beta}, \text{and} \quad \mathbf{C} = \mathbf{a} \tag{12.286}$$

The acceleration term of Eq. (12.284) becomes

$$\mathbf{a} \cdot \frac{\hat{\mathbf{R}}}{[1-\boldsymbol{\beta} \cdot \hat{\mathbf{R}}]^3} \frac{\hat{\mathbf{R}}-\boldsymbol{\beta}}{R} - \frac{1}{[1-\boldsymbol{\beta} \cdot \hat{\mathbf{R}}]^2} \frac{\mathbf{a}}{R} = \frac{1}{R} \frac{\hat{\mathbf{R}} \times (\hat{\mathbf{R}}-\boldsymbol{\beta}) \times \mathbf{a}}{[1-\boldsymbol{\beta} \cdot \hat{\mathbf{R}}]^3} \tag{12.287}$$

With substitution of Eq. (12.287) into Eq. (12.283), the electric field $\mathbf{E}(\mathbf{r}, t)$ at observer position $\mathbf{r}(t)$ at present time t becomes

$$\mathbf{E}(\mathbf{r}, t) = \frac{e}{4\pi\varepsilon_0} \left[\frac{1-\beta^2}{[1-\boldsymbol{\beta} \cdot \hat{\mathbf{R}}]^3} \frac{\hat{\mathbf{R}}-\boldsymbol{\beta}}{R^2} \right] + \frac{e}{4\pi\varepsilon_0 c^2} \left[\frac{1}{R} \frac{\hat{\mathbf{R}} \times (\hat{\mathbf{R}}-\boldsymbol{\beta}) \times \mathbf{a}}{[1-\boldsymbol{\beta} \cdot \hat{\mathbf{R}}]^3} \right] \tag{12.288}$$

The double cross product in the expression for the acceleration component of the electric field $\mathbf{E}(\mathbf{r}, t)$ of Eq. (12.288) demonstrates that the acceleration component of the electric field $\mathbf{E}_a(\mathbf{r}, t)$ is perpendicular to the direction $\hat{\mathbf{R}}$ of wave propagation from charge e at $\mathbf{r}_s(t_s)$ to the observer at $\mathbf{r}(t)$.

12.13 Derivation of the Magnetic Field of a Charged Particle in Arbitrary Motion from Lienard–Wiechert Potentials

In this section, the magnetic field produced by a charged particle in arbitrary motion is calculated using the Lienard–Wiechert potentials. The magnetic field $\mathbf{B}(t)$, which is measured by the observer at present time t, and arises from a charge e that moves at velocity $\mathbf{v}$ at distance R at earlier time t_s, is

$$\mathbf{B} = \nabla \times \mathbf{A} = [\nabla \times \mathbf{A}]_{t_s} - \frac{1}{c} \frac{\mathbf{R} \times (\partial \mathbf{A}/\partial t_s)}{R - (\mathbf{v} \cdot \mathbf{R}/c)} \tag{12.289}$$

Substitution into Eq. (12.289) of Eq. (12.233) for $\mathbf{A}(\mathbf{r}, t)$, gives

$$\mathbf{B} = \frac{e}{4\pi\varepsilon_0 c^2}\left[\nabla\times\frac{\mathbf{v}}{R-(\mathbf{v}\cdot\mathbf{R}/c)}\right]_{t_s} - \frac{e}{4\pi\varepsilon_0 c^2}\left[\frac{1}{R-(\mathbf{v}\cdot\mathbf{R}/c)}\right]\frac{\mathbf{R}}{c}\times\frac{\partial}{\partial t_s}\left[\frac{\mathbf{v}}{R-(\mathbf{v}\cdot\mathbf{R}/c)}\right] \tag{12.290}$$

To the first term of Eq. (12.290), the following vector identity is applied:

$$\nabla\times(f\mathbf{A}) = \nabla f\times\mathbf{A} + f\nabla\times\mathbf{A} \tag{12.291}$$

The chain rule of differentiation is then applied, and Eq. (12.290) becomes:

$$\begin{aligned}\mathbf{B} = {} & \frac{e}{4\pi\varepsilon_0 c^2}\left[\frac{\nabla\times\mathbf{v}}{R-(\mathbf{v}\cdot\mathbf{R}/c)} - \frac{\nabla[R-(\mathbf{v}\cdot\mathbf{R}/c)]\times\mathbf{v}}{[R-(\mathbf{v}\cdot\mathbf{R}/c)]^2}\right]_{t_s} \\ & - \frac{e}{4\pi\varepsilon_0 c^2}\frac{1}{[R-(\mathbf{v}\cdot\mathbf{R}/c)]^2}\frac{\mathbf{R}}{c}\times\left[\frac{\partial\mathbf{v}}{\partial t_s} - \frac{\mathbf{v}}{R-(\mathbf{v}\cdot\mathbf{R}/c)}\frac{\partial}{\partial t_s}(R-(\mathbf{v}\cdot\mathbf{R}/c))\right]\end{aligned} \tag{12.292}$$

Recall from Eqs. (12.262), (12.267), (12.270), and (12.271), the previously derived expressions

$$[\nabla R]_{t_s} = \hat{\mathbf{R}} \quad [\nabla(\mathbf{v}\cdot\mathbf{R})]_{t_s} = \mathbf{v} \tag{12.293}$$

and

$$\frac{\partial R}{\partial t_s} = -\frac{\mathbf{R}\cdot\mathbf{v}}{R} \quad \frac{\partial}{\partial t_s}(\mathbf{v}\cdot\mathbf{R}) = \mathbf{a}\cdot\mathbf{R} - \mathbf{v}\cdot\mathbf{v} \tag{12.294}$$

The velocity $\mathbf{v}$ is independent of the position $\mathbf{r}$ of the source charge, so

$$\nabla\times\mathbf{v} = 0 \tag{12.295}$$

With substitution of Eqs. (12.293–12.295) into Eq. (12.292), the observer magnetic field becomes

$$\begin{aligned}\mathbf{B} = {} & \frac{e}{4\pi\varepsilon_0 c^2}\left[-\frac{1}{[R-(\mathbf{v}\cdot\mathbf{R}/c)]^2}\left(\hat{\mathbf{R}} - \frac{\mathbf{v}}{c}\right)\times\mathbf{v}\right]_{t_s} \\ & - \frac{e}{4\pi\varepsilon_0 c^2}\frac{1}{[R-(\mathbf{v}\cdot\mathbf{R}/c)]^2}\left[\frac{\mathbf{R}\times\mathbf{a}}{c} - \frac{\mathbf{R}\times\mathbf{v}}{R-(\mathbf{v}\cdot\mathbf{R}/c)}\left(-\frac{\mathbf{R}\cdot\mathbf{v}}{cR} - \frac{\mathbf{a}\cdot\mathbf{R}-\mathbf{v}\cdot\mathbf{v}}{c^2}\right)\right]\end{aligned} \tag{12.296}$$

With substitution into Eq. (12.296) of the normalized velocity vector $\boldsymbol{\beta} = \mathbf{v}/c$, the expression for the magnetic field **B** is arranged with the acceleration vector **a** terms separated from the velocity vector $\boldsymbol{\beta} = \mathbf{v}/c$ terms, where

$$\mathbf{B} = \frac{e}{4\pi\varepsilon_0 c^2}\left[c\boldsymbol{\beta}\times\frac{\hat{\mathbf{R}}}{[R-\boldsymbol{\beta}\cdot\mathbf{R}]^2}\right]_{t_s} + \frac{e}{4\pi\varepsilon_0 c^2}c\boldsymbol{\beta}\times\frac{\hat{\mathbf{R}}}{[R-\boldsymbol{\beta}\cdot\mathbf{R}]^3}(\boldsymbol{\beta}\cdot\mathbf{R}-R\beta^2)$$
$$+\frac{e}{4\pi\varepsilon_0 c^2}\frac{1}{c}\frac{\mathbf{a}\times\mathbf{R}}{[R-\boldsymbol{\beta}\cdot\mathbf{R}]^2} + \frac{e}{4\pi\varepsilon_0 c^2}\frac{R}{[R-\boldsymbol{\beta}\cdot\mathbf{R}]^3}\frac{\mathbf{a}\cdot\mathbf{R}}{c}\boldsymbol{\beta}\times\hat{\mathbf{R}} \tag{12.297}$$

In the first two terms of Eq. (12.297), the velocity terms can be combined. The observer's magnetic $\mathbf{B}(\mathbf{r},t)$ field at the present time t is separated into acceleration terms with **a** and velocity terms with only **R** and **v**, where all terms on the right side of the equation are evaluated at the earlier time t_s, so

$$\mathbf{B} = \frac{e}{4\pi\varepsilon_0 c^2}\left[\frac{(1-\beta^2)Rc}{[R-\boldsymbol{\beta}\cdot\mathbf{R}]^3}(\boldsymbol{\beta}\times\hat{\mathbf{R}}) + \frac{1}{c}\left(\frac{\mathbf{a}\times\mathbf{R}}{[R-\boldsymbol{\beta}\cdot\mathbf{R}]^2} + \frac{(\mathbf{a}\cdot\mathbf{R})R}{[R-\boldsymbol{\beta}\cdot\mathbf{R}]^3}(\boldsymbol{\beta}\times\hat{\mathbf{R}})\right)\right] \tag{12.298}$$

Further simplification of Eq. (12.298), yields the explicit $1/R$ dependence of the acceleration terms, and the $1/R^2$ dependence of the velocity terms in the observer magnetic field $\mathbf{B}(\mathbf{r},t)$, where

$$\mathbf{B}(\mathbf{r},t) = \frac{e}{4\pi\varepsilon_0 c^2}\left[c\frac{1-\beta^2}{[1-\boldsymbol{\beta}\cdot\hat{\mathbf{R}}]^3}\boldsymbol{\beta}\times\frac{\hat{\mathbf{R}}}{R^2} + \frac{1}{c}\left(\frac{1}{[1-\boldsymbol{\beta}\cdot\hat{\mathbf{R}}]^2}\mathbf{a}\times\frac{\hat{\mathbf{R}}}{R} + \mathbf{a}\cdot\frac{\hat{\mathbf{R}}}{[1-\boldsymbol{\beta}\cdot\hat{\mathbf{R}}]^3}\boldsymbol{\beta}\times\frac{\hat{\mathbf{R}}}{R}\right)\right] \tag{12.299}$$

The magnetic field $\mathbf{B}(\mathbf{r},t)$ is perpendicular to the direction $\hat{\mathbf{R}}$ of wave propagation, and $\mathbf{B}(\mathbf{r},t)$ is perpendicular to the electric field $\mathbf{E}(\mathbf{r},t)$, where

$$c\mathbf{B} = \hat{\mathbf{R}}\times\mathbf{E} \tag{12.300}$$

By comparison, the electric field $\mathbf{E}(\mathbf{r},t)$ of Eq. (12.288) in the previous section is repeated here:

$$\mathbf{E}(\mathbf{r},t) = \frac{e}{4\pi\varepsilon_0}\left[\frac{1-\beta^2}{[1-\boldsymbol{\beta}\cdot\hat{\mathbf{R}}]^3}\frac{\hat{\mathbf{R}}-\boldsymbol{\beta}}{R^2}\right] + \frac{e}{4\pi\varepsilon_0 c^2}\left[\frac{1}{R}\frac{\hat{\mathbf{R}}\times(\hat{\mathbf{R}}-\boldsymbol{\beta})\times\mathbf{a}}{[1-\boldsymbol{\beta}\cdot\hat{\mathbf{R}}]^3}\right] \tag{12.301}$$

The velocity terms in the magnetic $\mathbf{B}(\mathbf{r},t)$ and electric $\mathbf{E}(\mathbf{r},t)$ fields vary inversely with the square of the distance $(1/R^2)$, whereas, the acceleration terms vary inversely with distance $(1/R)$. At large distances R between the source charge and the observer, the magnetic $\mathbf{B}(\mathbf{r},t)$ and electric $\mathbf{E}(\mathbf{r},t)$ fields at the observer are dominated by the $1/R$ acceleration terms.

12.14 Poynting Vector and Electromagnetic Energy Radiated from an Arbitrary Accelerated Charge per Solid Angle per Frequency Interval

In this section we follow chapter 14 in Jackson (1975). The Poynting vector **S** is the power density transmitted by the charge in the form of the electromagnetic field, which is caused by the charge acceleration **a**, where, in MKS units, and recalling Eq. (12.132),

$$\mathbf{S} = c^2\varepsilon_0 \mathbf{E}_a \times \mathbf{B}_a \tag{12.302}$$

From Eqs. (12.301) and (12.299), the components of the electric $\mathbf{E}_a(\mathbf{r}, t)$ and magnetic $\mathbf{B}_a(\mathbf{r}, t)$ fields that arise due to the charge acceleration are

$$\mathbf{E}_a = \frac{e}{4\pi\varepsilon_0 c^2}\left[\frac{1}{R}\frac{\hat{\mathbf{R}} \times (\hat{\mathbf{R}} - \boldsymbol{\beta}) \times \mathbf{a}}{[1 - \boldsymbol{\beta}\cdot\hat{\mathbf{R}}]^3}\right] \tag{12.303}$$

$$\mathbf{B}_a = \frac{e}{4\pi\varepsilon_0 c^2}\cdot\frac{1}{c}\left(\frac{1}{[1 - \boldsymbol{\beta}\cdot\hat{\mathbf{R}}]^2}\mathbf{a} \times \frac{\hat{\mathbf{R}}}{R} + \mathbf{a}\cdot\frac{\hat{\mathbf{R}}}{[1 - \boldsymbol{\beta}\cdot\hat{\mathbf{R}}]^3}\boldsymbol{\beta} \times \frac{\hat{\mathbf{R}}}{R}\right) \tag{12.304}$$

With substitution into Eq. (12.302) of Eq. (12.304) for the magnetic field $\mathbf{B}_a(\mathbf{r}, t)$, the Poynting vector **S** is shown to be a power density flow in the direction from the charge position $\mathbf{r}_s$ the prior, earlier time t_s to the observer position **r** at the present time t, where

$$\mathbf{S} = c^2\varepsilon_0 \mathbf{E}_a \times \left(\frac{\hat{\mathbf{R}} \times \mathbf{E}_a}{c}\right) = c\varepsilon_0 E_a^2 \hat{\mathbf{R}} \tag{12.305}$$

The BAC−CAB rule, applied to Eq. (12.306), shows that

$$\mathbf{E}_a \times (\hat{\mathbf{R}} \times \mathbf{E}_a) = \hat{\mathbf{R}}(\mathbf{E}_a\cdot\mathbf{E}_a) - \mathbf{E}_a(\mathbf{E}_a\cdot\hat{\mathbf{R}}) = E_a^2\hat{\mathbf{R}} \tag{12.306}$$

The charge acceleration component electric $\mathbf{E}_a$ and magnetic $\mathbf{B}_a$ fields are perpendicular to each other and the direction of propagation $\hat{\mathbf{R}}$, so

$$\mathbf{E}_a\cdot\hat{\mathbf{R}} = 0 \tag{12.307}$$

If the charge q is stationary with $\mathbf{v} = 0$ and $\mathbf{a} = 0$, then there arises only the $1/R^2$ static electric field $\mathbf{E}(\mathbf{r})$ of Coulomb's law. If the charge is moving at constant velocity $\mathbf{v} = const$ with no acceleration $\mathbf{a} = 0$, then the moving charge is a current $q\mathbf{v}$, and there arises only the $1/R^2$ static magnetic field $\mathbf{B}(\mathbf{r})$ of the Biot−Savart law. For zero charge acceleration **a**, the Poynting vector $\mathbf{S} = 0$, because there is no power flow with static electric and magnetic fields.

The emitted electromagnetic power per solid angle $\mathrm{d}P/\mathrm{d}\Omega$, which is produced by a charge undergoing acceleration **a**, is

$$\frac{\mathrm{d}P}{\mathrm{d}\Omega}(t) = |R^2\mathbf{S}| = c\varepsilon_0 |R\mathbf{E}_a|^2_{t_s} \tag{12.308}$$

The earlier t_s at the position $\mathbf{r}_s$ of the accelerating charge is related to the present, observer time t at the observer position $\mathbf{r}$ by

$$t_{ret} \equiv t_s = t - \frac{1}{c}R(t_s) \tag{12.309}$$

where

$$\mathrm{d}t_{ret} \equiv \mathrm{d}t_s = \mathrm{d}t - \frac{1}{c}\mathrm{d}R(t_s) \tag{12.310}$$

The earlier time t_s is termed the retarded time, which is denoted by t_{ret}.

Consider a distance-independent electric field vector $\mathbf{F}(t)$, which is the product of (1) the charge-to-observer distance R, evaluated at the present observer time t, and (2) the charge acceleration component of the electric field $\mathbf{E}_a(t)$, evaluated at the present observer time t, where

$$\mathbf{F}(t) = \sqrt{c\varepsilon_0}R(t)\mathbf{E}_a(t) \tag{12.311}$$

On the right side of the equation, one replaces the present observer time t by its retarded time-equivalent expression t_s, which shows the dependence of the observer time t upon the charge at earlier time t_s, where

$$\mathbf{F}(t) = \sqrt{c\varepsilon_0}R\left(t_s + \frac{1}{c}R(t_s)\right)\mathbf{E}_a\left(t_s + \frac{1}{c}R(t_s)\right) \tag{12.312}$$

The total energy radiated per solid angle $\mathrm{d}W/\mathrm{d}\Omega$ is the time integral of the emitted electromagnetic power per solid angle $\mathrm{d}P/\mathrm{d}\Omega$, where

$$\frac{\mathrm{d}W}{\mathrm{d}\Omega} = \int_{-\infty}^{\infty}\frac{\mathrm{d}P}{\mathrm{d}\Omega}\mathrm{d}t = \int_{-\infty}^{\infty}|\mathbf{F}(t)|^2\,\mathrm{d}t \tag{12.313}$$

Substitution into Eq. (12.313) of Eq. (12.312) for **F**, gives

$$\frac{\mathrm{d}W}{\mathrm{d}\Omega} = \int_{-\infty}^{\infty} c\varepsilon_0|R(t)E_a(t)|^2\,\mathrm{d}t \tag{12.314}$$

The time-dependent and distance-independent electric field $\mathbf{F}(t)$ can be related to its corresponding frequency-dependent form $\mathbf{F}(w)$ via the Fourier transform of $\mathbf{F}(t)$, where

$$\mathbf{F}(w) = \frac{1}{\sqrt{2\pi}} \int_{-\infty}^{\infty} dt \, \exp(iwt)\mathbf{F}(t) \tag{12.315}$$

The inverse Fourier transform of $\mathbf{F}(w)$ is

$$\mathbf{F}(t) = \frac{1}{\sqrt{2\pi}} \int_{-\infty}^{\infty} dw \, \exp(-iwt)\mathbf{F}(w) \tag{12.316}$$

The present observer time t is used in Eq. (12.316), due to convenience and practicality for the observer to consider a frequency spectrum $\mathbf{F}(w)$, which is based on the present observer time t, rather than considering a frequency spectrum $\mathbf{F}(w_s)$ that is based on an unknown earlier time t_s of the charge acceleration $\mathbf{a}(t_s)$.

With substitution into Eq. (12.313) of Eq. (12.316) $\mathbf{F}(t)$, the total energy radiated per solid angle $dW/d\Omega$ becomes

$$\frac{dW}{d\Omega} = \frac{1}{2\pi} \int_{-\infty}^{\infty} dt \int_{-\infty}^{\infty} dw \int_{-\infty}^{\infty} dw' \, [\mathbf{F}^*(w') \cdot \mathbf{F}(w)] \exp[-it(w' - w)] \tag{12.317}$$

The order of the time and frequency integrals in Eq. (12.317) can be reversed so that

$$\frac{dW}{d\Omega} = \frac{1}{2\pi} \int_{-\infty}^{\infty} dw \int_{-\infty}^{\infty} dw' \, [\mathbf{F}^*(w') \cdot \mathbf{F}(w)] \int_{-\infty}^{\infty} dt \, \exp[-it(w' - w)] \tag{12.318}$$

The last integral in Eq. (12.318) is the Dirac delta function, where

$$\int_{-\infty}^{\infty} dt \, \exp[-it(w' - w)] = \delta(w' - w) \tag{12.319}$$

With the Dirac delta function of Eq. (12.319) comes simplification, where substitution of Eq. (12.319) into Eq. (12.318), yields

$$\frac{dW}{d\Omega} = \frac{1}{2\pi} \int_{-\infty}^{\infty} |\mathbf{F}(w)|^2 \, dw \tag{12.320}$$

In this form, the total energy radiated per solid angle $dW/d\Omega$ suggests Parseval's theorem, which states that the summation (or integral) of a function equals the summation (or integral) of its transform (a Fourier transform in this case), where

$$\frac{\mathrm{d}W}{\mathrm{d}\Omega} = \frac{1}{2\pi}\int_{-\infty}^{\infty} |\mathbf{F}(w)|^2 \,\mathrm{d}w = \frac{1}{2\pi}\int_{-\infty}^{\infty} |\mathbf{F}(t)|^2 \,\mathrm{d}t \tag{12.321}$$

The positive or negative sign of the temporal frequencies w has no physical meaning; however, the positive or negative signs of the temporal (angular) frequency w and the spatial frequency k are used to indicate whether the wave is propagating in the forward or backward direction, relative to a fixed spatial and temporal reference point. Also, the distance-independent electric field vector $\mathbf{F}(t)$ has Fourier transform $\mathbf{F}(w)$, which is the energy emitted by the accelerated charge per solid angle per frequency interval $\mathrm{d}^2I/\mathrm{d}w\mathrm{d}\Omega$, where

$$|\mathbf{F}(w)|^2 = \frac{\mathrm{d}^2I(w,\hat{\mathbf{R}})}{\mathrm{d}w\mathrm{d}\Omega} \tag{12.322}$$

By convention, the differential energy emission $\mathrm{d}I$ per differential frequency $\mathrm{d}w$ and differential solid angle $\mathrm{d}\Omega$, which is denoted by $\mathrm{d}^2I/\mathrm{d}w\mathrm{d}\Omega$, is integrated over the positive frequency range, so that substitution of Eq. (12.322) into Eq. (12.320), yields

$$\frac{\mathrm{d}W}{\mathrm{d}\Omega} = \frac{1}{2\pi}\int_{0}^{\infty} \frac{\mathrm{d}^2I(w,\hat{\mathbf{R}})}{\mathrm{d}w\mathrm{d}\Omega}\mathrm{d}w \tag{12.323}$$

Comparison of the integral of Eq. (12.323) over positive frequencies, and the integral of Eq. (12.320) over positive and negative frequencies, then requires their integrands to have the following equality relation:

$$\frac{\mathrm{d}^2I(w,\hat{\mathbf{R}})}{\mathrm{d}w\mathrm{d}\Omega} = |F(w)|^2 + |F(-w)|^2 \tag{12.324}$$

One assumes that $\mathbf{F}(t_s)$ is real, so

$$F^*(w) = F(-w) \tag{12.325}$$

With substitution of Eq. (12.325) into Eq. (12.324), the observer-measured per solid angle frequency (energy) spectrum at observer position $\mathbf{r}$ at present time t for all the previous electron or charged particle accelerations $\mathbf{a}(t_s)$ at earlier times t_s is

$$\frac{\mathrm{d}^2I(w,\hat{\mathbf{R}})}{\mathrm{d}w\mathrm{d}\Omega} = 2|\mathbf{F}(w)|^2 \tag{12.326}$$

Where with Eq. (12.325), one obtains

$$[F(w)]^2 = F^*(w)F(w) = F(-w)F(w) \tag{12.327}$$

The Fourier transform of $\mathbf{F}(t)$ of Eq. (12.312) is $\mathbf{F}(w)$, where

$$\mathbf{F}(w) = \frac{1}{\sqrt{2\pi}} \int_{-\infty}^{\infty} \mathrm{d}t \exp(\mathrm{i}wt)\sqrt{c\varepsilon_0} R(t)\mathbf{E}_a(t) \tag{12.328}$$

The integral expression of $\mathbf{F}(w)$ must express this physical situation of electromagnetic wave emission from the accelerating charge at earlier time t_s from position $\mathbf{r}(t_s)$, transit time $R(t_s)/c$ for electromagnetic wave propagation over distance $R(t_s)$ between the sources, and then arrival at the observer position $\mathbf{r}(t)$ at present time t. That is, the integral of $\mathbf{F}(w)$ must express causality. The integral of Eq. (12.328) is changed from integration over the present time t to integration over the earlier time t_s. Also the present time t is replaced by its equivalent expression in terms of earlier time t_s and source to observer distance $R(t_s)$ at earlier time t_s, where

$$t = t_s + \frac{1}{c} R(t_s) \tag{12.329}$$

One takes the present time derivative of Eq. (12.329), where

$$\frac{\partial t_s}{\partial \mathrm{t}} = \frac{1}{1 - (\hat{\mathbf{R}} \cdot \mathbf{v}/c)} \tag{12.330}$$

One obtains the resulting differential expression from Eq. (12.330), which relates the differential present time of the observer to the differential earlier (retarded) time of the moving charge, where

$$\mathrm{d}t = (1 - \hat{\mathbf{R}} \cdot \boldsymbol{\beta})\mathrm{d}t_s \tag{12.331}$$

With substitution of Eqs. (12.329–12.331) into Eq. (12.328), one obtains the integral expression of $\mathbf{F}(w)$, which is integrated over the earlier (retarded) time t_s, where

$$\mathbf{F}(w) = \sqrt{\frac{c\varepsilon_0}{2\pi}} \int_{-\infty}^{\infty} \mathrm{d}t_s (1 - \hat{\mathbf{R}} \cdot \boldsymbol{\beta}) \exp\left[\mathrm{i}w\left(t_s + \frac{1}{c} R(t_s)\right)\right] R(t_s)\mathbf{E}_a(t_s) \tag{12.332}$$

With substitution into Eq. (12.332) of Eq. (12.303) for the acceleration-based component of the electric field $\mathbf{E}_a(t_s)$, the Fourier transform $\mathbf{F}(w)$ becomes

$$\mathbf{F}(w) = \frac{e}{4\pi\varepsilon_0 c^2} \sqrt{\frac{c\varepsilon_0}{2\pi}} \int_{-\infty}^{\infty} \mathrm{d}t_s \frac{\hat{\mathbf{R}} \times (\hat{\mathbf{R}} - \boldsymbol{\beta}) \times \mathbf{a}}{[1 - \boldsymbol{\beta} \cdot \hat{\mathbf{R}}]^2} \exp\left[\mathrm{i}w\left(t_s + \frac{1}{c} R(t_s)\right)\right] \tag{12.333}$$

The MKS and CGS units for energy are respectively joules and ergs, where joule $= 10^7$erg, and in MKS units, $4\pi\varepsilon_0 c^2 = 10^7$. With substitution of $\mathbf{F}(w)$ from Eq. (12.333) into 12.326 for $\mathrm{d}^2I/\mathrm{d}w\mathrm{d}\Omega$, the charged particle's emission of differential energy $\mathrm{d}I$ per differential frequency $\mathrm{d}w$ and solid angle $\mathrm{d}\Omega$, becomes

$$\frac{\mathrm{d}^2 I(w,\hat{\mathbf{R}})}{\mathrm{d}w\mathrm{d}\Omega} = 2|\mathbf{F}(w)|^2 = \frac{e^2}{16\pi^3\varepsilon_0 c^3}\left|\int_{-\infty}^{\infty}\mathrm{d}t_s \frac{\hat{\mathbf{R}}\times(\hat{\mathbf{R}}-\boldsymbol{\beta})\times\mathbf{a}}{[1-\boldsymbol{\beta}\cdot\hat{\mathbf{R}}]^2}\exp\left[\mathrm{i}w\left(t_s+\frac{1}{c}R(t_s)\right)\right]\right|^2 \tag{12.334}$$

Relative to a common reference point or origin, a simplification occurs if the distance $R(t_s)$ of the observer position $\mathbf{r}(t)$ at present time t from the charge position $\mathbf{r}_s(t_s)$ at earlier time t_s is much larger than the variation of distance ΔR over the integration of t_s from $[-\infty, \infty]$. That is, the long-distance approximation requires

$$R(t_s) \gg \Delta R \qquad -\infty \leq t_s \leq \infty \tag{12.335}$$

If the distance approximation is applicable, then

$$R(t_s) \cong |\mathbf{r}(t)| - \hat{\mathbf{r}}(t_s)\cdot\mathbf{r}_s(t_s) = r - \hat{\mathbf{r}}\cdot\mathbf{r}_s \tag{12.336}$$

The unit vector $\hat{\mathbf{r}}(t)$ points from the origin to the observer position $\mathbf{r}(t)$ at present time t, where

$$\hat{\mathbf{r}}(t) = \frac{\mathbf{r}(t)}{r(t)} \tag{12.337}$$

That is, $|\mathbf{r}(t)|$ or $r(t)$ is the distance of the observer from the origin at the present time t, and $|\mathbf{r}_s(t_s)|$ or $r_s(t_s)$ is the distance of the charge from the origin at the prior time t_s.

From Eq. (12.336), the distance $R(t_s)$ between the accelerating charge and the observer, as a function of earlier time t_s, is approximated as the component distance $\mathbf{r}_s\cdot\hat{\mathbf{r}}(t)$ of the charge, which points from the origin in the direction of the observer, where $\mathbf{r}_s\cdot\hat{\mathbf{r}}(t)$ is subtracted from the observer distance $\mathbf{r}(t)$ from the origin. In the long distance approximation, the unit vector $\hat{\mathbf{R}}(t_s)$, which points from the charge at prior time t_s to the observer at present time t, is replaced by the unit vector $\hat{\mathbf{r}}(t)$, where

$$\hat{\mathbf{R}}(t_s) \rightarrow \hat{\mathbf{r}}(t) \quad R(t_s) \gg \Delta R \tag{12.338}$$

In the long-distance approximation, with substitution of Eqs. (12.336−12.338) into Eq. (12.334), the energy emitted by the accelerated charge, per solid angle interval, per frequency interval, $\mathrm{d}^2I/\mathrm{d}w\mathrm{d}\Omega$, expressed in MKS units, then becomes

$$\frac{\mathrm{d}^2I(w,\hat{\mathbf{r}})}{\mathrm{d}w\mathrm{d}\Omega}=\frac{e^2}{16\pi^3\varepsilon_0c^3}\left|\int_{-\infty}^{\infty}\mathrm{d}t_s\frac{\hat{\mathbf{r}}\times(\hat{\mathbf{r}}-\boldsymbol{\beta})\times\mathbf{a}}{[1-\boldsymbol{\beta}\cdot\hat{\mathbf{r}}]^2}\exp\left[\mathrm{i}w\left(t_s+\frac{r-\hat{\mathbf{r}}\cdot\mathbf{r}_s}{c}\right)\right]\right|^2 \quad R(t_s)\gg\Delta R \tag{12.339}$$

The distance r from origin to observer is fixed in time. Thus, in the integrand, the phase-shift factor is fixed with a magnitude is unity, where

$$\left|\exp\left(\frac{\mathrm{i}wr}{c}\right)\right|=1 \tag{12.340}$$

As a result, the energy emitted by the accelerated charge, per solid angle interval, per frequency interval, $\mathrm{d}^2I/\mathrm{d}w\mathrm{d}\Omega$, expressed in MKS units, is simplified to

$$\frac{\mathrm{d}^2I(w,\hat{\mathbf{r}})}{\mathrm{d}w\mathrm{d}\Omega}=\frac{e^2}{16\pi^3\varepsilon_0c^3}\left|\int_{-\infty}^{\infty}\mathrm{d}t_s\frac{\hat{\mathbf{r}}\times(\hat{\mathbf{r}}-\boldsymbol{\beta})\times\mathbf{a}}{[1-\boldsymbol{\beta}\cdot\hat{\mathbf{r}}]^2}\exp\left[\mathrm{i}w\left(t_s-\frac{\hat{\mathbf{r}}\cdot\mathbf{r}_s}{c}\right)\right]\right|^2 \quad R(t_s)\gg\Delta R \tag{12.341}$$

The result of Eq. (12.341) applies to any other charged particle, where e is the charge of the particle. Also, $\mathrm{d}^2I/\mathrm{d}w\mathrm{d}\Omega$ is the per solid angle frequency spectrum, which is measured by the observer at observer position $\mathbf{r}$ at present time t, and was emitted by the charged particle at earlier time t_s, where

$$t=t_s+\frac{1}{c}R(t_s) \tag{12.342}$$

Causality with the finite speed of light c demands

$$t\geq t_s \tag{12.343}$$

Causality requires the electromagnetic field at present time t and position $\mathbf{r}$ of the stationary observer, where $\mathrm{d}\mathbf{r}/\mathrm{d}t=0$, is to be produced by acceleration $\mathbf{a}(t_s)$ of a charged particle at position $\mathbf{r}(t_s)$ and past time t_s, which occur within a "vacuum speed of light" distance $R=c(t-t_s)$, where $|\mathbf{r}(t)-\mathbf{r}(t_s)|\leq R$. The differential energy per differential frequency and solid angle, $\mathrm{d}^2I/\mathrm{d}w\mathrm{d}\Omega$, which is emitted by the accelerated charged particle, during the earlier time interval of interest T from initial time $t_s=0$ to final, later time $t_s=T$, where $T>0$, is measured by the stationary observer at the later, present position $\mathbf{r}$ at time t, where

$$\frac{\mathrm{d}^2I(w,\hat{\mathbf{r}})}{\mathrm{d}w\mathrm{d}\Omega}=\frac{e^2}{16\pi^3\varepsilon_0c^3}\left|\int_{0}^{T}\mathrm{d}t_s\frac{\hat{\mathbf{r}}\times(\hat{\mathbf{r}}-\boldsymbol{\beta})\times\mathbf{a}}{[1-\boldsymbol{\beta}\cdot\hat{\mathbf{r}}]^2}\exp\left[\mathrm{i}w\left(t_s-\frac{\hat{\mathbf{r}}\cdot\mathbf{r}_s}{c}\right)\right]\right|^2 \quad R(t_s)\gg\Delta R \tag{12.344}$$

and

$$t = T + \frac{1}{c}R(T) \tag{12.345}$$

Given a trajectory $\mathbf{r}_s(t_s)$ of a charge, one can compute the velocity $\mathbf{v}(t_s)$ and acceleration $\mathbf{a}(t_s)$ as a function of earlier time t_s. The resulting integrand is as a function of the electromagnetic wave frequency w and the unit vector direction $\hat{\mathbf{r}}(t)$, which points from the origin to the stationary observer position $\mathbf{r}(t)$ at the present time t. Performing the integration yields the energy, per solid angle increment, per frequency increment, as a function of direction $\hat{\mathbf{r}}(t)$ and electromagnetic wave frequency w. In CGS units, recalling section 12.6, the general expression for the differential energy emission from an accelerated charge per differential frequency and solid angle, $(\mathrm{d}^2 I/\mathrm{d}w\mathrm{d}\Omega)_{\mathrm{cgs}}$, is

$$\left(\frac{\mathrm{d}^2 I(w,\hat{\mathbf{r}})}{\mathrm{d}w\mathrm{d}\Omega}\right)_{\mathrm{cgs}} = \frac{e^2}{4\pi^2 c}\left|\int_{-\infty}^{\infty} \mathrm{d}t_s \frac{\hat{\mathbf{r}} \times (\hat{\mathbf{r}} - \boldsymbol{\beta}) \times \mathbf{a}}{[1 - \boldsymbol{\beta} \cdot \hat{\mathbf{r}}]^2} \exp\left[iw\left(t_s - \frac{\hat{\mathbf{r}} \cdot \mathbf{r}_s}{c}\right)\right]\right|^2 \quad R(t_s) \gg \Delta R \tag{12.346}$$

Coherent electromagnetic radiation emission occurs at wavelengths λ_{coh} for a localized bunch of N electrons with total charge e, which are confined to a radius r_b that is much smaller than the emitted wavelength λ_{coh}, where

$$\lambda_{\mathrm{coh}} \gg r_b \tag{12.347}$$

and

$$\lambda_{\mathrm{coh}} = 2\pi \frac{c}{w_{\mathrm{coh}}} \tag{12.348}$$

The corresponding angular frequencies w_{coh} for coherent electromagnetic emission are then

$$2\pi \frac{c}{r_b} \gg w_{\mathrm{coh}} \tag{12.349}$$

12.15 Calculation of the Electron Trajectory in Synchrotron Ring Insertion Devices

The electric $\mathbf{E}(\mathbf{r}, t)$ and magnetic $\mathbf{B}(\mathbf{r}, t)$ fields, and the differential energy emission per differential frequency and solid angle, $\mathrm{d}^2 I/\mathrm{d}w\mathrm{d}\Omega$ can be calculated from the position, velocity, and acceleration trajectory of a charged particle. This section discusses the calculation of the electron trajectories in synchrotron insertion devices, which then allow the calculation of the emitted X-ray electromagnetic field and power as a function of distance and angle and frequency. Of particular interest is the calculation of X-rays from wigglers, undulators, and micropole undulators.

In synchrotron radiation calculations for undulators, wigglers, and bending magnets, using the computer and numerical techniques, one can calculate the magnetic field at each point in the insertion device. The center axis of the insertion device is considered the z-axis, which gives the axial position of the electron in the linear or curved device. At each nth axial position z_n along the center, reference z-axis, one must calculate magnetic field $\mathbf{B}(\mathbf{r}_s(z_n))$, which acts on the electron at electron position $\mathbf{r}_s(z_n)$, where

$$\mathbf{B}(\mathbf{r}_s(z_n)) = B_x(\mathbf{r}_s(z_n))\hat{\mathbf{x}} + B_y(\mathbf{r}_s(z_n))\hat{\mathbf{y}} + B_z(\mathbf{r}_s(z_n))\hat{\mathbf{z}} \tag{12.350}$$

The magnetic field $\mathbf{B}(\mathbf{r}_s(z_n))$ is calculated at the electron position $\mathbf{r}(z_n)$ at axial position z_n. This allows one to calculate the electron acceleration $\mathbf{a}(z_n)$ at the electron position $\mathbf{r}_s(z_n)$ via the Lorentz force equation. At axial position z_n, the electron position $\mathbf{r}_s(z_n)$ and acceleration are

$$\mathbf{r}_s(z_n) = x(z_n)\hat{\mathbf{x}} + y(z_n)\hat{\mathbf{y}} + z(z_n)\hat{\mathbf{z}} \tag{12.351}$$

and

$$\mathbf{a}(z_n) = a_x(z_n)\hat{\mathbf{x}} + a_y(z_n)\hat{\mathbf{y}} + a_z(z_n)\hat{\mathbf{z}} \tag{12.352}$$

The calculated values for the electron vector position $\mathbf{r}_s(z_{n-1})$, vector velocity $\mathbf{v}(z_{n-1})$, vector acceleration $\mathbf{a}(z_{n-1})$, and magnetic field vector $\mathbf{B}(\mathbf{r}_s(z_{n-1}))$ values at the $(n-1)^{\text{th}}$ axial position z_{n-1} are used to calculate the electron position $\mathbf{r}_s(z_n)$, velocity $\mathbf{v}(z_n)$, and acceleration $\mathbf{a}(z_n)$ vectors at the next n^{th} axial position z_n.

First, one starts with the Lorentz force equation, where the acceleration vector $\mathbf{a}(z_{n-1})$ of the electron at vector position $\mathbf{r}_s(z_{n-1})$ is calculated from the cross product of the electron vector velocity $\mathbf{v}(z_{n-1})$ and vector magnetic field $\mathbf{B}(\mathbf{r}_s(z_{n-1}))$ for electron axial position z_{n-1}, with $n = 2, 3, 4, ..., N$ and electron mass m_e and charge e,

$$\mathbf{a}(z_{n-1}) = \frac{e}{m_e}\sqrt{1 - \frac{|\mathbf{v}(z_{n-1})|^2}{c^2}}[\mathbf{v}(z_{n-1}) \times \mathbf{B}(\mathbf{r}_s(z_{n-1}))] \tag{12.353}$$

Next, the velocity vector $\mathbf{v}(z_n)$ at axial position z_n is obtained from the acceleration vector $\mathbf{a}(z_{n-1})$ at axial position z_{n-1}, which is multiplied by time interval Δt_s between axial positions z_{n-1} and z_n. This product or velocity vector change $\Delta\mathbf{v}$ is then added to the velocity $\mathbf{v}(z_{n-1})$ at the previous axial position z_{n-1}, where

$$\mathbf{v}(z_n) = \mathbf{a}(z_{n-1})\Delta t_s + \mathbf{v}(z_{n-1}) \tag{12.354}$$

and

$$\mathbf{v}(z_n) = v_x(z_n)\hat{\mathbf{x}} + v_y(z_n)\hat{\mathbf{y}} + v_z(z_n)\hat{\mathbf{z}} \tag{12.355}$$

The electron vector position $\mathbf{r}_s(z_n)$ at axial point z_n is obtained from the electron velocity $\mathbf{v}(z_{n-1})$, which is multiplied by the time interval Δt_s between axial positions z_{n-1} and z_n. This product or position vector change $\Delta\mathbf{r}_s$ is then added to the electron vector position $\mathbf{r}_s(z_{n-1})$ at previous axial position z_{n-1}, where

$$\mathbf{r}_s(z_n) = \mathbf{v}(z_{n-1})\Delta t_s + \mathbf{r}_s(z_{n-1}) \tag{12.356}$$

The time interval Δt_s between axial positions z_{n-1} and z_n for $n = 2, 3, 4, \ldots, N$ is

$$\Delta t_s = \frac{\Delta z}{v_z(z_{n-1})} \tag{12.357}$$

The electron enters the insertion device at axial position z_1 and exits the device at axial position z_N, where there is a constant axial distance interval Δz between each pair of z_{n-1} and z_n axial points. The constant axial separation distance Δz is given by

$$\Delta z = z_n - z_{n-1} = \frac{z_N - z_1}{N - 1} \tag{12.358}$$

In an insertion device such as an undulator or wiggler, one chooses a sufficiently large number N of axial points $z_1, z_2, z_3, \ldots, z_N$ to ensure the change in the transverse position of the electron Δx and Δy are each very much smaller than the axial position change Δz between points z_{n-1} and z_n, where for $n = 2, 3, 4, \ldots, N$,

$$\Delta z \gg \Delta x = x(z_n) - x(z_{n-1}) \tag{12.359}$$

and

$$\Delta z \gg \Delta y = y(z_n) - y(z_{n-1}) \tag{12.360}$$

The very relativistic electrons in the insertion device have a total velocity, which is always less than speed of light c, but still is very close to c. The electron velocity $\mathbf{v}$ has mainly an axial component v_z, which is near the speed of light c, where $\mathbf{v} \cong c$ and $|\mathbf{v}| < c$. The axial velocity v_z of the electron in an insertion device is usually much larger than the transverse velocities v_x and v_y in the respective x- and y-directions, where

$$v_z \gg v_x \quad \text{and} \quad v_z \gg v_y \tag{12.361}$$

Therefore, to compute the trajectory of the very- or ultrarelativistic electron in a synchrotron ring insertion device accurately, one must have a computer, which can carry out the calculations to a sufficient number of decimal places, so as to allow accurate calculation of the near-zero $1/\gamma$ quantity for all points $n = 2, 3, 4, \ldots, N$, where

$$\frac{1}{\gamma} = \sqrt{1 - \frac{|\mathbf{v}(z_{n-1})|^2}{c^2}} \tag{12.362}$$

One should not let $1/\gamma$ go to zero at any point in the trajectory calculation due to rounding error. The time interval Δt_s between axial positions z_{n-1} and z_n in an insertion device with a very relativistic synchrotron electron beam is approximately constant, where $n = 2, 3, 4, \ldots, N$:

$$\Delta t_s = \frac{\Delta z}{v_z(z_{n-1})} \cong \frac{\Delta z}{c} \tag{12.363}$$

One obtains datasets of the electron position vector $\mathbf{r}(z_n)$, velocity vector $\mathbf{v}(z_n)$, and acceleration vector $\mathbf{a}(z_n)$ and time $t_s(z_n)$ as a function of axial position z_n of the electron as a function of time t_s in the electron frame, where for $n = 2, 3, 4, \ldots, N$,

$$t_s(z_n) = \Delta t_s + t_s(z_{n-1}) = \frac{\Delta z}{v_z(z_{n-1})} + t_s(z_{n-1}) \cong n\frac{\Delta z}{c} \tag{12.364}$$

The distance R of the observer from the electron in the insertion device, is assumed to be large, as compared to the variation of the distance ΔR from the electron to the observer positions, which occur during the electron trajectory through the insertion device, where $R \gg \Delta R$. The vector $\mathbf{R}$ from the moving electron to the fixed observer position is then approximated by the fixed observer position vector $\mathbf{r}$. Observer position vector $\mathbf{r}$ is measured from the origin at the center of the insertion device. The unit vector $\hat{\mathbf{r}}$, which points from insertion device center point to fixed observer position $\mathbf{r}$, and the above electron position $\mathbf{r}_s(z_n)$, velocity $\mathbf{v}(z_n)$, and acceleration $\mathbf{a}(z_n)$ vectors and the electron rest frame times $t_s(z_n)$ at axial positions z_n, are the needed electron trajectory datasets. The electron trajectory datasets then can be plugged into the previously derived expressions for the electric $\mathbf{E}(\mathbf{r}, t)$ and magnetic $\mathbf{B}(\mathbf{r}, t)$ fields and the per solid angle frequency spectrum $\mathrm{d}^2 I/\mathrm{d}w\mathrm{d}\Omega$ to determine X-ray emission as a function of distance, angle, and X-ray energy.

12.16 Simplified Velocity-Dependent Expression for the per Solid Angle Frequency Spectrum Emitted by Accelerated Charged Particles

In this section we follow chapter 14 in Jackson (1975). The above expression of Eq. (12.344) in MKS units (or Eq. (12.346) in CGS units) for the differential energy radiated per differential frequency and solid angle, $\mathrm{d}^2 I/\mathrm{d}w\mathrm{d}\Omega$, for an accelerated electron or other charged particle, is simplified to an expression, which requires knowledge only of the charged particle velocity trajectory $\mathbf{v}(t_s)$ from particle trajectory time $t_s = -\infty$ to $t_s = \infty$. This result will form the basis for the derivation of the energy radiated per solid angle for a moving magnetic moment $\boldsymbol{\mu}$, which is derived later in this chapter.

Simplification of $\mathrm{d}^2 I/\mathrm{d}w\mathrm{d}\Omega$ of Eq. (12.344) begins with simplification of the vector term in the integrand via integration by parts, where

$$\frac{\hat{\mathbf{r}}\times(\hat{\mathbf{r}}-\boldsymbol{\beta})\times\mathbf{a}}{[1-\boldsymbol{\beta}\cdot\hat{\mathbf{r}}]^2}=\frac{\mathrm{d}}{\mathrm{d}t_s}\left[\frac{\hat{\mathbf{r}}\times(\hat{\mathbf{r}}\times\boldsymbol{\beta})}{1-\boldsymbol{\beta}\cdot\hat{\mathbf{r}}}\right] \tag{12.365}$$

The simplified expression of Eq. (12.365) can be verified as follows. The observer position $\mathbf{r}$ is fixed relative to the origin, and thus $\hat{\mathbf{r}}$ is time independent, where

$$\mathrm{d}\frac{\hat{\mathbf{r}}}{\mathrm{d}t_s}=0 \tag{12.366}$$

On the right side of Eq. (12.365), performing the differentiation with respect to t_s, gives

$$\frac{\mathrm{d}}{\mathrm{d}t_s}\left[\frac{\hat{\mathbf{r}}\times(\hat{\mathbf{r}}\times\boldsymbol{\beta})}{1-\boldsymbol{\beta}\cdot\hat{\mathbf{r}}}\right]=\frac{1}{c}\frac{\hat{\mathbf{r}}\times(\hat{\mathbf{r}}\times\mathbf{a})}{1-\boldsymbol{\beta}\cdot\hat{\mathbf{r}}}+\frac{\hat{\mathbf{r}}\cdot\mathbf{a}}{c}\frac{\hat{\mathbf{r}}\times(\hat{\mathbf{r}}\times\boldsymbol{\beta})}{(1-\boldsymbol{\beta}\cdot\hat{\mathbf{r}})^2} \tag{12.367}$$

With $\hat{\mathbf{r}}\cdot\hat{\mathbf{r}}=1$, application of vector identity BAC−CAB to each of the two terms in Eq. (12.367), yields

$$\frac{\mathrm{d}}{\mathrm{d}t_s}\left[\frac{\hat{\mathbf{r}}\times(\hat{\mathbf{r}}\times\boldsymbol{\beta})}{1-\boldsymbol{\beta}\cdot\hat{\mathbf{r}}}\right]=\frac{1}{c}\frac{\hat{\mathbf{r}}(\hat{\mathbf{r}}\cdot\mathbf{a})-\mathbf{a}}{1-\boldsymbol{\beta}\cdot\hat{\mathbf{r}}}+\left(\frac{\hat{\mathbf{r}}\cdot\mathbf{a}}{c}\right)\frac{\hat{\mathbf{r}}(\hat{\mathbf{r}}\cdot\boldsymbol{\beta})-\boldsymbol{\beta}}{(1-\boldsymbol{\beta}\cdot\hat{\mathbf{r}})^2} \tag{12.368}$$

Rearrangement gives

$$\frac{\mathrm{d}}{\mathrm{d}t_s}\left[\frac{\hat{\mathbf{r}}\times(\hat{\mathbf{r}}\times\boldsymbol{\beta})}{1-\boldsymbol{\beta}\cdot\hat{\mathbf{r}}}\right]=\frac{1}{c}\frac{(\hat{\mathbf{r}}\cdot\mathbf{a})(\hat{\mathbf{r}}-\boldsymbol{\beta})-\mathbf{a}(1-\boldsymbol{\beta}\cdot\hat{\mathbf{r}})}{(1-\boldsymbol{\beta}\cdot\hat{\mathbf{r}})^2} \tag{12.369}$$

The BAC–CAB rule is again applied with

$$\mathbf{A} = \hat{\mathbf{r}}, \mathbf{B} = \hat{\mathbf{r}} - \boldsymbol{\beta}, \text{ and } \mathbf{C} = \mathbf{a} \tag{12.370}$$

Thus Eq. (12.367) becomes,

$$\frac{\mathrm{d}}{\mathrm{d}t_s}\left[\frac{\hat{\mathbf{r}} \times (\hat{\mathbf{r}} \times \boldsymbol{\beta})}{1 - \boldsymbol{\beta} \cdot \hat{\mathbf{r}}}\right] = \frac{\hat{\mathbf{r}} \times (\hat{\mathbf{r}} - \boldsymbol{\beta}) \times \mathbf{a}}{[1 - \boldsymbol{\beta} \cdot \hat{\mathbf{r}}]^2} \tag{12.371}$$

Next, Eq. (12.344) for the differential energy, which is emitted by the accelerated charge per differential solid angle and frequency interval, $\mathrm{d}^2 I/\mathrm{d}w\mathrm{d}\Omega$, expressed in MKS units, is set up for integration by parts, where

$$\begin{aligned}\frac{\mathrm{d}^2 I(w, \hat{\mathbf{r}})}{\mathrm{d}w\mathrm{d}\Omega} &= \frac{e^2}{16\pi^3 \varepsilon_0 c^3}\left|\int_{-\infty}^{\infty} \mathrm{d}t_s \frac{\hat{\mathbf{r}} \times (\hat{\mathbf{r}} - \boldsymbol{\beta}) \times \mathbf{a}}{[1 - \boldsymbol{\beta} \cdot \hat{\mathbf{r}}]^2} \exp\left[iw\left(t_s - \frac{\hat{\mathbf{r}} \cdot \mathbf{r}_s}{c}\right)\right]\right|^2 \\ &= \frac{e^2}{16\pi^3 \varepsilon_0 c^3}\left|\int_{-\infty}^{\infty} \mathrm{d}t_s \frac{\mathrm{d}}{\mathrm{d}t_s}\left[\frac{\hat{\mathbf{r}} \times (\hat{\mathbf{r}} \times \boldsymbol{\beta})}{1 - \boldsymbol{\beta} \cdot \hat{\mathbf{r}}}\right] \exp\left[iw\left(t_s - \frac{\hat{\mathbf{r}} \cdot \mathbf{r}_s}{c}\right)\right]\right|^2\end{aligned} \tag{12.372}$$

A finite solution requires the charge velocity $\mathbf{v}$ (i.e., $\boldsymbol{\beta} = \mathbf{v}/c$) to be zero at $t_s = \pm\infty$, so that

$$\boldsymbol{\beta}(\pm\infty) = \frac{\mathbf{v}(\pm\infty)}{c} = 0 \tag{12.373}$$

where

$$\left[\frac{\hat{\mathbf{r}} \times (\hat{\mathbf{r}} \times \boldsymbol{\beta})}{1 - \boldsymbol{\beta} \cdot \hat{\mathbf{r}}}\right]_{t_s = -\infty} = \left[\frac{\hat{\mathbf{r}} \times (\hat{\mathbf{r}} \times \boldsymbol{\beta})}{1 - \boldsymbol{\beta} \cdot \hat{\mathbf{r}}}\right]_{t_s = \infty} = 0 \tag{12.374}$$

and

$$\left[\frac{\hat{\mathbf{r}} \times (\hat{\mathbf{r}} \times \boldsymbol{\beta})}{1 - \boldsymbol{\beta} \cdot \hat{\mathbf{r}} \exp\left(iw\left(t_s - \frac{\hat{\mathbf{r}} \cdot \mathbf{r}_s}{c}\right)\right)}\right]_{t_s = -\infty}^{t_s = \infty} = 0 - 0 = 0 \tag{12.375}$$

The observer position vector $\mathbf{r}$ and unit vector $\hat{\mathbf{r}}$, which point from the origin to the observer position at distance r, are all fixed in time, so

$$\frac{\mathrm{d}}{\mathrm{d}t_s}\left\{\exp\left[iw\left(t_s - \frac{\hat{\mathbf{r}} \cdot \mathbf{r}_s}{c}\right)\right]\right\} = iw(1 - \boldsymbol{\beta} \cdot \hat{\mathbf{r}})\exp\left[iw\left(t_s - \frac{\hat{\mathbf{r}} \cdot \mathbf{r}_s}{c}\right)\right] \tag{12.376}$$

One can now apply integration of parts to Eq. (12.372), using Eqs. (12.373–12.376), where

$$\int_{-\infty}^{\infty} \mathrm{d}t_s \left[\frac{\hat{\mathbf{r}} \times (\hat{\mathbf{r}} \times \boldsymbol{\beta})}{1-\boldsymbol{\beta}\cdot\hat{\mathbf{r}}}\right] \frac{\mathrm{d}}{\mathrm{d}t_s} \exp\left[\mathrm{i}w\left(t_s - \frac{\hat{\mathbf{r}}\cdot\mathbf{r}_s}{c}\right)\right] =$$

$$w^2 \int_{-\infty}^{\infty} \mathrm{d}t_s (\hat{\mathbf{r}} \times (\hat{\mathbf{r}} \times \boldsymbol{\beta})) \exp\left[\mathrm{i}w\left(t_s - \frac{\hat{\mathbf{r}}\cdot\mathbf{r}_s}{c}\right)\right] \tag{12.377}$$

With substitution of Eq. (12.377) into Eq. (12.372), one obtains the useful expression for the differential energy emitted per differential frequency and solid angle from an accelerated charge, where in MKS units

$$\frac{\mathrm{d}^2 I(w,\hat{\mathbf{r}})}{\mathrm{d}w\mathrm{d}\Omega} = \frac{e^2 w^2}{16\pi^3 \varepsilon_0 c^3} \left| \int_{-\infty}^{\infty} \mathrm{d}t_s (\hat{\mathbf{r}} \times (\hat{\mathbf{r}} \times \boldsymbol{\beta})) \exp\left[\mathrm{i}w\left(t_s - \frac{\hat{\mathbf{r}}\cdot\mathbf{r}_s}{c}\right)\right] \right|^2 \tag{12.378}$$

This solution presumes the velocity $\mathbf{v}$ of charge e is zero for infinite time $t_s = \pm\infty$. If the velocity $\mathbf{v}$ of the charge is not zero at infinite time $t_s = \pm\infty$, but the charge acceleration $\mathbf{a}$ is zero $t_s = \pm\infty$, then one must use the previously derived integral expression $\mathrm{d}^2 I/\mathrm{d}w\mathrm{d}\Omega$, which is a function of acceleration $\mathbf{a}$ and velocity $\boldsymbol{\beta} = \mathbf{v}/c$. Causality requires the emitted energy per frequency and solid angle intervals, $\mathrm{d}^2 I/\mathrm{d}w\mathrm{d}\Omega$, which is measured by the observer at the present time t, to occur after the particle trajectory time "$t_s = -\infty$" to "$t_s = \infty$," so

$$\frac{\mathrm{d}^2 I(w,\hat{\mathbf{r}})}{\mathrm{d}w\mathrm{d}\Omega} = \frac{e^2 w^2}{16\pi^3 \varepsilon_0 c^3} \left| \int_{0}^{T} \mathrm{d}t_s (\hat{\mathbf{r}} \times (\hat{\mathbf{r}} \times \boldsymbol{\beta})) \exp\left[\mathrm{i}w\left(t_s - \frac{\hat{\mathbf{r}}\cdot\mathbf{r}_s}{c}\right)\right] \right|^2 \tag{12.379}$$

However, Eq. (12.379) produces unwanted, unintended contributions to the per solid angle frequency spectrum due to the infinitely abrupt acceleration of the charged particle at $t_s = 0$ and $t_s = T$. A uniform, white noise spectrum is produced by the "edges" at $t_s = 0$ and $t_s = T$, which adds to the "signal" spectrum produced by charged particle acceleration during the time interval defined by $t_s = 0$ to $t_s = T$. This noise artifact can be minimized by proper choice of integration limits and proper characterization of the integrand, such that the modeled physical situation is better represented by the velocity-only expression $\mathrm{d}^2 I/\mathrm{d}w\mathrm{d}\Omega$ of Eq. (12.379). This will be further discussed next.

Consider only the energy emitted by the accelerated charge during the charge's passage through a region of interest or a device such as a wiggler. Charge acceleration is assumed to occur only during a particular period of time between initial acceleration time T_{a1} and final acceleration time T_{a2}, where the charged particle propagates through the region of interest. The use of the acceleration- and velocity-based expression for $\mathrm{d}^2 I/\mathrm{d}w\mathrm{d}\Omega$ is straightforward. Any velocity change of the electron (charged particle), from $t_s = -\infty$ to $t_s = \infty$, produces contributions to the energy emitted per frequency per solid angle distribution, as seen in the acceleration- and velocity-dependent expression of Eq. (12.344), where

$$\frac{d^2 I(w,\hat{\mathbf{r}})}{dw d\Omega} = \frac{e^2}{16\pi^3 \varepsilon_0 c^3} \left| \int_{-\infty}^{\infty} dt_s \frac{\hat{\mathbf{r}} \times (\hat{\mathbf{r}} - \boldsymbol{\beta}) \times \mathbf{a}}{[1 - \boldsymbol{\beta} \cdot \hat{\mathbf{r}}]^2} \exp\left[iw\left(t_s + \frac{r - \hat{\mathbf{r}} \cdot \mathbf{r}_s}{c} \right) \right] \right|^2 \tag{12.380}$$

However, the unwanted acceleration contributions can be ignored by setting **a** to zero during the time periods within the integrand infinite time limits $t_s = [-\infty, \infty]$, where the particle acceleration is of no interest. That is, one may not be interested in the electromagnetic radiation and its contribution to the per solid angle energy frequency, which is produced by the electron during its acceleration or deceleration outside the region of interest.

The region of interest may be a synchrotron insertion device or some region of space in low Earth orbit, where one wishes to measure the emitted energy per unit frequency and per unit solid angle. The electromagnetic energy, which is produced by the charged particle at some previous time outside the region of interest, or at later time outside the region of interest is excluded from the integral by setting the acceleration to zero, where $\mathbf{a} = 0$. For electrons passing through a region of interest, such as a bending magnet, wiggler, or undulator in a synchrotron ring, the time interval of interest is from $t_s = T_{a1}$ to $t_s = T_{a2}$. The emitted energy per frequency and solid angle intervals, $d^2 I/dw d\Omega$, expressed in MKS units, is obtained by changing the limits of integration of Eq. (12.380), where

$$\frac{d^2 I(w,\hat{\mathbf{r}})}{dw d\Omega} = \frac{e^2}{16\pi^3 \varepsilon_0 c^3} \left| \int_{T_{a1}}^{T_{a2}} dt_s \frac{\hat{\mathbf{r}} \times (\hat{\mathbf{r}} - \boldsymbol{\beta}) \times \mathbf{a}}{[1 - \boldsymbol{\beta} \cdot \hat{\mathbf{r}}]^2} \exp\left[iw\left(t_s - \frac{\hat{\mathbf{r}} \cdot \mathbf{r}_s}{c} \right) \right] \right|^2 \tag{12.381}$$

and

$$\mathbf{a}(t_s > T_{a2}) = 0 \quad \text{and} \quad \mathbf{a}(t_s < T_{a1}) = 0 \tag{12.382}$$

However, the use of the simpler, velocity-only expression for $d^2 I/dw d\Omega$ requires special consideration. Consider Eq. (12.379), which is the flawed velocity-dependent expression for the differential emitted energy per differential frequency and solid angle, $d^2 I/dw d\Omega$, expressed in MKS units:

$$\frac{d^2 I(w,\hat{\mathbf{r}})}{dw d\Omega} = \frac{e^2 w^2}{16\pi^3 \varepsilon_0 c^3} \left| \int_{T_{a1}}^{T_{a2}} dt_s (\hat{\mathbf{r}} \times (\hat{\mathbf{r}} \times \boldsymbol{\beta})) \exp\left[iw\left(t_s - \frac{\hat{\mathbf{r}} \cdot \mathbf{r}_s}{c} \right) \right] \right|^2 \tag{12.383}$$

where

$$\boldsymbol{\beta}(t_s > T_{a2}) = 0 \quad \text{and} \quad \boldsymbol{\beta}(t_s < T_{a1}) = 0 \tag{12.384}$$

This expression assumes that the electron or charged particle accelerated from zero velocity $\mathbf{v}(t_s < T_{a1}) = 0$ to some possibly assumed, initial nonzero velocity $\mathbf{v}(T_{a1}) \neq 0$ in an infinitesimally small time interval, and the expression assumes that the charged particle decelerated from possibly nonzero velocity $\mathbf{v}(t_s = T_{a2}) \neq 0$ to zero velocity $\mathbf{v}(t_s > T_{a2}) = 0$ in an infinitesimally small time interval. The expression for $\mathrm{d}^2I/\mathrm{d}w\mathrm{d}\Omega$ is a Fourier transform of the time dependence of the velocity trajectory of the charged particle, and thus these two abrupt changes in charged particle velocity are two delta functions of velocity change or acceleration. The entrance and exit of the charged particle from the region of interest thus lead to an artifact of unwanted uniform spectrum of emitted electromagnetic energy.

In one possible real-life case, the particles enter and exit the region of interest with a constant velocity, and thus there is no electromagnetic radiation emission in the immediate vicinity before or after the region of interest. That is, the charged particle further back in time was created and accelerated to its initial constant velocity prior to its entrance into the region of interest. And at a much later point, after the particle exits the region of interest, the particle is accelerated again, or is decelerated to zero velocity.

Consider the time interval $T_{a1} - T_1$, where the charged particle travels at constant velocity $\mathbf{v}_1$ before entering the region of interest at time T_{a1}, so that

$$\mathbf{v}(T_1 < t_s \leq T_{a1}) = \mathbf{v}_1 \tag{12.385}$$

Next, consider the time period $T_2 - T_{a2}$, where the charged particle travels at constant velocity $\mathbf{v}_2$ after exiting the region of interest at time T_{a2}, so that

$$\mathbf{v}(T_{a2} < t_s \leq T_2) = \mathbf{v}_2 \tag{12.386}$$

Assume that time intervals $T_{a1} - T_1$ and $T_2 - T_{a2}$ of constant charged particle velocity, before and after the region of interest, are each long, as compared to the charged particle passage time through the region of interest, during time interval $T_{a2} - T_{a1}$, where charged particle acceleration $\mathbf{a}(T_{a1} \leq t_s \leq T_{a2})$ takes place. Then, the acceleration of the charged particle outside the region of interest $\mathbf{a}(t_s < T_1)$ and $\mathbf{a}(t_s > T_2)$, provides only low-frequency artifacts in the calculated charged particle energy emission per solid angle, per frequency distribution $\mathrm{d}^2I/\mathrm{d}w\mathrm{d}\Omega$, because of inverse relation of pulses between the time and frequency domains.

The angular frequency w in the Fourier transform is the inverse of time t_s. Hence, large time intervals Δt_s, which are centered about $t_s = 0$, produce narrow frequency intervals Δw about $w = 0$. For example, a unity amplitude, rectangular pulse of time duration T, which is centered about $t_s = 0$, is constructed by the integral sum of a continuum of sine waves. For the rectangular square pulse, the distribution of sine wave amplitudes, or Fourier transform $F(w)$, varies with frequency w as a Sinc function, where

$$F(w) = T\ Sinc(wT) \tag{12.387}$$

Frequency distribution $F(w)$ is centered about $w=0$, with the maximum center distribution of frequencies, which have positive amplitudes ranging from $w=-2\pi/T$ to $w=2\pi/T$. Increasing the rectangular pulse duration T decreases the primary range of frequencies $\pm 2\pi/T$ about $w=0$. As a result, the frequency spectrum per solid angle can be approximated by integrating over a sufficiently large time interval $[T_1, T_2]$, which contains the much smaller time interval of interest $[T_{1a}, T_{2a}]$, where

$$[T_1, T_2] \gg [T_{1a}, T_{2a}] \tag{12.388}$$

As a result of Eq. (12.388), the per solid angle, spectral distribution of charged particle energy emission of Eq. (12.378) can be approximated as,

$$\begin{aligned}\frac{\mathrm{d}^2 I(w, \hat{\mathbf{r}})}{\mathrm{d}w\mathrm{d}\Omega} &= \frac{e^2 w^2}{16\pi^3 \varepsilon_0 c^3}\left|\int_{-\infty}^{\infty} \mathrm{d}t_s (\hat{\mathbf{r}} \times (\hat{\mathbf{r}} \times \boldsymbol{\beta}))\exp\left[\mathrm{i}w\left(t_s - \frac{\hat{\mathbf{r}}\cdot\mathbf{r}_s}{c}\right)\right]\right|^2 \\ &\cong \frac{e^2 w^2}{16\pi^3 \varepsilon_0 c^3}\left|\int_{T_1}^{T_2} \mathrm{d}t_s (\hat{\mathbf{r}} \times (\hat{\mathbf{r}} \times \boldsymbol{\beta}))\exp\left[\mathrm{i}w\left(t_s - \frac{\hat{\mathbf{r}}\cdot\mathbf{r}_s}{c}\right)\right]\right|^2\end{aligned} \tag{12.389}$$

Next, one separates the integral of Eq. (12.389) into the sum of (1) an integral, whose time integration limits include the charged particle acceleration, and (2) the before and after integrals, whose time limits include only the charged particle constant velocity, where

$$\frac{\mathrm{d}^2 I(w, \hat{\mathbf{r}})}{\mathrm{d}w\mathrm{d}\Omega} \cong \frac{e^2 w^2}{16\pi^3 \varepsilon_0 c^3}\left|S_{1v} + \int_{T_{a1}}^{T_{a2}} \mathrm{d}t_s (\hat{\mathbf{r}} \times (\hat{\mathbf{r}} \times \boldsymbol{\beta}))\exp\left[\mathrm{i}w\left(t_s - \frac{\hat{\mathbf{r}}\cdot\mathbf{r}_s}{c}\right)\right] + S_{2v}\right|^2 \tag{12.390}$$

Velocities $\mathbf{v}(T_1 < t_s \le T_{a1}) = \mathbf{v}_1$ and $\mathbf{v}(T_{2a} < t_s \le T_2) = \mathbf{v}_2$ are constant, so

$$S_{1v} = \int_{T_1}^{T_{a1}} \mathrm{d}t_s \left(\hat{\mathbf{r}} \times \left(\hat{\mathbf{r}} \times \frac{\mathbf{v}_1}{c}\right)\right)\exp\left[\mathrm{i}w\left(t_s - \frac{\hat{\mathbf{r}}\cdot\mathbf{v}_1}{c}\right)\right] = \left(\hat{\mathbf{r}} \times \left(\hat{\mathbf{r}} \times \frac{\mathbf{v}_1}{c}\right)\right)\int_{T_1}^{T_{a1}} \mathrm{d}t_s \exp\left[\mathrm{i}\eta_1 t_s\right] \tag{12.391}$$

$$S_{2v} = \int_{T_{a2}}^{T_2} \mathrm{d}t_s \left(\hat{\mathbf{r}} \times \left(\hat{\mathbf{r}} \times \frac{\mathbf{v}_2}{c}\right)\right)\exp\left[\mathrm{i}w\left(t_s - \frac{\hat{\mathbf{r}}\cdot\mathbf{v}_2}{c}\right)\right] = \left(\hat{\mathbf{r}} \times \left(\hat{\mathbf{r}} \times \frac{\mathbf{v}_2}{c}\right)\right)\int_{T_{a2}}^{T_2} \mathrm{d}t_s \exp\left[\mathrm{i}\eta_2 t_s\right] \tag{12.392}$$

with

$$\eta_1 = w\left(1 - \frac{\hat{\mathbf{r}}\cdot\mathbf{v}_1}{c}\right) \quad \text{and} \quad \eta_2 = w\left(1 - \frac{\hat{\mathbf{r}}\cdot\mathbf{v}_2}{c}\right) \tag{12.393}$$

Now, let

$$T_1 = T_{a1} - \frac{2N_1\pi}{\eta_1} \quad N_1 = 0, 1, 2, 3, \ldots \text{and } N_1 \gg 1 \tag{12.394}$$

$$T_2 = T_{a2} + \frac{2N_1\pi}{\eta_1} \quad N_2 = 0, 1, 2, 3, \ldots \text{and } N_2 \gg 1 \tag{12.395}$$

One specifies N_1 and N_2 as sufficiently large positive integers, so that the abrupt velocity changes or acceleration at "$T_1 \to -\infty$" and "$T_2 \to \infty$", produce per solid angle spectral contributions only near the $w = 0$ frequency.

Given integers N_1 and N_2, where η_1 and η_2 are constant, and the origin-to-observer unit vector $\hat{\mathbf{r}}$ is constant, then the integrals of S_{1v} and S_{2v} are each zero, where

$$\int_{T_1}^{T_{a1}} dt_s \exp(\mathrm{i}\eta_1 t_s) = \left[\frac{\exp(\mathrm{i}\eta_1 t_s)}{\mathrm{i}\eta_1}\right]_{T_1}^{T_{a1}} = \frac{\exp(\mathrm{i}\eta_1 T_{a1}) - \exp(\mathrm{i}\eta_1 T_{a1})\exp(\mathrm{i}N_1 2\pi)}{\mathrm{i}\eta_1} = 0 \tag{12.396}$$

$$\int_{T_{a2}}^{T_2} dt_s \exp(\mathrm{i}\eta_2 t_s) = \left[\frac{\exp(\mathrm{i}\eta_2 t_s)}{\mathrm{i}\eta_2}\right]_{T_{a2}}^{T_2} = \frac{\exp(\mathrm{i}\eta_2 T_{a2})\exp(\mathrm{i}N_2 2\pi) - \exp(\mathrm{i}\eta_2 T_{a2})}{\mathrm{i}\eta_2} = 0 \tag{12.397}$$

and thus,

$$S_{1v} = S_{2v} = 0 \tag{12.398}$$

Hence, with substitution of Eqs. (12.398) into Eq. (12.390), one obtains the approximate expression for the differential energy emission per differential frequency and solid angle from a charge particle, where

$$\frac{d^2 I(w, \hat{\mathbf{r}})}{dw d\Omega} \cong \frac{e^2 w^2}{16\pi^3 \varepsilon_0 c^3} \left| \int_{T_{a1}}^{T_{a2}} dt_s (\hat{\mathbf{r}} \times (\hat{\mathbf{r}} \times \boldsymbol{\beta})) \exp\left[\mathrm{i}w\left(t_s - \frac{\hat{\mathbf{r}} \cdot \mathbf{r}_s}{c}\right)\right]\right|^2 \tag{12.399}$$

Finally, we consider the exact expression for the per solid angle, spectral distribution of energy emission from the charged particle, for large time intervals of constant particle velocity, which occur before $T_1 < t_s < T_{1a}$ and after $T_{2a} < t_s < T_2$ the time interval of interest (e.g. acceleration in the undulator) $T_{1a} < t_s < T_{2a}$. Assume the charged particle has zero velocity, $\mathbf{v} = 0$ from $t_s = -\infty$ to $t_s = -T_{01}$, where

$$\mathbf{v}(t_s \leq -T_{01}) = 0 \tag{12.400}$$

And assume that prior to the region of interest (e.g. undulator), the charged particle is accelerated from initial zero velocity $\mathbf{v}(-T_{01}) = 0$ to initial nonzero velocity $\mathbf{v}(T_1) = \mathbf{v}_1$ during the time interval $t_s = -T_{01}$ to later time $t_s = T_1$. The integral contribution of the acceleration S_{1a} of the charged particle from initial zero velocity to initial nonzero constant velocity $\mathbf{v}_1$, which is prior to the region of interest is

$$S_{1a} = \int_{-T_{01}}^{-T_1} dt_s(\hat{\mathbf{r}} \times (\hat{\mathbf{r}} \times \boldsymbol{\beta}))\exp\left[iw\left(t_s - \frac{\hat{\mathbf{r}} \cdot \mathbf{r}_s}{c}\right)\right] \tag{12.401}$$

Likewise, assume that the deceleration of the charged particle from region of interest (ROI) exit velocity $\mathbf{v}(T_2) = \mathbf{v}_2$ to final zero velocity $\mathbf{v}(T_{02}) = 0$ occurs after the ROI between time $t_s = T_1$ and later time $t_s = T_{02}$. The charged particle has zero velocity ($\mathbf{v} = 0$) from $t_s = T_{02}$ to $t_s = \infty$; i.e.,

$$\mathbf{v}(t_s \geq T_{02}) = 0 \tag{12.402}$$

The integral contribution of the deceleration S_{2a} of the charged particle from the nonzero, terminal, and constant velocity $\mathbf{v}_2$, after exiting the ROI, to final zero velocity, is then

$$S_{2a} = \int_{T_2}^{T_{02}} dt_s(\hat{\mathbf{r}} \times (\hat{\mathbf{r}} \times \boldsymbol{\beta}))\exp\left[iw\left(t_s - \frac{\hat{\mathbf{r}} \cdot \mathbf{r}_s}{c}\right)\right] \tag{12.403}$$

In this case, the exact expression for the differential energy emission per differential frequency and solid angle from an accelerated charged particle, which includes noise due to acceleration outside the ROI, is

$$\frac{d^2I(w, \hat{\mathbf{r}})}{dwd\Omega} \cong \frac{e^2w^2}{16\pi^3\varepsilon_0c^3}\left|S_{a1} + \int_{T_{a1}}^{T_{a2}} dt_s(\hat{\mathbf{r}} \times (\hat{\mathbf{r}} \times \boldsymbol{\beta}))\exp\left[iw\left(t_s - \frac{\hat{\mathbf{r}} \cdot \mathbf{r}_s}{c}\right)\right] + S_{a2}\right|^2 \tag{12.404}$$

The emitted energy per solid angle frequency spectrum $d^2I/dwd\Omega$ is measured by the observer at present time t, which occurs after the previous entire time history $t_s \in [-\infty, \infty]$ of the nonzero particle acceleration $\mathbf{a}(t_s)$; that is, the time-varying velocity trajectory $d\mathbf{v}(t_s)/dt_s$.

Again, assume that the charged particle velocity $\mathbf{v}(\pm\infty)$; i.e. $\boldsymbol{\beta}(\pm\infty) = 0$ is zero outside the region of interest at times much earlier or later than the time interval, where the charged particle is within the region of interest. In addition, assume a constant velocity before ($\mathbf{v}_1$) and after ($\mathbf{v}_2$) the region of interest over a time interval, which is large compared to the time of interest (e.g. particle passage time through the undulator), where the charged particle propagates through the region of interest (e.g. undulator). As stated in Jackson (1975), one can do the following mathematical trick to obtain the differential emitted energy per frequency and differential solid angle, $d^2I/dwd\Omega$, with minimum noise contribution from the charged particle accelerations outside the region of interest:

$$\frac{d^2I(w, \hat{\mathbf{r}})}{dwd\Omega} = \lim_{\varepsilon \to 0}\frac{e^2}{16\pi^3\varepsilon_0c^3}\left|\int_{-\infty}^{\infty} dt_s(\hat{\mathbf{r}} \times (\hat{\mathbf{r}} \times \boldsymbol{\beta}))\exp\left[iw\left(t_s - \frac{\hat{\mathbf{r}} \cdot \mathbf{r}_s}{c}\right)\right]\exp(-\varepsilon|t_s|)\right|^2 \tag{12.405}$$

12.17 Bremsstrahlung and Electromagnetic Wave Polarization

A charged particle q of charge state z with charge $q = ze$ propagates in a vacuum at velocity $\mathbf{v}$ with $\beta = v/c$, and then enters a material region and collides with the atoms, causing q to change speed and direction as it comes to a stop. The nuclear strong force or electromagnetic interactions of the positive nucleus and atomic electrons with the incident-charged particle q give rise to the change of speed and direction of q, as well as the electrons and nucleus of the struck atoms. As a result, all the charged articles in the collisions, which undergo acceleration, then emit electromagnetic radiation.

The electromagnetic amplitudes of the charged particle q and a struck charged particle Q, such as a nucleus and its associated electron cloud, sum to produce the net amplitude, whose squared magnitude is the emitted bremsstrahlung intensity from charged particle pair q and Q. Provided that incident-charged particle q, such as an electron, is much less massive than struck-charged particle Q, such as a nucleus or bound atomic electron cloud, then the bremsstrahlung emission is almost completely contributed from the less massive charged particle q (i.e., an electron). However, the sequence electromagnetic emission from the charged particle pair q with other struck-charged particles Q in the stopping material does not have a fixed phase relationship in time and space, which is due to the random nature of the collision process. Consequently, a distant observer measures the bremsstrahlung radiation as the sum of the bremsstrahlung intensities, which are emitted in the sequence of collisions between the incident-charged particle q with the set of struck-charged particles Q in the stopping material.

In many collision processes, the charged particle q, such as an electron or positron, has much less mass than the other charged particles Q, such as a nucleus or bound electron cloud. The change in speed and direction is much less for the more massive charged particle Q (nucleus), compared to the relatively large change in speed and direction of the lighter particle (electron). In spite of the larger charge state of the nucleus or bound atomic electron cloud Q, the bremsstrahlung emission from the highly accelerated electron or positron q almost completely accounts for the bremsstrahlung intensity at the distant observer. In the case of collisions between less massive charged particles q with more massive charged particles Q, the bremsstrahlung emission from the collision process can be approximated by that of the less massive charged particles q. In most collision processes, one can better measure or estimate the deceleration of the incident-charged particles q, such as electrons in an X-ray tube that accelerated over a voltage to strike a tungsten target. It is more difficult to measure or estimate the much smaller accelerations of the bound atomic electron cloud and nuclei Q of the tungsten atoms.

We now consider only the electromagnetic radiation, which is emitted from charged particle q as it collides with more massive atoms and undergoes deceleration. From the derivation from Section 12.16 in MKS units, the bremsstrahlung energy emitted per solid angle per frequency interval $\mathrm{d}^2I/\mathrm{d}w\mathrm{d}\Omega$ by charged particle q is given by Eq. (12.372), which is repeated below,

$$\frac{\mathrm{d}^2I(w,\hat{\mathbf{r}})}{\mathrm{d}w\mathrm{d}\Omega} = \frac{e^2}{16\pi^3\varepsilon_0c^3}\left|\int_{-\infty}^{\infty}\mathrm{d}t_s\frac{\hat{\mathbf{r}}\times(\hat{\mathbf{r}}-\boldsymbol{\beta})\times\mathbf{a}}{[1-\boldsymbol{\beta}\cdot\hat{\mathbf{r}}]^2}\exp\left[\mathrm{i}w\left(t_s-\frac{\hat{\mathbf{r}}\cdot\mathbf{r}_s}{c}\right)\right]\right|^2$$
$$= \frac{e^2}{16\pi^3\varepsilon_0c^3}\left|\int_{-\infty}^{\infty}\mathrm{d}t_s\frac{\mathrm{d}}{\mathrm{d}t_s}\left[\frac{\hat{\mathbf{r}}\times(\hat{\mathbf{r}}\times\boldsymbol{\beta})}{1-\boldsymbol{\beta}\cdot\hat{\mathbf{r}}}\right]\exp\left[\mathrm{i}w\left(t_s-\frac{\hat{\mathbf{r}}\cdot\mathbf{r}_s}{c}\right)\right]\right|^2 \tag{12.406}$$

Following Jackson (1975), we now examine the collision process in the low-frequency limit, where the spectrum can be determined by consideration of only the initial v_0 and final v velocities of charged particle q. The history of charged particle q deceleration determines its exact bremsstrahlung emission over time. As a simplifying approximation, we ignore the time history of the deceleration of charged particle q in the material-stopping medium. Also, one assumes that the significant deceleration of charged particle q occurs from initial velocity $v_0 = \beta_0 c$ to final velocity $v = \beta c$. As a result, in the limit of low frequencies as $w \to 0$, the argument of the exponential term in Eq. (12.406) goes to zero, where

$$\lim_{w\to 0}\exp\left[\mathrm{i}w\left(t_s-\frac{\hat{\mathbf{r}}\cdot\mathbf{r}_s}{c}\right)\right] \to 1 \tag{12.407}$$

The bremsstrahlung energy emitted per solid angle per frequency interval $\mathrm{d}^2I/\mathrm{d}w\mathrm{d}\Omega$ by charged particle q in the low-frequency limit becomes a perfect differential, where substitution of Eq. (12.407) into Eq. (12.406), yields

$$\lim_{w\to 0}\frac{\mathrm{d}^2I(w,\hat{\mathbf{r}})}{\mathrm{d}w\mathrm{d}\Omega} = \frac{e^2}{16\pi^3\varepsilon_0c^3}\left|\int_{-\infty}^{\infty}\mathrm{d}t_s\frac{\mathrm{d}}{\mathrm{d}t_s}\left[\frac{\hat{\mathbf{r}}\times(\hat{\mathbf{r}}\times\boldsymbol{\beta})}{1-\boldsymbol{\beta}\cdot\hat{\mathbf{r}}}\right]\right|^2 \tag{12.408}$$

The integration of the perfect differential of Eq. (12.408) is then the integrand evaluated at the initial time t_0, subtracted from the integrand evaluated at final time t, where

$$\lim_{w\to 0}\frac{\mathrm{d}^2I(w,\hat{\mathbf{r}})}{\mathrm{d}w\mathrm{d}\Omega} = \frac{e^2}{16\pi^3\varepsilon_0c^3}\left|\frac{\hat{\mathbf{r}}\times(\hat{\mathbf{r}}\times\boldsymbol{\beta})}{1-\boldsymbol{\beta}\cdot\hat{\mathbf{r}}}-\frac{\hat{\mathbf{r}}\times(\hat{\mathbf{r}}\times\boldsymbol{\beta}_0)}{1-\boldsymbol{\beta}_0\cdot\hat{\mathbf{r}}}\right|^2 \tag{12.409}$$

Dividing the bremsstrahlung energy emitted per solid angle per frequency interval $\mathrm{d}^2I/\mathrm{d}w\mathrm{d}\Omega$ by $\hbar w$, and using $\mathrm{d}w = \mathrm{d}E/\hbar$, one can convert to the differential number of bremsstrahlung photons per differential energy $\mathrm{d}E$ and solid angle $\mathrm{d}\Omega$, denoted by $\mathrm{d}^2N/\mathrm{d}E\mathrm{d}\Omega$, where

$$\frac{\mathrm{d}^2N(w,\hat{\mathbf{r}})}{\mathrm{d}E\mathrm{d}\Omega} = \lim_{w\to 0}\frac{1}{\hbar w}\frac{1}{\hbar}\frac{\mathrm{d}^2I(w,\hat{\mathbf{r}})}{\mathrm{d}w\mathrm{d}\Omega} \tag{12.410}$$

Also, one can introduce the complex and general polarization unit vector $\hat{\varepsilon}$ and its complex conjugate $\hat{\varepsilon}^*$ for expression of the polarization of electric field $\mathbf{E}_{\mathrm{br}}$ of the emitted bremsstrahlung, where

$$\frac{\hat{\mathbf{r}}\times(\hat{\mathbf{r}}\times\boldsymbol{\beta})}{1-\boldsymbol{\beta}\cdot\hat{\mathbf{r}}}-\frac{\hat{\mathbf{r}}\times(\hat{\mathbf{r}}\times\boldsymbol{\beta}_0)}{1-\boldsymbol{\beta}_0\cdot\hat{\mathbf{r}}=}\hat{\varepsilon}^*\cdot\left(\frac{\boldsymbol{\beta}}{1-\boldsymbol{\beta}\cdot\hat{\mathbf{r}}}-\frac{\boldsymbol{\beta}_0}{1-\boldsymbol{\beta}_0\cdot\hat{\mathbf{r}}}\right) \tag{12.411}$$

As a result, the differential number of bremsstrahlung photons per differential energy and solid angle, $\mathrm{d}^2N/\mathrm{d}E\mathrm{d}\Omega$, which are emitted by charged particle q in the low-frequency limit, becomes

$$\lim_{E\to 0}\frac{\mathrm{d}^2N(w,\hat{\mathbf{r}})}{\mathrm{d}E\mathrm{d}\Omega}=\frac{e^2}{16\pi^3\varepsilon_0\hbar c^3E}\left|\hat{\varepsilon}^*\cdot\left(\frac{\boldsymbol{\beta}}{1-\boldsymbol{\beta}\cdot\hat{\mathbf{r}}}-\frac{\boldsymbol{\beta}_0}{1-\boldsymbol{\beta}_0\cdot\hat{\mathbf{r}}}\right)\right|^2 \tag{12.412}$$

In the nonrelativistic limit, $\boldsymbol{\beta}\cdot\hat{\mathbf{r}}\ll 1$, Eq. (12.412) becomes

$$\lim_{E\to 0}\frac{\mathrm{d}^2N(w,\hat{\mathbf{r}})}{\mathrm{d}E\mathrm{d}\Omega}\cong\frac{e^2}{16\pi^3\varepsilon_0\hbar c^3E}\left|\hat{\varepsilon}^*\cdot(\boldsymbol{\beta}-\boldsymbol{\beta}_0)\right|^2 \tag{12.413}$$

However, in the relativistic limit, $\boldsymbol{\beta}\cdot\hat{\mathbf{r}}\gg 1$, one assumes only a small change in velocity, where $\boldsymbol{\beta}\cong\boldsymbol{\beta}_0$ and

$$\lim_{E\to 0}\frac{\mathrm{d}^2N(w,\hat{\mathbf{r}})}{\mathrm{d}E\mathrm{d}\Omega}\cong\frac{e^2}{16\pi^3\varepsilon_0\hbar c^3E}\left|\hat{\varepsilon}^*\cdot\left(\frac{\boldsymbol{\beta}-\boldsymbol{\beta}_0}{1-\boldsymbol{\beta}_0\cdot\hat{\mathbf{r}}}\right)\right|^2 \tag{12.414}$$

We now examine the generalized polarization unit vector $\hat{\varepsilon}$ for the electric field $\mathbf{E}_{\mathrm{br}}$ of the emitted bremsstrahlung. First, consider a z-directed bremsstrahlung electromagnetic plane wave, which is directed toward a distant observer from the decelerated charged particle q at the origin. The general bremsstrahlung electric field $\mathbf{E}_{br}$ is linearly polarized, if the transverse and linearly independent x- and y-component electric fields E_x and E_y have the same phase, where

$$\mathbf{E}_{\mathrm{br}}=(E_x\hat{\varepsilon}_x+E_y\hat{\varepsilon}_y)\exp[\mathrm{i}(\mathbf{k}\cdot\mathbf{r}-wt)] \tag{12.415}$$

The bremsstrahlung electromagnetic field, with unit wave vector $\hat{\mathbf{k}}=\mathbf{k}/k$, propagates through a medium characterized by electric permittivity $\varepsilon=\varepsilon_r\varepsilon_0$ and magnetic permeability $\mu=\mu_r\mu_0$ with relative ε_r and vacuum ε_0 electric permittivity, and with relative μ_r and vacuum μ_0 magnetic permeability. The cross product of the unit wave vector $\hat{\mathbf{k}}$ with the plane wave bremsstrahlung electric field $\mathbf{E}_{\mathrm{br}}$, and division by the wave phase velocity v_{ph}, gives the plane wave magnetic field $\mathbf{B}_{\mathrm{br}}$, where

$$\mathbf{B}_{\mathrm{br}}=\sqrt{\mu\varepsilon}(\hat{\mathbf{k}}\times\mathbf{E}_{\mathrm{br}})=\frac{1}{v_{\mathrm{ph}}}(\hat{\mathbf{k}}\times\mathbf{E}_{\mathrm{br}}) \tag{12.416}$$

The general bremsstrahlung electric field $\mathbf{E}_{\mathrm{br}}$ is elliptically polarized, if the transverse and linearly independent x- and y-component electric fields E_x and E_y have different phases, where

$$\mathbf{E}_{\mathrm{br}} = (E_{0x}\hat{\varepsilon}_x \pm \mathrm{i}E_{0y}\hat{\varepsilon}_y)\exp[\mathrm{i}(\mathbf{k}\cdot\mathbf{r} - wt)] \tag{12.417}$$

If the x and y transverse electric field amplitude components E_{0x} and E_{0y} equal amplitude E_0, then the electric field $\mathbf{E}_{\mathrm{br}}$ is circularly polarized, where

$$\mathbf{E}_{\mathrm{br}} = E_0(\hat{\varepsilon}_x \pm \mathrm{i}\hat{\varepsilon}_y)\exp[\mathrm{i}(\mathbf{k}\cdot\mathbf{r} - wt)] \tag{12.418}$$

Circular polarization introduces the complex unit polarization vectors $\hat{\varepsilon}$ and its complex conjugate $\hat{\varepsilon}^*$, where

$$\varepsilon = \hat{\varepsilon}_x + i\hat{\varepsilon}_y \quad \text{and} \quad \varepsilon^* = \hat{\varepsilon}_x - i\hat{\varepsilon}_y \tag{12.419}$$

The x- and y-component electric fields are the real parts of the complex electric field $\mathbf{E}_{\mathrm{br}}$, where

$$E_{\mathrm{brx}} = \mathrm{Re}[\mathbf{E}_{\mathrm{br}}]_x = E_0 \cos(\mathbf{k}\cdot\mathbf{r} - wt) \tag{12.420}$$

$$E_{\mathrm{bry}} = \mathrm{Re}[\mathbf{E}_{\mathrm{br}}]_y = \mp E_0 \sin(\mathbf{k}\cdot\mathbf{r} - wt) \tag{12.421}$$

The electric field vector $\mathbf{E}_{\mathrm{br}}$ rotates counterclockwise about the z-axis as seen by the observer for the case of left circular polarization, or positive helicity with a plus (+) sign. And as the observer looks at the incoming wave, a clockwise rotation of electric field vector $\mathbf{E}_{\mathrm{br}}$ is the case of right circular polarization, or negative helicity with a minus (−).

The left (positive helicity) $\hat{\varepsilon}_+$ and right (negative helicity) $\hat{\varepsilon}_-$ circular polarization of electromagnetic waves can form a dual set of polarizations, which can be used to describe the polarization of electromagnetic waves. The left, counterclockwise, and circular-polarized (positive helicity) unit vector $\hat{\varepsilon}_+$ is complex, and it is expressed as the sum of the x-component unit vector $\hat{\varepsilon}_x$ along the real axis, and the y-component unit vector $\hat{\varepsilon}_y$ along the positive imaginary axis, multiplied by the $1/\sqrt{2}$ normalization constant, where

$$\hat{\varepsilon}_+ = \frac{1}{\sqrt{2}}(\hat{\varepsilon}_x + i\hat{\varepsilon}_y) \tag{12.422}$$

The complex, right, clockwise, and circular-polarized (negative helicity) unit vector $\hat{\varepsilon}_-$ differs from $\hat{\varepsilon}_+$ only in that it is expressed as the y-component unit vector $\hat{\varepsilon}_y$ along the negative imaginary axis, where

$$\hat{\varepsilon}_- = \frac{1}{\sqrt{2}}(\hat{\varepsilon}_x - i\hat{\varepsilon}_y) \tag{12.423}$$

The following vector dot products of the positive helix $\hat{\varepsilon}_+$ unit vector and its complex conjugate $\hat{\varepsilon}_+{}^*$ are unity, as well as the negative helix $\hat{\varepsilon}_-$ unit vector and its complex conjugate $\hat{\varepsilon}_-{}^*$, where

$$\hat{\varepsilon}_+^* \cdot \hat{\varepsilon}_+ = \frac{1}{\sqrt{2}}(\hat{\varepsilon}_x + i\hat{\varepsilon}_y)^* \cdot \frac{1}{\sqrt{2}}(\hat{\varepsilon}_x + i\hat{\varepsilon}_y) = (1 + 0 - 0 + 1) = 1 \tag{12.424}$$

$$\hat{\varepsilon}_-^* \cdot \hat{\varepsilon}_- = \frac{1}{\sqrt{2}}(\hat{\varepsilon}_x - i\hat{\varepsilon}_y)^* \cdot \frac{1}{\sqrt{2}}(\hat{\varepsilon}_x - i\hat{\varepsilon}_y) = (1 + 0 - 0 + 1) = 1 \tag{12.425}$$

Also, the dot products of the orthogonal positive and negative helix vectors are zero:

$$\hat{\varepsilon}_+^* \cdot \hat{\varepsilon}_- = \frac{1}{\sqrt{2}}(\hat{\varepsilon}_x + i\hat{\varepsilon}_y)^* \cdot \frac{1}{\sqrt{2}}(\hat{\varepsilon}_x - i\hat{\varepsilon}_y) = (1 + 0 - 0 - 1) = 0 \tag{12.426}$$

$$\hat{\varepsilon}_-{}^* \cdot \hat{\varepsilon}_+ = \frac{1}{\sqrt{2}}(\hat{\varepsilon}_x - i\hat{\varepsilon}_y)^* \cdot \frac{1}{\sqrt{2}}(\hat{\varepsilon}_x + i\hat{\varepsilon}_y) = (1 + 0 - 0 - 1) = 0 \tag{12.427}$$

The complex conjugate, transverse unit vectors $\hat{\varepsilon}_+$ and $\hat{\varepsilon}_-$ are each perpendicular to the axial unit vector $\hat{\varepsilon}_z = \hat{\mathbf{k}}$ in the z-direction of the electromagnetic wave propagation, resulting in zero dot products

$$\hat{\varepsilon}_+^* \cdot \hat{\varepsilon}_z = \frac{1}{\sqrt{2}}(\hat{\varepsilon}_x + i\hat{\varepsilon}_y)^* \cdot \hat{\varepsilon}_z = (0 + 0) = 0 \tag{12.428}$$

$$\hat{\varepsilon}_-^* \cdot \hat{\varepsilon}_z = \frac{1}{\sqrt{2}}(\hat{\varepsilon}_x - i\hat{\varepsilon}_y)^* \cdot \hat{\varepsilon}_z = (0 + 0) = 0 \tag{12.429}$$

The bremsstrahlung electric field $\mathbf{E}_{\mathrm{br}}$, which is expressed in terms of unit positive $\hat{\varepsilon}_+$ and negative $\hat{\varepsilon}_-$ helical polarization vectors and complex amplitudes E_+ and E_-, becomes

$$\mathbf{E}_{\mathrm{br}} = (E_+\hat{\varepsilon}_+ + E_-\hat{\varepsilon}_-)\exp[\mathrm{i}(\mathbf{k} \cdot \mathbf{r} - wt)] \tag{12.430}$$

In general, the ratio of amplitude E_- divided by E_+ is

$$\frac{E_-}{E_+} = r\ \exp(i\alpha) \tag{12.431}$$

where

$$r = \pm 1\ \&\ \alpha = 0 \quad \text{linear polarization} \tag{12.432}$$

$$r \neq \pm 1\ \&\ \alpha = 0 \quad \text{elliptical polarization} \tag{12.433}$$

$$r \neq \pm 1\ \&\ \alpha \neq 0 \quad \text{elliptical polarization} - \text{rotated by angle } \alpha \tag{12.434}$$

The distant observer can measure the polarization of an incoming electromagnetic plane wave, such as a light wave, by experimental measurement of intensities using a linear polarizer and quarter wave plate and by employing the Stokes parameters, which were devised by Stokes in 1852. For an incoming z-directed electromagnetic wave with electric field amplitude **E**, the four Stokes parameters are linear-polarization scalar products in the x- and y-directions $\hat{\varepsilon}_x \cdot \mathbf{E}$ and $\hat{\varepsilon}_y \cdot \mathbf{E}$, as well as the positive and negative circular-polarization scalar products $\hat{\varepsilon}_+{}^* \cdot \mathbf{E}$ and $\hat{\varepsilon}_-{}^* \cdot \mathbf{E}$. The measurement of the electric field polarization of incident electromagnetic waves, using linear polarizers and quarter wave plates, allows one to obtain insight into the physical processes, which are involved in the electromagnetic source and/or the medium through which travels the electromagnetic wave from the source to the observer. The ability to measure the polarization of incoming electromagnetic waves can be an important tool in astronomy for understanding the nature of cosmic sources of radiation and the intervening space–gas–plasma medium, which is inaccessible to the observer.

Consider an incoming z-directed electromagnetic wave with electric field amplitude **E**, whose squared magnitude scalar product intensities are measured by an observer with a linear polarizer. Based on the x- and y-linear polarization unit vectors $\hat{\varepsilon}_x$ and $\hat{\varepsilon}_y$, the two scalar products form the basis of the four Stokes parameters or measurable intensities s_0, s_1, s_2, and s_3. The observer sees the incoming wave with linear polarization, which has x and y components, where

$$\hat{\varepsilon}_x \cdot \mathbf{E} \tag{12.435}$$

$$\hat{\varepsilon}_y \cdot \mathbf{E} \tag{12.436}$$

The x and y scalar coefficients for linear polarization, are the product of an amplitude, which is also the magnitude, and a phase factor, where

$$E_x = E_{0x} \exp(\mathrm{i}\alpha_x) \tag{12.437}$$

$$E_y = E_{0y} \exp(\mathrm{i}\alpha_y) \tag{12.438}$$

With Eqs. (12.435–12.438), one forms the 4 stokes parameters for linear polarization, where

$$s_0 = |\varepsilon_x \cdot \mathbf{E}|^2 \;+\; |\varepsilon_y \cdot \mathbf{E}|^2 = E_{0x}^2 + E_{0y}^2 \tag{12.439}$$

$$s_1 = |\varepsilon_x \cdot \mathbf{E}|^2 \;-\; |\varepsilon_y \cdot \mathbf{E}|^2 = E_{0x}^2 - E_{0y}^2 \tag{12.440}$$

$$s_2 = 2\mathrm{Re}[(\varepsilon_x \cdot \mathbf{E})^*(\varepsilon_y \cdot \mathbf{E})] = 2E_{0x}E_{0y} \cos(\alpha_y - \alpha_x) \tag{12.441}$$

$$s_3 = 2\mathrm{Im}[(\varepsilon_x \cdot \mathbf{E})^*(\varepsilon_y \cdot \mathbf{E})] = 2E_{0x}E_{0y} \sin(\alpha_y - \alpha_x) \tag{12.442}$$

Next, consider an incoming z-directed electromagnetic wave with electric field amplitude $\mathbf{E}$, whose squared magnitude scalar product intensities are measured by an observer with a quarter wave plate. Based on the basis of positive and negative circular-polarization unit vectors $\hat{\varepsilon}_+$ and $\hat{\varepsilon}_-$, the two scalar products form the basis of the four Stokes parameters or measurable intensities s_0, s_1, s_2, and s_3. The observer sees the incoming wave with circular polarization components, which are counter-clockwise (positive helicity) or clockwise (negative helicity), where

$$\hat{\varepsilon}_+{}^* \cdot \mathbf{E} \tag{12.443}$$

$$\hat{\varepsilon}_-{}^* \cdot \mathbf{E} \tag{12.444}$$

The positive helicity and negative helicity scalar coefficients for circular polarization, are the product of an amplitude, which is also the magnitude, and a phase factor, where

$$E_+ = E_{0+} \exp(i\alpha_+) \tag{12.445}$$

$$E_- = E_{0-} \exp(i\alpha_-) \tag{12.446}$$

With Eqs. (12.443–12.446), one forms the 4 stokes parameters for circular polarization, where

$$s_0 = |\varepsilon_+^* \cdot \mathbf{E}|^2 \ + \ |\varepsilon_-^* \cdot \mathbf{E}|^2 = E_{0+}^2 + E_{0-}^2 \tag{12.447}$$

$$s_1 = |\varepsilon_+^* \cdot \mathbf{E}|^2 - |\varepsilon_-^* \cdot \mathbf{E}|^2 = E_{0+}^2 - E_{0-}^2 \tag{12.448}$$

$$s_2 = 2\mathrm{Re}[(\varepsilon_+^* \cdot \mathbf{E})^*(\varepsilon_-^* \cdot \mathbf{E})] = 2E_{0+}E_{0-}\cos(\alpha_- - \alpha_-) \tag{12.449}$$

$$s_3 = 2\mathrm{Im}[(\varepsilon_+^* \cdot \mathbf{E})^*(\varepsilon_-^* \cdot \mathbf{E})] = 2E_{0+}E_{0-}\sin(\alpha_- - \alpha_-) \tag{12.450}$$

For the linear ($\hat{\varepsilon}_x$,$\hat{\varepsilon}_y$) or helicity ($\hat{\varepsilon}_+, \hat{\varepsilon}_-$) case, the Stokes parameters or measurable intensities s_0, s_1, s_2, and s_3 are not independent; rather, they are related. For a beam of monochromatic electromagnetic waves, which in general does not apply to bremsstrahlung, the Stokes parameter relation is

$$s_0^2 = s_1^2 + s_2^2 + s_3^2 \tag{12.451}$$

The s_0 Stokes parameter measures the relative intensity of the incoming electromagnetic wave. Stokes parameter s_1 measures the degree of polarization of the x-over y-linear polarization or negative-over-positive circular polarization. The Stokes parameters s_2 and s_3 yield the phase information between the two basis unit vector components. For a quasi-monochromatic beam (which also in general does not apply to bremsstrahlung), the Stokes parameter relation is

$$s_0^2 \geq s_1^2 + s_2^2 + s_3^2 \tag{12.452}$$

The observable Stokes parameters for quasi-monochromatic electromagnetic becomes macroscopic time averages, which is due to the beating of the superposition of a finite bandwidth of waves, where the phase-information Stokes parameters s_2 and s_3 become time-averaged $\langle\rangle$ expressions. The 4 time-averaged Stokes parameters for linear polarization are

$$s_0 = \langle E_{0x}^2 + E_{0y}^2 \rangle \qquad s_2 = 2\langle E_{0x}E_{0y}\cos(\alpha_y - \alpha_x)\rangle \tag{12.453}$$

$$s_1 = \langle E_{0x}^2 - E_{0y}^2 \rangle \qquad s_3 = 2\langle E_{0x}E_{0y}\sin(\alpha_y - \alpha_x)\rangle \tag{12.454}$$

The 4 time-averaged Stokes parameters for circular polarization are

$$s_0 = \langle E_{0+}^2 - E_{0-}^2 \rangle \quad s_2 = 2\langle E_{0+}E_{0-}\ \sin(\alpha_- - \alpha_-)\rangle \tag{12.455}$$

$$s_1 = \langle E_{0+}^2 - E_{0-}^2 \rangle \quad s_3 = 2\langle E_{0+}E_{0-}\ \sin(\alpha_- - \alpha_-)\rangle \tag{12.456}$$

In the case of visible light, and even monochromatic visible light, $s_1 = s_2 = s_3 = 0$. For more information, see Jackson (1975) and Born and Wolf (1970).

12.18 Net Neutron Magnetic Dipole Moment Produced in an Ultracold Neutron Population with Applied Magnetic Field, and Possibilities for Population Inversion

Section 12.16 derived a simplified expression for the emitted differential energy per differential frequency and solid angle from a moving charged particle, which required careful consideration of the temporal integration limits and charge particle velocity history before and after the region of interest. This simplified expression of Eq. (12.378) facilitates the derivation of the expression for the emitted differential energy per differential frequency and solid angle $\mathrm{d}^2I/\mathrm{d}w\mathrm{d}\Omega$ for a moving magnetic dipole moment $\boldsymbol{\mu}$. The moving magnetic dipole moment $\boldsymbol{\mu}$ applies to an electrically charged particle, such as an electron.

More significantly, the expression for $\mathrm{d}^2I/\mathrm{d}w\mathrm{d}\Omega$ will apply to neutrons, which have a magnetic dipole moment μ. In this section, we examine the condition of an ultracold neutron beam propagating in an applied magnetic field, which produces a larger fraction of neutrons in the lower-energy state (spin-down neutrons) than in the upper-energy state (spin-up neutrons). The ultracold neutron beam presents a net, moving magnetic dipole moment $\boldsymbol{\mu}$ by virtue of the excess number N_1 of spin-down neutrons with their longitudinal magnetic dipole moment $\mu_{\parallel}$ aligned parallel to the applied magnetic field $\mathbf{B}$ ($\mu_{\parallel}\uparrow B\uparrow$), compared to the smaller number N_2 of spin-up neutrons with their longitudinal magnetic dipole moment aligned antiparallel to the applied magnetic field ($\mu_{\parallel}\uparrow B\downarrow$). Cold and room-temperature neutron beams have almost no net magnetic dipole moment because the almost-equal populations of spin-up ($\mu_{\parallel}$ antiparallel to **B**) and spin-down ($\mu_{\parallel}$ parallel to **B**).

The longitudinal component $\mu_{\parallel}$ of the neutron magnetic dipole moment vector is aligned parallel or antiparallel with a z-directed applied magnetic field **B**. The resulting neutron potential energy U in the applied magnetic field **B** is

$$U = \boldsymbol{\mu} \cdot \mathbf{B} = |\mu||B|\cos\theta = \mu_{\parallel} B = \pm 6.01 \times 10^{-8} B \frac{\text{eV}}{\text{T}} \tag{12.457}$$

Given gyromagnetic ratio γ, neutron Landé g factor $g_n = -1.91$, nuclear magneton μ_n, and neutron spin vector **S**, where $|\mathbf{S}| = \hbar\sqrt{s(s+1)}$ and $|\mathbf{S}| = \hbar\sqrt{3}/2$ for neutron spin quantum number $s = 1/2$, the neutron magnetic dipole moment vector $\boldsymbol{\mu}$ is related to its oppositely directed spin vector **S** by

$$\boldsymbol{\mu} = \gamma \mathbf{S} = \frac{2 g_n \mu_n}{\hbar} \mathbf{S} \tag{12.458}$$

where

$$|\boldsymbol{\mu}| = \mu = g_n \mu_n \sqrt{3} = -1.91 \mu_n \sqrt{3} = \frac{\sqrt{3}}{2} \hbar \gamma_n = 1.04 \times 10^{-7} \frac{\text{ev}}{\text{T}} \tag{12.459}$$

In a beam of neutrons in a magnetic field, N_u spin-up neutrons are at the higher potential energy U_2. The spin-up neutrons' longitudinal component $\mu_{\parallel}$, which points from magnetic south S to north N, is aligned antiparallel to the applied magnetic field **B**, which points from magnetic north N to south S. Antiparallel alignment of $\mu_{\parallel}$ and **B** for spin-up neutrons results in a higher-energy state because the same-polarity poles of $\mu_{\parallel}$ and **B** are adjacent. The N_d spin-down neutrons at lower potential energy U_1 have their longitudinal component $\mu_{\parallel}$ aligned parallel to the applied magnetic field **B**. Parallel alignment of $\mu_{\parallel}$ and **B** for spin-down neutrons results in a lower-energy state because the opposite-polarity poles of $\mu_{\parallel}$ and **B** are adjacent. In thermal equilibrium at temperature T [K], the ratio r of the spin-up neutron population N_u at the higher potential energy U_2, divided by the spin-down neutron population N_d at the lower-energy state U_1, is given by the Boltzmann law, where

$$r = \frac{N_2}{N_1} = \exp\left(-\frac{U_2 - U_1}{kT}\right) = \exp\left(-\frac{2|\mu_{\parallel} B|}{kT}\right) \tag{12.460}$$

or

$$r = \frac{N_2}{N_1} = \exp\left(-\frac{U_2 - U_1}{kT}\right) = \exp\left(-\frac{1.40 \times 10^{-3}|B|}{T}\right) \tag{12.461}$$

where

$$U_2 = |\mu_{\parallel} B| \quad \text{and} \quad U_1 = -|\mu_{\parallel} B| \tag{12.462}$$

The Boltzmann constant $k = 8.62 \times 10^{-5}\text{eV}/K$ gives $kT = 0.025$ at room temperature 298°K. For laboratory magnetic fields $B[T]$, such as 2.0 T, the beam is almost evenly populated with magnetic spin vectors oriented parallel (lower-energy) and antiparallel to the magnetic vector, where the population ratio $\mu_{z0} = 0.99999$ shows a population difference of 1 part in 10^5. At sufficiently low temperatures and sufficiently high magnetic fields, almost all the neutrons will be in the spin-down state, residing in the lower-energy state with their magnetic moment vectors aligned parallel to the magnetic field vector. For example, at the very low temperature $T_{1/e} = 1.40 \times 10^{-3}\text{K}$ at $B = 1.0\ T$, the ratio of spin-up, higher-energy state neutrons, divided by the lower-energy, spin-down neutrons, is $r = 1/e$ or $r = 0.37$. The $1/e$ value of spin-up to spin-down neutron ratio r occurs at temperature $T_{1/e} = 1.40 \times 10^{-3}$ K for a 1 T applied magnetic field. This temperature $T_{1/e}$ corresponds to an ultracold neutron beam energy of $1.21 \times 10^{-7}\text{eV}$ (0.21 μeV) or an ultracold neutron beam wavelength of 824Å.

An ultracold neutron beam will present a large population fraction of lower-energy, spin-down neutrons whose longitudinal component $\mu_{\parallel}$ is aligned parallel to the applied magnetic field **B**. To produce a population-inverted beam of mostly spin-up, ultracold neutrons in the higher-energy state with $\mu_{\parallel}$ aligned antiparallel to **B** would require a negative temperature condition as in the case of the laser. One may speculate the following. Consider a population of ultracold neutrons, most of which are initially aligned with their longitudinal magnetic dipole moment $\mu_{\parallel}$ aligned parallel to an applied magnetic field **B**, where the neutrons thus reside in the lower-energy state.

Suppose the direction of the magnetic field is suddenly reversed from **B** to $-\mathbf{B}$ in a time interval Δt_{Bf}. A time interval $\Delta t_{\mu f}$ is required for the neutrons to flip their longitudinal magnetic dipole moment $\mu_{\parallel}$ in the opposite direction, so as to maintain parallel alignment with the new applied magnetic field $-\mathbf{B}$. Now assume the reversal of the magnetic field **B** to $-\mathbf{B}$ is so rapid, that the magnetic field flipping time interval Δt_{Bf} is less than the magnetic dipole realignment time interval $\Delta t_{\mu f}$, where

$$\Delta t_{Bf} < \Delta t_{\mu f} \tag{12.463}$$

Then, for a very short interval of population inversion time Δt_{in}, the majority of ultracold neutrons would be in the higher-energy state with their longitudinal magnetic dipole moment $\mu_{\parallel}$ aligned antiparallel to the flipped magnetic field, where

$$\Delta t_{\text{in}} = \Delta t_{\mu f} - \Delta t_{Bf} \tag{12.464}$$

The neutron realignment occurs within a Larmor precession period:

$$\Delta t_{\mu f} = t_{\Omega}[s] = \frac{2\pi}{\Omega} \tag{12.465}$$

The temporal variation of the magnetic field w must exceed the azimuthal, Lamor neutron precession frequency $\Omega = \gamma B$ (see Eq. (2.104) in Section 2.7 of Chapter 2), where

$$w = \frac{2\pi}{\Delta t_{Bf}} \gg \Omega = \gamma B \tag{12.466}$$

If magnetic field reversal time is much less than the magnetic dipole realignment time, then an inverted population occurs, where a majority of the neutrons are spin-up neutrons in the upper energy state, and the minority of neutrons are spin-down neutrons in the lower energy state, where

$$\frac{\Delta t_{Bf}}{\Delta t_{\mu f}} = \frac{(2\pi/w)}{(2\pi/\Omega)} = \frac{\Omega}{w} << 1 \tag{12.467}$$

The production of a high flux, ultracold neutron beams is presently not attainable. An applied magnetic field **B**, which is capable of ultra-fast switching, could create an inversion of ultracold neutrons. Initially the majority of ultracold neutrons are spin-down neutrons in the lower energy state, where their spins are in parallel alignment with the applied magnetic field **B**, where $(\mu_{||} \uparrow B \uparrow)$. If the direction of the magnetic field is suddenly reversed from **B** to $-\mathbf{B}$, the neutron spins would not have time to follow the change in the magnetic field direction. As a result, the ultracold neutrons would be left with their spins in anti-parallel alignment with the newly directed magnetic field $-\mathbf{B}$, where $(\mu_{||} \uparrow B \downarrow)$, which is the upper neutron energy state (spin-up). A population inversion of spin-up, ultracold neutrons would thus occur. The inverted population would consist of a majority of the ultracold neutrons in the upper energy state (spin-up), and a minority of ultracold neutrons in the lower energy state (spin-down). The inverted population of upper energy state, spin-up, ultracold neutrons, would then transition back to the lower energy state, and become spin-down, ultracold neutrons, whose spins are in parallel alignment with the newly directed magnetic field $-\mathbf{B}$, where $(\mu_{||} \uparrow B \uparrow)$. The flipping of the ultracold neutron spins is also accompanied by the flipping of the neutron magnetic dipole moments. The flipping or change of the direction of the neutron magnetic dipole moment is a spatial and temporal variation of the neutron magnetic dipole field, which creates electromagnetic radiation. If the spins of the inverted population of ultracold, spin-up neutrons flip together in unison, from antiparallel alignment to parallel alignment with $-\mathbf{B}$, then a laser-like emission of electromagnetic radiation is created, by the unison flipping of neutron magnetic dipole moments.

The Bloch equation of motion for the spin moment $\boldsymbol{\mu}$ in a magnetic field is the cross product of the neutron magnetic moment vector $\boldsymbol{\mu}$ and the magnetic field vector **B**, which is multiplied by the scalar gyromagnetic ratio γ, so that

$$\frac{d\boldsymbol{\mu}}{dt} = \gamma \boldsymbol{\mu} \times \mathbf{B} \tag{12.468}$$

The gyromagnetic ratio γ is proportional to the particle's Landé g factor, where

$$\gamma = \frac{2g\mu_n}{\hbar} = -1.83 \times 10^8 \text{rad/T} \cdot \text{s} \tag{12.469}$$

Independent of neutron wavelength $\lambda[\text{Å}]$, the neutron magnetic dipole moment vector $\boldsymbol{\mu}$ undergoes azimuthal precession at the Lamor frequency $\Omega = \gamma B$ at a fixed polar orientation angle θ about the magnetic field direction of $\theta = 125.3°$ for spin-up and $\theta = 54.7°$ for spin-down neutrons.

Similar to the Bloch equation is the Lorentz force equation:

$$\frac{d\mathbf{v}}{dt} = \frac{e}{m}\mathbf{v} \times \mathbf{B} \tag{12.470}$$

The solution for neutron magnetic moment vector $\boldsymbol{\mu}$ in the Bloch equation and the velocity vector $\mathbf{v}$ of a charged particle in the Lorentz equation are identical, if one sets the charge to mass ratio e/m of the charged particle in the Lorentz force equation, equal to the negative of the gyromagnetic ratio γ in the Bloch equation, where

$$\frac{e}{m} = -\gamma \tag{12.471}$$

For example, the ratio e/m in the Lorentz force equation solution, for the velocity vector $\mathbf{v}$ of a proton through a magnetic field, can be replaced by the proton gyromagnetic ratio $(-\gamma)$, which yields the trajectory of the proton magnetic dipole moment vector $\boldsymbol{\mu}$. From Liboff (2003), the magnitude of the electron magnetic dipole moment is $\mu_e = -\mu_b$, where the Bohr magneton is $\mu_b = 0.927 \times 10^{-20}$ $erg/gauss$. In comparison, the magnitude of the magnetic dipole moment for the proton is $\mu_p = 2.79\ \mu_N$, and for the neutron is $\mu_n = -1.91\mu_N$, where nuclear magneton is $\mu_N = 0.505 \times 10^{-23}$ $erg/gauss$.

Consider an ultracold neutron beam propagating through a long beam line, or a burst of neutrons from a neutron star, or neutrons, which are produced by nuclear spallation in a cosmic ray shower in the Earth's upper atmosphere. In all cases, these free neutrons disintegrate to protons after about 15 minutes in the frame of the moving neutron, or longer for the stationary observer in the case of relativistic neutrons. The mean free neutron lifetime is 885.7 s. A beam of free neutrons propagating through a magnetic field will undergo disintegration to become a proton beam via beta decay. In beta decay, a free neutron n^0, with a half-life of 613.9 s, decays by emission of an electron e^- and an electron antineutrino v_e to become a proton p^+, where

$$n^0 \to p^+ + e^- + \nu_e \tag{12.472}$$

The trajectory of the daughter proton in a periodic magnetic field can then be calculated from the trajectory of the magnetic dipole moment of the parent neutron in the same periodic magnetic field.

12.19 Energy Radiated per Solid Angle for a Moving Magnetic Dipole Moment of Neutrons and Charged Particles

Following Jackson (1975), the emitted differential energy per differential frequency and solid angle, $d^2I/dwd\Omega$ for a moving magnetic dipole moment $\boldsymbol{\mu}$, is now derived as follows. Assume N particles, each of charge e_i, which move at velocity $\mathbf{v}_i$ or $\boldsymbol{\beta}_i = \mathbf{v}_i/c$, and produce a current I given by

$$I = \sum_{i=1}^{N} e_i v_i \tag{12.473}$$

The differential energy per differential frequency and solid angle, which is emitted by the N charged particles, is obtained from Eq. (12.378), where

$$\frac{d^2I(w,\hat{\mathbf{r}})}{dwd\Omega} = \frac{w^2}{16\pi^3\varepsilon_0 c^3}\left|\int_{-\infty}^{\infty} dt_s\left(\hat{\mathbf{r}}\times\left(\hat{\mathbf{r}}\times\sum_{i=1}^{N} e_i\boldsymbol{\beta}_i\right)\right)\exp\left[iw\left(t_s - \frac{\hat{\mathbf{r}}\cdot\mathbf{r}_{s,i}}{c}\right)\right]\right|^2 \tag{12.474}$$

Now consider the case of a moving continuous charge distribution with charge density $\rho(\mathbf{r}',t)$:

$$\begin{gathered}\sum_{i=1}^{N} e_i\boldsymbol{\beta}_i \exp\left[iw\left(t_s - \frac{\hat{\mathbf{r}}\cdot\mathbf{r}_{s,i}}{c}\right)\right] \Rightarrow \\ \int_{-\infty}^{\infty} dx' \int_{-\infty}^{\infty} dy' \int_{-\infty}^{\infty} dz' \frac{\rho(\mathbf{r}',t_s)\mathbf{v}(\mathbf{r}',t_s)}{c}\exp\left[iw\left(t_s - \frac{\hat{\mathbf{r}}\cdot\mathbf{r}'}{c}\right)\right]\end{gathered} \tag{12.475}$$

The current density $\mathbf{J}(\mathbf{r}',t_s)$ is the product of the charge density $\rho(\mathbf{r}',t_s)$ and the charge density velocity $\mathbf{v}(\mathbf{r}',t_s)$, where

$$\mathbf{J}(\mathbf{r}',t_s) = \rho(\mathbf{r}',t_s)\mathbf{v}(\mathbf{r}',t_s) = c\rho(\mathbf{r}',t_s)\boldsymbol{\beta}(\mathbf{r}',t_s) \tag{12.476}$$

The current density $\mathbf{J}(\mathbf{r}',t_s)$ is the sum of the following two terms: (i) the curl of the magnetization vector $\mathbf{M}(\mathbf{r}',t_s)$; and (ii) the derivative of the polarization vector $d\mathbf{P}(\mathbf{r}',t_s)/dt_s$ with respect to the earlier time t_s, which results in radiation received by the observer at later time t, where

$$\mathbf{J}(\mathbf{r}',t_s) = \nabla\times\mathbf{M}(\mathbf{r}',t_s) + \frac{d\mathbf{P}(\mathbf{r}',t_s)}{dt_s} \tag{12.477}$$

The polarization vector $\mathbf{P}(\mathbf{r}', t_s)$ is the density of the electric dipole moments $\mathbf{p}_e(\mathbf{r}', t_s)$, where $\mathbf{p}_e(\mathbf{r}', t_s)$ is the product of (1) the separation vector $\mathbf{r}_e$ between a pair of equal and opposite electric charges $\pm e$, and (2) the magnitude of the electric charge e, where

$$\mathbf{p}_e(\mathbf{r}', t_s) = e\mathbf{r}_e(\mathbf{r}', t_s) \tag{12.478}$$

The separation vector $\mathbf{r}_e$ points from the negative charge $-e$ to the positive charge $+e$, and $|\mathbf{r}_e|$ is the distance between $-e$ and $+e$.

The magnetization vector $\mathbf{M}(\mathbf{r}', t_s)$ is the density of the magnetic dipoles $\boldsymbol{\mu}(\mathbf{r}', t_s)$. The magnetic dipole moment vector $\boldsymbol{\mu}(\mathbf{r}', t_s)$ points from the south S magnetic pole to the north N magnetic pole, where $\boldsymbol{\mu}(\mathbf{r}', t_s)$ is the product of the current I in a planar, current loop that encloses A. That is, using the right-hand rule, an azimuthal, counterclockwise positive loop current I defines the north N pole above the loop plane in the positive z-direction, and the south S pole below the loop plane. The two poles lie on the center z-axis on opposite sides of the current loop plane. Negative electrons flow in the opposite direction of the positive current. Thus, normal vector $\hat{\mathbf{z}}$ points in the positive direction from S to N magnetic poles, where $\hat{\mathbf{z}}$ is perpendicular to the loop area A, and the magnetic dipole moment vector $\boldsymbol{\mu}$ is

$$\boldsymbol{\mu}(\mathbf{r}', t_s) = IA(\mathbf{r}', t_s)\hat{\mathbf{z}} \tag{12.479}$$

With substitution of $\mathbf{J}(\mathbf{r}', t_s) = ec\boldsymbol{\beta}(\mathbf{r}', t_s)$ from Eq. (12.476) into Eq. (12.475), and then replacing the resulting $\mathbf{J}(\mathbf{r}', t_s)$ in the $\mathrm{d}^2 I/\mathrm{d}w\mathrm{d}\Omega$ expression by the alternative expression for $\mathbf{J}(\mathbf{r}', t_s)$ of Eq. (12.477), which is in terms of the magnetization $\mathbf{M}(\mathbf{r}', t_s)$ and polarization $\mathbf{P}(\mathbf{r}', t_s)$ vectors, one obtains

$$\frac{\mathrm{d}^2 I(w, \hat{\mathbf{r}})}{\mathrm{d}w\mathrm{d}\Omega} = \frac{w^2}{16\pi^3 \varepsilon_0 c^5} \times \left| \int_{-\infty}^{\infty} \mathrm{d}t_s \int_{-\infty}^{\infty} \mathrm{d}x' \int_{-\infty}^{\infty} \mathrm{d}y' \int_{-\infty}^{\infty} \mathrm{d}z' \exp\left[iw\left(t_s - \frac{\hat{\mathbf{r}}\cdot\mathbf{r}'}{c} \right) \right] \times \left\{ \hat{\mathbf{r}} \times \left(\hat{\mathbf{r}} \times \left[\nabla \times \mathbf{M}(\mathbf{r}', t_s) + \frac{\mathrm{d}\mathbf{P}(\mathbf{r}', t_s)}{\mathrm{d}t_s} \right] \right) \right\} \right|^2 \tag{12.480}$$

In Eq. (12.480), the integration of the first term, which is the double cross product of the observer position unit vector $\hat{\mathbf{r}}$ with the curl of the magnetization vector $\mathbf{M}(\mathbf{r}', t_s)$, or $\hat{\mathbf{r}} \times [\hat{\mathbf{r}} \times (\nabla \times \mathbf{M})]$, requires application of the vector identity

$$\nabla(\mathbf{A} \cdot \mathbf{B}) = (\mathbf{A} \cdot \nabla)\mathbf{B} + (\mathbf{B} \cdot \nabla)\mathbf{A} + \mathbf{A} \times (\nabla \times \mathbf{B}) + \mathbf{B} \times (\nabla \times \mathbf{A}) \tag{12.481}$$

where

$$\mathbf{A} = \hat{\mathbf{r}} \quad \text{and} \quad \mathbf{B} = \mathbf{M}(\mathbf{r}', t_s) \tag{12.482}$$

and

$$\hat{\mathbf{r}} = x\hat{\mathbf{x}} + y\hat{\mathbf{y}} + z\frac{\hat{\mathbf{z}}}{r} \tag{12.483}$$

One obtains

$$\hat{\mathbf{r}} \times (\nabla \times \mathbf{M}) = \nabla(\hat{\mathbf{r}}\cdot\mathbf{M}) - (\hat{\mathbf{r}}\cdot\nabla)\mathbf{M} - (\mathbf{M}\cdot\nabla)\hat{\mathbf{r}} - \mathbf{M} \times (\nabla \times \hat{\mathbf{r}}) \tag{12.484}$$

Constant unit vector $\hat{\mathbf{r}}$ points from the origin to the fixed observer position, where

$$\nabla \times \hat{\mathbf{r}} = 0 \tag{12.485}$$

After substitution of Eq. (12.485) into Eq. (12.484), the cross product of the resulting Eq. (12.484) with unit vector $\hat{\mathbf{r}}$ is performed, where $\hat{\mathbf{r}} \times \hat{\mathbf{r}} = 0$ and

$$\hat{\mathbf{r}} \times [\hat{\mathbf{r}} \times (\nabla \times \mathbf{M})] = \hat{\mathbf{r}} \times \nabla(\hat{\mathbf{r}}\cdot\mathbf{M}) - (\hat{\mathbf{r}} \times \mathbf{M})(\hat{\mathbf{r}}\cdot\nabla) \tag{12.486}$$

The gradient operator ∇ in rectangular coordinates is

$$\nabla = \hat{\mathbf{x}}\frac{\partial}{\partial x'} + \hat{\mathbf{y}}\frac{\partial}{\partial y'} + \hat{\mathbf{z}}\frac{\partial}{\partial z'} \tag{12.487}$$

In spherical coordinates, the dot product of the unit vector $\hat{\mathbf{r}}$, which points from the origin to the fixed observer position $\mathbf{r}$ with the gradient operator ∇, is

$$\hat{\mathbf{r}}\cdot\nabla = \hat{\mathbf{r}}\cdot\left(\hat{\mathbf{r}}\frac{\partial}{\partial r'} + \hat{\boldsymbol{\theta}}\frac{1}{r'}\frac{\partial}{\partial \theta'} + \hat{\boldsymbol{\varphi}}\frac{1}{r'}\sin\theta'\frac{\partial}{\partial \phi'}\right) = \frac{\partial}{\partial r'} \tag{12.488}$$

In spherical coordinates the magnetization vector is

$$\mathbf{M}(\mathbf{r}', t_s) = \hat{\mathbf{r}}M_r(\mathbf{r}', t_s) + \hat{\boldsymbol{\theta}}M_\theta(\mathbf{r}', t_s) + \hat{\boldsymbol{\varphi}}M_\phi(\mathbf{r}', t_s) \tag{12.489}$$

With Eq. (12.489), the first term of Eq. (12.486) becomes

$$\hat{\mathbf{r}} \times \nabla(\hat{\mathbf{r}}\cdot\mathbf{M}) = \hat{\mathbf{r}} \times \hat{\mathbf{r}}\frac{\partial M_r(\mathbf{r}', t_s)}{\partial r'} = 0 \tag{12.490}$$

With substitution of Eq. (12.490) into Eq. (12.486), one obtains

$$\hat{\mathbf{r}} \times [\hat{\mathbf{r}} \times (\nabla \times \mathbf{M})] = -(\hat{\mathbf{r}} \times \mathbf{M})(\hat{\mathbf{r}}\cdot\nabla) \tag{12.491}$$

After substitution of Eq. (12.491) for the first term of Eq. (12.480), one integrates by parts, where

$$\int_{-\infty}^{\infty} \mathrm{d}x' \int_{-\infty}^{\infty} \mathrm{d}y' \int_{-\infty}^{\infty} \mathrm{d}z'(-\hat{\mathbf{r}} \times \mathbf{M})(\hat{\mathbf{r}}\cdot\nabla)\exp\left[iw\left(t_s - \frac{\hat{\mathbf{r}}\cdot\mathbf{r}'}{c}\right)\right]$$

$$= \left[(-\hat{\mathbf{r}} \times \mathbf{M})(\hat{\mathbf{r}}\cdot\hat{\mathbf{r}})\exp\left[iw\left(t_s - \frac{\hat{\mathbf{r}}\cdot\mathbf{r}'}{c}\right)\right]\right]_{-\infty}^{\infty}$$

$$- \int_{-\infty}^{\infty} \mathrm{d}x' \int_{-\infty}^{\infty} \mathrm{d}y' \int_{-\infty}^{\infty} \mathrm{d}z'(-\hat{\mathbf{r}} \times \mathbf{M})(\hat{\mathbf{r}}\cdot\hat{\mathbf{r}})\frac{\partial}{\partial r'}\exp\left[iw\left(t_s - \frac{\hat{\mathbf{r}}\cdot\hat{\mathbf{r}}r'}{c}\right)\right], \tag{12.492}$$

and

$$\frac{\partial}{\partial r'}\exp\left[iw\left(t_s - \frac{\hat{\mathbf{r}}\cdot\hat{\mathbf{r}}r'}{c}\right)\right] = -\frac{iw}{c}\exp\left[iw\left(t_s - \frac{\hat{\mathbf{r}}\cdot\mathbf{r}'}{c}\right)\right] \tag{12.493}$$

In the first term on the right side of Eq. (12.492), the boundary conditions are zero at infinity $[M(\mathbf{r}' = \pm\infty, t_s) = 0]$,

$$\int_{-\infty}^{\infty} \mathrm{d}x' \int_{-\infty}^{\infty} \mathrm{d}y' \int_{-\infty}^{\infty} \mathrm{d}z'(-\hat{\mathbf{r}} \times \mathbf{M})(\hat{\mathbf{r}}\cdot\nabla)\exp\left[iw\left(t_s - \frac{\hat{\mathbf{r}}\cdot\mathbf{r}'}{c}\right)\right]$$

$$= -\int_{-\infty}^{\infty} \mathrm{d}x' \int_{-\infty}^{\infty} \mathrm{d}y' \int_{-\infty}^{\infty} \mathrm{d}z'\left(-\frac{iw}{c}\right)(-\hat{\mathbf{r}} \times \mathbf{M})\exp\left[iw\left(t_s - \frac{\hat{\mathbf{r}}\cdot\mathbf{r}'}{c}\right)\right] \tag{12.494}$$

Substitution of Eq. (12.494) for the first term in Eq. (12.480), yields

$$\frac{\mathrm{d}^2 I(w,\hat{\mathbf{r}})}{\mathrm{d}w\mathrm{d}\Omega} = \frac{w^2}{16\pi^3\varepsilon_0 c^5} \times$$

$$\left| \begin{array}{l} \int_{-\infty}^{\infty} \mathrm{d}t_s \int_{-\infty}^{\infty} \mathrm{d}x' \int_{-\infty}^{\infty} \mathrm{d}y' \int_{-\infty}^{\infty} \mathrm{d}z' \exp\left[iw\left(t_s - \frac{\hat{\mathbf{r}}\cdot\mathbf{r}'}{c}\right)\right] \times \\ \left\{\left(-\frac{iw}{c}\right)\hat{\mathbf{r}} \times \mathbf{M}(\mathbf{r}', t_s) + \hat{\mathbf{r}} \times \left(\hat{\mathbf{r}} \times \frac{\mathrm{d}\mathbf{P}(\mathbf{r}', t_s)}{\mathrm{d}t_s}\right)\right\} \end{array} \right|^2 \tag{12.495}$$

The integration of the second term of Eq. (12.495), requires reversing the order of spatial and time integration, and then performing the time integral, via integration by parts, where

$$\int_{-\infty}^{\infty} dt_s\, \hat{\mathbf{r}} \times \hat{\mathbf{r}} \times \frac{d\mathbf{P}(\mathbf{r}', t_s)}{dt_s} \exp\left[iw\left(t_s - \frac{\hat{\mathbf{r}}\cdot\mathbf{r}'}{c}\right)\right] =$$

$$\left[\hat{\mathbf{r}} \times \hat{\mathbf{r}} \times \mathbf{P}(\mathbf{r}', t_s) \exp\left[iw\left(t_s - \frac{\hat{\mathbf{r}}\cdot\mathbf{r}'}{c}\right)\right]\right]_{-\infty}^{\infty}$$

$$- \int_{-\infty}^{\infty} dt_s\, \hat{\mathbf{r}} \times \hat{\mathbf{r}} \times \mathbf{P}(\mathbf{r}', t_s) \frac{d}{dt_s} \exp\left[iw\left(t_s - \frac{\hat{\mathbf{r}}\cdot\mathbf{r}'}{c}\right)\right], \tag{12.496}$$

where

$$\frac{d}{dt_s} \exp\left[iw\left(t_s - \frac{\hat{\mathbf{r}}\cdot\mathbf{r}'}{c}\right)\right] = iw\left(1 - \frac{\hat{\mathbf{r}}\cdot\mathbf{v}'}{c}\right) \exp\left[iw\left(t_s - \frac{\hat{\mathbf{r}}\cdot\mathbf{r}'}{c}\right)\right] \tag{12.497}$$

In the first term on the right side of Eq. (12.496), one assumes zero boundary condition at infinity, where $\mathbf{P}(\mathbf{r}', t_s) = 0$ at $t_s = \pm\infty$, so

$$\int_{-\infty}^{\infty} dt_s\, \hat{\mathbf{r}} \times \hat{\mathbf{r}} \times \frac{d\mathbf{P}}{dt_s} \exp\left[iw\left(t_s - \frac{\hat{\mathbf{r}}\cdot\mathbf{r}'}{c}\right)\right] =$$

$$-iw(1 - \hat{\mathbf{r}}\cdot\boldsymbol{\beta}') \int_{-\infty}^{\infty} dt_s\, \hat{\mathbf{r}} \times \hat{\mathbf{r}} \times \mathbf{P}(\mathbf{r}', t_s) \exp\left[iw\left(t_s - \frac{\hat{\mathbf{r}}\cdot\mathbf{r}'}{c}\right)\right], \tag{12.498}$$

where

$$\boldsymbol{\beta}(t_s) = \frac{\mathbf{v}'(t_s)}{c} \tag{12.499}$$

The differential energy emission per differential frequency and solid angle $d^2I/dwd\Omega$ by current density $\mathbf{J}(\mathbf{r}', t_s)$, with substitution of Eqs. (12.494) and (12.498) into Eq. (12.480) for the first and second terms, becomes

$$\frac{d^2I(w, \hat{\mathbf{r}})}{dwd\Omega} = \frac{w^4}{16\pi^3\varepsilon_0 c^5} \times$$

$$\left| \int_{-\infty}^{\infty} dt_s \int_{-\infty}^{\infty} dx' \int_{-\infty}^{\infty} dy' \int_{-\infty}^{\infty} dz' \exp\left[iw\left(t_s - \frac{\hat{\mathbf{r}}\cdot\mathbf{r}'}{c}\right)\right] \times \left[\hat{\mathbf{r}} \times \frac{\mathbf{M}(\mathbf{r}', t_s)}{c} + (1 - \hat{\mathbf{r}}\cdot\boldsymbol{\beta}(t_s))\hat{\mathbf{r}} \times (\hat{\mathbf{r}} \times \mathbf{P}(\mathbf{r}', t_s))\right] \right|^2 \tag{12.500}$$

Using the Dirac delta function, the magnetization vector $\mathbf{M}(\mathbf{r}', t_s)$ in Eq. (12.500) is expressed as a function of the point-like magnetic dipole moments $\boldsymbol{\mu}(\mathbf{r}', t_s)$, and the polarization vector $\mathbf{P}(\mathbf{r}', t_s)$ in Eq. (12.500) is expressed as a function of the point-like electric dipole moments $\mathbf{p}_e(\mathbf{r}', t_s)$, where

$$\mathbf{M}(\mathbf{r}', t_s) = \boldsymbol{\mu}(t_s)\delta(\mathbf{r}' - \mathbf{r}_s(t_s)) \tag{12.501}$$

and

$$\mathbf{P}(\mathbf{r}', t_s) = \frac{\boldsymbol{\beta}(t_s) \times \boldsymbol{\mu}(t_s)\delta(\mathbf{r}' - \mathbf{r}_s(t_s))}{\sqrt{1-\beta^2}} \tag{12.502}$$

The magnetization vector $\mathbf{M}(\mathbf{r}', t_s)$ expression of Eq. (12.501) is expected, because the magnetization vector $\mathbf{M}(\mathbf{r}', t_s)$ is the density of the point-like, magnetic dipole moments $\boldsymbol{\mu}(\mathbf{r}', t_s)$. The magnetization vector $\mathbf{M}(\mathbf{r}', t_s)$ is measured in the frame of a stationary observer, who sees the charges move at nonzero velocity $\mathbf{v}'(\mathbf{r}', t_s)$. The polarization vector $\mathbf{P}(\mathbf{r}', t_s)$ of the moving charges, which is measured by the stationary observer, is then expressed in terms of the observer-measured, magnetization vector $\mathbf{M}(\mathbf{r}', t_s)$, where

$$\mathbf{P}(\mathbf{r}', t_s) = \frac{\boldsymbol{\beta}(t_s) \times \mathbf{M}(\mathbf{r}', t_s)}{\sqrt{1-\beta^2}} \tag{12.503}$$

With substitution of Eq. (12.503) into Eq. (12.500), one obtains

$$\frac{\mathrm{d}^2 I(w, \hat{\mathbf{r}})}{\mathrm{d}w\mathrm{d}\Omega} = \frac{w^4}{16\pi^3\varepsilon_0 c^5} \times \left| \int_{-\infty}^{\infty} \mathrm{d}t_s \int_{-\infty}^{\infty} \mathrm{d}x' \int_{-\infty}^{\infty} \mathrm{d}y' \int_{-\infty}^{\infty} \mathrm{d}z' [M_{\text{term}} + P_{\text{term}}] \exp\left[iw\left(t_s - \frac{\hat{\mathbf{r}}\cdot\mathbf{r}_s(t_s)}{c} \right) \right] \right|^2 \tag{12.504}$$

The magnetization term M_{term} and the polarization P_{term} term in Eq. (12.504) are

$$M_{\text{term}} = \frac{\hat{\mathbf{r}} \times \boldsymbol{\mu}(t_s)\delta(\mathbf{r}' - \mathbf{r}_s(t_s))}{c} \tag{12.505}$$

$$P_{\text{term}} = (1 - \hat{\mathbf{r}}\cdot\boldsymbol{\beta}(t_s))\hat{\mathbf{r}} \times \left(\hat{\mathbf{r}} \times \frac{\boldsymbol{\beta}(t_s) \times \boldsymbol{\mu}(t_s)\delta(\mathbf{r}' - \mathbf{r}_s(t_s))}{\sqrt{1-\beta^2}} \right) \tag{12.506}$$

In Eq. (12.504), spatial integration over all (x', y', z') space, yields the differential energy emission per differential frequency and solid angle $\mathrm{d}^2I/\mathrm{d}w\mathrm{d}\Omega$ from a moving magnetic dipole, where, expressed in MKS units:

$$\frac{\mathrm{d}^2I(w,\hat{\mathbf{r}})}{\mathrm{d}w\mathrm{d}\Omega} = \frac{w^4}{16\pi^3\varepsilon_0 c^5} \times$$

$$\left|\int_{-\infty}^{\infty} \mathrm{d}t_s \hat{\mathbf{r}} \times \left[\frac{\boldsymbol{\mu}(t_s)}{c} + [1-\hat{\mathbf{r}}\cdot\boldsymbol{\beta}(t_s)]\left(\hat{\mathbf{r}} \times \frac{\boldsymbol{\beta}(t_s)\times\boldsymbol{\mu}(t_s)}{\sqrt{1-\beta^2}}\right)\right] \exp\left[\mathrm{i}w\left(t_s - \frac{\hat{\mathbf{r}}\cdot\mathbf{r}_s(t_s)}{c}\right)\right]\right|^2 \tag{12.507}$$

In CGS units, recalling section 12.6, the differential energy emission per differential frequency and solid angle $(\mathrm{d}^2I/\mathrm{d}w\mathrm{d}\Omega)_{\mathrm{cgs}}$ from a moving magnetic dipole is

$$\left(\frac{\mathrm{d}^2I(w,\hat{\mathbf{r}})}{\mathrm{d}w\mathrm{d}\Omega}\right)_{\mathrm{cgs}} =$$

$$\frac{w^4}{4\pi^2c^3}\left|\int_{-\infty}^{\infty} \mathrm{d}t_s\, \hat{\mathbf{r}} \times \left[\boldsymbol{\mu}(t_s) + [1-\hat{\mathbf{r}}\cdot\boldsymbol{\beta}(t_s)]\left(\hat{\mathbf{r}} \times \frac{\boldsymbol{\beta}(t_s)\times\boldsymbol{\mu}(t_s)}{\sqrt{1-\beta^2}}\right)\right] \exp\left[\mathrm{i}w\left(t_s - \frac{\hat{\mathbf{r}}\cdot\mathbf{r}_s(t_s)}{c}\right)\right]\right|^2 \tag{12.508}$$

The differential energy emission per differential frequency and solid angle $\mathrm{d}^2I/\mathrm{d}w\mathrm{d}\Omega$ from a moving magnetic dipole $\boldsymbol{\mu}$, is a factor that is w^2 larger than that of an accelerated charge e. If the velocity of the magnetic dipoles is nonrelativistic, then $\mathrm{d}^2I/\mathrm{d}w\mathrm{d}\Omega$, expressed in CGS units, simplifies to

$$\left(\frac{\mathrm{d}^2I(w,\hat{\mathbf{r}})}{\mathrm{d}w\mathrm{d}\Omega}\right)_{\mathrm{cgs}} =$$
$$\frac{w^4}{4\pi^2c^3}\left|\int_{-\infty}^{\infty} \mathrm{d}t_s\;\; \hat{\mathbf{r}} \times [\boldsymbol{\mu}(t_s) + (\hat{\mathbf{r}} \times \boldsymbol{\beta}(t_s) \times \boldsymbol{\mu}(t_s))] \exp\left[\mathrm{i}w\left(t_s - \frac{\hat{\mathbf{r}}\cdot\mathbf{r}_s(t_s)}{c}\right)\right]\right|^2 \tag{12.509}$$

12.20 Relation of a Magnetization Vector of a Moving Charge with a Polarization Vector of a Stationary Charge

In this section, the magnetization vector **M** of a moving charge is related to the polarization vector $\mathbf{P}(\mathbf{r}', t_s)$ of a stationary charge. Both stationary and moving charges are measured relative to the stationary observer. To understand the expression for the polarization vector $\mathbf{P}(\mathbf{r}', t_s)$, consider the Lorentz force, which acts on a moving charge, relative to the applied electric **E** and magnetic **B** fields, measured in the stationary frame. The Lorentz force **F** acts on a charge (satellite) moving at velocity **v**, which is measured by a stationary observer (earth), who sees the charge (satellite) moving at velocity **v**. The stationary observer (earth) measures electric field **E**, magnetic field **B**, and force **F**, which acts on the charge (satellite), where

$$\mathbf{F} = e(\mathbf{E} + \mathbf{v} \times \mathbf{B}) \tag{12.510}$$

However, in the satellite rest frame of the moving charge, where $\mathbf{v} = 0$, only an electric field $\mathbf{E}'$ will be measured by the observer in the satellite, who is at rest, relative to the charge, where

$$\mathbf{E}' = \mathbf{E} + \mathbf{v} \times \mathbf{B} \tag{12.511}$$

The earth observer in the stationary earth frame, and the observer, who is stationary in the moving frame (satellite), each measure the same force **F**, where electric charge e is invariant in the Lorentz transformations, and

$$\mathbf{F} = e\mathbf{E}' = e(\mathbf{E} + \mathbf{v} \times \mathbf{B}) \tag{12.512}$$

An observer in the satellite, who is at rest, with $\mathbf{v} = 0$, in the moving frame of the charge, measures a current density $\mathbf{J}' = \sigma\mathbf{E}'$, which is proportional to the moving frame electric field $\mathbf{E}'$, via the intrinsic conductivity σ of the moving medium, where $\mathbf{E}' = \mathbf{E}$ from Eq. (12.511). The intrinsic conductivity σ of the moving or stationary medium is assumed to be a constant, which is independent of the electric and magnetic fields, and where σ is thus invariant under Lorentz transformations. Alternatively, a second observer on earth, who is at rest in the stationary frame of earth, sees the moving medium (satellite) with intrinsic conductivity σ, which propagates at nonzero velocity $\mathbf{v} \neq 0$. The observer on earth measures electric field **E**, magnetic field **B**, and current density $\mathbf{J}' = \sigma\mathbf{E}'$, where, $\mathbf{E}' = \mathbf{E} + \mathbf{v} \times \mathbf{B}$ from Eq. (12.511), and

$$\mathbf{J}' = \sigma\mathbf{E}' = \sigma(\mathbf{E} + \mathbf{v} \times \mathbf{B}) \tag{12.513}$$

Depending on the complexity of the media, the conductivity can be a constant σ or a tensor. From Eq. (12.513), the moving frame current density $\mathbf{J}'$, which is measured by an observer in the moving frame, can be calculated from the Lorentz invariant conductivity σ and the applied electric **E** and magnetic **B** fields, measured by an observer in the stationary frame. The magnetization vector **M** is related to the magnetic flux density **B** and applied field **H** via the vacuum permeability constant μ_0, where

$$\mathbf{H} = \frac{\mathbf{B}}{\mu_0} - \mathbf{M} \tag{12.514}$$

If the magnetic field arises only from magnetic dipoles in the case of zero applied magnetic field $\mathbf{H} = 0$, then the magnetic flux density $\mathbf{B}$ is produced by magnetic dipoles with magnetic moment $\boldsymbol{\mu}$, which form the magnetic dipole density or magnetization $\mathbf{M}$, where

$$\mathbf{B} = \mu_0 \mathbf{M} \tag{12.515}$$

The vacuum space permeability μ_0 and permittivity ε_0 constants are independent of the electric and magnetic fields, and thus μ_0 and ε_0 are invariant under Lorentz transformations. The current density $\mathbf{J}' = \sigma \mathbf{E}' = \sigma E$ is measured by an observer on the satellite, who is at rest in the moving frame of the charges, where the charges are stationary and $\mathbf{v} = 0$ in Eq. (12.513). However, the earth observer, who is at rest in the stationary frame of earth, where the charges are moving with nonzero velocity $\mathbf{v} \neq 0$, measures electric field $\mathbf{E}$, magnetization $\mathbf{M}$, via magnetic field $\mathbf{B} = \mu_0 \mathbf{M}$ of Eq. (12.515) , and current density $\mathbf{J}' = \sigma \mathbf{E}'$, where

$$\mathbf{J}' = \sigma \mathbf{E}' = \sigma(\mathbf{E} + \mathbf{v} \times \mu_0 \mathbf{M}) \tag{12.516}$$

The polarization vector $\mathbf{P}'$, which is measured by an observer in the moving frame, is related to the moving frame electric field $\mathbf{E}'$ via the vacuum permittivity constant ε_0 and Lorentz invariant relative permittivity ε_r by

$$\mathbf{P}' = \varepsilon_0 \varepsilon_r \mathbf{E}' = \varepsilon_0 \varepsilon_r (\mathbf{E} + \mathbf{v} + \mu_0 \mathrm{M}) \tag{12.517}$$

Measured by an observer in the moving frame, the electric displacement vector $\mathbf{D}'$ is obtained by substitution of Eq. (12.517) for $\mathbf{P}'$ followed by Eq. (12.511) for $\mathbf{E}'$, where

$$\mathbf{D}' = \varepsilon_0 \mathbf{E}' + \mathbf{P}' = (\varepsilon_0)(1 + \varepsilon_r)(\mathbf{E} + \mathbf{v} \times \mathbf{B}) \tag{12.518}$$

Since, the applied magnetic field intensity is zero, where $\mathbf{H} = 0$ via Eq. (12.514), the moving frame displacement vector $\mathbf{D}'$ of Eq. (12.518), becomes

$$\mathbf{D}' = \varepsilon_0 \mathbf{E}' + \mathbf{P}' = (\varepsilon_0)(1 + \varepsilon_r)(\mathbf{E} + \mathbf{v} \times \mu_0 \mathbf{M}) \tag{12.519}$$

With $c^2 = 1/\varepsilon_0 \mu_0$, Eq. (12.517) becomes

$$\mathbf{P}' = \varepsilon_0 \varepsilon_r \mathbf{E}' = \varepsilon_0 \varepsilon_r \mathbf{E} + \mathbf{v} \times \frac{\varepsilon_r \mathbf{M}}{c^2} \tag{12.520}$$

The polarization vector $\mathbf{P}$ and magnetization vector $\mathbf{M}$ are measured by an observer in the stationary frame, and together, with the relative velocity $\mathbf{v}$ of the moving frame, yield the polarization vector $\mathbf{P}'$, which is measured by an observer at rest with the charges in the moving frame, where $\boldsymbol{\beta} = \mathbf{v}/c$ and

$$\mathbf{P}' = \varepsilon_0\varepsilon_r\mathbf{E}' = \mathbf{P} + \varepsilon_r\frac{\boldsymbol{\beta}\times\mathbf{M}}{c} \tag{12.521}$$

Assume that the relative permittivity is approximately unity $\varepsilon_r \cong 1$, and assume zero applied electric field $\mathbf{E}$. As a result, the measured electric field $\mathbf{E}$ and polarization vector $\mathbf{P}$ in the stationary frame, are each zero, where $\mathbf{E} = 0$, and $\mathbf{P} = 0$ in Eq. (12.521), so that

$$\mathbf{P}' = \varepsilon_0\mathbf{E}' = \frac{\boldsymbol{\beta}\times\mathbf{M}}{c} \tag{12.522}$$

The moving charges in the moving frame, relative to the stationary frame, act as magnetic dipoles to produce a magnetization vector $\mathbf{M}$, which is viewed and measured in the stationary frame. However, an observer in the moving frame observes the charges as stationary and measures a polarization vector $\mathbf{P}'$, in contrast to the observer in the stationary frame, which observes the chargers as moving, and thus measures a magnetization vector $\mathbf{M}$.

References

Becker, R. (1982). *Electromagnetic fields and interactions*. New York, NY: Dover Publications.

Born, M., & Wolf, E. (1970). *Principles of optics* (4th ed., Chapter 10). New York, NY: Pergamon.

Eyges, L. (1980). *The classical electromagnetic field*. New York, NY: Dover Publications.

Feynman, R. P. (1964). *The Feynman lectures on physics* (Vol. 2, mainly electromagnetism and matter). Reading, MA: Addison-Wesley.

Jackson, J. D. (1975). *Classical electrodynamics*. New York, NY: Wiley.

Landau, L. D., & Lifshitz, E. M. (1975). *The classical theory of fields* (4th revised English edition). New York, NY: Pergamon Press.

Liboff, R. L. (2003). *Introductory quantum mechanics* (4th ed.). New York, NY: Addison Wesley.

Marion, J. B. (1965). *Classical electromagnetic radiation*. New York, NY: Academic Press.

Maxwell, J. C. (1954). *A treatise on electricity and magnetism* (Vols. 1 & 2). New York, NY: Dover.

Panofsky, W. K. H., & Phillips, M. (2005). *Classical electricity and magnetism*. New York, NY: Dover Publications.

Popović, B. D. (1971). *Introductory engineering electromagnetics*. Reading, MA: Addison-Wesley.

Purcell, E. M. (1965). *Electricity and magnetism*. New York, NY: McGraw-Hill.

Ramo, S., Whinnery, J. R., & Van Duzer, T. (1965). *Fields and waves in communication electronics*. New York, NY: Wiley.

Schwartz, M. (1972). *Principles of electrodynamics*. New York, NY: Dover Publications.
Schey, H. M. (2005). *Div grad curl and all that. An informal text on vector calculus*. New York, NY: W.W. Norton and Company.
Shadowitz, A. (1988a). *The electromagnetic field*. New York, NY: Dover Publications.
Shadowitz, A. (1988b). *Special relativity*. New York, NY: Dover Publications.
Shen, L. C., & Kong, J. A. (1983). *Applied electromagnetism*. Monterey, CA: Brooks.
Smith, R. J. (1966). *Circuits, devices, and systems*. New York, NY: Wiley.
Sommerfeld, A. (1964). *Electrodynamics*. New York, NY: Academic Press.
Stratton, J. A. (1941). *Electromagnetic theory*. New York, NY: McGraw-Hill.
Zahn, M. (1979). *Electromagnetic field theory: A problem solving approach*. New York, NY: Wiley.

13 Special Relativity, Electrodynamics, Least Action, and Hamiltonians

13.1 Special Relativity—Minkowski Space and Invariant Space–Time Distance

In this chapter, we cover special relativity, electrodynamics, least action, Euler–Lagrange equations, and Lagrangians and Hamiltonians. Topics covered include Minkowski space and the invariant space–time distance, length contraction, and time dilation. Then we discuss the Lorentz transformation of space–time position, four-vectors of position and velocity, and the four-vector and Lorentz transformation representation in Minkowski space. Velocity, mass, momentum, and energy are then examined in four-vector space. Next, we derive the radiated power from an accelerated charge moving at a nonrelativistic velocity, using the nonrelativistic Larmor formula and Lorentz transformation, applied at each instant of time. Covered next are electromagnetic four-vectors, electromagnetic fields and potentials as four-vectors, the electromagnetic field tensor, the Maxwell stress tensor, and force density. Then, the principle of least action and the Lagrangian are used to obtain the Euler–Lagrange equations, and the Hamiltonian is derived from the Lagrangian. Next, the Lagrangian for a relativistic charged particle in an electromagnetic field is derived. Finally, the Euler–Lagrange equation for the Langrangian density for real scalar fields, which is a starting point in quantum field theory, is derived. This chapter has drawn upon a number of sources, which include Adams (1997), Becker (1964), Feynman (1964), Foster and Nightingale (1995), French (1966), Jackson (1975), Kleppner and Kolenkow (1973), Landau and Lifshitz (1975), Lawson (1977), Marion (1965a,b), Panofsky and Phillips (2005), Pathria (2003), Pauli (1981), Purcell (1965), Schiff(1968), Schwartz (1972), Shadowitz (1988a,b), Sommerfeld (1964), Stratton (1941), and Reiser (1994).

In this chapter on special relativity, as a result of the finite speed of light limit for energy and matter, we will show a stationary observer (earth ground station) measures a moving, charged object (satellite) to be heavier (more massive), shorter in length, with a slower clock, and a higher charge density, than is measured by an observer at rest on the moving charged object (satellite). That is, stationary

Neutron and X-ray Optics. DOI: http://dx.doi.org/10.1016/B978-0-12-407164-3.00012-7

observers measure moving objects as heavier, shorter, and slower-clocked, than the observer at rest on the moving object.

Special relativity is a descendant of the Pythagorean Theorem (see Friedrichs, 1975), which is applied to the 4-dimensional space-time. Special relativity postulates that the speed of light c is constant in all inertial frames and that the laws of nature are the same in all inertial frames. An inertial frame has no net force, and thus the inertial frame moves at a constant velocity. The space–time distance interval, which is measured for a physical event, must produce the same measurement in all inertial frames, where

$$ds^2 = dx_0^2 + dy_0^2 + dz_0^2 - c^2\, dt_0^2 = dx^2 + dy^2 + dz^2 - c^2\, dt^2 \tag{13.1}$$

Examination of Eq. (13.1), suggests a 4-D space, and the use of a four-vector (x, y, z, ict) for space–time "distance", rather than the 3-D space of Newtonian mechanics, which uses time t and independent three-vector (x, y, z) for spatial distance.

In Newtonian mechanics, for an inertial frame moving at velocity v, the velocity of light relative to the frame is $\Delta v = c - v$, whereas in special relativity, the speed of light relative to all inertial frames is $\Delta v = c$. Newtonian mechanics also expected all the laws of nature to be the same in all inertial frames (as that was common experience and sense), time was invariant in all frames, and distance interval ds_N was invariant in all directions, where

$$ds_N^2 = dx_0^2 + dy_0^2 + dz_0^2 = dx^2 + dy^2 + dz^2 \tag{13.2}$$

To measure a proper time interval dt requires the observer to be at the exact same single spatial point (x, y, z) during the entire course of the time interval, and to measure a proper distance interval ds requires the observer to measure the entire length of ruler, or at least the two endpoints at the exact same single point in time t, where

$$ds = \sqrt{dx^2 + dy^2 + dz^2} \tag{13.3}$$

When two inertial frames are moving at different velocities, the two frames have a relative velocity, yet the speed of light c and the outcome of nature's laws must be exactly the same. An observer at rest in either frame can make the simultaneous measurement of the endpoints of a ruler at the same time to determine length. The observer can also measure an interval of time by performing the time measurement at a single spatial point.

Consider a pair of frames, moving (train) frame 1 and stationary (train platform) frame 2, which move at relative velocity v. The train-based observer, who is at rest on the moving frame 1 (train), measures length dL at a single time point t. However, the stationary (train platform) frame 2 observer, measures a nonzero time interval, where $dt_0 > 0$, during the course of the measurement of length dL by the frame 1 (train-based) observer. Next, the train-based observer of frame 1 measures time interval $\mathrm{d}t$ at a single spatial point (x, y, z). However, the stationary (train platform) frame 2 observer measures a nonzero length interval, where $\mathrm{d}L_0 > 0$, during the course of the measurement of time interval $\mathrm{d}t$ by the frame 1 (train-based) observer. In summary:

$$\mathrm{d}s = \mathrm{d}L = \sqrt{\mathrm{d}L_0^2 - c^2 t_0^2} = \mathrm{d}s_0 \quad \text{and} \quad \mathrm{d}s = \mathrm{i}ct = \sqrt{\mathrm{d}L_0^2 - c^2 t_0^2} = \mathrm{d}s_0 \tag{13.4}$$

In the Minkowski space, the magnitude of a 4-D space–time element of length $\mathrm{d}s_0$, or four-vector, is unchanged or invariant in a Lorentz transformation, or rotation about a point in 4-D space. This is similar to a 3-component spatial vector in Newtonian mechanics, which is invariant under rotation about a point in 3-D space, where time is independent and detached. The four-vector magnitude is also invariant under translation in the Lorentz transformation and Minkowski space, because the separation component distance between two points along each projection to an orthogonal axis is constant. The Lorentz transformation of spatial intervals (length) and time intervals (elapsed time) is obtained from the Minkowski space four-vector $(x, y, z, \mathrm{i}ct)$ for space–time "distance."

The space–time interval (distance) $\mathrm{d}s$ for a physical event is measured by an observer at rest on an inertial frame 1, which moves at constant velocity v relative to inertial frame 2. The observer at rest in inertial frame 2, which moves at constant velocity v relative to inertial frame 1, measures the space–time interval $\mathrm{d}s_0$ for the same physical event. The inertial stationary (train platform) frame observer and moving (train) frame observer, which move at constant relative velocity v, observe the same single physical event, and measure the same space–time distance $\mathrm{d}s_0 = \mathrm{d}s$. In terms of the 4-vector, space–time coordinates of the stationary (train platform) observer, $(x_0, y_0, z_0, \mathrm{i}ct_0)$, the invariance of the measurement of space–time distance for the same physical event by different inertial frame observers, is expressed as

$$\mathrm{d}s^2 = \mathrm{d}s_0^2 = \sqrt{\mathrm{d}x_0^2 + \mathrm{d}y_0^2 + \mathrm{d}z_0^2 - c^2\,\mathrm{d}t_0^2} \tag{13.5}$$

and

$$\mathrm{d}L_0^2 = \mathrm{d}x_0^2 + \mathrm{d}y_0^2 + \mathrm{d}z_0^2 \tag{13.6}$$

Substitution of Eq. (13.6) into Eq. (13.5), yields

$$\mathrm{d}s^2 = \mathrm{d}s_0^2 = \sqrt{\mathrm{d}L_0^2 - c^2\,\mathrm{d}t_0^2} \tag{13.7}$$

The moving (train) frame $(x, y, z, \mathrm{i}ct)$ moves at velocity v, which is measured relative to the stationary (train platform) frame. On the moving (train) frame, the train observer at rest on the moving frame (train) measures proper time interval $\mathrm{d}t$ and proper spatial interval $\mathrm{d}L$, where

$$\mathrm{d}s^2 = \mathrm{d}s_0^2 = \sqrt{\mathrm{d}x^2 + \mathrm{d}y^2 + \mathrm{d}z^2 - c^2\,\mathrm{d}t^2} \tag{13.8}$$

$$\mathrm{d}L^2 = \mathrm{d}x^2 + \mathrm{d}y^2 + \mathrm{d}z^2 \tag{13.9}$$

Substitution of Eq. (13.9) into Eq. (13.8), yields

$$\mathrm{d}s^2 = \mathrm{d}s_0^2 = \sqrt{\mathrm{d}L^2 - c^2\,\mathrm{d}t^2} \tag{13.10}$$

The invariant space–time intervals of the stationary (train platform) frame and moving (train) frame of Eqs. (13.7) and (13.10), are equated:

$$\mathrm{d}s_0^2 = \sqrt{\mathrm{d}L_0^2 - c^2\,\mathrm{d}t_0^2} = \sqrt{\mathrm{d}L^2 - c^2\,\mathrm{d}t^2} = \mathrm{d}s^2 \tag{13.11}$$

13.2 Special Relativity—Length Contraction

The time interval $\mathrm{d}t_0$ in a stationary frame (earth ground station), is measured by an observer at rest at a particular spatial point L_0 in the stationary frame, where $\mathrm{d}L_0 = 0$ in Eq. (13.11). On the moving frame (satellite), the proper time t and time interval $\mathrm{d}t$, and proper length interval L and length interval $\mathrm{d}L$ of objects at rest on the moving frame are measured by an satellite-based observer at rest on the moving frame. From Eq. (13.11), with $\mathrm{d}L_0 = 0$, one obtains the relation between the moving frame (satellite) proper time interval $\mathrm{d}t$ and the stationary (earth ground station) frame time interval $\mathrm{d}t_0$, where

$$\mathrm{i}c\mathrm{d}t_0 = \mathrm{d}t\sqrt{\frac{\mathrm{d}L^2}{\mathrm{d}t^2} - c^2} \tag{13.12}$$

The moving frame (satellite) observer, who at rest in the moving frame, measures the stationary frame (earth ground station) cross a spatial distance $\mathrm{d}L$ in time interval $\mathrm{d}t$. The moving frame (satellite) observer concludes that the stationary frame (earth ground station) is moving at velocity $\mathrm{d}L/dt$, which is equal to the relative velocity v between the stationary frame (earth ground station) and the moving frame (satellite), where

$$v = \frac{\mathrm{d}L}{\mathrm{d}t} \tag{13.13}$$

Substitution of Eq. (13.13) into Eq. (13.12), gives

$$\mathrm{i}c\mathrm{d}t_0 = \mathrm{d}t\sqrt{v^2 - c^2} = \mathrm{i}\,\mathrm{d}t\sqrt{c^2 - v^2} \tag{13.14}$$

Transposing Eq. (13.14), the proper time interval dt, which is measured by a clock at rest in the moving (satellite) frame, is then related to the stationary (earth ground station) frame time interval dt_0, which is measured by a clock at rest in the stationary (earth ground base) frame, where

$$\mathrm{d}t_0 = \mathrm{d}t\sqrt{1 - \frac{v^2}{c^2}} \tag{13.15}$$

One now multiplies both sides of Eq. (13.15) by speed of light c, where

$$c\,\mathrm{d}t_0 = c\,\mathrm{d}t\sqrt{1 - \frac{v^2}{c^2}} \tag{13.16}$$

In the moving (satellite) frame, the time interval dt and length interval dl of an object at rest in the moving frame (satellite) are measured by an observer and clock, which are at rest in the moving frame. The speed of light moves at the vacuum speed limit c, relative to the moving frame (satellite). The stationary observer on the moving frame (satellite) measures length dl of an object at rest in the moving frame, via the transit time dt of light required to cross the object length dl, where

$$\mathrm{d}l = c\mathrm{d}t \tag{13.17}$$

In the stationary (earth ground base) frame, the time interval dt_0 and length interval dl_0 of an object at rest in the stationary frame (earth ground station) are measured by an observer and clock, which are at rest in the stationary frame. The speed of light moves at the vacuum speed limit c, relative to the stationary frame (earth ground base). The stationary observer on the stationary frame (earth ground base) measures length dl_0 of an object at rest in the moving frame, via the transit time dt_0 of light required to cross the object length dl_0, where

$$\mathrm{d}l_0 = c\mathrm{d}t_0 \tag{13.18}$$

With substitution of Eqs. (13.17) and (13.18) into Eq. (13.16), the proper length interval dl, which is measured by an observer at rest in the moving (satellite) frame, is then related to the stationary (earth ground station) frame length interval dl_0, which is measured by an observer at rest in the stationary (earth ground base) frame, where

$$\mathrm{d}l_0 = \mathrm{d}l\sqrt{1 - \frac{v^2}{c^2}} \tag{13.19}$$

The observer in the stationary (earth ground station) frame measures a longer length dl_0 for a moving object (satellite), than the measured length dl by the moving frame, satellite-based observer, who is at rest on the moving object (satellite), where $dl_0 > dl$. Moving objects contract in length in their direction of motion relative to a stationary frame. Transposing Eq. (13.19), one obtains the measured or existing length in the rest frame in terms of the measured or existing length in the moving frame. One introduces the ratio β, which is the relative speed of the two frames v, divided by the speed of light c, where

$$\beta = \frac{v}{c} \tag{13.20}$$

A proper length interval dl in a moving frame (satellite), is measured by a moving frame observer (satellite-based observer) at a single time point t in the moving frame (satellite), where $dt = 0$, and substitution of Eq. (13.20) into Eq. (13.19), gives

$$dl_0 = dl\sqrt{1-\beta^2} = \frac{dl}{\gamma} \tag{13.21}$$

Here, we introduce the Lorentz factor γ:

$$\gamma = \frac{1}{\sqrt{1-\beta^2}} \tag{13.22}$$

13.3 Special Relativity—Time Dilation

We now return to the relationship between the measured time intervals in a stationary (earth ground station) frame and a moving (satellite) frame, which have relative velocity v. The proper time interval dt in the moving (satellite) frame, which is measured by a clock at rest in the moving frame, is related to the stationary (earth ground station) time interval dt_0, which is measured by a clock at rest in the stationary frame, where

$$dt_0 = dt\sqrt{1-\beta^2} = \frac{dt}{\gamma} \tag{13.23}$$

Transposing Eq. (13.23), the proper time interval dt, which is measured by a clock at rest in the moving (satellite) frame, is expressed in terms of the relative velocity v between the frames, via $\beta = v/c$ and the time interval dt_0, which is measured by a clock at rest in the stationary (earth ground station) frame, where

$$dt = \frac{dt_0}{\sqrt{1-\beta^2}} = \gamma \cdot dt_0 \tag{13.24}$$

The proper time interval dt, which is measured by the observer at rest in the moving frame (satellite), is larger than the time interval dt_0, which is measured by an observer at rest in the stationary (earth ground base) frame, where $\mathrm{d}t > \mathrm{d}t_0$. That is, moving clocks run slow (i.e., time dilation). From the perspective of an observer, who is at rest in either frame, the observer perceives the other frame as moving at velocity v, and the observer measures the elapsed time of the moving clock to be less than that of the stationary clock. The proper length interval $\mathrm{d}l$, and the proper time interval $\mathrm{d}t$ of the moving frame object (satellite), are both measured by the moving frame based observer (satellite-based observer), who is at rest in the moving frame (satellite).

13.4 Special Relativity—Lorentz Transformation of Space–Time Position

The length $\mathrm{d}l_0$ and time $\mathrm{d}t_0$ intervals in the stationary (earth-bound observer) frame, as exists or measured by an observer at rest in the stationary frame, are related to the length $\mathrm{d}l$ and time $\mathrm{d}t$ intervals in the moving frame, as exists or is measured by the satellite-bound observer, who is at rest on the moving frame (satellite), by the following two transformations:

$$\mathrm{d}l_0 = \gamma \cdot \mathrm{d}l \tag{13.25}$$

and

$$\mathrm{d}t_0 = \gamma \cdot \mathrm{d}t \tag{13.26}$$

If $v > 0$, then $\gamma > 1$, and

$$\mathrm{d}l_0 > \mathrm{d}l \quad \text{and} \quad \mathrm{d}t_0 > \mathrm{d}t \tag{13.27}$$

The time interval and spatial interval (length) of the moving frame object (satellite), are measured as $\mathrm{d}t_0$ and $\mathrm{d}l_0$ by the stationary (earth-based) frame observer, where $\mathrm{d}t_0$ and $\mathrm{d}l_0$ are each larger than the respective time interval $\mathrm{d}t$ and spatial interval (length) $\mathrm{d}l$ of the moving object (satellite), which is measured by the moving frame (satellite-based) observer, who is at rest on the moving frame (satellite). And vice versa by relativity of relative motion, which does not change the physical laws and observations in inertial frames, the "observers at rest" in a moving frame likewise measure larger time intervals and larger lengths than those measured by "other observers" moving relative to those "observers at rest."

Next, one considers the position transformation between the two frames, where the previously derived relation between the measured lengths is invoked, so that

$$l_0 = \gamma \cdot l \rightarrow z_0 = \gamma \cdot z \tag{13.28}$$

If one considers an elapsed time t in the moving (satellite-based observer) frame F, then the origin of the moving frame F moves to vt. Position z in the moving frame F is measured relative to the F frame origin $z = 0$. If the moving F frame origin translates to vt, then one translates $z \rightarrow z - vt$, so as to measure z from the F frame origin. The measured distance z_0 in the stationary (earth-bound observer) frame F_0, is obtained from moving frame (satellite-based observer) measurements of z and t, where

$$z_0 = \gamma \cdot (z - vt) \tag{13.29}$$

The distance z_0 measured in the stationary (earth-bound observer) frame F_0 is larger than that of z measured in the moving (satellite) frame F, which is evident if $t = 0$, because if $v > 0$, then $\gamma > 1$ and $z_0 > z$.

Alternatively, one can derive this result and obtain the expression for γ, via the requirement that the two frames have a common single relative velocity v in the z-direction, and invoke the simplest linear relations

$$z_0 = k \cdot (z - vt) \tag{13.30}$$

$$z = k_0 \cdot (z_0 + vt_0) \tag{13.31}$$

Substitution of Eq. (13.30) for z_0 into Eq. (13.31) for z gives

$$z = k_0 \cdot [k \cdot (z - vt) + vt_0] \tag{13.32}$$

Solve for t_0:

$$t_0 = k\left[t - \frac{z}{v}\left(1 - \frac{1}{kk_0}\right)\right] \tag{13.33}$$

The relative velocity v in the z-direction, the invariance of the four-vector length, and the speed of light in all frames, requires

$$z^2 - c^2t^2 = z_0^2 - c^2t_0^2 \tag{13.34}$$

Substitution into Eq. (13.34) of Eqs. (13.30) and (13.33) for the t_0 and z_0 organizing terms, gives

$$z^2\left[1 - k^2 + \frac{k^2c^2}{v^2}\left(1 - \frac{1}{kk_0}\right)^2\right] - t^2c^2\left(1 + k^2\frac{v^2}{c^2} - k^2\right) + 2zt\frac{k^2}{v}\left[v^2 - c^2\left(1 - \frac{1}{kk_0}\right)\right] = 0 \tag{13.35}$$

In Eq. (13.35), for arbitrary z and t, the coefficients of z^2, t^2, and zt must each be zero, where

$$1 + k^2\frac{v^2}{c^2} - k^2 \Rightarrow k = \gamma = \frac{1}{\sqrt{1 - (v/c)^2}} \tag{13.36}$$

$$v^2 - c^2\left(1 - \frac{1}{kk_0}\right) \Rightarrow k_0 = \gamma \tag{13.37}$$

$$1 - k^2 + \frac{k^2c^2}{v^2}\left(1 - \frac{1}{kk_0}\right)^2 = 0 \Rightarrow k_0 = \gamma \tag{13.38}$$

Substitution of k and k_0 of Eqs. (13.36) and (13.37) into Eqs. (13.30) and (13.31) for z_0 and z, gives

$$z_0 = \gamma(z - vt) \tag{13.39}$$

$$z = \gamma(z_0 + vt_0) \tag{13.40}$$

Next considered is the time transformation between the two frames. As expected, for relative motion between two frames, each is moving in the opposite direction of the other. Moving (satellite) frame F moves with speed v in the positive z-direction, relative to the stationary (earth ground station) frame F_0, and the stationary frame F_0 moves with speed v in the negative z-direction, relative to moving frame F. We now consider the position transformation from the moving frame F to the stationary frame F_0, and the transformation from stationary frame F_0 to moving frame F, where

$$z_0 = \gamma \cdot (z + vt) \tag{13.41}$$

and

$$z = \gamma \cdot (z_0 - vt_0) \tag{13.42}$$

Solving Eq. (13.42) for t_0, one obtains

$$t_0 = \frac{z_0}{v} - \frac{z}{v\gamma} \tag{13.43}$$

Substitution of z_0 from Eq. (13.41) into Eq. (13.43) for t_0, then gives the stationary frame (earth-bound observer) time t_0 in terms of the moving frame (satellite-bound observer) position z and time t, where

$$t_0 = \frac{\gamma \cdot (z + vt)}{v} - \frac{z}{v\gamma} \tag{13.44}$$

Rearrangement of Eq. (13.44), gives

$$t_0 = t\gamma + \frac{z(\gamma^2 - 1)}{v\gamma} \tag{13.45}$$

or

$$t_0 = t\gamma + \frac{z}{v}\gamma\beta^2 \tag{13.46}$$

With substitution of Eq. (13.20) into Eq. (13.46), one obtains

$$t_0 = \gamma\left(t + \frac{zv}{c^2}\right) \tag{13.47}$$

13.5 Special Relativity—Lorentz Transformation with Position and Velocity Four-Vectors

Lorentz showed that the form of Maxwell's equations were unchanged with the above-mentioned Lorentz transformations for position and time, and then Albert Einstein followed a suggestion by Poincaré, that all physical laws should be invariant under the above-mentioned Lorentz transformation for position and time. Since Newtonian mechanics was not invariant under the Lorentz transformation, Einstein developed special relativity to provide the correct laws of motion that were invariant under the Lorentz transformation. The 4×4 Lorentz transformation matrix Γ for transforming from the four-vector (x, y, z, ict) of the moving frame (satellite) to the four-vector (x_0, y_0, z_0, ict_0) of the observer in the stationary frame (earth-bound observer), is then

$$\begin{bmatrix} x_0 \\ y_0 \\ z_0 \\ ict_0 \end{bmatrix} = \Gamma \cdot \begin{bmatrix} x \\ y \\ z \\ ict \end{bmatrix} \tag{13.48}$$

The Lorentz transformation matrix is

$$\Gamma = \begin{bmatrix} 1 & 0 & 0 & 0 \\ 0 & 1 & 0 & 0 \\ 0 & 0 & \gamma & -i\beta\gamma \\ 0 & 0 & i\beta\gamma & \gamma \end{bmatrix} \tag{13.49}$$

where

$$\beta = \frac{v_z}{c} \tag{13.50}$$

One can obtain the relativistic transformation of 4-velocity $\mathbf{u}$ in the moving (satellite) frame to 4-velocity $\mathbf{u}_0$ in the stationary (earth ground base) frame, where the moving frame moves relative to the stationary frame at velocity v_z along the

z-direction. The differential for the transformation of the position-time 4-vector $\mathbf{s}$ in the moving frame to the position-time 4-vector $\mathbf{s}_0$ in the stationary frame is

$$\begin{bmatrix} \mathrm{d}x_0 \\ \mathrm{d}y_0 \\ \mathrm{d}z_0 \\ \mathrm{i}c\mathrm{d}t_0 \end{bmatrix} = \Gamma \cdot \begin{bmatrix} \mathrm{d}x \\ \mathrm{d}y \\ \mathrm{d}z \\ \mathrm{i}c\ \mathrm{d}t \end{bmatrix} \tag{13.51}$$

Dividing both sides of Eq. (13.51) by $\mathrm{d}t_0$, yields

$$\begin{bmatrix} u_{x0} \\ u_{y0} \\ u_{z0} \\ \mathrm{i}c \end{bmatrix} = \begin{bmatrix} \frac{\mathrm{d}x_0}{\mathrm{d}t_0} \\ \frac{\mathrm{d}y_0}{\mathrm{d}t_0} \\ \frac{\mathrm{d}z_0}{\mathrm{d}t_0} \\ \mathrm{i}c \end{bmatrix} = \Gamma \cdot \begin{bmatrix} \frac{\mathrm{d}x}{\mathrm{d}t_0} \\ \frac{\mathrm{d}y}{\mathrm{d}t_0} \\ \frac{\mathrm{d}z}{\mathrm{d}t_0} \\ c\frac{\mathrm{d}t}{\mathrm{d}t_0} \end{bmatrix} \tag{13.52}$$

Application of the chain rule to Eq. (13.52), gives

$$\begin{bmatrix} \frac{\mathrm{d}x}{\mathrm{d}t_0} \\ \frac{\mathrm{d}y}{\mathrm{d}t_0} \\ \frac{\mathrm{d}z}{\mathrm{d}t_0} \\ c\frac{\mathrm{d}t}{\mathrm{d}t_0} \end{bmatrix} = \begin{bmatrix} \frac{\mathrm{d}x}{\mathrm{d}t}\frac{\mathrm{d}t}{\mathrm{d}t_0} \\ \frac{\mathrm{d}y}{\mathrm{d}t}\frac{\mathrm{d}t}{\mathrm{d}t_0} \\ \frac{\mathrm{d}z}{\mathrm{d}t}\frac{\mathrm{d}t}{\mathrm{d}t_0} \\ c\frac{\mathrm{d}t}{\mathrm{d}t}\frac{\mathrm{d}t}{\mathrm{d}t_0} \end{bmatrix} \tag{13.53}$$

From Eq. (13.53), one obtains velocity transformation as follows:

$$\begin{bmatrix} u_{x0} \\ u_{y0} \\ u_{z0} \\ ic \end{bmatrix} = \Gamma \cdot \begin{bmatrix} u_x\frac{\mathrm{d}t}{\mathrm{d}t_0} \\ u_y\frac{\mathrm{d}t}{\mathrm{d}t_0} \\ u_z\frac{\mathrm{d}t}{\mathrm{d}t_0} \\ c\frac{\mathrm{d}t}{\mathrm{d}t_0} \end{bmatrix} \tag{13.54}$$

Given the relative velocity v between the stationary (earth ground station) and moving (satellite) frames, from Eq. (13.47), the differential expression dt_0 for the time t_0, which is measured by the stationary (earth ground station) frame observer, is expressed in terms of time t and axial position z, which is measured by the satellite-based observer, who is at rest on the moving frame (satellite), where

$$dt_0 = \gamma\left(dt + \frac{\beta}{c}dz\right) \tag{13.55}$$

Dividing both sides of Eq. (13.55) by dt, yields

$$\frac{dt_0}{dt} = \gamma\left(1 + \frac{\beta}{c}u_z\right) = \gamma\left(1 + \frac{v_z}{c^2}u_z\right) \tag{13.56}$$

where

$$u_z = \frac{dz}{dt} \tag{13.57}$$

With substitution of Eq. (13.56) into Eq. (13.54), the velocity transformation from $\mathbf{u}$ to $\mathbf{u}_0$ for two frames moving at relative velocity v_z along the z-direction becomes

$$\begin{bmatrix} u_{x0} \\ u_{y0} \\ u_{z0} \\ ic \end{bmatrix} = \frac{\Gamma}{\gamma\left(1 + \frac{v_z}{c^2}u_z\right)} \cdot \begin{bmatrix} u_x \\ u_y \\ u_z \\ ic \end{bmatrix} \tag{13.58}$$

13.6 Special Relativity—Four-Vector and Lorentz Transformation Representation in Minkowski Space

The Minkowski space four-vector for space−time is represented by space−time four-vector $\mathbf{s}$ and its differential $d\mathbf{s}$, which are specified by four orthogonal or "perpendicular" directions, where

$$\mathbf{s} = (x, y, z, ict) \tag{13.59}$$

and

$$d\mathbf{s} = (dx, dy, dz, ic\,dt) \tag{13.60}$$

The Minkowski space can be represented by the 2-D complex plane, where the Lorentz transformation represents a rotation of a four-vector about the origin of the Minkowski space. In the Minkowski complex plane, the vertical y-axis is the imaginary axis. That is, imaginary number i represents a 90° counterclockwise rotation from the real number, horizontal x-axis of the complex plane. In the 2-D Minkowski complex plane, the three-vector spatial component magnitude of the complex four-vector is projected onto the horizontal, real x-axis, and the single temporal component magnitude of the complex four-vector is projected onto the vertical y-axis.

The three-vector horizontal spatial component in the complex plane is obtained from the magnitude of the spatial three-vectors in 3-D Euclidian space, which represents our physically perceived 3-D space, where spatial position is represented by (x, y, z) for the perpendicular x-, y-, and z-axis. The Minkowski space is actually a 4-D space with three spatial dimensions bundled along the x-axis and the temporal component along the y-axis. The three dimensions of the object are represented by the horizontal x-coordinate, and the object propagates forward in time in the direction of the positive, vertical imaginary y-axis or time axis in the 2-D depiction of the Minkowski 4-D space.

The projection of a 3-D object onto the 2-D horizontal xy-plane, allows the vertical z-axis to represent the imaginary time axis in a 3-D depiction of Minkowski 4-D space. This 3-D depiction of the 4-D Minkowski space, allows good visualization of the trajectory of a particle or photon in 4-D space–time. This depiction shows the order of physical events, which is permitted by the speed of light limit for the transmission of energy and matter. If energy or matter is represented by a wave picture, then the speed of light is the limitation of the group velocity. The group velocity represents the propagation velocity of a modulation of a pure sine wave. If there is no modulation of the amplitude, frequency, or wavelength of the sine wave, which indicates an absence of the propagation of energy or matter, then the phase velocity of the sine wave can exceed the speed of light.

The space–time vector **s** measures the position in space–time relative to the origin of the complex plane. The differential space–time vector d**s** connects two points in the Minkowski four-space of space and time. The magnitudes of the space–time vector **s** and its differential d**s** are distances in the Minkowski four-space. These magnitudes or distances do not change upon translation or rotation. A different four-vector coordinate system in Minkowski space is obtained by either a trivial translation or less trivial rotation of an initial four-vector coordinate system. The magnitude or distance s and differential interval magnitude ds in the four-vector Minkowski space are

$$s = \sqrt{x^2 + y^2 + z^2 - c^2t^2} \tag{13.61}$$

and

$$\mathrm{d}s = \sqrt{\mathrm{d}x^2 + \mathrm{d}y^2 + \mathrm{d}z^2 - c^2\,\mathrm{d}t^2} \tag{13.62}$$

The 4-vector, differential time interval $d\tau$ is also invariant, because $d\tau$ is obtained from the differential, invariant, distance interval ds of Eq. (13.62), which is divided by the constant speed of light c, where

$$d\tau = \frac{ds}{c} = \frac{1}{c}\sqrt{dx^2 + dy^2 + dz^2 + (icdt)^2} \tag{13.63}$$

Rearrangement of Eq. (13.63) gives

$$d\tau = \sqrt{\frac{dx^2 + dy^2 + dz^2}{c^2} + (idt)^2} \tag{13.64}$$

The invariant distance interval ds, and the invariant time interval $d\tau$, represent four-vector magnitudes, which do not change upon a Lorentz transformation. The Lorentz transformation is a translation and/or rotation of the four-vector space–time coordinate system relative to a fixed origin. Translation is trivial, but rotation of axes to produce new coordinate systems is less so. The Lorentz transformation allows the separation distances between space–time points and measured and physical quantities in space–time to be invariant under translation and/or rotation of coordinate systems, such that the coordinate systems can be moving at constant velocities relative to one another.

Analogous to a circle, which is centered about the origin in the xy-plane, the radius of a circle about its origin in the complex, 2-D complex plane, represents the spatial and temporal components of the Minkowski four-vector. The complex plane radius is also the hypotenuse of a triangle, whose projection lengths on the real x-axis and imaginary y-axis, changes, as the radius rotates about the origin. However, the radius or hypotenuse is of constant or invariant length, as it rotates about the origin. The projection on the x-axis is analogous to the purely spatial component of the four-vector magnitude or distance, where

$$ict = 0 \quad \text{and} \quad ic\,dt = 0 \tag{13.65}$$

and

$$ds = \sqrt{x^2 + y^2 + z^2} \quad \text{and} \quad ds = \sqrt{dx^2 + dy^2 + dz^2} \tag{13.66}$$

The projection on the y-axis is analogous to the purely time component of the four-vector magnitude or distance, where

$$x = y = z = 0 \quad \text{and} \quad dx = dy = dz = 0 \tag{13.67}$$

$$\sqrt{x^2 + y^2 + z^2} = 0 \quad \text{and} \quad \sqrt{dx^2 + dy^2 + dz^2} = 0 \tag{13.68}$$

From Eq. (13.63), one obtains the imaginary time component, where

$$\tau = it \quad \text{and} \quad d\tau = i\,dt \tag{13.69}$$

Thus, the Pythagorean theorem analogously applies to special relativity, where the square of the hypotenuse magnitude equals the squared magnitudes of the x- and y-projections of the hypotenuse (i.e., squared lengths of the sides of the triangle equal the squared hypotenuse length).

13.7 Special Relativity—Velocity, Mass, Momentum, and Energy

We now determine the relation between (1) the measured mass and momentum of a moving object (e.g. satellite), which is measured in a stationary frame (e.g. an earth ground base) by the earth-based observer and (2) the measured mass and momentum of the moving object (satellite), which is measured by the moving frame observer (satellite-based observer), who is at rest in the moving frame (satellite). The stationary frame observer (earth ground base) measures velocity $\mathbf{v}_0$ for the moving frame (satellite). The satellite-based observer, who is at rest in the moving frame, measures velocity $\mathbf{v}$ of objects, which move relative to the moving frame (satellite), where

$$v = \sqrt{\left(\frac{dx}{dt}\right)^2 + \left(\frac{dy}{dt}\right)^2 + \left(\frac{dz}{dt}\right)^2} = \sqrt{v_x^2 + v_y^2 + v_z^2} \tag{13.70}$$

The stationary frame observer (earth-based observer) measures velocity $\mathbf{v}_0$ for the moving frame (satellite), where the satellite moves at velocity $\mathbf{v}_0$, relative to the earth-based observer, and

$$v_0 = \sqrt{\left(\frac{dx_0}{dt}\right)^2 + \left(\frac{dy_0}{dt}\right)^2 + \left(\frac{dz_0}{dt}\right)^2} = \sqrt{v_{x0}^2 + v_{y0}^2 + v_{z0}^2} \tag{13.71}$$

The invariant time interval $d\tau$ of Eq. (13.64), becomes

$$d\tau = dt\sqrt{1 - \frac{v_0^2}{c^2}} = \frac{dt}{\gamma} \tag{13.72}$$

The four-velocity $ds/d\tau$ is formed by substitution for ds of Eq. (13.60) and $d\tau$ of Eq. (13.72):

$$\frac{ds}{d\tau} = \left(\frac{dx}{d\tau}, \frac{dy}{d\tau}, \frac{dz}{d\tau}, ic\frac{dt}{d\tau}\right) = \gamma\left(\frac{dx}{dt}, \frac{dy}{dt}, \frac{dz}{dt}, ic\right) \tag{13.73}$$

In the stationary frame (e.g. earth ground base), with substitution from Eq. (13.72), the 4-vector velocity $\mathbf{v}_0$ of the moving object (e.g. satellite), which is measured by the stationary frame observer (earth-bound observer), is

$$\mathbf{v}_0 = \left(\frac{dx}{d\tau}, \frac{dy}{d\tau}, \frac{dz}{d\tau}, ic\frac{dt}{d\tau}\right) = \gamma\left(\frac{dx}{dt}, \frac{dy}{dt}, \frac{dz}{dt}, ic\right) \tag{13.74}$$

In the moving frame, the four-vector velocity $\mathbf{v}$, which is measured by the observer of the moving frame, is

$$\mathbf{v} = \left(\frac{dx}{dt}, \frac{dy}{dt}, \frac{dz}{dt}, ic\right) = (v_x, v_y, v_z, ic) \tag{13.75}$$

By substitution of Eqs. (13.74) and (13.75) into Eq. (13.73), the stationary frame velocity four-vector $\mathbf{v}_0$ is obtained from the moving-frame-velocity four-vector $\mathbf{v}$ by

$$\mathbf{v}_0 = \gamma\mathbf{v} = \gamma(v_x, v_y, v_z, ic) \tag{13.76}$$

In classical mechanics, momentum is the product of mass and velocity, and the same is true in relativity, provided that the property of mass is associated only with the body mass and does not include other parameters such as velocity. In relativity, the mass of a body, which is independent of velocity is the rest mass m_0. The rest mass or proper m_0 is measured in an inertial frame, where the mass m is at rest.

The moving frame moves at velocity v_0, which is measured relative to the stationary frame. The stationary frame momentum 4-vector $\mathbf{p}_0$ is obtained from the moving frame momentum 4-vector $\mathbf{p}$ by multiplication of Eq. (13.76) by rest mass m_0, where

$$m_0\mathbf{v}_0 = \gamma m_0\mathbf{v} = \frac{m_0\mathbf{v}}{\sqrt{1-(v_0/c)^2}} \tag{13.77}$$

The stationary frame momentum's four-vector is

$$\mathbf{p}_0 = (p_{x0}, p_{y0}, p_{z0}, i\gamma m_0 c) = \gamma m_0(v_x, v_y, v_z, ic) = m(v_x, v_y, v_z, ic) \tag{13.78}$$

The moving frame momentum's four-vector is

$$\mathbf{p} = (p_x, p_y, p_z, im_0c) = m_0(v_x, v_y, v_z, ic) = m_0\mathbf{v} \tag{13.79}$$

With substitution of Eqs. (13.78) and (13.79) into Eq. (13.77), the stationary frame momentum's four-vector $\mathbf{p}_0$ can be expressed in terms of the moving frame momentum's four-vector $\mathbf{p}$ by

$$\mathbf{p}_0 = \gamma(p_x, p_y, p_z, im_0c) = \gamma\mathbf{p} \tag{13.80}$$

A stationary observer (e.g. earth ground station) measures mass m and velocity v_0 for a moving object (e.g. satellite); whereas, an observer at rest on the moving mass (satellite) measures the satellite rest mass m_0, and from Eq. (13.77), one obtains the relation between the moving mass m and rest mass m_0, where

$$m = \gamma m_0 = \frac{m_0}{\sqrt{1-(v_0/c)^2}} = \frac{m_0}{\sqrt{1-\beta^2}} \tag{13.81}$$

The relation between the rest mass m_0, the mass total energy E_0, and the mass momentum, spatial component P_0 of a moving object (e.g. satellite), which is measured by a stationary observer (e.g. earth ground station), can be obtained in a simple manner, following Adams (1997). One starts with Eq. (13.81), which relates (1) the rest mass m_0 of the moving object (e.g. satellite), which is measured by a satellite-based observer, who is at rest in the moving frame (satellite) with (2) the moving object mass m, which is measured by an earth-bound observer, who is in the stationary frame, where $m = \gamma m_0$. Squaring this relation, gives $m^2 = \gamma^2 m_0^2$, where $\gamma^2 = 1/(1-\beta^2)$ and $\beta = v_0/c$. The mass momentum, spatial component $P_0 = mv_0$, is that of the moving object (satellite), which is measured by the stationary observer on earth, who measures velocity v_0 for the moving object (satellite). The rest mass energy E_{rm} of the moving object (satellite) is measured by the satellite-based observer, who is at rest in the moving frame of the object (satellite), where $E_{rm} = m_0c^2$. The total energy E_0 of the moving object (satellite), which is measured by the stationary observer (earth ground station) is $E_0 = mc^2$, where $E_0 = \gamma E_{rm}$. Finally, one multiplies the squared relation $m^2 = \gamma^2 m_0^2$ by c^2, expands the terms, and makes substitutions to obtain the relativistic conservation of energy and momentum, where

$$m^2c^2 = \gamma^2 m_0^2 c^2 \Rightarrow m^2c^2 - m^2v_0^2 = m_0^2c^2 \Rightarrow \frac{E_0^2}{c^2} - P_0^2 = m_0^2c^2 \tag{13.82}$$

The rotating earth and orbiting satellite are continuously changing direction, and thereby accelerating, and are thus non-inertial frames. However, if the satellite is in a circular orbit, which is concentric with the spherical surface of earth, then the instantaneous satellite velocity vector is tangent to the satellite circular orbit and is also parallel to the earth's surface velocity vector, which is tangent to the rotating earth surface. Hence, an instantaneous translation velocity difference exists between the parallel components of the instantaneous, tangential velocity vectors of the earth observer and overhead satellite.

The 4-vector force or Minkowski force is the time derivative of the 4-vector momentum. In the stationary frame, the 4-vector force $\mathbf{F}_0$ is the time derivative of the rest frame 4-vector momentum $\mathbf{p}_0$. The rest frame momentum $\mathbf{p}_0$ is related to the time derivative of the moving frame 4-vector momentum $\mathbf{p}$, via γ, where

$$\mathbf{F}_0 = \frac{d\mathbf{p}_0}{dt} = \frac{d}{dt}(\gamma\mathbf{p}) = \frac{d}{dt}\left(\frac{m_0\mathbf{v}}{\sqrt{1-(v_0/c)^2}}\right) = \frac{d}{d\tau}(m_0\mathbf{v}) \tag{13.83}$$

In the stationary frame, work W_0, which is done on a particle by the force per unit time or power S_0, is the dot product of the force F_0 that acts on the particle, where the particle moves at velocity $\mathbf{v}_0$, relative to the stationary frame. The power S_0 equals the time rate of change of the particle kinetic energy T_0, where

$$S_0 = \frac{\mathrm{d}W_0}{\mathrm{d}t} = \mathbf{F}_0 \cdot \mathbf{v}_0 = \frac{\mathrm{d}T_0}{\mathrm{d}t} \tag{13.84}$$

Substitution of Eq. (13.83) into Eq. (13.84), gives

$$\frac{\mathrm{d}T_0}{\mathrm{d}t} = \mathbf{v}_0 \cdot \frac{\mathrm{d}}{\mathrm{d}t}\left(\frac{m_0\mathbf{v}_0}{\sqrt{1-(v_0/c)^2}}\right) \tag{13.85}$$

One performs the integration of Eq. (13.85) over time t, where at initial time $t=0$, the particle velocity and kinetic energy are zero, $\mathbf{v}_0(0)=0$ and $T_0(0)=0$, and at final time t, the particle velocity and kinetic energy are $\mathbf{v}_0(t)=\mathbf{v}_0$ and $T_0(t)=T_0$,

$$\int_0^t \frac{\mathrm{d}T_0}{\mathrm{d}t} = T_0 = \int_0^t \mathbf{v}_0 \cdot \frac{\mathrm{d}}{\mathrm{d}t}\left(\frac{m_0\mathbf{v}_0}{\sqrt{1-(v_0/c)^2}}\right) = \int_0^t \frac{\mathrm{d}}{\mathrm{d}t}\left(\frac{m_0c^2}{\sqrt{1-(v_0/c)^2}}\right) = \left[\frac{m_0c^2}{\sqrt{1-(v_0/c)^2}}\right]_0^t \tag{13.86}$$

The stationary frame observer measures particle kinetic energy T_0 and mass γm_0, where the particle of rest mass m_0 moves at velocity v_0, relative to the stationary frame, so that

$$T_0 = \frac{m_0c^2}{\sqrt{1-(v/c)^2}} - m_0c^2 = mc^2 - m_0c^2 = (\gamma-1)m_0c^2 \tag{13.87}$$

The rest mass energy E_{rm} is

$$E_{\mathrm{rm}} = m_0c^2 \tag{13.88}$$

In the stationary frame, application of Eqs. (13.87) and (13.88), shows the total particle energy E_0 is the sum of its rest mass energy E_{rm} and its kinetic energy T_0, where

$$E_0 = E_{\mathrm{rm}} + T_0 = \gamma m_0c^2 \tag{13.89}$$

From Eqs. (13.78) and (13.89), the stationary frame observer (earth ground station) measures the 4-vector momentum $\mathbf{p}_0 = \gamma(p_{x0}, p_{y0}, p_{z0}, im_0c)$, which has invariant magnitude $|\mathbf{p}_0|$, and the earth-based observer measures the moving object (satellite) total energy $E_0 = \gamma E_{\mathrm{rm}}$. In contrast, from Eqs. (13.79) and (13.88), the observer, who is at rest on the moving object (satellite observer), measures the 4-vector momentum $\mathbf{p} = (0, 0, 0, im_0c)$, which has invariant magnitude $|\mathbf{p}|$, and the satellite-based observer measures the moving object (satellite) rest mass energy $E_{\text{rm}} = m_0c^2$. From Eq. (13.80), the stationary frame measured 4-vector momentum $\mathbf{p}_0$ of the moving object (satellite) is larger than the moving frame measured 4-vector momentum $\mathbf{p}$ by the Lorentz factor γ, where $\mathbf{p}_0 = \gamma\mathbf{p}$ and $\gamma \geq 1$. The relativistic conservation of momentum and energy is obtained from the invariance of the 4-vector momentum–energy in any inertial frame, which moves at some constant velocity v_0. As with the invariance of space–time distance of Eq. (13.1), one has invariant momentum–energy magnitude, where $|\mathbf{p}_0| = |\gamma\mathbf{p}|$, which yields the relativistic conservation of energy and momentum of Eq. (13.82). Using $|\mathbf{p}_0| = |\gamma\mathbf{p}|$, one relates (1) the rest mass energy $E_{\text{rm}} = m_0c^2$, which is measured by the satellite-bound observer in the moving frame, with (2) the moving object (satellite) spatial momentum component P and total energy $E_0 = \gamma m_0c^2$, which is measured by the earth-bound observer in the stationary frame, where again, one obtains the relativistic conservation of energy and momentum:

$$\sqrt{P_0^2 + m_0^2c^2} = \sqrt{p_{x0}^2 + p_{yp}^2 + p_{z0}^2 + m_0^2c^2} = |\mathbf{p}_0| = |\gamma\mathbf{p}| = \gamma m_0 c \Rightarrow E_0 = c\sqrt{p_0^2 + m_0^2c^2} \tag{13.90}$$

13.8 Radiated Power from an Accelerated Charge Moving at Nonrelativistic Velocity via the Nonrelativistic Larmor Formula and Lorentz Transformation at an Instant of Time

Following Marion (1965a), we now determine the equation for the electromagnetic power emission from an accelerating charge. One starts with the Larmor formula for radiated power P which, in terms of the charge acceleration $\mathbf{a}$, electric charge e, and speed of light c, is given in CGS (Gaussian) units by Eq. (14.22) of Jackson (1975) and Eq. (13.87) of Marion (1965a), and in MKS units by Eq. 20-19 of Panofsky and Phillips (2005), where

$$P_{\text{cgs}} = \frac{2e^2|\mathbf{a}|^2}{3c^3} \qquad P_{\text{mks}} = \frac{e^2|\mathbf{a}|^2}{6\pi\varepsilon_0c^3} \tag{13.91}$$

The Larmor formula of Eq. (13.91) in MKS units, was derived as Eq. (1.450) in section 1.5.2 of chapter 1. In this section, we now derive the Larmor formula, which uses special relativity, where this Larmor formula derivation is valid only for low relative velocities between the moving charge e and the observer. However, this derivation is especially valid at the instants of time, where the relative velocity between the observer and moving charge is near zero. Hence, one can consider a temporary inertial frame in which for one instant of time, the velocity of charge e is zero, where $\mathbf{v} = 0$. However, in this frame, the acceleration of the charge is generally nonzero, where $\mathbf{a} \neq 0$.

Consider the charge e undergoing acceleration $\mathbf{a}$. An accelerated reference frame, where the charge e is at rest ($\mathbf{v} = 0$), is not an inertial or Lorentz frame. However, at any instant of time t, there exists a temporary Lorentz frame, where the accelerating charge is at rest. Hence, for just that instant of time, one can derive an exact expression for the electromagnetic radiation in any observer frame Q_0, which moves with velocity $\mathbf{v}$ relative to the accelerated charge rest frame Q. The Larmor formula is exact in the instantaneous reference frame Q, where the accelerating charge is at rest.

Assume that the instantaneous rest frame of the moving charge Q, and the observer rest frame Q_0, has a relative velocity magnitude v_z along the z-axis. Recall that the rest-frame-velocity four-vector $\mathbf{v}_0$, as measured by the observer, is obtained from the moving frame of the accelerating charge with instantaneous-velocity four-vector $\mathbf{v}$, which is multiplied by the Lorentz factor γ, where

$$\mathbf{v}_0 = \gamma\mathbf{v} = \frac{\mathbf{v}}{\sqrt{1 - (v_z/c)^2}} \tag{13.92}$$

The velocity four-vector $\mathbf{v}$ in the instantaneous rest frame of the accelerating charge e is

$$\mathbf{v} = (v_x, v_y, v_z, \mathrm{i}c) \tag{13.93}$$

The acceleration four-vector $\mathbf{a}$ in the instantaneous rest frame of the accelerating charge e is obtained from the velocity four-vector $\mathbf{v}$ by taking the derivative with respect to the moving frame time t, where

$$\mathbf{a} = \left(\frac{\mathrm{d}v_x}{\mathrm{d}t}, \frac{\mathrm{d}v_y}{\mathrm{d}t}, \frac{\mathrm{d}v_z}{\mathrm{d}t}, 0\right) \tag{13.94}$$

The rest-frame-velocity four-vector $\mathbf{v}_0$ of the observer is given by

$$\mathbf{v}_0 = (v_{0x}, v_{0y}, v_{0z}, \mathrm{i}c) = \gamma\mathbf{v} \tag{13.95}$$

The acceleration four-vector $\mathbf{a}_0$ in the rest frame of the observer is obtained from the velocity four-vector $\mathbf{v}_0$ by taking the derivative of Eq. (13.95), with respect to the proper time τ of the observer frame, where

$$a_0 = \frac{\mathrm{d}}{\mathrm{d}\tau}(v_{0x}, v_{0y}, v_{0z}, \mathrm{i}c) = \frac{\mathrm{d}}{\mathrm{d}\tau}(\gamma\mathbf{v}) \tag{13.96}$$

Carrying out the differentiation in Eq. (13.96), with respect to proper time τ, on the equivalent expression $\gamma\mathbf{v}$ for the observer proper velocity $\mathbf{v}_0$, one obtains the proper charge acceleration $\mathbf{a}_0$, which is measured by the observer frame Q_0, where

$$a_0=\frac{\mathrm{d}}{\mathrm{d}\tau}(\gamma\mathbf{v})=\left(\left(\gamma\frac{\mathrm{d}v_x}{\mathrm{d}\tau}+v_x\frac{\mathrm{d}\gamma}{\mathrm{d}\tau}\right),\left(\gamma\frac{\mathrm{d}v_y}{\mathrm{d}\tau}+v_y\frac{\mathrm{d}\gamma}{\mathrm{d}\tau}\right),\left(\gamma\frac{\mathrm{d}v_z}{\mathrm{d}\tau}+v_z\frac{\mathrm{d}\gamma}{\mathrm{d}\tau}\right),\mathrm{i}c\frac{\mathrm{d}\gamma}{\mathrm{d}\tau}\right) \tag{13.97}$$

where

$$\frac{\mathrm{d}\gamma}{\mathrm{d}\tau}=\frac{\mathrm{d}}{\mathrm{d}\tau}\left[1-\left(\frac{v_z}{c}\right)^2\right]^{-1/2}=\frac{\gamma^3}{c^2}\left(v_z\frac{\mathrm{d}v_z}{\mathrm{d}\tau}\right) \tag{13.98}$$

The transverse velocities are zero, where

$$v_x=v_y=0 \tag{13.99}$$

As a result, Eq. (13.97) for the proper charge acceleration $\mathbf{a}_0$, which is measured by the observer in the observer frame Q_0, becomes

$$a_0=\frac{\mathrm{d}}{\mathrm{d}\tau}(\gamma\mathbf{v})=\left(\gamma\frac{\mathrm{d}v_x}{\mathrm{d}\tau},\gamma\frac{\mathrm{d}v_y}{\mathrm{d}\tau},\gamma\frac{\mathrm{d}v_z}{\mathrm{d}\tau}+v_z\frac{\mathrm{d}\gamma}{\mathrm{d}\tau},\mathrm{i}c\frac{\mathrm{d}\gamma}{\mathrm{d}\tau}\right) \tag{13.100}$$

From Eq. (13.72), the proper time τ of the stationary observer is related to the time t of the moving rest frame of the charge by

$$\mathrm{d}t=\gamma\mathrm{d}\tau\Rightarrow\mathrm{d}\tau=\frac{\mathrm{d}t}{\gamma} \tag{13.101}$$

Substitution of Eq. (13.101) for $\mathrm{d}\tau$ on the right side Eq. (13.98) for $\mathrm{d}\gamma/d\tau$, gives

$$\frac{\mathrm{d}\gamma}{\mathrm{d}\tau}=\frac{\gamma^4}{c^2}\left(v_z\frac{\mathrm{d}v_z}{\mathrm{d}t}\right) \tag{13.102}$$

Substitution of Eqs. (13.101) and (13.102) for $\mathrm{d}\tau$ and $\mathrm{d}\gamma/d\tau$ on the right side of Eq. (13.100), yields

$$\mathbf{a}_0=\begin{bmatrix}\dfrac{\mathrm{d}v_{0x}}{\mathrm{d}\tau}\\ \dfrac{\mathrm{d}v_{0y}}{\mathrm{d}\tau}\\ \dfrac{\mathrm{d}v_{0z}}{\mathrm{d}\tau}\\ 0\end{bmatrix}=\gamma^2\begin{bmatrix}\dfrac{\mathrm{d}v_x}{\mathrm{d}t}\\ \dfrac{\mathrm{d}v_y}{\mathrm{d}t}\\ \dfrac{\mathrm{d}v_z}{\mathrm{d}t}+\dfrac{v_z\gamma^2}{c^2}\left(v_z\dfrac{\mathrm{d}v_z}{\mathrm{d}t}\right)\\ \mathrm{i}\dfrac{\gamma^2}{c}\left(v_z\dfrac{\mathrm{d}v_z}{\mathrm{d}t}\right)\end{bmatrix} \tag{13.103}$$

The acceleration $\mathbf{a}_e$ in the moving frame Q of the charge e, where the charge e is instantaneously at rest, is transformed to the charge acceleration $\mathbf{a}_0$, which is measured by the observer in the observer frame Q_0, via the Lorentz transformation matrix, where

$$\mathbf{a}_0 = \Gamma \mathbf{a}_e \tag{13.104}$$

and

$$\Gamma = \begin{bmatrix} 1 & 0 & 0 & 0 \\ 0 & 1 & 0 & \\ 0 & 0 & \gamma & \mathrm{i}\beta\gamma \\ 0 & 0 & -\mathrm{i}\beta\gamma & \gamma \end{bmatrix} \tag{13.105}$$

$$\mathbf{a}_e = \left(\frac{\mathrm{d}v_{ex}}{\mathrm{d}t}, \frac{\mathrm{d}v_{ey}}{\mathrm{d}t}, \frac{\mathrm{d}v_{ez}}{\mathrm{d}t}, 0 \right) \tag{13.106}$$

and

$$\mathbf{a}_0 = \begin{bmatrix} 1 & 0 & 0 & 0 \\ 0 & 1 & 0 & \\ 0 & 0 & \gamma & \mathrm{i}\beta\gamma \\ 0 & 0 & -\mathrm{i}\beta\gamma & \gamma \end{bmatrix} \begin{bmatrix} \dfrac{\mathrm{d}v_{ex}}{\mathrm{d}t} \\ \dfrac{\mathrm{d}v_{ey}}{\mathrm{d}t} \\ \dfrac{\mathrm{d}v_{ez}}{\mathrm{d}t} \\ 0 \end{bmatrix} \tag{13.107}$$

Substitution into Eq. (13.107) of Eq. (13.103) for $\mathbf{a}_0$, gives

$$\mathbf{a}_0 = \gamma^2 \begin{bmatrix} \dfrac{\mathrm{d}v_x}{\mathrm{d}t} \\ \dfrac{\mathrm{d}v_y}{\mathrm{d}t} \\ \dfrac{\mathrm{d}v_z}{\mathrm{d}t} + \dfrac{v_z \gamma^2}{c^2}\left(v_z \dfrac{\mathrm{d}v_z}{\mathrm{d}t} \right) \\ \mathrm{i}\dfrac{\gamma^2}{c}\left(v_z \dfrac{\mathrm{d}v_z}{\mathrm{d}t} \right) \end{bmatrix} = \begin{bmatrix} 1 & 0 & 0 & 0 \\ 0 & 1 & 0 & \\ 0 & 0 & \gamma & \mathrm{i}\beta\gamma \\ 0 & 0 & -\mathrm{i}\beta\gamma & \gamma \end{bmatrix} \begin{bmatrix} \dfrac{\mathrm{d}v_{ex}}{\mathrm{d}t} \\ \dfrac{\mathrm{d}v_{ey}}{\mathrm{d}t} \\ \dfrac{\mathrm{d}v_{ez}}{\mathrm{d}t} \\ 0 \end{bmatrix} \tag{13.108}$$

where

$$\beta = \frac{v_z}{c} \tag{13.109}$$

Since

$$\frac{\mathrm{d}v_z}{\mathrm{d}t} + \frac{v_z\gamma^2}{c^2}\left(v_z\frac{\mathrm{d}v_z}{\mathrm{d}t}\right) = \frac{\mathrm{d}v_z}{\mathrm{d}t}(1+\gamma^2\beta^2) = \frac{1}{1-\beta^2}\frac{\mathrm{d}v_z}{\mathrm{d}t} \tag{13.110}$$

With substitution of Eq. (13.110) into Eq. (13.108), one obtains

$$\mathbf{a}_0 = \begin{bmatrix} \gamma^2\dfrac{\mathrm{d}v_x}{\mathrm{d}t} \\ \gamma^2\dfrac{\mathrm{d}v_y}{\mathrm{d}t} \\ \dfrac{\gamma^2}{1-\beta^2}\dfrac{\mathrm{d}v_z}{\mathrm{d}t} \\ \mathrm{i}\beta\gamma^4\dfrac{\mathrm{d}v_z}{\mathrm{d}t} \end{bmatrix} = \begin{bmatrix} 1 & 0 & 0 & 0 \\ 0 & 1 & 0 & 0 \\ 0 & 0 & \gamma & \mathrm{i}\beta\gamma \\ 0 & 0 & -\mathrm{i}\beta\gamma & \gamma \end{bmatrix} \begin{bmatrix} \dfrac{\mathrm{d}v_{ex}}{\mathrm{d}t} \\ \dfrac{\mathrm{d}v_{ey}}{\mathrm{d}t} \\ \dfrac{\mathrm{d}v_{ez}}{\mathrm{d}t} \\ 0 \end{bmatrix} \tag{13.111}$$

In the frame Q, where the accelerating charge e is instantaneously at rest, the acceleration $\mathbf{a}_e$ in the rest frame of the charge e, is calculated from the acceleration $\mathbf{a}_0$, which is measured by the observer in frame Q_0, where

$$\begin{bmatrix} \dfrac{\mathrm{d}v_x}{\mathrm{d}t} \\ \dfrac{\mathrm{d}v_y}{\mathrm{d}t} \\ \dfrac{\mathrm{d}v_z}{\mathrm{d}t} \\ \dfrac{\mathrm{d}v_z}{\mathrm{d}t} \end{bmatrix} = \begin{bmatrix} (1-\beta^2)\dfrac{\mathrm{d}v_{ex}}{\mathrm{d}t} \\ (1-\beta^2)\dfrac{\mathrm{d}v_{ey}}{\mathrm{d}t} \\ (1-\beta^2)^{3/2}\dfrac{\mathrm{d}v_{ez}}{\mathrm{d}t} \\ (1-\beta^2)^{3/2}\dfrac{\mathrm{d}v_{ez}}{\mathrm{d}t} \end{bmatrix} \tag{13.112}$$

The squared magnitude a^2 of the acceleration of the charge e, which is measured by the observer frame, where the electron has relative velocity v_z, is

$$a^2 = \mathbf{a}\cdot\mathbf{a} = \frac{\mathrm{d}\mathbf{v}}{\mathrm{d}t}\cdot\frac{\mathrm{d}\mathbf{v}}{\mathrm{d}t} = \left(\frac{\mathrm{d}v_x}{\mathrm{d}t}\right)^2 + \left(\frac{\mathrm{d}v_y}{\mathrm{d}t}\right)^2 + \left(\frac{\mathrm{d}v_z}{\mathrm{d}t}\right)^2 \tag{13.113}$$

The squared magnitude a_e^2 of the acceleration of the charge e, which is measured in the charge rest frame, is

$$a_e^2 = \mathbf{a}_e \cdot \mathbf{a}_e = \frac{d\mathbf{v}_e}{dt} \cdot \frac{d\mathbf{v}_e}{dt} = \left(\frac{dv_{ex}}{dt}\right)^2 + \left(\frac{dv_{ey}}{dt}\right)^2 + \left(\frac{dv_{ez}}{dt}\right)^2 \tag{13.114}$$

From the above matrix expression of Eq. (13.112), one obtains

$$\frac{dv_{ex}}{dt} = \frac{1}{1-\beta^2}\frac{dv_x}{dt} \tag{13.115}$$

$$\frac{dv_{ey}}{dt} = \frac{1}{1-\beta^2}\frac{dv_y}{dt} \tag{13.116}$$

$$\frac{dv_{ez}}{dt} = \frac{1}{(1-\beta^2)^{3/2}}\frac{dv_z}{dt} \tag{13.117}$$

Substitution into Eq. (13.114) of Eqs. (13.115–13.117), yields squared magnitude a_e^2 of the acceleration of the charge e, which is measured in its rest frame, but is expressed in terms of the x, y, and z acceleration components that are measured in the observer frame, where

$$a_e^2 = \frac{1}{(1-\beta^2)^2}\left[\left(\frac{dv_x}{dt}\right)^2 + \left(\frac{dv_y}{dt}\right)^2\right] + \frac{1}{(1-\beta^2)^3}\left(\frac{dv_z}{dt}\right)^2 \tag{13.118}$$

The expression Eq. (13.118) for the squared magnitude a_e^2 of the acceleration of the charge e, can be further simplified in three steps. The first step is

$$a_e^2 = \frac{1}{(1-\beta^2)^2}\left[\left(\frac{dv_x}{dt}\right)^2 + \left(\frac{dv_y}{dt}\right)^2 + \left(\frac{dv_z}{dt}\right)^2 - \left(\frac{dv_z}{dt}\right)^2\right] + \frac{1}{(1-\beta^2)^3}\left(\frac{dv_z}{dt}\right)^2 \tag{13.119}$$

The second step is

$$a_e^2 = \frac{1}{(1-\beta^2)^2}\left(\frac{d\mathbf{v}}{dt} \cdot \frac{d\mathbf{v}}{dt}\right) + \frac{1}{(1-\beta^2)^2}\left(\frac{dv_z}{dt}\right)^2\left(\frac{1}{1-\beta^2} - 1\right) \tag{13.120}$$

The third step is substitution of Eq. (13.121) below into Eq. (13.120),

$$\frac{1}{1-\beta^2} - 1 = \frac{\beta^2}{1-\beta^2} \quad \text{and} \quad \beta = \frac{v_z}{c} \tag{13.121}$$

The result of the third step yields a simplified expression for the squared magnitude a_e^2 of the acceleration of the charge e in the rest frame of the charge:

$$a_e^2 = \frac{1}{(1-\beta^2)^2}\left[\frac{d\mathbf{v}}{dt}\cdot\frac{d\mathbf{v}}{dt} + \frac{v_z^2}{c^2(1-\beta^2)}\left(\frac{dv_z}{dt}\right)^2\right] \tag{13.122}$$

The result of Eq. (13.122) applies to the case of an axial (z-directed), relative velocity v_z between the observer frame and the instantaneous rest frame of the charge e, where

$$\mathbf{v} = (0, 0, v_z) \tag{13.123}$$

Consider the more general case, where the relative velocity between the observer frame and the rest frame of charge e has components in the three directions of x, y, and z, where

$$\mathbf{v} = (v_x, v_y, v_z) \tag{13.124}$$

In the general case of a relative velocity $\mathbf{v}$, which has components in more than one direction, Eq. (13.122) for the squared magnitude a_e^2 of the acceleration of the charge e in the charge rest frame, becomes

$$a_e^2 = \frac{1}{(1-\beta^2)^2}\left[\frac{d\mathbf{v}}{dt}\cdot\frac{d\mathbf{v}}{dt} + \frac{1}{c^2(1-\beta^2)}\left(\mathbf{v}\cdot\frac{d\mathbf{v}}{dt}\right)^2\right] \tag{13.125}$$

Substitution for the a_e^2 of Eq. (13.125) into Eq. (13.91) in CGS units for radiated power P, gives the expression for the radiated power, as measured by the observer for a charge e, which moves at an arbitrary slow or fast velocity $\mathbf{v}$ that is less than the speed of light c, relative to the observer frame, where in CGS units,

$$P = \frac{2e^2a_e^2}{3c^3} = \frac{2e^2}{3c^3(1-\beta^2)^2}\left[\frac{d\mathbf{v}}{dt}\cdot\frac{d\mathbf{v}}{dt} + \frac{1}{c^2(1-\beta^2)}\left(\mathbf{v}\cdot\frac{d\mathbf{v}}{dt}\right)^2\right] \tag{13.126}$$

Division of Eq. (13.126) for the radiated power in CGS units by $4\pi\varepsilon_0$, yields the radiated power in MKS units. If the acceleration $d\mathbf{v}/dt$ of charge e is parallel, or approximately parallel to its velocity $\mathbf{v}$, relative to the observer, then Eq. (13.126) for the radiated power P from charge e at the observer, becomes

$$P = \frac{2e^2}{3c^3(1-\beta^2)^2}\left[\left|\frac{d\mathbf{v}}{dt}\right|^2 + \frac{|\mathbf{v}|^2}{c^2(1-\beta^2)}\left|\frac{d\mathbf{v}}{dt}\right|^2\right] \tag{13.127}$$

Simplifying Eq. (13.127), the radiated power P from the parallel acceleration $\mathbf{dv}/dt$ and velocity $\mathbf{v}$ of charge e, such as the transverse oscillation of an electron in an undulator, is then

$$P = \frac{2e^2}{3c^3(1-\beta^2)^3}\left|\frac{\mathrm{d}\mathbf{v}}{\mathrm{d}t}\right|^2 \tag{13.128}$$

If the acceleration $\mathbf{dv}/dt$ of charge e is perpendicular, or approximately perpendicular to its velocity $\mathbf{v}$, relative to the observer, such as an electron moving in a circle or an arc of a synchrotron ring bending magnet, then the radiated power P of Eq. (13.127) in CGS units, becomes (see Eq. (7.89) in Marion, 1965a),

$$P = \frac{2e^2}{3c^3(1-\beta^2)^2}\left|\frac{\mathrm{d}\mathbf{v}}{\mathrm{d}t}\right|^2 \tag{13.129}$$

There is minor acceleration in the direction of the electron velocity in a wiggler or undulator, but this acceleration is negligible compared to the transverse acceleration provided by their periodic magnetic field. In circular motion, the electron acceleration in a bending magnet is perpendicular to its velocity ($\mathbf{v}\perp\mathbf{a}$); whereas, in an undulator, the transverse electron velocity and acceleration are parallel ($\mathbf{v}||\mathbf{a}$). Jerry B. Marion (1965a) derives in his book, Eqs. (13.128) and (13.129) in CGS units, by the Lienard–Wiechert potentials in his chapter 7, which apply for all β, and in his chapter 13 by special relativity for $\beta << 1$. Comparison of Eq. (13.129) for a bending magnet, and Eq. (13.128) for an undulator, shows that given the same electron, squared magnitude of acceleration $|\mathbf{a}|^2$, the undulator emitted total power P_u, exceeds the bending magnet emitted total power P_b, by a factor of γ^2, where $P_\mathrm{u} = \gamma^2 P_\mathrm{b}$, for all β.

13.9 Special Relativity—Electromagnetic Four-Vectors

Electromagnetic potential (namely, the scalar potential and the magnetic vector potential) can be combined into a four-vector, and electromagnetic waves can be expressed via four-vectors and four-vector equations. The 3-D grad operator ∇ becomes the 4-D d'Alembertian operator ∇_4, where

$$\nabla_4 = \nabla + \frac{1}{\mathrm{i}c}\frac{\partial}{\partial t} = \hat{x}\frac{\partial}{\partial x} + \hat{y}\frac{\partial}{\partial y} + \hat{z}\frac{\partial}{\partial z} - \frac{\mathrm{i}}{c}\frac{\partial}{\partial t} \tag{13.130}$$

The wave equation is then expressed as

$$\nabla_4^2\Psi = \nabla^2\Psi - \frac{1}{c^2}\frac{\partial^2\Psi}{\partial t^2} \tag{13.131}$$

where

$$\nabla_4^2 = (\nabla_4 \cdot \nabla_4) = \nabla^2 - \frac{1}{c^2}\frac{\partial^2}{\partial t^2} \tag{13.132}$$

For the current-charge density four-vector $\mathbf{J}_4(\mathbf{J}, ic\rho)$, the continuity equation becomes

$$\nabla_4 \cdot \mathbf{J}_4 = \nabla_4 \cdot (\mathbf{J}, ic\rho) = \nabla \cdot \mathbf{J} + \frac{\partial \rho}{\partial t} = 0 \tag{13.133}$$

where

$$\nabla_4 \cdot \mathbf{J}_4 = \left(\nabla + \frac{1}{ic}\frac{\partial}{\partial t}\right) \cdot (\mathbf{J}, ic\rho) \tag{13.134}$$

In relativity, the charge is invariant, but the stationary frame measured charge density of a body can vary with the velocity of the body relative to the stationary frame. The charge density, which is independent of velocity, is the rest charge density ρ_0. The rest charge density or proper charge density ρ_0 is measured in an inertial frame, where the charge ρ is at rest. The stationary frame momentum four-vector $\mathbf{J}_{40}$ is obtained from the moving frame momentum four-vector $\mathbf{J}_4$ by multiplying Eq. (13.76) by rest charge density ρ_0:

$$\rho_0 \mathbf{v}_0 = \rho_0 \gamma \mathbf{v} = \frac{\rho_0 \mathbf{v}}{\sqrt{1 - (v_0/c)^2}} \tag{13.135}$$

The stationary frame current-charge density four-vector is

$$\mathbf{J}_{40} = (J_{x0}, J_{y0}, J_{z0}, ic\gamma\rho_0) = \rho_0(v_{x0}, v_{y0}, v_{z0}, ic\gamma) = \rho_0 \mathbf{v}_0 \tag{13.136}$$

The moving-frame current-charge density four-vector is

$$\mathbf{J}_4 = (J_x, J_y, J_z, i\rho_0 c) = \rho_0(J_x, J_y, J_z, ic) = \rho_0 \mathbf{v} \tag{13.137}$$

With substitution into Eq. (13.135) of Eqs. (13.136) and (13.137), the stationary frame current-charge density four-vector $\mathbf{J}_{40}$, can be expressed in terms of the moving-frame four-vector charge density $\mathbf{J}_4$ by

$$\mathbf{J}_{40} = \gamma(J_x, J_y, J_z, i\rho_0 c) = \gamma \mathbf{J}_4 \tag{13.138}$$

or

$$\mathbf{J}_{40} = \rho_0(v_{x0}, v_{y0}, v_{z0}, ic\gamma) = \rho_0\gamma(v_x, v_y, v_z, ic) = \rho(v_x, v_y, v_z, ic) \tag{13.139}$$

From Eq. (13.139), the stationary frame charge density ρ_0, can be expressed in terms of the moving frame charge density ρ by

$$\rho_0 = \frac{\rho}{\gamma} \tag{13.140}$$

13.10 Special Relativity—Electromagnetic Fields and Potentials as Four-Vectors

Application of the Lorentz gauge of Eq. (12.55) for regions of space without charge or currents, where$\rho = 0$ and $\mathbf{J} = 0$, one obtains from Eqs. (12.54) and (12.59), the wave equations of the scalar Φ and vector $\mathbf{A}$ potentials, which describe electromagnetic wave propagation in vacuum, where

$$\nabla^2\Phi - \frac{1}{c^2}\frac{\partial^2\Phi}{\partial t^2} = 0 \tag{13.141}$$

and

$$\nabla^2\mathbf{A} - \frac{1}{c^2}\frac{\partial^2\mathbf{A}}{\partial t^2} = 0 \tag{13.142}$$

From Eqs. (12.54) and (12.59) in chapter 12, in regions of space with nonzero charge density ρ and nonzero current density $\mathbf{J}$, the wave equations are

$$\nabla^2\Phi - \frac{1}{c^2}\frac{\partial^2\Phi}{\partial t^2} = -\frac{\rho}{\varepsilon_0} \tag{13.143}$$

and

$$\nabla^2\mathbf{A} - \frac{1}{c^2}\frac{\partial^2\mathbf{A}}{\partial t^2} = -\frac{1}{c^2}\frac{\mathbf{J}}{\varepsilon_0} \tag{13.144}$$

The electric $\mathbf{E}$ and magnetic $\mathbf{B}$ fields can be expressed in terms of the scalar Φ and vector $\mathbf{A}$ potentials, where

$$\mathbf{E} = -\nabla\Phi - \frac{\partial\mathbf{A}}{\partial t} \tag{13.145}$$

and

$$\mathbf{B} = \nabla \times \mathbf{A} \tag{13.146}$$

These two wave equations for the scalar Φ and vector $\mathbf{A}$ potentials with the source charge density ρ and current density $\mathbf{J}$ source terms, and the expressions for the electric $\mathbf{E}$ and magnetic $\mathbf{B}$ field in terms of the scalar and vector potentials, are equivalent to Maxwell's four equations.

One can apply the four-vector to represent compactly the source and propagation of electromagnetic fields and potentials with the d'Alembertian operator, which is represented by the symbol ∇_4, where

$$\nabla_4 = \nabla - \frac{\mathrm{i}}{c}\frac{\partial}{\partial t} \tag{13.147}$$

The scalar Φ and vector $\mathbf{A}$ are represented as a four-vector potential $\mathbf{A}_4$, where

$$\mathbf{A}_4 = \left(\mathbf{A}, \mathrm{i}\frac{\Phi}{c}\right) = \left(A_x, A_y, A_z, \mathrm{i}\frac{\Phi}{c}\right) \tag{13.148}$$

The Lorentz gauge is expressed by the d'Alembertian, which in four-vector notation, is

$$\nabla_4 \cdot \mathbf{A}_4 = \nabla_4 \cdot \left(\mathbf{A}, \mathrm{i}\frac{\Phi}{c}\right) = 0 \tag{13.149}$$

Substitution of Eqs. (13.147) and (13.148) into Eq. (13.149), yields

$$\nabla_4 \cdot (A_x, A_y, A_z, \mathrm{i}\Phi) = \left(\nabla - \frac{\mathrm{i}}{c}\frac{\partial}{\partial t}\right) \cdot \left(A_x, A_y, A_z, \mathrm{i}\frac{\Phi}{c}\right) = 0 \tag{13.150}$$

The result of Eq. (13.150) confirms Eq. (13.149) is the Lorentz gauge, which is

$$\nabla \cdot A + \frac{1}{c^2}\frac{\partial \Phi}{\partial t} = 0 \tag{13.151}$$

The pair of inhomogeneous wave equations of Eq. (13.143) and (13.144) for the scalar Φ and vector $\mathbf{A}$ potentials, which are expressed in terms of the source charge density ρ and current density $\mathbf{J}$ in four-vector notation, are now expressed in a single four-vector wave equation:

$$\nabla_4^2 \mathbf{A}_4 = -\frac{1}{c\varepsilon_0}\mathbf{J}_4 \tag{13.152}$$

Substitution of Eqs. 13.137, 13.147, and 13.148 into Eq. (13.152), and expansion of the resulting expression, gives

$$\nabla_4^2\left(A_x, A_y, A_z, \mathrm{i}\frac{\Phi}{c}\right) = \left(\nabla^2 - \frac{1}{c^2}\frac{\partial^2}{\partial t^2}\right) \cdot \left(A_x, A_y, A_z, \mathrm{i}\frac{\Phi}{c}\right) = -\frac{1}{c^2\varepsilon_0}(J_x, J_y, J_z, \mathrm{i}c\rho) \tag{13.153}$$

From Eq. (13.153), one extracts

$$\nabla^2 \Phi - \frac{1}{c^2}\frac{\partial^2 \Phi}{\partial t^2} = -\frac{\rho}{\varepsilon_0} \tag{13.154}$$

and

$$\nabla^2 \mathbf{A} - \frac{1}{c^2}\frac{\partial^2 \mathbf{A}}{\partial t^2} = -\frac{1}{c^2}\frac{\mathbf{J}}{\varepsilon_0} \tag{13.155}$$

13.11 Special Relativity—The Electromagnetic Field Tensor

The three components (E_x, E_y, E_z) of the electric field vector $\mathbf{E}(x, y, z)$, and the three components (B_x, B_y, B_z) of the magnetic field vector $\mathbf{B}(x, y, z)$, can be expressed in terms of an antisymmetric tensor called the *electromagnetic field tensor.*

The four-vector potential $\mathbf{A}_4$, is

$$\mathbf{A}_4 = \left(\mathbf{A}, \mathrm{i}\frac{\Phi}{c}\right) = \left(A_x, A_y, A_z, \mathrm{i}\frac{\Phi}{c}\right) \tag{13.156}$$

The elements F_{uv} of the electromagnetic tensor are defined in terms of the four-vector potential $\mathbf{A}_4$ and the four-vector position $\mathbf{x}_4$, where

$$\mathbf{A}_4 = \left(\mathbf{A}, \mathrm{i}\frac{\Phi}{c}\right) = \left(A_x, A_y, A_z, \mathrm{i}\frac{\Phi}{c}\right) \tag{13.157}$$

and

$$\mathbf{x}_4 = (x, y, z, \mathrm{i}ct) \tag{13.158}$$

The elements F_{uv} of the antisymmetric electromagnetic tensor are defined by

$$F_{uv} \equiv \frac{\partial A_v}{\partial x_u} - \frac{\partial A_u}{\partial x_v} \tag{13.159}$$

where

$$F_{uv} \equiv \frac{\partial A_v}{\partial x_u} - \frac{\partial A_u}{\partial x_v} = -F_{vu} \tag{13.160}$$

and

$$F_{uu} \equiv \frac{\partial A_u}{\partial x_u} - \frac{\partial A_u}{\partial x_u} = -F_{uu} = 0 \tag{13.161}$$

Eqs. (13.159−13.161) for the definition of F_{uv} allows one to generate, via Eqs. (13.157) and (13.158), the electric $\mathbf{E}$ and magnetic $\mathbf{B}$ fields in terms of the electromagnetic scalar Φ and vector $\mathbf{A}$ potentials, where

$$\mathbf{E} = -\nabla\Phi - \frac{\partial \mathbf{A}}{\partial t} \tag{13.162}$$

and

$$\mathbf{B} = \nabla \times \mathbf{A} \tag{13.163}$$

For example, application of Eqs. (13.157−13.161), yields

$$F_{12} \equiv \frac{\partial A_2}{\partial x_1} - \frac{\partial A_1}{\partial x_2} = \frac{\partial A_y}{\partial x_x} - \frac{\partial A_x}{\partial x_y} = B_z \quad u=1, \ \ v=2 \tag{13.164}$$

$$F_{14} \equiv \frac{\partial A_4}{\partial x_1} - \frac{\partial A_1}{\partial x_4} = \frac{\mathrm{i}}{c}\frac{\partial \Phi}{\partial x_x} - \frac{1}{\mathrm{i}c}\frac{\partial A_x}{\partial t} = \frac{\mathrm{i}}{c}E_x \quad u=1, \ \ v=4 \tag{13.165}$$

From Eqs. (13.157−13.163), the entire electromagnetic tensor F with all its elements F_{uv} in matrix form, is then

$$F = \begin{bmatrix} 0 & B_z & -B_y & -\frac{\mathrm{i}}{c}E_x \\ -B_z & 0 & B_x & -\frac{\mathrm{i}}{c}E_y \\ B_y & -B_x & 0 & -\frac{\mathrm{i}}{c}E_z \\ \frac{\mathrm{i}}{c}E_x & \frac{\mathrm{i}}{c}E_y & \frac{\mathrm{i}}{c}E_z & 0 \end{bmatrix} \tag{13.166}$$

One can express the continuity equation and Maxwell's equation compactly, via the electromagnetic tensor F_{uv}, where the current density four-vector $\mathbf{J}_4$ is expressed in terms of the x, y, and z spatial components of the current density vector (J_x, J_y, J_z) and the charge density ρ, where

$$\mathbf{J}_4 = (\mathbf{J}, \mathrm{i}c\rho) = (J_x, J_y, J_z, \mathrm{i}c\rho) \tag{13.167}$$

With Eqs. (13.167) and (13.169), the continuity equation is

$$\sum_{w=1}^{4} \frac{\partial J_w}{\partial q_w} = 0 \tag{13.168}$$

The position's four-vector $\mathbf{x}_4$ is also denoted by $\mathbf{q}$, where

$$\mathbf{x}_4 = \mathbf{q} = (x, y, z, \mathrm{i}ct) \tag{13.169}$$

With Eqs. 13.157 and Eq. (13.169), the Lorentz gauge is expressed by

$$\sum_{w=1}^{4} \frac{\partial A_w}{\partial q_w} = 0 \tag{13.170}$$

The two homogeneous Maxwell equations (i.e., without charge density or current density source terms) are expressed as

$$\nabla \cdot \mathbf{B} = 0 \quad \text{and} \quad \frac{\partial \mathbf{B}}{\partial t} + \nabla \times \mathbf{E} = 0 \tag{13.171}$$

With the electromagnetic field tensor F, the two homogeneous Maxwell equations can be expressed by

$$\frac{\partial F_{uv}}{\partial q_w} + \frac{\partial F_{vw}}{\partial q_u} + \frac{\partial F_{wu}}{\partial q_v} = 0 \quad u, v, w = 1, 2, 3, 4 \tag{13.172}$$

The two inhomogeneous Maxwell's equations with charge density or current density source terms are

$$\nabla \cdot \mathbf{E} = \rho \quad \text{and} \quad -\frac{\partial \mathbf{E}}{\partial t} + \nabla \times \mathbf{B} = \mathbf{J} \tag{13.173}$$

With the electromagnetic field tensor F, the two inhomogeneous Maxwell's equations can be expressed by

$$\sum_{w=1}^{4} \frac{\partial F_{uw}}{\partial q_w} = J_u \quad u, w = 1, 2, 3, 4 \tag{13.174}$$

Finally, the Lorentz force equation for force density **f** of electric field **E** and magnetic field **B** acting on charge density ρ is

$$\mathbf{f} = \rho[\mathbf{E} + (\mathbf{v} \times \mathbf{B})] \tag{13.175}$$

In the four-vector notation with the electromagnetic tensor F, the Lorentz force equation is expressed as

$$f_i = -\frac{e}{c}\sum_{k=1}^{4} F_{ik} v_k = -\frac{1}{c}\sum_{k=1}^{4} F_{ik} J_k \quad i, j, k = 1, 2, 3, 4 \tag{13.176}$$

Next, we determine the Lorentz transformation for the electromagnetic tensor F, via the self-similarity transformation with the Lorentz transformation matrix Γ and its inverse Γ^{-1}. The product of matrix Γ and its inverse Γ^{-1} is the unity matrix I, where

$$\Gamma\Gamma^{-1} = \begin{bmatrix} 1 & 0 & 0 & 0 \\ 0 & 1 & 0 & 0 \\ 0 & 0 & \gamma & -i\beta\gamma \\ 0 & 0 & i\beta\gamma & \gamma \end{bmatrix} \begin{bmatrix} 1 & 0 & 0 & 0 \\ 0 & 1 & 0 & 0 \\ 0 & 0 & \gamma & i\beta\gamma \\ 0 & 0 & -i\beta\gamma & \gamma \end{bmatrix} = I = \begin{bmatrix} 1 & 0 & 0 & 0 \\ 0 & 1 & 0 & 0 \\ 0 & 0 & 1 & 0 \\ 0 & 0 & 0 & 1 \end{bmatrix} \tag{13.177}$$

The electromagnetic tensor F is expressed by matrix equation

$$F' = \Gamma F \Gamma^{-1} \tag{13.178}$$

$$F' = \begin{bmatrix} 1 & 0 & 0 & 0 \\ 0 & 1 & 0 & 0 \\ 0 & 0 & \gamma & -\mathrm{i}\beta\gamma \\ 0 & 0 & \mathrm{i}\beta\gamma & \gamma \end{bmatrix} \begin{bmatrix} 0 & B_z & -B_y & -\frac{\mathrm{i}}{c}E_x \\ -B_z & 0 & B_x & -\frac{\mathrm{i}}{c}E_y \\ B_y & -B_x & 0 & -\frac{\mathrm{i}}{c}E_z \\ \frac{\mathrm{i}}{c}E_x & \frac{\mathrm{i}}{c}E_y & \frac{\mathrm{i}}{c}E_z & 0 \end{bmatrix} \begin{bmatrix} 1 & 0 & 0 & 0 \\ 0 & 1 & 0 & 0 \\ 0 & 0 & \gamma & \mathrm{i}\beta\gamma \\ 0 & 0 & -\mathrm{i}\beta\gamma & \gamma \end{bmatrix} \tag{13.179}$$

Carrying out the matrix multiplication in Eq. (13.179), gives

$$F' = \begin{bmatrix} 0 & B_z & -\gamma\left(B_y + \frac{\beta}{c}E_x\right) & -\mathrm{i}\gamma\left(\beta B_y + \frac{1}{c}E_x\right) \\ -B_z & 0 & \gamma\left(B_x - \frac{\beta}{c}E_y\right) & \mathrm{i}\gamma\left(\beta B_x - \frac{1}{c}E_y\right) \\ \gamma\left(B_y + \frac{\beta}{c}E_x\right) & -\gamma\left(B_x - \frac{\beta}{c}E_y\right) & 0 & -\frac{\mathrm{i}}{c}E_z \\ \mathrm{i}\gamma\left(\beta B_y + \frac{1}{c}E_x\right) & \mathrm{i}\gamma\left(\beta B_x - \frac{1}{c}E_y\right) & \frac{\mathrm{i}}{c}E_z & 0 \end{bmatrix} \tag{13.180}$$

$$F' = \begin{bmatrix} 0 & B'_z & -B'_y & -\frac{\mathrm{i}}{c}E'_x \\ -B'_z & 0 & B'_x & -\frac{\mathrm{i}}{c}E'_y \\ B'_y & -B'_x & 0 & -\frac{\mathrm{i}}{c}E'_z \\ \frac{\mathrm{i}}{c}E'_x & \frac{\mathrm{i}}{c}E'_y & \frac{\mathrm{i}}{c}E'_z & 0 \end{bmatrix} \tag{13.181}$$

Comparison of the two above expressions for F' in Eq. (13.181) allows one to summarize the transformation between (1) the electric **E** and magnetic **B** fields in the rest frame with electromagnetic tensor F, and (2) the electric $\mathbf{E}'$ and magnetic $\mathbf{B}'$ fields in the moving frame with electromagnetic tensor F'. The moving frame moves in the positive z-direction at velocity v, relative to the rest frame. By comparison of the corresponding matrix elements for the two expressions for F' in Eq. (13.181), one obtains

$$E'_z = E_z, \quad B'_z = B_z \tag{13.182}$$

$$E'_y = \gamma(E_y - vB_x), \quad B'_y = \gamma\left(B_y + \frac{\beta}{c}E_x\right) \tag{13.183}$$

$$E'_x = \gamma(E_x + vB_y), \quad B'_x = \gamma\left(B_x - \frac{\beta}{c}E_y\right) \tag{13.184}$$

The electric $\mathbf{E}$ and magnetic field $\mathbf{B}$ vectors can be resolved into parallel electric $E_\parallel$ and magnetic $B_\parallel$ field components, and perpendicular electric $E_\perp$ and magnetic $B_\perp$ field components, which are measured relative to the direction of the velocity between the two frames. One thus obtains more compact expressions that relate the stationary frame electric $\mathbf{E}$ and magnetic $\mathbf{B}$ fields to the moving frame electric $\mathbf{E}'$ and magnetic $\mathbf{B}'$ components, which are parallel $\parallel$ and perpendicular $\perp$ to the relative velocity $\mathbf{v}$, where

$$\mathbf{E}'_\perp = \gamma(\mathbf{E}_\perp + \mathbf{v} \times \mathbf{B}_\perp), \quad \mathbf{E}'_\parallel = \mathbf{E}_\parallel \tag{13.185}$$

and

$$\mathbf{B}'_\perp = \gamma\left(\mathbf{B}_\perp - \frac{1}{c^2}\mathbf{v} \times \mathbf{E}_\perp\right), \quad \mathbf{B}'_\parallel = \mathbf{B}_\parallel \tag{13.186}$$

If the magnetic field arises only from magnetic dipoles in the case of zero applied magnetic field ($\mathbf{H} = 0$), then the magnetic flux density $\mathbf{B}$ is produced by magnetic dipoles with magnetic moment $\boldsymbol{\mu}$ and magnetic dipole density or magnetization $\mathbf{M}$, where

$$\mathbf{B} = \mu_0 \mathbf{M} \tag{13.187}$$

The polarization vector $\mathbf{P}$ is related to the moving frame electric field $\mathbf{E}$, via the vacuum permittivity constant ε_0, and Lorentz invariant relative permittivity ε_r, where

$$\mathbf{P} = \varepsilon_0 \varepsilon_r \mathbf{E} \tag{13.188}$$

One next substitutes Eqs. (13.187) and (13.188) into Eqs. (13.185) and (13.186).

Using Eq. (13.188), the transformation relation of Eq. (13.185) for the moving frame parallel $\mathbf{E}'_{\parallel}$ and perpendicular $\mathbf{E}'_{\perp}$ electric field vector components, one finds the following. The parallel component of the polarization vector in the stationary frame $\mathbf{P}_{\parallel}$ and moving frame $\mathbf{P}'_{\parallel}$ are identical:

$$\mathbf{P}'_{\parallel} = \mathbf{P}_{\parallel} \tag{13.189}$$

However, the perpendicular $\perp$ components of the stationary frame polarization vector $\mathbf{P}_{\perp}$, are added to the cross product $\mathbf{v} \times \mathbf{M}_{\perp}$ of the relative velocity $\mathbf{v}$ and the stationary frame magnetization vector $\mathbf{M}_{\perp}$, which yields the perpendicular $\perp$ component of the moving frame polarization vector $\mathbf{P}'_{\perp}$, where $\boldsymbol{\beta} = \mathbf{v}/c$ and

$$\frac{\mathbf{P}'_{\perp}}{\varepsilon_r} = \gamma\left(\frac{\mathbf{P}_{\perp}}{\varepsilon_r\varepsilon_0} + \frac{\mathbf{v} \times \mathbf{M}_{\perp}}{c^2}\right) = \frac{\mathbf{P}_{\perp}}{\varepsilon_r\varepsilon_0\sqrt{1-\beta^2}} + \frac{1}{c}\frac{\boldsymbol{\beta} \times \mathbf{M}_{\perp}}{\sqrt{1-\beta^2}} \tag{13.190}$$

If the stationary electric field perpendicular component $\mathbf{E}_{\perp}$ (and hence $\mathbf{P}_{\perp}$) is zero in Eq. (13.190), then the magnetization perpendicular vector component in the stationary frame $\mathbf{M}_{\perp}$ produces a perpendicular polarization vector component in the moving frame $\mathbf{P}'_{\perp}$, where

$$\frac{\mathbf{P}'_{\perp}}{\varepsilon_r} = \frac{1}{c}\frac{\boldsymbol{\beta} \times \mathbf{M}_{\perp}}{\sqrt{1-\beta^2}}, \quad \mathbf{E}_{\perp} = \mathbf{P}_{\perp} = 0 \tag{13.191}$$

Using Eq. (13.187), the the transformation relation of Eq. (13.186) for the moving frame parallel $\mathbf{B}'_{\parallel}$ and perpendicular $\mathbf{B}'_{\perp}$ magnetic field vector components, one finds the following. The parallel components of the magnetization vector in the stationary frame $\mathbf{M}_{\parallel}$ and moving frame $\mathbf{M}'_{\parallel}$ are identical, where

$$\mathbf{M}'_{\parallel} = \mathbf{M}_{\parallel} \tag{13.192}$$

However, the perpendicular $\perp$ components of the stationary frame magnetization vector $\mathbf{M}_{\perp}$, are added to the cross product $\mathbf{v} \times \mathbf{P}_{\perp}$ of the relative velocity $\mathbf{v}$ and the stationary frame polarization vector $\mathbf{P}_{\perp}$, which yields the perpendicular $\perp$ component of the moving frame magnetization vector $\mathbf{M}'_{\perp}$, where

$$\mathbf{M}'_{\perp} = \gamma\left(\mathbf{M}_{\perp} - \frac{\mathbf{v} \times \mathbf{P}_{\perp}}{\varepsilon_r}\right) = \frac{\mathbf{M}_{\perp}}{\sqrt{1-\beta^2}} - \frac{c}{\varepsilon_r}\frac{\boldsymbol{\beta} \times \mathbf{P}_{\perp}}{\sqrt{1-\beta^2}} \tag{13.193}$$

If the stationary magnetic field perpendicular component $\mathbf{B}_{\perp}$ (and hence $\mathbf{M}_{\perp}$) is zero in Eq. (13.193), then the polarization perpendicular vector component in the stationary frame $\mathbf{P}_{\perp}$ produces a perpendicular magnetization vector component in the moving frame $\mathbf{M}'_{\perp}$, where

$$\mathbf{M}'_{\perp} = -\frac{c}{\varepsilon_r}\frac{\boldsymbol{\beta} \times \mathbf{P}_{\perp}}{\sqrt{1-\beta^2}}, \quad \mathbf{B}_{\perp} = \mathbf{M}_{\perp} = 0 \tag{13.194}$$

13.12 The Maxwell Stress Tensor

The conservation of energy is expressed by the integral form of the continuity equation, which includes (1) the time rate of change of energy per unit volume u in volume V; that is the temporal change of energy density $\mathrm{d}u/dt$, (2) the work Wdone by the electromagnetic field on the matter, which is contained inside volume V, and (3) the normal component $\hat{\mathbf{n}}\cdot\mathbf{S}$ of the energy current density or energy flow per unit area $\mathbf{S}$, which perpendicularly crosses the enclosing boundary surface Σ of volume V, where the surface unit normal vector is $\hat{\mathbf{n}}$. The integral form of the continuity equation is thus the change of energy within volume V, which is equal to the sum of (1) the work Wdone in volume V and (2) the energy that enters or exits volume V through the enclosing boundary surface Σ, where

$$-\frac{\partial}{\partial t}\iiint u\mathrm{d}V = \iint \mathbf{S}\cdot\hat{\mathbf{n}}\,\mathrm{d}\Sigma + W \tag{13.195}$$

The work W performed by the electromagnetic field on matter or the work performed by moving matter on the electromagnetic field inside volume V is the dot product of the electric field $\mathbf{E}$ and the electric current density $\mathbf{J}$. The work density w (work per unit volume) in volume V enclosed by surface Σ is

$$w = \mathbf{E}\cdot\mathbf{J} \tag{13.196}$$

where

$$W = \iiint w\mathrm{d}V = \iiint \mathbf{E}\cdot\mathbf{J}\mathrm{d}V \tag{13.197}$$

The Lorentz force equation is

$$\mathbf{F} = q(\mathbf{E} + \mathbf{v}\times\mathbf{B}) \tag{13.198}$$

Work W and power $P = \mathrm{d}W/dt$ are directed to a charge q by force acting in the direction of motion (velocity) of the charge q, where

$$P = \frac{\mathrm{d}W}{\mathrm{d}t} = \mathbf{F}\cdot\mathbf{v} = q\mathbf{E}\cdot\mathbf{v} \tag{13.199}$$

With substitution of Eq. (13.197) into Eq. (13.195), the energy conservation integral, becomes

$$-\frac{\partial}{\partial t}\iiint u\,\mathrm{d}V = \iint \mathbf{S}\cdot\hat{\mathbf{n}}\,\mathrm{d}\Sigma + \iiint \mathbf{E}\cdot\mathbf{J}\,\mathrm{d}V \tag{13.200}$$

The surface integral of Eq. (13.195) is converted to a volume integral by Gauss's theorem, so

$$-\frac{\partial}{\partial t}\iiint u\,\mathrm{d}V = \iiint \nabla\cdot\mathbf{S}\,\mathrm{d}V + \iiint \mathbf{E}\cdot\mathbf{J}\,\mathrm{d}V \tag{13.201}$$

or

$$\iiint\left(-\frac{\partial u}{\partial t} - \nabla\cdot\mathbf{S} - \mathbf{E}\cdot\mathbf{J}\right)\mathrm{d}V = 0 \tag{13.202}$$

The energy conservation expression of Eq. (13.202), is true for any volume, so that the integrand must be zero. Thus, the integrand becomes the differential equation, which expresses the conservation of energy, via the time rate of energy density, energy flux, and work performed on matter, where

$$-\frac{\partial u}{\partial t} - \nabla\cdot\mathbf{S} = \mathbf{E}\cdot\mathbf{J} \tag{13.203}$$

The current density $\mathbf{J}$ is obtained via the Maxwell equation of Ampere's law, where

$$\mathbf{J} = \varepsilon_0 c^2 \nabla\times\mathbf{B} - \varepsilon_0\frac{\partial\mathbf{E}}{\partial t} \tag{13.204}$$

The term $\mathbf{E}\cdot\mathbf{J}$ is then

$$\mathbf{E}\cdot\mathbf{J} = \varepsilon_0 c^2 \mathbf{E}\cdot(\nabla\times\mathbf{B}) - \varepsilon_0\mathbf{E}\cdot\frac{\partial\mathbf{E}}{\partial t} \tag{13.205}$$

where

$$\mathbf{E}\cdot\frac{\partial\mathbf{E}}{\partial t} = \frac{1}{2}\frac{\partial}{\partial t}(\mathbf{E}\cdot\mathbf{E}) \tag{13.206}$$

and

$$\mathbf{E}\cdot(\nabla\times\mathbf{B}) = \nabla\cdot(\mathbf{B}\times\mathbf{E}) + \mathbf{B}\cdot(\nabla\times\mathbf{E}) \tag{13.207}$$

The $\mathbf{E}\cdot\mathbf{J}$ work term becomes

$$\mathbf{E}\cdot\mathbf{J} = \varepsilon_0 c^2 \nabla\cdot(\mathbf{B}\times\mathbf{E}) + \varepsilon_0 c^2 \mathbf{B}\cdot(\nabla\times\mathbf{E}) - \frac{1}{2}\varepsilon_0\frac{\partial}{\partial t}(\mathbf{E}\cdot\mathbf{E}) \tag{13.208}$$

One now invokes the Maxwell equation of Faraday's law, where

$$\nabla\times\mathbf{E} = -\frac{\partial\mathbf{B}}{\partial t} \tag{13.209}$$

Substitution of Eq. (13.209) into Eq. (13.208), gives the work term $\mathbf{E}\cdot\mathbf{J}$ as

$$\mathbf{E}\cdot\mathbf{J} = -\varepsilon_0 c^2 \nabla\cdot(\mathbf{B}\times\mathbf{E}) - \frac{1}{2}\varepsilon_0 c^2 \frac{\partial}{\partial t}(\mathbf{B}\cdot\mathbf{B}) - \frac{1}{2}\varepsilon_0 \frac{\partial}{\partial t}(\mathbf{E}\cdot\mathbf{E}) \tag{13.210}$$

where

$$\mathbf{B}\cdot\frac{\partial \mathbf{B}}{\partial t} = \frac{1}{2}\frac{\partial}{\partial t}(\mathbf{B}\cdot\mathbf{B}) \tag{13.211}$$

Substitution of Eq. (13.210) for $\mathbf{E}\cdot\mathbf{J}$ into the differential equation of Eq. (13.203) for energy conservation, yields

$$-\frac{\partial u}{\partial t} - \nabla\cdot\mathbf{S} = -\varepsilon_0 c^2 \nabla\cdot(\mathbf{B}\times\mathbf{E}) - \frac{1}{2}\varepsilon_0 c^2 \frac{\partial}{\partial t}(\mathbf{B}\cdot\mathbf{B}) - \frac{1}{2}\varepsilon_0 \frac{\partial}{\partial t}(\mathbf{E}\cdot\mathbf{E}) \tag{13.212}$$

Comparison of the left and right sides of Eq. (13.212) defines the energy density u and power flux vector $\mathbf{S}$ $[W/m^2]$ or Poynting vector in terms of the electric $\mathbf{E}$ and magnetic $\mathbf{B}$ fields, where one also recalls section 1.5.1 in chapter 1:

$$u = \frac{1}{2}\varepsilon_0 c^2(\mathbf{B}\cdot\mathbf{B}) + \frac{1}{2}\varepsilon_0(\mathbf{E}\cdot\mathbf{E}) \tag{13.213}$$

and

$$\mathbf{S} = \varepsilon_0 c^2(\mathbf{E}\times\mathbf{B}) \tag{13.214}$$

Division of the time-averaged Poynting vector $\langle\mathbf{S}\rangle$ by the speed of light $c[m/s]$ yields the radiation pressure $\mathbf{P}[n/m^2]$, where

$$\mathbf{P} = \frac{\langle\mathbf{S}\rangle}{c} = \varepsilon_0 c\langle\mathbf{E}\times\mathbf{B}\rangle \tag{13.215}$$

The radiation pressure of the sun's electromagnetic waves on the tails of comets causes the deflection of the comet tails away from the sun. Ponderomotive pressure in laser-produced plasmas is another effect of radiation pressure. The intense electromagnetic fields of the focused laser, causes the plasma plume to be shaped into a convex, light-focusing lens, which further focuses the laser light onto the laser target. This increased focusing of laser light by the laser-produced plasma lens, leads to even larger focused laser electric and magnetic field densities at the target. These large electromagnetic fields create even greater radiation pressure and pondermotive forces. The ponderomotive forces, which arise from the intense focused laser electromagnetic field and plasma lens focusing, lead to much higher charge states, and larger ion energies, than would be associated with a normal Boltzmann energy distribution at the plasma electron and ion temperatures. See Chen (1974), Destler and Cremer (1983), and Hora (2000).

13.13 Force Density and the Maxwell Stress Tensor

In this section, we examine the relation of the electromagnetic force density of the electric and magnetic fields that act on charges, and the relation of the force density **f** to the Maxwell stress tensor T. Force density **f** is the force **F** per unit volume V that acts on a body of charge and/or mass. Force **F** [n], expressed in newtons, acts on volume V [m^3] and produces force density **f** [n/m^3] in units of newtons per cubic meter **f** [n/m^3], where

$$\mathbf{f} = \frac{\mathbf{F}}{V} \tag{13.216}$$

The electric field force density $\mathbf{f}^e$ for an electric field **E**, is first derived in terms of the Maxwell stress tensor T. Then, magnetic field force density $\mathbf{f}^m$ for a magnetic field **B**, is derived in terms of the Maxwell stress tensor T. The electromagnetic field force density $\mathbf{f}^{\mathrm{em}}$ is then the summation of the electric $\mathbf{f}^{\mathrm{e}}$ and magnetic $\mathbf{f}^{\mathrm{m}}$ field force densities, which is expressed by the Maxwell stress tensor T.

We start with relating the Maxwell stress tensor T to the electric field force density vector $\mathbf{f}^{\mathrm{e}}$. The electric force density vector $\mathbf{f}^{\mathrm{e}}$, which acts on charge density ρ, is obtained by invoking the Maxwell equation of Gauss's law for the electric field, where

$$\mathbf{f}^{\mathrm{e}} = \rho\mathbf{E} = (\varepsilon_0 \nabla \cdot \mathbf{E})\mathbf{E} \tag{13.217}$$

Recall, Gauss's Law for the electric field is

$$\rho = \varepsilon_0 \nabla \cdot \mathrm{E} \tag{13.218}$$

With expansion of Eq. (13.217), the x-, y-, and z-components f_x^{e}, f_y^{e}, and f_z^{e} of the electric force density vector $\mathbf{f}^{\mathrm{e}}$ are

$$f_x^{\mathrm{e}} = \varepsilon_0 \left(E_x \frac{\partial E_x}{\partial x} + E_x \frac{\partial E_y}{\partial y} + E_x \frac{\partial E_z}{\partial z} \right) \tag{13.219}$$

$$f_y^{\mathrm{e}} = \varepsilon_0 \left(E_y \frac{\partial E_x}{\partial x} + E_y \frac{\partial E_y}{\partial y} + E_y \frac{\partial E_z}{\partial z} \right) \tag{13.220}$$

$$f_z^{\mathrm{e}} = \varepsilon_0 \left(E_z \frac{\partial E_x}{\partial x} + E_z \frac{\partial E_y}{\partial y} + E_z \frac{\partial E_z}{\partial z} \right) \tag{13.221}$$

We now transform these expressions for x-, y-, and z-components f_x^{e}, f_y^{e}, and f_z^{e} of the electric field force density $\mathbf{f}^{\mathrm{e}}$ to an equivalent form, which can be better matched to the form of the expressions for the x-, y-, and z-components f_x^{m}, f_y^{m}, and f_z^{m} of the magnetic field force density vector $\mathbf{f}^{\mathrm{m}}$ that is to be subsequently derived.

In this manner, the results of the electric $\mathbf{f}^{\mathrm{e}}$ and magnetic $\mathbf{f}^{\mathrm{m}}$ force density vectors can be added to produce the electromagnetic force density vector $\mathbf{f}^{\mathrm{em}}$ in a convenient mathematical form, which best conveys the physicality. In the case of an electric field alone, the curl of the electric field is zero, since one assumes a zero magnetic field, where

$$\nabla \times \mathbf{E} = 0 \tag{13.222}$$

To aid evaluation of Eqs. (13.219–13.221), cyclic permutation of x, y, z gives the components of $\nabla \times \mathbf{E} = 0$, where

$$\frac{\partial E_x}{\partial y} = \frac{\partial E_y}{\partial x} \quad \frac{\partial E_y}{\partial z} = \frac{\partial E_z}{\partial y} \quad \frac{\partial E_z}{\partial x} = \frac{\partial E_x}{\partial z} \tag{13.223}$$

To evaluate the force density expressions of Eqs. (13.219–13.221), one uses the following useful relations:

$$E_x \frac{\partial E_x}{\partial x} = \frac{1}{2}\frac{\partial}{\partial x}(E_x^2), \quad E_y \frac{\partial E_y}{\partial y} = \frac{1}{2}\frac{\partial}{\partial y}(E_y^2), \quad E_y \frac{\partial E_y}{\partial y} = \frac{1}{2}\frac{\partial}{\partial y}(E_y^2) \tag{13.224}$$

$$E_x \frac{\partial E_y}{\partial y} = \frac{\partial}{\partial y}(E_x E_y) - E_y \frac{\partial E_x}{\partial y} = \frac{\partial}{\partial y}(E_x E_y) - E_y \frac{\partial E_y}{\partial x} = \frac{\partial}{\partial y}(E_x E_y) - \frac{1}{2}\frac{\partial}{\partial x}(E_y^2) \tag{13.225}$$

and

$$E_x \frac{\partial E_z}{\partial z} = \frac{\partial}{\partial z}(E_x E_z) - E_z \frac{\partial E_x}{\partial z} = \frac{\partial}{\partial z}(E_x E_z) - E_z \frac{\partial E_z}{\partial x} = \frac{\partial}{\partial z}(E_x E_z) - \frac{1}{2}\frac{\partial}{\partial x}(E_z^2) \tag{13.226}$$

Substitution of Eq. (13.224) into Eq. (13.219), gives the x-component of the electric force density vector $\mathbf{f}^{\mathrm{e}}$:

$$f_x^{\mathrm{e}} = \varepsilon_0 \left[\frac{1}{2}\frac{\partial}{\partial x}(E_x^2 - E_y^2 - E_z^2) + \frac{\partial}{\partial y}(E_x E_y) + \frac{\partial}{\partial z}(E_x E_z) \right] \tag{13.227}$$

Applied to Eq. (13.227), cyclic permutation of x, y, z gives the y- and z-components of $\mathbf{f}^{\mathrm{e}}$:

$$f_y^{\mathrm{e}} = \varepsilon_0 \left[\frac{\partial}{\partial x}(E_y E_x) + \frac{1}{2}\frac{\partial}{\partial y}(E_y^2 - E_z^2 - E_x^2) + \frac{\partial}{\partial z}(E_y E_z) \right] \tag{13.228}$$

and

$$f_z^{\mathrm{e}} = \varepsilon_0 \left[\frac{\partial}{\partial x}(E_z E_x) + \frac{\partial}{\partial y}(E_z E_y) + \frac{1}{2}\frac{\partial}{\partial z}(E_z^2 - E_x^2 - E_y^2) \right] \tag{13.229}$$

The electric force density $\mathbf{f}^e$ can be obtained from the electric field stress tensor T^e, where the x-, y-, and z-components f_x^e, f_y^e, and f_z^e of $\mathbf{f}^e$ are spatial derivatives of the components of 3×3 electric field tensor T^e, so that

$$f_x^e = \frac{\partial T_{xx}^e}{\partial x} + \frac{\partial T_{xy}^e}{\partial y} + \frac{\partial T_{xz}^e}{\partial z} \tag{13.230}$$

$$f_y^e = \frac{\partial T_{yx}^e}{\partial x} + \frac{\partial T_{yy}^e}{\partial y} + \frac{\partial T_{yz}^e}{\partial z} \tag{13.231}$$

$$f_z^e = \frac{\partial T_{zx}^e}{\partial x} + \frac{\partial T_{zy}^e}{\partial y} + \frac{\partial T_{zz}^e}{\partial z} \tag{13.232}$$

The 3×3 electric field tensor T^e is

$$T^e = \begin{bmatrix} T_{xx}^e & T_{xy}^e & T_{xz}^e \\ T_{yx}^e & T_{yy}^e & T_{yz}^e \\ T_{zx}^e & T_{zy}^e & T_{zz}^e \end{bmatrix} \tag{13.233}$$

Comparison of the three equations of Eqs. (13.227–13.229) for the x, y, and z electric force density components f_x^e, f_y^e, and f_z^e in terms of the electric field (E_x, E_y, E_z), and Eqs. (13.230–13.232) for the electric stress tensor components T_{ij}^e, then gives the electric stress tensor T_{ij}^e components, which are expressed by the electric field components (E_x, E_y, E_z), where $i, j = x, y, z$ and

$$T^e = \varepsilon_0 \begin{bmatrix} \frac{1}{2}(E_x^2 - E_y^2 - E_z^2) & E_x E_y & E_x E_z \\ E_y E_x & \frac{1}{2}(E_y^2 - E_z^2 - E_x^2) & E_y E_z \\ E_z E_x & E_z E_y & \frac{1}{2}(E_z^2 - E_x^2 - E_y^2) \end{bmatrix} \tag{13.234}$$

The electric stress tensor T^e is a symmetric matrix, where

$$T_{ij}^e = T_{ji}^e \quad i, j = x, y, z \tag{13.235}$$

This is because the products of the electric fields are commutative, where

$$E_i E_j = E_j E_i \quad i, j = x, y, z \tag{13.236}$$

We now derive the relation of the Maxwell stress tensor T to the magnetic field force density vector $\mathbf{f}^m$ in the absence of electric fields ($\mathbf{E} = 0$). The magnetic force density vector $\mathbf{f}^{\mathrm{m}}$ acts on current density $\mathbf{J}$, which propagates through magnetic field $\mathbf{B}$, where

$$\mathbf{f}^{\mathrm{m}} = \mathbf{J} \times \mathbf{B} \tag{13.237}$$

In the absence of electric fields, the Maxwell equation of Ampere's law shows that the curl of the magnetic field $\nabla \times \mathbf{B}$ is proportional to the current density $\mathbf{J}$ via the proportionality constant μ_0, the permeability of free space, where

$$\nabla \times \mathbf{B} = \mu_0 \mathbf{J} \tag{13.238}$$

Substitution into Eq. (13.237) of Eq. (13.238) for $\mathbf{J}$ in the force density equation, gives

$$\mathbf{f}^{\mathrm{m}} = \frac{1}{\mu_0}(\nabla \times \mathbf{B}) \times \mathbf{B} \tag{13.239}$$

Expansion of the cross products in Eq. (13.239), gives the x-component f_x^{m} of the force density vector $\mathbf{f}^{\mathrm{m}}$:

$$f_x^{\mathrm{m}} = \frac{1}{\mu_0}\left[B_z\left(\frac{\partial B_x}{\partial z} - \frac{\partial B_z}{\partial x}\right) - B_y\left(\frac{\partial B_y}{\partial x} - \frac{\partial B_x}{\partial y}\right)\right] \tag{13.240}$$

The other two y- and z-components, f_y^{m} and f_z^{m}, are obtained by cyclic permutation of the x, y, and z indices in Eq. (13.240), where

$$f_y^{\mathrm{m}} = \frac{1}{\mu_0}\left[B_x\left(\frac{\partial B_y}{\partial x} - \frac{\partial B_x}{\partial y}\right) - B_z\left(\frac{\partial B_z}{\partial y} - \frac{\partial B_y}{\partial z}\right)\right] \tag{13.241}$$

$$f_z^{\mathrm{m}} = \frac{1}{\mu_0}\left[B_y\left(\frac{\partial B_z}{\partial y} - \frac{\partial B_y}{\partial z}\right) - B_x\left(\frac{\partial B_x}{\partial z} - \frac{\partial B_z}{\partial x}\right)\right] \tag{13.242}$$

The expressions of Eqs. (13.240–13.241) for f_x^{m}, f_y^{m}, and f_z^{m} must be developed further to obtain a mathematically convenient and physically meaningful form, which can be added to its electric field counterpart. In particular, Eq. (13.240) for the x-component f_x^{m} can be expressed equivalently by

$$f_x^{\mathrm{m}} = \frac{1}{\mu_0}\left[\left(\frac{\partial}{\partial z}(B_z B_x) - B_x\frac{\partial B_z}{\partial z} - \frac{1}{2}\frac{\partial}{\partial x}B_z^2\right) + \left(\frac{\partial}{\partial y}(B_x B_y) - B_x\frac{\partial B_y}{\partial y} - \frac{1}{2}\frac{\partial}{\partial x}B_y^2\right)\right] \tag{13.243}$$

One can add to f_x^{m} an expression, which is zero, but yields a more mathematically convenient form. The zero expression to be added is

$$-B_x \frac{\partial B_x}{\partial x} + \frac{1}{2}\frac{\partial}{\partial x} B_x^2 = 0 \tag{13.244}$$

Eq. (13.243) for the x-component f_x^{m}, with the added zero expression of Eq. (13.244), becomes

$$f_x^{\mathrm{m}} = \frac{1}{\mu_0}\left[\frac{1}{2}\frac{\partial}{\partial x}(B_x^2 - B_y^2 - B_z^2) + \frac{\partial}{\partial y}(B_x B_y) + \frac{\partial}{\partial z}(B_z B_x) - B_x(\nabla \cdot \mathbf{B})\right] \tag{13.245}$$

Gauss's law for a magnetic field is

$$\nabla \cdot \mathbf{B} = 0 \tag{13.246}$$

With substitution of Eq. (13.246) into Eq. (13.245), the expression for f_x^{m}, becomes

$$f_x^{\mathrm{m}} = \frac{1}{\mu_0}\left[\frac{1}{2}\frac{\partial}{\partial x}(B_x^2 - B_y^2 - B_z^2) + \frac{\partial}{\partial y}(B_x B_y) + \frac{\partial}{\partial z}(B_z B_x)\right] \tag{13.247}$$

One obtains the expressions for f_y^{m} and f_z^{m}, via cyclic permutation of the indices of f_x^{m} of Eq. (13.247), where

$$f_y^{\mathrm{m}} = \frac{1}{\mu_0}\left[\frac{1}{2}\frac{\partial}{\partial y}(B_y^2 - B_z^2 - B_x^2) + \frac{\partial}{\partial z}(B_y B_z) + \frac{\partial}{\partial x}(B_x B_y)\right] \tag{13.248}$$

and

$$f_z^{\mathrm{m}} = \frac{1}{\mu_0}\left[\frac{1}{2}\frac{\partial}{\partial z}(B_z^2 - B_x^2 - B_y^2) + \frac{\partial}{\partial x}(B_z B_x) + \frac{\partial}{\partial y}(B_y B_z)\right] \tag{13.249}$$

The magnetic force density $\mathbf{f}^{\mathrm{m}}$ can be obtained from the magnetic field stress tensor T^{m}, where the x-, y-, and z-components f_x^{m}, f_y^{m}, and f_z^{m} of $\mathbf{f}^{\mathrm{m}}$ are spatial derivatives of the components of 3×3 magnetic field tensor T^{m}, so that

$$f_x^{\mathrm{m}} = \frac{\partial T_{xx}^{\mathrm{m}}}{\partial x} + \frac{\partial T_{xy}^{\mathrm{m}}}{\partial y} + \frac{\partial T_{xz}^{\mathrm{m}}}{\partial z} \tag{13.250}$$

$$f_y^{\mathrm{m}} = \frac{\partial T_{yx}^{\mathrm{m}}}{\partial x} + \frac{\partial T_{yy}^{\mathrm{m}}}{\partial y} + \frac{\partial T_{yz}^{\mathrm{m}}}{\partial z} \tag{13.251}$$

$$f_z^{\mathrm{m}} = \frac{\partial T_{zx}^{\mathrm{m}}}{\partial x} + \frac{\partial T_{zy}^{\mathrm{m}}}{\partial y} + \frac{\partial T_{zz}^{\mathrm{m}}}{\partial z} \tag{13.252}$$

The 3×3 magnetic field tensor T^{m} is

$$T^{\mathrm{m}} = \begin{bmatrix} T^{\mathrm{m}}_{xx} & T^{\mathrm{m}}_{xy} & T^{\mathrm{m}}_{xz} \\ T^{\mathrm{m}}_{yx} & T^{\mathrm{m}}_{yy} & T^{\mathrm{m}}_{yz} \\ T^{\mathrm{m}}_{zx} & T^{\mathrm{m}}_{zy} & T^{\mathrm{m}}_{zz} \end{bmatrix} \tag{13.253}$$

Comparison of the three equations of Eqs. (13.247–13.249) for the x, y, and z magnetic force density components f^{m}_x, f^{m}_y, and f^{m}_z, in terms of the magnetic field (B_x, B_y, B_z), and Eqs. (13.250–13.252) for the magnetic stress tensor components T^{m}_{ij}, then gives the magnetic stress tensor T^{m}_{ij} components, which are expressed by the magnetic field components (B_x, B_y, B_z), where $i, j = x, y, z$ and

$$T^{\mathrm{m}} = \frac{1}{\mu_0} \begin{bmatrix} \frac{1}{2}(B_x^2 - B_y^2 - B_z^2) & B_x B_y & B_x B_z \\ B_y B_x & \frac{1}{2}(B_y^2 - B_z^2 - B_x^2) & B_y B_z \\ B_z B_x & B_z B_y & \frac{1}{2}(B_z^2 - B_x^2 - B_y^2) \end{bmatrix} \tag{13.254}$$

The magnetic stress tensor T^{m} is a symmetric matrix, where

$$T^{\mathrm{m}}_{ij} = T^{\mathrm{m}}_{ji} \quad i, j = x, y, z \tag{13.255}$$

This is because the products of the magnetic fields are commutative, where

$$B_i B_j = B_j B_i \quad i, j = x, y, z \tag{13.256}$$

The electromagnetic force density $\mathbf{f}^{\mathrm{em}}$, which acts on the object charge density ρ and current density $\mathbf{J}$, is the sum of the electric force density $\mathbf{f}^{\mathrm{e}}$ and magnetic force density $\mathbf{f}^{\mathrm{m}}$, where

$$\mathbf{f}^{\mathrm{em}} = \mathbf{f}^{\mathrm{e}} + \mathbf{f}^{\mathrm{m}} \tag{13.257}$$

The electromagnetic force density $\mathbf{f}^{\mathrm{em}}$, which acts on the object, can be expressed in terms of the electric $\mathbf{E}$ and magnetic $\mathbf{B}$ fields that are applied to the object, via the electromagnetic tensor T^{em}. In particular the x-, y-, and z-components f^{em}_x, f^{em}_y, and f^{em}_z of the electromagnetic force density $\mathbf{f}^{\mathrm{em}}$ are obtained from the spatial derivatives of the electromagnetic tensor T^{em}:

$$f^{\mathrm{em}}_x = \frac{\partial T^{\mathrm{em}}_{xx}}{\partial x} + \frac{\partial T^{\mathrm{em}}_{xy}}{\partial y} + \frac{\partial T^{\mathrm{em}}_{xz}}{\partial z} \tag{13.258}$$

$$f^{\mathrm{em}}_y = \frac{\partial T^{\mathrm{em}}_{yx}}{\partial x} + \frac{\partial T^{\mathrm{em}}_{yy}}{\partial y} + \frac{\partial T^{\mathrm{em}}_{yz}}{\partial z} \tag{13.259}$$

$$f^{\mathrm{em}}_z = \frac{\partial T^{\mathrm{em}}_{zx}}{\partial x} + \frac{\partial T^{\mathrm{em}}_{zy}}{\partial y} + \frac{\partial T^{\mathrm{em}}_{zz}}{\partial z} \tag{13.260}$$

The 3×3 electromagnetic field tensor T^{em} is the sum of the 3×3 electric T^{e} and magnetic T^{m} field tensors, where

$$T^{\mathrm{em}} = \begin{bmatrix} T^{\mathrm{em}}_{xx} & T^{\mathrm{em}}_{xy} & T^{\mathrm{em}}_{xz} \\ T^{\mathrm{em}}_{yx} & T^{\mathrm{em}}_{yy} & T^{\mathrm{em}}_{yz} \\ T^{\mathrm{em}}_{zx} & T^{\mathrm{em}}_{zy} & T^{\mathrm{em}}_{zz} \end{bmatrix} = T^{\mathrm{e}} + T^{\mathrm{m}} = \begin{bmatrix} T^{\mathrm{e}}_{xx} & T^{\mathrm{e}}_{xy} & T^{\mathrm{e}}_{xz} \\ T^{\mathrm{e}}_{yx} & T^{\mathrm{e}}_{yy} & T^{\mathrm{e}}_{yz} \\ T^{\mathrm{e}}_{zx} & T^{\mathrm{e}}_{zy} & T^{\mathrm{e}}_{zz} \end{bmatrix} + \begin{bmatrix} T^{\mathrm{m}}_{xx} & T^{\mathrm{m}}_{xy} & T^{\mathrm{m}}_{xz} \\ T^{\mathrm{m}}_{yx} & T^{\mathrm{m}}_{yy} & T^{\mathrm{m}}_{yz} \\ T^{\mathrm{m}}_{zx} & T^{\mathrm{m}}_{zy} & T^{\mathrm{m}}_{zz} \end{bmatrix} \tag{13.261}$$

Substitution into Eqs. 13.234 and Eq. (13.254) into Eq. (13.261), gives the symmetric electromagnetic stress tensor T^{em}:

$$T^{\mathrm{em}} = \varepsilon_0 \begin{bmatrix} \varepsilon_0 E_x^2 + \dfrac{1}{\mu_0} B_x^2 - u & \varepsilon_0 E_x E_y + \dfrac{1}{\mu_0} B_x B_y & \varepsilon_0 E_x E_z + \dfrac{1}{\mu_0} B_x B_z \\ \varepsilon_0 E_y E_x + \dfrac{1}{\mu_0} B_y B_x & \varepsilon_0 E_y^2 + \dfrac{1}{\mu_0} B_y^2 - u & \varepsilon_0 E_y E_z + \dfrac{1}{\mu_0} B_y B_z \\ \varepsilon_0 E_x E_z + \dfrac{1}{\mu_0} B_x B_z & \varepsilon_0 E_y E_z + \dfrac{1}{\mu_0} B_y B_z & \varepsilon_0 E_z^2 + \dfrac{1}{\mu_0} B_z^2 - u \end{bmatrix} \tag{13.262}$$

where

$$T^{\mathrm{em}}_{ij} = T^{\mathrm{em}}_{ji} \quad i,j = 1,2,3,4 \tag{13.263}$$

In Eq. (13.262), the electromagnetic energy density u is given by

$$u = \frac{1}{2}\varepsilon_0 (E_x^2 + E_y^2 + E_z^2) + \frac{1}{2\mu_0}(B_x^2 + B_y^2 + B_z^2) = \frac{1}{2}\varepsilon_0 E^2 + \frac{1}{2\mu_0} B^2 \tag{13.264}$$

The Poynting vector $\mathbf{S}$ is energy flow per unit area per unit time or power flow per unit area with dimensional units $[W/\mathrm{cm}^2]$, where

$$\mathbf{S} = \frac{1}{\mu_0}(\mathbf{E} \times \mathbf{B}) = \sqrt{\frac{\varepsilon_0}{\mu_0}}[(E_y B_z - E_z B_y)\hat{\mathbf{x}} + (E_z B_x - E_x B_z)\hat{\mathbf{y}} + (E_x B_y - E_y B_x)\hat{\mathbf{z}}] \tag{13.265}$$

The electromagnetic momentum density $\mathbf{g}$ is obtained from the Poynting vector $\mathbf{S}$ via

$$\mathbf{g} = \frac{\mathbf{S}}{c^2} = \varepsilon_0 (\mathbf{E} \times \mathbf{B}) \tag{13.266}$$

The x-, y-, and z-components of electromagnetic momentum density vector $\mathbf{g}$ are

$$\mathbf{g} = \varepsilon_0 [(E_y B_z - E_z B_y)\hat{\mathbf{x}} + (E_z B_x - E_x B_z)\hat{\mathbf{y}} + (E_x B_y - E_y B_x)\hat{\mathbf{z}}] \tag{13.267}$$

The 4×4 electromagnetic momentum-energy tensor T, which is used with four-vectors of special relativity, is a generalization of the 3×3 Maxwell electromagnetic stress tensor T^{em}. The 4×4 electromagnetic momentum-energy tensor T is given by

$$T = \begin{bmatrix} T_{xx}^{\mathrm{em}} & T_{xy}^{\mathrm{em}} & T_{xz}^{\mathrm{em}} & cg_x \\ T_{yx}^{\mathrm{em}} & T_{yy}^{\mathrm{em}} & T_{yz}^{\mathrm{em}} & cg_y \\ T_{zx}^{\mathrm{em}} & T_{zy}^{\mathrm{em}} & T_{zz}^{\mathrm{em}} & cg_z \\ S_x/c & S_y/c & S_z/c & -u \end{bmatrix} \tag{13.268}$$

The three spatial x-, y-, and z-components and the temporal component t of the four-vector force density $\mathbf{f}$ can be obtained from the four-vector spatial derivatives $(x, y, z, \mathrm{i}ct)$ of the 4×4 electromagnetic momentum-energy tensor T. Alternatively, the four force density components f_p can be obtained by multiplying the 4×4 electromagnetic tensor F matrix with the column four-vector of the current density $\mathbf{J}$, where

$$f_p = \sum_{k=1}^{4} \frac{\partial T_{pk}}{\partial q_k} = -\frac{1}{c} \sum_{l=1}^{4} F_{pl} J^l \quad p, k, l = (1, 2, 3, 4) \tag{13.269}$$

Here, F_{pl} are the elements of the 4×4 electromagnetic tensor, and F and J^l are elements of the four-vector current density $\mathbf{J}$, where

$$F = \begin{bmatrix} 0 & B_z & -B_y & -\frac{\mathrm{i}}{c} E_x \\ -B_z & 0 & B_x & -\frac{\mathrm{i}}{c} E_y \\ B_y & -B_x & 0 & -\frac{\mathrm{i}}{c} E_z \\ \frac{\mathrm{i}}{c} E_x & \frac{\mathrm{i}}{c} E_y & \frac{\mathrm{i}}{c} E_z & 0 \end{bmatrix} \quad \text{and} \quad \mathbf{J} = \begin{bmatrix} J_x \\ J_y \\ J_z \\ \mathrm{i}c\rho \end{bmatrix} \tag{13.270}$$

13.14 The Principle of Least Action and the Lagrangian Yields Euler–Lagrange Equations

In this section, we derive the Euler–Lagrange equation from the principle of stationary or least action, also known as Hamilton's principle. In the nonrelativistic case, action G for a particle trajectory is the time integral of the difference of the particle kinetic energy W and potential energy U, which occurs during the time interval between the fixed initial and final times t_1 and t_2. The particle kinetic energy W minus the particle potential energy U is the Lagrange function L, where

$$G = \int_{t_1}^{t_2} (W - U)\,\mathrm{d}t = \int_{t_1}^{t_2} L\,\mathrm{d}t \tag{13.271}$$

The non-relativistic Lagrange function L, or Lagrangian L, is

$$L = W - U \tag{13.272}$$

Eq. (13.271) for the action G can apply to a relativistic particle or system of particles; however, the energy W is no longer the simple particle kinetic energy. Rather, the energy W is the nonrest mass portion of the total particle energy, as will be derived later, where

$$W = \frac{1}{2} m_0 v^2 \Rightarrow W = (1 - \beta^2) m_0 c^2 \tag{13.273}$$

The principle of least action states that a small variation of the particle trajectory about the true particle trajectory, yields the same time-integrated difference between the particle kinetic W and potential U energy, as that of the true path-integrated result, where

$$\delta G = \delta \int_{t_1}^{t_2} L\,\mathrm{d}t = 0 \tag{13.274}$$

The maxima or minima of a function occurs at the point, where the tangent line slope is zero, i.e., the differential of the function is zero. The minima–maxima principle is now extended to the case of a function, which is the action integral. The true particle trajectory path or ray path of a particle wave is the trajectory, where the action integral is minimum [Min(G)], and the differential of the action is zero, where

$$\mathrm{Min}(G) \rightarrow \delta G = 0 \tag{13.275}$$

Feynman (1964) discusses the relation of the principle of least action to quantum mechanics as follows. In the wave picture, a particle wave emitted from an origin point $\mathbf{r} = \mathbf{r}_1$ at initial time t_1 follows all possible paths to the terminal point $\mathbf{r} = \mathbf{r}_2$ at final time t_2. Each of the path amplitudes between the fixed initial and final point is proportional to $\exp(\mathrm{i}G/\hbar)$, where G is the action integral for that path. The action G of each path has dimensions of energy multiplied by time, as does the reduced Planck's constant $\hbar$. Thus, $G/\hbar$ represents the phase shift of the path. Assume that for all paths, the action G is much larger than the tiny Planck's constant. Then, for even a small change in the path between the fixed endpoints ΔG, the small change in the action G of the path, produces results in a large change in the phase shift $\Delta G/\hbar$ because $\hbar$ is so small.

The true path produces the minimum action Min(G). Paths that are not sufficiently close produce a uniform and complete range of phases. The resulting amplitudes have equal likelihood of being positive or negative and thus sum to zero. The only significant paths, are those paths close to the true path Min(G), which has the minimum action integral G. Those significant paths are on the order of Planck's constant or less, where specifically $G < h/4$, so that $G/\hbar \leq \pi/2$. That is, only paths that differ from the minimum action Min(G), by less than Planck's constant $[G - \text{Min}(G) < h/4]$, produce net, constructive summing of the path amplitudes. Thus, only those paths within a Planck's constant of action about the true path minimum action Min(G), produces a particle wave amplitude at the endpoint. That is, the particle follows the path of least action G, where $\delta G = 0$. Particle paths that are not close to the true trajectory path, violate the laws of energy conservation and motion. However, paths whose phase shifts do not differ by more than $\hbar/4$ from the true trajectory path, can occur without any violation of the laws of motion or energy.

The variation δG of the action integral for each ith spatial component for $i = x, y, x$ is then further developed by application of the chain rule for differentiation, where

$$\delta G = \delta \int_{t_1}^{t_2} L\text{d}t = \int_{t_1}^{t_2} \left[\frac{\partial L}{\partial q_i} \delta q_i + \left(\frac{\partial L}{\partial(\partial q_i/\partial t)} \right) \delta(\partial q_i/\partial t) \right] \text{d}t = 0 \quad i = x, y, x, \tag{13.276}$$

where

$$q_x = x(t), \quad q_y = y(t), \quad q_z = z(t), \quad q_y = y(t) \tag{13.277}$$

$$\frac{\partial q_x}{\partial t} = \frac{\partial x}{\partial t} = v_x(t), \quad \frac{\partial q_y}{\partial t} = \frac{\partial y}{\partial t} = v_y(t), \quad \frac{\partial q_z}{\partial t} = \frac{\partial z}{\partial t} = v_z(t) \tag{13.278}$$

Integration by parts is applied to the second term of Eq. (13.276) for the differential δG of the action integral G, where

$$\int_{t_1}^{t_2} \left(\frac{\partial L}{\partial(\partial q_i/\partial t)} \right) \delta(\partial q_i/\partial t)\text{d}t = \left[\frac{\partial L}{(\partial q_i/\partial t)} \delta q_i \right]_{t_1}^{t_2} - \int_{t_1}^{t_2} \text{d}t \frac{\text{d}}{\text{d}t} \left(\frac{\partial L}{\partial(\partial q_i/\partial t)} \right) \delta q_i \tag{13.279}$$

In Eq. (13.279), the time endpoints t_1 and t_2 are fixed, so that their variation is zero, where

$$\delta q_i(t_1) = \delta q_i(t_2) = 0 \quad i = x, y, x \tag{13.280}$$

and

$$\left[\frac{\partial L}{(\partial q_i/\partial t)} \delta q_i \right]_{t_1}^{t_2} = 0 - 0 = 0 \tag{13.281}$$

With substitution of Eqs. 13.279 with Eq. (13.281) into Eq. (13.276), the variation δG of the action integral G is then

$$\delta G = \int_{t_1}^{t_2} \left[\frac{\partial L}{\partial q_i} - \frac{\mathrm{d}}{\mathrm{d}t} \left(\frac{\partial L}{\partial(\partial q_i/\partial t)} \right) \right] \delta q_i \, \mathrm{d}t = 0 \quad i = x, y, x \tag{13.282}$$

The variation δG of the action integral will be zero for every possible variation δq_i of the ith spatial component ($i = x, y, x$), provided that the differential integrand expression of Eq. (13.282) of the Lagrange function is zero. The required zero integrand expression is the Euler–Lagrange equation, where

$$\frac{\partial L}{\partial q_i} - \frac{\mathrm{d}}{\mathrm{d}t} \left(\frac{\partial L}{\partial(\partial q_i/\partial t)} \right) = 0 \quad i = x, y, z \tag{13.283}$$

This Euler–Lagrange equation is in the stationary lab frame of the observer, who sees the trajectory of the moving particle, and calculates the action and variation of the action in terms of the particle position q_i, velocity $\partial q_i/\partial t$, and time t in the lab frame.

For a single particle with $i = x, y, z$, there are three simultaneous Euler–Lagrange equations for the x, y, and z spatial coordinates of the particle. The Lagrange function L has not been specifically expressed, and thus the burden remains to determine the relativistic correct expression L for the specific application. The Euler–Lagrange equation can be applied to systems of N particles with $3N$ spatial coordinates in x, y, and z, where the index ranges over $i = 1, 2, \ldots, 3N$, and one obtains $3N$ simultaneous Euler–Lagrange equations, where

$$\frac{\partial L}{\partial q_i} - \frac{\mathrm{d}}{\mathrm{d}t} \left(\frac{\partial L}{\partial(\partial q_i/\partial t)} \right) = 0 \quad i = 1, 2, \ldots, 3N \tag{13.284}$$

The $3N$ pair of variables q_i and $\partial q_i/\partial t$, can also denote other quantities, rather than particle position and trajectory, as a function of time t, and thus this general result can apply to a host of other modeling applications in science, engineering, and socio-politico-economics.

13.15 Derivation of the Hamiltonian from the Lagrangian

The Hamiltonian H can be derived from the Lagrange function L as follows. The particle velocity $\partial \mathbf{q}/\partial t$ is directly related to the particle momentum $\mathbf{p}$, via the relativistic correct expression,

$$\mathbf{p} = \gamma m_0 \frac{\mathrm{d}\mathbf{q}}{\mathrm{d}t} = \frac{m_0}{\sqrt{1 - (|d\mathbf{q}/dt|/c)^2}} \frac{\mathrm{d}\mathbf{q}}{\mathrm{d}t} \tag{13.285}$$

The nonrelativistic expression between momentum $\mathbf{p}$ and particle velocity $\partial\mathbf{q}/\partial t$ is even more suggestive of comparison with the Euler–Lagrange equation, where

$$\mathbf{p} = m_0 \frac{\mathrm{d}\mathbf{q}}{\mathrm{d}t} \tag{13.286}$$

The form of the Euler–Lagrange expression, suggests the $i = x, y, z$ components of general momentum p_i, can be related to the derivative of the Lagrange function, with respect to the particle velocity component general position $\mathrm{d}q_i/dt$, where

$$p_i = \frac{\partial L}{(\mathrm{d}q_i/dt)} \tag{13.287}$$

The particle energy E is obtained from the product of the particle momentum $\mathbf{p}$ and velocity $\mathbf{v}$, and this suggests the definition of the Hamiltonian H by

$$H(p_i, q_i, t) = \sum_{i=1}^{3} p_i \frac{\mathrm{d}q_i}{\mathrm{d}t} - L \tag{13.288}$$

With substitution into Eq. (13.288) of Eq. (13.272) for the nonrelativistic Lagrangian, the nonrelativistic expression of this Hamiltonian becomes

$$H(p_i, q_i, t) = \sum_{i=1}^{3} m_0 v_i^2 - (W - U) \tag{13.289}$$

The nonrelativistic charged particle has kinetic energy W, where

$$W = \frac{1}{2} \sum_{i=1}^{3} m_0 v_i^2 \tag{13.290}$$

With substitution of Eq. (13.290) into Eq. (13.289), this simple, single particle, nonrelativistic expression demonstrates that the Hamiltonian is the total particle energy, where

$$H(p_i, q_i, t) = \frac{1}{2} \sum_{i=1}^{3} m_0 v_i^2 + U = E \tag{13.291}$$

The total differential of the Hamiltonian $\mathrm{d}H$ is then

$$\mathrm{d}H(p_i,q_i,t) = \sum_{i=1}^{3} p_i \mathrm{d}\left(\frac{\mathrm{d}q_i}{\mathrm{d}t}\right) + \sum_{i=1}^{3} \mathrm{d}p_i \frac{\mathrm{d}q_i}{\mathrm{d}t} - \sum_{i=1}^{3} \mathrm{d}q_i \frac{\partial L}{\partial q_i} - \sum_{i=1}^{3} \mathrm{d}\left(\frac{\mathrm{d}q_i}{\mathrm{d}t}\right)_i \frac{\partial L}{\partial(\partial q_i/\partial t)} - \frac{\partial L}{\partial t}\mathrm{d}t \tag{13.292}$$

Evaluation of the first term in Eq. (13.292), yields

$$\sum_{i=1}^{3} p_i \mathrm{d}\left(\frac{\mathrm{d}q_i}{\mathrm{d}t}\right) = \sum_{i=1}^{3} \mathrm{d}\left(\frac{\mathrm{d}q_i}{\mathrm{d}t}\right)_i \frac{\partial L}{\partial(\partial q_i/\partial t)} \tag{13.293}$$

Evaluation of the derivative in the third term in Eq. (13.292), yields

$$\frac{\partial L}{\partial q_i} = \frac{\mathrm{d}}{\mathrm{d}t}\left(\frac{\partial L}{\partial(\partial q_i/\partial t)}\right) = \frac{\mathrm{d}p_i}{\mathrm{d}t} \tag{13.294}$$

With substitution of Eqs. (13.293) and (13.294) into Eq. (13.292), cancellation of the first and fourth terms in Eq. (13.292) occurs, so that

$$\mathrm{d}H(p_i, q_i, t) = \sum_{i=1}^{3} \frac{\mathrm{d}q_i}{\mathrm{d}t}\mathrm{d}p_i - \sum_{i=1}^{3} \frac{\mathrm{d}p_i}{\mathrm{d}t}\mathrm{d}q_i - \frac{\partial L}{\partial t}\mathrm{d}t \tag{13.295}$$

Next, the left side of Eq. (13.295) for dH is expanded, where

$$\mathrm{d}H(p_i, q_i, t) = \sum_{i=1}^{3} \frac{\mathrm{d}H}{\partial p_i}\mathrm{d}p_i + \sum_{i=1}^{3} \frac{\mathrm{d}H}{\partial q_i}\mathrm{d}q_i + \frac{\partial H}{\partial t}\mathrm{d}t \tag{13.296}$$

One equates Eqs. (13.295) and (13.296), and then transposes

$$\sum_{i=1}^{3}\left(\frac{\mathrm{d}q_i}{\mathrm{d}t} - \frac{\mathrm{d}H}{\partial p_i}\right)\mathrm{d}p_i + \sum_{i=1}^{3}\left(\frac{\mathrm{d}p_i}{\mathrm{d}t} + \frac{\mathrm{d}H}{\partial q_i}\right)\mathrm{d}q_i + \left(\frac{\partial L}{\partial t} + \frac{\partial H}{\partial t}\right)\mathrm{d}t = 0 \tag{13.297}$$

In Eq. (13.297), since dp_i, dq_i, and dt are arbitrary and can be zero or nonzero, one requires each of the three paired differences to equal zero, so that one obtains the following Hamiltonian relationships:

$$\frac{\mathrm{d}q_i}{\mathrm{d}t} = \frac{\mathrm{d}H}{\partial p_i}, \quad \frac{\mathrm{d}p_i}{\partial t} = -\frac{\mathrm{d}H}{\partial q_i}, \quad \frac{\partial L}{\partial t} = -\frac{\partial H}{\partial t} \tag{13.298}$$

In a conservative system, the Lagrangian L and Hamiltonian H do not depend explicitly on time t, and the Hamiltonian represents the total energy, which is a constant of motion, when

$$\frac{\partial L}{\partial t} = -\frac{\partial H}{\partial t} = 0 \tag{13.299}$$

If the Hamiltonian does not depend on one of the spatial coordinates q_i (e.g., $q_1 = x$), then that particular coordinate $q_1 = x$ is termed as a cyclic variable that is a constant of motion, when

$$\frac{\mathrm{d}p_1}{\partial t} = -\frac{\mathrm{d}H}{\partial q_1} = 0 \tag{13.300}$$

The spatial and velocity derivatives of the Lagrangian L and Hamiltonian H yield momentum and position, and hence, one can add a constant to the Lagrangian L or Hamiltonian H without changing the results.

13.16 Derivation of the Lagrangian for Relativistic Charged Particle in an Electromagnetic Field

So far, the expressions previously discussed can apply to either nonrelativistic or relativistic situations. However, to proceed with the goal of obtaining the Lagrange function L for the charged particle in an electromagnetic field requires explicit expressions for the force components F_x, F_y, and F_z. The magnetic vector potential is set to zero ($\mathbf{A} = 0$) to simplify the calculation of the correct relativistic form of the Lagrangian L, which will also be valid for nonzero vector potential ($\mathbf{A} \neq 0$).

With zero magnetic vector potential ($\mathbf{A} = 0$), the nonrelativistic Euler–Lagrange function L for charged particle with rest mass m_0 is

$$L = W - U = \frac{1}{2} m_0 \left[\left(\frac{\mathrm{d}x}{\mathrm{d}t}\right)^2 + \left(\frac{\mathrm{d}y}{\mathrm{d}t}\right)^2 + \left(\frac{\mathrm{d}z}{\mathrm{d}t}\right)^2 \right] - e\phi \tag{13.301}$$

Eq. (13.301) for the Lagrangian L is plugged into the Euler–Lagrange equation of Eq. (13.283), which then yields the x-, y-, and z-components of the nonrelativistic Lorentz force equation, where

$$m_0 \frac{\mathrm{d}}{\mathrm{d}t}\left(\frac{\mathrm{d}x}{\mathrm{d}t}\right) = -e\frac{\partial \phi}{\partial x} \Leftrightarrow \frac{\partial L}{\partial x} - \frac{\partial}{\partial t}\left(\frac{\partial L}{\partial(\partial x/\partial t)}\right) = 0 \tag{13.302}$$

$$m_0 \frac{\mathrm{d}}{\mathrm{d}t}\left(\frac{\mathrm{d}y}{\mathrm{d}t}\right) = -e\frac{\partial \phi}{\partial y} \Leftrightarrow \frac{\partial L}{\partial y} - \frac{\partial}{\partial t}\left(\frac{\partial L}{\partial(\partial y/\partial t)}\right) = 0 \tag{13.303}$$

and

$$m_0 \frac{\mathrm{d}}{\mathrm{d}t}\left(\frac{\mathrm{d}z}{\mathrm{d}t}\right) = -e\frac{\partial \phi}{\partial z} \Leftrightarrow \frac{\partial L}{\partial z} - \frac{\partial}{\partial t}\left(\frac{\partial L}{\partial(\partial z/\partial t)}\right) = 0 \tag{13.304}$$

The nonrelativistic form of the Lagrangian L of Eq. (13.301), has been confirmed to generate the Lorentz force equation, which has the x, y, and z components of Eqs. (13.302−13.304), via the Euler−Lagrange equations.

Next, we determine the relativistic correct Lagrangian L for the relativistic Lorentz force equations, whose x-, y-, and z-components are

$$\mathrm{F}_i = m_0 \frac{\mathrm{d}}{\mathrm{d}t}\left(\gamma \frac{\mathrm{d}q_i}{\mathrm{d}t}\right) = -e\frac{\partial \phi}{\partial q_i} \quad q_1, q_2, q_3 = x, y, z \tag{13.305}$$

Substitution of Eq. (13.307) into the second term of the Euler-Lagrange of Eq. (13.283), yields

$$\gamma = \left[1 - \frac{1}{c^2}\left(\frac{\mathrm{d}q_1}{\mathrm{d}t}\right)^2 - \frac{1}{c^2}\left(\frac{\mathrm{d}q_2}{\mathrm{d}t}\right)^2 - \frac{1}{c^2}\left(\frac{\mathrm{d}q_3}{\mathrm{d}t}\right)^2\right]^{-1/2} \tag{13.306}$$

The form of the x-, y-, and z-components of the relativistic Lorentz force equation and the relation $\mathrm{d}t = \gamma \mathrm{d}\tau$ between the proper, rest-frame time τ of the moving particle and the stationary, lab-frame time t suggest that the form of the Lagrangian L is

$$L = -\frac{m_0 c^2}{\gamma} - e\phi = -\sqrt{1-\beta^2} m_0 c^2 - e\phi \tag{13.307}$$

Substitution of Eq. (13.307) into the second term of the Euler-Lagrange of Eq. (13.283), yields

$$\frac{\mathrm{d}}{\mathrm{d}t}\left(\frac{\partial L}{\partial v_i}\right) = \frac{\mathrm{d}}{\mathrm{d}t}\left[\frac{\partial}{\partial v_i}\left(-m_0 c^2 \sqrt{1 - \frac{1}{c^2}(v_x^2 + v_y^2 + v_z^2)} - e\phi\right)\right] \quad i = x, y, z \tag{13.308}$$

For $i = x, y, z$, evaluation of Eq. (13.308), gives

$$\frac{\partial}{\partial v_i}\left(-m_0 c^2 \sqrt{1 - \frac{1}{c^2}(v_x^2 + v_y^2 + v_z^2)} - e\phi\right) = v_i m_0 \left[1 - \left(\frac{v_x^2 + v_y^2 + v_z^2}{c^2}\right)\right]^{-1/2} = \gamma m_0 v_i \tag{13.309}$$

Substitution of Eq. (13.307) into the first term of the Euler-Lagrange of Eq. (13.283), yields

$$\frac{\partial L}{\partial q_i} = \frac{\partial}{\partial q_i}\left(-m_0 c^2 \sqrt{1 - \frac{1}{c^2}(v_x^2 + v_y^2 + v_z^2)} - e\phi\right) = -e\frac{\partial \phi}{\partial q_i} \quad i = x, y, z \tag{13.310}$$

From Eqs. (13.308–13.310), the proposed Lagrangian L of Eq. (13.307), which is inserted into the Euler–Lagrange equation of Eq. (13.283), thus reproduces the relativistic Lorentz force equation for a charged particle in an applied electric field, where

$$\frac{\partial L}{\partial q_i} - \frac{\mathrm{d}}{\mathrm{d}t}\left(\frac{\partial L}{\partial(\partial q_i/\partial t)}\right) = 0 \Leftrightarrow m_0 \frac{\mathrm{d}}{\mathrm{d}t}\left(\gamma \frac{\mathrm{d}q_i}{\mathrm{d}t}\right) = -e\frac{\partial \phi}{\partial q_i} \quad i = 1, 2, \ldots, 3N \tag{13.311}$$

In what follows, the Euler–Lagrange equation will be used to identify the relativistic correct Lagrange function of an electron or other charged particle in an electromagnetic field. The scalar electric ϕ and magnetic vector potentials $\mathbf{A}$, and the resulting electric $\mathbf{E}$ and magnetic $\mathbf{B}$ fields, are required to satisfy the Lorentz force equation. The expression for the Lorentz force equation, which is then in terms of the applied scalar electric potential ϕ and the magnetic vector potential $\mathbf{A}$ that act on the electron, will be in the form of the Euler–Lagrange equation. With the Lagrange function L, we then obtain the Hamiltonian H for a charged particle, which propagates in an electric and magnetic field, where both the Lagrange function L and Hamiltonian H are calculated from the moving particle position, velocity, and time that is measured by an observer in the stationary, lab frame.

The Lorentz force equation for the force $\mathbf{F}$ of the electric $\mathbf{E}$ and magnetic $\mathbf{B}$ fields, which acts on an electron or other charged particle, is

$$\mathbf{F} = \frac{\mathrm{d}\mathbf{p}}{\mathrm{d}t} = e(\mathbf{E} + \mathbf{v} \times \mathbf{B}) \tag{13.312}$$

where

$$F_x = e(E_x + v_y B_z - v_z B_y) \tag{13.313}$$

$$F_y = e(E_y + v_z B_x - v_x B_z) \tag{13.314}$$

$$F_z = e(E_z + v_x B_y - v_y B_x) \tag{13.315}$$

With the x-component equation determined, one obtains the other two y- and z-component equations by inspection, using cyclic permutation of the x, y, and z vector components, where

$$\begin{array}{ll} + \text{ for cyclic clockwise permutation} & x \to y \to z \to x \\ - \text{ for cyclic counterclockwise permutation} & x \to z \to y \to x \end{array} \tag{13.316}$$

The electric **E** and magnetic **B** fields are expressed by the scalar electric potential ϕ and the magnetic vector potential **A** via

$$\mathbf{E} = -\nabla\phi - \frac{\partial \mathbf{A}}{\partial t} \tag{13.317}$$

$$\mathbf{B} = \nabla \times \mathbf{A} = \left(\frac{\partial A_z}{\partial y} - \frac{\partial A_y}{\partial z}\right)\hat{\mathbf{x}} + \left(\frac{\partial A_x}{\partial z} - \frac{\partial A_z}{\partial x}\right)\hat{\mathbf{y}} + \left(\frac{\partial A_y}{\partial x} - \frac{\partial A_x}{\partial y}\right)\hat{\mathbf{z}} \tag{13.318}$$

With substitution of Eqs. (13.317) and (13.318) into Eqs. (13.313–13.315), the resulting x-, y-, and z-components of the Lorentz force equation are

$$F_x = e\left[-\frac{\partial A_x}{\partial t} - \frac{\partial \phi}{\partial x} + v_y\left(\frac{\partial A_y}{\partial x} - \frac{\partial A_x}{\partial y}\right) - v_z\left(\frac{\partial A_x}{\partial z} - \frac{\partial A_z}{\partial x}\right)\right] \tag{13.319}$$

$$F_y = e\left[-\frac{\partial A_y}{\partial t} - \frac{\partial \phi}{\partial y} + v_z\left(\frac{\partial A_z}{\partial y} - \frac{\partial A_y}{\partial z}\right) - v_x\left(\frac{\partial A_y}{\partial x} - \frac{\partial A_x}{\partial y}\right)\right] \tag{13.320}$$

$$F_z = e\left[-\frac{\partial A_z}{\partial t} - \frac{\partial \phi}{\partial z} + v_x\left(\frac{\partial A_x}{\partial z} - \frac{\partial A_z}{\partial x}\right) - v_y\left(\frac{\partial A_z}{\partial y} - \frac{\partial A_y}{\partial z}\right)\right] \tag{13.321}$$

Rearranging Eqs. (13.319–13.321), we get

$$F_x = e\left[-\left(\frac{\partial A_x}{\partial t} + v_y\frac{\partial A_x}{\partial y} + v_z\frac{\partial A_x}{\partial z}\right) - \frac{\partial \phi}{\partial x} + \left(v_y\frac{\partial A_y}{\partial x} + v_z\frac{\partial A_z}{\partial x}\right)\right] \tag{13.322}$$

$$F_y = e\left[-\left(\frac{\partial A_y}{\partial t} + v_x\frac{\partial A_y}{\partial x} + v_z\frac{\partial A_y}{\partial z}\right) - \frac{\partial \phi}{\partial y} + \left(v_x\frac{\partial A_x}{\partial y} + v_z\frac{\partial A_z}{\partial y}\right)\right] \tag{13.323}$$

$$F_z = e\left[-\left(\frac{\partial A_z}{\partial t} + v_x\frac{\partial A_z}{\partial x} + v_y\frac{\partial A_z}{\partial y}\right) - \frac{\partial \phi}{\partial z} + \left(v_x\frac{\partial A_x}{\partial z} + v_y\frac{\partial A_y}{\partial z}\right)\right] \tag{13.324}$$

In Eqs. (13.322–13.324), addition and subtraction of the same term in each equation, give

$$F_x = e\left[-\left(\frac{\partial A_x}{\partial t} + v_x\frac{\partial A_x}{\partial x} + v_y\frac{\partial A_x}{\partial y} + v_z\frac{\partial A_x}{\partial z}\right) - \frac{\partial \phi}{\partial x} + \left(v_x\frac{\partial A_x}{\partial x} + v_y\frac{\partial A_y}{\partial x} + v_z\frac{\partial A_z}{\partial x}\right)\right] \tag{13.325}$$

$$F_y = e\left[-\left(\frac{\partial A_y}{\partial t} + v_x\frac{\partial A_y}{\partial x} + v_y\frac{\partial A_y}{\partial y} + v_z\frac{\partial A_y}{\partial z}\right) - \frac{\partial \phi}{\partial y} + \left(v_x\frac{\partial A_x}{\partial y} + v_y\frac{\partial A_y}{\partial y} + v_z\frac{\partial A_z}{\partial y}\right)\right] \tag{13.326}$$

$$F_z = e\left[-\left(\frac{\partial A_z}{\partial t} + v_x\frac{\partial A_z}{\partial x} + v_y\frac{\partial A_z}{\partial y} + v_z\frac{\partial A_z}{\partial z}\right) - \frac{\partial \phi}{\partial z} + \left(v_x\frac{\partial A_x}{\partial z} + v_y\frac{\partial A_y}{\partial z} + v_z\frac{\partial A_z}{\partial z}\right)\right] \tag{13.327}$$

The x-, y-, and z-components of the total time derivative $\mathbf{dA}/dt$ of vector potential $\mathbf{A}$ are

$$\frac{\mathrm{d}A_x}{\mathrm{d}t} = \frac{\partial A_x}{\partial t} + \left(\frac{\partial A_x}{\partial x}\frac{\partial x}{\partial t} + \frac{\partial A_x}{\partial y}\frac{\partial y}{\partial t} + \frac{\partial A_x}{\partial z}\frac{\partial z}{\partial t}\right) \tag{13.328}$$

$$\frac{\mathrm{d}A_y}{\mathrm{d}t} = \frac{\partial A_y}{\partial t} + \left(\frac{\partial A_y}{\partial x}\frac{\partial x}{\partial t} + \frac{\partial A_y}{\partial y}\frac{\partial y}{\partial t} + \frac{\partial A_y}{\partial z}\frac{\partial z}{\partial t}\right) \tag{13.329}$$

$$\frac{\mathrm{d}A_z}{\mathrm{d}t} = \frac{\partial A_z}{\partial t} + \left(\frac{\partial A_z}{\partial x}\frac{\partial x}{\partial t} + \frac{\partial A_z}{\partial y}\frac{\partial y}{\partial t} + \frac{\partial A_z}{\partial z}\frac{\partial z}{\partial t}\right) \tag{13.330}$$

With substitution of Eqs. (13.328–13.330) into Eqs. (13.325–13.327), the x-, y-, and z-components of the Lorentz force equation become

$$F_x = -e\frac{\mathrm{d}A_x}{\mathrm{d}t} + e\frac{\partial}{\partial x}[-\phi + \mathbf{v}\cdot\mathbf{A}] \tag{13.331}$$

$$F_y = -e\frac{\mathrm{d}A_y}{\mathrm{d}t} + e\frac{\partial}{\partial y}[-\phi + \mathbf{v}\cdot\mathbf{A}] \tag{13.332}$$

$$F_z = -e\frac{\mathrm{d}A_z}{\partial t} + e\frac{\partial}{\partial z}[-\phi + \mathbf{v}\cdot\mathbf{A}] \tag{13.333}$$

The particle position components q_j are independent of the velocity components $\partial q_i/\partial t$, which was applied to Eqs. (13.301–13.304), where

$$\frac{\partial}{\partial q_j}\left(\frac{\partial q_i}{\partial t}\right) = 0 \quad i,j = x,y,z \tag{13.334}$$

Substitution into Eqs. (13.331–13.333) of Eq. (13.305) for the x-, y-, and z-components of the force gives

$$\frac{\mathrm{d}}{\mathrm{d}t}(m_0 v_x \gamma + eA_x) = e\frac{\partial}{\partial x}[-\phi + \mathbf{v}\cdot\mathbf{A}] \tag{13.335}$$

$$\frac{\mathrm{d}}{\mathrm{d}t}(m_0 v_y \gamma + eA_y) = e\frac{\partial}{\partial y}[-\phi + \mathbf{v}\cdot\mathbf{A}] \tag{13.336}$$

$$\frac{\mathrm{d}}{\mathrm{d}t}(m_0 v_z \gamma + eA_z) = e\frac{\partial}{\partial z}[-\phi + \mathbf{v}\cdot\mathbf{A}] \tag{13.337}$$

Based on the results of Eqs. (13.335–13.337), the force equation can be expressed in terms of a canonical momentum vector $\mathbf{p}$ and a generalized potential U, where

$$\frac{\mathrm{d}\mathbf{p}}{\mathrm{d}t} = \frac{\partial}{\partial x}[-e\phi + e\mathbf{v}\cdot\mathbf{A}] \tag{13.338}$$

The canonical momentum $\mathbf{p}$ is the sum of the mechanical momentum $m_0\gamma(\mathrm{d}\mathbf{q}/dt)$ and the momentum contribution from the magnetic vector potential $\mathbf{A}$, where

$$\mathbf{p} = m_0\gamma\frac{\mathrm{d}\mathbf{q}}{\mathrm{d}t} + e\mathbf{A} \tag{13.339}$$

The generalized potential energy U is the sum of the potential energy contributions from the magnetic vector potential $\mathbf{A}$ and scalar electric potential ϕ, where

$$U = e\mathbf{v}\cdot\mathbf{A} - e\phi \tag{13.340}$$

The time derivatives of the three spatial components of the vector potential $\mathbf{A}$ can be expressed by

$$\frac{\mathrm{d}A_i}{\mathrm{d}t} = \frac{\mathrm{d}}{\mathrm{d}t}\left[\frac{\partial}{\partial v_i}(\mathbf{v}\cdot\mathbf{A})\right] \quad i = x, y, z \tag{13.341}$$

The scalar electric potential is independent of the particle velocity, where

$$\frac{\partial\phi}{\partial v_i} = 0 \quad i = x, y, z \tag{13.342}$$

Hence,

$$\frac{\mathrm{d}A_i}{\mathrm{d}t} = \frac{\mathrm{d}}{\mathrm{d}t}\left[\frac{\partial}{\partial v_i}(\mathbf{v}\cdot\mathbf{A} - \phi)\right] \quad i = x, y, z \tag{13.343}$$

From Eqs. (13.331–13.333), the Lorentz force equation is

$$F_i = -e\frac{\mathrm{d}A_i}{\mathrm{d}t} + e\frac{\partial}{\partial q_i}[-\phi + \mathbf{v}\cdot\mathbf{A}] \quad i = x, y, z \tag{13.344}$$

Substitution of Eq. (13.343) for $\mathrm{d}A_i/dt$ into the Lorentz force equation of Eq. (13.344), yields the generalized Lorentz force equation, which is expressed in terms of the generalized potential energy U, where

$$F_i = -e\frac{\mathrm{d}}{\mathrm{d}t}\left[\frac{\partial}{\partial(\partial q_i/\partial t)}(e\mathbf{v}\cdot\mathbf{A} - e\phi)\right] + \frac{\partial}{\partial q_i}[e\mathbf{v}\cdot\mathbf{A} - e\phi] \quad i = x, y, z \tag{13.345}$$

The generalized force equation has the form of the Euler–Lagrange equation of Eq. (13.283).

Without the magnetic vector potential ($\mathbf{A} = 0$), the relativistic Lagrangian L is

$$L = -\frac{m_0 c^2}{\gamma} - e\phi = -\sqrt{1-\beta^2}\, m_0 c^2 - e\phi \tag{13.346}$$

With the addition of nonzero magnetic vector potential ($\mathbf{A} \neq 0$), the Lagrangian for a relativistic charged particle in an electromagnetic field then becomes

$$L = -\sqrt{1-\beta^2}\, m_0 c^2 + e\mathbf{v}\cdot\mathbf{A} - e\phi \tag{13.347}$$

Recalling Eq. (13.288), the Hamiltonian H is

$$H(p_i, q_i, t) = \sum_{i=1}^{3} p_i \frac{\mathrm{d}q_i}{\mathrm{d}t} - L \tag{13.348}$$

where

$$\sum_{i=1}^{3} p_i \frac{\mathrm{d}q_i}{\mathrm{d}t} = \left(m_0 \gamma \frac{\mathrm{d}\mathbf{q}}{\mathrm{d}t} + e\mathbf{A} \right) \cdot \frac{\mathrm{d}\mathbf{q}}{\mathrm{d}t} = \gamma m_0 v^2 + e\mathbf{v}\cdot\mathbf{A} \tag{13.349}$$

and

$$\gamma m_0 v^2 = m_0 c^2 \frac{\beta^2}{\sqrt{1-\beta^2}} \tag{13.350}$$

Substitution of the Lagrangian L of Eq. (13.347) into the Hamiltonian H of Eq. (13.348), via Eqs. (13.349–13.350), gives

$$H = \left(m_0 c^2 \frac{\beta^2}{\sqrt{1-\beta^2}} + e\mathbf{v}\cdot\mathbf{A} \right) - \left(-\sqrt{1-\beta^2}\, m_0 c^2 + e\mathbf{v}\cdot\mathbf{A} - e\phi \right) \tag{13.351}$$

Converting from β to γ in Eq. (13.351), yields the Hamiltonian for a relativistic charged particle in an electromagnetic field, where

$$H = m_0 c^2 \gamma + e\phi \tag{13.352}$$

From Eq. (13.339), the relativistic mechanical momentum $\mathbf{P}$ is given by

$$\mathbf{P} = \gamma m_0 \mathbf{v} = \mathbf{p} - e\mathbf{A} \tag{13.353}$$

From Eq. (13.90), recall the invariant magnitude of the 4-vector for momentum-energy, expresses the relativistic conservation of energy and momentum, where

$$m_0c^2\gamma = c\sqrt{m_0^2c^2 + P^2} \tag{13.354}$$

With substitution of Eq. (13.354) into Eq. (13.352), the Hamiltonian becomes

$$H = c\sqrt{m_0^2c^2 + P^2} + e\phi \tag{13.355}$$

If one includes the magnetic vector potential **A**, then the mechanical momentum vector **P** of Eq. (13.355) is replaced by $\mathbf{p} - e\mathbf{A}$ of Eq. (13.355), so that the Hamiltonian for the charged particle in an electromagnetic field becomes

$$H = c\sqrt{m_0^2c^2 + (\mathbf{p} - e\mathbf{A})\cdot(\mathbf{p} - e\mathbf{A})} + e\phi \tag{13.356}$$

For information on the Lagrangian formulation for the electromagnetic field equations and the energy-momentum tensor of the electromagnetic field, see Marion (1965a).

13.17 Euler–Lagrange Equation for Langrangian Density for Real Scalar Fields

In this section, we derive the Euler–Lagrange equations for fields, which is a starting point in quantum field theory. Following Schiff (1968), the action is defined as the time integral of the Lagrange function L, where

$$G = \int_{t_1}^{t_2} L\,\mathrm{d}t \tag{13.357}$$

The Lagrangian L in the integrand is integrated over time. This implies the Lagrangian represents the temporal phase shift (angle) per unit time, and $L\,\mathrm{d}t$ is the phase shift (angle), which occurs in time $\mathrm{d}t$. In classical or quantum mechanics, an extended body or region can be described the Hamiltonian density $\tilde{H}$ and Lagrangian density $\tilde{L}$, where

$$H = \iiint \tilde{H}\,\mathrm{d}x\,\mathrm{d}y\,\mathrm{d}z \quad \text{and} \quad L = \iiint \tilde{L}\,\mathrm{d}x\,\mathrm{d}y\,\mathrm{d}z \tag{13.358}$$

The action can be expressed by the Lagrangian density $\tilde{L}$, where

$$G = \int_{t_1}^{t_2} \int_{-\infty}^{\infty} \mathrm{d}z \int_{-\infty}^{\infty} \mathrm{d}y \int_{-\infty}^{\infty} \mathrm{d}x \, \tilde{L} \, \mathrm{d}t \tag{13.359}$$

Between inertial frames with constant relative velocity, the Lagrangian L transforms by factor $1/\gamma$, the Hamiltonian transforms by factor γ, and the differential volume $\mathrm{d}x\,\mathrm{d}y\,\mathrm{d}z$ transforms (contracts) by factor $1/\gamma$. As a result, the Hamiltonian density $\tilde{H}$ transforms by factor γ^2, whereas the Lagrangian density $\tilde{L}$ transforms by a factor of unity or 1. That is, the Lagrangian density $\tilde{L}$ is the same in all reference frames and is termed a *Lorentz scalar*, and consequently, the Lagrangian density $\tilde{L}$ plays a fundamental role in quantum field theory and the standard model of physics. Lorentz invariant equations of motion can be expressed by the Lagrangian density $\tilde{L}$.

The Euler–Lagrange equations for fields in quantum field theory do not depend on the generalized coordinate q_i and velocity $\partial q_i/dt$ of particles; instead, they depend on a field amplitude $\psi(t, r)$. In quantum field theory, one employs the Lagrangian density $\tilde{L}$, which depends only upon field amplitude $\psi(t, r)$ and its first-order derivatives, where

$$\tilde{L} \equiv \tilde{L}\left(\psi, \frac{\partial \psi}{\partial t}, \frac{\partial \psi}{\partial x}, \frac{\partial \psi}{\partial y}, \frac{\partial \psi}{\partial z}, t\right) = \tilde{L}\left(\psi, \frac{\partial \psi}{\partial t}, \nabla\psi, t\right) \tag{13.360}$$

As is the case with the Lagrangian, the Euler–Lagrange equations for the Lagrangian density $\tilde{L}$ begin with the variation of the action set to zero, where

$$\delta G = \delta \int_{t_1}^{t_2} \mathrm{d}t \int_{z_1}^{z_2} \mathrm{d}z \int_{y_1}^{y_2} \mathrm{d}y \int_{x_1}^{x_2} \mathrm{d}x \, \tilde{L}\left(\psi, \frac{\partial \psi}{\partial t}, \nabla\psi, t\right) = 0 \tag{13.361}$$

The Lagrangian density $\tilde{L}$ is enclosed in a region with a fixed volume V and enclosure surface S, which are bounded by fixed spatial endpoints x_1 and x_2 along x, y_1 and y_2 along y, and z_1 and z_2 along z, as well as fixed initial and final time points t_1 and t_2.

Carrying out the variation on the right side of the Eq. (13.361), one obtains

$$\delta G = \int_{t_1}^{t_2} \mathrm{d}t \int_{z_1}^{z_2} \mathrm{d}z \int_{y_1}^{y_2} \mathrm{d}y \int_{x_1}^{x_2} \mathrm{d}x \left[\frac{\partial \tilde{L}}{\partial \psi}\delta\psi + \left(\frac{\partial \tilde{L}}{\partial \nabla\psi}\right) \cdot (\nabla\delta\psi) + \frac{\partial \tilde{L}}{\partial(\partial\psi/\partial t)}\delta\frac{\partial \psi}{\partial t}\right] = 0 \tag{13.362}$$

where

$$\frac{\partial \tilde{L}}{\partial \nabla\psi} = \frac{\partial \tilde{L}}{\partial(\partial\psi/\partial x)}\hat{\mathbf{x}} + \frac{\partial \tilde{L}}{\partial(\partial\psi/\partial y)}\hat{\mathbf{y}} + \frac{\partial \tilde{L}}{\partial(\partial\psi/\partial z)}\hat{\mathbf{z}} \tag{13.363}$$

One performs integration by parts on the the time integral of Eq. (10.362), where the last term evaluates to zero, because $\delta\psi$ is defined to be zero at the fixed time endpoints t_1 and t_2, and the fixed boundary surfaces at x_1 and x_2, y_1 and y_2, and z_1 and z_2, where

$$\int_{t_1}^{t_2} \mathrm{d}t \int_{z_1}^{z_2} \mathrm{d}z \int_{y_1}^{y_2} \mathrm{d}y \int_{x_1}^{x_2} \mathrm{d}x \, \frac{\partial \tilde{L}}{\partial(\partial\psi/\partial t)} \delta \frac{\partial \psi}{\partial t} = \int_{z_1}^{z_2} \mathrm{d}z \int_{y_1}^{y_2} \mathrm{d}y \int_{x_1}^{x_2} \mathrm{d}x \left[\frac{\partial \tilde{L}}{\partial(\partial\psi/\partial t)} \delta\psi \right]_{t_1}^{t_2} = 0 \tag{13.364}$$

The first three terms of the integrand remain of Eq. (13.362), and a second instance of integration by parts, but now over the spatial coordinates, yields

$$\delta G = \int_{t_1}^{t_2} \mathrm{d}t \int_{z_1}^{z_2} \mathrm{d}z \int_{y_1}^{y_2} \mathrm{d}y \int_{x_1}^{x_2} \mathrm{d}x \left[\frac{\partial \tilde{L}}{\partial\psi} \delta\psi - \nabla \cdot \left(\frac{\partial \tilde{L}}{\partial \nabla\psi} \right) \delta\psi - \frac{\partial}{\partial t} \left(\frac{\partial \tilde{L}}{\partial(\partial\psi/\partial t)} \right) \delta\psi \right] = 0 \tag{13.365}$$

The surface is fixed, and thus, the spatial variation of the endpoints of the x-, y-, and z-integrals are fixed, so that the second term in Eq. (13.365) integrates to zero, where

$$\int_{t_1}^{t_2} \mathrm{d}t \left(\frac{\partial \tilde{L}}{\partial \nabla\psi} \right)_{x_1}^{x_2} = \int_{t_1}^{t_2} \mathrm{d}t \left(\frac{\partial \tilde{L}}{\partial \nabla\psi} \right)_{y_1}^{y_2} = \int_{t_1}^{t_2} \mathrm{d}t \left(\frac{\partial \tilde{L}}{\partial \nabla\psi} \right)_{z_1}^{z_2} = 0 \tag{13.366}$$

The field variation $\delta\psi$ and the fixed boundary surface, which contains the Lagrangian density $\tilde{L}$, are both arbitrary. As a result, the integrand of Eq. (13.365) must be zero, since $\delta\psi$ can be nonzero, and thus one obtains the Euler–Lagrange equation for the Lagrangian density $\tilde{L}$, where

$$\frac{\partial \tilde{L}}{\partial\psi} - \nabla \cdot \left(\frac{\partial \tilde{L}}{\partial \nabla\psi} \right) - \frac{\partial}{\partial t} \left(\frac{\partial \tilde{L}}{\partial(\partial\psi/\partial t)} \right) = 0 \tag{13.367}$$

Expanding Eq. (13.367), the Euler-Lagrange equation for the Lagrangian density, becomes

$$\frac{\partial \tilde{L}}{\partial\psi} - \frac{\partial}{\partial x} \left(\frac{\partial \tilde{L}}{\partial(\partial\psi/\partial x)} \right) - \frac{\partial}{\partial y} \left(\frac{\partial \tilde{L}}{\partial(\partial\psi/\partial y)} \right) - \frac{\partial}{\partial z} \left(\frac{\partial \tilde{L}}{\partial(\partial\psi/\partial z)} \right) - \frac{\partial}{\partial t} \left(\frac{\partial \tilde{L}}{\partial(\partial\psi/\partial t)} \right) = 0 \tag{13.368}$$

One can express the Euler–Lagrange equation of Eq. (13.368) by the four-vector space–time derivative, where

$$\nabla_4 = \nabla + \frac{1}{\mathrm{i}c} \frac{\partial}{\partial t} = \hat{x} \frac{\partial}{\partial x} + \hat{y} \frac{\partial}{\partial y} + \hat{z} \frac{\partial}{\partial z} - \frac{\mathrm{i}}{c} \frac{\partial}{\partial t} \tag{13.369}$$

and

$$\sum_{i=1}^{4} \frac{\partial}{\partial \mu_i} \hat{\mathbf{u}}_i = \hat{x} \frac{\partial}{\partial x} + \hat{y} \frac{\partial}{\partial y} + \hat{z} \frac{\partial}{\partial z} - \frac{\mathrm{i}}{c} \frac{\partial}{\partial t} \tag{13.370}$$

With application of Eqs. (13.369) and (13.370) to Eq. (13.368), the Euler–Lagrange equation for the Lagrangian density $\tilde{L}$ in four-vector notation becomes

$$\frac{\partial \tilde{L}}{\partial \psi} - \sum_{i=1}^{4} \frac{\partial}{\partial \mu_i} \left(\frac{\partial \tilde{L}}{\partial(\partial \psi / \partial \mu_i)} \right) = 0 \tag{13.371}$$

References

Adams, S. (1997). *Relativity, an introduction to space–time physics*. London: Taylor & Francis.

Becker, R. (1964). *Electromagnetic fields and interactions*. New York, NY: Dover.

Chen, F. (1974). *Introduction to plasma physics*. New York, NY: Plenum Press.

Destler, W. W., & Cremer, J. T. (1983). Charge state measurements of collectively accelerated heavy ions. *Journal of Applied Physiology*, *54*(2), 636–639.

Feynman, R. P. (1964). *The Feynman lectures on physics* (Vol. II, (mainly electromagnetism and matter)). Reading, MA: Addison Wesley. (mainly electromagnetism and matter).

Foster, J., & Nightingale, J. D. (1995). *A short course in general relativity*. New York, NY: Springer Verlag.

French, A. P. (1966). *Special relativity*. New York, NY: W.W. Norton & Company.

Friedrichs, K. O. (1975). *From Pythagoras to Einstein*. Washington, DC: Mathematical Association of America.

Hora, H. (2000). *Laser plasma physics*. Bellingham, WA: SPIE Press.

Jackson, J. D. (1975). *Classical electrodynamics*. New York, NY: Wiley.

Kleppner, D., & Kolenkow, R. J. (1973). *An introduction to mechanics*. New York, NY: McGraw-Hill.

Landau, L. M., & Lifshitz, E. M. (1975). *The classical theory of fields* (4th ed.). New York, NY: Pergamon Press.

Lawson, J. D. (1977). *The physics of charged particle beams*. Oxford: Clarendon Press.

Marion, J. B. (1965a). *Classical electromagnetic radiation*. New York, NY: Academic Press.

Marion, J. B. (1965b). *Classical dynamics of particles and systems*. New York, NY: Academic Press.

Panofsky, W. K. H., & Phillips, M. (2005). *Classical electricity and magnetism*. New York, NY: Dover Publications.

Pathria, R. K. (2003). *The theory of relativity*. New York, NY: Dover Publications.

Pauli, W. (1981). *The theory of relativity*. New York, NY: Dover Publications.

Purcell, E. M. (1965). *Electricity and magnetism, Berkeley physics course* (Vol. 2). New York, NY: McGraw-Hill.

Reiser, M. P. (1994). *The theory and design of charged particle Beams*. New York, NY: John Wiley & Sons.

Schiff, L. I. (1968). *Quantum mechanics*. New York, NY: McGraw-Hill.

Schwartz, M. (1988a). *Principles of electrodynamics*. New York, NY: Dover.

Shadowitz, A. (1988a). *The electromagnetic field*. New York, NY: Dover.

Shadowitz, A. (1988b). *Special relativity*. New York, NY: Dover.

Sommerfeld, A. (1964). *Electrodynamics*. New York, NY: Academic Press.

Stratton, J. A. (1941). *Electromagnetic theory*. New York, NY: McGraw-Hill.

14 The Klein–Gordon and Dirac Equations

14.1 Relativistic Correct Schrödinger Wave Equation—the Klein–Gordon Equation

In this chapter, we cover the relativistic correct Schrödinger wave equation, namely, the Klein–Gordon equation and the Dirac wave equation for spin 1/2 particles. The general Dirac equation is presented and then solved for a free particle. Next, we derive the relativistic-based Dirac equation for the interaction of a bound electron interaction with a proton. The resulting Hamiltonian for the proton and bound electron is evaluated to yield the LS coupling term of magnetic neutron scatter and the hyperfine interaction terms for the atomic spectra of the hydrogen atom and hydrogen-like atoms. Also obtained are the equations of motion of the electron spin, orbital, and total angular momentum vectors of the bound electron.

Next, we derive the Dirac equation for the hydrogen and hydrogen-like atoms in the form of a pair of coupled Dirac eigenequations, which are then solved to yield asymptotic and then regular solutions. From the regular solution of the coupled Dirac eigenequation pair, we then obtain the electron energy levels formula for the hydrogen atom (and hydrogen-like atoms) and the quantum number relationships. Finally, we derive the nuclear electric quadrupole potential, which is obtained from the electric potential energy term of the Dirac Hamiltonian. This chapter follows the treatments of the Dirac equation in Liboff (2003), Schiff (1968), and Merzbacher (1970).

Schrödinger derived the relativistic wave equation in 1925, and independently Klein and Gordon published the same relativistic wave equation in 1927. For derivations, discussion, and application of the Schrödinger equation to X-ray and neutron optics, see Cremer (2012a, 2012b, and 2013). To obtain the relativistically correct Schrödinger equation, which is known as the *Klein–Gordon equation*, one starts with a Schrödinger equation, which has its operator form, equated with its equivalent differential form, and with potential energy set to zero $V = 0$, so that

$$H\Psi = E\Psi = i\hbar \frac{\partial}{\partial t}\Psi \tag{14.1}$$

Neutron and X-ray Optics. DOI: http://dx.doi.org/10.1016/B978-0-12-407164-3.00013-9

In Eq. (14.1), one is allowed to square the operator H and its corresponding eigenvalue E in their eigenequation for eigenfunction Ψ:

$$H^2\Psi = E^2\Psi = \left(\mathrm{i}\hbar\frac{\partial}{\partial t}\right)^2\Psi \tag{14.2}$$

Recall from section 13.7 and Eq. (13.90), conservation of energy requires the relativistic energy-momentum equation. One inserts the momentum operator $-\mathrm{i}\hbar\nabla$ in place of momentum p in the relativistic expression for E to obtain

$$E^2 = c^2(-\mathrm{i}\hbar\nabla)^2 + m^2c^2 \tag{14.3}$$

Here, m is the particle rest mass. Substitution of Eq. (14.3) into Eq. (14.2), yields

$$-\hbar^2c^2\nabla^2\Psi + m^2c^4\Psi = -\hbar^2\frac{\partial^2}{(\partial t)^2}\Psi \tag{14.4}$$

Transposing Eq. (14.4), one obtains the Klein–Gordon equation, which is the relativistically correct version of the Schrödinger wave equation, where

$$\frac{1}{c^2}\frac{\partial^2}{\partial t^2}\Psi - \nabla^2\Psi + \frac{m^2c^2}{\hbar^2}\Psi = 0 \tag{14.5}$$

In the course of normal neutron optics, the nonrelativistic Schrödinger equation is sufficient for most situations. However, the Lorentz invariant Maxwell's equations or special relativity is needed to understand spin–orbit magnetic scatter of neutrons or LS coupling, which leads to magnetic scatter of neutrons in magnetic materials. An example of this is in the spin–orbit interaction of the unpaired atomic electron electric field, which produces a magnetic field in the rest frame of a neutron moving at relative velocity v through the atomic electric field. The magnetic field in the neutron rest frame deflects the neutron via the neutron magnetic dipole moment of the neutron.

Another example is the LS coupling between the electric field of the nucleus and the electron, which moves with relative velocity v through the nuclear electric field in the course of the electron orbital motion. In the electron rest frame, the electron velocity, relative to the nucleus electric field, produces a magnetic field in the electron rest frames, which interacts with the electron spin magnetic dipole moment. This causes the electron to have a spin potential energy, which is based on the electron spin orientation in the magnetic field appearing in its rest frame. To calculate the spin–orbit potential energy in the lab frame, where the nucleus is at rest, and the electron is moving at velocity v, relative to the lab frame, one must perform a relativistic transformation of the velocities, that is, the Thomas precession.

14.2 The Dirac Wave Equation for Spin 1/2 Particles—Overview

To better understand the magnetic scatter of neutrons by unpaired atomic electrons, we present the derivation of the Dirac equation in the next few sections. The Dirac equation predicts LS coupling, which provides the effective magnetic field encountered by the magnetic dipole moment of an incident neutron that causes magnetic neutron scatter. The previously derived Klein–Gordon equation applies to the case of spinless particles. However, electrons, protons, and neutrons have spin, and as such, a relativistic correct wave equation accounting for spin is needed to describe the LS coupling and include the $1/2$ factor of Thomas precession. Also, any solution of the Dirac equation is also a solution to the Klein–Gordon equation, derived in Section 14.1.

The Dirac equation is a quantum mechanical wave equation, which is applied to spin 1/2 particles, such as electrons, protons, and neutrons. The Dirac equation includes the Lorentz transformation and total energy of special relativity. The solutions of the Dirac required the existence of antiparticles, such as the positron discovered by Carl Anderson in 1932 after Paul Dirac's formulation of the equation in 1928. Dirac discovered his equation while staring into a fireplace in Cambridge (see Zee, 2003 for details), and two years earlier, in 1926, Schrödinger formulated his wave equation after a vacation in a Swiss chalet, which presumably had at least one fireplace. The indoor or outdoor fireplace, being the source of warmth, light, and, cooking, has served as a meeting place for minds or a place to contemplate one's thoughts.

The solution of the Dirac equation for the hydrogen atom predicted the experimentally verified fine and hyperfine spectral lines of the atomic electron states measured by spectroscopy. The Dirac equation prediction of the experimentally verified spin, antiparticles, and the fine and hyperfine structure of the hydrogen atomic spectra, thus laid the foundations for quantum mechanics and its applications. The Dirac equation is normally applied to electrons. However, the Dirac equation can be applied to protons and neutrons, using proton mass and neutron mass in place of electron mass, and then further modified for neutrons by dropping the Coulomb term, although one may wish to retain the spin–orbit and Foldy neutron interactions with an imposed electric field. The Dirac equation is for particles of 1/2 spin, and similar to the Dirac equation, the Rarita–Schwinger equation is used to describe particles with 3/2 spin.

The Dirac equation does not include terms to create and remove particles, such as photon absorption and emission from electrons, and nor does the Dirac equation predict the Lamb shift, where the interaction of the electron with the vacuum leads to tiny shifts in the electron energy, which was measured by Lamb and Retherford (1947). They used microwaves to stimulate radio-frequency transitions between $^2S_{1/2}$ and $^2P_{1/2}$ levels of hydrogen, which showed a rise at 1.0 GHz of the $^2S_{1/2}$ above the $^2P_{1/2}$ level. At these long wavelengths, Doppler broadening, which is proportional to frequency, can be ignored.

Another effect not included in the Dirac equation is the anomalous magnetic moment of the electron. The Dirac equation predicts a g factor of 2 for the electron, which differs slightly from the experimentally measured value. The anomalous magnetic moment of the electron $a = \alpha/2\pi$ or $a = 0.001159$, was calculated by application of quantum electrodynamics by Julian Schwinger in 1948. The calculated anomalous magnetic moment of the electron is defined in terms of the fine-structure constant $\alpha = e^2/4\pi\varepsilon_0\hbar c$ or $\alpha = 137.036$, which is obtained from Planck's reduced constant $\hbar$, electron charge e, speed of light c, and the permittivity of free space ε_0.

The Dirac equation was then reformulated from a particle-based equation to field-based equation, where the electromagnetic field is quantized. This quantum field theory approach with the Dirac equation led to the development of quantum electrodynamics. In quantum field theory, the Lamb shift is explained by the virtual electric and magnetic fields in a vacuum that perturb Coulomb potential of the proton, which act on the electron in the hydrogen atom, and then creates a variation in the position of the electron, thus causing the observed Lamb shift. The quantized electromagnetic field of quantum field theory then predicts the Casimir–Polder force, which can be experimentally measured between two uncharged metallic plates in a vacuum with only a few microns or less separation in the absence of electromagnetic fields; see Bressi et al. (2002). The presence of the plates alters the expectation value of the quantized electromagnetic field of the vacuum. However, the Casimir force can also be interpreted as a van der Waals force; see Dzyaloshinskii, Lifshitz, and Pitaevskii (1961).

Zero-point energy is the lowest possible energy level or ground state in a quantized electromagnetic field, which interacts with physical system of particles, which are represented by de Broglie waves with quantized energy levels. That is, the zero-point energy is the energy of the ground state, where the ground state is positive and above zero. Zero-point energy, for example, occurs in the quantum mechanical simple harmonic oscillator wave functions. The quantized energy levels of the harmonic oscillator occur in a parabolic shape potential well, which is suspended above the zero-potential energy plane by the height of the zero-point energy. The concept of zero-point energy originated in 1910–1920 by the German school of scientists (Planck, Einstein, Stern, and Nernst) who postulated that all vacuum space is filled with a zero-point electromagnetic field. They postulated that the vacuum electromagnetic radiation arose due to fluctuations of the zero-point energy or ground state energy level of the quantized fields, which was caused by the Heisenberg uncertainty principle. The Heisenberg uncertainty principle states that all physical systems, including those at absolute zero temperature, have a zero-point energy greater than zero, such as the case of liquid helium-4 discovered by Kamerlingh Onnes in 1908. Zero-point energy is used to explain the experimentally observed liquid helium-4, which remains liquid at atmospheric pressure even at zero temperature and solidifies only at pressures above 25 atm.

In the case of the Casimir force between two closely spaced, uncharged metallic plates, the vacuum energy includes the electromagnetic radiation of all wavelengths, except wavelengths, which are excluded, due to boundary conditions imposed by the spacing between the two plates. As one brings the plates closer together, more wavelengths are excluded, 6+ and the vacuum energy between the plates decreases, relative to the vacuum outside the plates. The vacuum outside the plates is comprised of the full range of wavelengths. As the separation distance $\Delta x < 0$ between the two plates decreases, the potential energy $\Delta U < 0$ between the two plates decreases. As a result, there arises a force $F = -\nabla U$, which is in the negative x-direction, where $F = -\Delta U/\Delta x$ and $F < 0$. That is, the force acts in the direction to press the plates together. The Casimir force increases, with decreasing distance, between uncharged conductors or dielectrics. Whether due to the virtual vacuum electromagnetic fields, or van der Waal forces in the nanometer realm, the Casimir force is a very significant force, which has potential applications in nanotechnology.

14.3 Derivation of the Dirac Equation that Predicts Correct LS Coupling Term for Unpaired Atomic Electrons in Magnetic Neutron Scatter

The Dirac equation was formulated by Dirac under the demand that the equations of physics must have the same form under Lorentz transformations. Also, the equation should account for electron spin, which was demonstrated by the Stern–Gerlach experiment 7 years earlier, in 1922. The negative energy solutions, which were caused by the second derivative of time in the Klein Gordon equation, motivated Dirac to develop an equation with a single derivative of time. And due to the interchange of space and time in the Lorentz transformation, Dirac was motivated to develop the equation with single spatial and temporal derivatives.

Given the Schrödinger wave equation, the Klein–Gordon equation, and the Lorentz transformations, Dirac proposed a wave equation for a particle of rest mass m in the form

$$\left(i\hbar\frac{\partial}{\partial t} + i\hbar c\boldsymbol{\alpha}\cdot\nabla\right)\psi = \beta mc^2\psi \tag{14.6}$$

Vector parameter $\boldsymbol{\alpha}$ and scalar parameter β are to be determined by comparing this Dirac equation with the Klein–Gordon equation, which both apply to relativistic particles. With rearrangement and squaring of Eq. (14.6), the Dirac equation becomes

$$-\hbar^2\frac{\partial^2\psi}{\partial t^2} = (-i\hbar c\boldsymbol{\alpha}\cdot\nabla + \beta mc^2)^2\psi \tag{14.7}$$

The Dirac equation of Eq. (14.7), applies to relativistic particles of rest mass m with spin 1/2, such as electrons, protons, and neutrons.

Recall from section 13.7 and Eq. (13.90), conservation of energy requires the relativistic energy-momentum equation:

$$E^2\psi = (c^2(\mathbf{p}\cdot\mathbf{p}) + m^2c^4)\psi \tag{14.8}$$

Here, m is the electron rest mass and c is the speed of light. In the wave viewpoint, let total particle energy E and particle momentum $\mathbf{p}$ be expressed in their operator forms:

$$E \rightarrow \mathbf{ih}\frac{\partial}{\partial t} \quad \text{and} \quad \mathbf{p} \rightarrow -i\hbar\nabla \tag{14.9}$$

Substitution of the time derivative operator for E and gradient operator for $\mathbf{p}$ of Eq. (14.9) in the relativistic energy-momentum equation of Eq. (14.8), gives the Klein–Gordon equation:

$$-\hbar^2\frac{\partial^2\psi}{\partial t^2} = -\hbar^2c^2\nabla^2\psi + m^2c^4\psi \tag{14.10}$$

With substitution of Eq. (14.9) into Eq. (14.6), the time-independent Dirac equation is

$$E\psi = (c\boldsymbol{\alpha}\cdot\mathbf{p} + \beta mc^2)\psi \tag{14.11}$$

One expresses the eigenequation of Eq. (14.11) with the Dirac Hamiltonian H_{D} operator, where

$$H_{\mathrm{D}}\psi = E\psi \tag{14.12}$$

Comparison of Eqs. (14.11) and (14.12), yields the Dirac Hamiltonian operator H_D:

$$H_{\mathrm{D}} = c\boldsymbol{\alpha}\cdot\mathbf{p} + \beta mc^2 \tag{14.13}$$

The solution of the relativistic Dirac equation of Eq. (14.7) for a free particle, is required to also be a solution of the relativistic Klein Gordon equation of Eq. (14.10) for a free particle. This requires equating the relativistic energy-momentum expression of Eq. (14.8) for the Klein Gordon equation, with the squared relativistic energy-momentum expression of Eq. (14.13) for the Dirac equation.

$$c^2\mathbf{p}\cdot\mathbf{p} + m^2c^4 = (c\boldsymbol{\alpha}\cdot\mathbf{p} + \beta mc^2)^2 \tag{14.14}$$

The dot products are performed in Eq. (14.14), so that

$$c^2(p_x^2 + p_y^2 + p_z^2) + m^2c^4 = (c\alpha_x p_x + c\alpha_y p_y + c\alpha_z p_z + \beta mc^2)^2 \tag{14.15}$$

With expansion, Eq. (14.15) becomes

$$c^2(p_x^2+p_y^2+p_z^2)+m^2c^4=c^2(\alpha_x p_x+\alpha_y p_y+\alpha_z p_z)^2+(\alpha_x p_x+\alpha_y p_y+\alpha_z p_z)\beta mc^3 \\ +\beta mc^3(\alpha_x p_x+\alpha_y p_y+\alpha_z p_z)+(\beta mc^2)^2 \tag{14.16}$$

In Eq. (14.15), the components of momentum p_x, p_y, p_z commute with each other, where

$$p_x p_y = p_y p_x \quad p_y p_z = p_z p_y \quad p_z p_x = p_x p_z \tag{14.17}$$

Expansion of the first term on the right side of Eq. (14.16), yields

$$c^2(\alpha_x p_x+\alpha_y p_y+\alpha_z p_z)(\alpha_x p_x+\alpha_y p_y+\alpha_z p_z)=c^2(\alpha_x^2 p_x^2+\alpha_y^2 p_y^2+\alpha_z^2 p_z^2) \\ +c^2(\alpha_x\alpha_y+\alpha_y\alpha_x)p_x p_y+c^2(\alpha_x\alpha_z+\alpha_z\alpha_x)p_x p_z+c^2(\alpha_y\alpha_z+\alpha_z\alpha_y)p_y p_z \tag{14.18}$$

Since p_x, p_y, and p_z are generally nonzero, comparison of Eq. (14.18) for the first term on the right side of Eq. (14.16), with the first term on the left side of 14.16, yields

$$\alpha_x^2=\alpha_y^2=\alpha_z^2=1 \tag{14.19}$$

and

$$\alpha_x\alpha_y+\alpha_y\alpha_x=0 \tag{14.20}$$

$$\alpha_x\alpha_z+\alpha_z\alpha_x=0 \tag{14.21}$$

$$\alpha_y\alpha_z+\alpha_z\alpha_y=0 \tag{14.22}$$

Comparison of the last terms on each side of Eq. (14.16), yields

$$\beta^2=1 \tag{14.23}$$

Comparison of the two sides of Eq. (14.16), also requires

$$(\alpha_x p_x+\alpha_y p_y+\alpha_z p_z)\beta mc^3+\beta mc^3(\alpha_x p_x+\alpha_y p_y+\alpha_z p_z)=0 \tag{14.24}$$

Rearrangement of Eq. (14.24), yields

$$(\alpha_x\beta+\beta\alpha_x)mc^3 p_x+(\alpha_y\beta+\beta\alpha_y)mc^3 p_y+(\alpha_z\beta+\beta\alpha_z)mc^3 p_z=0 \tag{14.25}$$

Given m is nonzero, and p_x, p_y, and p_z are generally nonzero, one then extracts from 14.25 the 3 expressions:

$$\alpha_x\beta + \beta\alpha_x = 0 \quad (14.26)$$

$$\alpha_y\beta + \beta\alpha_y = 0 \quad (14.27)$$

$$\alpha_z\beta + \beta\alpha_z = 0 \quad (14.28)$$

The quantities α and β anticommute in pairs, and as a result, α and β are not numbers; instead, they are matrices, which must satisfy the above-mentioned constraints. The matrix for β, which satisfies these constraints of Eqs. (14.23) and (14.26–14.28), given Eqs. (14.19–14.22), must be a 4 × 4 matrix, where

$$\beta = \begin{bmatrix} 1 & 0 & 0 & 0 \\ 0 & 1 & 0 & 0 \\ 0 & 0 & -1 & 0 \\ 0 & 0 & 0 & -1 \end{bmatrix} \quad (14.29)$$

From Eqs. (14.23) and (14.29), matrix for β^2 is a unity matrix:

$$\beta^2 = \begin{bmatrix} 1 & 0 & 0 & 0 \\ 0 & 1 & 0 & 0 \\ 0 & 0 & -1 & 0 \\ 0 & 0 & 0 & -1 \end{bmatrix}\begin{bmatrix} 1 & 0 & 0 & 0 \\ 0 & 1 & 0 & 0 \\ 0 & 0 & -1 & 0 \\ 0 & 0 & 0 & -1 \end{bmatrix} = \begin{bmatrix} 1 & 0 & 0 & 0 \\ 0 & 1 & 0 & 0 \\ 0 & 0 & 1 & 0 \\ 0 & 0 & 0 & 1 \end{bmatrix} \quad (14.30)$$

The matrices for α_x, α_y, and α_z, which satisfy the constraints of Eqs. (14.19–14.22), given Eqs. (14.23) and (14.26–14.28), are required to be 4 × 4 matrices, where

$$\alpha = \alpha_x\hat{\mathbf{x}} + \alpha_y\hat{\mathbf{y}} + \alpha_z\hat{\mathbf{z}} \quad (14.31)$$

with

$$\alpha_x = \begin{bmatrix} 0 & 0 & 0 & 1 \\ 0 & 0 & 1 & 0 \\ 0 & 1 & 0 & 0 \\ 1 & 0 & 0 & 0 \end{bmatrix} = \begin{bmatrix} \begin{matrix} 0 & 0 \\ 0 & 0 \end{matrix} & \sigma_x \\ \sigma_x & \begin{matrix} 0 & 0 \\ 0 & 0 \end{matrix} \end{bmatrix} \quad (14.32)$$

$$\alpha_y = \begin{bmatrix} 0 & 0 & 0 & -i \\ 0 & 0 & i & 0 \\ 0 & -i & 0 & 0 \\ i & 0 & 0 & 0 \end{bmatrix} = \begin{bmatrix} \begin{matrix} 0 & 0 \\ 0 & 0 \end{matrix} & \sigma_y \\ \sigma_y & \begin{matrix} 0 & 0 \\ 0 & 0 \end{matrix} \end{bmatrix} \quad (14.33)$$

$$\alpha_z = \begin{bmatrix} 0 & 0 & 1 & 0 \\ 0 & 0 & 0 & -1 \\ 1 & 0 & 0 & 0 \\ 0 & -1 & 0 & 0 \end{bmatrix} = \begin{bmatrix} \begin{matrix} 0 & 0 \\ 0 & 0 \end{matrix} & \sigma_z \\ \sigma_z & \begin{matrix} 0 & 0 \\ 0 & 0 \end{matrix} \end{bmatrix} \quad (14.34)$$

From Liboff (2003), the Pauli spin matrix vector $\boldsymbol{\sigma}$ is

$$\boldsymbol{\sigma} = \sigma_x \hat{\mathbf{x}} + \sigma_y \hat{\mathbf{y}} + \sigma_z \hat{\mathbf{z}} \tag{14.35}$$

The x-, y-, and z-component Pauli spin matrices σ_x, σ_y, and σ_z are

$$\sigma_x = \begin{bmatrix} 0 & 1 \\ 1 & 0 \end{bmatrix} \quad \sigma_y = \begin{bmatrix} 0 & -\mathrm{i} \\ \mathrm{i} & 0 \end{bmatrix} \quad \sigma_z = \begin{bmatrix} 1 & 0 \\ 0 & -1 \end{bmatrix} \tag{14.36}$$

With substitution of Eq. (14.36) into Eq. (14.32–14.34), the matrix vector $\boldsymbol{\alpha}$ is expressed in terms of the x-, y-, and z-component spin matrices:

$$\boldsymbol{\alpha} = \begin{bmatrix} 0 & \sigma_x \\ \sigma_x & 0 \end{bmatrix} \hat{\mathbf{x}} + \begin{bmatrix} 0 & \sigma_y \\ \sigma_y & 0 \end{bmatrix} \hat{\mathbf{y}} + \begin{bmatrix} 0 & \sigma_z \\ \sigma_z & 0 \end{bmatrix} \hat{\mathbf{z}} \tag{14.37}$$

or simply

$$\boldsymbol{\alpha} = \begin{bmatrix} 0 & 1 \\ 1 & 0 \end{bmatrix} \boldsymbol{\sigma} = \begin{bmatrix} 0 & 0 & 0 & 1 \\ 0 & 0 & 1 & 0 \\ 0 & 1 & 0 & 0 \\ 1 & 0 & 0 & 0 \end{bmatrix} \hat{\mathbf{x}} + \begin{bmatrix} 0 & 0 & 0 & -\mathrm{i} \\ 0 & 0 & \mathrm{i} & 0 \\ 0 & -\mathrm{i} & 0 & 0 \\ \mathrm{i} & 0 & 0 & 0 \end{bmatrix} \hat{\mathbf{y}} + \begin{bmatrix} 0 & 0 & 1 & 0 \\ 0 & 0 & 0 & -1 \\ 1 & 0 & 0 & 0 \\ 0 & -1 & 0 & 0 \end{bmatrix} \hat{\mathbf{z}} \tag{14.38}$$

From Eq. (14.37), the cross product of the matrix vector $\boldsymbol{\alpha}$ with itself is not zero, because the Pauli spin matrices do not commute, where

$$\boldsymbol{\alpha} \times \boldsymbol{\alpha} = \begin{bmatrix} \sigma_y\sigma_z - \sigma_z\sigma_y & 0 \\ 0 & \sigma_y\sigma_z - \sigma_z\sigma_y \end{bmatrix} \hat{\mathbf{x}} + \begin{bmatrix} \sigma_z\sigma_x - \sigma_x\sigma_z & 0 \\ 0 & \sigma_z\sigma_x - \sigma_x\sigma_z \end{bmatrix} \hat{\mathbf{y}}$$

$$+ \begin{bmatrix} \sigma_x\sigma_y - \sigma_y\sigma_x & 0 \\ 0 & \sigma_x\sigma_y - \sigma_y\sigma_x \end{bmatrix} \hat{\mathbf{z}} \tag{14.39}$$

From Eq. (14.36), the commutation rules for the Pauli spin matrices are

$$\sigma_y\sigma_z - \sigma_z\sigma_y = \mathrm{i}\sigma_x \tag{14.40}$$

$$\sigma_z\sigma_x - \sigma_x\sigma_z = \mathrm{i}\sigma_y \tag{14.41}$$

$$\sigma_x\sigma_y - \sigma_y\sigma_x = \mathrm{i}\sigma_z \tag{14.42}$$

Substitution of Eqs. (14.40–14.42) into Eq. (14.39), gives

$$\boldsymbol{\alpha} \times \boldsymbol{\alpha} = \begin{bmatrix} \mathrm{i}\sigma_x & 0 \\ 0 & \mathrm{i}\sigma_x \end{bmatrix} \hat{\mathbf{x}} + \begin{bmatrix} \mathrm{i}\sigma_y & 0 \\ 0 & \mathrm{i}\sigma_y \end{bmatrix} \hat{\mathbf{y}} + \begin{bmatrix} \mathrm{i}\sigma_z & 0 \\ 0 & \mathrm{i}\sigma_z \end{bmatrix} \hat{\mathbf{z}} \tag{14.43}$$

The cross product of the matrix vector α with itself yields the Pauli spin matrix vector $\boldsymbol{\sigma}$ multiplied by i:

$$\alpha \times \alpha = \mathrm{i}\begin{bmatrix} 1 & 0 \\ 0 & 1 \end{bmatrix}\boldsymbol{\sigma} \tag{14.44}$$

The 4×4 matrix results require the wave function amplitude $\psi(\mathbf{r}, t)$ to be comprised of a 4×1 column vector, which is comprised of independent wave amplitude solutions. The four components $\psi_1(\mathbf{r}, t)$, $\psi_2(\mathbf{r}, t)$, $\psi_3(\mathbf{r}, t)$, and $\psi_4(\mathbf{r}, t)$ of the total wave function $\psi(\mathbf{r}, t)$ amplitude are orthogonal, where

$$\Psi_{\mathrm{T}}(\mathbf{r}, t) = [\psi_1(\mathbf{r}, t), \psi_2(\mathbf{r}, t), \psi_3(\mathbf{r}, t), \psi_4(\mathbf{r}, t)] \tag{14.45}$$

Recalling Eq. (14.11) and using Eq. (14.45), the Dirac equation can thus be expressed as

$$E\Psi_{\mathrm{T}}(\mathbf{r}, t) = (c\alpha \cdot \mathbf{p} + \beta mc^2)\Psi_{\mathrm{T}}(\mathbf{r}, t) \tag{14.46}$$

In matrix form, the Dirac equation of Eq. (14.46) becomes

$$\begin{bmatrix} E & 0 \\ 0 & E \end{bmatrix}\begin{bmatrix} \Phi_1 \\ \Phi_2 \end{bmatrix} = \begin{bmatrix} 0 & c\boldsymbol{\sigma} \cdot \mathbf{p} \\ c\boldsymbol{\sigma} \cdot \mathbf{p} & 0 \end{bmatrix}\begin{bmatrix} \Phi_1 \\ \Phi_2 \end{bmatrix} + \begin{bmatrix} mc^2 & 0 \\ 0 & -mc^2 \end{bmatrix}\begin{bmatrix} \Phi_1 \\ \Phi_2 \end{bmatrix} \tag{14.47}$$

The x-, y-, and z-component 2×2 Pauli spin matrices for Pauli spin vector matrix $\boldsymbol{\sigma}$, are

$$\boldsymbol{\sigma} = \hat{\mathbf{x}}\sigma_x + \hat{\mathbf{y}}\sigma_y + \hat{\mathbf{z}}z = \begin{bmatrix} 0 & 1 \\ 1 & 0 \end{bmatrix}\hat{\mathbf{x}} + \begin{bmatrix} 0 & -\mathrm{i} \\ \mathrm{i} & 0 \end{bmatrix}\hat{\mathbf{y}} + \begin{bmatrix} 1 & 0 \\ 0 & -1 \end{bmatrix}\hat{\mathbf{z}} \tag{14.48}$$

Wave functions $\Phi_1(\mathbf{r}, t)$ and $\Phi_2(\mathbf{r}, t)$ are orthogonal. The $\Phi_1(\mathbf{r}, t)$ orthogonal wave functions are $\psi_1(\mathbf{r}, t)$ and $\psi_2(\mathbf{r}, t)$, while $\Phi_2(\mathbf{r}, t)$ is comprised of orthogonal wave functions $\psi_3(\mathbf{r}, t)$ and $\psi_4(\mathbf{r}, t)$, where

$$\Psi_{\mathrm{T}}(\mathbf{r}, t) = [\Phi_1(\mathbf{r}, t), \Phi_2(\mathbf{r}, t)] \tag{14.49}$$

$$\Phi_1(\mathbf{r}, t) = [\psi_1(\mathbf{r}, t), \psi_2(\mathbf{r}, t)] \tag{14.50}$$

$$\Phi_2(\mathbf{r}, t) = [\psi_3(\mathbf{r}, t), \psi_4(\mathbf{r}, t)] \tag{14.51}$$

14.4 Solution of the Dirac Equation for a Free Particle

We now examine the Dirac solution for a free particle with spin 1/2 and mass m, which is expressed as a plane wave in the wave viewpoint. The Dirac equation of Eq. (14.47), is satisfied by four plane wave solutions of the form

$$\psi_1(\mathbf{r}, t) = u_1 \exp[\mathrm{i}(\mathbf{k} \cdot \mathbf{r} - wt)] \tag{14.52}$$

$$\psi_2(\mathbf{r}, t) = u_2 \exp[\mathrm{i}(\mathbf{k} \cdot \mathbf{r} - wt)] \tag{14.53}$$

$$\psi_3(\mathbf{r}, t) = u_3 \exp[\mathrm{i}(\mathbf{k} \cdot \mathbf{r} - wt)] \tag{14.54}$$

$$\psi_4(\mathbf{r}, t) = u_4 \exp[\mathrm{i}(\mathbf{k} \cdot \mathbf{r} - wt)] \tag{14.55}$$

The $\psi_1(\mathbf{r}, t)$, $\psi_2(\mathbf{r}, t)$, $\psi_3(\mathbf{r}, t)$, and $\psi_4(\mathbf{r}, t)$ functions are orthogonal. These functions are components, which comprise the 4-D, total wave amplitude $\Psi_\mathrm{T}(\mathbf{r}, t)$ with normalization constant N, where

$$\Psi_\mathrm{T}(\mathbf{r}, t) = [\psi_1(\mathbf{r}, t), \psi_2(\mathbf{r}, t), \psi_3(\mathbf{r}, t), \psi_4(\mathbf{r}, t)] \tag{14.56}$$

and

$$\begin{aligned}\Psi_\mathrm{T}(\mathbf{r}, t)(\Psi_\mathrm{T}(\mathbf{r}, t))^* = {} & N[\psi_1(\mathbf{r}, t), \psi_2(\mathbf{r}, t), \psi_3(\mathbf{r}, t), \psi_4(\mathbf{r}, t)] \\ & \times N^*[\psi_1(\mathbf{r}, t), \psi_2(\mathbf{r}, t), \psi_3(\mathbf{r}, t), \psi_4(\mathbf{r}, t)]^*\end{aligned} \tag{14.57}$$

Because $\psi_1(\mathbf{r}, t)$, $\psi_2(\mathbf{r}, t)$, $\psi_3(\mathbf{r}, t)$, and $\psi_4(\mathbf{r}, t)$ are orthogonal components of eigenvector $\Psi_\mathrm{T}(\mathbf{r}, t)$, the probability density of the eigenvector $\Psi_\mathrm{T}(\mathbf{r}, t)$ is the sum of the probability densities (i.e., squared magnitudes) of the four components, multiplied by the squared magnitude $|N|^2$ of the normalization constant, where

$$\Psi_\mathrm{T}(\mathbf{r}, t)(\Psi_\mathrm{T}(\mathbf{r}, t))^* = |N|^2(|\psi_1(\mathbf{r}, t)|^2 + |\psi_2(\mathbf{r}, t)|^2 + |\psi_3(\mathbf{r}, t)|^2 + |\psi_4(\mathbf{r}, t)|^2) \tag{14.58}$$

In Eq. (14.58), the normalization constant N is

$$N = \sqrt{u_1^2 + u_2^2 + u_3^2 + u_4^2} \tag{14.59}$$

With normalization constant N, the squared magnitude or probability density $|\Psi_\mathrm{T}(\mathbf{r}, t)|^2$, integrated over the volume of all space, then yields a probability of 1:

$$\int_{-\infty}^{\infty} \mathrm{d}z \int_{-\infty}^{\infty} \mathrm{d}y \int_{-\infty}^{\infty} \mathrm{d}x \, \Psi_\mathrm{T}(\mathbf{r}, t)(\Psi_\mathrm{T}(\mathbf{r}, t))^* = 1 \tag{14.60}$$

Energy E and momentum $\mathbf{p}$ are related to angular frequency w and wave vector $\mathbf{k}$, respectively, by the reduced Planck's constant $\hbar = h/2\pi$, where m is the particle rest mass and

$$E = \hbar w \quad \text{and} \quad \mathbf{p} = \hbar \mathbf{k} \tag{14.61}$$

From Eq. (14.6), the Dirac equation is

$$\frac{\partial \psi}{\partial t} = -c\boldsymbol{\alpha} \cdot \nabla \psi - \frac{\mathrm{i}}{\hbar} \beta mc^2 \psi \tag{14.62}$$

Recall Eq. (14.38) for α, and Eq. (14.29) for β, where

$$c\boldsymbol{\alpha} \cdot \nabla = c\left(\begin{bmatrix} 0 & 0 & 0 & 1 \\ 0 & 0 & 1 & 0 \\ 0 & 1 & 0 & 0 \\ 1 & 0 & 0 & 0 \end{bmatrix} \hat{\mathbf{x}} + \begin{bmatrix} 0 & 0 & 0 & -\mathrm{i} \\ 0 & 0 & \mathrm{i} & 0 \\ 0 & -\mathrm{i} & 0 & 0 \\ \mathrm{i} & 0 & 0 & 0 \end{bmatrix} \hat{\mathbf{y}} + \begin{bmatrix} 0 & 0 & 1 & 0 \\ 0 & 0 & 0 & -1 \\ 1 & 0 & 0 & 0 \\ 0 & -1 & 0 & 0 \end{bmatrix} \hat{\mathbf{z}} \right) \cdot \left[\frac{\partial}{\partial x} \hat{\mathbf{x}} + \frac{\partial}{\partial y} \hat{\mathbf{y}} + \frac{\partial}{\partial z} \hat{\mathbf{z}} \right] \tag{14.63}$$

or

$$c\boldsymbol{\alpha} \cdot \nabla = \begin{bmatrix} 0 & 0 & c\frac{\partial}{\partial z} & c\frac{\partial}{\partial x} - \mathrm{i}c\frac{\partial}{\partial y} \\ 0 & 0 & c\frac{\partial}{\partial x} + \mathrm{i}c\frac{\partial}{\partial y} & -c\frac{\partial}{\partial z} \\ c\frac{\partial}{\partial z} & c\frac{\partial}{\partial x} - \mathrm{i}c\frac{\partial}{\partial y} & 0 & 0 \\ c\frac{\partial}{\partial x} + \mathrm{i}c\frac{\partial}{\partial y} & -c\frac{\partial}{\partial z} & 0 & 0 \end{bmatrix} \tag{14.64}$$

and

$$\frac{\mathrm{i}}{\hbar} \beta mc^2 = \begin{bmatrix} \frac{\mathrm{i}}{\hbar} mc^2 & 0 & 0 & 0 \\ 0 & \frac{\mathrm{i}}{\hbar} mc^2 & 0 & 0 \\ 0 & 0 & -\frac{\mathrm{i}}{\hbar} mc^2 & 0 \\ 0 & 0 & 0 & -\frac{\mathrm{i}}{\hbar} mc^2 \end{bmatrix} \tag{14.65}$$

The Dirac equation of Eq. (14.62), in matrix form, is then

$$\begin{bmatrix} \dfrac{\partial\psi_1(\mathbf{r},t)}{\partial t} \\ \dfrac{\partial\psi_1(\mathbf{r},t)}{\partial t} \\ \dfrac{\partial\psi_1(\mathbf{r},t)}{\partial t} \\ \dfrac{\partial\psi_1(\mathbf{r},t)}{\partial t} \end{bmatrix} = \begin{bmatrix} -\dfrac{\mathrm{i}}{\hbar}mc^2 & 0 & c\dfrac{\partial}{\partial z} & -c\dfrac{\partial}{\partial x}+\mathrm{i}c\dfrac{\partial}{\partial y} \\ 0 & -\dfrac{\mathrm{i}}{\hbar}mc^2 & -c\dfrac{\partial}{\partial x}-\mathrm{i}c\dfrac{\partial}{\partial y} & c\dfrac{\partial}{\partial z} \\ -c\dfrac{\partial}{\partial z} & -c\dfrac{\partial}{\partial x}+\mathrm{i}c\dfrac{\partial}{\partial y} & \dfrac{\mathrm{i}}{\hbar}mc^2 & 0 \\ -c\dfrac{\partial}{\partial x}-\mathrm{i}c\dfrac{\partial}{\partial y} & c\dfrac{\partial}{\partial z} & 0 & \dfrac{\mathrm{i}}{\hbar}mc^2 \end{bmatrix} \begin{bmatrix} \psi_1(\mathbf{r},t) \\ \psi_2(\mathbf{r},t) \\ \psi_3(\mathbf{r},t) \\ \psi_4(\mathbf{r},t) \end{bmatrix} \tag{14.66}$$

From Eq. (14.9), the left side of Eq. (14.66), becomes

$$\begin{bmatrix} \dfrac{\partial\psi_1(\mathbf{r},t)}{\partial t} \\ \dfrac{\partial\psi_1(\mathbf{r},t)}{\partial t} \\ \dfrac{\partial\psi_1(\mathbf{r},t)}{\partial t} \\ \dfrac{\partial\psi_1(\mathbf{r},t)}{\partial t} \end{bmatrix} = \begin{bmatrix} -\dfrac{\mathrm{i}}{h}E & 0 & 0 & 0 \\ 0 & -\dfrac{\mathrm{i}}{h}E & 0 & 0 \\ 0 & 0 & -\dfrac{\mathrm{i}}{h}E & 0 \\ 0 & 0 & 0 & -\dfrac{\mathrm{i}}{h}E \end{bmatrix} \begin{bmatrix} \psi_1(\mathbf{r},t) \\ \psi_2(\mathbf{r},t) \\ \psi_3(\mathbf{r},t) \\ \psi_4(\mathbf{r},t) \end{bmatrix} \tag{14.67}$$

And from Eq. (14.9), the right side of Eq. (14.66), becomes

$$\begin{bmatrix} -\dfrac{\mathrm{i}}{\hbar}mc^2 & 0 & -c\dfrac{\mathrm{i}}{\hbar}p_z & -c\dfrac{\mathrm{i}}{\hbar}p_x+\mathrm{i}c\dfrac{\mathrm{i}}{\hbar}p_y \\ 0 & -\dfrac{\mathrm{i}}{\hbar}mc^2 & -c\dfrac{\mathrm{i}}{\hbar}p_x-\mathrm{i}c\dfrac{\mathrm{i}}{\hbar}p_y & c\dfrac{\mathrm{i}}{\hbar}p_z \\ -c\dfrac{\mathrm{i}}{\hbar}p_z & -c\dfrac{\mathrm{i}}{\hbar}p_x+\mathrm{i}c\dfrac{\mathrm{i}}{\hbar}p_y & \dfrac{\mathrm{i}}{\hbar}mc^2 & 0 \\ -c\dfrac{\mathrm{i}}{\hbar}p_x-\mathrm{i}c\dfrac{\mathrm{i}}{\hbar}p_y & c\dfrac{\mathrm{i}}{\hbar}p_z & 0 & -\dfrac{\mathrm{i}}{\hbar}mc^2 \end{bmatrix} \begin{bmatrix} \psi_1 \\ \psi_2 \\ \psi_3 \\ \psi_4 \end{bmatrix} \tag{14.68}$$

With substitution into Eq. (14.66) of Eq. (14.67) for the left side, and Eq. (14.68) for the right side, one obtains

$$\begin{bmatrix} E-mc^2 & 0 & -cp_z & -c(p_x-\mathrm{i}p_y) \\ 0 & E-mc^2 & -c(p_x+\mathrm{i}p_y) & cp_z \\ -cp_z & -c(p_x-\mathrm{i}p_y) & E+mc^2 & 0 \\ -c(p_x+\mathrm{i}p_y) & cp_z & 0 & E+mc^2 \end{bmatrix} \begin{bmatrix} u_1 \\ u_2 \\ u_3 \\ u_4 \end{bmatrix} = \begin{bmatrix} 0 \\ 0 \\ 0 \\ 0 \end{bmatrix} \tag{14.69}$$

To obtain a solution of Eq. (14.69), the determinant of coefficients of the four variables u_1, u_2, u_3, and u_4 in this set of four homogeneous equations must be zero, where

$$\begin{vmatrix} E-mc^2 & 0 & -cp_z & -c(p_x-ip_y) \\ 0 & E-mc^2 & -c(p_x+ip_y) & cp_z \\ -cp_z & -c(p_x-ip_y) & E+mc^2 & 0 \\ -c(p_x+ip_y) & cp_z & 0 & E+mc^2 \end{vmatrix} = 0 \tag{14.70}$$

Performing the diagonal multiplication in the determinant of Eq. (14.70), gives

$$(E^2-m^2c^4)^2 - 2(E^2-m^2c^4)c^2p_z^2 + c^4p_z^4 - c^4(p_x^2+p_y^2)^2 = 0 \tag{14.71}$$

or

$$[(E^2-m^2c^4)-c^2p_z^2]^2 - c^4(p_x^2+p_y^2)^2 = 0 \tag{14.72}$$

Rearrangement of Eq. (14.72), gives

$$E^2 - m^2c^4 - c^2p^2 = 0 \tag{14.73}$$

From Eq. (14.73), one obtains a positive E_+ and negative solution E_- for the total energy E, where

$$E_+ = +\sqrt{m^2c^4+c^2p^2} \quad \text{and} \quad E_- = -\sqrt{m^2c^4+c^2p^2} \tag{14.74}$$

The positive total energy E_+ is substituted into the matrix of Eq. (14.69), to solve for the four variables u_1, u_2, u_3, and u_4, where

$$\begin{bmatrix} E_+-mc^2 & 0 & -cp_z & -c(p_x-ip_y) \\ 0 & E_+-mc^2 & -c(p_x+ip_y) & cp_z \\ -cp_z & -c(p_x-ip_y) & E_++mc^2 & 0 \\ -c(p_x+ip_y) & cp_z & 0 & E_++mc^2 \end{bmatrix} \begin{bmatrix} u_1 \\ u_2 \\ u_3 \\ u_4 \end{bmatrix} = \begin{bmatrix} 0 \\ 0 \\ 0 \\ 0 \end{bmatrix} \tag{14.75}$$

Solving Eq. (14.75) for u_1, u_2, u_3, and u_4, one obtains for the positive total energy E_+ two linearly independent wave amplitude solutions $\Psi_a(E_+)$ and $\Psi_b(E_+)$. The first wave amplitude solution $\Psi_a(E_+)$ for positive total energy E_+ is

$$\Psi_a(E_+) = N_a[\psi_{1a}(\mathbf{r},t) + \psi_{2a}(\mathbf{r},t) + \psi_{3a}(\mathbf{r},t) + \psi_{4a}(\mathbf{r},t)] \tag{14.76}$$

For the positive energy first wave solution, the normalization constant N_a is

$$N_a = \sqrt{u_{1a}^2 + u_{2a}^2 + u_{3a}^2 + u_{4a}^2} = \sqrt{1 + \frac{p^2c^2}{(E_+ + mc^2)^2}}, \tag{14.77}$$

The 4 wave function components of the first wave solution for the positive energy are:

$$\psi_{1a}(\mathbf{r}, t) = \exp\left[\mathrm{i}\left(\mathbf{k}\cdot\mathbf{r} - \frac{E_+}{\hbar}t\right)\right] \tag{14.78}$$

$$\psi_{2a}(\mathbf{r}, t) = 0 \tag{14.79}$$

$$\psi_{3a}(\mathbf{r}, t) = \frac{cp_z}{E_+ + mc^2}\exp\left[\mathrm{i}\left(\mathbf{k}\cdot\mathbf{r} - \frac{E_+}{\hbar}t\right)\right] \tag{14.80}$$

$$\psi_{4a}(\mathbf{r}, t) = \frac{c(p_x + \mathrm{i}p_y)}{E_+ + mc^2}\exp\left[\mathrm{i}\left(\mathbf{k}\cdot\mathbf{r} - \frac{E_+}{\hbar}t\right)\right] \tag{14.81}$$

The second wave amplitude solution $\Psi_b(E_+)$ for positive total energy E_+ is

$$\Psi_b(E_+) = N_b[\psi_{1b}(\mathbf{r}, t) + \psi_{2b}(\mathbf{r}, t) + \psi_{3b}(\mathbf{r}, t) + \psi_{4b}(\mathbf{r}, t)] \tag{14.82}$$

For the second wave solution for the positive energy, the normalization constant N_b, again, is

$$N_b = \sqrt{u_{1b}^2 + u_{2b}^2 + u_{3b}^2 + u_{4b}^2} = \sqrt{1 + \frac{p^2c^2}{(E_+ + mc^2)^2}} \tag{14.83}$$

The 4 wave function components of the second wave solution for the positive energy are:

$$\psi_{1b}(\mathbf{r}, t) = 0 \tag{14.84}$$

$$\psi_{2b}(\mathbf{r}, t) = \exp\left[\mathrm{i}\left(\mathbf{k}\cdot\mathbf{r} - \frac{E_+}{\hbar}t\right)\right] \tag{14.85}$$

$$\psi_{3b}(\mathbf{r}, t) = \frac{c(p_x - \mathrm{i}p_y)}{E_+ + mc^2}\exp\left[\mathrm{i}\left(\mathbf{k}\cdot\mathbf{r} - \frac{E_+}{\hbar}t\right)\right] \tag{14.86}$$

$$\psi_{4b}(\mathbf{r}, t) = -\frac{cp_z}{E_+ + mc^2}\exp\left[\mathrm{i}\left(\mathbf{k}\cdot\mathbf{r} - \frac{E_+}{\hbar}t\right)\right] \tag{14.87}$$

From Eq. (14.74), the negative total energy E_- is substituted into the matrix of Eq. (14.69), to solve for the four variables u_1, u_2, u_3, and u_4, where

$$\begin{bmatrix} E_- - mc^2 & 0 & -cp_z & -c(p_x - ip_y) \\ 0 & E_- - mc^2 & -c(p_x + ip_y) & cp_z \\ -cp_z & -c(p_x - ip_y) & E_- + mc^2 & 0 \\ -c(p_x + ip_y) & cp_z & 0 & E_- + mc^2 \end{bmatrix} \begin{bmatrix} u_1 \\ u_2 \\ u_3 \\ u_4 \end{bmatrix} = \begin{bmatrix} 0 \\ 0 \\ 0 \\ 0 \end{bmatrix} \tag{14.88}$$

Solving Eq. (14.88) for u_1, u_2, u_3, and u_4, one obtains for the negative total energy E_- two linearly independent wave amplitude solutions $\Psi_c(E_-)$ and $\Psi_d(E_-)$. The first wave amplitude solution $\Psi_c(E_-)$ for negative total energy E_- is

$$\Psi_c(E_-) = N_c[\psi_{1c}(\mathbf{r}, t) + \psi_{2c}(\mathbf{r}, t) + \psi_{3c}(\mathbf{r}, t) + \psi_{4c}(\mathbf{r}, t)] \tag{14.89}$$

From Eq. (14.74), $E_- = -E_+$, and the normalization constant N_c for the negative energy first wave solution, is

$$N_c = \sqrt{u_{1c}^2 + u_{2c}^2 + u_{3c}^2 + u_{4c}^2} = \sqrt{1 + \frac{p^2c^2}{(E_- - mc^2)^2}} = \sqrt{1 + \frac{p^2c^2}{(E_+ + mc^2)^2}}, \tag{14.90}$$

The 4 wave function components of the first wave solution for the negative energy are:

$$\psi_{1c}(\mathbf{r}, t) = \frac{cp_z}{E_- - mc^2} \exp\left[i\left(\mathbf{k}\cdot\mathbf{r} - \frac{E_-}{\hbar}t\right)\right] \tag{14.91}$$

$$\psi_{2c}(\mathbf{r}, t) = \frac{c(p_x + ip_y)}{E_- - mc^2} \exp\left[i\left(\mathbf{k}\cdot\mathbf{r} - \frac{E_-}{\hbar}t\right)\right] \tag{14.92}$$

$$\psi_{3c}(\mathbf{r}, t) = 1 \tag{14.93}$$

$$\psi_{4c}(\mathbf{r}, t) = 0 \tag{14.94}$$

The second wave amplitude solution $\Psi_d(E_-)$ for negative total energy E_- is

$$\Psi_d(E_+) = N_d[\psi_{1d}(\mathbf{r}, t) + \psi_{2d}(\mathbf{r}, t) + \psi_{3d}(\mathbf{r}, t) + \psi_{4d}(\mathbf{r}, t)] \tag{14.95}$$

From Eq. (14.74), $E_- = -E_+$, and the normalization constant N_d for the negative energy second wave solution, is

$$N_d = \sqrt{u_{1d}^2 + u_{2d}^2 + u_{3d}^2 + u_{4d}^2} = \sqrt{1 + \frac{p^2c^2}{(E_- - mc^2)^2}} = \sqrt{1 + \frac{p^2c^2}{(E_+ + mc^2)^2}} \tag{14.96}$$

The 4 wave function components of the second wave solution for the negative energy are:

$$\psi_{1d}(\mathbf{r},t) = \frac{c(p_x - \mathrm{i}p_y)}{E_- - mc^2}\exp\left[\mathrm{i}\left(\mathbf{k}\cdot\mathbf{r} - \frac{E_-}{\hbar}t\right)\right] \tag{14.97}$$

$$\psi_{2d}(\mathbf{r},t) = -\frac{cp_z}{E_- - mc^2}\exp\left[\mathrm{i}\left(\mathbf{k}\cdot\mathbf{r} - \frac{E_-}{\hbar}t\right)\right] \tag{14.98}$$

$$\psi_{3d}(\mathbf{r},t) = 0 \tag{14.99}$$

$$\psi_{4d}(\mathbf{r},t) = 1 \tag{14.100}$$

These negative energy solutions for a negative electron correspond to a positive particle of the same mass of the electron behaving like an electron, which is called the *positron*.

14.5 A Useful Vector Property with the Pauli Spin Matrix

In this section, we derive the useful vector property of the Pauli spin vector matrix $\boldsymbol{\sigma}$ that is used in the Dirac equation. The expression to be derived is

$$(\boldsymbol{\sigma}\cdot\mathbf{a})(\boldsymbol{\sigma}\cdot\mathbf{b}) = (\mathbf{a}\cdot\mathbf{b})\begin{bmatrix}1 & 0\\ 0 & 1\end{bmatrix} + \mathrm{i}(\mathbf{a}\times\mathbf{b})\cdot\boldsymbol{\sigma} \tag{14.101}$$

The x-, y-, and z-component 2×2 Pauli spin matrices for Pauli spin vector matrix $\boldsymbol{\sigma}$ are

$$\boldsymbol{\sigma} = \hat{\mathbf{x}}\sigma_x + \hat{\mathbf{y}}\sigma_y + \hat{\mathbf{z}}z = \begin{bmatrix}0 & 1\\ 1 & 0\end{bmatrix}\hat{\mathbf{x}} + \begin{bmatrix}0 & -\mathrm{i}\\ \mathrm{i} & 0\end{bmatrix}\hat{\mathbf{y}} + \begin{bmatrix}1 & 0\\ 0 & -1\end{bmatrix}\hat{\mathbf{z}} \tag{14.102}$$

Given general vectors

$$\mathbf{a} = a_x\hat{\mathbf{x}} + a_y\hat{\mathbf{y}} + a_z\hat{\mathbf{z}} \quad \text{and} \quad \mathbf{b} = b_x\hat{\mathbf{x}} + b_y\hat{\mathbf{y}} + b_z\hat{\mathbf{z}} \tag{14.103}$$

one obtains

$$\begin{aligned}(\boldsymbol{\sigma}\cdot\mathbf{a})(\boldsymbol{\sigma}\cdot\mathbf{b}) = &\left(a_x\begin{bmatrix}0 & 1\\ 1 & 0\end{bmatrix} + a_y\begin{bmatrix}0 & -\mathrm{i}\\ \mathrm{i} & 0\end{bmatrix} + a_z\begin{bmatrix}1 & 0\\ 0 & -1\end{bmatrix}\right)\\ &\times\left(b_x\begin{bmatrix}0 & 1\\ 1 & 0\end{bmatrix} + b_y\begin{bmatrix}0 & -\mathrm{i}\\ \mathrm{i} & 0\end{bmatrix} + b_z\begin{bmatrix}1 & 0\\ 0 & -1\end{bmatrix}\right)\end{aligned} \tag{14.104}$$

Adding the matrices in Eq. (14.104), gives

$$(\boldsymbol{\sigma}\cdot\mathbf{a})(\boldsymbol{\sigma}\cdot\mathbf{b}) = \begin{bmatrix} a_z & a_x - \mathrm{i}a_y \\ a_x + \mathrm{i}a_y & -a_z \end{bmatrix}\begin{bmatrix} b_z & b_x - \mathrm{i}b_y \\ b_x + \mathrm{i}b_y & -b_z \end{bmatrix} \tag{14.105}$$

Matrix multiplication gives

$$(\boldsymbol{\sigma}\cdot\mathbf{a})(\boldsymbol{\sigma}\cdot\mathbf{b}) = \begin{bmatrix} (a_zb_z + a_yb_y + a_xb_x) + \mathrm{i}(a_xb_y - a_yb_x) & (a_zb_x - a_xb_z) + \mathrm{i}(a_yb_z - a_zb_y) \\ (a_xb_z - a_zb_x) + \mathrm{i}(a_yb_z - a_zb_y) & (a_xb_x + a_yb_y + a_zb_z) + \mathrm{i}(a_yb_x - a_xb_y) \end{bmatrix} \tag{14.106}$$

The vector representation is then

$$(\boldsymbol{\sigma}\cdot\mathbf{a})(\boldsymbol{\sigma}\cdot\mathbf{b}) = \begin{bmatrix} \mathbf{a}\cdot\mathbf{b} + \mathrm{i}(\mathbf{a}\times\mathbf{b})\cdot\hat{\mathbf{z}} & (\mathbf{a}\times\mathbf{b})\cdot\hat{\mathbf{y}} + \mathrm{i}(\mathbf{a}\times\mathbf{b})\cdot\hat{\mathbf{x}} \\ -(\mathbf{a}\times\mathbf{b})\cdot\hat{\mathbf{y}} + \mathrm{i}(\mathbf{a}\times\mathbf{b})\cdot\hat{\mathbf{x}} & \mathbf{a}\cdot\mathbf{b} - \mathrm{i}(\mathbf{a}\times\mathbf{b})\cdot\hat{\mathbf{z}} \end{bmatrix} \tag{14.107}$$

This expression can be decomposed into the following:

$$(\boldsymbol{\sigma}\cdot\mathbf{a})(\boldsymbol{\sigma}\cdot\mathbf{b}) = (\mathbf{a}\cdot\mathbf{b})\begin{bmatrix} 1 & 0 \\ 0 & 1 \end{bmatrix} + \mathrm{i}(\mathbf{a}\times\mathbf{b})\cdot\left(\begin{bmatrix} 0 & 1 \\ 1 & 0 \end{bmatrix}\hat{\mathbf{x}} + \begin{bmatrix} 0 & -\mathrm{i} \\ \mathrm{i} & 0 \end{bmatrix}\hat{\mathbf{y}} + \begin{bmatrix} 1 & 0 \\ 0 & -1 \end{bmatrix}\hat{\mathbf{z}}\right) \tag{14.108}$$

Substitution into Eq. (14.108) of Eq. (14.102) for Pauli spin vector matrix $\boldsymbol{\sigma}$, thus yields the useful identity

$$(\boldsymbol{\sigma}\cdot\mathbf{a})(\boldsymbol{\sigma}\cdot\mathbf{b}) = (\mathbf{a}\cdot\mathbf{b})\begin{bmatrix} 1 & 0 \\ 0 & 1 \end{bmatrix} + \mathrm{i}(\mathbf{a}\times\mathbf{b})\cdot\boldsymbol{\sigma} \tag{14.109}$$

14.6 Dirac Equation for Bound Electron–Proton Interaction—The Basis of LS Coupling

In this section, we derive the relativistic-based Dirac equation for the interaction of a bound electron with a nucleus, which is comprised of one proton, that is, the electron of the hydrogen atom. This electron–proton interaction also serves as the prototype for bound electron interaction in nuclei with more than one proton. The protons of a nucleus appear as a circulating current in the rest frame of the electron. In the electron rest frame, the electron magnetic dipole moment interacts with this magnetic field of the nuclear proton current, giving rise to LS coupling. The derivation and evaluation of the Dirac equation for the hydrogen atom will assume that the bound electron orbits the proton at nonrelativistic speeds.

The relativistic Dirac equation can then be examined for the nonrelativistic case to yield the fine-structure LS coupling and hyperfine-structure terms. These terms account for the interactions of the proton with the electric and magnetic fields, which arise from the electron charge, motion, and spin and the proton charge and spin. Beyond hydrogen, the more complicated nuclei and their atomic electrons form magnetic electric fields, which can be described in terms of charge, current, and electron, and nuclear magnetic dipole moments and nuclear electric quadrupole moments.

In this derivation of the Dirac equation for a bound electron in the hydrogen atom, the magnetic vector potential of the proton is included, which takes into account the proton spin and proton magnetic dipole moment. The time-independent Dirac equation and Hamiltonian, which were derived earlier in this chapter, are modified to include the electromagnetic field of the proton (or charged nucleus and aggregate proton spin) by including the scalar proton Coulomb potential ϕ and vector $\mathbf{A}$ potential of the proton magnetic dipole moment $\mathbf{m}$, where

$$H_{\mathrm{D}}\psi = E\psi \tag{14.110}$$

Given electron charge e, rest mass m, momentum $\mathbf{p}$, the scalar electric ϕ, and vector $\mathbf{A}$ potentials of the proton, the Dirac Hamiltonian operator of Eq. (14.13), becomes

$$H_{\mathrm{D}} = c\boldsymbol{\alpha}\cdot(\mathbf{p} - e\mathbf{A}) + \beta mc^2 + e\phi \tag{14.111}$$

The electron potential energy V is the product of the electron charge e and the scalar electric potential ϕ of the proton, where

$$V = e\phi \tag{14.112}$$

The Dirac equation of Eq. (14.46), is then expressed as

$$[c\boldsymbol{\alpha}\cdot(\mathbf{p} - e\mathbf{A}) + \beta mc^2 + e\phi - E]\Psi_{\mathrm{T}} = 0 \tag{14.113}$$

With Eq. (14.113), the Dirac equation in the matrix form of Eq. (14.47), becomes

$$\begin{bmatrix} 0 & c\boldsymbol{\sigma}\cdot(\mathbf{p} - e\mathbf{A}) \\ c\boldsymbol{\sigma}\cdot(\mathbf{p} - e\mathbf{A}) & 0 \end{bmatrix}\begin{bmatrix}\Phi_1 \\ \Phi_2\end{bmatrix} + \begin{bmatrix} mc^2 & 0 \\ 0 & -mc^2\end{bmatrix}\begin{bmatrix}\Phi_1 \\ \Phi_2\end{bmatrix} - \begin{bmatrix} E - e\phi & 0 \\ 0 & E - e\phi\end{bmatrix}\begin{bmatrix}\Phi_1 \\ \Phi_2\end{bmatrix} = \begin{bmatrix}0 \\ 0\end{bmatrix} \tag{14.114}$$

Wave functions $\Phi_1(\mathbf{r},t)$ and $\Phi_2(\mathbf{r},t)$ are orthogonal, and are comprised of orthogonal wave function pairs $\psi_1(\mathbf{r},t)$, $\psi_2(\mathbf{r},t)$ and $\psi_3(\mathbf{r},t)$, $\psi_4(\mathbf{r},t)$, respectively, where

$$\Phi_1(\mathbf{r},t) = [\psi_1(\mathbf{r},t), \psi_2(\mathbf{r},t)] \tag{14.115}$$

$$\Phi_2(\mathbf{r},t) = [\psi_3(\mathbf{r},t), \psi_4(\mathbf{r},t)] \tag{14.116}$$

The Pauli spin vector matrix $\boldsymbol{\sigma}$ is

$$\boldsymbol{\sigma} = \hat{\mathbf{x}}\sigma_x + \hat{\mathbf{y}}\sigma_y + \hat{\mathbf{z}}z = \begin{bmatrix} 0 & 1 \\ 1 & 0 \end{bmatrix}\hat{\mathbf{x}} + \begin{bmatrix} 0 & -\mathrm{i} \\ \mathrm{i} & 0 \end{bmatrix}\hat{\mathbf{y}} + \begin{bmatrix} 1 & 0 \\ 0 & -1 \end{bmatrix}\hat{\mathbf{z}} \tag{14.117}$$

From the matrix expression of Eq. (14.114), one obtains the pair of equations:

$$c\boldsymbol{\sigma} \cdot (\mathbf{p} - e\mathbf{A})\Phi_2 + (mc^2 + e\phi - E)\Phi_1 = 0 \tag{14.118}$$

and

$$c\boldsymbol{\sigma} \cdot (\mathbf{p} - e\mathbf{A})\Phi_1 + (-mc^2 + e\phi - E)\Phi_2 = 0 \tag{14.119}$$

The second equation of this pair, Eq. (14.119), is solved for wave function Φ_2, where

$$\Phi_2 = \frac{c\boldsymbol{\sigma} \cdot (\mathbf{p} - e\mathbf{A})}{E + mc^2 - e\phi}\Phi_1 \tag{14.120}$$

Substitution of Eq. (14.120) for Φ_2 into the first equation of this pair, Eq. (14.118), yields

$$c\boldsymbol{\sigma} \cdot (\mathbf{p} - q\mathbf{A})\frac{1}{E + mc^2 - e\phi}c\boldsymbol{\sigma} \cdot (\mathbf{p} - e\mathbf{A})\Phi_1 + (mc^2 + e\phi - E)\Phi_1 = 0 \tag{14.121}$$

Rearrangement of Eq. (14.121), gives

$$(E - mc^2)\Phi_1 = e\phi\Phi_1 + (\boldsymbol{\sigma} \cdot (\mathbf{p} - e\mathbf{A}))\frac{1}{E - mc^2 + 2mc^2 - e\phi}(\boldsymbol{\sigma} \cdot (\mathbf{p} - e\mathbf{A}))\Phi_1 \tag{14.122}$$

Let

$$E_1 = E - mc^2 \tag{14.123}$$

With substitution of Eq. (14.123) into Eq. (14.122), the Dirac eigenequation of Eq. (14.110) and the Dirac Hamiltonian H_D operator of Eq. (14.111), become

$$E_1\Phi_1 = e\phi\Phi_1 + \frac{1}{2m}(\boldsymbol{\sigma} \cdot (\mathbf{p} - e\mathbf{A}))\left(1 + \frac{E_1 - e\phi}{2mc^2}\right)^{-1}(\boldsymbol{\sigma} \cdot (\mathbf{p} - e\mathbf{A}))\Phi_1 \tag{14.124}$$

In Eq. (14.124), consider the nonrelativistic case, where

$$\left(1+\frac{E_1-e\phi}{2mc^2}\right)^{-1}=1-\frac{E_1-e\phi}{2mc^2} \tag{14.125}$$

With substitution of Eq. (14.125) into Eq. (14.124), the Dirac Hamiltonian expression H_D is then

$$E_1\Phi_1=e\phi\Phi_1+\frac{1}{2m}(\boldsymbol{\sigma}\cdot(\mathbf{p}-e\mathbf{A}))\left(1-\frac{E_1-e\phi}{2mc^2}\right)(\boldsymbol{\sigma}\cdot(\mathbf{p}-e\mathbf{A}))\Phi_1 \tag{14.126}$$

Insertion of the operator $-\mathrm{i}\hbar\nabla$ for electron momentum vector $\mathbf{p}$ gives

$$E_1\Phi_1=e\phi\Phi_1-\frac{1}{2m}(\boldsymbol{\sigma}\cdot(\mathrm{i}\hbar\nabla+e\mathbf{A}))\left(1-\frac{E_1-e\phi}{2mc^2}\right)(\boldsymbol{\sigma}\cdot(\mathbf{p}-e\mathbf{A}))\Phi_1 \tag{14.127}$$

Evaluation of the second term of Eq. (14.127), gives

$$\begin{aligned}&(\boldsymbol{\sigma}\cdot(\mathrm{i}\hbar\nabla+e\mathbf{A}))\left(1-\frac{E_1-e\phi}{2mc^2}\right)(\boldsymbol{\sigma}\cdot(\mathbf{p}-e\mathbf{A}))=\\&\left\{(\boldsymbol{\sigma}\cdot(\mathrm{i}\hbar\nabla+e\mathbf{A}))\left(1-\frac{E_1-e\phi}{2mc^2}\right)\right\}(\boldsymbol{\sigma}\cdot(\mathbf{p}-e\mathbf{A}))\\&+\left(1-\frac{E_1-e\phi}{2mc^2}\right)(\boldsymbol{\sigma}\cdot(\mathrm{i}\hbar\nabla+e\mathbf{A}))(\boldsymbol{\sigma}\cdot(\mathbf{p}-e\mathbf{A}))\end{aligned} \tag{14.128}$$

where

$$\mathrm{i}\hbar(\boldsymbol{\sigma}\cdot\nabla)\left(\frac{e\phi}{2mc^2}\right)=\mathrm{i}\frac{e\hbar(\boldsymbol{\sigma}\cdot\nabla\phi)}{2mc^2} \tag{14.129}$$

$$(\boldsymbol{\sigma}\cdot e\mathbf{A})\left(1-\frac{E_1-e\phi}{2mc^2}\right)=e\left(1-\frac{E_1-e\phi}{2mc^2}\right)(\boldsymbol{\sigma}\cdot\mathbf{A}) \tag{14.130}$$

$$\mathrm{i}\hbar(\boldsymbol{\sigma}\cdot\nabla)\left(1-\frac{E_1}{2mc^2}\right)=0 \tag{14.131}$$

With substitution of Eqs. (14.129–14.131) into Eq. (14.128), the second term of Eq. (14.127), becomes

$$(\boldsymbol{\sigma}\cdot(\mathrm{i}\hbar\nabla + e\mathbf{A}))\left(1-\frac{E_1-e\phi}{2mc^2}\right)(\boldsymbol{\sigma}\cdot(\mathbf{p}-e\mathbf{A})) = \\ \mathrm{i}\frac{e\hbar(\boldsymbol{\sigma}\cdot\nabla\phi)}{2mc^2}(\boldsymbol{\sigma}\cdot(\mathbf{p}-e\mathbf{A})) + e\left(1-\frac{E_1-e\phi}{2mc^2}\right)(\boldsymbol{\sigma}\cdot\mathbf{A})(\boldsymbol{\sigma}\cdot(\mathbf{p}-e\mathbf{A})) \tag{14.132}$$

With substitution of Eq. (14.132) into Eq. (14.127), the Dirac eigenequation, with the explicit expression for the Dirac Hamiltonian H_D operator, is then

$$E_1\Phi_1 = e\phi\Phi_1 - \mathrm{i}\frac{\hbar}{4m^2c^2}(\boldsymbol{\sigma}\cdot\nabla\phi)(\boldsymbol{\sigma}\cdot(\mathbf{p}-e\mathbf{A}))\Phi_1 \\ + \frac{1}{2m}\left(1-\frac{E_1-e\phi}{2mc^2}\right)(\boldsymbol{\sigma}\cdot(\mathbf{p}-e\mathbf{A}))(\boldsymbol{\sigma}\cdot(\mathbf{p}-e\mathbf{A}))\Phi_1 \tag{14.133}$$

The electron kinetic energy K is

$$K = E_1 - e\phi = \frac{p^2}{2m} \tag{14.134}$$

With substitution of Eq. (14.134) into Eq. (14.133), the Dirac Hamiltonian eigenequation, with explicit expression of the Dirac Hamiltonian operator H_D, is now in a form, which is ready for evaluation, where

$$E_1\Phi_1 = e\phi\Phi_1 - \mathrm{i}\frac{\hbar}{4m^2c^2}(\boldsymbol{\sigma}\cdot\nabla\phi)(\boldsymbol{\sigma}\cdot(\mathbf{p}-e\mathbf{A}))\Phi_1 \\ + \frac{1}{2m}\left(1-\frac{p^2}{4m^2c^2}\right)(\boldsymbol{\sigma}\cdot(\mathbf{p}-e\mathbf{A}))(\boldsymbol{\sigma}\cdot(\mathbf{p}-e\mathbf{A}))\Phi_1 \tag{14.135}$$

The Dirac eigenequation, with the explicit expression for the Dirac Hamiltonian H_D in Eq. (14.135), can be expressed in a compact form, which is the sum of three terms. The second term H_{D2} will be evaluated in Section 14.7, and the third term H_{D3} will be evaluated in Section 14.8, where

$$H_D\Phi_1 = E_1\Phi_1 = e\phi\Phi_1 - H_{D2}\Phi_1 + H_{D3}\Phi_1 \tag{14.136}$$

The second term H_{D2} is

$$H_{D2} = \mathrm{i}\frac{\hbar}{4m^2c^2}(\boldsymbol{\sigma}\cdot\nabla\phi)(\boldsymbol{\sigma}\cdot(\mathbf{p}-e\mathbf{A})) \tag{14.137}$$

and the third term H_{D3} is

$$H_{D3} = \frac{1}{2m}\left(1 - \frac{p^2}{4m^2c^2}\right)(\boldsymbol{\sigma}\cdot(\mathbf{p} - e\mathbf{A}))(\boldsymbol{\sigma}\cdot(\mathbf{p} - e\mathbf{A})) \tag{14.138}$$

14.7 Second Term H_{D2} of the Dirac Equation for Bound Electron–Proton Interaction that Includes an LS Coupling Term

The second term H_{D2} of Eq. (14.136) for the Dirac eigenequation and explicit expression of the Dirac Hamiltonian operator H_D for the hydrogen atom electron, which was derived in Section 14.6, is evaluated in this section, where Eq. (14.137) is repeated below:

$$H_{D2} = \mathrm{i}\frac{\hbar}{4m^2c^2}(\boldsymbol{\sigma}\cdot\nabla\phi)(\boldsymbol{\sigma}\cdot(\mathbf{p} - e\mathbf{A})) \tag{14.139}$$

First, we recall the identity for the paired dot products $(\boldsymbol{\sigma}\cdot\mathbf{a})(\boldsymbol{\sigma}\cdot\mathbf{b})$, which was derived in section 14.5 as Eq. (14.109), where $\mathbf{a}$ and $\mathbf{b}$ are

$$\mathbf{a} = \nabla\phi \quad \text{and} \quad \mathbf{b} = \mathbf{p} - e\mathbf{A} \tag{14.140}$$

The identity Eq. (14.109) is repeated below:

$$(\boldsymbol{\sigma}\cdot\mathbf{a})(\boldsymbol{\sigma}\cdot\mathbf{b}) = (\mathbf{a}\cdot\mathbf{b})\begin{bmatrix}1 & 0\\ 0 & 1\end{bmatrix} + \mathrm{i}(\mathbf{a}\times\mathbf{b})\cdot\boldsymbol{\sigma} \tag{14.141}$$

With substitution of Eq. (14.141) with Eq. (14.140), into Eq. (14.139), one obtains the second term H_{D2} expression

$$H_{D2} = \mathrm{i}\frac{\hbar}{4m^2c^2}(\nabla\phi\cdot(\mathbf{p} - e\mathbf{A}))\begin{bmatrix}1 & 0\\ 0 & 1\end{bmatrix} - \frac{\hbar}{4m^2c^2}[(\nabla\phi\times(\mathbf{p} - e\mathbf{A}))\cdot\boldsymbol{\sigma}] \tag{14.142}$$

Substitution of the operator expression $-\mathrm{i}\hbar\nabla$ for electron momentum vector $\mathbf{p}$ in the first term of Eq. (14.142) gives

$$H_{D2} = -\mathrm{i}\frac{\hbar}{4m^2c^2}(\nabla\varphi\cdot(\mathrm{i}\hbar\nabla + e\mathbf{A}))\begin{bmatrix}1 & 0\\ 0 & 1\end{bmatrix} - \frac{\hbar}{4m^2c^2}\boldsymbol{\sigma}\cdot[\nabla\phi\times(\mathbf{p} - e\mathbf{A})] \tag{14.143}$$

The electric potential ϕ of the protons of the nucleus is assumed to be spherically symmetric $\phi = \phi(r)$, so that

$$\nabla\phi = \mathbf{r}\frac{\partial\phi}{\partial r} \qquad \mathbf{p} = -\mathrm{i}\hbar\nabla \tag{14.144}$$

Substitution of Eq. (14.144) in the 2 terms of Eq. (14.143), gives

$$\nabla\phi\cdot(\mathrm{i}\hbar\nabla + e\mathbf{A}) = \frac{\partial\phi}{\partial r}\hat{\mathbf{r}}\cdot(\mathrm{i}\hbar\nabla + e\mathbf{A}) = \mathrm{i}\hbar\frac{\partial\phi}{\partial r}\frac{\partial}{\partial r} + e\frac{\partial\phi}{\partial r}A_r \tag{14.145}$$

and

$$\nabla\phi\times(\mathbf{p} - e\mathbf{A}) = \frac{\partial\phi}{\partial r}(\hat{\mathbf{r}}\times\mathbf{p}) - e\frac{\partial\phi}{\partial r}(\hat{\mathbf{r}}\times\mathbf{A}) \tag{14.146}$$

The electron angular momentum **L** is defined by

$$\mathbf{L} = \mathbf{r}\times\mathbf{p} = r\hat{\mathbf{r}}\times\mathbf{p} \tag{14.147}$$

With substitution of Eq. (14.147) into Eq. (14.146), one obtains

$$\nabla\phi\times(\mathbf{p} - e\mathbf{A}) = \frac{\partial\phi}{\partial r}\frac{\mathbf{L}}{r} - e\frac{\partial\phi}{\partial r}\hat{\mathbf{r}}\times\mathbf{A} \tag{14.148}$$

With substitution of Eq. (14.145) and (14.148) into Eq. (14.143), the second term H_{D2} of the Dirac Hamiltonian is then

$$H_{\mathrm{D2}} = \frac{\hbar}{4m^2c^2}\left(\hbar\frac{\partial\phi}{\partial r}\frac{\partial}{\partial r} - \mathrm{i}e\frac{\partial\phi}{\partial r}A_r\right)\begin{bmatrix}1 & 0\\ 0 & 1\end{bmatrix} - \frac{\hbar}{4m^2c^2}\frac{\partial\phi}{\partial r}\frac{(\boldsymbol{\sigma}\cdot\mathbf{L})}{r} + \frac{\hbar e}{4m^2c^2}\frac{\partial\phi}{\partial r}\boldsymbol{\sigma}\cdot(\hat{\mathbf{r}}\times\mathbf{A}) \tag{14.149}$$

The electron spin matrix vector **S** is related to the Pauli spin matrix vector $\boldsymbol{\sigma}$ by

$$\mathbf{S} = \frac{1}{2}\hbar\boldsymbol{\sigma} = \frac{1}{2}\hbar\left\{\begin{bmatrix}0 & 1\\ 1 & 0\end{bmatrix}\hat{\mathbf{x}} + \begin{bmatrix}0 & -\mathrm{i}\\ \mathrm{i} & 0\end{bmatrix}\hat{\mathbf{y}} + \begin{bmatrix}1 & 0\\ 0 & -1\end{bmatrix}\hat{\mathbf{z}}\right\} \tag{14.150}$$

Given the vacuum permeability μ_0, the magnetic vector potential **A** of the proton magnetic dipole moment **m**, is the cross product of **m** and the electron orbit radius unit vector $\hat{\mathbf{r}}$, where

$$\mathbf{A} = \frac{\mu_0}{4\pi}\left(\mathbf{m}\times\frac{\hat{\mathbf{r}}}{r^2}\right) \tag{14.151}$$

The BAC–CAB vector identity is invoked, where

$$\hat{\mathbf{r}} \times \mathbf{A} = \hat{\mathbf{r}} \times (\mathbf{m} \times \hat{\mathbf{r}}) = \mathbf{m}(\hat{\mathbf{r}} \cdot \hat{\mathbf{r}}) - \hat{\mathbf{r}}(\hat{\mathbf{r}} \cdot \mathbf{m}) = \mathbf{m} - \hat{\mathbf{r}}(\hat{\mathbf{r}} \cdot \mathbf{m}) \tag{14.152}$$

With substitution of Eq. (14.152) into Eq. (14.149), the second term H_{D2} of the Dirac Hamiltonian becomes

$$H_{\mathrm{D2}} = \frac{\hbar^2}{4m^2c^2}\left(\frac{\partial\phi}{\partial r}\frac{\partial}{\partial r} - \mathrm{i}\frac{e}{\hbar}\frac{\partial\phi}{\partial r}A_r\right)\begin{bmatrix} 1 & 0 \\ 0 & 1 \end{bmatrix} - \frac{1}{2m^2c^2}\frac{\partial\phi}{\partial r}\frac{\mathbf{S}\cdot\mathbf{L}}{r} + \frac{\mu_0 e}{8\pi m^2c^2}\frac{\partial\phi}{\partial r}\left[\mathbf{S}\cdot\frac{\mathbf{m} - \hat{\mathbf{r}}(\hat{\mathbf{r}}\cdot\mathbf{m})}{r^2}\right] \tag{14.153}$$

In Eq. (14.153), the electron spin **S** and electron angular momentum **L** commute, so that

$$\mathbf{S}\cdot\mathbf{L} = \mathbf{L}\cdot\mathbf{S} \tag{14.154}$$

The $\mathbf{L}\cdot\mathbf{S}$ term is the LS coupling term, which produces the net magnetic dipole moment vector of unpaired atomic electronics, and leads to magnetic scatter of neutrons via the neutron magnetic dipole moment vector. And the LS coupling term contains the correct factor of 1/2. This factor of 1/2 is not included in Schrödinger wave equation solutions that omit electron spin and relativity. However, Thomas precession does derive the correct factor of 1/2 in the LS coupling term. The electric field of the proton E_r is radial-directed due to spherical symmetry:

$$\mathbf{E} = E_r\hat{\mathbf{r}} = -\frac{\partial\phi}{\partial r}\hat{\mathbf{r}} \tag{14.155}$$

The magnetic vector potential **A** of the proton magnetic dipole moment is azimuthal with zero radial component $A_r = 0$, where

$$\mathbf{A} = A_\phi\hat{\boldsymbol{\phi}} \tag{14.156}$$

Hence, the second term H_{D2} of the Dirac Hamiltonian becomes

$$H_{\mathrm{D2}} = -\frac{\hbar^2 E_r}{4m^2c^2}\frac{\partial}{\partial r}\begin{bmatrix} 1 & 0 \\ 0 & 1 \end{bmatrix} + \frac{E_r}{2m^2c^2}\frac{\mathbf{S}\cdot\mathbf{L}}{r} - \frac{\mu_0 e}{8\pi m^2c^2}E_r\left[\mathbf{S}\cdot\frac{\mathbf{m} - \hat{\mathbf{r}}(\hat{\mathbf{r}}\cdot\mathbf{m})}{r^2}\right] \tag{14.157}$$

Fine structure terms, which appear in the Dirac equation, account for small interaction effects. These small interaction effects cause small energy shifts and splitting of electron energy levels in the hydrogen atom, as well as electron energy levels in other atoms, ions, and molecules. Fine structure arises from the interaction of the magnetic dipole moments of the electron spin and electron orbital motion. The hyperfine structure terms account for extremely small interaction effects, which are orders of magnitude smaller than the fine structure terms. The hyperfine structure terms are a result of the interaction of (1) the nucleus charge, nuclear magnetic dipole moment and nuclear electric quadrupole moment, with (2) the electric and magnetic fields, which are associated with the electron charge and the electron magnetic dipole moments (caused by electron spin and orbital motion). The nuclear magnetic dipole moment interaction, with the electron spin and orbital motion magnetic dipole moment, dominate over the much smaller nuclear electric quadrupole moment interactions with the electrons.

The H_{D2} expression includes two fine-structure terms: the LS coupling term and the Darwin $(\partial\phi/\partial r)(\partial/\partial r)$ term. The LS coupling term accounts for the interaction of the magnetic dipole moment of the electron spin matrix vector **S** with the electron magnetic dipole moment vector, which is associated with the electron angular momentum **L** that arises from the electron orbital motion. The electric field of the proton $E_r = -\partial\phi/\partial r$ is radial-directed due to spherical symmetry, and the Darwin fine-structure term $(\partial\phi/\partial r)(\partial/\partial r)$ is the product of the negated proton electric field $-E_r$ multiplied by the radial gradient $\partial/\partial r$ of the two wave function components Φ_{11} and Φ_{12} of the electron eigenvector Φ_1. The H_{D2} term also includes the hyperfine-structure term $\mathbf{S}\cdot\mathbf{m} - (\mathbf{S}\cdot\hat{\mathbf{r}})(\hat{\mathbf{r}}\cdot\mathbf{m})$, which accounts for the interaction of the electron spin matrix vector **S** with the proton or nuclear (multiproton nucleus) magnetic dipole moment vector **m**.

14.8 Third Term H_{D3} of the Dirac Equation for Bound Electron–Proton Interaction

The third term H_{D3} of the Dirac Hamiltonian H_D expression of Eq. (14.136) for the hydrogen atom, is evaluated in this section, where Eq. (14.138) is repeated below:

$$H_{D3} = \frac{1}{2m}\left(1 - \frac{p^2}{4m^2c^2}\right)(\boldsymbol{\sigma}\cdot(\mathbf{p} - e\mathbf{A}))(\boldsymbol{\sigma}\cdot(\mathbf{p} - e\mathbf{A})) \tag{14.158}$$

To evaluate Eq. (14.158), one applies the identity for the paired dot products $(\boldsymbol{\sigma}\cdot\mathbf{a})(\boldsymbol{\sigma}\cdot\mathbf{b})$, which was derived in section 14.5 as Eq. (14.109), where **a** and **b** are assigned as follows:

$$\mathbf{a} = \mathbf{b} = \mathbf{p} - e\mathbf{A} \tag{14.159}$$

The identity Eq. (14.109) is repeated below:

$$(\boldsymbol{\sigma}\cdot\mathbf{a})(\boldsymbol{\sigma}\cdot\mathbf{b}) = (\mathbf{a}\cdot\mathbf{b})\begin{bmatrix} 1 & 0 \\ 0 & 1 \end{bmatrix} + \mathrm{i}(\mathbf{a}\times\mathbf{b})\cdot\boldsymbol{\sigma} \tag{14.160}$$

With substitution of Eq. (14.159) into Eq. (14.160), one obtains:

$$(\boldsymbol{\sigma}\cdot(\mathbf{p}-e\mathbf{A}))(\boldsymbol{\sigma}\cdot(\mathbf{p}-e\mathbf{A}))=((\mathbf{p}-e\mathbf{A})\cdot(\mathbf{p}-e\mathbf{A}))\begin{bmatrix}1&0\\0&1\end{bmatrix}+\mathrm{i}\boldsymbol{\sigma}\cdot[(\mathbf{p}-e\mathbf{A})\times(\mathbf{p}-e\mathbf{A})] \tag{14.161}$$

This cross product of vector $(\mathbf{p}-e\mathbf{A})$ with $(\mathbf{p}-e\mathbf{A})$ generally is not zero:

$$(\mathbf{p}-e\mathbf{A})\times(\mathbf{p}-e\mathbf{A}) = -e(\mathbf{p}\times\mathbf{A}+\mathbf{A}\times\mathbf{p}) \tag{14.162}$$

This is due to the presence of operator $\mathbf{p} = -\mathrm{i}\hbar\nabla$, where

$$(\mathbf{p}-e\mathbf{A})\times(\mathbf{p}-e\mathbf{A}) = e\,\mathrm{i}\hbar(\nabla\times\mathbf{A}+\mathbf{A}\times\nabla) \tag{14.163}$$

Substitution of Eq. (14.163) into Eq. (14.161), gives

$$(\boldsymbol{\sigma}\cdot(\mathbf{p}-e\mathbf{A}))(\boldsymbol{\sigma}\cdot(\mathbf{p}-e\mathbf{A}))=((\mathbf{p}-e\mathbf{A})\cdot(\mathbf{p}-e\mathbf{A}))\begin{bmatrix}1&0\\0&1\end{bmatrix}-e\hbar\boldsymbol{\sigma}\cdot(\nabla\times\mathbf{A}+\mathbf{A}\times\nabla) \tag{14.164}$$

In the Dirac equation of Eq. (14.135), the second term of Eq. (14.164) operates on the electron wave function Φ_1, where

$$\boldsymbol{\sigma}\cdot(\nabla\times\mathbf{A}+\mathbf{A}\times\nabla)\Phi_1 \tag{14.165}$$

One invokes vector identity for magnetic vector potential $\mathbf{A}$ and scalar Φ_1:

$$\nabla\times\mathbf{A}\Phi_1 = \nabla\Phi_1\times\mathbf{A}+\Phi_1\nabla\times\mathbf{A} = -\mathbf{A}\times\nabla\Phi_1+\Phi_1\nabla\times\mathbf{A} \tag{14.166}$$

Application of the identity of Eq. (14.166) to Eq. (14.165), yields

$$(\nabla\times\mathbf{A}+\mathbf{A}\times\nabla)\Phi_1 = \nabla\times\mathbf{A}\Phi_1+\mathbf{A}\times\nabla\Phi_1 = \Phi_1(\nabla\times\mathbf{A}) \tag{14.167}$$

The ∇ operates only on the proton vector potential $\mathbf{A}$, so that

$$\Phi_1(\nabla\times\mathbf{A}) = (\nabla\times\mathbf{A})\Phi_1 \tag{14.168}$$

The proton magnetic field $\mathbf{B}$ is given by

$$\mathbf{B} = \nabla\times\mathbf{A} \tag{14.169}$$

With substitution of Eq. (14.167), with Eq. (14.168) into Eq. (14.164), one obtains

$$(\boldsymbol{\sigma}\cdot(\mathbf{p}-e\mathbf{A}))(\boldsymbol{\sigma}\cdot(\mathbf{p}-e\mathbf{A}))\Phi_1 = p^2\begin{bmatrix}1&0\\0&1\end{bmatrix}\Phi_1 - e\hbar(\boldsymbol{\sigma}\cdot\mathbf{B})\Phi_1 \tag{14.170}$$

In Eq. (14.170), the electron wave function Φ_1 can be dropped, and one obtains the dot product of electron spin vector matrix **S** and the proton magnetic dipole moment field **B**, where

$$(\boldsymbol{\sigma}\cdot(\mathbf{p}-e\mathbf{A}))(\boldsymbol{\sigma}\cdot(\mathbf{p}-e\mathbf{A}))=(\mathbf{p}-e\mathbf{A})^2\begin{bmatrix}1&0\\0&1\end{bmatrix}-2e(\mathbf{S}\cdot\mathbf{B}) \tag{14.171}$$

With substitution of Eq. (14.171) into Eq. (14.158), the third term H_{D3} of the Dirac Hamiltonian of Eq. (14.135), becomes

$$H_{D3}=\frac{1}{2m}\left(1-\frac{p^2}{4m^2c^2}\right)(\mathbf{p}-e\mathbf{A})^2\begin{bmatrix}1&0\\0&1\end{bmatrix}-\frac{e}{m}\left(1-\frac{p^2}{4m^2c^2}\right)(\mathbf{S}\cdot\mathbf{B}) \tag{14.172}$$

Upon expanding,

$$(\mathbf{p}-e\mathbf{A})^2=p^2-e\mathbf{p}\cdot\mathbf{A}-e\mathbf{A}\cdot\mathbf{p}+e^2A^2 \tag{14.173}$$

One next invokes the vector identity for the scalar product of a vector **C**, and the cross product $\mathbf{D}\times\mathbf{E}$ of 2 vectors, **D** and **E**, which apply to vector operators as well,

$$\mathbf{C}\cdot\mathbf{D}\times\mathbf{E}=\mathbf{D}\cdot\mathbf{E}\times\mathbf{C}=\mathbf{E}\cdot\mathbf{C}\times\mathbf{D} \tag{14.174}$$

with

$$\mathbf{A}=\frac{\mu_0}{4\pi}\left(\mathbf{m}\times\frac{\hat{\mathbf{r}}}{r^2}\right) \tag{14.175}$$

One then obtains

$$\mathbf{p}\cdot\mathbf{A}=\mathbf{A}\cdot\mathbf{p}=\frac{\mu_0}{4\pi}\left(\mathbf{m}\times\frac{\hat{\mathbf{r}}}{r^2}\right)\cdot\mathbf{p}=\frac{\mu_0}{4\pi}\left(\frac{\hat{\mathbf{r}}\times\mathbf{p}}{r^2}\right)\cdot\mathbf{m}=\frac{\mu_0}{4\pi}\frac{\mathbf{L}\cdot\mathbf{m}}{r^3} \tag{14.176}$$

Hence,

$$(\mathbf{p}-e\mathbf{A})^2=p^2-\frac{\mu_0 e}{2\pi}\frac{\mathbf{L}\cdot\mathbf{m}}{r^3}+e^2A^2 \tag{14.177}$$

With substitution of Eq. (14.177) into Eq. (14.172), the third term H_{D3} of the Dirac Hamiltonian becomes

$$H_{D3}=\left\{\frac{(\mathbf{p}-e\mathbf{A})^2}{2m}-\frac{p^4}{8m^3c^2}+\frac{\mu_0 e^2p^2}{16\pi m^3c^2}\frac{\mathbf{L}\cdot\mathbf{m}}{r^3}-\frac{p^2e^2A^2}{8m^3c^2}\right\}\begin{bmatrix}1&0\\0&1\end{bmatrix}-\frac{e}{m}(\mathbf{S}\cdot\mathbf{B}) \tag{14.178}$$

One can clarify the effects of the proton (nuclear) magnetic field and proton (nuclear) magnetic dipole interaction with the electron magnetic dipole moment in the electron rest frame as follows. The electron magnetic dipole moment vector $\boldsymbol{\mu}_s$ is obtained from the electron spin vector $\mathbf{S}$, via the Bohr magneton μ_b and electron spin g factor $g_s = 2$, where one recalls from section 2.5,

$$\boldsymbol{\mu}_s = -g_s \frac{\mu_b}{\hbar}\mathbf{S} = -\frac{2\mu_b}{\hbar}\mathbf{S} \tag{14.179}$$

and

$$\mathbf{S} = \frac{1}{2}\hbar\boldsymbol{\sigma} = \frac{1}{2}\hbar\left\{\begin{bmatrix} 0 & 1 \\ 1 & 0 \end{bmatrix}\hat{\mathbf{x}} + \begin{bmatrix} 0 & -\mathrm{i} \\ \mathrm{i} & 0 \end{bmatrix}\hat{\mathbf{y}} + \begin{bmatrix} 1 & 0 \\ 0 & -1 \end{bmatrix}\hat{\mathbf{z}}\right\} \tag{14.180}$$

The Bohr magneton μ_b for the electron of rest mass m_e, where $m_\mathrm{e} = m$ and charge e, is given by

$$\mu_b = \frac{\hbar e}{2m_e} \tag{14.181}$$

The electron spin vector $\mathbf{S}$ is an operator, which multiplies the electron wave function Φ_1 by $\hbar\boldsymbol{\sigma}/2$, where

$$\mathbf{S}\Phi_1 = \mathbf{S}\begin{bmatrix} \Phi_{11} \\ \Phi_{12} \end{bmatrix} = \frac{1}{2}\hbar\left\{\begin{bmatrix} 0 & 1 \\ 1 & 0 \end{bmatrix}\hat{\mathbf{x}} + \begin{bmatrix} 0 & -\mathrm{i} \\ \mathrm{i} & 0 \end{bmatrix}\hat{\mathbf{y}} + \begin{bmatrix} 1 & 0 \\ 0 & -1 \end{bmatrix}\hat{\mathbf{z}}\right\}\left[\Phi_{11}\ \Phi_{12}\right] \tag{14.182}$$

The electron orbital magnetic dipole moment vector $\boldsymbol{\mu}_L$ arises from the electron orbital motion, which produces an electron current loop around the proton (nucleus), and $\boldsymbol{\mu}_L$ is obtained from the electron orbital angular momentum vector $\mathbf{L}$ and orbital g factor $g_l = 1$ via

$$\boldsymbol{\mu}_L = -g_l \frac{\mu_b}{\hbar}\mathbf{L} = -\frac{\mu_b}{\hbar}\mathbf{L} \tag{14.183}$$

The electron orbital angular momentum vector $\mathbf{L}$ is an operator, which multiplies the electron wave function by $\sqrt{l(l+1)}$, where integer l is the electron angular momentum quantum number, and

$$\mathbf{L}\Phi_1 = \mathbf{L}\begin{bmatrix} \Phi_{11} \\ \Phi_{12} \end{bmatrix} = \hbar\sqrt{l(l+1)}\begin{bmatrix} \Phi_{11} \\ \Phi_{12} \end{bmatrix} \quad l = 0, 1, 2, \ldots \tag{14.184}$$

The magnetic dipole moment $\boldsymbol{\mu}_I$ for the proton (hydrogen nucleus) and other nuclei arises from the net spin $\mathbf{I}$ of the nucleus. In the case of the hydrogen atom, the net nuclear spin **I** is the proton spin. The magnetic dipole moment $\boldsymbol{\mu}_I$ of the proton (or nucleus) is obtained from the proton (or nucleus) spin **I** and the nuclear g factor g_I via

$$\mathbf{m} \equiv \boldsymbol{\mu}_I = -g_I \frac{\mu_n}{\hbar} \mathbf{I} \tag{14.185}$$

The magnetic dipole moment vector $\boldsymbol{\mu}_I$ of the proton or nucleus is also designated by **m**, where the nuclear magneton μ_n is expressed by proton rest mass m_p and elementary charge e, where

$$\mu_n = \frac{e\hbar}{2m_p} \tag{14.186}$$

The proton g factor is $g_p = 5.585$, and for the proton spin vector $\mathbf{I}_p$, the proton magnetic dipole moment vector **m** is

$$\mathbf{m} \equiv \boldsymbol{\mu}_I = -5.585 \frac{\mu_n}{\hbar} \mathbf{I} \tag{14.187}$$

With application of Eqs. (14.179–14.187) to Eq. (14.178), the third term H_{D3} of the Dirac Hamiltonian for the hydrogen atom is now expressed in terms of the electron magnetic dipole moment vectors due to spin $\boldsymbol{\mu}_s$ and electron orbital motion $\boldsymbol{\mu}_l$ and the nuclear (proton) magnetic dipole moment **m**:

$$H_{D3} = \left\{ \frac{(\mathbf{p} - e\mathbf{A})^2}{2m} - \frac{p^4}{8m^3c^2} + \frac{\mu_0 e p^2}{8\pi m^2 c^2} \frac{\boldsymbol{\mu}_L \cdot \mathbf{m}}{r^3} - \frac{p^2 e^2 A^2}{8m^3c^2} \right\} \begin{bmatrix} 1 & 0 \\ 0 & 1 \end{bmatrix} + (\boldsymbol{\mu}_s \cdot \mathbf{B}) \tag{14.188}$$

With rearrangement of Eq. (14.188), one obtains the expression for the third Hamiltonian term of Eq. (14.135), where

$$H_{D3} = \left\{ \frac{(\mathbf{p} - e\mathbf{A})^2}{2m} - \frac{p^4}{8m^3c^2} \left(1 + \frac{e^2 A^2}{p^2} \right) + \frac{\mu_0 e p^2}{8\pi m^2 c^2} \frac{\boldsymbol{\mu}_L \cdot \mathbf{m}}{r^3} \right\} \begin{bmatrix} 1 & 0 \\ 0 & 1 \end{bmatrix} + (\boldsymbol{\mu}_s \cdot \mathbf{B}) \tag{14.189}$$

By comparison, Eq. (14.157) for the second term H_{D2} of the Dirac Hamiltonian of Eq. (14.135), expressed in terms of the electron spin $\boldsymbol{\mu}_s$ and orbital motion $\boldsymbol{\mu}_L$ magnetic dipole moment vectors, is then

$$H_{D2} = -\frac{\hbar^2 E_r}{4m^2c^2} \frac{\partial}{\partial r} \begin{bmatrix} 1 & 0 \\ 0 & 1 \end{bmatrix} + \frac{e^2}{4m^4c^2} E_r \frac{(\boldsymbol{\mu}_s \cdot \boldsymbol{\mu}_L)}{r} + \frac{\mu_0 e^2}{8\pi m^3 c^2} E_r \left[\boldsymbol{\mu}_s \cdot \frac{\mathbf{m} - \hat{\mathbf{r}}(\hat{\mathbf{r}} \cdot \mathbf{m})}{r^2} \right] \tag{14.190}$$

where

$$\mu_s = -\frac{m}{e}\mathbf{S} \quad \text{and} \quad \mu_L = -\frac{2m}{e}\mathbf{L} \tag{14.191}$$

In Eq. (14.189) for the third term H_{D3} of the Dirac Hamiltonian, the total or canonical electron momentum **p** is the sum of the electron mechanical momentum **P** and the contribution to electron momentum due to the magnetic vector potential **A** of the proton magnetic dipole moment **m**, where

$$\mathbf{p} = \mathbf{P} + e\mathbf{A} = \gamma m\mathbf{v} + e\mathbf{A} \tag{14.192}$$

The relativistic electron mechanical momentum **P** is expressed in terms of the electron rest mass m, electron orbital velocity **v** about the proton nucleus, and the Lorentz factor γ. The electron velocity is assumed to be nonrelativistic, so

$$\mathbf{P} = \gamma m\mathbf{v} = \frac{m\mathbf{v}}{\sqrt{1-(v/c)^2}} \cong m\mathbf{v} \tag{14.193}$$

The electron kinetic energy W accounting for the electron orbital velocity **v** is the first term of the H_{D3} expression, where

$$W = \frac{P^2}{2m} = \frac{\mathbf{P}\cdot\mathbf{P}}{2m} = \frac{(\mathbf{p}-e\mathbf{A})\cdot(\mathbf{p}-e\mathbf{A})}{2m} = \frac{(\mathbf{p}-e\mathbf{A})^2}{2m} \tag{14.194}$$

The electron mechanical momentum **P** is assumed to be much larger than momentum contribution from the magnetic vector potential $e\mathbf{A}$ of the proton magnetic dipole moment vector **m**, so that $p \gg eA$ and

$$1 + \frac{e^2A^2}{p^2} \cong 1 \tag{14.195}$$

Since the electron of rest mass m is orbiting the proton at nonrelativistic speed v, then the electron kinetic energy W is related to its mechanical momentum P and speed v by

$$W = \frac{P^2}{2m} = \frac{mv^2}{2} \tag{14.196}$$

The electron rest mass energy E_{rm} is

$$E_{rm} = mc^2 \tag{14.197}$$

Hence,

$$\frac{p^4}{8m^3c^2} = \frac{1}{2mc^2}\left(\frac{p^2}{2m}\right)^2 \cong \frac{1}{E_{\rm rm}}\left(\frac{P^2}{2m}\right)^2 = \frac{W^2}{2E_{\rm rm}} \tag{14.198}$$

With substitution of Eq. (14.198) into Eq. (14.189), the third term H_{D3} of the Dirac Hamiltonian then becomes

$$H_{D3} = \left\{\frac{(\mathbf{p}-e\mathbf{A})^2}{2m} - \frac{W^2}{2E_{\rm rm}} + \frac{\mu_0 e^2 p^2}{16\pi m^3 c^2}\frac{\mathbf{L}\cdot\mathbf{m}}{r^3}\right\}\begin{bmatrix}1 & 0\\ 0 & 1\end{bmatrix} - \frac{e}{m}(\mathbf{S}\cdot\mathbf{B}) \tag{14.199}$$

Alternatively, in terms of electron spin $\boldsymbol{\mu}_s$ and orbital $\boldsymbol{\mu}_L$ magnetic dipole moment vectors, with substitution of Eq. (14.191) into Eq. (14.199), the third Hamiltonian term becomes

$$H_{D3} = \left\{\frac{(\mathbf{p}-e\mathbf{A})^2}{2m} - \frac{W^2}{2E_{\rm rm}} + \frac{\mu_0 e p^2}{8\pi m^2 c^2}\frac{\boldsymbol{\mu}_L\cdot\mathbf{m}}{r^3}\right\}\begin{bmatrix}1 & 0\\ 0 & 1\end{bmatrix} + (\boldsymbol{\mu}_s\cdot\mathbf{B}) \tag{14.200}$$

The H_{D3} expression contains the fine-structure term $p^4/8m^3c^2$, which is the relativistic correction term that has been expressed in its nonrelativistic form, $W^2/2E_{\rm rm}$. The hyperfine structure dot product term $\mathbf{S}\cdot\mathbf{B}$, accounts for the electron spin $\mathbf{S}$ interaction with the proton (or nucleus) magnetic field $\mathbf{B}$. In the electron rest frame, the electron spin matrix vector $\mathbf{S}$ interacts with the proton (nucleus) magnetic field $\mathbf{B}$, via the electron magnetic dipole moment vector $\boldsymbol{\mu}_s$, The dot product term $\boldsymbol{\mu}_L\cdot\mathbf{m}$ is also a hyperfine structure term, and accounts for the proton (or nucleus) magnetic dipole moment vector $\mathbf{m}$, which interacts with the electron magnetic dipole moment vector $\boldsymbol{\mu}_L$ that arises from electron orbital motion.

14.9 Magnetic Field of a Proton Current in an Electron Rest Frame Expressed by the Proton Magnetic Dipole Moment Vector and Electron Orbit Radius

The magnetic field $\mathbf{B}$ of the proton magnetic dipole moment $\mathbf{m}$ is derived as follows. Given the magnetic dipole moment $\mathbf{m}$ of the proton is constant,

$$\mathbf{A} = \frac{\mu_0}{4\pi}\left(\mathbf{m}\times\frac{\hat{\mathbf{r}}}{r^2}\right) = -\frac{\mu_0}{4\pi}\left(\mathbf{m}\times\nabla\frac{1}{r}\right) = \frac{\mu_0}{4\pi}\left(\nabla\frac{1}{r}\times\mathbf{m}\right) = \frac{\mu_0}{4\pi}\nabla\times\frac{\mathbf{m}}{r} \tag{14.201}$$

The resulting expression for the magnetic field $\mathbf{B}$ of the proton magnetic dipole moment $\mathbf{m}$ at the radial position r of the orbiting electron is

$$\mathbf{B} = \nabla\times\mathbf{A} = \frac{\mu_0}{4\pi}\nabla\times\nabla\times\frac{\mathbf{m}}{r} \tag{14.202}$$

Expanding Eq. (14.202), the magnetic field of the proton magnetic dipole moment, becomes

$$\mathbf{B} = \frac{\mu_0}{4\pi}\nabla\times\nabla\times\frac{\mathbf{m}}{r} = \frac{\mu_0}{4\pi}\nabla\left(\nabla\cdot\frac{\mathbf{m}}{r}\right) - \nabla^2\left(\frac{\mathbf{m}}{r}\right) \tag{14.203}$$

Since $\mathbf{m}$ is a constant, one obtains

$$\nabla\cdot\frac{\mathbf{m}}{r} = \mathbf{m}\cdot\nabla\left(\frac{1}{r}\right) + \frac{1}{r}\nabla\cdot\mathbf{m} = \mathbf{m}\cdot\nabla\left(\frac{1}{r}\right) = -\frac{\mathbf{m}\cdot\mathbf{r}}{r^3} \tag{14.204}$$

The first term for the proton magnetic field $\mathbf{B}$ is

$$\nabla\left(\nabla\cdot\frac{\mathbf{m}}{r}\right) = -\nabla\left(\frac{\mathbf{m}\cdot\mathbf{r}}{r^3}\right) = -\left[(\mathbf{m}\cdot\mathbf{r})\nabla\left(\frac{1}{r^3}\right) + \frac{1}{r^3}\nabla(\mathbf{m}\cdot\mathbf{r})\right] \tag{14.205}$$

Evaluation yields

$$\nabla\left(\nabla\cdot\frac{\mathbf{m}}{r}\right) = \frac{3\mathbf{r}}{r^5}(\mathbf{m}\cdot\mathbf{r}) - \frac{1}{r^3}(\mathbf{m}\cdot\nabla)\mathbf{r} \tag{14.206}$$

where

$$(\mathbf{m}\cdot\nabla)\mathbf{r} = \left(m_x\frac{\partial}{\partial x} + m_y\frac{\partial}{\partial y} + m_z\frac{\partial}{\partial z}\right)(x\hat{\mathbf{x}} + y\hat{\mathbf{y}} + z\hat{\mathbf{z}}) = \mathbf{m} \tag{14.207}$$

The first term on the right side of Eq. (14.203) is evaluated, where

$$\nabla\left(\nabla\cdot\frac{\mathbf{m}}{r}\right) = \frac{3\mathbf{r}}{r^5}(\mathbf{m}\cdot\mathbf{r}) - \frac{\mathbf{m}}{r^3} \tag{14.208}$$

Given the spherical symmetry, the second term in Eq. (14.203) for the proton magnetic field $\mathbf{B}$ is defined except at $r = 0$, where

$$\nabla^2\left(\frac{\mathbf{m}}{r}\right) = \mathbf{m}\nabla^2\left(\frac{1}{r}\right) = \mathbf{m}\frac{1}{r^2}\frac{\partial}{\partial r}\left(r^2\frac{\partial}{\partial r}\left(\frac{1}{r}\right)\right) \tag{14.209}$$

If $r \neq 0$, then

$$\frac{\partial}{\partial r}\left(r^2\frac{\partial}{\partial r}\left(\frac{1}{r}\right)\right) = \frac{\partial}{\partial r}\left(r^2\left(-\frac{1}{r^2}\right)\right) = \frac{\partial}{\partial r}(-1) = 0 \tag{14.210}$$

If $r = 0$, then there is a singularity, where

$$\frac{\partial}{\partial r}\left(r^2\frac{\partial}{\partial r}\left(\frac{1}{r}\right)\right) = \frac{\partial}{\partial r}\left(-\frac{0^2}{0^2}\right) \quad \text{not defined at } r = 0 \tag{14.211}$$

Hence, evaluation of the Laplacian $\nabla^2(1/r)$ requires care. In comparison, $\nabla(1/r)$ presents no problem because in the same spherically symmetric situation,

$$\nabla\frac{1}{r} = \hat{\mathbf{r}}\frac{\partial}{\partial r}\left(\frac{1}{r}\right) = -\frac{\hat{\mathbf{r}}}{r^2} = -\frac{\mathbf{r}}{r^3} \tag{14.212}$$

To evaluate the Laplacian $\nabla^2(1/r)$ requires introducing volume integration over a tiny sphere of radius r centered about the singularity at $r = 0$, where

$$\lim_{r\to 0}\int_0^{2\pi} \mathrm{d}\phi \int_{-\pi/2}^{\pi/2} \sin \mathrm{d}\theta \int_0^r r^2\nabla^2\frac{1}{r}\mathrm{d}r = \lim_{r\to 0}\int_0^{2\pi} \mathrm{d}\phi \int_{-\pi/2}^{\pi/2} \mathrm{d}\theta \sin\theta \int_0^r r^2\nabla\cdot\nabla\frac{1}{r}\mathrm{d}r \tag{14.213}$$

Application of Gauss's law to equate the volume integral to integration over the spherical surface is invoked, where

$$\lim_{r\to 0}\int_0^{2\pi} \mathrm{d}\phi \int_0^{\pi} \mathrm{d}\theta \sin\theta \int_0^r r^2\nabla\cdot\nabla\frac{1}{r}\mathrm{d}r = \lim_{r\to 0}\int_0^{2\pi} r\,\mathrm{d}\phi \int_0^{\pi} \mathrm{d}\theta r \sin\theta\left(\hat{\mathbf{r}}\cdot\nabla\frac{1}{r}\right) \tag{14.214}$$

One performs the gradient operation in the integrand on the right side of Eq. (14.214), and then integrates, so that

$$\lim_{r\to 0}\int_0^{2\pi} r\,\mathrm{d}\phi \int_0^{\pi} \mathrm{d}\theta r \sin\theta\left(\hat{\mathbf{r}}\cdot\nabla\frac{1}{r}\right) = \lim_{r\to 0}\int_0^{2\pi} \mathrm{d}\phi \int_0^{\pi} \mathrm{d}\theta r^2 \sin\theta\left(-\frac{1}{r^2}\right) = -4\pi \tag{14.215}$$

Thus the result of integration in Eq. (14.213), is

$$\lim_{r\to 0}\int_0^{2\pi} \mathrm{d}\phi \int_{-\pi/2}^{\pi/2} \sin \mathrm{d}\theta \int_0^r r^2\nabla^2\frac{1}{r}\mathrm{d}r = -4\pi \tag{14.216}$$

Examination of the integrated result of Eq. (14.216), suggests the integrand is a Dirac delta function $\delta(\mathbf{r})$, where

$$\nabla^2\frac{1}{r} = -4\pi\delta(\mathbf{r}) \tag{14.217}$$

Substitution of Eq. (14.217) back into the volume integral of Eq. (14.213), confirms the following:

$$\lim_{r\to 0}\int_0^{2\pi} \mathrm{d}\phi \int_{-\pi/2}^{\pi/2} \sin \mathrm{d}\theta \int_0^r r^2[-4\pi\delta(r)]\mathrm{d}r = -4\pi \tag{14.218}$$

By the definition of the Dirac delta function $\delta(\mathbf{r})$, which is a result of the limiting operation $\mathbf{r} \to 0$ in the volume integral of Eq. (14.213), if one performs the volume integration of an integrand that is comprised of the function $f(\mathbf{r})$, multiplied by the Dirac delta function $\delta(\mathbf{r})$, then one obtains the function evaluated at $\mathbf{r} = 0$; i.e.,$f(0)$, where

$$\int_0^{2\pi} \mathrm{d}\phi \int_{-\pi/2}^{\pi/2} \sin \mathrm{d}\theta \int_0^r r^2 f(r)\delta(r)\mathrm{d}r = f(0) \tag{14.219}$$

where

$$f(r) = f(0) = -4\pi \tag{14.220}$$

With substitution of Eq. (14.217) into Eq. (14.209), the second term for the proton magnetic field $\mathbf{B}$ is

$$\nabla^2\left(\frac{\mathbf{m}}{r}\right) = \mathbf{m}\nabla^2\left(\frac{1}{r}\right) = -\mathbf{m}4\pi\delta(\mathbf{r}) \tag{14.221}$$

In the laboratory frame, the proton is at rest, and the electron orbits at radius r, and experiences magnetic field $\mathbf{B}$ of the proton magnetic dipole moment vector $\mathbf{m}$. The magnetic field $\mathbf{B}$ of the proton magnetic dipole moment $\mathbf{m}$ is thus obtained by substitution of Eqs. (14.221) and (14.208) into Eq. (14.203), where

$$\mathbf{B} = \frac{3\mathbf{r}}{r^5}(\mathbf{m} \cdot \mathbf{r}) - \frac{\mathbf{m}}{r^3} + 4\pi\delta(\mathbf{r})\mathbf{m} \tag{14.222}$$

Equation (14.222) for the magnetic field $\mathbf{B}$, which is imposed on the electron by the proton magnetic dipole moment $\mathbf{m}$, can be expressed in terms of the unit radius vector $\hat{\mathbf{r}}$, which points from the proton to the electron at radius r, where

$$\mathbf{B} = 3\frac{\hat{\mathbf{r}}}{r^3}(\mathbf{m} \cdot \hat{\mathbf{r}}) - \frac{\mathbf{m}}{|r\hat{\mathbf{r}}|^3} + 4\pi\delta(r\hat{\mathbf{r}})\mathbf{m} \tag{14.223}$$

This magnetic field $\mathbf{B}$ is derived for only the proton or nucleus magnetic dipole moment $\mathbf{m}$. Recall in chapter 9, was derived the spin–orbit interaction of incident neutrons of velocity $\mathbf{v}$ with the atomic electric field $\mathbf{E}$, which is the sum of the electric fields of all the atomic electrons (paired and unpaired), as well as the nucleus electric field. The atomic electric field $\mathbf{E}$ produces magnetic field $\mathbf{B} = \mathbf{E} \times \mathbf{v}/c^2$ in the neutron rest frame. The magnetic field $\mathbf{B}$ acts on the neutron magnetic dipole moment $\boldsymbol{\mu}$ in the neutron rest frame, thereby causing neutron scatter via the total electric field $\mathbf{E}$ of the atom.

14.10 Evaluated Dirac Equation for Bound Electron–Proton Interaction that Predicts LS Coupling and Hyperfine Interactions

The second H_{D2} and third H_{D3} operator terms of the Dirac Hamiltonian operator H_D in the Dirac eigenequation expression of Eqs. (14.135) and (14.136), which describe the bound electron interaction with the proton in a hydrogen atom, can also approximate the interaction of an outer, single valence electron in other hydrogen-like atoms, such as lithium, which have more than one proton. That is, the inner electron cloud shields an equal and opposite amount of nuclear charge, so that the net nuclear charge which interacts with the outer valence single electron is effectively one proton. In the hydrogen atom or hydrogen-like atoms, the electron wave amplitude eigenvector is comprised of two orthogonal components Φ_{11} and Φ_{12}, where

$$\Phi_1 = \begin{bmatrix} \Phi_{11} \\ \Phi_{12} \end{bmatrix} = \begin{bmatrix} u_1 \exp[\mathrm{i}(\mathbf{k} \cdot \mathbf{r} - wt)] \\ u_2 \exp[\mathrm{i}(\mathbf{k} \cdot \mathbf{r} - wt)] \end{bmatrix} \tag{14.224}$$

The fully evaluated Dirac Hamiltonian H_D from Eq. (14.135) is

$$H_D\Phi_1 = E_1\Phi_1 = [e\phi - H_{D2} + H_{D3}]\Phi_1 \tag{14.225}$$

From Eqs. (14.157) and (14.200), the second H_{D2} and third H_{D3} terms of the Dirac Hamiltonian H_D are

$$H_{D2} = -\frac{\hbar^2 E_r}{4m^2c^2}\frac{\partial}{\partial r}\begin{bmatrix} 1 & 0 \\ 0 & 1 \end{bmatrix} + \frac{E_r}{2m^2c^2}\frac{\mathbf{L}\cdot\mathbf{S}}{r} - \frac{\mu_0 e E_r}{8\pi m^2 c^2}\left[\mathbf{S}\cdot\frac{\mathbf{m} - \hat{\mathbf{r}}(\hat{\mathbf{r}}\cdot\mathbf{m})}{r^2}\right] \tag{14.226}$$

and

$$H_{D3} = \left\{\frac{(\mathbf{p} - e\mathbf{A})^2}{2m} - \frac{W^2}{2E_{\mathrm{rm}}} + \frac{\mu_0 e^2 p^2}{16\pi m^3 c^2}\frac{\mathbf{L}\cdot\mathbf{m}}{r^3}\right\}\begin{bmatrix} 1 & 0 \\ 0 & 1 \end{bmatrix} - \frac{e}{m}(\mathbf{S}\cdot\mathbf{B}) \tag{14.227}$$

The eigenequation for the electron orbital angular momentum vector operator $\mathbf{L}$ is

$$\mathbf{L}\Phi_1 = \mathbf{L}\begin{bmatrix} \Phi_{11} \\ \Phi_{12} \end{bmatrix} = \hbar\sqrt{l(l+1)}\begin{bmatrix} \Phi_{11} \\ \Phi_{12} \end{bmatrix} \quad l = 0, 1, 2, \ldots \tag{14.228}$$

With $\mathbf{S} = \boldsymbol{\sigma}(\hbar/2)$, one has the electron spin vector operator eigenequations

$$\mathbf{S}\Phi_1 = \mathbf{S}\begin{bmatrix} \Phi_{11} \\ \Phi_{12} \end{bmatrix} = \frac{1}{2}\hbar\left\{\begin{bmatrix} 0 & 1 \\ 1 & 0 \end{bmatrix}\hat{\mathbf{x}} + \begin{bmatrix} 0 & -\mathrm{i} \\ \mathrm{i} & 0 \end{bmatrix}\hat{\mathbf{y}} + \begin{bmatrix} 1 & 0 \\ 0 & -1 \end{bmatrix}\hat{\mathbf{z}}\right\}[\Phi_{11}\ \Phi_{12}] \tag{14.229}$$

Substitution of Eqs. (14.224) and (14.226–14.229) into Eq. (14.225), yields the fully evaluated Dirac Hamiltonian:

$$H_{\mathrm{D}}\begin{bmatrix}\Phi_{11}\\ \Phi_{12}\end{bmatrix}=E_1\begin{bmatrix}\Phi_{11}\\ \Phi_{12}\end{bmatrix} \tag{14.230}$$

or

$$E_1\begin{bmatrix}\Phi_{11}\\ \Phi_{12}\end{bmatrix}=\left(e\phi+\frac{\mu_0 eE_r}{8\pi m^2c^2}\left[\mathbf{S}\cdot\frac{\mathbf{m}-\hat{\mathbf{r}}(\hat{\mathbf{r}}\cdot\mathbf{m})}{r^2}\right]-\frac{E_r}{2m^2c^2}\frac{\mathbf{L}\cdot\mathbf{S}}{r}-2\mu_b\frac{\mathbf{S}\cdot\mathbf{B}}{\hbar}\right)\begin{bmatrix}\Phi_{11}\\ \Phi_{12}\end{bmatrix}+$$
$$\left(\frac{(\mathbf{p}-e\mathbf{A})^2}{2m}-\frac{W^2}{2E_{rm}}+\frac{\hbar^2E_r}{4m^2c^2}\frac{\partial}{\partial r}+\frac{\mu_0e^2p^2}{16\pi m^3c^2}\frac{\mathbf{L}\cdot\mathbf{m}}{r^3}\right)\begin{bmatrix}1&0\\0&1\end{bmatrix}\begin{bmatrix}\Phi_{11}\\ \Phi_{12}\end{bmatrix} \tag{14.231}$$

The magnetic field **B** of the proton magnetic dipole moment **m**, which interacts with the orbiting electron at radius vector **r** (and distance $r=|\mathbf{r}|$) in the hydrogen atom, includes the Dirac delta function $\delta(\mathbf{r})$, where

$$\mathbf{B}=\frac{3\mathbf{r}}{r^5}(\mathbf{m}\cdot\mathbf{r})-\frac{\mathbf{m}}{r^3}+4\pi\delta(\mathbf{r})\mathbf{m} \tag{14.232}$$

In terms of the magnetic dipole moment vectors for electron spin $\boldsymbol{\mu}_s$ and electron orbital motion $\boldsymbol{\mu}_L$ and the proton magnetic dipole vector **m**, the second H_{D2} and third H_{D3} terms of the Dirac Hamiltonian H_{D} are

$$H_{\mathrm{D2}}=-\frac{\hbar^2E_r}{4m^2c^2}\frac{\partial}{\partial r}\begin{bmatrix}1&0\\0&1\end{bmatrix}+\frac{e^2E_r}{4m^4c^2}\frac{\boldsymbol{\mu}_s\cdot\boldsymbol{\mu}_L}{r}+\frac{\mu_0e^2E_r}{8\pi m^3c^2}\left[\boldsymbol{\mu}_s\cdot\frac{\mathbf{m}-\hat{\mathbf{r}}(\hat{\mathbf{r}}\cdot\mathbf{m})}{r^2}\right] \tag{14.233}$$

and

$$H_{\mathrm{D3}}=\left\{\frac{(\mathbf{p}-e\mathbf{A})^2}{2m}-\frac{W^2}{2E_{\mathrm{rm}}}+\frac{\mu_0ep^2}{8\pi m^2c^2}\frac{\boldsymbol{\mu}_L\cdot\mathbf{m}}{r^3}\right\}\begin{bmatrix}1&0\\0&1\end{bmatrix}+(\boldsymbol{\mu}_s\cdot\mathbf{B}) \tag{14.234}$$

The magnetic dipole moment vector operator $\boldsymbol{\mu}_{\mathbf{S}}$ for the electron spin angular momentum vector operator **S** and the magnetic dipole moment vector operator $\boldsymbol{\mu}_{\mathbf{L}}$ for the electron orbital angular momentum vector operator **L** are

$$\boldsymbol{\mu}_s=-\frac{m}{e}\mathbf{S}\quad\text{and}\quad\boldsymbol{\mu}_L=-\frac{2m}{e}\mathbf{L} \tag{14.235}$$

The Dirac Hamiltonian and wave equation here account for electron spin and yield the experimentally correct LS coupling term $\mathbf{L} \cdot \mathbf{S}$ with the needed multiplicative factor of $1/2$. This factor of $1/2$ is missing in the ordinary Schrödinger equation and Hamiltonian, which do not account for spin. However, the Thomas precession analysis also yields the factor of $1/2$, which accounts for the electron acceleration due to its circular orbit around the proton, albeit at constant speed—the electron continuously changes direction. The LS coupling term derived in the Dirac equation is independent of the magnetic vector potential $\mathbf{A}$ of the proton magnetic dipole moment $\mathbf{m}$. That is, the LS coupling term is also obtained when one ignores the proton magnetic dipole moment with $\mathbf{A}=0$. This is expected, in that the LS coupling term is a result of the interaction of the magnetic dipole moments of the electron spin $\boldsymbol{\mu}_s$ and orbital motion $\boldsymbol{\mu}_L$.

In Eq. (14.231), one converts total energy E_1 to its operator, partial time derivative form, where

$$E_1 \rightarrow -i\hbar\frac{\partial}{\partial t} \tag{14.236}$$

This Dirac Hamiltonian of Eq. (14.231) becomes the Dirac version of the Schrödinger equation, which is now relativistically correct and accounts for electron spin, where

$$-i\hbar\frac{\partial}{\partial t}\begin{bmatrix}\Phi_{11}\\ \Phi_{12}\end{bmatrix} = \left(e\phi + \frac{\mu_0 eE_r}{8\pi m^2c^2}\left[\mathbf{S}\cdot\frac{\mathbf{m}-\hat{\mathbf{r}}(\hat{\mathbf{r}}\cdot\mathbf{m})}{r^2}\right] - \frac{E_r}{2m^2c^2}\frac{\mathbf{L}\cdot\mathbf{S}}{r} - 2\mu_b\frac{\mathbf{S}\cdot\mathbf{B}}{\hbar}\right)\begin{bmatrix}\Phi_{11}\\ \Phi_{12}\end{bmatrix} + \left(\frac{(\mathbf{p}-e\mathbf{A})^2}{2m} - \frac{W^2}{2E_{\mathrm{rm}}} + \frac{\hbar^2E_r}{4m^2c^2}\frac{\partial}{\partial r} + \frac{\mu_0e^2p^2}{16\pi m^3c^2}\frac{\mathbf{L}\cdot\mathbf{m}}{r^3}\right)\begin{bmatrix}1 & 0\\ 0 & 1\end{bmatrix}\begin{bmatrix}\Phi_{11}\\ \Phi_{12}\end{bmatrix} \tag{14.237}$$

Given stationary wave functions Φ_{11} and Φ_{12}, the time derivative extracts angular frequency w:

$$\frac{\partial}{\partial t}\Phi_1 = \frac{\partial}{\partial t}\begin{bmatrix}\Phi_{11}\\ \Phi_{12}\end{bmatrix} = \frac{\partial}{\partial t}\begin{bmatrix}u_1\exp[i(\mathbf{k}\cdot\mathbf{r}-wt)]\\ u_2\exp[i(\mathbf{k}\cdot\mathbf{r}-wt)]\end{bmatrix} = -iw\begin{bmatrix}\Phi_{11}\\ \Phi_{12}\end{bmatrix} \tag{14.238}$$

With substitution of Eq. (14.238) into Eq. (14.237), the Dirac wave equation becomes a time-independent expression, which states the conservation of energy, where

$$\begin{bmatrix}0\\ 0\end{bmatrix} = \left(\hbar w + e\phi + \frac{\mu_0 eE_r}{8\pi m^2c^2}\left[\mathbf{S}\cdot\frac{\mathbf{m}-\hat{\mathbf{r}}(\hat{\mathbf{r}}\cdot\mathbf{m})}{r^2}\right] - \frac{E_r}{2m^2c^2}\frac{\mathbf{L}\cdot\mathbf{S}}{r} - 2\mu_b\frac{\mathbf{S}\cdot\mathbf{B}}{\hbar}\right)\begin{bmatrix}\Phi_{11}\\ \Phi_{12}\end{bmatrix} + \left(\frac{(\mathbf{p}-e\mathbf{A})^2}{2m} - \frac{W^2}{2E_{\mathrm{rm}}} + \frac{\hbar^2E_r}{4m^2c^2}\frac{\partial}{\partial r} + \frac{\mu_0e^2p^2}{16\pi m^3c^2}\frac{\mathbf{L}\cdot\mathbf{m}}{r^3}\right)\begin{bmatrix}1 & 0\\ 0 & 1\end{bmatrix}\begin{bmatrix}\Phi_{11}\\ \Phi_{12}\end{bmatrix} \tag{14.239}$$

14.11 Equation of Motion for Electron Spin, Orbital, and Total Angular Momentum Vectors

The equation of motions of the electron spin **S**, orbital **L**, and total angular momentum $\mathbf{J} = \mathbf{L} + \mathbf{S}$ in the Dirac wave equation are examined in the Heisenberg picture. The goal is to determine if **S**, **L**, or $\mathbf{J} = \mathbf{L} + \mathbf{S}$ is time invariant, and thus is a constant of motion. As will be shown later in this chapter, only the sum $\mathbf{J} = \mathbf{L} + \mathbf{S}$ of the spin and orbital angular momentum (i.e., total angular momentum) of the electron is independent of time, and thus a constant of electron motion in the Dirac wave equation.

In the Heisenberg picture (see chapter 3 in Cremer, 2012b), a time-independent operator **C**, which is used in the Schrödinger equation with time-varying state functions $\psi(t)$, is converted into a time-dependent operator $\mathbf{C}(t)$ of the Heisenberg picture that uses time-independent state functions ϕ. This conversion is obtained via the transformation involving the Hamiltonian operator H, where

$$A(t) = \exp\left(\mathrm{i}\frac{Ht}{\hbar}\right)A\exp\left(-\mathrm{i}\frac{Ht}{\hbar}\right) \tag{14.240}$$

The time derivative is taken so that

$$\frac{\mathrm{d}A(t)}{\mathrm{d}t} = \frac{\mathrm{d}}{\mathrm{d}t}\left[\exp\left(\mathrm{i}\frac{Ht}{\hbar}\right)A\exp\left(-\mathrm{i}\frac{Ht}{\hbar}\right)\right] \tag{14.241}$$

$$\begin{aligned}\frac{\mathrm{d}A(t)}{\mathrm{d}t} = {}& \frac{\mathrm{i}}{\hbar}\left[\exp\left(\mathrm{i}\frac{Ht}{\hbar}\right)HA\exp\left(-\mathrm{i}\frac{Ht}{\hbar}\right) - \exp\left(\mathrm{i}\frac{Ht}{\hbar}\right)AH\exp\left(-\mathrm{i}\frac{Ht}{\hbar}\right)\right]\\ & + \exp\left(\mathrm{i}\frac{Ht}{\hbar}\right)\frac{\partial A}{\partial t}\exp\left(-\mathrm{i}\frac{Ht}{\hbar}\right)\end{aligned} \tag{14.242}$$

Assume that $A(t)$ does not have explicit time dependence, where

$$\frac{\partial A}{\partial t} = 0 \tag{14.243}$$

One substitutes Eq. (14.243) into Eq. (14.242) and uses commutation brackets "[ ]" to express the commutation between operators H and $A(t)$, where $[H, A] = HA - AH$ and

$$\frac{\mathrm{d}A(t)}{\mathrm{d}t} = \frac{\mathrm{i}}{\hbar}[HA(t) - A(t)H] \tag{14.244}$$

The operator $A(t)$ can be a vector operator, such as the electron spin matrix angular momentum vector **S** and the electron orbital angular moment vector **L**. However, as will be shown later, the angular momentum vectors for electron spin **S** and electron orbital motion **L** are each not constants of motion in the Dirac wave equation because **S** and **L** do not commute with the Dirac Hamiltonian H_D, where

$$H_{\mathrm{D}} = c\boldsymbol{\alpha}\cdot(\mathbf{p}-e\mathbf{A})+\beta mc^2+e\phi \tag{14.245}$$

The electron spin $\mathbf{S}(t)$, and electron orbital angular $\mathbf{L}(t)$ momentum vector operators are dependent on time t, and therefore $\mathbf{S}(t)$ and $\mathbf{L}(t)$ do not commute with the time-independent, Dirac equation Hamiltonian operator H_D, where $[H_D, \mathbf{S}] \neq 0$ and $[H_D, \mathbf{L}] \neq 0$.

Consider the case of the Heisenberg picture equation of motion for the electron orbital angular momentum matrix vector **L**, where

$$\frac{d\mathbf{L}}{dt} = \frac{i}{\hbar}(H_{\mathrm{D}}\mathbf{L}-\mathbf{L}H_{\mathrm{D}}) \tag{14.246}$$

Insertion into Eq. (14.246) of Eq. (14.245) for the Dirac Hamiltonian H_D, gives

$$\frac{d\mathbf{L}}{dt} = \frac{ic}{\hbar}((\boldsymbol{\alpha}\cdot\mathbf{p})\mathbf{L}-\mathbf{L}(\boldsymbol{\alpha}\cdot\mathbf{p}))+\frac{ic}{\hbar}(\beta mc^2\mathbf{L}-\mathbf{L}\beta mc^2)+\frac{ic}{\hbar}(e\phi\mathbf{L}-\mathbf{L}e\phi) \tag{14.247}$$

The product constant of the electron charge e and the proton scalar electric potential ϕ commute with the electron orbital angular momentum vector operator **L**, and **L** commutes with constant β, so that last two terms of Eq. (14.247) are zero, where

$$\beta mc^2\mathbf{L}-\mathbf{L}\beta mc^2 = 0 \quad \text{and} \quad e\phi\mathbf{L}-\mathbf{L}e\phi = 0 \tag{14.248}$$

The electron orbit position **r**, linear momentum **p**, and angular momentum **L** relative to the origin at the center of the nucleus are

$$\mathbf{r} = \hat{\mathbf{x}}x+\hat{\mathbf{y}}y+\hat{\mathbf{z}}z \tag{14.249}$$

$$\mathbf{p} = \hat{\mathbf{x}}p_x+\hat{\mathbf{y}}p_y+\hat{\mathbf{z}}p_z \tag{14.250}$$

$$\mathbf{L} = \mathbf{r}\times\mathbf{p} = \hat{\mathbf{x}}(yp_z-zp_y)+\hat{\mathbf{y}}(zp_x-xp_z)+\hat{\mathbf{z}}(xp_y-yp_x) \tag{14.251}$$

With Eqs. (12.249−12.251), one evaluates the first term of Eq. (14.247) for the equation of motion for **L**:

$$\begin{aligned}\frac{ic}{\hbar}((\boldsymbol{\alpha}\cdot\mathbf{p})\mathbf{L}-\mathbf{L}(\boldsymbol{\alpha}\cdot\mathbf{p})) = {} & (\alpha_x p_x+\alpha_y p_y+\alpha_z p_z)\big[\hat{\mathbf{x}}(yp_z-zp_y)+\hat{\mathbf{y}}(zp_x-xp_z)\\ & +\hat{\mathbf{z}}(xp_y-yp_x)\big]-\big[\hat{\mathbf{x}}(yp_z-zp_y)+\hat{\mathbf{y}}(zp_x-xp_z)\\ & +\hat{\mathbf{z}}(xp_y-yp_x)\big](\alpha_x p_x+\alpha_y p_y+\alpha_z p_z)\end{aligned} \tag{14.252}$$

The matrix vector $\boldsymbol{\alpha}$ is a constant, and the electron momentum components (p_x, p_y, p_z) are independent of the electron position components (x, y, z) so that the second term $\mathbf{L}(\boldsymbol{\alpha}\cdot\mathbf{p})$ is zero, where

$$\mathbf{L}(\boldsymbol{\alpha}\cdot\mathbf{p}) = -i\hbar h\left[\hat{\mathbf{x}}\left(y\frac{\partial}{\partial z} - z\frac{\partial}{\partial y}\right) + \hat{\mathbf{y}}\left(z\frac{\partial}{\partial x} - x\frac{\partial}{\partial z}\right)\right.$$
$$\left. + \hat{\mathbf{z}}\left(x\frac{\partial}{\partial y} - y\frac{\partial}{\partial x}\right)\right](\alpha_x p_x + \alpha_y p_y + \alpha_z p_z) = 0 \tag{14.253}$$

However, the first term $(\boldsymbol{\alpha}\cdot\mathbf{p})\mathbf{L}$ is nonzero, where

$$(\boldsymbol{\alpha}\cdot\mathbf{p})\mathbf{L} = -i\hbar\left(\alpha_x\frac{\partial}{\partial x} + \alpha_y\frac{\partial}{\partial y} + \alpha_z\frac{\partial}{\partial z}\right)[\hat{\mathbf{x}}(yp_z - zp_y) + \hat{\mathbf{y}}(zp_x - xp_z) + \hat{\mathbf{z}}(xp_y - yp_x)] \tag{14.254}$$

Carrying out the differentiation yields

$$(\boldsymbol{\alpha}\cdot\mathbf{p})\mathbf{L} = -i\hbar[\hat{\mathbf{x}}(\alpha_y p_z - \alpha_z p_y) + \hat{\mathbf{y}}(\alpha_z p_x - \alpha_x p_z) + \hat{\mathbf{z}}(\alpha_x p_y - \alpha_y p_x)] = -i\hbar(\boldsymbol{\alpha}\times\mathbf{p}) \tag{14.255}$$

The electron orbital momentum $\mathbf{L}$ is thus not time invariant and thus is not a constant of motion in the Heisenberg picture for the equation of motion of Eq. (14.247) for $\mathbf{L}$, where

$$\frac{d\mathbf{L}}{dt} = c(\boldsymbol{\alpha}\times\mathbf{p}) \tag{14.256}$$

Next, consider the electron spin angular momentum $\mathbf{S}$ expressed via the Pauli spin vector matrix $\boldsymbol{\sigma}$, where

$$\frac{d\boldsymbol{\sigma}}{dt} = \frac{i}{\hbar}(H_D\boldsymbol{\sigma} - \boldsymbol{\sigma}H_D) \tag{14.257}$$

and

$$\mathbf{S} = \frac{\hbar}{2}\boldsymbol{\sigma} \tag{14.258}$$

Insertion into Eq. (14.258) of Eq. (14.245) for the Dirac Hamiltonian H_D gives

$$\frac{d\boldsymbol{\sigma}}{dt} = \frac{ic}{\hbar}((\boldsymbol{\alpha}\cdot\mathbf{p})\boldsymbol{\sigma} - \boldsymbol{\sigma}(\boldsymbol{\alpha}\cdot\mathbf{p})) + \frac{ic}{\hbar}(\beta mc^2\boldsymbol{\sigma} - \boldsymbol{\sigma}\beta mc^2) + \frac{ic}{\hbar}(e\phi\boldsymbol{\sigma} - \boldsymbol{\sigma}e\phi) \tag{14.259}$$

The product constant of the electron charge e and the proton scalar electric potential ϕ commute with the electron spin angular momentum vector operator $\mathbf{S}$ (and Pauli electron spin vector operator $\boldsymbol{\sigma}$), and $\mathbf{S}$ (and $\boldsymbol{\sigma}$) commutes with constant β, so that last two terms of Eq. (14.259) are zero, where

$$\beta mc^2\boldsymbol{\sigma} - \boldsymbol{\sigma}\beta mc^2 = 0 \quad \text{and} \quad e\phi\boldsymbol{\sigma} - \boldsymbol{\sigma}e\phi = 0 \tag{14.260}$$

With substitution of Eq. (14.260) into Eq. (14.259), one obtains

$$\frac{\mathrm{d}\boldsymbol{\sigma}}{\mathrm{d}t} = \frac{\mathrm{i}c}{\hbar}((\boldsymbol{\alpha}\cdot\mathbf{p})\boldsymbol{\sigma} - \boldsymbol{\sigma}(\boldsymbol{\alpha}\cdot\mathbf{p})) \tag{14.261}$$

where

$$\boldsymbol{\alpha} = \begin{bmatrix} 0 & 1 \\ 1 & 0 \end{bmatrix} \boldsymbol{\sigma} = \begin{bmatrix} 0 & \sigma_x \\ \sigma_x & 0 \end{bmatrix} \hat{\mathbf{x}} + \begin{bmatrix} 0 & \sigma_y \\ \sigma_y & 0 \end{bmatrix} \hat{\mathbf{y}} + \begin{bmatrix} 0 & \sigma_z \\ \sigma_z & 0 \end{bmatrix} \hat{\mathbf{z}} \tag{14.262}$$

and

$$\boldsymbol{\alpha}\cdot\mathbf{p} = \begin{bmatrix} 0 & \sigma_x p_x + \sigma_y p_y + \sigma_z p_z \\ \sigma_x p_x + \sigma_y p_y + \sigma_z p_z & 0 \end{bmatrix} \tag{14.263}$$

Substitution of Eq. (14.263) into the right side of Eq. (14.261), gives

$$\begin{gathered}(\boldsymbol{\alpha}\cdot\mathbf{p})\boldsymbol{\sigma} - \boldsymbol{\sigma}(\boldsymbol{\alpha}\cdot\mathbf{p}) = \begin{bmatrix} 0 & \sigma_x p_x + \sigma_y p_y + \sigma_z p_z \\ \sigma_x p_x + \sigma_y p_y + \sigma_z p_z & 0 \end{bmatrix} (\sigma_x\hat{\mathbf{x}} + \sigma_y\hat{\mathbf{y}} + \sigma_z\hat{\mathbf{z}}) - \\ (\sigma_x\hat{\mathbf{x}} + \sigma_y\hat{\mathbf{y}} + \sigma_z\hat{\mathbf{z}}) \begin{bmatrix} 0 & \sigma_x p_x + \sigma_y p_y + \sigma_z p_z \\ \sigma_x p_x + \sigma_y p_y + \sigma_z p_z & 0 \end{bmatrix}\end{gathered} \tag{14.264}$$

Evaluation of the x-component in Eq. (14.264), gives

$$\begin{gathered}(\boldsymbol{\alpha}\cdot\mathbf{p})\sigma_x - \sigma_x(\boldsymbol{\alpha}\cdot\mathbf{p}) = \\ \begin{bmatrix} 0 & (\sigma_y\sigma_x - \sigma_x\sigma_y)p_y + (\sigma_z\sigma_x - \sigma_x\sigma_z)p_z \\ (\sigma_y\sigma_x - \sigma_x\sigma_y)p_y + (\sigma_z\sigma_x - \sigma_x\sigma_z)p_z & 0 \end{bmatrix}\end{gathered} \tag{14.265}$$

Recall that the commutation rules for the Pauli spin matrices are

$$\sigma_y\sigma_z - \sigma_z\sigma_y = \mathrm{i}\sigma_x \tag{14.266}$$

$$\sigma_z\sigma_x - \sigma_x\sigma_z = \mathrm{i}\sigma_y \tag{14.267}$$

$$\sigma_x\sigma_y - \sigma_y\sigma_x = \mathrm{i}\sigma_z \tag{14.268}$$

With the Pauli spin matrix commutation rules, the x-component becomes

$$(\boldsymbol{\alpha}\cdot\mathbf{p})\sigma_x-\sigma_x(\boldsymbol{\alpha}\cdot\mathbf{p})=\mathrm{i}\begin{bmatrix}0 & -\sigma_z p_y+\sigma_y p_z\\ -\sigma_z p_y+\sigma_y p_z & 0\end{bmatrix}=\mathrm{i}(\boldsymbol{\sigma}\times\mathbf{p})_x\begin{bmatrix}0 & 1\\ 1 & 0\end{bmatrix} \tag{14.269}$$

Evaluation of the y-component in Eq. (14.264), gives

$$(\boldsymbol{\alpha}\cdot\mathbf{p})\sigma_y-\sigma_y(\boldsymbol{\alpha}\cdot\mathbf{p})= \begin{bmatrix}0 & (\sigma_x\sigma_y-\sigma_y\sigma_x)p_x+(\sigma_z\sigma_y-\sigma_y\sigma_z)p_z\\ (\sigma_x\sigma_y-\sigma_y\sigma_x)p_x+(\sigma_z\sigma_y-\sigma_y\sigma_z)p_z & 0\end{bmatrix} \tag{14.270}$$

With the Pauli spin matrix commutation rules, the y-component becomes

$$(\boldsymbol{\alpha}\cdot\mathbf{p})\sigma_y-\sigma_y(\boldsymbol{\alpha}\cdot\mathbf{p})=\mathrm{i}\begin{bmatrix}0 & \sigma_z p_x-\sigma_x p_z\\ \sigma_z p_x+\sigma_x p_z & 0\end{bmatrix}=\mathrm{i}(\boldsymbol{\sigma}\times\mathbf{p})_y\begin{bmatrix}0 & 1\\ 1 & 0\end{bmatrix} \tag{14.271}$$

Evaluation of the z-component in Eq. (14.264), gives

$$(\boldsymbol{\alpha}\cdot\mathbf{p})\sigma_z-\sigma_z(\boldsymbol{\alpha}\cdot\mathbf{p})= \begin{bmatrix}0 & (\sigma_x\sigma_z-\sigma_z\sigma_x)p_x+(\sigma_y\sigma_z-\sigma_z\sigma_y)p_y\\ (\sigma_x\sigma_z-\sigma_z\sigma_x)p_x+(\sigma_y\sigma_z-\sigma_z\sigma_y)p_y & 0\end{bmatrix} \tag{14.272}$$

With the Pauli spin matrix commutation rules, the z-component becomes

$$(\boldsymbol{\alpha}\cdot\mathbf{p})\sigma_z-\sigma_z(\boldsymbol{\alpha}\cdot\mathbf{p})=\mathrm{i}\begin{bmatrix}0 & -\sigma_y p_x+\sigma_x p_y\\ -\sigma_y p_x+\sigma_x p_y & 0\end{bmatrix}=\mathrm{i}(\boldsymbol{\sigma}\times\mathbf{p})_z\begin{bmatrix}0 & 1\\ 1 & 0\end{bmatrix} \tag{14.273}$$

Substitution into Eq. (14.261) of Eqs. (14.269), (14.271), and (14.273), yields the electron spin vector $\boldsymbol{\sigma}$ equation of motion:

$$\frac{\mathrm{d}\boldsymbol{\sigma}}{\mathrm{d}t}=\frac{\mathrm{i}c}{\hbar}((\boldsymbol{\alpha}\cdot\mathbf{p})\boldsymbol{\sigma}-\boldsymbol{\sigma}(\boldsymbol{\alpha}\cdot\mathbf{p}))=-\frac{c}{\hbar}(\boldsymbol{\sigma}\times\mathbf{p})\begin{bmatrix}0 & 1\\ 1 & 0\end{bmatrix} \tag{14.274}$$

Recall Eq. (14.38), where

$$\boldsymbol{\sigma}\begin{bmatrix}0 & 1\\ 1 & 0\end{bmatrix}=\boldsymbol{\alpha} \tag{14.275}$$

With substitution of Eq. (14.275) into Eq. (14.274), the $\boldsymbol{\sigma}$ equation of motion becomes

$$\frac{\mathrm{d}\boldsymbol{\sigma}}{\mathrm{d}t} = -2\frac{c}{\hbar}(\alpha \times \mathbf{p}) \tag{14.276}$$

Replacement of the Pauli spin vector matrix $\boldsymbol{\sigma}$ for the electron-by-electron spin vector matrix $\mathbf{S}$ then gives

$$\frac{\mathrm{d}\mathbf{S}}{\mathrm{d}t} = \frac{\hbar}{2}\frac{\mathrm{d}\boldsymbol{\sigma}}{\mathrm{d}t} = -c(\alpha \times \mathbf{p}) \tag{14.277}$$

Hence, the nonzero result of equation of motion of Eq. (14.277) for electron spin vector $\mathbf{S}$ demonstrates that $\mathbf{S}$ varies with time and thus is not a constant of motion.

However, the total electron angular momentum $\mathbf{J}$, which is the sum of the spin $\mathbf{S}$ and orbital angular momentum $\mathbf{L}$ components, is a constant of motion and is time invariant, where

$$\mathbf{J} = \mathbf{L} + \mathbf{S} \tag{14.278}$$

The total electron angular momentum $\mathbf{J}$ commutes with the Dirac Hamiltonian H_{D}, where

$$\frac{\mathrm{d}\mathbf{J}}{\mathrm{d}t} = -\frac{\mathrm{i}}{\hbar}(\mathbf{J}H_{\mathrm{D}} - H_{\mathrm{D}}\mathbf{J}) = -\frac{\mathrm{i}c}{\hbar}(\mathbf{J}(\alpha \cdot \mathbf{p}) - (\alpha \cdot \mathbf{p})\mathbf{J}) = \mathrm{i}\hbar(\alpha \times \mathbf{p}) - \mathrm{i}\hbar(\alpha \times \mathbf{p}) = 0 \tag{14.279}$$

Hence, the total electron angular momentum vector $\mathbf{J}$, which accounts for the electron spin and orbital motion, remains constant in time throughout the electron trajectory.

Associated with symmetry is a conservation law. For example, if electron energy is constant, then the translation of the electron in time is invariant (the electron state does not change with time; i.e., symmetry with time). Symmetry of the object under an imposed condition or operation means that the object looks unchanged to the observer before, during, and after the imposed conditions or operation. If the electron total angular does not change with time, then rotation of the electron is invariant (the electron rotation exhibits spherical symmetry—electron state does not change upon rotation; i.e., symmetry with rotation). If the electrons were free instead of being bound in an atom, and the electron is moving in a straight line at constant speed (constant velocity and thus linear momentum), then the electron linear momentum is constant with time or conserved. Associated with constant linear momentum is the invariance of the electron in translation through space (the electron state is unchanged with constant linear motion; i.e., linear symmetry).

14.12 Derivation of the Dirac Equation for the Hydrogen Atom—Step 1: Evaluation of $\alpha \cdot \mathbf{p}$ in the Dirac Hamiltonian

In this section, we derive the Dirac Hamiltonian H_D and Dirac equation for the hydrogen atom. The result of this section is the expression of the Dirac equation in the form of a pair of eigenequations, which are in a form that is ready for solution by a power series approach (which will be discussed in the next section). The resulting pair of eigenvector solutions, and their associated energy eigenvalue solutions for the hydrogen atom, provides an archetype of the atomic electron properties and interactions with and nucleus for other isotopes. This is especially true for one-electron atoms and ions, whose outer electron (ions with only one electron), allows a hydrogen atom approximation for the single outer electron interaction with the net sum of the inner electron charge and nuclear charge. We follow the approach to the derivation of the coupled Dirac equations and the resulting solution of the total electron energy E as presented by Schiff (1968) and Merzbacher (1970).

The derived expression for the total electron energy E has a similar form to that of the nonrelativistic expression for electron energy obtained by Sommerfeld (1916) with the old quantum mechanics, which occurred prior to the Schrödinger equation in 1926 and the Dirac equation in 1928. The energy expression E is also similar to results obtained by the solution of the relativistic Schrödinger equation (i.e., the previously derived Klein–Gordon equation) obtained by Schrödinger (1925) and Klein and Gordon (1927).

In the hydrogen atom, one assumes spherical symmetry, where the proton electric potential is radial $\phi(r)$, and the proton electric field E_r, which acts on the electron, is radial-directed, where

$$\mathbf{E} = -\nabla\phi = \frac{1}{r}\frac{\partial\phi}{\partial r}\hat{\mathbf{r}} \tag{14.280}$$

The magnetic vector potential of the proton magnetic dipole moment is ignored, where

$$\mathbf{A} = 0 \tag{14.281}$$

The zero magnetic vector potential of the proton is required, so there is no momentum transfer to the electron. This allows the total electron angular momentum to be conserved, which is associated with a spherically symmetric central field that acts on the electron.

The Dirac Hamiltonian H_D for the electron with rest mass m and charge e, which is acted upon by the central electric field of the proton via electric potential ϕ, is given by

$$H_D = c\alpha \cdot \mathbf{p} + \beta mc^2 + e\phi \tag{14.282}$$

where

$$\mathbf{p} = \hat{\mathbf{x}}p_x + \hat{\mathbf{y}}p_y + \hat{\mathbf{z}}p_z \tag{14.283}$$

and

$$\alpha = \begin{bmatrix} 0 & 1 \\ 1 & 0 \end{bmatrix}\boldsymbol{\sigma} = \begin{bmatrix} 0 & \sigma_x \\ \sigma_x & 0 \end{bmatrix}\hat{\mathbf{x}} + \begin{bmatrix} 0 & \sigma_y \\ \sigma_y & 0 \end{bmatrix}\hat{\mathbf{y}} + \begin{bmatrix} 0 & \sigma_z \\ \sigma_z & 0 \end{bmatrix}\hat{\mathbf{z}} \tag{14.284}$$

With substitution of Eq. (14.36) into Eq. (14.284), one obtains

$$\alpha = \begin{bmatrix} 0 & 1 \\ 1 & 0 \end{bmatrix}\boldsymbol{\sigma} = \begin{bmatrix} 0 & 0 & 0 & 1 \\ 0 & 0 & 1 & 0 \\ 0 & 1 & 0 & 0 \\ 1 & 0 & 0 & 0 \end{bmatrix}\hat{\mathbf{x}} + \begin{bmatrix} 0 & 0 & 0 & -i \\ 0 & 0 & i & 0 \\ 0 & -i & 0 & 0 \\ i & 0 & 0 & 0 \end{bmatrix}\hat{\mathbf{y}} + \begin{bmatrix} 0 & 0 & 1 & 0 \\ 0 & 0 & 0 & -1 \\ 1 & 0 & 0 & 0 \\ 0 & -1 & 0 & 0 \end{bmatrix}\hat{\mathbf{z}} \tag{14.285}$$

Recall from Eq. (14.29),

$$\beta = \begin{bmatrix} 1 & 0 & 0 & 0 \\ 0 & 1 & 0 & 0 \\ 0 & 0 & -1 & 0 \\ 0 & 0 & 0 & -1 \end{bmatrix} \tag{14.286}$$

Next, one evaluates $\alpha \cdot \mathbf{p}$ in Eq. (14.282) by starting with vector identity

$$\nabla = \hat{\mathbf{r}}(\hat{\mathbf{r}}\cdot\nabla) - \hat{\mathbf{r}} \times (\hat{\mathbf{r}} \times \nabla) \tag{14.287}$$

With $\mathbf{p} = -\,\mathrm{i}h\nabla$, the expression for the electron orbital angular momentum $\mathbf{L}$ is

$$\mathbf{L} = \mathbf{r} \times \mathbf{p} = -\,\mathrm{i}\hbar\mathbf{r} \times \nabla \tag{14.288}$$

and

$$\hat{\mathbf{r}} \times \nabla = \frac{-\,\mathrm{i}\hbar}{-\,\mathrm{i}\hbar}\hat{\mathbf{r}} \times \nabla = \frac{\mathrm{i}}{\hbar}\hat{\mathbf{r}} \times \mathbf{p} = \frac{\mathrm{i}}{\hbar}\mathbf{L} \tag{14.289}$$

Hence,

$$\nabla = \hat{\mathbf{r}}\frac{\partial}{\partial r} - \frac{\mathrm{i}}{\hbar}\frac{\hat{\mathbf{r}}}{r} \times \mathbf{L} \tag{14.290}$$

and

$$\alpha \cdot \mathbf{p} = -\,\mathrm{i}\hbar\alpha \cdot \nabla = -\,\mathrm{i}\hbar(\alpha \cdot \hat{\mathbf{r}})\frac{\partial}{\partial r} - \frac{\alpha \cdot (\hat{\mathbf{r}} \times \mathbf{L})}{r} \tag{14.291}$$

Next, one evaluates $\alpha \cdot (\hat{\mathbf{r}} \times \mathbf{L})$ by invoking the Pauli spin vector matrix identity of Eq. (14.101), with $\mathbf{a} = \hat{\mathbf{r}}$ and $\mathbf{b} = \mathbf{L}$, where

$$(\sigma \cdot \mathbf{a})(\sigma \cdot \mathbf{b}) = (\mathbf{a} \cdot \mathbf{b})\begin{bmatrix} 1 & 0 \\ 0 & 1 \end{bmatrix} + \mathrm{i}(\mathbf{a} \times \mathbf{b}) \cdot \sigma \quad (14.292)$$

Application of this identity yields

$$(\sigma \cdot \hat{\mathbf{r}})(\sigma \cdot \mathbf{L}) = (\hat{\mathbf{r}} \cdot \mathbf{L})\begin{bmatrix} 1 & 0 \\ 0 & 1 \end{bmatrix} + \mathrm{i}(\hat{\mathbf{r}} \times \mathbf{L}) \cdot \sigma \quad (14.293)$$

The electron orbital angular momentum vector $\mathbf{L}$ is perpendicular to the radial unit vector $\hat{\mathbf{r}}$. For example, if the electron orbits in the xy-plane, then $\mathbf{L}$ is z-directed, perpendicular to the xy-plane that contains $\hat{\mathbf{r}}$, so

$$\hat{\mathbf{r}} \cdot \mathbf{L} = 0 \quad (14.294)$$

With substitution of Eq. (14.294) in Eq. (14.293), one obtains,

$$\mathrm{i}(\hat{\mathbf{r}} \times \mathbf{L}) \cdot \sigma = (\sigma \cdot \hat{\mathbf{r}})(\sigma \cdot \mathbf{L}) \quad (14.295)$$

One then invokes the relation of Eq. (14.38) between α and σ, where

$$\alpha\begin{bmatrix} 0 & 1 \\ 1 & 0 \end{bmatrix} = \sigma \quad \text{and} \quad \sigma\begin{bmatrix} 0 & 1 \\ 1 & 0 \end{bmatrix} = \alpha \quad (14.296)$$

Post multiplication of both sides of Eq. (14.295), is performed with the 2×2 flip matrix, where

$$\mathrm{i}(\hat{\mathbf{r}} \times \mathbf{L}) \cdot \sigma\begin{bmatrix} 0 & 1 \\ 1 & 0 \end{bmatrix} = (\sigma \cdot \hat{\mathbf{r}})(\sigma \cdot \mathbf{L})\begin{bmatrix} 0 & 1 \\ 1 & 0 \end{bmatrix} \quad (14.297)$$

In Eq. (14.297), the reversal of order of scalar product operators $(\sigma \cdot \hat{\mathbf{r}})$ and $(\sigma \cdot \mathbf{L})$ on the right-hand side introduces a minus sign, because of the reversed order of the $\hat{\mathbf{r}}$ and $\mathbf{L}$ operators in the cross product $(\hat{\mathbf{r}} \times \mathbf{L})$on the left-hand side, where

$$(\sigma \cdot \hat{\mathrm{r}})(\sigma \cdot L)\begin{bmatrix} 0 & 1 \\ 1 & 0 \end{bmatrix} = -(\sigma \cdot L)(\sigma \cdot \hat{\mathrm{r}})\begin{bmatrix} 0 & 1 \\ 1 & 0 \end{bmatrix} \quad (14.298)$$

Application of Eq. (14.296) and the transitive property of equality to Eqs. (14.297) and (14.298), yields

$$\mathrm{i}(\hat{\mathbf{r}} \times \mathbf{L}) \cdot \alpha = (\alpha \cdot \hat{\mathbf{r}})(\sigma \cdot \mathbf{L}) \quad (14.299)$$

where

$$(\hat{\mathbf{r}} \times \mathbf{L}) \cdot \boldsymbol{\alpha} = \boldsymbol{\alpha} \cdot (\hat{\mathbf{r}} \times \mathbf{L}) \tag{14.300}$$

With application of Eq. (14.300) to Eq. (14.299), one obtains

$$\boldsymbol{\alpha} \cdot (\hat{\mathbf{r}} \times \mathbf{L}) = -i(\boldsymbol{\alpha} \cdot \hat{\mathbf{r}})(\boldsymbol{\sigma} \cdot \mathbf{L}) \tag{14.301}$$

Substitution of Eq. (14.301) into Eq. (14.291) for $\boldsymbol{\alpha} \cdot \mathbf{p}$, gives

$$\boldsymbol{\alpha} \cdot \mathbf{p} = -\mathrm{i}\hbar(\boldsymbol{\alpha} \cdot \hat{\mathbf{r}})\frac{\partial}{\partial r} + \mathrm{i}\frac{(\boldsymbol{\alpha} \cdot \hat{\mathbf{r}})(\boldsymbol{\sigma} \cdot \mathbf{L})}{r} \tag{14.302}$$

This expression for $\boldsymbol{\alpha} \cdot \mathbf{p}$ will be plugged into the Dirac Hamiltonian H_D and then evaluated in the next section.

14.13 Derivation of the Dirac Equation for the Hydrogen Atom—Step 2: Evaluation of $\boldsymbol{\sigma} \cdot \mathbf{L}$ in a Dirac Hamiltonian $\boldsymbol{\alpha} \cdot \mathbf{p}$ Term

The expression for $\boldsymbol{\alpha} \cdot \mathbf{p}$ of Eq. (14.302), which was derived in Section 14.12, is now substituted in the Dirac Hamiltonian H_D of Eq. (14.282), so that

$$H_\mathrm{D} = -\mathrm{i}\hbar c(\boldsymbol{\alpha} \cdot \hat{\mathbf{r}})\frac{\partial}{\partial r} + \mathrm{i}c\frac{(\boldsymbol{\alpha} \cdot \hat{\mathbf{r}})(\boldsymbol{\sigma} \cdot \mathbf{L})}{r} + \beta mc^2 + e\phi \tag{14.303}$$

Next, one evaluates the dot product $\boldsymbol{\sigma} \cdot \mathbf{L}$ of the Pauli electron spin vector matrix $\boldsymbol{\sigma}$ with the orbital angular momentum vector $\mathbf{L}$ by expressing $\boldsymbol{\sigma} \cdot \mathbf{L}$ in terms of the total electron angular momentum $\mathbf{J} = \mathbf{L} + \mathbf{S}$, which is a constant of motion in the electron trajectory in the central field of the proton. Recall from the last section that neither $\boldsymbol{\sigma}$ [and $\mathbf{S} = (\hbar/2)\boldsymbol{\sigma}$] nor $\mathbf{L}$ is a constant of electron motion. The total electron angular momentum $\mathbf{J}$ commutes with the Dirac Hamiltonian H_D because $\mathbf{J}$ is time invariant due to the central force electric potential $\varphi(r)$ of the nucleus.

The term $\boldsymbol{\sigma} \cdot \mathbf{L}$ suggests expression in terms of J^2, where

$$J^2 = (\mathbf{L} + \mathbf{S}) \cdot (\mathbf{L} + \mathbf{S}) = L^2 + \mathbf{S} \cdot \mathbf{L} + \mathbf{L} \cdot \mathbf{S} + S^2 \tag{14.304}$$

The squared magnitude of the total electron angular momentum J^2 has associated positive quantum numbers j, where

$$J^2 = \hbar^2 \mathrm{j}(\mathrm{j} + 1) \tag{14.305}$$

The total electron angular momentum quantum numbers j are half-integers, where

$$j = \frac{1}{2}, \frac{3}{2}, \frac{5}{2}, \ldots \tag{14.306}$$

Electron spin **S** and orbital angular momentum **L** commute, so

$$\mathbf{S} \cdot \mathbf{L} + \mathbf{L} \cdot \mathbf{S} = 2\mathbf{S} \cdot \mathbf{L} = \hbar \boldsymbol{\sigma} \cdot \mathbf{L} \tag{14.307}$$

The squared magnitude S^2 of the electron spin vector **S** is

$$\begin{aligned} S2 &= \mathbf{S} \cdot \mathbf{S} = \frac{\hbar^2}{4} \boldsymbol{\sigma} \cdot \boldsymbol{\sigma} \\ &= \frac{\hbar^2}{4} \left\{ \begin{bmatrix} 0 & 1 \\ 1 & 0 \end{bmatrix} \hat{\mathbf{x}} + \begin{bmatrix} 0 & -i \\ i & 0 \end{bmatrix} \hat{\mathbf{y}} + \begin{bmatrix} 1 & 0 \\ 0 & -1 \end{bmatrix} \hat{\mathbf{z}} \right\} \cdot \left\{ \begin{bmatrix} 0 & 1 \\ 1 & 0 \end{bmatrix} \hat{\mathbf{x}} + \begin{bmatrix} 0 & -i \\ i & 0 \end{bmatrix} \hat{\mathbf{y}} + \begin{bmatrix} 1 & 0 \\ 0 & -1 \end{bmatrix} \hat{\mathbf{z}} \right\} \end{aligned} \tag{14.308}$$

Evaluation of the S^2 expression gives

$$S^2 = \frac{\hbar^2}{4} \left\{ \begin{bmatrix} 1 & 0 \\ 0 & 1 \end{bmatrix} + \begin{bmatrix} 1 & 0 \\ 0 & 1 \end{bmatrix} + \begin{bmatrix} 1 & 0 \\ 0 & 1 \end{bmatrix} \right\} = \frac{3\hbar^2}{4} \begin{bmatrix} 1 & 0 \\ 0 & 1 \end{bmatrix} \tag{14.309}$$

With substitution of Eq. (14.307) and Eq. (14.309) into Eq. (14.304), J^2 becomes

$$J^2 = L^2 + \hbar \boldsymbol{\sigma} \cdot \mathbf{L} + \frac{3\hbar^2}{4} \begin{bmatrix} 1 & 0 \\ 0 & 1 \end{bmatrix} \tag{14.310}$$

To evaluate $\boldsymbol{\sigma} \cdot \mathbf{L}$ in the J^2 expression, one again invokes the Pauli spin vector matrix identity of Eq. (14.101) with $\mathbf{a} = \mathbf{L}$ and $\mathbf{b} = \mathbf{L}$, where

$$(\boldsymbol{\sigma} \cdot \mathbf{a})(\boldsymbol{\sigma} \cdot \mathbf{b}) = (\mathbf{a} \cdot \mathbf{b}) \begin{bmatrix} 1 & 0 \\ 0 & 1 \end{bmatrix} + i(\mathbf{a} \times \mathbf{b}) \cdot \boldsymbol{\sigma} \tag{14.311}$$

Evaluation of Eq. (14.311) with the substituted parameters for **a** and **b**, gives

$$(\boldsymbol{\sigma} \cdot \mathbf{L})(\boldsymbol{\sigma} \cdot \mathbf{L}) = L^2 \begin{bmatrix} 1 & 0 \\ 0 & 1 \end{bmatrix} + i(\mathbf{L} \times \mathbf{L}) \cdot \boldsymbol{\sigma} \tag{14.312}$$

The cross product $\mathbf{L} \times \mathbf{L}$ is not zero because $\mathbf{L} = \mathbf{r} \times \mathbf{p}$ and via the Heisenberg uncertainty principle, $\mathbf{r}$ and $\mathbf{p}$ do not commute. One starts with electron angular momentum $\mathbf{L}$ and position $\mathbf{r}$ operators, where

$$\mathbf{L} = \hat{\mathbf{x}}L_x + \hat{\mathbf{y}}L_y + \hat{\mathbf{z}}L_z = \mathbf{r} \times \mathbf{p} = \begin{bmatrix} \hat{\mathbf{x}} & \hat{\mathbf{y}} & \hat{\mathbf{z}} \\ x & y & z \\ p_x & p_y & p_z \end{bmatrix} \tag{14.313}$$

Evaluation of the matrix determinant of Eq. (14.313) for the electron orbital angular momentum operator $\mathbf{L}$, gives

$$\mathbf{L} = \hat{\mathbf{x}}(yp_z - zp_y) + \hat{\mathbf{y}}(zp_x - xp_z) + \hat{\mathbf{z}}(xp_y - yp_x) \tag{14.314}$$

The cross product of $\mathbf{L} \times \mathbf{L}$ is then

$$\mathbf{L} \times \mathbf{L} = \begin{bmatrix} \hat{\mathbf{x}} & \hat{\mathbf{y}} & \hat{\mathbf{z}} \\ yp_z - zp_y & zp_x - xp_z & xp_y - yp_x \\ yp_z - zp_y & zp_x - xp_z & xp_y - yp_x \end{bmatrix} \tag{14.315}$$

Evaluation of the matrix determinant of Eq. (14.315), gives

$$(\mathbf{L} \times \mathbf{L})_x = zp_y(p_x x - xp_x) - yp_z(p_x x - xp_x) + yp_x(zp_x - zp_x) + xp_y(xp_z - xp_z) \tag{14.316}$$

$$(\mathbf{L} \times \mathbf{L})_y = xp_z(p_y y - yp_y) - zp_x(p_y y - yp_y) + zp_y(xp_y - xp_y) + yp_z(yp_x - yp_x) \tag{14.317}$$

$$(\mathbf{L} \times \mathbf{L})_z = yp_x(p_z z - zp_z) - xp_y(p_z z - zp_z) + xp_z(yp_z - yp_z) + zp_x(zp_y - zp_y) \tag{14.318}$$

where

$$zp_x - zp_x = xp_z - xp_z = xp_y - xp_y = yp_x - yp_x = yp_z - yp_z = zp_y - zp_y = 0 \tag{14.319}$$

The Heisenberg uncertainty principle commutation relations for electron position $\mathbf{r}$ and linear momentum $\mathbf{p}$ are

$$[\mathbf{r}, \mathbf{p}] = \mathbf{r} \cdot \mathbf{p} - \mathbf{p} \cdot \mathbf{r} = -[\mathbf{p}, \mathbf{r}] = i\hbar \tag{14.320}$$

where

$$p_x x - xp_x = p_y y - yp_y = p_z z - zp_z = -i\hbar \tag{14.321}$$

and

$$\begin{array}{ll} \text{Clockwise } x \to y \to z \to x \text{ Positive} \\ \text{Counterclockwise } x \to z \to y \to x \text{ Negative} \end{array} \tag{14.322}$$

Application of Eqs. (14.320–14.322) to Eqs. (14.316–14.318), yields

$$(\mathbf{L} \times \mathbf{L})_x = -\,\mathrm{i}\hbar(zp_y - yp_z) = \mathrm{i}\hbar(yp_z - zp_y) = \mathrm{i}\hbar(\mathbf{r} \times \mathbf{p})_x = \mathrm{i}hL_x \tag{14.323}$$

$$(\mathbf{L} \times \mathbf{L})_y = -\,\mathrm{i}\hbar(xp_z - zp_x) = \mathrm{i}\hbar(zp_x - xp_z) = \mathrm{i}\hbar(\mathbf{r} \times \mathbf{p})_y = \mathrm{i}hL_y \tag{14.324}$$

$$(\mathbf{L} \times \mathbf{L})_z = -\,\mathrm{i}\hbar(yp_x - xp_y) = \mathrm{i}\hbar(xp_y - yp_x) = \mathrm{i}\hbar(\mathbf{r} \times \mathbf{p})_z = \mathrm{i}hL_z \tag{14.325}$$

Hence,

$$\mathbf{L} \times \mathbf{L} = \mathrm{i}\hbar\mathbf{L} \tag{14.326}$$

Substitution of Eq. (14.326) into Eq. (14.312), yields

$$(\boldsymbol{\sigma} \cdot \mathbf{L})(\boldsymbol{\sigma} \cdot \mathbf{L}) = L^2 \begin{bmatrix} 1 & 0 \\ 0 & 1 \end{bmatrix} - \hbar(\mathbf{L} \cdot \boldsymbol{\sigma}) \tag{14.327}$$

The scalar (dot) product is independent of the order of the vectors, where

$$\mathbf{L} \cdot \boldsymbol{\sigma} = \boldsymbol{\sigma} \cdot \mathbf{L} \tag{14.328}$$

Application of Eq. (14.328) to Eq. (14.327), gives

$$(\boldsymbol{\sigma} \cdot \mathbf{L})(\boldsymbol{\sigma} \cdot \mathbf{L}) = L^2 \begin{bmatrix} 1 & 0 \\ 0 & 1 \end{bmatrix} - \hbar(\boldsymbol{\sigma} \cdot \mathbf{L}) \tag{14.329}$$

One transposes Eq. (14.329),

$$\hbar(\boldsymbol{\sigma} \cdot \mathbf{L}) = L^2 \begin{bmatrix} 1 & 0 \\ 0 & 1 \end{bmatrix} - (\boldsymbol{\sigma} \cdot \mathbf{L})(\boldsymbol{\sigma} \cdot \mathbf{L}) \tag{14.330}$$

To recap, the scalar operator expression $(\boldsymbol{\sigma} \cdot \mathbf{L})(\boldsymbol{\sigma} \cdot \mathbf{L})$, which involves the scalar product of the Pauli, electron spin, angular momentum, vector matrix operator $\boldsymbol{\sigma}$, and the electron orbital angular momentum vector operator $\mathbf{L}$, result from expansion of the scalar operator J^2. The scalar operator J^2 is the squared magnitude of the total electron angular momentum vector operator $\mathbf{J} = \mathbf{L} + \mathbf{S}$, where the electron spin, angular momentum, vector matrix operator $\mathbf{S}$, is expressed by the Pauli electron spin vector matrix operator $\boldsymbol{\sigma}$. The scalar product $(\boldsymbol{\sigma} \cdot \mathbf{L})$, which is multiplied with itself, can now be simply expressed by the squared, scalar product $(\boldsymbol{\sigma} \cdot \mathbf{L})^2$ in Eq. (14.330), which is transposed to obtain:

$$L^2 \begin{bmatrix} 1 & 0 \\ 0 & 1 \end{bmatrix} = (\boldsymbol{\sigma} \cdot \mathbf{L})^2 + \hbar(\boldsymbol{\sigma} \cdot \mathbf{L}) \tag{14.331}$$

With substitution into Eq. (14.310) of Eq. (14.330) for $\hbar(\boldsymbol{\sigma} \cdot \mathbf{L})$, the J^2 operator becomes

$$J^2 = L^2 + L^2 \begin{bmatrix} 1 & 0 \\ 0 & 1 \end{bmatrix} - (\boldsymbol{\sigma} \cdot \mathbf{L})(\boldsymbol{\sigma} \cdot \mathbf{L}) + \frac{3\hbar^2}{4} \begin{bmatrix} 1 & 0 \\ 0 & 1 \end{bmatrix} \tag{14.332}$$

One adds $\hbar^2/4$ to both sides of the equation, so

$$J^2 + \frac{\hbar^2}{4} = 2L^2 \begin{bmatrix} 1 & 0 \\ 0 & 1 \end{bmatrix} - (\boldsymbol{\sigma} \cdot \mathbf{L})^2 + \hbar^2 \begin{bmatrix} 1 & 0 \\ 0 & 1 \end{bmatrix} \tag{14.333}$$

Substitution into Eq. (14.333) of Eq. (14.331) for the $2L^2$ term gives

$$J^2 + \frac{\hbar^2}{4} = (\boldsymbol{\sigma} \cdot \mathbf{L})^2 + 2\hbar(\boldsymbol{\sigma} \cdot \mathbf{L}) + \hbar^2 \begin{bmatrix} 1 & 0 \\ 0 & 1 \end{bmatrix} \tag{14.334}$$

14.14 Derivation of the Dirac Equation for a Hydrogen Atom—Step 3: Introduction of the Squared Angular Momentum Operator K^2 in a Dirac Hamiltonian

Next, one defines an electron total angular momentum squared operator K^2, which is expressed in terms of the total angular momentum squared operator J^2, where

$$K^2 = J^2 + \frac{\hbar^2}{4} = \left(\boldsymbol{\sigma} \cdot \mathbf{L} + \hbar \begin{bmatrix} 1 & 0 \\ 0 & 1 \end{bmatrix}\right)^2 \tag{14.335}$$

and

$$K = \pm\sqrt{J^2 + \frac{\hbar^2}{4}} = \pm\left(\boldsymbol{\sigma} \cdot \mathbf{L} + \hbar \begin{bmatrix} 1 & 0 \\ 0 & 1 \end{bmatrix}\right) \tag{14.336}$$

Scalar operator K^2 differs from J^2 by the constant $\hbar^2/4$. Since the total squared electron angular momentum magnitude J^2 and the electron total angular momentum vector $\mathbf{J}$ are time invariant, the operator K^2 and scalar operator K are time-invariant constants of electron motion. As a result, the scalar operator K commutes with the electron Dirac Hamiltonian H_D, squared angular momentum magnitude J^2, and z-component J_z of $\mathbf{J}$, where

$$[K, H_D] = [K, J^2] = [K, J_z] = 0 \tag{14.337}$$

The square root of K^2 gives the pair of $\pm$ solutions for K, with two of the four wave function solutions of the Dirac equation for the hydrogen atom taking the positive solution $K > 0$, and the other two wave function solutions taking the negative solution $K < 0$, which can thus be expressed β, where from Eq. (14.29),

$$\beta = \begin{bmatrix} 1 & 0 \\ 0 & -1 \end{bmatrix} \tag{14.338}$$

With substitution of Eq. (14.338) into Eq. (14.336), the scalar operator K can be thus expressed by

$$K = \beta\left(\boldsymbol{\sigma} \cdot \mathbf{L} + \hbar \begin{bmatrix} 1 & 0 \\ 0 & 1 \end{bmatrix}\right) \tag{14.339}$$

Since $\beta^2 = 1$, one can solve for $\boldsymbol{\sigma} \cdot \mathbf{L}$ in Eq. (14.339) in terms of matrix β and K, where

$$\boldsymbol{\sigma} \cdot \mathbf{L} = \beta K - \hbar \begin{bmatrix} 1 & 0 \\ 0 & 1 \end{bmatrix} \tag{14.340}$$

Substitution of Eq. (14.340) for $\boldsymbol{\sigma} \cdot \mathbf{L}$ in the Dirac Hamiltonian H_D of Eq. (14.303), yields

$$H_D = -i\hbar c(\boldsymbol{\alpha} \cdot \hat{\mathbf{r}})\frac{\partial}{\partial r} + ic\frac{(\boldsymbol{\alpha} \cdot \hat{\mathbf{r}})}{r}\left\{\beta K - \hbar \begin{bmatrix} 1 & 0 \\ 0 & 1 \end{bmatrix}\right\} + \beta mc^2 + e\phi \tag{14.341}$$

14.15 Derivation of the Dirac Equation for a Hydrogen Atom—Step 4: Obtain a Pair of Coupled First-Order Differential Equations from the Developed Dirac Hamiltonian

One is now ready to express the Dirac Hamiltonian in terms of three Hermitian operators. From the resulting Hamiltonian, one can then obtain a pair of coupled, first-order, differential equations for the pair of electron eigenfunctions for the hydrogen atom. These equations can then be solved by a pair of power series expressions.

Recall that a Hermitian operator acts on an eigenfunction to yield a real eigenvalue, which corresponds to a physically measurable value such as linear or angular momentum, energy, position, or electric potential. The first operator K is related to the total electron angular momentum (spin and orbital angular momentum). The scalar operator for the total electron angular momentum operator K is Hermitian, because it is comprised of Hermitian operators $\mathbf{L}$ and $\boldsymbol{\sigma}$, where Eq. (14.336) for K is repeated below,

$$K = \pm\sqrt{J^2 + \frac{\hbar^2}{4}} = \pm\left(\boldsymbol{\sigma}\cdot\mathbf{L} + \hbar\begin{bmatrix} 1 & 0 \\ 0 & 1 \end{bmatrix}\right) \tag{14.342}$$

Recall from Eq. (14.305), that the squared magnitude J^2 of the total electron angular momentum is

$$J^2 = \hbar^2 j(j+1) \tag{14.343}$$

The electron total angular momentum quantum numbers j are half-integers, where

$$j = \frac{1}{2}, \frac{3}{2}, \frac{5}{2}, \cdots \tag{14.344}$$

With substitution of Eq. (14.343) into Eq. (14.342), the scalar operator K for the total electron angular momentum is thus expressed in terms of the quantum number j of the total angular momentum of the electron, where

$$K = \pm\hbar\sqrt{j(j+1) + \frac{1}{4}} \tag{14.345}$$

Spherical symmetry demands the position vector $\mathbf{r}$ to have only a radial component:

$$\mathbf{r} = r\hat{\mathbf{r}} \quad \text{and} \quad \hat{\mathbf{r}} = \frac{\mathbf{r}}{|\mathbf{r}|} = \frac{\mathbf{r}}{r} \tag{14.346}$$

Via multiplication of the speed of light c by $\boldsymbol{\alpha}\cdot\hat{\mathbf{r}}$, a second Hermitian operator $c\alpha_r$, which is comprised only of constants c and α_r, acts as an electron radial velocity operator, where

$$c\boldsymbol{\alpha}\cdot\hat{\mathbf{r}} = c\alpha_r \tag{14.347}$$

Finally, the third Hermitian operator p_r is the radial-directed electron linear momentum operator p_r, which is defined by

$$p_r = -i\hbar\left(\frac{\partial}{\partial r} + \frac{1}{r}\hbar\begin{bmatrix} 1 & 0 \\ 0 & 1 \end{bmatrix}\right) \tag{14.348}$$

With substitution of Eqs. (14.347) and (14.448) into Eq. (14.341), the Dirac Hamiltonian operator becomes

$$H_{\mathrm{D}} = c\alpha_r p_r + \mathrm{i}\frac{1}{r}c\alpha_r\beta K + \beta mc^2 + e\phi \tag{14.349}$$

From Eq. (14.29), the Dirac parameter β is

$$\beta = \begin{bmatrix} 1 & 0 \\ 0 & -1 \end{bmatrix} \tag{14.350}$$

The Dirac equation must satisfy the relativistic energy and momentum relation, which is set by the Klein–Gordon equation, and requires from Eqs. (14.19), (14.23), and (14.26–14.28):

$$\alpha_r^2 = \beta^2 = 1 \quad \text{and} \quad \alpha_r\beta + \beta\alpha_r = 0 \tag{14.351}$$

These requirements are met with α_r set to

$$\alpha_r = \begin{bmatrix} 0 & -\mathrm{i} \\ \mathrm{i} & 0 \end{bmatrix} \tag{14.352}$$

The Dirac equation is

$$H_{\mathrm{D}}\begin{bmatrix} \Phi_1 \\ \Phi_2 \end{bmatrix} = E\begin{bmatrix} \Phi_1 \\ \Phi_2 \end{bmatrix} \tag{14.353}$$

With operators p_r, α_r, and β of Eqs. (14.348), (14.352), and (14.350) explicitly expressed, the Dirac Hamiltonian of Eq. (14.349), becomes

$$H_{\mathrm{D}} = -\mathrm{i}\hbar c\begin{bmatrix} 0 & -\mathrm{i} \\ \mathrm{i} & 0 \end{bmatrix}\left(\frac{\partial}{\partial r} + \frac{1}{r}\right) + \mathrm{i}\frac{1}{r}c\begin{bmatrix} 0 & -\mathrm{i} \\ \mathrm{i} & 0 \end{bmatrix}\begin{bmatrix} 1 & 0 \\ 0 & -1 \end{bmatrix}K + \begin{bmatrix} 1 & 0 \\ 0 & -1 \end{bmatrix}mc^2 + \begin{bmatrix} 1 & 0 \\ 0 & 1 \end{bmatrix}e\phi \tag{14.354}$$

Consolidating,

$$H_{\mathrm{D}} = \begin{bmatrix} e\phi + mc^2 & -\frac{1}{r}(cK + \hbar c) - \hbar c\frac{\partial}{\partial r} \\ -\frac{1}{r}(cK - \hbar c) + \hbar c\frac{\partial}{\partial r} & e\phi - mc^2 \end{bmatrix} \tag{14.355}$$

where

$$(H_{\mathrm{D}} - E)\begin{bmatrix} \Phi_1 \\ \Phi_2 \end{bmatrix} = \begin{bmatrix} 0 \\ 0 \end{bmatrix} \tag{14.356}$$

Substitution of Eq. (14.355) for the Dirac Hamiltonian operator into the Dirac eigenequation matrix of Eq. (14.356), yields,

$$\begin{bmatrix} e\phi + mc^2 - E & -\frac{1}{r}(cK + \hbar c) - \hbar c\frac{\partial}{\partial r} \\ -\frac{1}{r}(cK - \hbar c) + \hbar c\frac{\partial}{\partial r} & e\phi - mc^2 - E \end{bmatrix}\begin{bmatrix} \Phi_1 \\ \Phi_2 \end{bmatrix} = \begin{bmatrix} 0 \\ 0 \end{bmatrix} \tag{14.357}$$

This Dirac equation has two solutions associated with eigenvalues E_1 and E_2 and their associated eigenfunctions Φ_1 and Φ_1, which are radial-dependent solutions with assumed form

$$\begin{bmatrix} \Phi_1 \\ \Phi_2 \end{bmatrix} = \frac{1}{r}\begin{bmatrix} F(r) \\ G(r) \end{bmatrix} \tag{14.358}$$

Substitution of the eigenfunctions of Eq. (14.358) into the Dirac equation of Eq. (14.357), yields a pair of equations for $F(r)$ and $G(r)$, where

$$(e\phi + mc^2 - E)\frac{F}{r} - \frac{1}{r}(cK + \hbar c)\frac{G}{r} - \hbar c\frac{\partial}{\partial r}\left(\frac{G}{r}\right) = 0 \tag{14.359}$$

$$-\frac{1}{r}(cK - \hbar c)\frac{F}{r} + \hbar c\frac{\partial}{\partial r}\left(\frac{F}{r}\right) + (e\phi - mc^2 - E)\frac{G}{r} = 0 \tag{14.360}$$

Carrying out the derivatives in Eqs. (14.359) and (14.360), gives

$$(e\phi + mc^2 - E)F - cK\frac{G}{r} - \hbar c\frac{\mathrm{d}G}{\mathrm{d}r} = 0 \tag{14.361}$$

$$-cK\frac{F}{r} + \hbar c\frac{\mathrm{d}F}{\mathrm{d}r} + (e\phi - mc^2 - E)G = 0 \tag{14.362}$$

To solve the coupled Dirac equations, one makes the following substitutions of variables to simplify the mathematics, where, for electron rest mass m,

$$\alpha_1 = \frac{mc^2 + E}{\hbar c} \qquad \alpha_2 = \frac{mc^2 - E}{\hbar c} \qquad \rho = r\alpha \tag{14.363}$$

where

$$\alpha = +\sqrt{\alpha_1\alpha_2} = \frac{+\sqrt{m^2c^4 - E^2}}{\hbar c} \tag{14.364}$$

The scalar operator K of Eq. (14.345) for the total electron angular momentum, is divided by $\hbar$, and yields constant k, which is thus a quantum number associated with the electron total angular momentum quantum number j, where

$$k = \frac{K}{\hbar} = \pm\sqrt{j(j+1) + \frac{1}{4}} \tag{14.365}$$

Quantum number k will act as one of the parameters in the Dirac equation solution. With rearrangement of Eqs. (14.361) and (14.362), and substitution of Eq. (14.365), the pair of coupled Dirac equations for the hydrogen atom becomes

$$\left(\frac{\mathrm{d}}{\mathrm{d}\rho} + \frac{k}{\rho}\right)G - \left(\frac{\alpha_2}{\alpha} + \frac{e\phi}{\hbar c\alpha}\right)F = 0 \tag{14.366}$$

$$\left(\frac{\mathrm{d}}{\mathrm{d}\rho} - \frac{k}{\rho}\right)F - \left(\frac{\alpha_1}{\alpha} - \frac{e\phi}{\hbar c\alpha}\right)G = 0 \tag{14.367}$$

14.16 Asymptotic Solution—A Coupled, Dirac Eigenequation Pair for the Hydrogen Atom

In this section, the coupled Dirac equation, which was derived in Section 14.15, is solved for the hydrogen atom, and follows Schiff (1968). The solution uses a power series in ρ, which converges to an exponential for the asymptotic case, where $\rho = r\alpha$. In the asymptotic case, the solution is expressed by the summation of the terms of an infinite power series in ρ^v with large-exponent, index integers v, which converge to $\exp(2\rho)$. The solution also applies to ions with atomic numbers greater than 1 where the ion has one electron only. Also, the Dirac solution can be applied to atoms with atomic numbers greater than 1 for the outermost electron. In this manner, the charge of the inner electrons and nuclear charge, sums to effectively produce a field, which is attributable to a positive, point-like, nuclear charge that acts on the single outermost electron.

The Dirac solution predicts the experimentally measured fine and hyperfine structure of the hydrogen atomic spectra, thus helping to validate the quantum mechanical approach to the theory of magnetic neutron scatter by unpaired atomic electrons. In particular, the energy eigenvalues will be determined for the central, radial-dependent coulomb electric field of the nucleus that acts on the electron. The nuclear electric potential energy $\phi(r)$ is a function of radius r from the nuclear center, where

$$\phi(r) = -\frac{Ze^2}{r} \tag{14.368}$$

The charge of the hydrogen nucleus is generalized to Z protons, so the results of this derivation will apply to all single-electron elements with higher atomic number Z.

Given the radial electric potential energy $\phi(r)$ of the nucleus, the solution of the Dirac equation for the electron interaction in a single-electron atom is derived as follows. A parameter γ is defined as

$$\gamma = \frac{Ze^2}{\hbar c} \tag{14.369}$$

The dimensionless fine-structure constant α_f is expressed in CGS units by

$$\alpha_f = \frac{e^2}{\hbar c} = \frac{1}{137.0359} \tag{14.370}$$

Hence, parameter γ is much less than 1 for low and moderate atomic number Z nuclei, where

$$\gamma = \frac{Ze^2}{\hbar c} = Z\alpha_f = \frac{Z}{137.0359} \ll 1 \tag{14.371}$$

As a result, one obtains

$$\frac{\phi(r)}{\hbar c\alpha} = -\frac{\gamma}{\rho} \tag{14.372}$$

where

$$\alpha = \frac{+\sqrt{m^2c^4 - E^2}}{\hbar c} \quad \text{and} \quad \rho = r\alpha \tag{14.373}$$

One starts with Eqs. (14.366) and (14.367) for the previously derived pair of Dirac equations for the hydrogen atom:

$$\left(\frac{\mathrm{d}}{\mathrm{d}\rho} + \frac{k}{\rho}\right)G - \left(\frac{\alpha_2}{\alpha} + \frac{e\phi}{\hbar c\alpha}\right)F = 0 \tag{14.374}$$

$$\left(\frac{\mathrm{d}}{\mathrm{d}\rho} - \frac{k}{\rho}\right)F - \left(\frac{\alpha_1}{\alpha} - \frac{e\phi}{\hbar c\alpha}\right)G = 0 \tag{14.375}$$

Next, one proposes a pair of series solutions for $F(\rho)$ and $G(\rho)$ that have asymptotic solutions, which satisfy the boundary conditions at infinity, where for $\rho = r\alpha \to \infty$, the functions go to zero [$F(\infty) = 0$ and $G(\infty) = 0$]. The pair of series solutions for $F(\rho)$ and $G(\rho)$ are

$$F(\rho) = f(\rho)\exp(-\rho) \tag{14.376}$$

$$G(\rho) = g(\rho)\exp(-\rho) \tag{14.377}$$

Substitution into Eqs. (14.374–14.375) of Eqs. (14.376–14.377) for $F(\rho)$ and $G(\rho)$, yields a pair of coupled differential equations, which are dependent on f and g, where

$$\frac{dg}{d\rho} - g + \frac{kg}{\rho} - \left(\frac{\alpha_2}{\alpha} - \frac{\gamma}{\rho}\right)f = 0 \tag{14.378}$$

$$\frac{df}{d\rho} - f + \frac{kf}{\rho} - \left(\frac{\alpha_1}{\alpha} + \frac{\gamma}{\rho}\right)g = 0 \tag{14.379}$$

One then specifies the solutions f and g in the form of a power series, where

$$f = \rho^s(a_0 + a_1\rho + a_2\rho^2 + \cdots + a_v\rho^v + \cdots) \quad a_0 \neq 0 \tag{14.380}$$

$$g = \rho^s(b_0 + b_1\rho + b_2\rho^2 + \cdots + b_v\rho^v + \cdots) \quad b_0 \neq 0 \tag{14.381}$$

Since the solutions $F(\rho)$ and $G(\rho)$ are to be finite at $r = 0$, then $s \geq 1$ or s must be close to 1 if s is less than 1.

The power series expressions for f and g of Eqs. (14.380–14.381) are substituted into Eqs. (14.378–14.379) for the pair of coupled differential equations, and the coefficients of ρ^{s+v-1} are equated to zero. For integer $v > 0$, the pair of iteration equations for a_v and b_v is

$$(s + v + k)b_v - b_{v-1} + a_v\gamma - \frac{\alpha_2}{\alpha}a_{v-1} = 0 \quad v = 1, 2, 3, \ldots \tag{14.382}$$

$$(s + v - k)a_v - a_{v-1} - b_v\gamma - \frac{\alpha_1}{\alpha}b_{v-1} = 0 \quad v = 1, 2, 3, \ldots \tag{14.383}$$

One can obtain a relation between a_v and b_v by multiplying the first equation of this pair by α and the second equation by α_2, and then subtracting the resulting second equation from the first, where one recalls that $\alpha^2 = \alpha_1\alpha_2$. The resulting equation, which directly relates a_v to b_v is then

$$b_v[\alpha(s + v + k) + \gamma\alpha_2] = a_v[\alpha_2(s + v - k) - \alpha\gamma] \quad v = 1, 2, 3, \ldots \tag{14.384}$$

With Eq. (14.384) for the relation between a_v and b_v, the form of the solution is examined for large index integer v, where v becomes much larger than any of the constants α, α_1, α_2, s, and k:

$$vb_v - b_{v-1} + a_v\gamma - \frac{\alpha_2}{\alpha}a_{v-1} = 0 \quad v \gg s, k, \alpha, \alpha_1, \alpha_2 \tag{14.385}$$

$$va_v - a_{v-1} - b_v\gamma - \frac{\alpha_1}{\alpha}b_{v-1} = 0 \quad v \gg s, k, \alpha, \alpha_1, \alpha_2 \tag{14.386}$$

The relation between a_ν and b_ν for large ν yields

$$\nu b_\nu \alpha = \nu a_\nu \alpha_2 \quad \nu \gg s, k, \alpha, \alpha_1, \alpha_2 \tag{14.387}$$

Solving for b_ν,

$$b_\nu = \frac{a_\nu \alpha_2}{\alpha} \quad \nu \gg s, k, \alpha, \alpha_1, \alpha_2 \tag{14.388}$$

Substituting b_ν and $b_{\nu-1}$ in this pair of equations for large ν yields a pair of expressions in terms of a_ν and $a_{\nu-1}$ only, where

$$a_\nu\left(\gamma + \nu\frac{\alpha_2}{\alpha}\right) - 2\frac{\alpha_2}{\alpha}a_{\nu-1} = 0 \quad \nu \gg s, k, \alpha, \alpha_1, \alpha_2 \tag{14.389}$$

$$-a_{\nu-1} + a_\nu\left(\nu - \frac{\alpha_2}{\alpha}\gamma\right) - \frac{\alpha_1}{\alpha}\frac{a_{\nu-1}\alpha_2}{\alpha} = 0 \quad \nu \gg s, k, \alpha, \alpha_1, \alpha_2 \tag{14.390}$$

and

$$\alpha^2 = \alpha_1\alpha_2 \quad \nu \gg \frac{\alpha_2}{\alpha}\gamma \quad \nu\frac{\alpha_2}{\alpha} \gg \gamma \tag{14.391}$$

Eq. (14.389) and Eq. (14.390) have the same recursion relation for a_ν:

$$a_\nu = \frac{2}{\nu}a_{\nu-1} \quad \nu \gg s, k, \alpha, \alpha_1, \alpha_2 \tag{14.392}$$

The alternative but equivalent relation between a_ν and b_ν for large ν is

$$a_\nu = \frac{\alpha}{\alpha_2}b_\nu \quad \nu \gg s, k, \alpha, \alpha_1, \alpha_2 \tag{14.393}$$

Substitution of Eqs. (14.392–14.393) into Eqs. (14.385–14.386) for a_ν and b_ν, yields a pair of expressions in terms of b_ν and $b_{\nu-1}$ only, where

$$b_\nu\left(\nu + \frac{\alpha}{\alpha_2}\gamma\right) - 2b_{\nu-1} = 0 \quad \nu \gg s, k, \alpha, \alpha_1, \alpha_2 \tag{14.394}$$

$$b_\nu(\nu\alpha - \alpha_2\gamma) - 2\alpha b_{\nu-1} = 0 \quad \nu \gg s, k, \alpha, \alpha_1, \alpha_2 \tag{14.395}$$

Eq. (14.394) and Eq. (14.395) produce the same recursion relation for b_ν:

$$b_\nu = \frac{2}{\nu}b_{\nu-1} \quad \nu \gg s, k, \alpha, \alpha_1, \alpha_2 \tag{14.396}$$

One substitutes the recursion relation for a_v of Eq. (14.392) into the asymptotic form of the power series solution for f of Eq. (14.380), for large v, so that

$$f = \rho^s\left(2a_0 + \frac{a_1}{2}\rho + \frac{2a_2}{3}\rho^2 + \cdots + \frac{2a_{v-1}}{v}\rho^v + \cdots\right) \quad v \gg s, k, \alpha, \alpha_1, \alpha_2 \tag{14.397}$$

One substitutes the recursion relation for b_v of Eq. (14.396) into the asymptotic form of the power series solution for g of Eq. (14.381), for large v, so that

$$g = \rho^s\left(2b_0 + \frac{b_1}{2}\rho + \frac{2b_2}{3}\rho^2 + \cdots + \frac{2b_{v-1}}{v}\rho^v + \cdots\right) \quad v \gg s, k, \alpha, \alpha_1, \alpha_2 \tag{14.398}$$

For large v, application of Eq. (14.392) for the a_v recursion in the power series for f in Eq. (14.397), and application of Eq. (14.396) for the b_v recursion in the power series for g in Eq. (14.398), yield

$$f = \rho^s a_0\left(2 + 2\rho + \frac{2^2}{2!}\rho^2 + \frac{2^3}{3!}\rho^2 + \cdots + \frac{2^v}{v!}\rho^v \cdots\right) = \rho^s a_0 \exp(2\rho) \quad v \gg s, k, \alpha, \alpha_1, \alpha_2 \tag{14.399}$$

and

$$g = \rho^s b_0\left(2 + 2\rho + \frac{2^2}{2!}\rho^2 + \frac{2^3}{3!}\rho^2 + \cdots + \frac{2^v}{v!}\rho^v \cdots\right) = \rho^s b_0 \exp(2\rho) \quad v \gg s, k, \alpha, \alpha_1, \alpha_2 \tag{14.400}$$

14.17 Regular Solution of the Coupled, Dirac Eigenequation Pair and Electron Energy Formula for the Hydrogen Atom

In this section, we obtain the regular solution of the Dirac equation from the power series solution derived in Section 14.16. In that discussion, we obtained asymptotic solutions of f and g for large-integer indexes v of ρ^v, where $\rho = r\alpha$. That is, the power series solutions f and g, in terms of $a_v\rho^v$ and $b_v\rho^v$, each converged to an exponential solution $\exp(2\rho)$ for summation of the infinite series of terms with large-integer index v, where $v \gg s, k, \alpha, \alpha_1, \alpha_2$. However, a regular solution will require the power series to terminate at a moderate or low value of integer index v, which is discussed next.

Assume that the series terminates at a particular integer index q, where $v = q$, so that in Eqs. (14.380–14.381):

$$a_{q+1} = 0 \quad \text{and} \quad b_{q+1} = 0 \tag{14.401}$$

Recall Eqs. (14.380) and (14.381) for the infinite power series solutions f and g of the coupled pair of Dirac eigenequations, which describe the relativistic, bound electron motion (including electron spin) in the spherically symmetric electron field of the proton nucleus of the hydrogen atom, or point-like, positive nuclear charge in a hydrogen-like atom. The infinite series solutions of f and g in Eqs. (14.380) and (14.381) are now converted to finite power series solutions f and g, by terminating the infinite series at integer index $v = q$, where

$$f = \rho^s(a_0 + a_1\rho + a_2\rho^2 + \cdots + a_q\rho^q) \quad a_0 \neq 0 \tag{14.402}$$

$$g = \rho^s(b_0 + b_1\rho + b_2\rho^2 + \cdots + b_q\rho^q) \quad b_0 \neq 0 \tag{14.403}$$

The index integer q is also called the radial quantum number. For a positive index integer $q > 0$, one obtains the finite power series in f and g in Eqs. (14.402) and (14.403), by application of the termination condition of Eq. (14.401) to the initial pair of recursion equations for a_q and b_q from Eqs. (14.382) and (14.383), where

$$\alpha b_q = -\alpha_2 a_q \quad q = 0, 1, 2, 3\ldots \tag{14.404}$$

$$\alpha a_q = -\alpha_1 b_q \quad q = 0, 1, 2, 3\ldots \tag{14.405}$$

With $\alpha^2 = \alpha_1\alpha_2$ from Eq. (14.391), the second recursion equation of Eq. (14.405), becomes identical to the first recursion equation of Eq. (14.404), so that both recursion equations for a_q and $b_q$are identical, and the single recursion relation between a_q and b_q is then

$$\alpha b_q = -\alpha_2 a_q \quad q = 0, 1, 2, 3\ldots \tag{14.406}$$

Equation (14.406), for the recursion relation between a_v and b_v for the finite power series equations for f and g of Eqs. (14.402) and (14.403) with $v = q$, is now applied to Eq. (14.384). Equation (14.384) is the recursion relation between a_v and b_v, which was obtained from the initial pair of recursion equations (14.382) and (14.383) that were obtained from the infinite power series equations for f and g of Eqs. (14.380) and (14.381). The recursion relation between a_v and b_v of Eq. (14.384), for the infinite power series, is repeated below, where

$$b_q[\alpha(s + q + k) + \gamma\alpha_2] = a_q[\alpha_2(s + q - k) - \alpha\gamma] \quad q = 0, 1, 2, 3\ldots \tag{14.407}$$

where

$$a_0 \neq 0 \quad \text{and} \quad b_0 \neq 0 \tag{14.408}$$

Substitution of Eq. (14.406) into Eq. (14.407) gives

$$\gamma(\alpha_1 - \alpha_2) = 2\alpha(s + q) \quad q = 0, 1, 2, 3\ldots, \tag{14.409}$$

Given electron rest mass m, we now recall Eq. (14.363) for α_1 and α_2, and recall Eq. (14.364) for α, which are repeated below:

$$\alpha = \frac{+\sqrt{m^2c^4 - E^2}}{\hbar c} \qquad \alpha_1 = \frac{mc^2 + E}{\hbar c} \qquad \alpha_2 = \frac{mc^2 - E}{\hbar c} \tag{14.410}$$

Substitution into Eq. (14.409) of Eq. (14.140) for α, α_1, and α_2 gives

$$\gamma^2 E^2 = (m^2c^4 - E^2)(s+q)^2 \quad q = 0, 1, 2, 3\ldots \tag{14.411}$$

One solves Eq. (14.411) for the total electron energy E to obtain

$$E = mc^2\left[1 + \frac{\gamma^2}{(s+q)^2}\right]^{-1/2} \quad q = 0, 1, 2, 3\ldots \tag{14.412}$$

In Eq. (14.412) for the total electron energy E, s is the exponent of the multiplicative term ρ^s in the infinite and finite power series solutions f and g of Eqs. (14.380) and (14.381) and Eqs. (14.402) and (14.403) for the coupled pair of Dirac eigenequations. One solves for exponent s in Eq. (14.412) by examination of the case for $v = 0$, which is applied to the pair of recursion equations of Eqs. (14.382) and (14.383) for a_v and b_v, where

$$(s + k)b_0 + a_0\gamma = 0 \quad v = 0 \tag{14.413}$$

$$(s - k)a_0 - b_0\gamma = 0 \quad v = 0 \tag{14.414}$$

One can solve for s via the matrix equation, which is formed from Eqs. (14.413) and (14.414), where

$$\begin{bmatrix} \gamma & s+k \\ s-k & -\gamma \end{bmatrix}\begin{bmatrix} a_0 \\ b_0 \end{bmatrix} = \begin{bmatrix} 0 \\ 0 \end{bmatrix} \tag{14.415}$$

The determinant of the matrix must be zero, where

$$-\gamma^2 - (s+k)(s-k) = 0 \tag{14.416}$$

One solves for s and obtains a double-valued solution:

$$s = \pm\sqrt{k^2 - \gamma^2} \tag{14.417}$$

The positive solution for s is used because of the boundary condition at the origin with the term ρ^s common to both solutions f and g, so that

$$s = +\sqrt{k^2 - \gamma^2} > 0 \tag{14.418}$$

where

$$\gamma^2 = \left(\frac{Z}{137.0359}\right)^2 \ll 1 \tag{14.419}$$

Recalling Eq. (14.345), quantum number k, which is obtained from scalar operator K for the total electron angular momentum via $k = K/\hbar$, is associated with the electron total angular momentum quantum number j by

$$k = \frac{K}{\hbar} = \pm\sqrt{j(j+1) + \frac{1}{4}} = \pm\left(j + \frac{1}{2}\right) \tag{14.420}$$

Since the square root expression for quantum number k has a positive and negative solution, we now must determine which solution of k to use. To choose the solution for k, we start by comparing the relations between a_0 and b_0 that were obtained from (1) the $v = 0$ case with the initial pair of iteration equations and (2) the $q = 0$ case when one terminates the power series at an integer index v equal to integer radial quantum number q, where $q = v$.

The zero radial quantum number $q = 0$ case gives

$$\alpha b_0 = -\alpha_2 a_0 \tag{14.421}$$

Since parameters α_2 and α are positive ($\alpha_2 > 0$ and $\alpha > 0$), then the ratio b_0/a_0 is negative:

$$\frac{b_0}{a_0} = -\frac{\alpha_2}{\alpha} < 0 \tag{14.422}$$

The zero-integer index $v = 0$ case gives

$$(s + k)b_0 + a_0\gamma = 0 \tag{14.423}$$

and

$$(s - k)a_0 - b_0\gamma = 0 \tag{14.424}$$

Solving for b_0/a_0 in Eqs. (14.423) and (14.424), yield

$$\frac{b_0}{a_0} = -\frac{\gamma}{s + k} \tag{14.425}$$

and

$$\frac{b_0}{a_0} = \frac{s - k}{\gamma} \tag{14.426}$$

Multiplying Eqs. (14.425) and (14.426) gives

$$\left(\frac{b_0}{a_0}\right)^2 = -\frac{s-k}{s+k} > 0 \tag{14.427}$$

To fulfill the condition of positive $(b_0/a_0)^2$ requires that

$$|k| > s > 0 \tag{14.428}$$

That is, since ρ^s must not be a singularity at $\rho = 0$, then $s > 0$. Also, γ is positive ($\gamma > 0$).

Consistency for the zero-integer indexes $v = 0$ and $q = 0$ requires

$$\frac{b_0}{a_0} = -\frac{\gamma}{s+k} = \frac{s-k}{\gamma} < 0 \tag{14.429}$$

This requirement is then met by requiring k be positive ($k > 0$) when $v = q = 0$, where

$$\text{if } \ v = q = 0 \quad \text{then } \ k > 0 \tag{14.430}$$

Hence, one uses the positive solution for the total electron angular momentum quantum number k, where $k > 0$ only if $v = q = 0$, where

$$k = \frac{K}{\hbar} = +\left(j + \frac{1}{2}\right) > 0 \quad v = q = 0 \tag{14.431}$$

If the integer index v and radial quantum number q are each nonzero, where $v \neq 0$ and $q \neq 0$, then the total electron angular momentum quantum number k can be positive or negative, where

$$k = \frac{K}{\hbar} = \pm\left(j + \frac{1}{2}\right) > 0 \quad v = q \neq 0 \tag{14.432}$$

The positive and negative values of k for nonzero integer index $v \neq 0$, and thus nonzero radial quantum number $q \neq 0$, yield two sets of quantum numbers for each radial quantum number n, as will be shown later in this chapter.

The squaring of k gives a single positive value for k^2. With substitution for quantum number k and parameter $\gamma = Z\alpha_f$ in the expression for exponent s of ρ^s, where $\rho = r\alpha$, one then obtains the single, positive solution for exponent s for the case of $v = q = 0$ or $v = q \neq 0$, where

$$s = +\sqrt{\mathrm{j}(\mathrm{j}+1) + \frac{1}{4} - (z\alpha_\mathrm{f})^2} \tag{14.433}$$

Recalling Eq. (14.420), one can combine $j(j+1)$ and $1/4$ so that the single positive solution for s becomes

$$s = +\sqrt{\left(j+\frac{1}{2}\right)^2 - (Z\alpha_f)^2} \tag{14.434}$$

In Eq. (14.412) for the total electron energy E, we substitute Eq. (14.434) for the positive solution of exponent s in the multiplicative factor ρ^s in the power series solutions f and g of the coupled pair of Dirac eigenequations. The coupled pair of Dirac eigenequations describes the relativistic, bound electron motion (including the electron spin) in the spherically symmetric, electric field of the proton nucleus in the hydrogen atom, or a point-like, nuclear positive charge in a hydrogen-like atom. With electron rest mass m, Dirac equation parameter $\gamma = Z\alpha_{\mathrm{f}}$, and the fine structure constant $\alpha_{\mathrm{f}} \cong 1/137$, one obtains the total energy E for a bound electron in the hydrogen atom, or a hydrogen-like atom with atomic number Z. Thus, with substitution of Eq. (14.434) for exponent s into Eq. (14.412) for total electron energy E, one obtains the total electron energy E as a function of the total electron angular momentum quantum number j and the electron radial quantum number q, where

$$E = mc^2\left[1 + (Z\alpha_f)^2\left(\sqrt{\left(j+\frac{1}{2}\right)^2 - (Z\alpha_f)^2} + q\right)^{-2}\right]^{-1/2} \tag{14.435}$$

The electron total angular momentum j is a positive half-integer, where

$$j = l + \frac{1}{2} = \frac{1}{2}, \frac{3}{2}, \frac{5}{2}, \ldots \tag{14.436}$$

The power series index integer q (i.e., the radial quantum number q) is zero or a positive integer ($q \geq 0$), where

$$q = 0, 1, 2, 3\ldots \tag{14.437}$$

This formula for total electron energy E obtained as a solution to the Dirac equation accounts for the fine-structure lines for hydrogen and hydrogen-like atoms in atomic spectroscopy. The matching of measured fine-structure atomic spectra data, with the energy predictions of this energy formula, is one way to validate quantum mechanics. In the next section, the quantum numbers of the relativistic-based formula for the electron energy E, are related to the quantum numbers and energy level notation of the nonrelativistic hydrogen atom. The Dirac equation and its solution are also stepping stones to quantum electrodynamics and relativistic quantum field theories.

One can see the fine-structure terms in Eq. (14.435) for the electron energy E, via a binomial series expansion, where

$$(1+x)^{-1/2} = 1 - \frac{x}{2} + \frac{3}{8}x^2 - \frac{5}{16}x^3 + \frac{35}{128}x^4 + \cdots \quad -1 < x < 1 \tag{14.438}$$

First we examine the total electron energy E of Eq. (14.412) before substitution for s and γ:

$$E = mc^2\left[1 + \frac{\gamma^2}{(s+q)^2}\right]^{-1/2} \quad q = 0, 1, 2, 3\ldots \tag{14.439}$$

From Eq. (14.434), we obtain the single, positive solution for exponent s, where

$$s = +\sqrt{\left(j+\frac{1}{2}\right)^2 - (Z\alpha_f)^2} \tag{14.440}$$

and

$$x = \frac{\gamma^2}{(s+q)^2} \ll 1 \tag{14.441}$$

with

$$\gamma = Z\alpha_f = \frac{Z}{137.0359} \ll 1 \tag{14.442}$$

One can then expand the radical in Eq. (14.439) into a binomial series to obtain the total energy E of an electron in a hydrogen atom or hydrogen-like atom or ion:

$$E = mc^2\left[1 - \frac{\gamma^2}{2(s+q)^2} + \frac{3}{8}\frac{\gamma^4}{(s+q)^4} - \frac{5}{16}\frac{\gamma^6}{(s+q)^6}\right] \quad q = 0, 1, 2, 3\ldots \tag{14.443}$$

Rearrangement gives the total electron energy E as

$$E = mc^2\left[1 - \frac{\gamma^2}{2(s+q)^2} - \frac{\gamma^4}{2(s+q)^4}\left(\frac{5}{8}\frac{\gamma^2}{(s+q)^2} - \frac{3}{4}\right)\right] \quad q = 0, 1, 2, 3\ldots \tag{14.444}$$

The fine-structure terms of the atomic spectra of hydrogen are expressed by the powers of $[\gamma/(s+q)]^2$ in the formula for total electron energy E, as is made evident in the binomial series expansion of the radical of E.

14.18 Quantum Number Relationships in the Dirac Electron Energy Formula for a Hydrogen Atom

The total electron angular momentum quantum numbers j and k in the relativistic expression for electron energy E, which was derived from the Dirac equation that was applied to the hydrogen atom (or hydrogen-like atoms or ions), are now related to the electron orbital quantum number l and the radial quantum number n. The electron total quantum number n is defined by

$$n = k + q = j + \frac{1}{2} + q > 0 \tag{14.445}$$

where

$$n = 1, 2, 3, \ldots \tag{14.446}$$

$$q = 0, 1, 2, 3\ldots \tag{14.447}$$

The electron total quantum number n is the principal quantum number for the nonrelativistic theory of the hydrogen atom. In this section, the radial q and total angular momentum j quantum numbers of the relativistic electron energy E formula of Eq. (14.435), which is based on the Dirac equation for the hydrogen atom, are related to the principal quantum number n for electron energy and the electron orbital l and spin s angular momentum quantum numbers. The quantum numbers n, l, and s were developed in nonrelativistic quantum mechanics to explain the atomic spectra of hydrogen.

The total electron angular momentum quantum number k is positive ($k > 0$) when the radial quantum number q is zero ($q = 0$), and when the radial quantum number q is greater than zero ($q > 0$), then k can be positive ($k > 0$) or negative ($k < 0$). The total angular momentum quantum numbers j and orbital angular momentum quantum numbers l are given by

$$j = l + \frac{1}{2} = \frac{1}{2}, \frac{3}{2}, \frac{5}{2}, \ldots \tag{14.448}$$

and

$$l = 0, 1, 2, 3, \ldots \tag{14.449}$$

The total electron quantum number j is the sum of the electron orbital angular momentum quantum number l and the total electron spin angular momentum quantum number $S = 1/2$. The range of possible quantum numbers j for the total electron angular momentum is

$$j = l + S, l + S - 1, l + S - 2, \ldots, |l - S| + 1, |l - S| \tag{14.450}$$

Since the total electron spin angular momentum quantum number $S = 1/2$, there are just two values of electron total angular momentum j, namely,

$$j = l \pm \frac{1}{2} \tag{14.451}$$

The electron total spin quantum number S has the range of azimuthal component spin quantum numbers s, given by

$$s = S, S-1, S-2, \ldots, -S+1, -S \tag{14.452}$$

With electron total spin quantum number $S = 1/2$, there are just two values for the electron azimuthal component spin quantum numbers s, namely,

$$s = \pm 1/2 \tag{14.453}$$

Recall from Eq. (14.435), that for an electron of rest mass m, the total electron energy E in a one-electron atom, or ion with atomic number Z in the Dirac result, is given by

$$E = mc^2\left[1+(Z\alpha_f)^2\left(\sqrt{\left(j+\frac{1}{2}\right)^2-(Z\alpha_f)^2}+q\right)^{-2}\right]^{-1/2} \tag{14.454}$$

where

$$(Z\alpha_f)^2 \gg \left(j+\frac{1}{2}\right)^2 \tag{14.455}$$

If one ignores $Z\alpha_f$ compared to $j + (1/2)$ inside the radical, then one obtains the electron energy E for the nonrelativistic hydrogen atom in terms of the principal quantum number n, where $n = j + (1/2) + q$. This result was originally derived by Arnold Sommerfeld in the old Bohr quantum theory, and was applied to explain the atomic spectra of hydrogen, where

$$E = \frac{mc^2}{\sqrt{1+\left[Z\alpha_f/\left(j+\frac{1}{2}+q\right)\right]^2}} = \frac{mc^2}{\sqrt{1+\left(\frac{Z\alpha_f}{n}\right)^2}} \tag{14.456}$$

In 1916, the rightmost expression for the electron total energy E, as a function of just the principal quantum number n, was derived theoretically by Sommerfeld, who based this work on the Bohr atomic model. However, with the experimentally measured wavelength spectra of the hydrogen atom, Johannes Rydberg derived in 1888 the rightmost expression empirically for spectral wavelengths λ. In the Rydberg formula, the wavelength λ of a spectral line of hydrogen resulted from the electron transition from an initial energy level to a final energy level, with respective principal quantum numbers n_i and n_f. Associated with the Rydberg formula was the Rydberg constant, where $R_y = 1.097 \times 10^7\ \text{m}^{-1}$. The Rydberg formula was

$$\frac{1}{\lambda} = R_y\left(\frac{1}{n_f^2} - \frac{1}{n_i^2}\right) \tag{14.457}$$

Then, in 1900, Planck introduced his postulate (and Planck's constant h) for the energy E of oscillators in a radiating blackbody as a function of frequency v, which was quantized via integer n, where $E = nhv$. The Rydberg constant R_y, multiplied by hc, then gives $hcR_y = 13.6$ eV or 1 Ry of energy, which is the energy of the electron in its lowest state in the hydrogen atom. Generalized to hydrogen-like atoms or ions of atomic number Z, the Rydberg formula under the Bohr–Sommerfeld model became

$$\frac{1}{\lambda} = R_y Z^2\left(\frac{1}{n_f^2} - \frac{1}{n_i^2}\right) \tag{14.458}$$

The Bohr–Sommerfeld model identified the empirical Rydberg constant R_y with the fine-structure constant α_f and one-half the rest mass energy of the electron $mc^2/2$, where

$$R_y = \frac{mc^2}{2}\alpha_f^2 \tag{14.459}$$

and

$$\alpha_f = \frac{e^2}{4\pi\varepsilon_0 \hbar c} \cong \frac{1}{137} \tag{14.460}$$

The energy E of the photon is related to its wavelength λ by

$$E = \frac{hc}{\lambda} \tag{14.461}$$

An initial, unbound electron at an energy level of infinity ($n_i = \infty$), which becomes bonded with a proton in final energy level $n_f = n$ to form a hydrogen atom, attains bound energy E, where

$$E = -R_y\left(\frac{Z}{n}\right)^2 = -\frac{mc^2}{2}\left(\frac{\alpha_f Z}{n}\right)^2 = -\frac{mc^2}{2}\left(\frac{\gamma}{n}\right)^2 \tag{14.462}$$

In comparison, the Dirac formula for the total bound, electron energy E, where $Z\alpha_f$ is ignored compared to $j + (1/2)$, can then undergo a binomial expansion of the radical and thereby yield the same result as the Bohr–Sommerfeld model, but with the addition of the electron rest mass energy mc^2, where

$$E = \frac{mc^2}{\sqrt{1 + \left(\frac{Z\alpha_f}{n}\right)^2}} = mc^2\left[1 - \frac{1}{2}\left(\frac{Z\alpha_f}{n}\right)^2\right] = mc^2\left[1 - \frac{1}{2}\left(\frac{\gamma}{n}\right)^2\right] \tag{14.463}$$

The Dirac energy formula for total electron energy E in its approximate form, yields the Bohr–Sommerfeld result, and the atomic spectra wavelength of the empirical Rydberg equation of hydrogen spectra. In its exact form, the Dirac energy formula extends the approximate result to predict the fine-structure atomic spectra, and detailed interactions of the electron with the nucleus in the hydrogen atom and in hydrogen-like atoms and ions.

We now obtain the relation between the quantum numbers for the electron principal energy levels, and the quantum numbers for the total, orbital, and azimuthal angular momentum. Then, the quantum number sets for the electron orbitals are listed and expressed in the notation of nonrelativistic electron orbitals in the hydrogen atom and hydrogen-like atoms and ions. Recall the defining equation for scalar operator K^2 of Eq. (14.335) in the derivation of the coupled pair of Dirac equations, where

$$K = \hbar k = \pm\sqrt{J^2 + \frac{\hbar^2}{4}} = \pm\left(\boldsymbol{\sigma}\cdot\mathbf{L} + \hbar\begin{bmatrix}1 & 0\\ 0 & 1\end{bmatrix}\right) \tag{14.464}$$

Also, recall in Eq. (14.464), that the squared magnitude of the total electron angular momentum J^2 of Eq. (14.310) is given by

$$J^2 = L^2 + \hbar\boldsymbol{\sigma}\cdot\mathbf{L} + \frac{3\hbar^2}{4}\begin{bmatrix}1 & 0\\ 0 & 1\end{bmatrix} \tag{14.465}$$

Substitution of $\hbar^2 j(j+1)$ for J^2 and $\hbar^2 l(l+1)$ for L^2, gives the relation between the electron total angular momentum quantum number j, the orbital angular momentum quantum number l, and the spin angular momentum quantum number s, via the Pauli spin matrix $\boldsymbol{\sigma}$, and the orbital angular momentum matrix $\mathbf{L}$, where

$$j(j+1) = l(l+1) + \frac{1}{\hbar}\boldsymbol{\sigma}\cdot\mathbf{L} + \frac{3}{4}\begin{bmatrix}1 & 0\\ 0 & 1\end{bmatrix} \tag{14.466}$$

Solving for $\boldsymbol{\sigma}\cdot\mathbf{L}$, one obtains

$$\frac{1}{\hbar}\boldsymbol{\sigma}\cdot\mathbf{L} = j(j+1) - l(l+1) - \frac{3}{4}\begin{bmatrix}1 & 0\\ 0 & 1\end{bmatrix} \tag{14.467}$$

Also, recall in the derivation that the expression for $\boldsymbol{\sigma} \cdot \mathbf{L}$ is given by

$$\boldsymbol{\sigma} \cdot \mathbf{L} = \beta K - \hbar h \begin{bmatrix} 1 & 0 \\ 0 & 1 \end{bmatrix} \tag{14.468}$$

where

$$\beta = \begin{bmatrix} 1 & 0 \\ 0 & -1 \end{bmatrix} \tag{14.469}$$

Substitution of $\boldsymbol{\sigma} \cdot \mathbf{L}$ in the equation for $j(j+1)$, gives

$$j(j+1) = l(l+1) + \begin{bmatrix} 1 & 0 \\ 0 & -1 \end{bmatrix} k - \frac{1}{4} \begin{bmatrix} 1 & 0 \\ 0 & 1 \end{bmatrix} \tag{14.470}$$

One then solves for total angular momentum quantum number k to obtain

$$k = j(j+1) - l(l+1) + \frac{1}{4} \tag{14.471}$$

$$k = -j(j+1) + l(l+1) - \frac{1}{4} \tag{14.472}$$

One requires the total quantum number n to be greater than zero $(n > 0)$ for the set of radial quantum numbers $q = 0, 1, 2, 3\ldots$. Also, the total angular momentum k must be positive $(k > 0)$, if the radial quantum number q is zero $(q = 0)$. The quantum numbers must obey the defining equation for the total quantum number n, where n is also the principal quantum number in the nonrelativistic theory of the hydrogen atom, and was introduced earlier, where

$$n = j + \frac{1}{2} + q = k + q > 0 \quad q = 0, 1, 2, 3\ldots \tag{14.473}$$

The minimum total $j = 1/2$ and minimum orbital $l = 0$ angular momentum quantum numbers can occur when the total quantum number n equals 1 $(n = 1)$, which requires the radial quantum number q to be zero $(q = 0)$. The second of the pair of equations for the total electron angular momentum k then yields a negative k value $(k < 0)$, which contradicts the requirement that $k > 0$ if $q = 0$. Thus, the second expression for k is eliminated from consideration. Hence, only the first equation of the pair of k equations satisfies the required range of quantum numbers and index integers without contradiction. The nonzero, positive, or negative integer quantum number k for the total electron angular momentum is then given by the single relation

$$k = j(j+1) - l(l+1) + \frac{1}{4} \tag{14.474}$$

The two total angular quantum numbers $j = l + S$ and $j = l - S$ are then substituted into Eq. (14.474), and yield a pair of equations that relate the quantum numbers

$$k = j(j+1) - l(l+1) + \frac{1}{4} = \left(l + \frac{1}{2}\right)\left(l + \frac{1}{2} + 1\right) - l(l+1) + \frac{1}{4} = l + 1 \quad j = l + \frac{1}{2} \tag{14.475}$$

and

$$k = j(j+1) - l(l+1) + \frac{1}{4} = \left(l - \frac{1}{2}\right)\left(l - \frac{1}{2} + 1\right) - l(l+1) + \frac{1}{4} = -l \quad j = l - \frac{1}{2} \tag{14.476}$$

The electron radial quantum number n, total angular momentum quantum numbers k and j, orbital angular momentum l, and total spin quantum number $S = 1/2$, with its two associated spin-up and spin-down azimuthal spin component quantum numbers $s = \pm 1/2$, yield the following set of electron orbitals in the hydrogen atom.

For zero radial quantum number $q = 0$, there is only the positive electron quantum number $k = 3$, which is associated with the electron total angular momentum quantum number $j = 5/2$, and orbital angular momentum quantum number $l = 2$. These quantum numbers represent the ${}^2D_{5/2}$ electron orbital in hydrogen, where, for azimuthal component spin quantum numbers $s = \pm 1/2$, the single quantum number set is

$$q = 0 \text{ and } k = 3 \Leftrightarrow l = 2 \text{ and } j = \frac{5}{2} \Leftrightarrow {}^2D_{5/2} \tag{14.477}$$

For radial quantum number $q = 1$, there is the positive and negative electron quantum number $k = \pm 2$, which is associated with the electron total angular momentum quantum number $j = 3/2$, and orbital angular momentum quantum numbers $l = 2$ for $k = -2$, and $l = 1$ for $k = 2$. These quantum numbers represent the ${}^2D_{3/2}$ for $k = -2$ orbital and ${}^2P_{3/2}$ orbital for $k = 2$ in hydrogen, where, for azimuthal component spin quantum numbers $s = \pm 1/2$, the two quantum number sets are

$$q = 1 \text{ and } k = -2 \Leftrightarrow l = 2 \text{ and } j = \frac{3}{2} \Leftrightarrow {}^2D_{3/2} \tag{14.478}$$

and

$$q = 1 \text{ and } k = 2 \Leftrightarrow l = 1 \text{ and } j = \frac{3}{2} \Leftrightarrow {}^2P_{3/2} \tag{14.479}$$

For radial quantum number $q = 2$, there are the positive and negative electron quantum numbers $k = \pm 1$, which is associated with the electron total angular momentum quantum number $j = 1/2$, and orbital angular momentum quantum numbers $l = 1$ for $k = -1$ and $l = 0$ for $k = 1$. These quantum numbers represent the ${}^2P_{1/2}$ for $k = -1$ orbital and ${}^2S_{1/2}$ orbital for $k = 1$ in hydrogen, where, for azimuthal component spin quantum numbers $s = \pm 1/2$, the two quantum number sets are

$$q = 2 \text{ and } k = -1 \Leftrightarrow l = 1 \text{ and } j = \frac{1}{2} \Leftrightarrow {}^2P_{1/2} \tag{14.480}$$

and

$$q = 2 \text{ and } k = 1 \Leftrightarrow l = 0 \text{ and } j = \frac{1}{2} \Leftrightarrow {}^2S_{1/2} \tag{14.481}$$

14.19 The Nuclear Electric Quadrupole Potential

The hyperfine interaction, which is due to the electric quadrupole moment of the nucleus, can be obtained from the electric potential energy term $e\phi$ of the Dirac Hamiltonian and wave equations. Electric quadrupole moments arise, and thus are nonzero only in those nuclei, whose charge distribution is not spherically symmetric, such as the deuterium nucleus and nitrogen-14, oxygen-17, and chlorine-35. Relative to the origin at the center of the nucleus, the nuclear charge density $\rho(\mathbf{r}_s)$ is at position vector $\mathbf{r}_s$, and the observer is at position vector $\mathbf{r}$ with angle χ between charge density and observer vectors. Given the vacuum permittivity ε_0, the electric potential $\phi(\mathbf{r})$ of the nuclear charge distribution, which is experienced by an electron at position $\mathbf{r}$, is found by volume integration over the nuclear charge density $\rho(\mathbf{r}_s)$, where

$$\phi(\mathbf{r}) = \frac{1}{4\pi\varepsilon_0}\int_0^{2\pi} \mathrm{d}\phi_s \int_0^{\pi} \mathrm{d}\theta_s \int_0^{\infty} \mathrm{d}r_s \frac{\rho(\mathbf{r}_s)}{|\mathbf{r} - \mathbf{r}_s|} \tag{14.482}$$

The law of cosines is

$$|\mathbf{r} - \mathbf{r}_s| = \sqrt{r^2 + r_s^2 - 2rr_s \cos\chi} \tag{14.483}$$

Observer and charge density vectors point outward from the origin at the center of the nucleus, with angle χ between them. Angle χ is expressed in terms of the polar and azimuthal angles for the charge density (θ_s, ϕ_s) and observer (θ, ϕ) by

$$\cos\chi = \cos\theta\cos\theta_s + \sin\theta\sin\theta_s\cos(\phi - \phi_s) \tag{14.484}$$

The denominator term is prepared for series expansion, where

$$\frac{1}{|\mathbf{r}-\mathbf{r}_s|}=r^{-1}\left[1+\left(\frac{r_s}{r}\right)^2-2\frac{r_s}{r}\cos\chi\right]^{-1/2} \tag{14.485}$$

The electron radius r about the center of the nucleus is assumed to be much larger than the maximum distance r_s of nuclear charge from the center, where $r_s/r \ll 1$, so that the Taylor series formula is

$$(1+x)^n = 1+nx+\frac{n(n-1)}{2!}x^2+\cdots \tag{14.486}$$

where

$$n=-\frac{1}{2} \quad \text{and} \quad x=\left(\frac{r_s}{r}\right)^2-2\frac{r_s}{r}\cos\chi \tag{14.487}$$

Application of the Taylor series formula of Eq. (14.486) to Eq. (14.485), yields

$$\frac{1}{|\mathbf{r}-\mathbf{r}_s|}=\frac{1}{r}\left[1+\frac{r_s}{r}\cos\chi-\frac{1}{2}\left(\frac{r_s}{r}\right)^2+\frac{3}{8}\left(\frac{r_s}{r}\right)^4-\frac{12}{8}\left(\frac{r_s}{r}\right)^3\cos\chi+\frac{12}{8}\left(\frac{r_s}{r}\right)^2\cos^2\chi+\cdots\right] \tag{14.488}$$

Terms higher than quadratic are ignored, so

$$\frac{1}{|\mathbf{r}-\mathbf{r}_s|}=\frac{1}{r}\left[1+\frac{r_s}{r}\cos\chi-\frac{1}{2}\left(\frac{r_s}{r}\right)^2(3\cos^2\chi-1)\right] \tag{14.489}$$

Substitution of Eq. (14.489) into the integral of Eq. (14.482) for the nuclear electric potential $\phi(\mathbf{r})$ at the electron, thus yields the multipole expansion of the nuclear potential, where

$$\phi(\mathbf{r})=\frac{1}{4\pi\varepsilon_0}\int_0^{2\pi}d\phi_s\int_0^{\pi}d\theta_s\int_0^{\infty}dr_s\frac{\rho(\mathbf{r}_s)}{r}\left[1+\frac{r_s}{r}\cos\chi-\frac{1}{2}\left(\frac{r_s}{r}\right)^2(3\cos^2\chi-1)\right] \tag{14.490}$$

In Eq. (14.490), the first term is the electric monopole moment, the second term is the electric dipole moment, and the third term is the electric quadrupole moment $Q(\mathbf{r},\mathbf{r}_s,\chi)$, where

$$Q(\mathbf{r},\mathbf{r}_s,\chi)=\frac{\rho(\mathbf{r}_s)}{r}\left[1+\frac{r_s}{r}\cos\chi-\frac{1}{2}\left(\frac{r_s}{r}\right)^2(3\cos^2\chi-1)\right] \tag{14.491}$$

For atoms heavier than hydrogen, one can calculate the effective monopole, dipole, and quadrupole components of the electric potential, which are experienced by the outer electron. The distribution of charge to be integrated, is then the sum of the positive nuclear charge distribution, and the negative electron charge distribution for the electrons inside the orbit of the outer electron.

References

Bressi, G., Carugno, G., Onofrio, R., & Ruoso, G. (2002). Measurement of the Casimir force between parallel metallic surfaces. *Physical Review Letters*, *88*(4), 041804.

Cremer, J. T. (2012a). In Hawkes P. W. (Ed.), *Neutron and X-ray microscopy—Part 1. Advances in imaging and electron physics* (Vol. 172). London: Academic Press, an imprint of Elsevier Press.

Cremer, J. T. (2012b). In Hawkes P. W. (Ed.), *Neutron and X-ray microscopy—Part 2. Advances in imaging and electron physics* (Vol. 173). London: Academic Press, an imprint of Elsevier Press.

Cremer, J. T. (2013). In Hawkes P. W. (Ed.), *Chapters 1 and 2. Advances in imaging and electron physics* (Vol. 175). London: Academic Press, an imprint of Elsevier Press.

Dzyaloshinskii, I. E., Lifshitz, E. M., & Pitaevskii, L. P. (1961). General theory of van der Waals' forces. *Soviet Physics Uspekhi*, *4*(2), 153–176.

Lamb, W. E, & Retherford, R. C. (1947). Fine structure of the hydrogen atom by a microwave method. *Physical Review*, *72*(3), 241–243.

Liboff, R. L. (2003). *Introductory quantum mechanics*. New York, NY: Addison-Wesley.

Merzbacher, E. (1970). *Quantum mechanics*. New York, NY: John Wiley & Sons.

Schiff, L. I. (1968). *Quantum mechanics*. New York, NY: McGraw-Hill.

Schiff, L. I. (1968). *Quantum Mechanics* (3rd ed.). New York, NY: McGraw Hill Book Company.

Sommerfeld, A. (1916). Zur Quantentheorie der Spektrallinien. *Annalen der Physik, Bd*, *51*(1–94), 125–167.

Zee, A. (2003). *Quantum field theory in a nutshell*. Princeton, NJ: Princeton University Press.

15 Neutron and X-ray Optics in General Relativity and Cosmology

15.1 Special and General Relativity—History and Relation to Neutron and X-ray Optics

In cosmology, many processes involve neutron and X-ray optical phenomena, which occur within the realm of the special and general relativistic setting. In this chapter, we present general relativity and its occurrence in cosmology for the purpose of facilitating insight and observations from the point of view of naturally occuring neutron and X-ray optical processes. Many cosmological experimental results and observations of cosmological processes involve X-ray and neutron sources and their corresponding processes of scatter, refraction, diffraction, and reflection. General relativity is one of the foundations of modern cosmology that is used to interpret and theorize the experimental and observational findings cosmology. The purpose of this chapter is to give the reader an understanding of general relativity that can be applied to neutron and X-ray optical investigations of cosmological processes.

The development of special and general relativity is an example attributed to an African proverb: "It takes a village to raise a child," in which the "child" is special and general relativity as can be seen by the multitude of Wikipedia entries (http://en.wikipedia.org) regarding the history of relativity and its foundations. Consider the histroy of the Lorentz transformation prior to Albert Einstein's independent derivation of the Lorentz transformation using the speed of light invariance in all frames and physical law invariance in all intertial frames. In 1887, assuming an incompressible medium and the Doppler effect (proposed by Christian Doppler in 1842), Voigt developed the transformations of space and time between inertial frames in relative motion at constant veclocity. In an attempt to explain the Michelson–Morley ether wind experiment of 1887, G. Fitzgerald in 1889 and H. Lorentz in 1892 proposed the length contraction formula. Lorentz published several papers on the transformations of distance and time between inertial frames in relative motion in 1889, 1895, and 1904.

Finally, in 1904, H. Poincairé developed the modern-day Lorentz transfomations and developed the mathematical framework that showed the Lorentz transformations were a subset of the Poincairé group of symmetry transformations. Poincairé

Neutron and X-ray Optics. DOI: http://dx.doi.org/10.1016/B978-0-12-407164-3.00014-0

had presented two papers on the principle of relative motion in 1900, and later in 1904, Poincairé developed the term *relativity*, in which he asserted no experiment can discriminate between a state of uniform motion and rest. In 1905–1906, Poincairé realized that using $\sqrt{-1}$ in the Lorentz transformation allowed the Lorentz transfomation to be realized as a rotation in 4-D space, with time as the fourth dimension. Minlowski used Poincairé's result to restate Maxwell's equations in four dimensions and show their invariance under the Lorentz transformation and reformulated special relativity in four-vectors, thereby developing the Minkowski space and diagram.

Based on these two assumptions, the following is the case: (i) the speed of light has a constant value whether observed in a rest frame or a moving frame; and (ii) the physical laws and their equations should be the same in inertial frames that move at constant velocities relative to one another. From these two principles, Einstein independently derived the Lorentz transfomations in his special relativity paper of 1905 that described space and time and the equivalence of mass and energy. With special relativity, Einstein demonstrated that the rates of clocks depend on their position in a gravitational field and that the difference in the clock rate is proportional to the gravitational potential to first approximation.

In 1907, Einstein realized that the principle of relativity could be extended to gravitational fields. That is, a uniformly accelerated box that is not in a gravitational field is equivalent to a box at rest in a constant gravitational field. Einstein realized that the gravitational field is completely equivalent to the acceleration of a reference frame but in a sufficiently small region of space such that the gravitational field is uniform. In the intervening years (1907–1912), Einstein studied gravitational fields and applied special relativity to produce papers on the gravitational redshift of light and the gravity deflection of light.

To progress further, Einstein would need the mathematics principles developed in the 1800s, specifically the mathematics of vectors and tensors (H. Hamilton, H. Grassmann, J. W. Gibbs, O. Heaviside), curved surfaces (Euler and C. Gauss), and differential geometry and Riemannian space (H. Grassmann, W. Clifford, C. Jacobi, B. Riemann, N. Lobachesky, F. Klein, E. Beltrami, E.B. Christofel, R. Lipshitz, T. Levi-Cita, and G. Ricci-Cubastro). Prior to Einstein, Clifford introduced geometric algebra and first suggested gravity may be a manifestation of the underlying geometry of space. In 1900, T. Levi-Cita and G. Ricci-Cubastro published a book on tensors called *Méthodes de calcul différentiel absolu et leurs applications,* which was used by Einstein and others to learn about tensor calculus, including the parallel transport of vectors and covariant differentiation. Following Einstein's 1915 publication on general relativity and additional subsequent publications, Levi-Cita and Einstein began corresponding, and Levi-Cita corrected some tensor calculus errors in Einstein's papers.

Upon Einstein's return to Switzerland in 1912 to teach at ETH (Eidgenössische Technische Hochschule) in Zürich, Switzerland, a colleague, Marcel Grossmann, introduced Einstein to Riemannian geometry and differential geometry. Also, Italian mathematician Tullio Levi-Civita introduced Einstein to general covariance and the use of tensors. Through his colleague and friend Herman Minkowski (who developed the 4-vector, Minkowski space-time of special relativity), David Hilbert, originally a pure mathematician, began investigating mathematical physics, and following the progress of physics, especially after 1911. By late 1915, Einstein had published many papers on general relativity that culminated with the correct field equations of gravitation. Almost immediately afterward, David Hilbert independently derived the field equations of gravity via the principle of least action, in which he fully credited Einstein as the originator of general relativity and the gravitational field equations.

With the correct field equations, Karl Schwarzschild published a paper in 1916 providing the solution of the correct Einstein gravitational equations applied to the gravitational field of a spherically symmetric, massive sphere like the sun in our solar system. Also, Schwarzschild did the first study of constant-density stars and black holes. However, black holes were first predicted by J. Michell in 1783 in a letter written to the H. Cavendish of the Royal Society of London. For a detailed review of the correspondence and interaction of Hilbert and Einstein, see Logunov, Mestvirishvili, and Petrov (2004). This chapter draws upon many sources, which include Arfken (1970), Graustein (1962), Guggenheimer (1977), Wrede (1972), Lawden (2002), Friedrichs (1975), Dutton (1986), Joos and Freeman (1986), Soper (2008), Greenwood (1977), Sommerfeld (1964), Shadowitz (1988a,b), Sen and Powers (2011), Brannon (2004), Foster and Nightingale (1995), Pathria (2003), Einstein (2004), Landau and Lifschitz (1975), Pauli (1981), Misner, Thorne, and Wheeler (1973), McMahon (2006), and Zee (2003).

15.2 Equivalence Principle, Manifolds, Parallel Vector Translation, and Covariant Derivatives

As Landau points out in his book, the accelerating or noninertial frames of reference are not completely identical to the actual gravitational fields because at an infinite distance from the source of gravity, the gravitational field goes to zero and the inertial frame of reference is unchanged. See page 226 in Landau and Lifshitz (1975), in which Landau points out that the centrifugal force of a rotating system increases with the radius from the axis of rotation in which the centrifugal force at infinity would be a nonphysical infinity. The local, rotating, equivalent noninertial frame of reference would thus also increase in its acceleration with radial distance from the axis of rotation. Thus, the equivalence principle is limited to very localized regions of space with uniform gravitation and acceleration, and noninertial frames vary from one local region of space to the next.

Einstein realized that if all accelerated systems are equivalent, and the force of gravity can be equivalent to an accelerated system, then trajectories on curved surfaces absent forces are forms of acceleration in that one is changing direction even while maintaining constant speed. As a result, Einstein realized that the geometry that describes these curved surfaces has physical significance. In general relativity, Einstein assumed that the physical quantitites, processes, and laws are assumed to be independent of the coordinate system, and no single or unique reference point in the universe is assumed to exist. Also, in general relativity, gravity acts on the 4-D space of the universe (three spatial dimensions and one temporal dimension) to warp or curve the otherwise flat space.

In the parallel transport of a vector around a loop on a spherical surface, the vector is pointing in a different direction than its original direction. The change of the pointing direction of the vector is proportional to the area enclosed by the loop. If the surface were flat, the vector would point in the same original direction after one parallel transport cycle about the loop. Parallel transport of a vector about a loop requires that the vector maintain a constant angle with the tangent vector from point-to-point around the loop. Parallel transport forms the basis of covariant differentiation, in which the unit vectors of a coordinate system generally point in different directions from point-to-point in the curvilinear coordinate system. Covariant differentiation is the generalization of the directional derivative of vector calculus, which is the gradient of a scalar function.

General relativity requires the covariant derivative, which is the gradient of a vector. One can extend covariant differentiation to the gradient of a tensor or to the scalar quantities of a tensor. The covariant derivative must be expressed in a manner that makes the covariant derivative independent of the coordinate system. A vector is a list of numbers relative to a set of basis vectors, which are linearly independent vectors, which in linear combination can span or represent every vector in the vector space or coordinate system. Independent of the basis vectors, the vector is an invariant geometric object in space; that is, the vector is an arrow pointing in a particular direction in space, unchanged by the choice of coordinate system and the associated coordinate system basis vectors.

Covariant transformation of the vector is required, so the physical quantity, which is represented by a vector magnitude and direction in physical space that is unchanged by the choice of basis vectors and coordinate system. The covariant derivative, which represents a variation of a physical quantity in space due to a change in position or time, is required to change under a change (transformation) of coordinate systems in the same manner as a vector. That is, the change of the physical quantity with a change in position or time should be independent of the coordinate system. The covariant derivative must change by covariant transformation under a change of coordinate systems. There are two sets of basis vectors for each curvilinear coordinate system in general; namely, the natural basis vectors and dual basis vectors. In the spherical and cylindrical coordinate systems, the natural and dual basis vectors are different; however, in the Cartesian or rectangular system, the natural and normal basis vectors are identical.

A geodesic in Euclidean space is a straight line, which is the shortest distance between two points. On a manifold or curved surface or curved space, the shortest distance L between two points a and b is a geodesic curve, and the distance between a and b measured along the geodesic curve is determined by a metric tensor $g(\mathbf{r})$ that varies with vector position $\mathbf{r}$. A manifold describes a global space, which locally can be approximated as a Euclidean space of specific dimension termed the *dimension of the manifold.* A line or circle is a 1-D manifold, while a plane or sphere is a 2-D manifold.

The immediate vicinity around each point on a manifold is the approximately flat Euclidean space. However, on the global level, the space can be curved and have a complicated structure. The Earth appears flat beneath our feet, but on a very tall structure or mountaintop, one can see the curvature of the Earth. That is, any point on the 2-D spherical surface is contained within the small radius of a circular region (disk) that is approximately flat. The set of small flat disks corresponding to a large set of points on the spherical surface of the Earth can be joined to form the familiar flat geographical map of the Earth.

In general relativity, the manifold surfaces are assumed to be continuous, allow differentiation, and are endowed with a Riemannian metric, which allows the measurement of distances and angles. In particular, the 4-D Lorentzian manifold accounts for the three spatial dimensions and the fourth dimension of time. With a Riemannian metric, geometric entities such as angles, lengths of curves, areas, volumes, curvatures, gradients of scalar and vector fields, and divergences of vector fields and parallel transport of vectors can be described on a manifold (i.e., a 1-, 2-, 3-, or n-D surface embedded in a higher-dimensional space). The parallel transport is a method of connecting the local geometry of one point to another nearby point.

15.3 Surfaces and Gauss's Remarkable Theorem, Gaussian Curvature, and Metric Coefficients

In 1828, C. Gauss proposed and proved one of the important tenets of differential geometry. The Theorema Egregium, or "remarkable theorem" states that the *Gaussian curvature* of a *surface* can be determined entirely by measuring angles, distances, and their rates on the surface itself. One does not require knowledge of how the 2-D surface is embedded in the higher 3-D space. That is, Gaussian curvature is an intrinsic invariant of the surface in which the 2-D surface curvature does not depend on the 3-D space in which the surface is embedded. However, the curve has only extrinsic curvature κ, which depends on how the 1-D curve is embedded in the 2-D plane.

The extrinsic curvature at each point P of the 1-D curve, which lies in the 2-D (x, y) plane, is defined as the derivative $\mathrm{d}\theta/ds$ of the curve's tangent angle θ with respect to its position on the curve. If the position of point P on the curve is measured as a function of path length s and $\theta(s)$, the curve makes a fixed reference to the x-axis, then the extrinsic curvature κ is

$$\kappa = \frac{\mathrm{d}\theta}{\mathrm{d}s} \tag{15.1}$$

For each point $P(x, y)$ on the curve, the tangent line is defined by

$$\tan\theta = \frac{\mathrm{d}y}{\mathrm{d}x} \tag{15.2}$$

where

$$ds^2 = dx^2 + dy^2 \tag{15.3}$$

By the chain rule, the curvature κ of Eq. (15.1), can be expressed in terms of the first and second derivatives of curve coordinate y with respect to x, where

$$\kappa = \frac{d\theta}{ds} = \frac{dx}{ds}\frac{d\tan^{-1}(dy/dx)}{dx} = \frac{dx}{ds}\frac{d^2y/dx^2}{1+(dy/dx)^2} = \frac{d^2y/dx^2}{[1+(dy/dx)^2]^{3/2}} \tag{15.4}$$

and

$$\frac{ds}{dx} = \sqrt{1+(dy/dx)^2} \tag{15.5}$$

Now we examine the intrinsic Gaussian curvature of a 2-D surface. The radius R of an osculating circle measures the curvature C of a curve at each point P. At point P, the osculating circle and curve share the same tangent line and curvature. The tangent is the best line that approximates the curve at point P, and the osculating circle best approximates the curvature of the curve at point P. The curvature C of the curve at point P is the reciprocal of the radius R [length] of the osculating circle in which curvature C has units of [1/length], where

$$C = \frac{1}{R} \tag{15.6}$$

A straight line has zero curvature because its radius of curvature is infinity.

The Gaussian curvature K at each point P on a surface is defined as the product of curvatures C_1 and C_2 of the two principal curves that intersect at point P. The Gaussian curvature K [1/area] has units of inverse area, where

$$K = \frac{1}{R_1 R_2} \tag{15.7}$$

The use of surfaces embeded via circles to define the Gaussian curvature is extrinsic; however, the Gaussian curvature is an intrinsic property of the surface because the calculation of the Gaussian curvature does not depend on the particular embedding surface. Homogeneous and isotropic spaces can be flat, spherical, or hyperbolic as defined by the Gaussian curvature, where

$$K = \frac{1}{R_1 R_2} = 0 \quad R_1 = R_2 = \infty \leftrightarrow \text{flat space} \tag{15.8}$$

A plane ($R_1 = R_2 = \infty$) and a cylindrical surface ($R_1 = R, R_2 = \infty$) each have zero Gaussian curvature. If the radius-defining osculating circles, which define the curvature of the two principal curves at each surface point P, lie on the same side of the surface, then one has a spherical space, where

$$K = \frac{1}{R_1 R_2} > 0 \quad R_1 = R_2 = R \leftrightarrow \text{spherical space} \tag{15.9}$$

If the radius-defining osculating circles lie on opposite sides of the surface, then one has a hyperbolic space, where R_1 and R_2 have opposite signs, where

$$K = \frac{1}{R_1 R_2} < 0 \quad R_1 = R_2 = R \leftrightarrow \text{hyperbolic space} \tag{15.10}$$

One simple measurement of curvature of a surface by observers confined to the surface is summing the angles of a triangle that lies on the surface. The sum θ of the interior angles α, β, and γ of a triangle on a flat surface is 180°, exceeds 180° on a spherical surface, and is less than 180° on a hyperbolic surface, where $\theta = \alpha + \beta + \gamma$:

$$\theta - \pi = 0 \quad \text{flat or cylindrical surface} \tag{15.11}$$

$$\theta - \pi > 0 \quad \text{spherical surface} \tag{15.12}$$

$$\theta - \pi < 0 \quad \text{hyperbolic surface} \tag{15.13}$$

and

$$\theta = \alpha + \beta + \gamma \tag{15.14}$$

The Girard (Harriot) theorem for a triangle that lies on a spherical surface of radius R states the triangle area A is the sum θ of the interior angles of a triangle minus 180°, where

$$A = (\theta - \pi)R^2 = \frac{1}{K^2}(\theta - \pi) \tag{15.15}$$

Spherical trigonometry has practical importance in aeronautics and international flight-path planning. The Lambert theorem for a triangle, which lies on a hyperbolic surface with Gaussian curvature K, states the triangle area A is 180° minus the sum θ of the interior angles of a triangle, where

$$A = -\frac{1}{K^2}(\pi - \theta) \tag{15.16}$$

The Girard and Lambert theorems are special cases of the Gauss–Bonnet theorem, which relates the geometry of the surface, expressed by the surface curvature, to the topology of the surface, expressed by the Euler characteristic.

In topology, surfaces can be categorized by the Euler characteristic χ, which is originally applied by Euler to polyhedra, but can be extended to any surface by the polyhedralization of the surface; that is, the surface is comprised of polyhedra with a sufficient number of faces to approximate the surface. The Euler characteristic χ is calculated from the number of edges E, faces F, and vertices V of the surface of the polyhedra solid, where

$$\chi = V - E + F \tag{15.17}$$

Any convex polyhedra has a Euler characteristic $\chi = 2$, and for some common geometric entities, the Euler characteristic is

$$\chi = 0 \quad \text{torus, circle, Möbius strip} \tag{15.18}$$

$$\chi = 1 \quad \text{line segment and disk} \tag{15.19}$$

$$\chi = 2 \quad \text{sphere} \tag{15.20}$$

Riemann extended Gauss's remarkable theorem to higher-dimensional spaces or manifolds such that the distances, angles, and curvature of an $n-1$ dimensional surface were intrinisc to the surface, independent of the n-D or higher-order embedding spaces. Einstein used the Riemannian theory of manifolds to develop general relativity, in which the curvature of space is determined by the gravitational fields of masses embedded in the space. Having a surface completely defined by its own parameters, without reference to the embedding space, was the "remarkable" aspect of Gauss's remarkable theorem for 2-D surfaces embedded in 3-D spaces, which was then extended by Riemann to n-D surfaces embedded in $n+1$ and higher-dimensional spaces.

A distance interval ds on a flat 2-D surface or Euclidean surface with coordinates x and y is

$$\mathrm{d}s^2 = \mathrm{d}x^2 + \mathrm{d}y^2 \tag{15.21}$$

On a curved surface, 2-D surface, or non-Euclidean surface, the distance interval ds becomes

$$\mathrm{d}s^2 = F\,\mathrm{d}x^2 + G\,\mathrm{d}x\,\mathrm{d}y + H\,\mathrm{d}y^2 \tag{15.22}$$

The metric coefficients $F(x,y)$, $G(x,y)$, and $H(x,y)$ generally depend on the surface position (x,y). The metric coefficients have 3 independent components out of 4 total components in two dimensions, 6 independent components out of 9 components in three dimensions, and 9 independent components out of 16 components in four dimensions. For example, in two dimensions, symmetry gives the same coefficient G for the equivalent dx dy and dy dx terms. For more details on differential geometry, see Guggenheimer (1977), and for vectors and tensors, see Wrede (1972).

15.4 Curvilinear Coordinate Systems

A vector is a list or tuple of numbers that depends on a coordinate system or reference frame. If the coordinates are transformed by rotation or stretching, then the vector components also transform so as to maintain the vector magnitude and direction. The vector is covariant or contravariant, depending on the relation between the vector components and the coordinates. *Contravariant vectors* are regular vectors with units of distance (such as position, velocity, and acceleration). *Covariant vectors* have units of inverse distance as in the gradient, where the gradient of the electric and gravitational potential yields covariant electric field and gravitational field vectors.

In an ordinary rectangular or Cartesian coordinate system, the coordinate lines are perpendicular to each other at every point, thus forming a rectangular set of grid lines. At each point in the Cartesian coordinate system, the basis vectors point in the same direction. That is, at each point, the coordinate lines are mutually perpendicular and point in the same direction. However, in curvilinear coordinate systems, the angles between the axes can change from point-to-point. In curvilinear coordinate systems, the coordinate lines have curvature (i.e., they are coordinate curves rather than coordinate lines), and the angles between the coordinate curves are not necessarily mutually perpendicular, and also the angles between the coordinate curves can change from point-to-point. That is, in the curvilinear coordinate systems, the direction of the basis vectors changes from point-to-point.

The familiar rectangular, cylindrical, and spherical coordinate systems are simple curvilinear coordinate systems in which the coordinate curves are perpendicular at each point. However, the cylindrical and spherical coordinate systems have coordinate curves, and the rectangular or Cartesian coordinate system has only coordinate lines. Most important, however, is the basis vectors for the spherical and cylindrical coordinate systems do not point in the same physical direction from point to point. In the Cartesian coordinate system, at each point, the unit vectors in the x-, y-, and z-directions remain aligned in the same x, y, and z physical directions at all other points.

In the cylindrical coordinate system, the azimuthal unit vector points in the azimuthal direction at each point in a circle about the z-axis. However, the azimuthal direction points in a physically different direction at each point on the circle. That is, the azimuthal unit vector points in a physically different direction at each point along a circle about the z-axis. Also, the unit radial vector points in a different physical direction at each point in a circle about the z-axis. In contrast, the axial unit vector has the same direction from point-to-point.

In the spherical coordinate system, at each point, the radial, polar, and azimuthal vectors each point in different physical directions from point-to-point. In parallel vector transport, the vector maintains the same angle with respect to the tangent vector along a curve on a surface as the vector is parallel-transported along the curve. Starting at a point A on a loop, which lies on the surface of a sphere, after the parallel transport of the vector around the loop, the vector returns to point A, but the vector now points in a different physical direction. In contrast, a vector parallel-transported around a loop, which lies on a flat surface, returns to starting point A physically pointing in the same direction.

Physical quantities and the flow of physical quantities that are not confined to curved surfaces, but rather are allowed to propagate or be defined at points in a rectangular volume or flat surface, can be depicted mathematically by simple Cartesian coordinates. In general, the parallel transport of vector **A** around closed curve C, where constant angle θ is maintained between vector **A**, and the tangent vector $\mathbf{T}(p)$ at each curve point p, as vector **A** is parallel-transported around closed curve C, results in vector **A** pointing in a different direction, after one round trip. As a result of parallel vector transport, if the flow of physical quantities occurs on curved surface, then the description of the physical processes on the curved surface by surface tangent vectors (planes) and normal vectors (planes) differs, unlike a planar surface. On a planar surface, the tangent vectors (planes) overlap with the normal vectors (planes). The tangent vectors (planes) are obtained by taking the derivative of surface curves and the normal vectors (planes) are obtained by taking the gradient vector of the set of coordinate points that lie on the surface. On a planar surface, the tangent and normal vectors (planes) remain fixed in direction, measured relative to three fixed reference coordinate axes, as one moves from one point to the next on the planar surface. However, in general, on a curved surface, the tangent and normal vectors vary in direction, measured relative to three fixed reference coordinate axes, as one moves from one point to the next on the curved surface. As a result, the analysis of physical processes on curved surfaces, uses curvilinear coordinate systems that have two different basis vector sets (natural and dual), and employs covariant differentiation, which is based on parallel vector transport and has two sets of vector and tensor derivatives (covariant and contravariant). This is the case not only in general relativity, but also in fluid mechanics and meteorology, surface physics, chemical, mechanical, civil, and aeronautical engineering, and other branches of science and engineering that involved tensors and vectors confined to curved surfaces and curved volumes.

In this section, we examine coordinate systems in particular curvilinear coordinates and transformations of differential and integral expressions of length, area, and volume from one coordinate system to the next. We also examine the use of generalized coordinates to express physical laws and quantizes that are independent of the chosen coordinate system. We start by examining a most versatile and amazing equation; namely,

$$\nabla^2\phi + k^2\phi = 0 \tag{15.23}$$

Here, k^2 is a constant, where

$$k^2 = 0 \quad \text{Laplace's equation} \tag{15.24}$$

$$k^2 > 0 \quad \text{Helmholtz equation} \tag{15.25}$$

$$k^2 < 0 \quad \text{Diffusion equation} \tag{15.26}$$

$$k^2 < \text{const} \times \text{K.E.} \quad \text{Schrodinger equation} \tag{15.27}$$

These physical laws should maintain their form in other coordinate systems and should be obeyed by observers in all the frames in which the frames are moving at constant velocity relative to one another, or all the frames are moving with acceleration relative to each other. Special relativity handles the case of inertial frames that move at constant velocity relative to one another. General relativity handles the case of frames that have acceleration relative to other frames. To characterize the frames that are accelerating relative to one another via general relativity, one must first examine the transformation between coordinate systems, and also examine how length, surface area, and volume are characterized and length, area, and volume are transformed from one coordinate system to the next. Then, one must examine how these transformations are affected by velocity and acceleration.

There are 11 coordinate systems in which the above equation is separable, all of which are particular cases of the confocal ellipsoidal system; see Arfken (1970) and Eisenhart (1934). In rectangular coordinates (x, y, z), there are three mutually perpendicular planes, which are described by the simple equations given constants (c_1, c_2, c_3), where

$$x = c_1 \quad yz\text{-plane} \tag{15.28}$$

$$y = c_2 \quad xz\text{-plane} \tag{15.29}$$

$$z = c_3 \quad xy\text{-plane} \tag{15.30}$$

Two perpendicular planes intersect to form a single straight line, and the three pairs of intersecting planes produce three mutually perpendicular straight line axes. The three mutually perpendicular planes intersect to form a single point.

Next, consider curvilinear coordinate systems in which one can form three mutually perpendicular, 2-D surfaces. In Cartesian coordinates, the three unique, 2-D surfaces are planes with constant x, y, and z. In spherical coordinates, the three unique, 2-D surfaces are surfaces with constant r, θ, or ϕ. The intersection of each pair of unique 2-D surfaces produces a 1-D curve along each of the three spatial coordinates. And the three unique and mutually perpendicular surfaces intersect at a single point. The three unique pairings of these unique, 2-D surfaces leads to three mutually perpendicular curves. These three mutually perpendicular surfaces and the three mutually perpendicular curves intersect at the same single point.

For example, cylindrical and spherical coordinates are curvilinear coordinate systems whose three spatial directions are mutually perpendicular curves (and lines). In the case of a spherical coordinate system (r, θ, ϕ), given constants (c_1, c_2, c_3), the three mutually perpendicular surfaces are given by

$$r = c_1 \quad \text{spherical surface} \tag{15.31}$$

$$\theta = c_2 \quad \text{conical surface} \tag{15.32}$$

$$\phi = c_3 \quad \text{planar surface} \tag{15.33}$$

In the rectangular coordinate system, the lines that lie along curves or parallel to the respective differential coordinate intervals $(\mathrm{d}x, \mathrm{d}y, \mathrm{d}z)$ are mutually perpendicular. Similarly, curves along or parallel to the cylindrical differential coordinate interval $(\mathrm{d}r, r\,\mathrm{d}\phi, \mathrm{d}z)$ are mutually parallel, and curves along or parallel to the spherical differential coordinate interval $(\mathrm{d}r, r\,\mathrm{d}\theta, r \sin\theta\,\mathrm{d}\phi)$ are mutually parallel. The rectangular, cylindrical, and spherical spatial coordinate systems and differential coordinate intervals are

$$x, y, z\ \mathrm{d}x, \mathrm{d}y, \mathrm{d}z \quad \text{rectangular coordinates} \tag{15.34}$$

$$\rho, \phi, z\rho, \mathrm{d}\phi, \mathrm{d}z \quad \text{cylindrical coordinates} \tag{15.35}$$

$$r, r\theta, r\phi \sin\theta \quad \mathrm{d}r, r\,\mathrm{d}\theta, r\,\mathrm{d}\phi \sin\theta \quad \text{spherical coordinates} \tag{15.36}$$

Now consider a curvilinear coordinate system (q_1, q_2, q_3), in which the three spatial axes are not perpendicular. With constants (c_1, c_2, c_3), one can form three surfaces, which are not mutually perpendicular, where

$$q_1 = c_1 \quad \text{spherical surface} \tag{15.37}$$

$$q_2 = c_2 \quad \text{conical surface} \tag{15.38}$$

$$q_3 = c_3 \quad \text{planar surface} \tag{15.39}$$

Next, let us express these nonperpendicular, curvilinear coordinates (q_1, q_2, q_3) in the very familiar and comfortable rectangular coordinates (x, y, z), where

$$q_1 = q_1(x, y, z) = c_1 \tag{15.40}$$

$$q_2 = q_2(x, y, z) = c_2 \tag{15.41}$$

$$q_3 = q_3(x, y, z) = c_3 \tag{15.42}$$

For $i = 1, 2, 3$, the ith surface $q_i = c_i$ has its unit, normal vector $\mathbf{n}_i$ pointing in the direction of increasing q_i. The cylindrical coordinates (ρ, ϕ, z) are obtained from the rectangular coordinates (x, y, z) via

$$q_1 = \rho(x, y, z) = \sqrt{x^2 + y^2} \tag{15.43}$$

$$q_2 = \phi(x, y, z) = \tan^{-1}\frac{y}{x} \tag{15.44}$$

$$q_3 = z(x, y, z) = z \tag{15.45}$$

The spherical coordinates (r, θ, ϕ) are obtained from the rectangular coordinates (x, y, z) via

$$q_1 = r(x, y, z) = \sqrt{x^2 + y^2 + z^2} \tag{15.46}$$

$$q_2 = \theta(x, y, z) = \tan^{-1}\left(\frac{\sqrt{x^2 + y^2}}{z}\right) \tag{15.47}$$

$$q_3 = z(x, y, z) = \tan^{-1}\left(\frac{y}{x}\right) \tag{15.48}$$

Each point P on the ith surface with its associated normal vector $\mathbf{n}_i$ is formed by the intersection of two coordinate curves $C_j(q_j)$ and $C_k(q_k)$ for the coordinates q_j and q_k for $j = 1, 2, 3$ and $k = 1, 2, 3$. The coordinate curves are not perpendicular at point P; however, $C_j(q_j)$ and $C_k(q_k)$ would be perpendicular in the case of curvilinear coordinate systems with three mutually perpendicular spatial axes such as the rectangular, cylindrical, and spherical coordinate systems.

The coordinate curves $C_j(q_j)$ and $C_k(q_k)$ can be perpendicular or not perpendicular. In either case, at each point P, the intersecting pair of coordinate curves $C_j(q_j)$ and $C_k(q_k)$ has unit, tangent vectors $\mathbf{t}_j$ and $\mathbf{t}_k$. The pair of coordinate curve tangent vectors $\mathbf{t}_j$ and $\mathbf{t}_k$ at each point P on the surface describes a tangent plane to the surface at point P. The surface at each point P can thus be characterized by normal vector $\mathbf{n}_i$ or the tangent plane at P, or the surface can be characterized by the pair of orthogonal tangent vectors $\mathbf{t}_j$ and $\mathbf{t}_k$ of the two intersecting orthogonal coordinate curves at point P.

15.5 Metric Tensor and Invariant Distance in Coordinate System Transformations

The spatial distance between two points should be a constant that is independent of the coordinate system in nonrelativistic systems, and the space–time distance between two points should be a constant in relativistic systems. Also, scalar quantizes should be invariant of coordinate systems in nonrelativistic and relativistic systems. In the nonrelativistic systems, the spatial distance between two points is invariant. Given differential distance ds, the square of the differential distance ds^2 between two points in the rectangular and curvilinear coordinate systems is given by

$$\mathrm{d}s^2 = \mathrm{d}x^2 + \mathrm{d}y^2 + \mathrm{d}z^2 = \sum_{i=1}^{3}\sum_{j=1}^{3} g_{ij}^2 \,\mathrm{d}q_i \,\mathrm{d}q_j \tag{15.49}$$

The expression for ds^2 in terms of the curvilinear coordinates (q_1, q_2, q_3) has six cross terms, comprised of the symmetric pair $g_{12}^2 q_1 q_2$ and $g_{21}^2 q_2 q_1$, as well as three squared, diagonal terms: $g_{11}^2 q_1^2$, $g_{22}^2 q_2^2$, and $g_{33}^2 q_3^2$. The coefficients g_{ij}, which are called the metric coefficients of the metric tensor g, characterize the coordinate system (q_1, q_2, q_3). The metric coefficients g_{ij} form a 3×3 metric matrix for Galilean, nonrelativistic systems.

One can determine the metric g_{ij} for a curvilinear coordinate system (q_1, q_2, q_3) as follows. The rectangular coordinates (x, y, z) can be expressed in terms of curvilinear coordinates (q_1, q_2, q_3), where

$$x = x(q_1, q_2, q_3) \tag{15.50}$$

$$y = y(q_1, q_2, q_3) \tag{15.51}$$

$$z = z(q_1, q_2, q_3) \tag{15.52}$$

In cylindrical coordinates (ρ, ϕ, z), the rectangular coordinates (x, y, z) are

$$x = x(\rho, \phi, z) = \rho \cos \phi \tag{15.53}$$

$$y = y(\rho, \phi, z) = \rho \sin \phi \tag{15.54}$$

$$z = z(\rho, \phi, z) = z \tag{15.55}$$

In spherical coordinates (r, θ, ϕ), the rectangular coordinates (x, y, z) are

$$x = x(r, \theta, \phi) = r \sin \theta \cos \phi \tag{15.56}$$

$$y = y(r, \theta, \phi) = r \sin \theta \sin \phi \tag{15.57}$$

$$z = z(r, \theta, \phi) = r \cos \phi \tag{15.58}$$

In Eqs. (15.56–15.58), one performs the partial differentiation of the three generalized Cartesian equations, which are expressed in terms of curvilinear coordinates (q_1, q_2, q_3); namely $x = x(q_1, q_2, q_3)$, $y = y(q_1, q_2, q_3)$, and $z = z(q_1, q_2, q_3)$ to obtain

$$\mathrm{d}x = \frac{\partial x}{\partial q_1}\mathrm{d}q_1 + \frac{\partial x}{\partial q_2}\mathrm{d}q_2 + \frac{\partial x}{\partial q_3}\mathrm{d}q_3 \tag{15.59}$$

$$\mathrm{d}y = \frac{\partial y}{\partial q_1}\mathrm{d}q_1 + \frac{\partial y}{\partial q_2}\mathrm{d}q_2 + \frac{\partial y}{\partial q_3}\mathrm{d}q_3 \tag{15.60}$$

$$\mathrm{d}z = \frac{\partial z}{\partial q_1}\mathrm{d}q_1 + \frac{\partial z}{\partial q_2}\mathrm{d}q_2 + \frac{\partial z}{\partial q_3}\mathrm{d}q_3 \tag{15.61}$$

One now substitutes Eqs. (15.59−15.61) for the differentials dx, dy, and dz in Eq. (15.49) for squared differential distance ds^2 to obtain

$$
\begin{aligned}
\mathrm{d}s^2 &= \sum_{i=1}^{3}\sum_{j=1}^{3} g_{ij}^2\,\mathrm{d}q_i\,\mathrm{d}q_j \\
&= \left(\frac{\partial x}{\partial q_1}\mathrm{d}q_1 + \frac{\partial x}{\partial q_2}\mathrm{d}q_2 + \frac{\partial x}{\partial q_3}\mathrm{d}q_3\right)^2 + \left(\frac{\partial y}{\partial q_1}\mathrm{d}q_1 + \frac{\partial y}{\partial q_2}\mathrm{d}q_2 + \frac{\partial y}{\partial q_3}\mathrm{d}q_3\right)^2 \\
&\quad + \left(\frac{\partial z}{\partial q_1}\mathrm{d}q_1 + \frac{\partial z}{\partial q_2}\mathrm{d}q_2 + \frac{\partial z}{\partial q_3}\mathrm{d}q_3\right)^2
\end{aligned} \tag{15.62}
$$

Consolidation of terms in Eq. (15.62), gives the expression of the metric coefficients g_{ij} for a curvilinear coordinate system (q_1, q_2, q_3) in terms of derivates of Cartesian variables x, y, and z with respect to q_1, q_2, and q_3, where

$$
g_{ij}^2 = \frac{\partial x}{\partial q_i}\frac{\partial x}{\partial q_j} + \frac{\partial y}{\partial q_i}\frac{\partial y}{\partial q_j} + \frac{\partial z}{\partial q_i}\frac{\partial z}{\partial q_j} \tag{15.63}
$$

with

$$
\mathrm{d}s^2 = \mathrm{d}x^2 + \mathrm{d}y^2 + \mathrm{d}z^2 = \sum_{i=1}^{3}\sum_{j=1}^{3}\left(\frac{\partial x}{\partial q_i}\frac{\partial x}{\partial q_j} + \frac{\partial y}{\partial q_i}\frac{\partial y}{\partial q_j} + \frac{\partial z}{\partial q_i}\frac{\partial z}{\partial q_j}\right)\mathrm{d}q_i\,\mathrm{d}q_j \tag{15.64}
$$

where

$$
i,j = 1,2,3 \Leftrightarrow x, y, z \quad \text{rectangular coordinates} \tag{15.65}
$$

$$
i,j = 1,2,3 \Leftrightarrow r, \theta, z \quad \text{cylindrical coordinates} \tag{15.66}
$$

$$
i,j = 1,2,3 \Leftrightarrow r, \theta, \phi \quad \text{spherical coordinates} \tag{15.67}
$$

The differential length ds is then

$$
\mathrm{d}s = \sqrt{\mathrm{d}x^2 + \mathrm{d}y^2 + \mathrm{d}z^2} = \sqrt{\sum_{i=1}^{3}\sum_{j=1}^{3}\left(\frac{\partial x}{\partial q_i}\frac{\partial x}{\partial q_j} + \frac{\partial y}{\partial q_i}\frac{\partial y}{\partial q_j} + \frac{\partial z}{\partial q_i}\frac{\partial z}{\partial q_j}\right)\mathrm{d}q_i\,\mathrm{d}q_j} \tag{15.68}
$$

or

$$
\mathrm{d}s = \sqrt{\mathrm{d}x^2 + \mathrm{d}y^2 + \mathrm{d}z^2} = \sqrt{\sum_{i=1}^{3}\sum_{j=1}^{3} g_{ij}^2\,\mathrm{d}q_i\,\mathrm{d}q_j} \tag{15.69}
$$

15.6 Metric Tensor and Invariant Area and Volume in Coordinate System Transformations

Let us now consider the transformations between the curvilinear coordinate systems (x, y, z) and (q_1, q_2, q_3) for area and volume. The differential length ds_k in the kth direction in the (x, y, z) coordinate system is invariant of the coordinate system and is expressed in the curvilinear coordinate system (q_1, q_2, q_3) by

$$ds_k = \sqrt{dx_k^2 + dy_k^2 + dz_k^2} = \sqrt{\sum_{i=1}^{3}\sum_{j=1}^{3} g_{ijk}^2\, dq_i\, dq_j} \tag{15.70}$$

The differential area is also invariant of the choice of coordinate system. With $k, l = (1, 2, 3)$, the relation between differential area $dx_k\, dx_l$ in the (x, y, z) coordinate system, where $(1, 2, 3) = (x, y, z)$, and differential area $dq_k\, dq_l$ in the curvilinear coordinate system (q_1, q_2, q_3), is then the product of (i) the differential length in the kth direction along the direction q_k and (ii) the differential length in the lth direction along the direction q_l, where

$$dx_k\, dx_l = ds_k\, ds_l = \sqrt{\sum_{i=1}^{3}\sum_{j=1}^{3} g_{ijk}^2\, dq_i\, dq_j}\sqrt{\sum_{i=1}^{3}\sum_{j=1}^{3} g_{ijl}^2\, dq_i\, dq_j} \tag{15.71}$$

The differential volume dV is independent of the coordinate system. In a similar fashion to the treatment of area, the relation of volume in the two curvilinear coordinate systems is

$$\begin{aligned} dx_1\, dx_2\, dx_3 &= ds_1\, ds_2\, ds_3 \\ &= \sqrt{\sum_{i=1}^{3}\sum_{j=1}^{3} g_{ij1}^2\, dq_i\, dq_j}\sqrt{\sum_{i=1}^{3}\sum_{j=1}^{3} g_{ij2}^2\, dq_i\, dq_j}\sqrt{\sum_{i=1}^{3}\sum_{j=1}^{3} g_{ij3}^2\, dq_i\, dq_j} \end{aligned} \tag{15.72}$$

If the curvilinear coordinate system has perpendicular coordinate axes or curves, then the metric coefficient cross terms are all zero, where the metric coefficient is zero, and when $i \neq j$, where

$$g_{ij} = 0 \quad \text{if} \quad i \neq j \quad \text{perpendicular coordinate axes} \tag{15.73}$$

and

$$ds^2 = dx^2 + dy^2 + dz^2 = \sum_{i=1}^{3} g_{ii}^2\, dq_i^2 = \sum_{i=1}^{3}\left[\left(\frac{\partial x}{\partial q_i}\right)^2 + \left(\frac{\partial y}{\partial q_i}\right)^2 + \left(\frac{\partial z}{\partial q_i}\right)^2\right] dq_i^2 \tag{15.74}$$

For a curvilinear coordinate system with perpendicular, coordinate axes or curves, the spatial length $\mathrm{d}s_i$ of a line (curve) segment in each of the q_i directions is

$$\mathrm{d}s_i = g_{ii}\,\mathrm{d}q_i \tag{15.75}$$

That is, the differential line or curve segment $\mathrm{d}s_i$ has units of length, where

$$\mathrm{d}s_i = \mathrm{d}q_i\sqrt{\left(\frac{\partial x}{\partial q_i}\right)^2 + \left(\frac{\partial y}{\partial q_i}\right)^2 + \left(\frac{\partial z}{\partial q_i}\right)^2} \quad i = 1, 2, 3 \tag{15.76}$$

For perpendicular coordinate curves and axes, the metric coefficients or scale factors g_{ii}, which depend on the three curvilinear coordinates (q_1, q_2, q_3), can have units of length that are obtained from the first derivatives of the coordinates (x, y, z) of the (x, y, z) coordinate system with respect to the curvilinear coordinates (q_1, q_2, q_3) in the (q_1, q_2, q_3) coordinate system. The differential area $\mathrm{d}A_{ij}$ in the curvilinear coordinate system (q_1, q_2, q_3) is the product of perpendicular differential line (curve) segments $\mathrm{d}s_i$ and $\mathrm{d}s_j$, where

$$\begin{aligned}
\mathrm{d}A_{ij} &= \mathrm{d}s_i\,\mathrm{d}s_j \\
&= g_{ii}\,g_{jj}\,\mathrm{d}q_i\,\mathrm{d}q_j \\
&= \mathrm{d}q_i\,\mathrm{d}q_j\sqrt{\left[\left(\frac{\partial x}{\partial q_i}\right)^2 + \left(\frac{\partial y}{\partial q_i}\right)^2 + \left(\frac{\partial z}{\partial q_i}\right)^2\right]\left[\left(\frac{\partial x}{\partial q_j}\right)^2 + \left(\frac{\partial y}{\partial q_j}\right)^2 + \left(\frac{\partial z}{\partial q_j}\right)^2\right]} \\
&\quad i,j = 1, 2, 3
\end{aligned} \tag{15.77}$$

The differential volume $\mathrm{d}V$ in the curvilinear coordinate system (q_1, q_2, q_3) is the product of orthogonal and perpendicular differential line or curve segments $\mathrm{d}s_i$, $\mathrm{d}s_j$, and $\mathrm{d}s_k$, where

$$\mathrm{d}V = \mathrm{d}s_i\,\mathrm{d}s_j\,\mathrm{d}s_k = g_{ii}g_{jj}g_{kk}\,\mathrm{d}q_i\,\mathrm{d}q_j\,\mathrm{d}q_k \quad i,j,k = 1, 2, 3 \tag{15.78}$$

15.7 Curvilinear Coordinate Systems and the Jacobian Determinant

Now we return to the case of a general curvilinear coordinate system (q_1, q_2, q_3), in which the axes are generally not perpendicular. One can relate differential length, area, and volume in curvilinear coordinate system (q_1, q_2, q_3) to that of the curvilinear coordinate system (x, y, z) by the Jacobian determinant as follows. A single differential length $\mathrm{d}x$ can be related to curvilinear coordinate $\mathrm{d}q_1$ by

$$\mathrm{d}x = \frac{\mathrm{d}x}{\mathrm{d}q_1}\mathrm{d}q_1 \tag{15.79}$$

The Jacobian is the derivative dx/dq_1. The differential area $dx\,dy$ is related to curvilinear coordinate differential area $dq_1\,dq_2$ by result of the Jacobian determinant

$$dx\,dy = J dq_1\,dq_2 = \begin{vmatrix} \dfrac{\partial x}{\partial q_1} & \dfrac{\partial x}{\partial q_2} \\ \dfrac{\partial y}{\partial q_1} & \dfrac{\partial y}{\partial q_2} \end{vmatrix} dq_1\,dq_2 \tag{15.80}$$

The Jacobian matrix J is given by the determinant

$$J = \begin{vmatrix} \dfrac{\partial x}{\partial q_1} & \dfrac{\partial x}{\partial q_2} \\ \dfrac{\partial y}{\partial q_1} & \dfrac{\partial y}{\partial q_2} \end{vmatrix} \tag{15.81}$$

Evaluation of the Jacobian determinant gives the relation between the areas $dx\,dy$ and $dq_1\,dq_2$, where

$$dx\,dy = \left(\frac{\partial x}{\partial q_1}\frac{\partial y}{\partial q_2} - \frac{\partial y}{\partial q_1}\frac{\partial x}{\partial q_2}\right) dq_1\,dq_2 \tag{15.82}$$

The use of the Jacobian determinant relating the area $dx\,dy$ to area $dq_1\,dq_2$ is derived as follows. The differential area $dx\,dy$ is the result of the cross product of the differential changes in vectors $\mathbf{dx}$ and $\mathbf{dy}$ in the (x, y) coordinate system expressed in terms of changes in vectors $\mathbf{dq}_1$ and $\mathbf{dq}_2$ of the (q_1, q_2) coordinate system. Consider the position vector $\mathbf{r}$ with x- and y-components $x(q_1, q_2)$ and $y(q_1, q_2)$ expressed in terms of coordinates (q_1, q_2), where $\hat{\mathbf{x}}$ and $\hat{\mathbf{y}}$ are unit vectors:

$$\mathbf{r} = x(q_1, q_2)\hat{\mathbf{x}} + y(q_1, q_2)\hat{\mathbf{y}} \tag{15.83}$$

A variation or length differential dq_1 corresponds to vector length differential $d\mathbf{r}$:

$$\frac{d\mathbf{r}}{dq_1} = \frac{\partial x}{\partial q_1}\hat{\mathbf{x}} + \frac{\partial y}{\partial q_1}\hat{\mathbf{y}} \tag{15.84}$$

The differential length $(d\mathbf{r})_{dq_1}$ corresponds to differential length dq_1:

$$(d\mathbf{r})_{dq_1} = \frac{\partial x}{\partial q_1} dq_1\hat{\mathbf{x}} + \frac{\partial y}{\partial q_1} dq_1\hat{\mathbf{y}} \tag{15.85}$$

Similarly, the differential dq_2 corresponds to differential $d\mathbf{r}$:

$$\frac{d\mathbf{r}}{dq_2} = \frac{\partial x}{\partial q_2}\hat{\mathbf{x}} + \frac{\partial y}{\partial q_2}\hat{\mathbf{y}} \tag{15.86}$$

The differential length $(\mathbf{dr})_{dq_2}$ corresponds to differential length dq_2:

$$(\mathbf{dr})_{dq_2} = \frac{\partial x}{\partial q_2} dq_2 \hat{\mathbf{x}} + \frac{\partial y}{\partial q_2} dq_2 \hat{\mathbf{y}} \tag{15.87}$$

The differential vectors $(\mathbf{dr})_{dq_1}$ and $(\mathbf{dr})_{dq_2}$ are perpendicular to and equal differential lengths dx and dy. And differential lengths dx and dy correspond to differential lengths dq_1 and dq_2. The parallelogram formed by the cross product of the differential vectors $(\mathbf{dr})_{dq_1}$ and $(\mathbf{dr})_{dq_2}$ defines the area $(\mathbf{dr})_{dq_1}(\mathbf{dr})_{dq_2}$, where

$$(\mathbf{dr})_{dq_1} \times (\mathbf{dr})_{dq_2} = dx\, dy \hat{\mathbf{z}} = \left(\frac{\partial x}{\partial q_1} dq_1 \hat{\mathbf{x}} + \frac{\partial y}{\partial q_1} dq_1 \hat{\mathbf{y}} \right) \times \left(\frac{\partial x}{\partial q_2} dq_2 \hat{\mathbf{x}} + \frac{\partial y}{\partial q_2} dq_2 \hat{\mathbf{y}} \right) \tag{15.88}$$

Evaluation of the cross product yields the

$$(\mathbf{dr})_{dq_1} \times (\mathbf{dr})_{dq_2} = dx\, dy \hat{\mathbf{z}} = \left(\frac{\partial x}{\partial q_1} \frac{\partial y}{\partial q_2} - \frac{\partial x}{\partial q_2} \frac{\partial y}{\partial q_1} \right) dq_1\, dq_2\, \hat{\mathbf{z}} \tag{15.89}$$

In rectangular coordinates (x, y), the unit vector $\hat{\mathbf{z}}$ is perpendicular to the differential and oriented area $dx\, dy\hat{\mathbf{z}}$, in which the orientation of area $dx\, dy$ is defined by the perpendicular or normal vector $\hat{\mathbf{z}}$. Since normal vector $\hat{\mathbf{z}}$ is common to both sides of Eq. (15.89), one obtains the differential area $dx\, dy$ as the product of the differential area $dq_1\, dq_2$ in the curvilinear coordinate system (q_1, q_2), multiplied by the evaluated determinant of the Jacobian:

$$dx\, dy = \left(\frac{\partial x}{\partial q_1} \frac{\partial y}{\partial q_2} - \frac{\partial x}{\partial q_2} \frac{\partial y}{\partial q_1} \right) dq_1\, dq_2 \tag{15.90}$$

One can similarly relate differential area $dy\, dz$ to differential area $dq_2\, dq_3$ and relate differential area $dx\, dz$ to curvilinear differential area $dq_1\, dq_3$. Thus, one can obtain the general result for relating differential area $dx_i\, dx_j$ to differential area $dq_i\, dq_j$, where $i, j = 1, 2, 3$, via the Jacobian determinant for the general case:

$$dx_i\, dx_j = J dq_i\, dq_j = \begin{vmatrix} \dfrac{\partial x_i}{\partial q_i} & \dfrac{\partial x_i}{\partial q_j} \\ \dfrac{\partial x_j}{\partial q_i} & \dfrac{\partial x_j}{\partial q_j} \end{vmatrix} dq_i\, dq_j \tag{15.91}$$

The Cartesian coordinates (x, y, z) are denoted by (x_1, x_2, x_3), where

$$(x_1, x_2, x_3) = (x, y, z) \tag{15.92}$$

The Jacobian matrix J for the general case is given by the determinant

$$J = \begin{vmatrix} \dfrac{\partial x_i}{\partial q_i} & \dfrac{\partial x_i}{\partial q_j} \\ \dfrac{\partial x_j}{\partial q_i} & \dfrac{\partial x_j}{\partial q_j} \end{vmatrix} \quad i,j = 1,2,3 \tag{15.93}$$

For example, for $i = 2$ and $j = 3$,

$$\mathrm{d}x_2\ \mathrm{d}x_3 = \mathrm{d}y\ \mathrm{d}z = J\ \mathrm{d}q_2\ \mathrm{d}q_3 = \begin{vmatrix} \dfrac{\partial y}{\partial q_2} & \dfrac{\partial y}{\partial q_3} \\ \dfrac{\partial z}{\partial q_2} & \dfrac{\partial z}{\partial q_3} \end{vmatrix} \mathrm{d}q_2\ \mathrm{d}q_3 \tag{15.94}$$

This Jacobian determinant result extends beyond the comparison of the curvilinear coordinate system (q_1, q_2, q_3) and rectangular coordinates (x, y, z), and applies to any pair of orthogonal or nonorthogonal curvilinear coordinate systems (x_1, x_2, x_3) and (q_1, q_2, q_3).

Next, we consider the transformation between differential volumes in rectangular (x, y, z) and curvilinear (q_1, q_2, q_3) coordinate systems in which the curves or lines along or parallel to the respective q_1, q_2, and q_3 axes are not necessarily perpendicular. Again, we shall find that the Jacobian determinant describes the transformation. One starts with differential rectangular coordinate lengths $(\mathbf{dr})_{\mathrm{d}q_1}$, $(\mathbf{dr})_{\mathrm{d}q_2}$, and $(\mathbf{dr})_{\mathrm{d}q_3}$ along x, y, and z that correspond to differential curvilinear lengths $\mathrm{d}q_1$, $\mathrm{d}q_2$, and $\mathrm{d}q_3$, which are given by

$$(\mathbf{dr})_{\mathrm{d}q_1} = \frac{\partial x}{\partial q_1}\mathrm{d}q_1\ \hat{\mathbf{x}} + \frac{\partial y}{\partial q_1}\mathrm{d}q_1\ \hat{\mathbf{y}} + \frac{\partial z}{\partial q_1}\mathrm{d}q_1\ \hat{\mathbf{z}} \tag{15.95}$$

$$(\mathbf{dr})_{\mathrm{d}q_2} = \frac{\partial x}{\partial q_2}\mathrm{d}q_2\ \hat{\mathbf{x}} + \frac{\partial y}{\partial q_2}\mathrm{d}q_2\ \hat{\mathbf{y}} + \frac{\partial z}{\partial q_2}\mathrm{d}q_1\ \hat{\mathbf{z}} \tag{15.96}$$

and

$$(\mathbf{dr})_{\mathrm{d}q_3} = \frac{\partial x}{\partial q_3}\mathrm{d}q_3\ \hat{\mathbf{x}} + \frac{\partial y}{\partial q_3}\mathrm{d}q_3\ \hat{\mathbf{y}} + \frac{\partial z}{\partial q_3}\mathrm{d}q_3\ \hat{\mathbf{z}} \tag{15.97}$$

The triple scalar product of three vectors **A**, **B**, and **C** forms a parallel-piped scalar volume. The base vectors **B** and **C** are the two sides that form the base area, and the base area is defined by the cross product $\mathbf{B} \times \mathbf{C}$. The scalar product of height vector **A** with the base area vector $\mathbf{B} \times \mathbf{C}$, which is normal to **B** and **C**, then forms the parallel-piped volume defined by the scalar triple product. The scalar triple product and the parallel-piped volume is expressed by the determinant of the 3×3 matrix that is formed by the vectors **A**, **B**, and **C**. Also, with a square matrix, the determinant of the matrix equals the determinant of the transpose of the matrix, where

$$\mathbf{A}\cdot(\mathbf{B}\times\mathbf{C})=\begin{vmatrix} A_x & A_y & A_z \\ B_x & B_y & B_z \\ C_x & C_y & C_z \end{vmatrix}=\begin{vmatrix} A_x & B_x & C_x \\ A_y & B_y & C_y \\ A_z & B_z & C_z \end{vmatrix} \tag{15.98}$$

and

$$\det(A^{\mathrm{T}})=\det(A) \tag{15.99}$$

The differential volume $\mathrm{d}V$ of a parallel-piped formed with rectangular differential vector lengths $(\mathbf{dr})_{\mathrm{d}q_1}$, $(\mathbf{dr})_{\mathrm{d}q_2}$, and $(\mathbf{dr})_{\mathrm{d}q_3}$ along the x-, y-, and z-directions is then given by the triple scalar product of the lengths, where

$$(\mathbf{dr})_{\mathrm{d}q_1}\cdot\left[(\mathbf{dr})_{\mathrm{d}q_2}\times(\mathbf{dr})_{\mathrm{d}q_3}\right]=\begin{vmatrix} \frac{\partial x}{\partial q_1}\mathrm{d}q_1 & \frac{\partial y}{\partial q_1}\mathrm{d}q_1 & \frac{\partial z}{\partial q_1}\mathrm{d}q_1 \\ \frac{\partial x}{\partial q_2}\mathrm{d}q_2 & \frac{\partial y}{\partial q_2}\mathrm{d}q_2 & \frac{\partial z}{\partial q_2}\mathrm{d}q_2 \\ \frac{\partial x}{\partial q_3}\mathrm{d}q_3 & \frac{\partial y}{\partial q_3}\mathrm{d}q_3 & \frac{\partial z}{\partial q_3}\mathrm{d}q_3 \end{vmatrix} \tag{15.100}$$

Since the determinant of a square matrix equals the determinant of the transpose of the matrix, the triple scalar product can be represented by the Jacobian matrix, where

$$(\mathbf{dr})_{\mathrm{d}q_1}\cdot\left[(\mathbf{dr})_{\mathrm{d}q_2}\times(\mathbf{dr})_{\mathrm{d}q_3}\right]=\begin{vmatrix} \frac{\partial x}{\partial q_1}\mathrm{d}q_1 & \frac{\partial x}{\partial q_2}\mathrm{d}q_2 & \frac{\partial x}{\partial q_3}\mathrm{d}q_3 \\ \frac{\partial y}{\partial q_1}\mathrm{d}q_1 & \frac{\partial y}{\partial q_2}\mathrm{d}q_2 & \frac{\partial y}{\partial q_3}\mathrm{d}q_3 \\ \frac{\partial z}{\partial q_1}\mathrm{d}q_1 & \frac{\partial z}{\partial q_2}\mathrm{d}q_2 & \frac{\partial z}{\partial q_3}\mathrm{d}q_3 \end{vmatrix} \tag{15.101}$$

By inspection, one can bring the common factor $\mathrm{d}q_1\,\mathrm{d}q_2\,\mathrm{d}q_3$ outside the determinant, and the differential volume $\mathrm{d}V$ in Cartesian coordinates $\mathrm{d}x\,\mathrm{d}y\,\mathrm{d}z$ is related to the differential volume in curvilinear coordinates $\mathrm{d}q_1\,\mathrm{d}q_2\,\mathrm{d}q_3$ by the Jacobian determinant, where

$$\mathrm{d}V=\mathrm{d}x\,\mathrm{d}y\,\mathrm{d}z=J(q_1,q_2,q_3)\mathrm{d}q_1\,\mathrm{d}q_2\,\mathrm{d}q_3=\begin{vmatrix} \frac{\partial x}{\partial q_1} & \frac{\partial x}{\partial q_2} & \frac{\partial x}{\partial q_3} \\ \frac{\partial y}{\partial q_1} & \frac{\partial y}{\partial q_2} & \frac{\partial y}{\partial q_3} \\ \frac{\partial z}{\partial q_1} & \frac{\partial z}{\partial q_2} & \frac{\partial z}{\partial q_3} \end{vmatrix}\mathrm{d}q_1\,\mathrm{d}q_2\,\mathrm{d}q_3 \tag{15.102}$$

In general, the Jacobian can apply to transformations between two N-D curvilinear coordinate systems $(x_1, x_2, \ldots, x_N)$ and $(q_1, q_2, \ldots, \mathrm{d}q_N)$, in which the differential hyper-volume $\mathrm{d}U$ is invariant with the choice of coordinate systems, where

$$\begin{aligned}\mathrm{d}U &= \mathrm{d}x_1\,\mathrm{d}x_2\ldots\mathrm{d}x_N\\ &= J(q_1, q_2, \ldots, q_N)\mathrm{d}q_1\,\mathrm{d}q_2\ldots\mathrm{d}q_N\\ &= \begin{vmatrix}\dfrac{\partial x_1}{\partial q_1} & \cdots & \dfrac{\partial x_1}{\partial q_N}\\ \vdots & \ddots & \vdots\\ \dfrac{\partial x_N}{\partial q_1} & \cdots & \dfrac{\partial x_N}{\partial q_N}\end{vmatrix}\mathrm{d}q_1\,\mathrm{d}q_2\ldots\mathrm{d}q_N\end{aligned} \tag{15.103}$$

The physical lengths, areas, and volumes should be independent of the coordinate system, and we now have derived Jacobian-determinant-based expressions and metric-coefficient-based expressions that transform these invariant physical quantities from a Cartesian coordinate system to a curvilinear coordinate system.

15.8 Metric Coefficients, Jacobian Determinant, and Cartesian-to-Curvilinear Coordinate System Transformation

In this section, we will further discuss the metric coefficients, Jacobian determinant, and the transformation between rectangular and other curvilinear coordinate systems, as well as between pairs of curvilinear coordinate systems. Consider the differential change $\mathrm{d}\boldsymbol{\lambda}$ of a vector $\boldsymbol{\lambda}(x, y, z)$ given vector displacement $\mathrm{d}\mathbf{s}$ of vector $\mathbf{s}(x, y, z)$, where

$$\boldsymbol{\lambda} = \hat{\mathbf{x}}\lambda_x + \hat{\mathbf{y}}\lambda_y + \hat{\mathbf{z}}\lambda_z \tag{15.104}$$

and

$$\mathbf{s} = \hat{\mathbf{x}}x + \hat{\mathbf{y}}y + \hat{\mathbf{z}}z \tag{15.105}$$

The gradient operator is defined for scalar arguments; hence, one has

$$\mathrm{d}\lambda_x = \mathrm{d}\mathbf{s}\cdot\nabla\lambda_x = \mathrm{d}x\frac{\partial\lambda_x}{\partial x} + \mathrm{d}y\frac{\partial\lambda_x}{\partial y} + \mathrm{d}z\frac{\partial\lambda_x}{\partial z} \tag{15.106}$$

$$\mathrm{d}\lambda_y = \mathrm{d}\mathbf{s}\cdot\nabla\lambda_y = \mathrm{d}x\frac{\partial\lambda_y}{\partial x} + \mathrm{d}y\frac{\partial\lambda_y}{\partial y} + \mathrm{d}z\frac{\partial\lambda_y}{\partial z} \tag{15.107}$$

and

$$d\lambda_z = \mathbf{ds} \cdot \nabla\lambda_z = dx\frac{\partial\lambda_z}{\partial x} + dy\frac{\partial\lambda_z}{\partial y} + dz\frac{\partial\lambda_z}{\partial z} \qquad (15.108)$$

The differential change vector $\mathbf{d\lambda}(x, y, z)$ can be expressed as the dot product of a Jacobian vector matrix $\mathbf{J}$ and vector displacement $\mathbf{ds}(x, y, z)$, where

$$\mathbf{d\lambda} = \mathbf{J} \cdot \mathbf{ds} = \begin{bmatrix} \frac{\partial\lambda_x}{\partial x} & \frac{\partial\lambda_x}{\partial y} & \frac{\partial\lambda_x}{\partial z} \\ \frac{\partial\lambda_y}{\partial x} & \frac{\partial\lambda_y}{\partial y} & \frac{\partial\lambda_y}{\partial z} \\ \frac{\partial\lambda_z}{\partial x} & \frac{\partial\lambda_z}{\partial y} & \frac{\partial\lambda_z}{\partial z} \end{bmatrix} \begin{bmatrix} dx \\ dy \\ dz \end{bmatrix} = \begin{bmatrix} dx\frac{\partial\lambda_x}{\partial x} + dy\frac{\partial\lambda_x}{\partial y} + dz\frac{\partial\lambda_x}{\partial z} \\ dx\frac{\partial\lambda_y}{\partial x} + dy\frac{\partial\lambda_y}{\partial y} + dz\frac{\partial\lambda_y}{\partial z} \\ dx\frac{\partial\lambda_z}{\partial x} + dy\frac{\partial\lambda_z}{\partial y} + dz\frac{\partial\lambda_z}{\partial z} \end{bmatrix} \qquad (15.109)$$

and

$$\mathbf{d\lambda} = \begin{bmatrix} d\lambda_x \\ d\lambda_y \\ d\lambda_z \end{bmatrix} \quad \mathbf{J} = \begin{bmatrix} \nabla\lambda_x \\ \nabla\lambda_y \\ \nabla\lambda_z \end{bmatrix} = \begin{bmatrix} \frac{\partial\lambda_x}{\partial x} & \frac{\partial\lambda_x}{\partial y} & \frac{\partial\lambda_x}{\partial z} \\ \frac{\partial\lambda_y}{\partial x} & \frac{\partial\lambda_y}{\partial y} & \frac{\partial\lambda_y}{\partial z} \\ \frac{\partial\lambda_z}{\partial x} & \frac{\partial\lambda_z}{\partial y} & \frac{\partial\lambda_z}{\partial z} \end{bmatrix} \quad \mathbf{ds} = \begin{bmatrix} dx \\ dy \\ dz \end{bmatrix} \qquad (15.110)$$

The Jacobian of a function gives the orientation of the tangent plane to the function at a given point. Also, the Jacobian is the gradient or first derivative of a vector function of several variables, and the gradient is the first derivative of a scalar function of several variables. The squared differential length (namely, $d\lambda^2$ or $\mathbf{d\lambda} \cdot \mathbf{d\lambda}$) is given by

$$(d\lambda)^2 = \mathbf{d\lambda}^{\mathrm{T}} \cdot \mathbf{d\lambda} = (\mathbf{J} \cdot \mathbf{ds})^{\mathrm{T}} \cdot (\mathbf{J} \cdot \mathbf{ds}) \qquad (15.111)$$

Since the dot product operation is commutative, associative, and distributive, we obtain

$$(d\lambda)^2 = \mathbf{J}^{\mathrm{T}} \cdot \mathbf{ds}^{\mathrm{T}} \cdot \mathbf{J} \cdot \mathbf{ds} = \mathbf{ds}^{\mathrm{T}} \cdot \mathbf{J}^{\mathrm{T}} \cdot \mathbf{J} \cdot \mathbf{ds} \qquad (15.112)$$

The metric tensor $\mathbf{G}$, which was derived in a previous section, is the dot product of the Jacobian matrix $\mathbf{J}$ and its transpose $\mathbf{J}^{\mathrm{T}}$, where

$$\mathbf{G} = \mathbf{J}^{\mathrm{T}} \cdot \mathbf{J} \qquad (15.113)$$

The matrix elements g_{kl} of the metric tensor $\mathbf{G}$ are

$$g_{kl} = \frac{\partial \lambda_i}{\partial q_k}\frac{\partial \lambda_i}{\partial q_l} \quad i,k,l = x,y,z \text{ and } q_x, q_y, q_z = x,y,z \tag{15.114}$$

Recall from Eq. (15.69) that

$$\mathrm{d}\lambda = \sqrt{\mathrm{d}\lambda_x^2 + \mathrm{d}\lambda_y^2 + \mathrm{d}\lambda_z^2} = \sqrt{\sum_{k=1}^{3}\sum_{l=1}^{3} g_{kl}^2 \,\mathrm{d}q_k \,\mathrm{d}q_l} \quad q_1, q_2, q_3 = x,y,z \tag{15.115}$$

The squared differential length $\mathrm{d}\lambda^2$ is then expressed in terms of the metric tensor $\mathbf{G}$:

$$(\mathrm{d}\lambda)^2 = \mathbf{ds}^{\mathrm{T}} \cdot \mathbf{G} \cdot \mathbf{ds} \tag{15.116}$$

where

$$\mathbf{G} = \mathbf{J}^{\mathrm{T}} \cdot \mathbf{J} = \begin{bmatrix} \frac{\partial \lambda_x}{\partial x} & \frac{\partial \lambda_y}{\partial x} & \frac{\partial \lambda_z}{\partial x} \\ \frac{\partial \lambda_x}{\partial y} & \frac{\partial \lambda_y}{\partial y} & \frac{\partial \lambda_z}{\partial z} \\ \frac{\partial \lambda_x}{\partial z} & \frac{\partial \lambda_y}{\partial z} & \frac{\partial \lambda_z}{\partial z} \end{bmatrix} \begin{bmatrix} \frac{\partial \lambda_x}{\partial x} & \frac{\partial \lambda_x}{\partial y} & \frac{\partial \lambda_x}{\partial z} \\ \frac{\partial \lambda_y}{\partial x} & \frac{\partial \lambda_y}{\partial y} & \frac{\partial \lambda_y}{\partial z} \\ \frac{\partial \lambda_z}{\partial x} & \frac{\partial \lambda_z}{\partial y} & \frac{\partial \lambda_z}{\partial z} \end{bmatrix} \tag{15.117}$$

One can generalize the squared differential length $\mathrm{d}\lambda^2$ expression to curvilinear coordinates (q_1, q_2, q_3). The differential change $\mathbf{d\lambda}$ of a vector $\boldsymbol{\lambda}(\lambda_1, \lambda_2, \lambda_3)$ in one curvilinear coordinate system is expressed by the vector displacement $\mathbf{ds}$ of vector $\mathbf{s}(q_1, q_2, q_3)$ in another curvilinear coordinate system, where

$$\boldsymbol{\lambda} = \hat{\mathbf{q}}_1\lambda_1 + \hat{\mathbf{q}}_1\lambda_1 + \hat{\mathbf{q}}_3\lambda_3 \tag{15.118}$$

and

$$\mathbf{s} = \hat{\mathbf{q}}_1 s_1 + \hat{\mathbf{q}}_2 s_2 + \hat{\mathbf{q}}_3 s_3 \tag{15.119}$$

The metric tensor $\mathbf{G}$ relates the squared differential length expressed by $\mathrm{d}\lambda^2$ in one curvilinear coordinate system to the squared differential length expression $\mathrm{d}s^2$ in another curvilinear coordinate system:

$$\mathbf{G} = \mathbf{J}^{\mathrm{T}} \cdot \mathbf{J} = \begin{bmatrix} \frac{\partial \lambda_1}{\partial q_1} & \frac{\partial \lambda_2}{\partial q_1} & \frac{\partial \lambda_3}{\partial q_1} \\ \frac{\partial \lambda_1}{\partial q_2} & \frac{\partial \lambda_2}{\partial q_2} & \frac{\partial \lambda_3}{\partial q_2} \\ \frac{\partial \lambda_1}{\partial q_3} & \frac{\partial \lambda_2}{\partial q_3} & \frac{\partial \lambda_3}{\partial q_3} \end{bmatrix} \begin{bmatrix} \frac{\partial \lambda_1}{\partial q_1} & \frac{\partial \lambda_1}{\partial q_2} & \frac{\partial \lambda_1}{\partial q_3} \\ \frac{\partial \lambda_2}{\partial q_1} & \frac{\partial \lambda_2}{\partial q_2} & \frac{\partial \lambda_2}{\partial q_3} \\ \frac{\partial \lambda_3}{\partial q_1} & \frac{\partial \lambda_3}{\partial q_2} & \frac{\partial \lambda_3}{\partial q_3} \end{bmatrix} \tag{15.120}$$

As an example, consider the differential expressions of a squared length in the spherical and rectangular coordinate systems, where

$$\lambda_1 = x = r \sin\theta \cos\phi \quad \lambda_2 = y = r \sin\theta \sin\phi \quad \lambda_3 = z = r \cos\theta \tag{15.121}$$

and

$$q_1 = r = \sqrt{x^2 + y^2 + z^2} \quad q_2 = \theta = \tan^{-1}\left(\frac{\sqrt{x^2 + y^2}}{z}\right) \quad q_3 = \phi = \tan^{-1}\left(\frac{y}{x}\right) \tag{15.122}$$

The metric tensor **G** relating a squared differential length in rectangular and spherical coordinate systems is then

$$\begin{aligned}
\mathbf{G} &= \mathbf{J}^{\mathrm{T}} \cdot \mathbf{J} \\
&= \begin{bmatrix} \sin\theta\cos\phi & \sin\theta\sin\phi & \cos\theta \\ r\cos\theta\cos\phi & r\cos\theta\sin\phi & -r\sin\theta \\ -r\sin\theta\sin\phi & r\cos\theta\cos\phi & 0 \end{bmatrix} \\
&\times \begin{bmatrix} \sin\theta\cos\phi & r\cos\theta\cos\phi & -r\sin\theta\sin\phi \\ \sin\theta\sin\phi & r\cos\theta\sin\phi & r\cos\theta\cos\phi \\ \cos\theta & -r\sin\theta & 0 \end{bmatrix}
\end{aligned} \tag{15.123}$$

Carrying out the matrix multiplication yields the diagonal metric tensor **G** relating rectangular and spherical coordinate systems, where

$$\mathbf{G} = \mathbf{J}^{\mathrm{T}} \cdot \mathbf{J} = \begin{bmatrix} 1 & 0 & 0 \\ 0 & r^2 & 0 \\ 0 & 0 & r^2\sin^2\theta \end{bmatrix} \tag{15.124}$$

The determinant g of the metric tensor **G** is

$$g = \det|\mathbf{G}| = \det|\mathbf{J}^{\mathrm{T}} \cdot \mathbf{J}| \tag{15.125}$$

The differential volumes $\mathrm{d}\lambda_x\,\mathrm{d}\lambda_y\,\mathrm{d}\lambda_z$ is proportional to the differential volume $\mathrm{d}x\,\mathrm{d}y\,\mathrm{d}z$ by the proportionality constant. The square root of the determinant of the metric tensor $\sqrt{g}$, where

$$\mathrm{d}\lambda_x\,\mathrm{d}\lambda_y\,\mathrm{d}\lambda_z = \sqrt{g}\,\mathrm{d}x\,\mathrm{d}y\,\mathrm{d}z \tag{15.126}$$

Provided that the determinant of the metric tensor is nonzero ($g \neq 0$), and thus the determinant of the Jacobian matrix **J** is nonzero, where $\det|\mathbf{J}| \neq 0$, one has the inverse relation for the squared differential distance $\mathrm{d}s^2$:

$$(\mathrm{d}s)^2 = \mathbf{ds}^{\mathrm{T}} \cdot \mathbf{ds} = (\mathbf{J}^{-1} \cdot \mathrm{d}\boldsymbol{\lambda})^{\mathrm{T}} \cdot (\mathbf{J}^{-1} \cdot \mathrm{d}\boldsymbol{\lambda}) \tag{15.127}$$

The squared differential distance ds^2 can be expressed in terms of the inverse metric tensor $\mathbf{G}^{-1}$, where

$$(ds)^2 = d\boldsymbol{\lambda}^T \cdot \mathbf{G}^{-1} \cdot d\boldsymbol{\lambda} \tag{15.128}$$

where

$$\mathbf{G}^{-1} = (\mathbf{J}^{\mathrm{T}})^{-1} \cdot \mathbf{J}^{-1} = \begin{bmatrix} \frac{\partial \lambda_x}{\partial x} & \frac{\partial \lambda_y}{\partial x} & \frac{\partial \lambda_z}{\partial x} \\ \frac{\partial \lambda_x}{\partial y} & \frac{\partial \lambda_y}{\partial y} & \frac{\partial \lambda_z}{\partial z} \\ \frac{\partial \lambda_x}{\partial z} & \frac{\partial \lambda_y}{\partial z} & \frac{\partial \lambda_z}{\partial z} \end{bmatrix}^{-1} \begin{bmatrix} \frac{\partial \lambda_x}{\partial x} & \frac{\partial \lambda_x}{\partial y} & \frac{\partial \lambda_x}{\partial z} \\ \frac{\partial \lambda_y}{\partial x} & \frac{\partial \lambda_y}{\partial y} & \frac{\partial \lambda_y}{\partial z} \\ \frac{\partial \lambda_z}{\partial x} & \frac{\partial \lambda_z}{\partial y} & \frac{\partial \lambda_z}{\partial z} \end{bmatrix}^{-1} \tag{15.129}$$

The inverse metric tensor $\mathbf{G}^{-1}$ relating a squared differential length in rectangular and spherical coordinate systems can be obtained from the metric tensor $\mathbf{G}$, via the identity relation, where the product of the matrix $\mathbf{G}$ and its inverse $\mathbf{G}^{-1}$ gives the unity, diagonal matrix I, provided that the determinant of $\mathbf{G}$ is nonzero, where

$$\mathbf{G}^{-1}\mathbf{G} = 1 \tag{15.130}$$

By inspection, one obtains the inverse metric tensor $\mathbf{G}^{-1}$ from the reciprocal of metric tensor $\mathbf{G}$:

$$\mathbf{G}^{-1} = (\mathbf{J}^{st})^{-1} \cdot \mathbf{J}^{-1} = \begin{bmatrix} 1 & 0 & 0 \\ 0 & r^2 & 0 \\ 0 & 0 & r^2 \sin^2\theta \end{bmatrix}^{-1} = \begin{bmatrix} 1 & 0 & 0 \\ 0 & r^{-2} & 0 \\ 0 & 0 & r^{-2} \sin^{-2}\theta \end{bmatrix} \tag{15.131}$$

where

$$\mathbf{G}^{-1}\mathbf{G} = \begin{bmatrix} 1 & 0 & 0 \\ 0 & r^{-2} & 0 \\ 0 & 0 & r^{-2} \sin^{-2}\theta \end{bmatrix} \begin{bmatrix} 1 & 0 & 0 \\ 0 & r^2 & 0 \\ 0 & 0 & r^2 \sin^2\theta \end{bmatrix} = \begin{bmatrix} 1 & 0 & 0 \\ 0 & 1 & 0 \\ 0 & 0 & 1 \end{bmatrix} \tag{15.132}$$

15.9 Variation of Physical Quantity by Displacement—Scalar and Vector Fields

Joos (1986) considers the questions "How does a scalar point function $\phi(s)$ change with displacement ds?" and "How does vector $\mathbf{v}(s)$ change with displacement ds?" These questions are discussed in this section and serve as an introduction to the transformation of properties, vectors, and tensors under a change of coordinate systems, where the transformations are required to maintain the invariance of the total magnitudes of the vectors and tensors. The total magnitudes of the vector or tensor represent basic physical quantities, properties, or relations that are intrinsic to the fabric of space–time and independent of coordinate systems. However, vector or tensor components may vary with coordinate system or inertial reference frame in special relativity or accelerated frames in general relativity. As a result, observers in different coordinate systems or frames of references may measure and observe different components. However, the observers in all the coordinate systems and frames of reference should measure or observe the same total magnitude of a fundamental physical quantity, or observe the same fundamental physical relations.

We start with a displacement ds in the scalar field $\phi(s)$. A scalar quantity, such as charge q, mass m, or temperature T, is expressed as a function $f(x, y, z)$ of points (x, y, z) in a region and forms a scalar field. One obtains scalar field $q(x, y, z)$, $m(x, y, z)$, or $T(x, y, z)$. The change in the value of the function $\mathrm{d}f(x, y, z)$, which corresponds to a displacement ds, depends on the direction of displacement and thus depends on the differential displacement vector d**s**, where

$$\mathrm{d}\mathbf{s} = \hat{\mathbf{x}}\,\mathrm{d}x + \hat{\mathbf{y}}\,\mathrm{d}y + \hat{\mathbf{z}}\,\mathrm{d}z \tag{15.133}$$

The unit vectors $(\hat{\mathbf{x}}, \hat{\mathbf{y}}, \hat{\mathbf{z}})$ define the three directions for the (x, y, z) coordinate system. The change in the function $\mathrm{d}f(x, y, z)$ by the chain rule is

$$\mathrm{d}f = \frac{\partial f}{\partial x}\mathrm{d}x + \frac{\partial f}{\partial y}\mathrm{d}y + \frac{\partial f}{\partial z}\mathrm{d}z \tag{15.134}$$

A change in the function $\mathrm{d}f(x, y, z)$ can be considered as the scalar product of differential displacement vector d**s** and the gradient of the function $f(x, y, z)$, where

$$\mathrm{d}f = \mathrm{d}\mathbf{s} \cdot \nabla f \tag{15.135}$$

Now consider a vector quantity which has a total constant magnitude but is comprised of spatial components whose individual magnitudes can vary. Harking back to the Pythagorean theorem, for example, the total squared distance between two points, connecting the two endpoints of a displacement vector, equals the sum of the squares of its spatial component distances in each of the spatial directions, which can be generalized from two or three dimensions to n dimensions; see Friedrichs (1975).

Assume a vector field comprised of a set of points with attached vectors, which contain a scalar quantity (the vector magnitude) with an associated direction of action. Now consider the differential change $\mathrm{d}\boldsymbol{\lambda}$ of a vector $\boldsymbol{\lambda}(x, y, z)$ given a vector displacement $\mathrm{d}\mathbf{s}(x, y, z)$, where

$$\boldsymbol{\lambda} = \hat{\mathbf{x}}\lambda_x + \hat{\mathbf{y}}\lambda_y + \hat{\mathbf{z}}\lambda_z \tag{15.136}$$

The gradient operator is defined for scalar arguments, and hence, one has

$$d\lambda_x = \mathbf{ds}\cdot\nabla\lambda_x = dx\frac{\partial\lambda_x}{\partial x} + dy\frac{\partial\lambda_x}{\partial y} + dz\frac{\partial\lambda_x}{\partial z} \tag{15.137}$$

$$d\lambda_y = \mathbf{ds}\cdot\nabla\lambda_y = dx\frac{\partial\lambda_y}{\partial x} + dy\frac{\partial\lambda_y}{\partial y} + dz\frac{\partial\lambda_y}{\partial z} \tag{15.138}$$

$$d\lambda_z = \mathbf{ds}\cdot\nabla\lambda_z = dx\frac{\partial\lambda_z}{\partial x} + dy\frac{\partial\lambda_z}{\partial y} + dz\frac{\partial\lambda_z}{\partial z} \tag{15.139}$$

The differential change $d\boldsymbol{\lambda}(x,y,z)$ in coordinate system (x,y,z) can be expressed in terms of the differential displacement $d\mathbf{q}$ of the position vector $\mathbf{q}(u,v,w)$ of another coordinate system $(u,v,w)u,v,w$, where

$$d\mathbf{q}(u,v,w) = \hat{\mathbf{u}}\,du + \hat{\mathbf{v}}\,dv + \hat{\mathbf{w}}\,dw \tag{15.140}$$

The differential change $d\boldsymbol{\lambda}(x,y,z)$ with respect to the unit vector $\hat{\mathbf{u}}$ component of differential displacement vector $d\mathbf{q}$ in the other (u,v,w) coordinate system (namely, du) is

$$\frac{d\lambda_x}{du} = \frac{\mathbf{ds}}{du}\cdot\nabla\lambda_x = \frac{dx}{du}\frac{\partial\lambda_x}{\partial x} + \frac{dy}{du}\frac{\partial\lambda_x}{\partial y} + \frac{dz}{du}\frac{\partial\lambda_x}{\partial z} \tag{15.141}$$

$$\frac{d\lambda_y}{du} = \frac{\mathbf{ds}}{du}\cdot\nabla\lambda_y = \frac{dx}{du}\frac{\partial\lambda_y}{\partial x} + \frac{dy}{du}\frac{\partial\lambda_y}{\partial y} + \frac{dz}{du}\frac{\partial\lambda_y}{\partial z} \tag{15.142}$$

$$\frac{d\lambda_z}{du} = \frac{\mathbf{ds}}{du}\cdot\nabla\lambda_z = \frac{dx}{du}\frac{\partial\lambda_z}{\partial x} + \frac{dy}{du}\frac{\partial\lambda_z}{\partial y} + \frac{dz}{du}\frac{\partial\lambda_z}{\partial z} \tag{15.143}$$

The differential change $d\boldsymbol{\lambda}(x,y,z)$ with respect to the unit vector $\hat{\mathbf{v}}$ component of differential displacement vector $d\mathbf{q}$ in the other (u,v,w) coordinate system (namely, dv) is

$$\frac{d\lambda_x}{dv} = \frac{\mathbf{ds}}{dv}\cdot\nabla\lambda_x = \frac{dx}{dv}\frac{\partial\lambda_x}{\partial x} + \frac{dy}{dv}\frac{\partial\lambda_x}{\partial y} + \frac{dz}{dv}\frac{\partial\lambda_x}{\partial z} \tag{15.144}$$

$$\frac{d\lambda_y}{dv} = \frac{\mathbf{ds}}{dv}\cdot\nabla\lambda_y = \frac{dx}{dv}\frac{\partial\lambda_y}{\partial x} + \frac{dy}{dv}\frac{\partial\lambda_y}{\partial y} + \frac{dz}{dv}\frac{\partial\lambda_y}{\partial z} \tag{15.145}$$

$$\frac{d\lambda_z}{dv} = \frac{\mathbf{ds}}{dv}\cdot\nabla\lambda_z = \frac{dx}{dv}\frac{\partial\lambda_z}{\partial x} + \frac{dy}{dv}\frac{\partial\lambda_z}{\partial y} + \frac{dz}{dv}\frac{\partial\lambda_z}{\partial z} \tag{15.146}$$

The differential change $d\boldsymbol{\lambda}(x,y,z)$ with respect to the unit vector $\hat{\mathbf{w}}$ component of differential displacement vector $\mathbf{dq}$ in the other (u,v,w) coordinate system (namely, dv) is

$$\frac{d\lambda_x}{dw} = \frac{\mathbf{ds}}{dw}\cdot\nabla\lambda_x = \frac{dx}{dw}\frac{\partial\lambda_x}{\partial x} + \frac{dy}{dw}\frac{\partial\lambda_x}{\partial y} + \frac{dz}{dw}\frac{\partial\lambda_x}{\partial z} \tag{15.147}$$

$$\frac{d\lambda_y}{dw} = \frac{\mathbf{ds}}{dw}\cdot\nabla\lambda_y = \frac{dx}{dw}\frac{\partial\lambda_y}{\partial x} + \frac{dy}{dw}\frac{\partial\lambda_y}{\partial y} + \frac{dz}{dw}\frac{\partial\lambda_y}{\partial z} \tag{15.148}$$

$$\frac{d\lambda_z}{dw} = \frac{\mathbf{ds}}{dw}\cdot\nabla\lambda_z = \frac{dx}{dw}\frac{\partial\lambda_z}{\partial x} + \frac{dy}{dw}\frac{\partial\lambda_z}{\partial y} + \frac{dz}{dw}\frac{\partial\lambda_z}{\partial z} \tag{15.149}$$

In short, one can compactly express the differential change $d\boldsymbol{\lambda}(x,y,z)$ in coordinate system (x,y,z) in terms of the differential displacement $\mathbf{dq}$ of the position vector $\mathbf{q}(u,v,w)$ in another coordinate system (u,v,w):

$$\frac{d\lambda_i}{dq_k} = \sum_{j=1}^{3}\frac{\mathbf{ds}_j}{dq_k}\cdot\nabla\lambda_j = \sum_{j=1}^{3}\frac{\partial\lambda_i}{\partial x_j}\frac{\partial x_j}{\partial q_k} = \frac{\partial\lambda_i}{\partial x_j}\frac{\partial x_j}{\partial q_k} \quad k=1,2,3 \tag{15.150}$$

A vector space V of finite dimension n is defined by n basis vectors $\mathbf{e}_1,\mathbf{e}_2,\mathbf{e}_3,\ldots,\mathbf{e}_n$. A vector $\mathbf{v}$ in V has coordinates $v_1,v_2,v_3,\ldots,v_n$ relative to the basis vectors, where

$$\mathbf{v} = v^i\mathbf{e}_i \equiv \sum_{i=1}^{n} v^i\mathbf{e}_i \quad i=1,2,3,\ldots,n \tag{15.151}$$

This expression of the vector $\mathbf{v}$ uses the Einstein summation convention, which compresses extensive, tedious expressions into compact expressions. The integer index i, which is summed over, is the summation index; or, alternatively, integer index i is termed the *dummy index*, since the result does not depend on index i because one could have used j, k, l, or some other index in place of i. If the same index i appears twice in a single term $v^i\mathbf{e}_i$ (as superscript v^i and as subscript $\mathbf{e}_i$), then summation over i for $\sum_i v^i\mathbf{e}_i$ is compactly expressed by $v^i\mathbf{e}_i$, as shown in Eq. (15.151).

In some modeling of physical processes, the axes used to define the space characterizing the physical process do not have the same physical dimensions. For example, the thermodynamic state space are comprised of the defining axes of the pressure, volume, and temperature of a fluid. The axes do not have the same physical dimensions (e.g., length), and thus, one cannot compare the axes, nor can one use the lengths and angles between two points in the thermodynamic state space to do this. Such spaces are termed *nonmetric* or *affine* spaces. In affine spaces, one is thus limited to properties that remain invariant under the change of scale and angle of the axes. The discussions that follow are limited to physical spaces in which each axis has a dimension of length. In the four-vectors, the imaginary fourth component based on time ict is given a spatial dimension of length ct via the multiplication of time t by the speed of light c.

15.10 Invariance, Summation Convention, and Contravariant Vectors

Dutton (1986) shows that if the coordinate axes are no longer orthogonal (perpendicular), then one must employ two types of vectors, contravariant and covariant vectors, which each have their own set of basis vectors and vector components that together specify a vector. A vector expressed in contravariant form or its dual covariant form is required in order for scalar products of the vectors to be formed in the nonorthogonal coordinate systems. Scalar quantities result from the vector dot product of the contravariant (covariant) vectors with itself (the squared magnitude of the vector) or with another like contravariant (covariant) vector.

The scalar quantities, which represent physical properties, must be the same in all the coordinate systems, whether orthogonal or nonorthogonal. In addition, there is no longer a need for or advantage to the basis vectors having unit length. In fact, the lengths of the basis vectors carry the spatial and temporal distance (ct) scaling information. In ordinary rectangular coordinates that describe planer or flat surfaces and rectangular volumes, the vector components and basis vectors of the contravariant and covariant vectors are identical. However, in the case of curved surfaces and spaces, the vector components and basis vectors differ between the contravariant and covariant vectors.

A vector must be coordinate system–invariant, in which the vector components must vary inversely with a change of basis vectors. That is, the vector must vary in the opposite manner to the imposed change of basis vectors. The position vector and the time derivatives of the position vector $\mathbf{r}$; namely, velocity $\mathbf{v}$, acceleration $\mathbf{a}$, jerk $\mathbf{j}$ (the third time derivative) are contravariant vectors. In Einstein summation convention, the contravariant component (e.g., v^i) has upper indices and the basis vector $\mathbf{e}_i$ has the lower indices. In the case of four-vectors, the position vector is

$$\mathbf{r} = r^i \mathbf{e}_i \equiv \sum_{i=1}^{4} r^i \mathbf{e}_i \tag{15.152}$$

Basis vectors $\mathbf{e}_i$ are termed *natural basis vectors.* The velocity $\mathbf{v}$, acceleration $\mathbf{a}$, and jerk $\mathbf{j}$ four-vectors are expressed in Einstein summation convention as

$$\mathbf{v} = \frac{d\mathbf{r}}{dt} = v^i \mathbf{e}_i \quad \mathbf{a} = \frac{d^2\mathbf{r}}{dt^2} = a^i \mathbf{e}_i \quad \mathbf{j} = \frac{d^3\mathbf{r}}{dt^3} = j^i \mathbf{e}_i \tag{15.153}$$

In general relativity of curved surfaces and spaces, the basis vectors $\mathbf{e}_i$ for a contravariant vector are not unit vectors, and in general, the basis vectors $\mathbf{e}_i$ are not mutually perpendicular.

In a change from coordinate system $(y_1, y_2, y_3, \ldots, y_n)$ to coordinate system $(x_1, x_2, x_3, \ldots, x_n)$, the contravariant components W^j of vector $\mathbf{W}$ transform to the contravariant components w^i of vector $\mathbf{w}$, where

$$w^i = \frac{\partial y^i}{\partial x^j} W^j \equiv \sum_{j=1}^{n} \frac{\partial y^i}{\partial x^j} W^j \quad \text{contravariant vector} \tag{15.154}$$

with

$$\mathbf{w} = \mathbf{e}_i w^i \equiv \sum_{i=1}^{n} \mathbf{e}_i w^i \tag{15.155}$$

and

$$\mathbf{W} = \mathbf{e}_j W^j \equiv \sum_{j=1}^{n} \mathbf{e}_j W^j \tag{15.156}$$

The contravariant vector components transform in the same manner as coordinate differentials. On the change of coordinate systems, as the components W^j of the contravariant vector $\mathbf{W}$ change to components w^i of the contravariant vector $\mathbf{w}$, and the basis vectors $\mathbf{e}_j$ of $\mathbf{W}$ change to basis vector $\mathbf{e}_i$ of $\mathbf{w}$. The change from contravariant component W^j to w^i is offset by the opposite change from natural basis vector $\mathbf{e}_j$ to $\mathbf{e}_i$. This causes the magnitudes of contravariant vectors $\mathbf{W}$ and $\mathbf{w}$ to be identical, where

$$|\mathbf{W}| = |\mathbf{w}| \tag{15.157}$$

The squared magnitudes are equal as well, where

$$|\mathbf{w}|^2 = |\mathbf{W}|^2 \tag{15.158}$$

Hence, the fundamental physical quantity, which is represented by the invariant vector magnitude, is also invariant under the change of coordinate systems or frames of reference.

A physical quantity is represented by vector $\mathbf{w}$ in coordinate system $(y_1, y_2, y_3, \ldots, y_n)$ and by vector $\mathbf{W}$ in another coordinate system $(x_1, x_2, x_3, \ldots, x_n)$. The physical quantity is assumed to be independent of the coordinate system. Consequently, the magnitude of vector $\mathbf{w}$ and $\mathbf{W}$, which is an invariant physical quantity, are equal, where

$$w^2 = \mathbf{w} \cdot \mathbf{w} = |\mathbf{w}|^2 = \sum_{i=1}^{n} (w^i)^2 \tag{15.159}$$

and

$$W^2 = \mathbf{W} \cdot \mathbf{W} = |\mathbf{W}|^2 = \sum_{j=1}^{n} (W^j)^2 \tag{15.160}$$

Recall that in special relativity, one employs $i = \sqrt{-1}$ as the fourth coordinate of time in the space–time four-component or four-vector, where $n = 4$ and coordinates are $(q_1, q_2, q_3, ictq_4)$. The space–time distance interval is a physical quantity that must be the same in all directions and be invariant upon a change of coordinate systems from $(q^1, q^2, q^3, ictq^4)$ to $(\lambda^1, \lambda^2, \lambda^3, ict\lambda^4)$, where

$$|\mathrm{d}\boldsymbol{\lambda}|^2 = \sum_{l=1}^{4} (\mathrm{d}\lambda^i)^2 = \sum_{k=1}^{4} \sum_{l=1}^{4} g_{kl}^2 \, \mathrm{d}q^k \, \mathrm{d}q^l \tag{15.161}$$

15.11 Natural Basis Vectors are Tangent to Surfaces in a Contravariant Vector Space

Consider a point on a surface specified by position vector $\mathbf{r}(x, y, z)$, whose rectangular coordinates (x, y, z) are expressed in terms of general, curvilinear coordinates (u, v, w) via

$$\mathbf{r}(x, y, z) = x(u, v, w)\hat{\mathbf{x}} + y(u, v, w)\hat{\mathbf{y}} + z(u, v, w)\hat{\mathbf{z}} \tag{15.162}$$

In a curvilinear coordinate system (u, v, w), the natural basis vectors $(\mathbf{e}_u, \mathbf{e}_v, \mathbf{e}_w)$ for contravariant vectors form the tangent vector $\mathbf{t}(x, y, z)$ for each point P on a surface, where

$$\mathbf{t}(x, y, z) = \mathbf{e}_u + \mathbf{e}_v + \mathbf{e}_w \tag{15.163}$$

The ith natural basis vector $\mathbf{e}_i$ of coordinate system (u, v, w), where $(i = u, v, w)$, is obtained from the derivative of a vector $\mathbf{r}$, which is expressed by $\mathbf{r}(x, y, z)$ in the (x, y, z) coordinate system, with respect to the ith spatial coordinate of the coordinate system (u, v, w), where

$$\mathbf{e}_u = \frac{\mathrm{d}\mathbf{r}(x, y, z)}{\mathrm{d}u} \tag{15.164}$$

$$\mathbf{e}_v = \frac{\mathrm{d}\mathbf{r}(x, y, z)}{\mathrm{d}v} \tag{15.165}$$

$$\mathbf{e}_w = \frac{\mathrm{d}\mathbf{r}(x, y, z)}{\mathrm{d}w} \tag{15.166}$$

Recall that the intersection of each pair of unique 2-D surfaces produces a 1-D curve along each of the three spatial coordinates. And the three unique and mutually perpendicular surfaces intersect at a single point P. The three unique pairings of these unique, 2-D surfaces leads to three mutually perpendicular curves. These three mutually perpendicular surfaces and the three mutually perpendicular curves intersect at the same single point P. Each point P on a surface is the intersection of the three coordinate curves, and each curve has its own tangent line at point P.

The tangent lines along each of the coordinate curves thus form the components of the tangent vector that describes the tangent plane at point P. Dual to the tangent vectors is the normal vectors, in which at point P, each tangent line of a coordinate curve has a perpendicular or normal line at point P. Thus, a surface comprised of a set of points P can be described by the normal vector at the points P in which the normal vector components are perpendicular to the tangent line components of the tangent vector associated with each point P. The dual basis vectors, which are normal vectors to the surface, describe surfaces and their enclosed volumes in a space of covariant (dual) vectors to be discussed in the next section. The natural basis vectors, which are tangent vectors to surfaces, describe surfaces and their enclosed volumes in a space of contravariant vectors described in this section.

We show how one calculates the natural basis vectors for contravariant vectors in a curvilinear coordinate system in which we use the spherical coordinate system as a specific example. Consider a spherical surface $\mathbf{r}(x, y, z)$ that is expressed in rectangular (x, y, z) and spherical coordinates (r, θ, ϕ), respectively, by

$$\mathbf{r}(x, y, z) = x\hat{\mathbf{x}} + y\hat{\mathbf{y}} + z\hat{\mathbf{z}} \tag{15.167}$$

and

$$\mathbf{r}(x, y, z) = r \sin\theta \cos\phi\, \hat{\mathbf{x}} + r \sin\theta \sin\phi\, \hat{\mathbf{y}} + r \cos\theta\, \hat{\mathbf{z}} \tag{15.168}$$

where

$$u = r \quad v = \theta \quad w = \phi \tag{15.169}$$

The expressions for (x, y, z) coordinates $x(u, v, w)$, $y(u, v, w)$, and $z(u, v, w)$ are

$$x(u, v, w) = r \sin\theta \cos\phi \tag{15.170}$$

$$y(u, v, w) = r \sin\theta \sin\phi \tag{15.171}$$

$$z(u, v, w) = r \cos\theta \tag{15.172}$$

The tangent vector $\mathbf{t}(x, y, z)$ in rectangular coordinates (x, y, z) is obtained from the tangent vector $\mathbf{t}(r, \theta, \phi)$ in spherical coordinates (r, θ, ϕ) via the natural basis vectors $(\mathbf{e}_r, \mathbf{e}_\theta, \mathbf{e}_\phi)$ that relate the rectangular (x, y, z) to spherical coordinates (r, θ, ϕ) via

$$\mathbf{t}(x, y, z) = \mathbf{e}_r + \mathbf{e}_\theta + \mathbf{e}_\phi \tag{15.173}$$

The three natural basis vectors ($\hat{\mathbf{e}}_r$,$\hat{\mathbf{e}}_\theta$,$\hat{\mathbf{e}}_\phi$), which are obtained from position vector $\mathbf{r}$, are tangential unit vectors to their respective coordinate curves. Here, $\hat{\mathbf{e}}_r$ is the unit vector tangent to the coordinate curve, formed by the intersection of two constant coordinate surfaces; namely, constant polar angle coordinate surface $\theta=\theta_0$ and the constant azimuth angle coordinate surface $\phi=\phi_0$. The radial contravariant basis vector $\hat{\mathbf{e}}_r$ points outward from point P on the spherical surface. Likewise, the polar basis vector $\hat{\mathbf{e}}_\theta$ is a unit vector tangent to the coordinate curve, formed by the intersection of the constant radial coordinate surface $r=r_0$ and constant azimuth angle coordinate surface $\phi=\phi_0$. And finally, the azimuth basis vector $\hat{\mathbf{e}}_\phi$ is a unit vector tangent to the coordinate curve, formed by the intersection of the constant radial coordinate surface $r=r_0$ and the constant polar angle coordinate surface $\theta=\theta_0$.

The position vector $\mathbf{r}(x,y,z)$ is given with its rectangular coordinates (x,y,z) expressed in terms of spherical coordinates (r,θ,ϕ):

$$\mathbf{r}(x,y,z)=r\sin\theta\cos\phi\hat{\mathbf{x}}+r\sin\theta\sin\phi\hat{\mathbf{y}}+r\cos\theta\hat{\mathbf{z}} \tag{15.174}$$

The three natural basis vectors $(\mathbf{e}_r,\mathbf{e}_\theta,\mathbf{e}_\phi)$ are thus directly obtained by differentiation:

$$\mathbf{e}_r=\frac{\partial\mathbf{r}}{\partial r}=\sin\theta\cos\phi\hat{\mathbf{x}}+\sin\theta\sin\phi\hat{\mathbf{y}}+\cos\theta\,\hat{\mathbf{z}} \tag{15.175}$$

$$\mathbf{e}_\theta=\frac{\partial\mathbf{r}}{\partial\theta}=r\cos\theta\cos\phi\hat{\mathbf{x}}+r\cos\theta\sin\phi\hat{\mathbf{y}}-r\sin\theta\hat{\mathbf{z}} \tag{15.176}$$

$$\mathbf{e}_\phi=\frac{\partial\mathbf{r}}{\partial\phi}=-r\sin\theta\sin\phi\hat{\mathbf{x}}+r\sin\theta\cos\phi\hat{\mathbf{y}} \tag{15.177}$$

These natural basis vectors are not normalized.

If the vector position is given in rectangular coordinates only, where

$$\mathbf{r}(x,y,z)=x\hat{\mathbf{x}}+y\hat{\mathbf{y}}+z\hat{\mathbf{z}} \tag{15.178}$$

then one must apply the chain rule to obtain three natural basis vectors $(\mathbf{e}_r,\mathbf{e}_\theta,\mathbf{e}_\phi)$ in the spherical coordinate system:

$$\mathbf{e}_u=\mathbf{e}_r=\frac{\partial\mathbf{r}(x,y,z)}{\partial r}=\frac{\partial\mathbf{r}(x,y,z)}{\partial x}\frac{\partial x}{\partial r}+\frac{\partial\mathbf{r}(x,y,z)}{\partial y}\frac{\partial y}{\partial r}+\frac{\mathrm{d}\mathbf{r}(x,y,z)}{\mathrm{d}z}\frac{\partial z}{\partial r} \tag{15.179}$$

$$\mathbf{e}_v=\mathbf{e}_\theta=\frac{\partial\mathbf{r}(x,y,z)}{\partial\theta}=\frac{\partial\mathbf{r}(x,y,z)}{\partial x}\frac{\partial x}{\partial\theta}+\frac{\partial\mathbf{r}(x,y,z)}{\partial y}\frac{\partial y}{\partial\theta}+\frac{\mathrm{d}\mathbf{r}(x,y,z)}{\mathrm{d}z}\frac{\partial z}{\partial\theta} \tag{15.180}$$

$$\mathbf{e}_w=\mathbf{e}_\phi=\frac{\partial\mathbf{r}(x,y,z)}{\partial\phi}=\frac{\partial\mathbf{r}(x,y,z)}{\partial x}\frac{\partial x}{\partial\phi}+\frac{\partial\mathbf{r}(x,y,z)}{\partial y}\frac{\partial y}{\partial\phi}+\frac{\mathrm{d}\mathbf{r}(x,y,z)}{\mathrm{d}z}\frac{\partial z}{\partial\phi} \tag{15.181}$$

The natural basis vectors $\hat{\mathbf{x}}, \hat{\mathbf{y}}, \hat{\mathbf{z}}$ for the rectangular coordinate system (x, y, z) are obtained via

$$\frac{\partial \mathbf{r}(x, y, z)}{\partial x} = \hat{\mathbf{x}} \tag{15.182}$$

$$\frac{\partial \mathbf{r}(x, y, z)}{\partial y} = \hat{\mathbf{y}} \tag{15.183}$$

$$\frac{\partial \mathbf{r}(x, y, z)}{\partial z} = \hat{\mathbf{z}} \tag{15.184}$$

Substitution gives the same unnormalized expressions for the natural basis vectors $(\mathbf{e}_r, \mathbf{e}_\theta, \mathbf{e}_\phi)$ in the spherical coordinate system in terms of the (x, y, z) rectangular coordinate system's natural basis vectors $(\hat{\mathbf{x}}, \hat{\mathbf{y}}, \hat{\mathbf{z}})$ as obtained previously. The normalized spherical coordinate system's natural basis vectors $(\hat{\mathbf{e}}_u, \hat{\mathbf{e}}_v, \hat{\mathbf{e}}_w)$ are obtained from the unnormalized natural basis vectors $(\mathbf{e}_r, \mathbf{e}_\theta, \mathbf{e}_\varphi)$ by division of the ith natural basis vector $\mathbf{e}_i$ by its magnitude $|\mathbf{e}_i|$, where

$$\hat{\mathbf{e}}_u = \frac{\mathbf{e}_u}{|\mathbf{e}_u|} \Rightarrow |\mathbf{e}_r| = 1 \text{ and } \hat{\mathbf{e}}_r = \mathbf{e}_r \tag{15.185}$$

$$\hat{\mathbf{e}}_v = \frac{\mathbf{e}_v}{|\mathbf{e}_v|} \Rightarrow |\mathbf{e}_\theta| = r \text{ and } \hat{\mathbf{e}}_\theta = \frac{\mathbf{e}_\theta}{r} \tag{15.186}$$

$$\hat{\mathbf{e}}_w = \frac{\mathbf{e}_w}{|\mathbf{e}_w|} \Rightarrow |\mathbf{e}_\phi| = r \sin\theta \text{ and } \hat{\mathbf{e}}_\phi = \frac{\mathbf{e}_\phi}{r \sin\theta} \tag{15.187}$$

Finally, the natural basis vectors for the spherical coordinate system are mutually orthogonal because the dot product between the different components is zero, where

$$\hat{\mathbf{e}}_r \cdot \hat{\mathbf{e}}_\theta = \hat{\mathbf{e}}_\theta \cdot \hat{\mathbf{e}}_\phi = \hat{\mathbf{e}}_\phi \cdot \hat{\mathbf{e}}_r = 0 \tag{15.188}$$

15.12 Geometric View of Contravariant Representation of a Vector

We now present a geometric interpretation of the contravariant representation of vectors and tensors. Consider the 2-D case in which a contravariant vector **A** extends from the origin O to a point P in a nonorthogonal coordinate system defined by the skewed (i.e., tilted rather than perpendicularly vertical) X^1 and X^2 axes. The components A^1 and A^2 of vector **A** are parallel to skewed axes X^1 and X^2. As an example, the tilted X^1 and X^2 coordinate axes are not perpendicular in the Minkowski diagram for four-vector space, where the spatial and time components of the four-vector are parallel to the untilted axes of the rest frame, and are also parallel to the tilted axes of the moving frame.

If the contravariant vector **A** increases incrementally in length from its endpoint P by differential change d**A**, then the components of d**A** are dA^1 and dA^2 are also parallel to the skewed axes X^1 and X^2. The point P is a shared vertex point for two similar triangles. The first triangle is formed by vector **A** and components A^1 and A^2, and the second triangle is formed by differential change d**A** and components dA^1 and dA^2. The components dA^1 and dA^2 extend over the respective intervals dX^1 and dX^2 on the X^1 and X^2 axes, so that

$$\frac{A^1}{A} = \frac{dA^1}{dA} = \frac{\partial X^1}{\partial A} \quad \text{and} \quad \frac{A^2}{A} = \frac{dA^2}{dA} = \frac{\partial X^2}{\partial A} \tag{15.189}$$

Rearrangement of Eq. (15.189), confirms the transformation relations of a contravariant vector, where

$$A^1 = \frac{\partial X^1}{\partial A}A \quad \text{and} \quad A^2 = \frac{dX^2}{dA}A \tag{15.190}$$

If contravariant vector **A** is now represented by a different set of oblique axes (x^1, x^2), then the components A^1 and A^2 of vector **A** in the coordinate system with oblique axes (X^1, X^2) transform to components a^1 and a^2 in the new coordinate system with axes (x^1, x^2), where

$$a^1 = \frac{\partial x^1}{\partial A}A \quad \text{and} \quad a^2 = \frac{\partial x^2}{\partial A}A \tag{15.191}$$

Application of the chain rule to Eq. (15.191), gives

$$a^1 = \left(\frac{\partial x^1}{\partial X^1}\frac{\partial X^1}{\partial A} + \frac{\partial x^1}{\partial X^2}\frac{\partial X^2}{\partial A}\right)A \tag{15.192}$$

and

$$a^2 = \left(\frac{\partial x^2}{\partial X^1}\frac{\partial X^1}{\partial A} + \frac{\partial x^2}{\partial X^2}\frac{\partial X^2}{\partial A}\right)A \tag{15.193}$$

Substitution into Eq. (15.192) and Eq. (15.193) of Eq. (15.190) for $(\partial X^j/\partial A)A$ with $j = 1, 2$ then yields the rule for transformation of the components of contravariant vectors under a change of coordinate systems and their associated basis vectors, where

$$a^1 = \frac{\partial x^1}{\partial X^1}A^1 + \frac{\partial x^1}{\partial X^2}A^2 \tag{15.194}$$

and

$$a^2 = \frac{\partial x^2}{\partial X^1} A^1 + \frac{\partial x^2}{\partial X^2} A^2 \tag{15.195}$$

This geometric description of the contravariant vector can be applied to 3-, 4-, or n-D coordinate systems. With $i = 1, 2, 3, \ldots, n$ components dA^i of the n-D contravariant vector **A** parallel to the coordinate axes, this geometric analysis demonstrates that the components of the contravariant vector transform in the same manner as coordinate differentials, where

$$a^i = \frac{\partial x^i}{\partial X^j} A^j \quad i, j = 1, 2, 3, \ldots, n \tag{15.196}$$

15.13 Geometric View of Covariant (Dual) Representation of a Vector

We now examine the dual or covariant vector counterpart to the contravariant vector. In review, the contravariant vector is a vector whose components transform in the same manner as coordinate differentials, where the components are parallel to the axes of a coordinate system that has perpendicular or skewed (tilted non-perpendicular) axes. And the contravariant vectors have basis vectors, which are the tangent vectors at the surface points, and are defined by the contravariant vectors. Also, contravariant vectors commonly have units of distance. Finally, the contravariant vector components transform inversely from old to new basis vectors.

Surfaces and the volume enclosed are defined by either tangential vectors at each surface point or by normal vectors at each surface point. And the components of the vector can be formed by components parallel to the coordinate system axes, or the vector components can be formed by perpendicular projections onto the coordinate system axes. Also, vector components can vary inversely with a change of basis vectors and coordinate systems, or the vector components could vary in the same manner as the change of base vectors. These dual possibilities suggest the existence of a dual vector to the vector. Indeed, corresponding to vectors or contravariant vectors are dual vectors or covariant vectors. The normalized covariant basis vectors are identical to the normalized contravariant basis vectors for the rectangular coordinate system, so in either representation, the normalized basis vectors for the rectangular coordinate system are $(\hat{\mathbf{x}}, \hat{\mathbf{y}}, \hat{\mathbf{z}})$.

As the name *covariant vector* suggests, the components of this vector must covary with a change or variance in the basis vectors that accompany a change of coordinate systems. Yet the magnitude of the vector must remain constant in that the vector magnitude represents a physical quantity. The dual vector has a set of basis vectors that are the normal vectors at the surface points defined by this vector. And the components of the dual vector would be formed by perpendicular projections onto the coordinate system axes. Also, the dual vector would have or contain units of inverse distance (spatial wave number). Finally, the dual vector components transform in the same manner as the transformation of the basis vectors. As in the case of contravariant vectors, generally the covariant basis vectors $\mathbf{e}_i$ are not unit vectors and the covariant basis vectors $\mathbf{e}_i$ are not mutually perpendicular. The covariant vector components (e.g., E_i) have lower indices, and the basis vectors $\mathbf{e}^i$ for the covariant vector have upper indices and follow Einstein summation, where

$$\mathbf{E} = E_i\mathbf{e}^i \equiv \sum_{i=1}^{n} E_i\mathbf{e}^i \quad i = 1, 2, 3, \ldots, n \tag{15.197}$$

The basis vector $\mathbf{e}^i$ vectors for the covariant vector are also termed *dual basis vectors.*

We now present a geometric interpretation of the covariant representation of vectors and tensors. Consider again the 2-D case in which a contravariant vector **A** extends from the origin O to a point P in a nonorthogonal coordinate system defined by the skewed X^1 and the X^2 axes. The components A_1 and A_2 of vector **A**, however, are no longer parallel to skewed axes X^1 and X^2, but rather, the components A_1 and A_2 are the projections of vector **A** onto the X^1 and the X^2 axes. The projection of vector **A** onto an axis X^1 is formed by a straight line $L_{\perp 1}$ extending from the tip of the vector **A** to the axis X^1 and perpendicular to axis X^1. In a similar manner, all the components of covariant vector **A** are formed by projections onto the set of skewed or unskewed coordinate axes. Given the tilted X^1 and X^2 coordinate axes of the Minkowski diagram for four-vector space, the spatial and time components of the four-vector would be formed by projections onto the untilted axes of the rest frame, and projections onto the tilted axes of the moving frame.

If the contravariant vector **A** increases incrementally in length from its endpoint P by differential change d**A**, then the components of d**A** are $\mathrm{d}A_1$ and $\mathrm{d}A_2$ are also projections to the skewed axes X^1 and X^2, which are obtained by dropping lines perpendicular to the axes from the tip of the incremental vector d**A** onto the coordinate axes. The point P is a shared vertex point for only of the two similar triangles. The first triangle, whose vertex is at the origin O, is formed by vector **A** and components A_1 that lies and extends along the X^1 axis and component A_2 that is perpendicular to the same axis X^1. There is an angle θ between vector **A** and its component A_1 that lies on the X^1 axis. The second triangle whose vertex is at point P is formed by differential change d**A** and component $\mathrm{d}A_1$ that is parallel to the X^1 axis and component $\mathrm{d}A_2$ that is perpendicular to the same axis X^1.

There is the same angle θ between differential vector d**A** and its component $\mathrm{d}A_1$ that is parallel to the X^1 axis. One thus can form the relations

$$\cos\theta = \frac{A_1}{A} = \frac{\mathrm{d}A}{\mathrm{d}A_1} = \frac{\partial A}{\partial X^1} \tag{15.198}$$

and

$$\sin\theta = \frac{A_2}{A} = \frac{\mathrm{d}A}{\mathrm{d}A_2} = \frac{\partial A}{\partial X^2} \tag{15.199}$$

Rearrangement of Eq. (15.198) and Eq. (15.199), confirms the transformation relations of a covariant vector, where

$$A_1 = \frac{\partial A}{\partial X^1}A \quad \text{and} \quad A_2 = \frac{\mathrm{d}A}{\mathrm{d}X^2}A \tag{15.200}$$

If covariant vector **A** is now represented by a different set of oblique axes (x^1, x^2), then the components A_1 and A_2 of vector **A** in the coordinate system with oblique axes (X^1, X^2) transform to components a_1 and a_2 in the new coordinate system with axes (x^1, x^2), where

$$a_1 = \frac{\partial A}{\partial x_1}A \quad \text{and} \quad a_2 = \frac{\partial A}{\partial x_2}A \tag{15.201}$$

Application of the chain rule to Eq. (15.201), gives

$$a_1 = \left(\frac{\partial A}{\partial X^1}\frac{\partial X^1}{\partial x^1} + \frac{\partial A}{\partial X^2}\frac{\partial X^2}{\partial x^1}\right)A \tag{15.202}$$

and

$$a_2 = \left(\frac{\partial A}{\partial X^1}\frac{\partial X^1}{\partial x^2} + \frac{\partial A}{\partial X^2}\frac{\partial X^2}{\partial x^2}\right)A \tag{15.203}$$

Substitution for $(\partial X^j/\partial x^i)A$ with $i, j = 1, 2$ then yields the rule for transformation of the components of covariant vectors under a change of coordinate systems and their associated basis vectors, where

$$a_1 = \frac{\partial X^1}{\partial x^1}A_1 + \frac{\partial X^2}{\partial x^1}A_2 \tag{15.204}$$

and

$$a_2 = \frac{\partial X^1}{\partial x^2}A_1 + \frac{\partial X^2}{\partial x^2}A_2 \tag{15.205}$$

This geometric description of the contravariant vector can be applied to 3-, 4-, or n-D coordinate systems. With $i = 1, 2, 3, \ldots, n$ components $\mathrm{d}A_i$ of the n-D covariant vector **A** as projections onto the coordinate axes, this geometric analysis shows that the components of the covariant vector transform in the same manner as the gradient of a scalar function, where

$$a_i = \frac{\partial X^j}{\partial x_i} A_j \quad i, j = 1, 2, 3, \ldots, n \tag{15.206}$$

In special relativity, the Loedel diagram is a variant of the Minkowski diagram applied to contravariant four-vectors, and the Brehme diagram is a variant of the Minkowski diagram applied to covariant four-vectors. Further details on these diagrammatic methods of visualizing four-vectors in special relativity can be found in Shadowitz (1988a, b).

15.14 Covariant Vectors Transformation Is Similar to the Gradient of a Scalar Function

The gradient of a scalar function, such as an electric field $\mathbf{E} = -\nabla\phi$, which is obtained from the scalar electric potential, is an example of a covariant/dual vector. *Covariant* denotes a vector that varies in the variance or gradient. The gradient ∇ is found by "dividing" the scalar electric potential $\phi(X, Y, Z)$ by the differential vector $\mathrm{d}\mathbf{R}$, where

$$\nabla = \hat{\mathbf{X}}\frac{\partial}{\partial X} + \hat{\mathbf{Y}}\frac{\partial}{\partial Y} + \hat{\mathbf{Z}}\frac{\partial}{\partial Z} \tag{15.207}$$

and

$$\mathrm{d}\mathbf{R} = \mathrm{d}X\,\hat{\mathbf{X}} + \mathrm{d}Y\,\hat{\mathbf{Y}} + \mathrm{d}Z\,\hat{\mathbf{Z}} \tag{15.208}$$

The covariant electric field vector in the (X, Y, Z) coordinate system is then

$$\mathbf{E} = E_X\hat{\mathbf{X}} + E_Y\hat{\mathbf{Y}} + E_Z\hat{\mathbf{Z}} = -\nabla\phi = -\left(\hat{\mathbf{X}}\frac{\partial\phi}{\partial X} + \hat{\mathbf{Y}}\frac{\partial\phi}{\partial Y} + \hat{\mathbf{Z}}\frac{\partial\phi}{\partial Z}\right) \tag{15.209}$$

The covariant electric field vector contains units of inverse length, as expected for a covariant/dual vector. In a different coordinate system (x, y, z), which is denoted by lowercase letters, the scalar electric potential $\phi(x, y, z)$ is expressed by

$$\mathbf{E} = E_x\hat{\mathbf{x}} + E_y\hat{\mathbf{y}} + E_z\hat{\mathbf{z}} = -\nabla\phi = -\left(\hat{\mathbf{x}}\frac{\partial\phi}{\partial x} + \hat{\mathbf{y}}\frac{\partial\phi}{\partial y} + \hat{\mathbf{z}}\frac{\partial\phi}{\partial z}\right) \tag{15.210}$$

where

$$\frac{\partial \phi}{\partial x} = \frac{\partial \phi}{\partial X}\frac{\partial X}{\partial x} + \frac{\partial \phi}{\partial Y}\frac{\partial Y}{\partial x} + \frac{\partial \phi}{\partial Z}\frac{\partial Z}{\partial x} \tag{15.211}$$

$$\frac{\partial \phi}{\partial y} = \frac{\partial \phi}{\partial X}\frac{\partial X}{\partial y} + \frac{\partial \phi}{\partial Y}\frac{\partial Y}{\partial y} + \frac{\partial \phi}{\partial Z}\frac{\partial Z}{\partial y} \tag{15.212}$$

$$\frac{\partial \phi}{\partial z} = \frac{\partial \phi}{\partial X}\frac{\partial X}{\partial z} + \frac{\partial \phi}{\partial Y}\frac{\partial Y}{\partial z} + \frac{\partial \phi}{\partial Z}\frac{\partial Z}{\partial z} \tag{15.213}$$

In the compactness of the Einstein summation convention with integer indices i and j, the transformation of the electric potential from the (X, Y, Z) coordinate system to the (x, y, z) coordinate system is given by

$$\frac{\partial \phi}{\partial x^i} = \frac{\partial \phi}{\partial X^j}\frac{\partial X^j}{\partial x^i} \quad i, j = 1, 2, 3 \tag{15.214}$$

where

$$i, k = (1, 2, 3) = (x, y, z) \quad \text{and} \quad j = (1, 2, 3) = (X, Y, Z) \tag{15.215}$$

Let scalar quantity ϕ be a component x^k of position vector $\mathbf{r}$ so that

$$\frac{\partial x^k}{\partial x^i} = \frac{\partial x^k}{\partial X^j}\frac{\partial X^j}{\partial x^i} \tag{15.216}$$

For all three components of position vector $\mathbf{r}$, one obtains

$$\nabla_{xyz} x^k = \frac{\partial x^k}{\partial X^j} \nabla_{xyz} X^j \tag{15.217}$$

The transformation relation from basis vectors $\mathbf{e}^j$ in the curvilinear coordinate system (X, Y, Z) to dual basis vectors $\mathbf{e}^k$ in curvilinear coordinate system (x, y, z), where

$$\mathbf{e}^k = \frac{\partial x^k}{\partial X^j} \mathbf{e}^j \tag{15.218}$$

The dual basis vectors $\mathbf{e}^k$ of the covariant vector in the (x, y, z) coordinate system with $k = x, y, z$, and the dual basis vectors $\mathbf{e}^j$ of the covariant vector in the (X, Y, Z) coordinate system with $j = X, Y, Z$ are

$$\mathbf{e}^k = \nabla_{xyz} x^k \quad \text{and} \quad \mathbf{e}^j = \nabla_{xyz} X^j \tag{15.219}$$

The gradient ∇_{xyz} with respect to the (x, y, z) coordinate system is

$$\nabla_{xyz} = \hat{\mathbf{x}}\frac{\partial}{\partial x} + \hat{\mathbf{y}}\frac{\partial}{\partial y} + \hat{\mathbf{z}}\frac{\partial}{\partial z} \tag{15.220}$$

The covariant basis vector transformation relation can be generalized to covariant four-vectors and covariant n-D vectors from dual basis vectors in the curvilinear coordinate system $(X^1, X^2, X^3, \ldots, X^n)$ to curvilinear coordinate system $(x^1, x^2, x^3, \ldots, x^n)$, where $i, j, k = 1, 2, 3, \ldots, n$.

In short, the covariant or dual vector $\mathbf{A}$ in coordinate system $(X^1, X^2, X^3, \ldots, X^n)$ is expressed by dual basis vectors $\mathbf{e}^j$ and covariant (dual) vector components A_j:

$$\mathbf{A} = \mathbf{e}^j A_j \equiv \sum_{j=1}^{n} \mathbf{e}^j A_j \tag{15.221}$$

Covariant vector $\mathbf{a}$ in coordinate system $(x^1, x^2, x^3, \ldots, n)$ is expressed by dual basis vectors $\mathbf{e}^j$ and covariant vector components a_j:

$$\mathbf{a} = a_i \mathbf{e}^i \equiv \sum_{i=1}^{n} a_i \mathbf{e}^i \tag{15.222}$$

The transformation between covariant components of vectors $\mathbf{A}$ and $\mathbf{a}$ is

$$a_i = \frac{\partial X^j}{\partial x_i} A_j \quad i, j = 1, 2, 3, \ldots, n \tag{15.223}$$

The transformation between the dual basis vectors of covariant vectors $\mathbf{A}$ and $\mathbf{a}$ is

$$\mathbf{e}^k = \frac{\partial x^k}{\partial X^j} \mathbf{e}^j \quad i, j = 1, 2, 3, \ldots, n \tag{15.224}$$

One obtains the dual basis vectors $\mathbf{e}^k$ in the (x, y, z) coordinate system and the dual basis vectors $\mathbf{e}^j$ in the (X, Y, Z) coordinate system by

$$\mathbf{e}^k = \nabla_{xyz} x^k \quad \text{and} \quad \mathbf{e}^j = \nabla_{XYZ} X^j \tag{15.225}$$

where

$$\nabla_{xyz} = \frac{\partial}{\partial x_1}\hat{\mathbf{x}}_1 + \frac{\partial}{\partial x_2}\hat{\mathbf{x}}_2 + \frac{\partial}{\partial x_3}\hat{\mathbf{x}}_3 + \cdots + \frac{\partial}{\partial x_n}\hat{\mathbf{x}}_n \tag{15.226}$$

and

$$\nabla_{XYZ} = \frac{\partial}{\partial X_1}\hat{\mathbf{X}}_1 + \frac{\partial}{\partial X_2}\hat{\mathbf{X}}_2 + \frac{\partial}{\partial X_3}\hat{\mathbf{X}}_3 + \cdots + \frac{\partial}{\partial X_n}\hat{\mathbf{X}}_n \tag{15.227}$$

On a change of coordinate systems, as the components A_j of the covariant vector $\mathbf{A}$ change to components a_i of the covariant vector $\mathbf{a}$, and the dual basis vectors $\mathbf{e}^j$ of $\mathbf{A}$ change to dual basis vector $\mathbf{e}^i$ of $\mathbf{a}$. In the change of coordinate systems for covariant vectors, the component change from A_j to a_i changes in the same manner as the basis vector change from $\mathbf{e}^j$ to $\mathbf{e}^i$. This allows the magnitudes of covariant vectors $\mathbf{A}$ and $\mathbf{a}$ to be identical, where

$$|\mathbf{A}| = |\mathbf{a}| \tag{15.228}$$

In the covariant vector representation, the space–time four-vector has components $(q_1, q_2, q_3, ictq_4)$. The space–time distance interval is a physical quantity that must be the same in all directions and invariant upon a change of coordinate systems from $(q_1, q_2, q_3, ictq_4)$ to $(\lambda_1, \lambda_2, \lambda_3, ict\lambda_4)$, where

$$|\mathrm{d}\boldsymbol{\lambda}|^2 = \sum_{l=1}^{4} (\mathrm{d}\lambda_i)^2 = \sum_{k=1}^{4}\sum_{l=1}^{4} (g^{kl})^2 \mathrm{d}q_k\, \mathrm{d}q_l \tag{15.229}$$

15.15 Dual Basis Vectors are Normal to Surfaces in the Covariant Vector Space

We now derive the covariant basis vectors $(\mathbf{e}^r, \mathbf{e}^\theta, \mathbf{e}^\phi)$ in spherical coordinates. In spherical coordinates, the covariant representation of vector $\mathbf{V}$ uses the dual basis vectors $(\mathbf{e}^r, \mathbf{e}^\theta, \mathbf{e}^\phi)$ with upper indices and covariant vector components with lower indices (V_r, V_θ, V_ϕ). The dual basis vectors $(\mathbf{e}^r, \mathbf{e}^\theta, \mathbf{e}^\phi)$ are the normal vectors at each point on a surface, where the surface is formed by vector $\mathbf{V}$. Alternatively, the contravariant representation of vector $\mathbf{V}$ uses the natural basis vectors and contravariant vectors $(\mathbf{e}_r, \mathbf{e}_\theta, \mathbf{e}_\phi)$ with lower indices and contravariant vector components with upper indices (V^r, V^θ, V^ϕ). The natural basis vectors $(\mathbf{e}_r, \mathbf{e}_\theta, \mathbf{e}_\phi)$ are the tangent vectors at each point on a surface, where the surface is formed by vector $\mathbf{V}$.

The dual basis vectors $(\mathbf{e}^r, \mathbf{e}^\theta, \mathbf{e}^\phi)$ are obtained from the rectangular coordinate system gradient ∇_{xyz} of the spherical coordinate $r(x, y, z)$, $\theta(x, y, z)$, and $\phi(x, y, z)$, which are expressed as a function of the rectangular coordinates (x, y, z), where

$$\mathbf{e}^r = \nabla_{xyz} r \quad \mathbf{e}^\theta = \nabla_{xyz}\theta \quad \mathbf{e}^\phi = \nabla_{xyz}\phi \tag{15.230}$$

and

$$\nabla_{xyz} = \hat{\mathbf{x}}\frac{\partial}{\partial x} + \hat{\mathbf{y}}\frac{\partial}{\partial y} + \hat{\mathbf{z}}\frac{\partial}{\partial z} \tag{15.231}$$

with

$$r(x,y,z) = \sqrt{x^2 + y^2 + z^2} \tag{15.232}$$

$$\theta(x,y,z) = \tan^{-1}\left(\frac{\sqrt{x^2 + y^2}}{z}\right) \tag{15.233}$$

$$\phi(x,y,z) = \tan^{-1}\left(\frac{y}{x}\right) \tag{15.234}$$

Carrying out the gradient operation on Eqs. (15.232–15.234), yields dual basis vectors $(\mathbf{e}^r, \mathbf{e}^\theta, \mathbf{e}^\phi)$ in terms of the rectangular coordinates (x, y, z), where

$$\mathbf{e}^r = \nabla_{xyz} r = \left(\hat{\mathbf{x}}\frac{\partial}{\partial x} + \hat{\mathbf{y}}\frac{\partial}{\partial y} + \hat{\mathbf{z}}\frac{\partial}{\partial z}\right)\sqrt{x^2 + y^2 + z^2} = \hat{\mathbf{x}}x + \hat{\mathbf{y}}y + \frac{\hat{\mathbf{z}}z}{r} \tag{15.235}$$

$$\begin{aligned}
\mathbf{e}^\theta &= \nabla_{xyz}\theta \\
&= \left(\hat{\mathbf{x}}\frac{\partial}{\partial x} + \hat{\mathbf{y}}\frac{\partial}{\partial y} + \hat{\mathbf{z}}\frac{\partial}{\partial z}\right)\tan^{-1}\left(\frac{\sqrt{x^2 + y^2}}{z}\right) \\
&= \left[\hat{\mathbf{x}}x + \frac{\hat{\mathbf{y}}y}{z\sqrt{x^2 + y^2}} - \frac{\hat{\mathbf{z}}\sqrt{x^2 + y^2}}{z^2}\right]\left[1 + \frac{x^2 + y^2}{z^2}\right]^{-1}
\end{aligned} \tag{15.236}$$

and

$$\mathbf{e}^\phi = \nabla_{xyz}\phi = \left(\hat{\mathbf{x}}\frac{\partial}{\partial x} + \hat{\mathbf{y}}\frac{\partial}{\partial y} + \hat{\mathbf{z}}\frac{\partial}{\partial z}\right)\tan^{-1}\left(\frac{y}{x}\right) = \left(-\hat{\mathbf{x}}\frac{y}{x^2} + \hat{\mathbf{y}}\frac{1}{x}\right)\left(1 + \frac{y^2}{x^2}\right)^{-1} \tag{15.237}$$

The rectangular coordinates (x, y, z) are then expressed in terms of (r, θ, ϕ) to obtain the spherical coordinate system, and covariant basis vectors $(\mathbf{e}^r, \mathbf{e}^\theta, \mathbf{e}^\phi)$ expressed by (r, θ, ϕ), where

$$\mathbf{e}^r = \nabla_{xyz} r = \hat{\mathbf{x}}\sin\theta\cos\phi + \hat{\mathbf{y}}\sin\theta\sin\phi + \hat{\mathbf{z}}\cos\theta \tag{15.238}$$

$$\mathbf{e}^\theta = \nabla_{xyz}\theta = \hat{\mathbf{x}}\frac{\cos\theta\cos\phi}{r} + \hat{\mathbf{y}}\frac{\cos\theta\sin\phi}{r} - \hat{\mathbf{z}}\frac{\sin\theta}{r} \tag{15.239}$$

and

$$\mathbf{e}^\phi = \nabla_{xyz}\phi = -\hat{\mathbf{x}}\frac{\sin\phi}{r\sin\theta} + \hat{\mathbf{y}}\frac{\cos\phi}{r\sin\theta} \tag{15.240}$$

These dual basis vectors are not normalized.

The normalized spherical coordinate system's dual basis vectors ($\hat{\mathbf{e}}^u$,$\hat{\mathbf{e}}^v$,$\hat{\mathbf{e}}^w$) are obtained from the unnormalized dual basis vectors ($\mathbf{e}^u, \mathbf{e}^v, \mathbf{e}^w$), by dividing the ith dual basis vector $\mathbf{e}^i$ by its magnitude $|\mathbf{e}^i|$, where

$$\hat{\mathbf{e}}^u = \frac{\mathbf{e}^u}{|\mathbf{e}^u|} \Rightarrow |\mathbf{e}^r| = 1 \text{ and } \hat{\mathbf{e}}^r = \mathbf{e}^r \tag{15.241}$$

$$\hat{\mathbf{e}}^v = \frac{\mathbf{e}^v}{|\mathbf{e}^v|} \Rightarrow |\mathbf{e}^\theta| = \frac{1}{r} \text{ and } \hat{\mathbf{e}}^\theta = r\mathbf{e}^\theta \tag{15.242}$$

$$\hat{\mathbf{e}}^w = \frac{\mathbf{e}^w}{|\mathbf{e}^w|} \Rightarrow |\mathbf{e}^\phi| = \frac{1}{r\sin\theta} \text{ and } \hat{\mathbf{e}}_\phi = r\sin\theta\mathbf{e}_\phi \tag{15.243}$$

The dual basis vectors for the spherical coordinate system are mutually orthogonal because the dot product between the different components is zero, where

$$\hat{\mathbf{e}}^r \cdot \hat{\mathbf{e}}^\theta = \hat{\mathbf{e}}^\theta \cdot \hat{\mathbf{e}}^\phi = \hat{\mathbf{e}}^\phi \cdot \hat{\mathbf{e}}^r = 0 \tag{15.244}$$

The dual basis vectors are generally not unit vectors, as demonstrated by the dual basis vectors ($\mathbf{e}^r, \mathbf{e}^\theta, \mathbf{e}^\phi$) for the spherical coordinate system. The dot product of the unnormalized ith dual basis component vectors $\mathbf{e}^i$ and itself ($\mathbf{e}^i \cdot \mathbf{e}^i$) is 1 or some other nonzero value:

$$\mathbf{e}^r \cdot \mathbf{e}^r = 1 \quad \mathbf{e}^\theta \cdot \mathbf{e}^\theta = \frac{1}{r^2} \quad \mathbf{e}^\phi \cdot \mathbf{e}^r = \frac{1}{r^2\sin^2\theta} \qquad 15.245)$$

However, the dot product of the normalized i^{th} dual basis component vectors $\hat{\mathbf{e}}^i$ and itself ($\hat{\mathbf{e}}^i \cdot \hat{\mathbf{e}}^i$) is just 1:

$$\hat{\mathbf{e}}^r \cdot \hat{\mathbf{e}}^r = 1 \quad \hat{\mathbf{e}}^\theta \cdot \hat{\mathbf{e}}^\theta = 1 \quad \hat{\mathbf{e}}^\phi \cdot \hat{\mathbf{e}}^r = 1 \tag{15.246}$$

15.16 Covariant and Contravariant Tensors, Pseudo and Polar Scalars, Vectors, and Tensors

In this section, the concept of contravariant and covariant vectors is extended to tensors. Also, the contravariant (covariant) forms of the metric tensor are expressed as the dot product of a pair of contravariant (covariant) basis vectors. Two vectors may be multiplied in the manner of a dot product, which produces a scalar, or in the manner of a cross product that produces another vector. Two vectors can also be multiplied in the manner that leads to a tensor. That is, an n-D row vector V can be multiplied from the left by an n-D column vector U to produce an $n \times n$ matrix that represents an n-D tensor T.

For example, consider the 3-D row vector V multiplied from its left side by a 3-D column vector U to produce the 3×3 matrix representation of tensor T, where

$$T = \mathbf{UV} = \begin{bmatrix} U_1 \\ U_2 \\ U_3 \end{bmatrix} \begin{bmatrix} V_1 & V_2 & V_3 \end{bmatrix} = \begin{bmatrix} U_1V_1 & U_1V_2 & U_1V_3 \\ U_2V_1 & U_2V_2 & U_2V_3 \\ U_3V_1 & U_3V_2 & U_3V_3 \end{bmatrix} \tag{15.247}$$

Let covariant vectors U and V be obtained from covariant vectors u and v, respectively, via the covariant transformation of vector components, where

$$U_j = \frac{\partial x^i}{\partial X^j} u_i \equiv \sum_{i=1}^{3} \frac{\partial x^i}{\partial X^j} u_i \tag{15.248}$$

and

$$V_k = \frac{\partial x^l}{\partial X^k} v_l = \sum_{l=1}^{3} \frac{\partial x^l}{\partial X^k} v_l \tag{15.249}$$

Substitution of Eq. (15.248) and Eq. (15.249) into Eq. (15.247) for tensor $T = UV$, yields the transformation from a covariant tensor with components T_{il} to a covariant tensor with components T_{jk}, where

$$T_{jk} = U_j V_k = \frac{\partial x^i}{\partial X^j} \frac{\partial x^l}{\partial X^k} u_i v_l = \frac{\partial x^i}{\partial X^j} \frac{\partial x^l}{\partial X^k} T_{il} \tag{15.250}$$

and

$$T_{il} = u_i v_l \tag{15.251}$$

That is, the transformation between covariant tensors T_{il} and T_{jk} is

$$T_{jk} = \frac{\partial x^i}{\partial X^j} \frac{\partial x^l}{\partial X^k} T_{il} \tag{15.252}$$

Now let contravariant vectors U and V be obtained from contravariant vectors u and v, respectively, via the contravariant transformation of vector components, where

$$U^j = \frac{\partial X^j}{\partial x^i} u^i \equiv \sum_{i=1}^{3} \frac{\partial X^j}{\partial x^i} u^i \tag{15.253}$$

and

$$V^k = \frac{\partial X^k}{\partial x^l} v^l = \sum_{l=1}^{3} \frac{\partial X^k}{\partial x^l} v^l \tag{15.254}$$

Substitution of Eqs. (15.253) and (15.254) into the expression for tensor $T = UV$, yields the transformation from a covariant tensor with components T^{il} to a covariant tensor with components T^{jk}, where

$$T^{jk} = U^j V^k = \frac{\partial X^j}{\partial x^i}\frac{\partial X^k}{\partial x^l} u^i v^l = \frac{\partial X^j}{\partial x^i}\frac{\partial X^k}{\partial x^l} T^{il} \tag{15.255}$$

and

$$T^{il} = u^i v^l \tag{15.256}$$

That is, the transformation between contravariant tensors T^{il} and T^{jk} is

$$T^{jk} = \frac{\partial X^j}{\partial x^i}\frac{\partial X^k}{\partial x^l} T^{il} \tag{15.257}$$

A polar vector, also termed a *contravariant vector* or *vector*, such as the position vector **r**, reverses sign when the coordinate axes are reversed. That is, the coordinate axes reverse direction when the conventional right-handed coordinate system becomes a left-handed coordinate system, where

$$(x, y, z) \to (-x,\ -y,\ -z) \Rightarrow \mathbf{r} \to -\mathbf{r} \tag{15.258}$$

The abbreviation *polar* for *polar vector* refers to the magnittude and direction angle θ for a vector in a polar coordinate system, where $0 \le \theta \le 2\pi$.

A vector is required to have its components transform in a specified manner under the operation of proper rotation. Proper rotation occurs about an axis at an arbitrary angle, while improper rotation is the mirror image of the vector and its rotation axis followed by proper rotation of the mirror-image vector about the mirror-image axis. Improper rotation is also equivalently described as inversion through the origin followed by proper rotation. In summary, polar or contravariant vector changes sign upon an improper rotation, and the mirror image of a polar vector has its longitudinal component reversed, but its transverse components are unchanged. Polar vector reflection across a plane can be observed by looking at oneself in a mirror.

In contrast to polar vectors is the pseudovector, which also is termed an *axial vector* or *covariant vector*. The pseudovector or axial vector does not reverse sign upon reversal of the coordinate systems. Equivalently stated, the axial vector does not change sign under an improper rotation. The mirror image of a pseudovector has an unchanged longitudinal component, but its transverse components are reversed. Examples of pseudovectors include the cross products of vectors, which include angular momentum, torque, and the magnetic field, where

$$(x, y, z) \to (-x,\ -y,\ -z) \Rightarrow \mathbf{c} = \mathbf{a} \times \mathbf{b} \to \mathbf{c} = \mathbf{a} \times \mathbf{b} \tag{15.259}$$

and

$$\mathbf{a} \to -\mathbf{a} \quad \mathbf{b} \to -\mathbf{b} \tag{15.260}$$

In the reversal of polar vectors **a** and **b**, each vector reverses signs ($\mathbf{a} \to -\mathbf{a}$ and $\mathbf{b} \to -\mathbf{b}$) upon change of coordinate axes $(x, y, z) \to (-x, -y, -z)$. The resulting axial vector, cross product $\mathbf{c} = \mathbf{a} \times \mathbf{b}$ of two negative vectors $-\mathbf{a}$ and $-\mathbf{b}$, yields a positive cross product **c**, where $\mathbf{c} \to \mathbf{c}$. Hence, the axial cross product vector **c** does not change sign upon reversal of the coordinate axes. The concept of polar and axial vectors can be extended to scalars and tensors, where upon inversion of the coordinate axes, a pseudoscalar and pseudotensor do not change sign.

The cross product of two vectors $\mathbf{a} \times \mathbf{b}$ defines a plane or area, where the resulting vector **c** is perpendicular to the plane, and thus perpendicular to the two vectors **a** and **b**. By convention, the right-handed rule is employed to pick a single, consistent, particular direction of resulting vector $\mathbf{c} = \mathbf{a} \times \mathbf{b}$. The right-handed rule is based on a right-handed screw. Vector arrows **a** and **b** are mounted and extend from the same point on the threads of a right-handed screw like the minute and hour hands of a clock. Application of torque (i.e., force to the tip of vector arrow **a** in the clockwise direction toward vector arrow **b**) moves the right-handed threads of the screw in the direction of the resulting cross product vector **c**.

The vector cross product **c** of **a** and **b** can be expressed as the product of an antisymmetric matrix with a column vector, where

$$\mathbf{c} = \mathbf{a} \times \mathbf{b} = \begin{bmatrix} 0 & -a_3 & a_2 \\ a_3 & 0 & -a_1 \\ -a_2 & a_1 & 0 \end{bmatrix} \begin{bmatrix} b_1 \\ b_2 \\ b_3 \end{bmatrix} = \begin{bmatrix} 0 & -b_3 & b_2 \\ b_3 & 0 & -b_1 \\ -b_2 & b_1 & 0 \end{bmatrix} \begin{bmatrix} a_1 \\ a_2 \\ a_3 \end{bmatrix} \tag{15.261}$$

and

$$\mathbf{a} = a_1\hat{\mathbf{x}}_1 + a_2\hat{\mathbf{x}}_2 + a_3\hat{\mathbf{x}}_3 \quad \text{and} \quad \mathbf{b} = b_1\hat{\mathbf{x}}_1 + b_2\hat{\mathbf{x}}_2 + b_3\hat{\mathbf{x}}_3 \tag{15.262}$$

Clockwise (counterclockwise) cyclic permutation of indexes $(1, 2, 3)$ for the positive (negative) spatial components of $(\hat{\mathbf{x}}_1, \hat{\mathbf{x}}_2, \hat{\mathbf{x}}_3)$ yields the components of cross product **c**, where

$$\mathbf{c} = (a_2b_3 - a_3b_2)\hat{\mathbf{x}}_1 + (a_3b_1 - a_1b_3)\hat{\mathbf{x}}_2 + (a_1b_2 - a_2b_1)\hat{\mathbf{x}}_3 \tag{15.263}$$

The commutation $[\mathbf{ab}, \mathbf{ba}]$ of the tensor products of vectors **a** and **b** yields the antisymmetric tensor that describes the cross product $\mathbf{c} = \mathbf{a} \times \mathbf{b}$ of the two vectors, as shown by

$$\mathbf{ab} - \mathbf{ba} = \begin{bmatrix} 0 & a_1b_2 - b_1a_2 & a_1b_3 - b_1a_3 \\ a_2b_1 - b_2a_1 & 0 & a_2b_3 - b_2a_3 \\ a_3b_1 - b_3a_1 & a_3b_2 - b_3a_2 & 0 \end{bmatrix} \tag{15.264}$$

where

$$\mathbf{ab} = \begin{bmatrix} a_1b_1 & a_1b_2 & a_1b_3 \\ a_2b_1 & a_2b_2 & a_2b_3 \\ a_3b_1 & a_3b_2 & a_3b_3 \end{bmatrix} \quad \text{and} \quad \mathbf{ba} = \begin{bmatrix} b_1a_1 & b_1a_2 & b_1a_3 \\ b_2a_1 & b_2a_2 & b_2a_3 \\ b_3a_1 & b_3a_2 & b_3a_3 \end{bmatrix} \tag{15.265}$$

Thus, the commutation $[\mathbf{ab}, \mathbf{ba}]$ yields the components of the vector product $\mathbf{c}$, where

$$\mathbf{ab} - \mathbf{ba} = \begin{bmatrix} 0 & c_3 & -c_2 \\ -c_3 & 0 & c_1 \\ c_2 & -c_1 & 0 \end{bmatrix} \tag{15.266}$$

The axial (covariant) vector cross product $\mathbf{c}$ of two polar vectors $\mathbf{a}$ and $\mathbf{b}$ can thus be obtained from the commutation $[\mathbf{ab}, \mathbf{ba}]$ of their tensor product $\mathbf{ab}$.

A mixed tensor has an upper contravariant index and a lower covariant index. The multiplication of a covariant tensor T_{ij} with a contravariant tensor T^{ik} can lead to a mixed tensor, if the upper index in one tensor is identical to the lower index in the other tensor, which allows summation over the pair of identical or dummy indices (Einstein summation convention). This summation leads to contraction of the tensor product, where

$$T_j^k = T_{ij}T^{ik} \equiv \sum_{i=1}^{n} T_{ij}T^{ik} \tag{15.267}$$

15.17 Natural and Dual Basis Vectors are Reciprocals

Consider position vector $\mathbf{r}(q_1, q_2, q_3)$ whose curvilinear coordinates (q_1, q_2, q_3) are expressed in terms of general, curvilinear coordinates (u_1, u_2, u_3) via

$$\mathbf{r}(q_1, q_2, q_3) = q_1(u_1, u_2, u_3)\hat{\mathbf{q}}_1 + q_2(u_1, u_2, u_3)\hat{\mathbf{q}}_2 + q_3(u_1, u_2, u_3)\hat{\mathbf{q}}_3 \tag{15.268}$$

The orthonormality between the natural basis vectors $\mathbf{e}_j$ and the dual basis unit vectors $\mathbf{e}^i$ is obtained via their dot product, where

$$\mathbf{e}^i \cdot \mathbf{e}_j = \nabla u^i \cdot \frac{\partial \mathbf{r}}{\partial u^j} = \frac{\partial u^i}{\partial q_1}\frac{\partial q_1}{\partial u^j} + \frac{\partial u^i}{\partial q_2}\frac{\partial q_2}{\partial u^j} + \frac{\partial u^i}{\partial q_3}\frac{\partial q_3}{\partial u^j} = \frac{\partial u^i}{\partial u^j} = \delta_j^i \tag{15.269}$$

Here, the Kronnecker delta function δ_j^i is defined by

$$\delta_j^i = \begin{matrix} 0 & i \neq j \\ 1 & i = j \end{matrix} \tag{15.270}$$

The dot product of $\mathbf{e}_j$ and $\mathbf{e}^i$ gives three nonzero values of unity and six zero values. The ith dual basis vector $\mathbf{e}^i$ component of dual basis vector $\mathbf{e}^d$ is the ith component of the normal vector $\mathbf{e}^d$ to the surface described by vector $\mathbf{r}$. And the ith dual basis vector $\mathbf{e}^i$ component is the gradient ∇_q of the ith spatial component (u_1, u_2, u_3) with (q_1, q_2, q_3) spatial derivatives, where

$$\mathbf{e}^i = \nabla_q u^i \tag{15.271}$$

The jth natural basis vector $\mathbf{e}_j$ component of natural basis vector $\mathbf{e}$ is the jth component of the tangent vector $\mathbf{e}$ to the surface described by vector $\mathbf{r}$. And the jth natural basis vector $\mathbf{e}_j$ component is the spatial derivative of vector $\mathbf{r}$ with respect to the jth spatial component (u_1, u_2, u_3), where

$$\mathbf{e}_j = \frac{\partial \mathbf{r}}{\partial u^j} \tag{15.272}$$

In spherical coordinates, the dot product of the covariant basis vectors $(\mathbf{e}^r, \mathbf{e}^\theta, \mathbf{e}^\phi)$ and contravariant basis vectors $(\mathbf{e}_r, \mathbf{e}_\theta, \mathbf{e}_\phi)$ is a case of orthonormality between the natural basis vectors $\mathbf{e}_j$ and dual basis unit vectors $\mathbf{e}^i$. The three nonzero, unity dot products $\mathbf{e}^i \cdot \mathbf{e}_j = 1$ for identical indices $i = j = r, \theta, \phi$ are

$$\begin{aligned} \mathbf{e}^r \cdot \mathbf{e}_r &= (\nabla_{xyz} r) \cdot \left(\frac{\partial \mathbf{r}}{\partial r}\right) \\ &= (\hat{\mathbf{x}} \sin\theta \cos\phi + \hat{\mathbf{y}} \sin\theta \sin\phi + \hat{\mathbf{z}} \cos\theta) \\ &\quad \cdot (\sin\theta \cos\phi \hat{\mathbf{x}} + \sin\theta \sin\phi \hat{\mathbf{y}} + \cos\theta \hat{\mathbf{z}}) = 1 \end{aligned} \tag{15.273}$$

$$\begin{aligned} \mathbf{e}^\theta \cdot \mathbf{e}_\theta &= (\nabla_{xyz} \theta) \cdot \left(\frac{\partial \mathbf{r}}{\partial \theta}\right) \\ &= \left(\hat{\mathbf{x}} \frac{\cos\theta \cos\phi}{r} + \hat{\mathbf{y}} \frac{\cos\theta \sin\phi}{r} - \hat{\mathbf{z}} \frac{\sin\theta}{r}\right) \\ &\quad \cdot (r \cos\theta \cos\phi \hat{\mathbf{x}} + r \cos\theta \sin\phi \hat{\mathbf{y}} - r \sin\theta \hat{\mathbf{z}}) = 1 \end{aligned} \tag{15.274}$$

and

$$\begin{aligned} \mathbf{e}^\phi \cdot \mathbf{e}_\phi &= (\nabla_{xyz} \phi) \cdot \left(\frac{\partial \mathbf{r}}{\partial \varphi}\right) \\ &= \left(-\hat{\mathbf{x}} \frac{\sin\phi}{r \sin\theta} + \hat{\mathbf{y}} \frac{\cos\phi}{r \sin\theta}\right) \cdot (-r \sin\theta \sin\phi \hat{\mathbf{x}} + r \sin\theta \cos\phi \hat{\mathbf{y}}) = 1 \end{aligned} \tag{15.275}$$

The other six terms are zero ($\mathbf{e}^i \cdot \mathbf{e}_j = 0$) for different indices $i \neq j = r, \theta, \phi$, where

$$(\nabla_{xyz} r) \cdot \left(\frac{\partial \mathbf{r}}{\partial \theta}\right) = (\nabla_{xyz} r) \cdot \left(\frac{\partial \mathbf{r}}{\partial \phi}\right) = 0 \tag{15.276}$$

$$(\nabla_{xyz} \theta) \cdot \left(\frac{\partial \mathbf{r}}{\partial r}\right) = (\nabla_{xyz} \theta) \cdot \left(\frac{\partial \mathbf{r}}{\partial \phi}\right) = 0 \tag{15.277}$$

$$(\nabla_{xyz} \phi) \cdot \left(\frac{\partial \mathbf{r}}{\partial r}\right) = (\nabla_{xyz} \phi) \cdot \left(\frac{\partial \mathbf{r}}{\partial \theta}\right) = 0 \tag{15.278}$$

The dot product of dual basis vector component $\mathbf{e}^j$ and vector $\boldsymbol{\lambda}$ with contravariant components λ^i and natural basis vector components $\mathbf{e}_i$ gives the contravariant components λ^j, where

$$\boldsymbol{\lambda} \cdot \mathbf{e}^j = \lambda^i \mathbf{e}_i \cdot \mathbf{e}^j = \lambda^i \delta_i^j = \lambda^j \tag{15.279}$$

Similarly, the dot product of natural basis vector component $\mathbf{e}_j$ and vector $\boldsymbol{\lambda}$ with covariant components λ_i and dual basis vector components $\mathbf{e}^i$ gives covariant components λ_j, where

$$\boldsymbol{\lambda} \cdot \mathbf{e}_j = \lambda_i \mathbf{e}^i \cdot \mathbf{e}_j = \lambda_i \delta_j^i = \lambda_j \tag{15.280}$$

15.18 Natural and Dual Basis Vectors and the Metric Tensor

We now examine the metric tensor g_{ij} formed by the dot product of a pair of natural basis vectors due to the dot product of two contravariant vectors. Also, we will examine the metric tensor g^{ij} formed by the dot product of a pair of dual basis vectors due to the dot product of two contravariant vectors.

The dot product of two vectors $\boldsymbol{\lambda}$ and $\boldsymbol{\mu}$ in their contravariant representation is

$$\boldsymbol{\lambda} \cdot \boldsymbol{\mu} = \lambda^i \mathbf{e}_i \cdot \mu^j \mathbf{e}_j = g_{ij} \lambda^i \mu^j \tag{15.281}$$

The Einstein summation convention is assumed in the dot product of the two n-D vectors, where

$$\boldsymbol{\lambda} \cdot \boldsymbol{\mu} = \left(\sum_{i=1}^{n} \lambda^i \mathbf{e}_i\right) \cdot \left(\sum_{j=1}^{n} \mu^j \mathbf{e}_j\right) = \sum_{i=1}^{n} \sum_{j=1}^{n} g_{ij} \lambda^i \mu^j \tag{15.282}$$

In the case of four-vectors ($n = 4$), and Einstein summation convention is assumed henceforth, in which summation over index i is assumed if index i occurs as both a lower and an upper index in the same term. The natural basis metric tensor components g_{ij} is the dot product of the natural basis vectors $\mathbf{e}_i$ and $\mathbf{e}_j$ for vectors $\boldsymbol{\lambda}$ and $\boldsymbol{\mu}$ in their contravariant representation, where

$$g_{ij} = \mathbf{e}_i \cdot \mathbf{e}_j = \frac{\partial \mathbf{r}}{\partial u^i} \cdot \frac{\partial \mathbf{r}}{\partial u^j} \tag{15.283}$$

Next, one considers the dot product of two vectors $\boldsymbol{\lambda}$ and $\boldsymbol{\mu}$ in their covariant representation, where

$$\boldsymbol{\lambda} \cdot \boldsymbol{\mu} = \lambda_i \mathbf{e}^i \cdot \mu_j \mathbf{e}^j = g^{ij} \lambda_i \mu_j \tag{15.284}$$

The dual basis metric tensor components g^{ij} is the dot product of the natural basis vectors $\mathbf{e}^i$ and $\mathbf{e}^j$ for vectors $\boldsymbol{\lambda}$ and $\boldsymbol{\mu}$ in their covariant representation, where

$$g^{ij} = \mathbf{e}^i \cdot \mathbf{e}^j = \nabla_q u^i \cdot \nabla_q u^j \tag{15.285}$$

Finally, one considers the dot product of vector $\boldsymbol{\lambda}$ in its contravariant representation and vector $\boldsymbol{\mu}$ in its covariant representation gives

$$\boldsymbol{\lambda} \cdot \boldsymbol{\mu} = \lambda^i \mathbf{e}_i \cdot \mu_j \mathbf{e}^j = \lambda^i \mu_j \delta^i_j \tag{15.286}$$

Alternatively, the dot product of vector $\boldsymbol{\lambda}$ in its covariant representation and vector $\boldsymbol{\mu}$ in its contravariant representation gives

$$\boldsymbol{\lambda} \cdot \boldsymbol{\mu} = \lambda_i \mathbf{e}^i \cdot \mu^j \mathbf{e}_j = \lambda_i \mu^j \delta^j_i \tag{15.287}$$

In summary, the dot product of vectors $\boldsymbol{\lambda}$ in its covariant or contravariant forms with vector $\boldsymbol{\mu}$ in its covariant or contravariant form, gives rise to four different ways of expressing the cross product, where

$$\boldsymbol{\lambda} \cdot \boldsymbol{\mu} = g_{ij} \lambda^i \mu^j = g^{ij} \lambda_i \mu_j = \lambda^i \mu_j \delta^i_j = \lambda_i \mu^j \delta^j_i \tag{15.288}$$

Due to the dot product, reversing the order of the indices i and j in metric tensor components g_{ij} and g^{ij} do not change their values, where

$$g_{ij} = g_{ji} \quad \text{and} \quad g^{ij} = g^{ji} \tag{15.289}$$

Application of the Kronnecker delta function of Eq. (15.270), gives

$$\boldsymbol{\lambda} \cdot \boldsymbol{\mu} = g_{ij} \lambda^i \mu^j = g^{ij} \lambda_i \mu_j = \lambda^i \mu_i = \lambda_i \mu^i \tag{15.290}$$

For arbitrary vector components λ_i and λ^i, the above relation shows

$$g^{ij} \mu_j \lambda_i = \mu^i \lambda_i \quad \text{and} \quad g_{ij} \mu^j \lambda^i = \mu_i \lambda^i \tag{15.291}$$

One can drop arbitrary λ_i and λ^i, which shows that the dual basis metric g^{ij} can be used to raise the index, and therefore obtain the contravariant components μ^i of vector $\boldsymbol{\mu}$ from its covariant components μ_j, where

$$g^{ij}\mu_j = \mu^i \tag{15.292}$$

And the natural basis metric g_{ij} can be used to lower the index and therefore obtain the covariant components μ_i of vector $\boldsymbol{\mu}$ from its contravariant components μ^j, where

$$g_{ij}\mu^j = \mu_i \tag{15.293}$$

Combining Eqs. (15.292) and (15.293) gives

$$g^{ij}g_{jk}\mu^k = g^{ij}\mu_j = \mu^i \quad \text{and} \quad g_{ij}g^{jk}\mu_k = g_{ij}\mu^j = \mu_i \tag{15.294}$$

Equation (15.294) holds for arbitrary μ^i and μ_i, so that

$$g^{ij}g_{jk} = \delta^i_k \quad \text{and} \quad g_{ij}g^{jk} = \delta^i_k \tag{15.295}$$

A mixed tensor is contravariant with rank s and covariant with rank p, and has s upper indices and p lower indices. Contraction of the tensor occurs, if the upper and lower indices are identical, and summation then occurs over the identical index k. The contracted tensor has a contravariant rank, $s-1$ and covariant rank, $p-1$. As an example of contraction, consider the mixed tensor with contravariant rank 2 and covariant rank 3, where a pair of upper and lower indices are identical. One then obtains a tensor with contravariant rank 1 and covariant rank 2, where

$$T^{ab}_{bcd} \equiv \sum_{b=1}^{4} T^{ab}_{bcd} = T^{a}_{cd} \tag{15.296}$$

The dot product or inner product of vector $\boldsymbol{\lambda}$ with itself; that is, $\boldsymbol{\lambda}\cdot\boldsymbol{\lambda}$ is assumed to be positive definite, where

$$g_{ij}\lambda^i\lambda^j \geq 0 \tag{15.297}$$

The inner product (vector λ^i multiplied by itself) is zero, if the vector is zero, where $\lambda^i = 0$ and

$$g_{ij}\lambda^i\lambda^j = 0 \tag{15.298}$$

However, to provide a model for the space–time of general relativity, we must relax this condition and require the metric tensor be nonsingular in the sense that the matrix $|g_{ij}|$ has an inverse. A manifold or surface that possesses a positive definite metric tensor is Riemannian. If the metric tensor is indefinite, then the description is pseudo-Riemannian (or semi-Riemannian).

The length s of the n-D vector in general, or a four-vector (in particular λ^a), is

$$s = \left|g_{ab}\lambda^a\lambda^b\right|^{1/2} = \left|g^{ab}\lambda_a\lambda_b\right|^{1/2} = \left|\lambda_a\lambda^a\right|^{1/2} \tag{15.299}$$

In particular, the indexes are $a, b = x, y, z, t$ for four-vector coordinates (x, y, z, ict). If vector $\boldsymbol{\lambda}$ has nonzero components, where $\lambda^a \neq 0$, but the dot product of the vector $\boldsymbol{\lambda}$ with itself is zero, where $\boldsymbol{\lambda} \cdot \boldsymbol{\lambda}$, then the metric tensor g_{ab} is indefinite, where

$$|\lambda_a\lambda^a|^{1/2} \equiv \sum_{a=1}^{n} |\lambda_a\lambda^a|^{1/2} = 0 \tag{15.300}$$

In this case, the vector λ^a is described as null. The angle θ between two nonnull vectors λ^a, μ^b is given by

$$\cos\theta = \frac{g_{ab}\lambda^a\mu^b}{\left|g_{cd}\lambda^c\mu^d\right|^{1/2}\left|g_{ef}\lambda^e\mu^f\right|^{1/2}} \tag{15.301}$$

This generalizes the formula

$$\cos\theta = \frac{(\boldsymbol{\lambda}\cdot\boldsymbol{\mu})}{|\boldsymbol{\lambda}||\boldsymbol{\mu}|} \tag{15.302}$$

If the metric tensor is indefinite, this formula can lead to

$$|\cos\theta| > 1 \tag{15.303}$$

This results in a nonreal value of θ. Two vectors are orthogonal or perpendicular, if their inner product is zero. This definition makes sense even if one or both vectors are null. In fact, a null vector is a nonzero vector, which is perpendicular to itself.

A further aspect of indefiniteness in the metric tensor is that we may have a curve whose tangent vector at each point is null, so that at every point,

$$g_{ab}\frac{dx^a}{ds}\frac{dx^b}{ds} = 0 \quad \text{null or light-like} \tag{15.304}$$

This yields a curve of zero length, called a *null curve*. Photons follow a null path. If the tangents to the curve are time-like, then we describe the curve or that part of the curve as time-like; a particle with mass follows a time-like path:

$$g_{ab}\frac{dx^a}{ds}\frac{dx^b}{ds} > 0 \quad \text{time-like} \tag{15.305}$$

A space-like curve is described by a negative product, where

$$g_{ab}\frac{dx^a}{ds}\frac{dx^b}{ds} < 0 \quad \text{space-like} \tag{15.306}$$

15.19 Differentials in Curvilinear Coordinates, Curved Surfaces, and Parallel Transport

In this section, we examine the differentiation of vectors on a curved surface (that is, covariant differentiation), and the need to define parallel transport of vectors along the paths on which the derivatives are defined. The following discussion applies to both the contravariant and covariant forms of a vector. However, in this section, we consider only the contravariant vector representation. Differentiation of a vector or tensor on curved surfaces and spaces, is considerably more complicated than on planar surfaces and rectangular spaces of ordinary Cartesian or Galilean systems of Newtonian and classical mechanics.

Consider a vector field **A**, which is expressed by covariant components A_i at point x^i and components $A_i + dA_i$ at $x^i + dx^i$. In Euclidean, Cartesian, or flat space, the differential dA_i is a vector and the derivative $\partial A_i/\partial x^k$ is a tensor. Vector **A** has $k = 1, 2, 3, 4$ covariant components $A'_k(x^1, x^2, x^3, x^4)$ expressed in a primed Euclidean frame coordinate system by coordinates (x'^1, x'^2, x'^3, x'^4). In the unprimed Euclidean coordinate system (x^1, x^2, x^3, x^4), vector **A** is expressed by covariant components $A_i(x^1, x^2, x^3, x^4)$, and the covariant vector components of vector **A** in the two Euclidean coordinate systems are related by the vector transformation

$$A_i = \frac{\partial x'^k}{\partial x^i}A'_k \tag{15.307}$$

The transformation of the covariant components vector differential d**A** between the primed and unprimed Euclidean coordinate systems, is also a vector transformation, where

$$dA_i = \frac{\partial x'^k}{\partial x^i}dA'_k \tag{15.308}$$

The covariant and contravariant components and transformations are identical in the flat spaces of Euclidean (Cartesian) coordinate systems.

Differentiation of vector and tensor quantities in curvilinear coordinate systems is complicated by the variation in magnitude and direction of basis vectors with positions in the curvilinear space. In Euclidean space, natural and dual basis vectors are identical, and the basis vectors maintain their magnitude and direction from point to point. In curvilinear coordinates, the natural and dual basis vectors differ, and the basis vectors vary in magnitude and direction from point to point on curved surfaces and lines in curvilinear coordinate systems such as cylindrical and spherical coordinate systems. The transformation of the covariant components of the vector differential **dA** between the primed and unprimed curvilinear coordinate systems is given by

$$\mathrm{d}A_i = \frac{\partial x'^k}{\partial x^i}\mathrm{d}A'_k + A'_k \mathrm{d}\left(\frac{\partial x'^k}{\partial x^i}\right) \tag{15.309}$$

Performing the differentiation in Eq. (15.309), yields

$$\mathrm{d}A_i = \frac{\partial x'^k}{\partial x^i}\mathrm{d}A'_k + A'_k \frac{\partial^2 x'^k}{\partial x^i \partial x^j}\mathrm{d}x^j \tag{15.310}$$

or

$$\frac{\mathrm{d}A_i}{\mathrm{d}x^j} = \frac{\partial x'^k}{\partial x^i}\frac{\mathrm{d}A'_k}{\mathrm{d}x^j} + A'_k \frac{\partial^2 x'^k}{\partial x^i \partial x^j} \tag{15.311}$$

In curvilinear coordinates, $\mathrm{d}A^i$ is not a vector unless the primed coordinates x'^k are linear functions of the unprimed coordinates x^k. If the primed coordinates x'^k are linear functions of the unprimed coordinates x^k as in Galilean (Cartesian) space, then the second derivative term is zero, and one obtains the usual vector transformation, where

$$\frac{\partial^2 x'^k}{\partial x^i \partial x^j} = 0 \tag{15.312}$$

and

$$\frac{\mathrm{d}A_i}{\mathrm{d}x^j} = \frac{\partial x'^k}{\partial x^i}\frac{\mathrm{d}A'_k}{dx^j} \tag{15.313}$$

If the differential of a covariant vector $\mathrm{d}A_i$ is to transform like a vector, then one requires the vector covariant components A_i at point x^i and covariant components $A_i + \mathrm{d}A_i$ at $x^i + \mathrm{d}x^i$ be subtracted from each other at the same point in space. In the case of Cartesian coordinates, the parallel translation of the covariant vector components A_i from point x^i to point $x^i + \mathrm{d}x^i$ cause complete overlap of the covariant components A_i and $A_i + \mathrm{d}A_i$. In parallel translation, the covariant vector components A_i are translated parallel to themselves from point to point along a curve connecting points x^i and $x^i + \mathrm{d}x^i$.

In flat Cartesian space, the covariant vector components A_i and $A_i + \mathrm{d}A_i$ at point $x^i + \mathrm{d}x^i$ are identical. Hence, subtraction of covariant vector A_i from covariant vector $A_i + \mathrm{d}A_i$ at point $x^i + \mathrm{d}x^i$ yields differential $\mathrm{d}A_i$. Parallel transport of a covariant vector in rectilinear or Cartesian coordinates, thereby ensures the covariant vector components A_i do not change, due to translation from point x^i to point

$x^i + \mathrm{d}x^i$. In the same manner, if the differential of a contravariant vector $\mathrm{d}A^i$ is to transform like a vector, then one requires the vector contravariant components A^i at point x^i and contravariant components $A^i + \mathrm{d}A^i$ at $x^i + \mathrm{d}x^i$, to be subtracted from each other at the same point in space. Contravariant vector components A^i at point x^i are parallel-transported to point $x^i + \mathrm{d}x^i$ for subtraction from contravariant vector components $A^i + \mathrm{d}A^i$, which yields contravariant vector differential $\mathrm{d}A^i$.

Parallel transport of a vector around a simple closed curve or loop on a flat or curved surface, requires the vector be shifted parallel to itself from point-to-point. On a flat surface, the components of the vector do not change upon parallel transport. However, from point-to-point along a curve, which is embedded on a curved surface, the components of the vector do change. In order to define parallel transport of a vector prom point to point on a loop on a flat or curved surface, one uses the tangent vector at each point on the loop as a reference line.

Consider a simple closed curve or loop, where the position vector $\mathbf{r}(s)$ at each point along the loop depends on the distance s along the loop from some arbitrary starting point $s = 0$ on the loop. The tangent vector to each point on the loop is $\mathbf{dr}/\mathrm{d}s$. In parallel transport of vector $\mathbf{A}$ from point to point along the loop, the angle θ between vector $\mathbf{A}$ and tangent vector $\mathbf{dr}/\mathrm{d}s$ at each point along the curve, remains constant. The loop on the flat or curve surface is assumed to be continuous, but the curve may not necessarily be continuously differentiable. That is, the curve can be a polygon on a flat or curved surface. At the vertex points, where there is an abrupt change in the tangent vector $\mathbf{dr}/\mathrm{d}s$, the vector $\mathbf{A}$ transported along the curve through a vertex causes an abrupt change in the direction of vector $\mathbf{A}$ under parallel transport.

Consider parallel transport of vector $\mathbf{A}$ around a polygon perimeter or a circle, which is embedded on a flat surface. Vector $\mathbf{A}$ situated at starting point $\mathbf{r}(s = 0)$ points at angle θ_0 with respect to the tangent vector $(\mathbf{dr}/\mathrm{d}s)_{s=0}$. As vector $\mathbf{A}$ is parallel-transported from point to point around the polygon or circle, a constant angle θ_0 is maintained with respect to the tangent vector $(\mathbf{dr}/\mathrm{d}s)_s$. After completing one lap around polygon perimeter or circle, vector $\mathbf{A}$ returns to starting point $\mathbf{r}(s = 0)$, with the final angle identical to its initial angle θ_0, which is measured relative to the tangent vector $(\mathbf{dr}/\mathrm{d}s)_{s=0}$.

Next, consider the parallel transport of a vector $\mathbf{A}$ around a polygon perimeter or a circle, which is embedded on a curved, spherical surface. Vector $\mathbf{A}$ situated at starting point $\mathbf{r}(s = 0)$ points at angle θ_0 with respect to the tangent vector $(\mathbf{dr}/\mathrm{d}s)_{s=0}$. As vector $\mathbf{A}$ is parallel-transported from point to point around the polygon or circle, a constant angle θ_0 is maintained with respect to the tangent vector $(\mathbf{dr}/\mathrm{d}s)_s$. After completing one lap around the polygon perimeter or circle, vector $\mathbf{A}$ returns to starting point $\mathbf{r}(s = 0)$ with final angle θ_f, which is measured relative to the tangent vector $(\mathbf{dr}/\mathrm{d}s)_{s=0}$. The final angle θ_f of vector $\mathbf{A}$, relative to the tangent vector $(\mathbf{dr}/\mathrm{d}s)_{s=0}$ after one parallel-transport revolution around the polygon perimeter or circle, has changed from the initial angle θ_0 of vector $\mathbf{A}$, relative to the tangent vector $(\mathbf{dr}/\mathrm{d}s)_{s=0}$ at starting point $\mathbf{r}(s = 0)$. The change in direction of vector $\mathbf{A}$ after one parallel-transport revolution around the polygon perimeter or circle on a curved surface, such as a spherical surface, indicates that the surface is curved.

Let us consider in more detail the surface of a sphere. A great circle or Riemannian circle of a sphere is defined as the intersection of the sphere and a plane that intercepts the center point of the sphere. The great circle shares the same center point as the sphere, and the great circle is the largest circle, which can be drawn on a sphere. The diameter and the circumference of any great circle is equal the diameter and circumference of the sphere. The shortest distance between two points on a plane is a straight line, and the shortest distance between two points on a spherical surface is the arc length along a great circle on which lie the two points. A small circle of a sphere is defined as the intersection of a plane and a sphere, where the plane does not contain the center point of the sphere. Small circles have smaller diameters (and thus larger curvatures) than the sphere and great circles. Since the small-circle curvature is larger than that of a great circle, the distance along the arc of a small circle that connects two points is longer than the arc of a great circle that connects the two points.

Now we consider parallel transport of vector **A** around a great circle on a spherical surface such as the Earth, where vector **A** maintains the same angle θ_0 relative to the tangent vector at each point on the great circle. To allow easier visualization, suppose that the vector **A** angle is zero $\theta_0 = 0$, so the direction of vector **A** is parallel to the tangent vector at each point on the great circle. Thus, vector **A** starts with angle $\theta_0 = 0$ relative to the prime meridian line of longitude at the Earth's equator. Vector **A** is parallel-transported along the prime meridian to the north pole. At the north pole, vector **A** continues on the great circle, which consists of the prime meridian and the 180° line of longitude. Vector **A** is parallel-transported along the 180° line of longitude past the equator and onto the south pole. At the south pole, vector **A** continues along the same line of longitude, which is now again the prime meridian, and returns to its starting point at the equator. At the equator, the final pointing direction of vector **A**, after one revolution of parallel transport around the great circle, is the same as the initial vector **A** pointing direction prior to parallel transport.

Next, we consider the parallel transport of vector **A** on a spherical surface, around a small circle or a loop, which is not a great circle. As an example of a loop that is not a great circle, one particular loop that can be easily visualized, is the parallel transport of vector **A** around a triangle, which lies on the Earth's surface. Parallel transport of vector **A** starts at a point on the equator on the prime meridian. The parallel transport of vector **A** around the triangle loop, which is embedded on the Earth's spherical surface, requires a constant angle θ_0 between vector **A** and the tangent vector $(d\mathbf{r}/ds)_s$ from point to point around the triangle. Again to aid visualization, assume that the angle is zero $\theta_0 = 0$, so vector **A** is parallel to the tangent vector at each point on the triangle path. Vector **A** is parallel-transported along the prime meridian to the north pole. However, at the north pole, vector **A** and the tangent vector $(d\mathbf{r}/ds)_s$ change direction and turn a corner at the top vertex of the triangle toward the south. Vector **A** travels southward down the 90° line of longitude, maintaining the same $\theta_0 = 0$ relative to the tangent vector at each point until the vector reaches the equator. At the equator, vector **A** and tangent vector $(d\mathbf{r}/ds)_s$ again change direction and turn a corner at the right vertex of the triangle toward the west. Vector **A** is parallel-transported westward along the equator line to its original starting point at the prime meridian. Vector **A** maintained a constant angle $\theta_0 = 0$ with respect to the tangent vector $(d\mathbf{r}/ds)_s$ at each point of the triangle loop on the spherical surface. However, in the one revolution of parallel transport of vector **A** around the triangle loop, the tangent vector $(d\mathbf{r}/ds)_s$ and vector **A** point in a different direction than their initial directions.

Examination of other examples of curved segments and closed curves on spherical surfaces, which shows the change in direction of vector **A**, and the tangent vector $(d\mathbf{r}/ds)_s$ on a curve segment, or closed simple or complicated curve, or spherical surface, depends on the path of the curve segment or closed curve. Consider examples of curve segments and closed curves on curved surfaces. The curve segment can be the portion of a curve or a parabola or cycloid curve. The closed curve can be a simple closed curve, which is a nonself-intersecting continuous loop, such as a circle, a bicuspid curve, or a curve on a Möbius strip surface. However, the curve can also be a complicated closed curve, which is a self-intersecting curve, such as a Lissajous curve of complex harmonic motion.

Consider two people who have only 2-D perception of left–right and forward–backward, with no height perception. The two people are assumed to have no physical perception, that they are on a spherical surface. Instead, they perceive themselves as positioned on a planar surface, which is the local tangent plane at the points where they stand. Starting from two separate points at the equator, the two people move along parallel lines of longitude to the north pole. The lines of longitude start as separate parallel lines at the equator and then converge to a single point at the north pole.

The two people can consider either (1) there is an attractive force between them as they move in their separate paths from the equator to the single point on the north pole or (2) they are moving on converging, curved paths on a spherical surface, which causes a continuous change in their direction (acceleration), as their paths converge to a single point on the north pole. From these two possibilities, the people can conclude the force between them is real or conclude the force is a result of their propagation on curved paths on a spherical surface. The apple fell from the tree onto Newton, who produced the laws of motion and gravity. Einstein observed the falling (accelerating) apple and concluded the curvature of the apple surface, like the curvature of space, is caused by the mass, which is contained within the apple or spatial volume. And thus Einstein concluded that (1) the acceleration of the apple, which is caused by the gravity force of the earth's matter, is equivalent to (2) the acceleration of the apple, which is caused by the change in direction of the apple trajectory path along a curved surface on a four-dimensional, spatial volume. As presented in elementary and graphic detail by Adams (1997), matter bends space, and space determines particle trajectory; that is, the space-time metric is warped by matter (and thus warped by energy as well, via $E = \gamma m_0 c^2$), and the particle trajectory is then determined by the warped space-time metric.

15.20 Absolute and Covariant Derivatives of Vectors with Contravariant Representation

In this section, we examine absolute and covariant derivatives of vectors and tensors with contravariant representation of the vector and tensor components. General relativity postulates that space–time is a smooth 4-D Riemannian manifold M, where points are termed *events*. Locally, the manifold M is the Minkowski 4-D space–time in which local relativity is valid locally. That is, special relativity is obeyed at each point and the small flat space region surrounding each point on the global manifold M. Freely falling particles move on time-like geodesics of manifold M, where

$$g_{ab}\frac{dx^a}{ds}\frac{dx^b}{ds} > 0 \quad \text{time-like} \tag{15.314}$$

The strong equivalence principle of general relativity states that all physical laws that hold in the flat, Minkowski space (special relativity) can be expressed in terms of vectors and tensors. These vector and tensors can be expressed as functions of the points on the manifold M and hold in every reference frame or coordinate system, provided that the ordinary derivatives are replaced by absolute derivatives.

In the case of curvilinear coordinates and curved surfaces and spaces, one must transform $\partial A_i/\partial x^j$ from Cartesian to curvilinear coordinates in such a manner that the physical quantity represented by $\partial A_i/\partial x^j$ is independent of the coordinate system, which requires $\partial A_i/\partial x^j$ transforms as a vector for changes of curvilinear coordinate systems. In the curvilinear coordinate system, the vector $\mathbf{A}$ is parallel-transported from points x^i to $x^i + dx^i$. Consider vector $\mathbf{A}$ expressed in its contravariant form, in which its contravariant vector components A^i are parallel-transported from point x^i to point $x^i + dx^i$. At point , the contravariant components of vector $\mathbf{A}$ are A^i, and at an infinitesimally close point at $x^i + dx^i$, the components of vector $\mathbf{A}$ are $A^i + dA^i$. However, in the parallel transport of vector $\mathbf{A}$ from point x^i to the infinitesimally close point at $x^i + dx^i$, the change δA^i in contravariant components of vector $\mathbf{A}$ can differ from the evaluated difference dA^i of $\mathbf{A}(x^i)$ and $\mathbf{A}(x^i + dx^i)$.

The difference between the evaluated vector $\mathbf{A}$ at points x^i and $x^i + dx^i$ and the change of vector $\mathbf{A}$ upon parallel translation from points x^i and $x^i + dx^i$ is then defined as the absolute contravariant differential DA^i in terms of the contravariant vector components A^i of vector $\mathbf{A}$, where

$$DA^i = dA^i - \delta A^i \tag{15.315}$$

The change δA^i in the contravariant components of vector $\mathbf{A}$ upon parallel transport between infinitesimally close points is linearly dependent on the values of the contravariant components A^i. If the transformation of the sum of two vectors is identical to the transformation of the sum of the vector components, then one requires a linear dependence between δA^i and A^i.

The linear dependence between change δA^i due to vector $\mathbf{A}$ translation between two infinitesimally close points x^i and $x^i + dx^i$ and the evaluated difference dA^i of vector $\mathbf{A}$ calculated at points x^i and $x^i + dx^i$ can be expressed by the Christoffel function of the second kind Γ^i_{kj} acting as a proportionality factor. Given the differential vector $\mathbf{dx}$ between points x^i and $x^i + dx^i$ and the contravariant vector components dx^j of $\mathbf{dx}$, one obtains the linear relation between δA^i and A^i as

$$\delta A^i = -\Gamma^i_{kj}A^k\, dx^j \tag{15.316}$$

The Christoffel function of the second kind or connection coefficient Γ^i_{kj} is a function of the coordinates x^i of position vector $\mathbf{x}$ in the general curvilinear coordinate system. In Galilean or Cartesian coordinate systems of flat surfaces and spaces, Γ^i_{kj} is zero:

$$\Gamma^i_{kj} = 0 \quad \text{Cartesian (Galilean) coordinate systems} \tag{15.317}$$

In curvilinear coordinate systems, such as the spherical and cylindrical coordinate systems, Γ^i_{kj} is nonzero in general.

If a vector or tensor, which is independent of coordinate systems, is zero in one coordinate system, then the vector or tensor is zero in all the coordinate systems. The connection coefficient Γ^i_{kj} terms are not tensors because in general, if all the terms Γ^i_{kj} can be made to be zero at all points in one particular curvilinear coordinate system, Γ^i_{kj} is nonzero in other curvilinear coordinate systems. For example, the connection coefficient Γ^i_{kj} acts as the gravitational strength terms in general relativity, and Γ^i_{kj} can be set to be zero over an infinitesimal region of space via the application of the equivalence principle. However, in other regions of space or other curvilinear coordinate systems, the gravity term Γ^i_{kj} is nonzero.

Substitution of Eq. (15.316) for δA^i in the contravariant absolute differential expression of Eq. (15.315), gives

$$\mathrm{D}A^i = \mathrm{d}A^i + \Gamma^i_{kj}A^k\,\mathrm{d}x^j \tag{15.318}$$

In Eq. (15.318), expansion of the evaluated difference of the contravariant components $\mathrm{d}A^i$ of vector A gives the expression for the absolute differential of vector $\mathbf{A}$ in terms of contravariant components $\mathrm{D}A^i$, where

$$\mathrm{D}A^i = \frac{\mathrm{d}A^i}{\mathrm{d}x^j}\mathrm{d}x^j + \Gamma^i_{kj}A^k\,\mathrm{d}x^j = \left(\frac{\mathrm{d}A^i}{\mathrm{d}x^j} + \Gamma^i_{kj}A^k\right)\mathrm{d}x^j \tag{15.319}$$

The contravariant, absolute differential vector components $\mathrm{D}A^i$ is a product of the contravariant vector $\mathrm{d}x^j$ and the contravariant tensor expression in parentheses, which is the covariant derivative of contravariant vector components A^i, denoted by $A^i_{;j}$ with semicolon notation, where

$$A^i_{;j} = \frac{\mathrm{d}A^i}{\mathrm{d}x^j} + \Gamma^i_{kj}A^k \tag{15.320}$$

Thus, the contravariant absolute differential components $\mathrm{D}A^i$ of vector $\mathbf{A}$ is the product of the covariant derivative, which is a mixed tensor $A^i_{;j}$, and the contravariant differential components $\mathrm{d}x^j$ of position vector $\mathbf{x}$, where

$$\mathrm{D}A^i = A^i_{;j}\mathrm{d}x^j = \left(\frac{\mathrm{d}A^i}{\mathrm{d}x^j} + \Gamma^i_{kj}A^k\right)\mathrm{d}x^j \tag{15.321}$$

15.21 Absolute and Covariant Derivatives of Vectors and Tensors with Covariant Representation

We now examine absolute and covariant derivatives of vectors and tensors with covariant representation of the vector and tensor components. If vector **A** undergoes parallel transport between infinitesimally close points x^i and $x^i + \mathrm{d}x^i$, then the change δA_i in the covariant components of vector **A** are linearly dependent on the values of the covariant components A_i. This can be shown as described next. A scalar quantity is unchanged by parallel transport, and thus the scalar vector product of vectors **A** and **B** is unchanged by parallel transport, where

$$\delta(A_i B^i) \equiv \delta\left(\sum_{i=1}^{4} A_i B^i\right) = 0 \tag{15.322}$$

The dot product of two vectors **A** and **B**, which represents a physical quantity that has magnitude but no direction, is assumed not to change its value from point to point. In this case,

$$\delta(A_i B^i) = B^i \delta A_i + A_i \delta B^i = 0 \tag{15.323}$$

where

$$B^i \delta A_i = -A_i \delta B^i \tag{15.324}$$

One assumes a linear dependence between δB^i and B^i via the connection coefficient Γ^i_{kj}:

$$A_i \delta B^i = -\Gamma^i_{kj} B^k A_i \, \mathrm{d}x^j \tag{15.325}$$

Substitution of $A_i \delta B^i$ by $-B^i \delta A_i$ in Eq. (15.325), gives

$$B^i \delta A_i = \Gamma^i_{kj} B^k A_i \, \mathrm{d}x^j \tag{15.326}$$

Changing dummy variable i to k on the left side of Eq. (15.326) gives

$$(\delta A_k - \Gamma^i_{kj} A_i \, \mathrm{d}x^j) B^k = 0 \tag{15.327}$$

Since B^k is arbitrary, one obtains linear dependence between the covariant components A_i and the change δA_i in the covariant components of vector **A** upon parallel transport between infinitesimally close points x^i and $x^i + \mathrm{d}x^i$, where

$$\delta A_k = \Gamma^i_{kj} A_i \, \mathrm{d}x^j \tag{15.328}$$

The difference between the evaluated vector **A** at points x^i and $x^i + dx^i$, and the change of vector **A** upon parallel translation from points x^i and $x^i + dx^i$, are defined as the absolute covariant differential DA_i in terms of the covariant vector components A_i of vector **A**, where

$$DA_i = dA_i - \delta A_i \tag{15.329}$$

Substitution into Eq. (15.329) of Eq. (15.328) for δA_i gives

$$DA_i = dA_i - \Gamma^i_{kj} A_i \, dx^j \tag{15.330}$$

In Eq. (15.330), expansion of the evaluated difference of the covariant components dA_i of vector **A**, gives the expression for the absolute differential of vector **A** in terms of covariant components DA_i, where

$$DA_i = \frac{dA_i}{dx^j} dx^j - \Gamma^k_{ij} A_k \, dx^j = \left(\frac{dA_i}{dx^j} + \Gamma^k_{ij} A_k \right) dx^j \tag{15.331}$$

The covariant, absolute differential vector components DA_i are a product of the contravariant vector dx^j and the contravariant tensor expression in parentheses, which is the covariant derivative of covariant vector components A^i and denoted by $A_{i;j}$ with semicolon notation, where

$$A_{i;j} = \frac{dA_i}{dx^j} + \Gamma^k_{ij} A_k \tag{15.332}$$

Thus, the covariant absolute differential components DA_i of vector **A** is the product of the covariant derivative, which is a covariant tensor $A_{i;j}$, and the contravariant differential components dx^j of position vector **x**, where

$$DA_i = A_{i;j} \, dx^j = \left(\frac{dA_i}{dx^j} + \Gamma^k_{ij} A_k \right) dx^j \tag{15.333}$$

In a Cartesian (Galilean) coordinate system of flat surfaces and rectangular volumes, $\Gamma^i_{kj} = 0$ and the contravariant DA^i and covariant DA_i absolute differential components of vector **A**, reduce to the respective ordinary contravariant dA^i and covariant dA_i differential components, which are identical, where

$$\text{Galilean} \quad \Gamma^i_{kj} = 0 \quad DA^i = DA_i = dA^i = dA_i \tag{15.334}$$

One can extend the covariant and absolute derivatives of vectors in curvilinear coordinate systems for curved surfaces and spaces to tensors. The calculation of the covariant derivative of a tensor, considers the change of the tensor under an infinitesimal parallel displacement. Consider the case of a tensor expressed in its contravariant form. The contravariant tensor is expressed as the product of two contravariant vectors **A** and **B** via the product of their contravariant components A^iB^k. In parallel transport, the differential of the product of contravariant tensor components $\delta(A^iB^k)$ is given by

$$\delta\left(A^iB^k\right) = A^i\delta B^k + B^k\delta A^i \tag{15.335}$$

Substitution into Eq. (15.335) the linear expressions for δB^k and δA^i in terms of the connection coefficients, using Eq. (15.316), gives the differential vector product expression

$$\delta(A^iB^k) = -(A^i\Gamma^k_{jm}B^j + B^k\Gamma^i_{jm}A^j)\mathrm{d}x^m \tag{15.336}$$

This form of the vector product A^iB^k or tensor suggests the form of the differential tensor δA^{ik} due to parallel tensor transport of arbitrary contravariant tensor components A^{ik}, where

$$\delta A^{ik} = -(A^{im}\Gamma^k_{mj} + A^{mk}\Gamma^i_{mj})\mathrm{d}x^j \tag{15.337}$$

The absolute differential $\mathrm{D}A^{ik}$ of contravariant tensor components A^{ik} is

$$\mathrm{D}A^{ik} = \mathrm{d}A^{ik} - \delta A^{ik} = \frac{\partial A^{ik}}{\partial x^j}\mathrm{d}x^j - \delta A^{ik} \tag{15.338}$$

Substitution into Eq. (15.338) of Eq. (15.337) for δA^{ik} gives the absolute differential $\mathrm{D}A^{ik}$ of the contravariant tensor components A^{ik}, where

$$\mathrm{D}A^{ik} = \left(\frac{\partial A^{ik}}{\partial x^j} + A^{im}\Gamma^k_{mj} + A^{mk}\Gamma^i_{mj}\right)\mathrm{d}x^j \tag{15.339}$$

The covariant derivative $A^{ik}_{;j}$ of the contravariant tensor components A^{ik} is the quantity in parentheses:

$$A^{ik}_{;j} = \frac{\partial A^{ik}}{\partial x^j} + \Gamma^i_{mj}A^{mk} + \Gamma^k_{mj}A^{im} \tag{15.340}$$

In the same fashion, one obtains the absolute differential $\mathrm{D}A_{ik}$ and the covariant derivative $A_{ik;j}$ of the covariant tensor components A_{ik} as

$$\mathrm{D}A_{ik} = \left(\frac{\partial A_{ik}}{\partial x^j} - \Gamma^m_{ij}A_{mk} - \Gamma^m_{kj}A_{im}\right)\mathrm{d}x^j \tag{15.341}$$

and

$$A_{ik;j} = \frac{\partial A_{ik}}{\partial x^j} - \Gamma^m_{ij}A_{mk} - \Gamma^m_{kj}A_{im} \tag{15.342}$$

Similarly, the absolute differential DA_{ik} and the covariant derivative $A^i_{k;j}$ of the mixed tensor components A_{ik} as

$$DA^i_k = \left(\frac{\partial A^i_k}{\partial x^j} - \Gamma^m_{kl}A^i_m + \Gamma^i_{mj}A^m_k\right)dx^j \tag{15.343}$$

$$A^i_{k;j} = \frac{\partial A^i_k}{\partial x^j} - \Gamma^m_{kl}A^i_m + \Gamma^i_{mj}A^m_k \tag{15.344}$$

The covariant derivative of a product follows the same rule as the ordinary derivative of a product, where

$$(A_iB_j)_{;k} = A_{i;k}B_j + A_iB_{j;k} \tag{15.345}$$

and

$$(A^iB^j)_{;k} = A^i_{;k}B_j + A_iB^j_{;k} \tag{15.346}$$

15.22 Geodesic Curves and Connection Coefficients Expressed by a Metric Tensor

The metric tensor components are the building blocks used to construct the connection coefficients and represent trajectories and geodesic curves in general relativity. A straight line can be defined as a line in which the tangent vectors at each point all point in the same direction. Consider vector $\mathbf{r}(s)$ as a function of arc length s measured from some reference point on the line used as a parameter, where

$$\mathbf{r}(s) = x(s)\hat{\mathbf{x}} + y(s)\hat{\mathbf{y}} + z(s)\mathbf{z} \tag{15.347}$$

The arc length s along the curve is given by

$$s = \int_0^s ds\sqrt{\left(\frac{dx}{ds}\right)^2 + \left(\frac{dy}{ds}\right)^2 + \left(\frac{dz}{ds}\right)^2} \tag{15.348}$$

where

$$ds = \sqrt{(dx)^2 + (dy)^2 + (dz)^2} \tag{15.349}$$

The tangent vector $\boldsymbol{\lambda}$ to the curve $\mathbf{r}(s)$ is given by

$$\boldsymbol{\lambda} = \frac{d\mathbf{r}}{ds} = \frac{dx(s)}{ds}\hat{\mathbf{x}} + \frac{dy(s)}{ds}\hat{\mathbf{y}} + \frac{dz(s)}{ds}\hat{\mathbf{z}} \tag{15.350}$$

Substitution into Eq. (15.350) of Eq. (15.349) for ds gives

$$\boldsymbol{\lambda} = \frac{\mathrm{d}\mathbf{r}}{\mathrm{d}s} = \frac{\hat{\mathbf{x}}\,\mathrm{d}x + \hat{\mathbf{y}}\,\mathrm{d}y + \hat{\mathbf{z}}\,\mathrm{d}z}{\sqrt{(\mathrm{d}x)^2 + (\mathrm{d}y)^2 + (\mathrm{d}z)^2}} \tag{15.351}$$

The magnitude $|\boldsymbol{\lambda}|$ of the tangent vector $\boldsymbol{\lambda}$ is 1, where

$$|\boldsymbol{\lambda}| = \left|\frac{\mathrm{d}\mathbf{r}}{\mathrm{d}s}\right| = \left|\frac{\hat{\mathbf{x}}\,\mathrm{d}x + \hat{\mathbf{y}}\,\mathrm{d}y + \hat{\mathbf{z}}\,\mathrm{d}z}{\sqrt{(\mathrm{d}x)^2 + (\mathrm{d}y)^2 + (\mathrm{d}z)^2}}\right| = 1 \tag{15.352}$$

The tangent vector $\boldsymbol{\lambda}$ is thus a unit vector with a constant length of 1. As a result, the derivative of the tangent unit vector $\boldsymbol{\lambda}$ with respect to arc length s is zero, where

$$\frac{\mathrm{d}\boldsymbol{\lambda}}{\mathrm{d}s} = \frac{\mathrm{d}^2\mathbf{r}}{\mathrm{d}s^2} = 0 \tag{15.353}$$

One way of characterizing a straight line in Euclidean space and (by extension) a geodesic in a manifold is related to the idea of parallel transport of a vector along a curve. Let C be a curve in 3-D Euclidean space given parametrically by $\mathrm{d}u^i(t)/\mathrm{d}t$ as a function of time t, and let P_0 be the initial point on curve C to which is attached vector $\boldsymbol{\lambda}_0$ at initial time t_0. The initial vector $\boldsymbol{\lambda}_0$ at initial time t_0 and point P_0 is parallel-transported along curve C without any change in the vector length and direction. The result is a parallel field of vectors along the curve C, which is generated by the parallel transport of $\boldsymbol{\lambda}_0$ along C. Since there is no change in length or direction of vector $\boldsymbol{\lambda}(t)$ along C, then vector $\boldsymbol{\lambda}(t)$ satisfies the differential equation

$$\frac{\mathrm{d}\boldsymbol{\lambda}}{\mathrm{d}t} = \frac{\mathrm{d}\boldsymbol{\lambda}}{\mathrm{d}s}\frac{\mathrm{d}s}{\mathrm{d}t} = 0 \tag{15.354}$$

Here $\boldsymbol{\lambda}(t_0) = \boldsymbol{\lambda}_0$ is the initial vector condition and $s(t)$ is the distance along the curve C from starting point P_0 is a function of time t.

The tangent vector $\boldsymbol{\lambda}$ can be expressed by the natural basis unit vectors $\hat{\mathbf{e}}_i$ and contravariant components λ^i, where

$$\boldsymbol{\lambda} = \lambda^i\hat{\mathbf{e}}_i = \sum_{i=1}^{3}\lambda^i\hat{\mathbf{e}}_i \tag{15.355}$$

The zero derivative of the tangent unit vector $\boldsymbol{\lambda}$ with respect to arc length s becomes

$$0 = \frac{\mathrm{d}\boldsymbol{\lambda}}{\mathrm{d}s} = \frac{\mathrm{d}}{\mathrm{d}s}(\lambda^i\hat{\mathbf{e}}_i) = \frac{\mathrm{d}\lambda^i}{\mathrm{d}s}\hat{\mathbf{e}}_i + \lambda^i\frac{\mathrm{d}\hat{\mathbf{e}}_i}{\mathrm{d}s} \tag{15.356}$$

where

$$\lambda^i = \frac{\mathrm{d}u^i}{\mathrm{d}s} \tag{15.357}$$

By the chain rule, one has

$$\frac{\mathrm{d}\hat{\mathbf{e}}_i}{\mathrm{d}s} = \frac{\mathrm{d}\hat{\mathbf{e}}_i}{\mathrm{d}u^j}\frac{\mathrm{d}u^j}{\mathrm{d}s} \tag{15.358}$$

At each point in space, the derivative $\mathrm{d}\hat{\mathbf{e}}_i/du^j$ is linearly related to the basis unit vectors $\hat{\mathbf{e}}_k$ by the connection coefficient Γ^i_{kj}, where

$$\frac{\mathrm{d}\hat{\mathbf{e}}_i}{du^j} = \Gamma^k_{ij}\hat{\mathbf{e}}_k \tag{15.359}$$

Substitution into Eq. (13.356) of Eq. (15.358), followed by substitution of Eq. (15.359), gives

$$0 = \frac{\mathrm{d}\boldsymbol{\lambda}}{\mathrm{d}s} = \frac{\mathrm{d}}{\mathrm{d}s}(\lambda^i\hat{\mathbf{e}}_i) = \frac{\mathrm{d}\lambda^i}{\mathrm{d}s}\hat{\mathbf{e}}_i + \lambda^i\Gamma^k_{ij}\frac{\mathrm{d}u^j}{\mathrm{d}s}\hat{\mathbf{e}}_k \tag{15.360}$$

Exchanging dummy indices i and k in the second term of Eq. (15.360), yields

$$0 = \frac{\mathrm{d}\boldsymbol{\lambda}}{\mathrm{d}s} = \frac{\mathrm{d}}{\mathrm{d}s}(\lambda^i\hat{\mathbf{e}}_i) = \left(\frac{\mathrm{d}\lambda^i}{\mathrm{d}s} + \lambda^k\Gamma^i_{kj}\frac{\mathrm{d}u^j}{\mathrm{d}s}\right)\hat{\mathbf{e}}_i \tag{15.361}$$

Since natural basis vector $\hat{\mathbf{e}}_i$ is arbitrary,

$$\frac{\mathrm{d}\lambda^i}{\mathrm{d}s} + \lambda^k\Gamma^i_{kj}\frac{\mathrm{d}u^j}{\mathrm{d}s} = 0 \tag{15.362}$$

Substitution into Eq. (15.362) of (15.357) for λ^i and λ^k, gives

$$\frac{\mathrm{d}^2u^i}{\mathrm{d}s^2} + \Gamma^i_{kj}\frac{\mathrm{d}u^j}{\mathrm{d}s}\frac{\mathrm{d}u^k}{\mathrm{d}s} = 0 \tag{15.363}$$

The connection coefficients Γ^i_{jk} can be expressed in terms of the metric tensor, which is shown as described next. One starts by noting that

$$\frac{\partial\hat{\mathbf{e}}_i}{\partial u^j} = \frac{\partial^2\mathbf{r}}{\partial u^j\partial u^i} = \frac{\partial^2\mathbf{r}}{\partial u^i\partial u^j} = \frac{\partial\hat{\mathbf{e}}_j}{\partial u^i} \tag{15.364}$$

so that

$$\Gamma^k_{ij}\hat{\mathbf{e}}_k = \Gamma^k_{ji}\hat{\mathbf{e}}_k \tag{15.365}$$

In Eq. (15.365), forming the dot product with dual basis vector $\hat{\mathbf{e}}^m$ then gives the symmetry property between the exchange of indices i and j, where

$$\Gamma_{ij}^m = \Gamma_{ji}^m \tag{15.366}$$

The covariant metric tensor component g_{ij} is obtained by the dot products of the natural basis vectors for coordinates i and j, where

$$g_{ij} = \hat{\mathbf{e}}_i \cdot \hat{\mathbf{e}}_j \tag{15.367}$$

In Eq. (15.367), the chain rule applied to the differentiation of covariant metric tensor component g_{ij} with respect to contravariant position vector component u^k gives

$$\frac{\partial g_{ij}}{\partial u^k} = \frac{\partial \hat{\mathbf{e}}_i}{\partial u^k} \cdot \hat{\mathbf{e}}_j + \hat{\mathbf{e}}_i \cdot \frac{\partial \hat{\mathbf{e}}_j}{\partial u^k} \tag{15.368}$$

Natural basis vectors $\hat{\mathbf{e}}_i$ and $\hat{\mathbf{e}}_j$ are parallel-transported between two infinitesimally close points connected by vector differential $\mathrm{d}\mathbf{u}$ with contravariant components $\mathrm{d}u^k$. The variation of the component natural basis vectors $\mathrm{d}\hat{\mathbf{e}}_i$ and $\mathrm{d}\hat{\mathbf{e}}_j$ with respect to the vector differential components $\mathrm{d}u^k$ are linearly related to the other natural basis vectors $\hat{\mathbf{e}}_i$, $\hat{\mathbf{e}}_j$, and $\hat{\mathbf{e}}_m$ via the connection coefficients Γ_{ik}^m and Γ_{jk}^m, where substitution gives

$$\frac{\partial g_{ij}}{\partial u^k} = \Gamma_{ik}^m \hat{\mathbf{e}}_m \cdot \hat{\mathbf{e}}_j + \hat{\mathbf{e}}_i \cdot \Gamma_{jk}^m \hat{\mathbf{e}}_m \tag{15.369}$$

In Eq. (15.369), using the definition $g_{ij} = \hat{\mathbf{e}}_i \cdot \hat{\mathbf{e}}_j$ again yields

$$\frac{\partial g_{ij}}{\partial u^k} = \Gamma_{ik}^m g_{mj} + \Gamma_{jk}^m g_{im} \tag{15.370}$$

Clockwise permutation of the indices ijk in Eq. (15.370) yields

$$\frac{g_{jk}}{\partial u^i} = \Gamma_{ji}^m g_{mk} + \Gamma_{ki}^m g_{jm} \tag{15.371}$$

and, again, clockwise permutation of the indices ijk in Eq. (15.371) yields

$$\frac{g_{ki}}{\partial u^j} = \Gamma_{kj}^m g_{mi} + \Gamma_{ij}^m g_{km} \tag{15.372}$$

Subtracting the last equation, Eq. (15.372), from the sum of the previous two equations, Eq. (15.370) and Eq. (15.371), and using the symmetry of the ij indices in both Γ_{ij}^m and g_{ij} yields

$$2\Gamma_{ki}^m g_{mj} = \frac{\partial g_{ij}}{\partial u^k} + \frac{\partial g_{jk}}{\partial u^i} - \frac{\partial g_{ki}}{\partial u^j} \tag{15.373}$$

In Eq. (15.373), contraction with $\frac{1}{2}g^{lj}$ yields the connection coefficient:

$$\Gamma^{l}_{ki} = \frac{1}{2}g^{lj}\left(\frac{\partial g_{ij}}{\partial u^k} + \frac{\partial g_{jk}}{\partial u^i} - \frac{\partial g_{ki}}{\partial u^j}\right) \tag{15.374}$$

Christoffel terms of the first kind $\Gamma_{i,kj}$ are obtained by multiplication of components of the connection coefficient Γ^{i}_{kj} by the components of the metric tensor g_{im}, where

$$\Gamma_{ikj} = g_{im}\Gamma^{m}_{kj} \tag{15.375}$$

The inverse is

$$\Gamma^{i}_{kj} = g^{im}\Gamma_{ikj} \tag{15.376}$$

The Christoffel symbol of the first kind can be expanded into metric tensor terms, where

$$\Gamma_{abc} = \frac{1}{2}\left(\frac{\partial g_{ac}}{\partial u^b} + \frac{\partial g_{ba}}{\partial u^c} - \frac{\partial g_{bc}}{\partial u^a}\right) \tag{15.377}$$

and

$$\Gamma^{a}_{bc} = g^{ad}\Gamma_{dbc} \tag{15.378}$$

If we use a general parameter t to parameterize a straight line, then the geodesic equation is a more complicated form. However, for parameters related to the arc length s by an equation of the form

$$t = As + B \tag{15.379}$$

Here A, B are constants ($A \neq 0$), and the geodesic equation has basically the same form as when s is used:

$$\frac{\mathrm{d}^2u^i}{\mathrm{d}t^2} + \Gamma^{i}_{jk}\frac{\mathrm{d}u^j}{\mathrm{d}t}\frac{\mathrm{d}u^k}{\mathrm{d}t} = 0 \tag{15.380}$$

Such privileged parameters are called *affine parameters*, and for an affine parameter, $\mathrm{d}s/\mathrm{d}t$ is a constant, so one is taken along the geodesic at a constant sort of rate (if one thinks of t as time, then the geodesic is traversed at a constant speed). If a general parameter $t = f(s)$ is used to parameterize a straight line in Euclidean space, then the geodesic takes the form

$$\frac{\mathrm{d}^2u^i}{\mathrm{d}t^2} + \Gamma^{i}_{jk}\frac{\mathrm{d}u^j}{\mathrm{d}t}\frac{\mathrm{d}u^k}{\mathrm{d}t} = h(s)\frac{\mathrm{d}u^i}{\mathrm{d}t} \tag{15.381}$$

where

$$h(s) = -\frac{d^2t}{ds^2}\left(\frac{dt}{ds}\right)^{-2} \tag{15.382}$$

15.23 Gradient of a Vector in Curvilinear Coordinates

We now derive the gradient of a vector in curvilinear coordinates that are applicable to the curved surfaces of general relativity or meterology and fluid mechanics. Consider a vector field $\mathbf{v}$ in curvilinear coodinates (q^1, q^2, q^3, q^4) that is expressed in its contravariant vector form $\mathbf{v}(q^1, q^2, q^3, q^4)$ with contravariant components $v^i(q^1, q^2, q^3, q^4)$ and natural basis vectors $\mathbf{e}_i(q^1, q^2, q^3, q^4)$, where

$$\mathbf{v} = v^i \mathbf{e}_i \equiv \sum_{i=1}^{4} v^i \mathbf{e}_i \tag{15.383}$$

The vector differential $d\mathbf{v}(q^1, q^2, q^3, q^4)$, which is the product of the contravariant components $r^i(q^1, q^2, q^3, q^4)$ and the natural basis vectors $\mathbf{e}_i(q^1, q^2, q^3, q^4)$, are then obtained by applying the product rule of differentiation to Eq. (15.383):

$$d\mathbf{v} = dv^i \, \mathbf{e}_i + v^i \, d\mathbf{e}_i \tag{15.384}$$

In Eq. (15.384), one can apply the chain rule to the differentials of the contravariant components dv^i and to the natural basis vectors $d\mathbf{e}_i$, where

$$dv^i = \frac{\partial v^i}{\partial q^j} dq^j \equiv \sum_{j=1}^{4} \frac{\partial v^i}{\partial q^j} dq^j \tag{15.385}$$

and

$$d\mathbf{e}_i = \frac{\partial \mathbf{e}_i}{\partial q^j} dq^j \equiv \sum_{i=1}^{4} \frac{\partial \mathbf{e}_i}{\partial q^j} dq^j \tag{15.386}$$

Next, these contravariant differential coordinates dq^j of curvilinear coordinate system (q^1, q^2, q^3, q^4) in the expressions for dv^i and $d\mathbf{e}^i$ can be related to the covariant differential vector $d\mathbf{r}(x_1, x_2, x_3, x_4)$ in another curvilinear coordinate system (x_1, x_2, x_3, x_4), where

$$d\mathbf{r} = dq^j \frac{\partial \mathbf{r}}{\partial q^j} \equiv \sum_{j=1}^{4} dq^j \frac{\partial \mathbf{r}}{\partial q^j} \tag{15.387}$$

The dot product of covariant derivative $\partial\mathbf{r}/\partial q_j$ is taken on both sides of Eq. (15.387) for covariant differential vector $\mathbf{dr}(x_1, x_2, x_3, x_4)$, where

$$\mathbf{dr}\cdot\frac{\partial\mathbf{r}}{\partial q_j} = \mathrm{d}q^j\left(\frac{\partial\mathbf{r}}{\partial q^j}\cdot\frac{\partial\mathbf{r}}{\partial q_j}\right) \tag{15.388}$$

Recall from section 15.17, that $\partial\mathbf{r}/\partial q^j$ is the dual basis vector $\mathbf{e}^j$, and $\partial\mathbf{r}/\partial q_j$ is natural basis vector $\mathbf{e}_j$:

$$\mathbf{e}^j = \frac{\partial\mathbf{r}}{\partial q^j} \quad \text{and} \quad \mathbf{e}_j = \frac{\partial\mathbf{r}}{\partial q_j} \tag{15.389}$$

Also, recall that the dual basis vectors $\mathbf{e}^k$ are orthogonal to the natural basis vector $\mathbf{e}_j$, and their dot product is expressed by the Kronnecker delta function δ_j^k, where

$$\mathbf{e}^k\cdot\mathbf{e}_j = \nabla x^k\cdot\frac{\partial\mathbf{r}}{\partial x^j} = \frac{\partial x^k}{\partial q_1}\frac{\partial q_1}{\partial x^j} + \frac{\partial x^k}{\partial q_2}\frac{\partial q_2}{\partial x^j} + \frac{\partial x^k}{\partial q_3}\frac{\partial q_3}{\partial x^j} + \frac{\partial x^k}{\partial q_4}\frac{\partial q_4}{\partial x^j} = \frac{\partial x^k}{\partial x^j} = \delta_j^k \tag{15.390}$$

As indicated by the Kronnecker delta function δ_j^k, like coefficients $k = 1, 2, 3, 4$ yield a dot product of unity, where

$$\mathbf{e}^k\cdot\mathbf{e}_k = \mathbf{e}^j\cdot\mathbf{e}_j = \frac{\partial\mathbf{r}}{\partial q^j}\cdot\frac{\partial\mathbf{r}}{\partial q_j} = 1 \tag{15.391}$$

The above is an example of dummy index k that can be replaced by dummy index j in this summation that is expressed with the Einstein summation convention. The dummy index is the summation index integer in which the result of the summation does not depend on which index integer k or j is used. That is,

$$\mathbf{e}^k\cdot\mathbf{e}_k \equiv \sum_{k=1}^{4}\mathbf{e}^k\cdot\mathbf{e}_k = \sum_{k=1}^{4}\mathbf{e}^j\cdot\mathbf{e}_j = 1 \tag{15.392}$$

Thus, the result of application of the dot product to the equation for the covariant differential vector $\mathbf{dr}$ is then

$$\mathrm{d}q^j = \mathbf{dr}\cdot\frac{\partial\mathbf{r}}{\partial q^j} \tag{15.393}$$

With substitution into Eqs. (15.385) and (15.386) of Eq. (15.393) for contravariant coordinate differentials $\mathrm{d}q^j$, the contravariant vector components $\mathrm{d}v^i$ and natural basis vectors $\mathbf{de}_i$ become

$$\mathrm{d}v^i = \frac{\partial v^i}{\partial q^j}\left(\mathbf{dr}\cdot\frac{\partial\mathbf{r}}{\partial q^j}\right) = \frac{\partial v^i}{\partial q^j}(\mathbf{e}^j\cdot\mathbf{dr}) \tag{15.394}$$

and

$$\mathbf{de}_i = \frac{\partial \mathbf{e}_i}{\partial q^j}\left(\mathbf{dr}\cdot\frac{\partial \mathbf{r}}{\partial q^j}\right) = \frac{\partial \mathbf{e}_i}{\partial q^j}(\mathbf{e}^j\cdot\mathbf{dr}) \tag{15.395}$$

In this result, recall that the dot product is commutative, where

$$\mathbf{dr}\cdot\mathbf{e}^j = \mathbf{e}^j\cdot\mathbf{dr} \tag{15.396}$$

With the substitution into Eq. (15.384) of Eqs. (15.394) and (15.395) for $\mathrm{d}v^i$ and $\mathbf{de}_i$, the vector differential $\mathbf{dv}(q^1, q^2, q^3, q^4)$ becomes

$$\mathbf{dv} = \mathrm{d}v^i\mathbf{e}_i + v^i\mathbf{de}_i = \frac{\partial v^i}{\partial q^j}(\mathbf{e}^j\cdot\mathbf{dr})\mathbf{e}_i + v^i\frac{\partial \mathbf{e}_i}{\partial q^j}(\mathbf{e}^j\cdot\mathbf{dr}) \tag{15.397}$$

Rearrangement of Eq. (15.397), gives the vector differential $\mathbf{dv}$ for all $\mathbf{dr}$, where

$$\mathbf{dv} = \left(\frac{\partial v^i}{\partial q^j}\mathbf{e}_i\mathbf{e}^j + v^i\frac{\partial \mathbf{e}_i}{\partial q^j}\mathbf{e}^j\right)\cdot\mathbf{dr} \quad \text{for all } \mathbf{dr} \tag{15.398}$$

The gradient of a vector is defined as

$$\mathbf{dv} \equiv \frac{\mathbf{dv}}{\mathbf{dr}}\cdot\mathbf{dr} \tag{15.399}$$

Comparison of Eq. (15.398) and Eq. (15.399) for the vector differential $\mathbf{dv}$ for arbitrary vector differential $\mathbf{dr}$ yields the following expression for the gradient of vector $\mathbf{v}$:

$$\frac{\mathbf{dv}}{\mathbf{dr}} = \frac{\partial v^i}{\partial q^j}\mathbf{e}_i\mathbf{e}^j + v^i\frac{\partial \mathbf{e}_i}{\partial q^j}\mathbf{e}^j \tag{15.400}$$

where

$$\frac{\partial \mathbf{e}_i}{\partial q^j} = \frac{\partial \mathbf{r}}{\partial q^j\partial q^i} = \frac{\partial \mathbf{r}}{\partial q^i\partial q^j} \tag{15.401}$$

The Christoffel symbol of the second kind Γ^k_{ij} is defined as the dot product between the derivative of the contravariant derivative natural basis $\partial\mathbf{e}_i/\partial q^j$ and the dual basis vector $\mathbf{e}^k$, where

$$\Gamma^k_{ij} = \mathbf{e}^k\cdot\frac{\partial \mathbf{e}_i}{\partial q^j} = \mathbf{e}^k\cdot\frac{\partial \mathbf{r}}{\partial q^j\partial q^i} = \mathbf{e}^k\cdot\frac{\partial \mathbf{r}}{\partial q^i\partial q^j} = \Gamma^k_{ji} \tag{15.402}$$

Taking the dot product of the natural basis vector $\mathbf{e}_k$ on both sides of Eq. (15.402) gives

$$\mathbf{e}_k \cdot \Gamma^k_{ij} = (\mathbf{e}_k \cdot \mathbf{e}^k) \cdot \frac{\partial \mathbf{e}_i}{\partial q^j} \tag{15.403}$$

The dot product of the covariant form of the vector with the contravariant form of the same vector is equivalent to the summation index k that appears as both a lower and an upper index in the same product term, where

$$\mathbf{e}_k \cdot \mathbf{e}^k = 1 \quad \mathbf{e}_k \cdot \Gamma^k_{ij} = \sum_{k=1}^{4} \mathbf{e}_k \Gamma^k_{ij} \equiv \mathbf{e}_k \Gamma^k_{ij} \tag{15.404}$$

Hence, the contravariant derivative $\partial \mathbf{e}_i / \partial q^j$ of the natural basis component vector $\mathbf{e}_i$ is expressed in terms of the summed products of the natural basis vector $\mathbf{e}_k$ with the connection coefficient Γ^k_{ij}, where

$$\frac{\partial \mathbf{e}_i}{\partial q^j} = \mathbf{e}_k \Gamma^k_{ij} \tag{15.405}$$

The Christoffel function of the second kind Γ^k_{ij} is also called the *affinity term* or *connection coefficient* because Γ^k_{ij} characterizes how the base vectors vary from point-to-point in space. By the chain rule, the covariant basis vector increment $\mathrm{d}\mathbf{e}_i$ is

$$\mathrm{d}\mathbf{e}_i = \frac{\partial \mathbf{e}_i}{\partial q^j} \mathrm{d}q^j \tag{15.406}$$

Substitution into Eq. (15.406) of Eq. (15.405) for the $\partial \mathbf{e}_i / \partial q^j$ then yields the covariant basis vector differential $\mathrm{d}\mathbf{e}_i$ expressed in terms of the connection coefficients Γ^k_{ij}, where

$$\mathrm{d}\mathbf{e}_i = \mathbf{e}_k \Gamma^k_{ij} \mathrm{d}q^j \tag{15.407}$$

Next, Eq. (15.405) for the contravariant derivative $\partial \mathbf{e}_i / \partial q^j$, which is expressed in terms of the connection coefficients Γ^k_{ij}, is substituted into Eq. (15.400) for the gradient $\mathrm{d}\mathbf{v}/d\mathbf{r}$ of vector $\mathbf{v}$ to obtain

$$\frac{\mathrm{d}\mathbf{v}}{\mathrm{d}\mathbf{r}} = \frac{\partial v^i}{\partial q^j} \mathbf{e}_i \mathbf{e}^j + v^i \Gamma^k_{ij} \mathbf{e}_k \mathbf{e}^j \tag{15.408}$$

The dummy index k in the second term, which is a repeated upper and lower index indicating summation over k, is exchanged with index i. This allows the indices of the basis vectors in the first and second terms on the right side of Eq. (15.408) to be matched, where $k \leftrightarrow i$, so that

$$\frac{\mathbf{dv}}{\mathbf{dr}} = \frac{\partial v^i}{\partial q^j}\mathbf{e}_i\mathbf{e}^j + v^k\Gamma^i_{kj}\mathbf{e}_i\mathbf{e}^j \tag{15.409}$$

The distributive property can now be applied to Eq. (15.409), where

$$\frac{\mathbf{dv}}{\mathbf{dr}} = \left(\frac{\partial v^i}{\partial q^j} + v^k\Gamma^i_{kj}\right)\mathbf{e}_i\mathbf{e}^j \tag{15.410}$$

Since the product term of the natural and dual basis vectors is arbitrary and independent of the expression the parentheses. The expression in parentheses is defined as the covariant derivative with respect to spatial component q^j along a curve in the vector space $\mathbf{v}$, where

$$v^i_{;j} \equiv \frac{\partial v^i}{\partial q^j} + v^k\Gamma^i_{kj} \quad \text{covariant derivative} \tag{15.411}$$

The gradient $\mathbf{dv}/d\mathbf{r}$ of vector $\mathbf{v}$ is the absolute derivative, where

$$\frac{\mathbf{dv}}{\mathbf{dr}} \equiv \frac{\mathbf{Dv}}{\mathbf{dr}} \equiv v^i_{;j}\mathbf{e}_i\mathbf{e}^j \quad \text{absolute derivative} \tag{15.412}$$

15.24 Cylindrical Coordinate System—Covariant and Contravariant Metric Coefficients and Connection Coefficients

In this section, we calculate the covariant and contravariant metric tensor components in cylindrical coordinates for the case of three dimensions, as well as the Christoffel functions of the second kind in cylindrical coordinates. In cylindrical coordinates, one starts with position vector $\mathbf{q}(r, \theta, z)$ expressed by the natural basis vectors $(\hat{\mathbf{x}}, \hat{\mathbf{y}}, \hat{\mathbf{z}})$ of the Cartesian coordinate system, where

$$\mathbf{q}(r, \theta, z) = \hat{\mathbf{x}}r\cos\theta + \hat{\mathbf{y}}r\sin\theta + \hat{\mathbf{z}}z \tag{15.413}$$

with

$$x = r\cos\theta \quad y = r\sin\theta \quad z = z \tag{15.414}$$

The natural basis vectors $(\mathbf{e}_r, \mathbf{e}_\theta, \mathbf{e}_z)$ in the cylindrical coordinate system are obtained via

$$\mathbf{e}_r = \frac{\partial \mathbf{q}(r,\theta,z)}{\partial r} = \frac{\partial}{\partial r}(\hat{\mathbf{x}} r \cos\theta + \hat{\mathbf{y}} r \sin\theta + \hat{\mathbf{z}} z) = \hat{\mathbf{x}} \cos\theta + \hat{\mathbf{y}} \sin\theta \qquad (15.415)$$

$$\mathbf{e}_\theta = \frac{\partial \mathbf{q}(r,\theta,z)}{\partial \theta} = \frac{\partial}{\partial \theta}(\hat{\mathbf{x}} r \cos\theta + \hat{\mathbf{y}} r \sin\theta + \hat{\mathbf{z}} z) = -\hat{\mathbf{x}} r \sin\theta + \hat{\mathbf{y}} r \cos\theta \qquad (15.416)$$

$$\mathbf{e}_z = \frac{\partial \mathbf{q}(r,\theta,z)}{\partial z} = \frac{\partial}{\partial z}(\hat{\mathbf{x}} r \cos\theta + \hat{\mathbf{y}} r \sin\theta + \hat{\mathbf{z}} z) = \hat{\mathbf{z}} \qquad (15.417)$$

In cylindrical coordinates, the scalar components g_{ij} of covariant metric tensors are obtained via the dot product of the natural basis vector components, where

$$g_{11} = g_{rr} = \mathbf{e}_r \cdot \mathbf{e}_r = 1 \quad g_{12} = g_{r\theta} = \mathbf{e}_r \cdot \mathbf{e}_\theta = 0 \quad g_{13} = g_{rz} = \mathbf{e}_r \cdot \mathbf{e}_z = 0 \qquad (15.418)$$

$$g_{21} = g_{\theta r} = \mathbf{e}_\theta \cdot \mathbf{e}_r = 0 \quad g_{22} = g_{\theta\theta} = \mathbf{e}_r \cdot \mathbf{e}_\theta = r^2 \quad g_{23} = g_{\theta z} = \mathbf{e}_r \cdot \mathbf{e}_z = 0 \qquad (15.419)$$

$$g_{31} = g_{zr} = \mathbf{e}_z \cdot \mathbf{e}_r = 0 \quad g_{32} = g_{z\theta} = \mathbf{e}_z \cdot \mathbf{e}_\theta = 0 \quad g_{33} = g_{zz} = \mathbf{e}_r \cdot \mathbf{e}_z = 1 \qquad (15.420)$$

In matrix form, the covariant metric tensor $[g_{ij}]$ in cylindrical coordinates is

$$[g_{ij}] = \begin{bmatrix} 1 & 0 & 0 \\ 0 & r^2 & 0 \\ 0 & 0 & 1 \end{bmatrix} \qquad (15.421)$$

In cylindrical coordinates, the contravariant metric tensor $[g^{ij}]$ is the inverse of the covariant metric tensor $[g_{ij}]$, where

$$[g^{ij}] = \begin{bmatrix} 1 & 0 & 0 \\ 0 & r^{-2} & 0 \\ 0 & 0 & 1 \end{bmatrix} \qquad (15.422)$$

Alternatively, the contravariant metric tensor $[g^{ij}]$ can be obtained from transformation from cylindrical coordinates (r, θ, z) to Cartesian coordinates (x, y, z), where

$$\mathbf{q}(x,y,z) = \hat{\mathbf{r}} r + \hat{\boldsymbol{\theta}} r\theta + \hat{\mathbf{z}} z \qquad (15.423)$$

with

$$r = \sqrt{x^2 + y^2} \quad \theta = \tan^{-1}\frac{y}{x} \quad z = z \qquad (15.424)$$

The three dual basis vectors $\mathbf{e}^r$, $\mathbf{e}^\theta$, and $\mathbf{e}^z$ for the contravariant vector representation in cylindrical coordinates are obtained via the Cartesian coordinate gradient of the three cylindrical component coordinates (r, θ, z), where

$$\mathbf{e}^r = \nabla r = \nabla\sqrt{x^2+y^2} = \hat{\mathbf{x}}\frac{x}{r} + \hat{\mathbf{y}}\frac{y}{r} = \hat{\mathbf{x}}\cos\theta + \hat{\mathbf{y}}\sin\theta \tag{15.425}$$

$$\mathbf{e}^\theta = \nabla\theta = \nabla\tan\frac{y}{x} = -\hat{\mathbf{x}}\frac{y}{x^2}\cos^2\theta + \hat{\mathbf{y}}\frac{1}{x}\cos^2\theta = -\hat{\mathbf{x}}\frac{\sin\theta}{r} + \hat{\mathbf{y}}\frac{1}{r}\cos\theta \tag{15.426}$$

$$\mathbf{e}^z = \nabla z = \hat{\mathbf{z}} \tag{15.427}$$

The relation between the normal and the dual basis vectors in cylindrical coordinates is then

$$\mathbf{e}^r = \mathbf{e}_r \quad \mathbf{e}^\theta = \frac{1}{r}\mathbf{e}_\theta \quad \mathbf{e}^z = \mathbf{e}_z \tag{15.428}$$

In cylindrical coordinates, the scalar components g^{ij} of contravariant metric tensors are obtained via the dot product of the dual basis vector components, where

$$g^{11} = g^{rr} = \mathbf{e}^r\cdot\mathbf{e}^r = 1 \quad g^{12} = g^{r\theta} = \mathbf{e}^r\cdot\mathbf{e}^\theta = 0 \quad g^{13} = g^{rz} = \mathbf{e}^r\cdot\mathbf{e}^z = 0 \tag{15.429}$$

$$g^{21} = g^{\theta r} = \mathbf{e}^\theta\cdot\mathbf{e}^r = 0 \quad g^{22} = g^{\theta\theta} = \mathbf{e}^\theta\cdot\mathbf{e}^\theta = \frac{1}{r^2} \quad g^{23} = g^{\theta z} = \mathbf{e}^\theta\cdot\mathbf{e}^z = 0 \tag{15.430}$$

$$g^{31} = g^{zr} = \mathbf{e}^z\cdot\mathbf{e}^r = 0 \quad g^{32} = g^{z\theta} = \mathbf{e}^z\cdot\mathbf{e}^\theta = 0 \quad g^{33} = g^{zz} = \mathbf{e}^z\cdot\mathbf{e}^z = 1 \tag{15.431}$$

Recall that the Christoffel function of the second kind or connection coefficients Γ^k_{ij} are defined by

$$\Gamma^k_{ij} = \mathbf{e}^k\cdot\frac{\partial\mathbf{e}_i}{\partial q^j} = \frac{\partial g_{ki}}{\partial q^j} = \frac{\partial g_{kj}}{\partial q^i} = \mathbf{e}^k\cdot\frac{\partial\mathbf{e}_j}{\partial q^i} = \Gamma^k_{ji} \quad i,j,k = 1,2,3 \tag{15.432}$$

where

$$\mathbf{e}_i = \frac{\partial\mathbf{r}}{\partial q^i} \tag{15.433}$$

In cylindrical coordinates, only 3 of the 27 Christoffel functions are nonzero; namely, $\Gamma^2_{12} = \Gamma^2_{21}$ and Γ^1_{22}, where

$$\Gamma^\theta_{\theta r} = \Gamma^\theta_{r\theta} = \mathbf{e}^\theta\cdot\frac{\partial\mathbf{e}_r}{\partial q^\theta} = \left(-\hat{\mathbf{x}}\frac{\sin\theta}{r} + \hat{\mathbf{y}}\frac{1}{r}\cos\theta\right)\cdot\frac{\partial}{\partial\theta}(\hat{\mathbf{x}}\cos\theta + \hat{\mathbf{y}}\sin\theta) = \frac{1}{r} \tag{15.434}$$

and

$$\Gamma^r_{\theta\theta} = \mathbf{e}^r\cdot\frac{\partial\mathbf{e}_\theta}{\partial q^\theta} = (\hat{\mathbf{x}}\cos\theta + \hat{\mathbf{y}}\sin\theta)\cdot\frac{\partial}{\partial\theta}(-\hat{\mathbf{x}}r\sin\theta + \hat{\mathbf{y}}r\cos\theta) = -r \tag{15.435}$$

15.25 Spherical Coordinate System—Covariant and Contravariant Metric Coefficients and Connection Coefficients

In this section, we calculate the covariant and contravariant metric tensor components in spherical coordinates for the case of three dimensions, as well as the Christoffel functions of the second kind. Recall that the spherical coordinate natural basis vectors expressed in terms of Cartesian unit vectors are

$$\mathbf{e}_r = \frac{\partial \mathbf{r}}{\partial r} = \sin\theta\cos\phi\hat{\mathbf{x}} + \sin\theta\sin\phi\hat{\mathbf{y}} + \cos\theta\,\hat{\mathbf{z}} \tag{15.436}$$

$$\mathbf{e}_\theta = \frac{\partial \mathbf{r}}{\partial \theta} = r\cos\theta\cos\phi\hat{\mathbf{x}} + r\cos\theta\sin\phi\hat{\mathbf{y}} - r\sin\theta\hat{\mathbf{z}} \tag{15.437}$$

$$\mathbf{e}_\phi = \frac{\partial \mathbf{r}}{\partial \phi} = -r\sin\theta\sin\phi\hat{\mathbf{x}} + r\sin\theta\cos\phi\hat{\mathbf{y}} \tag{15.438}$$

In spherical coordinates, the scalar components g_{ij} of covariant metric tensors are obtained via the dot product of the natural basis vector components, where

$$g_{11} = g_{rr} = \mathbf{e}_r \cdot \mathbf{e}_r = 1 \quad g_{12} = g_{r\theta} = \mathbf{e}_r \cdot \mathbf{e}_\theta = 0 \quad g_{13} = g_{r\phi} = \mathbf{e}_r \cdot \mathbf{e}_\phi = 0 \tag{15.439}$$

$$g_{21} = g_{\theta r} = \mathbf{e}_\theta \cdot \mathbf{e}_r = 0 \quad g_{22} = g_{\theta\theta} = \mathbf{e}_\theta \cdot \mathbf{e}_\theta = r^2 \quad g_{23} = g_{\theta\phi} = \mathbf{e}_\theta \cdot \mathbf{e}_\phi = 0 \tag{15.440}$$

$$g_{31} = g_{\phi r} = \mathbf{e}_\phi \cdot \mathbf{e}_r = 0 \quad g_{32} = g_{\phi\theta} = \mathbf{e}_\phi \cdot \mathbf{e}_\theta = 0 \quad g_{33} = g_{\phi\phi} = \mathbf{e}_\phi \cdot \mathbf{e}_\phi = r^2\sin^2\theta \tag{15.441}$$

In matrix form, the covariant metric tensor $[g_{ij}]$ in spherical coordinates is

$$[g_{ij}] = \begin{bmatrix} 1 & 0 & 0 \\ 0 & r^2 & 0 \\ 0 & 0 & r^2\sin^2\theta \end{bmatrix} \tag{15.442}$$

The dual basis vectors for spherical coordinates expressed in terms of the Cartesian unit vectors are

$$\mathbf{e}^r = \nabla_{xyz} r = \hat{\mathbf{x}}\sin\theta\cos\phi + \hat{\mathbf{y}}\sin\theta\sin\phi + \hat{\mathbf{z}}\cos\theta \tag{15.443}$$

$$\mathbf{e}^\theta = \nabla_{xyz}\theta = \hat{\mathbf{x}}\frac{\cos\theta\cos\phi}{r} + \hat{\mathbf{y}}\frac{\cos\theta\sin\phi}{r} - \hat{\mathbf{z}}\frac{\sin\theta}{r} \tag{15.444}$$

$$\mathbf{e}^\phi = \nabla_{xyz}\phi = -\hat{\mathbf{x}}\frac{\sin\phi}{r\sin\theta} + \hat{\mathbf{y}}\frac{\cos\phi}{r\sin\theta} \tag{15.445}$$

In spherical coordinates, the scalar components g^{ij} of contravariant metric tensors are obtained via the dot product of the dual basis vector components, where

$$g^{11} = g^{rr} = \mathbf{e}^r \cdot \mathbf{e}^r = 1 \quad g^{12} = g^{r\theta} = \mathbf{e}^r \cdot \mathbf{e}^\theta = 0 \quad g^{13} = g^{r\phi} = \mathbf{e}^r \cdot \mathbf{e}^\phi = 0 \tag{15.446}$$

$$g^{21} = g^{\theta r} = \mathbf{e}^\theta \cdot \mathbf{e}^r = 0 \quad g^{22} = g^{\theta\theta} = \mathbf{e}^\theta \cdot \mathbf{e}^\theta = \frac{1}{r^2} \quad g^{23} = g^{\theta\phi} = \mathbf{e}^\theta \cdot \mathbf{e}^\phi = 0 \tag{15.447}$$

$$g^{31} = g^{\phi r} = \mathbf{e}^\phi \cdot \mathbf{e}^r = 0 \quad g^{32} = g^{\phi\theta} = \mathbf{e}^\phi \cdot \mathbf{e}^\theta = 0 \quad g^{33} = g^{\phi} = \mathbf{e}^\phi \cdot \mathbf{e}^\phi = 1/r^2 \sin^2\theta \tag{15.448}$$

In spherical coordinates, the contravariant metric tensor $[g^{ij}]$, which is also the inverse of the covariant metric tensor $[g_{ij}]$, is

$$\left[g^{ij}\right] = \begin{bmatrix} 1 & 0 & 0 \\ 0 & \dfrac{1}{r^2} & 0 \\ 0 & 0 & \dfrac{1}{r^2 \sin^2\theta} \end{bmatrix} \tag{15.449}$$

The Christoffel function of the second kind or connection coefficients Γ^k_{ij} are defined by

$$\Gamma^k_{ij} = \mathbf{e}^k \cdot \frac{\partial \mathbf{e}_i}{\partial q^j} = \frac{\partial g_{ki}}{\partial q^j} = \frac{\partial g_{kj}}{\partial q^i} = \mathbf{e}^k \cdot \frac{\partial \mathbf{e}_j}{\partial q^i} = \Gamma^k_{ji} \quad i, j, k = 1, 2, 3 \tag{15.450}$$

where

$$\mathbf{e}_i = \frac{\partial \mathbf{r}}{\partial q^i} \tag{15.451}$$

In spherical coordinates, only 9 of the 27 Christoffel functions are nonzero; namely, $\Gamma^2_{12} = \Gamma^2_{21}$ and Γ^1_{22} and $\Gamma^3_{13} = \Gamma^3_{31}$ and $\Gamma^3_{23} = \Gamma^3_{32}$ and Γ^1_{33} and Γ^2_{33}, where

$$\begin{aligned}\Gamma^\theta_{r\theta} &= \mathbf{e}^\theta \cdot \frac{\partial \mathbf{e}_r}{\partial q^\theta} \\ &= \left(\hat{\mathbf{x}}\frac{\cos\theta\cos\phi}{r} + \hat{\mathbf{y}}\frac{\cos\theta\sin\phi}{r} - \hat{\mathbf{z}}\frac{\sin\theta}{r}\right) \\ &\quad \cdot \frac{\partial}{\partial\theta}(\hat{\mathbf{x}}\sin\theta\cos\phi + \hat{\mathbf{y}}\sin\theta\sin\phi + \hat{\mathbf{z}}\cos\theta) = \frac{1}{r}\end{aligned} \tag{15.452}$$

$$\begin{aligned}\Gamma^r_{\theta\theta} &= \mathbf{e}^r \cdot \frac{\partial \mathbf{e}_\theta}{\partial q^\theta} \\ &= (\hat{\mathbf{x}}\sin\theta\cos\phi + \hat{\mathbf{y}}\sin\theta\sin\phi + \hat{\mathbf{z}}\cos\theta) \\ &\quad \cdot \frac{\partial}{\partial\theta}(r\cos\theta\cos\phi\hat{\mathbf{x}} + r\cos\theta\sin\phi\hat{\mathbf{y}} - r\sin\theta\hat{\mathbf{z}}) = -r\end{aligned} \tag{15.453}$$

$$\Gamma^{\phi}_{r\phi} = \mathbf{e}^{\phi} \cdot \frac{\partial \mathbf{e}_r}{\partial q^{\phi}}$$
$$= \left(-\hat{\mathbf{x}} \frac{\sin \phi}{r \sin \theta} + \hat{\mathbf{y}} \frac{\cos \phi}{r \sin \theta} \right) \cdot \frac{\partial}{\partial \phi} (\sin \theta \cos \phi \hat{\mathbf{x}} + \sin \theta \sin \phi \hat{\mathbf{y}} + \cos \theta \hat{\mathbf{z}}) = \frac{1}{r} \tag{15.454}$$

$$\Gamma^{\phi}_{\theta\phi} = \mathbf{e}^{\phi} \cdot \frac{\partial \mathbf{e}_{\theta}}{\partial q^{\phi}}$$
$$= \left(-\hat{\mathbf{x}} \frac{\sin \phi}{r \sin \theta} + \hat{\mathbf{y}} \frac{\cos \phi}{r \sin \theta} \right) \cdot \frac{\partial}{\partial \phi} (r \cos \theta \cos \phi \hat{\mathbf{x}} + r \cos \theta \sin \phi \hat{\mathbf{y}} - r \sin \theta \hat{\mathbf{z}}) = \cot \theta \tag{15.455}$$

$$\Gamma^{r}_{\phi\phi} = \mathbf{e}^{r} \cdot \frac{\partial \mathbf{e}_{\phi}}{\partial q^{\phi}}$$
$$= (\hat{\mathbf{x}} \sin \theta \cos \phi + \hat{\mathbf{y}} \sin \theta \sin \phi + \hat{\mathbf{z}} \cos \theta) \cdot \frac{\partial}{\partial \phi} (-r \sin \theta \sin \phi \, \hat{\mathbf{x}} + r \sin \theta \cos \phi \hat{\mathbf{y}}) = -r \sin^2 \theta \tag{15.456}$$

and

$$\Gamma^{\theta}_{\phi\phi} = \mathbf{e}^{\theta} \cdot \frac{\partial \mathbf{e}_{\phi}}{\partial q^{\phi}}$$
$$= \left(\hat{\mathbf{x}} \frac{\cos \theta \cos \phi}{r} + \hat{\mathbf{y}} \frac{\cos \theta \sin \phi}{r} - \hat{\mathbf{z}} \frac{\sin \theta}{r} \right) \cdot \frac{\partial}{\partial \phi} (-r \sin \theta \sin \phi \hat{\mathbf{x}} + r \sin \theta \cos \phi \hat{\mathbf{y}}) = -\sin \theta \cos \theta \tag{15.457}$$

15.26 Newton's Laws and Maxwell's Equations in General Relativity

In this section, Newton's three laws of motion and Maxwell's equations for electromagnetic waves are expressed in their general relativistic form that is applicable to processes in curved space–time. Cosmology is a prime example of a setting in which general relativity has a significant presence and thereby requires modification of these laws for curved space–time. We first consider Newton's three laws of motion and then Maxwell's equations, including the general relativistic expression of the Lorentz gauge.

In Galilean relativity or classical mechanics, Newton's second law is

$$\mathbf{F} = m_0 \frac{\mathrm{d}^2\mathbf{x}}{\mathrm{d}t^2} \tag{15.458}$$

The rest mass m_0 remains unchanged with particle velocity in Newtonian mechanics so that the moving mass m equals the rest mass, where $m = m_0$. In Newton's second law of special relativity, the moving mass m is a function of its velocity, where $m = \gamma m_0$. In the stationary frame, the four-vector force $\mathbf{F}_0$ is the time derivative of the rest frame four-vector momentum $\mathbf{p}_0$. And the rest frame momentum $\mathbf{p}_0$ is related to the time derivative of the moving frame four-vector momentum $\mathbf{p}$ via γ, so that

$$\mathbf{F}_0 = \frac{d\mathbf{p}_0}{dt} = \frac{\mathrm{d}}{\mathrm{d}t}(\gamma\mathbf{p}) = \frac{\mathrm{d}}{\mathrm{d}t}\left(\frac{m_0\mathbf{v}}{\sqrt{1-(v/c)^2}}\right) = \frac{\mathrm{d}}{\mathrm{d}\tau}(m_0\mathbf{v}) \tag{15.459}$$

In general relativity, recalling the contravariant absolute derivative of Eq. (15.321), Newton's second law becomes

$$f^\mu = m_0\left(\frac{\mathrm{d}^2x^\mu}{\mathrm{d}\tau^2} + \Gamma^\mu_{\nu\sigma}\frac{\mathrm{d}x^\nu}{\mathrm{d}\tau}\frac{\mathrm{d}x^\sigma}{\mathrm{d}\tau}\right) = \frac{\mathrm{D}p^\mu}{\mathrm{d}\tau} \tag{15.460}$$

That is, given particle rest mass m_0, the absolute derivative of the particle four-momentum vector components p^μ, with respect to the proper time τ, gives the external four-force vector components f^μ, which act on the particle. In the case of Euclidean space and special relativity; i.e., when

$$\Gamma^\mu_{\nu\sigma} = 0 \tag{15.461}$$

and

$$g_{\mu\nu}\lambda^{\mu\nu} = c^2\mathrm{d}t^2 - \mathrm{d}x^2 - \mathrm{d}y^2 - \mathrm{d}z^2 \tag{15.462}$$

Newton's first law states that free particles move in straight lines through space. In general relativity, free particles follow geodesics in space–time. When no force acts on a particle rest mass m_0, where the four-force vector components are zero ($f^\mu = 0$), then one obtains the geodesic equation for the free particle trajectory along a geodesic, where from Eq. (15.460),

$$\frac{\mathrm{D}(\mathrm{d}x^\mu/d\tau)}{\mathrm{d}\tau} = \frac{\mathrm{d}^2x^\mu}{\mathrm{d}\tau^2} + \Gamma^\mu_{\nu\sigma}\frac{\mathrm{d}x^\nu}{\mathrm{d}\tau}\frac{\mathrm{d}x^\sigma}{\mathrm{d}\tau} = 0 \tag{15.463}$$

That is, the absolute derivative of the four-velocity vector components $dx^{\mu}/d\tau$ of a free particle, with respect to the proper time τ, equals zero. In the case of no external forces acting on the particle, the particle moves as a free body on a geodesic line, with only gravity influencing its motion.

In general relativity, gravity is no longer considered a force; rather, it is incorporated into the curvature of the space–time by the connection coefficients $\Gamma^{\mu}_{\nu\sigma}$. The connection coefficients $\Gamma^{\mu}_{\nu\sigma}$ are expressed in terms of the covariant g_{ab} and contravariant g^{ab} metric tensor components, and the metric tensor components of a specified space–time model contain the gravity influencing parameters. For example, the Schwarzschild solution of the Einstein field equations, which is valid for regions outside a spherical body of mass M, employs metric tensor components that contains the mass M of the body and the gravitational constant G.

Newton's law of gravity and the force of gravity can be derived from general relativity, as a limiting case, when the curved space–time of general relativity becomes the flat space–time of special relativity and Newtonian mechanics. This limiting case occurs as the distance r of a test particle of rest mass m_0, which for a large body of mass M goes to infinity ($r \to \infty$), and/or the mass M of the large body goes to zero ($M \to 0$). The limiting case can be simply specified by the condition $GM/rc^2 \ll 1$, in the Schwarzschild solution of Einstein's field equation, where G is the gravitational constant and c is the speed of light. Finally, note that in general relativity, Newton's third law, which states that every action has an opposed and equal reaction, applies to nongravitational forces.

The curvature of space–time affects Maxwell's equations. In curved space–time, the derivatives of the flat space–time of special relativity is replaced by the covariant derivatives of general relativity. In CGS units, recall from Eq. (13.174), the two Maxwell source (inhomogeneous) equations, which are Ampere's law and Gauss's law for electric field. In general relativity, recalling the covariant derivative of Eq. (15.340), the two Maxwell source equations are expressed by

$$\frac{4\pi}{c} J^b = \frac{\partial F^{ab}}{\partial q^a} + \Gamma^{a}_{\mu a} F^{\mu b} + \Gamma^{b}_{\mu a} F^{a\mu} \tag{15.464}$$

where

$$\nabla \times \mathbf{B} = \mu_0 \mathbf{J} + \mu_0 \varepsilon_0 \frac{\partial \mathbf{E}}{\partial t} \quad \text{and} \quad \nabla \cdot \mathbf{E} = \frac{\rho}{\varepsilon_0} \tag{15.465}$$

The position four-vector is $\mathbf{q} = (x, y, z, ict)$, and the q^a spatial derivative component $\partial F^{ab}/\partial q^a$ of the electromagnetic tensor F^{ab} is

$$\frac{\partial F^{ab}}{\partial q^a} \equiv \sum_{a=1}^{4} \frac{\partial F^{ab}}{\partial q^a} = \frac{\partial F^{xb}}{\partial x} + \frac{\partial F^{yb}}{\partial y} + \frac{\partial F^{zb}}{\partial z} - \frac{i}{c}\frac{\partial F^{tb}}{\partial t} \tag{15.466}$$

Recalling Eq. (15.340), the covariant derivative $F^{ab}_{;a}$ for the electromagnetic tensor F^{ab} is

$$F^{ab}_{;a} \equiv \frac{\partial F^{ab}}{\partial q^a} + \Gamma^a_{\mu a} F^{\mu b} + \Gamma^b_{\mu a} F^{a\mu} \tag{15.467}$$

In CGS units, recalling Eq. (13.172), the two homogeneous Maxwell's equations, which are Gauss's law for the magnetic field and Faraday's law. In general relativity, using the covariant derivative of Eq. (15.340), the source-free Maxwell equations are expressed by

$$F^{ab}_{;c} + F^{ca}_{;b} + F^{bc}_{;a} = 0 \tag{15.468}$$

where

$$\nabla \cdot \mathbf{B} = 0 \quad \text{and} \quad \nabla \times E = -\frac{\partial \mathbf{B}}{\partial t} \tag{15.469}$$

Finally, the Lorentz gauge in general relativity uses the covariant derivative of Eq. (15.340), such that

$$A^a_{;a} \equiv \frac{\partial A^a}{\partial q^a} + \Gamma^a_{ca} A^c = 0 \tag{15.470}$$

The four-vector potential $\mathbf{A}$ is given by

$$\mathbf{A} = \left(A_x, A_y, A_z, \mathrm{i}\frac{\Phi}{c} \right) \tag{15.471}$$

15.27 The Curvature, Ricci, and Einstein Tensors and the Curvature or Ricci Scalar

In this section, we derive the curvature tensor of a surface by calculating the change $\Delta\mathbf{A}$ in a vector $\mathbf{A}$ after parallel transport around an arbitrary, infinitesimally small, closed loop on a curved surface. First, we consider the covariant form of vector $\mathbf{A}$ and derivation of the curvature or Riemann tensor $\mathbf{R}$ with matrix elements R^i_{kjm}. That is, we determine the change $\Delta\mathbf{A}$ in the contravariant components of vector $\mathbf{A}$ after parallel transport around an arbitrary, infinitesimal, closed loop on the surface that leads a solution expressed by the curvature tensor components R^i_{kjm}. One integrates the covariant form of the vector $\mathbf{A}$ around the closed loop in which $\mathbf{A}$ varies from point-by-point on the loop, where

$$\Delta A_k = \oint \delta A_k = \oint \Gamma^i_{kj} A_i \, \mathrm{d}x^j \tag{15.472}$$

The values of the covariant components A_i of vector **A** at points inside the infinitesimally small loop vary, and depend on the path, which is used to approach an interior point.

For an infinitesimally small loop and enclosed area, the values of the covariant components A_i at the enclosed points are approximated as unique (single-valued). These approximately single-valued covariant components A_i in the interior of the infinitesimally small loop have values, which are specified by the values of A_i at points on the loop. Recall the prior derived relation

$$\delta A_k = \Gamma^i_{kj} A_i \, \mathrm{d}x^j \tag{15.473}$$

This variation δA_k of vector **A** depends on the connection coefficient Γ^i_{kj} and values of A_i, given a differential change $\mathbf{dx}$ in vector position $\mathbf{x}$.

Variation δA_k is evaluated at points on the loop and inside the loop. The values of δA_k normally depend on the direction of $\mathbf{dx}$, where the values of A_i and δA_k vary depending on the path direction of $\mathbf{dx}$. However, for the case of the infinitesimally small loop, the variation of δA_k can be approximated as being independent of the direction of the differential change $\mathbf{dx}$. Hence, one can equate δA_k with $\mathrm{d}A_k$, where

$$\delta A_k \cong \mathrm{d}A_k \tag{15.474}$$

As a result, one can directly relate the ordinary partial derivatives $\partial A_k / \partial x^j$ of the covariant components A_i of vector **A** components to A_i via the proportionality connection term Γ^i_{kj}:

$$\frac{\partial A_k}{\partial x^j} = \Gamma^i_{kj} A_i \tag{15.475}$$

The loop on the curved surface encloses an area, in which the closed loop line integral for vector **A** can be related to the surface integral of the curl of vector **A** over the loop-enclosed surface by Stoke's theorem, where

$$\Delta A_k = \frac{1}{2}\left[\frac{\partial(\Gamma^i_{km} A_i)}{\partial x^j} - \frac{\partial(\Gamma^i_{kj} A_i)}{\partial x^j}\right] \Delta S^{jm} \tag{15.476}$$

Here, ΔS^{jm} is the differential area enclosed by the loop. Carrying out the derivatives in Eq. (15.476), gives

$$\Delta A_k = \frac{1}{2}\left[\frac{\partial \Gamma^i_{km}}{\partial x^j} A_i - \frac{\partial \Gamma^i_{kj}}{\partial x^m} A_i + \Gamma^i_{km} \frac{\partial A_i}{\partial x^j} - \Gamma^i_{kj} \frac{\partial A_i}{\partial x^m}\right] \Delta S^{jm} \tag{15.477}$$

With dummy variable p and n,

$$\frac{\partial A_i}{\partial x^j} = \Gamma^n_{ij} A_n = \Gamma^p_{ij} A_p \quad \text{and} \quad \frac{\partial A_i}{\partial x^m} = \Gamma^n_{im} A_n = \Gamma^p_{im} A_p \tag{15.478}$$

Substitution into Eq. (15.477) of Eq. (15.478) for the derivatives $\partial A_i/\partial x^j$ and $\partial A_i/\partial x^m$ gives

$$\Delta A_k = \frac{1}{2} R^i_{kjm} A_i \Delta S^{jm} \tag{15.479}$$

In Eq. (15.479), the covariant form of the curvature tensor or Riemann tensor R^i_{kjm} is then

$$R^i_{kjm} = \frac{\partial \Gamma^i_{km}}{\partial x^j} A_i - \frac{\partial \Gamma^i_{kj}}{\partial x^m} A_i + \Gamma^n_{km}\Gamma^i_{nj} - \Gamma^n_{kj}\Gamma^i_{nm} \tag{15.480}$$

At each point on a flat surface, the Christoffel components of a second kind Γ^i_{kj} are zero, and thus, the curvature tensor components R^i_{kjm} are also zero, where

$$\Gamma^i_{kj} = 0 \leftrightarrow R^i_{kjm} = 0 \quad \text{flat surface} \tag{15.481}$$

On a flat surface, the vector **A** undergoes parallel transport around a closed loop that is independent of the loop path. Thus, vector **A** returns to its original orientation after one cycle around any closed loop path. However, on a curved surface, Γ^i_{kj} and R^i_{kjm} are each nonzero in general at each point, where

$$\Gamma^i_{kj} \neq 0 \leftrightarrow R^i_{kjm} \neq 0 \quad \text{curved surface} \tag{15.482}$$

On a curved surface, the vector **A**, which undergoes parallel transport around a closed loop, depends on the loop path, and thus it generally returns to a different orientation after one cycle around the loop in which the changed vector orientation depends on the loop path. Parallel transport around a small, closed loop allows calculation of the curvature tensor, which then determines if the surface is curved ($R^i_{kjm} \neq 0$) or flat ($R^i_{kjm} = 0$).

Next, we determine the change $\Delta\mathbf{A}$ in the contravariant components of vector **A** after parallel transport around an arbitrary, infinitesimal, closed loop on the surface. One integrates the covariant form of vector **A** around the closed loop in which **A** varies from point by point on the loop, where

$$\Delta A^k = \oint \delta A^k = \oint \Gamma^i_{kj} A^i \, \mathrm{d}x^j \tag{15.483}$$

The j remains a superscript on the components x^j of the displacement vector **x** to indicate that summation over j is required by the Einstein summation convention. One then invokes the invariance of a scalar $\mathbf{A} \cdot \mathbf{B}$ or $A^k B_k$ under parallel displacement, where

$$\Delta(A^k B_k) = \Delta A^k B_k + A^k \Delta B_k = 0 \tag{15.484}$$

With dummy indices, one has the equivalent scalar invariance expression

$$\Delta A^i B_i + A^k \Delta B_k = 0 \tag{15.485}$$

Here, A^k are the contravariant vector components of vector **A**, and B_k are the components of any covariant vector **B**, where

$$\Delta B_k = \frac{1}{2} R^i_{kjm} B_i \Delta S^{jm} \tag{15.486}$$

Substitution of Eq. (15.486) for ΔB_k in the scalar invariance expression of Eq. (15.485), gives

$$\left(\Delta A^i + \frac{1}{2} A^k R^i_{kjm} \Delta S^{jm} \right) B_i = 0 \tag{15.487}$$

The covariant components B_i of vector **B** is arbitrary, so that the expression inside the parentheses is set to zero, where

$$\Delta A^i = -\frac{1}{2} A^k R^i_{kjm} \Delta S^{jm} \tag{15.488}$$

As a result, the change $\Delta \mathbf{A}$ of vector **A**, which is expressed by the change ΔA^i in the contravariant components A^i after parallel transport around an arbitrary, infinitesimally small, closed loop on a curved surface, is again expressed in terms of the curvature tensor components R^i_{kjm}.

Unlike ordinary differentiation, covariant differentiation of the vector **A** with respect to x^k and x^j generally depends on the order of differentiation; that is,

$$A_{i;j;k} \neq A_{i;k;j} \tag{15.489}$$

where

$$A_{i;j} = \frac{dA_i}{dx^j} + \Gamma^k_{ij} A_k \tag{15.490}$$

However, the commutation of the derivatives with respect to x^k and x^j of the covariant components A_i of vector **A** yields the curvature tensor components R^i_{kjm}, where

$$A_{i;j;k} - A_{i;k;j} = A_m R^m_{ijk} \tag{15.491}$$

And the commutation of the derivatives with respect to x^k and x^j of the contravariant components A^i of vector **A** yields the curvature tensor components R^i_{kjm}:

$$A^i_{;j;k} - A^i_{;k;j} = -A^m R^i_{mjk} \tag{15.492}$$

The associated type (0,4) tensor components R_{abcd} is defined by contracting the curvature tensor R^e_{bcd} with the covariant components of the metric tensor g_{ae}, where

$$R_{abcd} \equiv g_{ae} R^e_{bcd} \tag{15.493}$$

where

$$R_{abcd} = -R_{bacd} \quad R_{abcd} = -R_{abdc} \quad R_{abcd} = R_{cdab} \tag{15.494}$$

The cyclic identity of the curvature tensor R^a_{bcd} is given by

$$R^a_{bcd} + R^a_{cdb} + R^a_{dbc} = 0 \tag{15.495}$$

The Bianchi identity for the covariant derivatives $R^a_{bcd;e}$ of the curvature tensor R^a_{bcd} is

$$R^a_{bcd;e} + R^a_{bde;c} + R^a_{bec;d} = 0 \tag{15.496}$$

Note that the contraction of the curvature tensor, via R^a_{acd}, is zero:

$$R^a_{acd} = 0 \tag{15.497}$$

However, the contraction of curvature tensor, via R^c_{abc}, is nonzero and produces the symmetric Ricci tensor R_{ab} defined by

$$R_{ab} \equiv R^c_{abc} \tag{15.498}$$

where

$$R_{ab} = R_{ba} \tag{15.499}$$

The Ricci tensor R_{ab} is given by

$$R_{ab} = \frac{\partial \Gamma^c_{ab}}{\partial x^c} - \frac{\partial \Gamma^c_{ac}}{\partial x^b} + \Gamma^c_{ab}\Gamma^d_{cd} - \Gamma^d_{ac}\Gamma^c_{bd} \tag{15.500}$$

where

$$\Gamma^a_{\mu a} = \frac{1}{2} g^{-1} \frac{\partial g}{\partial x^b} = \frac{1}{2} \frac{\partial}{\partial x^b} \ln|g| \tag{15.501}$$

The modulus sign is needed because g is not necessarily positive in the indefinite case. Also, the determinant of the covariant metric tensor components $|g_{ab}|$ is denoted by g. Here, the connection coefficients Γ^a_{bc} are given by

$$\Gamma^a_{bc} = \frac{1}{2} g^{ad} \left(\frac{g_{dc}}{\partial x^b} + \frac{g_{bd}}{\partial x^c} - \frac{\partial g_{bc}}{\partial x^d} \right) \tag{15.502}$$

Further contraction of the Ricci tensor components R_{ab} with the contravariant metric tensor components g^{ab} gives the curvature scalar or Ricci scalar R, which is defined as

$$R \equiv g^{ab} R_{ab} = R_a^a \tag{15.503}$$

The curvature scalar R is invariant, where $\delta R = 0$. For a 2-D surface, the curvature scalar R equals twice the Gaussian curvature K of the surface defined in terms of the two principal radii of curvature ρ_1 and ρ_2, where

$$R = 2K = \frac{2}{\rho_1 \rho_2} \tag{15.504}$$

The principal radii of curvature ρ_1 and ρ_2 have the same sign if the corresponding centers of curvature are on the same side of the surface ($K > 0$), and they have opposite signs if the centers are on opposite sides of the surface ($K < 0$), where $K = 1/\rho_1 \rho_2$.

Finally, the symmetric Einstein tensor components G_{ab} is defined in terms of the scalar curvature R and the covariant components of the metric tensor g_{ab} and the Ricci tensor R_{ab}, where

$$G_{ab} \equiv R_{ab} - \frac{1}{2} R g_{ab} \tag{15.505}$$

and

$$G_{ab} = G_{ba} \tag{15.506}$$

Using the Bianchi identity, one can show the Einstein tensor G_{ab} has a single, zero divergence $G^{ab}_{;a}$, in which

$$G^{ab}_{;a} = 0 \tag{15.507}$$

15.28 The Stress Tensor, Universe Fluid Model, and the Equation of Continuity and Motion

In this section, we examine the stress tensor for the universe modeled as a fluid, and derive the equations of fluid continuity and the fluid equation of motion from the divergence of the stress tensor of the fluid, which is zero. We develop the stress tensor and the universe fluid model, which was employed by Einstein to develop the Einstein field equations for gravity, which we cover in the next section.

The universe is modeled as simple macroscopic fluid, devoid of shear viscous, bulk viscous, and heat-conductive properties where the symmetric, contravariant stress tensor T^{ab} is defined by

$$T^{ab} = (p + \rho/c^2)u^a u^b - pg^{ab} \tag{15.508}$$

where

$$T^{ab} = T^{ba} \tag{15.509}$$

The energy−momentum−stress tensor is a function of the fluid pressure p, the fluid density ρ, the speed of light in a vacuum c, and the world velocity four-vectors u^a and u^b are vectors with a = 0, 1, 2, 3 and $b = 0, 1, 2, 3$ expressed in terms of the proper time derivative of the position vector $\mathbf{x}^0$ with proper time coordinate $x^4 = c\tau$, radial coordinate $x^0 = r$, polar angle coordinate $x^1 = \theta$, and azimuth angle coordinate $x^2 = \phi$; such that

$$u^a = \frac{\mathrm{d}x^a}{\mathrm{d}\tau} = \frac{\mathrm{d}}{\mathrm{d}\tau}(r^0, \theta^0, \phi^0, c\tau) = (0, c) \tag{15.510}$$

The stress tensor T is given in matrix form by

$$\begin{aligned} T &= [T^{ab}] \\ &= \begin{bmatrix} T^{00} & T^{01} & T^{02} & T^{03} \\ T^{10} & T^{11} & T^{12} & T^{13} \\ T^{20} & T^{21} & T^{22} & T^{23} \\ T^{30} & T^{31} & T^{32} & T^{33} \end{bmatrix} \\ &= \begin{bmatrix} \begin{bmatrix} T^{00} & T^{01} & T^{02} \\ T^{10} & T^{11} & T^{12} \\ T^{20} & T^{21} & T^{22} \end{bmatrix} & \begin{bmatrix} T^{03} \\ T^{13} \\ T^{23} \end{bmatrix} \\ \begin{bmatrix} T^{30} & T^{31} & T^{32} \end{bmatrix} & [T^{33}] \end{bmatrix} \end{aligned} \tag{15.511}$$

Stress tensor component T^{ab} is the flux of the ath component of the four-momentum vector $(p_1, p_2, p_3, -E)$, which crosses a surface of constant x^b. The momentum flux T^{ij} is the flow of the ith momentum component in the jth direction, where

$$T^{ij} = \begin{bmatrix} T^{00} & T^{01} & T^{02} \\ T^{10} & T^{11} & T^{12} \\ T^{20} & T^{21} & T^{22} \end{bmatrix} \quad i, j = 0, 1, 2 \tag{15.512}$$

The pressure in the x-, y-, and z-directions are respectively given by the diagonal elements T^{00}, T^{11}, and T^{22}, and the energy density is given by diagonal element T^{33}. The energy flux in the x-, y-, and z-directions is given by the row elements T^{30}, T^{31}, and T^{32}, respectively. The momentum density in the x-, y-, and z-directions is given by the column elements T^{03}, T^{13}, and T^{23}, respectively.

The divergence of the stress tensor components T^{ab} is zero:

$$\frac{\partial T^{ab}}{\partial x^a} = 0 \tag{15.513}$$

Substitution into Eq. (15.513) of Eq. (15.508) for T^{ab}, gives

$$\frac{\partial(\rho u^a)}{\partial x^a} u^b + \rho u^a \frac{\partial u^b}{\partial x^a} + \frac{p}{c^2}\frac{\partial u^a}{\partial x^a} u^b + \frac{p}{c^2} u^a \frac{\partial u^b}{\partial x^a} + \frac{1}{c^2}\frac{\partial p}{\partial x^a} u^a u^b - \frac{\partial p}{\partial x^a} g^{ab} = 0 \tag{15.514}$$

One recalls Eq. (15.510), where

$$u^b u_b = c^2 \tag{15.515}$$

Differentiation of Eq. (15.515) gives

$$\frac{\partial u^b}{\partial x^a} u_b + u^b \frac{\partial u_b}{\partial x^a} = 0 \tag{15.516}$$

where

$$\frac{\partial u^b}{\partial x^a} u_b = u^b \frac{\partial u_b}{\partial x^a} = 0 \tag{15.517}$$

Contraction of the components of $\partial T^{ab}/\partial x^a$ in Eq. (15.514) with the covariant velocity components u_b in the divergence equation of Eqs. (15.513–15.514), gives

$$\frac{\partial T^{ab}}{\partial x^a} u_b = 0 \tag{15.518}$$

with

$$g^{ab} u_b = u^a \tag{15.519}$$

Dividing Eq. (15.518) by c^2, gives the continuity equation

$$\frac{\partial(\rho u^a)}{\partial x^a} + \frac{p}{c^2}\frac{\partial u^a}{\partial x^a} = 0 \tag{15.520}$$

Hence,

$$\frac{\partial(\rho u^a)}{\partial x^a} = -\frac{p}{c^2}\frac{\partial u^a}{\partial x^a} \tag{15.521}$$

One substitutes the result of Eq. (15.521) from the continuity equation into Eq. (15.514) for the divergence $\partial T^{ab}/\partial x^a$ of the stress tensor, where the first and third terms cancel, yielding the equation of motion

$$\left(\rho + \frac{p}{c^2}\right)\frac{\partial u^b}{\partial x^a}u^a = \left(g^{ab} - \frac{u^a u^b}{c^2}\right)\frac{\partial p}{\partial x^a} \tag{15.522}$$

15.29 Einstein Field Equations

In this section, we present the Einstein field equations, which employ the stress tensor and universe fluid model discussed in Section 15.28. We first consider the metric tensor. The contravariant metric tensor components g^{ab} depend on the specific coordinate system parameters and the gravitational potential energy. A proposed relativistic field equation should reduce to Poisson's equation in the nonrelativistic, Newtonian limit, where given gravitational constant G, gravitational potential energy U_g, and mass density ρ, is

$$\nabla^2 V_g = 4\pi G\rho \tag{15.523}$$

In 1915, Einstein first published the proposed field equation in which the Ricci tensor components R^{ab} are proportional to the stress tensor T^{ab} via a coupling constant κ, where

$$R^{ab} = \kappa T^{ab} \tag{15.524}$$

where

$$\kappa = -\frac{8\pi G}{c^4} \tag{15.525}$$

The coupling constant κ is expressed by -8π multiplied by the ratio of the gravitational constant G, divided by c^4, for speed of light c. The expression of the coupling constant κ was chosen, so that the Newtonian limit of this field equation would reduce to Poisson's equation.

At this point, the field equation suffers the same flaw as Ampere's law in a vacuum for the magnetic flux density $\mathbf{B}$, with only the free electric current density vector $\mathbf{J}_f$. Maxwell realized the vacuum displacement current $\partial\mathbf{E}/\partial t$, which multiplied by the constant $1/c$, and added to the free electric current density vector $\mathbf{J}_f$ in Ampere's law, allowed the divergence of both sides of the Ampere Law equation to be zero. The complete Ampere's law, expressed in CGS–Gaussian units, becomes

$$\nabla \times \mathbf{B} = \frac{4\pi}{c}\mathbf{J}_f + \frac{1}{c}\frac{\partial \mathbf{E}}{\partial t} \tag{15.526}$$

Taking the divergence of both sides of Eq. (15.526), gives zero on both sides, where now one invokes Gauss's law for the electric field on the right side of Eq. (15.526):

$$\nabla\cdot\left(\frac{4\pi}{c}\mathbf{J}_f+\frac{1}{c}\frac{\partial\mathbf{E}}{\partial t}\right)=\frac{4\pi}{c}\nabla\cdot\mathbf{J}_f+\frac{\partial}{\partial t}(\nabla\cdot\mathbf{E})=\frac{4\pi}{c}\left(\nabla\cdot\mathbf{J}_f+\frac{\partial\rho_f}{\partial t}\right)=0 \tag{15.527}$$

Gauss's law for the electric field $\mathbf{E}$ expressed in CGS–Gaussian units for free charge density ρ_f is

$$\nabla\cdot\mathbf{E}=4\pi\rho_f \tag{15.528}$$

Charge conservation requires the continuity equation

$$\nabla\cdot\mathbf{J}_f+\frac{\partial\rho_f}{\partial t}=0 \tag{15.529}$$

The divergence of the left side of Eq. (15.287), gives zero, because the divergence of the curl of a vector $\mathbf{B}$ is zero:

$$\nabla\cdot(\nabla\times\mathbf{B})=0 \tag{15.530}$$

Einstein encountered a similar mathematical contraction of his original field equation where the divergence of the stress tensor T^{ab} on the right side of Eq. (15.524) is zero, but the divergence of the Ricci tensor R^{ab} is generally nonzero, so that

$$\nabla\cdot R^{ab}=\nabla\cdot(\kappa T^{ab})=\kappa\nabla\cdot T^{ab} \tag{15.531}$$

where

$$\nabla\cdot T^{ab}=0 \quad \text{and} \quad \nabla\cdot R^{ab}\neq 0 \tag{15.532}$$

As a result, later in 1915, Einstein modified the field equation of Eq. (15.524) by subtracting the product of the Ricci scalar and contravariant metric tensor $Rg^{ab}/2$ from the Ricci tensor R^{ab} on the left-hand side of Eq. (15.524) and obtained the corrected field equation, which had zero divergence on both sides of the equation, where

$$R^{ab}-\frac{1}{2}Rg^{ab}=\kappa T^{ab} \tag{15.533}$$

The mixed tensor component of the corrected Einstein field equation is

$$R^{a}_{b}-\frac{1}{2}Rg^{a}_{b}=\kappa T^{a}_{b} \tag{15.534}$$

Contraction of the mixed tensor form of the field equation in Eq. (15.334), gives

$$R_a^a - \frac{1}{2}Rg_a^a = \kappa T_a^a \tag{15.535}$$

Performing the scalar product in Eq. (15.535), one obtains the Ricci scalar R in terms of the stress scalar T:

$$R = -\kappa T \tag{15.536}$$

where

$$R = R_a^a, \quad g_a^a = \delta_a^a = 4, \quad \text{and} \quad T = T_a^a \tag{15.537}$$

Consequently, the corrected Einstein field equation can be expressed by the contravariant components of the stress tensor T^{ab} and scalar T and the Ricci tensor R^{ab}, where

$$R^{ab} = \kappa\left(T^{ab} - \frac{1}{2}Tg^{ab}\right) \tag{15.538}$$

In terms of the Einstein gravity tensor G^{ab}, the corrected field equation is

$$G^{ab} = \kappa T^{ab} \tag{15.539}$$

The Einstein gravity tensor contravariant components G^{ab} are expressed in terms of the contravariant components of the Ricci tensor R^{ab} and metric tensor g^{ab} and the contracted Ricci tensor, which is the Ricci scalar, where

$$G^{ab} \equiv R^{ab} - \frac{1}{2}Rg^{ab} \tag{15.540}$$

In 1917, Einstein applied the field equation to a stable and cosmic gas model universe in which the dimensionless scale factor $R(t)$ as a function of time t is used to express the spatial extent of the universe via the Hubble ratio $H(t)$. The Hubble ratio $H(t)$ is the ratio of the time derivative $\mathrm{d}R(t)/dt$ divided by dimensionless scale factor $R(t)$. In the case of a stable universe, as Einstein assumed in 1917,

$$H(t) = \frac{1}{R}\frac{\mathrm{d}R}{\mathrm{d}t} = 0 \tag{15.541}$$

The stable universe model with $H(t) = 0$ requires a counteracting force to oppose the gravitational-induced collapse of the cosmic gas. Einstein improvised by adding to the left side of the field equation a term to counteract the gravitational collapse. This improvised term was the product of a cosmological constant Λ, multiplied by the contravariant components of the metric tensor g^{ab}. The improvised Einstein field equation for a stable universe was then

$$G^{ab} + \Lambda g^{ab} = \kappa T^{ab} \tag{15.542}$$

The detection of the Doppler red-shifted, receding galaxies by Hubble suggested the universe was expanding, which eliminated the need for the cosmological constant term Λg^{ab} and Einstein returned to his original field equation:

$$G^{ab} = R^{ab} - \frac{1}{2} R g^{ab} = \kappa T^{ab} \tag{15.543}$$

This form of the Einstein field equation is expressed by the contravariant tensor components of the Ricci R^{ab} and stress T^{ab} tensors. One can equivalently express the Einstein equations with the covariant tensor components of the Ricci R_{ab} and stress T_{ab} tensors.

15.30 Hilbert Derived Einstein Field Equations from the Principle of Least Action

In 1915, just after Einstein's published version of the above field equation, Hilbert published a derivation of the same field expression using the least action principle. The action integral S is given by

$$S = \int \mathrm{d}x^0 \int \mathrm{d}x^1 \int \mathrm{d}x^2 \int \mathrm{d}x^3 \left[\frac{R}{2\kappa} + L_{\mathrm{M}}\right] \sqrt{-g} \tag{15.544}$$

With coupling constant κ, the term L_{M} describes matter field, R is the Ricci scalar, and g is the determinant of the metric tensor in which integration is preformed over the 4-D volume of the three spatial dimensions and the fourth dimension of time. The variation of the action integral is set to zero, where

$$\delta S = \int \mathrm{d}x^0 \int \mathrm{d}x^1 \int \mathrm{d}x^2 \int \mathrm{d}x^3 \left[\frac{1}{2\kappa} \frac{\delta(R\sqrt{-g})}{\delta g^{ab}} + \frac{\delta(\sqrt{-g} L_{\mathrm{M}})}{\delta g^{ab}}\right] \delta g^{ab} = 0 \tag{15.545}$$

Expanding the first term of the integrand,

$$\delta S = \int \mathrm{d}x^0 \int \mathrm{d}x^1 \int \mathrm{d}x^2 \int \mathrm{d}x^3 \left[\frac{1}{2\kappa} \left(\frac{\delta R}{\delta g^{ab}} + \frac{R}{\sqrt{-g}} \frac{\delta\sqrt{-g}}{\delta g^{ab}}\right) \right. \tag{15.546}$$
$$\left. + \frac{\delta(\sqrt{-g} L_{\mathrm{M}})}{\delta g^{ab}\sqrt{-g}} \right] \delta g^{ab} \sqrt{-g} = 0$$

The portion integrand in brackets must be zero for arbitrary nonzero δg^{ab}:

$$\frac{1}{2\kappa}\left(\frac{\delta R}{\delta g^{ab}}+\frac{R}{\sqrt{-g}}\frac{\delta\sqrt{-g}}{\delta g^{ab}}\right)+\frac{1}{\sqrt{-g}}\frac{\delta\left(\sqrt{-g}L_{\mathrm{M}}\right)}{\delta g^{ab}}=0 \tag{15.547}$$

The last term in the left side of Eq. (15.547) is identified with the stress tensor T^{ab}:

$$\frac{\delta R}{\delta g^{ab}}+\frac{R}{\sqrt{-g}}\frac{\delta\sqrt{-g}}{\delta g^{ab}}=\kappa T^{ab} \tag{15.548}$$

where

$$T_{ab}=-\frac{2}{\sqrt{-g}}\frac{\delta\left(\sqrt{-g}L_{\mathrm{M}}\right)}{\delta g^{ab}}=-2\frac{\partial L_{\mathrm{M}}}{\delta g^{ab}}+g_{ab}L_{\mathrm{M}} \tag{15.549}$$

The variation of the Ricci tensor δR is derived as follows. One starts with the variation of the curvature tensor R^i_{jab}, where

$$R^i_{jab}=\frac{\partial}{\partial x^a}\Gamma^i_{bj}-\frac{\partial}{\partial x^b}\Gamma^i_{aj}+\Gamma^i_{ak}\Gamma^k_{bj}-\Gamma^i_{bk}\Gamma^k_{aj} \tag{15.550}$$

The variation of the curvature tensor δR^i_{jab} in Eq. (15.550), is given by

$$\delta R^i_{jab}=\frac{\partial}{\partial x^a}\delta\Gamma^i_{bj}-\frac{\partial}{\partial x^b}\delta\Gamma^i_{aj}+\delta\Gamma^i_{ak}\Gamma^k_{bj}+\Gamma^i_{ak}\delta\Gamma^k_{bj}-\delta\Gamma^i_{bk}\Gamma^k_{aj}-\Gamma^i_{bk}\Gamma^k_{aj} \tag{15.551}$$

The variation of the Christoffel function of a second kind is a tensor, which thus has a covariant derivative given by

$$\nabla_\lambda(\delta\Gamma^i_{ba})=\frac{\partial}{\partial x^\lambda}(\delta\Gamma^i_{ba})+\Gamma^i_{j\lambda}\delta\Gamma^j_{ba}-\Gamma^j_{b\lambda}\delta\Gamma^i_{ja}-\Gamma^j_{a\lambda}\delta\Gamma^i_{bj} \tag{15.552}$$

The variation of the curvature tensor δR^i_{jab} of Eq. (15.551), equals the difference of the two covariant derivatives of Christoffel function variations of Eq. (15.552), where

$$\delta R^i_{jab}=\nabla_a(\delta T^i_{bj})-\nabla_b(\delta T^i_{bj}) \tag{15.553}$$

From Eq. (15.553), one obtains the variation of the Ricci tensor δR_{ab} by contracting two indices of the curvature tensor δR^i_{aib}, where

$$\delta R_{ab}\equiv\delta R^i_{aib}=\nabla_i(\delta T^i_{ba})-\nabla_b(\delta T^i_{ia}) \tag{15.554}$$

The derivation of the variation of the Ricci scalar δR starts with the definition of the Ricci scalar R as the contraction of the covariant Ricci tensor components R_{ab} with the contravariant components of the metric tensor g^{ab}, where

$$R = g^{ab} R_{ab} \tag{15.555}$$

In Eq. (15.555), the variation of the Ricci scalar δR is then

$$\delta R = R_{ab}\delta g^{ab} + g^{ab}\delta R_{ab} \tag{15.556}$$

Substitution into Eq. (15.556) of Eq. (15.554) for the variation of the Ricci tensor δR_{ab} gives

$$\delta R = R_{ab}\delta g^{ab} + g^{ab}[\nabla_i(\delta T^i_{ba}) - \nabla_b(\delta T^i_{ia})] \tag{15.557}$$

Covariant vector component variables x_j are independent of the contravariant vector component variables x^a and x^b, so that the covariant vector gradient ∇_j, which is based on derivatives with respect to x_j, where ∇_j is applied to the contravariant metric tensor g^{ab}, yields zero, where

$$\nabla_j g^{ab} = 0 \tag{15.558}$$

In Eq. (15.557), one can bring the metric tensor g^{ab} inside the gradient operator, where

$$\delta R = R_{ab}\delta g^{ab} + \nabla_j(g^{ab}\delta T^j_{ba} - g^{aj}\delta T^i_{ia}) \tag{15.559}$$

The last term in Eq. (15.559) is a total derivative, which yields boundary terms when integrated. As a result, the last term can be dropped because it does not contribute to the variation of the action, and the variation of the Ricci scalar δR becomes

$$\delta R = R_{ab}\delta g^{ab} \tag{15.560}$$

The variation of the contravariant metric tensor components δg^{ab} is now derived as follows. One starts with Jacobi's formula for differentiating a determinant, where, applied to the determinant g of the metric tensor, gives

$$\delta g = g^{ab} g \delta g_{ab} \tag{15.561}$$

With substitution of Eq. (15.561) for δg, variation of $\delta\sqrt{-g}$ gives

$$\delta\sqrt{-g} = -\frac{\delta g}{2\sqrt{-g}} = \frac{1}{2}\sqrt{-g}\left(g^{ab}\delta g_{ab}\right) = -\frac{1}{2}\sqrt{-g}\left(g_{ab}\delta g^{ab}\right) \tag{15.562}$$

Rearrangement of Eq. (15.562), gives

$$\frac{\delta\sqrt{-g}}{\sqrt{-g}}\frac{1}{\delta g^{ab}} = -\frac{1}{2}g_{ab} \tag{15.563}$$

In summary, given the action S,

$$S = \int \mathrm{d}x^0 \int \mathrm{d}x^1 \int \mathrm{d}x^2 \int \mathrm{d}x^3 \left[\frac{R}{2\kappa} + L_{\mathrm{M}}\right]\sqrt{-g} \tag{15.564}$$

The variation δS of the action in Eq. (15.564), then yields

$$\frac{\delta R}{\delta g^{ab}} + \frac{R}{\sqrt{-g}}\frac{\delta\sqrt{-g}}{\delta g^{ab}} = \kappa T^{ab} \tag{15.565}$$

where

$$\delta R = R_{ab}\delta g^{ab} \tag{15.566}$$

and

$$\frac{\delta\sqrt{-g}}{\sqrt{-g}}\frac{1}{\delta g^{ab}} = -\frac{1}{2}g_{ab} \tag{15.567}$$

Substitution of Eqs. (15.566–15.567) into Eq. (15.565), gives the Einstein field equations, which are derived from the least action principle:

$$R_{ab} - \frac{1}{2}Rg_{ab} = \kappa T_{ab} \tag{15.568}$$

This form of the Einstein field equation is expressed by the covariant tensor components of the Ricci R_{ab} and stress T_{ab} tensors compared to the form discussed in previous Section 15.29, which uses the contravariant components R^{ab} and T^{ab}.

15.31 Gravity Waves Derived from Linearized Einstein Field Equations for the Weak Gravity Condition

In 1916, Einstein derived a weak field solution to his field equations that predicted gravity waves. The gravity waves, which mediate the gravitational force, propagate through empty space–time at the speed of light c, thereby negating "instant action at a distance." The weak field solution assumed that the covariant metric tensor components g_{ab} were the sum of the flat space, Minkowski metric components n_{ab}

of special relativity added to a small perturbation comprised of the curved space metric component h_{ab}, which is much less than 1:

$$g_{ab} = n_{ab} + \varepsilon h_{ab} \quad |h_{ab}| \ll 1 \tag{15.569}$$

Parameter ε is a constant, where

$$|\varepsilon| \ll 1 \tag{15.570}$$

The covariant components n_{ab} of the flat space, Minkowski metric tensor are

$$[n_{ab}] = \begin{bmatrix} -1 & 0 & 0 & 0 \\ 0 & -1 & 0 & 0 \\ 0 & 0 & -1 & 0 \\ 0 & 0 & 0 & 1 \end{bmatrix} \tag{15.571}$$

The building block of the curvature or Riemann mixed tensor components R^a_{bcd}, the covariant Ricci tensor components R_{ab}, and the scalar R have the connection coefficients Γ^a_{bc}:

$$\Gamma^a_{bc} = \frac{1}{2} g^{ad} \left(\frac{\partial g_{dc}}{\partial x^b} + \frac{\partial g_{bd}}{\partial x^c} - \frac{\partial g_{bc}}{\partial x^d} \right) \tag{15.572}$$

The weak field metric tensor components derivatives, such as $\partial g_{bc}/\partial x^d$, are expressed by

$$\frac{\partial g_{bc}}{\partial x^d} = \frac{\partial}{\partial x^d}(n_{bc} + \varepsilon h_{bc}) = \frac{\partial n_{bc}}{\partial x^d} + \varepsilon \frac{\partial h_{bc}}{\partial x^d} \tag{15.573}$$

To obtain the contravariant perturbations component h^{ab} from the covariant perturbation components h_{ab}, one can raise the indices with the Minkowski metric, where

$$h^{ab} = n^{ac} n^{bd} h_{cd} \tag{15.574}$$

The contraction of the covariant g_{ab} and contravariant g^{bc} metric tensor components, yields the Dirac delta mixed tensor components δ^c_a, where

$$g_{ab} g^{bc} = \delta^c_a \tag{15.575}$$

One assumes the linearized form of the contravariant metric tensor component g^{bc} in Eq. (15.575) to be:

$$g_{ab} = n_{ab} + \varepsilon h_{ab} \qquad g^{bc} = n^{bc} + r\varepsilon h^{bc} \tag{15.576}$$

In Eq. (15.576), constant r is solved as follows. Substitution of Eq. (15.576) into Eq. (15.575) for the linearized forms of for g_{ab} and g^{bc}, yields

$$(n_{ab} + \varepsilon h_{ab})(n^{bc} + r\varepsilon h^{bc}) = \delta^c_a \tag{15.577}$$

Expansion of Eq. (15.577), yields

$$n_{ab}n^{bc} + \varepsilon h_{ab}n^{bc} + r\varepsilon h^{bc}n_{ab} + r\varepsilon^2 h_{ab}h^{bc} = \delta^c_a \tag{15.578}$$

In Eq. (15.578), one can neglect the ε^2 quadratic term to obtain

$$\delta^c_a + \varepsilon\left(h_{ab}n^{bc} + rh^{bc}n_{ab}\right) = \delta^c_a \tag{15.579}$$

Equation (15.579) requires the expression in parentheses to be zero, where

$$-n^{bc}h_{ab} = rn_{ab}h^{bc} \tag{15.580}$$

On the right side of Eq. (15.580), the covariant and contravariant components can be transformed to their respective contravariant and covariant components, where

$$rn_{ab}h^{bc} = rn_{ab}n^{be}n^{cf}h_{ef} = r\delta^e_a n^{cf}h_{ef} = rn^{cf}h_{af} \tag{15.581}$$

The dummy index f in the last term of Eq. (15.581) can be replaced by index b, so that Eq. (15.580) becomes

$$-n^{bc}h_{ab} = rn^{bc}h_{ab} \tag{15.582}$$

In Eq. (15.582), the constant r is solved, where $r = -1$, so that the linearized form of the contravariant g^{bc} metric tensor components become

$$g^{bc} = n^{bc} - \varepsilon h^{bc} \tag{15.583}$$

Given perturbation constant ε, the spatial derivatives of the covariant metric tensor g_{ab} components yield the perturbation components h_{ab}, where

$$\frac{\partial g_{dc}}{\partial x^b} = \varepsilon\frac{\partial h_{dc}}{\partial x^b}, \quad \frac{\partial g_{bd}}{\partial x^c} = \varepsilon\frac{\partial h_{bd}}{\partial x^c}, \quad \frac{\partial g_{bc}}{\partial x^d} = \varepsilon\frac{\partial h_{bc}}{\partial x^d} \tag{15.584}$$

With substitution into Eq. (15.572) of the linearized form of the contravariant metric tensor g^{ad} of Eq. (15.583) and the perturbation components of the covariant metric tensors g_{dc}, g_{bd}, and g_{bc} of Eq. (15.584), the connection coefficient Γ^a_{bc}, becomes

$$\Gamma^a_{bc} = \frac{1}{2}(\varepsilon n^{ad} - \varepsilon^2 h^{ad})\left(\frac{\partial h_{dc}}{\partial x^b} + \frac{\partial h_{bd}}{\partial x^c} - \frac{\partial h_{bc}}{\partial x^d}\right) \tag{15.585}$$

In Eq. (15.585), the second-order terms can be dropped, so the linearized connection coefficient Γ^a_{bc} becomes

$$\Gamma^a_{bc} = \frac{1}{2}\varepsilon n^{ad}\left(\frac{\partial h_{dc}}{\partial x^b} + \frac{\partial h_{bd}}{\partial x^c} - \frac{\partial h_{bc}}{\partial x^d}\right) \tag{15.586}$$

From Eq. (15.480) for the curvature or Riemann tensor components R^a_{bcd}, with the substitution of Eq. (15.586) for the linearized connection coefficient Γ^a_{bc}, the curvature tensor becomes

$$R^a_{bcd} = \frac{\partial \Gamma^a_{bd}}{\partial x^c} - \frac{\partial \Gamma^a_{bc}}{\partial x^d} + \Gamma^e_{bd}\Gamma^a_{ec} - \Gamma^e_{bc}\Gamma^a_{ed} \tag{15.587}$$

where

$$\Gamma^e_{bd}\Gamma^a_{ec} \propto \varepsilon^2(n^{ed} - \varepsilon h^{ed})(n^{ad} - \varepsilon h^{ad}) \tag{15.588}$$

$$\Gamma^e_{bc}\Gamma^a_{ed} \propto \varepsilon^2(n^{ed} - \varepsilon h^{ed})(n^{ad} - \varepsilon h^{ad}) \tag{15.589}$$

In Eq. (15.587), one drops the ε^2 quadratic terms, and the curvature tensor R^a_{bcd} components become

$$R^a_{bcd} = \frac{\partial \Gamma^a_{bd}}{\partial x^c} - \frac{\partial \Gamma^a_{bc}}{\partial x^d} \tag{15.590}$$

Using Eq. (15.586), substitution for the connection coefficients Γ^a_{bd} and Γ^a_{bc} gives

$$R^a_{bcd} = \frac{\varepsilon}{2}\frac{\partial}{\partial x^c}\left[n^{ae}\left(\frac{\partial h_{ed}}{\partial x^b} + \frac{\partial h_{be}}{\partial x^d} - \frac{\partial h_{bd}}{\partial x^e}\right)\right] - \frac{\varepsilon}{2}\frac{\partial}{\partial x^d}\left[n^{af}\left(\frac{\partial h_{fc}}{\partial x^b} + \frac{\partial h_{bf}}{\partial x^c} - \frac{\partial h_{bc}}{\partial x^f}\right)\right] \tag{15.591}$$

In Eq. (15.591), the first term in brackets, e is a dummy variable that can be replaced by f, which shows that the middle terms in the first and second brackets cancel. Taking the second derivative then gives the curvature mixed tensor R^a_{bcd} components:

$$R^a_{bcd} = \frac{\varepsilon}{2}n^{af}\left(\frac{\partial^2 h_{fd}}{\partial x^c \partial x^b} + \frac{\partial^2 h_{bc}}{\partial x^d \partial x^f} - \frac{\partial^2 h_{bd}}{\partial x^c \partial x^f} - \frac{\partial^2 h_{fc}}{\partial x^d \partial x^b}\right) \tag{15.592}$$

The covariant Ricci tensor components R_{bd} are obtained by contraction of the mixed tensor components of the curvature tensor R^a_{bcd}, where

$$R_{bd} = R^c_{bcd} \tag{15.593}$$

and

$$R^a_{bcd} = \frac{\varepsilon}{2} n^{cf} \left(\frac{\partial^2 h_{fd}}{\partial x^c \partial x^b} + \frac{\partial^2 h_{bc}}{\partial x^d \partial x^f} - \frac{\partial^2 h_{bd}}{\partial x^c \partial x^f} - \frac{\partial^2 h_{fc}}{\partial x^d \partial x^b} \right) \tag{15.594}$$

Since n^{cf} is a constant, one rearranges Eq. (15.594), where

$$R^a_{bcd} = \frac{\varepsilon}{2} \left(\frac{\partial^2}{\partial x^c \partial x^b} h_{fd} n^{cf} + \frac{\partial^2}{\partial x^d \partial x^f} h_{bc} n^{cf} - \frac{\partial^2}{\partial x^c \partial x^f} h_{bd} n^{cf} - \frac{\partial^2}{\partial x^d \partial x^b} h_{fc} n^{cf} \right) \tag{15.595}$$

We now contract the metric tensor terms in Eq. (15.595).

$$h^c_d = h_{fd} n^{cf}, \quad h^f_b = h_{bc} n^{cf}, \quad h = h_{fc} n^{cf} \tag{15.596}$$

In Eq. (15.596), contraction of the Minkowski tensor contravariant tensor components (e.g., n^{cd}) with the metric perturbation tensor component (e.g., h_{bd}) gives the Ricci tensor covariant components R_{bd}, where the contravariant derivative of scalar h is zero, and

$$R^a_{bcd} = \frac{\varepsilon}{2} \left(\frac{\partial^2 h^c_d}{\partial x^c \partial x^b} + \frac{\partial^2 h^f_b}{\partial x^d \partial x^f} - n^{cf} \frac{\partial^2 h_{bd}}{\partial x^c \partial x^f} - \frac{\partial^2 h}{\partial x^d \partial x^b} \right) = \frac{\varepsilon}{2} \left(\frac{\partial^2 h^c_d}{\partial x^c \partial x^b} + \frac{\partial^2 h^f_b}{\partial x^d \partial x^f} - n^{cf} \frac{\partial^2 h_{bd}}{\partial x^c \partial x^f} \right) \tag{15.597}$$

The d'Alembertian operator Λ is defined by

$$\Lambda = \frac{\partial^2}{\partial t^2} - \left(\frac{\partial^2}{\partial x^2} + \frac{\partial^2}{\partial y^2} + \frac{\partial^2}{\partial z^2} \right) = n^{cf} \frac{\partial}{\partial x^c} \frac{\partial}{\partial x^f} \tag{15.598}$$

From Eq. (15.571) for the covariant metric tensor n_{cf}, the contravariant metric tensor n^{cf} is a constant, so that

$$\Lambda h_{bd} = \frac{\partial^2}{\partial x^c \partial x^f} \left(n^{cf} h_{bd} \right) \tag{15.599}$$

The metric perturbation scalar h is obtained by contracting the Minkowski tensor contravariant component n^{cd} with the metric perturbation tensor component h_{cd}, where

$$h = n^{cd} h_{cd} \tag{15.600}$$

With substitution of Eq. (15.599) into Eq. (15.597) and with the dummy variable f set to d, the covariant Ricci tensor R_{bd} becomes

$$R_{bd} = R^c_{bcd} = \frac{\varepsilon}{2} \left(\frac{\partial^2 h^c_d}{\partial x^c \partial x^b} + \frac{\partial^2 h^f_b}{\partial x^d \partial x^f} - \Lambda h_{bd} \right) = \frac{\varepsilon}{2} \left(\frac{\partial^2 h^c_d}{\partial x^c \partial x^b} + \frac{\partial^2 h^c_b}{\partial x^d \partial x^c} - \Lambda h_{bd} \right) \tag{15.601}$$

We now obtain the Ricci scalar R with $g_{bd} \cong n_{bd}$. Contraction of the contravariant metric tensor components g^{bd} with the covariant Ricci tensor components R_{bd}, gives the Ricci scalar R, where

$$R = g^{bd} R_{bd} \cong \frac{\varepsilon}{2} n^{bd} \left(\frac{\partial^2 h^c_d}{\partial x^c \partial x^b} + \frac{\partial^2 h^c_b}{\partial x^d \partial x^c} - \Lambda h_{bd} \right) \tag{15.602}$$

Note that scalar h is a constant, so differentiation yields zero, where

$$\frac{\partial^2 h}{\partial x^d \partial x^b} = 0 \tag{15.603}$$

Carrying out the contractions in Eq. (15.602), gives

$$R \cong \frac{\varepsilon}{2} \left(\frac{\partial^2 h^{bc}}{\partial x^c \partial x^b} + \frac{\partial^2 h^{dc}}{\partial x^d \partial x^c} - \Lambda h_{bd} \right) \tag{15.604}$$

In the second term of Eq. (15.604), the dummy variable d can be changed to b, so that

$$R = \frac{\varepsilon}{2} \left(\frac{\partial^2 h^{bc}}{\partial x^c \partial x^b} + \frac{\partial^2 h^{dc}}{\partial x^d \partial x^c} - \Lambda h_{bd} \right) = \varepsilon \left(\frac{\partial^2 h^{bc}}{\partial x^c \partial x^b} - \frac{\Lambda h_{bd}}{2} \right) \tag{15.605}$$

With substitution of Eq. (15.605) for the Ricci scalar R and Eq. (15.601) for the covariant components R_{bd} of the Ricci tensor, the Einstein field equation of Eq. (15.568) becomes

$$\kappa T_{bd} = R_{bd} - \frac{1}{2} R g_{bd} \cong \frac{\varepsilon}{2} \left(\frac{\partial^2 h^c_d}{\partial x^c \partial x^b} + \frac{\partial^2 h^c_b}{\partial x^d \partial x^c} - \Lambda h_{bd} \right) - \frac{\varepsilon}{2} \left(\frac{\partial^2 h^{bc}}{\partial x^c \partial x^b} - \frac{\Lambda h_{bd}}{2} \right) g_{bd} \tag{15.606}$$

With $g_{bd} \cong n_{bd}$, where $n_{bd} h^{bc} = h^c_d$, the field equation of Eq. (15.606), becomes

$$\begin{aligned} 2\kappa T_{bd} &\cong \varepsilon \left[\frac{\partial^2 h^c_d}{\partial x^c \partial x^b} + \frac{\partial^2 h^c_b}{\partial x^d \partial x^c} - \Lambda \left(h_{bd} - \frac{n_{bd} h_{bd}}{2} \right) - n_{bd} \frac{\partial^2 h^{bc}}{\partial x^c \partial x^b} \right] \\ &= \varepsilon \left[\frac{\partial^2 h^c_b}{\partial x^d \partial x^c} - \Lambda \left(h_{bd} - \frac{n_{bd} h_{bd}}{2} \right) \right] \end{aligned} \tag{15.607}$$

Through the d'Alembertian operator Λ, Eq. (15.607) is the precursor to the simplified gravity equation for weak field gravity using a perturbed Minkowski metric that leads to linearized connection coefficients.

In Eq. (15.607), on the right side, there are four terms, all of which are derivatives. As with the Lorentz gauge and the electromagnetic field differential equations, one can find an additive quantity, which when added simplifies the expression, but does not change the solution because the derivatives of the additive quantity is zero. The simplified wave equation is obtained via a gauge transformation, which is a coordinate transformation that leaves unchanged the mixed curvature tensor components R^a_{bcd}, Ricci covariant tensor components R_{ab}, and Ricci scalar R to the first order in constant parameter ε. The gauge transformation is

$$x^{a'} = x^a + f^a \tag{15.608}$$

Here, function $f^a(x^b)$ is a function of position x^b, in which the spatial derivative components of function f^a are much less than 1:

$$\left|\frac{\partial f^a}{\partial x^b}\right| \ll 1 \tag{15.609}$$

The coordinate transformation between the perturbed metric tensor elements h_{bd} and $h_{b'd'}$ is given by

$$h_{b'd'} = \frac{\partial x^a}{\partial x^{b'}}\frac{\partial x^c}{\partial x^{d'}}h_{ac} \tag{15.610}$$

Like Eqs. (15.608) and (15.609), the gauge transformation with functions $f_a(x^b)$ and $f_b(x^d)$, where $\partial f_a(x^b)/\partial x^b \ll 1$ and $\partial f_b(x^d)/\partial x^d \ll 1$, are given by:

$$x^b + f^b = x^{b'} \quad \text{and} \quad x^d + f^d = x^{d'} \tag{15.611}$$

Given the Kronnecker delta function δ^a_b, substitution of Eq. (15.611) into Eq. (15.610), then gives

$$h_{b'd'} = \frac{\partial x^a}{\partial(x^b + f^b)}\frac{\partial x^c}{\partial(x^d + f^d)}h_{ac} = \left(\delta^a_b - \frac{\partial f^a}{\partial x^b}\right)\left(\delta^c_d - \frac{\partial f^c}{\partial x^d}\right)h_{ac} \tag{15.612}$$

Multiplying and dropping the derivative product term in Eq. (15.612), yields

$$h_{b'd'} = h_{ac}\delta^a_b\delta^c_d - \frac{\partial f^a}{\partial x^b}\delta^c_d h_{ac} - \frac{\partial f^c}{\partial x^d}\delta^a_b h_{ac} \tag{15.613}$$

Application of the Kronnecker delta function in Eq. (15.613), gives

$$h_{b'd'} = h_{bd} - \frac{\partial f^a}{\partial x^b}h_{ad} - \frac{\partial f^c}{\partial x^d}h_{bc} \tag{15.614}$$

Contraction of the metric tensor with the last two terms in Eq. (15.614), gives

$$h_{b'd'} = h_{bd} - \frac{\partial f_d}{\partial x^b} - \frac{\partial f_b}{\partial x^d} \tag{15.615}$$

This covariant tensor component relation also holds for the contravariant tensor components, where

$$h^{b'd'} = h^{bd} - \frac{\partial f^d}{\partial x^b} - \frac{\partial f^b}{\partial x^d} \tag{15.616}$$

Next, one defines h' in terms of the contraction of the constant part of the metric tensor covariant components n_{bd} with the contravariant tensor components of the variable, perturbed portion of the metric tensor $h^{b'd'}$, or alternatively, by the contraction of constant, contravariant metric component n^{bd} with covariant tensor components of the variable, perturbed portion of the metric tensor $h_{b'd'}$, where

$$h' \equiv n_{bd}h^{b'd'} = n^{bd}h_{b'd'} \tag{15.617}$$

Substitution into the middle term of Eq. (15.617) for h' of Eq. (15.616) for $h^{b'd'}$, and after subsequent multiplication and contraction, one sets dummy variable $b = d$ to obtain

$$h' = n_{bd}h^{b'd'} = n_{bd}h^{bd} - n_{bd}\frac{\partial f^d}{\partial x^b} - n_{bd}\frac{\partial f^b}{\partial x^d} = h - \frac{\partial f_b}{\partial x^b} - \frac{\partial f_d}{\partial x^d} = h - 2\frac{\partial f_b}{\partial x^b} \tag{15.618}$$

Substitution into the right term of Eq. (15.617) for h' of Eq. (15.615) for $h_{b'd'}$ and after subsequent multiplication and contraction, one sets dummy variable $b = d$ to obtain

$$h' = n^{bd}h_{b'd'} = n^{bd}h_{bd} - n^{bd}\frac{\partial f_d}{\partial x^b} - n^{bd}\frac{\partial f_b}{\partial x^d} = h - \frac{\partial f^b}{\partial x^b} - \frac{\partial f^d}{\partial x^d} = h - 2\frac{\partial f^b}{\partial x^b} \tag{15.619}$$

Next, in the Einstein field equation of Eq. (15.607), we can express the transformation of the second term in parenthesis. With the covariant tensor $h_{b'd'}$ of Eq. (15.615), the contravariant tensor $h^{b'd'}$of Eq. (15.616), scalar constant h' of Eq. (15.618) for covariant vector f_b, and scalar constant h' of Eq. (15.619) for contravariant vector f^b, one expresses the covariant and contravariant transformations of the second term of the Einstein field equation (15.607), where

$$h_{b'd'} - \frac{h'n_{b'd'}}{2} = h_{bd} - \frac{\partial f_d}{\partial x^b} - \frac{\partial f_b}{\partial x^d} - \frac{hn_{bd}}{2} + n_{bd}\frac{\partial f_b}{\partial x^b} \tag{15.620}$$

and

$$h^{b'd'} - \frac{h'n^{b'd'}}{2} = h^{bd} - \frac{\partial f^d}{\partial x^b} - \frac{\partial f^b}{\partial x^d} - \frac{hn^{bd}}{2} + n^{bd}\frac{\partial f^b}{\partial x^b} \tag{15.621}$$

Recall that the metric tensor components n_{bd} are constant, so that

$$n_{b'd'} = n_{bd} \tag{15.622}$$

The inverse matrix element $\partial x^b/\partial x'^d$ is obtained via the binomial approximation for the contravariant and covariant versions, where

$$\frac{\partial x^b}{\partial x'^d} = \frac{\partial x^b}{\partial(x^d + f^d)} = \frac{\partial x^b}{\partial x^d}\left(1 + \frac{\partial f^d}{\partial x^d}\right)^{-1} \cong \frac{\partial x^b}{\partial x^d}\left(1 - \frac{\partial f^d}{\partial x^d}\right) = \delta^b_d - \frac{\partial f^b}{\partial x^d} \tag{15.623}$$

and

$$\frac{\partial x_b}{\partial x^{d'}} = \delta^b_d - \frac{\partial f_b}{\partial x^d} \tag{15.624}$$

Next, one takes the derivative of $h^{b'd'}$ relative to $x^{d'}$, and substitution for $\partial x^b/\partial x^{d'}$ of Eq. (15.623) gives

$$\frac{\partial h^{b'd'}}{\partial x^{d'}} = \frac{\partial h^{b'd'}}{\partial x^b}\frac{\partial x^b}{\partial x^{d'}} = \frac{\partial h^{b'd'}}{\partial x^b}\left(\delta^b_d - \frac{\partial f^b}{\partial x^d}\right) \tag{15.625}$$

Recall $\partial f_b(x^d)/\partial x^d \ll 1$ and for $b = d$ in the Kronecker delta function of Eq. (15.625), one has

$$\delta^b_d \gg \frac{\partial f^b}{\partial x^d} \tag{15.626}$$

Application of Eq. (15.626), where the Kronecker delta function $\delta^b_d = 1$ for $b = d$, and $\delta^b_d = 0$ for $b \neq d$, gives

$$\frac{\partial h^{b'd'}}{\partial x^{d'}} \cong \frac{\partial h^{b'd'}}{\partial x^d} \tag{15.627}$$

In Eq. (15.627), application of the contravariant derivative $\partial/\partial x^d$ to the contravariant tensor $h^{b'd'}$ of Eq. (15.616), where $\partial f^b/\partial x^b = 0$ for $f_b(x^d)$, gives

$$\frac{h^{b'd'}}{\partial x^d} = \frac{\partial h^{bd}}{\partial x^d} - \frac{\partial^2 f^d}{\partial x^d \partial x^b} - \frac{\partial^2 f^b}{\partial x^d \partial x^d} = \frac{\partial h^{bd}}{\partial x^d} - \frac{\partial^2 f^b}{\partial x^d \partial x^d} \tag{15.628}$$

Substitution of Eq. (15.628) into Eq. (15.627) and recalling Eq. (15.598) for the d'Alembertian operator Λ, one obtains the gauge condition for the gravitational waves, where

$$\frac{\partial h^{b'd'}}{\partial x^{d'}} \cong \frac{\partial h^{bd}}{\partial x^{d}} - \frac{\partial^2 f^b}{\partial x^d \partial x^d} = \frac{\partial h^{bd}}{\partial x^{d}} - \Lambda f^b \tag{15.629}$$

The gauge condition for gravity waves is thus similar to the Lorentz gauge of electromagnetic waves.

In a manner similar to the Lorentz gauge condition of Eq. (12.55) and Eq. (13.149) for electromagnetic waves, where the electromagnetic 4-vector potential $\mathbf{A}_4$ has zero divergence, in the case of gravity waves, one is free to choose the form of the gauge f^b, as long as the Riemann curvature tensor R^a_{bcd} is unchanged. Hence, we make the simplified mathematical choice of choosing gauge f^b, where the variable portion $h^{b'd'}$ of the contravariant metric tensor $g^{b'd'}$ for the gravity waves has zero divergence, so that the gravity wave gauge condition is then

$$\frac{\partial h^{b'd'}}{\partial x^{d'}} = 0 \tag{15.630}$$

As a result of Eq. (15.630), one then requires the gauge choice f^b to be a solution of the following equation:

$$\frac{\partial h^{bd}}{\partial x^{d}} = \Lambda f^b = \frac{\partial^2 f^b}{\partial t^2} - \left(\frac{\partial^2 f^b}{\partial x^2} + \frac{\partial^2 f^b}{\partial y^2} + \frac{\partial^2 f^b}{\partial z^2}\right) \tag{15.631}$$

We now recall the field equation of Eq. (15.607), and apply the gauge choice f_c, where

$$2\kappa T_{bd} = \varepsilon\left[\frac{\partial^2 h^c_b}{\partial x^d \partial x^c} - \Lambda\left(h_{bd} - \frac{n_{bd} h_{bd}}{2}\right)\right] \tag{15.632}$$

As in Eq. (15.630), the gauge is chosen so that the first term is zero in Eq. (15.632), where

$$\frac{\partial^2 h^c_b}{\partial x^d \partial x^c} = 0 \tag{15.633}$$

With substitution of Eq. (15.633) into Eq. (15.632), application of the gauge choice to the Einstein field equation for the condition of weak gravity, yields the simplified field equation, which is now the explicit gravitational wave equation, where

$$-\kappa T_{bd} = \frac{\varepsilon\Lambda}{2}\left(h_{bd} - \frac{h n_{bd}}{2}\right) \tag{15.634}$$

or

$$-\kappa T_{bd} = \frac{\varepsilon}{2}\left(\frac{\partial^2}{\partial t^2} - \frac{\partial^2}{\partial x^2} - \frac{\partial^2}{\partial y^2} - \frac{\partial^2}{\partial z^2}\right)\left(h_{bd} - \frac{h n_{bd}}{2}\right) \tag{15.635}$$

In empty space–time, the stress tensor is zero ($T_{bd} = 0$), and the gravity wave equation of Eq. (15.635), then clearly shows that gravity waves travel at the speed of light c, where

$$\left(\frac{\partial^2}{\partial x^2} + \frac{\partial^2}{\partial y^2} + \frac{\partial^2}{\partial z^2} - \frac{1}{c^2}\frac{\partial^2}{\partial t^2}\right)\left(h_{bd} - \frac{1}{2}hn_{bd}\right) = 0 \tag{15.636}$$

One can solve Eq. (15.636) for the empty space–time wave equation. The solution h_{bd} of the empty space–time gravitational wave equation is specified by amplitude A_{bd}, the four-vector, space–time position x^{μ}, and the constant four-vector gravity wave number k_{μ}, where

$$h_{bd} = A_{bd} \exp(\mathrm{i}k_{\mu}x^{\mu}) \tag{15.637}$$

Much experimental effort has been and is being directed toward the detection of the gravity waves. The experimental verification of gravity waves is another test of general relativity.

15.32 Schwarzschild Solution to Einstein Field Equations and Experimental Predictions

The first solution to the Einstein field equations was derived by Schwarzschild in the foxhole as a soldier fighting in World War I. He derived a metric tensor for the case of a static, spherically symmetric, gravitational field in the empty space–time that surrounds a massive spherical body, such as a star. Schwarzschild assumed that the curved space–time surrounding the massive body asymptotically approached a flat space–time as the radial distance from the body increased to infinity. He also assumed that the resulting field equation in the empty space–time, which surrounded the massive body, was given by the Ricci tensor R_{ab} of Eq. (15.500), set to zero, where

$$R_{ab} = \frac{\partial \Gamma^{c}_{ac}}{\partial x^{b}} - \frac{\partial \Gamma^{c}_{ab}}{\partial x^{c}} + \Gamma^{d}_{ac}\Gamma^{c}_{db} - \Gamma^{d}_{ab}\Gamma^{c}_{dc} = 0 \tag{15.638}$$

In the empty space–time outside the massive body, the stress tensor T_{ab} is assumed to be zero, and the Ricci scalar R is assumed to be zero, where

$$R = 0 \quad T_{ab} = 0 \tag{15.639}$$

The connection coefficients Γ^{a}_{bc} are obtained from the metric tensor components and its derivatives:

$$\Gamma^{a}_{bc} = \frac{1}{2}g^{ad}\left(\frac{g_{dc}}{\partial x^{b}} + \frac{g_{bd}}{\partial x^{c}} - \frac{\partial g_{bc}}{\partial x^{d}}\right) \tag{15.640}$$

This Schwarzschild foxhole formulation is the simplest form of the field equations, yet it provides much valued insight into the space–time about spherical symmetric, massive bodies such as the sun in the static case. The Schwarzschild formulation also allows the relatively simple calculation and design of many experiments, which can be used to test general relativity, such as black holes, the bending of light by a massive object, spectral shifts by gravity, radar sounding, the geodesic effect, and the perihelion advance of planets in their orbit about the sun in our solar system.

Schwarzschild assumed a line element based on a diagonal metric tensor of the form

$$c^2\,d\tau^2 = A(r)dt^2 - B(r)dr^2 - r^2\,d\theta^2 - r^2\sin^2\theta\,d\phi^2 \tag{15.641}$$

where

$$\lim_{r\to\infty} A(r) = c^2 \qquad \lim_{r\to\infty} B(r) = 1 \tag{15.642}$$

and

$$g_{00} = B(r) \quad g_{11} = -r^2 \quad g_{22} = -r^2\sin^2\theta \quad g_{33} = A(r) \tag{15.643}$$

Plugging this form of the covariant metric tensor components into the field equation $R_{ab} = 0$, and evaluating the resulting equations, one solves for $A(r)$ and $B(r)$ and thereby obtains the covariant, diagonal, Schwarzschild metric tensor g_{sc}, where

$$g_{sc} = [g_{ab}] = \begin{bmatrix} -\left(1-\dfrac{2MG}{c^2 r}\right)^{-1} & 0 & 0 & 0 \\ 0 & -r^2 & & 0 \\ 0 & 0 & -r^2\sin^2\theta & 0 \\ 0 & 0 & 0 & \left(1-\dfrac{2MG}{c^2 r}\right)c^2 \end{bmatrix} \tag{15.644}$$

Here, M is the mass of the spherical body, G the gravitational constant, c the speed of light, r the radial coordinate measured from the center of mass M, and θ the polar angle coordinate. The Schwarzschild line element is thus

$$c^2\,d\tau^2 = \left(1-\frac{2MG}{c^2 r}\right)c^2\,dt^2 - \left(1-\frac{2MG}{c^2 r}\right)^{-1} dr^2 - r^2\,d\theta^2 - r^2\,d\phi^2\sin^2\theta \tag{15.645}$$

The radial coordinate r is measured from the center of the spherical body of mass M, whose surface is at radial coordinate r_M. If the radial coordinate decreases from infinity $r = \infty$ to radial coordinate r_{bh}, which is exterior to the body, where $r_{bh} = 2MG/c^2$, where $r_{bh} > r_M$, then the body is a black hole, where

$$r_{bh} = \frac{2MG}{c^2} > r_M \quad \text{black hole condition} \tag{15.646}$$

The black-hole-defining radial coordinate $r = r_{bh}$ is a singularity of the Schwarzschild metric and line element that causes them to become infinite, where

$$\left(1 - \frac{2MG}{c^2 r_{bh}}\right)^{-1} = \infty \tag{15.647}$$

A photon that originates at radial coordinate r is less than the black-hole-defining radial coordinate r_{bh}; that is, $r < r_{bh}$ cannot escape the black-hole body. Thus, the black-hole body appears dark because no light can be emitted to the outside observer at $r > r_{bh}$.

The Schwarzschild metric provides many important insights into general relativity and suggests many experiments that can be used to test general relativity. In particular, consider the measurement of radial distances near the sun or about the massive, static, spherically symmetric body in the proper time frame τ, where $\mathrm{d}t = 0$ in which a radial line along a constant polar θ and azimuth ϕ angle is chosen, where $\mathrm{d}\theta = 0$ and $\mathrm{d}\phi = 0$. The Schwarzschild metric of Eq. (15.645), becomes

$$-c^2\,\mathrm{d}\tau^2 = \mathrm{d}R^2 \equiv \left(1 - \frac{2MG}{c^2 r}\right)^{-1} \mathrm{d}r^2 \tag{15.648}$$

Variable r is no longer a radial distance; rather, variable R becomes the radial distance in the Schwarzschild solution to the Einstein field equations. For a clock at a fixed position in space (r, θ, ϕ), where $\mathrm{d}r = 0$, $\mathrm{d}\theta = 0$, and $\mathrm{d}\phi = 0$, the Schwarzschild metric of Eq. (15.645), becomes

$$\mathrm{d}\tau^2 = \left(1 - \frac{2MG}{c^2 r}\right)\mathrm{d}t^2 \tag{15.649}$$

The time interval and length interval relations between inertial frames, with constant relative velocity v of special relativity, can be compared to the time and length relations in the Schwarzschild solution of Einstein's equations. In special relativity, the length interval $\mathrm{d}l_0$ and time interval $\mathrm{d}\tau$ of objects and events in the proper frame, and the proper frame-measured length $\mathrm{d}l$ and time intervals $\mathrm{d}t$ for the same objects and events an inertial frame, which moves at constant velocity v, relative to the proper frame, are related by the Lorentz contraction and time dilation relations, and are independent of position r:

$$\mathrm{d}l = \mathrm{d}l_0\sqrt{1 - \frac{v^2}{c^2}} \quad \text{and} \quad \mathrm{d}t = \mathrm{d}\tau \Big/ \sqrt{1 - \frac{v^2}{c^2}} \tag{15.650}$$

In comparison, in the Schwarzschild solution to the Einstein field equations, the time and length interval relations for proper radial distance interval $\mathrm{d}R$ and proper time interval $\mathrm{d}\tau$ in curved space, are related to variation $\mathrm{d}r$ in radial coordinate r and variation $\mathrm{d}t$ in time coordinate t in flat space by the following relations, which depend on position r, where

$$\mathrm{d}r = \mathrm{d}R\sqrt{1-\frac{2MG}{c^2r}} \quad \text{and} \quad \mathrm{d}t = \mathrm{d}\tau \Big/ \sqrt{1-\frac{2MG}{c^2r}} \tag{15.651}$$

For example, the radial distance R between two points at positions r_1 and r_2 with $r_2 > r_1$ on a radial line is found by integrating $\mathrm{d}R$, where

$$\begin{aligned} R &= R_2 - R_1 \\ &= \int_{r_1}^{r_2}\left(1-\frac{2MG}{c^2r}\right)^{-1/2}\mathrm{d}r = \left[\sqrt{r\left(r-\frac{2GM}{c^2}\right)} + \frac{2GM}{c^2}\ln\left\{\sqrt{r}+\sqrt{r-\frac{2GM}{c^2}}\right\}\right]_{r_1}^{r_2} \end{aligned} \tag{15.652}$$

and

$$R = R_2 - R_1 = r_2 - r_1 \quad \frac{2MG}{c^2r} \ll 1 \tag{15.653}$$

Another prediction of the Schwarzschild solution of the Einstein field equations is the rate of clocks in a gravitational field and the shift of spectral lines. Consider the proper time interval $\mathrm{d}\tau_1$ recorded by a clock C_1 at radial coordinate r_1, and the proper time interval $\mathrm{d}\tau_2$ recorded by another clock C_2 at radial coordinate r_2, where

$$\mathrm{d}\tau_1 = \frac{\mathrm{d}s_1}{ic} = \mathrm{d}t\sqrt{1-\frac{2MG}{c^2r_1}} \quad \text{and} \quad \mathrm{d}\tau_2 = \frac{\mathrm{d}s_2}{ic} = \mathrm{d}t\sqrt{1-\frac{2MG}{c^2r_2}} \tag{15.654}$$

In Eq. (15.654), taking the ratio of $\mathrm{d}\tau_1$ divided by $\mathrm{d}\tau_2$ gives

$$\frac{\mathrm{d}\tau_1}{\mathrm{d}\tau_2} = \sqrt{1-\frac{2MG}{c^2r_1}} \Big/ \sqrt{1-\frac{2MG}{c^2r_2}} \tag{15.655}$$

Now suppose that at the radial coordinate r_2 (relative to the center of a massive body), a transmitter emits N pulses in the proper time interval $\mathrm{d}\tau_2$ recorded clock C_2 with associated transmission frequency $f_2 = N/\mathrm{d}\tau_2$. The N pulses are received at radial coordinate r_1 (relative to the center of a massive body) in the proper time interval $\mathrm{d}\tau_1$ recorded by clock C_1 with associated reception frequency $f_1 = N/\mathrm{d}\tau_1$.

As a result, the ratio of the received frequency f_1 divided by the transmitted frequency f_2 is

$$\frac{f_1}{f_2} = \sqrt{1 - \frac{2MG}{c^2 r_2}} \Big/ \sqrt{1 - \frac{2MG}{c^2 r_1}} \tag{15.656}$$

where

$$\frac{f_1}{f_2} \cong \left(1 - \frac{MG}{c^2 r_2}\right)\left(1 + \frac{MG}{c^2 r_1}\right) \cong 1 + \frac{MG}{c^2}\left(\frac{1}{r_1} - \frac{1}{r_2}\right) \quad \frac{2MG}{c^2 r_1}, \frac{2MG}{c^2 r_2} \ll 1 \tag{15.657}$$

The fractional shift of frequency $\Delta f / f_2$ at the receiver due to the gravitational field is then

$$\frac{\Delta f}{f_2} = \frac{f_1 - f_2}{f_2} \cong \frac{MG}{c^2}\left(\frac{1}{r_1} - \frac{1}{r_2}\right) \quad \frac{2MG}{c^2 r_1}, \frac{2MG}{c^2 r_2} \ll 1 \tag{15.658}$$

If $r_1 < r_2$, in which the receiver is closer to the massive body than the transmitter is, then the gravitational shift $\Delta f / f_2$ at the receiver is toward the blue, higher end of the spectrum. Conversely, if $r_1 > r_2$, with the transmitter nearer the massive body, then the gravitational shift is toward the red, lower end of the spectrum.

The Schwarzschild solution of the Einstein field equations also yields an equation for general particle motion that includes photons at radial coordinate r, which is measured from the center of a large spherical mass M. The equation of motion can be derived using the four Euler equations of a Lagrangian, which is comprised of the x, y, and z spatial and proper time τ components, where the Lagrangian L describes the space–time interval of the Schwarzschild metric, where

$$\frac{d}{d\tau}\left(\frac{\partial L}{\partial x^a / \partial \tau}\right) - \frac{\partial L}{\partial x^a} = 0 \tag{15.659}$$

and

$$L \equiv \frac{1}{2} g_{ab} \frac{\partial x^a}{\partial \tau} \frac{\partial x^b}{\partial \tau} \tag{15.660}$$

Substitution of the Schwarzschild metric tensor components of Eq. (15.644) into Eq. (15.660), gives

$$L = \frac{1}{2}\left[c^2\left(1 - \frac{2MG}{c^2 r}\right)\left(\frac{\partial t}{\partial \tau}\right)^2 - \left(1 - \frac{2MG}{c^2 r}\right)^{-1}\left(\frac{\partial r}{\partial \tau}\right)^2 - r^2\left(\left(\frac{\partial \theta}{\partial \tau}\right)^2 + \left(\frac{\partial \phi}{\partial \tau}\right)^2 \sin^2\theta\right)\right] \tag{15.661}$$

Because of the spherical symmetry of the Schwarzschild solution, one can set the polar angle $\theta = \pi/2$, so that one obtains by substitution of Eq. (15.661) into Eq. (15.659), the four Euler equations for the three spatial and one temporal component.

The Euler equation for the time coordinate t is expressed via integration constant k as

$$\left(1 - \frac{2MG}{c^2 r}\right)\left(\frac{\mathrm{d}t}{\mathrm{d}\tau}\right) = k \tag{15.662}$$

Since $\theta = \pi/2$, with $\mathrm{d}\theta = 0$, there is no Euler equation for polar coordinate θ. The Euler equation for the azimuth coordinate ϕ is expressed via integration constant h as

$$r^2 \frac{\mathrm{d}\phi}{\mathrm{d}\tau} = h \tag{15.663}$$

The Euler equation for the radial coordinate r is

$$\left(1 - \frac{2MG}{c^2 r}\right)^{-1} \frac{\partial^2 r}{\partial \tau^2} + \frac{MG}{r^2}\left(\frac{\partial t}{\partial \tau}\right)^2 - \left(1 - \frac{2MG}{c^2 r}\right)^{-2} \frac{MG}{c^2 r^2}\left(\frac{\partial r}{\partial \tau}\right)^2 - r\left(\frac{\partial \phi}{\partial \tau}\right)^2 = 0 \tag{15.664}$$

The proper time is defined with the Schwarzschild line element, which is evaluated at $\theta = \pi/2$. The resulting equation is simpler, and can be used in place of the Euler equation for the radial coordinate r, where

$$c^2 \mathrm{d}\tau^2 = g_{ab} \frac{\mathrm{d}x^a}{\mathrm{d}\tau} \frac{\mathrm{d}x^b}{\mathrm{d}\tau} \tag{15.665}$$

Substitution into Eq. (15.665) of Eq. (15.644) for the covariant Schwarzschild metric tensor components, gives

$$c^2 = c^2\left(1 - \frac{2MG}{c^2 r}\right)\left(\frac{\mathrm{d}t}{\mathrm{d}\tau}\right)^2 - \left(1 - \frac{2MG}{c^2 r}\right)^{-1}\left(\frac{\mathrm{d}r}{\mathrm{d}\tau}\right)^2 - r^2\left(\frac{\mathrm{d}\phi}{\mathrm{d}\tau}\right)^2 \tag{15.666}$$

Division of both sides by $(\mathrm{d}\phi/d\tau)^2$ gives

$$c^2\left(\frac{\mathrm{d}\phi}{\mathrm{d}\tau}\right)^{-2} = c^2\left(1 - \frac{2MG}{c^2 r}\right)\left(\frac{\mathrm{d}t}{\mathrm{d}\tau}\right)^2\left(\frac{\mathrm{d}\phi}{\mathrm{d}\tau}\right)^{-2} - \left(1 - \frac{2MG}{c^2 r}\right)^{-1}\left(\frac{\mathrm{d}r}{\mathrm{d}\phi}\right)^2 - r^2 \tag{15.667}$$

Substitution the for $\mathrm{d}t/\mathrm{d}\tau$ and $\mathrm{d}\phi/\mathrm{d}\tau$ from the Euler equations for coordinates t and ϕ, yields the energy equation

$$\left(\frac{\mathrm{d}r}{\mathrm{d}\phi}\right)^2 + r^2\left(1 + \frac{c^2 r^2}{h^2}\right)\left(1 - \frac{2MG}{c^2 r}\right) - \frac{c^2 k^2 r^4}{h^2} = 0 \tag{15.668}$$

Equation (15.668) can be put into more compact form by setting $u = 1/r$, so that

$$\left(\frac{du}{d\phi}\right)^2 + u^2 = E + \frac{2MG}{h^2}u + \frac{2MG}{c^2}u^3 \tag{15.669}$$

where

$$E = \frac{c^2(k^2 - 1)}{h^2} \tag{15.670}$$

Equation (15.670) is identical to the energy equation in the flat space of Newtonian mechanics except for the last term in u^3, which is the general relativistic correction. In Newtonian mechanics, the conservation of energy and angular momentum are applied to a particle trajectory in an equatorial plane $\theta = \pi/2$, where the particle moves under the gravitational influence of a spherical body of mass M at distance r from the body center in which $u = 1/r$, where

$$\left(\frac{du}{d\phi}\right)^2 + u^2 = E + \frac{2MG}{h^2}u \tag{15.671}$$

Here h is angular momentum per mass and again, E is a constant.

These Euler equations for motion can be applied to a particle moving in a circular orbit of constant radial coordinate r about a massive object of mass M, where $dr/d\tau = 0$ and $d^2r/d\tau^2 = 0$. The Euler equation for the radial coordinate r reduces to

$$MG\left(\frac{dt}{d\tau}\right)^2 = r^3\left(\frac{d\phi}{d\tau}\right)^2 \tag{15.672}$$

Recall Eq. (15.662) for the Euler equation for time coordinate t:

$$\left(1 - \frac{2MG}{c^2 r}\right)\left(\frac{dt}{d\tau}\right) = k \tag{15.673}$$

Equations (15.672) and (15.673) can be combined to obtain

$$\left(\frac{d\phi}{d\tau}\right)^2 = \frac{k^2 MG}{r}\left(r - \frac{2MG}{c^2}\right)^{-2} \tag{15.674}$$

and

$$\left(\frac{dt}{d\tau}\right)^2 = r^2k^2\left(r - \frac{2MG}{c^2}\right)^{-2} \tag{15.675}$$

One then recalls Eq. (15.667) for the general Euler equation for the radial coordinate r:

$$c^2\left(\frac{d\phi}{d\tau}\right)^{-2} = c^2\left(1-\frac{2MG}{c^2r}\right)\left(\frac{dt}{d\tau}\right)^2\left(\frac{d\phi}{d\tau}\right)^{-2} - \left(1-\frac{2MG}{c^2r}\right)^{-1}\left(\frac{dr}{d\phi}\right)^2 - r^2 \tag{15.676}$$

Since the constant orbit radius does not vary with azimuth angle, then $dr/d\phi = 0$. Substitution into Eq. (15.676) of Eq. (15.675) for $(dt/d\tau)^2$ and Eq. (15.674) for $(d\phi/d\tau)^2$, gives an expression for the constant k^2, where

$$k^2 = \frac{1}{r}\left(r-\frac{2MG}{c^2}\right)^2\left(r-\frac{3MG}{c^2}\right)^{-1} \tag{15.677}$$

Again, recall from Eq. (15.662), the Euler equation for time coordinate t:

$$\left(1-\frac{2MG}{c^2r}\right)\left(\frac{dt}{d\tau}\right) = k \tag{15.678}$$

Substitution for k, gives the relation between the proper time interval $\Delta\tau$, which is measured by an orbiting observer in the Schwarzschild curved space region, who is located at radial coordinate r, which is measured from the center of the massive body with mass M. Given the time interval Δt, which is measured by a fixed observer in a flat space region at infinity, the proper time interval $\Delta\tau$ measured by an orbiting observer is

$$\Delta\tau = \Delta t\sqrt{1-\frac{3MG}{rc^2}} \tag{15.679}$$

Transposing Eq. (15.679), one obtains,

$$\frac{dt}{d\tau} = \left(1-\frac{3MG}{rc^2}\right)^{-1/2} \tag{15.680}$$

Recall the Euler equation for the particle in orbit with radial coordinate r, where from Eq. (15.672),

$$MG\left(\frac{dt}{d\tau}\right)^2 = r^3\left(\frac{d\phi}{d\tau}\right)^2 \tag{15.681}$$

Multiplying both sides of Eq. (15.681) by $d\tau^2$, and then dividing both sides by dt^2, gives the relativistic version of Kepler's third law:

$$\left(\frac{d\phi}{dt}\right)^2 = \frac{MG}{r^3} \tag{15.682}$$

The period of the particle orbit is the change in the coordinate time Δt for one revolution $\Delta\phi = 2\pi$, where the relativistic version of Kepler's third law becomes

$$\Delta t = 2\pi\sqrt{\frac{r^3}{MG}} \tag{15.683}$$

Substitution of Eq. (15.683) for Δt in Eq. (15.679) relating $\Delta\tau$ and Δt, gives an expression for the proper time interval $\Delta\tau$, which is measured by an observer in a circular orbit at a constant radial position r, measured relative to the center of the spherical body with mass M, where

$$\Delta\tau = 2\pi\sqrt{\frac{r^3}{MG}\left(1 - \frac{3MG}{rc^2}\right)} \tag{15.684}$$

Equation (15.684) gives the condition for the circular orbit to occur, which is that the radial coordinate r must exceed a minimum value, where

$$r > \frac{3MG}{c^2} \tag{15.685}$$

In the case of a photon that orbits at the speed of light c, and whose proper time interval approaches zero, where $\Delta\tau \to 0$, then the photon has an orbit whose radial coordinate approaches this minimum limit, where

$$r \to \frac{3MG}{c^2} \tag{15.686}$$

We now consider the geodesic effect of the parallel transport of a vector, which moves in circular orbit at radial coordinate r in the equatorial plane with polar coordinate $\theta = \pi/2$ about a spherical body with mass M. The vector is attached to a freely moving particle, which propagates along a geodesic line in the curved space–time of the Schwarzschild solution of the Einstein field equations. For example, the vector could be the axis of a gyroscope, which is mounted on a satellite, and orbits in a circle of constant radius about the center of the Earth.

In general relativity, a freely moving particle propagates along a geodesic line in which the free particle is acted on gravity via the curvature of space. Gravity is incorporated into the connection coefficients Γ^a_{bc}. The connection coefficients Γ^a_{bc} are expressed by the covariant g_{ab} and contravariant g^{ab} metric tensor components. And the metric tensor components contain the relevant parameters of the spherical body mass M and the gravitational constant G. That is, gravity is no longer treated as a force in general relativity. A free particle, which moves on a geodesic line, is acted upon by gravity, via the curvature of the space–time, rather than by gravity as a force. The geodesic equation for the parallel-transported vector, which represents the axis of a gyroscope in circular orbit about the Earth with constant radial coordinate r and constant polar angle coordinate $\theta = \pi/2$, is given by Eq. (15.463), where

$$\frac{\mathrm{D}(\mathrm{d}x^a/\mathrm{d}\tau)}{\mathrm{d}\tau} = \frac{\mathrm{d}^2x^a}{\mathrm{d}\tau^2} + \Gamma^a_{bc}\frac{\mathrm{d}x^b}{\mathrm{d}\tau}\frac{\mathrm{d}x^c}{\mathrm{d}\tau} = 0 \tag{15.687}$$

where

$$\frac{\mathrm{d}r}{\mathrm{d}\tau} = 0 \quad \text{and} \quad \frac{\mathrm{d}\theta}{\mathrm{d}\tau} = 0 \tag{15.688}$$

Let λ^a be the vector components of gyroscope vector axis $\boldsymbol{\lambda}$, which is parallel-transported in the circular orbit about the Earth, where

$$\lambda^a = \frac{\mathrm{d}x^a}{\mathrm{d}\tau} \tag{15.689}$$

With substitution of Eq. (15.689) into Eq. (15.687), the geodesic equation for the freely moving (orbiting) gyroscope axis vector $\boldsymbol{\lambda}$ is then

$$\frac{\mathrm{D}\lambda^a}{\mathrm{d}\tau} = \frac{\mathrm{d}\lambda^a}{\mathrm{d}\tau} + \Gamma^a_{bc}\lambda^b\frac{\mathrm{d}x^c}{\mathrm{d}\tau} = 0 \tag{15.690}$$

The parallel-transported vector λ^a is perpendicular to the tangent vector $\mathrm{d}x^b/\mathrm{d}\tau$ of the circular trajectory so that the line element, which is the dot product of these two vectors, is zero, and the orthogonality condition is obtained by recalling the Schwarzschild metric tensor g_{ab} of Eq. (15.644), where

$$g_{ab}\lambda^a\frac{\mathrm{d}x^b}{\mathrm{d}\tau} = \left(1 - \frac{2MG}{c^2r}\right)c^2\lambda^t\frac{\mathrm{d}x^t}{\mathrm{d}\tau} - r^2\lambda^\phi\frac{\mathrm{d}x^\phi}{\mathrm{d}\tau} = 0 \tag{15.691}$$

From Eq. (15.690), the radial r and polar angle θ geodesic equations are

$$\frac{\mathrm{d}\lambda^r}{\mathrm{d}\tau} + \Gamma^r_{tt}\lambda^t\frac{\mathrm{d}x^t}{\mathrm{d}\tau} + \Gamma^r_{\phi\phi}\lambda^\phi\frac{\mathrm{d}x^\phi}{\mathrm{d}\tau} = 0 \tag{15.692}$$

$$\frac{\mathrm{d}\lambda^\theta}{\mathrm{d}\tau} = 0 \tag{15.693}$$

From Eq. (15.690), the geodesic equations for azimuth angle ϕ and time t are equivalent, as can be seen by setting $\phi \to t$ in the equation for ϕ or setting $t \to \phi$ in the equation for t, where

$$\frac{\mathrm{d}\lambda^\phi}{\mathrm{d}\tau} + \Gamma^\phi_{r\phi}\lambda^r\frac{\mathrm{d}x^\phi}{\mathrm{d}\tau} = 0 \tag{15.694}$$

$$\frac{\mathrm{d}\lambda^t}{\mathrm{d}\tau} + \Gamma^t_{rt}\lambda^r\frac{\mathrm{d}x^t}{\mathrm{d}\tau} = 0 \tag{15.695}$$

Substitution of Eq. (15.644) for the Schwarzschild metric tensor component g_{ab} into Eq. (15.374) for the connection coefficients Γ^a_{bc}, yields

$$\Gamma^r_{tt} = \frac{GM}{r^2c^2}\left(1 - \frac{2GM}{rc^2}\right) \quad \Gamma^r_{\phi\phi} = -r\left(1 - \frac{2GM}{rc^2}\right) \tag{15.696}$$

$$\Gamma^\phi_{r\phi} = \frac{1}{r} \quad \Gamma^t_{rt} = \frac{GM}{r^2c^2}\left(1 - \frac{2GM}{rc^2}\right)^{-1} \tag{15.697}$$

Recall from Eqs. (15.680) and (15.682) for the Euler equation results that

$$a \equiv \frac{\mathrm{d}t}{\mathrm{d}\tau} = \left(1 - \frac{3MG}{rc^2}\right)^{-1/2} \quad \text{and} \quad \Omega \equiv \frac{\mathrm{d}\phi}{\mathrm{d}\tau} = \sqrt{\frac{MG}{r^3}} \tag{15.698}$$

Both the defined a and the angular coordinate speed Ω are constants. The four-velocity vector components are then given by

$$\left(\frac{\mathrm{d}x^r}{\mathrm{d}\tau}, \frac{\mathrm{d}x^\theta}{\mathrm{d}\tau}, \frac{\mathrm{d}x^\phi}{\mathrm{d}\tau}, \frac{\mathrm{d}x^t}{\mathrm{d}\tau}\right) = (0, 0, \Omega a, a) \tag{15.699}$$

With substitution of Eqs. (15.696–15.697) into Eqs. (15.694–15.695), using Eq. (15.699), the line element orthogonality condition allows expression of λ^t in terms of λ^ϕ, where

$$\lambda^t = \frac{\Omega r^2}{c^2}\left(1 - \frac{2GM}{rc^2}\right)^{-1} \lambda^\phi \tag{15.700}$$

With substitution of Eqs. (15.696–15.697) into Eqs. (15.692–15.695), using Eq. (15.699), one obtains"

$$\frac{\mathrm{d}\lambda^r}{\mathrm{d}\tau} - \frac{r\Omega}{a}\lambda^\phi = 0 \tag{15.701}$$

$$\frac{\mathrm{d}\lambda^\theta}{\mathrm{d}\tau} = 0 \tag{15.702}$$

$$\frac{\mathrm{d}\lambda^\phi}{\mathrm{d}\tau} + \frac{a\Omega}{r}\lambda^r = 0 \tag{15.703}$$

Solving Eqs. (15.701–15.703), one obtains:

$$\lambda^r = \frac{A}{a}\cos(\phi_0 - \Omega\tau) \tag{15.704}$$

$$\lambda^\theta = B \tag{15.705}$$

$$\lambda^\phi = \frac{A}{r}\sin(\phi_0 - \Omega\tau) \tag{15.706}$$

Here, A, B, and ϕ_0 are constants of integration.

If the gyroscope axis vector $\boldsymbol{\lambda}$ is initially pointing in the radial direction, then $B = 0$, and if the time origin occurs when $t = \tau = 0$, then $\phi_0 = 0$, so that Eqs. (15.704–15.706) become

$$\lambda^r = \frac{A}{a}\cos(\Omega\tau) \tag{15.707}$$

$$\lambda^\theta = 0 \tag{15.708}$$

$$\lambda^\phi = -\frac{A}{r}\sin(\Omega\tau) \tag{15.709}$$

One parallel-transport orbit of the gyroscope axis vector $\boldsymbol{\lambda}$ around the Earth occurs in time coordinate Δt and proper time $\Delta\tau$ intervals of

$$\Delta t = \frac{2\pi}{\Omega} \tag{15.710}$$

and from Eq. (15.679)

$$\Delta\tau = \Delta t\sqrt{1 - \frac{3MG}{rc^2}} = \frac{2\pi}{\Omega}\sqrt{1 - \frac{3MG}{rc^2}} \tag{15.711}$$

The final azimuth direction angle ϕ_λ of the gyroscope axis vector $\boldsymbol{\lambda}$ after parallel transport around the Earth for one orbit is less than 2π radians, which demonstrates the geodesic effect of parallel transport about a massive body that curves space–time, where

$$\phi_\lambda = \Omega\Delta t - \Omega\Delta\tau = 2\pi\left(1 - \sqrt{1 - \frac{3MG}{rc^2}}\right) \tag{15.712}$$

The final azimuth direction angle ϕ_λ, which is attained in a single parallel-transport orbit about the Earth, is thus the difference between the aziumuth angular coordinate interval $\Omega\Delta t$ of a fixed observer, and the proper azimuth angular interval $\Omega\Delta\tau$ of an observer orbiting with the gyroscope.

This result is the example of the parallel transport of a vector around a closed loop embedded on a curved surface, where the vector points in a different direction than its original direction after one parallel-transport revolution around the loop. The space–time in the vicinity of the Earth is slightly curved due to the significant mass of the Earth. The absence of other forces allows the free body satellite to move in a geodesic orbit about the Earth. In general relativity, gravity acts on the satellite via the curvature of the space–time, which causes the satellite embedded in this curved space–time to move in a geodesic orbit about the Earth. In Newtonian mechanics, the orbit of the satellite is caused by gravity, which acts as a radial inward force, and adds a radial inward velocity component to the tangential velocity component of the satellite, which then causes the satellite to orbit the Earth.

15.33 The Friedmann Equation and Cosmological Issues

In this section, we present the Friedmann equation, which is a solution of the Einstein field equations and provides an equation of motion of the universe as a whole. Also, the Friedmann equation is derived without general relativity using only Euler's continuity and motion equations with the empirical-based Hubble's law from observational astronomy. That is, one could speculate that the Friedmann equation could have been derived prior to general relativity and Hubble's law by using the Euler equations of continuity and motion, and hypothesizing the three scenarios, the expansion speed of the universe increases, decreases, or remains constant with distance from the Earth observer.

The Friedmann equation is another solution of Einstein's field equations that uses the Robertson–Walker metric, which is based on the universe being spatially homogeneous and isotropic. The Walker–Robertson line element is given by

$$\mathrm{d}\tau^2 = \mathrm{d}t^2 - \left(\frac{\mathrm{d}r^2}{1-kr^2} + r^2\mathrm{d}\theta^2 + r^2\mathrm{d}\phi^2\sin^2\theta\right)R^2(t) \tag{15.713}$$

The Walker–Robertson metric tensor is a diagonal tensor with covariant diagonal components g_{aa}, given by

$$g_{00} = -\frac{R^2}{1-kr^2} \quad g_{11} = R^2r^2 \quad g_{22} = R^2r^2\sin^2\theta \quad g_{44} = 1 \tag{15.714}$$

Here $R(r)$ is a dimensionless scale factor that depends on time t. Constant k is related to the spatial curvature, where a flat space has $k = 0$, a spherical-like space of positive constant curvature and finite volume has $k = 1$, and a space of constant negative curvature and infinite volume has $k = -1$. In the Friedmann equation, the universe is modeled as a fluid with pressure p, proper frame fluid density ρ, and fluid particle velocity u_a, where the covariant stress tensor T_{ab} is obtained from Eq. (15.508) for the contravariant stress tensor,

$$T_{ab} = \left(\rho + \frac{p}{c^2}\right)u_a u_b - p g_{ab} \tag{15.715}$$

One then substitutes the diagonal metric tensor coefficients g_{aa} of Eq. (15.714) and stress tensor T_{ab} of Eq. (15.715) into the Einstein field equation of Eq. (15.568), and obtains the Friedmann equation:

$$\left(\frac{\mathrm{d}R}{\mathrm{d}t}\right)^2 + k = \left(\frac{8\pi G}{3c^4}\right)\rho R^2 \tag{15.716}$$

The Friedmann equation is effectively the equation of motion of the universe, where the fluid pressure p dropped out in the course of derivation.

The Friedmann equation can be derived without general relativity using only Newtonian mechanics, which uses only Euler's equation of motion and the continuity equation combined with Hubble's law. This derivation follows Chapter 6 of Foster and Nightingale (1995). The Friedmann equation derived from Newtonian mechanics accounts only for the gravity contribution to energy and neglects the pressure contribution to the total energy. Furthermore, in principle, the derivation does not take into account the relativistic speed of the particles or the universe speed limit for light, energy, and matter in this simplistic fluid model universe.

The Friedmann equation result is now derived by Newtonian mechanics and experimental-based Hubble's law. Assume a Newtonian dust, which is a fluid with zero pressure, moving according to Newton's laws of motion and gravitation, of uniform density $\rho(t)$ as a function of time t. This Newtonian dust is in a state of uniform expansion, with gravity being the only force acting on it. Consequently, the position vector $\mathbf{r}$ of a fluid particle at any time t is given by

$$\mathbf{r} = R(t)\mathbf{d} \tag{15.717}$$

where $\mathbf{d}$ is a constant vector, which is determined by the initial position of a fluid particle, and $R(t)$ is a dimensionless scale factor depending only on time t. This model assumes (neglects) that fluid pressure contributions to total energy and to the pressure effects on the gravitational field are negligible compared to the fluid density, and neglects relativistic effects of high-speed particles of the fluid.

Differentiation of position vector $\mathbf{r}$ of a fluid particle with respect to time gives

$$\frac{\mathrm{d}\mathbf{r}}{\mathrm{d}t} = \mathbf{r}H(t) \tag{15.718}$$

Here, $H(t)$ is Hubble's constant, defined by

$$H(t) = \frac{1}{R}\frac{\mathrm{d}R}{\mathrm{d}t} \tag{15.719}$$

The dust particle thus has a Hubble constant associated with the scale factor $R(t)$.

Next, the Newtonian continuity equation is given by

$$\frac{\partial \rho}{\partial t} + \nabla \cdot \left(\rho \frac{\mathrm{d}\mathbf{r}}{\mathrm{d}t} \right) = 0 \tag{15.720}$$

One assumes spherical coordinates with radial dependence only for the Hubble expression $\mathbf{dr}/dt = \mathbf{r}H(t)$ of Eq. (15.718), which is then substituted into the continuity equation of Eq. (15.720), and gives

$$\frac{\mathrm{d}\rho}{\mathrm{d}t} + 3\rho \frac{\partial r}{\partial t} = \frac{\mathrm{d}\rho}{\mathrm{d}t} + 3\rho H r = 0 \tag{15.721}$$

The result in Eq. (15.721) was obtained by setting $\mathbf{A} = \mathbf{r}H(t)$, where in spherical coordinates, the divergence of a vector $\mathbf{A} = A_r \hat{\mathbf{r}}$ with radial dependence only is

$$(\nabla \cdot \mathbf{A})_r = \frac{1}{r^2} \frac{\partial (r^2 A_r)}{\partial r} = \frac{1}{r^2} \frac{\partial (r^3)}{\partial r} = 3 \frac{\partial r}{\partial t} \tag{15.722}$$

With n equal to a constant, integration of Eq. (15.722) with respect to time t yields the integrated form of the continuity equation, where

$$\rho R^3 = n \tag{15.723}$$

Let us consider present-day values (i.e., time $t = t_0$) in the Hubble expansion of the universe. The present time t_0 values of fluid density ρ_0, dimensionless expansion parameter R_0, and Hubble's constant H_0 are then denoted by

$$\rho_0 = \rho(t_0) \quad R_0 = R(t_0) \quad H_0 = H(t_0) \tag{15.724}$$

The constant n at present time t_0 can be expressed as

$$n = \rho_0 R_0^3 \tag{15.725}$$

Substitution into the continuity equation of Eq. (15.723) for the constant n of Eq. (15.725), and solving for density $\rho(t)$ at time t in the resulting form of the continuity equation, yields

$$\rho(t) = \frac{\rho_0 R_0^3}{R^3(t)} \tag{15.726}$$

Next, Euler's equation of motion is given by

$$\left(\frac{\partial}{\partial t} + \frac{\mathrm{d}\mathbf{r}}{\mathrm{d}t} \cdot \nabla \right) \frac{\mathrm{d}\mathbf{r}}{\mathrm{d}t} = \mathbf{F} \tag{15.727}$$

where

$$\frac{d\mathbf{r}}{dt} = \mathbf{r}H(t) \tag{15.728}$$

With application of Eq. (15.722) to Eq. (15.727), and substitution of Eq. (15.728), Euler's equation of motion becomes

$$\left(\frac{dH}{dt} + H^2\right)\mathbf{r} = \mathbf{F} \tag{15.729}$$

Here $\mathbf{F}$ is the body force per unit mass. Force $\mathbf{F}$ is due to gravity and the divergence of the force $\nabla \cdot \mathbf{F}$ is proportional to the mass density ρ via the gravitational constant G, where

$$\nabla \cdot \mathbf{F} = -4\pi G\rho \tag{15.730}$$

Taking the divergence of both sides of the Euler equation of Eq. (15.729), yields

$$\nabla \cdot \left(\frac{dH}{dt} + H^2\right)\mathbf{r} = \nabla \cdot \mathbf{F} \tag{15.731}$$

Below are three useful relations, which will be used in the derivation that follows:

$$\nabla \cdot \mathbf{r} = 3 \qquad \frac{d}{dt}\left(\frac{1}{R}\frac{dR}{dt}\right) + \frac{1}{R^2}\left(\frac{dR}{dt}\right)^2 = \frac{1}{R}\frac{d^2R}{dt^2} \qquad \frac{dR}{dt}\frac{\partial^2 R}{\partial t^2} = \frac{1}{2}\frac{d}{dt}\left(\frac{dR}{dt}\right)^2 \tag{15.732}$$

Carrying out the divergence operation in Eq. (15.731), using the left-most relation in Eq. (15.732), yields

$$3\left(\frac{dH}{dt} + H^2\right) = -4\pi G\rho \tag{15.733}$$

On substitutes Eq. (15.719) for Hubble's constant $H(t)$ and Eq. (15.726) for $\rho(t)$ into Eq. (15.733) and applies the middle relation of Eq. (15.732) to obtain the second-order differential equation for the scale factor $R(t)$, where

$$\frac{d}{dt}\left(\frac{1}{R}\frac{dR}{dt}\right) + \frac{1}{R^2}\left(\frac{dR}{dt}\right)^2 = -\frac{4\pi}{3}\frac{G\rho_0 R_0^3}{R^3} \Rightarrow \frac{d^2R}{dt^2} = -\frac{4\pi}{3}\frac{G\rho_0 R_0^3}{R^2} \tag{15.734}$$

One multiplies both sides of the right-hand side of Eq. (15.734) by dR/dt, applies the right-most relation in Eq. (15.732), then integrates over time, and thus obtains the differential equation for energy in the universe, which is expressed by the scale factor $R(t)$ and its squared time derivative $(\mathrm{d}R/dt)^2$, where

$$\left(\frac{\mathrm{d}R}{\mathrm{d}t}\right)^2 - \frac{1}{R}\left(\frac{8\pi G\rho_0 R_0^3}{3}\right) = -k \quad \text{total energy} \tag{15.735}$$

Here, k is a constant of integration in which $-k$ represents the total energy of the universe. The total universe energy k can have a continuous range of values, including the range from $k=-1$ to $k=0$ to $k=1$. The kinetic energy of the universe is represented by

$$\left(\frac{\mathrm{d}R}{\mathrm{d}t}\right)^2 \quad \text{kinetic energy} \tag{15.736}$$

The gravitational potential energy of the universe is represented by

$$-\frac{1}{R}\left(\frac{8\pi G\rho_0 R_0^3}{3}\right) \quad \text{potential energy} \tag{15.737}$$

If $k=-1$, the total energy of the universe is positive, and the universe has an excess of kinetic energy that allows it to keep expanding at an ultimately constant rate. As $t \to \infty$, $R \to \infty$, and

$$\lim_{R\to\infty} \frac{\mathrm{d}R}{\mathrm{d}t} \to 1 \quad t \to \infty \tag{15.738}$$

If $k=0$, the total energy of the universe is zero, and the universe has just enough kinetic energy to allow its expansion at a decreasing rate. As $t \to \infty$, $R \to \infty$, and

$$\lim_{R\to\infty} \frac{\mathrm{d}R}{\mathrm{d}t} \to 0 \quad t \to \infty \tag{15.739}$$

If $k=1$, the total energy of the universe is negative, and the universe does not have enough kinetic energy for continued expansion. The universe expands until

$$\frac{\mathrm{d}R}{\mathrm{d}t} = 0 \quad \text{and} \quad R = \frac{8\pi G\rho_0 R_0^3}{3} \tag{15.740}$$

The universe then subsequently contracts after reaching a finite radius R, where the rate of increase goes to zero ($\mathrm{d}R/\mathrm{d}t=0$). This simple model was derived without special or general relativity, in which the cosmological constant k is allowed to vary continuously from -1 to $+1$.

At present, the basic view of cosmology, which is based on experimental observation, is that the universe is homogeneous, with a constant density of galaxies and mass, where the number of galaxies per solid angle is equal in all directions, and the galaxies are red-shifted, where $\Delta\lambda/\lambda$ increases with distance. The universe does not appear to be infinite, which is based on Olber's paradox that the night sky is not as bright as the day sky, and radioactive dating of meteorites gives the age of the universe as at least 4 billion years. Finally, in all directions, the background microwave radiation corresponds to black-body radiation of about 2.7 K.

Warping of space occurs, due to the presence of matter, and thus energy. Dark energies are to be found in space as simply expanded energy, producing a volume so great that dark energy is difficult to measure on the scale of an Earth-based lab, but can be measured readily when weighing galaxies. Current estimates of dark energy and dark matter is 95%; with dark energy taking up 65% and dark matter taking up 30% of energy in the universe. The growing body of experimental results and measurements in astrophysics and the cosmological models make for fascinating developments in our view and observation of the universe. For example, see Cheng (2010), Dodelson (2003), Peebles (1993), Penrose (2004), Shu (1982), and Weinberg (2008).

One interesting model, which serves as a counterpoint to parallel universes and universes with extra dimensions, is the Orr space and equidynamic model of the universe conceived by Orr (2003, 2004). In Orr space, the universe is composed of one fabric, known as *spacenergy*. The basic unit is not the particle, but a combined space and energy unit, which comprise two aspects of one whole. Neither spatial relationships nor energy exists without the other. That is, a finite amount of energy cannot exist at a single point, which in that case produces a singularity of infinite energy density. Instead, a finite amount of energy can exist only in a finite spatial region. A finite region of space cannot exist unless the space contains a finite amount of energy. The differential within the single spacenergy unit is the motive principle of change resulting in fluctuation.

From Orr (2003, 2004), the motive principle is the differential between the space and the energy aspects of spacenergy. Nature must possess an inherent motive property. This imperfect state is what makes movement an inherent condition. Fluctuation is the inherent movement of the basic unit, which is caused by the differential between space and energy components, and interaction with adjoining spacenergy units. The number of spacenergy units is constant, and the spacenergy units can be neither created nor destroyed, and the differential capacity of the spacenergy unit remains invariant. This fundamental spacenergy unit possesses a variable volume capacity of near singularity to astronomically large, in which the spacenergy can expand into a space-like field, or compress to de Broglie waves and solitons of matter (particles) and to wave packets of energy (photons). Volume is an attribute of spacenergy units, not its identity. The spacenergy unit is the physical entity, and volume is the measurement of the stretch or compression of one or more spacenergy units.

From Orr (2003, 2004), all the invariance or conservation of spacenergy leads to a redistributive, cyclic, and closed universe that obeys causality, with only forward-directed time in which extra dimensions and parallel universes are excluded. The Orr space universe is a closed system, which has no surroundings, and thus, there is no exchange or orrtropy across the closed universe boundary. *Orrtropy* is the flow of all summed attributes of energy, space, and their relationships in the closed universe system. Thermodynamics accounts for the flow of energy, matter, and entropy in a system that interacts with an outside surrounding environment. However, equidynamics and orrtropy, which includes entropy, is a description of the closed Orr space universe system, which has neither an interacting nor a surrounding environment.

From Orr (2003, 2004), all the interactions are a manifestation of the equidynamic principle. The differential and resolution of the differential are the conditions for energy to exist. *Equidynamics* is the act of resolving a differential within a spacenergy unit and between spacenergy units to create a local balance. One movement to balance invites the next movement to balance. The average condition of the universe is a dimpled fabric of spacenergy. The universe is mostly an expanded continuum of spacenergy units, dotted minimally with spacenergy convergences to energy and matter. The energy–matter actions on each local level propagate out at the speed of light, and the spacenergy interactions on the local level propagate at the speed of space to surrounding local regions. The flow of each of the local spacenergy fluctuations out into global expanse, and their resulting interactions, create the complexity of the universe, with causality obeyed.

From Orr (2003, 2004), the Heisenberg uncertainty principle is a form of an average, stating that the product of the standard deviation of the particle position multiplied by the standard deviation of the particle velocity is greater than or equal to Planck's constant. The amplitude of a sine wave averages to zero, but the standard deviation or average variation (deviation) of the sine wave amplitude is nonzero. Variations around the average, allow the possibility for spacenergy fluctuations to average constructively their high variations together, or average constructively their low variations together, which produces a new average higher or lower state. Destructive interference between high and low fluctuations of spacenergy units leads to the average between high and low. Equidynamics allows fluctuations of spacenergy to sum non-preferentially to higher or lower states. Also, equidynamics allows for the nonpreferentiality of the convergence of spacenergy to matter-energy manifestations, or divergence to space-like manifestations.

From Orr (2003, 2004), the restriction of the speed of light is not how fast particles and energy packets can move, but rather the speed that the spacenergy continuum permits. The spacenergy continuum can move faster than the speed of light, but the converged spacenergy packets of matter and energy cannot. The phase velocity of an unmodulated sine wave amplitude of an electromagnetic or de Broglie wave can be faster than the speed of light, but the group velocity of a phase-, amplitude-, or frequency-modulated electromagnetic or de Broglie wave is limited to the speed of light or less because modulation represents energy propagation. In the same manner, the fluctuations of the spacenergy continuum is limited to the speed of spacenergy, and the converged spacenergy in the form of packets of matter or energy is limited to the speed of light. The expanded space-field is placed on the energy spectrum with an astronomically long wavelength of a diffuse energy value, which is difficult to measure in an Earth-based laboratory, but is readily measured as dark matter and dark energy when weighing galaxies.

From Orr (2003, 2004), if the continuum of spacenergy is averaged locally and globally so as to resolve the maximal differential, then the maximally averaged attributes of space and energy would be a contiguous continuum called the *spacenergy sea.* The spacenergy sea is the original and preferred condition that is the average of the maximally expanded and maximally converged universe. During its existence, the spacenergy sea has no net modulation or net fluctuations on a local and global level, and time ceases. Existing in the spacenergy sea is the domain of chance or causal probability in which anything that might be possible is then physically possible.

From Orr (2003, 2004), a sufficient local sequence of fluctuations of the space and energy aspects, within single spacenergy units and their neighbors, can lead to an imbalance or modulation in the energy and space aspects of the universe, which leads to a global imbalance or modulation. With this, the spacenergy sea condition ends, and time begins anew. As the casual chains of imbalance and modulation commence, the former large realm of possibilities of the spacenergy sea becomes irretrievably foreclosed, and what remains is the much smaller set of possibilities, which are associated with the particular cascade of imbalance that has occurred and is evolving. In contrast to the preferred condition of the spacenergy sea, are the extreme conditions of the big bang, when the universe is maximally converged, and the opposite extreme condition of the void, where the universe is maximally expanded.

References

Adams, S. (1997). *Relativity. An introduction to space-time physics.* London, UK: Taylor and Francis, Ltd.

Arfken, G. (1970). *Mathematical methods for physicists.* New York, NY: Academic Press.

Brannon, R. M. (2004). *Curvilinear analysis in Euclidean space.* Albuquerque, NM: UNM Supplemental Book Draft, University of New Mexico. website: <http://www.mech.utah.edu/~brannon/public/curvilinear.pdf>.

Cheng, T.-P. (2010). *Relativity, Gravitation and Cosmology: A Basic Introduction.* New York: Oxford University Press, Inc.

Dodelson, S. (2003). Moderncosmology. Oxford: Academic Press.

Dutton, J. A. (1986). *Dynamics of atmospheric motion.* New York, NY: Dover Publications.

Einstein, A. (2004). *The meaning of relativity.* Princeton, NJ: Princeton University Press.

Eisenhart, L. P. (1934). Separable systems in Euclidean 3-space. *Physical Review, 45*, 427–428.

Foster, J., & Nightingale, J. D. (1995). *A short course in general relativity.* New York, NY: Springer.

Friedrichs, K. O. (1975). *From Pythagoras to Einstein.* Washington, DC: Mathematical Association of America.

Graustein, W. C. (1962). *Differential geometry.* New York, NY: Dover Publications.

Greenwood, D. T. (1977). *Classical dynamics.* New York, NY: Dover Publications.

Guggenheimer, H. W. (1977). *Differential geometry.* New York, NY: Dover Publication.

Joos, G. (1986). *Theoretical physics* (3rd Edition). New York, NY: Dover Publications.

Landau, L. M., & Lifshitz, E. M. (1975). *The classical theory of fields* (4th ed.). New York, NY: Pergamon Press.

Lawden, D. F. (2002). *Introduction to tensor calculus, relativity, and cosmology*. New York, NY: Dover Publications.

Logunov, A. A., Mestvirishvili, M. A., & Petrov, V. A. (2004). How were the Hilbert–Einstein equations discovered? *Physical-Uspekhi*, *47*(6), 607–621.

McMahon, D. (2006). *Relativity demystified*. New York, NY: McGraw-Hill.

Misner, C. W., Thorne, K. S., & Wheeler, J. A. (1973). *Gravitation*. New York, NY: W.H. Freeman.

Orr, N. B. (2003). A theoretical paradigm of spacetime in cosmological modeling. *Frontier in Astroparticle Physics and Cosmology (University Academy Press) University of Tokyo*, pp. 471–476. Available from <http://adsabs.harvard.edu/abs/2004fapc.conf..471O>. The 6th RESCEU International Symposium, *Frontier in Astroparticle Physics and Cosmology*, November 4–7, 2003, Tokyo, Japan.

Orr, N. B. (2004). Galactic starbursts as a manifestation of a general Equipoise dynamic. *Starbursts 2004: Institute of Astronomy Cambridge University (Astrophysics & Space Science Library)*. Paper on CD. Available from: <http://adsabs.harvard.edu/abs/2005sdlb.procP.88O>. Starbursts International Conference, Institute of Astronomy, University of Cambridge (UK), September 6–10, 2004.

Pathria, R. K. (2003). *The theory of relativity*. New York, NY: Dover Publications.

Pauli, W. (1981). *The theory of relativity*. New York, NY: Dover Publications.

Peebles, P. J. E. (1993). *Principles of physical cosmology*. Princeton, NJ: Princeton University Press.

Penrose, R. (2004). *The road to reality*. New York, NY: Vintage Books.

Sen, M., & Powers, J. M. (2011). *Lecture notes on mathematical methods*. IN: Department of Aerospace and Mechanical Engineering, University of Notre Dame. Available from: <http://www.nd.edu/%7Epowers/ame.60611/M.pdf>.

Shadowitz, A. (1988a). *The electromagnetic field*. New York, NY: Dover Publicaitons.

Shadowitz, A. (1988b). *Special relativity*. New York, NY: Dover Publications.

Shu, F. H. (1982). *The physical universe. An introduction to astronomy*. Sausalito, CA: University Science Books.

Sommerfeld, A. (1964). *Electrodynamics*. New York, NY: Academic Press.

Soper, D. (2008). *Classical field theory*. New York, NY: Dover Publications.

Weinberg, S. (2008). *Cosmology*. Oxford: Oxford University Press.

Wrede, R. C. (1972). *Introduction to vector and tensor analysis*. New York, NY: Dover Publications.

Zee, A. (2003). *Quantum field theory in a nutshell*. Princeton, NJ: Princeton University Press.

16 Radiation Imaging Systems and Performance

16.1 Liouville's Theorem, the Vlasov Equation, and the Continuity Equation

In this section, we derive Liouville's theorem, which is a statement of conservation of phase space, an important consideration in imaging and focusing systems. For example, take an ensemble of N noninteracting particles with the particle phase-space density $f(q_x, q_y, q_z, P_x, P_y, P_z)$ as a function of the x, y, and z spatial position components q_x, q_y, and q_z, and the mechanical momentum x, y, and z components P_x, P_y, and P_z. The N particles at the source are contained in the source phase-space volume. The conservation of phase-space volume, states that the integration over the entire range of spatial coordinates in x, y, and z mechanical momentum coordinates P_x, P_y and P_z, yields the same number of particles N throughout the particle trajectories, where

$$\int \mathrm{d}q_x \int \mathrm{d}q_y \int \mathrm{d}q_z \int \mathrm{d}P_x \int \mathrm{d}P_y \int \mathrm{d}P_z f(q_x, q_y, q_z, P_x, P_y, P_z) = N \tag{16.1}$$

The number of particles N remains constant, but the particle density and shape of the volume can vary, so long as the integration over the phase-space volume yields the invariant particle number N.

Assume that an ensemble of particles each has constant total momentum P. The total momentum P of the particle of rest mass m, relativistic mass γm, and velocity βc is

$$P = \gamma m \beta c \tag{16.2}$$

The trace-space area A_i for $i = x, y, x$ is the projection of the phase-space volume onto the $q_i - P_i$ plane, divided by the total mechanical momentum P per particle, where

$$A_i = \frac{1}{P} \int \mathrm{d}q_i \int \mathrm{d}P_i \tag{16.3}$$

Neutron and X-ray Optics. DOI: http://dx.doi.org/10.1016/B978-0-12-407164-3.00015-2

If there is no coupling between the x, y, and z dimensions, then each trace-space area A_x for x, A_y for y, and A_z for z remains constant, but the shape of the area can vary, so that

$$A_x = \frac{1}{P}\int \mathrm{d}x \int \mathrm{d}P_x, \quad A_y = \frac{1}{P}\int \mathrm{d}y \int \mathrm{d}P_y, \quad A_z = \frac{1}{P}\int \mathrm{d}z \int \mathrm{d}P_z \tag{16.4}$$

If there is an energy change to the ensemble of particles, via an external force acting in the direction of particle motion, then the particle momentum changes, and the conservation of phase-space volume no longer applies.

The differential of phase-space volume is zero, because the number of particles N is fixed, where $\mathrm{d}N = 0$ and

$$\frac{\mathrm{d}}{\mathrm{d}t}\iiint\iiint f(q_x, q_y, q_z, P_x, P_y, P_z)\mathrm{d}q_x\,\mathrm{d}q_y\,\mathrm{d}q_z\,\mathrm{d}P_x\,\mathrm{d}P_y\,\mathrm{d}P_z = \frac{\mathrm{d}N}{\mathrm{d}t} = 0 \tag{16.5}$$

One can bring the time derivative, which is independent of the spatial and momentum coordinates inside the integrand:

$$\iiint\iiint \frac{\mathrm{d}}{\mathrm{d}t}\left[f(q_x, q_y, q_z, P_x, P_y, P_z)\mathrm{d}q_x\,\mathrm{d}q_y\,\mathrm{d}q_z\,\mathrm{d}P_x\,\mathrm{d}P_y\,\mathrm{d}P_z\right] = \frac{\mathrm{d}N}{\mathrm{d}t} = 0 \tag{16.6}$$

Application of the time derivative in Eq. (16.6), gives

$$\begin{aligned}&\iiint\iiint \mathrm{d}q_x\,\mathrm{d}q_y\,\mathrm{d}q_z\,\mathrm{d}P_x\,\mathrm{d}P_y\,\mathrm{d}P_z\frac{\mathrm{d}}{\mathrm{d}t}\left[f(q_x, q_y, q_z, P_x, P_y, P_z)\right] + \\ &f(q_x, q_y, q_z, P_x, P_y, P_z)\frac{\mathrm{d}}{\mathrm{d}t}\left[\mathrm{d}q_x\,\mathrm{d}q_y\,\mathrm{d}q_z\,\mathrm{d}P_x\,\mathrm{d}P_y\,\mathrm{d}P_z\right] = \frac{\mathrm{d}N}{\mathrm{d}t} = 0\end{aligned} \tag{16.7}$$

The conservation phase-space density f requires its derived time to be zero, so the first term of Eq. (16.7) is zero, where $\mathrm{d}f/dt = 0$. From Eq. (16.7), one thus obtains Liouville's theorem, where

$$\int \mathrm{d}q_x \int \mathrm{d}q_y \int \mathrm{d}q_z \int \mathrm{d}P_x \int \mathrm{d}P_y \int \mathrm{d}P_z \frac{\mathrm{d}f(q_x, q_y, q_z, P_x, P_y, P_z)}{\mathrm{d}t} = 0 \tag{16.8}$$

With substitution of Eq. (16.8) into Eq. (16.7), the second term of Eq. (16.7) is also zero, and the resulting expression is a second version of Liouville's theorem, where

$$f(q_x, q_y, q_z, P_x, P_y, P_z)\frac{\mathrm{d}}{\mathrm{d}t}\left[\mathrm{d}q_x\,\mathrm{d}q_y\,\mathrm{d}q_z\,\mathrm{d}P_x\,\mathrm{d}P_y\,\mathrm{d}P_z\right] = 0 \tag{16.9}$$

For particle phase-space density $f \neq 0$, the second version of Liouville's theorem in Eq. (16.9), states that the differential phase-space volume $\mathrm{d}^3q\mathrm{d}^3P$ is independent of time, where

$$\frac{\mathrm{d}}{\mathrm{d}t}\left[\mathrm{d}q_x\,\mathrm{d}q_y\,\mathrm{d}q_z\,\mathrm{d}P_x\,\mathrm{d}P_y\,\mathrm{d}P_z\right] = 0 \tag{16.10}$$

We now further examine the first version of Liouville's theorem of Eq. (16.8). The total number N of particles must remain constant, as well as the total number ($6N$) of their associated coordinate points. The particle phase-space density f also must be constant in time, where

$$\frac{\mathrm{d}f(q_x, q_y, q_z, P_x, P_y, P_z)}{\mathrm{d}t} = 0 \tag{16.11}$$

Application of the chain rule to Eq. (16.11), gives Liouville's theorem, which is expressed by the Boltzmann equation without collisions. This is also known as the Vlasov equation, where

$$\frac{\mathrm{d}f}{\mathrm{d}t} = \frac{\partial f}{\partial t} + \sum_{i=1}^{3} \frac{\partial f}{\partial q_i}\frac{\mathrm{d}q_i}{\mathrm{d}t} + \sum_{i=1}^{3} \frac{\partial f}{\partial P_i}\frac{\mathrm{d}P_i}{\mathrm{d}t} = 0 \tag{16.12}$$

In the second summation term in Eq. (16.12), the change of momentum with time $\mathrm{d}\mathbf{P}/dt$ is force $\mathbf{F}$, where

$$\mathbf{F} = \frac{\mathrm{d}\mathbf{P}}{\mathrm{d}t} = \hat{\mathbf{x}}\frac{\mathrm{d}P_x}{\mathrm{d}t} + \hat{\mathbf{y}}\frac{\mathrm{d}P_y}{\mathrm{d}t} + \hat{\mathbf{z}}\frac{\mathrm{d}P_z}{\mathrm{d}t} \tag{16.13}$$

In the first summation term in Eq. (16.12), the phase-space velocity vector $\mathbf{v}(\dot{\mathbf{q}},\dot{\mathbf{P}})$ is the sum of the time rate of change of the position $\dot{\mathbf{q}}$ and momentum $\dot{\mathbf{P}}$, where

$$\mathbf{v}(\dot{\mathbf{q}},\dot{\mathbf{P}}) = \dot{\mathbf{q}} + \dot{\mathbf{P}} = \dot{\mathbf{q}} + \mathbf{F} \tag{16.14}$$

and

$$\mathbf{v}(\dot{\mathbf{q}},\dot{\mathbf{P}}) = \left(\frac{\mathrm{d}q_x}{\mathrm{d}t} + \frac{\mathrm{d}P_x}{\mathrm{d}t}\right)\hat{\mathbf{x}} + \left(\frac{\mathrm{d}q_y}{\mathrm{d}t} + \frac{\mathrm{d}P_y}{\mathrm{d}t}\right)\hat{\mathbf{y}} + \left(\frac{\mathrm{d}q_z}{\mathrm{d}t} + \frac{\mathrm{d}P_z}{\mathrm{d}t}\right)\hat{\mathbf{z}} \tag{16.15}$$

The Vlasov equation of Eq. (16.12) can be expressed by the pair of vector dot products of the particle phase-space velocity vector $\mathbf{v}(\dot{\mathbf{q}},\dot{\mathbf{P}})$ with (1) the spatial gradient $\nabla_q f$ of particle phase-space density $f(\mathbf{q},\mathbf{P})$, and (2) the momentum gradient $\nabla_p f$ of particle phase-space density $f(\mathbf{q},\mathbf{P})$, where

$$\frac{\mathrm{d}f}{\mathrm{d}t} = \frac{\partial f}{\partial t}\mathrm{d}t + \dot{\mathbf{q}}\cdot\nabla_q f + \mathbf{F}\cdot\nabla_p f = 0 \tag{16.16}$$

and

$$\nabla_q = \hat{\mathbf{x}}\frac{\partial}{\partial x} + \hat{\mathbf{y}}\frac{\partial}{\partial y} + \hat{\mathbf{z}}\frac{\partial}{\partial z} \tag{16.17}$$

$$\nabla_p = \hat{\mathbf{x}}\frac{\partial}{\partial P_x} + \hat{\mathbf{y}}\frac{\partial}{\partial P_y} + \hat{\mathbf{z}}\frac{\partial}{\partial P_z} \tag{16.18}$$

In Eq. (16.6), evaluation of the dot product $\mathbf{v} \cdot \nabla_q f$ of the phase-space velocity $\mathbf{v}(\dot{q}_i, \dot{P}_i)$ with the spatial gradient of the phase-space density $\nabla_q f$ gives

$$\sum_{i=1}^{3} \frac{\partial f}{\partial q_i} \frac{\mathrm{d}q_i}{\mathrm{d}t} = \frac{\partial f}{\partial q_x} \frac{\mathrm{d}q_x}{\mathrm{d}t} + \frac{\partial f}{\partial q_y} \frac{\mathrm{d}q_y}{\mathrm{d}t} + \frac{\partial f}{\partial q_z} \frac{\mathrm{d}q_z}{\mathrm{d}t} = \dot{\mathbf{q}} \cdot \nabla_q f \tag{16.19}$$

In Eq. (16.6), evaluation of the dot product $\mathbf{F} \cdot \nabla_p f$ of the phase-space velocity $\mathbf{v}(\dot{q}_i, \dot{P}_i)$ with the momentum gradient of the phase-space density $\nabla_p f$ gives

$$\sum_{i=1}^{3} \frac{\partial f}{\partial P_i} \frac{\mathrm{d}P_i}{\mathrm{d}t} = \frac{\partial f}{\partial P_x} \frac{\mathrm{d}P_x}{\mathrm{d}t} + \frac{\partial f}{\partial P_y} \frac{\mathrm{d}P_y}{\mathrm{d}t} + \frac{\partial f}{\partial P_z} \frac{\mathrm{d}P_z}{\mathrm{d}t} = \mathbf{F} \cdot \nabla_p f \tag{16.20}$$

We now examine the relation between Liouville's theorem and the continuity equation in phase space. The constant number of particles N requires the phase-space density $f = f(\mathbf{q}, \mathbf{P})$ and current density $f\mathbf{v}$ to obey the continuity equation, given that the particle velocity $\mathbf{v}(\dot{\mathbf{q}}, \dot{\mathbf{P}})$ is a function of the change of position $\dot{q}_i$ and the force F_i or change of momentum $\dot{P}_i$ with time t, where $i = x, y, z$. The dot indicates differentiation with respect to time t, and the spatial q_i, momentum P_i, velocity $\dot{q}_i$, and force $\dot{P}_i$ are independent variables in each of the three i^{th} directions (x, y, and z). With the spatial ∇_q and momentum ∇_p gradients, which sums to the phase-space gradient ∇, the continuity equation for phase space is

$$\frac{\partial f}{\partial t} + \nabla \cdot (f\mathbf{v}) = \frac{\partial f}{\partial t} + (\nabla_q + \nabla_p)(f(\dot{\mathbf{q}} + \dot{\mathbf{P}})) = 0 \tag{16.21}$$

where

$$\nabla = \nabla_q + \nabla_p = \left(\frac{\partial}{\partial x} + \frac{\partial}{\partial P_x}\right)\hat{\mathbf{x}} + \left(\frac{\partial}{\partial y} + \frac{\partial}{\partial P_y}\right)\hat{\mathbf{y}} + \left(\frac{\partial}{\partial z} + \frac{\partial}{\partial P_z}\right)\hat{\mathbf{z}} \tag{16.22}$$

With application of the combined spatial and momentum gradient ∇ to the phase-space density current $f\mathbf{v}$ in Eq. (16.21), the continuity equation becomes

$$\frac{\partial f}{\partial t} + (\dot{\mathbf{q}} + \dot{\mathbf{P}}) \cdot (\nabla_q + \nabla_p)f + f(\nabla_q + \nabla_p) \cdot \dot{\mathbf{q}} + f(\nabla_q + \nabla_p) \cdot \dot{\mathbf{P}} = 0 \tag{16.23}$$

where

$$\begin{aligned} \nabla \cdot (\dot{\mathbf{q}} + \dot{\mathbf{P}}) &= \nabla_q + \nabla_p \\ &= \left[\left(\frac{\partial}{\partial x} + \frac{\partial}{\partial P_x}\right)\hat{\mathbf{x}} + \left(\frac{\partial}{\partial y} + \frac{\partial}{\partial P_y}\right)\hat{\mathbf{y}} + \left(\frac{\partial}{\partial z} + \frac{\partial}{\partial P_z}\right)\hat{\mathbf{z}}\right] \\ &\quad \cdot \left[\left(\frac{\mathrm{d}q_x}{\mathrm{d}t} + \frac{\mathrm{d}P_x}{\mathrm{d}t}\right)\hat{\mathbf{x}} + \left(\frac{\mathrm{d}q_y}{\mathrm{d}t} + \frac{\mathrm{d}P_y}{\mathrm{d}t}\right)\hat{\mathbf{y}} + \left(\frac{\mathrm{d}q_z}{\mathrm{d}t} + \frac{\mathrm{d}P_z}{\mathrm{d}t}\right)\hat{\mathbf{z}}\right] \end{aligned} \tag{16.24}$$

The position **q** and momentum **P** vectors are independent variables, so evaluation of the dot product gives

$$\nabla \cdot (\dot{\mathbf{q}} + \dot{\mathbf{P}}) = (\nabla_q + \nabla_p)(\dot{\mathbf{q}} + \dot{\mathbf{P}}) = \sum_{i=1}^{3} \frac{\partial \dot{q}_i}{\partial q_i} + \sum_{i=1}^{3} \frac{\partial \dot{P}_i}{\partial P_i} \tag{16.25}$$

Assume that the system of particles maintain a constant energy, which can be described by a Hamiltonian. The system of particles of constant energy can be described by a Hamiltonian $H(\mathbf{q}, \mathbf{P}, t)$, where one can apply Hamilton's equations, namely,

$$\dot{q}_i = \frac{\partial H}{\partial P_i} \quad \text{and} \quad \dot{P}_i = -\frac{\partial H}{\partial q_i} \tag{16.26}$$

where

$$\frac{\partial \dot{q}_i}{\partial q_i} = \frac{\partial H}{\partial q_i \partial P_i} \quad \text{and} \quad \frac{\partial \dot{P}_i}{\partial P_i} = -\frac{\partial H}{\partial q_i \partial P_i} \tag{16.27}$$

Application of Hamilton's equations gives

$$\nabla \cdot \mathbf{v} = (\nabla_q + \nabla_p) \cdot (\dot{\mathbf{q}} + \dot{\mathbf{P}}) = \sum_{i=1}^{3} \frac{\partial \dot{q}_i}{\partial q_i} + \sum_{i=1}^{3} \frac{\partial \dot{P}_i}{\partial P_i} = \frac{\partial H}{\partial q_i \partial P_i} - \frac{\partial H}{\partial q_i \partial P_i} = 0 \tag{16.28}$$

From Eq. (16.28), since the phase-space divergence of the phase-space velocity is zero, where $\nabla \cdot \mathbf{v} = 0$, then the spatial divergence of the time rate of change of position $\nabla_q \cdot \dot{\mathbf{q}}$ is equal to and in opposition to the momentum divergence of the time rate of change of momentum (force) $\nabla_p \cdot \dot{\mathbf{P}}$, where

$$\nabla_q \cdot \dot{\mathbf{q}} = -\nabla_p \cdot \dot{\mathbf{P}} \tag{16.29}$$

Given $\nabla \cdot \mathbf{v} = 0$, the phase-space continuity equation for phase-space current density $f\mathbf{v}$ then states Liouville's theorem:

$$\frac{\mathrm{d}f}{\mathrm{d}t} = \frac{\partial f}{\partial t} + \mathbf{v} \cdot \nabla f = \frac{\partial f}{\partial t} + (\dot{\mathbf{q}} + \mathbf{F}) \cdot (\nabla_q + \nabla_p) f = 0 \tag{16.30}$$

The total time derivative of the phase-space density is zero, where $\mathrm{d}f/dt = 0$. This confirms that the phase-space density f is constant in time, and thus the density of points anywhere in phase space is a constant f_0, where

$$f(q_x, q_y, q_z, P_x, P_y, P_z) = f_0 = \text{const} \tag{16.31}$$

Consider particles and photons, which are emitted from an extended source. As the group of particles and photons propagate outward in all directions, the extent of the physical space occupied by the photons and particles increases, whereas the range of transverse momentum at each position in the physical space decreases. In phase space, as distance z from the source is increased, the increase of the spatial component $\Delta x \Delta y > 0$ in the z-direction is exactly offset by the decrease of the transverse momentum component $\Delta p_x \Delta p_y < 0$. At large distances from the source, the divergence angle between a local group of particles and photons decreases, and one measures approximately parallel trajectory paths. If no particles or photons are lost by absorption, conversion, or collisions, which change momentum magnitude, then the phase-space density is constant along the particle or photon trajectory.

Manifestations of Liouville's theorem occur with galactic cosmic rays, which are measured by ground-based and Earth-based orbit observatories. In the vast empty tracts of galactic space with gravitational and magnetic fields, the phase-space density of the very energetic charged and neutral particles and photons remains approximately constant, which is caused by the near-total lack of collisions due to the very small particle densities, and the very small interaction cross sections at high energies. As a result, the phase invariance of the cosmic rays in their trajectory from their distant origins to the Earth, allows Earth-based measurements to provide direct information on the distant processes. See Bradt and Olbert (2008) for more information on this topic.

Scatter, refraction, reflection, and diffraction of neutrons, charged and neutral particles, gamma rays, and X-rays, occur in environments with very large gravitational and magnetic fields in astrophysical processes. Recall that the mean lifetime of free but unstable neutrons is about 885.7 s. A burst of neutrons from an astrophysical source becomes a burst of protons, electrons, and electron antineutrinos by beta decay after about a 15 min transit time, which is measured in the neutron rest frame. Astrophysical processes involve the scatter and the optics of neutrons, X-rays, gamma rays, and charged and neutral particle beams, which interact in mediums of varying density, composition, and shape, with imposed electromagnetic and gravitational fields.

The study of neutron and X-ray optics and scatter extends from nanostructures in the laboratory to the cosmos. For example, many X-ray and gamma bursts are associated with neutron stars and black holes. Neutron stars are old, massive stars, whose fusion processes are sufficiently diminished, so that gravitational collapse overcomes the outward pressure, and compresses the star to a minimum radius, which is limited by the Pauli exclusion principle. The protons and electrons are compressed so closely that electron–proton pairs are converted into neutrons, where collapse to a singularity, is prevented by the Pauli exclusion principle, which requires that no two neutrons (or fermions in general) can occupy the same place and quantum state at the same time. The conservation of angular momentum causes the collapsed neutron star to rotate at a much faster rate than the original star at its former larger radius.

The magnetic dipole moment of the rapidly rotating neutron star produces an intense magnetic field, which can accelerate charged particles in plasmas and beams. The rapid rotation of the neutron star and the periodic acceleration of the charged particles within its vicinity, lead to bursts of electromagnetic fields from radio to gamma frequencies. The highly magnetized, rotating neutron stars or pulsars, emit bursts of photons with periods, which can range from milliseconds to seconds, and is termed the *lighthouse effect*, because of the similarity to the rotating light beam of a lighthouse.

These wide-band, pulsar, photon beams pass through the plasma interstellar medium, where the frequency-dependent medium dispersion, which is due to the plasma electrons, causes the photon velocity through the medium to vary according to photon frequency. For example, low-frequency radio waves travel at a slower velocity than the higher frequency radio waves. The resulting time delay between the arrivals of pulse components of different frequencies, allows one to make a measurement of the plasma electron density between the source and observer. Turbulence in the interstellar plasma can cause scatter of the pulsar-emitted radio waves, which leads to intermittent radio wave pulse amplitudes, and is similar to the twinkling of stars seen by observers on Earth, due to disturbances in the atmosphere.

The black hole was predicted by John Michell in 1783 in a paper, which he presented to the British Royal Society (see Ellis, 1999). Michell was an English geologist and "the father of seismology," who constructed a torsion balance for measuring gravitational forces that was used by Henry Cavendish to measure the gravitational force between two test masses. A black hole of 3–20 times the mass of the sun, but shrunk to a radius of about 30 km, can be formed by the collapse of a neutron star whose mass exceeds the Tolman–Oppenheimer–Volkoff limit. Space-time can be warped by massive objects, and photons can thus be deflected, so that the black hole acts as a gravity lens. For more astrophysical information, including black holes, see the NASA web site (http://www.nasa.gov), as well as Wikipedia.

Finally, the continuity equation is examined without the assumption of a Hamiltonian to describe the system of particles. One then determines what type of forces can be imposed upon the particles that allow Liouville's theorem to be applicable, where the particle energy does not change. That is, which external forces can interact with the particles, and still allow a constant phase-space density throughout the trajectory paths of the particles. One must exclude those forces, which cause the momentum and energy of the particles to change. That is, particle radiation processes and collisions of the particles with the medium, which reduce the particle total energy and momentum, are excluded from Liouville's theorem. Also excluded from Liouville's theorem, are collisions between the particles, which would change the particle energy and momentum.

If the force $\mathbf{F}$ is not dependent on momentum, such as gravity, then with $\nabla \cdot \mathbf{v} = 0$, the momentum gradient of the force $\nabla_p \cdot \mathbf{F}$ and the spatial gradient of the velocity $\nabla_q \cdot \dot{\mathbf{q}}$ are each zero, where using Eq. (16.29), one obtains

$$\nabla_q \cdot \dot{\mathbf{q}} = -\nabla_p \cdot \dot{\mathbf{P}} = -\nabla_p \cdot \mathbf{F} = 0 \tag{16.32}$$

The magnetic force $\mathbf{F}_B$, which acts on a charged particle, also has a zero momentum divergence of the time rate of change of momentum (force) $\nabla_p \cdot \mathbf{F}_B = 0$, which is shown next. The Lorentz or magnetic force $\mathbf{F}_B$ acts perpendicular to the direction of the charged particle momentum, so that the particle energy is unchanged, where

$$\mathbf{F}_B = e\dot{\mathbf{q}} \times \mathbf{B} \tag{16.33}$$

Vector identity shows

$$\nabla_p \cdot \mathbf{F}_B = e\nabla_p \cdot (\dot{\mathbf{q}} \times \mathbf{B}) = e\mathbf{B} \cdot (\nabla_p \times \dot{\mathbf{q}}) - e\dot{\mathbf{q}} \cdot (\nabla_p \times \mathbf{B}) \tag{16.34}$$

Magnetic field $\mathbf{B(q)}$ is dependent upon position $\mathbf{q}$ and independent of momentum $\mathbf{P}$, so in the last term of Eq. (16.34),

$$\nabla_p \times \mathbf{B} = 0 \tag{16.35}$$

Velocity $\dot{\mathbf{q}}$ is obtained from the momentum gradient ∇_p of the total, relativistic particle energy E and rest mass m, where

$$\dot{\mathbf{q}} = \nabla_p E = c^2 \frac{\mathbf{P}}{E} \tag{16.36}$$

The total, relativistic particle energy E is

$$E = c\sqrt{P_x^2 + P_y^2 + P_z^2 + m^2c^2} \tag{16.37}$$

From Eq. (16.36), the curl of a gradient is zero:

$$\nabla_p \times \dot{\mathbf{q}} = \nabla_p \times \nabla_p E = 0 \tag{16.38}$$

Consequently, the momentum gradient ∇_p of the magnetic force $\mathbf{F}_B$ is zero in Eq. (16.34), where

$$\nabla_q \cdot \dot{\mathbf{q}} = -\nabla_p \cdot \mathbf{F}_B = -\nabla_p \cdot \dot{\mathbf{P}}_B = 0 \tag{16.39}$$

Without assumption of a Hamiltonian, recall that the continuity equation of Eq. (16.23) is

$$\frac{\partial f}{\partial t} + (\dot{\mathbf{q}} + \dot{\mathbf{P}}) \cdot (\nabla_q + \nabla_p)f + f(\nabla_q + \nabla_p) \cdot \dot{\mathbf{q}} + f(\nabla_q + \nabla_p) \cdot \dot{\mathbf{P}} = 0 \tag{16.40}$$

Since $\mathbf{P}$ and $\mathbf{q}$ are independent variables, then

$$\nabla_q \cdot \dot{\mathbf{P}} = 0 \qquad \nabla_p \cdot \dot{\mathbf{q}} = 0 \tag{16.41}$$

If the force is not dependent on momentum, then with substitution into Eq. (16.40) of Eqs. (16.32) and (16.41), the continuity equation becomes the collision-less Boltzmann equation:

$$\frac{\partial f}{\partial t} + (\dot{\mathbf{q}} + \dot{\mathbf{P}}) \cdot (\nabla_q + \nabla_p) f = 0 \tag{16.42}$$

The Vlasov equation is an equivalent statement of Liouville's theorem, which applies to particles that are acted upon by gravity and magnetic fields. In terms of the phase-space velocity $\mathbf{v}$ and gradient ∇, the continuity equation for momentum-independent forces of gravity and magnetism is obtained by substitution into Eq. (16.42) of Eqs. (16.14) and (16.22), where

$$\frac{\partial f}{\partial t} + \mathbf{v} \cdot \nabla f = 0 \tag{16.43}$$

16.2 Fraction of Source Intensity Intercepted and Focused by Lens

Next, we derive a useful formula for calculating the fraction of source-emitted light, which is intercepted by a lens, and focused onto an image plane (see McFarland, Lanza, & Poulos, 1991). A lens at object distance s_0 from a source intercepts a fraction ε_l of the isotropic emitted source light, and then focuses that light onto an image plane, where

$$\varepsilon_l = \frac{\pi \cdot r_l^2}{4\pi \cdot s_0^2} = \left(\frac{r_l}{2s_0}\right)^2 \tag{16.44}$$

The fraction ε_l of light, which is intercepted by the lens from the source and focused onto the image plane at distance s_i, can be expressed in terms of the lens transverse magnification M, which is also the ratio of the image radius r_{img}, divided by the source radius r_{sou}, where

$$M = \frac{s_i}{s_0} = \frac{r_{\text{img}}}{r_{\text{sou}}} \tag{16.45}$$

The magnification M is also determined by dividing image area A_{img} by source area A_{sou}, where, for a cylindrical lens geometry,

$$M = \frac{\pi (r_{\text{img}})^2}{\pi (r_{\text{sou}})^2} = \frac{A_{\text{img}}}{A_{\text{sou}}} \tag{16.46}$$

The lens-maker equation is invoked:

$$\frac{1}{s_0}+\frac{1}{s_i}=\frac{1}{f} \tag{16.47}$$

One then solves Eq. (16.47) for the object distance s_0 in terms of the lens focal length f and magnification M for image distance s_i, where

$$s_0=\left(1+\frac{s_0}{s_i}\right)f=(1+M^{-1})f \tag{16.48}$$

Substitution for s_0 of Eq. (16.48) into Eq. (16.44) for ε_l, yields the fraction of light, which is intercepted by the lens from the source, and transferred to the image:

$$\varepsilon_l=\left[\frac{r_l}{2(1+M^{-1})\cdot f}\right]^2 \tag{16.49}$$

Given the lens radius r_l, the lens f-stop $f^{\#}$ is the ratio of the lens focal length f, divided by the lens diameter $2r_l$, where $f^{\#}=f/2r_l$. The fraction ε_l of source light, which is transmitted by the lens in vacuum (or air) to the image, can be expressed in terms of the lens f-stop $f^{\#}$ and the image magnification M, where

$$\varepsilon_l=\left[\frac{1}{4(1+M^{-1})f^{\#}}\right]^2 \quad \text{vacuum (air) between source and image} \tag{16.50}$$

If the lens is immersed in a medium with refractive index n_{med}, which extends from the source to the image, then the fraction ε_l of source light, which is intercepted by the lens, and transferred to the image, can be expressed in terms of the magnification of the image over the source object, the lens numerical aperture NA, and the medium refractive index, where

$$\varepsilon_l=\left[\frac{\text{NA}}{2n_{\text{med}}(1+M^{-1})}\right]^2 \quad \text{material medium between source and image} \tag{16.51}$$

16.3 Depth of Field of a Lens

The depth of field of a lens determines the depth of the object that remains in focus, which in our case could be an axial thick object. The depth of field or depth of object (DO) of a lens can be calculated from the lens radius r_l, focal length f, f-stop $f^{\#}=f/2r_l$, the near and far point object distances s_{01} and s_{02} from the lens center plane, the allowable image blur diameter σ_b, and the divergence angle α, which is subtended by the lens diameter from the on-axis image point at r_{im}. From Pedrotti and Pedrotti (1993), the depth of field DO of the lens is defined by

$$\text{DO} = s_{02} - s_{01} \tag{16.52}$$

Measured relative to the thin lens center principle plane, the lens-maker formula relates the pair of near object s_{01} and image s_{i1} point distances and the pair of far object s_{02} and image s_{i2} point distances, where

$$s_{01} = f\left(\frac{s_{i1}}{s_{i1} - f}\right) \tag{16.53}$$

and

$$s_{02} = f\left(\frac{s_{i2}}{s_{i2} - f}\right) \tag{16.54}$$

With substitution of Eqs. (16.53) and (16.54) into Eq. (16.52) the depth of field DO becomes

$$\text{DO} = \left(\frac{s_{i2}}{s_{i2} - f} - \frac{s_{i1}}{s_{i1} - f}\right)f \tag{16.55}$$

The image point midway between the near and far image points is at axial distance s_{im} from the lens center plane, and corresponds to the axial object distance s_{0m} halfway between the corresponding near- and far-object points. A diagonal ray from a point on the lens perimeter of radius r_l, intercepts the center z-axis at the middle image point at axial distance s_{im}, and forms angle α. The diagonal ray then propagates axial distance z_{b} to a point on the image plane at radius σ_{b} from the center z-axis, where

$$\tan\alpha = \frac{2r_l}{s_{\text{im}}} = \frac{\sigma_b}{z_b} \tag{16.56}$$

Hence, the axial distance z_{b} is

$$z = \frac{\sigma_{\text{b}} s_{\text{im}}}{2r_l} = \frac{\sigma_{\text{b}} s_{\text{im}} f^{\#}}{f} \tag{16.57}$$

The near image s_{i1} and far image s_{i2} distances are related to the middle image distance s_{im} via the maximum tolerated distances $\pm z$ for an image blur radius σ_{b}, where

$$s_{i1} = s_{\text{im}} - z \tag{16.58}$$

and

$$s_{i2} = s_{\text{im}} + z \tag{16.59}$$

With substitution of Eqs. (16.58) and (16.59) into Eq. (16.55), the depth of field DO is

$$\mathrm{DO} = \left(\frac{s_{\mathrm{im}} + z}{s_{\mathrm{im}} + z - f} - \frac{s_{\mathrm{im}} - z}{s_{\mathrm{im}} - z - f}\right) f \tag{16.60}$$

After factoring, one obtains

$$\mathrm{DO} = \frac{2zf^2}{(s_{\mathrm{im}} - f)^2 - z^2} \tag{16.61}$$

Substitution into Eq. (16.61) of Eq. (16.57) for z, gives the depth of field DO, where

$$\mathrm{DO} = \frac{2\sigma_{\mathrm{b}} s_{\mathrm{im}} f^{\#} f^3}{f^2 (s_{\mathrm{im}} - f)^2 - (\sigma_{\mathrm{b}} s_{\mathrm{im}} f^{\#})^2} \tag{16.62}$$

Via the lens-maker formula of Eq. (16.47), the middle image distance s_{im} is expressed in terms of the middle object distance s_{0m} (or magnification M) and lens focal length f by

$$s_{\mathrm{im}} = \frac{s_{0m} f}{s_{0m} - f} = f(1 + M) \tag{16.63}$$

Substitution for s_{im} (expressed by s_{0m} and f) in the depth of field DO expression gives

$$\mathrm{DO} = \frac{2\sigma_{\mathrm{b}} s_{0m} (s_{0m} - f) f^{\#} f^2}{f^4 - (\sigma_{\mathrm{b}} s_{0m} f^{\#})^2} \tag{16.64}$$

16.4 Radiation Dose Rate and Radiation Shielding

Neutrons are indirectly ionizing radiation, but generally neutrons have more biological effectiveness per unit of absorbed dose than gamma rays and X-rays. Neutrons deliver the radiation dose to imaged specimens, via the energy transfer from neutrons to charged particles, such as electrons, protons, and nuclei. The radiation kerma is an acronym for "kinetic energy released in material", and is the kinetic energy of radiation, which is released per unit of mass of matter. The kerma K is the sum of the initial kinetic energies E_K of all the charged ionizing particles released by the interaction of the indirectly ionizing neutrons in the specimen mass m or

$$K = \frac{\mathrm{d}E_K}{\mathrm{d}m} \tag{16.65}$$

The kerma K [rad] can be calculated via

$$K = 9.64 \times 10^{15}\Phi \sum_{i=1}^{P} \frac{w_i}{A_i}(E^i_{\mathrm{abs}}\sigma^i_{\mathrm{abs}} + E^i_{\mathrm{incoh}}\sigma^i_{\mathrm{incoh}} + E^i_{\mathrm{coh}}\sigma^i_{\mathrm{coh}}) \tag{16.66}$$

Here,

$$\mathrm{rad} = 100\frac{\mathrm{erg}}{\mathrm{gram}} \tag{16.67}$$

The ith isotope of the irradiated specimen has weight fraction w_i, atomic weight A_i, and total kinetic energy transfer to charged particles for the thermal neutron absorption E^i_{abs}, incoherent neutron scatter E^i_{incoh}, and coherent neutron scatter E^i_{coh}. The kerma K is proportional to the incident neutron fluence Φ [n/cm^2].

The thermal neutron dose rate equivalent H [millirem/hour], which is delivered to the imaged specimen of thickness T and mass density ρ, is given by

$$H = \frac{CQ\Phi_0}{T\rho}\int_0^{E_m} E(1 - \exp\{-T(\mu_{\mathrm{abs}} + \mu_{\mathrm{incoh}} + \mu_{\mathrm{coh}})\})\mathrm{d}E \tag{16.68}$$

The specimen linear absorption coefficient $\mu_{\mathrm{abs}}(E)$ varies with neutron energy E [eV] and the incident neutron flux is $\Phi_0\,[\mathrm{n/cm^2}/s]$. The constant $C = 1.6 \times 10^{-12}$ eV/erg is the conversion factor from eV to ergs, and Q is the quality factor where $Q = 2.3$ for thermal neutrons. The dose rate D [millirem/hour] is multiplied by the quality factor Q, which gives the dose rate equivalent H [millirem/hour] that is delivered to the specimen, where

$$H = QD \tag{16.69}$$

The dose D in rads, which is expressed in units of [energy/gram], is a measure of the absorbed energy per gram of specimen material.

There is an increase of biological effects per unit dose D [milli − rad] with linear energy transfer (LET). The LET is measured in units of energy per neutron propagation distance in the specimen keV/μm. The quality factor Q increases with increasing LET, and the LET changes with the depth of the sample. The dose D [rad] is equal to the kerma K [rad], only if the charged particle equilibrium (CPE) is achieved during irradiation.

When a thermal neutron is scattered elastically and isotropically by a nucleus in the sample, the energy transferred E_{tr} to the struck nucleus of atomic mass M_{A} is given by

$$E_{\mathrm{tr}} = E_n\frac{2M_{\mathrm{A}}M_n}{(M_{\mathrm{A}}+M_n)^2} \tag{16.70}$$

The maximal energy transfer occurs when the neutron mass is equal to the nuclear mass, which is the case for hydrogen (H_1^1 isotope). For hydrogen recoils, the mean energy transfer $\overline{E_{tr}} = 0.5$. For nuclear masses close to the neutron mass, the mean energy transfer is significant: $\overline{E_{tr}} = 0.142$ for carbon nuclei, $\overline{E_{tr}} = 0.124$ for nitrogen nuclei, and $\overline{E_{tr}} = 0.083$ for oxygen nuclei. However, more relevant and prevalent to energy transferred by irradiation of a specimen by thermal neutrons, is the thermal neutron capture by hydrogen and nitrogen, as well as strongly absorbing contrast agents such as boron, lithium, and gadolinium.

The four most prevalent biological elements, namely, hydrogen, carbon, nitrogen, and oxygen, receive more energy transfer per neutron elastic collision, than the less prevalent biological elements of sodium, chlorine, potassium, phosphorous, calcium, and iron. The energy carried away by each recoil nucleus, are absorbed locally, because the maximum ranges of these particles are much less than 1.0 mm. Tissue from a mouse may have 10.4% H, 18.8% C, 3.0% N, and 67.7% O. As such, the absorbed dose will be dominated by elastic scattering from the hydrogen for higher energy nonthermal neutrons (<2.5 MeV), but for thermal neutrons (0.025 eV) the dose rate D and the dose rate equivalent will be dominated by thermal neutron absorption in hydrogen $H^1(n,\gamma)H^2$ and nitrogen $N^{14}(n,p)C^{14}$.

The gamma and neutron cross sections of elements, isotopes, and common compounds and mixtures, as a function of energy, can be found in several online and published sources, which include the following:

1. Thermal neutron capture and capture gamma emission: Lawrence Berkeley National Laboratory, Isotopes Project (http://ie.lbl.gov/ng.html).
2. Coherent, incoherent, absorption cross sections, scatter lengths, and scatter length densities for thermal neutrons in isotopes: National Institute of Science and Technology, Center of Neutron Research, Neutron Scattering Lengths and Cross Sections (http://www.ncnr.nist.gov/).
3. Thermal neutron imaging and radiography: Neutra, Paul Scherrer Institute (PSI), Switzerland (http://neutra.web.psi.ch/facility/index.html).
4. Neutron cross sections of all sorts as a function of energy: Los Alamos Laboratory T2 database ENDF-neutron-ACE (http://t2.lanl.gov).
5. Neutron Cross Section database: Brookhaven National Laboratory CINDA database (http://www.nndc.bnl.gov/nndc/cinda/).
6. Lawrence Berkeley National Laboratory, The Center for X-ray Optics database (http://henke.lbl.gov/optical_constants/).
7. Tabulation of nuclear energy levels and gamma/particle emission for nuclear excited states: Firestone R. B., Shirley, V. S. (1996). *Table of isotopes*, New York, NY: John Wiley & Sons.
8. National Institute of Science and Technology, Physical Measurement Laboratory, M. J. Berger et al., "Stopping-power and range tables for electrons, protons, and helium ions, ESTAR, PSTAR, and ASTAR (http://www.nist.gov/pml/data/star/index.cfm).
9. National Institute of Science and Technology, Physical Measurement Laboratory, M. J. Berger et al., XCOM: photon cross sections database (http://www.nist.gov/pml/data/xcom/index.cfm).

The prompt gamma noise, which is due to thermal neutron absorption in the neutron shielding and neutron moderators, can be reduced by using lithium carbonate (Li_2CO_3) with 2.11 g/cm^3 density) as the moderator, deuterium oxide (D_2O)

for cooling water, and Li_2CO_3 shielding for neutrons and gammas, supplemented by bismuth for gamma shielding, and the use of Teflon (C_2F_4 polymer) in place of hydrogen-based plastics that are exposed to thermal neutrons.

The deuterium (0.5 mb), carbon-12 (3.5 mb), fluorine-19 (9.6 mb), oxygen-16 (0.01 mb), lithium-6 (5.5 mb), and lithium-7 (46 mb) contribute very little to capture gammas, because their cross sections for emission of capture gammas are very small. Deuterium oxide (heavy water) can be used as the inner moderator layer, such that the average neutron energy, which reaches the outer layer of lithium carbonate, is reduced to well below the 1.75 MeV threshold in Li-6 and the 0.55 MeV threshold in Li-7 for inelastic neutron scatter and gamma emission.

Inelastic neutron scatter, where the nucleus is excited from its ground state to a higher nuclear state does not occur in hydrogen and deuterium, and inelastic scatter of fast neutrons below about 2.5 MeV does not occur in carbon, nitrogen, and oxygen, because their first excited states are respectively 4.44 MeV, 2.31 MeV, and 6.05 eV, which are above or too close to 2.5 MeV. Hence, hydrogen (H), deuterium (D), carbon (C), nitrogen (N), and oxygen (O) do not undergo inelastic scatter with 2.5 MeV neutrons, which are produced from the nuclear fusion of two deuterium nuclei (DD fusion produces 2.5 MeV DD neutrons). In fluorine, the threshold for inelastic neutron scatter and gamma emission is about 0.16 MeV. These materials have a very small thermal neutron absorption cross sections and produce almost no prompt gamma emission, with only minor gamma emission due to inelastic fast neutron scatter in the Teflon fluorine (0.9 barn inelastic scatter cross section for the first five fluorine excited nuclear states below 2.5 MeV). Fluorine has only 0.01 barn cross section for thermal neutron absorption. Below the 2.5 MeV DD neutron energy, lithium-7 has a small inelastic cross section of 0.19 barn for gamma emission from only its first excited nuclear state, and lithium-6 has a very small inelastic cross section of 0.07 barn for gamma emission from only its first two excited nuclear states.

Thin-walled Teflon tubes of D_2O can be coiled around odd-shaped structures of fast neutron generators, which produce fast neutrons by nuclear fusion reactions. That is, these fast neutron generators produce fast neutrons by the nuclear fusion of two nuclei, where (1) deuterium (D) and deuterium (D) fuse to produce 2.5 MeV neutrons, (2) deuterium (D) and tritium (T) fuse to produce 14 MeV neutrons, and (3) tritium (T) and tritium (T) fuse to produce a uniform distribution of 0 to 9 MeV neutrons, see Figure 147 in section 1.5 in Cremer (2012a). These coiled Teflon tubes, which contain D_2O, act as a fast neutron moderator and absorber of thermal neutrons with almost zero prompt gamma emission. One can also use deuterated benzene as a moderator, which is normally used as a scintillant. The Li_2CO_3 powder can be compressed into neutron shielding bricks, which are enclosed in Teflon bags, and packed like sandbags for neutron shielding.

Bismuth (2.8×10^{22} atom/cm^3) with cross section 15.6 barn and 5.2 cm one-tenth thickness for absorption for 2.2 MeV gammas, is about eightfold more expensive than lead (3.3×10^{22} atom/cm^3) with cross section 15.3 barn and 4.6 cm one-tenth thickness for absorption of 2.2 MeV gammas. Bismuth has about 88% of the gamma shielding capability as lead. However, the environmental cost and toxicity of lead, provides the incentive to choose bismuth, rather than lead as a gamma shielding alternative overall. The author observed this directly as former engineering manager of the former Reactor Experiments subsidiary of Thermo-Fisher, which began replacing lead with bismuth in some of its shielding products, and faced costly issues with the cleanup of lead in its manufacturing facility. Bismuth is a nontoxic metal used in casting precise molds because of its low melting point (271°C) and expansion upon cooling. The bismuth can be easily cast into bricks and/or custom-shaped shielding. Also, one can use bismuth bead packed in Teflon bags and packed like sandbags for use as gamma shielding.

16.5 Imaging System Modulation Transfer Function (MTF)

Next, we determine the modulation transfer function MTF(f), which is a popular image-quality metric that reflects the imaging system's amplitude response to sinusoidal patterns of spatial frequency f. The spatial frequency f characterizes contrast variation, or the spatial detail of an object, and the reciprocal of the spatial frequency, is the spatial period $T = 1/f$, which corresponds to the object spatial detail. The MTF(f) describes the range of detail, expressed as contrast variation, which can be transferred from the input of the detector to its output. For spatial frequencies greater than zero ($f > 0$), the MTF(f) describes the range of detail of the object, which is expressed as contrast variation that is transferred from the input of the detector to its output. The MTF(f) is the ratio of the modulation at the detector input (scintillator), divided by the corresponding modulation at the detector output (the charged coupled device, or CCD):

$$\mathrm{MTF}(f) = \frac{M_{\mathrm{out}}(f)}{M_{\mathrm{in}}(f)} \tag{16.71}$$

The modulation $M(f)$ is defined as the ratio of the neutron fluence variation $\Delta\Phi$ about its mean, divided by the mean neutron fluence Φ_0, i.e.,

$$M(f) = \frac{\Delta\Phi(f)}{\Phi_0(f)} \tag{16.72}$$

At zero spatial frequency $f = 0$, the MTF(f) is maximized, with a value of 1:

$$\mathrm{MTF}(0) = 1 \tag{16.73}$$

As the spatial frequency increases, the MTF(f) monotonically decreases to zero, where

$$\lim_{f \to \infty} \mathrm{MTF}(f) \to 0 \tag{16.74}$$

The line spread function LSF(x) is the image intensity profile at the detector output as a function of horizontal distance x, which is produced by a vertical (y-direction) neutron or an X-ray line source. The y-vertical neutron or X-ray line source in the image is smeared transversely in the x-direction to produce a bandlike diffraction pattern at the output. The detector MTF(f) can be calculated from the magnitude of the Fourier transform line spread function LSF in the image output. One can eliminate the need for a line source of neutrons or X-rays with a knife-edge test.

In this test, a plate opaque to the incident z-directed neutron or photon beam blocks half the beam from striking the detector input surface, such that the half-plane at the detector input surface is covered ($x < 0$), and the other half-plane is uncovered ($x \geq 0$). For example, a thin sheet of gadolinium can be used for thermal neutrons, and a thin sheet of tungsten for X-rays. For fast neutrons, one must use a thick sheet of polyethylene, and for energetic gammas, a thick sheet of tungsten. This produces an image with sharp contrast between the dark and light image regions. The spatial derivative $\mathrm{d}I/\mathrm{d}x$ of the line intensity profile $I(x)$, which is across the light and dark image regions, along the horizontal direction x, is the horizontal line spread function LSF(x). The magnitude of the Fourier transform of the horizontal line spread function LSF(x), yields the horizontal modulation transfer function MTF(f_x), as a function of spatial frequency f_x along x.

As the spatial frequency increases from zero, the MTF(f) decreases from 1 to 0. The sample feature gain $G \cdot$ MTF(f) rolls off toward zero, as the sample spatial frequency f increases to the detector limited spatial frequency, which is defined by the largest pixel diameter in any portion of the detector chain (e.g., scintillator pixel resolution or CCD pixel size). The resolution σ of the imaging system is conservatively defined as the reciprocal of the spatial frequency f_σ, where the modulation transfer function (MTF) falls to 0.15, so that

$$\mathrm{MTF}(f_\sigma) = 0.15 \tag{16.75}$$

The imaging system resolution σ [mm] is the inverse of f_σ [mm^{-1}], where

$$\sigma = \frac{1}{f_\sigma} \tag{16.76}$$

Let us calculate the MTF of a single module of the detector, characterized by a pixel, which can be the resolution element of the scintillator, the diameter of a tapered fiber-optic transferring light from the scintillator to the CCD, or the size of the CCD pixel. Assume a square pixel with edge dimension p_k at detector stage k. The sharpest response of the detector stage k is as wide as 1 pixel of the detector stage, which is modeled as a square pixel with width p_k and pulse intensity $1/p_k$ for the unity area. Assume that a plane wave with an imposed spatial pattern with spatial frequency f is incident on the kth detector stage. The MTF$_k$(f) of the pixilated detector stage k is then

$$\mathrm{MTF}_k(f) = \frac{1}{p_k}\left|\int_{-p_k/2}^{p_k/2} \exp(i \cdot 2\pi \cdot f \cdot x)\mathrm{d}x\right| = \frac{\sin(\pi \cdot f \cdot p_k)}{\pi \cdot f \cdot p_k} \tag{16.77}$$

Suppose that we use a lens in place of a tapered fiber-optic to transfer light from the scintillator to the CCD. We will not derive the MTF of a lens, but rather will only present its expression, as described next. The MTF of a uniformly illuminated circular lens $\mathrm{MTF_L}$ can be expressed in terms of the lens numerical aperture $\mathrm{NA_L}$, which for a lens of diameter D and focal length F_l, is given by

$$\mathrm{NA_L} = \frac{D}{2F_l} \tag{16.78}$$

With wavelength λ of the scintillator-emitted light, the circular lens MTF is expressed as

$$\mathrm{MTF_L}(f) = \left(\frac{2}{\pi}\right) \cdot \left[\cos^{-1}(\phi) - \cos(\varphi)\sin(\phi)\right] \tag{16.79}$$

For more information on circular lens MTF, see Spring and Davidson (2011) and the Nikon Microscopy U web site (http://www.microscopyu.com). Parameter $\phi(f)$ is given by

$$\phi(f) = \frac{\lambda \cdot f}{2 \cdot \mathrm{NA_L}} \tag{16.80}$$

The resulting cutoff spatial frequency f_c for the lens is

$$f_c = \frac{2 \cdot \mathrm{NA_L}}{\lambda} \tag{16.81}$$

The overall theoretical line spread function of the detector LSF(x) is the convolution of line spread functions of all the K detector stages or, equivalently, the product of the stage MTFs, such that the overall detector MTF is

$$\mathrm{MTF}(f) = \mathrm{MTF_L}(f) \prod_{k=1}^{K} \mathrm{MTF}_k(f) \tag{16.82}$$

Consider the range of spatial frequencies f, where the scintillator detector fibers and CCD pixel MTFs decrease from 1 to near 0, that is

$$0 < \mathrm{MTF}_k(f) \ll 1 \tag{16.83}$$

In the same frequency range, the lens MTF is still unity:

$$\mathrm{MTF}_L = 1 \tag{16.84}$$

The overall system MTF is determined by the scintillator fiber detector and the CCD chip. Hence, the effective system MTF can omit the lens MTF, such that

$$\mathrm{MTF}(f) = \prod_{k=1}^{K} \mathrm{MTF}_k(f) \tag{16.85}$$

Assuming three detector stages (the scintillator, the lens, and CCD), the overall detector MTF is

$$\mathrm{MTF}(f) = \frac{\sin(\pi \cdot f \cdot p_i)}{\pi \cdot f \cdot p_i} \cdot \frac{\sin(\pi \cdot f \cdot p_o)}{\pi \cdot f \cdot p_o} \tag{16.86}$$

If a tapered fiber-optic with fiber diameter p_l is used in place of the lens, the system MTF(f) is

$$\mathrm{MTF}(f) = \frac{\sin(\pi \cdot f \cdot p_i)}{\pi \cdot f \cdot p_i} \cdot \frac{\sin(\pi \cdot f \cdot p_l)}{\pi \cdot f \cdot p_l} \cdot \frac{\sin(\pi \cdot f \cdot p_o)}{\pi \cdot f \cdot p_o} \tag{16.87}$$

16.6 Rose Model, Contrast, Detective Quantum Efficiency (DQE), and Noise Power Spectrum (NPS)

In this section, the limitations of the imaging system are discussed in the context of the Rose model. The Rose model provides criteria for the identification of an object feature in the presence of background noise, which will be expressed by the detective quantum efficiency DQE(f). The DQE(f) depends on the both the object and detector, via the contrast of the object feature, surrounding object background, and upon the spatial frequency-dependent detector gain and quantum noise.

The Rose model, developed by Albert Rose of the Radio Corporation of America during the 1940s and 1950s, is a probabilistic model of low-contrast threshold detection between two regions of an image, the "target" and "background". Starting in World War II in the early 1940s into the 1950s, H.R. Blackwell, a U.S. Navy scientist, who was engaged in visual perception studies for sailor identification of enemy vessels at night, developed a database of visual acuity. For example, is a large dim light easier to see than a small bright light, and what level of light and what size of the light source is required? Rose used Blackwell's data in his model.

The Rose model assumes that the number of photons used to image the target and background are Poisson-distributed, an identical number of photons are emitted from background and target regions of the same area A, and contrast C is the ratio of the difference ΔN between the number of photons emitted from the target and background divided by the total number N of detected photons, where

$$C = \frac{\Delta N}{N} \tag{16.88}$$

In the Rose model, the signal-to-noise ratio (SNR) is the ratio of the signal $\Delta N = CN$ divided by the noise $\sqrt{N}$, so that

$$\mathrm{SNR} = C\sqrt{N} \tag{16.89}$$

If Φ is the photon fluence or photons per unit area (flux is fluence per unit time) that arrives at the detector input used to form an image, then

$$N = \Phi A \tag{16.90}$$

where A is the area of the target embedded in the background of the object. The SNR of the target photon signal that arrives at the detector input is

$$\mathrm{SNR} = C\sqrt{\Phi A} \tag{16.91}$$

Rose examined Blackwell's empirical visual acuity data and found that $\mathrm{SNR} = 5 - 7$ was required for a 50% probability of distinguishing the target from its background. That is, in the Rose model, an observer has a probability of 50% for detection of a low-contrast target in a noisy background, if the SNR of the light reaching the observer is 5–7. For further information, see Hasegawa (1991). The Rose model can be applied to thermal or fast neutrons, X-rays, or gammas that arrive at the detector input, which is comprised of a detection medium appropriate for the radiation.

For example, a gadolinium-doped phosphor can image thermal neutrons, X-rays, and gammas. A plastic scintillator sheet or fiber array, or a liquid scintillant capillary array, can detect fast neutrons and gammas, and gadolinium-based thermoluminescent plate or a boron-doped microchannel plate can image thermal neutrons. With the exception of the thermoluminescent plate and film, the other light-emission imaging detectors are usually viewed indirectly via a 45° mirror by a thermoelectric-cooled CCD camera with image intensifier. The CCD camera stacks a large number of image frames of short exposure to limit the dark noise in each frame. Furthermore, image processing can be performed on each image frame before stacking the frames that produce the final image that quantum noise limited. That is, the noise is limited by the object, not the CCD camera electronics.

Consider a fast neutron fluence Φ_0, which is incident on the input plane of the object, and is transmitted through the background-only portion of the object, and transmitted through the portion of the object with the embedded feature. The image contrast of the object feature is determined by the fluence of neutrons transmitted through the background-only portion Φ_b and feature portions Φ_f of the object. At the input face of the scintillator, the total object-transmitted fluence Φ_T is the sum of the object-transmitted, background-only fluence Φ_b and feature fluence Φ_f, where

$$\Phi_T = \Phi_b + \Phi_f \tag{16.92}$$

The contrast C is defined as the difference between (1) the background fluence Φ_b, transmitted through the portion of the object, which does not contain the feature, and (2) the feature fluence Φ_f, transmitted through the portion of the object, which contains the feature, where the difference is divided by the object background fluence Φ_b, so that

$$C = \frac{\Phi_f - \Phi_b}{\Phi_b} = \frac{\Phi_f}{\Phi_b} - 1 \qquad (16.93)$$

The input neutrons from a feature in the object and the surrounding nonfeature portion of the object, are assumed to be uncorrelated. The input noise at the detector is assumed to arise only from the object feature or the background portions of the object surrounding the feature. Neglected are scattered neutrons incident on the detector, which did not originate from the object and its embedded feature. These assumptions allow the detection system to be characterized by a DQE, which directly relates the input SNR SNR_{in} to the output SNR SNR_{out} at the output image of the detector. The detector adds noise to the overall imaging system and degrades the input signal SNR as a function of spatial frequency f [mm^{-1}] of an object feature.

The detective quantum efficiency DQE(f) expresses the decrease of a detected signal SNR_{out} from the detector input SNR_{in}, and is defined as

$$\mathrm{DQE}(f) = \left(\frac{\mathrm{SNR}_{out}}{\mathrm{SNR}_{in}}\right)^2 \qquad (16.94)$$

DQE(f) measures the efficiency of the detector to transfer the sinusoidal spatial variation of a detected signal from the detector input to output, as a function of the spatial frequency f [mm^{-1}]. An ideal detector transfers 100% of the SNR over the range of detected spatial frequencies f to the detector output yielding DQE(f) = 1. For nonideal detectors at a zero spatial frequency, the detected signal SNR is reduced from the detector input to output, because of the detector quantum and dark noise. In real detectors, the DQE(f) is less than 1 for all spatial frequencies f, where

$$\mathrm{DQE}(f) < \mathrm{DQE}(0) = 1 \quad f > 0 \qquad (16.95)$$

The noise power spectrum NPS(f) is the size of the noise signal, as a function of spatial frequency f [mm^{-1}] in units of line-pairs per millimeter. For white noise, the NPS(f) is a constant, as a function of spatial frequency f. The detector noise, which occurs without the presence of an object, must be measured to determine the noise characteristics of the detection system, as a function of spatial frequency, i.e., the detector noise power spectrum NPS(f). Noise is the uncertainty with which a signal of spatial frequency f is reproduced, and it can be characterized by the standard deviation of the intensity measurement obtained at the imaging detector. The Wiener spectrum describes the noise amplitude as a function of spatial frequency f, and the Wiener spectrum equals the Fourier transform of the autocorrelation function of the intensity variations in a uniformly exposed detector output.

The noise power spectrum $\mathrm{NPS}(u, v)$ or the Wiener spectrum, is experimentally measured by recording an image, which is obtained without an object between the neutron (X-ray) source and detector. The resulting image will have variations due to detector nonuniformities and inherent statistical fluctuations in the neutron (X-ray) source beam. If $D(x, y)$ represents the per-pixel electron counts at the CCD camera detector output, then the fluctuations $\Delta D(x, y)$ can be calculated by subtracting the mean per-pixel electron counts D_{avg}, so that

$$\Delta D(x, y) = D(x, y) - D_{\mathrm{avg}} \tag{16.96}$$

The Wiener spectrum is then obtained as the 2-D Fourier transform of the autocorrelation function $R(X, Y)$ of the CCD per-pixel electron counts $D(x, y)$, i.e.,

$$\mathrm{NPS}(u, v) = W(u, v) = \iint R(X, Y)\exp(-2\pi \cdot \mathrm{i}uX)\exp(-2\pi \cdot \mathrm{i}vY)\mathrm{d}X\,\mathrm{d}Y \tag{16.97}$$

The 2-D autocorrelation function $R(X, Y)$ is given by

$$R(X, Y) = \iint \Delta D(x, y)\Delta D(x + X)\Delta D(y + Y)\mathrm{d}x\,\mathrm{d}y \tag{16.98}$$

As a quick and crude approximation of the $\mathrm{NPS}(u, v)$, one can take a series of images with no object, subtract the average background level, and then take the Fourier transform magnitude to provide the noise power as a function of the spatial frequencies u and v in the x- and y-directions. For further information on DQE, NPS, and imaging system performance, see Cunningham and Shaw (1999).

16.7 Derivation of DQE from NPS, Contrast, Gain, and MTF

The expression of the $\mathrm{DQE}(f)$ for a neutron or X-ray imaging system is derived in this section. One assumes light emission from a radiation-to-light-emission transducer (e.g., phosphor, microchannel plate, and plastic scintillator) that is viewed by a CCD camera with an image intensifier (a low noise, high-gain light amplifier). This analysis or its approach to derive the $\mathrm{DQE}(f)$ can be applied to thermal or fast neutron, gamma, or X-ray imaging refractive or radiographic systems. In refractive X-ray or thermal neutron imaging or in X-ray or thermal neutron radiography, one can use a gadolinium-based phosphor as the radiation-to-light-emission transducer, which is imaged by a CCD camera. In a fast neutron radiographic imaging system, a plastic sheet scintillator or a pixilated scintillator (comprised of an array-solid fiber-optic scintillators, or an array of hollow capillary arrays filled with liquid scintillant) is viewed with a CCD camera.

One assumes that a source of X-rays or neutrons illuminates the target and background regions of the object. The transmitted, scattered, refracted, or diffracted neutrons or X-rays then travel to the detector input. At the detector input is a transducer, which converts the incident radiation to light, and the emitted light is imaged by a CCD camera with an image intensifier. Given the detector system gain, MTF, the noise or uncertainty of the detector transduction of X-ray or neutron radiation to light, and the contrast of the target with the background, one can obtain the DQE(f) as a function of the spatial frequency f of the target features as follows.

The detected object feature neutron signal S_{in} at the detector input face of area A_{d} is

$$S_{\mathrm{in}} = (\Phi_{\mathrm{f}} - \Phi_{\mathrm{b}}) \cdot A_{\mathrm{d}} \tag{16.99}$$

Associated with the object feature signal S_{in}, is the input Poisson noise N_{in} at the detector input face given by

$$N_{\mathrm{in}} = \sqrt{\Phi_{\mathrm{b}} A_{\mathrm{d}}} \tag{16.100}$$

The input Poisson noise at the input surface of the detector N_{in} is proportional to the square root of the fluence transmitted by the background portion of the object $\sqrt{\Phi_{\mathrm{b}}}$ that arrives at the detector input. The object feature $\mathrm{SNR_{in}}$ at the detector input surface , is the ratio of the input signal $\mathrm{S_{in}}$ of Eq. (16.99), divided by the input noise $\mathrm{N_{in}}$ of Eq. (16.100), where

$$\mathrm{SNR_{in}} = \frac{\Phi_{\mathrm{f}} - \Phi_{\mathrm{b}}}{\sqrt{\Phi_{\mathrm{b}}}} \sqrt{A_{\mathrm{d}}} \tag{16.101}$$

In terms of contrast C, the object feature SNR at the input surface $\mathrm{SNR_{in}}$ of the detector, with substitution of Eq. (16.93) into Eq. (16.101), becomes

$$\mathrm{SNR_{in}} = C\sqrt{\Phi_{\mathrm{b}} A_{\mathrm{d}}} \tag{16.102}$$

The signal neutrons are amplified by the frequency-dependent detector gain $G_{\mathrm{s}}(f)$, which is a product of the frequency-independent gain G_{d} and the frequency-dependent modulation transfer function MTF(f) of Eq. (16.71). The gain $G_{\mathrm{s}}(f)$ of the object feature details as a function of spatial frequency f is then

$$G_{\mathrm{s}}(f) = G_{\mathrm{d}} \cdot \mathrm{MTF}(f) \tag{16.103}$$

The frequency-independent (constant or DC) gain G_{d} is a product of the frequency-independent gains of the detector modules between the input and output of the detection system. The output image signal electrons S_{out} are produced by the neutrons from the object feature, where

$$S_{\mathrm{out}} = G_{\mathrm{d}} \cdot \mathrm{MTF}(f) \cdot \varepsilon_n (\Phi_{\mathrm{f}} - \Phi_{\mathrm{b}}) \cdot A_{\mathrm{d}} \tag{16.104}$$

The output image electron noise N_{out} is equal to the square root of the spatial frequency-dependent noise power spectrum NPS(f), where

$$N_{\text{out}} = \sqrt{\text{NPS}} \tag{16.105}$$

Added in quadrature are the three noise components at the output image, which sum up to equal NPS(f):

$$\text{NPS}(f) = \sigma_{\text{nn}}^2 + \sigma_{\text{np}}^2 + \sigma_{\text{nd}}^2 \tag{16.106}$$

The first noise contribution σ_{det}, which is the associated Poisson uncertainty of the detected neutrons, is

$$\sigma_{\text{det}} = \sqrt{\varepsilon_{\text{n}} \Phi_{\text{T}} A_{\text{d}}} \tag{16.107}$$

The probability that a thermal or fast neutron or X-ray or gamma photon will interact with the detection medium is ε_n. The Poisson uncertainty σ_{det} of incident neutrons is amplified by the overall frequency-dependent gain of the detection system $G_s(f)$ of Eq. (16.103). The resultant frequency-dependent noise contribution $\sigma_{\text{nn}}(f)$ to the electron noise in the output image is due to the uncertainty of neutron arrival at the detector input:

$$\sigma_{\text{nn}}(f) = \sigma_{\text{det}} G_s(f) = G_{\text{d}} \cdot \text{MTF}(f) \cdot \sqrt{\frac{\varepsilon_{\text{n}} \Phi_{\text{T}} A_{\text{d}}}{n_{\text{p}}}} \tag{16.108}$$

The second noise contribution results in the Poisson uncertainty of the light emission from the radiation-to-light transducer at the detector input. The Poisson uncertainty of detected neutrons per-pixel σ_{det} is multiplied by the Poisson uncertainty of the transducer photon amplification gain σ_{pho}, where

$$\sigma_{\text{pho}} = \sqrt{G_{\text{d}}} \tag{16.109}$$

The output image noise contribution σ_{np} from the transducer light emission is independent of the object and thus is independent of the spatial frequency f of the object details. The electron noise contribution to the output image σ_{np} from the detector amplification of the statistical fluctuations of the neutron-generated light photons in the transducer is then obtained via Eqs. (16.107–16.108), where

$$\sigma_{\text{np}} = \sigma_{\text{det}} \sigma_{\text{pho}} = \sqrt{G_{\text{d}} \varepsilon_{\text{n}} \Phi_{\text{T}} A_{\text{d}}} \tag{16.110}$$

The third contribution to the electron noise in the output image is from the electron dark noise σ_{nd} of the image intensifier and the CCD sensor (chip). The electron dark noise σ_{nd} is independent of the incident neutrons and the detection system gain. Dark noise σ_{nd} is the dark noise electrons per output image frame n_r, and is the sum of the dark noise electrons from the light amplifier (dual multichannel plate or dual MCP) and the CCD chip, where

$$\sigma_{nd} = n_r \tag{16.111}$$

The detector is assumed to be quantum noise limited, such that the noise σ_{nd} from the n_r dark noise electrons in an output image frame is much smaller than the sum of the two quantum noise components of statistical fluctuation of the detected neutrons σ_{nn} and transducer light emission σ_{np}, where

$$\sigma_{nd} \ll \sigma_{nn} + \sigma_{np} \tag{16.112}$$

With application of Eq. (16.112), and substitution of Eqs. (16.108) and (16.110) into Eq. (16.106), the quantum noise limited output noise $N_{out}(f)$ is then approximated as

$$N_{out}(f) = \sqrt{\varepsilon_n \Phi_T A_d \cdot (G_d^2 \cdot \mathrm{MTF}^2(f) + G_d)} \tag{16.113}$$

From Eq. (16.104) for the output signal $S_{out}(f)$, and Eq. (16.113) for the output noise $N_{out}(f)$, the output signal to noise $\mathrm{SNR}_{out}(f)$ in the quantum noise limited regime is

$$\mathrm{SNR}_{out}(f) = \frac{S_{out}}{N_{out}} = \frac{G_d \cdot \mathrm{MTF}(f)}{\sqrt{G_d^2 \cdot \mathrm{MTF}^2(f) + G_d}} \cdot \frac{\Phi_f - \Phi_b}{\sqrt{\Phi_T}} \cdot \sqrt{\varepsilon_n A_d} \tag{16.114}$$

With substitution of Eq. (16.101) and Eq. (16.114), the detection system's unnormalized DQE(f) is

$$\mathrm{DQE}(f) = \frac{\mathrm{SNR}_{out}^2(f)}{\mathrm{SNR}_{in}^2} = \frac{G_d \mathrm{MTF}^2(f)}{G_d \mathrm{MTF}^2(f) + 1} \cdot \frac{\Phi_b}{\Phi_T} \cdot \varepsilon_n \tag{16.115}$$

Recall Eq. (16.93) for contrast C. The probability of neutron or X-ray scatter in the transducer is ε_n, and the ratio of the object-transmitted background fluence Φ_b, divided by the total object-transmitted fluence Φ_T is

$$\frac{\Phi_b}{\Phi_T} = \frac{\Phi_b}{\Phi_f + \Phi_b} = \frac{1}{C+2} \tag{16.116}$$

With substitution of Eq. (16.116) into Eq. (16.115), the un-normalized DQE(f) can be expressed in terms of the C between the object feature and background, where

$$\mathrm{DQE}(f) = \frac{\mathrm{SNR}_{out}^2(f)}{\mathrm{SNR}_{in}^2} = \frac{G_d \mathrm{MTF}^2(f)}{G_d \mathrm{MTF}^2(f) + 1} \cdot \frac{\varepsilon_n}{C+2} \tag{16.117}$$

From Eq. (16.93), one obtains the range of contrast values, where contrast $C = -1$ for zero-transmitted sample feature fluence $\Phi_f = 0$, with non-zero background transmitted fluence $\Phi_b \neq 0$, and contrast $C = \infty$ for zero-transmitted background fluence $\Phi_b = 0$, with nonzero sample feature fluence $\Phi_f \neq 0$. Hence,

$$-1 \leq C \leq \infty \tag{16.118}$$

The case of maximum negative contrast $C = -1$ corresponds to a different choice of the noise N_{in} at the detector input. In the alternative choice, the object noise, which arrives at the detector input face, is assumed to be proportional to the square root of the total object-transmitted neutron or X-ray fluence $\sqrt{\Phi_T}$ from both the feature portion of the object and the background portions of the object, rather than from only the background portion, where

$$N_{in} = \sqrt{\Phi_T A_d} = \sqrt{(\Phi_f + \Phi_s) A_d} \tag{16.119}$$

The resulting object feature's SNR_{in}, which arrives at the input surface of the detector, is then

$$SNR_{in} = \frac{\Phi_f - \Phi_b}{\sqrt{\Phi_T}} \sqrt{A_d} \tag{16.120}$$

With substitution of Eq. (16.114) and Eq. (16.120), the resulting un-normalized DQE(f) becomes

$$DQE(f) = \frac{SNR_{out}^2(f)}{SNR_{in}^2} = \frac{G_d MTF^2(f)}{G_d MTF^2(f) + 1} \cdot \varepsilon_n \tag{16.121}$$

The normalized DQE is now determined, which applies to either choice of detector input noise. At zero spatial frequency, the MTF(0) is unity, such that Eq. (16.121) becomes

$$DQE(0) = \frac{G_d^2 \cdot \varepsilon_n}{G_d^2 + G_d} = \frac{G_d}{G_d + 1} \cdot \frac{\varepsilon_n}{C + 2} \tag{16.122}$$

As a result of Eqs. (16.120) and (16.121), the normalized $DQE_n(f)$ is defined by

$$DQE_n(f) = \frac{(G_d + 1) MTF^2(f)}{G_d MTF^2(f) + 1} \tag{16.123}$$

For zero spatial frequency $f = 0$, since MTF(0) = 1, then the normalized $DQE_n(0)$ is unity, where

$$DQE_n(0) = 1 \tag{16.124}$$

Given $\mathrm{MTF}(f) \leq 1$, for all spatial frequencies $f \geq 0$, the normalized $\mathrm{DQE}_n(f)$ is equal to the squared $\mathrm{MTF}(f)$, if the zero-spatial-frequency gain G_d; that is, the DC gain is sufficiently small, where $G_\mathrm{d} \ll 1$, so that

$$\mathrm{DQE_n}(f) = \mathrm{MTF}^2(f) \tag{16.125}$$

when

$$G_\mathrm{d} \ll 1 \tag{16.126}$$

Given $\mathrm{MTF}(f) \leq 1$, for all spatial frequencies $f \geq 0$, the normalized $\mathrm{DQE}_n(f)$ is unity for the detector-resolvable frequencies $f \leq f_r$, where $0.15 = \mathrm{MTF}(f_r)$, if the DC gain is sufficiently large, where $G_\mathrm{d} \gg 1$, so that

$$\mathrm{DQE_n}(f < f_\mathrm{r}) = 1 \tag{16.127}$$

when

$$0.15 = \mathrm{MTF}(f_\mathrm{r}) \gg \sqrt{\frac{1}{G_\mathrm{d}}} \tag{16.128}$$

However, the un-normalized $\mathrm{DQE}(f)$ is less than unity because of its dependence on the probability of neutron interaction (scatter) in the transducer, where

$$\mathrm{DQE}(f \leq f_r) = \frac{\varepsilon_n}{C+2} \qquad \text{for } 0.15 \gg \sqrt{\frac{1}{G_\mathrm{d}}} \tag{16.129}$$

16.8 Criteria for Sample Feature Imaging

In this section, we examine the range of source parameters and detector parameters, which is necessary to image features of interest in the sample object. The criteria for imaging sample features are expressed in terms of the neutron source yield Y, exposure time τ, detector distance from source r_0, sample background thickness t and linear attenuation μ, feature contrast C and spatial frequency f, and the imaging detection system $\mathrm{DQE}(f)$ and detector area A_d.

The sample is modeled as a slab of material with neutron linear attenuation μ_b [cm^{-1}] and thickness t_b. The sample slab serves as a background for a small embedded feature, where the feature has thickness t_f and linear attenuation μ_f [cm^{-1}]. Given exposure time τ [s], the neutron (X-ray) fluence Φ_0 [n/cm^2], which is incident on the detector area A_d that is at distance r_0 from the isotropic source of yield of $Y\ [n/s]$, is given by

$$\Phi_0 = \frac{Y\tau}{4\pi \cdot r_0^2} \tag{16.130}$$

The fluence Φ_f, transmitted through the feature portion of the sample to the input surface of the detector, is

$$\Phi_f = \Phi_0 \exp[-\mu_b(t_b - t_f)] \cdot \exp(-\mu_f t_f) \tag{16.131}$$

The fluence Φ_b, transmitted through the sample without the feature to the input surface of the detector, is

$$\Phi_b = \Phi_0 \exp(-\mu_b t_b) \tag{16.132}$$

Substitution for sample background Φ_b of Eq. (16.132), and feature fluence Φ_f of Eq. (16.131) into Eq. (16.93) for contrast C, yields

$$C = \exp[t_f(\mu_b - \mu_f)] - 1 \tag{16.133}$$

where

$$C = t_f(\mu_b - \mu_f) \quad \text{for} \quad t_f(\mu_b - \mu_f) \ll 1 \tag{16.134}$$

Identification of a sample feature in the Rose model requires SNR_{out} in the output image to exceed 5, where

$$\text{SNR}_{\text{out}} \geq 5 \tag{16.135}$$

The minimum source Ψ_0 yield, which is required to detect a feature in the output image, can be calculated from the detector unnormalized DQE(f) and sample SNR_{in} that arrives at the detector, where

$$\text{SNR}_{\text{in}} \geq \frac{5}{\sqrt{\text{DQE}(f)}} \tag{16.136}$$

The input SNR_{in} is expressed in terms of the contrast C between the sample feature and the background, and the transmitted fluence Φ_b from the background portion of the sample, which arrives at the input detection surface area A_d:

$$\text{SNR}_{\text{in}} = C\sqrt{\Phi_b A_d} \tag{16.137}$$

Substitution into Eq. (16.137) of Eq. (16.132) for sample background fluence Φ_b, yields the SNR_{in} at the detector input:

$$\text{SNR}_{\text{in}} = C\sqrt{\frac{Y\tau}{4\pi \cdot r_0^2} \cdot \exp(-\mu_b t_b) A_d} \tag{16.138}$$

The thermal neutron source yield Y and exposure time τ, the sample feature contrast C, the neutron transmission through the sample background $\exp(-\mu_b t_b)$, and the transducer (e.g., phosphor or scintillator) area A_d at the detector input, together determine the conditions for resolving sample feature details at spatial frequency f. With substitution of Eq. (16.138) into Eq. (16.136), the sample feature detection condition is

$$\frac{C^2}{25} \cdot \frac{Y\tau}{4\pi \cdot r_0^2} \cdot \exp(-\mu_b t_b) \cdot A_d \cdot \mathrm{DQE}(f) \geq 1 \tag{16.139}$$

Equation (16.139) can be transposed to obtain the yield-exposure $Y\tau$ product required for imaging features in a sample via neutron or X-ray microscopy. Given an imaging detection system with a specified or measured DQE(f), as a function of spatial frequency f, one can calculate the minimum product $Y\tau$ of source yield Y and exposure time τ, which is required to form an image of area A_d at detector distance r_0 from the source with a neutron or X-ray lens. One must also specify the sample feature contrast C and spatial frequency f and the linear attenuation μ_b and thickness t_b of the sample background, where the feature is embedded. The minimum yield-exposure-time product $Y\tau$ is then

$$Y\tau \geq \frac{100\pi}{\mathrm{DQE}(f)} \cdot \frac{r_0^2}{\mathrm{A_d}} \cdot \frac{\exp(\mu_b t_b)}{C^2} \tag{16.140}$$

Next, assume a sufficiently large zero-frequency gain G_d. With substitution Eq. (16.129) into Eq. (16.140), and then transposing, the detection of a sample feature with spatial frequency f, which is within the resolution capability of the detection system ($f \leq f_r$), must then satisfy the condition

$$\frac{Y\tau}{4\pi \cdot r_0^2} \cdot \exp(-\mu_b t_b) \cdot A_d \cdot \frac{C^2 \varepsilon_n}{C+2} \geq 25 \quad \text{for} \quad 0.15 \gg \sqrt{\frac{1}{G_d}} \tag{16.141}$$

16.9 Imaging System Performance—Receiver Operating Characteristic (ROC), Positive and Negative Likelihood, and Positive and Negative Predictive Value Curves

Performance testing of imaging detectors includes the comparison of the image features with the actual object features. That is, does the image correctly show the actual object features, or are there image artifacts that provide misleading information regarding the object structures? Or, does the image fail to include important object features due to noise, lack of contrast, or lack of resolution? To test the performance of the imaging system, one can employ Rose model test objects, which contain various ranges of hole thicknesses and hole diameters. A Rose model test object is a wedge with several rows of holes. Each row of holes has the same mean thickness, but the hole diameters vary as fractions or multiples of the row thickness. The mean thickness of each column of the Rose phantom varies from thin to thick (see Hasegawa, 1991).

For example, for fast or thermal neutron radiographic imaging or thermal neutron refractive imaging, the Rose model test object will be a wedge of hydrogen-rich polyethylene with holes of various diameters and thicknesses. In a row of holes with varying diameters (i.e., detail) but with a constant thickness (i.e., contrast), a range of detail is presented for a set contrast. In a column of holes with the same diameter (i.e., detail) but varying thickness (contrast), a range of contrast is presented for a set detail. Detail can be quantified by the spatial frequency (line per millimeter). For X-ray or gamma radiographic or X-ray refractive imaging, the Rose phantom "slice of Swiss cheese" is made of lead or bismuth. Stepped wedges composed of different materials such as Stepped phantoms, which are comprised of a staggered stack of different materials, such as steel, lead, Lucite, aluminum, and polyethylene, can also be used to demonstrate contrast in X-ray/gamma or fast/thermal neutron radiography, or X-ray and thermal neutron microscopy.

Now that the test object is obtained, let the tests of imaging fidelity begin. The perfect discrimination or resolution of an imaged object feature from the imaged object background rarely occurs. There is usually an overlap or blurring of the imaged object feature with its surrounding background via noise from the object, its surroundings, as well as the imaging detection system. A diagnostic of an imaging detection system performance is the receiver operating characteristic (ROC), which was developed in World War II for the detection of enemy targets in the background of the battlefield. The ROC is used to determine the fidelity of an imaging detection system to discriminate an imaged object feature from the imaged object background. In particular, the ROC is simply a plot of the true positive rate TPR (fraction of "hits" out of the sum of "hits" and "miss") vs. the false positive rate FPR (the fraction of "false alarms" out of the total of "false alarms" and "correct rejections").

For an imaging detection system, the TPR or sensitivity is defined as the probability that an identification or evaluation of image feature is a true representation of the true object feature. The TP is the fraction of "hits" or successful identifications of imaged object features that are correct, and FN is the fraction of "misses" or failures to identify object features that are actually present in the object. The FN is the type II error of failing to image object features, or the failure to identify actual object features in the imaging system. Sensitivity or the TPR, is the fraction of true positives TP ("hits") out of the total of the true positives ("hits") TP and false negatives ("misses") FN:

$$\mathrm{TPR} = \frac{\mathrm{TP}}{\mathrm{TP} + \mathrm{FN}} \quad \text{sensitivity} \tag{16.142}$$

FPR is the fraction of false positives FP out of the sum of false positives FP and true negatives TN, where

$$\mathrm{FPR} = \frac{\mathrm{FP}}{\mathrm{FP} + \mathrm{TN}} \tag{16.143}$$

The specificity S of an imaging system is the probability of successful discrimination, i.e., successful rejection of false object features that are not present in the object. The specificity S is the fraction of true negatives TN out of the total of false positives FP and true negatives TN, where

$$S = 1 - \mathrm{FPR} = \frac{\mathrm{TN}}{\mathrm{FP} + \mathrm{TN}} \quad \text{specificity} \tag{16.144}$$

The false positive FP is a type I error. Specificity is an indicator of noise or false signals from the detector, object, or surroundings that are mistaken as a real object feature.

ROC is a plot of TPR vs. FPR, where

$$\mathrm{ROC} \Rightarrow \mathrm{TPR} = \frac{\mathrm{TP}}{\mathrm{TP} + \mathrm{FN}} \quad \text{vs.} \quad \mathrm{FPR} = \frac{\mathrm{FP}}{\mathrm{FP} + \mathrm{TN}} \tag{16.145}$$

That is, ROC measures (1) the TPR or the number of correct identifications of actual object features ("hits" TP) in the detector image, per number of actual sample features, which were identified ("hits" TP) or not identified ("misses" FN), as a function of (2) the FPR or the number of incorrect positive identifications ("false alarms" FP) of nonexistent sample features in the detector image, per number of nonexistent sample features, which were incorrectly identified ("false alarms"FP) or not identified ("correctly rejected" TN). The TPR is plotted on the ordinate or vertical, y-axis, and the FPR is plotted on the abscissa, or horizontal x-axis.

The ROC curve is the relative operating characteristic curve for TPR vs. FPR. Equivalently, the ROC curve is a plot of TPR, divided by the specificity S. Each point on the ROC curve corresponds to a decision threshold DT, which is defined by the ratio of the TPR or sensitivity, divided by the specificity S, where

$$\mathrm{DT} = \frac{\text{sensitivity}}{\text{specificity}} = \frac{\mathrm{TPR}}{S} = \frac{\mathrm{TPR}}{1 - \mathrm{FPR}} \tag{16.146}$$

In the case of perfect detector discrimination, where there are only "hits" (TP) with "correct rejections" (TN) accompanied by no "misses" ($\mathrm{FN} = 0$), and no "false alarms" ($\mathrm{FP} = 0$), $\mathrm{TPR} = 1$ and $\mathrm{FPR} = 0$ result. This point ($\mathrm{TPR} = 1, \mathrm{FPR} = 0$) appears at the upper-left corner of the ROC plot, where a zero false positive rate ($\mathrm{FPR} = 0$) corresponds to a specificity of 1 ($S = 1$). That is, perfect discrimination for a detection system has a detector sensitivity of 1 ($\mathrm{TPR} = 1$) and a detection system specificity of 1 ($S = 1$), and the decision threshold for the detection system is 1 ($\mathrm{DT} = 1$). The detection system includes the source, the sample, the imaging detector, and the surroundings. Hypothetically, particular configurations of the source and detector and surroundings, which include particular samples and their embedded targets, can lead to perfect discrimination.

Consider now the realistic case of imperfect discrimination for a detection system. Each point on the ROC curve represents a specific ratio of sensitivity TPR, divided by specificity S. Changing the sample or embedded targets in the sample, or reducing the source output, changing the detector radiation to a light transducer or the introduction of a noisy light amplifier in the detector, or introducing a noisy source, environment, or sample, can lead to a detection system with imperfect discrimination. In the case of an imperfect detection system, the ROC curve moves toward that of a 45° positive-slope diagonal line, which corresponds to random guessing:

$$\text{TPR} = \text{FPR} \quad \text{random guess} \tag{16.147}$$

An excellent diagnostic performance for the detection system, occurs for those detection system ROC curves, which occupy the upper-left corner of the plot, above the 45°, positive-slope diagonal line. A poor diagnostic performance for the detection system, occurs for those detection system ROC curves, which occupy the lower-right corner, below the 45°, positive-slope diagonal line of random guess.

The positive likelihood ratio PLR is TPR, divided by FPR (or, equivalently, $1 - S$), where

$$\text{PLR} = \frac{\text{TPR}}{\text{FPR}} = \frac{\text{TPR}}{1 - S} = \frac{\text{TP}}{\text{FN} + \text{TP}} \Big/ \frac{\text{FP}}{\text{FP} + \text{TN}} \tag{16.148}$$

The positive likelihood ratio (PLR) is the ratio of the fraction of TP ("hits") out of the total of TP ("hits") and FN ("misses"), divided by the fraction of FP ("false alarms") out of the total of FP ("false alarms") and TN ("correct rejections"). That is, PLR is the number of correct identifications of actual object features ("hits," TP), per number of actual object features, which were identified correctly ("hits," TP) or were not identified ("misses," FN), divided by the number of incorrect identifications of nonexistent object features by the detection system ("false alarms," FP), per number of nonexistent object features, which were incorrectly identified ("false alarms," FP) or were correctly not identified ("correct rejections," TN).

FNR is the fraction of FN "misses" out of the total of FN "misses" and TP "hits", where

$$\text{FNR} = \frac{\text{FN}}{\text{TP} + \text{FN}} \tag{16.149}$$

The negative likelihood ratio (NLR) is the FNR divided by S:

$$\text{NLR} = \frac{1 - \text{TPR}}{1 - \text{FPR}} = \frac{\text{FNR}}{S} = \frac{\text{FN}}{\text{TP} + \text{FN}} \Big/ \frac{\text{TN}}{\text{FP} + \text{TN}} \tag{16.150}$$

NLR is the fraction of false negatives FN ("misses") out of the sum of true positives TP ("hits") and FN ("misses"), divided by the fraction of true negatives TN ("correct rejections") out of the sum of false positives FP ("false alarms") and TN ("correct rejections"). That is, NLR is the number of failed identifications of actual object features ("misses," FN), per number of actual object features, which were identified correctly ("hits," TP) or were not identified ("misses," FN) divided by the number of correct rejections of nonexistent object features by the detection system ("correct rejections," TN), per number of nonexistent object features, which were correctly rejected ("correct rejections," TN) or were incorrectly identified ("false alarms," FP).

The positive predictive value (PPV) is the fraction of true positives TP ("hits") out of the total of the true positives ("hits," TP) and false positives ("false alarms," FP):

$$\mathrm{PPV} = \frac{\mathrm{TP}}{\mathrm{TP} + \mathrm{FP}} \tag{16.151}$$

PPV is the probability that one images and detects a sample feature ("hit," TP), provided that a positive detection has occurred, either as a true positive ("hit," TP) or as a false positive ("false alarm," FP).

NPV is the fraction of true negatives ("correct rejections," TN) out of the total of the true negatives ("correct rejections," TN) and false negatives ("misses," FN), where

$$\mathrm{NPV} = \frac{\mathrm{TN}}{\mathrm{TN} + \mathrm{FN}} \tag{16.152}$$

NPV is the probability that one correctly rejects an image feature ("correct rejection," TN), provided that a negative detection has occurred, either as a true negative ("correct rejection," TN) or as a false negative ("miss," FN). For further information on statistics, see Kirkwood and Sterne (2003). Also, see the online MedCalc statistics software manual from MedCalc of Mariakerke, Belgium (http://www.medcalc.org).

16.10 Reliable Detection of a Signal in the Presence of Background and Noise

The accurate detection of real and weak radiation sources, which are in the presence of natural noise and background, is considered in this section. The reliable detectable activity (RDA) is the amount of gamma activity, which a detector or system (combination of the detectors) can be expected to detect reliably. The detector or system of detectors measures gross count rates, which is the sum of the background count rate and the net count rate. The background counts are emitted from natural occurring cosmic and terrestrial radioactive sources. The RDA equation and concepts for gamma detection also apply to detection systems for X-rays and fast and thermal neutrons.

One can solve for the minimum detectable count rate R_m for a specified background, count time, false alarm rate, and confidence by the solution of an implicit expression for the minimum detectable count rate, via iteration or by the intersection of a pair of plotted curves. The following implicit RDA equation for R_m assumes that the photon noise background flux, which reaches the detector from the test sample, is a Gaussian probability distribution. The following equation for the minimum detectable count rate R_m can be solved by iteration or graphically:

$$R_m = F_A\sqrt{\frac{B}{T}} + Z\sqrt{\frac{R_m + B}{T}} \tag{16.153}$$

The variables are defined as follows. Here, T is the count time (sec) for the radiation measurement by the detector or system; R_m is the minimum counts per second, which are emitted by the minimum detectable radiation test source; and B is the mean of the Gaussian distributed background radiation, measured in units of counts per second. The sum $R_m + B$ is the mean of the Gaussian distributed background radiation, plus the test source in units of counts per second.

Here, F_A is the false alarm sigma, which is the number of background count standard deviations above the average background count, where a gross count recording will result in a false detection or false acceptance of radiation. For example, $F_A = 2$ corresponds to a false alarm probability of 2.3%. Here, Z is the confidence sigma, which is the number of gross count standard deviations below the average gross count, above which a gross count recording will result in a successful detection or true acceptance of radiation. For example, $Z = 1.28$ corresponds to a probability of detection of 90%. And for example, the minimum detectable count rate of 10 signal photons per second, corresponds to a count time of 10 s in a background of 89 noise photons per second, which has a false alarm probability of 2.3%, and a 90% confidence of detection.

The following is a derivation for this RDA equation, which is used in many radiation monitoring systems to determine the presence of radioactive material, given a specified false alarm rate F_A, confidence Z, and measured background count rate B for count time T. The background count is measured prior to the measurement (for the same count time T), of persons or objects that may be radioactive. The probability of naturally occurring radiation, which produce count numbers in the detector that exceed the alarm level $N > A_{FA}$ (i.e., a "false alarm"), is given by the area under the Gaussian curve, which is to the right of the alarm level A count value. The alarm level A_{FA} (counts) for detector count time T is thus set by specification of the false alarm probability P_{FA}, which corresponds to a specified number of standard deviations F_A that are above the mean counts μ of the Gaussian background radiation, via the relation

$$A_{FA} = \mu + F_A\sqrt{\mu} \tag{16.154}$$

That is, A_{FA} [counts] is equal to the false alarm level, which is the minimum number of counts, which are needed to set off the alarm. The parameter μ [counts] is the average number of background counts, which are recorded by the detector in count time T. One standard deviation σ [counts] of the Gaussian distribution is given by

$$\sigma = \sqrt{\mu} \tag{16.155}$$

Table 16.1 False Alarm Rate F_A

F_A	P_{FA}
Number of standard deviations to the right of the mean M of the Gaussian distribution of detector count numbers	Percent of total area under the Gaussian distribution curve of count numbers that is between the following two points: 1. F_A standard deviations to the right of the mean detector count number M 2. The rightmost point of infinite counts
0.0	50%
1.0	16%
2.0	2.3%
3.0	0.15%
4.0	0.003%
5.0	0.00005%

The false alarm rate F_A can then be expressed as the number of standard deviations above the mean counts μ of the Gaussian distributed background radiation. See Table 16.1, which lists the values for F_A and the corresponding P_{FA}.

Test radiation sources, which produce count numbers that exceed the confidence alarm level A_Z counts for count time T, will be detected with a success probability of P_{det}, which corresponds to confidence level Z, where

$$A_Z = M - Z\sqrt{M} \tag{16.156}$$

The mean counts M is the sum of the background and test radiation. That is, mean counts M is the sum of the background radiation mean μ and the single-valued radiation counts N from the test source, for count time T, where

$$M = N + \mu \tag{16.157}$$

Substitution into Eq. (16.156) of (16.157) for M for the confidence alarm count A_Z, yields

$$A_Z = (N + \mu) - Z\sqrt{N + \mu} \tag{16.158}$$

See Table 16.2 which shows the values of Z and the corresponding P_{det}.

Next, we seek to determine the minimum source strength N_{sou} count, which produces N_{det} counts at the detector in count time T for a specified source-to-detector distance r. This requires that both the confidence alarm level count A_Z and the false alarm level count A_{FA} are equal to the N_{det} counts, i.e.,

$$N_{det} = A_Z = A_{FA} \tag{16.159}$$

Table 16.2 Confidence Level Z

Z	$\boldsymbol{P_{det}}$
Number of standard deviations to the left of the mean value *M* of Gaussian distribution of detector count numbers	Confidence of detection (percent probability of successful detection)
0	50%
0.47	68%
1.28	90%
1.64	95%
2.33	99%

Then, one equates A_{FA} of Eq. (16.154) to A_Z of Eq. (16.158) to obtain

$$\mu + F_A\sqrt{\mu} = (N + \mu) - Z\sqrt{N + \mu} \tag{16.160}$$

Next, one solves for the measured detector counts N_{det}, measured in count time T, to obtain the transcendental equation

$$N = F_A\sqrt{\mu} + Z\sqrt{N + \mu} \tag{16.161}$$

Finally, dividing both sides of Eq. (16.161) by count time T, one obtains R_m, the minimum counts per second, which are measured by the detector, and emitted by the minimum detectable radiation test source:

$$R_m = F_A\sqrt{\frac{B}{T}} + Z\sqrt{\frac{R_m + B}{T}} \tag{16.162}$$

In Eq. (16.163), the detector measured background count rate B is the mean, detector measured counts μ of the Gaussian distributed background radiation, divided by detector count time T. The minimum detection count rate R_m of the detector is the measured radiation counts N from the test radiation source, divided by detector count time T, where

$$B = \frac{\mu}{T} \qquad R_m = \frac{N}{T} \tag{16.163}$$

The minimum detectable gamma count rate R_m for a 2.3% false alarm rate ($F_A = 2$), but with 95% confidence ($Z = 1.64$) for a count time of $T = 10$ seconds, is

$$R_m = 2\sqrt{\frac{B}{10}} + 1.64\sqrt{\frac{R_m + B}{10}} \tag{16.164}$$

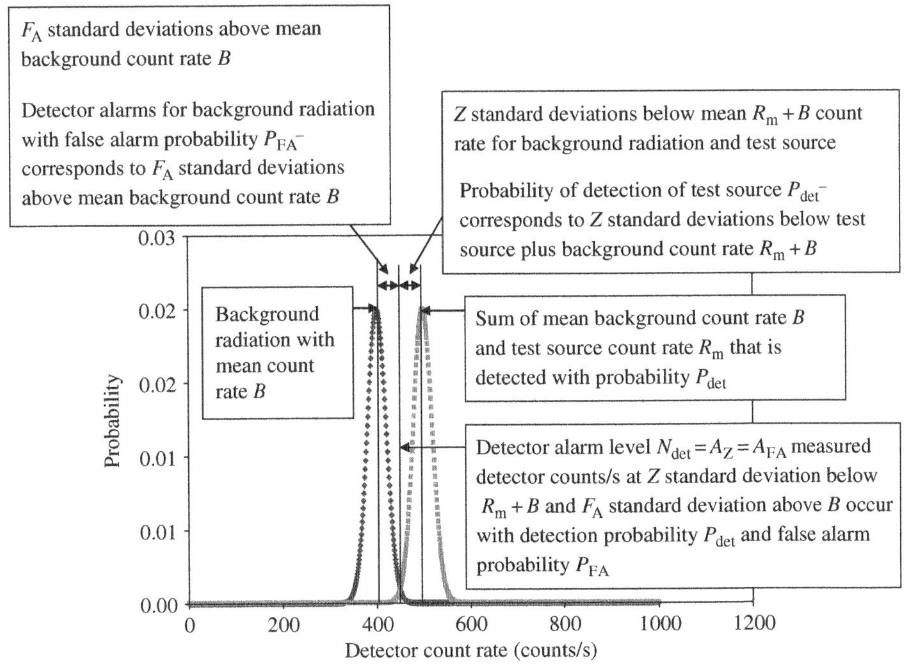

Figure 16.1 The probability of detector counts, which are shown as (1) the Gaussian curve for the detector background count distribution with mean B (counts/s) averaged for count time T (s), and (2) the Gaussian curve for the detector count distribution, which is the sum of a minimum-strength, radiation test source of R_m (counts/s) and the same Gaussian background count distribution with mean B (counts/s) averaged for the same count time T (s). The alarm level N_{det} (counts) is set for a specified false alarm rate F_A (and probability of false alarm P_{FA}) and a specified confidence level Z (and detection probability of a minimum-strength test source P_{det}) within the background for detector count time T.

See Figure 16.1. For example, a background of $B = 10^5\ c/s$ requires a minimum detectable gamma signal count rate of $R_m = 364$ counts$/s$.

References

Bradt, H., & Olbert, S. (2008). *Astrophysical processes*. Cambridge UK: Cambridge University Press.

Cremer, J. T. (2012). In Hawkes P.W. (Ed.), Neutron and X-ray microscopy—Part 1. *Advances in imaging and electron physics* (Vol. 172). London: Academic Press, an imprint of Elsevier Press.

Cunningham, I. A., & Shaw, R. (1999). Signal-to-noise optimization of medical imaging systems. *Journal of the Optical Society of America, A16*(3), 621–632.

Ellis, A. (1999). Black Holes—Part 1—History. *Journal of the Astronomical Society of Edinburgh, 39*, 349.

Hasegawa, B. H. (2001). *The physics of medical X-ray imaging* (2nd ed.). Madison, WI: Medical Physics Publishing.

Kirkwood, B., & Sterne, J. (2003). *Essentials of medical statistics*. Oxford, UK: Blackwell Science.

McFarland, E. W., Lanza, R. C., & Poulos, G. W. (1991). Multi-dimensional neutron computed tomography using cooled charge-coupled devices. *IEEE Transactions on Nuclear Science*, *38*(2), 612–622.

Pedrotti, L., & Pedrotti, L. S. (1993). *Introduction to optics*. Englewood Cliffs, NJ: Prentice Hall.

Shultis, J. K., & Faw, R. E. (1996). *Radiation shielding*. New York, NY: Prentice Hall.

Spring, K. R., & Davidson, M. W. (2011). Modulation transfer function. Nikon Microscopy U, Molecular Expressions, Florida State University, and Nikon, Inc. 2011. <http://www.microscopyu.com>.

17 Neutron and Charged Particle Magnetic Optics

17.1 Charged Particle Motion in Axial Symmetric Magnetic Field—Busch's Theorem

In this chapter, we cover the topics of magnetic optics for neutrons and charged particles. Presented first is charged particle motion in an axial symmetric magnetic field and Busch's theorem, followed by the paraxial ray equation and focusing of charged particles by an axial symmetric magnetic field. Next, the paraxial ray equation for azimuth symmetric magnetic and electric fields, using Busch's theorem, are derived from the least-action principle. Covered next are multipole magnetic fields and lenses for charged particles, including the focusing–defocusing (FODO) quadrupole magnetic lens pair for charged particles. Next, we derive the neutron trajectory in an applied sextupole magnetic field, which focuses neutrons via the magnetic field gradient. Presented next is a neutron 1-D magnetic lens, which is a magnetic compound refractive lens that consists of a single row of N cylindrical magnetic fields. Next, the use of paired sheets of opposite magnetic charges are shown to model magnetic fields for rectangular, permanent magnets. We then show a 1-D periodic magnetic field can act as a gravitational trap for ultracold spin-up neutrons. Finally, the electron trajectory in wigglers and undulator photon sources is derived, and the emitted X-ray wavelength, bandwidth, cone angle, and power from an undulator are presented.

The following several sections on charged particle beam optics follow the treatment of electric and magnetic lenses by Lawson (1977) and Reiser (1994). In lens design, one usually wants to maximize refractive power to obtain short, practical focal lengths, which facilitates relatively easy measurements and reduces the size, complexity, and cost of the experiment. For example, a focal length of 50 m requires a large vacuum chamber and building, where simple and convenient tabletop measurements do not apply. As a result, the shorter focal lengths of multipole magnetic lenses, such as the quadrupole magnetic lens and FODO pairs of quadrupole lenses, are used in place of solenoidal magnetic lenses for energetic charged particle beams.

We start by deriving the paraxial ray equation for the trajectory of charged particles in a static, axial symmetric magnetic field $\mathbf{B}$, using cylindrical coordinates, where

$$\mathbf{B} = B_r\hat{\mathbf{r}} + B_z\hat{\mathbf{z}} \tag{17.1}$$

Neutron and X-ray Optics. DOI: http://dx.doi.org/10.1016/B978-0-12-407164-3.00016-4

The axial component B_z of the magnetic field is independent of radial position r, where

$$\frac{\mathrm{d}B_z}{\mathrm{d}r} = 0 \tag{17.2}$$

A solenoid magnetic field is the typical example of the axial symmetric magnetic field. The azimuth angle θ component of the Lorentz force, which acts on a charged particle of rest mass m_0 and charge e, is then

$$F_\theta = \frac{1}{r}\frac{\mathrm{d}}{\mathrm{d}t}\left(\gamma m_0 r^2 \frac{\mathrm{d}\theta}{\mathrm{d}t}\right) = -e\left(\frac{\mathrm{d}r}{\mathrm{d}t}B_z - \frac{\mathrm{d}z}{\mathrm{d}t}B_r\right) \tag{17.3}$$

Rearrangement of Eq. (17.3), gives the azimuth angle component of the cross product $\mathbf{v} \times \mathbf{B}$ of the charged particle velocity $\mathbf{v} = \mathrm{d}\mathbf{r}/\mathrm{d}t$ and applied magnetic field $\mathbf{B}$, where

$$(\mathbf{v} \times \mathbf{B})_\theta = \frac{\mathrm{d}r}{\mathrm{d}t}B_z - \frac{\mathrm{d}z}{\mathrm{d}t}B_r = -\frac{1}{er}\frac{\mathrm{d}}{\mathrm{d}t}\left(\gamma m_0 r^2 \frac{\mathrm{d}\theta}{\mathrm{d}t}\right) \tag{17.4}$$

In the applied solenoid magnetic field, a charged particle moves in circular orbit of radius r about a center z-axis. The magnetic flux ψ, which passes through the plane of the charged particle circular orbit, is enclosed by the orbit radius r, which is given by

$$\psi = \int_0^{2\pi} \mathrm{d}\theta \int_0^r B_z r \, \mathrm{d}r = 2\pi \int_0^r B_z r \, \mathrm{d}r \tag{17.5}$$

The temporal rate of change of magnetic flux $\mathrm{d}\psi/\mathrm{d}t$, which is enclosed by the circular orbit of the charged particle, changes as the particle orbit radius increases from r to $r + \mathrm{d}r$, and the particle axial position changes from z to $z + \mathrm{d}z$, where

$$\frac{\mathrm{d}\psi}{\mathrm{d}t} = \frac{\mathrm{d}r}{\mathrm{d}t}\frac{d\psi}{\mathrm{d}r} + \frac{\mathrm{d}z}{\mathrm{d}t}\frac{\mathrm{d}\psi}{\mathrm{d}z} = 2\pi\frac{\mathrm{d}r}{\mathrm{d}t}\frac{\mathrm{d}}{\mathrm{d}r}\int_0^r B_z r \, \mathrm{d}r + 2\pi\frac{\mathrm{d}z}{\mathrm{d}t}\frac{\mathrm{d}}{\mathrm{d}z}\int_0^r B_z r \, \mathrm{d}r \tag{17.6}$$

Evaluation of the first and second integral terms of Eq. (17.6), yields

$$2\pi\frac{\mathrm{d}r}{\mathrm{d}t}\frac{\mathrm{d}}{\mathrm{d}r}\int_0^r B_z r \, \mathrm{d}r = 2\pi\frac{\mathrm{d}r}{\mathrm{d}t}B_z r \tag{17.7}$$

and

$$2\pi\frac{\mathrm{d}z}{\mathrm{d}t}\frac{\mathrm{d}}{\mathrm{d}z}\int_0^r B_z r \, \mathrm{d}r = 2\pi\frac{\mathrm{d}z}{\mathrm{d}t}\int_0^r \frac{\mathrm{d}B_z}{\mathrm{d}z} r \, \mathrm{d}r \tag{17.8}$$

From Maxwell equations, one applies Gauss's law for the magnetic field to Eq. (17.8), where

$$\nabla \cdot \mathbf{B} = \frac{1}{r}\frac{\mathrm{d}}{\mathrm{d}r}(rB_r) + \frac{\mathrm{d}B_z}{\mathrm{d}z} = 0 \tag{17.9}$$

With substitution into Eq. (17.6) of Eq. (17.7) and Eq. (17.8), via the results of Eq. (17.9), the expression for the time derivative of the magnetic flux, which is enclosed by the charged particle orbit, where

$$\frac{\mathrm{d}\psi}{\mathrm{d}t} = 2r\pi\frac{\mathrm{d}r}{\mathrm{d}t}B_z - 2\pi\frac{\mathrm{d}z}{\mathrm{d}t}\int_0^r \frac{r\mathrm{d}r}{r}\frac{\mathrm{d}}{\mathrm{d}r}(rB_r) \tag{17.10}$$

Evaluation of the integral in Eq. (17.10), yields

$$\frac{\mathrm{d}\psi}{\mathrm{d}t} = 2r\pi\left(\frac{\mathrm{d}r}{\mathrm{d}t}B_z - \frac{\mathrm{d}z}{\mathrm{d}t}B_r\right) \tag{17.11}$$

Substitution into Eq. (17.11) of Eq. (17.4), yields

$$\frac{\mathrm{d}\psi}{\mathrm{d}t} = 2\pi\frac{1}{e}\frac{\mathrm{d}}{\mathrm{d}t}\left(\gamma m_0 r^2\frac{\mathrm{d}\theta}{\mathrm{d}t}\right) \tag{17.12}$$

Integration of Eq. (17.12) yields

$$\psi - \psi_0 = -\frac{2\pi\gamma m_0 r^2}{e}\frac{\mathrm{d}\theta}{\mathrm{d}t} \tag{17.13}$$

In Eq. (17.13), at initial time $t = 0$, an initial magnetic flux $\psi(0) = \psi_0$ is enclosed by the initial charged particle orbit radius $r(0) = r_0$, where the initial charged particle azimuth velocity $\mathrm{d}\theta/\mathrm{d}t$ is zero; that is,

$$\frac{\mathrm{d}\theta}{\mathrm{d}t}(t = 0) = 0 \quad \text{and} \quad \psi(t = 0) = \psi_0 \tag{17.14}$$

With initial conditions of Eq. (17.14), one solves Eq. (17.13) for the charged particle azimuth velocity $\mathrm{d}\theta/\mathrm{d}t$, where

$$\frac{\mathrm{d}\theta}{\mathrm{d}t} = -\frac{1}{2\pi}\frac{1}{r^2\gamma}\frac{e}{m_0}(\psi - \psi_0) \tag{17.15}$$

In Eq. (17.15), the initial magnetic flux ψ_0, which was enclosed by the initial radius r_0 of the charged particle circular orbit, and the final magnetic flux ψ, which was enclosed by the final radius r of the charged particle circular orbit, are

$$\psi = \pi \cdot r^2 B_z \quad \text{and} \quad \psi_0 = \pi \cdot r_0^2 B_{z0} \tag{17.16}$$

If the magnetic field is static, then the charged particle azimuth velocity $d\theta/dt$ is constant with time. Eq. (17.15) is Busch's theorem, which expresses the conservation of canonical angular momentum P_θ, where, given the azimuth component A_θ of the magnetic vector potential,

$$P_\theta = erA_\theta + \gamma m_0 r^2 \frac{\mathrm{d}\theta}{\mathrm{d}t} \tag{17.17}$$

Equation (17.15) for $\mathrm{d}\theta/\mathrm{d}t$ can be obtained by substitution of Eqs. (17.18) and (17.19) into the expression for P_θ in Eq. (17.17), where

$$A_\theta = \frac{1}{2\pi}\frac{\psi}{r} = \frac{1}{2} rB_z \tag{17.18}$$

and

$$P_\theta = \frac{e}{2\pi}\psi_0 = \frac{e}{2} r_0^2 B_{z0} \tag{17.19}$$

If the charged particle enters an axially symmetric magnetic field from a field-free region, where $B_{z0} = 0$ and $\psi_0 = 0$, then the charged particle rotates about the center z-axis at the Larmor frequency w_L, which is obtained by substitution of Eqs. (17.18) and (17.19) into Eq. (17.17), where

$$w_L = \frac{\mathrm{d}\theta}{\mathrm{d}t} = -\frac{eB_z}{2\gamma m_0} \tag{17.20}$$

The charged particle also rotates about individual magnetic field lines at the cyclotron frequency w_c, which is twice the charged particle Larmor frequency w_L about the center z-axis, where

$$w_c = 2w_L \tag{17.21}$$

17.2 Paraxial Ray Equation and Focusing of Charged Particles by an Axial Symmetric Magnetic Field

In this section, we derive the paraxial ray equation for charged particles in a static, axial symmetric magnetic field, where the z-component of the magnetic field is approximated as independent of the radius in the vicinity of the particle beam. That is, the paraxial ray equation for the charged particle trajectory in the circular symmetric solenoid-type magnetic field, is derived from the Lorentz force equations. The radial component of the Lorentz force equation axial symmetric magnetic field is then

$$\frac{\mathrm{d}}{\mathrm{d}t}\left(\gamma m_0 \frac{\mathrm{d}r}{\mathrm{d}t}\right) - \gamma m_0 r\left(\frac{\mathrm{d}\theta}{\mathrm{d}t}\right)^2 = er\frac{\mathrm{d}\theta}{\mathrm{d}t}B_z \tag{17.22}$$

The charged particle axial velocity $\mathrm{d}z/dt$ is assumed to be approximately constant, and much greater than its radial velocity $\mathrm{d}r/dt$ in this approximation, so that

$$\frac{\mathrm{d}^2 r}{\mathrm{d}t^2} - r\left(\frac{\mathrm{d}\theta}{\mathrm{d}t}\right)^2 = \frac{er}{\gamma m_0}\frac{\mathrm{d}\theta}{\mathrm{d}t}B_z \tag{17.23}$$

where

$$\frac{\mathrm{d}z}{\mathrm{d}t} \gg \frac{\mathrm{d}r}{\mathrm{d}t} \quad \text{and} \quad \frac{\mathrm{d}z}{\mathrm{d}t} \cong \beta c \tag{17.24}$$

with

$$\frac{\mathrm{d}\gamma}{\mathrm{d}t} \cong 0 \quad \text{and} \quad \frac{\mathrm{d}\beta}{\mathrm{d}t} = 0 \tag{17.25}$$

With substitution into Eq. (17.23) of Eq. (17.20) for $\mathrm{d}\theta/\mathrm{d}t$, the radial component of the Lorentz force equation as a function of axial position z is then

$$\frac{\mathrm{d}^2 r}{\mathrm{d}t^2} + \frac{r}{4}\left(\frac{eB_z}{\gamma m_0}\right)^2 = 0 \tag{17.26}$$

One converts Eq. (17.26) from time dependence t to axial position dependence z via

$$\frac{\mathrm{d}r}{\mathrm{d}t} = \frac{\mathrm{d}r}{\mathrm{d}z}\frac{\mathrm{d}z}{\mathrm{d}t} = \frac{\mathrm{d}r}{\mathrm{d}z}\beta c \tag{17.27}$$

and

$$\frac{\mathrm{d}^2 r}{\mathrm{d}t^2} = \frac{\mathrm{d}^2 r}{\mathrm{d}z^2}\left(\frac{\mathrm{d}z}{\mathrm{d}t}\right)^2 + \left(\frac{\mathrm{d}r}{\mathrm{d}z}\right)\frac{\mathrm{d}^2 z}{\mathrm{d}t^2} \cong \frac{\mathrm{d}^2 r}{\mathrm{d}z^2}\beta^2 c^2 + \frac{\mathrm{d}r}{\mathrm{d}z}\frac{\mathrm{d}\beta}{\mathrm{d}t} \cong \frac{\mathrm{d}^2 r}{\mathrm{d}z^2}\beta^2 c^2 \tag{17.28}$$

With substitution of Eq. (17.28) into Eq. (17.26), the radial component of the Lorentz force equation as a function of axial position z is then

$$\frac{\mathrm{d}^2 r}{\mathrm{d}z^2} + \frac{r}{4}\left(\frac{eB_z}{\gamma \beta c m_0}\right)^2 = 0 \tag{17.29}$$

From Eq. (17.20), the azimuth component of the Lorentz force equation becomes

$$w_L = \frac{\mathrm{d}\theta}{\mathrm{d}t} = \frac{\mathrm{d}\theta}{\mathrm{d}z}\frac{\mathrm{d}z}{\mathrm{d}t} = -\frac{eB_z}{2\gamma m_0} \tag{17.30}$$

Acting on the charged particle in the axial symmetric magnetic field are the radial and azimuth force components, which are expressed by the Larmor frequency w_L, where

$$\frac{d^2r}{dz^2} = -r\left(\frac{w_L}{\beta c}\right)^2 \tag{17.31}$$

and

$$\theta = \theta_0 + \frac{1}{\beta c}\int_0^z w_L \, dz \tag{17.32}$$

Eqs. (17.31) and (17.32) are the paraxial ray equations for a charged particle in a static, axial symmetric magnetic field.

An axial symmetric magnetic field, such as a solenoid of short axial length, can be used to focus charged particles, where the z-component of the magnetic field is approximated as independent of the radius in the vicinity of the particle beam. Given that the charged particle is at the initial nonzero, radial position $r > 0$ in the axial symmetric magnetic field, one can determine the axial distance or focal length f, within which the charged particle is magnetically deflected to the center z-axis at zero, radial position $r = 0$. The charged particle enters the magnetic field from the left side at $z = 0$ with nonzero radius $r > 0$. The z-directed magnetic field is assumed to have limited axial extent L_z, which is much smaller than the focal length f or axial distance $f = \Delta z$ that is required for the charged particle trajectory to intercept the z-axis at $r = 0$, where

$$f \gg L_z \tag{17.33}$$

The magnetic field is assumed to act as a thin lens extending from $z = 0$ to $z = L_z$, and has a single principal plane at $z = L_z/2$. One invokes the paraxial approximation in which the radial position r of the charged particle inside the axial symmetric magnetic field is assumed to be approximately constant from $z = 0$ to $z = L_z$; however, the slope dr/dz of the particle trajectory is assumed to be nonzero. Inside the thin magnetic lens, the radial position r is assumed to be constant so that the radial Lorentz force equation can be integrated over axial distance z, where

$$\frac{dr}{dz} = r_0 - r\int_0^{L_z}\left(\frac{w_L}{\beta c}\right)^2 dz \tag{17.34}$$

The charged particle enters the magnetic lens from the left, and it has a zero initial slope ($z = 0$):

$$\frac{dr}{dz}(z = 0) = r_0 = 0 \tag{17.35}$$

The center transverse plane of the magnetic lens or principal plane is separated from the lens focal plane by axial, focal distance f. From an off-axis point at radius r in the principal plane of the magnetic lens, and in the positive z-direction, a diagonal ray converges from radius r to the lens focal point on the z-axis at $r = 0$. The diagonal ray subtends angle θ, relative to the negative z-direction, where

$$\tan\theta = \frac{\mathrm{d}r}{\mathrm{d}z} = \frac{r}{f} \tag{17.36}$$

The focal length f of the magnetic lens is then

$$f = \frac{r}{\mathrm{d}r/dz} \tag{17.37}$$

or

$$\frac{1}{f} = \frac{1}{r}\frac{\mathrm{d}r}{\mathrm{d}z} \tag{17.38}$$

With substitution into Eq. (17.38) of Eq. (17.34), using Eq. (17.35), one obtains the magnetic lens focal length f, where

$$\frac{1}{f} = \frac{1}{r}\frac{\mathrm{d}r}{\mathrm{d}z} = -\int_0^{L_z}\left(\frac{w_L}{\beta c}\right)^2 \mathrm{d}z \tag{17.39}$$

For the thin-lens paraxial approximation, the charged particle trajectory is thus assumed to be a straight line, which starts from nonzero radius r at the center, transverse plane of the magnetic lens at $z = L_z/2$, and passes through the z-axis ($r = 0$) at focal length f or axial position z_{f+}:

$$z_{f+} = f + \frac{L_z}{2} \tag{17.40}$$

The magnetic lens is symmetric. A charged particle, which is incident at $z = L_z$ and nonzero radius r from the right, is focused to the left ($z < 0$) at focal distance f_- from the principal, center plane at axial position z_{f-}, where

$$z_{f-} = \frac{L_z}{2} - f \tag{17.41}$$

and

$$\frac{1}{f_-} = \frac{1}{r}\frac{\mathrm{d}r}{\mathrm{d}z} = -\int_{L_z}^{0}\left(\frac{w_L}{\beta c}\right)^2 \mathrm{d}z = \int_0^{L_z}\left(\frac{w_L}{\beta c}\right)^2 \mathrm{d}z \tag{17.42}$$

The sign convention then has focal lengths with opposite signs depending on the direction of the incident-charged particles:

$$\frac{1}{f_-} = -\frac{1}{f} \tag{17.43}$$

At the ends of the solenoid magnetic field lens are the radial magnetic field components, which, along with the Larmor precession of the charged particles about the axial magnetic field in the lens interior, lead to rotation of the electrons as well as to aberrations. Aberration also occurs due to radial variation in the imperfect solenoid magnetic field. To minimize aberrations, the electron beam trajectory should be very close to the center z-axis to maintain the paraxial approximation. Higher-order terms outside the paraxial approximation (i.e., variation of the radial position of the electron within the lens) then cause lens aberrations. Also, the lens focal length should be much larger than the lens axial length, to maintain the thin-lens approximation of a single center lens principal plane that one uses to calculate the thin-lens focal length.

17.3 Paraxial Ray Equation for Azimuthally Symmetric Magnetic and Electric Fields and Busch's Theorem Derived from the Least-Action Principle

The paraxial ray equation, which is for a cylindrically symmetric system with electric and magnetic fields, can also be derived from the least-action principle for a charged particle beam, where the Hamiltonian H is conserved. The derivation given next follows the derivations given in the books by Lawson (1977), and by Panofsky and Phillips (1962). The canonical angular momentum vector $\mathbf{P}$ of a charged particle, is integrated along a trajectory path, which occurs between fixed endpoints r_1 and r_2, where all paths are possible, and each path has a different time of flight and value of H. The variation of the action integral of the canonical angular momentum P is zero, where

$$\delta \int_{r_1}^{r_2} \mathbf{P} \cdot \mathbf{dr} = 0 \tag{17.44}$$

Given charged particle velocity vector $\mathbf{v}$ and the magnetic vector potential $\mathbf{A}$, the canonical momentum $\mathbf{P}$ is

$$\mathbf{P} = \frac{\mathbf{v}}{c} \gamma m_0 c + e\mathbf{A} \tag{17.45}$$

With cylindrically symmetric electric and magnetic fields, there is only magnetic vector potential $\mathbf{A}$, with azimuth component A_θ:

$$\mathbf{A} = A_\theta \hat{\boldsymbol{\theta}} \tag{17.46}$$

In terms of expression as a function of axial position z rather than time t, the variation of the action integral of Eq. (17.44), with substitution of Eqs. (17.45) and (17.46), becomes integration along the z-axis from axial position z_1 to axial position z_2:

$$\delta\int_{r_1}^{r_2}\mathbf{P}\cdot\mathrm{d}\mathbf{r}=m_0c\delta\int_{z_1}^{z_2}\mathrm{d}z\left(\beta\gamma\sqrt{\left(\frac{\mathrm{d}r}{\mathrm{d}z}\right)^2+r^2\left(\frac{\mathrm{d}\theta}{\mathrm{d}z}\right)^2+1}+\frac{eA_\theta}{m_0c}r\frac{\mathrm{d}\theta}{\mathrm{d}t}\right)=0 \qquad (17.47)$$

Recall section 13.14, which derives the Euler equations from the least action principle, and is applied to Eq. (17.47), so that one obtains the Euler equations for the radial and azimuth directions:

$$\frac{\mathrm{d}}{\mathrm{d}z}\frac{\partial}{\partial r/\partial z}\left(\frac{\mathbf{v}}{c}\gamma m_0c+e\mathbf{A}\right)-\frac{\partial}{\partial r}\left(\frac{\mathbf{v}}{c}\gamma m_0c+e\mathbf{A}\right)=0 \qquad (17.48)$$

and

$$\frac{\mathrm{d}}{\mathrm{d}z}\frac{\partial}{\partial\theta/\partial z}\left(\frac{\mathbf{v}}{c}\gamma m_0c+e\mathbf{A}\right)-\frac{\partial}{\partial\theta}\left(\frac{\mathbf{v}}{c}\gamma m_0c+e\mathbf{A}\right)=0 \qquad (17.49)$$

Integration of the Euler equation of Eq. (17.49) for the azimuth component gives

$$\frac{\mathrm{d}\theta}{\mathrm{d}z}\frac{r^2\gamma\beta m_0c}{\sqrt{(\mathrm{d}r/\mathrm{d}z)^2+r^2(\mathrm{d}\theta/\mathrm{d}z)^2+1}}+eA_\theta r=\mathrm{const}=P_\theta \qquad (17.50)$$

The above Euler equation of Eq. (17.50), is Busch's theorem for the canonical angular momentum P_θ, which is constant for the charged particle trajectory through the azimuthally symmetric arrangement of electric and magnetic fields. The Euler equation for the radial component of the charged particle trajectory of Eq. (17.48), is more complicated, where

$$\begin{aligned}&\frac{\partial}{\partial z}\frac{\mathrm{d}r}{\mathrm{d}z}\frac{\gamma\beta}{\sqrt{(\mathrm{d}r/\mathrm{d}z)^2+r^2(\mathrm{d}\theta/\mathrm{d}z)^2+1}}-\sqrt{\left(\frac{\mathrm{d}r}{\mathrm{d}z}\right)^2+r^2\left(\frac{\mathrm{d}\theta}{\mathrm{d}z}\right)^2+1}\cdot\frac{1}{2\beta\gamma}\frac{\partial}{\partial r}(\beta\gamma)\\&-\frac{r\gamma\beta}{\sqrt{(\mathrm{d}r/\mathrm{d}z)^2+r^2(\mathrm{d}\theta/\mathrm{d}z)^2+1}}\left(\frac{\mathrm{d}\theta}{\mathrm{d}z}\right)^2+\frac{e}{m_0c}\frac{\mathrm{d}\theta}{\mathrm{d}z}\frac{\partial}{\partial r}(rA_\theta)=0\end{aligned} \qquad (17.51)$$

The expression of Eq. (17.50) for $\partial\theta/\partial z$, which is obtained from the azimuth Euler equation of Eq. (17.49), is inserted into the radial Euler equation of Eq. (17.51).

This yields the radial trajectory component of the charged particle in the cylindrical symmetric electric and magnetic field systems, where

$$\frac{d^2r}{dz^2} + \frac{1}{\lambda^2}\left[1 + \left(\frac{dr}{dz}\right)^2\right] \cdot \left\{\left[1 + \left(\frac{dr}{dz}\right)^2\right]\frac{\partial\lambda}{\partial r} - \frac{dr}{dz}\frac{\partial\lambda}{\partial z}\right\} = 0 \tag{17.52}$$

with

$$\lambda = \beta^2\gamma^2(1 - \eta^2) \tag{17.53}$$

and

$$\eta = \frac{1}{\beta\gamma m_0 c}\left(\frac{P_\theta}{r} - eA_\theta\right) = \frac{r}{\beta c}\frac{\partial\theta}{\partial t} \tag{17.54}$$

17.4 Multipole Magnetic Fields and Lenses for Charged Particles

There is a need for magnetic lenses for charged particles with a short focal length. The axial symmetric magnetic field has relatively weak focusing power, and thus relatively long focal length compared to multipole magnetic fields. Recall from Eq. (17.39), and dropping the minus sign, that in a static, axial symmetric magnetic lens of length L_z, the focal length f is given by

$$f = \frac{r}{dr/dz} = \left[\int_0^{L_z}\left(\frac{w_L}{\beta c}\right)^2 dz\right]^{-1} \tag{17.55}$$

Focal length f increases with the charged particle axial velocity βc, and in the laboratory magnetic fields of 1 to a few tesla, the Larmor frequency w_L of Eq. (17.20), is

$$w_L = \frac{eB_z}{2\gamma m_0} = \sqrt{1-\beta^2}\,\frac{eB_z}{2m_0} \tag{17.56}$$

For the case of electrons and magnetic fields of a few tesla or less, the ratio $w_L/c\beta$ in the integrand of Eq. (17.55), is typically much less than 1 for the relativistic electron beams, where $\beta \cong 1$ and

$$\frac{w_L}{\beta c} = \frac{\sqrt{1-\beta^2}}{\beta c}\frac{eB_z}{2m_0} = 293\frac{\sqrt{1-\beta^2}}{\beta}B_z\left[\frac{1}{T\cdot m}\right] < 1 \tag{17.57}$$

With relativistic electron beams, the integral of Eq. (17.55) is much less than 1, and one obtains long focal lengths of many meters with the use of a static, axial symmetric magnetic field.

In synchrotron rings and other facilities, which use relativistic electron beams, and require relatively short focal length lenses, one must use more powerful magnetic lenses, than the short solenoid-type magnetic lenses that are used in low-voltage electron microscopes and vacuum tube electronics. The large electric fields for focusing the relativistic electron beams are not practical, and the large electric fields also have the disadvantage of altering the electron beam energy. Also, electric fields are not well suited for producing recirculating electron beams, which move in closed orbits, around the relatively compact space of a synchrotron ring. The need arose for magnetic lenses with more focusing power. One can curve the z-axis, via a toroidal magnetic field, and thus produce closed orbits. Where the magnetic field is constant in both the circular axial direction and the radial direction. Alternatively, one can employ magnetic fields, which do not have circular symmetry.

Cylindrical multipole magnetic fields, such as quadrupole, sextupole, or octupole magnetic fields, provide first-order focusing, as opposed to second-order focusing by circular symmetric magnetic fields. That is, second-order focusing of charged particles, with the solenoid magnetic lenses, provides focal lengths, which are inversely proportional to the square of the magnetic field amplitude. In comparison, magnetic multipole lenses, which have first-order focusing, provide shorter focal lengths that are inversely proportional to the magnetic field amplitude. In a multipole magnetic or electric lens, the radial field magnetic B_r or radial electric field E_r in vacuum space satisfies the static Maxwell equations:

$$\nabla \cdot \mathbf{E} = 0 \quad \nabla \cdot \mathbf{B} = 0 \quad \nabla \times \mathbf{B} = 0 \quad \nabla \times \mathbf{E} = 0 \tag{17.58}$$

In a $2n$-pole multipole magnetic or electric field, the radial field magnetic B_r and radial electric field E_r have the relation

$$E_r, B_r \propto f(z) r^{n-1} \cos[2(n-1)\theta] \tag{17.59}$$

Equation (17.59) satisfies the static, vacuum Maxwell equations given in Eq. (17.58).

Quadrupole ($2n = 4$), sextupole ($2n = 6$), and octupole ($2n = 8$) magnetic or electric fields, have four, six, and eight poles of alternating polarity, which are evenly spaced in a circle centered about the z-axis. The interior magnetic or electric pole faces are at radial distance a from the center z-axis. For example, a pure quadrupole electric field has four evenly spaced, alternating polarity, poles, where the radial E_r and azimuth E_θ electric field components in Cartesian and cylindrical coordinates are

$$E_x = E_0 \frac{x}{a} \quad \text{and} \quad E_y = -E_0 \frac{y}{a} \tag{17.60}$$

and

$$E_r = \frac{E_0 r}{a}\cos 2\theta \quad \text{and} \quad E_\theta = -\frac{E_0 r}{a}\sin 2\theta \tag{17.61}$$

Along the center z-axis, the electric field is zero, and at the surface of the electrode at $r = a$, the electric field has a maximum E_0, where

$$E_r^2 + E_\theta^2 = E_0^2 \tag{17.62}$$

Similarly, a pure quadrupole magnetic field has four evenly spaced, alternating-polarity poles, where the radial B_r and azimuth B_θ magnetic field components in Cartesian and cylindrical coordinates are

$$B_x = B_0\frac{x}{a} \quad \text{and} \quad B_y = -B_0\frac{y}{a} \tag{17.63}$$

and

$$B_r = \frac{B_0 r}{a}\cos 2\theta \quad \text{and} \quad B_\theta = -\frac{B_0 r}{a}\sin 2\theta \tag{17.64}$$

Along the center z-axis, the magnetic field is zero, and at the surface of the electrode at $r = a$, the electric field is at a maximum value of B_0, where

$$B_r^2 + B_\theta^2 = B_0^2 \tag{17.65}$$

The quadrupole magnetic lens and its aberrations are covered in Hawkes (1970).

The charged particle trajectory along the z-axis inside a magnetic quadrupole lens in Cartesian coordinates, as a function of propagation time t, is

$$\gamma m_0 \frac{\mathrm{d}^2 x}{\mathrm{d}t^2} + e\beta c B_0 \frac{x}{a} = 0 \tag{17.66}$$

and

$$\gamma m_0 \frac{\mathrm{d}^2 y}{\mathrm{d}t^2} + e\beta c B_0 \frac{y}{a} = 0 \tag{17.67}$$

And from Eqs. (17.66) and (17.67), as a function of position z along the z-axis, the trajectory of a charged particle inside a quadrupole magnetic lens is expressed as

$$\frac{\mathrm{d}^2 x}{\mathrm{d}z^2} = -\frac{eB_0}{\gamma m_0 \beta c}\frac{x}{a} \tag{17.68}$$

and

$$\frac{d^2y}{dz^2} = + \frac{eB_0}{\gamma m_0 \beta c} \frac{y}{a} \tag{17.69}$$

The negative sign for the x-component of the charged particle trajectory, indicates focusing or deflection toward the center z-axis, and the positive sign for the y-component, indicates defocusing or deflection away from the z-axis. The solutions for the pair of equations in lens matrix form are

$$\begin{bmatrix} x(L_z) \\ \dfrac{dx}{dz}(L_z) \end{bmatrix} = \begin{bmatrix} \cos\left(L_z\sqrt{\kappa}\right) & \dfrac{1}{\sqrt{\kappa}}\sin\left(L_z\sqrt{\kappa}\right) \\ -\sqrt{\kappa}\sin\left(L_z\sqrt{\kappa}\right) & \cos\left(L_z\sqrt{\kappa}\right) \end{bmatrix} \begin{bmatrix} x(0) \\ \dfrac{dx}{dz}(0) \end{bmatrix} \tag{17.70}$$

and

$$\begin{bmatrix} y(L_z) \\ \dfrac{dy}{dz}(L_z) \end{bmatrix} = \begin{bmatrix} \cosh\left(L_z\sqrt{\kappa}\right) & \dfrac{1}{\sqrt{\kappa}}\sinh\left(L_z\sqrt{\kappa}\right) \\ \sqrt{\kappa}\sinh\left(L_z\sqrt{\kappa}\right) & \cosh\left(L_z\sqrt{\kappa}\right) \end{bmatrix} \begin{bmatrix} x(0) \\ \dfrac{dx}{dz}(0) \end{bmatrix} \tag{17.71}$$

Here, the constant $\sqrt{\kappa}$ acts as the spatial frequency of the charged particle oscillation and deflection in the applied magnetic field, and κ is proportional to the Larmor frequency w_L or cyclotron frequency w_c, and inversely proportional to the inner radius a of the pole faces of the quadrupole magnets, and the axial particle velocity βc, where

$$\kappa = \frac{eB_0}{a\gamma m_0 \beta c} = \frac{2w_L}{a\beta c} = \frac{w_c}{a\beta c} \tag{17.72}$$

Application of the small-angle approximation to Eqs. (17.70) and (17.71), where $L_z\sqrt{\kappa} \ll 1$, gives

$$\begin{bmatrix} x(L_z) \\ \dfrac{dx}{dz}(L_z) \end{bmatrix} = \begin{bmatrix} 1 & L_z \\ -\kappa L_z & 1 \end{bmatrix} \begin{bmatrix} x(0) \\ \dfrac{dx}{dz}(0) \end{bmatrix} \tag{17.73}$$

and

$$\begin{bmatrix} y(L_z) \\ \dfrac{dy}{dz}(L_z) \end{bmatrix} = \begin{bmatrix} 1 & L_z \\ -\kappa L_z & 1 \end{bmatrix} \begin{bmatrix} x(0) \\ \dfrac{dx}{dz}(0) \end{bmatrix} \tag{17.74}$$

As with the solenoid lens, one assumes that the quadrupole magnetic lens is a thin lens with axial length L_z, which is much smaller than focal length f. The thin lens approximation assumes that the charged particle trajectory in the quadrupole magnetic lens maintains a constant radial distance from the center z-axis ($r = \text{const}$), but the slope dr/dz of the trajectory changes at the quadrupole lens principal plane. Focusing of charged particles occurs in the transverse x-direction, and defocusing occurs in the transverse y-direction. In the x-direction, the focal length f_{quad} of the magnetic quadrupole lens, is obtained by application of Eq. (17.55) and Eq. (17.73), where

$$f_{\text{quad}} = \frac{x}{dx/dz} = \left[\frac{e}{\gamma m_0 \beta c}\int_0^{L_z}\frac{B_0}{a}dz\right]^{-1} = \left[\frac{2}{a}\int_0^{L_z}\frac{w_L}{\beta c}dz\right]^{-1} \tag{17.75}$$

For laboratory magnetic fields of less than a few tesla, and relativistic electron beams with $\beta \cong 1$, the ratio $w_L/\beta c$ of the charged particle Larmor frequency w_L, divided by its axial velocity βc, is much less than 1:

$$\frac{w_L}{\beta c} = \frac{\sqrt{1-\beta^2}}{\beta c}\frac{eB_0}{2m_0} = 293\frac{\sqrt{1-\beta^2}}{\beta}B_0\left[\frac{1}{T\cdot m}\right] \ll 1 \tag{17.76}$$

With Eq. (17.76), comparison of Eq. (17.75) with Eq. (17.55) shows, that for relativistic electron beams, the focal length of the quadrupole magnetic lens f_{quad} is much less than the focal length f of the circular symmetric magnetic field, where

$$f_{\text{quad}} = \left[\frac{2}{a}\int_0^{L_z}\frac{w_L}{\beta c}dz\right]^{-1} \ll f = \left[\int_0^{L_z}\left(\frac{w_L}{\beta c}\right)^2 dz\right]^{-1} \tag{17.77}$$

Sextupoles and octupoles are used for controlling nonlinear effects, which are caused by focusing forces that are not proportional to the charged particle displacement. The sextupole magnetic lens is used for the magnetic focusing of charged particles or uncharged particles, such as the neutron, which have a net magnetic dipole moment. The magnetic dipole moment of a charged or uncharged particle, which is acted upon by the quadrupole, sextupole, or octupole magnetic field produces deflection of the particle in the plane that is parallel to the particle magnetic dipole moment vector. The x and y component sextupole magnetic fields in Cartesian and cylindrical coordinates are given by Reiser (1994),

$$B_x = \left(\frac{B_0}{a}\right)^2 xy \quad \text{and} \quad B_y = \frac{1}{2}\left(\frac{B_0}{a}\right)^2 (x^2 - y^2) \tag{17.78}$$

and

$$B_x = \left(\frac{B_0}{a}\right)^2 r^2 \sin 3\theta \quad \text{and} \quad B_y = \left(\frac{B_0}{a}\right)^2 r^2 \cos 3\theta \tag{17.79}$$

The pole surfaces of the sextupole magnetic are at radial distance a and have surface magnetic field B_0.

17.5 FODO Quadrupole Magnetic Lens Pair for Charged Particles

The F quadrupole magnetic lens focuses z-directed charged particles in the transverse, horizontal direction, but defocuses in the vertical, transverse direction. The D quadrupole magnetic lens focuses vertically but defocuses horizontally. Separation of the F and D quadrupole lenses by a sufficient distance, which prevents cancellation of the F and D magnetic fields, thereby forms a lens pair that produces strong focusing in both horizontal and vertical, transverse directions, hence the term FODO quadrupole lens. In this section is derived the focal length and trajectory of charged particles, which propagate through a pair of quadrupole thin lenses, each of axial length L_z and separated by axial distance D_z between their respective center, principal planes. The identical quadrupole lenses are azimuthally rotated relative to one another by 90°. In this manner, the pair of quadrupole lenses provide overall focusing of the charged particle in the transverse x- and y-directions. The focusing in x and defocusing in y of the first quadrupole lens, is followed by a second quadrupole lens, where defocusing occurs in the x-direction and focusing occurs in the y-direction.

The charged particle is farther from the center z-axis upon entering the first quadrupole magnetic lens, which focuses the particle closer to the z-axis in the x-direction, and defocuses farther from the z-axis in the y-direction. Recall that the magnetic field is zero at the center axis, and increases with the radius to the magnetic pole surfaces at radius a. The charged particle, which has been focused in the x-direction closer to the z-axis by the first quadrupole lens, is less defocused in the x-direction by the second quadrupole lens. And the charged particle, which has been defocused in the y-direction farther from the z-axis by the first quadrupole lens, is then more focused in the y-direction by the second quadrupole lens. Consequently, the pair of FODO quadrupole magnetic lenses provide overall focusing of the charged particle beam in both x- and y-transverse directions.

One can describe the trajectory of the charged particles through the FODO by transfer matrices. The axial drift space distance D_z, which is between the pair of quadrupole lenses in the FODO, gives the drift space matrix:

$$\begin{bmatrix} y(L_z) \\ \dfrac{\mathrm{d}y}{\mathrm{d}z}(L_z) \end{bmatrix} = \begin{bmatrix} 1 & D_z \\ 0 & 1 \end{bmatrix} \begin{bmatrix} x(0) \\ \dfrac{\mathrm{d}x}{\mathrm{d}z}(0) \end{bmatrix} \tag{17.80}$$

Recalling Eq. (17.72), with application of the small-angle approximation $L_z\sqrt{\kappa} \ll 1$, the charged particle position $x(z)$, $y(z)$, and slope $\mathrm{d}x/dz(z)$, $\mathrm{d}y/dz(z)$ in the x- and y-directions at the exit of the FODO, is expressed by the transfer matrix in terms of the incident x- and y-positions $x(0)$, $y(0)$ and slopes $\mathrm{d}x/dz(0)$, $\mathrm{d}y/dz(0)$ at the FODO entrance, where

$$\begin{bmatrix} x(L_z) \\ \dfrac{\mathrm{d}x}{\mathrm{d}z}(L_z) \end{bmatrix} = \begin{bmatrix} 1 & L_z \\ \kappa L_z & 1 \end{bmatrix} \begin{bmatrix} 1 & D_z \\ 0 & 1 \end{bmatrix} \begin{bmatrix} 1 & L_z \\ -\kappa L_z & 1 \end{bmatrix} \begin{bmatrix} x(0) \\ \dfrac{\mathrm{d}x}{\mathrm{d}z}(0) \end{bmatrix} \tag{17.81}$$

and

$$\begin{bmatrix} y(L_z) \\ \dfrac{dy}{dz}(L_z) \end{bmatrix} = \begin{bmatrix} 1 & L_z \\ -\kappa L_z & 1 \end{bmatrix} \begin{bmatrix} 1 & D_z \\ 0 & 1 \end{bmatrix} \begin{bmatrix} 1 & L_z \\ \kappa L_z & 1 \end{bmatrix} \begin{bmatrix} y(0) \\ \dfrac{dy}{dz}(0) \end{bmatrix} \tag{17.82}$$

To simplify matrix multiplication, the axial length L_z of each quadrupole magnetic lens is assumed to be much smaller than the drift space distance D_z between the two lenses, where

$$L_z \ll D_z \tag{17.83}$$

The matrix multiplication gives the charged particle position and slope at the exit of the FODO in terms of its incident position and slope, the axial length L_z of each quadrupole magnetic lens, the drift space distance D_z, the magnetic field magnitude B_0 at the quadrupole pole faces, and the charged particle axial velocity βc and velocity-dependent relativistic mass factor γ contained in constant κ, where

$$\begin{bmatrix} x(z) \\ \dfrac{dx}{dz}(z) \end{bmatrix} = \begin{bmatrix} 1-\kappa L_z D_z & D_z \\ -\kappa^2 D_z L_z^2 & 1+\kappa L_z D_z \end{bmatrix} \begin{bmatrix} x(0) \\ \dfrac{dx}{dz}(0) \end{bmatrix} \tag{17.84}$$

and

$$\begin{bmatrix} y(z) \\ \dfrac{dy}{dz}(z) \end{bmatrix} = \begin{bmatrix} 1+\kappa L_z D_z & D_z \\ -\kappa^2 D_z L_z^2 & 1-\kappa L_z D_z \end{bmatrix} \begin{bmatrix} y(0) \\ \dfrac{dy}{dz}(0) \end{bmatrix} \tag{17.85}$$

Consider the general transfer matrix for a symmetric thin lens, which is expressed by the lens focal length f, where

$$\begin{bmatrix} r(z) \\ \dfrac{dr}{dz}(z) \end{bmatrix} = \begin{bmatrix} 1 & 0 \\ -\dfrac{1}{f} & 1 \end{bmatrix} \begin{bmatrix} r(0) \\ \dfrac{dr}{dz}(0) \end{bmatrix} \tag{17.86}$$

A pair of thin lenses, which are separated by drift distance D, are expressed by the product of the pair of general transfer matrices for the thin lenses and the drift space transfer matrix, where

$$\begin{bmatrix} r(z) \\ \dfrac{dr}{dz}(z) \end{bmatrix} = \begin{bmatrix} 1 & 0 \\ -\dfrac{1}{f} & 1 \end{bmatrix} \begin{bmatrix} 1 & D \\ 0 & 1 \end{bmatrix} \begin{bmatrix} 1 & 0 \\ -\dfrac{1}{f} & 1 \end{bmatrix} \begin{bmatrix} r(0) \\ \dfrac{dr}{dz}(0) \end{bmatrix} \tag{17.87}$$

Matrix multiplication is then performed. One assumes that the focal length f of each thin lens is much greater than the lens separation distance D. The general transfer matrix representation of the pair of separated, thin lenses is

$$\begin{bmatrix} r(z) \\ \frac{dr}{dz}(z) \end{bmatrix} = \begin{bmatrix} 1 & D \\ -\frac{2}{f} & 1 \end{bmatrix} \begin{bmatrix} r(0) \\ \frac{dr}{dz}(0) \end{bmatrix} \quad f \gg D \tag{17.88}$$

The FODO lens is comprised of two quadrupole lenses, each of length L_z, and separated by drift distance D_z, with magnetic field magnitude B_0 at the surface. Comparison of the transfer matrix for the FODO and pair of separated lenses, yields the focal length of the f_{fodo}, where

$$f_{\text{fodo}} = \frac{2}{\kappa^2 D_z L_z^2} = \frac{a^2(\beta c)^2}{2w_L^2 D_z L_z^2} \tag{17.89}$$

The FODO pair of quadrupole lenses provides second-order focusing, via the inverse quadratic dependence of f_{fodo} on the pole surface magnetic field magnitude B_0:

$$\left(\frac{w_L}{\beta c}\right)^2 = \frac{1-\beta^2}{\beta^2 c^2}\left(\frac{eB_z}{2m_0}\right)^2 < 1 \tag{17.90}$$

One can compare (1) the focal length f_{fodo} of the FODO quadrupole magnetic lens, to (2) the focal length f_{sol} of a solenoid magnetic lens of length $2L_z$, with axial-directed magnetic field amplitude B_0. The ratio of the FODO focal length, divided by the solenoid lens focal length f_{sol}, is then

$$\frac{f_{\text{fodo}}}{f_{\text{sol}}} = \frac{a^2}{2L_z D_z} \ll 1 \tag{17.91}$$

The pole face radius a of the FODO is much smaller than the individual quadrupole lengths L_z and separation distance D_z. Hence, the FODO focal length is much shorter than the equivalent solenoid lens. If the two quadruple lenses are adjacent and separated by distance $D_z = L_z$ between their principal planes, then the ratio becomes

$$\frac{f_{\text{fodo}}}{f_{\text{sol}}} = \frac{a^2}{2L_z^2} \ll 1 \tag{17.92}$$

17.6 Neutron Trajectory in Magnetic Fields and Magnetic Field Gradient Focusing of Neutrons

Excellent coverage of practical theory, experiments, and devices in magnetic neutron optics is found by Williams (1988). For more updated coverage, see Anderson et al. (2009) and Furrer et al. (2009). The Shimizu group at Riken in Japan has pioneered development of magnetic quadrupole and sextupole (including superconducting sextupole) magnets for focusing neutrons, and the group has also fabricated arrays of magnesium fluoride neutron prisms; see Shimizu (2002). The sextupole magnet is not only an important focusing device for charged particle beams, but it also is a very useful polarizing, focusing, and imaging neutron optic. The neutrons are focused via the gradient of the sextupole magnetic field. Following Shimizu, Suzuki, and Oku (2010) and recalling Chapter 2, the neutron position and spin trajectory in an applied magnetic field, is given by the Lorentz force equation for the neutron position vector **r** and the Bloch equation for the neutron magnetic dipole moment vector $\boldsymbol{\mu}$, via the oppositely directed, neutron spin unit vector $\boldsymbol{\sigma}$, where

$$\frac{d^2\mathbf{r}}{dt^2} = -\alpha\nabla(\boldsymbol{\sigma}\cdot\mathbf{B}) \tag{17.93}$$

and

$$\frac{d\boldsymbol{\sigma}}{dt} = \gamma_n(\boldsymbol{\sigma}\times\mathbf{B}) \tag{17.94}$$

The Lorentz force equation parameter α is

$$\alpha = \left|\frac{\mu_n}{m_n}\right| = 5.77\ \mathrm{m^2 s^{-2} T^{-1}} \tag{17.95}$$

Here, m_n is the neutron rest mass. With the reduced Planck's constant $\hbar = h/2\pi$, the neutron gyromagnetic ratio γ_n in the Bloch equation is given by

$$\gamma_n = \frac{2\mu_n}{\hbar} = -1.83\times10^8\ \mathrm{s^{-1}T^{-1}} \tag{17.96}$$

In the Lorentz force equation of Eq. (17.93), the neutron magnetic dipole moment $\boldsymbol{\mu}$ is accelerated by the force of the gradient $\nabla|\mathbf{B}|$ of the applied magnetic field **B**. In the Bloch equation of Eq. (17.94), the neutron magnetic dipole moment $\boldsymbol{\mu}$ and its oppositely directed neutron spin unit vector $\boldsymbol{\sigma}$ undergo precession about the applied magnetic field **B** at the Larmor frequency w_L, via $\boldsymbol{\mu}\times\mathbf{B}$. The associated "current loop of moving charge" of the neutron magnetic dipole moment $\boldsymbol{\mu}$ interacts with the applied magnetic field **B**, via the Lorentz force, which is expressed in terms of the neutron spin unit vector $\boldsymbol{\sigma}$ by $\boldsymbol{\sigma}\times\mathbf{B}$. The Larmor precession frequency w_L of the neutron spin unit vector $\boldsymbol{\sigma}$ is given by

$$w_L = -\gamma_n|\mathbf{B}| \tag{17.97}$$

The local magnetic field at the neutron position varies as the neutron travels through the inhomogeneous magnetic field. The angular frequency w_B of the magnetic field direction $\hat{\mathbf{B}}$ in the neutron rest frame is

$$w_B = \frac{\mathrm{ds}}{\mathrm{d}t}\left|\frac{\partial\hat{\mathbf{B}}}{\partial\mathbf{s}}\right| \tag{17.98}$$

Here, **s** is the coordinate along the neutron trajectory curve, and the unit magnetic field vector $\hat{\mathbf{B}}$ or magnetic field direction is

$$\hat{\mathbf{B}} = \frac{\mathbf{B}}{|\mathbf{B}|} \tag{17.99}$$

Consider that the local magnetic field **B** is sufficiently strong that the ratio is

$$\frac{w_L}{w_B} \gg 1 \tag{17.100}$$

Then the neutron polarization along the magnetic field is adiabatically transported and almost conserved. In this case, the Lorentz force Eq. (17.93) for $\mathrm{d}^2\mathbf{r}/dt^2$ becomes

$$\frac{\mathrm{d}^2\mathbf{r}}{\mathrm{d}t^2} = \mp\,\alpha\nabla|\mathbf{B}| \tag{17.101}$$

Here, $\mp$ corresponds to the cases of the neutron spin, which are parallel and anti-parallel to the local magnetic field, respectively.

Consider the case of neutrons with spin, which is parallel to the local magnetic field that pass through a sextupole magnet, where the magnitude of the magnetic field is given by

$$|\mathbf{B}| = \left(\frac{G}{2}\right)(x^2 + y^2) \tag{17.102}$$

Here, G is a positive constant that indicates the strength of the magnetic field gradient. The z-axis is parallel to the neutron beam axis, and the x- and y-axes are each perpendicular to the beam axis. Inside the sextupole magnet, the x-, y-, and z-direction neutron trajectory equations are obtained, via substitution of sextupole magnetic field Eq. (17.102) into Lorentz force Eq. (17.101), where

$$\frac{\mathrm{d}^2x}{\mathrm{d}t^2} = -\,w^2x \tag{17.103}$$

$$\frac{\mathrm{d}^2y}{\mathrm{d}t^2} = -\,w^2y \tag{17.104}$$

and

$$\frac{d^2z}{dt^2} = 0 \tag{17.105}$$

where

$$w^2 = G\alpha \tag{17.106}$$

Given neutron speed v, wavelength λ, and wave number k, the neutron wave angular frequency w is

$$w = kv = \frac{2\pi}{\lambda}v \tag{17.107}$$

With substitution of Eq. (17.107) into Eq. (17.106), one determines the magnetic field gradient parameter G, where

$$G \propto \frac{1}{\lambda^2} \tag{17.108}$$

Note that for a pulsed neutron source, the neutron wavelength λ is proportional to the neutron time of flight t, so that the neutron gradient parameter G is inversely proportional to the square of the time of flight t, where

$$G \propto \frac{1}{t^2} \tag{17.109}$$

Outside the sextupole, the neutrons proceed in straight-line trajectories through a downstream focal point on the z-axis at focal distance f, measured from the sextupole entrance:

$$f = z_m + \frac{h}{wm_n\lambda}\cot\left(\frac{wm_n\lambda}{h}z_m\right) \tag{17.110}$$

Here z_m is the length of the sextupole magnet (along the z-axis), and w is the neutron angular frequency. Alternatively, the sextupole magnet focal length, expressed in terms of neutron speed v, is

$$f = z_m + \frac{h}{2\pi m_n v}\cot\left(\frac{2\pi m_n v}{h}z_m\right) \tag{17.111}$$

The chromatic aberration can be removed, if the magnetic field gradient can be varied in time, as neutrons propagate though the sextupole. Removal of chromatic aberration, requires that the magnetic field gradient is decreased, with increasing arrival times of the longer wavelength neutrons, which occur after the beginning of each neutron burst. The gradient can be altered by the current I in an electromagnetic, pulsed sextupole, such that the current I is decreased with longer neutron time of flight (longer neutron wavelengths), where

$$I \propto G \propto \frac{1}{t^2} \tag{17.112}$$

17.7 Neutron Compound Cylindrical Magnetic Lens

At the Argonne National Laboratory, Littrell et al. (2007) fabricated and tested a magnetic compound refractive lens for focusing and polarizing cold neutron beams in one dimension. The magnetic compound refractive lens was 157.5 cm long and was comprised of a row of 99 pairs of cylindrical magnets that produced a 1.01 T magnetic field in a 3-mm gap. Each nickel-plated, NdFeB cylindrical magnet was 1.27 cm in diameter and 0.95 cm thick, with a magnetic field of 1.37 T. The refractive index n for the cylindrical magnetic lens is obtained from the general expression for the refractive index.

The compound refractive, cylindrical magnetic lens was tested on the polarized reflectometer ASTERIX at the Los Alamos Neutron Scattering Center (LANSCE) of the Los Alamos National Laboratory in New Mexico. A spallation source produced pulsed neutrons over the range of 4–13 Å, which were polarized, and a radiofrequency spin flipper placed after the polarizer, allowed the neutrons to have their spins remain parallel or be flipped antiparallel to the direction of the magnetic field, along the stack of 99 cylindrical magnetic lenses.

Recall from Section 2.1, the total neutron energy E is equal to the neutron kinetic energy W in the no-field vacuum region, where air can be approximated as vacuum. In vacuum (air), one assumes the neutron maintains constant total energy E, after the neutron crosses into the magnetic field region from the no-field region. However, in the magnetic field region, the neutron kinetic energy is reduced, as the neutron acquires magnetic potential energy U. The acquired neutron potential energy is assumed to be much smaller than the neutron total energy E, where $U \ll E$. Based on this, one can obtain the neutron refractive index in the magnetic field region, where one recalls Eq. (2.4),

$$n^2 = 1 - \frac{U}{E} \Rightarrow n = 1 - \frac{U}{2W} \quad U \ll E = W \tag{17.113}$$

The neutron kinetic energy W outside the magnetic field, corresponding to the vacuum, field-free wavelength λ, is

$$E = W = \frac{h^2}{2m_n \lambda^2} \tag{17.114}$$

The magnetic potential energy of interaction U of the neutron with the row of 99 cylindrical magnetic fields, each with an approximate diameter of 1.27 cm (neglecting fringing fields), and 3-mm transverse width relative to the incident 3-mm (or smaller) neutron beam, is

$$U = -\mu \cdot B = \pm |\mu_{||}| B \tag{17.115}$$

The focal length f_1 of a single cylindrical magnetic lens, is expressed in terms of the radius of curvature R and refractive index n:

$$f_1 = \frac{R}{2(n-1)} \tag{17.116}$$

With substitution into Eq. (17.113) for neutron refractive index n of Eq. (17.114) for total neutron energy E and Eq. (17.115) for neutron acquired potential energy in the magnetic field U, one then obtains focal length f_1 of a single cylindrical magnetic lens, by substitution of the resulting expression for refractive index n into Eq. (17.116), where

$$f_1 = \pm \frac{Rh^2}{2|\mu| B m_n \lambda^2} \tag{17.117}$$

The magnetic lens is birefringent. Recall from chapter 2, spin-down neutrons, with their magnetic dipole longitudinal moment component $\mu_{||}$ parallel (azimuthal spin vector component S_z antiparallel) to the direction of the imposed magnetic field vector **B**, are in the lower spin-state energy, and are focused by the magnetic lens (+sign). The spin-up neutrons at the higher spin-state energy, with their magnetic dipole longitudinal moment component $\mu_{||}$ antiparallel (azimuthal spin vector component S_z parallel) to the direction of **B**, are defocused (−sign).

A cylindrical volume with radius (and radius of curvature) R, which contains an approximately constant 2.5 T magnetic field, causes a spin-down neutron to experiences the same refractive focusing (and focal length f_1), as a biconcave, cylindrical MgF_2 lens, with the same radius of curvature R. Material lenses refract the spin-down and spin-up neutrons in the same direction, whereas magnetic lenses refract spin-up and spin-down neutrons in the opposite directions.

Recall from chapter 2, the refractive index n for a material is given by the neutron wavelength λ and the scatter length density $\rho_n b$, which is the product of the number density ρ of scatter units (single nucleus or set of nuclei comprising a molecule) and the scatter length b of the scatter unit (single nucleus or molecule). In terms of decrement δ for the material, the material refractive index n is

$$n = 1 - \delta = 1 - \frac{\lambda^2 \rho b}{2\pi} o \quad \delta \ll 1 \tag{17.118}$$

The decrement δ is much less than 1. For example, at cold neutron wavelength $\lambda = 4\,\mathring{A}$, the MgF_2 scatter length density is $\rho b = 5.06 \times 10^{-10}\ \mathrm{cm}^{-2}$ and the decrement is $\delta = 1.3 \times 10^{-5}$.

As a result of the very small decrements at cold neutron wavelengths, a relatively short focal length f, requires stacking N cylindrical magnetic lenses, each of focal length f_1, to produce a focal length f that is N-fold shorter than a single lens, where

$$f = \frac{f_1}{N} \tag{17.119}$$

A radius of curvature $R = 1.0$ cm with a 1.0 T magnetic field and $N = 100$ cylindrical, magnetic lenses, gives a focal length of $f = 210$ cm for $\lambda = 8\,\mathring{A}$ neutrons. However, the actual, stacked 99 cylindrical lenses produced a thick, compound refractive lens of length (thickness), $t = 157.5$ cm. The thick lens focal length f_{thk}, given thin-lens focal length f for the N-lens CRL of length t, is from Cremer (2012) and Pantell et al. (2003):

$$f_{\mathrm{thk}} = f \frac{\sqrt{t/f}}{\sin\left(\sqrt{t/f}\right)} \tag{17.120}$$

For $\lambda = 12.6\,\mathring{A}$ neutrons, Littrell et al. (2007) calculated a thin-lens focal length $f = 48$ cm, and obtained a measured, thick lens focal length of $f_{\mathrm{thk}} = 101$ cm.

Littrell et al. (2007) compared their magnetic compound refractive lens (MCRL) with the sextupole lens, developed by Suzuki et al. (2003). The sextupole in the Halbach arrangement was developed by Suzuki et al. (2003) for the JAEA JRR-3 small-angle scattering instrument (SANS) in Japan. The Japanese sextupole lens has an aperture diameter of 25 mm and a length of 300 mm; the magnitude of the magnetic field at 10 mm from the center is 0.98 T. Their sextupole lens focused 9.75 $\mathring{A}$ neutrons in a symmetric arrangement, with equal object and image distances of 9.85 m.

The sextupole lens magnetic field **B** is radial-dependent, where r_{ap} is the sextupole aperture radius over, which has a magentic field gradient, and L is the total sextupole length. In cylindrical coordinates with radius r and azimuth angle ϕ, the sextupole lens magnetic field **B** in the analysis by Littrell et al. (2007) is

$$\mathbf{B} = \left(\frac{r}{r_{\mathrm{ap}}}\right)^2 B_{\max}[\sin 3\phi\,\hat{\mathbf{x}} + \cos 3\phi\,\hat{\mathbf{y}}] \tag{17.121}$$

In Cartesian coordinates, the sextupole lens magnetic field **B** is

$$\mathbf{B} = \frac{1}{r_{\mathrm{ap}}^2} B_{\max}\left[xy\hat{\mathbf{x}} + (x^2 - y^2)\hat{\mathbf{y}}\right] \tag{17.122}$$

The equation of motion of the neutron in the sextupole magnetic field is from Newton's second law, which assumes no change in the neutron energy (adiabatic), where

$$\frac{d^2 r}{dt^2} = \mp \frac{|\mu|}{m_n} \nabla \mathbf{B} \tag{17.123}$$

The sextupole magnetic field **B** increases from zero at the center ($r = 0$) to a maximum magnetic field B_{max} at the bore radius r_{ap}. The focal length of the sextupole f_s is given by

$$f_s = \frac{h^2 r_{ap}^2}{2|\mu| B_{max} m_n \lambda^2 L} \tag{17.124}$$

With the same bore radius, the sextupole has twofold more refractive power per unit length than the MCRL. However, the MCRL is a simpler magnetic refractive lens than the sextupole lens. The CMRL MCRL can be matched more easily to the requirements of a 1-D slit geometry for applications, which require intensity and polarization to be maximized at the sample, such as in neutron reflectometry.

17.8 Calculation of Magnetic Fields of Planar Magnets by Paired Magnetic Charge Sheets

In this section, we derive the magnetic field **B** for a magnetic charge sheet, which can be used as the basis of modeling permanent rectangular magnets with simple analytic formulas. One can represent arrays of permanent magnets by magnetic charge sheets. For charge sheet representation of permanent magnets, see Tatchyn (1993, 1995) and Tatchyn and Csonka (1987). For information on integration of line charges, see Zahn (1979), and concerning scalar magnetic potentials, see Popović (1971) and Jackson (1975).

Consider a permanent magnet, which is modeled as a linear homogeneous medium with a static magnetic field that has no net macroscopic currents. Ampere's law in this case is

$$\nabla \times \mathbf{B} = \nabla \times \mu \mathbf{H} = \mu \nabla \times \mathbf{H} = 0 \tag{17.125}$$

As in the electrostatic case with no net charge density, one can then define a scalar magnetic potential V_m, where

$$\mathbf{H} = -\nabla V_m \tag{17.126}$$

The curl of the gradient is zero:

$$\nabla \times \nabla V_m = 0 \tag{17.127}$$

And the divergence of the gradient is the Laplacian:

$$\nabla \cdot \nabla V_m = \nabla^2 V_m = 0 \tag{17.128}$$

Recall that for the electrostatic case, one can define the electric scalar potential V_e, where

$$\mathbf{E} = -\nabla V_e \tag{17.129}$$

$$\nabla \times \nabla V_e = 0 \tag{17.130}$$

and

$$\nabla \cdot \nabla V_e = \nabla^2 V_e = 0 \tag{17.131}$$

And recall that for the electric scalar potential of a polarized dielectric with zero net charge, the superposition or summation of potentials from electric dipoles exists. The scalar potential of a single electric dipole V_{ed} is expressed by the dipole moment vector $\mathbf{p}$, which is directed from the negative to positive charge. The unit vector $\hat{\mathbf{r}}$ points from the center point (x_c, y_c, z_c), which is centered between the negative and positive charges, and terminates on the observer point (x, y, z), where

$$V_{\mathrm{ed}} = \mathbf{p} \cdot \frac{\hat{\mathbf{r}}}{4\pi\varepsilon_0} \tag{17.132}$$

The electric scalar potential of a polarized dielectric V_d is produced by N electric dipoles, which are contained in volume V that, and are enclosed by surface S, where V_d is expressed in terms of the polarization vector $\mathbf{P}$ (polarization per unit volume), so that

$$\mathbf{P} = \frac{N\mathbf{p}}{V} \tag{17.133}$$

The electric scalar potential of a polarized dielectric V_d, is the sum of the net dipole electric field contributions from (1) the electric dipole charge density ρ_p in the dielectric volume, and (2) the electric dipole charge density σ_p on the dielectric surface, where

$$V_d = \frac{1}{4\pi\varepsilon_0} \iiint_V \mathrm{d}V \frac{\rho_p}{|\mathbf{r} - \mathbf{r}_p|} + \frac{1}{4\pi\varepsilon_0} \iint_S \mathrm{d}S \frac{\sigma_p}{|\mathbf{r} - \mathbf{r}_p|} \tag{17.134}$$

and

$$\rho_p = -\nabla \cdot \mathbf{P} \quad \sigma_p = \mathbf{P} \cdot \hat{\mathbf{r}} \tag{17.135}$$

The minus sign occurs in the polarized charge density expression ρ_p, because (1) the positive-directed electric dipole moment vector **p** is defined to point from the negative to the positive charge, and (2) the positive-directed electric field is defined to point oppositely from the positive to the negative charge.

In a similar manner, the magnetic scalar potential of a single current loop or separated pair of magnetic charges is expressed in terms of the magnetic dipole moment vector **m**, which points from the magnetic south pole (south or negative magnetic charge) to the magnetic north pole (north or positive magnetic charge). One assumes that magnetic monopoles do not exist physically, and that the current loop is used as the physical defining cause of the magnetic dipole moment vector. However, for magnetostatic cases where $\nabla \times \mathbf{H} = 0$, such as permanent magnets, one can employ the mathematical construct of magnetic monopole charges to solve for the static magnetic fields in a similar manner as with electrostatic cases, where $\nabla \times \mathbf{E} = 0$ holds and one solves for static electric fields of a dielectric.

In both the electrostatic and magnetostatic cases, one solves Laplace's equation, which also must satisfy the imposed boundary conditions for the potential or its spatial derivative. The magnetostatic field solution in a vacuum outside a permanent magnet, which is modeled by physical current loops or unphysical magnetic monopole charges, yield the lines of magnetic field intensity **H** and magnetic flux density **B** that point in the same direction, where $\mathbf{H} = \mathbf{B}/\mu_0$. However, inside a permanent magnet, the magnetic intensity **H** points in the opposite direction of the resulting magnetic flux density **B**. The correct calculation of magnetic fields inside a permanent magnet of permeability μ, requires the use of physically viable current loops to model and calculate the magnetic flux density **B** and magnetic field intensity $\mathbf{H} = \mathbf{B}/\mu$.

The magnetic scalar potential of a linear, homogeneous permanent magnet V_m, is produced by N magnetic dipoles, which are contained in volume V, and enclosed by surface S, where V_m is expressed in terms of the magnetization vector **M** (polarization per unit volume), so that

$$\mathbf{M} = \frac{N\mathbf{m}}{V} \tag{17.136}$$

The magnetic scalar potential of the permanent magnet V_m, is the sum of the net dipole magnetic field contributions from (1) the magnetic dipole charge density ρ_m in the magnet volume, and (2) the magnetic dipole charge density σ_m on the magnet surface, where

$$V_m = \frac{1}{4\pi}\iiint_V \mathrm{d}V \frac{\rho_m}{|\mathbf{r} - \mathbf{r}_m|} + \frac{1}{4\pi}\iint_S \mathrm{d}S \frac{\sigma_m}{|\mathbf{r} - \mathbf{r}_m|} \tag{17.137}$$

and

$$\rho_m = -\nabla \cdot \mathbf{M} \quad \sigma_m = \mathbf{M} \cdot \hat{\mathbf{r}} \tag{17.138}$$

The minus sign occurs in magnetic charge density expression ρ_m, because (1) the positive-directed magnetic dipole vector **m** is defined to point from the south to the north magnetic pole, and (2) the positive-directed magnetic field is defined to point oppositely from the north to the south magnetic pole.

The magnetic field of a rectangular permanent magnet, which has its north and magnetic poles of area A separated by distance d, can be modeled by a positive and negative magnetic charge sheet pair of area A separated by distance d. The magnetic charge sheet of positive magnetic charge of area A represents the north magnetic pole face, which is separated by distance d from a parallel sheet of negative magnetic charge that represents the south magnetic pole face. Between the two charge sheets is a vacuum. Linear transformation of coordinates can allow rotation of the planar charge sheet pairs and their associated magnetic fields, so as to represent permanent magnets aligned in the xy, yz, and xz planes.

Planar magnets, which are represented by pairs of planar charge sheets, can model rectangular permanent magnets that can be arrayed into configurations, such as planar-magnet-based dipole, quadrupole, sextupole, and octupole magnets. Consider a set of N rectangular magnets, which are arrayed to construct an N-pole multipole magnet, such as a quadrupole magnet ($N = 4$). The inside planar pole faces are at a minimum distance r_{min} from the center z-axis of the multipole magnet, which is comprised of planar magnets. Within a small radius r about the center z-axis, the magnetic field, where $r \ll r_{\text{min}}$, very closely approximates the multipole magnetic field, such as the quadrupole field for $N = 4$.

An undulator, which is constructed of a line of alternating dipole magnet pairs, is easily modeled using magnetic charge sheets. Corrections to the multipole magnetic field are provided by the addition of outer planar magnets, whose charge sheet representation can guide the design to produce very accurate multipole fields in the center region of the planar-magnet-based multipole magnet or undulator; see Cobb and Tatchyn (1994), Tatchyn (1993), Tatchyn and Cremer (1997), and Tatchyn et al. (1993).

Consider a magnetic charge sheet in the xy-plane, which is bounded by interval $[x_1, x_2]$ along the x-direction, and by interval $[y_1, y_2]$ along the y-direction, and centered at point (x_c, y_c, z_0) in the xy-plane at $z = z_0$, where

$$x_c = \frac{1}{2}(x_1 + x_2) \quad \text{and} \quad x_c = \frac{1}{2}(x_1 + x_2) \tag{17.139}$$

The magnetic scalar potential for a single magnetic charge sheet is found by integration over the surface magnetic charge density $\rho_m(x_m, y_m)$, which has a nonzero, constant value of ρ_m on the charge sheet and is zero elsewhere:

$$V_m(x, y, z) = \frac{1}{4\pi}\int_{y_1}^{y_2} \mathrm{d}y_m \int_{x_1}^{x_2} \mathrm{d}x_m \frac{\rho_m}{\sqrt{(x - x_m)^2 + (x - x_m)^2 + (z - z_0)^2}} \tag{17.140}$$

and

$$\rho_m(x_m, y_m) = \rho_m \quad x_1 \leq x_m \leq x_2 \text{ and } y_1 \leq y_m \leq y_2 \tag{17.141}$$

$$\rho_m(x_m, y_m) = 0 \quad \text{otherwise} \tag{17.142}$$

The magnetic field $\mathbf{B}(x, y, z)$ is the negative gradient of the magnetic scalar potential V_m:

$$\mathbf{B} = -\nabla V_m = -\left(\frac{\partial V_m}{\partial x}\hat{\mathbf{x}} + \frac{\partial V_m}{\partial y}\hat{\mathbf{y}} + \frac{\partial V_m}{\partial z}\hat{\mathbf{z}}\right) \tag{17.143}$$

where

$$B_x = -\frac{\partial V_m}{\partial x} = \frac{1}{4\pi}\int_{y_1}^{y_2} \mathrm{d}y_m \int_{x_1}^{x_2} \mathrm{d}x_m \left(-\frac{\partial}{\partial x}\right) \frac{\rho_m}{\sqrt{(x-x_m)^2 + (y-y_m)^2 + (z-z_0)^2}} \tag{17.144}$$

$$B_y = -\frac{\partial V_m}{\partial y} = \frac{1}{4\pi}\int_{x_1}^{x_2} \mathrm{d}x_m \int_{y_1}^{y_2} \mathrm{d}y_m \left(-\frac{\partial}{\partial y}\right) \frac{\rho_m}{\sqrt{(x-x_m)^2 + (y-y_m)^2 + (z-z_0)^2}} \tag{17.145}$$

$$B_z = -\frac{\partial V_m}{\partial z} = \frac{1}{4\pi}\int_{y_1}^{y_2} \mathrm{d}y_m \int_{x_1}^{x_2} \mathrm{d}x_m \left(-\frac{\partial}{\partial z}\right) \frac{\rho_m}{\sqrt{(x-x_m)^2 + (y-y_m)^2 + (z-z_0)^2}} \tag{17.146}$$

Performing the first derivatives in Eqs. (17.144–17.146) for the x- and y-magnetic field components B_x, B_y, and B_z, give

$$B_x = -\frac{\rho_m}{4\pi}\int_{y_1}^{y_2} \mathrm{d}y_m \int_{x_1}^{x_2} \mathrm{d}x_m \frac{x - x_m}{[(x-x_m)^2 + (y-y_m)^2 + (z-z_0)^2]^{3/2}} \tag{17.147}$$

$$B_y = -\frac{\rho_m}{4\pi}\int_{x_1}^{x_2} \mathrm{d}x_m \int_{y_1}^{y_2} \mathrm{d}y_m \frac{y - y_m}{[(x-x_m)^2 + (y-y_m)^2 + (z-z_0)^2]^{3/2}} \tag{17.148}$$

$$B_z = -\frac{\rho_m}{4\pi}\int_{y_1}^{y_2} \mathrm{d}y_m \int_{x_1}^{x_2} \mathrm{d}x_m \frac{z - z_0}{[(x-x_m)^2 + (y-y_m)^2 + (z-z_0)^2]^{3/2}} \tag{17.149}$$

Performing the second integration in Eqs. (17.147–17.149) for the x-, y-, and z-magnetic field components B_x, B_y, and B_z, give

$$B_x = -\int_{y_1}^{y_2} dy_m \left(\frac{\rho_m/4\pi}{\sqrt{(x-x_2)^2+(y-y_m)^2+(z-z_0)^2}} - \frac{\rho_m/4\pi}{\sqrt{(x-x_1)^2+(y-y_m)^2+(z-z_0)^2}} \right) \tag{17.150}$$

$$B_y = -\int_{x_1}^{x_2} dy_m \left(\frac{\rho_m/4\pi}{\sqrt{(x-x_m)^2+(y-y_2)^2+(z-z_0)^2}} - \frac{\rho_m/4\pi}{\sqrt{(x-x_m)^2+(y-y_1)^2+(z-z_0)^2}} \right) \tag{17.151}$$

and

$$\begin{aligned} B_z = {} & \frac{\rho_m(z-z_0)}{4\pi} \int_{y_1}^{y_2} dy_m \frac{x-x_2}{[(y-y_m)^2+(z-z_0)^2]\sqrt{(x-x_2)^2+(y-y_m)^2+(z-z_0)^2}} \\ & - \frac{\rho_m(z-z_0)}{4\pi} \int_{y_1}^{y_2} dy_m \frac{x-x_1}{[(y-y_m)^2+(z-z_0)^2]\sqrt{(x-x_1)^2+(y-y_m)^2+(z-z_0)^2}} \end{aligned} \tag{17.152}$$

Performing the third integration in Eqs. (17.150–17.152) for the x-, y-, and z-magnetic field components B_x, B_y, and B_z, for a rectangular charge sheet set in the xy-plane, give

$$\begin{aligned} B_x = {} & -\frac{\rho_m}{4\pi} \times \ln \left[\frac{y-y_2-\sqrt{(x-x_2)^2+(y-y_2)^2+(z-z_0)^2}}{y-y_1-\sqrt{(x-x_2)^2+(y-y_1)^2+(z-z_0)^2}} \right. \\ & \left. \times \frac{y-y_1-\sqrt{(x-x_1)^2+(y-y_1)^2+(z-z_0)^2}}{y-y_2-\sqrt{(x-x_1)^2+(y-y_2)^2+(z-z_0)^2}} \right] \end{aligned} \tag{17.153}$$

$$\begin{aligned} B_y = {} & -\frac{\rho_m}{4\pi} \times \ln \left[\frac{x-x_2-\sqrt{(x-x_2)^2+(y-y_2)^2+(z-z_0)^2}}{x-x_1-\sqrt{(x-x_1)^2+(y-y_2)^2+(z-z_0)^2}} \right. \\ & \left. \times \frac{x-x_1-\sqrt{(x-x_1)^2+(y-y_1)^2+(z-z_0)^2}}{x-x_2-\sqrt{(x-x_1)^2+(y-y_1)^2+(z-z_0)^2}} \right] \end{aligned} \tag{17.154}$$

and

$$
\begin{aligned}
B_z = {} & \frac{\rho_m}{4\pi}\tan^{-1}\left[\frac{(x-x_1)(y-y_1)}{(z-z_0)\sqrt{(x-x_1)^2+(y-y_1)^2+(z-z_0)^2}}\right] \\
& -\frac{\rho_m}{4\pi}\tan^{-1}\left[\frac{(x-x_1)(y-y_2)}{(z-z_0)\sqrt{(x-x_1)^2+(y-y_2)^2+(z-z_0)^2}}\right] \\
& -\frac{\rho_m}{4\pi}\tan^{-1}\left[\frac{(x-x_2)(y-y_1)}{(z-z_0)\sqrt{(x-x_2)^2+(y-y_2)^2+(z-z_0)^2}}\right] \\
& +\frac{\rho_m}{4\pi}\tan^{-1}\left[\frac{(x-x_2)(y-y_2)}{(z-z_0)\sqrt{(x-x_2)^2+(y-y_2)^2+(z-z_0)^2}}\right]
\end{aligned}
\tag{17.155}
$$

One can readily perform these integrations using integral tables or the Wolfram Mathematica software and, even more conveniently, the Wolfram Mathematica Online Integrator at http://integrals.wolfram.com. In particular, the integration over y_m to obtain the z-component magnetic field B_z, was performed using the integral result, which is in Eq. (17.156) below, where $r_m = y - y_m$. Since the observer position (x, y, z) is constant y, then $\mathrm{d}r_m = -\,\mathrm{d}y_m$ and

$$
\int \frac{\mathrm{d}r}{(a^2+r^2)\sqrt{b^2+r^2}} = \frac{1}{a\sqrt{b^2-a^2}}\tan^{-1}\left(\frac{r\sqrt{b^2-a^2}}{a\sqrt{b^2+r^2}}\right) \tag{17.156}
$$

where

$$
a^2 = (z-z_0)^2 \tag{17.157}
$$

and

$$
b^2 = (x-x_2)^2 + (z-z_0)^2 \quad \text{and} \quad b^2 = (x-x_1)^2 + (z-z_0)^2 \tag{17.158}
$$

The magnetic field at the observer position (x, y, z), which is produced by a permanent magnet of thickness d with pole face areas A, is obtained by the superposition of the x-, y-, and z-component magnetic field contributions from a pair of charge sheets of area A. The charge sheets are parallel to the xy-plane, and separated by distance d, with their center points at (x_c, y_c, z_0) and $(x_c, y_c, z_0 + d)$. Each permanent, planar magnet is represented by a pair of separated charge sheets, and the observer magnetic field outside the permanent magnets is the superposition or summation of the magnetic fields from all the paired charged sheets.

Cyclic permutation of x, y, and, z in these formulas for the magnetic field components, yields the observer magnetic field, which are produced by charge sheets parallel to the yz- and xz-planes. For example, the magnetic field at observer point (x, y, z), which are produced by a charge sheet parallel to the yz-plane, is obtained from these magnetic field component equations by letting $x \rightarrow y$, $y \rightarrow z$, and $z \rightarrow x$. From the resulting magnetic field component expressions for the yz-plane charge sheet, one obtains the observer magnetic field for a charge sheet parallel to the zx-plane by letting $y \rightarrow z$, $z \rightarrow x$, and $x \rightarrow y$.

To calculate the magnetic field of an arbitrarily oriented charge sheet, one determines the Euler angle rotation angles (α, β, γ), which are required to rotate an xy-plane charge sheet about its center point (x_c, y_c, z_c) to the prescribed, arbitrary orientation. The magnetic field components B_x, B_y, and B_z at fixed observer points (x, y, z), are calculated for an identical charge sheet lying in the xy-plane, using these derived expressions. Then, the x-, y-, and z-magnetic field components B_x, B_y, and B_z for the rectangular xy-plane charge sheet, are subjected to a coordinate transformation, which use the same Euler angles (α, β, γ). This yields the magnetic field components B'_x, B'_y, and B'_z at (x, y, z), which are produced by the charge sheet oriented at the prescribed, arbitrary orientation. An arbitrarily oriented permanent magnet is thus comprised of a pair of xy-plane charge sheets, which undergo the same Euler angle rotations about their respective center points. However, one of the charge sheets must be translated parallel to the other charge sheet, until the two charge sheets are completely eclipsed relative to the observer point. That is, a perpendicular line from the observer point, passes through the center points of both charge sheets, after the operation of rotation and translation.

17.9 1-D Magnetic Gravitational Trap for Ultracold Neutrons

This section presents the results found by Frank and Nosov (2004). They propose a magnetic gravitational trap for ultracold neutrons that uses a Vladimirskiĭ mirror. The trap consists of alternating currents, which flow along the y-axis, with period d in the $z = 0$ plane, where the current density J is given by

$$J = \frac{c}{2\pi} B_m \sin\left(\frac{2\pi}{d} x\right) \tag{17.159}$$

The resulting magnetic field has x- and z-components B_x and B_z at or above the plane $(z \geq 0)$, given by

$$B_x = B_m \exp\left(-\frac{2\pi}{d} z\right) \sin\left(\frac{2\pi}{d} x\right) \tag{17.160}$$

$$B_z = B_m \exp\left(-\frac{2\pi}{d} z\right) \cos\left(\frac{2\pi}{d} x\right) \tag{17.161}$$

where

$$\exp\left(-\frac{2\pi}{d}z\right) \ll 1 \tag{17.162}$$

The potential energy $U(z)$ of a neutron of rest mass m_n depends on the neutron height z above the $z = 0$ plane, and is the sum of magnetic U_m and gravitational U_g contributions:

$$U(z) = \pm \mu_{\parallel} B_m \exp\left(-\frac{2\pi}{d}z\right) + m_n g z \tag{17.163}$$

The magnetic potential U_m for the spin-up $(+)$ and spin-down $(-)$ neutrons is the product of the magnetic field magnitude and the component $\mu_{\parallel}$ of the neutron dipole moment vector $\boldsymbol{\mu}$, which is parallel or antiparallel to the direction of the magnetic field $\mathbf{B}$, where

$$U_m = \boldsymbol{\mu} \cdot \mathbf{B} = \pm \mu_{\parallel} |\mathbf{B}| = \pm \mu_{\parallel} B_m \exp\left(-\frac{2\pi}{d}z\right) \tag{17.164}$$

The gravitational potential energy U_g depends on height z and gravitational acceleration g:

$$U_g = m_n g z \tag{17.165}$$

The magnetic-gravitational potential energy $U(z) = U_{\mathrm{m}}(z) + U_{\mathrm{g}}(z)$ has a minimum and thus forms a potential well, but only for spin-up (+) neutrons, where the sum of the z-derivatives of Eqs. (17.164) and (17.65) forms the total z-derivative $\mathrm{d}U/\mathrm{d}z$, given by

$$\frac{\mathrm{d}U}{\mathrm{d}z} = -\mu_{\parallel} B_m \frac{2\pi}{d} \exp\left(-\frac{2\pi}{d}z\right) + m_n g = 0 \tag{17.166}$$

For the spin-up $(+)$ neutrons, one obtains by transposing Eq. (17.166),

$$\mu_{\parallel} B_m \frac{2\pi}{d} \exp\left(-\frac{2\pi}{d}z\right) = m_n g \tag{17.167}$$

At a distance $z = L$ above the plane, one solves for $2\pi/d$ in Eq. (17.167) to obtain the required period d of the magnetic field to trap the neutrons, where

$$\frac{2\pi}{d} = \frac{m_n g}{\mu_{\parallel} B_m} \exp\left(\frac{2\pi}{d}L\right) \tag{17.168}$$

The presence of the magnetic-gravitation potential well allows the trapping of spin-up (+) neutrons, which have sufficiently small vertical velocities. Such neutrons oscillate with a small amplitude $z-L$ about the equilibrium plane $z=L$, with resulting potential energy $U(z)$ given by the Taylor series expansion of U(z) about equilibrium point L:

$$U(z)=\mu_{\parallel}B_m\exp\left(-\frac{2\pi}{d}L\right)+m_n gL+\frac{1}{2}(z-L)^2\left(\frac{2\pi}{d}\right)^2\mu_{\parallel}B_m\exp\left(-\frac{2\pi}{d}L\right) \tag{17.169}$$

The variation of the potential $U(z)$ with height z is given by the last term, which is denoted by $u(z)$, where

$$u(z)=C(z-L)^2 \tag{17.170}$$

The elastic constant C for the neutron oscillation in the magnetic gravitational potential well, is obtained by comparison of Eqs. (17.169) and (17.170), where

$$C=\frac{1}{2}\left(\frac{2\pi}{d}\right)^2\mu_{\parallel}B_m\exp\left(-\frac{2\pi}{d}L\right) \tag{17.171}$$

Substitution into Eq. (17.171) of Eq. (17.168) for $2\pi/d$, gives

$$C=\frac{1}{2}\frac{(m_n g)^2}{\mu_{\parallel}B_m}\exp\left(\frac{2\pi}{d}L\right) \tag{17.172}$$

The neutron oscillation frequency near the equilibrium point is given by

$$w=\sqrt{\frac{C}{m_n}}=\sqrt{\frac{m_n g^2}{\mu_{\parallel}B_m}\exp\left(\frac{2\pi}{d}L\right)} \tag{17.173}$$

The vertical amplitude of the neutron motion from the equilibrium point $z=L$ to the highest point $z=z_0>L$ is associated with energy E_z, which is assumed to be much less than the neutron potential energy $U(L)$ at $z=L$, where

$$E_z=\frac{1}{2}C(z_0-L)^2\ll U(L)=\mu_{\parallel}B_m\exp\left(-\frac{2\pi}{d}L\right) \tag{17.174}$$

The classical description is valid if

$$E_z\gg\hbar w=\varepsilon \tag{17.175}$$

where from Eq. (17.173),

$$\varepsilon = \hbar\sqrt{\frac{m_n g^2}{\mu_{\|} B_m}\exp\left(\frac{2\pi}{d}L\right)} \tag{17.176}$$

In the case of lower neutron energies and smaller oscillation amplitudes, the quantum description is required, where the neutron is considered a linear oscillator with the nth quantum state energy E_n, where

$$E_n = \left(n+\frac{1}{2}\right)\varepsilon = \hbar\left(n+\frac{1}{2}\right)\sqrt{\frac{m_n g^2}{\mu_{\|} B_m}\exp\left(\frac{2\pi}{d}L\right)} \tag{17.177}$$

The neutron wave function ψ_n for the nth quantum state is

$$\psi_n(z-z_0) = \left(\frac{m_n w}{\hbar\pi}\right)^{1/4}\frac{1}{\sqrt{2^n n!}}\exp\left[-\frac{m_n w}{2\hbar}(z-L)^2\right]H_n\left[(z-L)\sqrt{\frac{m_n w}{\hbar}}\right] \tag{17.178}$$

Here, H_n is the Hermite polynomial. The spatial distribution of neutrons stored on a magnetic mirror were measured, and thus the possibility of observing the quantum properties of the magnetic gravitational trap was shown by Nesvizhevsky et al. (2002, 2003).

One recently developed neutron optic by Cremer et al. (2010), which is similar to the Vladimirskiĭ mirror previously discussed, is the PMF neutron optic, which can focus and separate neutrons axially by wavelength and vertically by spin. As a result, the PMF has applications as a neutron spectrometer, monochromator, and polarizer. The PMF is an arrangement of alternating north–south poles, that forms a half-wiggler magnet configuration, similar to the full wigglers of synchrotron X-ray sources. Also, the relatively inexpensive PMF can provide strong, transverse, magnetic field gradients over relatively long axial lengths (2–3 m), which causes relatively large vertical neutron deflections and separation of spin states.

Spin-down neutrons in the PMF are deflected downward and unfocused toward the PMF pole faces. The spin-up neutrons (higher-energy spin state) are deflected upward and are focused into horizontal lines at a fixed vertical distance in a parallel plane above the PMF surface plane. This vertical distance is proportional to the period of the PMF, but is independent of the neutron wavelength. The axial focal distance of the line-focused neutrons, which is measured from the PMF horizontal entrance slit, is proportional to the neutron energy and inversely proportional to the square of the neutron wavelength.

The PMF magnetic field lines lie in vertical planes above the PMF surface, and the spin-up neutrons exit the PMF with their longitudinal magnetic moments aligned antiparallel to the direction of the last-encountered PMF magnetic field lines. Thus, a detector can scan along the axial direction at a fixed vertical distance above the pole faces to record the spectral distribution of the aligned (polarized) spin-up neutrons.

The PMF consists of an axial (z-axis) row of N rectangular magnets with alternating magnetic polarity in the vertical (y-axis) direction. Along the z-axis, the magnets of axial width Δz_m are separated by aluminum spacers of axial width Δz_s. Neutrons propagate in the axial z-direction through the sinusoidal magnetic field lines, which lie in parallel, vertical yz-planes above the horizontal xz-plane of the PMF surface. The magnets can be permanent, or they can be continuous or pulsed electromagnets.

The PMF magnetic field component in the transverse x-direction is zero, but it varies periodically with spatial period p in the axial z-direction and vertical y-direction. The periodic PMF magnetic field along z is achieved by a line of magnets, whose surface polarity alternates (N–S–N–S–N–S...N–S, where the dashes indicate the aluminum spacer between each pair of magnets). A single period p consists of two magnets with surface poles N and S, separated by an aluminum spacer and followed by another aluminum spacer. The magnetic field amplitude at the magnet surface is determined by the vertical magnetization vector **M** of the magnet, and in this case, they are rectangular, neodymium (NdFeB, grade 42) permanent magnets. For more details of the PMF, see Cremer et al. (2010).

17.10 Electron Trajectory in Wigglers and Undulator X-ray Sources

Synchrotron X-ray sources are typically bending magnets, wigglers, and undulators. Wigglers and undulators are comprised of planar magnets. Paired magnetic charged sheets can be readily applied to model the planar magnets, which comprise wigglers and produce more broadband radiation than the undulator. In this section, we examine the sinusoidal, transverse undulations of the axial, relativistic electron beam trajectory in a transverse, sinusoidal magnetic field that comprises wigglers and undulators. The transverse sinusoidal electron trajectory causes the electron to act as an oscillating electric dipole, which propagates at relativistic speeds in the axial direction through the transverse magnetic field. The transverse magnetic field causes the electron to oscillate in another transverse direction, due to the Lorentz force.

In a wiggler and an undulator, electrons, which propagate in the axial, z-direction, are sinusoidally wiggled in the transverse y-direction by an axially dependent, static magnetic field that has sinusoidal amplitude variation in the x-direction $B_x(z)$, where

$$\mathbf{B} = \hat{\mathbf{x}} B_{0x} \sin k_z z \tag{17.179}$$

The sinusoidal wiggler magnetic field has spatial period p and wave k_z, where

$$k_z = \frac{2\pi}{p} \tag{17.180}$$

The Lorentz force equation for the trajectory of an electron of rest mass m_e and charge e through the magnetic field **B** is

$$\mathbf{F} = \frac{\mathrm{d}\mathbf{p}}{\mathrm{d}t} = \frac{\mathrm{d}}{\mathrm{d}t}(\gamma m_e \mathbf{v}) = e(\mathbf{v} \times \mathbf{B}) \tag{17.181}$$

The magnetostatic field is perpendicular to the electrons and thus performs no work, which assumes negligible radiated energy compared to the electron beam energy. As a result, the total electron velocity magnitude $|\mathbf{v}|$ and relativistic mass factor γ are constant, so

$$\gamma m_e \frac{\mathrm{d}\mathbf{v}}{\mathrm{d}t} = e(\mathbf{v} \times \mathbf{B}) \tag{17.182}$$

The x- and z-component electron velocity trajectory equations are

$$\frac{\mathrm{d}v_x}{\mathrm{d}t} = -\frac{v_z \Omega}{\gamma} \sin k_z z \tag{17.183}$$

and

$$\frac{\mathrm{d}v_z}{\mathrm{d}t} = \frac{v_x \Omega}{\gamma} \sin k_z z \tag{17.184}$$

The electron cyclotron frequency Ω rad/s, in terms of magnetic field amplitude B_{0x} [T], is

$$\Omega = \frac{eB_{0x}}{m_e} = 1.758 \times 10^{11} B_{0x} \tag{17.185}$$

The electron transverse component velocities are assumed to be much smaller than the axial z-electron velocity components $v_x(t)$ and $v_z(t)$, where

$$v_x, v_y \ll v_z \tag{17.186}$$

The electron enters the wiggler at initial time $t = 0$, and the axial position z of the electron in the wiggler is the product of time t and the axial electron velocity $v_z(t)$, which is approximated to be constant in this case:

$$v_z(t) \cong v_{\parallel} \tag{17.187}$$

and

$$z \cong v_z t = v_{\parallel} t \tag{17.188}$$

In the lab frame, the time variation of the magnetic field, which the electron encounters, is then

$$\mathbf{B} = \hat{\mathbf{x}} B_{0x} \sin(k_z v_{\parallel} t) = \hat{\mathbf{x}} B_{0x} \sin(wt) \tag{17.189}$$

In the lab frame, the temporal wiggle frequency w of the electron, which is caused by the imposed sinusoidal magnetic field with spatial period p, is

$$w = k_z v_{\parallel} = \frac{2\pi}{p} v_{\parallel} \tag{17.190}$$

With these approximations, one can solve for the electron trajectory in the wiggler subject to initial conditions:

$$x(t=0) = y(t=0) = z(t=0) = 0 \tag{17.191}$$

and

$$v_x(t=0) = v_y(t=0) = 0 \quad \text{and} \quad v_z(t=0) = v_{\parallel} \tag{17.192}$$

With substitution of Eq. (17.187) into Eq. (17.183), the x-component electron velocity trajectory equation, becomes

$$\frac{\mathrm{d}v_x}{\mathrm{d}t} = -\frac{v_{\parallel}\Omega}{\gamma}\sin wt \tag{17.193}$$

Integration of Eq. (17.193) with initial conditions of Eq. (17.192), yields

$$v_x = \frac{v_{\parallel}\Omega}{\gamma w}\cos wt = \frac{v_{\parallel}K}{\gamma}\cos wt \tag{17.194}$$

In Eq. (17.194), the dimensionless, magnetic deflection parameter K is obtained from the electron cyclotron frequency Ω, divided by the electron wiggle frequency $w = k_z v_{\parallel}$, where

$$K = \frac{\Omega}{w} = \frac{1.758 \times 10^{11} B_{0x} p}{2\pi \cdot v_{\parallel}} \tag{17.195}$$

If the electrons are relativistic, then one can approximate the electron velocity as being equal to the speed of light, where $v_{\parallel} \cong c$. The deflection parameter K is then expressed by magnetic field amplitude B_{0x} [T] and wiggler period p [cm], where

$$K = 0.934 B_{0x} p \tag{17.196}$$

Substitution of the transverse electron velocity $v_x(t)$ solution of Eq. (17.194) into the trajectory Eq. (17.184) for axial electron velocity $v_z(t)$, with $z \cong v_{\parallel} t$ and $w = k_z v_{\parallel}$, gives

$$\frac{\mathrm{d}v_z}{\mathrm{d}t} = \frac{v_x \Omega}{\gamma}\sin k_z z = \frac{v_{\parallel} w K^2}{\gamma^2}\cos wt \sin wt = \frac{v_{\parallel} w K^2}{2\gamma^2}\sin 2wt \tag{17.197}$$

Integration of Eq. (17.197) with initial conditions of Eq. (17.192), yields the time-dependent axial electron velocity $v_z(t)$ as a perturbation about its initial axial velocity $v_{\parallel}$, which is required in order for the total electron velocity magnitude to be constant $v_{\parallel}$, and yet allow the relatively small, transverse wiggle velocity $v_x(t)$, where

$$v_z = v_{\parallel} - \frac{v_{\parallel} K^2}{4\gamma^2}\cos 2wt \tag{17.198}$$

The electron transverse $x(t)$ and axial position $z(t)$, as a function of elapsed time t in the wiggler, is obtained by integrating the electron velocity $v_x(t)$ and $v_z(t)$ expressions of Eqs. (17.194) and (17.197), with initial conditions of Eq. (17.191):

$$x = \frac{Kv_{\parallel}}{\gamma w}\sin wt \cong \frac{Kp}{\gamma 2\pi}\sin\frac{2\pi}{p}ct \tag{17.199}$$

and

$$z = v_{\parallel}t - \frac{v_{\parallel}K^2}{8w\gamma^2}\sin 2wt = ct - \frac{K^2p}{16\pi\gamma^2}\sin\frac{4\pi}{p}ct \tag{17.200}$$

where

$$\gamma = \left[1 - \frac{v_x^2}{c^2} - \frac{v_z^2}{c^2}\right]^{-1/2} \tag{17.201}$$

One approximates $v_{\parallel} \cong c$.

An undulator is distinguished from a wiggler by the deflection parameter K, where

$$K < 1 \quad \text{undulator} \tag{17.202}$$

$$K \gg 1 \quad \text{wiggler} \tag{17.203}$$

The electron, with its x- and z-component velocities, oscillates at angle θ_e, with respect to the z-axis, where substitution of Eq. (17.194) and Eq. (17.187), yield

$$\tan\theta_e = \frac{v_x}{v_z} \cong \frac{K}{\gamma}\cos wt \tag{17.204}$$

From Attwod (1999), the characteristic half-angle θ_{rad} of radiation emission is

$$\theta_{\text{rad}} \cong \frac{1}{2\gamma} \tag{17.205}$$

In the case of the undulator, which has a weak magnetic field $B_x(z)$, the deflection parameter K is less than 1, where $K < 1$. The maximum angle of oscillating electron $\theta_e = K/\gamma$ is contained within θ_{rad}. As a result, interference effects between the spatially overlapping, emitted electromagnetic radiation causes the radiation to be narrowband. In the case of the wiggler with a strong magnetic field $B_x(z)$, the deflection parameter is much greater than 1, where $K \gg 1$. For $K \gg 1$, the electron excursion angle θ_e is no longer contained within θ_{rad}, and hence, the interference effects do not occur between the emitted radiation, because the emitted radiation does not spatially overlap. However, the emitted photon flux and energies are larger in a wiggler in which a continuum of radiation wavelengths is emitted.

17.11 Derivation of Emitted X-ray Wavelength, Bandwidth, and Cone Angle for the Undulator

Following Attwood (1999), in this section, we derive the emitted X-ray wavelength, bandwidth, and cone angle for an undulator. We consider the case of the undulator in which the deflection parameter is less than 1, where $K < 1$. One can solve for v_z in Eq. (17.198) for γ, where $v_{\parallel} \cong c$, so that

$$\frac{v_z}{c} = \sqrt{1 - \frac{1}{\gamma^2} + \frac{K^2}{\gamma^2}\cos^2 wt} \tag{17.206}$$

Application of the binomial approximation for the square root gives

$$\frac{v_z}{c} = 1 - \frac{1}{2\gamma^2} + \frac{K^2}{2\gamma^2}\cos^2 wt \tag{17.207}$$

Given

$$\cos^2 wt = \frac{1}{2}(1 + \cos 2wt) \tag{17.208}$$

one obtains

$$\frac{v_z}{c} = 1 - \frac{1}{2\gamma^2}\left(1 + \frac{K^2}{2}\right) + \frac{K^2}{4\gamma^2}\cos 2wt \tag{17.209}$$

The average of $\bar{v}_z/c$ over one period of the undulator gives

$$\frac{\bar{v}_z}{c} = 1 - \frac{1}{2\gamma^2}\left(1 + \frac{K^2}{2}\right) \tag{17.210}$$

or

$$\left(\frac{\bar{v}_z}{c}\right)^2 \cong 1 - \frac{1}{\gamma^2}\left(1 + \frac{K^2}{2}\right) \tag{17.211}$$

One can thus define the effective axial or z-component relativistic factor γ_z of the relativistic factor γ, where

$$\gamma_z = \frac{\gamma}{\sqrt{1 + (K^2/2)}} \tag{17.212}$$

In an undulator or a wiggler, the period of the magnetic field p is Lorentz-contracted in the rest frame of the electron to a shorter period p_e, where

$$p_e = \frac{p}{\gamma} \tag{17.213}$$

In an undulator, the amplitude of the electron oscillation is less than the radiated wavelength in which the wiggled electron is a radiating electric dipole. In the rest frame of the electron, the frequency f_e of the emitted radiation is based on the Lorentz-contracted period p_e of the undulator magnetic field:

$$f_e = \frac{c}{p_e} = \frac{c\gamma}{p} \tag{17.214}$$

In the lab frame, the emitted radiation from the wiggled electron is Doppler shifted. The frequency f of the Doppler-shifted radiation, which is measured by an observer in the lab frame, depends on the relative velocity between the electron rest frame and the observer in the lab frame. As a result, the frequency f depends on angle θ between the axial direction of the transverse wiggling electron and the laboratory observer, where

$$f = \frac{f_e}{\gamma(1 - \beta \cos \theta)} = \frac{c}{p(1 - \beta \cos \theta)} \tag{17.215}$$

The electron beam is relativistic, where $\beta \cong 1$, so that

$$\gamma^2 = \frac{1}{1 - \beta^2} = \frac{1}{1 - \beta} \cdot \frac{1}{1 + \beta} = \frac{1}{2} \frac{1}{1 - \beta} \tag{17.216}$$

For small angles θ, relative to the electron beam propagation direction, one can use the small-angle approximation

$$\cos \theta \cong 1 - \frac{1}{2}\theta^2 \quad \theta \ll 1 \tag{17.217}$$

With β approximately 1, and with substitution of Eq. (17.217) into Eq. (17.215), the frequency f of the Doppler-shifted radiation, which is measured by an observer in the lab frame at small angles θ, becomes

$$f = \frac{c}{p} \cdot \frac{1}{1 - \beta + \theta^2/2} = \frac{2c\gamma^2}{p(1 + \gamma^2\theta^2)} \tag{17.218}$$

The observed wavelength λ in the lab frame is

$$\lambda = \frac{c}{f} = \frac{p}{2\gamma^2}(1 + \gamma^2\theta^2) \tag{17.219}$$

Since $\theta \ll 1$, one can replace the relativistic factor γ in Eq. (17.219) by the axial relativistic factor γ_z of Eq. (17.212), and thus obtain the undulator equation or wavelength λ_n for the nth harmonic radiation:

$$\lambda_n = \frac{c}{f_n} + \frac{p}{2\gamma^2 n}\left(1 + \frac{K^2}{2} + \gamma^2\theta^2\right) \tag{17.220}$$

In an undulator, as K increases toward 1 and higher, there occur higher-order harmonics in which the integer harmonic number is denoted by n. The even harmonic wavelengths $n = 2, 4, 6, \ldots$ occur off-axis, whereas the odd harmonic wavelengths $n = 1, 3, 5, \ldots$ are emitted as narrowband radiation in a forward-directed, narrow cone angle about the z-axis. The bandwidth is set by the number of undulator oscillation periods N, where

$$\frac{\Delta\lambda}{\lambda} = \frac{1}{nN} \tag{17.221}$$

Consider the fundamental or first ($n = 1$) harmonic, which is contained within a small cone angle θ_c, relative to the z-axis, where $\theta_c \ll 1$. The first harmonic wavelength λ_c at angle θ_c relative to the z-axis ($\theta = 0$) is given by

$$\lambda_1(\theta = \theta_c) = \frac{p}{2\gamma^2}\left(1 + \frac{K^2}{2} + \gamma^2\theta_c^2\right) \tag{17.222}$$

The bandwidth contained within angle θ_c for the first harmonic is

$$\Delta\lambda_1 = \lambda_1(\theta = \theta_c) - \lambda_1(\theta = 0) = \frac{p\theta_c^2}{2} \tag{17.223}$$

The fractional bandwidth contained within cone angle θ_c for the first harmonic is

$$\frac{\Delta\lambda_1(\theta_c)}{\lambda_1(0)} = \frac{1}{N} = \frac{p\theta_c^2}{2}\left[\frac{p}{2\gamma^2}\left(1+\frac{K^2}{2}\right)\right]^{-1} \tag{17.224}$$

The cone angle θ_c containing the first harmonic is then

$$\theta_c = \frac{\sqrt{1 + (K^2/2)}}{\gamma\sqrt{N}} \tag{17.225}$$

For the electron rest mass m_n, the relativistic factor γ for electrons propagating in a beam at energy E_e is given by

$$\gamma = \frac{E_e}{m_e c^2} = 1957\, E_e \text{ Gev} \tag{17.226}$$

From the very handy *X-ray Data Booklet* of the Lawrence Berkeley Laboratory, CXRO (2009) in section 2.B.2 by K. J. Kim, the following useful formula for low K undulators is applied. This formula refers to the lab frame wavelength λ_1 [Å] for the fundamental frequency of forward-emitted X-rays, which are forward-scattered in a narrow cone about the z-axis. X-rays emitted by the wiggled electron beam of energy E [Gev], which passes through an undulator with period p [cm], have a fundamental wavelength λ_1 [Å] given by

$$\lambda_1\left[\mathring{A}\right] = \frac{13.06p\,[\mathrm{cm}]}{E^2\,[\mathrm{GeV}]}\left(1+\frac{K^2}{2}\right) \tag{17.227}$$

17.12 X-ray Power and Differential Power Emission per Solid Angle from an Undulator

Following Attwood (1999) and Marshall (1985), in this section, we derive the differential power per solid angle, which is emitted from an undulator for the fundamental or first harmonic frequency, and is forward directed in a narrow cone, about the z-axis, in the direction of the axial propagating, and transversely wiggled relativistic electron beam. The narrow bandwidth, narrow emission angle cone, and intensity of the undulator X-ray source yields many research and commercial applications.

From Eq. (1.447) of section 1.5.2, the differential power per solid angle $\mathrm{d}P_e/\mathrm{d}\Omega_e$ irradiated by a transverse, oscillating, single electron in the electron rest frame is

$$\frac{\mathrm{d}P_e}{\mathrm{d}\Omega_e} = \frac{e^2}{16\pi^2\varepsilon_0 c^3}\left(\frac{\mathrm{d}v_x}{\mathrm{d}t}\right)_e^2 \sin^2\Theta_e \tag{17.228}$$

where

$$\sin^2\Theta_e = 1-\cos^2\Theta_e \tag{17.229}$$

In the spherical coordinate system,

$$\cos\Theta_e = \cos\theta_{ek}\cos\theta_{ea} + \sin\theta_{ek}\sin\theta_{ea}\cos(\phi_{ek}-\phi_{ea}) \tag{17.230}$$

Here, polar angle Θ_e is measured between the direction of electron acceleration $(\mathrm{d}v_x/\mathrm{d}t)_e$ and the direction of the wave vector $\mathbf{k}_0$ of the emitted radiation in the electron rest frame. The polar angles θ_{ek} and θ_{ea} are the angles in the xz-plane of the wave vector and electron acceleration vector, respectively, which are measured from the z-axis in the electron rest frame. In the yz-plane, the wave vector azimuth angle ϕ_{ek} and electron acceleration vector azimuth angle ϕ_{ea} are each measured from the transverse x-axis in the electron rest frame.

We consider only the power per solid angle emitted by the first harmonic along the tiny cone angle about the z-axis, where $\theta_{ea} = \pi/2$ and $\phi_{ea} = 0$, and polar angle Θ_e is given by

$$\sin^2 \Theta_e = 1 - \sin^2 \theta_{ek} \cos^2 \phi_{ek} \tag{17.231}$$

One invokes the Lorentz transformation between angle θ_e in the electron rest frame, and the angle θ in the lab frame. This Lorentz transformation is obtained from appendix F.2 in Atwood (1999), or from section 17 of chapter 4 in Cremer (2012), where

$$\sin \theta_e = \frac{\sin \theta}{\gamma(1 - \beta \cos \theta)} \tag{17.232}$$

Equation (17.232) is obtained from the relativistic Doppler equation for the electromagnetic wave frequency of the electron rest frame f_{ze}, which is expressed in terms of the lab frame frequency f_z. Angle θ is the angle between the wave vector **k**, and the direction of the relative velocity $v = \beta c$ between the two frames with relativistic factor γ. The relativistic Doppler equation can be obtained from appendix F.1 in Atwood (1999), or Kleppner & Kolenkow (1973), where

$$f_e = f_\gamma(1 - \beta \cos \theta) \tag{17.233}$$

With the x-directed magnetic field with z-directed electron beam, the transverse y-directed electromagnetic frequency component f_{ey} in the electron rest frame, which is perpendicular to the z-direction of relative motion between the two frames, is identical to the lab frame y-component frequency f_y:

$$f_{ye} = f_y = f \sin \theta \tag{17.234}$$

Finally,

$$\sin \theta_e = \frac{f_{ye}}{f_e} = \frac{f \sin \theta}{f_\gamma(1 - \beta \cos \theta)} = \frac{\sin \theta}{\gamma(1 - \beta \cos \theta)} \tag{17.235}$$

To Eq. (17.235) for $\sin \theta_e$, one applies the binomial approximation to β, and represents $\cos \theta$ by the first two terms of its series representation, where $\beta \cong 1$ and $\theta \ll 1$, so that

$$\beta \cong 1 - \frac{1}{2\gamma^2} \tag{17.236}$$

and

$$1 - \beta \cos \cong 1 - \left(1 - \frac{1}{2\gamma^2}\right)\left(1 - \frac{\theta^2}{2}\right) \tag{17.237}$$

Substitution of Eq. (17.237) into Eq. (17.235), gives the approximation

$$\sin\theta_e \cong \frac{2\gamma\theta}{1+\gamma^2\theta^2} \tag{17.238}$$

Taking the differential of both sides, one obtains

$$d(\sin\theta_e) \cong d\left(\frac{2\gamma\theta}{1+\gamma^2\theta^2}\right) \tag{17.239}$$

or

$$\cos\theta_e\, d\theta_e \cong \frac{2\gamma(1-\gamma^2\theta^2)}{(1+\gamma^2\theta^2)^2}\, d\theta \tag{17.240}$$

Alternatively,

$$\cos^2\theta_e = 1-\sin^2\theta_e \cong 1-\left(\frac{2\gamma\theta}{1+\gamma^2\theta^2}\right)^2 \tag{17.241}$$

or

$$\cos\theta_e = \sqrt{\frac{(1+\gamma^2\theta^2)^2-(2\gamma\theta)^2}{(1+\gamma^2\theta^2)^2}} = \sqrt{\frac{(1-\gamma^2\theta^2)^2}{(1+\gamma^2\theta^2)^2}} = \frac{1-\gamma^2\theta^2}{1+\gamma^2\theta^2} \tag{17.242}$$

Substitution of Eq. (17.242) for $\cos\theta_e$ into Eq. (17.240) for $\cos\theta_e\, d\theta_e$, gives

$$\frac{1-\gamma^2\theta^2}{1+\gamma^2\theta^2}\, d\theta_e \cong \frac{2\gamma(1-\gamma^2\theta^2)}{(1+\gamma^2\theta^2)^2}\, d\theta \tag{17.243}$$

From Eq. (17.243), one then obtains the relation between the lab frame polar differential angle $d\theta$, and the electron rest frame differential polar angle $d\theta_e$, where

$$d\theta_e \cong \frac{2\gamma}{1+\gamma^2\theta^2}\, d\theta \tag{17.244}$$

Since the azimuth angle is perpendicular to the relativistic axial motion of the electron, the azimuth is unchanged in the Lorentz transformation, where

$$\phi_{ek} = \phi_k = \phi \tag{17.245}$$

One drops the "k" subscript from azimuth angle ϕ_k. With substitution of Eqs. (17.238), (17.244), and (17.245), the equation for the solid angle differential $d\Omega_e$ in the electron rest frame, in terms of angles in the lab frame, is then

$$d\Omega_e = \sin\theta_e\, d\theta_e\, d\phi_e \cong \frac{4\gamma^2}{(1+\gamma^2\theta^2)}\theta\cdot d\theta\cdot d\phi \tag{17.246}$$

For small angles $\sin\theta \cong \theta$, one obtains the transformation from the differential solid angle $d\Omega$ in the lab frame to the differential solid angle $d\Omega_e$ in the electron frame, where

$$d\Omega_e \cong \frac{4\gamma^2}{(1+\gamma^2\theta^2)^2}d\Omega \tag{17.247}$$

Finally, substitution of Eq. (17.238) with Eq. (17.245) into Eq. (17.231), gives the expression for polar angle Θ_e as

$$\sin^2\Theta_e = 1 - \left(\frac{2\gamma\theta}{1+\gamma^2\theta^2}\right)^2\cos^2\phi \tag{17.248}$$

Substitution into Eq. (17.228) of Eq. (17.248) for $\sin^2\Theta_e$, yields the power per solid angle $dP_e/d\Omega_e$ per electron in the electron rest frame:

$$\frac{dP_e}{d\Omega_e} = \frac{e^2}{16\pi^2\varepsilon_0 c^3}\left(\frac{dv_x}{dt}\right)_e^2\left[1-\left(\frac{2\gamma\theta}{1+\gamma^2\theta^2}\right)^2\cos^2\phi\right] \tag{17.249}$$

For radiation emission close to the z-axis of electron beam propagation through the wiggler, one replaces the relativistic factor γ by its z-component γ_z and then approximates

$$1-\left(\frac{2\gamma_z\theta}{1+\gamma_z^2\theta^2}\right)\cos^2\phi = \frac{1+2\gamma_z^2\theta^2(1-2\cos^2\phi)+\gamma_z^4\theta^4}{(1+\gamma_z^2\theta^2)^2} \tag{17.250}$$

With substitution of Eq. (17.250) into Eq. (17.249), the differential power per solid angle $dP_e/d\Omega_e$, which is emitted along a narrow cone about the z-axis in the forward direction of electron beam propagation in the electron rest frame, and then becomes

$$\frac{dP_e}{d\Omega_e} = \frac{e^2}{16\pi^2\varepsilon_0 c^3}\left(\frac{dv_x}{dt}\right)_e^2\left[\frac{1+2\gamma_z^2\theta^2(1-2\cos^2\phi)+\gamma_z^4\theta^4}{(1+\gamma_z^2\theta^2)^2}\right] \tag{17.251}$$

We now determine the x-directed electron acceleration $(\mathrm{d}v_x/dt)_e$ in the electron rest frame, given the x-directed acceleration in the lab frame derived previously. The electron x-component position in the lab frame with $v_{\|} \cong c$ is obtained from Eq. (17.199), where

$$x = \frac{Kc}{\gamma w} \sin wt \tag{17.252}$$

To transform to the rest frame of the electron, one employs the relativistic transformation with the z-component relativistic factor γ_z, since we are concerned with emission only along the z-axis, where the average axial electron velocity relative to the lab frame is $\bar{v}_z \cong c$. The Lorentz transformations of Eq. (13.48) from section 13.5 are recalled, where

$$t = \gamma_z\left(t_e + \frac{\bar{v}_z}{c}\frac{z_e}{c}\right) \cong \gamma_z\left(t_e + \frac{z_e}{c}\right) \cong \gamma_z t_x \quad \frac{z_e}{c} \ll t_e \tag{17.253}$$

and

$$x = x_e \tag{17.254}$$

In the previous section 17.11, we derived the z-component relativistic factor γ_z, which was based on the total relativistic factor γ and magnetic deflection parameter K, where Eq. (17.212) is repeated below

$$\gamma_z = \frac{\gamma}{\sqrt{1 + (K^2/2)}} \tag{17.255}$$

Here, z_e is a small axial variation about the average electron position in the electron rest frame. Substitution for t of Eq. (17.253) into Eq. (17.252) for $x = x_e$, gives the electron x-component position x_e in the electron rest frame, as follows:

$$x_e = \frac{Kc}{\gamma w} \sin \omega\gamma_z t_e \tag{17.256}$$

The x-component acceleration in the electron rest frame is

$$\frac{\mathrm{d}v_{ex}}{\mathrm{d}t_e} = \frac{\mathrm{d}^2 x_e}{\mathrm{d}t_e^2} = -\frac{Kcw\gamma_z^2}{\gamma} \sin w\gamma_z t_e \tag{17.257}$$

where

$$w = k_z v_{\|} = \frac{2\pi}{p} c \tag{17.258}$$

The root mean square (RMS) acceleration $(dv_x/dt)_e^{\text{rms}}$ of the electron in the x-direction in the electron rest frame is

$$\left(\frac{dv_x}{dt}\right)_e^{\text{rms}} = \sqrt{\frac{1}{T}\int_0^T dt_e \left(\frac{dv_x}{dt}\right)_e^2} = \frac{Kcw\gamma_z^2}{\gamma}\sqrt{\frac{1}{T}\int_0^T dz \sin^2\frac{2\pi}{T}t_e} = \frac{Kcw\gamma_z^2}{\gamma\sqrt{2}} \quad (17.259)$$

where

$$T = \frac{1}{w\gamma_z} \quad (17.260)$$

Substitution into Eq. (17.259) of Eq. (17.255) for γ_z, gives the x-directed RMS acceleration as

$$\left(\frac{dv_x}{dt}\right)_e^{\text{rms}} = \frac{1}{\sqrt{2}}\frac{Kcw\gamma}{1+(K^2/2)} = \frac{2\pi}{p\sqrt{2}}\frac{Kc^2\gamma}{1+(K^2/2)} \quad (17.261)$$

The RMS differential power per solid angle $(dP/d\Omega)_{\text{rms}}$ irradiated by a single transverse-oscillating electron in the electron rest frame is obtained, where

$$\left(\frac{dP}{d\Omega}\right)_e^{\text{rms}} = \frac{e^2}{16\pi^2\varepsilon_0 c^3}\left[\left(\frac{dv_x}{dy}\right)_e^{\text{rms}}\right]^2 \sin^2\Theta \quad (17.262)$$

Substitution of Eq. (17.248), via Eq. (17.250), and Eq. (17.261) into Eq. (17.262), gives the RMS power emission per solid angle per electron $(dP/d\Omega)_{\text{rms}}$ in the electron rest frame:

$$\left(\frac{dP}{d\Omega}\right)_e^{\text{rms}} = \frac{e^2c\gamma^2}{8\varepsilon_0 p^2}\cdot\frac{K^2}{[1+(K^2/2)]^2}\cdot\left[\frac{1+2\gamma_z^2\theta^2(1-2\cos^2\phi)+\gamma_z^4\theta^4}{(1+\gamma_z^2\theta^2)^2}\right] \quad (17.263)$$

We now only need to find the relation between the differential power dP_e in the electron rest frame and the differential power in the lab frame dP, in order to obtain the differential power per solid angle $dP/d\Omega$ in the lab frame. This is done by recalling Eq. (17.233) for the angle-dependent photon energy Doppler shift from the electron rest frame to the lab frame, where

$$hf = \frac{hf_e}{\gamma_z(1-\beta\cos\theta)} \quad (17.264)$$

The z-component Lorentz factor γ_z is used instead of the total Lorentz factor γ, since the emission is along the z-axis. The polar angle θ with respect to the z-axis is assumed to be very small, and with the z-component of $\beta = v/c$, allows the following approximations:

$$\beta_z = \frac{v_z}{c} \cong 1 - \frac{1}{2\gamma_z^2} \quad \text{and} \quad \cos\theta = 1 - \frac{\theta^2}{2} \tag{17.265}$$

Substitution of Eq. (17.265) into Eq. (17.264), gives

$$hf \cong \frac{2\gamma}{1+\gamma_z^2\theta^2} hf_e \tag{17.266}$$

The Lorentz relation between time intervals in the electron rest frame dt_e and lab frame dt is given by the z-component Lorentz factor γ_z:

$$dt = \gamma_z dt_e \tag{17.267}$$

Given M photons, the differential radiated power in the electron rest frame dP_e and the lab frame dP are

$$dP_e = \frac{Mhf_e}{dt_e} \tag{17.268}$$

and

$$dP = \frac{Mhf}{dt} \tag{17.269}$$

Substitution of Eqs. (17.266) and (17.267) into Eq. (17.269) for dP, gives

$$dP = \frac{Mhf}{dt} = \frac{Mhf_e}{\gamma_z dt_e} \frac{2\gamma_z}{1+\gamma_z^2\theta^2} \tag{17.270}$$

Substitution of Eq. (17.268) into Eq. (17.270), yields the transformation from the differential power dP_e in electron frame to the differential power in the lab dP frame, where

$$dP = \frac{2}{1+\gamma_z^2\theta^2} dP_e \tag{17.271}$$

The previously derived Lorentz relation between $d\Omega$ and $d\Omega_e$ of Eq. (17.247) is flipped, with the z-component relativistic gamma factor γ_z used in place of γ, so that

$$d\Omega \cong \frac{(1+\gamma_z^2\theta^2)^2}{4\gamma_z^2} d\Omega_e \tag{17.272}$$

Combining the results for the Lorentz relations between dP and dP_e of Eq. (17.271), and between dΩ and dΩ_e of Eq. (17.272), one obtains the Lorentz relation between the differential power per solid angle in the lab frame dP/dΩ and in the electron rest frame (dP/dΩ)$_e$, where

$$\frac{\mathrm{d}P}{\mathrm{d}\Omega} = \frac{8\gamma_z^2}{(1+\gamma_z^2\theta^2)^3}\frac{\mathrm{d}P_e}{\mathrm{d}\Omega_e} \tag{17.273}$$

Since emission is in a tiny cone about the z-axis, the relativistic factor γ has been replaced by the z-component relativistic factor γ_z. Substitution into Eq. (17.273) of Eq. (17.263) for dP_e/dΩ_e of the electron rest frame, then yields the lab frame RMS differential power per solid angle dP/dΩ emitted per electron for the first harmonic along the z-axis, where

$$\left(\frac{\mathrm{d}P}{\mathrm{d}\Omega}\right)^{\mathrm{rms}} = \frac{e^2c\gamma^4}{\varepsilon_0 p^2}\cdot\frac{K^2}{[1+(K^2/2)]^3}\cdot\left[\frac{1+2\gamma_z^2\theta^2(1-2\cos^2\phi)+\gamma_z^4\theta^4}{(1+\gamma_z^2\theta^2)^5}\right] \tag{17.274}$$

The RMS power per electron P^{rms} of the first harmonic can be calculated, via integrating Eq. (17.274) over azimuth angle ϕ and polar angle θ, where one starts with

$$\int_0^{2\pi}\mathrm{d}\phi(1-2\cos^2\phi) = \int_0^{2\pi}\mathrm{d}\phi\left[1-2\left(\frac{1+\cos 2\phi}{2}\right)\right] = -\frac{1}{2}[\sin 2\phi]_0^{2\pi} = 0 \tag{17.275}$$

To obtain the RMS power per electron of the first harmonic, one is left with the integration

$$\begin{aligned} P^{\mathrm{rms}} &= \int_0^{\pi}\mathrm{d}\theta\,\sin\theta\int_0^{2\pi}\mathrm{d}\phi\left(\frac{\mathrm{d}P}{\mathrm{d}\Omega}\right)^{\mathrm{rms}} \\ &= \frac{e^2c\gamma^4}{\varepsilon_0 p^2}\cdot\frac{K^2}{[1+(K^2/2)]^3}\cdot\int_0^{\pi}\mathrm{d}\theta\,\sin\theta\int_0^{2\pi}\mathrm{d}\phi\frac{1+\gamma_z^4\theta^4}{(1+\gamma_z^2\theta^2)^5} \end{aligned} \tag{17.276}$$

After integration over ϕ in Eq. (17.276), one is left with integration over θ, where

$$P^{\mathrm{rms}} = \frac{e^2c\gamma^4}{\varepsilon_0 p^2}\cdot\frac{2\pi K^2}{[1+(K^2/2)]^3}\cdot\int_0^{\pi}\mathrm{d}\theta\frac{\theta+\gamma_z^4\theta^5}{(1+\gamma_z^2\theta^2)^5} \tag{17.277}$$

As an approximation, $\sin\theta$ is replaced by θ because $\gamma_z \gg 1$, and the integrand becomes small for values of $\theta > 1$ with a $1/\theta^5$ dependence. However, the integrand is significantly larger for small values of $\theta < 1$ with the $1/\theta^5$ dependence. Variable substitution is done in Eq. (17.277) to perform the integration over θ, where

$$r = \gamma_z\theta \quad x = 1 + r^2 \quad \mathrm{d}x = 2r\,\mathrm{d}r \tag{17.278}$$

and

$$1 + r^4 = 1 + (r^2)^2 = 1 + (x-1)^2 = x^2 - 2x + 2 \tag{17.279}$$

The integral of Eq. (17.277) becomes

$$P^{\mathrm{rms}} = \frac{e^2 c\gamma^4}{\varepsilon_0 p^2} \cdot \frac{2\pi K^2}{[1+(K^2/2)]^3} \cdot \frac{2\pi}{\gamma_z^2} \cdot \frac{1}{2} \cdot \int_1^{1+\gamma_z^2\pi^2} \mathrm{d}x \frac{x^2 - 2x + 2}{x^5} \tag{17.280}$$

Since the integrand has a x^{-3} dependence, integration from $1 + \gamma_z^2\pi^2$ to ∞ is small compared to the integration from 1 to $1 + \gamma_z^2\pi^2$, and one uses the more mathematically convenient integration limits of 1 to ∞, where

$$\int_1^{1+\gamma_z^2\pi^2} \mathrm{d}x \frac{x^2 - 2x + 2}{x^5} \cong \int_1^{\infty} \mathrm{d}x(x^{-3} - 2x^{-4} + 2x^{-5}) = \frac{1}{3} \tag{17.281}$$

With substitution of Eq. (17.281) into Eq. (17.280), integration yields the RMS total power P^{rms} per electron of the first harmonic electromagnetic radiation, which is emitted predominantly in the forward z-direction within a $1/2\gamma$ angle, where

$$P^{\mathrm{rms}} = \frac{e^2 c\gamma^2\pi}{3\varepsilon_0 p^2} \cdot \frac{K^2}{[1+(K^2/2)]^2} \tag{17.282}$$

The total, forward-emitted RMS power P^{rms} of the first harmonic can be expressed in terms of the electron beam current I and the number of undulator periods N. Consider the electrons of charge e, axial velocity $v \cong c$, and electron number per axial length n_l for undulator length L, which contains an average of $\overline{n}$ electrons over L, where

$$I = evn_l \cong ec\frac{\overline{n}}{L} \tag{17.283}$$

The undulator length L equals the product of the undulator period p and number of periods N, where

$$L = Np \tag{17.284}$$

One obtains the average number of electrons $\overline{n}$ in the undulator for electron beam current I from Eqs. (17.283) and (17.284):

$$\overline{n} = \frac{LI}{ec} = \frac{NpI}{ec} \tag{17.285}$$

The RMS first harmonic power P_I^{rms} for undulator beam current I, is the product of the per-electron RMS first harmonic power P^{rms}, and the average number of electrons $\bar{n}$ in the undulator, where

$$P_I^{\mathrm{rms}} = \bar{n}P^{\mathrm{rms}} = \frac{NpI}{ec}P^{\mathrm{rms}} \tag{17.286}$$

For an undulator of N periods through which passes electron current I, the first harmonic power, forward directed in a narrow cone in the direction of electron beam propagation, is

$$P_I^{\mathrm{rms}} = NI \cdot \frac{e\gamma^2\pi}{3\varepsilon_0 p} \cdot \frac{K^2}{[1+(K^2/2)]^2} \tag{17.287}$$

or

$$P_I^{\mathrm{rms}}\,[\mathrm{watt}] = 1.90 \times 10^{-6}\frac{N\gamma^2 I\,[\mathrm{amp}]}{p\,[\mathrm{cm}]} \cdot \frac{K^2}{[1+(K^2/2)]^2} \tag{17.288}$$

The power $P_{I,\mathrm{wig}}^{\mathrm{rms}}$, which is emitted by a wiggler, and integrated over all angles and wavelengths, for a specified K, is identical to Eq. (17.287) for the undulator first harmonic, except that the wiggler expression drops the denominator term $1 + (K^2/2)$. Hence, for a wiggler, the power $P_{I,\mathrm{wig}}^{\mathrm{rms}}$ emitted over all angles and wavelengths is

$$P_{I,\mathrm{wig}}^{\mathrm{rms}}\,[\mathrm{watt}] = 1.90 \times 10^{-6}\frac{K^2 N\gamma^2 I\,[\mathrm{amp}]}{p\,[\mathrm{cm}]} \tag{17.289}$$

For more information, see Kim (1989).

References

Anderson, I. S., McGreevy, R. L., & Bilheux, H. Z. (Eds.), (2009). *Neutron imaging and its applications* New York, NY: Springer.

Attwood, D. (1999). *Soft X-rays and extreme ultraviolet radiation. Principles and applications*. Cambridge: Cambridge University Press.

Center for X-Ray Optics and Advanced Light Source. (2009). X-Ray Data Booklet. *LBNL/ Pub 490 Rev. 3*, October 2009.

Cobb, J., & Tatchyn, R. (1994). Tests of planar permanent magnet multipole focusing elements. *Nuclear Instruments and Methods*, ***A****347*, 562.

Cremer, J. T., Williams, D. L., Fuller, M. J., Gary, C. K., Piestrup, M. A., & Pantell, R. H., et al. (2010). Periodic magnetic field as a polarized and focusing thermal neutron spectrometer and monochromator. *Review of Scientific Instruments*, *81*(1), 13902.

Cremer, J. T. (2012). In P. W. Hawkes (Ed.), *Neutron and X-ray microscopy—Part 1. Advances in imaging and electron physics* (Vol. 172). London: Academic Press, an imprint of Elsevier Press.

Frank, A. I., & Nosov, V. G. (2004). Quantum effects in a one-dimensional magnetic gravitational trap for ultracold neutrons. *JETP Letters*, *70*(7), 313–315.

Furrer, A., Mesot, J., & Strässle, T. (2009). *Neutron scattering in condensed matter physics*. Hackensack, NJ: World Scientific.

Hawkes, P. W. (1970). *Quadrupoles in electron lens design*. New York, NY: Academic Press.

Jackson, J. D. (1975). *Classical electrodynamics*. New York, NY: John Wiley & Sons.

Kim, K. J. (1989). Characteristics of synchrotron radiation. *American Institute of Physics—Physics of Particle Accelerators, AIP Conference Proceedings*, *184*, 565–632.

Kleppner, D., & Kolenkow, R. J. (1973). *An introduction to mechanics*. New York, NY: McGraw-Hill.

Lawson, J. D. (1977). *The physics of charged particle beams*. Oxford: Oxford University Press.

Littrell, K. C., te Velthuis, S. G. E., Felcher, G. P., Park, S., Kirby, B. J., & Fitzsimmons, M. R. (2007). Magnetic compound refractive lens for focusing and polarizing cold neutron beams. *Review of Scientific Instruments*, *78*(3), 035101–035108.

Marshall, T. G. (1985). *Free electron lasers*. New York, NY: Macmillan.

Nesvizhevsky, V. V., Börner, H., Petukhov, A. K., Abele, H., Bäßler, S., & Rueß, F. J., et al. (2002). Quantum states of neutrons in the earth gravitational field. *Nature*, *415*, 297–299.

Nesvizhevsky, V. V., Börner, H., Gagarski, A. M., Petoukhov, A. K., Petrov, G. A., & Abele, H., et al. (2003). Measurement of quantum states of neutrons in the Earth's gravitational field. *Physical Review D*, *67*, 102002.

Panofsky, W. K. H., & Phillips, M. (1962). *Classical electricity and magnetism*. Reading, MA: Addison-Wesley.

Pantell, R. H., Feinstein, J., Beguiristain, H. R., Piestrup, M. A., Gary, C. K., & Cremer, J. T. (2003). Characteristics of the thick, compound refractive lens. *Applied Optics*, *42*(4), 719–723.

Popović, B. D. (1971). *Introductory engineering electromagnetics*. Reading, MA: Addison-Wesley.

Reiser, M. (1994). *Theory and design of charged particle beams*. New York, NY: John Wiley & Sons.

Shimizu, H. (2002). Development of neutron optical devices. *Applied Physics A*, *74*(Suppl.), S326–S328.

Shimizu, H., Suzuki, J., & Oku, T. (2010). Neutron polarization apparatus, U.S. Patent No.2010/0032554 A1, February 11.

Suzuki, J., Oku, T., Adashi, T., Shimizu, H. M., Hirumachi, T., & Tsuchihashi, I., et al. (2003). Cold neutron beam focusing by a superconducting sextupole magnet. *Journal of Applied Crystallography*, *36*, 795–799.

Tatchyn, R. (1993). Permanent magnet edge-field quadrupoles as compact focusing elements for single pass particle accelerators. *SLAC-PUB-6058*, Stanford Linear Accelerator Center, Menlo Park, California: Center Publications.

Tatchyn, R. (1995). Image tuning techniques for enhancing the performance of pure permanent magnet undulators with small gap/period. *SLAC-Pub-95-7035*, Stanford Linear Accelerator Center Publications, Stanford Linear Accelerator Center, Menlo Park, California.

Tatchyn, R., Boyce, R., Halbach, K., Nuhn, H.-D., Seeman, J., Winick, H., et al. (1993). Design considerations for a 60 meter pure permanent magnet undulator for the SLAC Linac Coherent Light Source (LCLS). Institute of Electrical and Electronics Engineers, IEEE Catalog No. 93CH3279-7. *Proceedings of the 1993 Particle Accelerator Conference*, Washington, DC, May 17–20, 1993 (pp. 1608–1616).

Tatchyn, R., & Cremer, T. (1997). Analytical undulator model for relativistic high-energy FEL simulations. *Nuclear Instruments and Methods*, *393*(1–3), 114–118.

Tatchyn, R., & Csonka, P. (1987). Modelling and characterization of a micropole undulator with a 726 μm period. *Journal of Physics D: Applied Physics*, *20*, 394.

Williams, W. G. (1988). *Polarized neutrons.* Oxford: Clarendon Press.

Zahn, M. (1979). *Electromagnetic field theory, a problem solving approach.* New York, NY: John Wiley & Sons.

CPSIA information can be obtained at www.ICGtesting.com
Printed in the USA
BVOW08*0331070714

358074BV00004B/10/P

9 780124 071643